MARINE OMICS
Principles and Applications

MARINE OMICS
Principles and Applications

edited by
Se-Kwon Kim

CRC Press
Taylor & Francis Group
Boca Raton London New York

CRC Press is an imprint of the
Taylor & Francis Group, an **informa** business

CRC Press
Taylor & Francis Group
6000 Broken Sound Parkway NW, Suite 300
Boca Raton, FL 33487-2742

First issued in paperback 2019

ISBN-13: 978-1-4822-5820-2 (hbk)
ISBN-13: 978-0-367-87081-2 (pbk)

Library of Congress Cataloging-in-Publication Data

Names: Kim, Se-Kwon, editor.
Title: Marine OMICS : principles and applications / editor, Se-Kwon Kim.
Description: Boca Raton : Taylor & Francis, 2016. | Includes bibliographical
references and index. | "This book provides comprehensive coverage on
current trends in marine omics of various relevant topics such as
genomics, lipidomics, proteomics, foodomics, transcriptomics,
metabolomics, nutrigenomics, pharmacogenomics and toxicogenomics as
related to and applied to marine biotechnology, molecular biology, marine
biology, marine microbiology, environmental biotechnology, environmental
science, aquaculture, pharmaceutical science and bioprocess
engineering."--Provided by publisher.
Identifiers: LCCN 2016008915
Subjects: LCSH: Marine biotechnology. | MESH: Marine Biology | Biotechnology
| Genomics | Molecular Biology | Computational Biology
Classification: LCC TP248.27.M37 M387 2016 | DDC 578.77--dc23
LC record available at https://lccn.loc.gov/2016008915

**Visit the Taylor & Francis Web site at
http://www.taylorandfrancis.com**

**and the CRC Press Web site at
http://www.crcpress.com**

Contents

Section I Introduction to Marine Omics

Section II Marine Genomics

Section III Marine Metagenomics

Section IV   Marine Glycomics

Section V   Marine Transcriptomics

Section VI   Marine Metabolomics

Section VII   Marine Nutrigenomics

Section VIII   Marine Pharmacogenomics

Section IX   Marine Bioinformatics

Section X   Marine Omics and Its Application in Nanotechnology

Section XI   Marine Lipidomics

Section XII   Marine Biocatalysts: Approach and Applications

Section XIII   Marine Foodomics

Section XIV   Marine Toxicogenomics

Preface

This book provides a comprehensive coverage on the current trends in marine omics of various relevant topics such as genomics, lipidomics, proteomics, foodomics, transcriptomics, metabolomics, nutrigenomics, pharmacogenomics, and toxicogenomics as related and applied to marine biotechnology, molecular biology, marine biology, marine microbiology, environmental biotechnology, environmental science, aquaculture, pharmaceutical science, and bioprocess engineering.

This book contains 32 chapters.

Chapter 1 provides the general introduction to the topics covered in this book and deals with marine omics in detail. Chapter 2 deals with algal omics. Chapters 3 through 7 deal with marine genomics and described genomics advances in biosynthetic pathways, fungal genomics, genomics perspective of marine plants, recent advancement and wide-area applications in marine genomics, and puffer genome for tracking human genes.

Furthermore, this book also deals with marine metagenomics. Chapters 8 through 11 discuss metagenomics of sponge and coral holobionts, metagenomics of marine actinomycetes, marine microbial diversity, and gene-targeted metagenomics.

Chapters 12 through 22 discuss marine glycomics, marine transcriptomics, marine metabolomics, marine nutrigenomics, marine pharmacogenomics, and marine bioinformatics. Various topics such as marine invertebrates genomic approach, algal transcriptomics, *de novo* transcriptomics, untargeted metabolomics, and bioinformatics are discussed in detail.

The application part of this book (Chapters 22 through 28) deals with the nanotechnology approach in omics. Finally, Chapters 29 through 32 discuss lipidomics, foodomics, and toxicogenomics in detail.

Se-Kwon Kim
Busan, Republic of Korea

Acknowledgments

I thank CRC Press for their continuous encouragement and suggestions to get this wonderful compilation published. I also extend my sincere gratitude to all the contributors for providing their help, support, and advice to accomplish this task. Further, I thank Dr. Panchanathan Manivasagan and Dr. Jayachandran Venkatesan who worked with me throughout the course of this book project. I strongly recommend this book to marine biotechnologist and omics researchers/students/industrialists and hope that this will help enhance their understanding in this field.

Prof. Se-Kwon Kim
Pukyong National University
Busan, South Korea

Editor

Se-Kwon Kim, PhD, is working as a distinguished professor in the Department of Marine Bio Convergence Science and Technology and the director of Marine Bioprocess Research Center (MBPRC) at Pukyong National University, Busan, South Korea.

He earned his MSc and PhD from Pukyong National University and conducted his postdoctoral studies at the Laboratory of Biochemical Engineering, University of Illinois, Urbana–Champaign, Illinois. Later, he became a visiting scientist at the Memorial University of Newfoundland and University of British Columbia in Canada.

Dr. Kim served as president of the Korean Society of Chitin and Chitosan in 1986–1990 and the Korean Society of Marine Biotechnology in 2006–2007. As credit to his research, he won the best paper award from the American Oil Chemists' Society in 2002. Dr. Kim was also the chairman of the 7th Asia-Pacific Chitin and Chitosan Symposium, which was held in South Korea in 2006. He was the chief editor of the Korean Society of Fisheries and Aquatic Science during 2008–2009. He is also a board member of the International Marine Biotechnology Association (IMBA) and the International Society for Nutraceuticals and Functional Foods (ISNFF).

His major research interests are the investigation and development of bioactive substances from marine resources. His experience in marine bioprocessing and mass-production technologies for the marine bio-industry is the key asset of holding majorly funded marine bio projects in Korea. Furthermore, he expanded his research fields up to the development of bioactive materials from marine organisms for their applications in oriental medicine, cosmeceuticals, and nutraceuticals. To date, he has authored around 700 research papers and 70 books and holds 130 patents.

Contributors

Shiek S.S.J. Ahmed
Faculty of Allied Health Sciences
Chettinad Hospital and Research Institute
Kelambakkam, Tamil Nadu, India

Kumarappan Alagappan
AccuVis Bio
Abu Dhabi, United Arab Emirates

V. Alexandar
Faculty of Allied Health Sciences
Chettinad Academy of Research Education
Kelambakkam, Tamil Nadu, India

Nabeel M. Alikunhi
Coastal and Marine Resources Core Lab
King Abdullah University of Science and
    Technology
Thuwal, Kingdom of Saudi Arabia

Natália Alvarenga
Laboratory of Organic Chemistry and
    Biocatalysis
Institute of Chemistry of São Carlos
University of São Paulo
São Carlos, São Paulo, Brazil

P. Anantharaman
Faculty of Marine Sciences
Annamalai University
Parangipettai, Tamil Nadu, India

Jie Bai
Shenzhen Key Lab of Marine Genomics
Guangdong Provincial Key Lab of
    Molecular Breeding in Marine Economic
    Animals
BGI Academy of Marine Sciences
and
Shenzhen BGI Fisheries Sci & Tech Co., Ltd
Shenzhen, Guangdong, People's Republic
    of China

Chao Bian
Shenzhen Key Lab of Marine Genomics
Guangdong Provincial Key Lab of
    Molecular Breeding in Marine Economic
    Animals
BGI Academy of Marine Sciences
and
Shenzhen BGI Fisheries Sci & Tech Co., Ltd
Shenzhen, Guangdong, People's Republic
    of China

Willian G. Birolli
Laboratory of Organic Chemistry and
    Biocatalysis
Institute of Chemistry of São Carlos
University of São Paulo
São Carlos, São Paulo, Brazil

P.V. Bramhachari
Department of Biotechnology and Botany
Krishna University
Machilipatnam, Andhra Pradesh, India

Pranjal Chandra
Amity Institute of Biotechnology
Amity University
Noida, Uttar Pradesh, India

Jie Chen
State Key Laboratory of Microbial
    Metabolism
Shanghai Jiao Tong University
Minhang, Shanghai, People's Republic of
    China

Maria J. Chen
Department of Molecular and Cell Biology
University of Connecticut
Storrs, Connecticut

Thomas T. Chen
Department of Molecular and Cell Biology
University of Connecticut
Storrs, Connecticut

Katarzyna Chojnacka
Department of Chemistry
Wroclaw University of Technology
Wroclaw, Poland

E. Dilipan
Faculty of Marine Sciences
Annamalai University
Parangipettai, Tamil Nadu, India

Jacinta S. D'Souza
Department of Biology
UM DAE Centre for Excellence in basic
    Sciences
University of Mumbai
Mumbai, India

Xiaodong Fang
BGI Genomics Co., Ltd
Shenzhen, Guangdong, People's Republic
    of China

Irlon M. Ferreira
Laboratory of Organic Chemistry and
    Biocatalysis
Institute of Chemistry of São Carlos
University of São Paulo
São Carlos, São Paulo, Brazil

Yuki Fujii
Laboratory of Glycobiology and Marine
    Biochemistry
Yokohama City University
Yokohama, Japan

Jiharu Hamako
Laboratory of Glycobiology and Marine
    Biochemistry
Yokohama City University
Yokohama, Japan

Imtiaj Hasan
Laboratory of Glycobiology and Marine
    Biochemistry
Yokohama City University
Yokohama, Japan

Masahiro Hosono
Laboratory of Glycobiology and Marine
    Biochemistry
Yokohama City University
Yokohama, Japan

Dhinakarasamy Inbakandan
Centre of Ocean Research
Sathyabama University
Chennai, Tamil Nadu, India

Sougata Jana
Department of Pharmaceutics
Gupta College of Technological Sciences
Asansol, West Bengal, India

Subrata Jana
Department of Chemistry
V.E.C. Sarguja University
Ambikapur, Chhattisgarh, India

Bhavanath Jha
Division of Marine Biotechnology and
    Ecology
Central Salt & Marine Chemicals Research
    Institute
Bhavnagar, Gujarat, India

Pavan P. Jutur
Centre for Advanced Bioenergy Research
International Centre for Genetic
    Engineering and Biotechnology
New Delhi, India

Robert A. Kanaly
Laboratory of Glycobiology and Marine
    Biochemistry
Yokohama City University
Yokohama, Japan

Sivakumar Kannan
Faculty of Marine Sciences
Annamalai University
Parangipettai, Tamil Nadu, India

Ramachandran Karthik
Department of Medical Biotechnology
Chettinad Hospital and Research Institute
Kelambakkam, Tamil Nadu, India

Valliappan Karuppiah
Marine Biotechnology Laboratory
Shanghai Jiao Tong University
Minhang, Shanghai, People's Republic of
    China

K. Kathiresan
Faculty of Marine Sciences
Annamalai University
Parangipettai, Tamil Nadu, India

Sarkar M.A. Kawsar
Laboratory of Glycobiology and Marine
    Biochemistry
Yokohama City University
Yokohama, Japan

S.S. Khora
Division of Medical Biotechnology
VIT University
Vellore, Tamil Nadu, India

Se-Kwon Kim
Department of Marine-Bio Convergence
    Science
and
Marine Bioprocess Research Center
Pukyong National University
Busan, South Korea

Yasuhiro Koide
Laboratory of Glycobiology and Marine
    Biochemistry
Yokohama City University
Yokohama, Japan

Ozcan Konur
Department of Materials Engineering
Yildirim Beyazit University
Ankara, Turkey

D. İpek Kurtböke
Faculty of Science, Health, Education and
    Engineering
University of the Sunshine Coast
Maroochydore DC, Queensland,
    Australia

Zhiyong Li
Marine Biotechnology Laboratory
Shanghai Jiao Tong University
Minhang Shanghai, People's Republic
    of China

Yue Liang
Department of Biotechnology and Life
    Science
Tokyo University of Agriculture and
    Technology
Tokyo, Japan

Chun-Mean Lin
Department of Molecular and Cell Biology
University of Connecticut
Storrs, Connecticut

Jay H. Lo
Department of Molecular and Cell Biology
University of Connecticut
Storrs, Connecticut

Ming-Wei Lu
Department of Aquaculture
National Taiwan Ocean University
Keelung, Taiwan, Republic of China

Yoshiaki Maeda
Department of Biotechnology and Life
    Science
Tokyo University of Agriculture and
    Technology
Tokyo, Japan

Taei Matsui
Laboratory of Glycobiology and Marine
    Biochemistry
Yokohama City University
Yokohama, Japan

Izabela Michalak
Department of Chemistry
Wroclaw University of Technology
Wroclaw, Poland

Avinash Mishra
Division of Marine Biotechnology and
    Ecology
Central Salt & Marine Chemicals Research
    Institute
Bhavnagar, Gujarat, India

Ana M. Mouad
Laboratory of Organic Chemistry and
    Biocatalysis
Institute of Chemistry of São Carlos
University of São Paulo
São Carlos, São Paulo, Brazil

Pagolu Navya
Division of Medical Biotechnology
VIT University
Vellore, Tamil Nadu, India

Asha A. Nesamma
Center for Advanced Bioenergy
    Research
International Centre for Genetic
    Engineering and Biotechnology
New Delhi, India

E.P. Nobi
Ministry of Environment, Forest and
    Climate Change
New Delhi, India

Yukiko Ogawa
Laboratory of Glycobiology and Marine
    Biochemistry
Yokohama City University
Yokohama, Japan

Yasuhiro Ozeki
Laboratory of Glycobiology and Marine
    Biochemistry
Yokohama City University
Yokohama, Japan

Sivasankar Palaniappan
Faculty of Marine Sciences
Annamalai University
Parangipettai, Tamil Nadu, India

J. Paniagua-Michel
Department of Marine Biotechnology
Centro de Investigaction Cientifica y de
    Educacion Superior de Ensenada
Ensenada, Mexico

Manish Kumar Patel
Division of Marine Biotechnology and
    Ecology
Central Salt & Marine Chemicals Research
    Institute
Bhavnagar, Gujarat, India

Chao Peng
Shenzhen Key Lab of Marine Genomics
Guangdong Provincial Key Lab of
    Molecular Breeding in Marine Economic
    Animals
BGI Academy of Marine Sciences
and
Shenzhen BGI Fisheries Sci & Tech Co., Ltd
Shenzhen, Guangdong, People's Republic
    of China

Leonel Pereira
Department of Life Sciences/MARE UC
    and IMAR
Faculty of Sciences and Technology
University of Coimbra
Coimbra, Portugal

Thirunavukkarasu Periyasamy
Department of Aquaculture
National Taiwan Ocean University
Keelung, Taiwan, Republic of China

André L.M. Porto
Laboratory of Organic Chemistry and
    Biocatalysis
Institute of Chemistry of São Carlos
University of São Paulo
São Carlos, São Paulo, Brazil

Munish Puri
Centre for Chemistry and Biotechnology
Deakin University
Geelong, Victoria, Australia

Sultana Rajia
Laboratory of Glycobiology and Marine
    Biochemistry
Yokohama City University
Yokohama, Japan

G.P.C.N. Raju
Emory University
Atlanta, Georgia

Avinesh R. Reddy
Centre for Chemistry and
    Biotechnology
Deakin University
Geelong, Victoria, Australia

Balamurugan Sadaiappan
Department of Biotechnology
Sri Sankara Arts and Science College
Enathur, Kanchipuram, Tamil Nadu, India

Kandasamy Saravanakumar
Department of Environmental and
    Resource
Shanghai Jiao Tong University
Minhang, Shanghai, People's Republic
    of China

Ramachandran Saravanan
Faculty of Allied Health Sciences
Chettinad Hospital and Research
    Institute
Kelambakkam, Tamil Nadu, India

Olga N. Sekurova
Department of Pharmacognosy
University of Vienna
Vienna, Austria

Kashif M. Shaikh
Center for Advanced Bioenergy
    Research
International Centre for Genetic
    Engineering and Biotechnology
New Delhi, India

Sushrut Sharma
Amity Institute of Biotechnology
Amity University Uttar Pradesh
Noida, Uttar Pradesh, India

Qiong Shi
Shenzhen Key Lab of Marine Genomics
Guangdong Provincial Key Lab of
    Molecular Breeding in Marine Economic
    Animals
BGI Academy of Marine Sciences
and
Shenzhen BGI Fisheries Sci & Tech Co., Ltd
and
BGI Education Center
Graduate University of Chinese Academy
    of Sciences
Shenzhen, Guangdong, People's Republic
    of China

V.L. Sirisha
Department of Biology
UM DAE Centre for Excellence in Basic
    Sciences
University of Mumbai
Mumbai, India

Ananya Srivastava
Department of Chemistry
Indian Institute of Technology
New Delhi, India

Renesha Srivastava
Amity Institute of Biotechnology
Amity University Uttar Pradesh
Noida, Uttar Pradesh, India

Poongodi Subramaniam
Faculty of Marine Sciences
Annamalai University
Parangipettai, Tamil Nadu, India

Vasuki Subramanian
Faculty of Marine Sciences
Annamalai University
Parangipettai, Tamil Nadu, India

Venkataramanan Subramanian
Biosciences Center
National Renewable Energy Laboratory
Golden, Colorado

Shigeki Sugawara
Laboratory of Glycobiology and Marine
    Biochemistry
Yokohama City University
Yokohama, Japan

Pankaj Suman
Amity Institute of Biotechnology
Amity University Uttar Pradesh
Noida, Uttar Pradesh, India

Ying Sun
Section on Marine BioBank
China National GeneBank
Shenzhen, Guangdong, People's Republic
    of China

Tsuyoshi Tanaka
Department of Biotechnology and Life
    Science
Tokyo University of Agriculture and
    Technology
Tokyo, Japan

T. Thangaradjou
Department of Science & Technology
Science and Engineering Research Board
New Delhi, India

Antonio Trincone
National Research Council
Institute of Biomolecular Chemistry
Naples, Italy

Chun-Hsi Tso
Department of Aquaculture
National Taiwan Ocean University
Keelung, Taiwan, Republic of China

Jayachandran Venkatesan
Department of Marine-Bio Convergence
    Science
and
Marine Bioprocess Research Center
Pukyong National University
Busan, South Korea

Xinxin You
Shenzhen Key Lab of Marine Genomics
Guangdong Provincial Key Lab of
    Molecular Breeding in Marine Economic
    Animals
BGI Academy of Marine Sciences
and
Shenzhen BGI Fisheries Sci & Tech Co., Ltd
Shenzhen, Guangdong, People's Republic
    of China

Basit Yousuf
Division of Marine Biotechnology and
    Ecology
Central Salt & Marine Chemicals Research
    Institute
Bhavnagar, Gujarat, India

Xinhui Zhang
Shenzhen Key Lab of Marine Genomics
Guangdong Provincial Key Lab of
    Molecular Breeding in Marine Economic
    Animals
BGI Academy of Marine Sciences
and
Shenzhen BGI Fisheries Sci & Tech Co., Ltd
Shenzhen, Guangdong, People's Republic
    of China

Sergey B. Zotchev
Department of Pharmacognosy
University of Vienna
Vienna, Austria

Section I

Introduction to Marine Omics

1

Introduction to Marine Omics

Se-Kwon Kim and Jayachandran Venkatesan

CONTENTS

1.1 Overview of the Chapter Contents in This Book

Marine biotechnology has been an emerging area of research in the last five decades, which provides abundant resource materials for human consumption. Omics in marine technology changed several aspects of research at the molecular level. Different kinds of omics, such as nutrigenomics, proteomics, transcriptomics, genomics, and metabolomics, are important for the development of marine biotechnology industries (Qin et al. 2012). In recent years, significant research have been conducted on omics, which are presented by Dr. Ozcan Konur in Chapter 2. The main points of the chapter are as follows. The development of algal biofuels as a third-generation biofuel has been considered as a major solution for global problems. The design and production of algal biofuels, such as algal biohydrogen and algal biodiesel, and algal biocompounds, such as algal drugs and algal nutritional ingredients, have also been an important research area in recent years. Based on the results from studies on algal omics, the issues relating to the optimization of algal production through omic analysis have also emerged as an important research area in particular. Although there have been many studies on algal omics and on omics and there have been a limited number of scientometric studies on algal biofuels (Konur 2011). In Chapter 3, Professor J. Paniagua-Michel discusses the increasing need for alternatives to fossil hydrocarbon-based chemicals, and the recent advances in omics, and metabolic engineering, are promissory options for the development of biobased solutions for bioproducts and cell factories. Recent advances in omics in microalgae are now orientated toward the optimization and diversification of the production and application of bioproducts. Omic approaches contribute to redefine metabolism, functional diversity, and biosynthetic capacity of fluxes of compounds of strains into a desired biosynthetic pathway. During the last couple of decades, microalgae have been considered as a suitable and valuable feedstock for biofuels, pharmaceuticals, nutraceuticals, biomedicals, and material applications (Guarnieri and Pienkos 2015; Paniagua-Michel et al. 2015). Fungal genomics is explained further by Saravanan et al. in Chapter 4. In Chapter 5, Dr. Dilipan et al. explained the genomic perspective of marine plants with special reference to seagrass, reviewing the molecular taxonomy and evolutionary lineages of seagrass as different molecular markers that could resolve taxonomic plasticity. Although their use in these studies seems convincing, the origin and evolutionary lineage of seagrass, salt marsh, and freshwater should be studied in detail. Furthermore, studies on different molecular markers need to be performed from different coastal areas to know the origin of seagrass.

Chapter 6 describes the recent advancement of whole genome sequencing of marine vertebrates, including teleost fishes, marine reptiles, seabirds, and marine mammals, and overviews the recent advancement and wide-area applications of whole genome sequencing in marine vertebrates, ranging from fishes, reptiles, and birds to mammals. In the past 3 years, genome sequencing has been extensively applied to a large number of marine species, although only around 10 reports are available, and also the chapter deals with few unpublished genomic information of several marine animals, for example, mudskippers (amphibious fishes) and penguins (an attractive example of the seabirds), for a wide range of comparison. So far, marine genomics has been utilized mainly for marker-assisted molecular breeding, development of marine medicine, and conservation of marine species. In Chapter 7, Professor S.S. Khora explains about the puffer fish genome that can provide a more distant evolutionary comparison (400 million years versus 100 million years for mouse and rat), which permits a more accurate triangulation of genome function than a mouse or rat alone. Genomic features that are common to puffer fish, rodents, and human will focus on the essential core genes that define being a vertebrate. Further, Karuppiah et al. explained about coral holobiont omics: microbes and dinoflagellates. Coral holobiont refers to the collective community of coral host and its metazoan, protist, and microbial symbionts. The significance of association between symbionts and their invertebrate hosts is becoming increasingly apparent. Meta-analysis determines the presence of symbionts associated with healthy, diseased, or bleached coral as well as whether coral reef habitats harbor clusters of distinct taxa. The relationship between coral and its symbionts represents a valuable model that can be applied to the broader discipline of invertebrate–symbionts. This chapter focuses on the omics of microbes, viruses, and dinoflagellates in coral holobionts, including background, techniques, application and latest progress, and future perspectives (Lee et al. 2012). In Chapter 9, Professor İpek Kurtböke explains about metagenomics of marine actinomycetes. Metagenomics proved to be a powerful approach to both the analysis of microbial communities and the discovery of novel enzymes and bioactive compounds. However, the success of the latter is often hampered by the large size of libraries needed to be screened, fragmentation of DNA encoding large biosynthetic gene clusters, and absence of functional expression of genes in heterologous hosts. Targeting metagenomes of marine actinomycetes, which have emerged as a promising source of new bioactive compounds, may circumvent some of these problems. The development of novel approaches to the actinomycete-targeted metagenomics coupled with the state-of-the-art technologies (e.g., synthetic biology) is, however, needed to fully explore and exploit this potential. This chapter provides an overview of marine actinomycetes and their importance in biodiscovery of new pharmaceuticals. It also provides examples on how metagenomics can be used to reveal the true biosynthetic potential of marine actinomycetes. Furthermore, it highlights recent advances in the discovery of novel bioactive molecules based on metagenomics and suggests how this process can be streamlined. The latter includes a brief overview of DNA assembly methods and construction of specialized hosts for the heterologous expression of biosynthetic genes for bioactive compounds. The last section offers an insight into future perspectives of metagenomics targeted to marine actinomycetes for biodiscovery of new drug leads. In the consecutive chapter, Dr. Sadaiappan et al. explain in detail marine microbial diversity and its pyrosequencing.

Chapter 11 explains gene-targeted metagenomics for the study of biogeochemical cycling from coastal–saline ecosystems as well as the detail about carbon, nitrogen, and sulphur metabolism. In Chapter 14, Dr. Karthik, with Prof. Saravanan et al., describes the study of Indian marine molluscs in glycomic approach.

> The marine glycome is highly beneficial to mankind in one way by possessing economic importance and nutritive value or the other by producing molecules along with proteins and lipids of medicinal significance. The economic value of these glycans (GLY) is evident by their consumption across the globe and their functioning as an active ingredient of the cosmetic industry. The medicinal importance of the marine molluscan GLY is expanding with time, as numerous mollusc derived drugs against conditions of pain, cancer are already in the pharmaceutical market. The structural characterization of marine GLY has been a constant barrier towards the quest of developing fast glycome based drugs, primarily due to high structural complexity attributed by the heavy branching of sugars. Further, technological access and undersophistication of instrumentation in the Indian context has largely restricted the glycome characterization and analysis to the western World. Thus the current chapter holds importance, where molluscan proteoglycans (PG) from the coast of South India has been studied and characterized structurally using Fourier Transform Infrared (FT-IR), Circular Dichroism (CD) spectroscopy and Matrix Assisted Laser Desorption Ionization (MALDI-TOF) mass spectroscopy (MS). The crude PG of the posterior salivary gland of *S. prashadii* was fractionated using Sephadex G-50 gel filtration chromatography. The fractionated PG was purified using reverse phase high performance liquid chromatography (RP-HPLC) over a retention time of 30 minutes. The total protein and neutral sugar concentration of the active purified fraction was 7.85 mg/g and 69.52 μg/g of gland respectively, whereas CD spectrum manifested 59.78% of α-helix and 4.38% of β-sheet in PG secondary structure. MALDI-TOF MS analysis of the PG in positive ion mode recorded molecular mass up ≥2000 Da. Further studies are required to authorize the economic and medicinal importance of low molecular weight PG from cuttlefish.

Chapter 15 deals with transcriptomics in marine macroalgae. The main approach of the chapter is the next-generation renewable biofuels (Jutur and Nesamma 2015; Pavan and Asha 2015). In Chapter 16, Dr. Lu et al. use the transcriptomic approaches to study the pathogenesis of the nervous necrosis virus.

In Section VI, Dr. Patel et al. (Chapter 17) describe in detail the untargeted metabolomics of halophytes and their future perspective. Metabolome analysis of halophytes has become an invaluable tool for the study of metabolic changes occurring during abiotic stresses. Tools, such as mass spectrometry, allow the identification of metabolites and pathways. Integration of "omics" data, such as genomics, transcriptomics, and proteomics with metabolomics, reveals the biochemical pathways from gene to metabolites. These studies are necessary to fill the gap of cellular processes for the metabolic regulatory network. The database of metabolites from halophytes is expected to integrate with transcriptomics and proteomics for the discovery of novel metabolic pathways in the future.

Chapter 21 explains the approach of lectin genomics, and Wu et al. (2014) showed that an sea urchin egg lectin/rhamnose-binding lectins (SUEL/RBL) with triple tandem repeating domains isolated from American purple sea urchin (*Strongylocentrotus purpuratus*) and a fucose-binding lectin isolated from European sea bass (*Dicentrarchus labrax*) both cause apoptosis of cancer cells through protein methyltransferase-5 and transcription factor E2F. These lectins, exogenously induced through a replication-deficient adenovirus vector engineered to carry their respective genes, significantly suppressed levels of anti-apoptosis factors such as Bcl-2 and X-linked inhibitor of apoptosis protein. However, the lectins did not enhance caspase activation (Wu et al. 2014).

Bioinformatics is an important tool in marine biotechnology and is explained in detail in Chapter 22. In recent years, continuous development has been achieved in nanotechnology in various fields, concerning marine biotechnology, and nanoapproach is recently growing and will have a huge impact in the future. Dr. Nabeel et al. explain the nanotechnology approach for marine omics in Chapter 23. Ranesha Srivastava et al. explain the application of nanotechnology in omics in nanodiagnostics.

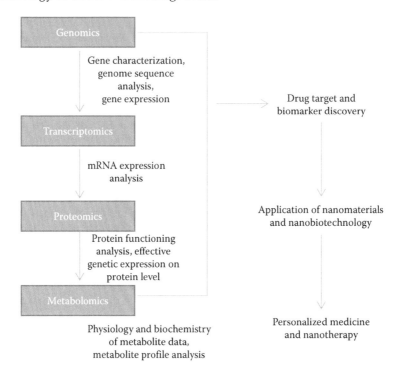

Dr. Tsuyoshi Tanaka et al. describe the general marine microalgal lipid compounds and their function were first introduced. Then, the concept and development of lipidomics, the common techniques used for lipidomic studies, and the basic application of each technique were reviewed. In the same section, several newly developed MS and chromatography for lipid analyses were also briefly introduced. Finally, the recent development of microalgal lipidomics was summarized. Due to technological difficulties, the profiling and quantification of the overall microalgal lipidome are still under development. However, the current technology can still provide high-quality data to facilitate new discoveries on the metabolism and function of microalgal lipids (Yamada et al. 2013). Environmental problem is a huge concern in humans, including global warming. Dr. Inbakandan explains about predicting the impact of ocean acidification by omics perspective in detail in Chapter 28.

Dr. P.V. Brahmachari et al. describe current advances in biotechnology-driven marine microbial metagenomics. The huge potential in diversity of marine life is still not fully exploited due to the complexity in culturing microorganisms under laboratory conditions. They emphasized the importance of using a metagenomic technique to access the uncultured majority of marine microbial communities. They also report the current biotechnological advances of several novel biocatalysts, enzymes, proteins, peptides, and natural products isolated by sequence-based and function-screening strategies assisted by next-generation sequencing technology. The potential future perspectives of metagenomics in bioprospecting for novel gene clusters, novel biocatalysts and bioactive compounds, and uncharacterized metabolism increased the complexity of biogeochemical pathways.

References

Guarnieri, M.T. and P.T. Pienkos. 2015. Algal omics: Unlocking bioproduct diversity in algae cell factories. *Photosynthesis Research* 123(3):255–263.

Jutur, P.P. and A.A. Nesamma. 2015. Chapter 24—Genetic engineering of marine microalgae to optimize bioenergy production. In *Handbook of Marine Microalgae*, S.-K. Kim (ed.). Boston, MA: Academic Press.

Konur, O. 2011. The scientometric evaluation of the research on the algae and bio-energy. *Applied Energy* 88(10):3532–3540.

Lee, O.O., J. Yang, S. Bougouffa et al. 2012. Spatial and species variations in bacterial communities associated with corals from the Red Sea as revealed by pyrosequencing. *Applied and Environmental Microbiology* 78(20):7173–7184.

Paniagua-Michel, J., J.O. Soto, and E. Morales-Guerrero. 2015. Drugs and leads from the ocean through biotechnology. In *Springer Handbook of Marine Biotechnology*, S.-K. Kim (ed.). Berlin, Germany: Springer, pp. 711–729.

Pavan, P.J. and A.N. Asha. 2015. Marine microalgae: Exploring the systems through an omics approach for biofuel production. In *Marine Bioenergy*, S.-K. Kim (ed.). CRC Press Editor. CRC Press, Boca Raton, FL.

Qin, S., S. Watabe, and H.Z. Lin. 2012. Omics in marine biotechnology. *Chinese Science Bulletin* 57(25):3251–3252.

Wu, L., X. Yang, X. Duan, L. Cui, and G. Li. 2014. Exogenous expression of marine lectins DlFBL and SpRBL induces cancer cell apoptosis possibly through PRMT5-E2F-1 pathway. *Scientific Reports* 4:4505.

Yamada, T., T. Uchikata, S. Sakamoto et al. 2013. Supercritical fluid chromatography/Orbitrap mass spectrometry based lipidomics platform coupled with automated lipid identification software for accurate lipid profiling. *Journal of Chromatography A* 1301:237–242.

2

Algal Omics: The Most-Cited Papers

Ozcan Konur

CONTENTS

2.1 Overview

2.1.1 Issues

Some of the most important public policy issues in recent years relate to global warming, air pollution, and energy security (e.g., Jacobson 2009, Wang et al. 2008, Yergin 2006). With the increasing global population, food security has also become a major public policy issue (e.g., Lal 2004). The development of biofuels generated from the biomass has been a long-awaited solution to these global problems (e.g., Demirbas 2007, Goldemberg 2007, Lynd et al. 1991). However, the development of the early biofuels produced from the agricultural plants, such as sugarcane (e.g., Goldemberg 2007), and agricultural wastes, such as corn stovers (e.g., Bothast and Schlicher 2005), has resulted in a series of substantial concerns about the food security (e.g., Godfray et al. 2010).

Therefore, the development of algal biofuels as a third-generation biofuel has been considered as a major solution for the global problems of global warming, air pollution, energy security, and food security (e.g., Chisti 2007, Demirbas 2007, Kapdan and Kargi 2006, Spolaore et al. 2006, Volesky and Holan 1995).

The design of production of algal biofuels, such as algal biohydrogen and algal biodiesel, and algal biocompounds, such as algal drugs and algal food ingredients, has also been an important research area in recent years (e.g., Cardozo et al. 2007, Grima et al. 2003,

Pulz 2001). Based on the results from the studies on the algal omics, the issues relating to the optimization of the algal biocompound and algal biofuel production through omic analysis have also emerged as an important research area in particular (e.g., Badger and Price 2003, Badger et al. 2006, Fulda et al. 2000, Hackett et al. 2007, Keller et al. 2005, 2007, Lemaire et al. 2004, Lindell et al. 2004, Palenik et al. 2006, Price et al. 2008).

For example, Borowitzka (1999) argues that the "main problem facing the commercialization of new microalgae and microalgal products is the need for closed culture systems and the fact that these are very capital intensive. The high cost of microalgal culture systems relates to the need for light and the relatively slow growth rate of the algae." Similarly, Pulz (2001) argues that the design of the technical and technological basis for photobioreactors for the production of algal biofuels and biocompounds is the "most important issue for economic success in the field of phototrophic biotechnology."

Li et al. (2008) argue that the production of algal biofuels can be coupled with flue gas CO_2 mitigation, wastewater treatment, and the production of high-value chemicals. They assert that the developments in microalgal cultivation and downstream processing (e.g., harvesting, drying, and thermochemical processing) would further "enhance the cost-effectiveness of the biofuel from microalgae strategy." Similarly, Greenwell et al. (2010) argue that "simulations that incorporate financial elements, along with fluid dynamics and algae growth models, are likely to be increasingly useful for predicting reactor design efficiency and life cycle analysis to determine the viability of the various options for large-scale culture." They argue that the greatest potential for cost reduction and increased yields most probably lies within closed or hybrid closed-open production systems.

Although there have been many studies on the algal omics (e.g., Anderson et al. 2012, Badger and Price 2003, Badger et al. 2006, Fulda et al. 2000, Goodstein et al. 2012, Hackett et al. 2007, Hagemann 2011, Keller et al. 2005, 2007, Leliaert et al. 2012, Lemaire et al. 2004, Lindell et al. 2004, Palenik et al. 2006, Price et al. 2008, Wu et al. 2011) and on the omics as a whole (e.g., Arumugam et al. 2011, Bartel 2004, Cox et al. 2011, Gentleman et al. 2004, Gerlinger et al. 2012, Heid et al. 1996, Knox et al. 2011, Kumar et al. 2004, Schwanhausser et al. 2011, Tamura et al. 2007) and there have been a limited number of scientometric studies on the algal biofuels (Konur 2011), there has not been any study on the citation classics in the algal omics as in the other research fields as highly cited papers (e.g., Baltussen and Kindler 2004a,b, Dubin et al. 1993, Gehanno et al. 2007, Konur 2012a,b, 2013, Paladugu et al. 2002, Wrigley and Matthews 1986).

As North's New Institutional Theory suggests, it is important to have up-to-date information about the current public policy issues to develop a set of viable solutions to satisfy the needs of all the key stakeholders (North 1994, Konur 2000, 2002a,b,c, 2006a,b, 2007a,b, 2012c,d).

Therefore, following a scientometric overview of the research in omics in general and algal omics in particular, the brief information on a selected set of 25 citation classics in the field of the algal omics is presented in this chapter to inform the key stakeholders about omic design of algae for the efficient production of algal biofuels and biocompounds, such as algal drugs and algal food ingredients, for the solution of these problems in the long run, thus complementing a number of recent scientometric and review studies on the biofuels and global energy research (Konur 2011, 2012e–p, 2013, 2015a–m, 2016a–b). The scientometric overview of the research in omics at large and algal omics in particular was necessary since there have been limited number of scientometric studies in omics in the literature as indexed by the SSCI or SCIE as of July 2014 (Kuzhabekova and Kuzma 2014, Mansiaux and Carrat 2012, Wonkam et al. 2011).

2.1.2 Methodology

A two-stage search strategy was applied in this chapter. At the first stage, a search on the omics in great detail was carried out. For this purpose, a search was carried out in the SCIE and SSCI databases (version 5.14) in July 2014 to locate the papers relating to the omics at length using the keyword and journal set of Source = (*Annual Review of Genomics and Human Genetics* or *Briefings in Functional Genomics* or *Public Health Genomics* or *Metabolomics* or *Pharmacogenetics and Genomics* or *Journal of Proteomics* or *Journal of Nutrigenetics and Nutrigenomics* or *Expert Review of Proteomics* or *BMC Medical Genomics* or *Metallomics* or *Proteomics Clinical Applications* or *Comparative Biochemistry and Physiology D-Genomics & Proteomics* or *Current Proteomics* or *Journal of Genetics and Genomics* or *Marine Genomics* or *Plant Omics* or *Genes & Genomics* or *Molecular & Cellular Proteomics* or *Pharmacogenomics Journal* or *BMC Genomics* or *Pharmacogenomics* or *Genomics* or *Physiological Genomics* or *Proteomics* or *Biochimica et Biophysica Acta—Proteins and Proteomics* or *Functional & Integrative Genomics* or *Molecular Genetics and Genomics* or *Current Genomics* or *Comparative and Functional Genomics* or *Genome Biology* or *Genome Research* or *Tree Genetics & Genomes*) or Topic = (*omics) and not Topic = (economics or ergonomics) in the abstract pages of the papers. The key bibliometric data were extracted from this search for the overview of the omic literature in a general way.

At the second stage, a search on the algal omics in particular was carried out. For this purpose, a search was carried out in the SCIE and SSCI databases (version 5.14) in July 2014 to locate the papers relating to the algal omics using the keyword set of Topic = ((alga* or microalga* or macroalga* or cyanobacter*) and (*omics*)) in the abstract pages of the papers. The key bibliometric data were extracted from this search for the overview of the algal omic literature at large. It was necessary to focus on the key references by selecting articles and reviews.

The located highly cited 25 papers were arranged in the order of the decreasing number of citations. The summary information about the located citation classics is presented in the order of the decreasing number of citations for each topical area.

The information relating to the document type, affiliation of the authors, the number and gender of the authors, the country of the authors, the journal where the paper was published, the subject area of the journal where the paper was indexed, the concise topic of the paper, and the total number of citations received for the paper for the Web of Science and Google Scholar is given in a table for each paper.

2.1.3 Research on the Omics: Overview

It is essential first to consider the broader omic research before analyzing the research on the algal omics, which is a subset of the broader literature, to explore the research trends in the field of omics.

Using the keywords and the selected journal set related to the omics, 151,829 references were located in omics. Of these references, 129,858 were articles and reviews comprising 85.5% of the sample. Meeting abstracts, editorial materials, and other types of references formed the remaining part of the sample. The three most prolific authors, Y. Zhang, Y. Wang, and J. Wang, produced between 392 and 522 papers each. The list of the most prolific authors was dominated by the Asian and the US authors.

The most prolific country in terms of the number of publications is the United States with 54,149 papers forming 41.7% of the sample. Germany, England, and China followed the United States with 9.1%, 9.0%, and 8.5% of the sample, respectively. English was the dominant language of scientific communication in omics comprising 98.6% of the sample.

The most prolific institution was the *National Institutes of Health* (NIH) of the United States with 3405 papers, comprising 2.6% of the sample. *Harvard University* of the United States, *National Scientific Research Center* (CNRS) of France, and *Chinese Academy of Sciences* of China followed the top institution with 3405, 3177, and 2477 papers, respectively.

Similar to the most recent dynamic research areas like nanoresearch or bioinformatic research, the omic research has been a solely recent phenomenon as 92.6% of the papers were published after 2000. The share of the research in the 1980s, 1990s, 2000s, and 2010s was 0.9%, 6.4%, 46.5%, and 46.1%, respectively. These findings suggest that the omic research has been a recent research area as the research boomed in the 2000s and 2010s with the steady increase in the number of publications over time jumping in 2000 from 1337 to 1847, although the research has been carried out since the 1980s.

The most prolific journal in terms of the number of publications was *Genomics*, publishing only 5.2% of the sample. *BMC Genomics*, *Proteomics*, and *Genome Research* followed the top journal with 5.0%, 3.8%, and 2.7% of the sample, respectively. It is notable that the contribution of the omic journals to the research at the top of the list in omics was significant.

The most prolific subject category in terms of the number of publications was *Genetics Heredity* with 34,257 papers forming 26.4% of the sample. *Biochemistry Molecular Biology*, *Biotechnology Applied Microbiology*, and *Biochemical Research Methods* followed the top subject category with 23.0%, 21.3%, and 16.3% of the sample, respectively. These findings suggest that these top 4 subject categories share a common set of journals where a journal was indexed under more than one subject category.

The most-cited papers in omics were mostly related to the development of omics and its application in a number of areas in general, such as the development of the computer software programs and quantitative methods for the genomic analysis (Bartel 2004, Gentleman et al. 2004, Heid et al. 1996, Kumar et al. 2004, Tamura et al. 2007). For example, the most-cited paper with 17,993 citations was related to discussion of molecular evolutionary genetics analysis (MEGA) software version 4.0 (Tamura et al. 2007). Bartel (2004) discusses the genomics, biogenesis, mechanism, and function of the micro-RNAs in a review paper with 9338 citations. Kumar et al. (2004) discuss the integrated software for MEGA and sequence alignment in a paper with 9317 citations. Gentleman et al. (2004) discuss the bioconductor, the open software development for computational biology, and bioinformatics in a paper with 4174 citations. Heid et al. (1996) discuss the development of the real-time quantitative polymerase chain reaction method in a paper with 3632 citations. It is notable that these top citation classics in omics were published in the high-impact genomic journals.

Contrary to the citation classics, the hottest papers in omics published in the last 3 years between 2011 and 2013 were related mostly to the application of the omics in the medicine (Arumugam et al. 2011, Cox et al. 2011, Gerlinger et al. 2012, Knox et al. 2011, Schwanhausser et al. 2011). For example, the hottest paper with 934 citations is related to the intratumor heterogeneity and branched evolution revealed by multiregion sequencing (Gerlinger et al. 2012). Arumugam et al. (2011) investigate the enterotypes of the human gut microbiome in a paper with 739 citations. Schwanhausser et al. (2011) investigate the global quantification of mammalian gene expression control in a paper with 642 citations. Knox et al. (2011) discuss the DrugBank 3.0, a comprehensive resource for *omic* research on drugs in a paper with 453 citations. Cox et al. (2011) discuss the Andromeda, a peptide search engine integrated into the MaxQuant environment, in a paper with 361 citations. It is further notable that some of these top hottest papers in omics were published in the high-impact multidisciplinary journals like *Nature* and other high-impact journals like *Journal of Proteome Research*.

2.1.4 Research on the Algal Omics: Overview

Using the keywords related to algal omics, only 1158 references were located in algal omics. Of these references, 1107 were articles and reviews. Meeting abstracts, editorial materials, and other references formed the remaining part of the sample. Considering the fact that there were 129,858 papers in the parent omic research field, the research on the algal omics formed 0.9% of the larger parental data set. This finding suggests that the field of algal omics was a highly specialized field of research with relatively a small sample size and with a specific set of shareholders such as authors, institutions, and countries.

Another three most prolific authors, P. C. Wright, C. Bowler, and S. Bhattacharya, each produced between 12 and 16 papers. The list of the most prolific authors for algal omics was dominated by U.S. Asian, and European authors.

The most prolific country in terms of the number of publications is the United States with 434 papers forming 39.2% of the sample. Germany, France, and China followed the United States with 13.3%, 9.7%, and 8.8% of the sample, respectively. English was the dominant language of scientific communication in algal omics with 99.7% of the sample in English.

The most prolific institution was the *Department of Energy* of the United States with 55 papers. CNRS of France; *Pierre and Marie Curie University, or University Paris VI,* of France; and *Max Planck Society* of Germany followed the top institution with 53, 41, and 37 papers, respectively.

Like the nanoresearch and the parent omic research, the algal omic research has been a recent phenomenon as 97.4% of the papers were published after 2000. The research in algal omics in the 1990s, 2000s, and 2010s formed 2.4%, 36.7%, and 60.7% of the sample, respectively. There was a general increasing trend in the number of papers over time, especially in the 2010s. These findings support the earlier findings that algal omics received public attention in the 2000s and 2010s and would continue to do so in the near future as it meets the needs of the society for the omic research related to the algae.

The most prolific journal in terms of the number of publications was *PLoS One* publishing 2.8% of the sample with 31 papers. *BMC Genomics, Molecular Biology and Evolution,* and *Proceedings of the National Academy of Sciences of the United States of America* followed the top journal with 2.2% each of the sample, respectively. It is notable that the contribution of the journals in omics to the research at the top of the journal list in algal omics was insignificant.

The most prolific subject category in terms of the number of publications was *Plant Sciences* with 189 papers forming 17.1% of the sample. *Biochemistry Molecular Biology, Biotechnology Applied Microbiology,* and *Biochemical Research Methods* followed the top subject category with 25.1%, 14.8%, and 9.9% of the sample, respectively. These findings suggest that these top 4 subject categories share a common set of journals, where a journal was indexed under more than one subject category.

There were 26,789 citations received for 1,107 papers in algal omics. The average citation rate per paper was 24.2 and the H-index was 77. These findings suggest that considering the fact that this research field was a relatively new, the citation impact of 1107 papers in algal omics was relatively significant. The number of citations rose over time, but especially, spectacularly in the 2010s, as 60.1% of the citations were received in the 2010s.

The most-cited papers in algal omics were heavy in the omic analysis of the algae (e.g., Badger and Price 2003, Badger et al. 2006, Fulda et al. 2000, Hackett et al. 2007, Keller et al. 2005, 2007, Lemaire et al. 2004, Lindell et al. 2004, Palenik et al. 2006, Price et al. 2008).

For example, the most-cited paper with 264 citations was related to the CO_2-concentrating mechanisms (CCMs) in cyanobacteria (Badger and Price 2003).

There was a similar trend for the hottest papers published in the last 3 years between 2011 and 2013 (Anderson et al. 2012, Goodstein et al. 2012, Hagemann 2011, Leliaert et al. 2012, Wu et al. 2011). For example, the hottest paper with 229 citations was related to the phytozome, a comparative platform for green plant genomics (Goodstein et al. 2012). Leliaert et al. (2012) discuss the phylogeny and molecular evolution of the green algae in a paper with 67 citations. Anderson et al. (2012) discuss the globally distributed genus *Alexandrium* with the emphasis on the multifaceted roles in marine ecosystems and impacts on human health in a review paper with 64 citations. Hagemann (2011) discusses the molecular biology of cyanobacterial salt acclimation in a review paper with 58 citations. Wu et al. (2011) investigate the in vivo lipidomics of algae using single-cell Raman spectroscopy in a paper with 56 citations.

In the following part, the brief information on the most-cited papers will be provided.

2.2 Research on Algal Omics

2.2.1 Introduction

The research on the omics in general and algal omics in particular has been one of the most dynamic research areas in recent years. Twenty-five citation classics in the field of algal omics with more than 81 citations were located, and the key emerging issues from these papers were presented below in the decreasing order of the number of citations (Table 2.1). These papers give strong hints about the determinants of the optimal design of algae through the algal omics and emphasize that algal omics offer the valuable information about the omic design of the algae for the optimal production of algal biofuels, such as algal biohydrogen and algal diesel, and algal biocompounds, such as algal drugs and algal food ingredients.

The papers were dominated by the researchers from at least 10 countries, usually through the intracountry institutional collaboration, and they were multiauthored. The number of the authors for the papers arranged from 2 to 25. The United States (13 papers), Germany (8 papers), Australia (5 papers), and France and Sweden (2 papers each) were the most prolific countries.

Similarly, all these papers were published in the journals indexed by the SCI and/or SCIE. There was also no paper indexed by the SSCI. The number of citations ranged from 79 to 264 for the Web of Science and from 106 to 380 for the Google Scholar databases. Twenty of the papers were articles, while only 5 of them were reviews.

The papers were published between 1996 and 2009, while 1 and 24 of them were published in the 1990s and 2000s, respectively. This finding supports the earlier findings that the research on the algal omics boomed especially in the 2000s and 2010s, but older papers had more citations.

There was a significant gender deficit among the most-cited papers in algal omics as there were 4 female first authors of 19 papers with known author gender. This issue merits further research.

The most prolific journals were *Proceedings of the National Academy of Sciences of the United States of America* (seven papers), *Journal of Experimental Botany* and *Plant Physiology* (three papers each), and *Molecular Biology and Evolution* and *Molecular and Cellular Proteomics*

TABLE 2.1

Research on the Algal Omics

No.	Paper Ref.	Year	Doc.	Affil.	Country	No. of Authors	M/F	Journal	Subject Area	Topic	Total No. of Citations WN	Total No. of Citations GS
1.	Badger and Price	2003	R	Australian Natl. Univ.	Australia	2	M	*J. Exp. Bot.*	Plant Sci.	CCMs in cyanobacteria	264	380
2.	Lindell et al.	2004	A	MIT, San Diego State Univ.	United States	6	F	*Proc. Natl. Acad. Sci. United States*	Mult. Sci.	Transfer of photosynthesis genes to and from *Prochlorococcus* viruses	213	299
3.	Kettler et al.	2007	A	MIT, Harvard Univ. +4	United States, Germany	14	M	*PLoS Genet.*	Gen. Her.	Patterns and implications of gene gain and loss in the evolution of *Prochlorococcus*	190	215
4.	Keller et al.	2005	A	Univ. Calif San Francisco	United States	5		*Curr. Biol.*	Bioch. Mol. Biol, Cell Biol.	Proteomic analysis of isolated *Chlamydomonas* centrioles	168	215
5.	Price et al.	2008	A	Australian Natl. Univ.	Australia	4	M	*J. Exp. Bot.*	Plant Sci.	Cyanobacterial CCM	137	202
6.	Hackett et al.	2007	A	Univ. Iowa +1	United States	6	M	*Mol. Biol. Evol.*	Bioch. Mol. Biol., Evol. Biol., Gen. Her.	Monophyly of cryptophytes and haptophytes	131	189
7.	Badger et al.	2006	A	Australian Natl. Univ.	Australia	4	M	*J. Exp. Bot.*	Plant Sci.	Environmental plasticity and ecological genomics of the cyanobacterial CCM	131	178

(Continued)

TABLE 2.1 (*Continued*)

Research on the Algal Omics

No.	Paper Ref.	Year	Doc.	Affil.	Country	No. of Authors	M/F	Journal	Subject Area	Topic	Total No. of Citations WN	Total No. of Citations GS
8.	Lemaire et al.	2004	A	Univ. Paris 11	France	6		*Proc. Natl. Acad. Sci. United States*	Mult. Sci.	New thioredoxin targets in the unicellular photosynthetic eukaryote *C. reinhardtii*	128	161
9.	Palenik et al.	2006	A	Univ. Calif., San Diego +1	United States	17	M	*Proc. Natl. Acad. Sci. United States*	Mult. Sci.	Genome sequence of *Synechococcus* CC9311	115	159
10.	Fulda et al.	2000	A	Univ. Rostock, Stockholm Univ. +1	Sweden, Germany	5	F	*Eur. J. Biochem.*	Bioch. Mol. Biol.	Proteomics of *Synechocystis* sp. strain PCC 6803	110	167
11.	Van Mooy et al.	2006	a	Univ. Washington +1	United States	5	M	*Proc. Natl. Acad. Sci. United States*	Mult. Sci.	Impact of sulfolipids on phosphorus demand by picocyanobacteria in oligotrophic marine environments	104	135
12.	Vaulot et al.	2008	R	Univ. Paris 06, Norwegian Inst. Water. Res.	France, Norway	4	M	*Fems Microbiol. Rev.*	Microbiol.	Diversity of small eukaryotic phytoplankton in marine ecosystems	100	148
13.	Herranen et al.	2004	A	Turku Univ.	Finland	7		*Plant Physiol.*	Plant Sci.	Functional proteomics of membrane protein complexes in *Synechocystis* sp. PCC 6803	100	140

(*Continued*)

TABLE 2.1 (*Continued*)

Research on the Algal Omics

No.	Paper Ref.	Year	Doc.	Affil.	Country	No. of Authors	M/F	Journal	Subject Area	Topic	Total No. of Citations WN	Total No. of Citations GS
14.	Gilson and McFadden	1996	A	Univ. Melbourne	Australia	2	M	*Proc. Natl. Acad. Sci. United States*	Mult. Sci.	Miniaturized nuclear genome of eukaryotic endosymbiont	96	129
15.	Molnar et al.	2009	A	Univ. Cambridge, Univ. E Anglia +2	England, Germany	8	M	*Plant J.*	Plant Sci.	Gene silencing by artificial micro-RNAs in the unicellular alga *C. reinhardtii*	94	156
16.	Bohnert et al.	2001	R	Univ. Arizona, Purdue Univ. +2	United States	25	M	*Plant Physiol. Biochem.*	Plant Sci.	Genomics approach toward salt stress tolerance	93	182
17.	Swingley et al.	2008	A	Hokkaido Univ., Univ. Sydney +6	Japan, Australia, United States	25	M	*Proc. Natl. Acad. Sci. United States*	Mult. Sci.	Niche adaptation and genome expansion in the chlorophyll d-producing cyanobacterium *A. marina*	93	110
18.	Hess et al.	2001	R	Humboldt Univ., MIT	United States, Germany	7	M	*Photosynth. Res.*	Plant Sci.	Photosynthetic apparatus of *Prochlorococcus*	91	121
19.	Turmel et al.	2006	A	Univ. Laval	Canada	3	F	*Mol. Biol. Evol.*	Bioch. Mol. Biol., Evol. Biol., Gen. Her	Chloroplast genome sequence of *C. vulgaris*	89	110
20.	Whitelegge et al.	2002	A	Univ. Calif Los Angeles, Purdue Univ. +2	United States	5	M	*Mol. Cell. Proteomics*	Bioch. Res. Meth.	Oligomeric membrane protein cytochrome b_6f complex from cyanobacterium *M. laminosus*	89	139

(*Continued*)

TABLE 2.1 (*Continued*)

Research on the Algal Omics

No.	Paper Ref.	Year	Doc.	Affil.	Country	No. of Authors	M/F	Journal	Subject Area	Topic	Total No. of Citations WN	Total No. of Citations GS
21.	Liska et al.	2004	A	Weizmann Inst. Sci., Max Planck Inst.	Israel, Germany	4	M	*Plant Physiol.*	Plant. Sci.	Salinity tolerance in *Dunaliella* as revealed by homology-based proteomics.	88	138
22.	Barbier et al.	2005	R	Michigan State Univ., Univ. Potsdam	United States, Germany	8	M	*Plant Physiol.*	Plant. Sci.	Comparative genomics of two closely related unicellular thermoacidophilic red algae, *G. sulphuraria* and *C. merolae*	82	116
23.	Huang et al.	2002	A	Stockholm Univ., Washington Univ. +2	Sweden, United States	7		*Mol. Cell. Proteomics*	Bioch. Res. Meth.	Proteomics of *Synechocystis* sp. strain PCC 6803 identification of plasma membrane proteins	82	127
24.	Duhring et al.	2006	A	Univ. Freiburg, Humboldt Univ.	Germany	4	M	*Proc. Natl. Acad. Sci. United States*	Mult. Sci.	Expression of the photosynthesis gene *isiA*	80	128
25.	Naumann et al.	2006	A	Univ. Munster, Univ. Penn	Germany, United States	7	F	*Proteomics*	Bioch. Res. Meth., Bioch. Mol. Biol.	Remodeling of bioenergetic pathways under iron deficiency in *C. reinhardtii*	79	106

A, article; R, review; M, male; F, female; WN, Web of Knowledge; GS, Google Scholar.

(two papers each). It is notable that there are seven papers published in ultrahigh-impact multidisciplinary journals like *Proceedings of the National Academy of Sciences of the United States of America*. These findings suggest that the research on the algal omics has been considered most significant by the shareholders in the research community.

2.2.2 Most-Cited Papers in Algal Omics

Badger and Price (2003) discuss the CCMs in cyanobacteria with the emphasis on the molecular components, their diversity, and evolution in a review paper originating from Australia with 246 citations. They note that the CCM components include at least four modes of active inorganic carbon uptake, including two bicarbonate transporters and two CO_2 uptake systems associated with the operation of specialized type I NAD(P) H dehydrogenase (NDH-1) complexes. All these uptake systems serve to accumulate HCO_3^- in the cytosol of the cell, which is subsequently used by the Rubisco-containing carboxysome protein microcompartment within the cell to elevate CO_2 around Rubisco. They argue that cyanobacteria have evolved two types of carboxysomes, correlated with the form of Rubisco present (forms 1A and 1B). The two HCO_3^- and CO_2 transport systems are distributed variably, with some cyanobacteria (*Prochlorococcus mannus* species) appearing to lack CO_2 uptake systems entirely. There are multiple carbonic anhydrases in many cyanobacteria, but, surprisingly, several cyanobacterial genomes lack any identifiable CA genes.

Lindell et al. (2004) discuss the transfer of photosynthesis genes to and from *Prochlorococcus* viruses in a paper originating from the United States with 213 citations. They report the presence of genes central to oxygenic photosynthesis in the genomes of three phages from two viral families (*Myoviridae* and *Podoviridae*) that infect the marine cyanobacterium *Prochlorococcus*. The genes that encode the photosystem II (PSII) core reaction center protein D1 (*psbA*) and a high-light-inducible protein (HLIP) (*hli*) are present in all three genomes. Both myoviruses contain additional *hli* gene types, and one of them encodes the second PSII core reaction center protein D2 (*psbD*), whereas the other encodes the photosynthetic electron transport proteins plastocyanin (*petE*) and ferredoxin (*petF*). These uninterrupted, full-length genes are conserved in their amino acid sequence, suggesting that they encode functional proteins that may help maintain photosynthetic activity during infection. Phylogenetic analyses show that phage D1, D2, and HLIP proteins cluster with those from *Prochlorococcus*, indicating that they are of cyanobacterial origin. Their distribution among several *Prochlorococcus* clades further suggests that the genes encoding these proteins were transferred from host to phage multiple times.

Kettler et al. (2007) investigate the patterns and implications of gene gain and loss in the evolution of *Prochlorococcus* in a paper originating from Germany and the United States with 190 citations. They describe the genomes of eight newly sequenced isolates and combine them with the first four genomes for a comprehensive analysis of the core (shared by all isolates) and flexible genes of the *Prochlorococcus* group and the patterns of loss and gain of the flexible genes over the course of evolution. They find that there are 1273 genes that represent the core shared by all 12 genomes. They describe a phylogeny for all 12 isolates by subjecting their complete proteomes to three different phylogenetic analyses. They find many of the genetic differences among isolates, especially for genes involved in outer membrane synthesis and nutrient transport, within the same clade. Nevertheless, they identify some genes defining high-light (HL) and low-light (LL) ecotypes and clades within these broad ecotypes, helping to demonstrate the basis of HL and LL adaptations in *Prochlorococcus*. They argue that besides identifying islands and demonstrating their role throughout the history of *Prochlorococcus*, reconstruction of past gene gains and losses shows that much of the variability exists at the "leaves of the tree" between the most closely related strains.

Keller et al. (2005) investigate the proteomic analysis of isolated *Chlamydomonas* centrioles to explore orthologs of ciliary-disease genes in a paper originating from the United States with 168 citations. By combining proteomic data with information about gene expression and comparative genomics, they identify 45 cross-validated centriole candidate proteins in two classes. Members of the first class of proteins (BUG1–BUG27) are encoded by genes whose expression correlates with flagellar assembly and which therefore may play a role in ciliogenesis-related functions of basal bodies. Members of the second class (POC1–POC18) are implicated by comparative genomic and comparative proteomic studies to be conserved components of the centriole. They confirm centriolar localization for the human homologs of four candidate proteins. Three of the cross-validated centriole candidate proteins are encoded by orthologs of genes (*OFD1*, *NPHP-4*, and *PACRG*) implicated in mammalian ciliary function and disease, suggesting that oral–facial–digital syndrome and nephronophthisis may involve a dysfunction of centrioles and/or basal bodies.

Price et al. (2008) discuss the advances in understanding the cyanobacterial CCM with the emphasis on the functional components, Ci transporters, diversity, genetic regulation, and prospects for engineering into plants in a paper originating from Australia with 137 citations. They note that cyanobacteria can possess up to five distinct transport systems for Ci uptake. Through database analysis of some 33 complete genomic DNA sequences for cyanobacteria, they find that considerable diversity exists in the composition of transporters employed, although in many species this diversity is yet to be confirmed by comparative phenomics. In addition, two types of carboxysomes are known within the cyanobacteria that have apparently arisen by parallel evolution, and considerable progress has been made toward understanding the proteins responsible for carboxysome assembly and function. Progress has also been made toward identifying the primary signal for the induction of the subset of CCM genes known as CO_2-responsive genes, and transcriptional regulators CcmR and CmpR have been shown to regulate these genes.

Hackett et al. (2007) discuss whether the phylogenomic analysis supports the monophyly of cryptophytes and haptophytes and the association of rhizaria with chromalveolates in a paper originating from the United States with 131 citations. They assess chromalveolate monophyly using a multigene data set of nuclear genes that includes members of all 6 eukaryotic supergroups. They use an automated phylogenomics pipeline followed by targeted database searches to assemble a 16-protein data set (6735 amino acids) from 46 taxa for tree inference. Maximum likelihood and Bayesian analyses of these data support the monophyly of haptophytes and cryptophytes. They argue that this relationship is consistent with a gene replacement via horizontal gene transfer of plastid-encoded *rpl36* that is uniquely shared by these taxa. They further argue that the haptophytes + cryptophytes are sister to a clade that includes all other chromalveolates and, surprisingly, two members of the rhizaria, *Reticulomyxa filosa* and *Bigelowiella natans*.

Badger et al. (2006) investigate the environmental plasticity and ecological genomics of the cyanobacterial CCM in a paper originating from Australia with 131 citations. There are two primary functional elements of this CCM: first, the containment of Rubisco in carboxysome protein microbodies within the cell (the sites of CO_2 elevation) and, second, the presence of several inorganic carbon (Ci) transporters that deliver HCO_3^- intracellularly. Cyanobacteria show both species adaptation and acclimation of this mechanism. Between species, they argue that there are differences in the suites of Ci transporters in each genome, the nature of the carboxysome structures, and the functional roles of carbonic anhydrases. Within a species, different CCM activities can be induced depending on the Ci availability in the environment. This acclimation is largely based on the induction of multiple Ci transporters with different affinities and specificities for either CO_2 or HCO_3^- as substrates.

Lemaire et al. (2004) discuss the new thioredoxin targets in the unicellular photosynthetic eukaryote *Chlamydomonas reinhardtii* in a paper originating from France with 128 citations. These proteins were retained specifically on a thioredoxin affinity column made of a monocysteinic thioredoxin mutant able to form mixed disulfides with its targets. Of a total of 55 identified targets, 29 had been found previously in higher plants or *Synechocystis*, but 26 were new targets. Biochemical tests were performed on three of them, showing a thioredoxin-dependent activation of isocitrate lyase and isopropylmalate dehydrogenase and a thioredoxin-dependent deactivation of catalase that is redox insensitive in *Arabidopsis*.

Palenik et al. (2006) investigate the genome sequence of *Synechococcus* CC9311 to gain insights into adaptation to a coastal environment in a paper originating from the United States with 115 citations. They find that the genome of a coastal cyanobacterium, *Synechococcus* sp. strain CC9311, has significant differences from an open-ocean strain, *Synechococcus* sp. strain WH8102, and these are consistent with the differences between their respective environments. CC9311 has a greater capacity to sense and respond to changes in its (coastal) environment. It has a much larger capacity to transport, store, use, or export metals, especially iron and copper. In contrast, phosphate acquisition is less important, consistent with the higher concentration of phosphate in coastal environments. They predict that CC9311 has differences in its outer membrane lipopolysaccharide, and this may be the characteristic of the speciation of some cyanobacterial groups. In addition, they argue that the types of potentially horizontally transferred genes are markedly different between the coastal and open-ocean genomes and suggest a more prominent role for phages in horizontal gene transfer in oligotrophic environments.

Fulda et al. (2000) investigate the proteomics of *Synechocystis* sp. strain PCC 6803 in a paper originating from Germany and Sweden with 110 citations. Periplasmic proteins isolated by cold osmotic shock of *Synechocystis* sp. PCC 6803 cells were identified using 2D PAGE, MS, and genome analysis. They find that most of the periplasmic proteins represent *hypothetical proteins* with unknown function. They also find a number of proteases of different specificities and several enzymes involved in cell wall biosynthesis. In salt-adapted cells, six proteins were greatly enhanced and three proteins were newly induced. Most of the salt-enhanced proteins are involved in the alteration of cell wall structure of salt-adapted cells. The precursors of all 57 periplasmic proteins identified have a signal peptide. Forty-seven of which contain a typical Sec-dependent signal peptide, whereas 10 contain a putative twin-arginine signal peptide.

Van Mooy et al. (2006) investigate whether the sulfolipids dramatically decrease phosphorus demand by picocyanobacteria in oligotrophic marine environments in a paper originating from the United States with 104 citations. They find that *Prochlorococcus*, the cyanobacterium that dominates North Pacific Subtropical Gyre (NPSG) phytoplankton, primarily synthesizes sulfoquinovosyldiacylglycerol (SQDG), a lipid that contains sulfur and sugar instead of phosphate. In axenic cultures of *Prochlorococcus*, it was observed that <1% of the total PO_4^{3-} uptake was incorporated into membrane lipids. Liquid chromatography/mass spectrometry of planktonic lipids in the NPSG confirmed that SQDG was the dominant membrane lipid. Furthermore, the analyses of SQDG synthesis genes from the Sargasso Sea environmental genome showed that the use of sulfolipids in subtropical gyres was confined primarily to picocyanobacteria; no sequences related to known heterotrophic bacterial SQDG lineages were found.

Vaulot et al. (2008) discuss the diversity of small eukaryotic phytoplankton (≤3 μm) in marine ecosystems in a review paper originating from Norway and France with 100 citations. They argue that while the prokaryotic component of picophytoplankton is dominated by two genera, *Prochlorococcus* and *Synechococcus*, the eukaryotic fraction is much more diverse. Since the discovery of the ubiquitous *Micromonas pusilla* in the early 1950s, just over 70 species that can be <3 μm have been described. In the last couple of years, the first genomes of photosynthetic picoplankton have become available, providing key information on their physiological capabilities. They discuss the range of methods that can be used to assess small phytoplankton diversity, present the species described to date, review the existing molecular data obtained on field populations, and end up by looking at the promises offered by genomics.

Herranen et al. (2004) investigate the functional proteomics of membrane protein complexes in *Synechocystis* sp. PCC 6803 in a paper originating from Finland with 100 citations. Besides the protein complexes involved in linear photosynthetic electron flow and ATP synthesis (photosystem [PSI, PSII], cytochrome b_6f, and ATP synthase), they identify four distinct complexes containing NDH-1 subunits, as well as several novel, still uncharacterized, protein complexes. They find that the most distinct modulation observed in PSs occurred in iron-depleted conditions, which induced an accumulation of CP43' protein associated with PSI trimers. The NDH-1 complexes, on the other hand, responded readily to changes in the CO_2 concentration and the growth mode of the cells and represented an extremely dynamic group of membrane protein complexes. They conclude that the NdhF3, NdhD3, and CupA proteins assemble together to form a small low CO_2-induced protein complex and further demonstrate the presence of a fourth subunit, Sll1735, in this complex.

Gilson and McFadden (1996) investigate whether the miniaturized nuclear genome of eukaryotic endosymbiont contains genes that overlap, genes that are cotranscribed, and the smallest known spliceosomal introns in a paper originating from Australia with 96 citations. They analyze nucleotide sequence data from a subtelomeric fragment of chromosome III as a preliminary investigation of the coding capacity of this vestigial genome. They identify several housekeeping genes, including U6 small nuclear RNA, ribosomal proteins S4 and S13, a core protein of the spliceosome (small nuclear ribonucleoprotein [snRNP] E), and a clp-like protease (clpP). They find that the protein-encoding genes are typically eukaryotic in overall structure and their messenger RNAs are polyadenylylated. A novel feature is the abundance of 18-, 19-, or 20-nucleotide introns, the smallest spliceosomal introns known. Two of the genes, U6 and S13, overlap, while another two genes, snRNP E and clpP, are cotranscribed in a single mRNA. They conclude that the overall gene organization is extraordinarily compact, making the nucleomorph a unique model for eukaryotic genomics.

Molnar et al. (2009) investigate the highly specific gene silencing by artificial micro-RNAs in the unicellular alga *C. reinhardtii* in a paper originating from England and Germany with 94 citations. They demonstrate that the unicellular alga *C. reinhardtii*, like diverse multicellular organisms, contains miRNAs. These RNAs resemble the miRNAs of land plants in that they direct site-specific cleavage of target mRNA with miRNA-complementary motifs and, presumably, act as regulatory molecules in growth and development. They develop a novel artificial miRNA system based on ligation of DNA oligonucleotides that can be used for specific high-throughput gene silencing in green algae.

Bohnert et al. (2001) discuss a genomics approach toward salt stress tolerance in a review paper originating from the United States with 93 citations. They analyze global gene expression profiles monitored under salt stress conditions through abundance profiles in several species: in the cyanobacterium *Synechocystis* PCC6803, in unicellular (*Saccharomyces cerevisiae*) and multicellular (*Aspergillus nidulans*) fungi, the eukaryotic alga *Dunaliella salina*, the halophytic land plant *Mesembryanthemum crystallinum*, the glycophytic *Oryza sativa*, and the genetic model *Arabidopsis thaliana*. They find that expanding the gene count, stress brings about a significant increase of transcripts for which no function is known. More than 400,000 T-DNA-tagged lines of *A. thaliana* have been generated, and lines with altered salt stress responses have been obtained.

Swingley et al. (2008) investigate the niche adaptation and genome expansion in the chlorophyll d-producing cyanobacterium *Acaryochloris marina* in a paper originating from Australia, Japan, and the United States with 93 citations. They find that the DNA content of *A. marina* is composed of 8.3 million base pairs, which is among the largest bacterial genomes sequenced thus far. This large array of genomic data are distributed into nine single-copy plasmids that code for >25% of the putative ORFs. They argue that heavy duplication of genes related to DNA repair and recombination (primarily recA) and transposable elements could account for genetic mobility and genome expansion. They also find that *A. marina* carries a unique complement of genes for these phycobiliproteins in relation to those coding for antenna proteins related to those in *Prochlorococcus* species. They conclude that the genus *Acaryochloris* is a fitting candidate for understanding genome expansion, gene acquisition, ecological adaptation, and photosystem modification in the cyanobacteria.

Hess et al. (2001) discuss the photosynthetic apparatus of *Prochlorococcus* to gain insights through comparative genomics in a review paper originating from Germany and the United States with 91 citations. They compare information obtained from the complete genome sequences of two *Prochlorococcus* strains, with special emphasis on genes for the photosynthetic apparatus. These two strains, *Prochlorococcus* MED4 and MIT 9313, are representatives of HL- and LL-adapted ecotypes, characterized by their low or high Chl b/a ratio, respectively. They find that both genomes are significantly smaller (1700 and 2400 kbp) than those of other cyanobacteria, and the LL-adapted strain has significantly more genes than its HL counterpart. In keeping with their comparative light-dependent physiologies, MED4 has many more genes encoding putative HLIPs and photolyases to repair UV-induced DNA damage, whereas MIT 9313 possesses more genes associated with the photosynthetic apparatus. These include two *pcb* genes encoding Chl-binding proteins and a second copy of the gene *psbA*, encoding the PSII reaction center protein D1. In addition, MIT 9313 contains a gene cluster to produce chromophorylated phycoerythrin. The latter represents an intermediate form between the phycobiliproteins of non-Chl *b*-containing cyanobacteria and an extremely modified β phycoerythrin as the sole derivative of phycobiliproteins still present in MED4.

Turmel et al. (2006) investigate whether the chloroplast genome sequence of *Chara vulgaris* sheds new light into the closest green algal relatives of land plants in a paper originating from Canada with 89 citations. They determine the complete chloroplast genome sequence (184,933 bp) of a representative of the Charales, *C. vulgaris*, and compared this genome to those of *Mesostigma* (Mesostigmatales), *Chlorokybus* (Chlorokybales), *Staurastrum* and *Zygnema* (Zygnematales), *Chaetosphaeridium* (Coleochaetales), and selected land plants. They argue that *Charales* diverged before the Coleochaetales and Zygnematales. *Chara* remained at the same basal position in trees including more land plant taxa and inferred from 56 proteins/genes. They find support for T1 than for the topology of the four-gene tree.

They argue that many of the features conserved in land plant cpDNAs were inherited from their green algal ancestors. The intron content data predicted that at least 15 of the 21 land plant group II introns were gained early during the evolution of streptophytes and that a single intron was acquired during the transition from charophycean green algae to land plants.

Whitelegge et al. (2002) investigate the full subunit coverage liquid chromatography electrospray ionization mass spectrometry (LCMS+) of an oligomeric membrane protein cytochrome b$_6$f complex from spinach and the cyanobacterium *Mastigocladus laminosus* in a paper originating from the United States with 89 citations. They detect the products of *petA*, *petB*, *petC*, *petD*, *petG*, *petL*, *petM*, and *petN* in complexes from both spinach and *M. laminosus*, while the spinach complex also contained ferredoxin-NADP+ oxidoreductase. Products of the spinach chloroplast genome, *PetG*, *PetL*, and *PetN*, all retained their initiating formylmethionine, while the nuclear encoded *PetM* was cleaved after import from the cytoplasm. While the sequences of *PetG* and *PetN* revealed no discrepancy with translations of the spinach chloroplast genome, *Phe* was detected at position 2 of *PetL*. The spinach chloroplast genome reports a codon for Ser at position 2 implying the presence of a DNA sequencing error or a previously undiscovered RNA editing event. Full subunit coverage of an oligomeric intrinsic membrane protein complex by LCMS+ presents a new facet to intact mass proteomics.

Liska et al. (2004) investigate whether the enhanced photosynthesis and redox energy production contribute to salinity tolerance in *Dunaliella* as revealed by homology-based proteomics in a paper originating from Israel and Germany with 88 citations. They identified 80% of the salt-induced proteins found that salinity stress upregulated key enzymes in the Calvin cycle, starch mobilization, and redox energy production, regulatory factors in protein biosynthesis and degradation, and a homolog of a bacterial Na$^+$-redox transporters. They argue that *Dunaliella* responds to high salinity by the enhancement of photosynthetic CO$_2$ assimilation and by diversion of carbon and energy resources for synthesis of glycerol, the osmotic element in *Dunaliella*. They argue that the ability of *Dunaliella* to enhance photosynthetic activity at high salinity is remarkable because, in most plants and cyanobacteria, salt stress inhibits photosynthesis.

Barbier et al. (2005) discuss the comparative genomics of two closely related unicellular thermoacidophilic red algae, *Galdieria sulphuraria* and *Cyanidioschyzon merolae*, to explore the molecular basis of the metabolic flexibility of *G. sulphuraria* and significant differences in carbohydrate metabolism of both algae in a review paper originating from Germany and the United States with 82 citations. They note that a compelling strategy for gene identification is the comparison of similarly sized genomes of related organisms with different physiologies. Using this approach, they identify candidate genes that are critical to the metabolic versatility of *Galdieria*. They find that more than 30% of the *Galdieria* sequences did not relate to any of the *Cyanidioschyzon* genes. At a closer inspection of these sequences, they find a large number of membrane transporters and enzymes of carbohydrate metabolism that are unique to *Galdieria*. They argue that genes involved in the uptake of reduced carbon compounds and enzymes involved in their metabolism are crucial to the metabolic flexibility of *G. sulphuraria*.

Huang et al. (2002) investigate the proteomics of *Synechocystis* sp. strain PCC 6803 with the emphasis on the identification of plasma membrane proteins in a paper originating from Sweden and the United States with 82 citations. They find a total of 57 different membrane proteins, of which 17 are integral membrane spanning proteins. Among the 40 peripheral proteins, 20 are located on the periplasmic side of the membrane, while 20 are on the cytoplasmic side. Among the proteins identified are subunits of the two photosystems and Vipp1, which has been suggested to be involved in vesicular transport between plasma and thylakoid membranes and is thus relevant to the possibility that plasma membranes are the initial site for photosystem biogenesis. They also identify four subunits of the Pilus complex responsible for cell motility as well as several subunits of the TolC and TonB transport systems.

Duhring et al. (2006) investigate whether internal antisense RNA regulates expression of the photosynthesis gene *isiA* in a paper originating from Germany with 80 citations. They note that cyanobacteria respond to iron deficiency by expressing *IsiA* (iron stress–induced protein A), which forms a giant ring structure around PSI. They show that this process is controlled by *IsrR* (iron stress–repressed RNA), a cis-encoded antisense RNA transcribed from the *isiA* noncoding strand. They find that artificial overexpression of *IsrR* under iron stress causes a strongly diminished number of *IsiA*–PSI supercomplexes, whereas *IsrR* depletion results in premature expression of *IsiA*. They argue that the coupled degradation of *IsrR/isiA* mRNA duplexes is a reversible switch that can respond to environmental changes.

Naumann et al. (2007) investigate the comparative quantitative proteomics to investigate the remodeling of bioenergetic pathways under iron deficiency in *C. reinhardtii* in a paper originating from Germany and the United States with 79 citations. They find that although photosynthetic electron transfer is largely compromised under iron deficiency, the functional antenna size of PSII significantly increased. Concomitantly, stress-related chloroplast polypeptides, like 2-cys peroxiredoxin and a stress-inducible light-harvesting protein, LhcSR3, as well as a novel light-harvesting protein and several proteins of unknown function, were induced under iron deprivation. Respiratory oxygen consumption did not decrease, and accordingly, polypeptides of respiratory complexes, harboring numerous iron–sulfur clusters, were only slightly diminished or even increased under low iron. Consequently, they argue that iron deprivation induces a transition from photoheterotrophic to primarily heterotrophic metabolism, indicating that a hierarchy for iron allocations within organelles of a single cell exists, which is closely linked with the metabolic state of the cell.

2.2.3 Conclusion

The research on the omics in general and algal omics in particular has been one of the most dynamic research areas in recent years. Twenty-five citation classics in the field of algal omics with more than 81 citations were located, and the key emerging issues from these papers were presented earlier in the decreasing order of the number of citations (Table 2.1). These papers give strong hints about the determinants of the optimal design of algae through the algal omics and emphasize that algal omics offer the valuable information about the omic design of the algae for the optimal production of algal biofuels, such as algal biohydrogen and algal diesel, and algal biocompounds, such as algal drugs and algal food ingredients.

2.3 Conclusion

Global warming, air pollution, and energy security contributed to some of the most important public policy issues in recent years (e.g., Jacobson 2009, Wang et al. 2008, Yergin 2006). With the increasing global population, food security has also become a major public policy issue (e.g., Lal 2004). The development of biofuels generated from the biomass has been a long-awaited solution to these global problems (e.g., Demirbas 2007, Goldemberg 2007, Lynd et al. 1991). However, the development of the early biofuels produced from the agricultural plants, such as sugarcane (e.g., Goldemberg 2007), and agricultural wastes, such as corn stovers (e.g., Bothast and Schlicher 2005), has resulted in a series of substantial concerns about the food security (e.g., Godfray et al. 2010).

Therefore, the development of algal biofuels as a third-generation biofuel has been considered as a major solution for the global problems of global warming, air pollution, energy security, and food security (e.g., Chisti, 2007, Demirbas 2007, Kapdan and Kargi 2006, Spolaore et al. 2006, Volesky and Holan 1995).

The design of production of algal biofuels, such as algal biohydrogen and algal biodiesel, and algal biocompounds, such as algal drugs and algal food ingredients, has also been an important research area in recent years (e.g., Cardozo et al. 2007, Grima et al. 2003, Pulz 2001). Based on the results from the studies on the algal omics, the issues relating to the optimization of the algal biocompound and algal biofuel production through omic analysis have also emerged as an important research area in particular (e.g., Badger and Price 2003, Badger et al. 2006, Fulda et al. 2000, Hackett et al. 2007, Keller et al. 2005, 2007, Lemaire et al. 2004, Lindell et al. 2004, Palenik et al. 2006, Price et al. 2008).

Although there have been many studies on the algal omics (e.g., Anderson et al. 2012, Badger and Price 2003, Badger et al. 2006, Fulda et al. 2000, Goodstein et al. 2012, Hackett et al. 2007, Hagemann 2011, Keller et al. 2005, 2007, Leliaert et al. 2012, Lemaire et al. 2004, Lindell et al. 2004, Palenik et al. 2006, Price et al. 2008, Wu et al. 2011) and on the omics as a whole (e.g., Arumugam et al. 2011, Bartel 2004, Cox et al. 2011, Gentleman et al. 2004, Gerlinger et al. 2012, Heid et al. 1996, Knox et al. 2011, Kumar et al. 2004, Schwanhausser et al. 2011, Tamura et al. 2007) and there have been a limited number of scientometric studies on the algal biofuels (Konur 2011), there has not been any study on the citation classics in the algal omics as in the other research fields as highly cited papers (e.g., Baltussen and Kindler 2004a,b, Dubin et al. 1993, Gehanno et al. 2007, Konur 2012a,b, 2013, Paladugu et al. 2002, Wrigley and Matthews 1986).

As North's New Institutional Theory suggests, it is important to have up-to-date information about the current public policy issues to develop a set of viable solutions to satisfy the needs of all the key stakeholders (North 1994, Konur 2000, 2002a,b,c, 2006a,b, 2007a,b, 2012c,d).

Therefore, following a scientometric overview of the research in omics at large and algal omics in particular, the brief information on a selected set of 25 citation classics in the field of the algal omics is presented in this chapter to inform the key stakeholders about omic design of algae for the efficient production of algal biofuels and biocompounds such as algal drugs and algal food ingredients for the solution of these problems in the long run, thus complementing a number of recent scientometric and review studies on the biofuels and global energy research (Konur 2011, 2012e–p, 2013, 2015a–m, 2016a,b). The scientometric overview of the research in omics in general and algal omics in particular was necessary since there have been a limited number of scientometric studies in omics in the literature as indexed by the SSCI or SCIE as of July 2014 (Kuzhabekova and Kuzma 2014, Mansiaux and Carrat 2012, Wonkam et al. 2011).

Using the keywords and the selected journal set related to the omics, 151,829 references were located in omics. Of these references, 129,858 were articles and reviews comprising 85.5% of the sample. Meeting abstracts, editorial materials, and other types of references formed the remaining part of the sample.

The most prolific country in terms of the number of publications was the United States with 54,149 papers forming 41.7% of the sample. Germany, England, and China followed the United States with 9.1%, 9.0%, and 8.5% of the sample, respectively. English was the dominant language of scientific communication in omics comprising 98.6% of the sample.

Similar to the most recent dynamic research areas like nanoresearch or bioinformatic research, the omic research has been a solely recent phenomenon as 92.6% of the papers were published after 2000. The share of the research in the 1980s, 1990s, 2000s, and 2010s was 0.9%, 6.4%, 46.5%, and 46.1%, respectively. These findings suggest that the omic research has been a recent research area as the research boomed in the 2000s and 2010s with the steady increase in the number of publications over time jumping in 2000 from 1337 to 1847, although the research has been carried out since the 1980s.

The most prolific subject category in terms of the number of publications was *Genetics Heredity* with 34,257 papers forming 26.4% of the sample. *Biochemistry Molecular Biology*, *Biotechnology Applied Microbiology*, and *Biochemical Research Methods* followed the top subject category with 23.0%, 21.3%, and 16.3% of the sample, respectively. These findings suggest that these top 4 subject categories share a common set of journals where a journal was indexed under more than one subject category.

The most-cited papers in omics were mostly related to the development of omics and its application in a number of areas in general such as the development of the computer software programs and quantitative methods for the genomic analysis (Bartel 2004, Gentleman et al. 2004, Heid et al. 1996, Kumar et al. 2004, Tamura et al. 2007). It is notable that these top citation classics in omics were published in the high-impact genomic journals.

Contrary to the citation classics, the hottest papers in omics published in the last 3 years between 2011 and 2013 were related mostly to the application of the omics in the medicine (Arumugam et al. 2011, Cox et al. 2011, Gerlinger et al. 2012, Knox et al. 2011, Schwanhausser et al. 2011). It is further notable that some of these top hottest papers in omics were published in the high-impact multidisciplinary journals like *Nature* and high-impact other journals like *Journal of Proteome Research*.

Using the keywords related to algal omics, only 1158 references were located in algal omics. Of these references, 1107 were articles and reviews. Meeting abstracts, editorial materials, and other references formed the remaining part of the sample. Considering the fact that there were 129,858 papers in the parent omic research field, the research on the algal omics formed 0.9% of the larger parental data set. This finding suggests that the field of algal omics was a highly specialized field of research with relatively a small sample size and with a specific set of shareholders such as authors, institutions, and countries.

The most prolific country in terms of the number of publications was the United States with 434 papers forming 39.2% of the sample. Germany, France, and China followed the United States with 13.3%, 9.7%, and 8.8% of the sample, respectively. English was the dominant language of scientific communication in algal omics with 99.7% of the sample in English.

Like the nanoresearch and the parent omic research, the algal omic research has been a recent phenomenon as 97.4% of the papers were published after 2000. The research in algal omics in the 1990s, 2000s, and 2010s formed 2.4%, 36.7%, and 60.7% of the sample, respectively. There was a general increasing trend in the number of papers over time especially in the 2010s. These findings support the earlier findings that algal omics received public attention in the 2000s and 2010s and would continue to do so in the near future as it meets the needs of the society for the omic research related to the algae.

The most prolific subject category in terms of the number of publications was *Plant Sciences* with 189 papers forming 17.1% of the sample. *Biochemistry Molecular Biology, Biotechnology Applied Microbiology,* and *Biochemical Research Methods* followed the top subject category with 25.1%, 14.8%, and 9.9% of the sample, respectively. These findings suggest that these top four subject categories share a common set of journals where a journal was indexed under more than one subject category.

There were 26,789 citations received for 1107 papers in algal omics. The average citation rate per paper was 24.2 and the H-index was 77. These findings suggest that considering the fact that this research field was a relatively new, the citation impact of 1107 papers in algal omics was relatively significant. The number of citations rose over time, but especially spectacularly in the 2010s as 60.1% of the citations, was received in the 2010s.

The most-cited papers in algal omics were heavy in the omic analysis of the algae (e.g., Badger and Price 2003, Badger et al. 2006, Fulda et al. 2000, Hackett et al. 2007, Keller et al. 2005, 2007, Lemaire et al. 2004, Lindell et al. 2004, Palenik et al. 2006, Price et al. 2008).

There was a similar trend for the hottest papers published in the last 3 years between 2011 and 2013 (Anderson et al. 2012, Goodstein et al. 2012, Hagemann 2011, Leliaert et al. 2012, Wu et al. 2011).

The research on the omics in general and algal omics in particular has been one of the most dynamic research areas in recent years. Twenty-five citation classics in the field of algal omics with more than 81 citations were located, and the key emerging issues from these papers were presented earlier in the decreasing order of the number of citations (Table 2.1). These papers give strong hints about the determinants of the optimal design of algae through the algal omics and emphasize that algal omics offer the valuable information about the omic design of the algae for the optimal production of algal biofuels, such as algal biohydrogen and algal diesel, and algal biocompounds, such as algal drugs and algal food ingredients.

The papers were dominated by the researchers from at least 10 countries, usually through the intracountry institutional collaboration, and they were multiauthored. The number of the authors for the papers arranged from 2 to 25. The United States (13 papers), Germany (8 papers), Australia (5 papers), and France and Sweden (2 papers each) were the most prolific countries.

Similarly, all these papers were published in the journals indexed by the SCI and/or SCIE. There was also no paper indexed by the SSCI. The number of citations ranged from 79 to 264 for the Web of Science and from 106 to 380 for the Google Scholar databases. Twenty of the papers were articles, while only 5 of them were reviews.

The papers were published between 1996 and 2009, while 1 and 24 of them were published in the 1990s and 2000s, respectively. This finding supports the earlier findings that the research on the algal omics boomed especially in the 2000s and 2010s, but older papers had more citations.

There was a significant gender deficit among the most-cited papers in algal omics as there were 4 female first authors of 19 papers with known author gender. This issue merits further research.

The most prolific journals were *Proceedings of the National Academy of Sciences of the United States of America* (seven papers), *Journal of Experimental Botany* and *Plant Physiology* (three papers each), and *Molecular Biology and Evolution* and *Molecular and Cellular Proteomics* (two papers each). It is notable that there are seven papers published in ultrahigh-impact multidisciplinary journals like *Proceedings of the National Academy of Sciences of the United States of America*. These findings suggest that the research on the algal omics has been considered most significant by the shareholders in the research community.

The citation classics presented in this chapter confirm the predictions that the marine algae have a significant potential to serve as a major solution for the global problems of warming, air pollution, energy security, and food security in the form of algal biofuels and algal biocompounds based on the omic design of algae to meet such high demands.

Further research is recommended for the detailed studies including scientometric studies and citation classic studies to inform the key stakeholders about the potential of the marine algae for the solution of the global problems of warming, air pollution, energy security, and food security in the form of algal biofuels and algal biocompounds through the omic design of algae.

References

Anderson, D.M., T.J. Alpermann, A.D. Cembella, Y. Collos, E. Masseret, and M. Montresor. 2012. The globally distributed genus Alexandrium: Multifaceted roles in marine ecosystems and impacts on human health. *Harmful Algae* 14:10–35.

Arumugam, M., J. Raes, E. Pelletier, D. Le Paslier, T. Yamada, D.R. Mende et al. 2011. Enterotypes of the human gut microbiome. *Nature* 473:174–180.

Badger, M.R. and P.D. Price. 2003. CO_2 concentrating mechanisms in cyanobacteria: Molecular components, their diversity and evolution. *Journal of Experimental Botany* 54:609–622.

Badger, M.R., G.D. Price, B.M. Long, and F.J. Woodger. 2006. The environmental plasticity and ecological genomics of the cyanobacterial CO_2 concentrating mechanism. *Journal of Experimental Botany* 57:249–265.

Baltussen, A. and C.H. Kindler. 2004a. Citation classics in anesthetic journals. *Anesthesia & Analgesia* 98:443–451.

Baltussen, A. and C.H. Kindler. 2004b. Citation classics in critical care medicine. *Intensive Care Medicine* 30:902–910.

Barbier, G., C. Oesterhelt, M.D. Larson, R.G. Halgren, C. Wilkerson, R.M. Garavito et al. 2005. Comparative genomics of two closely related unicellular thermo-acidophilic red algae, *Galdieria sulphuraria* and *Cyanidioschyzon merolae*, reveals the molecular basis of the metabolic flexibility of *Galdieria sulphur*aria and significant differences in carbohydrate metabolism of both algae. *Plant physiology* 137:460–474.

Bartel, D.P. 2004. MicroRNAs: Genomics, biogenesis, mechanism, and function. *Cell* 116:281–297.

Bohnert, H.J., P. Ayoubi, C. Borchert, R.A. Bressan, R.L. Burnap, J.C. Cushman et al. 2001. A genomics approach towards salt stress tolerance. *Plant Physiology and Biochemistry* 39:295–311.

Borowitzka, M.A. 1999. Commercial production of microalgae: Ponds, tanks, tubes and fermenters. *Journal of Biotechnology* 70:313–321.

Bothast, R.J. and M.A. Schlicher. 2005. Biotechnological processes for conversion of corn into ethanol. *Applied Microbiology and Biotechnology* 67:19–25.

Cardozo, K.H.M., T. Guaratini, M.P. Barros, V.R. Falcão, A.P. Tonon, N.P. Lopes et al. 2007. Metabolites from algae with economical impact. *Comparative Biochemistry and Physiology C—Toxicology & Pharmacology* 146:60–78.

Chisti, Y. 2007. Biodiesel from microalgae. *Biotechnology Advances* 25:294–306.

Cox, J., N. Neuhauser, A. Michalski, R.A. Scheltema, J.V. Olsen, and M. Mann. 2011. Andromeda: A peptide search engine integrated into the MaxQuant environment. *Journal of Proteome Research* 10:1794–1805.

Demirbas, A. 2007. Progress and recent trends in biofuels. *Progress in Energy and Combustion Science* 33:1–18.

Dubin, D., A.W. Hafner, and K.A. Arndt. 1993. Citation-classics in clinical dermatological journals—Citation analysis, biomedical journals, and landmark articles, 1945–1990. *Archives of Dermatology* 129:1121–1129.

Duhring, U., I.M. Axmann, W.R. Hess, and A. Wilde. 2006. An internal antisense RNA regulates expression of the photosynthesis gene *isiA*. *Proceedings of the National Academy of Sciences of the United States of America* 103:7054–7058.

Fulda, S., F. Huang, F. Nilsson, M. Hagemann, and B. Norling. 2000. Proteomics of *Synechocystis* sp. strain PCC 6803. *European Journal of Biochemistry* 267:5900–5907.

Gehanno, J.F., K. Takahashi, S. Darmoni, and J. Weber. 2007. Citation classics in occupational medicine journals. *Scandinavian Journal of Work, Environment & Health* 33:245–251.

Gentleman, R.C., V.J. Carey, D.M. Bates, B. Bolstad, M. Dettling, S. Dudoit et al. 2004. Bioconductor: Open software development for computational biology and bioinformatics. *Genome Biology* 5:Art.No.R80.

Gerlinger, M., A.J. Rowan, S. Horswell, J. Larkin, D. Endesfelder, E. Gronroos et al. 2012. Intratumor heterogeneity and branched evolution revealed by multiregion sequencing. *New England Journal of Medicine* 366:883–892.

Gilson, P.R. and G.I. McFadden. 1996. The miniaturized nuclear genome of eukaryotic endosymbiont contains genes that overlap, genes that are cotranscribed, and the smallest known spliceosomal introns. *Proceedings of the National Academy of Sciences of the United States of America* 93:7737–7742.

Godfray, H.C.J., J.R. Beddington, I.R. Crute, L. Haddad, D. Lawrence, J.F. Muir et al. 2010. Food security: The challenge of feeding 9 billion people. *Science* 327:812–818.

Goldemberg, J. 2007. Ethanol for a sustainable energy future. *Science* 315:808–810.

Goodstein, D.M., S.Q. Shu, R. Howson, R. Neupane, R.D. Hayes, J. Fazo et al. 2012. Phytozome: A comparative platform for green plant genomics. *Nucleic Acids Research* 40:D1178–D1186.

Greenwell, H.C., L.M.L. Laurens, R.J. Shields, R.W. Lovitt, and K.J. Flynn. 2010. Placing microalgae on the biofuels priority list: A review of the technological challenges. *Journal of the Royal Society Interface* 7:703–726.

Grima, E.M., E.H. Belarbi, F.G.A. Fernandez, A.R. Medina, and Y. Chisti. 2003. Recovery of microalgal biomass and metabolites: Process options and economics. *Biotechnology Advances* 20:491–515.

Hackett, J.D., H.S. Yoon, S. Li, A. Reyes-Prieto, S.E. Rummele, and D. Bhattacharya. 2007. Phylogenomic analysis supports the monophyly of cryptophytes and haptophytes and the association of rhizaria with chromalveolates. *Molecular Biology and Evolution* 24:1702–1713.

Hagemann, M. 2011. Molecular biology of cyanobacterial salt acclimation. *FEMS Microbiology Reviews* 35:87–123.

Heid, C.A., J. Stevens, K.J. Livak, and P.M. Williams. 1996. Real time quantitative PCR. *Genome Research* 6:986–994.

Herranen, M., N. Battchikova, P. Zhang, A. Graf, S. Sirpio, V. Paakkarinen et al. 2004. Towards functional proteomics of membrane protein complexes in *Synechocystis* sp. PCC 6803. *Plant Physiology* 134:470–481.

Hess, W.R., G. Rocap, C.S. Ting, F. Larimer, S. Stilwagen, J. Lamerdin et al. 2001. The photosynthetic apparatus of *Prochlorococcus*: Insights through comparative genomics. *Photosynthesis Research* 70:53–71.

Huang, F., I. Parmryd, F. Nilsson, A.L. Persson, H.B. Pakrasi, B. Andersson et al. 2002. Proteomics of *Synechocystis* sp. strain PCC 6803 identification of plasma membrane proteins. *Molecular & Cellular Proteomics* 1:956–966.

Jacobson, M.Z. 2009. Review of solutions to global warming, air pollution, and energy security. *Energy & Environmental Science* 2:148–173.

Kapdan, I.K. and F. Kargi. 2006. Bio-hydrogen production from waste materials. *Enzyme and Microbial Technology* 38:569–582.

Keller, L.C., E.P. Romijn, I. Zamora, J.R. Yates, and W.F. Marshall. 2005. Proteomic analysis of isolated *Chlamydomonas* centrioles reveals orthologs of ciliary-disease genes. *Current Biology* 15:1090–1098.

Kettler, G.C., A.C. Martiny, K. Huang, J. Zucker, M.L. Coleman, S. Rodrigue et al. 2007. Patterns and implications of gene gain and loss in the evolution of *Prochlorococcus*. *PLOS Genetics* 3:e231.

Knox, C., V. Law, T. Jewison, P. Liu, S. Ly, A. Frolkis et al. 2011. DrugBank 3.0: A comprehensive resource for 'Omics' research on drugs. *Nucleic Acids Research* 39:D1035–D1041.

Konur, O. 2000. Creating enforceable civil rights for disabled students in higher education: An institutional theory perspective. *Disability & Society* 15:1041–1063.

Konur, O. 2002a. Assessment of disabled students in higher education: Current public policy issues. *Assessment and Evaluation in Higher Education* 27:131–152.

Konur, O. 2002b. Access to employment by disabled people in the UK: Is the Disability Discrimination Act working? *International Journal of Discrimination and the Law* 5:247–279.

Konur, O. 2002c. Access to Nursing Education by disabled students: Rights and duties of Nursing programs. *Nurse Education Today* 22:364–374.

Konur, O. 2006a. Participation of children with dyslexia in compulsory education: Current public policy issues. *Dyslexia* 12:51–67.

Konur, O. 2006b. Teaching disabled students in Higher Education. *Teaching in Higher Education* 11:351–363.

Konur, O. 2007a. A judicial outcome analysis of the Disability Discrimination Act: A windfall for the employers? *Disability & Society* 22:187–204.

Konur, O. 2007b. Computer-assisted teaching and assessment of disabled students in Higher Education: The interface between academic standards and disability rights. *Journal of Computer Assisted Learning* 23:207–219.

Konur, O. 2011. The scientometric evaluation of the research on the algae and bio-energy. *Applied Energy* 88:3532–3540.

Konur, O. 2012a. 100 citation classics in Energy and Fuels. *Energy Education Science and Technology Part A—Energy Science and Research* 30(si 1):319–332.

Konur, O. 2012b. What have we learned from the citation classics in Energy and Fuels: A mixed study. *Energy Education Science and Technology Part A—Energy Science and Research* 30(si 1):255–268.

Konur, O. 2012c. The gradual improvement of disability rights for the disabled tenants in the UK: The promising road is still ahead. *Social Political Economic and Cultural Research* 4:71–112.

Konur, O. 2012d. The policies and practices for the academic assessment of blind students in higher education and professions. *Energy Education Science and Technology Part B: Social and Educational Studies* 4(si 1):240–244.

Konur, O. 2012e. Prof. Dr. Ayhan Demirbas' scientometric biography. *Energy Education Science and Technology Part A—Energy Science and Research* 28:727–738.

Konur, O. 2012f. The evaluation of the research on the biofuels: A scientometric approach. *Energy Education Science and Technology Part A—Energy Science and Research* 28:903–916.

Konur, O. 2012g. The evaluation of the research on the biodiesel: A scientometric approach. *Energy Education Science and Technology Part A—Energy Science and Research* 28:1003–1014.

Konur, O. 2012h. The evaluation of the research on the bioethanol: A scientometric approach. *Energy Education Science and Technology Part A—Energy Science and Research* 28:1051–1064.

Konur, O. 2012i. The evaluation of the research on the microbial fuel cells: A scientometric approach. *Energy Education Science and Technology Part A—Energy Science and Research* 29:309–322.

Konur, O. 2012j. The evaluation of the research on the biohydrogen: A scientometric approach. *Energy Education Science and Technology Part A—Energy Science and Research* 29:323–338.

Konur, O. 2012k. The evaluation of the biogas research: A scientometric approach. *Energy Education Science and Technology Part A—Energy Science and Research* 29:1277–1292.

Konur, O. 2012l. The scientometric evaluation of the research on the production of bio-energy from biomass. *Biomass and Bioenergy* 47:504–515.

Konur, O. 2012m. The evaluation of the global energy and fuels research: A scientometric approach. *Energy Education Science and Technology Part A—Energy Science and Research* 30:613–628.

Konur, O. 2012n. The evaluation of the biorefinery research: A scientometric approach. *Energy Education Science and Technology Part A—Energy Science and Research* 30(si 1):347–358.

Konur, O. 2012o. The evaluation of the bio-oil research: A scientometric approach. *Energy Education Science and Technology Part A—Energy Science and Research* 30(si 1):379–392.

Konur, O. 2012p. The evaluation of the research on the biofuels: A scientometric approach. *Energy Education Science and Technology Part A—Energy Science and Research* 28:903–916.

Konur, O. 2013. What have we learned from the research on the International Financial Reporting Standards (IFRS)? A mixed study. *Energy Education Science and Technology Part D: Social Political Economic and Cultural Research* 5:29–40.

Konur, O. 2015a. Algal economics and optimization. In: Kim, S.K. and Lee, C.G. (Eds.). *Marine Bioenergy: Trends and Developments*, pp. 691–716. Boca Raton, FL: CRC Press.

Konur, O. 2015b. Algal high-value consumer products. In: Kim, S.K. and Lee, C.G. (Eds.). *Marine Bioenergy: Trends and Developments*, pp. 653–682. Boca Raton, FL: CRC Press.

Konur, O. 2015c. Algal photobioreactors. In: Kim, S.K. and Lee, C.G. (Eds.). *Marine Bioenergy: Trends and Developments*, pp. 81–108. Boca Raton, FL: CRC Press.

Konur, O. 2015d. Algal biosorption of heavy metals from wastes. In: Kim, S.K. and Lee, C.G. (Eds.). *Marine Bioenergy: Trends and Developments*, pp. 597–626. Boca Raton, FL: CRC Press.

Konur, O. 2015e. Current state of research on algal bioelectricity and algal microbial fuel cells. In: Kim, S.K. and Lee, C.G. (Eds.). *Marine Bioenergy: Trends and Developments*, pp. 527–556. Boca Raton, FL: CRC Press.

Konur, O. 2015f. Current state of research on algal biodiesel. In: Kim, S.K. and Lee, C.G. (Eds.). *Marine Bioenergy: Trends and Developments*, pp. 487–512. Boca Raton, FL: CRC Press.

Konur, O. 2015g. Current state of research on algal biohydrogen. In: Kim, S.K. and Lee, C.G. (Eds.). *Marine Bioenergy: Trends and Developments*, pp. 393–422. Boca Raton, FL: CRC Press. Publication no. MB/4.

Konur, O. 2015h. Current state of research on algal biomethanol. In: Kim, S.K. and Lee, C.G. (Eds.). *Marine Bioenergy: Trends and Developments*, pp. 327–370. Boca Raton, FL: CRC Press.

Konur, O. 2015i. Current state of research on algal biomethane. In: Kim, S.K. and Lee, C.G. (Eds.). *Marine Bioenergy: Trends and Developments*, pp. 273–302. Boca Raton, FL: CRC Press.

Konur, O. 2015j. Current state of research on algal bioethanol. In: Kim, S.K. and Lee, C.G. (Eds.). *Marine Bioenergy: Trends and Developments*, pp. 217–244. Boca Raton, FL: CRC Press.

Konur, O. 2015k. Algal photosynthesis, biosorption, biotechnology, and biofuels. In: Kim, S.K. (Eds.). *Springer Handbook of Marine Biotechnology*, pp. 1131–1161. Berlin, Germany: Springer.

Konur, O. 2015l. The scientometric study of the global energy research. In: Prasad, R., Sivakumar, S. and Sharma, U.C. (Eds.). *Energy Science and Technology*. Vol.1, Opportunities and Challenges, pp. 475–489. Houston, TX: Studium Press LLC.

Konur, O. 2015m. The review of citation classics on the global energy research. In: Prasad, R., Sivakumar, S., and Sharma, U.C. (Eds.). *Energy Science and Technology*. Vol.1, Opportunities and Challenges, pp. 490–526. Houston, TX: Studium Press LLC.

Konur, O. 2016a. Scientometric overview regarding the surface chemistry of nanobiomaterials. In: Grumezescu, A.M. (Ed.). *Surface Chemistry of Nanobiomaterials, Applications of Nanobiomaterials*. Vol. 3, pp. 463–486. Amsterdam, the Netherlands: Elsevier.

Konur, O. 2016b. Scientometric overview regarding the nanobiomaterials in antimicrobial therapy. In: Grumezescu, A.M. (Ed.). *NanoBioMaterials in Antimicrobial Therapy, Applications of Nanobiomaterials*, pp. 511–535. Amsterdam, the Netherlands: Elsevier.

Kumar, S., K. Tamura, and N. Nei. 2004. MEGA3: Integrated software for molecular evolutionary genetics analysis and sequence alignment. *Briefings in Bioinformatics* 5:150–163.

Kuzhabekova, A. and J. Kuzma. 2014. Mapping the emerging field of genome editing. *Technology Analysis & Strategic Management* 26:321–352.

Lal, R. 2004. Soil carbon sequestration impacts on global climate change and food security. *Science* 304:1623–1627.

Leliaert, F., D.R. Smith, H. Moreau, M.D. Herron, H. Verbruggen, and C.F. Delwiche. 2012. Phylogeny and molecular evolution of the green algae. *Critical Reviews in Plant Sciences* 31:1–46.

Lemaire, S.D., B. Guillon, P. Le Marechal, E. Keryer, M. Miginiac-Maslow, and P. Decottignies. 2004. New thioredoxin targets in the unicellular photosynthetic eukaryote *Chlamydomonas reinhardtii*. *Proceedings of the National Academy of Sciences of the United States of America* 101:7475–7480.

Li, Y., M. Horsman, N. Wu, C.Q. Lan, and N. Dubois-Calero. 2008. Biofuels from microalgae. *Biotechnology Progress* 24(4):815–820.

Lindell, D., M.B. Sullivan, Z.I. Johnson, A.C. Tolonen, F. Rohwer, and S.W. Chisholm. 2004. Transfer of photosynthesis genes to and from *Prochlorococcus* viruses. *Proceedings of the National Academy of Sciences of the United States of America* 101:11013–11018.

Liska, A.J., A. Shevchenko, U. Pick, and A. Katz. 2004. Enhanced photosynthesis and redox energy production contribute to salinity tolerance in *Dunaliella* as revealed by homology-based proteomics. *Plant physiology* 136:2806–2817.

Lynd, L.R., J.H. Cushman, R.J. Nichols, and C.E. Wyman. 1991. Fuel ethanol from cellulosic biomass. *Science* 251:1318–1323.

Mansiaux, Y. and F. Carrat. 2012. Contribution of genome-wide association studies to scientific research: A bibliometric survey of the citation impacts of GWAS and candidate gene studies published during the same period and in the same journals. *PLOS ONE* 7:e51408.

Molnar, A., A. Bassett, E. Thuenemann, F. Schwach, S. Karkare, S. Ossowski et al. 2009. Highly specific gene silencing by artificial microRNAs in the unicellular alga *Chlamydomonas reinhardtii*. *Plant Journal* 58:165–174.

Naumann, B., A. Busch, J. Allmer, E. Ostendorf, M. Zeller, and H. Kirchhoff. 2007. Comparative quantitative proteomics to investigate the remodeling of bioenergetic pathways under iron deficiency in *Chlamydomonas reinhardtii*. *Proteomics* 7:3964–3979.

North, D. 1994. Economic-performance through time. *American Economic Review* 84:359–368.

Paladugu, R., M.S. Chein, S. Gardezi, and L. Wise. 2002. One hundred citation classics in general surgical journals. *World Journal of Surgery* 26:1099–1105.

Palenik, B., Q. Ren, C.L. Dupont, G.S. Myers, J.F. Heidelberg, J.H. Badger et al. 2006. Genome sequence of *Synechococcus* CC9311: Insights into adaptation to a coastal environment. *Proceedings of the National Academy of Sciences of the United States of America* 103:13555–13559.

Price, G.D., M.R. Badger, F.J. Woodger, and B.M. Long. 2008. Advances in understanding the cyanobacterial CO_2-concentrating-mechanism (CCM): Functional components, Ci transporters, diversity, genetic regulation and prospects for engineering into plants. *Journal of Experimental Botany* 59:1441–1461.

Pulz, O. 2001. Photobioreactors: Production systems for phototrophic microorganisms. *Applied Microbiology and Biotechnology* 57:287–293.

Schwanhausser, B., D. Busse, N. Li, G. Dittmar, J. Schuchhardt, J. Wolf et al. 2011. Global quantification of mammalian gene expression control. *Nature* 473:337–342.

Spolaore, P., C. Joannis-Cassan, E. Duran, and A. Isambert. 2006. Commercial applications of microalgae. *Journal of Bioscience and Bioengineering* 101:87–96.

Swingley, W.D., M. Chen, P.C. Cheung, A.L. Conrad, L.C. Dejesa, J. Hao et al. 2008. Niche adaptation and genome expansion in the chlorophyll *d*-producing cyanobacterium *Acaryochloris marina*. *Proceedings of the National Academy of Sciences of the United States of America* 105:2005–2010.

Tamura, K., J. Dudley, M. Nei, and S. Kumar. 2007. MEGA4: Molecular evolutionary genetics analysis (MEGA) software version 4.0. *Molecular Biology and Evolution* 24:1596–1599.

Turmel, M., C. Otis, and C. Lemieux. 2006. The chloroplast genome sequence of *Chara vulgaris* sheds new light into the closest green algal relatives of land plants. *Molecular Biology and Evolution* 23:1324–1338.

Van Mooy, B.A., G. Rocap, H.F. Fredricks, C.T. Evans, and A.H. Devol. 2006. Sulfolipids dramatically decrease phosphorus demand by picocyanobacteria in oligotrophic marine environments. *Proceedings of the National Academy of Sciences of the United States of America* 103:8607–8612.

Vaulot, D., W. Eikrem, M. Viprey, and H. Moreau. 2008. The diversity of small eukaryotic phytoplankton (≤ 3 μm) in marine ecosystems. *FEMS Microbiology Reviews* 32:795–820.

Volesky, B. and Z.R. Holan. 1995. Biosorption of heavy-metals. *Biotechnology Progress* 11:235–250.

Wang, B., Y.Q. Li, N. Wu, and C.Q. Lan. 2008. CO_2 bio-mitigation using microalgae. *Applied Microbiology and Biotechnology* 79:707–718.

Whitelegge, J.P., H. Zhang, R. Aguilera, R.M. Taylor, and W.A. Cramer. 2002. Full subunit coverage liquid chromatography electrospray ionization mass spectrometry (LCMS+) of an oligomeric membrane protein cytochrome b_6f complex from spinach and the cyanobacterium *Mastigocladus laminosus*. *Molecular & Cellular Proteomics* 1:816–827.

Wonkam, A., M.A. Kenfack, W.F.T. Muna, and O. Ouwe-Missi-Oukem-Boyer. 2011. Ethics of human genetic studies in Sub-Saharan Africa: The case of Cameroon through a bibliometric analysis. *Developing World Bioethics* 11:120–127.

Wrigley, N. and S. Matthews. 1986. Citation-classics and citation levels in geography. *Area* 18:185–194.

Wu, H.W., J.V. Volponi, A.E. Oliver, A.N. Parikh, B.A. Simmons, and S. Singh. 2011. In vivo lipidomics using single-cell Raman spectroscopy. *Proceedings of the National Academy of Sciences of the United States of America* 108:3809–3814.

Yergin, D. 2006. Ensuring energy security. *Foreign Affairs* 85:69–82.

Section II

Marine Genomics

3

Omics Advances of Biosynthetic Pathways of Isoprenoid Production in Microalgae

J. Paniagua-Michel and Venkataramanan Subramanian

CONTENTS

3.1 Introduction

The increasing need for alternatives to fossil hydrocarbon-based chemicals and the recent advances in omics and metabolic engineering are promissory options for the development of biobased solutions for bioproducts and cell factories. Recent advances in omics in microalgae are now orientated toward the optimization and diversification of the production and applications of bioproducts. Moreover, omics approaches contribute to redefine metabolism, functional diversity, and biosynthetic capacity of fluxes of compounds of strains into a desired biosynthetic pathway (Guarnieri and Pienkos 2014). During the last couple of decades, microalgae have been considered as a suitable and valuable feedstock for biofuel, pharmaceuticals, nutraceutical, biomedical, and materials applications (Paniagua-Michel et al. 2014). Among the ubiquitous metabolites in microalgae, isoprenoids are one of the largest classes of small molecules found on earth with a wide variety of important functions and applications. Isoprenoids (also referred to as terpenes) constitute one of the most diverse groups of natural products in nature and derived from the universal precursors isopentenyl diphosphate (IPP) and its isomer, dimethylallyl diphosphate (DMAPP) (Paniagua-Michel et al. 2012; Subramanian et al. 2012). As a rule, isoprenoids hold a skeleton containing between 5 and 30 carbons, and their respective modified compounds, usually referred to as hemi-, mono-, sesqui-, di-, and triterpenes, composed of 5, 10, 15, 20, and 30 carbons, respectively (Withers and Keasling 2007). The biosynthesis of isoprenoids is ubiquitous among the three domains of life (eubacteria, archaebacteria, and eukaryotes).

Isoprenoids are often as intriguingly complicated as bioactive natural products, making chemical synthesis and supply uneconomic (Klein-Marcuschamer et al. 2007), which is why much effort has been directed toward their production in microbial hosts. It is expected that in the following decades, the large challenges and barriers that prevent present technology from producing a significant fraction of pharmaceuticals, fine chemicals, and polymers could be overcome through biotechnological processes and omics technologies (Hatti-Kaul et al. 2007). Moreover, supply, target identification, toxicity, and complex structure are among other equally important obstacles. The recent advances in "omics" technologies in life sciences are contributing to elucidate the full potential of microalgae as valuable feedstocks, with utility in industrial biotechnology and in a variety of applications. In this chapter, the current status of microalgal isoprenoids and the role of omics technologies, or otherwise specified, in bioproduct optimization and applications are reviewed. Emphasis is focused in the metabolic pathways of microalgae involved in the production of commercially important products, namely, hydrocarbons and biofuels, nutraceuticals, and pharmaceuticals.

3.2 Isoprenoids: Functions and Applications

Isoprenoids exhibit important biochemical functions, for instance, they play an important role as photosynthetic pigments (carotenoids, side chain of chlorophyll) and as quinones in electron transport chains and are also integral parts of membranes (prenyllipids in archaebacteria and sterols in eubacteria and eukaryotes) and in regulation (prenylation of proteins). They are also hormones (gibberellins, brassinosteroids, abscisic acid) and plant defense compounds (monoterpenes, sesquiterpenes, diterpenes) (Lange et al. 2000). Among the different classes of isoprenoids, the site of synthesis varies, which is reflected in the conformational structure of products. Thus, sesquiterpenes are typically synthesized in the cytosol, while carotenoid pigments and sterols in plastids. Under these conditions, several derivative branches of the isoprenoid pathway are responsible to the biosynthesis and storage of structurally and functionally diverse compounds in the different cellular compartments (Jørgensen et al. 2005).

Actually, there are a varied number of compounds derived from isoprenoids, which can be used in a diverse number of products and holds great potential for a rich source of commercial products, such as therapeutics, nutraceuticals, and fine chemicals (Klein-Marcuschamer et al. 2007). At present, there are isoprenoid products used in cancer therapy, the treatment of infectious diseases, and crop protection (Withers and Keasling 2007), as well as in hydrocarbon biosynthesis and biofuel production. During the past decade, biosynthesis of isoprenoids has focused mostly on carotenoids, which can be used as food colorants, nutraceuticals, cosmetic products, and pharmaceuticals. The properties and characteristics of the carotenoid molecules place them as versatile markers for new tools and methods for metabolic engineering (see Figure 3.3). Some of these strategies have been specifically designed to improve yield; others have been used as proof of concept of diverse isoprenoid overproduction and applications. The carotenoids are a class of tetraterpenoid isoprenoid ubiquitous in all photosynthetic organisms as in microalgae. They serve a number of uses ranging from therapeutics, biofuels, pharmaceuticals, nutraceuticals, and fine chemicals to coloring agents for agriculture and aquaculture. They are also used as precursors of vitamin A in humans (Yeum and Russell 2002; Muntendam et al. 2009) and applied in prevention of macular degeneration (Landrum and Bone 2001; and a useful anticancer agents (Bhuvaneswari and Nagini 2005). Particularly, algal β-carotene is preferred by the health market because it is a mixture of *trans* and *cis* isomers and widely used as anticancer preventive medicine; such a mixture could be difficult to be obtained by chemical synthesis. Some carotenoids, such as canthaxanthin, astaxanthin, and lutein, are common source of pigments and have accordingly been included as ingredients of feed for salmonid fish and trout, as well as poultry. Recently, β-carotene has been the object of intense demand as provitamin A (retinol) in multivitamin preparations; it is usually included in the formulation of healthy foods, mainly as antioxidant. Recently, the uses of isoprenoids as alternative substrates for generation of biofuels have been intensified. The interplay of isoprenoid precursors and regulation of major pathways of biosynthesis in algae is under intense research for novel biofuels by carbon partitioning between the isoprenoids and fatty acids (Subramanian et al. 2012; Hildebrand et al. 2013).

Moreover, many carotenoid enzymes are catalytically promiscuous and accept substrates other than their natural ones, enabling synthesis of even more diverse molecular architectures (Klein-Marcuschamer et al. 2007). Because all isoprenoids share common metabolic precursors, they are also considered suitable genetic platforms for the biosynthesis of carotenoids and other valuable metabolites. The recent developments on genomics and other approaches from systems biology, namely, proteomics, transcriptomics, metabolomics, computational biology, and synthetic biology, have contributed with new tools for manipulating metabolic pathways of isoprenoids. Figure 3.1 exemplifies the strategy of functional genomics in eliciting metabolism of isoprenoids.

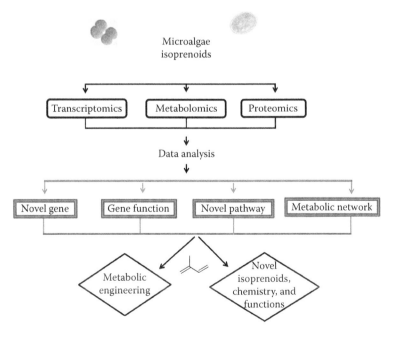

FIGURE 3.1
Scheme of the strategy for the use of functional genomics in the metabolic engineering of isoprenoids from microalgae. (Modified after Yang, D. et al., *Eng. Life Sci.*, 14, 456, 2014.)

The developments and discovery on new and diversified applications of isoprenoid products have placed microalgae as a valuable feedstock as biofuel, nutraceutical, biomedical, and materials science applications (Guarnieri and Pienkos 2014). Thus, the discovery of novel compounds or pathways with the recent advances in omics technologies may contribute to elucidate new important aspects of isoprenoids and carotenoids for many human benefits.

3.3 Pathways of Isoprenoid Biosynthesis in Microalgae: Omics Advances

Isoprenoids are typically synthesized via two different metabolic pathways: the mevalonate (MVA) pathway and the deoxyxylulose (DXP) pathway. The DXP pathway is also called as the mevalonate-independent pathway (MVI) or the methylerythritol pathway (MEP). Both these pathways yield the five-carbon (5C) monomer IPP and its isomer, DMAPP. These two precursor compounds are further condensed to yield longer chain isoprenoids that can serve as alternative fuels. Whereas the synthesis of sterols occurs in the cytosol of higher plants via the MVA pathway, synthesis of carotenoids and phytols occurs in the plastids via the DXP pathway (Lichtenthaler 1999). In contrast, the synthesis of terpenoids occurs via the DXP pathway in green algae such as *Botryococcus braunii*, *Scenedesmus obliquus*, *Chlamydomonas* sp., *Chlorella fusca* (Disch et al. 1998; Rager and Metzger 2000; Metzger et al. 2002, 2003; Sato et al. 2003; Achitouv et al. 2004), and *Dunaliella salina* (Capa-Robles et al. 2009; Paniagua-Michel et al. 2009). The red alga *Cyanidium caldarium* exhibits a dichotomy in that the cytosolic MVA pathway is used to synthesize sterols, while phytols are synthesized via the chloroplastic DXP pathway (Schwender et al. 1997; Disch et al. 1998; Lichtenthaler 1999). The phytoflagellate *Euglena gracilis* is known to produce both phytols and sterols via the MVA pathway (Schwender et al. 1997; Disch et al. 1998), and cyanobacteria synthesize phytol and β-carotene via the DXP pathway (Disch et al. 1998).

3.3.1 DXP Pathway

The DXP pathway is initiated by the formation of 1-deoxyxylulose-5-phosphate (DXP) from the glycolysis intermediates, glyceraldehyde-3-phosphate and pyruvate (Rohmer et al. 1996; Paniagua-Michel et al. 2012; Figure 3.2). DXP is reduced to 2-C-methylerythritol-4-phosphate (MEP) and then undergoes a conjugation reaction with cytidine-5-phosphate (CTP) to form 4-(cytidine 5′-diphospho)-2-C-methylerythritol (CDP-ME), which further gets phosphorylated to form 4-(cytidine 5′-diphospho)-2-C-methylerythritol 2-phosphate (CDP-MEP). CDP-MEP then gets converted to 2-C-methylerythritol 2,4-cyclodiphosphate (ME-cPP) and to hydroxymethylbutenyl 4-diphosphate (HMBPP). The latter is reduced by NADPH, yielding a mixture of IPP and DMAPP.

3.3.2 MVA Pathway

The mevalonate pathway is initiated by the condensation of two molecules of acetyl-CoA to produce acetoacetyl-CoA (Figure 3.2). This product undergoes another round of condensation to give 3-hydroxy-3-methylglutaryl-CoA (HMG-CoA), which is subsequently

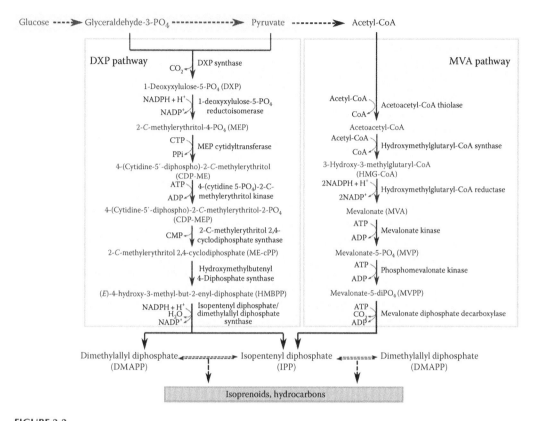

FIGURE 3.2

Isoprenoid biosynthesis pathways. DXP pathway (left panel)—the pathway originates from the glycolysis inter-mediates, glyceraldehyde-3-phosphate and pyruvate. The final products of the DXP pathway are a mixture of isopentenyl diphosphate (IPP) and dimethylallyl diphosphate (DMAPP). MVA pathway (right panel)—the pathway is initiated by a Claisen condensation reaction of two acetyl-CoA molecules yielding acetoacetyl-CoA, followed by another round of condensation, reduction by NADPH, two phosphorylation reactions, and a final ATP-dependent carboxylation reaction generating IPP, which undergoes isomerization to yield DMAPP. The green box represents the glycolytic pathway.

reduced by NADPH to form mevalonic acid (MVA). Two sequential phosphorylation reactions follow to yield mevalonate 5-phosphate (MVP) and mevalonate 5-diphosphate (MVPP). MVPP is then carboxylated in an ATP-dependent reaction to yield IPP, which is further converted to DMAPP via an isomerization reaction.

3.3.3 Omics of Isoprenoid Biosynthesis in Algae and Cyanobacteria

With the increasing demand for alternate sources of fuels to replace petroleum, focus has shifted toward biofuel research. Biofuels have advanced from being pro-duced from crop-based sources (first-generation biofuel) to lignocellulosic materi-als (second-generation biofuels) and most recently toward the use of photosynthetic microorganisms; the latter can harvest carbon dioxide (CO_2) from the environment, thereby eliminating the conflict with food/fuel and land use (advanced fuels). In fact, CO_2 can also be obtained from industries that produce it as a pollutant waste, thereby making the technology an environmentally friendly means of biofuel generation.

Among the different sources of liquid biofuels, this chapter discusses hydrocarbon-based biofuels. Liquid fuels comprise C4–C12-containing hydrocarbons (i.e., gasoline), C9–C23-containing hydrocarbons (i.e., diesel), and C8–C16-containing hydrocarbons (i.e., jet fuel). All these molecules arise from a common isoprenoid pathway, from the 5C precursors IPP and DMAPP, which polymerize to longer chains that serve as substitutes for fuels (Figure 3.2). Omics tools in relation to isoprenoid biosynthesis from algae and cyanobacteria have only recently been explored. Due to the limited knowledge in this field, we have expanded this section to cover other hydrocarbon-based biofuels, including alkanes and triacylglycerols (TAGs).

3.3.4 Genomics of Algae and Cyanobacteria in Relation to Hydrocarbon Synthesis

The recent sequencing of the red alga, *Porphyridium purpureum*, revealed the identity of many genes involved in the MEP pathway (Bhattacharya et al. 2013). Although the two genes, acetoacetyl-CoA thiolase and 3-hydroxy-3-methylglutaryl-CoA synthase, were identified in the genome, they are known to be shared by other biochemical pathways, and hence their identification does not prove the existence of the MVA pathway in this organism. Moreover, the gene encoding the enzyme, 3-hydroxy-3-methylglutaryl-CoA reductase, which is the first and committing step for the MVA pathway, was not detected (Bhattacharya et al. 2013), suggesting that the MVA pathway is absent in *P. purpureum*. In contrast, all the genes required for the MEP pathway were identified, thus suggesting that this organism has the necessary genetic information for synthesizing the 5C isoprenoid precursors. Furthermore, four genes that code for prenyltransferases were identified (Bhattacharya et al. 2013). These enzymes catalyze the condensation of the 5C precursors to yield longer chain isoprenoid compounds. Another red alga, *Pyropia yezoensis*, was recently sequenced, but its genome has not yet been analyzed with respect to biofuel production capability (Nakamura et al. 2013).

Other algal genomes, such as those of the model unicellular alga, *Chlamydomonas reinhardtii*, and the model multicellular alga, *Volvox carteri*, have been sequenced and annotated (Merchant et al. 2007; Prochnik et al. 2010). However, these genomes were primarily used toward understanding selected photosynthetic processes, cell motility, and hydrogen (H_2) and lipid production. *Chlorella variabilis* NC64A was sequenced and studied with respect to virus/algal interactions, although this genus is also known to accumulate higher amounts of lipids that serve as a potential source of biofuel (Blanc et al. 2010; Guarnieri et al. 2013). The halo-tolerant green alga *D. salina* was sequenced and analyzed due to its tolerance to saline environments (Polle et al. 2009) as well as its ability to produce β-carotene (Ben-Amotz et al. 1982; Katz et al. 1995) and its potential for production of lipids as biofuel precursors (Smith et al. 2010).

Cyanobacteria have been primarily studied for their ability to produce biofuel compounds/precursors, which include H_2, isoprene, ethanol, ethylene, and hexane. Whereas isoprene is synthesized from the MEP pathway, alkanes originate from intermediates of the fatty acid metabolism (Schirmer et al. 2010). It has been shown that *Synechocystis* sp., PCC6803, is capable of producing the 5C compound isoprene via the MEP pathway, by the introduction of the nonnative isoprene synthase gene, *IspS*, from the vine *Pueraria montana* (kudzu, Lindberg et al. 2010). Ever since the first sequencing report of the photosynthetic autotroph, cyanobacteria *Synechocystis* sp. PCC6803, a huge number of cyanobacterial genomes have been sequenced (Kaneko and Tabata 1997; Shih et al. 2013), thus providing a comprehensive view of the gene organization in this class of biofuel-relevant organisms.

3.3.5 Transcriptomics of Algae and Cyanobacteria in Relation to Hydrocarbon Synthesis

The most studied microalga toward hydrocarbon production is the green colonial microalga *B. braunii*. This alga is known to differ in its ability to produce different types of hydrocarbons and is classified accordingly into three races, A, B, and L. Whereas n-alkadiene and triene hydrocarbons (C23–C33) are produced by race A strains (Metzger et al. 1985a), triterpenoid hydrocarbons/botryococcenes (C30–C37, Metzger et al. 1985b) and methylated squalenes (C31–C34) (Huang and Poulter 1989; Achitouv et al. 2004) are produced by race B strains, and lycopadiene, a tetraterpenoid hydrocarbon (C40, Metzger and Casadevall 1987; Metzger et al. 1990), is produced by race L strains. Figure 3.3 shows a schematic representation of the potential uses of botryococcene from *B. braunii* as a likely precursor of biofuels, according to Niehaus et al. (2011).

Although the genome of *B. braunii* has not been sequenced, there have been a few reports on transcriptome analyses of *B. braunii* races A and B strains (Baba et al. 2012; Ioki et al. 2012a,b; Molnar et al. 2012). The race A strain, BOT-88-2, and the race B strain, BOT-22, were sequenced using the 454 pyrosequencing technique in a Genome Sequencer-FLX system (Baba et al. 2012; Ioki et al. 2012a). This yielded a total of 29,038 and 27,427 nonredundant sequences of over 2 kb for BOT-88-2 (Baba et al. 2012) and BOT-22 (Ioki et al. 2012a) strains, respectively, of which 964 and 725 sequences were annotated successfully based on sequence similarity. Whereas putative genes involved in very-long-chain fatty acid (VLCFA) synthesis were identified in BOT-88-2, identification of genes involved in the conversion of unsaturated VLCFA to hydrocarbons was not identified (Baba et al. 2012). In contrast, BOT-22 strain revealed the presence of 17 nonredundant sequences

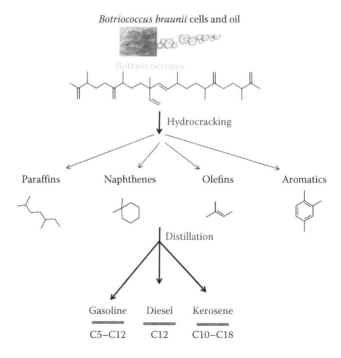

FIGURE 3.3

Schematic pathway of the isoprenoid botryococcene from *Botryococcus braunii* as a likely potential precursor of combustible fuels. (Modified after Niehaus, T.D. et al., *Proc. Natl. Acad. Sci. USA*, 108, 12260, 2011.)

associated with botryococcene synthesis (Ioki et al. 2012a). Furthermore, transcriptome analysis suggested that botryococcene oil biosynthesis occurs via the MVI pathway in the race B strain, BOT-22. Transcriptome of another race B strain, BOT-70, was analyzed by sequencing the 5′ ends of a full-length enriched cDNA library using the dye terminator method (Ioki et al. 2012b). In contrast to the pyrosequencing methods, only 9345 high-quality sequences were obtained, of which 509 sequences were annotated and functionally categorized. Consistent with the race B strain, a large number of lipid metabolism–related genes were associated with terpenoid biosynthesis. In particular, 10 nonredundant sequences associated with botryococcene synthesis were discovered (Ioki et al. 2012b). These included four out of eight genes involved in MVI pathway and four of the five genes involved in triterpene biosynthesis from IPP (Ioki et al. 2012b). Thus, the transcriptome profile indicated that the synthesis of botryococcene in the race B strain most likely occurs through the MVI pathway (Ioki et al. 2012b). The BOT-70 transcriptome also revealed the presence of an alternate entry route into the MVI pathway, that is, via xylulose 5-phosphate, which is similar to cyanobacteria, suggesting that photosynthetic products from the reductive pentose phosphate pathway could be directed toward IPP production (Ioki et al. 2012b). This study also suggested that the initial reactions of botryococcene biosynthesis in BOT-70 occur in the chloroplast, whereas the final reactions are catalyzed outside the chloroplast (Ioki et al. 2012b).

These studies were followed by another in-depth transcriptomic study that provided a global view of terpenome biosynthesis in the *B. braunii* race B strain, Showa (Molnar et al. 2012). The authors used *de novo* assembly of the transcriptome, annotated and curated the genes for the enzymes, and reconstructed pathways involved in the biosynthesis of liquid terpenoid compounds that originate from the photosynthetic precursor 3-phosphoglycerate in addition to competing storage compounds like fatty acids, TAGs, and starch. They further went on to determine the possible routes for export of the hydrocarbon compounds into the extracellular environment (Molnar et al. 2012).

Transcriptomic analysis using pyrosequencing has also been applied to the nonmodel algal species, *Dunaliella tertiolecta*, which accumulates up to 67% lipids (Takagi et al. 2006; Rismani-Yazdi et al. 2011). This study revealed the identity of most of the enzymes involved in fatty acid synthesis, desaturation, elongation, and catabolism, as well as enzymes involved in TAG synthesis (Rismani-Yazdi et al. 2011). Starch biosynthesis was also confirmed in this organism, as evidenced by the presence of the respective enzymes (Rismani-Yazdi et al. 2011). Starch is known to serve as a carbon storage molecule in algae, which can be directed toward the production of other biofuels such as H_2 and ethanol (Gfeller and Gibbs 1984; Gibbs et al. 1986; Mus et al. 2007; Dubini et al. 2009; Rismani-Yazdi et al. 2011). Another oleaginous microalga, *Neochloris oleoabundans*, which accumulates up to 55% lipids, was analyzed using RNA-Seq technology (Tornabene et al. 1983; Li et al. 2008; Rismani-Yazdi et al. 2012). Consistent with its lipid-accumulating capacity, the fatty acid biosynthesis pathway in this organism was found to be upregulated along with a simultaneous downregulation of fatty acid degradation genes, under nitrogen limitation (i.e., lipid-accumulating conditions). In addition, a few genes involved in TAG formation were also found to be upregulated, indicating TAG accumulation under the same conditions (Rismani-Yazdi et al. 2012). Finally, using a combination of *de novo* assembled transcriptome and proteome analyses, Guarnieri et al. were able to identify fatty acid and TAG synthesis pathways in the high-lipid (~60% fatty acids) producing strain, *Chlorella vulgaris* (Guarnieri et al. 2011).

A selection strategy was used in conjunction with transcriptome analysis (using RNA-Seq) in the microalga *Tisochrysis lutea* (Tiso; Carrier et al. 2014), which identified the genomic-level changes that led to improved lipid accumulation. Briefly, the UV mutation–induced high-lipid-producing Tiso strain (Tiso-S2M2) was compared with its parental wild-type strain (Tiso-Wt), which led to the identification of two transcripts, coding for long-chain fatty acid ligase and GDSL lipase, both of which were downregulated in the Tiso-S2M2 strain. These transcript-coding proteins are known to be involved in fatty acid catabolism, thereby suggesting their role in the observed increased accumulation of fatty acids in these strains (Carrier et al. 2014).

Several transcriptomic analyses of cyanobacteria have been published, albeit with emphasis on different organism adaptations. These include the impact of environmental stresses like temperature, light, and nutrient changes. The nutrient-based stresses include nitrogen starvation, saline stress, and mixotrophic versus photoheterotrophic conditions (Huang et al. 2013; Qiao et al. 2013; Nakajima et al. 2014). In relation to biofuel production, it was determined that under mixotrophic conditions, the oxidative pentose phosphate and the glycolytic pathways are activated, in comparison to autotrophic conditions (Yoshikawa et al. 2013). Another study focused in understanding how to improve free fatty acid production by cyanobacteria. Although cyanobacteria are known to excrete free fatty acids, their overall productivity is hampered by product toxicity. This toxicity of free fatty acids has been attributed to the disruption of the electron transport chain and oxidative phosphorylation, thereby affecting cellular energy production (Ruffing 2013). Using RNA-Seq analysis, 15 specific gene targets were identified as being differentially regulated in response to the presence of free fatty acids. These genes included hypothetical proteins, reactive-oxygen species–degrading proteins, and free fatty acid transporters (Ruffing 2013). Deletions of these genes resulted in improved growth, physiology, and production of free fatty acids. This study proved to be a classic example of the use of an omics technique to improve biofuel production in a photosynthetic organism. Similar studies have been published in relation to improvement of ethanol and butanol resistance in cyanobacteria, thereby increasing the productivity of these alcohols by the organism (Wang et al. 2012; Zhu et al. 2013).

3.3.6 Proteomics of Algae and Cyanobacteria in Relation to Hydrocarbon Synthesis

Whereas genomics and transcriptomics have been used more commonly in studies related to hydrocarbon production by algae and cyanobacteria, instances of proteomics techniques are still very limited. In one liquid chromatography (LC)–mass spectrometry (MS)-based proteomics study of the model green alga, *C. reinhardtii*, it was observed that three enzymes responsible for the synthesis of farnesyl pyrophosphate, geranylgeranyl pyrophosphate, and solanesyl pyrophosphate, which belong to the isoprenoid biosynthesis pathway, were downregulated in response to copper (Cu) deprivation (Hsieh et al. 2013). However, the role of Cu deprivation and its relevance to isoprenoid metabolism are unknown.

In another study, the cyanobacteria *Synechocystis* sp. PCC6803 proteome was investigated to reveal its cellular response to hexane (Liu et al. 2012). This study was carried out toward improving the knowledge of the mechanisms of alkane tolerance in cyanobacteria and thereby aimed at helping the development of a more alkane-tolerant and a robust alkane-producing strain. A quantitative approach using isobaric tag for relative and absolute quantification (iTRAQ) technique, combined with the LC–MS/MS technique, was used, which revealed molecular targets that can be engineered to produce better hexane-tolerant strains of cyanobacteria. iTRAQ proteomics has also been used in order to understand the response of cyanobacteria to nitrogen starvation (Huang et al. 2013).

Another proteomics technique, using 2D electrophoresis, was utilized to study the impact of UV-B exposure on cyanobacteria. This work revealed downregulation of proteins involved in the MVI pathway in this organism, and this result was attributed to their role in carotenoids and chlorophyll biosynthesis, thereby affecting the overall photosynthetic ability of the organism (Gao et al. 2009).

Proteomics techniques have been more prevalent in studies related to lipid production by green algae. Comparative proteomics was undertaken recently to study mutants that show increased and decreased lipid content with respect to wild-type *Chlamydomonas* cells (Choi et al. 2013). The 2D polyacrylamide gel electrophoresis (PAGE) combined with MALDI-TOF analysis led to the identification of 16 proteins that were exclusively detected in the wild-type cells. These proteins were suggested to be involved in algal growth and/or cell division (Choi et al. 2013). In addition, 17 proteins that showed increases in the low-lipid strain as opposed to the WT and the high-lipid-producing strain were also detected, suggesting their probable involvement in negatively regulating lipid synthesis (Choi et al. 2013). Nguyen et al. published an LC–MS/MS-based proteomics study on purified *C. reinhardtii* oil bodies, in order to understand the mechanisms of oil accumulation in algae (Nguyen et al. 2011b). Their study revealed the identity of 33 proteins that are involved in metabolism of lipids in this alga under nitrogen-deprivation conditions (Nguyen et al. 2011b). This study also showed that oil bodies are also involved in oil synthesis, degradation, and lipid homeostasis, in addition to just being lipid storage compartments (Nguyen et al. 2011). Prior to this study, Moellering and Benning used an MS approach to identify proteins enriched in lipid droplets (Moellering and Benning 2010). They identified a protein called the major lipid droplet protein (MLDP), which was found to be the most abundant protein in the lipid droplet fraction. This protein showed sequence similarity exclusively to proteins encoded in other green algal genomes such as in *Chlorella* sp. NC64A, *C. vulgaris*, and *V. carteri*. MDLP gene expression was downregulated, using an RNAi approach, which resulted in increased lipid droplet size (Moellering and Benning 2010).

There have been few reports of proteomics approaches applied to the high-lipid-producing algal genus *Chlorella*. Li et al. studied the effects of using a two-stage cultivation process on lipid production by *Chlorella protothecoides*, where algal cultivation in the first stage was done in nutrient-rich medium to initiate biomass accumulation and the second stage of cultivation was done in Cu-stress medium; Cu stress was used as a trigger for lipid accumulation (Li et al. 2013). Using 2D analysis, they showed that proteins involved in hydrolysis of starch to glucose and its further conversion to the lipid precursors, pyruvate and acetyl-CoA (via glycolysis), were upregulated under Cu-stress conditions, suggesting the importance of starch as a possible contributor to lipid synthesis. Additionally, proteins involved in the lipid synthesis pathway were upregulated, which was consistent with the increased rate of fatty acid synthesis (Li et al. 2013). Similar proteomics studies were undertaken in the microalga *Isochrysis galbana* under *N*-starvation conditions, further suggesting the role of glycolysis and the citrate transport system as the main route for lipid synthesis in this organism (Song et al. 2013). The *C. vulgaris* UTEX 395 strain has been investigated intensively using transcriptomics and proteomics techniques, due to its ability to accumulate large quantities of fatty acids and TAGs (Guarnieri et al. 2011, 2013). Pienko's group used GeLC/MS techniques to identify cell cycle regulator protein homologs such as right open reading frame (RIO) and cell untimely torn (cut4), as well as cell signaling regulator proteins such as constitutive triple response (ctr1), which is involved in ethylene signaling in plants (Chen et al. 2005), under nitrogen-replete conditions (Guarnieri et al. 2013). Additionally, they reported a rapid increase in abundance of the fatty acid synthesis protein, heteromeric acetyl-CoA carboxylase (ACCase), and the TAG biosynthesis–related protein, diacylglyceride acyltransferase, in response to nitrogen starvation, which was consistent with the observed fatty acid and TAG accumulation under these conditions (Guarnieri et al. 2013). Moreover, they also identified transcription factors such as the universal stress protein (USP), pTAC14, and SNF2, which are known to play roles in cell survival (via USP), replication, transcription, and plasticidal metabolism (via pTAC14) and lipid metabolism (via SNF2) in other organisms (Kerk et al. 2003; Kamisaka et al. 2007; Gonzalez-Ballester et al. 2010; Gao et al. 2011).

In conjunction with the transcriptomics study on Tiso strains (see Section 3.3.5), Cadoret's lab also conducted a proteomics study to compare the proteome of the mutant strain, Tiso-S2M2, to the wild-type Tiso strain (Garnier et al. 2014), in response to the lipid-producing, nitrogen starvation conditions. Using 2D electrophoresis, they identified 17 proteins that were regulated by nitrogen starvation during the lipid accumulation phase in both the strains. Additionally, they identified 33 proteins that were differentially regulated in the Tiso-S2M2 strain exclusively under nitrogen starvation conditions. In particular, proteins involved in the breakdown of carbohydrates to acetyl-CoA were found to be upregulated, which was consistent with the increased TAG accumulation by this organism, similar to observations in *C. reinhardtii* and *I. galbana* (Li et al. 2010; Song et al. 2013). This study also led to the identification of four genetic targets, the plastid beta-ketoacyl-ACP reductase, coccolith scale–associated protein, and two glycoside hydrolases, which could be involved in the overaccumulation of lipids under nitrogen starvation (Garnier et al. 2014).

3.3.7 Metabolomics of Algae and Cyanobacteria in Relation to Hydrocarbon Synthesis

Although metabolomics studies have been conducted in algae, they have been primarily confined to the model green alga *C. reinhardtii.* The focus of these studies has been to optimize the necessary metabolite extraction conditions, to understand biochemical pathways under different nutrient regimes, to assist genome annotation, and to study H_2 production (Bolling and Fiehn 2005; Lee do and Fiehn 2008; May et al. 2008; Matthew et al. 2009; Wienkoop et al. 2010; Nguyen et al. 2011a; Jamers et al. 2013). Other instances of metabolomics approach usage have been applied to algae such as *C. vulgaris* and *Scenedesmus vacuolatus* and cyanobacteria, such as *Fischerella ambigua, Synechocystis* sp. PCC6803, and *Nostoc* sp. PCC7120 (Lin et al. 2008; Krall et al. 2009; Sans-Piche et al. 2010; Zhang et al. 2014). To our knowledge, there have been no published metabolomics studies with respect to isoprenoid biosynthesis, although there have been few reports related to the production of other hydrocarbons such as lipids. For instance, Ito et al. quantified more than 300 metabolites from the green alga *Pseudochoricystis ellipsoidea* MBIC 11204 under nitrogen-rich (+N) and nitrogen-deprived (−N) conditions (Ito et al. 2013). Using two techniques, namely, capillary electrophoresis–MS and LC–MS techniques, they analyzed metabolites involved in central metabolism (sugars, organic acids, amino acids, etc.), as well as lipids (fatty acids, phospholipids, glycolipids, neutral lipids) and pigments (Ito et al. 2013). This study identified 16 metabolites that decreased dramatically under −N conditions, some of which are involved in N assimilation and N transport. Similarly, 34 metabolites increased under −N conditions, with 4 of them >20-fold (Ito et al. 2013). These four metabolites were molecular species of TAGs that are major components of oil bodies (Ito et al. 2013). Starch synthesis–related metabolites increased, which are metabolites involved in carbon fixation and/or pentose phosphate pathways (Ito et al. 2013). Interestingly, the levels of citrate, fructose 1,6-diphosphate, 6-phosphogluconate, and pyruvate were decreased under −N conditions in their study, in contrast to increased levels of these metabolites reported in

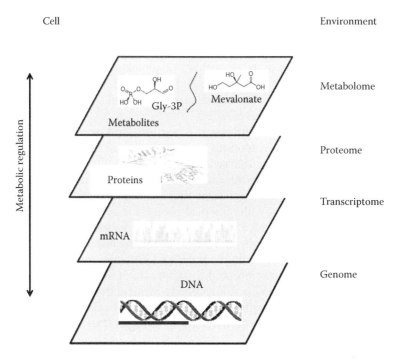

FIGURE 3.4
Representation of the expected integration of the different -omes interacting in a cell in a hierarchical order from the bottom to the top, which can be exemplified for the metabolic regulation of isoprenoid biosynthesis of the figure. (Modified after Nielsen, J. and Oliver, S., *Trends Biotechnol.*, 23, 544, 2005.)

C. reinhardtii under similar conditions (Bolling and Fiehn 2005; Ito et al. 2013). Overall, this study displayed snapshots of the metabolome of this organism at a single time point under N-starvation conditions (Ito et al. 2013). Recently, a lipidomics study of high-lipid-producing algae *Chlorella* sp. and *Nannochloropsis* sp. under N-deprivation condition was recently published (Martin et al. 2014). This study revealed the similarity between the two algal species in the remodeling intracellular lipid pools in order to increase TAG synthesis. For example, decreases in the chloroplast glycolipids, namely, monogalactosyldiacylglycerol, digalactosyldiacylglycerol, and sulfoquinovosyldiacylglycerol, and in phosphoglycerolipids, namely, phosphatidylcholine and phosphatidylethanolamine, and increases in the non-nitrogen-containing phosphatidylglycerol levels were observed (Martin et al. 2014). A schematic representation of the interaction of the different omics for metabolic regulation of isoprenoids is exemplified in Figure 3.4.

3.4 Isoprenoids, Metabolic Engineering, and Genomics of Microalgae

Since ancient times and more recently, society has benefited tremendously from isoprenoids. Recent biotechnological tools have contributed into the understanding of genes functions in isoprenoid production in photosynthetic systems, namely, microalgae. Metabolic engineering has the potential to manipulate and to reconstitute biosynthetic pathways of desirable compounds in hosts. The complete genomes of photosynthetic algae (about a dozen; http://genome.jgi-psf.org; jgi.doe.gov), represented by microalgal model organisms such as *C. reinhardtii, C. variabilis,* and *V. carteri,* have been registered to study the basic biology of these microalgae (Molnar et al. 2012) and to optimize yields of industrially important compounds. Actually, several genome projects are currently in progress for other algae. The genome of *C. reinhardtii* was one of the earliest algal genomes sequenced and is a useful reference for most molecular analyses (Grossman et al. 2007; Merchant et al. 2007; Ramos et al. 2011). It is expected that complete biotechnological developments based on these approaches shall contribute for new products and optimize algal commercial exploitation (Ramos et al. 2011).

The discovery and study of the genes that regulate the clever processes of the pathways of biosynthesis of metabolites from microalgae have been possible by the recent sequencing of transcripts from microalgal cultures grown under stressing and normal conditions. That next-generation sequencing has been a useful tool to discover pathways involved in TAG biosynthesis by assembling the complete transcriptome of the microalgae *D. tertiolecta* and *C. vulgaris* (Molnar et al. 2012). The new approaches, such as Algal Functional Annotation Tool, allow access to transcripts on pathway maps, including the statistically rigorous functional analysis of large sets of transcripts. Besides, data mining has facilitated the public analysis of large- and functional-scale datasets for algae. The possibility of the reassembling of biosynthetic-relevant genes from different biological sources in a heterologous microorganism may be exemplified by engineering isoprenoids into microbial hosts. In the case of microalgae, engineering of isoprenoids into microbial hosts is a strategic approach for the production of isoprenoid-derived compounds such as carotenoids. This strategy could provide more benefits compared to traditional approaches of complex isoprenoid compounds, mainly in aspects as stereochemically complex compounds. *D. salina* and *Haematococcus pluvialis* are the main sources of beta-carotene and astaxanthin, respectively, under commercial production. However, the developments on metabolic engineering for improved carotenoid production in algae are very scarce, and it is expected that in the following decades, commercial exploitation of engineered strains could be a reality, mainly in aspects related with carotenoid overproduction. The modern revolution in genomic banks and in the available well-documented and well-characterized libraries is an essential supporting tool to accomplish successful programs in metabolic engineering and synthetic biology (Anderson et al. 2010; Redding-Johanson et al. 2011). The advent of metagenomics-based approaches is regarded as an attractive option to screen large clone libraries for the discovery and functional activities of natural products from uncultured microorganisms, namely, microalgae and isoprenoids (Figure 3.5).

3.4.1 Genomics and Carotenoid Production in Microalgae

The design of microorganisms expressing specific features is possible with the basic key issue aspects: the introduction of heterologous pathways, the elimination of native reactions that are unnecessary or not physiologically significant, and the reengineering of the regulatory networks through rational or random approaches (Klein-Marcuschamer 2007). Due to the photosynthetic nature of microalgae, these engineering approaches in the pathway of biosynthesis of isoprenoid carotenoids must often be complemented by suitable photobioreactors aiming to optimize the full potential of the microbial catalyst for sustainable production (Kiss and Stephanopoulos 1991; Klein-Marcuschamer 2007). The stable transformation of an important number of microalgal species, including several Chlorophyta (e.g., *C. reinhardtii, V. carteri, Chlorella kessleri* (Radakovits et al. 2010), has been possible by genetic engineering. In the case of the halophilic *D. salina,* the lack of efficient

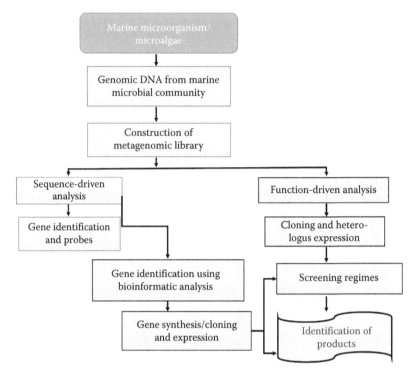

FIGURE 3.5

Omics-based strategies to identify novel compounds and biocatalysts from marine ecosystems. (Modified after Kennedy, J. et al., *Microb. Cell Fact.*, 7, 27, 2008.)

and highly active promoters, selective markers, reporter genes, and stable transformation methods (Coll 2006; Hallmann 2007) is an obstacle for the development of a stable and reproducible transformation system. Different *Dunaliella* species and strains, namely, *D. salina*, *D. tertiolecta*, and *D. viridis*, have been the main focus of genetic transformation experiments (Feng et al. 2009; Polle et al. 2009). Early engineering attempts performed by Geng et al. (2004) reported the nuclear transformation of *D. salina* and stable foreign expression of hepatitis B surface antigen (HbsAg) under the control of the maize ubiquitin (Ubil-Ω) promoter as well as the chloramphenicol acetyltransferase (CAT), a selectable gene under the control of the simian virus 40 (SV40) promoter (Ramos et al. 2011).

In general, carotenoid production and yield face difficulties because of the limited availability of the substrates, the accumulation of intermediates, and a limited storage capacity for end products (Marcuschamer 2007). It is known that in isoprenoids, metabolic fluxes can differ among families and species, as in the case of the production of astaxanthin by various hosts. The several studies performed on the production capacities of microalgae showed that wild-type production levels of astaxanthin differ dramatically between hosts (Muntendam et al. 2009). The highest production levels of astaxanthin were reported in a wild type of the Chlorophyte microalgae *H. pluvialis*, around 3% of dry weight (Bhosale and Bernstein 2005; Muntendam et al. 2009), while *Chlorella zofingiensis* produces about 0.1%, and *Xanthophyllomyces dendrorhous* (an industrial high producer), between 0.4% and 2.5% of dry weight (An et al. 1989).

Concerning other commercially produced carotenoids, β-carotene is a useful metabolite in the health food, chemical, and pharmaceutical industries. The Chlorophyte microalga

D. salina Teod is the richest natural source of β-carotene, which accumulates up to about 10%–14% on dry weight basis. In fact, the carotenogenic capacity of commercial *Dunaliella* (50 mg/g cell dry weight) is obviously larger than that of recombinant *E. coli* (1.5 mg/g cell dry weight). This Chlorophyte is one of a few microalgae with economical interest currently cultivated on an industrial scale in several countries since the 1980s and presently is the focus of intense fundamental and applied research. *D. salina* has the capacity to overcome stress or growth-limiting conditions (e.g., salt, temperature, light, and nutrient deficiencies) by synthesizing and accumulating β-carotene in lipid globules between the interthylakoid spaces of the chloroplasts (Jin and Polle 2009; Ramos et al. 2011). The isoprenoid β-carotene exhibits all-*trans* and 9-*cis* stereoisomers, with a wide range of important economical applications such as human health, cosmetic, and aquaculture industries (Raja et al. 2007; Paniagua-Michel et al. 2012). Genomic sequencing initiative in *D. salina* strain CCAP 19/18 (USDA-DOE Joint Genome Institute [JGI]) is under way toward a better understanding of the response of *D. salina* to abiotic stress and approaches its nuclear and plastid transformation. The molecular information related to the isoprenoid biosynthesis pathway in *D. salina* is still very limited and incomplete (Ye et al. 2008). The need for an integrated genomics, proteomics, and metabolomics approach in order to elucidate the stress-induced overaccumulation of carotenoids in this species is an issue under intense research. At this time, highly efficient transformation systems are not available; the advent of massively sequenced genomes will enhance the transformation of *D. salina*, which needs more genomic tools when compared to development in *C. reinhardtii*. On the other hand, contradictory findings remain on the factors involved in the transcriptional regulation of DXS and HDR (GenBank accession numbers: FJ469276 and FJ040210) (Ramos et al. 2011). Actually, only few sequences have been completed for genes that code for enzymes involved in the carotenoid biosynthesis pathway in *D. salina*. Among these genes, only the following have been published: PSY (GenBank accession number: AY601075) PDS (GenBank accession number: AY954517) (Zhu et al. 2005), ZDS (GenBank accession number: HM754265) (Ye and Jiang 2010), and LCY-b (EU327876) (Ramos et al. 2008). That information emphasized the differential expression of multiple PSY genes that might regulate the biosynthesis of carotenoids in *Dunaliella*. It is expected that in the next decade, advances in several remaining pathway genes will lead to a better understanding of the current genome and on the molecular basis of carotenogenesis in this Chlorophyte. According to Ramos et al. (2011), the overproducing carotenoid conditions are not enhanced at the transcriptional and translational levels by the carotenoid biosynthetic enzymes, including PSY or PDS, but a dramatic increase was observed by the activity of a key lipid biosynthesis regulatory enzyme, ACCase. It is recognized that by the synthetic biology, isoprenoids can be efficiently produced for useful market applications. In that process, the natural biosynthetic pathway is transferred to a suited host, and the economic feasibility of the products will depend on the final titers of the pathway rationally designed, built, and optimized according to the desired product (Klein-Marcuschamer et al. 2007; Anthony et al. 2009; Immethun et al. 2013). It is expected that genomics of natural functions will enhance the rate of intermediaries and new compound discoveries. The successful incorporation of synthetic biology and omics would lead to transform microorganisms, including microalgae, for industrial-scale production of industrially important compounds or diversification of the already existed (Figure 3.6).

Recently comparative genomics has been developed as a mechanism to take advantage of the amount of genomic data available, as in the case of the reconstruction of biosynthetic pathway of carotenoids in cyanobacteria (Liang et al. 2006). Concerning the genes involved

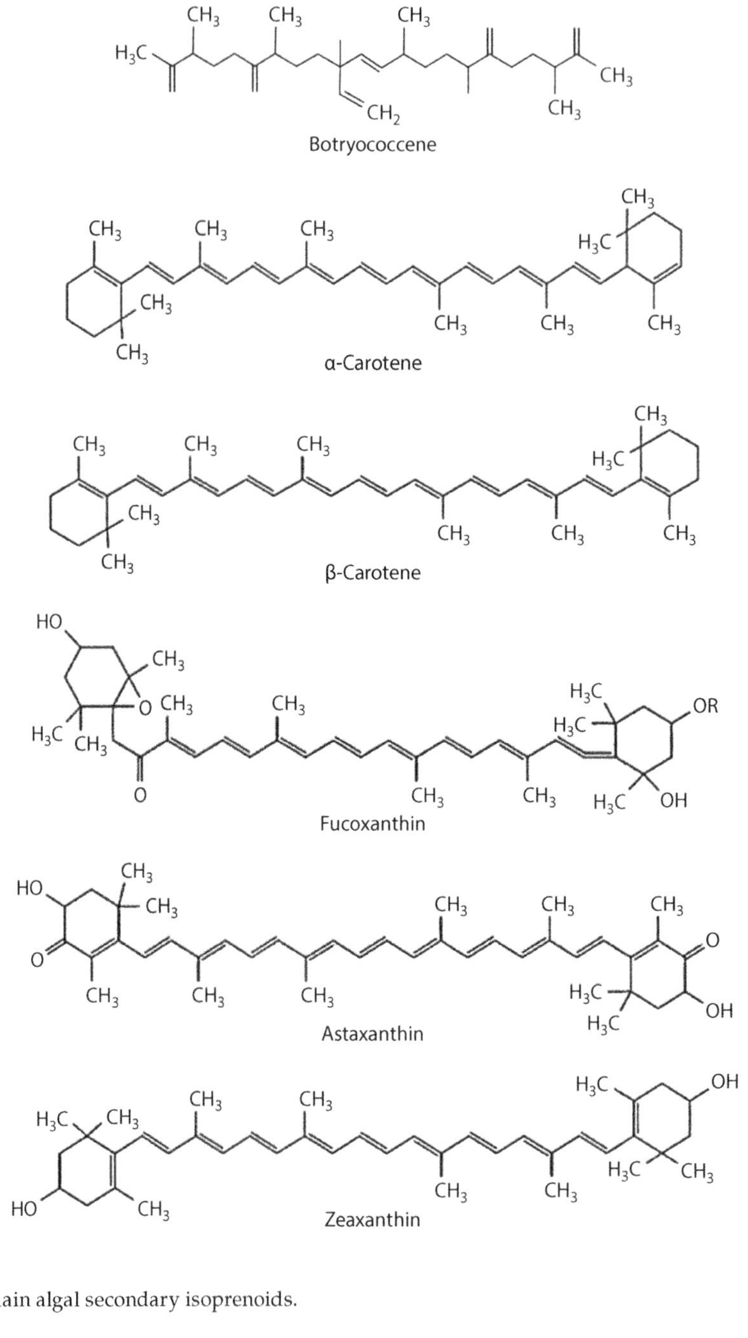

FIGURE 3.6
Examples of main algal secondary isoprenoids.

in carotenoid biosynthesis across cyanobacterial genomes, geranylgeranyl pyrophosphate synthase and phytoene synthase genes were widely distributed among 18 cyanobacterial strains. Cyanobacteria share the same carotenoid biosynthesis pathway to lycopene and beta-carotene. A remarkable feature was that in unicellular cyanobacteria, each step of the carotenogenic pathway is usually catalyzed by one gene product, whereas multiple ketolase genes are common in filamentous bacteria (Liang et al. 2006).

3.5 Summary

Recent advent of "omics" technologies and emerging algal biotechnology will contribute to harness the full potential for value-added compounds, such as isoprenoids. These important molecules actually are facing obstacles that impede their diversification and full exploitation, which has been part of the dirty chemistry. The need for sustainable biobased biotechnology for multiuse feedstocks, with utility in an array of industrial developments such as pharmaceuticals, nutraceuticals, biomedical compounds, biofuels, and derivative biorefineries, are highly promissory approaches for the development and discoveries of new functionalities of isoprenoids and algal products coupled with omics technologies.

References

Achitouv, E., Metzger, P., Rager, M.N., and Largeau, C. (2004). C31–C34 methylated squalenes from a Bolivian strain of *Botryococcus braunii*. *Phytochemistry* 65: 3159–3165.

An, G.H., Schuman, D.B., and Johnson, E.A. (1989). Isolation of *Phaffia rhodozyma* mutants with increased astaxanthin content. *Appl Environ Microbiol* 55: 116–124.

Anthony, J.R., Anthony, L.C., Nowroozi, F., Kwon, G., Newman, J.D., and Keasling, J.D. (2009). Optimization of the mevalonate-based isoprenoid biosynthetic pathway in *Escherichia coli* for production of the anti-malarial drug precursor amorpha-4,11-diene. *Metab Eng* 11: 13–19.

Baba, M., Ioki, M., Nakajima, N., Shiraiwa, Y., and Watanabe, M.M. (2012). Transcriptome analysis of an oil-rich race A strain of *Botryococcus braunii* (BOT-88-2) by *de novo* assembly of pyrosequencing cDNA reads. *Bioresour Technol* 109: 282–286.

Ben-Amotz, A., Katz, A., and Avron, M. (1982). Accumulation of B-carotene-rich globules from *Dunaliella bardawil* (Chlorophyceae). *J Phycol* 18: 529–537.

Bhattacharya, D., Price, D.C., Chan, C.X., Qiu, H., Rose, N., Ball, S., Weber, A.P. et al. (2013). Genome of the red alga *Porphyridium purpureum*. *Nat Commun* 4: 1941.

Bhosale, P. and Bernstein, P.S. (2005). Microbial xanthophylls. *Appl Microbiol Biotechnol* 68: 445–455.

Bhuvaneswari, V. and Nagini, S. (2005). Lycopene: A review of its potential as an anticancer agent. *Curr Med Chem Anticancer Agents* 5: 627–635.

Blanc, G., Duncan, G., Agarkova, I., Borodovsky, M., Gurnon, J., Kuo, A., Lindquist, E. et al. (2010). The *Chlorella variabilis* NC64A genome reveals adaptation to photosymbiosis, coevolution with viruses, and cryptic sex. *Plant Cell* 22: 2943–2955.

Bolling, C. and Fiehn, O. (2005). Metabolite profiling of *Chlamydomonas reinhardtii* under nutrient deprivation. *Plant Physiol* 139: 1995–2005.

Capa-Robles, W., Paniagua-Michel, J., and Olmos Soto, J. (2009). The biosynthesis and accumulation of -carotene in *Dunaliella salina* proceed via the glyceraldehyde 3-phosphate/pyruvate pathway. *Nat Prod Res* 23: 1021–1028.

Carrier, G., Garnier, M., Le Cunff, L., Bougaran, G., Probert, I., De Vargas, C., Corre, E., Cadoret, J.P., and Saint-Jean, B. (2014). Comparative transcriptome of wild type and selected strains of the microalgae *Tisochrysis lutea* provides insights into the genetic basis, lipid metabolism and the life cycle. *PLoS One* 9: e86889.

Chen, Y.F., Etheridge, N., and Schaller, G.E. (2005). Ethylene signal transduction. *Ann Bot* 95: 901–915.

Choi, Y.E., Hwang, H., Kim, H.S., Ahn, J.W., Jeong, W.J., and Yang, J.W. (2013). Comparative proteomics using lipid over-producing or less-producing mutants unravels lipid metabolisms in *Chlamydomonas reinhardtii*. *Bioresour Technol* 145: 108–115.

Coll, J.M. (2006). Methodologies for transferring DNA into eukaryotic microalgae. *Span J Agric Res* 4: 316–330.

Disch, A., J. Schwender, C. Muller, H. K. Lichtenthaler, and M. Rohmer (1998). Distribution of the mevalonate and glyceraldehyde phosphate/pyruvate pathways for isoprenoid biosynthesis in unicellular algae and the cyanobacterium Synechocystis PCC 6714. *Biochem J* 333(Pt 2): 381–388.

Dubini, A., Mus, F., Seibert, M., Grossman, A.R., and Posewitz, M.C. (2009). Flexibility in anaerobic metabolism as revealed in a mutant of *Chlamydomonas reinhardtii* lacking hydrogenase activity. *J Biol Chem* 284: 7201–7213.

Feng, S., Xue, L., Liu, H., and Lu, P. (2009). Improvement of efficiency of genetic transformation for *Dunaliella salina* by glass beads method. *Mol Biol Rep* 36: 1433–1439.

Gao, Y., Xiong, W., Li, X.B., Gao, C.F., Zhang, Y.L., Li, H., and Wu, Q.Y. (2009). Identification of the proteomic changes in *Synechocystis* sp. PCC 6803 following prolonged UV-B irradiation. *J Exp Bot* 60: 1141–1154.

Gao, Z.P., Yu, Q.B., Zhao, T.T., Ma, Q., Chen, G.X., and Yang, Z.N. (2011). A functional component of the transcriptionally active chromosome complex, Arabidopsis pTAC14, interacts with pTAC12/HEMERA and regulates plastid gene expression. *Plant Physiol* 157: 1733–1745.

Garnier, M., Carrier, G., Rogniaux, H., Nicolau, E., Bougaran, G., Saint-Jean, B., and Cadoret, J.P. (2014). Comparative proteomics reveals proteins impacted by nitrogen deprivation in wild-type and high lipid-accumulating mutant strains of *Tisochrysis lutea. J Proteomics* 105: 107–120.

Geng, D.G., Han, Y., Wang, Y., Wang, P., Zhang, L., Li, W., and Sun, Y.R. (2004). Construction of a system for the stable expression of foreign genes in *Dunaliella salina. Acta Bot Sin* 46: 342–346.

Gfeller, R.P. and Gibbs, M. (1984). Fermentative metabolism of *Chlamydomonas reinhardtii*: I. Analysis of fermentative products from starch in dark and light. *Plant Physiol* 75: 212–218.

Gibbs, M., Gfeller, R.P., and Chen, C. (1986). Fermentative metabolism of *Chlamydomonas reinhardtii*: III. Photoassimilation of acetate. *Plant Physiol* 82: 160–166.

Gonzalez-Ballester, D., Casero, D., Cokus, S., Pellegrini, M., Merchant, S.S., and Grossman, A.R. (2010). RNA-seq analysis of sulfur-deprived *Chlamydomonas* cells reveals aspects of acclimation critical for cell survival. *Plant Cell* 22: 2058–2084.

Grossman, A.R., Croft, M., Gladyshev, V.N., Merchant, S.S., Posewitz, M.C., Prochnik, S., and Spalding, M.H. (2007). Novel metabolism in *Chlamydomonas* through the lens of genomics. *Curr Opin Plant Biol* 10: 190–198.

Guarnieri, M.T., Nag, A., Smolinski, S.L., Darzins, A., Seibert, M., and Pienkos, P.T. (2011). Examination of triacylglycerol biosynthetic pathways via *de novo* transcriptomic and proteomic analyses in an unsequenced microalga. *PLoS One* 6: e25851.

Guarnieri, M.T., Nag, A., Yang, S., and Pienkos, P.T. (2013). Proteomic analysis of *Chlorella vulgaris*: Potential targets for enhanced lipid accumulation. *J Proteomics* 93: 245–253.

Guarnieri, M.T. and Pienkos, P.T. 2015. Algal omics: Unlocking bioproduct diversity in algae cell factories. *Photosynth Res.* 123:255–263.

Hallmann, A. (2007). Algal transgenics and biotechnology. *Transgenic Plant J* 1: 81–98.

Hatti-Kaul, R., Törnvall, U., Gustafsson, L., and Börjesson, P. (2007). Industrial biotechnology for the production of bio-based chemicals—A cradle-to-grave perspective. *Trends Biotechnol* 25: 119–124.

Hildebrand, M., Abbriano, R.M., Polle, J.E.W., Traller, J.C., Trentacoste, E.M., Smith, S.R., and Davis, A.K. (2013). Metabolic and cellular organization in evolutionarily diverse microalgae as related to biofuels production. *Curr Opin Chem Biol* 17: 506–514.

Hsieh, S.I., Castruita, M., Malasarn, D., Urzica, E., Erde, J., Page, M.D., Yamasaki, H. et al. (2013). The proteome of copper, iron, zinc, and manganese micronutrient deficiency in *Chlamydomonas reinhardtii. Mol Cell Proteomics* 12: 65–86.

Huang, S., Chen, L., Te, R., Qiao, J., Wang, J., and Zhang, W. (2013). Complementary iTRAQ proteomics and RNA-seq transcriptomics reveal multiple levels of regulation in response to nitrogen starvation in *Synechocystis* sp. PCC 6803. *Mol Biosyst* 9: 2565–2574.

Huang, Z. and Poulter, C.D. (1989). Tetramethylsqualene, a triterpene from *Botryococcus braunii* var. showa. *Phytochemistry* 28: 1467–1470.

Immethun, C.M., Hoynes-O'Connor, A.G., Balassy, A., and Moon, T.S. (2013). Microbial production of isoprenoids enabled by synthetic biology. *Front Microbiol* 4: 1–8.

Ioki, M., Baba, M., Nakajima, N., Shiraiwa, Y., and Watanabe, M.M. (2012a). Transcriptome analysis of an oil-rich race B strain of *Botryococcus braunii* (BOT-22) by de novo assembly of pyrosequencing cDNA reads. *Bioresour Technol* 109: 292–296.

Ioki, M., Baba, M., Nakajima, N., Shiraiwa, Y., and Watanabe, M.M. (2012b). Transcriptome analysis of an oil-rich race B strain of *Botryococcus braunii* (BOT-70) by de novo assembly of 5′-end sequences of full-length cDNA clones. *Bioresour Technol* 109: 277–281.

Ito, T., Tanaka, M., Shinkawa, H., Nakada, T., Ano, Y., Kurano, N., Soga, T., and Tomita, M. (2013). Metabolic and morphological changes of an oil accumulating trebouxiophycean alga in nitrogen-deficient conditions. *Metabolomics* 9: 178–187.

Jamers, A., Blust, R., De Coen, W., Griffin, J.L., and Jones, O.A. (2013). Copper toxicity in the microalga *Chlamydomonas reinhardtii*: An integrated approach. *Biometals* 26: 731–740.

Jin, E. and Polle, J.E.W. (2009). Carotenoid biosynthesis in *Dunaliella* (Chlorophyta). In Ben-Amotz, A., Polle, J.E.W., and Subba Rao, D.V. (eds.) *The Alga Dunaliella: Biodiversity, Physiology, Genomics and Biotechnology*. Science Publishers, Enfield, NH, pp. 147–172.

Jørgensen, K., Rasmussen, A.V., Morant, M., Nielsen, A.H., Bjarnholt, N., Zagrobelny, M., Bak, S., and Møller, B.L. (2005). Metabolon formation and metabolic channeling in the biosynthesis of plant natural products. *Curr Opin Plant Biol* 8: 280–291.

Kamisaka, Y., Tomita, N., Kimura, K., Kainou, K., and Uemura, H. (2007). DGA1 (diacylglycerol acyltransferase gene) overexpression and leucine biosynthesis significantly increase lipid accumulation in the Deltasnf2 disruptant of *Saccharomyces cerevisiae*. *Biochem J* 408: 61–68.

Kaneko, T. and Tabata, S. (1997). Complete genome structure of the unicellular cyanobacterium *Synechocystis* sp. PCC6803. *Plant Cell Physiol* 38: 1171–1176.

Katz, A., Jimenez, C., and Pick, U. (1995). Isolation and characterization of a protein associated with carotene globules in the alga *Dunaliella bardawil*. *Plant Physiol* 108: 1657–1664.

Kennedy, J., Marchesi, J.R., and Dobson, A.D. (2008). Marine metagenomics: Strategies for the discovery of novel enzymes with biotechnological applications from marine environments. *Microb Cell Fact* 7: 27.

Kerk, D., Bulgrien, J., Smith, D.W., and Gribskov, M. (2003). Arabidopsis proteins containing similarity to the universal stress protein domain of bacteria. *Plant Physiol* 131: 1209–1219.

Kiss, R.D. and Stephanopoulos, G. (1991). Metabolic activity control of the L-lysine fermentation by restrained growth fed-batch strategies. *Biotechnol Prog* 7: 501–509.

Klein-Marcuschamer, D., Ajikumar, P.K., and Stephanopoulos, G. (2007). Engineering microbial cell factories for biosynthesis of isoprenoid molecules: Beyond lycopene. *Trends Biotechnol* 25: 417–424.

Krall, L., Huege, J., Catchpole, G., Steinhauser, D., and Willmitzer, L. (2009). Assessment of sampling strategies for gas chromatography-mass spectrometry (GC-MS) based metabolomics of cyanobacteria. *J Chromatogr B Analyt Technol Biomed Life Sci* 877: 2952–2960.

Landrum, J.T. and Bone, R.A. (2001). Lutein, zeaxanthin, and the macular pigment. *Arch Biochem Biophys* 385: 28–40.

Lange, B.M., Tamas, R., William, M., and Croteau, R. (2000). Isoprenoid biosynthesis: The evolution of two ancient and distinct pathways across genomes. *Proc Natl Acad Sci USA* 97: 13172–13177.

Lee Do, Y. and Fiehn, O. (2008). High quality metabolomic data for *Chlamydomonas reinhardtii*. *Plant Methods* 4: 7.

Li, Y., Han, D., Hu, G., Dauvillee, D., Sommerfeld, M., Ball, S., and Hu, Q. (2010). *Chlamydomonas* starchless mutant defective in ADP-glucose pyrophosphorylase hyper-accumulates triacylglycerol. *Metab Eng* 12: 387–391.

Li, Y., Horsman, M., Wang, B., Wu, N., and Lan, C.Q. (2008) Effects of nitrogen sources on cell growth and lipid accumulation of green alga *Neochloris oleoabundans*. *Appl Microbiol Biotechnol* 81: 629–636.

Li, Y., Mu, J., Chen, D., Han, F., Xu, H., Kong, F., Xie, F., and Feng, B. (2013). Production of biomass and lipid by the microalgae *Chlorella protothecoides* with heterotrophic-Cu(II) stressed (HCuS) coupling cultivation. *Bioresour Technol* 148: 283–292.

Liang, C., Zhao, F., Wei, W., Wen, Z., and Qin, S. (2006). Carotenoid biosynthesis in Cyanobacteria: Structural and evolutionary scenarios on comparative genomics. *Int J Biol Sci* 2: 197–207.

Lichtenthaler, H. K. (1999). The 1-deoxy-D-xylulose-5-phosphate pathway of isoprenoid biosynthesis in plants. *Annu Rev Plant Physiol Plant Mol Biol* 50: 47–65.

Lin, Y., Schiavo, S., Orjala, J., Vouros, P., and Kautz, R. (2008). Microscale LC-MS-NMR platform applied to the identification of active cyanobacterial metabolites. *Anal Chem* 80: 8045–8054.

Lindberg, P., Park, S. and Melis, A. (2010). Engineering a platform for photosynthetic isoprene production in cyanobacteria, using *Synechocystis* as the model organism. *Metab Eng* 12: 70–79.

Liu, J., Chen, L., Wang, J., Qiao, J., and Zhang, W. (2012). Proteomic analysis reveals resistance mechanism against biofuel hexane in *Synechocystis* sp. PCC 6803. *Biotechnol Biofuels* 5: 68.

Martin, G.J., Hill, D.R., Olmstead, I.L., Bergamin, A., Shears, M.J., Dias, D.A., Kentish, S.E., Scales, P.J., Botte, C.Y., and Callahan, D.L. (2014). Lipid profile remodeling in response to nitrogen deprivation in the microalgae *Chlorella* sp. (Trebouxiophyceae) and *Nannochloropsis* sp. (Eustigmatophyceae). *PLoS One* 9: e103389.

Matthew, T., Zhou, W., Rupprecht, J., Lim, L., Thomas-Hall, S.R., Doebbe, A., Kruse, O. et al. (2009). The metabolome of *Chlamydomonas reinhardtii* following induction of anaerobic H_2 production by sulfur depletion. *J Biol Chem* 284: 23415–23425.

May, P., Wienkoop, S., Kempa, S., Usadel, B., Christian, N., Rupprecht, J., Weiss, J. et al. (2008). Metabolomics- and proteomics-assisted genome annotation and analysis of the draft metabolic network of *Chlamydomonas reinhardtii*. *Genetics* 179: 157–166.

Merchant, S.S., Prochnik, S.E., Vallon, O., Harris, E.H., Karpowicz, S.J., Witman, G.B., Terry, A. et al. (2007). The *Chlamydomonas* genome reveals the evolution of key animal and plant functions. *Science* 318: 245–250.

Metzger, P., Allard, B., Casadevall, E., Berkaloff, C., and Coute, A. (1990). Structure and chemistry of a new chemical race of *Botryococcus braunii* that produces lycopadiene, a tetraterpenoid hydrocarbon. *J Phycol* 26: 258–266.

Metzger, P., Berkaloff, C., Coute, A., and Casadevall, E. (1985a). Alkadiene- and botryococcene-producing races of wild strains of *Botryococcus braunii*. *Phytochemistry* 24: 2305–2312.

Metzger, P. and Casadevall, E. (1987). Lycopadiene, a tetraterpenoid hydrocarbon from new strains of the green alga *Botryococcus braunii*. *Tetrahedron Lett* 28: 3931–3934.

Metzger, P., M. N. Rager, and C. Largeau (2002). Botryolins A and B, two tetramethylsqualene triethers from the green microalga Botryococcus braunii. *Phytochemistry* 59(8): 839–843.

Metzger, P., M. N. Rager, N. Sellier, and C. Largeau (2003). Lycopanerols I-L, four new tetraterpenoid ethers from Botryococcus braunii. *J Nat Prod* 66(6): 772–778.

Metzger, P., Casadevall, E., Pouet, M.-J., and Pouet, Y. (1985b). Structures of some botryococcenes: Branched hydrocarbons from the B race of the green alga *Botryococcus braunii*. *Phytochemistry* 24: 2995–3002.

Moellering, E.R. and Benning, C. (2010). RNA interference silencing of a major lipid droplet protein affects lipid droplet size in *Chlamydomonas reinhardtii*. *Eukaryot Cell* 9: 97–106.

Molnar, I., Lopez, D., Wisecaver, J.H., Devarenne, T.P., Weiss, T.L., Pellegrini, M., and Hackett, J.D. (2012). Bio-crude transcriptomics: Gene discovery and metabolic network reconstruction for the biosynthesis of the terpenome of the hydrocarbon oil-producing green alga, *Botryococcus braunii* race B (Showa). *BMC Genomics* 13: 576.

Muntendam, R., Melillo, E., Ryden, A.M., and Kayser, O. (2009). Perspectives and limits of engineering the isoprenoid metabolism in heterologous hosts. *Appl Microbiol Biotechnol* 73: 980–990.

Mus, F., Dubini, A., Seibert, M., Posewitz, M.C., and Grossman, A.R. (2007). Anaerobic acclimation in *Chlamydomonas reinhardtii*: Anoxic gene expression, hydrogenase induction, and metabolic pathways. *J Biol Chem* 282: 25475–25486.

Nakajima, T., Kajihata, S., Yoshikawa, K., Matsuda, F., Furusawa, C., Hirasawa, T., and Shimizu, H. (2014). Integrated metabolic flux and omics analysis of *Synechocystis* sp. PCC 6803 under mixotrophic and photoheterotrophic conditions. *Plant Cell Physiol* 55: 1605–1612.

Nakamura, Y., Sasaki, N., Kobayashi, M., Ojima, N., Yasuike, M., Shigenobu, Y., Satomi, M. et al. (2013). The first symbiont-free genome sequence of marine red alga, Susabi-nori (*Pyropia yezoensis*). *PLoS One* 8: e57122.

Nguyen, A.V., Toepel, J., Burgess, S., Uhmeyer, A., Blifernez, O., Doebbe, A., Hankamer, B., Nixon, P., Wobbe, L., and Kruse, O. (2011a). Time-course global expression profiles of *Chlamydomonas reinhardtii* during photo-biological H(2) production. *PLoS One* 6: e29364.

Nguyen, H.M., Baudet, M., Cuine, S., Adriano, J.M., Barthe, D., Billon, E., Bruley, C. et al. (2011b). Proteomic profiling of oil bodies isolated from the unicellular green microalga *Chlamydomonas reinhardtii*: With focus on proteins involved in lipid metabolism. *Proteomics* 11: 4266–4273.

Niehaus, T.D., Shigeru, O., Devarenne, T.P., Watt, D.S., Sviripa, V., and Chappella, J. (2011). Identification of unique mechanisms for triterpene biosynthesis in *Botryococcus braunii*. *Proc Natl Acad Sci USA* 108: 12260–12265.

Nielsen, J. and Oliver, S. (2005). The next wave in metabolome analysis. *Trends Biotechnol* 23: 544–546.

Paniagua-Michel, J., Capa-Robles, W., Olmos-Soto, J., and Gutierrez-Millan, L.E. (2009). The carotenogenesis pathway via the isoprenoid-β-carotene interference approach in a new strain of *Dunaliella salina* isolated from Baja California Mexico. *Mar Drugs* 7: 45–56.

Paniagua-Michel, J., Olmos-Soto, J., and Acosta, M. (2012). Pathways of carotenoid biosynthesis in bacteria and microalgae. In Barredo, J.-L. (ed.) *Microbial Carotenoids from Bacteria and Microalgae: Methods and Protocols*, Methods in Molecular Biology, Vol. 892. Springer–Humana Press, Holanda, the Netherlands, pp. 1–12.

Paniagua-Michel, J., Olmos-Soto, J., and Morales-Guerrero, E. (2014). Drugs and leads from the ocean. In Kim, S.K. (ed.) *Handbook of Marine Biotechnology*. Springer, New York.

Polle, J.E.W., Tran, D., and Ben-Amotz, A. (2009). History, distribution, and habitats of algae of the genus *Dunaliella Teodoresco* (Chlorophyceae). In Ben-Amotz, A., Polle, J.E.W., and Rao, S. (eds.) *The Alga Dunaliella: Biodiversity, Physiology, Genomics & Biotechnology*. Science Publishers, Enfield, NH, pp. 15–44.

Prochnik, S.E., Umen, J., Nedelcu, A.M., Hallmann, A., Miller, S.M., Nishii, I., Ferris, P. et al. (2010). Genomic analysis of organismal complexity in the multicellular green alga *Volvox carteri*. *Science* 329: 223–226.

Qiao, J., Huang, S., Te, R., Wang, J., Chen, L., and Zhang, W. (2013). Integrated proteomic and transcriptomic analysis reveals novel genes and regulatory mechanisms involved in salt stress responses in *Synechocystis* sp. PCC 6803. *Appl Microbiol Biotechnol* 97: 8253–8264.

Radakovits, R., Jinkerson, R.E., Darzins, A., and Posewitz, M.C. (2010). Genetic engineering of algae for enhanced biofuel production. *Eukaryot Cell* 9: 486–501.

Rager, M. N. and P. Metzger (2000). Six novel tetraterpenoid ethers, lycopanerols B-G, and some other constituents from the green microalga Botryococcus braunii. *Phytochemistry* 54(4): 427–437.

Raja, R., Hemaiswarya, S., and Rengasamy, R. (2007). Exploitation of *Dunaliella* for β-carotene production. *Appl Microbiol Biotechnol* 74: 517–523.

Ramos, A., Coesel, S., Marques, A., Rodrigues, M., Baumgartner, A., Noronha, J., Rauter, A., Brenig, B., and Varela, J. (2008). Isolation and characterization of a stress-inducible *Dunaliella salina* Lcy-β gene encoding a functional lycopene β-cyclase. *Appl Microbiol Biotechnol* 79: 819–828.

Ramos, A., Polle, J., Tran, D., Cushman, J.C., Jin, E.S., and Varela, J.C. (2011). The unicellular green alga *Dunaliella salina* Teod. as a model for abiotic stress tolerance: Genetic advances and future perspectives. *Algae* 26: 3–20.

Redding-Johanson, A.M., Tanveer, S., Rossana, B., Krupa, C.R., Szmidt, H.L., Adams, P.D., Keasling, J.D., Lee, T.S., Mukhopadhyay, A., and Petzold, C.J. (2011). Targeted proteomics for metabolic pathway optimization: Application to terpene production. *Metab Eng* 13: 194–203.

Rismani-Yazdi, H., Haznedaroglu, B.Z., Bibby, K., and Peccia, J. (2011). Transcriptome sequencing and annotation of the microalgae *Dunaliella tertiolecta*: Pathway description and gene discovery for production of next-generation biofuels. *BMC Genomics* 12: 148.

Rismani-Yazdi, H., Haznedaroglu, B.Z., Hsin, C., and Peccia, J. (2012). Transcriptomic analysis of the oleaginous microalga *Neochloris oleoabundans* reveals metabolic insights into triacylglyceride accumulation. *Biotechnol Biofuels* 5: 74.

Rager, M. N. and P. Metzger (2000). Six novel tetraterpenoid ethers, lycopanerols B-G, and some other constituents from the green microalga Botryococcus braunii. *Phytochemistry* 54(4): 427–437.

Ruffing, A.M. (2013). RNA-Seq analysis and targeted mutagenesis for improved free fatty acid production in an engineered cyanobacterium. *Biotechnol Biofuels* 6: 113.

Sans-Piche, F., Kluender, C., Altenburger, R., and Schmitt-Jansen, M. (2010). Anchoring metabolic changes to phenotypic effects in the chlorophyte *Scenedesmus vacuolatus* under chemical exposure. *Mar Environ Res* 69(Suppl): S28–S30.

Sato, Y., Y. Ito, S. Okada, K. Murakami, and H. Abe (2003). Biosynthesis of the triterpenoids, botryo-coccenes and tetramethylsqualene in the B race of Botryococcus braunii via the non-mevalon-ate pathway. *Tetrahedron Lett* 44: 7035–7037.

Schirmer, A., Rude, M.A., Li, X., Popova, E., and del Cardayre, S.B. (2010). Microbial biosynthesis of alkanes. *Science* 329: 559–562.

Schwender, J., J. Zeidler, R. Groner, C. Muller, M. Focke, S. Braun, F. W. Lichtenthaler, and H. K. Lichtenthaler (1997). Incorporation of 1-deoxy-D-xylulose into isoprene and phytol by higher plants and algae. *FEBS Lett* 414(1): 129–134.

Shih, P.M., Wu, D., Latifi, A., Axen, S.D., Fewer, D.P., Talla, E., Calteau, A. et al. (2013). Improving the coverage of the cyanobacterial phylum using diversity-driven genome sequencing. *Proc Natl Acad Sci USA* 110: 1053–1058.

Smith, D.R., Lee, R.W., Cushman, J.C., Magnuson, J.K., Tran, D., and Polle, J.E. (2010). The *Dunaliella salina* organelle genomes: Large sequences, inflated with intronic and intergenic DNA. *BMC Plant Biol* 10: 83.

Song, P., Li, L., and Liu, J. (2013). Proteomic analysis in nitrogen-deprived *Isochrysis galbana* during lipid accumulation. *PLoS One* 8: e82188.

Subramanian, V., Dubini, A., and Seibert, M. (2012). Metabolic pathways in green algae with poten-tial value for biofuel production. In: Gordon, R. and Seckbach, J. (eds.) *The Science of Algal Fuels*. Springer, Dordrecht, pp. 399–422.

Takagi, M., Karseno and Yoshida, T. (2006). Effect of salt concentration on intracellular accumulation of lipids and triacylglyceride in marine microalgae *Dunaliella* cells. *J Biosci Bioeng* 101: 223–226.

Tornabene, T.G., Holzer, G., Lien, S., and Burris, N. (1983). Lipid composition of the nitrogen starved green alga *Neochloris oleoabundans. Enzyme Microb Technol* 5: 435–440.

Wang, J., Chen, L., Huang, S., Liu, J., Ren, X., Tian, X., Qiao, J., and Zhang, W. (2012). RNA-seq based identification and mutant validation of gene targets related to ethanol resistance in cyanobac-terial *Synechocystis* sp. PCC 6803. *Biotechnol Biofuels* 5: 89.

Wienkoop, S., Weiss, J., May, P., Kempa, S., Irgang, S., Recuenco-Munoz, L., Pietzke, M. et al. (2010). Targeted proteomics for *Chlamydomonas reinhardtii* combined with rapid subcellular protein fractionation, metabolomics and metabolic flux analyses. *Mol Biosyst* 6: 1018–1031.

Withers, S.T. and Keasling, J.D. (2007). Biosynthesis and engineering of isoprenoid small molecules. *Appl Microbiol Biotechnol* 73: 980–990.

Yang, D., Du, X., Yang, Z., Liang, Z., Guo, Z., and Liu, Y. (2014). Transcriptomics, proteomics, and metabolomics to reveal mechanisms underlying plant secondary metabolism. *Eng Life Sci* 14: 456–466.

Ye, Z.W. and Jiang, J.G. (2010). Analysis of an essential carotenogenic enzyme: ζ-carotene desaturase from unicellular alga *Dunaliella salina. J Agric Food Chem* 58: 11477–11482.

Ye, Z.W., Jiang, J.G., and Wu, G.H. (2008). Biosynthesis and regulation of carotenoids in *Dunaliella*: Progresses and prospects. *Biotechnol Adv* 26: 352–360.

Yeum, K.J. and Russell, R.M. (2002). Carotenoid bioavailability and bioconversion. *Annu Rev Nutr* 22: 483–504.

Yoshikawa, K., Hirasawa, T., Ogawa, K., Hidaka, Y., Nakajima, T., Furusawa, C., and Shimizu, H. (2013). Integrated transcriptomic and metabolomic analysis of the central metabolism of *Synechocystis* sp. PCC 6803 under different trophic conditions. *Biotechnol J* 8: 571–580.

Zhang, W., Tan, N.G., and Li, S.F. (2014). NMR-based metabolomics and LC-MS/MS quantification reveal metal-specific tolerance and redox homeostasis in *Chlorella vulgaris. Mol Biosyst* 10: 149–160.

Zhu, H., Ren, X., Wang, J., Song, Z., Shi, M., Qiao, J., Tian, X., Liu, J., Chen, L., and Zhang, W. (2013). Integrated OMICS guided engineering of biofuel butanol-tolerance in photosynthetic *Synechocystis* sp. PCC 6803. *Biotechnol Biofuels* 6: 106.

Zhu, Y.H., Jiang, J.G., Yan, Y., and Chen, X. (2005). Isolation and characterization of phytoene desatu-rase cDNA involved in the β-carotene biosynthetic pathway in *Dunaliella salina. J Agric Food Chem* 53: 5593–5597.

4

Marine Fungal Genomics: Trichoderma

K. Saravanakumar, K. Kathiresan, and Jie Chen

CONTENTS

4.1 Introduction

Fungi are a large group of the eukaryotic organisms, occurring in various habitats either in free forms or in association with other microbes, plants, or animals. The fungi act largely as decomposer of the organic matters and extensively involved in the organic and nutrient cycles (Thorn 1997). They are used in various beneficial applications such as direct source of the food (mushrooms), fermentation process, antibiotics, enzymes, industrial detergents, biological pesticides, and bioactive compounds of mycotoxins such as alkaloids and polyketides (Furbino et al. 2014; Suryanarayanan et al. 2009).

Marine ecosystem is the largest system on earth and the fascinating source of novel microorganisms, including fungi due to the presence of vagaries of physical and chemical conditions of the ocean (Liu et al. 2010). Marine fungi, like terrestrial ones, do have potential for applications (Lee et al. 2010; Saravanakumar et al. 2012; Ziemert et al. 2014). In particular, marine *Trichoderma* species have potential for controlling plant and soil pathogens, biofertilizer effects, biofuel production, novel secondary metabolites synthesis, bioremediation process, and nanoparticle (NP) synthesis (Saravanakumar 2012). However, molecular genomics of marine *Trichoderma* is far clear, and hence this chapter deals with the basics of omics for the fungus *Trichoderma*.

4.2 Fungi: Marine *Trichoderma*

Fungus research increased the traditional interest in genetics to genomics studies of complementation, mitotic recombination, extrachromosomal inheritance, host–parasite interactions, and plant and fungal pathogen interactions. The classification of fungi is proposed by Ainsworth (1973), and over 60,000 species of fungi are known. The classification and genetics of the fungus vary with genes present in the fungi. Fungi diversity encompasses four major groups of organisms such as ascomycetes, basidiomycetes, zygomycetes, and chytrids. Fungal diversity has been extensively studied for biology and omics; however, genomic studies on marine *Trichoderma* are only limited, despite of diverse genetics and physiology of the fungus. Genomic studies in relation to the growth, photoreception, gene silencing, gene activation, gene tolerance, molecular basis of conidiation, and biological pathways of sexual reproduction are required to resolve the biological issues of marine *Trichoderma*.

4.2.1 Biology

Microbial colonization in several habitats is crucially dependent on its potential to defend its ecological niche and to thrive and prosper despite competition for nutrients, space, and light (Schuster and Schmoll 2010). Many of the fungi, especially those of the genus *Trichoderma*, are isolated from the marine environment. Marine *Trichodermas* extensively grow and sporulate in marine or estuarine environment, as in facultative or obligate forms. Marine *Trichodermas* are ecologically, physiologically, and genetically defined clearly, but taxonomically they are prevalent components of diversified ecosystems in a wide range of climate zones (Kubicek et al. 2008; Meincke et al. 2010). *Trichoderma/ Hypocrea* (Ascomycetes) has been isolated from the decaying coastal or marine litter seawater, deep sea soil, and marine animals and plants (Gams and Bissett 1998; Harman et al. 2004; Kathiresan et al. 2011; Kohlmeyer 1974). Marine *Trichoderma* acts as mycorrhizal and saprophytic fungi to provide a wide range of the phosphate and nitrogen and is capable of degrading the lignin (Saravanakumar et al. 2013; Saravanakumar and Kathiresan 2013).

Ecological divergences of marine *Trichoderma* vary according to abiotic and biotic factors, including the intra- and intraspecific interactions. Unlike terrestrial system, marine ecosystems with varied physicochemical conditions can support novel *Trichoderma* species with higher bioprospecting potential. Abiotic factors, such as temperature and alkaline condition, induce the genetic differences in marine *Trichoderma*; however, enzymatic genes that confer resistance to the temperature, salinity, and alkalinity require due attention for the understanding of the researchers.

4.2.2 History and Nomenclature

The fungal genus *Trichoderma* was originally described by Persoon in 1974, and the biocontrol activity of these fungi was discovered in 1930s (Weindling 1932). This genus was considered to have only one species, *T. viride*, for several decades (Bisby 1939). The morphological classification of this genus was first described by Rifai (1969). Subsequently, the genus was classified into 5 sections, that is, *Longibrachiatum, Trichoderma, Hypocrea* anamorphs, *Pachybasium B*, and *Gliocladium*, and 14 clades, that is, *Rufa, Pachybasium A, Pachybasioides, Hypocreanum, Chlorospora, Lixii/catoptron, Virens, Semiorbis, Strictipillis, Stromatica, Ceramica, Luta, Psychrophila*, and *lone lineages* (Bissett 1991; Druzhinina and Kubicek 2005). Over 100 *Trichoderma* species have been physiologically discovered from different isolation sources of terrestrial region (Druzhinina et al. 2006). Marine *Trichoderma* species are also frequently isolated from marine or estuarine ecosystems of all latitudes (Table 4.1). This fungus was first described by Kohlmeyer (1974) in a marine environment and subsequently described to be found from Canadian shellfish and mussels (Brewer et al. 1993). A number of researchers have reported marine *Trichoderma* in various marine ecological systems such as marine water (Brewer et al. 1993; Kis-Papo et al. 2001; Saravanakumar 2012; Song et al. 2010), sediments (Paz et al. 2010a; Ren et al. 2009; Saravanakumar 2012; Sun et al. 2006, 2008), aquatic plants, mangroves (Immaculatejeyasantha 2011; Karthik Raja and Nirmala 2011; Saravanakumar 2012), mangrove detritus (Kathiresan et al. 2011), aquatic animals, sponges (Amagata et al. 1998; Khamthong et al. 2012; Kobayashi et al. 1993; Paz et al. 2010b; Ruiz et al. 2007; Sallenave et al. 1999; Sperry et al. 1998; Wang 2008; Wang et al. 2008; Wiese et al. 2011), mollusks (Brewer et al. 1993; Grishkan et al. 2003), Cnidaria (Da Silva et al. 2008), algae (Brewer et al. 1993; Garo et al. 2003; El-Kassas and El-Taher 2009; Wang et al. 2008; Yamada et al. 2014), echinoderms (Devi et al. 2012), and tunicates (Ruiz et al. 2013; Sallenave et al. 1999).

4.3 Molecular Taxonomy

Earlier, *Trichoderma* received a lot of the taxonomic debate in the identification. In some cases, misidentification of certain species occurred. The detection and isolation of this fungus from soil by using the traditional microbiological techniques are quite laborious and lead the bias in isolation (Thornton 2004). However, it is difficult to overcome these biases without the proper understanding of the strains. Hence, the molecular oligonucleotide database of *Trichoderma* significantly facilitates the identification of *Trichoderma*. In addition, till date, there are no specific genetic databases or taxonomical tools for the identification of marine *Trichoderma* from soil. Hence, the molecular monitoring and isolation of *Trichoderma* species from marine ecosystem also follows the methods that are commonly used for the genetic diversity analysis of *Trichoderma* from terrestrial region. A quick molecular identification of marine *Trichoderma* assessed by oligonucleotide sequences (TrichOkey) and a customized similarity search tool (TrichoBLAST) at www.isth.info (Druzhinina et al. 2005; Kopchinskiy et al. 2005), BLAST interface in the National Center for Biotechnology Information (NCBI) (http://blast.ncbi.nlm.nih.gov/), and phenotype microarray has been used for the characterization of the novel *Trichoderma* isolated from the marine and terrestrial resources, which facilitate the 96 carbon utilization patterns of the 96 carbon sources (Bochner et al. 2001; Druzhinina et al. 2006). Molecular techniques are emphasizing the advanced research on the fungal diversity in various ecological systems. The development of new molecular techniques can establish the database combined with other environmental factors, which facilitate metagenomic analysis.

TABLE 4.1

Diversity and Biological Function of Marine *Trichoderma*

Species	Isolation Sources	Chemicals	Biological Activity	References
T. hamatum	Mollusk (*Mytilus edulis*)	—	—	Brewer et al. (1993)
Trichoderma sp.	Marine sponge, seaweed	Unknown antimicrobial compounds	Antimicrobial activity	Masuma et al. (2001)
T. longibrachiatum	*Cerastoderma edule* (mollusk)	Longibrachin-like peptaibols	Acute toxicity	Landreau et al. (2002)
Trichoderma sp.	Mollusk and sediments			Brewer et al. (1993); Grishkan et al. (2003)
T. viride	Rhizosphere soil of *Avicennia marina*	—	Nematicidal activity	Syed et al. (2003)
T. virens	Ascidian (*Didemnum molle*) green alga (*Halimeda*)	Trichodermamides A and B	*In vitro* cytotoxicity against HCT-116 cell lines, moderate antimicrobial activity	Garo et al. (2003)
T. virens	Marine ascidian	Trichodermamide	—	Eliane et al. (2003)
T. virens	Alga (*Halimeda* sp.) and tunicate (*Didemnum molle*)	Heptelidic acid chlorohydrin	—	Garo et al. (2003)
T. aureiviride	Sea sediments (South China)	Laminarinase, amylase, *N*-acetyl-β-ᴅ-glucosaminidase, β-ᴅ-gluco- and galactosidase, cellulose		Burtseva et al. (2006)
T. longibrachiatum	Mussels (*Mytilus edulis*)	Peptaibols (acetyl) group, aminoisobutyric acid (Aib), and *N*-terminal acyl (most often)	—	Mohamed-Benkada et al. (2006)
T. longibrachiatum	Mussels (*Mytilus edulis*)	Polyunsaturated fatty acids		Ruiz et al. (2007b)
Trichoderma sp. strain H1-7	Marine sediment	—	Tyrosinase inhibitor-producing mechanism	Katsuhisa Yamada et al. (2008)
Trichoderma sp.	Sponge (*Agelas dispar*)	Vertinolide	Antioxidant activity	Neumann et al. (2007)

(Continued)

TABLE 4.1 (*Continued*)

Diversity and Biological Function of Marine *Trichoderma*

Species	Isolation Sources	Chemicals	Biological Activity	References
T. tomentosum	Sponge (*Gelliodes fibrosa*)	—	—	Wang (2008)
T. reesei	Sediment	Trichodermatide B–D	Cytotoxicity activity	Sun et al. (2006, 2008)
T. reesei	Marine mud	Trichodermatides A–D	Cytotoxicity against A375-S2 human melanoma cell line	Sun et al. (2008)
T. asperellum	Sediments	Aspereline A–F	Antimicrobial activity	Ren et al. (2009)
T. viride	Marine polluted soil	—	Biodegradation of chromium	El-Kassas and El-Taher (2009)
Trichoderma sp.Gc1	Marine sponges (*Geodia corticostylifera* and *Chelonaplysylla erecta*)	—	Reduction reaction	Rocha et al. (2009)
Trichoderma sp.	Sediments	Sorbicillin, sorbicillin (A), 2′,3′ dihydrosorbicillin	Cytotoxicity activity	Du et al. (2009)
T. atroviride	Sediments (root of mangroves *Ceriops tagal*) and sponge (*Psammocinia* sp.)	—	—	Sun et al. (2009); Paz et al. (2010a)
T. longibrachiatum	Mollusk (*Cerastoderma edule and Mytilus edulis*) and sponge (*Psammocinia* sp. and *Holothuria* sp.)	—	Cytotoxicity activity	Sperry et al. (1998); Sallenave et al. (1999); Ruiz et al. (2007b); Paz (2010b)
Trichoderma sp.	Marine sediment	Trichoderone, a novel cytotoxic cyclopentenone and cholesta-7, 22-diene-3b, 5a, 6b-triol	—	Yi et al. (2010)
T. koningii	Sediments, mollusk, seawater	—	—	Brewer et al. (1993); Kis-Papo et al. (2001); Song et al. (2010)

(Continued)

TABLE 4.1 (*Continued*)

Diversity and Biological Function of Marine *Trichoderma*

Species	Isolation Sources	Chemicals	Biological Activity	References
Trichoderma sp.	Sponge (*Agelas dispar*)	Trichodimerol, bislongiquinolide, and bisvertinol	—	Abdel-Latif et al. (2010)
Trichoderma sp.	Sponge	Trichoderins, novel aminolipopeptides	Antimycobacterial substances with activity	Pruksakorn et al. (2010)
Trichoderma sp. *T. cerinum*	Marine sediments Sponge (*Tethya aurantium*)	Trichoderone —	Cytotoxicity activity	You et al. (2010) Wiese et al. (2011)
T. longibrachiatum, T. atroviride, T. harzianum, T. asperelloides	Sponges (*Psammocinia* sp.)	Soluble and volatile metabolites	Biocontrol agents	Gal-Hemed et al. (2011)
T. harzianum	Marine sponge *Gelliodes fibrosa, Suberites zeteki, Tethya aurantium, Halichondria okadai, Psammocinia* sp., *Mycale cecilia* Cnidarian (*Echinogorgia rebekka*)	Trichoharzin, trichodermaketone C	Antimicrobial acidity	Kobayashi et al. (1993); Amagata et al. (1998); Wang (2008); Paz (2010b); Wiese et al. (2011b)
Hypocrea lixii *T. virens*	Seawater and drift wood	—	Antimicrobial activity	Immaculatejeyasanta et al. (2011); Karthik Raja and Nirmala (2011)
Trichoderma sp. Gc1	Marine sponges (*Geodia corticostylifera* and *Chelonaplysylla erecta*)	—	Biodegradation of pesticide DDT (1,1-dichloro-2,2-*bis*-(4-chlorophenyl) ethane)	Ortega et al. (2011)
Trichoderma sp.	Saline land soil	(E)-6-(2,4-dihydroxyl-5-methylphenyl)-6-oxo-2-hexenoic acid	Cytotoxicity activity	Ma et al. (2011)
T. estonicum	Mangrove sediments	Protease production	Blood stain removal	Saravanakumar and Kathiresan (2012)

(Continued)

TABLE 4.1 (*Continued*)

Diversity and Biological Function of Marine *Trichoderma*

Species	Isolation Sources	Chemicals	Biological Activity	References
Trichoderma sp. Gc1	Sponge (*G. corticostylifera, A. sydowii*, and *Bionectria* sp.)	—	Bioreduction of 1-(4-methoxyphenyl) ethanone	Lenilson et al. (2012)
T. pseudokoningii	Echinoderm (*Holothuria* sp.)	—	—	Devi et al. (2012)
T. aereoviride	Marine sponge (*Gelliodes fibrosa*), Cnidarian (*Annella* sp.), seawater and sediments	—	Antimicrobial activity	Wang et al. (2008); Khamthong et al. (2012)
T. aureoviride PSU-F95	Marine environment	Tetrahydroanthraquinone and xanthone derivatives	Antimycobacterial activity	Khamthong et al. (2012)
T. harzianum	Marine ascidian	Fungicide tyrosol 2-(4-hydroxyphenyl)	Antagonistic activity	Devi et al. (2013)
T. viride	Sediments, alga (*Gracilaria verrucosa*), mollusk (*Mytilus edulis*), sponge (*Suberites zeteki*), seawater	—	—	Brewer et al. (1993); Wang et al. (2008); El-Kassas and El-Taher (2009); Yamada et al. (2014)
T. asperellum, T. arundinaceum, T. brevicompactum, T. ghanense, T. aggressivum, H. viridescens, T. hamatum, T. harzianum/H. lixii, T. atroviride, T. koningii (TSK10), *T. estonicum/ H. estonica,* and *T. viride/H. rufa*	Sediments, seawater	—	Biotechnological applications	Saravanakumar (2012); Saravanakumar et al. (2013)

4.3.1 DNA Extraction

The fundamental requirement of *Trichoderma* for assessing the molecular diversity is DNA or RNA isolated from soil or other isolation sources (culture-independent method) and microbial purified *Trichoderma* culture (culture-dependent method). Many protocols have been developed for the isolation of the fungal DNA/RNA from soil or other isolation sources (Robe et al. 2003). Fungal DNA extraction is attributed by the chemical and/or enzymatic treatments and physical procedures of bead beating (Miller et al. 1999). Recently, scientists are interested to use the commercial hit for the DNA extraction (Borneman et al. 1996). Cetyltrimethylammonium bromide DNA extraction protocol is predominantly used in the extraction of nucleic acids from microbial culture of the *Trichoderma* (Doohan et al. 1998). However, the proper standardization and optimization of the protocol is necessary for the molecular analysis of *Trichoderma* from soil.

4.3.2 Genetic Markers

It is worth mentioning that the identification and deposition of the *Trichoderma* strains in the culture collection center based only on the morphological key characteristics is 50% acceptable; however, these techniques are proved to be unacceptable (Druzhinina and Kubicek 2005). Hence, wise researchers are focused on the *Trichoderma* diversity assessment by molecular markers. Generally, two kinds of molecular techniques are used for the community analysis of *Trichoderma* (Figure 4.1): (1) culture-independent molecular analysis,

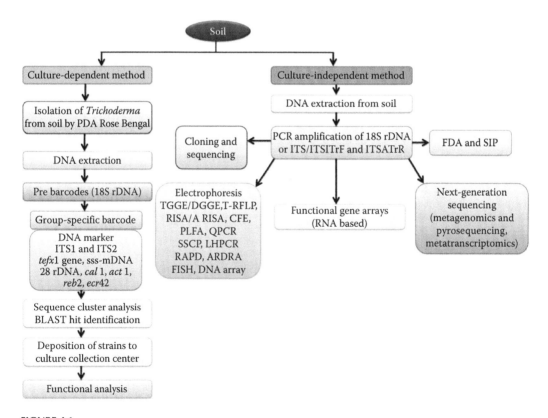

FIGURE 4.1

Experimental scheme: A molecular diversity analysis of marine *Trichoderma*.

where DNA is directly extracted from the soil or any kind of isolation source and the targeted specific *Trichoderma* DNA marker genes are amplified and analyzed by polymerase chain reaction–denaturing gradient gel electrophoresis (PCR-DGGE) as metagenomic or pyrosequencing approach (Meincke et al. 2010; Wu et al. 2013), and (2) culture-dependent fungal diversity analysis, where *Trichoderma* from the soil is isolated by using the microbiological techniques and targeted DNA marker genes are extracted from purified culture and amplified gene by using the PCR-based techniques (Sun et al. 2012).

Trichoderma diversity was analyzed by culture-independent molecular approach. In this approach, a specific forward primer ITS1*Tr*F (5′-ACTCCCAAACCCAATGTGAA-3′) and a reverse primer ITS4*Tr*R (5′-TGTGCAAACTACTGCGCA-3′) are designed for *Trichoderma* gene amplification in soil DNA or any isolation source of DNA by *in silico* primer development methods (Meincke et al. 2010). For the culture-dependent approach, genetic classification and molecular identification of marine *Trichoderma* have been made using phylogenetic inferences such as internal transcribed spacers 1 and 2 (ITS1 and ITS2), which are located between the 18S rDNA and 28S rDNA (Kullnig-Gradinger et al. 2002; White et al. 1990). Small subunit RNA-coding genes (ssu-mDNA) of the mitochondrial ribosomal DNA and 28S rDNA gene (Kullnig-Gradinger et al. 2002), protein-coding genes such as (*cal*1) calmodulin and (*act*1) actin (Chaverri et al. 2003), RNA polymerase B subunit 2 (*rpb*2), and the last large exon endochitinase 42 (*ech*42) (Kullnig-Gradinger et al. 2002). However, this phylogenetic inference of *Trichoderma* gene cluster of the sequences is not consistent with very similar species. Hence, the multigene approaches (translation elongation factor EF-1a [*tef*1 gene]) are encouraged for the ideal identification of these fungi (Druzhinina and Kubicek 2005; Kullnig-Gradinger et al. 2002). RNA polymerase gene barcodes (*rpb*2 exon) are also used for the identification by TrichOkey, TrichoBLAST, and TrichoCHIT (Druzhinina et al. 2005; Nagy et al. 2007). Currently, the status of molecular identification of *Trichoderma* by using DNA marker is quite ideal, but still a large number of new species could not be detected due to occasional dearth of available sequence match.

Also, review the genetic DNA markers for the identification of the *Trichoderma* from both the isolation sources of marine and their terrestrial origin. On the other hand, it is important to discuss the molecular threat; generally, some workers recognized species identity based on their similitude with BLAST analysis, wherein criticisms of these issues are being clearly described in a review on species concepts and biodiversity of *Trichoderma* and *Hypocrea*: from aggregate species to species clusters (Druzhinina and Kubicek 2005). These issues had been resolved by the new *Trichoderma* barcoding system that defines the differences between nucleotide sequences (Druzhinina et al. 2004). Finally, the proper comparison of biochemical, physiological, and molecular aspects can result in the proper identification of *Trichoderma*.

4.3.3 Molecular Profiling

The molecular profiling/differentiation of *Trichoderma* has been obtained on the basis of targeted nucleotide genes from the PCR product of mixed target genes, such as rDNA and ITS region of various *Trichoderma* species by analyzing the PCR products through molecular community profiling methods, including some electrophoresis techniques such as denaturing gel electrophoresis (PCR-DGGE), temperature gradient gel electrophoresis (TGGE), random amplified polymorphism DNA (RAPD), thermal restriction fragment length polymorphism, and sequence-based techniques such as cloning sequences and pyrosequencing.

Genetic diversity profiles of *Trichoderma* are accessed by the electrophoresis techniques of PCR-DGGE and TGGE (Meincke et al. 2010; Wu et al. 2013), which are advanced molecular tools used for studying the microbial community in the soil, and PCR-RAPD (Zimand et al. 1994) and RFLP (random fragment length polymorphism) (Sadfi-Zouaoui et al. 2009), which are frequently used for the assessment of genetic variations within individual species. However, the current microbial ecological assessments by PCR-based electrophoresis techniques are not frequently accepted due to the reproducibility of the results. Hence, the more advanced techniques such as cloning sequence and pyrosequencing techniques are encouraged in the microbial community studies. Culture-independent molecular approach of pyrosequencing or metagenomic analysis provides novel insights and significant research advances in *Trichoderma* ecology. The study on the microbial community analysis of *Trichoderma* from both terrestrial and marine ecosystem is sparse due to the lack of molecular ecological information about the influence of various experimental factors, especially selection of PCR primer for genetic diversity studies.

4.4 Omics

The fungal genus *Trichoderma* has multifunctional genomics in a variety of applications. Identification of the most important bioprospecting strains is necessary based on gene expression studies. The biological activity of *Trichoderma* has increased according to their molecular mechanisms and different interaction conditions in both inside (pathogen resistance, photosynthesis, promotion of plant growth) and outside (mycoparasitism, toxin degradation, antimicrobial activity, or biocontrol of phytopathogens). Furthermore, the understanding of the omics of *Trichoderma* can facilitate various applications of agriculture, industrial, and pharmaceutical research. However, the omics information of marine *Trichoderma* is limited.

4.4.1 *Trichoderma* Genome

The geographical occurrence, diversity, and genome of *Trichoderma* have been documented in the Europe and worldwide in a database found in the website http://nt.ars-grin.gov/taxadescriptions/keys/TrichodermaIndex.cfm (Chavarri and Samuels 2003; Samuels et al. 2014). The Index Fungorum database could facilitate in assessing the list of *Trichoderma* genomic records (http://www.indexfungorum.org/Names/Names.asp), and the International Subcommission of Trichoderma and Hypocrea also provides information of this fungus, which is reported in different resources (http://www.isth.info/biodiversity/index.php).

A number of fungal genome projects are being investigated by the US Department of Energy Joint Genome Institute (JGI, http:/www.jgi.doe.gov). Over 155 projects are involved in the barcoding of the fungi from 71 different genera with the objective of analysis of the entire genome and resequencing of existing genomes (Lorito et al. 2010). However, some JGI projects are focusing on different *Trichoderma* species: two programs investigating the genome of mycoparasitic biocontrol fungi *T. virens* (http://genome.jgidoe.gov/TriviGv29_8_2/TriviGv29_8_2.home.html), four of *T. reesei* (Martinez et al. 2008) (http://genome.jgi.doe.gov/Trire2/Trire2.home.html), three of *T. longibrachiatum* (http://genome.jgi.doe.gov/Trilo3/Trilo3.home.html), two of *T. harzianum* (http://genome.jgi.doe.gov/Triha1/Triha1.home.html), and one each of *T. reesei* RUT-30 (http://genome.jgi.doe.gov/TrireRUTC30_1/TrireRUTC30_1.home.html),

T. atroviride (http://genome.jgi-psf.org/Triat1/Triat1.home.html), *T. harzianum* (http://genome.jgi.doe.gov/Triha1/Triha1.home.html), and *T. asperellum* (http://genome.jgi.doe.gov/Trias1/Trias1.home.html).

Marine *Trichoderma* has been ecologically proved but not yet taxonomically described. Further, its biological systems and total homologous, heterologous, and stress tolerance gene expressions are poorly studied in comparison to the *Trichoderma* derived from terrestrial origin. Hence, there is a clear need to explore the phenotype and biological applications underlying the genetic systems. According to ecological description, *Trichoderma* could be divided into two main groups such as gene expression of marine *Trichoderma* and terrestrial *Trichoderma*.

The first *Trichoderma* expressed sequence tag (EST) libraries were made from *T. reesei* QM6a (Diener et al. 2004; Lorito et al. 2010). The redundant clones include ITS, translation elongation factor, exoglucanases, hydrolytic enzymes, heat shock proteins, hydrophobins, and ABC transporter, representing various biological activities. The available molecular genetic studies of *Trichoderma* genome data until August 2014, in accordance with the NCBI database, cover a total of 9,183 gene sequences, and ESTs of 147,390 are documented with respective applications of polymer degradation enzyme and biocontrol efficiency. Among the *Trichoderma* species, *T. reesei* and *T. harzianum* have received major attention due to their remarkable applications in the biochemical engineering industries and agricultural applications. A fundamental biological efficiency of *Trichoderma* encodes a variety of gene expressions that lead to the enzyme activities, including endochitinases, β-*N*-acetylhexosaminidase (*N*-acetyl-β-D-glucosaminidases), chitin-1,4-β-chitobiosidases, proteases, cellulases, amylases, pactinases, exo- and endo-β-1,3 glucanases, lipase, xylanases, mannanases, phytases, and phospholipidases (Ait-Lahsen et al. 2001; de la Cruz et al. 1992; Saravanakumar et al. 2013; Viterbo et al. 2001, 2002). The platelet-activating factor acetylhydrolase (PAF-AH) enzyme has the vital role in the antagonism and stress tolerance of *T. harzianum* in diverse ecological systems, thus revealing that the activation of PAF-AH may help develop the fertilizer for crop cultivation in different agriculture farming systems (Yu et al. 2014).

Trichoderma-derived cell wall–degrading enzymes (CWDEs) play a significant role in the control of phytopathogens (Harman 2000). A new biocontrol gene *taabc2* has been identified from the *T. atroviride* P1 (ATCC 74058) and elucidated its role in the ABC transporter membrane interactions with the various plant pathogenic fungi, and the upregulation of this gene enhances the production of the CWDEs and metabolites that lead to antagonistic activity against plant pathogens (Ruocco et al. 2009). The expression of *ech 42* and *nag1* genes of the biocontrol strain *T. atroviride* has been triggered for various signals and cell wall degradation of the pathogen (Mach et al. 1999). *T. reesei* is a great industrial source for cellulase enzyme, which is used for the conversion of the biomass into simple sugar and fuels, and the genomic analysis provides extraordinary capacity of *T. reesei* in the industrial application for bioethanol production (Martinez et al. 2008). A number of transcriptomes of *Trichoderma* have been reported (Lorito et al. 2010).

The defensive actions of *Trichoderma* species use lytic enzymes (Kubicek et al. 2001; Viterbo et al. 2002), proteolytic enzymes (Suarez et al. 2007), ABC transporter pumps (Ruocco et al. 2009), and biocontrol metabolites (Eziashi et al. 2007; Reino et al. 2008). However, the growth and biocontrol efficiency of *Trichoderma* are not independent of the temperature (Mukherjee and Raghu 1997). *Trichoderma* and its host interactions are involved in the stress response, nitrogen storage, cross-pathway control, lipid metabolism, and cell signaling process (Seidl et al. 2009). Besides regulatory mechanisms triggering the defense of *Trichoderma*, signal transduction triggers biocontrol and mycoparasitism genes, which especially include the mitogen-activated protein kinase, cascades, and cAMP pathways (Zeilinger and Omann

2007). These signaling mechanisms in *Trichoderma* target efficient biocontrol and interactions with plant root (Viterbo et al. 2005). The beneficial action of this fungus is not only to control the plant pathogens but also to protect and enhance the plant growth. Hence, these fungi are the promising source for agriculture and industrial applications.

4.4.2 Functional Genomics

Microbial functional genomics is the field of molecular biology that studies the basics of genetics of targeted microorganisms based on genomic sequences. A *Trichoderma* functional genomics study was first started in 2001 to explore their genetics, molecular diversity, and commercial applications (Rey et al. 2004). Initially, this project has been supported by the New BioTechnic and academic groups in Spain. Recently, it has extended to all agricultural fungal biological laboratories worldwide with the special attention in the study of the functional genomics of the potential biocontrol strains. This project has been attempted with the following specific objectives:

(1) To identify the strains isolated from different ecological systems, with molecular approaches to their taxonomical diversity, and also their biotechnological applications

(2) To define diversified EST sequences and also different ecological stress, antagonism, plant–pathogen interactions, biocontrol enzyme production, CWDE production, etc.

(3) To specify gene functional genomics of *Trichoderma* that focuses on the wide applications of biocontrol activities, mycoparasitism, enzyme production, and agro-food industrial commercial utilization (Ray et al. 2004)

A researcher has even reported a highly potent biocontrol of *Trichoderma* with the multifunctional effects on the agro-food application. The question here is whether the strains are stable for the environmental conditions. This can be answered by accomplishing proper studies of the functional genomics of the selected genome. However, functional genomic analyses vary according to their respective applications (Vizcaíno et al. 2006). Figure 4.2 illustrates the general schematic diagram of a gene functional analysis of *Trichoderma* fungi (Rey et al. 2004).

4.4.3 Gene Expression

The gene expressions of *Trichoderma* sp. vary based on different ecological conditions. In order to understand the engineering of marine *Trichoderma* with enhanced biotechnological applications, a great interest has been elucidated on the molecular mechanism operating at the transcriptional network that controls the gene expression in marine *Trichoderma*, and this provides the advanced genetic information for biotechnological utilization. However, the current scenario of gene expression studies on marine *Trichoderma* is limited. Hence, the recent gene expression studies of *Trichoderma* derived from terrestrial origin are here dealt with. The preliminary gene expression of each clone varies according to the interactions and combined effects of the environmental factors on *Trichoderma* growth. The virulence of selected clones from microarray has been tested in respect to the specific application of RNAs by Northern experiments. Different types of gene expressions are reported from *Trichoderma* with respect to biological applications.

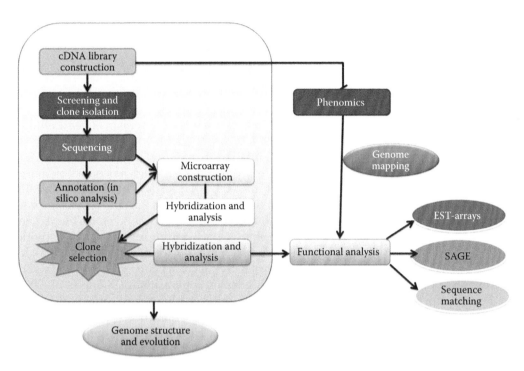

FIGURE 4.2
Experimental scheme: Gene functional analysis of marine *Trichoderma*. (Modified from Ray, M. et al., *Appl. Mycol. Biotechnol. Fungal Genom.*, 4, 225, 2004.)

Trichoderma species reduce the severity of plant diseases by inhibiting phytopathogens through antagonistic and mycoparasitism activities (Hermosa et al. 2012). Hydrolytic enzymes and CWDEs, such as 1,3-β-glucanases, proteases, and chitinases, have remarkable role in the mycoparasitism of *Trichoderma*. Hydrolytic enzyme activities vary according to gene expressions and functions. The gene expression of the biocontrol *Trichoderma* species has been successfully analyzed by high-density oligonucleotide microarray methods based on *Trichoderma* EST sequences (Samolski et al. 2009). A lytic enzyme has an important role to play in the biocontrol and mycoparasitism of *Trichoderma* on plant pathogenic fungi.

Several biocontrol gene expressions are reported in *Trichoderma* species such as *Tga*1 gene with the higher antifungal activity that has been reported from *T. atroviride* P1ATCC 74058 (Reithner et al. 2005). The expression of gene encoding the lytic enzyme, such as NAGases (exc1 and exc2), chitinases (chit42 and chit33), proteases (prb1), and β-glucanases (bgn13.1), has been reported in the biocontrol of *T. harzianum* T-32 and T-78 (Lopez-Mondejar et al. 2011). The RNA interference of endochitinase of *T. virens* has the biocontrol properties against pineapple and sugarcane diseases (Dumaresq et al. 2012). The expression of genes WRKY18 and WRKY40 of *Trichoderma* stimulates JA signals through JAZ receptor and negatively regulates the expression of defense genes FMO1, PAD3, and CYP71A13 in the roots of *Arabidopsis thaliana*, and colonization is increased in the FMO1 knockout mutant (Brotman et al. 2013). MDAR gene expression upregulates the monodehydroascorbate reductase in *Trichoderma* treatment, which affects the osmoprotection and oxidative stress–related gene expression in *Arabidopsis* and cucumber (*Cucumis sativus* L.) and significantly increases germination (Brotman et al. 2013).

Endochitinase (*ech-42*) gene expression promotes the ability of lytic enzyme chitinase production by the biocontrol strain *T. harzianum*, which improves the mycoparasitism against plant pathogen of *Botrytis cinerea* (Carsolio et al. 1994). Recently, it is proved that the *cis*-regulatory elements of XYR1 and CRE1 in the *T. reesei* enhance the cellulase gene expression for the production of cellulases and hemicellulases for utilization in paper and fuel industries (Rocha et al. 2009). Further, cellulase-encoding gene expressions, such as β-xylanases (*xyn*, *xyn*2) and β-mannanases (*man*1), *acetyl xylan* esterase (*axe*1), β-glucuronidase (*glr*1), β-L-arabinofuranosidase (abf1), cr-galactosidases (*agl*1, *agl*2, and *agl*3), β-glucosidase (bgl1), and β-xylosidase (*bxl*1), regulate the conversion of hemicellulose polymer materials (Margolles-Clark et al. 1997).

The gene expressions of ThPTR2, Qid74, T34hsp70, and *Erg*1 in *T. harzianum* CECT 2413 induce peptide transport and enhance mycoparasitism and biocontrol activities (Cardoza et al. 2006; Montero-Barrientos et al. 2008, 2011; Rosado et al. 2007; Vizcaino et al. 2006). The gene expressions of TmkA, TvBgn2, TvBgn3, and Sm1 in *T. virens* improve biocontrol activities (Dzonovic et al. 2007; Viterbo et al. 2005). *Beta tubulin*, Thke 11, and serine protease gene expression in *T. harzianum* shows significant biocontrol activities against phytopathogens (Hermosa et al. 2011; Li et al. 2007). Monooxygenase expression in *T. hamatum* LU 593 also increases biocontrol activities (Carpenter et al. 2008). TrCCD1 gene expression in *T. reesei* QM9414 shows higher mycoparasitism (Zhong et al. 2009). Tag 3 gene expression in *T. asperellum* shows significant biocontrol effects (Marcello et al. 2010). The gene expression of Tri5 in *T. brevicompactum* increases the biocontrol effects against plant pathogens (Tijerino et al. 2011). However, the targeted beneficial gene expression has been improved with respect to biotechnological applications by the more advanced approach of reverse genetic techniques, which extensively targeted gene disruptions/replacement (knockout) of gene function in *Trichoderma* (Bhadauria et al. 2009; Timberlake and Marshall 1989).

4.5 Biological Properties

The genes of *Trichoderma* are a remarkable source for biological and industrial applications with a variety of biological functions. Marine-derived *Trichoderma* received recent attention because of its remarkable tolerance to alkalinity and salinity (Saravanakumar et al. 2012). In addition, the activation of the multiple genes in marine *Trichoderma* can emphasize the diverse applications in biocontrol activities, pharmaceutical industry, enzyme production, biofuels, bioremediation, and NP synthesis.

4.5.1 Biocontrol Mechanisms

Plant diseases and soil nutrient deficiency cause crop yield losses, and this emphasizes a need for potent microbial fertilizers with multiple properties of biocontrol of phytopathogen and fungicidal activities and also plant growth promotion and soil nutrient solubilization. Microbial fertilizers are environment friendly and economically viable. Globally, *Trichoderma*-derived biocontrol agents are proved to be a remarkable source for agriculture utilization with multifunctions (Figure 4.3).

T. viride, isolated from the rhizosphere of the mangrove plant species *Avicennia marine*, has been reported as the biocontrol agent to control the root-knot nematode *Meloidogyne javanica* (Syed et al. 2003). Marine *T. harzianum* TSK8 and *T. estonicum* SKS1 isolated from the mangrove rhizosphere showed high soil phosphate solubilization and promoted

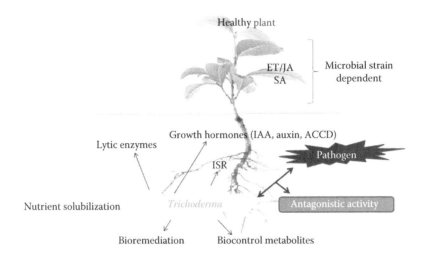

FIGURE 4.3
Multifunction *Trichoderma* (ET, ethylene; JA, jasmonic acid; SA, salicylic acid; ISR, induced systemic resistance; IAA, indole-3-acetic acid; ACCD, ACC deaminase). (Modified from Hermosa, R. et al., *Microbiology*, 158, 17, 2012.)

nutrient availability, thereby improving the growth of mangroves *Rhizophora apiculata* (Blume), *Rhizophora mucronata* (Lam.), and *Avicennia marina* (Forsk.) Vierh (Saravanakumar 2012; Saravanakumar et al. 2013). The increasing global human population needs a higher quantity of food products, thus encouraging saline water irrigation for arid-zone agriculture, or desert agriculture. The saline water supply can increase the soil salinity and increase the severity of diseases caused by plant pathogens (Hasegawa et al. 2000). These issues lead to finding new microorganisms to tolerate higher salinity and to control plant pathogens or diseases, which then promotes the search for the novel saline-tolerant *Trichoderma* as a biocontrol agent or plant growth promoter for arid-zone agriculture. Thus, marine sponges–derived *T. longibrachiatum*, *T. atroviride*, *T. harzianum*, and *T. asperelloides* are efficient in controlling cucumber pathogens, such as *Pseudomonas syringae* pv. lachrymans and *Rhizoctonia solani* in arid-zone agriculture (Hamed et al. 2011). Marine Ascidian–associated *T. harzianum* is known to produce the fungicide tyrosol [2-(4-hydroxyphenyl) to control plant pathogens such as *Penicillium piceum* and *Sclerotium rolfsii* and tomato fungus such as *Chalara* sp. (Devi et al. 2013).

4.5.2 Enzymes and Bioenergy

Trichoderma species produce novel enzymes of industrial and biocontrol importance (cellulase, lipase, amylase, laccase, xylanase, etc.) as well as their use in leather, detergent, and food preservation. However, only few studies are available on the enzyme production by marine-derived *Trichoderma*. Salt-tolerant laccase-secreting strain of *Trichoderma* sp. NFCCI-2745, isolated from coconut husk retting pile in the estuarine waters of North Kerala, India, is efficient to produce the laccase enzyme to treat saline phenolic effluents (Divya et al. 2013). *T. aureviride* isolated from seafloor is efficient to produce enzyme such as laminarinase, amylase, *N*-acetyl-β-D-glucosaminidase, β-D-gluco- and galactosidase, and cellulase (Burtseva et al. 2006). *T. estonicum* isolated from mangrove soil produces protease with a potential of removing blood stain (Saravanakumar and Kathiresan 2012). The potential of terrestrial *Trichoderma* species on bioenergy conversation is well established, but for marine-derived *Trichoderma* species, only limited studies are available. Marine strains of *Trichoderma* have recently been proved for their ability in cellulase production for pretreatment of waste lignocellulosic material and fermentation for ethanol production (Ahmed and Mervat 2012; Saravanakumar et al. 2013).

4.5.3 Bioremediation

The continuous increases of chemical industries result in the discharge of industrial effluents throughout the coastal area and agriculture lands, which challenge safe disposal of industrial effluents (Donmez and Aksu 2001). Continuous discharge of the effluents into the environment induces several changes in the environments. This pollution seriously affects the supply of microbial nutrients in agriculture lands, leading to great loss of crop cultivation. In this regard, *Trichoderma* species are sole source of fertilizer with biocontrol and bioremediation functions. Marine *T. viride* is a chromium-resistant strain isolated from marine resource and is known to remove chromium in industrial effluents (El-Kassas and El-Taher 2009). *Trichoderma* sp. Gc1 is known to degrade the pesticide (DDT) 1,1-dichloro-2,2-*bis*-(4-chlorophenyl) ethane (Ortega et al. 2011) and to reduce the contamination of 1-(4-methoxyphenyl) ethanone (Lenilson et al. 2009). *T. asperellum* is efficient to remove Cu(II) from aqueous solution (Tang and Ting 2012). Marine-derived strains of *Trichoderma* may be better efficient for bioremediation than the terrestrial ones due to the saline, thermo, and alkaline tolerance of marine strains.

4.5.4 Metabolites

The unexpected increases of new human diseases and infections have triggered to search for novel bioactive compounds from the biological resources with higher medicinal values. In this regard, over the past several decades, researchers have extensively studied the biocompounds from plants and microbes derived from terrestrial regions. Marine ecosystem is biologically diverse with a great promise for discovery of biocompounds with higher industrial applications (Sithranga Boopathy and Kathiresan 2010). Many researchers have focused on all groups of marine fungi, but few studies are available on the secondary metabolites of marine-derived *Trichoderma*. *T. virens* is a producer of trichodermamides A and B with high pharmaceutical value (Garo et al. 2003). *Trichoderma* is a producer of trichoderins, a novel aminolipopeptide, trichodimerol, bislongiquinolide, bisvertinol, trichoderone, a novel cytotoxic cyclopentenone, cholesta-7, 22-diene-3b, 5a, 6b-triol, didemnidones, and trichoderone 35, 36 with antimycobacterial activity and pharmaceutical applications (Abdel-Latif et al. 2010; Pruksakorn et al. 2010; You et al. 2010). *T. longibrachiatum* is a producer of peptaibols—acetyl group, aminoisobutyric acid (Aib), and an *N*-terminal acyl with unknown pharmaceutical utilization (Mohamed-Benkada et al. 2006). *Trichoderma* strain H1-7 is a producer of tyrosinase-inhibiting unknown compound (Yamada et al. 2008). *T. aureoviride* PSU-F95 is a producer of tetrahydroanthraquinone and xanthone derivatives with strong antimicrobial activity (Khamthong et al. 2012). Four new polyketide derivatives, trichodermatides A–D (1–4), extracted from the marine-derived fungus *T. reesei* exhibit strong cytotoxicity against A375-S2 human melanoma cell line (Sun et al. 2009). Tandyukisin (1), a novel decalin derivative with an enolic b-ketoaldehyde, is extracted from marine sponge–derived *T. harzianum* and exhibits moderate cytotoxicity against human cancer cell lines (Yamada et al. 2014). Marine *T. koningii* isolated from a shellfish (the cockle *Cerastoderma edule*) is known to produce mycotoxin containing Pheol and Aib, two amino-acid characteristics of peptaibols (Landreau et al. 2002).

Longibrachin-A-I has been identified from a culture of *T. longibrachiatum* (Leclerc et al. 1998). It had also been isolated from *Gliocladium deliquescens* with the name of Gliodesquin A (Bruckner and Przybylski 1984) and from *T. koningii* with the name of Trichokonin-VI (Huang et al. 1996). *T. longibrachiatum* is also known to produce the new trichobrachins, 11-residue peptaibols (Ruiz et al. 2007a). Marine *Trichoderma* species, *Hypocrea lixii* TSK8 and *Hypocrea rufa* SKS2 isolated from marine biotope, are known to produce heptadecanoic acid and octadecadienoic acid with remarkable anti–skin cancer activity (Saravanakumar 2012; Saravanakumar et al. 2012). A marine strain of *T. longibrachiatum* isolated from mussels produces fatty acids, predominantly oleic (up to 15.3% of the total fatty acid mixture for submerged fermentation and 33.9% for agar surface fermentation), linoleic (46.1% for submerged fermentation and 40.3% for agar surface fermentation), and palmitic acids (28.1% for agar surface fermentation and 19.1% for agar surface fermentation) (Ruiz et al. 2007a). The endophytic fungus *Trichoderma* with high antioxidant potentials has been isolated from the mangrove leaves collected from Andaman and Nicobar Island, India (Saravanakumar and Kathiresan 2014). Recently, marine fungus *Trichoderma* is found to be the considerable source for the secondary metabolites with the higher bioactivity in agriculture and human disease control. *T. longibrachiatum* is the source for isolation of potential antifungal agents. In addition, some *Trichoderma* species such as *T. citrinoviride*, *T. harzianum*, *T. longibrachiatum*, and *H. orientalis* are reportedly opportunistic human pathogenic fungi that can cause respiratory problems (Kredics et al. 2003; Schuster and Schmoll 2010).

4.5.5 Nanoparticle Synthesis

Nanobiotechnology is the convergence of molecular biology and bioengineering and is a fast-growing field of research with enormous applications (Asmathunisha and Kathiresan 2013; Fortina et al. 2005). NPs have been used in bioremediation, food preservation, cotton finishing, drug delivery, etc. (Hebeish et al. 2011; Piotrowska et al. 2009). There are several biological and chemical techniques for the synthesis of NPs, but they are costly and not environment friendly. This emphasizes the importance of biological synthesis of NPs from various biological resources (Asmathunisha and Kathiresan 2013). *Trichoderma* species of terrestrial origin are known to produce silver NPs (Fayaz et al. 2010; Gajbhiye et al. 2009; Huang et al. 2013); subsequently, several *Trichoderma* species derived from the terrestrial region have been utilized for various metal NP synthesis. The cellulolytic *T. reesei/Hypocrea jecorina* is proved to be a remarkable resource for large-scale synthesis of silver NPs (Vahabi et al. 2011). *T. koningiopsis* is known to synthesize copper (Cu) NPs, and it also removes Cu in wastewater by adsorption mechanism (Salvadori et al. 2014). The biocontrol strain *T. koningii* is known to produce gold and silver NPs with higher cytotoxicity and antimicrobial activity (Maliszewska 2013; Tripathi et al. 2013). Currently, green-synthesized silver NPs are widely used in the clinical research of drug delivery due to their dispersion, uniformity in size, and similarity to molecular molecules. Relatively, silver NPs synthesized by *T. harzianum* can improve the activity of anti-fascioliasis drug triclabendazole (Gherbawy et al. 2013), and it also has antimicrobial activity against bacterial pathogens such as *Staphylococcus aureus* and *Klebsiella pneumoniae* (Ahluwalia et al. 2014). The bioengineered silver nanobowls by *T. viride* show a significant antimicrobial activity against gram-negative and gram-positive bacteria (Chitra and Annadurai 2013). However, very recently, marine *Trichoderma* isolated from mangrove biotope is proved to synthesize silver and gold NPs applied in bioelectricity production (Saravanakumar 2012).

4.6 Conclusions and Future Prospectus

Although *Trichoderma* species of terrestrial origin are well known for their genomic studies and applications in agriculture and industries, only a few such studies are available for marine *Trichoderma* species. As marine-derived *Trichoderma* are likely to have better bioprospecting potential than their terrestrial counterparts, this chapter calls for a greater research attention on marine *Trichoderma* for its gene expression in multibiological functions of applications in bioremediation, biofuels, biofertilizer, and biocontrol effects as well as in the production of high-value metabolites in the field medicine. Advanced studies on marine *Trichoderma* are required for agricultural engineering, nanotechnological applications, and gene signaling involved in microbial interactions.

Acknowledgments

The authors are thankful to Shanghai Jiao Tong University and Annamalai University for providing the necessary facility to complete this chapter. The first author thanks the National Science and Technology Basic Work Project (2014FY120900), China, for providing a postdoctoral fellowship.

References

Abdel-Latif, H., A. Mohammed, and H. Wafaa Mohamed. 2010. Mutagenesis and inter-specific protoplast fusion between *Trichoderma koningii* and *Trichoderma reesei* for biocontrol improvement. *American Journal of Scientific and Industrial Research* 1:504–515.

Ahluwalia, V., J. Kumar, R. Sisodia, N. A. Shakil, and S. Walia. 2014. Green synthesis of silver nanoparticles by *Trichoderma harzianum* and their bio-efficacy evaluation against *Staphylococcus aureus* and *Klebsiella pneumonia*. *Industrial Crops and Products* 55:202–206.

Ahmed, M., A. El-Bondkly, and M. M. A. El-Gendy. 2012. Cellulase production from agricultural residues by recombinant fusant strain of a fungal endophyte of the marine sponge *Latrunculia corticata* for production of ethanol. *Antonie van Leeuwenhoek* 101:331–346.

Ainsworth, G. C. 1973. Introduction and keys to higher taxa. In *The Fungi: An Advanced Treatise, IVB*, G. C. Ainsworth, F. K. Sparrow, and A. S. Sussman (eds.). New York: Academic Press, pp. 1–7.

Ait-Lahsen, H., A. Soler, M. Rey, J. de la Cruz, E. Monte, and A. Llobell. 2001. An antifungal exo-α (1→3)-glucanase (AGN13.1) from the biocontrol fungus *T. harzianum*. *Applied and Environmental Microbiology* 67:5833–5839.

Amagata, T., Y. Usami, K. Minoura, T. Ito, and A. Numata. 1998. Cytotoxic substances produced by a fungal strain from a sponge: Physico-chemical properties and structures. *Journal of Antibiotics* 51:33–40.

Asmathunisha, N. and K. Kathiresan. 2013. A review on biosynthesis of nanoparticles by marine organisms. *Colloids and Surfaces B: Biointerfaces* 103:283–287.

Bhadauria, V., S. Banniza, Y. Wei, and Y. L. Peng. 2009. Reverse genetics for functional genomics of phytopathogenic fungi and oomycetes. *Comparative and Functional Genomics*, 2009, Article ID 380719, 11pp. http//dx.doi.org/10.1155/2009/380719.

Bisby, G. R. 1939. *Trichoderma viride* pers. ex Fries, and notes on *Hypocrea. Transactions of the British Mycological Society* 23:149–168.

Bissett, J. 1991. A revision of the genus *Trichoderma*. II. Infrageneric classification. *Canadian Journal of Botany* 69:2357–2372.

Bochner, B. R., P. Gadzinski, and E. Panomitros. 2001. Phenotype microarrays for high-throughput phenotypic testing and assay of gene function. *Genome Research* 11:1246–1255.

Borneman, J., P. W. Skroch, K. M. O'Sullivan, J. A. Palus, N. G. Rumjanek, J. L. Jansen, J. Nienhuis et al. 1996. Molecular microbial diversity of an agricultural soil in Wisconsin. *Applied and Environmental Microbiology* 62:1935–1943.

Brewer, D., M. Greenwel, and A. Taylor. 1993. Studies of *Trichoderma* sp. isolated from *Mytilus eludis* collected on shores of Capbreton and Prince Edward islands. *Proceedings of the Nova Scotian Institute of Science* 40:29–40.

Brotman, Y., U. Landau, A. Cuadros-Inostroza, T. Takayuki, A. R. Femie, I. Chet, A. Viterbor, and L. Willmitzer. 2013. *Trichoderma*-plant root colonization: Escaping early plant defense responses and activation of the antioxidant machinery for saline stress tolerance. *PLoS Pathogens* 9(3):e11003221.

Bruckner, H. and M. Przybylski. 1984. Isolation and structure characterization of polypeptide antibiotics of the peptaibol class by high-performance liquid chromatography with field desorption and fast atom bombardment mass spectrometry. *Journal of Chromatography* 296:263–275.

Burtseva, Y., N. Verigina, V. Sova, M. Pivkin, and T. Zvyagintseva. 2006. Comparative characterization of laminarinases from the filamentous marine fungi *Chatomium indicum* Corda and *Trichoderma aureviride* Rifai. *Journal of Applied Phycology* 18:375–380.

Cardoza, R. E., Vizcaíno, J. A., Hermosa, M. R., Sousa, S., González, F. J., Llobell, A. et al. (2006). Cloning and characterization of the erg1 gene of *Trichoderma harzianum*: Effect of the erg1 silencing on ergosterol biosynthesis and resistance to terbinafine. *Fungal Genetics and Biology* 43:164–178.

Carpenter, M. A., H. J. Ridgway, A. M. Stringer, A. J. Hay, and A. Stewart. 2008. Characterization of a *Trichoderma hamatum* monooxygenase gene involved in antagonistic activity against fungal plant pathogens. *Current Genetics* 53:193–205.

Carsolio, C., A. Gutierrez, B. Jimenez, M. Van Montagu, and A. Herrera-Estrella. 1994. Primary structure and expression pattern of the 33-kDa chitinase gene from the mycoparasitic fungus *Trichoderma harzianum. Proceedings of the National Academy of Sciences of the United States of America* 91(23):10903–10907.

Chaverri, P., L. A. Castleburry, G. J. Samuels, and D. M. Geiser. 2003. Multilocus phylogenetic structure of *Trichoderma harzianum/Hypocrea lixii* complex. *Molecular Phylogenetics Evolution* 27:302–313.

Chaverri, P. and G. J. Samuels. 2003. *Hypocrea/Trichoderma* (Acomycota, Hypocreals, Hypocreaceae): Species with green ascospores. *Studies in Mycology* 48:1–116.

Chitra, K. and G. Annadurai. 2013. Bioengineered silver nanobowls using *Trichoderma viride* and its antibacterial activity against gram-positive and gram-negative bacteria. *Journal of Nanostructure in Chemistry* 3:9.

Da Silva, M., M. R. Z. Passarini, R. C. Bonugli, and L. D. Sette. 2008. Cnidarian-derived filamentous fungi from Brazil: Isolation, characterisation and RBBR decolourisation screening. *Environmental Technology* 29:1331–1339.

de la Cruz, J., A. Hidalgo-Gallego, J. M. Lora, T. Benitez, J. A. Pintor-Toro, and A. Llobell. 1992. Isolation and characterization of three chitinases from *Trichoderma harzianum. European Journal of Biochemistry* 206:859–867.

Devi, T. P., S. Kulanthaivel, D. Kamil, J. L. Borah, N. Prabhakaran, and N. Srinivasa. 2013. Biosynthesis of silver nanoparticles from Trichoderma species. *Indian Journal of Experimental Biology* 51:543–547.

Devi, T. P., N. Prabhakaran, K. Deeba, J. L. Borah, and P. Pandey. 2012. Development of species specific markers for detection of *Trichoderma* species. *Vegetos* 25:207–217.

Diener, S. E., M. K. Chellappan, T. K. Mitchell, N. Dunn-Coleman, M. Ward, and R. A. Dean. 2004. Insight into *Trichoderma reesei's* genome content, organization and evolution revealed through BAC library characterization. *Fungal Genetics and Biology* 41:1077–1087.

Divya, L. M., G. K. Prasanth, and C. Sadasivan. 2013. Isolation of a salt tolerant laccase secreting strain of *Trichoderma* sp. NFCCI-2745 and optimization of culture conditions and assessing its effectiveness in treating saline phenolic effluents. *Journal of Environmental Sciences* 25:2410–2416.

Djonovic, S., W. A. Vargas, M. V. Kolomiets, M. Horndeski, A. Wiest, and C. M. Kenerley. 2007. A proteinaceous elicitor Sm1 from the beneficial fungus *Trichoderma virens* is required for induced systemic resistance in maize. *Plant Physiology* 145:875–889.

Donmez, G. and Z. Aksu. 2001. Bioaccumulation of copper (II) and nickel (10 by the non-adapted and adapted growing *Cundidu* sp. *Journal of Water Research* 35:1425–1434.

Doohan, F. M., D. W. Parry, P. Jenkinson, and P. Nicholson. 1998. The use of species-specific PCR-based assays to analyse *Fusarium ear* blight of wheat. *Plant Pathology* 47:197–205.

Druzhinina, I. S., P. Chaverri, P. Fallah, C. P. Kubicek, and G. J. Samuels. 2004. *Hypocrea flaviconidia*, a new species from Costa Rica with yellow conidia. *Studies in Mycology* 50:401–407.

Druzhinina, I. S., A. G. Koptchinski, M. Komon, J. Bissett, G. Szakacs, and C. P. Kubicek. 2005. An oligonucleotide barcode for species identification in *Trichoderma* and *Hypocrea*. *Fungal Genetic Biology* 42:813–828.

Druzhinina, I. S., A. G. Kopchinskiy, and C. P. Kubicek. 2006. The first 100 *Trichoderma* species characterized by molecular data. *Mycoscience* 47:55–64.

Druzhinina, I. and C. P. Kubicek. 2005. Species concepts and biodiversity in *Trichoderma* and *Hypocrea*: From aggregate species to species clusters? *Journal of Zhejiang University Science B* 6:100–112.

Du, L., D. Li, T. Zhu, S. Cai, F. Wang, X. Xiao, and Q. Gu. 2009. New alkaloids and diterpenes from a deep ocean sediment derived fungus *Penicillium* sp. *Tetrahedron* 65:1033–1039.

Dumaresq, A. S. R., W. L. de Araujo, N. J. Talbot, and C. R. Thornton. 2012. RNA interference of endo-chitinases in the sugarcane endophyte *Trichoderma virens* 223 reduces its fitness as a biocontrol agent of pineapple disease. *PLoS ONE* 7(10):e47888.

Eliane, G., M. S. Courtney, R. J. Paul, W. Fenical, E. Lobkovsky, and J. Clardy. 2003. Trichodermamides A and B, cytotoxic modified dipeptides from the marine-derived fungus *Trichoderma virens*. *J Nat Prod* 66:423–426.

El-Kassas, H. Y. and E. M. El-Taher. 2009. Optimization of batch process parameters by response surface methodology for mycoremediation of chrome-VI by a chromium resistant strain of marine *Trichoderma Viride*. *American-Eurasian Journal of Agricultural and Environmental Science* 5:676–681.

Eziashi, E. I., I. B. Omamor, and E. E. Odigie. 2007. Antagonism of *Trichoderma viride* and effects of extracted water soluble compounds from *Trichoderma* species and benlate solution on *Ceratocystis paradoxa*. *African Journal of Biotechnology* 6:388–392.

Fayaz, A. M., K. Balaji, M. Girilal, R. Yadav, P. T. Kalaichelvan, and R. Venketesan. 2010. Biogenic synthesis of silver nanoparticles and their synergistic effect with antibiotics: A study against gram-positive and gram-negative bacteria. *Nanomedicine: Nanotechnology, Biology and Medicine* 6:103–109.

Fortina, P., L. J. Kricka, S. Surrey, and P. Grodzinski. 2005. Nanobiotechnology: The promise and reality of new approaches to molecular recognition. *Trends in Biotechnology* 23:168–173.

Furbino, L. E., V. M. Godinho, I. F. Santiago, F. M. Pellizari, T. M. A. Alves, C. L. Zani, P. A. S. Junior et al. 2014. Diversity patterns, ecology and biological activities of fungal communities associated with the endemic macroalgae across the Antarctic Peninsula. *Microbial Ecology* 67:775–787.

Gajbhiye, M., J. Kesharwani, A. Ingle, A. Gade, and M. Rai. 2009. Fungus mediated synthesis of silver nanoparticles and their activity against pathogenic fungi in combination with fluconazole. *Nanomedicine: Nanotechnology, Biology and Medicine* 5:382–386.

Gal-Hemed, I., L. Atanasova, M. Komon-Zelazowska, I. S. Druzhinina, A. Viterbo, and O. Yarden. 2011. Marine isolates of *Trichoderma* spp. as potential halotolerant agents of biological control for Arid-zone agriculture. *Applied and Environmental Microbiology* 77:5100–5109.

Gams, W. and J. Bissett (eds.). 1998. *Morphology and Identification of Trichoderma*. London, U.K.: Taylor & Francis Ltd.

Garo, E., C. M. Starks, P. R. Jensen, W. Fenical, E. Lobkovsky, and J. Clardy. 2003. Trichodermamides A and B, cytotoxic modified dipeptides from the marine-derived fungus *Trichoderma virens*. *Journal of Natural Products* 66:423–426.

Gherbawy, Y. A., I. M. Shalaby, M. S. Abd El-sadek, H. M. Elhariry, and A. A. Banaja. 2013. The anti-fascioliasis properties of silver nanoparticles produced by *Trichoderma harzianum* and their improvement of the anti-fascioliasis drug *Triclabendazole*. *International Journal of Molecular Sciences* 14:21887–21898.

Grishkan, I., A. B. Korol, E. Nevo, and S. P. Wasser. 2003. Ecological stress and sex evolution in soil microfungi. *Proceedings of the Royal Society of London* 270:13–18.

Harman, G. E. 2000. Myths and dogmas of biocontrol. Changes in perceptions derived from research on *Trichoderma harzianum* T-22. *Plant Disease* 84:377–393.

Harman, G. E., C. R. Howell, A. Viterbo, I. Chet, and M. Lorito. 2004. Trichoderma species—Opportunistic, avirulent plant symbionts. *Nature Reviews Microbiology* 2:43–56.

Hasegawa, P. M., R. A. Bressan, J. K. Zhu, and H. J. Bohnert. 2000. Plant cellular and molecular responses to high salinity. *Annual Review of Plant Physiology and Plant Molecular Biology* 51:463–499.

Hebeish, A., A. El-Shafei, S. Sharaf, and S. Zaghloul. 2011. Novel precursors for green synthesis and application of silver nanoparticles in the realm of cotton finishing. *Carbohydrate Polymers* 84:605–613.

Hermosa, R., L. Botella, E. Keck, J. A. Jimenez, M. Montero-Barrientos, V. Arbona, A. Gomez Cadenas, E. Monte, and C. Nicolas. 2011. The overexpression in *Arabidopsis thaliana* of a *Trichoderma harzianum* gene that modulates glucosidase activity, and enhances tolerance to salt and osmotic stresses. *Journal of Plant Physiology* 168:1295–1302.

Hermosa, R., A. Viterbo, I. Chet, and E. Monte. 2012. Plant-beneficial effects of *Trichoderma* and of its genes. *Microbiology* 158:17–25.

Huang, Q., Y. Tezuka, Y. Hatanaka, T. Kikuchi, A. Nishi, and K. Tubaki. 1996. Studies on metabolites of mycoparasitic fungi. V. Ion-spray ionization mass spectrometric analysis of trichokonin II, a peptaibol mixture obtained from the culture broth of *Trichoderma koningii*. *Chemical Pharmaceutical Bulletin* 44:590–593.

Huang, X., J. Ge, J. Fan, X. Chen, X. Xu, J. Li, Y. Zhang et al. 2013. Characterization and optimization of xylanases and endoglucanase production by *Trichoderma viride* HG 623 using response surface methodology (RSM). *African Journal of Microbiological Research* 7:4521–4532.

Immaculatejeyasanta, K., P. Madhanraj, J. Patterson, and A. Panneerselvam. 2011. Case study on the extra cellular enzyme of marine fungi associated with mangrove driftwood of Muthupet Mangrove, Tamil Nadu, India. *Journal of Pharmacy Research* 4:1385–1387.

Karthik Raja, N. and D. Nirmala. 2011. Enhanced production of alpha amylase using vegetable wastes by *Aspegilllus niger* strain SK01 marine isolate. *Indian Journal of Geo-Marine Sciences* 40:130–133.

Kathiresan, K., K. Saravanakumar, R. Anburaj, V. Gomathi, G. Abirami, S. K. Sahu, and S. Anandhan. 2011. Microbial enzyme activity in decomposing leaves of mangroves. *International Journal of Advanced Biotechnology and Research* 2:382–389.

Khamthong, N., V. Rukachaisirikul, K. Tadpetch, M. Kaewpet, S. Phongpaichit, S. Preedanon, and J. Sakayaroj. 2012. Tetrahydroanthraquinone and xanthone derivatives from the marine-derived fungus *Trichoderma aureoviride* PSU-F95. *Archives of Pharmacal Research* 35:461–468.

Kis-Papo, T., I. Grishkan, A. Oren, S. P. Wasser, and E. Nevo. 2001. Spatiotemporal diversity of filamentous fungi in the hypersaline Dead Sea. *Mycological Research* 105:749–756.

Kobayashi, M., H. Uehara, K. Matsunami, S. Aoki, and I. Kitagawa. 1993. Trichoharzin, a new polyketide produced by the imperfect fungus *Trichoderma harzianum* separated from the marine sponge *Micale cecilia*. *Tetrahedron Letters* 34:7925–7928.

Kohlmeyer, J. 1974. On the definition and taxonomy of higher marine fungi. *Veroffentlichnedne s Instituts fur Meeresforschung Bremerhaven Supplement* 5:263–286.

Kopchinskiy, A. G., M. Komon, C. P. Kubicek, and I. S. Druzhinina. 2005. *Tricho* BLAST: A multiloci database of phylogenetic markers for *Trichoderma* and *Hypocrea* powered by sequence diagnosis and similarity search tools. *Mycological Research* 109:658–660.

Kredics, L., Z. Antal, I. Doczi, L. Manczinger, F. Kevei, and E. Nagy. 2003. Clinical importance of the genus *Trichoderma*. A review. *Acta Microbiologica et Immunologica Hungarica* 50:105–117.

Kubicek, C. P., M. Komon-Zelazowska, and I. S. Druzhinina. 2008. Fungal genus *Hypocrea/Trichoderma*: From barcodes to biodiversity. *Journal of Zhejiang University Sciences B* 9:753–763.

Kubicek, C. P., R. L. Mach, C. K. Peterbauer, and M. Lorito. 2001. *Trichoderma*: From genes to biocontrol. *Journal of Plant Pathology* 83:11–23.

Kullnig-Gradinger, C. M., G. Szakacs, and C. P. Kubicek. 2002. Phylogeny and evolution of the fungal genus *Trichoderma*: A multigene approach. *Mycological Research* 106:757–767.

Landreau, A., Y. F. Pouchus, C. Sallenave-Namont, J. F. Biard, M. C. Boumard, T. R. du Pont, F. Mondeguer et al. 2002. Combined use of LC/MS and a biological test for rapid identification of marine mycotoxins produced by *Trichoderma koningii*. *Journal of Microbiological Methods* 48:181–194.

Leclerc, G., S. Rebuffat, C. Goulard, and B. Bodo. 1998. Directed biosynthesis of peptaibol antibiotics in two *Trichoderma* strains I. Fermentation and isolation. *Journal of Antibiotics* 51:170–177.

Lee, H. I., M. H. Kim, K. Y. Kim, and J. S. So. 2010. Screening and selection of stress resistant *Lactobacillus* spp. isolated from the marine oyster *Crassostrea gigas*. *Anaerobe* 16:522–526.

Lenilson, C. R., H. V. Ferreira, E. F. Pimenta, R. G. S. Berlinck, M. H. R. Seleghim, D. C. D. Javaroti, D. C. D. Javaroti et al. 2009. Bioreduction of a-chloroacetophenone by whole cells of marine fungi. *Biotechnology Letters* 31:1559–1563.

Lenilson, C. R., V. F. Hercules, F. L. Rodrigo, D. S. Lara, and Andre, L. M. P. 2012. Stereoselective bioreduction of 1-(4-Methoxyphenyl) ethanone by whole cells of marine-derived fungi. *Mar Biotechnol* 14:358–362.

Li, C. et al. 2014. Two Antarctic penguins reveal insights into their evolutionary history and molecular changes related to the Antarctic environment. *GigaScience* 3:27.

Li, L., S. J. Wright, S. Krystofova, G. Park, and K. V. Borkovich. 2007. Heterotrimeric G Protein signaling in filamentous fungi. *Annual Review of Microbiology* 61:423–452.

Liu, X., E. Ashforth, B. Ren, F. Song, H. Dai, M. Liu, J. Wang et al. 2010. Bioprospecting microbial natural product libraries from the marine environment for drug discovery. *Journal of Antibiotics* 63:415–422.

Lopez-Mondejar, R., M. Ros, and J. A. Pascual. 2011. Mycoparasitism-related genes expression of *Trichoderma harzianum* isolates to evaluate their efficacy as biological control agent. *Biological Control* 56:59–66.

Lorito, M., S. L. Woo, G. E. Harman, and E. Monte. 2010. Translational research on *Trichoderma*: From 'omics to the field. *Annual Review of Phytopathology* 48:395–417.

Ma, L., W. Liu, Y. Huang, and X. Rong. 2011. Two acid sorbicillin analogues from saline lands-derived fungus *Trichoderma* sp. *The Journal of Antibiotics* 64:645–647.

Mach, R. L., C. K. Peterbauer, K. Payer, S. Jaksits, S. L. Woo, S. Zeilinger, C. M. Kullnig et al. 1999. Expression of two major chitinase genes of *Trichoderma atroviride* (*T. harzianum* P1) is triggered by different regulatory signals. *Applied Journal of Environmental Microbiology* 65:1858–1863.

Maliszewska, I. 2013. Microbial mediated synthesis of gold nanoparticles: Preparation, characterization and cytotoxicity studies. *Digest Journal of Nanomaterials and Biostructures* 8:1123–1131.

Marcello, C. M., A. S. Steindorff, S. P. Silva, R. N. Silva, and L. A. M. Bataus. 2010. Expression analysis of the exo-β-1, 3-glucanase from the mycoparasitic fungus *Trichoderma asperellum*. *Microbiological Research* 165:75–81.

Margolles-Clark, E., M. Ihnen, and M. Penttila. 1997. Expression patterns of ten hemicellulase genes of the filamentous fungus *Trichoderma reesei* on various carbon sources. *Journal of Biotechnology* 57:167–179.

Martinez, D., R. M. Berka, B. Henrissat, M. Saloheimo, M. Arvas, S. E. Baker, J. Chapman et al. 2008. Genome sequencing and analysis of the biomass-degrading fungus *Trichoderma reesei* (syn. *Hypocrea jecorina*). *Nature Biotechnology* 26:553–560.

Masuma, R., Y. Yamaguchi, M. Noumi, O. S. Mura, and M. Namikoshi. 2001. Effect of sea water on hyphal growth and antimicrobial metabolite production in a marine fungi. *Mycoscience* 42:455–459.

Meincke, R., N. Weinert, V. Radl, M. Schloter, K. Smalla, and G. Berg. 2010. Development of a molecular approach to describe the composition of *Trichoderma* communities. *Journal of Microbiological Methods* 80:63–69.

Miller, D. N., J. E. Bryant, E. L. Madsen, and W. C. Ghiorse. 1999. Evaluation and optimization of DNA extraction and purification procedures for soil and sediment samples. *Applied and Environmental Microbiology* 65:4715–4724.

Mohamed-Benkada, M., M. Montagu, J. F. Biard, F. Mondeguer, P. Verite, M. Dalgalarrondo, J. Bissett et al. 2006. New short peptaibols from a marine *Trichoderma* strain. *Rapid Communications in Mass Spectrometry* 20:1176–1180.

Montero-Barrientos, M., R. Hermosa, R. E. Cardoza, S. Gutierrez, and E. Monte. 2011. Functional analysis of the *Trichoderma harzianum* nox1 gene, encoding an NADPH oxidase, relates production of reactive oxygen species to specific biocontrol activity against *Pythium ultimum*. *Applied and Environmental Microbiology* 77:3009–3016.

Montero-Barrientos, M., R. Hermosa, C. Nicolás, R. E. Cardoza, S. Gutiérrez, and E. Monte. 2008. Overexpression of a *Trichoderma* HSP70 gene increases fungal resistance to heat and other abiotic stresses. *Fungal Genetic Biology* 45:1506–1513.

Mukherjee, P. K. and K. Raghu. 1997. Effect of temperature on antagonistic and biocontrol potential of *Trichoderma* sp. on *Sclerotium rolfsii*. *Mycopathologia* 139:151–155.

Nagy, V., V. Seidl, G. Szakacs, M. Komon-Zelazowska, C. P. Kubicek, and I. S. Druzhinina. 2007. Application of DNA bar codes for screening of industrially important fungi: The haplotype of *Trichoderma harzianum* sensu stricto indicates superior chitinase formation. *Applied Environmental Microbiology* 73:7048–7058.

Neumann K., A. Abdel-Lateff, A. D. Wright, S. Kehraus, A. Krick, and G. M. König. 2007. Novel sorbicillin derivatives with an unprecedented carbon skeleton from the sponge-derived fungus *Trichoderma* sp. *Eur J Org Chem* 2:2268–2275.

Ortega, S. N., M. Nitschke, A. M. Mouad, M. D. Landgraf, M. O. O. Rezende, M. H. R. Seleghim, L. D. Sette et al. 2011. Isolation of Brazilian marine fungi capable of growing on DDD pesticide. *Biodegradation* 22:43–50.

Paz, Z., M. Komon-Zepazowska, I. S. Druzginina, M. M. Aveskamp, A. Shnaiderman, Y. Akuma, S. Carneli et al. 2010a. Diversity and potential antifungal properties of fungi associated with a Mediterranean sponge. *Botanica Marina* 42:17–26.

Paz, Z., M. Komon-Zelazowska, I. S. Druzhinina, M. M. Aveskamp, A. Schnaiderman, and A. Aluma. 2010b. Diversity and potential antifungal properties of fungi associated with a Mediterranean sponge. *Fungal Diversity* 42:17–26.

Persoon, C. H. 1974. Dipositio methodica fungorum. *Romer's Neues Magazin Botanische* 1:81–128.

Piotrowska, G. B., J. Golimowski, and L. Pawel. 2009. Urban nanoparticles: Their potential toxicity, waste and environmental management. *Waste Management* 29:2587–2595.

Pruksakorn, P., M. Arai, N. Kotoku, C. Vilcheze, A. D. Baughn, P. Moodley, W. R. Jacobs, and M. Kobayashi. 2010. Trichoderins, novel aminolipopeptides from a marine sponge-derived *Trichoderma* sp., are active against dormant mycobacteria. *Bioorganic and Medicinal Chemistry Letters* 20:3658–3663.

Ray, M., A. Libell, E. Monte, F. Scala, and M. Lorito. 2004. Genomics of *Trichoderma*. *Applied Mycology and Biotechnology, Fungal Genomics* 4:225–248.

Reino, J. L., R. F. Guerrero, R. Hernandez-Galan, and I. G. Collado. 2008. Secondary metabolites from species of the biocontrol agent *Trichoderma*. *Phytochemistry* 7:89–123.

Reithner, B., K. Brunner, R. Schuhmacher, I. Peissl, V. Seidl, R. Krska, and S. Zeilinger. 2005. The G protein alpha subunit Tga1 of *Trichoderma atroviride* is involved in chitinase formation and differential production of antifungal metabolites. *Fungal Genetic Biology* 42:749–760.

Ren, J., C. Xue, L. Tian, M. Xu, J. Chen, and Z. Deng. 2009. Asperelines A-F, peptaibols from the marine-derived fungus *Trichoderma asperellum*. *Journal of Natural Products* 72:1036–1044.

Rifai, M. A. 1969. A revision of the genus *Trichoderma*. *Mycological Papers* 116:1–56.

Robe, P., R. Nalin, C. Capellano, T. A. Vogel, and P. Simonet. 2003. Extraction of DNA from soil. *European Journal of Soil Biology* 39:183–190.

Rocha, C. L., H. V. Ferreira, E. F. Pimenta, R. G. S. Berlinck, M. H. R. Seleghim, D. C. D. Javaroti, L. D. Sette et al. 2009. Bioreduction of alpha-chloroacetophenone by whole cells of marine fungi. *Biotechnology Letters* 31:1559–1563.

Rosado, I., M. Rey, A. Codon, J. Gonavites, M.A. Moreno-Mateos, and Benitez, T. 2007. QID74 Cell wall protein of *Trichoderma harzianum* is involved in cell protection and adherence to hydrophobic surfaces. *Fungal Genetics and Biology* 44(10):950–964.

Ruiz, N., N. Dubois, G. Wielgosz-Collin, T. R. du Pont, J. P. Berge, Y. F. Pouchus, and G. Barnathan. 2007a. Lipid content and fatty acid composition of a marine-derived *Trichoderma longibrachiatum* strain cultured by agar surface and submerged fermentations. *Process Biochemistry* 42:676–680.

Ruiz, N., C. Roullier, K. Petit, C. Sallenave-Namont, O. Grovel, and Y. F. Pouchus. 2013. Marine-derived *Trichoderma*: A source of novel bioactive metabolites. In *Trichoderma: Biology and Applications*, P. K. Mukherjee et al. (eds.). Wallingford, U.K.: CABI, pp. 247–279.

Ruiz, N., G. Wielgosz-Collin, L. Poirier, O. Grovel, K. E. Petit, M. Mohamed-Benkada, T. R. du Pont et al. 2007b. New Trichobrachins, 11-residue peptaibols from a marine strain of *Trichoderma longibrachiatum*. *Peptides* 28:1351–1358.

Ruocco, M., S. Lanzuise, F. Vinale, R. Marra, D. Turra, S. L. Woo, and M. Lorito. 2009. Identification of a new biocontrol gene in *Trichoderma atroviride*: The role of an ABC transporter membrane pump in the interaction with different plant–pathogenic fungi. *Molecular Plant Microbe Interactions* 22:291–301.

Sadfi-Zouaoui, N., I. Hannachi, M. Rouaissi, M. Hajlaoui, M. Rubio, E. Monte, A. Boudabous, and M. Hermosa. 2009. Biodiversity of *Trichoderma* strains in Tunisia. *Canadian Journal of Microbiology* 55:154–162.

Sallenave, C., Y. F. Pouchus, M. Bardouil, P. Lassus, M. F. Roquebert, and J. F. Verbist. 1999. Bioaccumulation of mycotoxins by shellfish: Contamination of mussels by metabolites of a *Trichoderma koningii* strain isolated in the marine environment. *Toxicon* 37:77–83.

Salvadori, M. R., R. A. Ando, C. A. O. Do Nascimento, and B. Correa. 2014. Bioremediation from wastewater and extracellular synthesis of copper nanoparticles by the fungus *Trichoderma koningiopsis*. *Journal of Environmental Science and Health, Part A Toxic/Hazardous Substances and Environmental Engineering* 49:1286–1295.

Samolski, I., A. de Luis, J. A. Vizcaino, E. Monte, and M. B. Suarez. 2009. Gene expression analysis of the biocontrol fungus *Trichoderma harzianum* in the presence of tomato plants, chitin, or glucose using a high-density oligonucleotide microarray. *BMC Microbiology* 9:217.

Samuels, G. J., P. Chaverri, D. F. Farr, and E. B. McCray. 2014. *Trichoderma* online, Systematic Mycology and Microbiology Laboratory, ARS, USDA, Washington, DC. Retrieved September 28, 2014 from /taxadescriptions/keys/TrichodermaIndex.cfm.

Saravanakumar, K. 2012. Studies on mangroves derived *Trichoderma* and their biotechnological applications. PhD thesis, Annamalai University, Chidambaram, Tamil Nadu, India, 290pp.

Saravanakumar, K. and K. Kandasamy. 2014. Antioxidant activity of the mangrove endophytic fungus *Trichoderma* sp. *Journal of Coastal Life Medicine* 2:566–570.

Saravanakumar, K. and K. Kathiresan. 2012. Statistical optimization of protease production by mangrove-derived *Trichoderma estonicum* and its potential on blood stain removal. *International Journal for Biotechnology and Molecular Biology Research* 3:15–21.

Saravanakumar, K., S. K. Sahu, and K. Kandasamy. 2012. *In silico* studies on fungal metabolite against skin cancer protein (4, 5-diarylisoxazole HSP90 chaperone). *International Scholarly Research Network ISRN Dermatology*, article ID 626214, 5pp. doi:10.5402/2012/626214.

Saravanakumar, K., V. Shanmuga Arasu, and K. Kathiresan. 2013. Effect of *Trichoderma* on soil phosphate solubilization and growth improvement of *Avicennia marina*. *Aquatic Botany* 104:101–105.

Schuster, A. and M. Schmoll. 2010. Biology and biotechnology of *Trichoderma*. *Applied Microbiological and Biotechnology* 87:787–799.

Seidl, V., C. Seibel, C. P. Kubicek, and M. Schmoll. 2009. Sexual development in the industrial workhorse *Trichoderma reesei*. *Proceedings of the National Academy of Sciences of the United States of America* 106:13909–13914.

Sithranga Boopathy, N. and K. Kathiresan. 2010. Anticancer drugs from marine flora: An overview. *Journal of Oncology* article ID 214186, 18pp. http://dx.doi.org/10.1155/2010/214186.

Song, F., H. Dai, Y. Tony, B. Ren, C. Chen, N. Sun, X. Liu et al. 2010. Trichodermaketones A-D and 7-*O* methylkoninginin D from the marine fungus *Trichoderma koningii*. *Journal of Natural Products* 73:806–810.

Sperry, S., G. J. Samuels, and P. Crews. 1998. Vertinoid polyketides from the saltwater culture fungus *Trichoderma longibrachiatum* separated from a *Haliclona marine* sponge. *Journal of Organic Chemistry* 63:10011–10014.

Suarez, M. B., J. A. Vizcaino, A. Llobell, and E. Monte. 2007. Characterization of genes encoding novel peptidases in the biocontrol 20 fungus *Trichoderma harzianum* CECT 2413 using the Tricho EST functional genomics approach. *Current Genetics* 51:331–342.

Sun, R.-Y., Z. C. Liu, K. Fu, L. Fan, and J. Chen. 2012. *Trichoderma* biodiversity in China. *Journal of Applied Genetics* 53:343–354.

Sun, S., L. Tian, Y. N. Wang, H. H. Wu, X. Lu, and Y. H. Pei. 2009. A novel natural product from the fermentation liquid of marine fungus *Trichoderma atroviride* G20–12. *Asian Journal of Traditional Medicines* 4:123–127.

Sun, Y., L. Tian, J. Huang, H. Y. Ma, Z. Zheng, A. L. Lv, K. Yasukawa, and Y. H. Pei. 2008. Trichodermatides A-D, novel polyketides from the marine derived fungus *Trichoderma reesei*. *Organic Letters* 10:393–396.

Sun, Y., L. Tian, Y. F. Huang, Y. Sha, and Y. H. Pei. 2006. A new cyclotetrapeptide from marine fungus *Trichoderma reesei*. *Pharmazie* 61:809–810.

Suryanarayanan, T. S., N. Thirunavukkarasu, M. B. Govindarajulu, F. Sasse, R. Jansen, and T. S. Murali. 2009. Fungal endophytes and bioprospecting. *Fungal Biology Reviews* 23:9–19.

Syed, S. S., A. S. Imran, and S. M. Fatima. 2003. *Avicenna marina* (mangrove) soil amendment changes the fungal community in the rhizosphere and root tissue of mungbean and contributes to control of root-knot nematodes. *Phytopathologia Mediterranea* 42:135–140.

Tan, W. S. and A. S. Y. Ting. 2012. Efficacy and reusability of alginate-immobilized live and heat-inactivated *Trichoderma asperellum* cells for Cu (II) removal from aqueous solution. *Bioresource Technology* 123:290–295.

Thorn, R. G. 1997. The fungi in soil. In *Modern Soil Microbiology*, J. D. Elsas, J. T. Trevors, and E. M. H. Wellington (eds.). New York: Marcel Dekker, Inc., pp. 63–127.

Thornton, C. R. 2004. An immunological approach to quantifying the saprotrophic growth dynamics of *Trichoderma* species during antagonistic interactions with *Rhizoctonia solani* in a soil-less mix. *Environmental Microbiology* 6:323–334.

Tijerino, A., R. E. Cardoza, J. Moraga, M. G. Malmierca, F. Vincente, J. Aleu, I. G. Collado et al. 2011. Overexpression of the trichodiene synthase gene Tbtri5 increases trichodermin production and antimicrobial activity in *Trichoderma brevicompactum*. *Fungal Genetics and Biology* 48:285–296.

Timberlake, W. E. and M. A. Marshall. 1989. Genetic engineering of filamentous fungi. *Science* 244:1313–1317.

Tripathi, R. M., R. K. Gupta, A. Shrivastav, M. P. Singh, B. R. Shrivastav, and P. Singh. 2013. *Trichoderma koningii* assisted biogenic synthesis of silver nanoparticles and evaluation of their antibacterial activity. *Advances in Natural Sciences: Nanosciences and Nanotechnology* 4:035005.

Uhen, M. D., 2007. Evolution of marine mammals: Back to the sea after 300 million years. *Anat. Rec.* 290:514–522.

Vahabi, K., G. Mansoori, and V. Karimi. 2011. Biosynthesis of silver nanoparticles by fungus *Trichoderma reesei*. *Insciences Journal* 1:65–79.

Viterbo, A., S. Haran, D. Friesem, O. Ramot, and I. Chet. 2001. Antifungal activity of a novel endochitinase gene (chit36) from *Trichoderma harzianum* Rifai TM. *FEMS Microbiology Letters* 200:169–174.

Viterbo, A., M. Harel, B. A. Horwitz, I. Chet, and P. K. Mukherjee. 2005. *Trichoderma* MAP-kinase signaling is involved in induction of plant systemic resistance. *Applied and Environmental Microbiology* 71:6241–6246.

Viterbo, A., M. Montero, O. Ramot, D. Friesem, E. Monte, A. Llobell, and I. Chet. 2002. Expression regulation of the endochitinase chit36 from *Trichoderma asperellum* (*T. harzianum* T-203). *Current Genetics* 42:114–122.

Vizcaino, J. A., R. E. Cardoza, M. Hauser, R. Hermosa, M. Rey, A. Llobell, J. M. Becker et al. 2006. *ThPTR2*, a di/tri-peptide transporter gene from *Trichoderma harzianum*. *Fungal Genetics and Biology* 43:234–246.

Wang, S. L., J. H. Peng, T. W. Liang, and Liu, K. C. 2008. Purification and characterization of a chitosanase from *Serratia marcescens* TKU011. *Carbohydrates Research* 343:1316–1323.

Wang, G., Q. Li, and P. Zhu. 2008. Phylogenetic diversity of culturable fungi associated with the Hawaiian sponges *Suberites zeteki* and *Gelliodes fibrosa*. *Antonie van Leeuwenhoek* 93:163–174.

Weindling, R. 1932. *Trichoderma lignorum* as a parasite of other soil fungi. *Phytopathology* 22:837–845.

White, T. J., T. Bruns, S. Lee, and J. W. Taylor. 1990. Amplification and direct sequencing of fungal ribosomal RNA genes for phylogenetics. In *PCR Protocols: A Guide to Methods and Applications*, M. A. Innis, D. H. Gelfand, J. J. Shinsky, T. J. White (eds.). San Diego, CA: Academic Press, pp. 315–322.

Wiese, J., B. Ohlender, M. Blumel, R. Schmaljohann, and J. F. Imhoff. 2011. Phylogenetic identification of marine fungi isolated from the marine sponge *Tethya aurantium* and identification of their secondary metabolites. *Marine Drugs* 9:561–585.

Wu, F., W. Wang, Y. Ma, Y. Liu, X. Ma, L. An, and H. Feng. 2013. Prospect of beneficial microorganisms applied in potato cultivation for sustainable agriculture. *African Journal of Microbiology Research* 7:2150–2158.

Yamada, K., C. Imada, M. Uchino, T. Kobayanshi, N. Hamada-sato, and K. Takano. 2008. Phenotypic characteristics and cultivation conditions of inhibitor-producing fungi isolated form marine sediment. *Fisheries Sciences* 74:662–669.

Yamada, T., Y. Mizutani, Y. Umebayashi, N. Inno, M. Kawashima, T. Kikuchi, and R. Tanaka. 2014. Tandyukisin, a novel ketoaldehyde decalin derivative, produced by a marine sponge-derived *Trichoderma harzianum*. *Tetrahedron Letters* 55:662–664.

Yi, Y., J. Huo, W. Zhang, L. Li, P. Sun, B. Liu, and H. Tang. 2010. Reducing anthraquione-like compounds with antimicrobial activity. Patent 101885676, China.

You, J., H. Dai, Z. Chen, G. Liu, Z. He, F. Song, X. Yang et al. 2010. Trichoderone, a novel cytotoxic cyclopentenone and cholesta-7, 22-diene-3β, 5α, 6β-triol, with new activities from the marine-derived fungus *Trichoderma* sp. *Journal of Industrial Microbiology and Biotechnology* 37:245–252.

You, X. et al., 2014. Mudskipper genomes provide insights into terrestrial adaptation of amphibious fishes. *Nat. Commun.* 5:5594.

Yu, C., L. Fan, Q. Wu, K. Fu, and S. Gao. 2014. Biological role of *Trichoderma harzianum*-derived platelet-activating factor Acetylhydrolase (PAF-AH) on stress response and antagonism. *PLoS ONE* 9:e100367.

Zeilinger, S. and M. Omann. 2007. Trichoderma biocontrol: Signal transduction pathways involved in host sensing and mycoparasitism. *Gene Regulation and Systems Biology* 1:227–234.

Zhong, Y. H., T. H. Wang, X. L. Wang, G. T. Zhang, and H. N. Yu. 2009. Identification and characterization of a novel gene, TrCCD1, and its possible function in hyphal growth and conidiospore development of *Trichoderma reesei*. *Fungal Genetic Biology* 46:255–263.

Ziemert, N., A. Lechner, M. Wietz, N. Millan-Aguinaga, K. L. Chavarria, and P. R. Jensen. 2014. Diversity and evolution of secondary metabolism in the marine actinomycete genus *Salinispora*. *Proceedings of the National Academy of Sciences of the United States of America* 111: E1130–E1139.

Zimand, G., L. Valinsky, and Y. Elad. 1994. Use of the RAPD procedure for the identification of *Trichoderma* strains. *Mycological Research* 98:531–534.

5

Genomic Perspective of Marine Plants with Special Reference to Seagrasses

E. Dilipan, E.P. Nobi, and T. Thangaradjou

CONTENTS

5.1 Introduction

Marine plants are generally considered to be of two types: marine algae (seaweed) and marine angiosperms (seagrasses). These seagrasses constitute about 0.01% of over 3 lakh flowering plants. Though few in numbers, these monocots form extensive meadow in tropical and temperate seas. Seagrass research gained momentum only in the last two decades and mostly in the aspects of taxonomy, ecology, nutrient requirements, secondary metabolites, floral–faunal assemblages, etc. But all these knowledge parallel exposed taxonomy uncertainties in seagrass molecular taxonomy and evolutionary aspects. The uncertainty invited to focus on genomic outlook of seagrasses in this chapter is based on reports from various parts of the world. This chapter on genomic perspective of seagrass is extensive and includes the topics on distribution, taxonomic uncertainty of seagrasses, genetic diversity, and species diversity.

5.2 Distribution

Seagrasses are rhizomatous marine angiosperms or flowering plants, adapted to live and reproduce under submerged saline environment and are closely related to terrestrial lilies and gingers than to true grasses. Seagrasses grow in shallow, sheltered soft-bottomed coastal habitats such as coastlines, estuaries, and lakes. Seagrasses have returned to the sea after acquiring numerous physiological and morphological characteristics (including functional vascular systems) typical of many advanced terrestrial plant species, which have ribbonlike, grassy leaves (CEN, 2005). Morphologically, the size of seagrasses varies from a fingernail to plants with leaves as long as one meter. Shapes vary from oval, fern, long spaghetti-like leaf, and ribbon (McKenzie, 2008).

Seagrasses are unique among flowering plants, in which all genera can live entirely immersed in seawater except *Enhalus* plants, which can emerge to the surface to reproduce, while all others can flower and be pollinated under water (Phillips and Menez, 1988). The pollination systems of seagrasses are well adapted for aquatic environment. Seagrass forms tiny flowers, fruits, and seeds, and hydrophilic pollination can occur above water surface (*Enhalus*), on water surface (*Halodule*), and beneath water surface (*Thalassia*), which have separate male and female plants (dioecious) or have both sexes on the same plant (monoecious). For example, the Posidoniaceae are exclusively monoecious, the Hydrocharitaceae and Zosteraceae have both monoecious and dioecious species, and the Cymodoceaceae are exclusively dioecious; overall about 75% of all seagrass species are dioecious (Waycott and Les, 1996). One facet of sterility in dioecious hydrophiles is the low percentage of flowering in several species, and seagrasses are rare flowering according to Hartog (1970). McMillan (1980) noted that environmental conditions have wide effects on the reproductive biology of many seagrasses. Therefore, outcrossing plays a key role to sustain their life forms in marine. The predominance of dicliny in hydrophiles serves as an evidence of an inevitable association with outcrossing and production of genetically variant progeny in seagrasses (Hartog, 1970). Les (1988) hypothesized three components: hydrophiles are characterized by outcrossing, which leads to the production of genetically variable offspring, the selective advantage of outcrossing in hydrophiles offsets the evolutionary costs associated with transitions to dicliny, and dicliny has evolved in hydrophiles as a mechanism for promoting outcrossing. Furthermore, it is difficult to rationalize the slow evolutionary diversification of this group with a supposed history of outcrossing and prolific genetic variability.

There are an estimated 72 named species of seagrasses worldwide, represented by 12 genera in five families (Hydrocharitaceae, Cymodoceaceae, Posidoniaceae, Zosteraceae, and Ruppiaceae); the greatest diversity is found in tropical regions. The species share a fundamentally similar architecture and physiology and perform similar ecosystem functions (Hemminga and Duarte, 2000; Spalding et al., 2003). They occur on a variety of substrata, from mud to sand and even bedrock, generally in areas sheltered from wave action and strong currents. Seagrasses grow from intertidal to a depth of 15 m dominantly forming extensive meadows, where light intensity is greatest, but in some areas, they can occur down to depths of 50–60 m (Coles et al., 2003; Spalding et al., 2003). Though there are different thoughts on global seagrass distribution, Short et al. (2007) described six biogeographical regions based on seagrass assemblage of different taxonomic groups in temperate and tropical areas and physical separations of world's oceans. It is generally agreed that *Enhalus, Halodule, Halophila, Syringodium, Thalassia,* and *Thalassodendron* are tropical genera, but some of the species of these genera are also represented in temperate regions, whereas *Amphibolis, Heterozostera, Phyllospadix, Posidonia, Pseudalthenia,* and *Zostera* are typical temperate species (Hartog and Yang, 1988).

In India, there are 14 species of seagrasses recorded along the east and west coasts (Kannan and Thangaradjou, 2006). Seagrass species, namely, *Enhalus acoroides, Thalassia hemprichii, Cymodocea* spp., and *Halodule* spp., contribute more biomass, and their photosynthetic productivity is also higher as compared to the other most productive seagrass regions of the world (Kannan, 2005). The Gulf of Mannar, Palk Bay, Andaman and Nicobar Islands (Thangaradjou et al., 2010), and Lakshadweep islands (Nobi et al., 2011) are identified as seagrass hot spots of India.

5.3 Taxonomic Uncertainty in Seagrasses

Hartog (1970) identified six species throughout the world by means of leaf tip morphology and leaf width, whereas Phillips and Menez (1988) identified three species using only leaf tip morphology. Taxonomic confusion has been especially prominent in the identification of *H. uninervis*, and some researchers divided plants with narrow and wide leaves into *H. tridentata* and *H. uninervis*, respectively (Hartog, 1964; Ohba and Miyata, 2007), whereas others considered all plants with varying leaf widths to be *H. uninervis* (Hartog, 1970; Phillips and Menez, 1988).

Though a number of different morphological variations is observed in *Z. marina* from different environmental conditions (Hartog, 1972), it is still considered as the same species. The field identification of *Zostera* species in Australia continues to be difficult using the characters described in taxonomic treatments (Sainty and Jacobs, 1981). Taxonomic distinctions in *Zostera* have been attained primarily on the basis of leaf morphology (Hartog, 1970), which are environmentally labile (Phillips, 1972) or differentiated ecotypically (Backman, 1991), at least with respect to length and width characters. The extent of variation in *Zostera* was not adequately understood, as evidenced by recent discoveries of *Z. caulescens* having leaves an order of magnitude larger than previously reported for the species (Aioi et al., 1998). In particular, species recognition in *Zostera* relies strongly on leaf tip and retinacular morphology, which exhibit extensive variation within species (Phillips and Menez, 1988). In spite of that, there is no indication that this typical variety is identical with the material named by Linnaeus (1753) as *Z. marina*. However, Phillips and Willey-Echeverria (1990) proposed that the wide leaf morphology of *Z. marina* from pacific coast of America should be considered as *Z. asiatica*. Whether these different morphologies represent different varieties of the same species or contain different taxa requires further critical investigation with all available techniques on a worldwide scale.

The generic distinction between *Zostera* and *Heterozostera* has become increasingly unsettled due to uncertainty in the reliability of key taxonomic characters. Taxonomists have found it difficult to separate the morphologically similar *Zostera* and *Heterozostera* (Aston, 1973; Jacobs and Williams, 1980). Aston (1973) and Phillips and Menez (1988) followed Hartog (1970) who distinguished between monopodial (former) and sympodial (latter) rhizomes to separate the genera. However, Tomlinson (1982) and Soros-Pottruff and Posluszny (1995) showed that this often-cited feature (sympodial, unbranched rhizome) is erroneous and should not be used to distinguish the genera.

Despite the limited size of the seagrass flora, there is no generally accepted number of species, and an examination of the literature indicates that there is 20% uncertainty about the total number of seagrass species (Waycott, 1999). Current identification keys are based at the species level on differences in the shape of margins and venation of leaves as diagnostic criteria (Hartog, 1970; Phillips and Menez, 1988). Another phenomenon is the "trait" of morphological plasticity within the same species, hypothesized to enable survival in different ecological niches (Bricker et al., 2011). All these factors make correct assignment of species based alone on conventional identification keys difficult or even impossible. For these reasons, closely related species, for example, *Halophila* spp. (Hartog and Kuo, 2006), still form a taxonomically unresolved complex without correct assignment of species and subspecies. Even molecular phylogenies using matK, *rbcL*, and *trnK* and internal transcribed spacer (*ITS*) regions yielded different results resolving the genus complexes. More exclusive criteria such as the structure of reproductive organs are of limited use for the identification at all times, since they are highly reduced in most seagrass species, very difficult to find in some species, and not available throughout the year. Hence, there is a great need to quantify genetic plasticity of seagrass species and conduct genetic studies across large scales to resolve the uncertainties on the taxonomic status of seagrass.

5.4 Genetic Diversity

Random amplified polymorphic DNA (RAPD) and amplified fragment length polymorphism (AFLP) markers, in fact, also confirm high population connectivity with higher but still low within-site genetic variation. In both RAPD and AFLP analyses, the partitioning of genetic diversity was distributed similarly with a higher percentage of variation within seagrass meadows (Waycott et al., 2006). In general, the genetic diversity of different seagrasses from different locations was studied to assess the population diversity, in which high levels of RAPD genetic variability were revealed in species such as *Halophila johnsonii* (Freshwater et al., 2003; Smith et al., 1997), *Thalassia testudinum* (Kirsten et al., 1998), *C. nodosa* (Procaccini and Mazzela, 1996), *Thalassodendron ciliatum* (Banderia and Nilson, 2001), *P. australis* (Waycott, 1995), *P. oceanica* (Jover et al., 2003), and *Z. muelleri* (Jones et al., 2008) other than hydrophillous species.

DNA-based molecular markers (restriction fragment length polymorphisms [RFLPs], microsatellites) have found more genetic and phenotypic substructuring both within and among patches of seagrass (Alberte et al., 1994; Fain et al., 1992; Laushman, 1993; Reusch et al., 1999; Ruckelshaus, 1996; Williams and Orth, 1998). Fain et al. (1992) used comparative restriction analyses (RFLPs) of nuclear DNA encoding RNA (rDNA) to demonstrate genetic distinctions between geographically disjunct eelgrass populations from California that were correlated with leaf morphology and habitat depth. Examination of sequence variations between the four eelgrass populations examined ranged from 0.00 to 0.69, and three of the populations could be distinguished by unique RFLPs. These studies verified genetic diversity between seagrass populations (de Heij and Nienhuis, 1992) but did not reveal significant within-population genetic diversity. Arano et al. (2003) reported a complete lack of detectable RFLP variation among *E. acoroides* populations from different geographic regions, which would support the existence of a single *Enhalus* species.

Waycott and Barnes (2001) used AFLP and found very low genetic differentiation among *T. testudinum* from three regions of the Caribbean and North Atlantic regions with high levels of gene flow between all sites, which indicates that the long-distance vegetative fragment dispersal is highly probable and genetic uniformity may be related to a long-term environmental change over geographic range.

In contrast to isozyme and allozyme studies, other molecular DNA studies (RAPD) found a higher level of genetic diversity within population of *T. testudinum* of the Florida Keys rather than between geographically distinct population of the Steinhatchee River (Kirsten et al., 1998). It is also confirmed that Jamaican population and the two Florida Keys sites were distinct genetic individuals within and between the three populations. This is in contrast to the northern population that had the fewest RAPD phenotypes. The lower within-community genetic diversity of northern populations of *T. testudinum* may reflect the limited introduction of new genets (e.g., drift plants) or a low level of flowering and seed production perhaps because of less-than optimum water temperatures or water transparencies.

Waycott (1995) identified 16 four-primer genotypes from 22 samples and reported that the levels of genetic variation in *Posidonia australis* are equivalent to or greater than that observed in other hydrophilous species and also specified that the intrapopulation genetic variation of *P. australis* meadows need not be monoclonal in all cases. Smith et al. (1997) studied DNA banding patterns of *H. johnsonii* from the three northern American species that showed considerable variation both within and between species. Though the similarity values both within and between the species are higher than that reported by Alberte et al. (1994) for *Zostera marina* L., the *Halophila* data, however, do not represent the extensive sampling. The introduced species *Halophila stipulacea* has high RAPD diversity in meadows sampled along the Sicilian coasts and high population connectivity between distinct localities (Procaccini et al., 1999). DNA fingerprinting techniques found that the clonal variation in *T. testidinum* populations in Florida Bay are discrete beds that were not genetically uniform (Davis et al., 1999) and confirmed the role of sexual reproduction in maintaining population genetic variation.

Larkin et al. (2006) reported that two *T. testudinum* populations from the south Texas Gulf Coast showed genetic diversity and genetic distance between the two populations was low, and the majority of the genetic variation was attributed to differences between individuals of within populations.

Procaccini and Mazzela (1996) found high levels of RAPD genetic variability in *Cymodocea nodosa* from the island of Ischia (Gulf of Naples, Italy), a population known to reproduce sexually. In contrast to these results, Alberto et al. (2001) found low genetic variability from the Ria Formosa National Park, Portugal, and interpreted this with that of unsuccessful sexual reproduction and recovery of population only by horizontal vegetative propagation of *Cymodocea nodosa*. Genetic population structure of the seagrass *Thalassodendron ciliatum* of sandy and rocky habitats of southern Mozambique found 71.6% of genetic variation within populations (Bandeira and Nilsson, 2001). Angel (2002) made RAPD analysis of three populations of *H. wrightii* from Texas (Christmas Bay, Corpus Christi) and Florida (Florida Bay) and recorded that 72% of the loci were polymorphic and reproducible. All individuals appeared to have unique genotypes, with plants from Corpus Christi more closely clustered to those of Florida Bay, and were due to similar habitats that acted as a selective force on both the population. Recently, Nguyen et al. (2013) studied AFLP among the subspecies of *H. ovalis* subsp. *ovalis*, *H. ovata*, and *H. ovalis* subsp. *ramamurthiana*, and they concluded that *H. ovalis* subsp. *ramamurthiana* was genetically different between species and geographies. The genetic diversity of seagrasses based on microsatellite markers revealed that the genetic variability in the Western Pacific is higher than in the Eastern Indian Ocean (Nguyen et al., 2014).

The congeneric *H. johnsonii* shows very high clonality and genetic homogeneity in south Florida, with the same RAPD phenotype present in more than 50% of the samples (Freshwater et al., 2003). Jover et al. (2003) assessed the genetic diversity in western Mediterranean population of the *Posidonia oceanica* and found the lowest clonal diversity in the less structured and youngest prairies. Conversely, a high genotypic diversity was found in the highly structured meadows. RAPD analysis showed genetic differences among Mediterranean Sea sampling stations, with a decrease in genetic diversity along an anthropogenic disturbance gradient (Micheli et al., 2005). New Zealand seagrass showed the highest genetic similarity within each of the sites, indicating a low degree of gene flow between populations in *Zostera muelleri* (Jones et al., 2008) and *P. oceanica* (Rotini et al., 2011). Dilipan (2012) reported that RAPD technique is a useful tool for the analysis of genetic diversity among Indian seagrasses and can be used as the tool to resolve the taxonomy issues.

5.5 Species Diversity

Only limited work on the 18S rDNA sequencing of seagrasses has been carried out elsewhere, and no conspicuous work was carried out on Indian seagrasses. The position of the Alismatales at the base of the monocotyledons supports the notion that this is an old group. More rapidly evolving DNA sequences have been shown to be very useful at unraveling relationships among more closely related taxa. Among these, the *ITS* region of nuclear ribosomal DNA (nrDNA) has been utilized extensively in phylogenetic studies in many angiosperm families and has been particularly useful in improving our understanding of species relationships in these families (Baldwin et al., 1995). Molecular data (18S ribosomal DNA) of *Zostera* have not shown much evolutionary relationship due to limited taxa (Uchiyama, 1996). Phylogenetic studies using *rbcL* and *matK* gene sequence data in the family of Hydrocharitaceae having 15 genera and 80 species found reduction of Hydrocharitaceae-epihydrophily from four times to three times (Tanaka et al., 1997). The phylogenetic analysis of the *ITS* region of the nrDNA for the seagrass genus *Halophila* (Hydrocharitaceae) showed that widespread Pacific species *H. ovalis* is paraphyletic and may contain cryptic species (Waycott et al., 2002). The molecular phylogeny of 11 species of *Halophila* found that *H. decipiens* and *H. engelmannii* were distinct species and that there was a 100% overlap between populations of the former species with the populations from Australia, Caribbean, and Florida (Waycott et al., 2002).

The molecular phylogenetic analyses have begun to clarify some finer details of seagrass relationships. First, the emerging pattern of phylogenetic relationships indicates that seagrasses should be divided among six separate families (Les et al., 1997). The *rbcL* survey by Les et al. (1997) indicated that marine angiosperms have evolved in at least three separate lineages. Another independent lineage (Zannichelliaceae), which contains *Lepilaena marina*, represents a fourth marine angiosperm origin, given that this species is known to occupy marine habitats (Womersley, 1984). Procaccini et al. (1999) used *trnL* intron sequence data to study relationships among six marine genera in five families and recovered distinct clades containing (1) *Cymodocea/Posidonia/Ruppia*, (2) *Zostera/Phyllospadix*, and (3) *Halophila*.

Incorporation of molecular data in cladistic analyses provided the next major refinement in phylogenetic reconstructions of seagrass relationships. Les et al. (1993) conducted preliminary cladistic analyses of Alismatidae using *rbcL* gene sequence data from 8 families but included only a few seagrass genera, which do not provide much insight into their relationships. Cladistic analyses of molecular and nonmolecular data have been applied to questions of seagrass species relationships, but this approach only has been taken relatively recently and many species remain unstudied (Waycott et al., 2006). When population distributions are viewed using a genetic perspective, higher genetic differentiation among populations reflects lower genetic exchange (gene flow) and indicates long-term isolation. The application of molecular markers to study the levels of population connectivity potentially can provide significant insights into the factors influencing present-day distribution of seagrass species, particularly when applied across broad spatial scales (Waycott et al., 2006). Yet, genetic studies in populations of *Cymodocea nodosa*, *Posidonia oceanica*, *Zostera noltii*, and *Z. marina* have reported high differentiation along their distributional ranges, although species and site dependent (Alberto et al., 2008; Arnaud-Haond et al., 2007; Becheler et al., 2010). Recently, Wissler et al. (2011) constructed orthologous gene clusters shared between two seagrass species such as *Z. marina* and *P. oceanica*, and they found that seagrass genes have diverged from their terrestrial counterparts via an initial aquatic stage characteristic of the order and to the fully derived marine stage characteristics of seagrasses.

Nuclear (ITS) and plastid (*trnK* intron, *rbcL*) genomes of Zosteraceae genera from Australia and New Zealand waters found two major clades with high divergence and two subclass with low divergence at both morphological and molecular levels (Les et al., 2002). Analysis of matK sequences of all 11 species in *Zostera*, in addition to *Heterozostera tasmanica* and *Phyllospadix iwatensis*, found monophyly in *H. tasmanica* and subgenus *Zosterella*, and Zosteraceae consists of three taxa of genus *Phyllospadix*, subgenus *Zosteralla*, and *Heterozostera* (Tanaka et al., 2003). Based upon these findings, they suggested some refinements in taxonomic classification of Zosteraceae. However, Tanaka et al. (2003) did not set the time frame for the origin and evolution of the family. Similarly, Kato et al. (2003) examined the phylogeny of *Zostera* species based on *rbcL* and matK nucleotide sequences with respect to the origin and diversification of seagrasses in Japanese waters, which showed age-related classification of Zosteraceae to reveal the origin and evolution. Li and Zhou (2009) conducted separate and combined analyses of 103 morphological characters and 52 *rbcL* sequences to explore the controversial phylogenies of the families Cymodoceaceae, Najadaceae, and Hydrocharitaceae, and they found that the monophyly of the Alismatales is strongly supported with a bootstrap value of 97%. However, different markers were used for different species: *rbcL* and matK of the Hydrocharitaceae (Tanaka et al., 1997), nuclear (18SrRNA, *ITS*) and other cpDNA (*trnL*) loci in *Thalassia*, nuclear (*ITS*) and cpDNA (*rbcL*, *trnL*) in *Halodule* (Les et al., 1997), chloroplast *trnL* intron and *rbcL* of several seagrasses (Procaccini et al., 2007), nuclear (ITS) and cpDNA (*rbcL*, *trnL*) loci of *Posidonia*, nuclear *ITS* of *Halophila* (Waycott et al., 2002), *ITS* (*ITS*-1 and *ITS*-2 regions including the 5.8S rDNA gene), *trnK* introns and *rbcL* of *Zostera* (Les et al., 2002), *rbcL* and *matK* of *Zostera* (Kato et al., 2003), *ITS*1, 5.8S rDNA, and *ITS*2 of *Halophila* (Uchimura et al., 2008), and *ITS*1, 5.8S rDNA, and *ITS*2 of *Halophila* (Short et al., 2010) and chloroplastic markers such as *trnL-F, matK/ trnK* in *Posidonia* (Aires et al., 2011) and *rbcL*/matK/*psbA*/*trnH* of Indian seagrasses (Lucas et al., 2012), and there is no generally agreed consensus on conserved molecular region useful for seagrass taxonomy and evolutionary history. Similarly, Dilipan (2012) analyzed molecular (18S rRNA) and nonmolecular data of tropical and temperate seagrass species and concluded that the seagrass might be of independent origin. Of these, *Halophila* genus could be the intermediate group for both regions. Vy et al. (2015) have proposed a new DNA bar-coding system for seagrasses using ITS sequences.

5.6 Summary

In this chapter, the molecular taxonomy and evolutionary lineages of seagrasses have been reviewed as different molecular markers that could resolve the taxonomic plasticity. Although their use in these studies seems convincing, the origin and evolutionary lineage between seagrass, salt marsh, and freshwater should be studied in detail. Furthermore, the studies on different molecular markers need to be performed from different coastal areas to know the origin of seagrass.

References

Aioi, K., T. Komatsu, and K. Morita, 1998. The world's longest seagrass, *Zostera caulescens* from Northeastern Japan. *Aquat. Bot.*, 61: 87–93.

Aires, T., N. Marba, R.L. Cunha, G.A. Kendrick, and D.I. Walker, 2011. Evolutionary history of the seagrass genus *Posidonia*. *Mar. Ecol. Prog. Ser.*, 421: 117–130.

Alberte, R.S., G.K. Suba, G. Procaccini, R.C. Zimmerman, and S.R. Fain, 1994. Assessment of genetic diversity of seagrass populations using DNA fingerprinting: Implications for population stability and management. *Proc. Natl. Acad. Sci. U.S.A.*, 91: 1049–1053.

Alberto, F., P. Massa, E. Manent, S. Diaz-Almela, C.M. Arnaud-Haond, C.M. Duarte, and E.A. Serrão, 2008. Genetic differentiation and secondary contact zone in the seagrass *Cymodocea nodosa* across the Mediterranean-Atlantic transition region. *J. Biogeogr.*, 35: 1270–1294.

Alberto, F., L. Mata, and R. Santos, 2001. Genetic homogeneity in the seagrass *Cymodocea nodosa* at its northern Atlantic limit revealed through RAPD. *Mar. Ecol. Prog. Ser.*, 221: 299–301.

Angel, R., 2002. Genetic diversity in *Halodule wrightii* using Random Amplified Polymorphic DNA. *Aquat. Bot.*, 74: 165–174.

Arano, K.G., J.N. Ouborg, and E.D.R.V. Steveninck, 2003. Chloroplast DNA phylogeography of Indo-Pacific seagrasses (Abstract). *Gulf. Mex. Sci.*, 21: 126pp.

Arnaud-Haond, S., M. Migliaccio, E. Diaz-Almela, S. Teixeira, M.S. Van de Vliet, F. Alberto, G. Procaccini, C.M. Duarte, and E. Serrao, 2007. Vicariance patterns in the Mediterranean Sea: West-east cleavage and low dispersal in the endemic seagrass *Posidonia oceanica*. *J. Biogeogr.*, 34: 963–976.

Aston, H.I., 1973. *Aquatic Plants of Australia*. Melbourne University Press, Carleton, Victoria, Australia. 368pp.

Backman, T.W.H., 1991. Genotypic and phenotypic variability of *Zostera marina* on the west coast of North America. *Can. J. Bot.*, 69: 1361–1371.

Baldwin, B.G., M.J. Sanderson, J.M. Porter, M.F. Wojceishowski, C.S. Campbell, and M.J. Donoghue, 1995. The ITS region of nuclear ribosomal DNA: A valuable source of evidence on angiosperm phylogeny. *Ann. Missouri Bot. Gard.*, 82: 247–277.

Bandeira, S.O. and P.G. Nilsson, 2001. Genetic population structure of the seagrass *Thalassodendron ciliatum* in sandy and rocky habitats in southern Mozambique. *Mar. Biol.*, 139: 1007–1012.

Becheler, R., O. Diekmann, C. Hily, Y. Moalic, and S. Arnaud-Haond, 2010. The concept of population in clonal organisms: Mosaics of temporally colonized patches are forming highly diverse meadows of *Zostera marina* in Brittany. *Mol. Ecol.*, 19: 2394–2407.

Bricker, E., M. Waycott, A. Calladine, and J.C. Zieman, 2011. High connectivity across environmental gradients and implications for phenotypic plasticity in a marine plant. *Mar. Ecol. Prog. Ser.*, 423: 57–67.

CEN, 2005. *Watching the Seagrass Grow—A Guide for Community Seagrass Monitoring in NSW*, 2nd edn. The Community Environment Network, Ourimbah, New South Wales, Australia. 68pp.

Coles, R., L. McKenzie, and S. Campbell, 2003. The seagrasses of eastern Australia. In: *World Atlas of Seagrasses*, Green, E.P. and F.T. Short (eds.). University of California Press, Berkeley, CA, pp. 119–133.

Davis, J.L., D.L. Childers, and D.N. Kuhn, 1999. Clonal variation in a Florida Bay *Thalassia testudinum* meadow: Molecular genetic assessment of population structure. *Mar. Ecol. Prog. Ser.*, 186: 127–136.

de Heij, H. and P.H. Nienhuis, 1992. Intraspecific variation on isozyme patterns of phenotypically separated populations of *Zostera marina* L. in the south-western Netherlands. *J. Exp. Mar. Biol. Ecol.*, 161: 1–14.

den Hartog, C., 1964. An approach to the taxonomy of the seagrass *Halodule* Endl. (Potamogetonaceae). *Blumea*, 12: 289–312.

den Hartog, C., 1970. *The Seagrasses of the World*. North-Holland Publishing, Amsterdam, the Netherlands, 275pp.

den Hartog, C., 1972. The seagrasses of Brazil. *Acta Bot. Neerl.*, 21: 512–516.

den Hartog, C. and J. Kuo, 2006. Taxonomy and biogeography in seagrasses. In: *Seagrasses: Biology, Ecology and Conservation*, Larkum, A.W.D., R.J. Orth, and C.M. Duarte (eds.). Springer-Verlag, Berlin, Germany, pp. 1–23.

den Hartog, C. and Z.D. Yang, 1988. Occurrence of the seagrass *Halodule pinifolia* (Miki) den Hartog in the Indian Ocean. *Curr. Sci.*, 57: 1172–1174.

Dilipan, E., 2012. Taxonomy of Indian seagrasses: Molecular (RAPD, 18S rDNA) and Bioinformatic (SNDI) tools for seagrass identification. PhD, thesis, Annamalai University, Annamalainagar, India, 107pp.

Fain, S.R., A. DeTomasko, and R.S. Alberte, 1992. Characterisation of disjunct populations of *Zostera marina* (eelgrass) from California: Genetic differences resolved by restriction-fragment length polymorphisms. *Mar. Biol.*, 112: 683–689.

Freshwater, D.W., R.A. York, W.J. Kenworthy, and M. Waycott, 2003. Multi-locus genotyping of the threatened seagrass, *Halophila johnsonii* Eiseman reveals a high level of clonality. *Aquat. Bot.* (unpublished).

Hemminga, M.A. and C.M. Duarte, 2000. *Seagrass Ecology: An Introduction*. Cambridge University Press, Cambridge, U.K., 298pp.

Jacobs, S.W.L. and A. Williams, 1980. Notes on the genus *Zostera* s. lat. in New South Wales. *Telopea*, 1: 451–455.

Jones, C.T., C.E.C. Gemmil, and C.A. Pilditch, 2008. Genetic variability of New Zealand seagrass (*Zostera muelleri*) assessed at multiple spatial scales. *Aquat. Bot.*, 88: 39–46.

Jover, M.A., L. del Castillo-Agudo, M. Garcia-Carrascosa, and J. Segura, 2003. Random amplified polymorphic DNA assessment of diversity in western Mediterranean populations of the seagrass *Posidonia oceanica*. *Am. J. Bot.*, 90: 364–369.

Kannan, L., 2005. Seagrasses of India: Eco-biology and conservation. In: *The Proceedings of Nature Symposium Marine Plants, their Chemistry and Utilization*, June 23–25, 2005, Seaweed Research and Utilization Association, Tuticorin, India, 13pp.

Kannan, L. and T. Thangaradjou, 2006. Identification and assessment of biomass and productivity of seagrasses. In: *The Proceedings of the National Training Workshop on Marine and Coastal Biodiversity Assessment for Conservation and Sustainable Utilization*, SDMRI Publication, Vol. 10, pp. 9–15. Tuticorin, India.

Kato, K., K. Aioi, Y. Omori, N. Takahata, and Y. Satta, 2003. Phylogenetic analyses of *Zostera* species based on rbcL and matK nucleotide sequences: Implications for the origin and diversification of seagrasses in Japanese water. *Genes Genet. Syst.*, 78: 329–342.

Kirsten, J.H., C.J. Dawes, and B.J. Cochrane. 1998. Randomly amplified polymorphism detection (RAPD) reveals high genetic diversity in *Thalassia testudinum* Banks ex König (Turtlegrass). *Aquat. Bot.*, 61: 269–287.

Larkin, P., E. Quevedo, S. Salinas, J. Parker, K. Storey, and B. Hardegree, 2006. Genetic structure of two *Thalassia testudinum* populations from the south Texas Gulf coast. *Aquat. Bot.*, 85: 198–202.

Laushman, R.H., 1993. Population genetics of hydrophilous angiosperms. *Aquat. Bot.*, 44: 147–158.

Les, D.H., 1988. Breeding systems, population structure and evolution in hydrophylous angiosperms. *Ann. Missouri Bot. Gard.*, 75: 819–835.

Les, D.H., M.A. Cleland, and M. Waycott, 1997. Phylogenetic studies in Alismatidae, II—Evolution of marine angiosperms (seagrasses) and hydrophily. *Syst. Bot.*, 22(3): 443–463.

Les, D.H., D.K. Garvin, and C.F. Wimpee, 1993. Phylogenetic studies in the monocot subclass Alismatidae: Evidence for a reappraisal of the aquatic order Najadales. *Mol. Phylogenet. Evol.*, 2: 304–314.

Les, D.H., M.L. Moody, S.W.L. Jacobs, and R.J. Bayer, 2002. Systematics of seagrasses (Zosteraceae) in Australia and New Zealand. *Syst. Bot.*, 27: 468–484.

Li, X. and Z. Zhou, 2009. Phylogenetic studies of the core Alismatales inferred from morphology and rbcL sequences. *Prog. Nat. Sci.*, 19: 931–945.

Linnaeus, C., 1753. *Species Plantarum*, Vol. 1. Laurenti Salvii, Stockholm, Sweden, 1200pp.

Lucas, C., T. Thangaradjou, and J. Papenbrock, 2012. Development of a DNA barcoding system for seagrasses: Successful but not simple. *PLoS ONE*, 7(1): 1–12. Australia.

McKenzie, L.J., 2008. *Seagrass Educators Handbook*. Seagrass Watch Head Quarters Department of Primary Industries and Fisheries, 20pp. Australia.

McMillan, C., 1980. Isozymes of tropical seagrasses from the Indo-Pacific and the Gulf of Mexico-Carribean. *Aquat. Bot.*, 8: 163–172.

Micheli, C., P. Paganin, A. Peirano, G. Caye, A. Meinesz, and C.N. Bianchi, 2005. Genetic variability of *Posidonia oceanica* (L.) Delile in relation to local factors and biogeographic patterns. *Aquat. Bot.*, 82: 210–221.

Nguyen, X.V., M. Detcharoen, P. Tuntiprapas, U. Soe-Htun, S.B. Japar, H.Z. Muta, A. Prathep, and J. Papenbrock, 2014. Genetic species identification and population structure of Halophila (Hydrocharitaceae) from the Western Pacific to the Eastern Indian Ocean. *BMC Evol. Biol.*, 14: 92.

Nguyen, X.V., S. Höfler, Y. Glasenapp, T. Thangaradjou, C. Lucas, and J. Papenbrock, 2015. New insights into DNA barcoding of seagrasses. *Syst. Biodivers.* 13(5):496–508.

Nguyen, X.V., T. Thangaradjou, and J. Papenbrock, 2013. Genetic variation among Halophila ovalis (Hydrocharitaceae) and closely related seagrass species from the coast of Tamil Nadu, India—An AFLP fingerprint approach. *Syst. Biodivers.*, 11: 467–476.

Nobi, E.P., E. Dilipan, K. Sivakumar, and T. Thangaradjou, 2011. Distribution and biology of seagrass resources of Lakshadweep group of Islands, India. *Indian J. Geo-Mar. Sci.*, 40(5): 624–634.

Ohba, T. and M. Miyata, 2007. *Seagrasses of Japan*. Hokkaido University Press, Sapporo, Japan, 114pp.

Phillips, R.C., 1972. The ecological life history of *Zostera marina* L. (eelgrass) in Puget Sound, Washington. PhD, dissertation, University of Washington, Seattle, WA, 154pp.

Phillips, R.C. and E.G. Menez, 1988. Seagrasses, Smithsonian Contributions to the Marine Sciences, No. 34. Smithsonian Institution Press, Washington, DC, 104pp.

Phillips, R.C. and S. Willey-Echeverria, 1990. *Zostera asiatica* Miki on the Pacific coast of North America. *Pacific Sci.*, 44: 130–134.

Procaccini, G. and L. Mazzella, 1996. Genetic variability and reproduction in two Mediterranean seagrasses. In: *Seagrass Biology: Proceedings of an International Workshop*, Kuo, J., R.C. Phillips, D.I. Walker, and H. Kirkman (eds.). SCIENCES UWA, Rottnest Island, Western Australia, Australia, pp. 85–92.

Procaccini, G., L. Mazzella, R.S. Alberte, and D.H. Les, 1999. Chloroplast tRNA Leu (UAA) intron sequences provide phylogenetic resolution of seagrass relationships. *Aquat. Bot.*, 62: 269–283.

Procaccini, G., J.L. Olsen, and T.B.H. Reusch, 2007. Contribution of genetics and genomics to seagrass biology and conservation. *J. Exp. Mar. Biol. Ecol.*, 350: 234–259.

Reusch, T.B.H., W.T. Stam, and J.L. Olsen, 1999. Microsatellite loci in eelgrass *Zostera marina* reveal marked polymorphism within and among populations. *Mol. Ecol.*, 8: 317–321.

Rotini, A., C. Micheli, L. Valiante, and L. Migliore, 2011. Assessment of *Posidonia oceanica* (L.) Delile conservation status by standard and putative approaches: The case study of Santa Marinella meadow (Italy, W Mediterranean). *Open J. Ecol.*, 1(2): 48–56.

Ruckelshaus, M.H., 1996. Estimation of genetic neighborhood parameters from pollen and seed dispersal in the marine angiosperm *Zostera marina* L. *Evolution*, 50: 856–864.

Sainty, G.R. and S.W.L. Jacobs, 1981. *Waterplants of New South Wales*. Water Resources Commission, Sydney, New South Wales, Australia, 550pp.

Short, F.T., W.C. Dennison, T.J.B. Carruthers, and M. Waycott, 2007. Global seagrass distribution and diversity: A bioregional model. *J. Exp. Mar. Biol. Ecol.*, 350: 3–20.

Short, F.T., G.E. Moore, and K.A. Peyton, 2010. *Halophila ovalis* in the Tropical Atlantic Ocean. *Aquat. Bot.*, 93: 141–146.

Smith, J.J., C. McMillan, W.J. Kenworthy, and K. Bird, 1997. Flowering and genetic banding patterns of *Halophila johnsonii* and conspecifics. *Aquat. Bot.*, 59: 323–331.

Soros-Pottruff, C.L. and U. Posluszny, 1995. Developmental morphology of reproductive structures of *Zostera* and a reconsideration of *Heterozostera* (Zosteraceae). *Int. J. Plant Sci.*, 156: 143–158.

Spalding, M., M. Taylor, C. Ravilious, F.T. Short, and E. Green, 2003. Global overview: The distribution and status of seagrasses. In: *World Atlas of Seagrasses*, Green, E.P. and F.T. Short (eds.). University of California Press, Berkley, CA, pp. 5–26.

Tanaka, N., J. Kuo, Y. Omori, and K.N. Aioi, 2003. Phylogenetic relationships in the genera *Zostera* and *Heterozostera* (Zosteraceae) based on matK sequence data. *J. Plant Res.*, 116: 273–279.

Tanaka, N., H. Setoguchi, and J. Murata, 1997. Phylogeny of the family of Hydrocharitaceae inferred from rbcL and matK gene sequence data. *J. Plant. Res.*, 110: 329–337.

Thangaradjou, T., E.P. Nobi, E. Dilipan, K. Sivakumar, and S. Susila, 2010. Heavy metal enrichment in seagrasses of Andaman Islands and its implication to the health of the coastal ecosystem. *Indian J. Geo-Mar. Sci.*, 39(1): 85–91.

Tomlinson, P.B., 1982. Helobiae (Alismatidae). In: *Anatomy of the Monocotyledons*, Metcalfe, C.R. (ed.). Clarendon Press, Oxford, U.K, 559pp.

Uchimura, M., E.J. Faye, S. Shimada, T. Inoue, and Y. Nakamura, 2008. A reassessment of *Halophila* species (Hydrocharitaceae) diversity with special reference to Japanese representatives. *Bot. Mar.*, 51: 258–268.

Uchiyama, H., 1996. An easy method for investigating molecular systematic relationships in the genus *Zostera*, Zosteraceae. In: *The Proceedings of an International Workshop Seagrass Biology*, Kuo, J., R.C. Phillips, D.I. Walker, and H. Kirkman (eds.). The University of Western Australia, Faculty of Sciences, Nedlands, Western Australia, Australia, pp. 79–84.

Waycott, M., 1995. Assessment of genetic variation and clonality in the seagrass *Posidonia australis* using RAPD and allozyme analysis. *Mar. Ecol. Prog. Ser.*, 116: 289–295.

Waycott, M., 1999. Mating systems and population genetics of marine angiosperms (seagrasses). In: *Systematics and Evolution of Monocots, Proceedings of the Second International Conference on Comparative Biology of Monocotyledons*, Vol. 1, Wilson, K.L. and D. Morrison (eds.). CSIRO Publishing, Sydney, New South Wales, Australia, pp. 277–285.

Waycott, M. and P.A.G. Barnes, 2001. AFLP diversity within and between populations of the Caribbean seagrass *Thalassia testudinum* (Hydrocharitaceae). *Mar. Biol.*, 139: 1021–1028.

Waycott, M., D.W. Freshwater, R.A. York, A. Calladine, and W.J. Kenworthy, 2002. Evolutionary trends in the seagrass genus *Halophila* (Thouars): Insights from molecular phylogeny. *Bull. Mar. Sci.*, 71: 1299–1308.

Waycott, M. and D.H. Les, 1996. An integrated approach to the evolutionary study of seagrasses. In: *Seagrass Biology: Proceedings of an International Workshop*, Kuo, J., R.C. Phillips, D.I. Walker, and H. Kirkman (eds.). SCIENCES UWA, Rottnest Island, Western Australia, Australia, pp. 71–78.

Waycott, M., G. Procaccini, D.H. Les, and T.B.H. Reusch, 2006. Seagrass evolution and conservation: A genetic perspective. In: *Seagrasses: Biology, Ecology and Conservation*, Larkum, A.W.D., R.J. Orth, and C.M. Duarte (eds.). Springer Verlag, Berlin, Germany, pp. 25–50.

Williams, S.L. and R.J. Orth, 1998. Genetic diversity and structure of natural and transplanted Eelgrass populations in the Chesapeake and Chincoteague Bays. *Estuaries*, 21: 118–128.

Wissler, L., F.M. Codoner, J. Gu, T.B.H. Reusch, J.L. Olsen, G. Procaccini, and E.B. Bauer, 2011. Back to the sea twice: Identifying candidate plant genes for molecular evolution to marine life. *BMC Evol. Biol.*, 11(8): 1–12.

Womersley, H.B.S., 1984. In: *The Marine Benthic Flora of Southern Australia. Part 1*, Woolman, D.J. (ed.). Adelaide, South Australia, Australia, 329pp. South Australian Government Printing Division, Adelaide.

6

Marine Genomics:
Recent Advancement and Wide-Area Applications

Xinxin You, Chao Bian, Xiaodong Fang, Ying Sun, Jie Bai,
Chao Peng, Xinhui Zhang, and Qiong Shi

CONTENTS

6.1 Introduction

Genomics studies the structure, function, and diversity of genomes (genome being the collective term for all the genetic information contained in a particular organism). As a discipline, genomics began with the first attempts to obtain large-scale sequence data for individual organism by sequencing the genomic DNA. Genome sequencing has ushered in a new era of investigation in the biological sciences, allowing us to embark for the first time on a truly comprehensive study of vertebrate evolution, the results of which will touch nearly every aspect of vertebrate biological enquiry (Cock et al., 2010).

Genome sequencing was initially a very costly process, and consequently early efforts were focused on laboratory model organisms such as the bacteria *Escherichia coli* and the yeast *Saccharomyces cerevisiae* followed by the animal species *Caenorhabditis elegans* and *Drosophila melanogaster* and the plant *Arabidopsis thaliana*. Marine organisms were very poorly represented among these early genomic models and were to some extent left behind as the application of genomic approaches to several terrestrial models allowed an acceleration of our understanding of the biology, ecology, and evolutionary history of these species (Cock et al., 2010). With the rapid development of sequencing technologies, the sequencing capacity has been dramatically increased, while the sequencing cost has been drastically reduced. Accordingly, whole genome sequencing has become available for marine vertebrates.

Marine vertebrates are mainly living in a marine environment. They include marine fishes, marine reptiles, seabirds, and marine mammals. These animals have an internal skeleton and make up about 4% of animal population in the sea. Fugu, commonly known as "torafugu," is a teleost fish with the first published marine genome (Aparicio et al., 2002). Its small genome size (392 Mb) is primarily due to differences in repeated sequence content. The compact genome of Fugu has been sequenced to over 95% coverage, and more than 80% of the assembly is in multigene-sized scaffolds. In the 365 Mb (no gaps) genome assembly, repetitive DNA accounts for less than one-sixth of the sequence, and gene loci occupy about one-third of the genome. As with the human genome, gene loci are not evenly distributed but are clustered into sparse and dense regions. Some "giant" genes were observed that had average coding sequence sizes but were spread over genomic lengths significantly larger than those of their human orthologs (Aparicio et al., 2002).

In the past 3 years, whole genome sequencing has been explosively applied into large amount of marine species. BGI, the largest genomic research center in the world, has been leading almost all areas of genome sequencing. As an important research platform for marine genomics, our lab (Shenzhen Key Lab of Marine Genomics in the BGI) has been involved into studies on over 30 marine organisms, such as groupers, mudskippers, Chinese large yellow croaker, Chinese pearl oyster, seahorses, Asian arowana, cavefishes, Chinese sturgeon, cone snails, sea cucumbers Chinese mitten crab, Chinese white dolphin, and penguins. Here, we provide a minireview of these published genomes in the past 3 years. Table 6.1 outlines these recently published genome data of marine vertebrates. This chapter also deals with a few unpublished genomic information of several marine animals, that is, mudskippers and penguins, for a wide range of comparison.

TABLE 6.1

Published Genome Data of Recently Sequenced Marine Vertebrates

Species	Genome Size (Mb)	Contig N50/Scaffold N50 (kb)	References
Atlantic cod	824	2.3/393	Star et al. (2011)
Sea lamprey	886	13/185	Smith et al. (2013)
Coelacanth	2860	12.7/924	Amemiya et al. (2013)
Green sea turtle	2240	20.4/3780	Wang et al. (2013)
Pacific bluefin tuna	800	7.5/136	Nakamura et al. (2013)
Common minke whale	2760	22.6/12800	Yin et al. (2014)
Elephant shark	974	46.6/4522	Venkatesh et al. (2014)
Half-smooth tongue sole	470	27/510	Chen et al. (2014)
Rainbow trout	1900	7.7/384	Berthelot et al. (2014)

6.2 Recent Advancement in Whole Genome Sequencing of Marine Vertebrates

6.2.1 Marine Fishes

Marine fishes are abundant in oceans. Although no exact species number is available, they exhibit great species diversity (in the total fish >32,000 species) and high economic importance. Therefore, most of the sequenced marine organisms are fishes.

6.2.1.1 Teleost Fishes

Most living fishes are members of this group. Here, we also present recently published data of mudskippers, which are amphibious fishes and provide novel insights into terrestrial adaptation. By presenting the first genome sequencing of amphibious fishes, we also offer a new model for understanding the adaptive evolution of animals from water to land.

6.2.1.1.1 *Atlantic Cod* (Gadus morhua)

The Atlantic cod is a well-known benthopelagic fish belonging to the family Gadidae. It is a large, cold-adapted teleost that sustains long-standing commercial fisheries and incipient aquaculture. The Atlantic cod genome was published in 2011 and provided insights into complex thermal adaptations in its hemoglobin gene cluster and an unusual immune architecture compared to the other sequenced vertebrates (Star et al., 2011). The genome assembly was obtained exclusively by 454 sequencing of shotgun and paired-end libraries and automated annotation identified 22,154 genes. The major histocompatibility complex (MHC) II is a conserved feature of the adaptive immune system of jawed vertebrates, but Atlantic cod has lost the genes for MHCII, CD4, and invariant chain that are essential for the function of this pathway. Nevertheless, Atlantic cod is not exceptionally susceptible to disease under natural conditions. A highly expanded number of MHCI genes and a unique composition of its Toll-like receptor (TLR) families were existent in the Atlantic cod genome. This indicates how the Atlantic cod immune system has evolved compensatory mechanisms in both adaptive and innate immunity in the absence of MHCII (Star et al., 2011). These observations affect fundamental assumptions about the evolution of the adaptive immune system and its components in vertebrates.

6.2.1.1.2 *Coelacanth* (Latimeria chalumnae)

Coelacanths are a group of deep-sea fishes once thought to be extinct. They are lobe-finned fish more closely related to tetrapods than to ray-finned fish. Only two living species are known: the Cormoran coelacanth (*Latimeria chalumnae*) is native to the coastal regions of the Indian Ocean, while the Sulawesi coelacanth (*Latimeria menadoensis*) is native to the Celebes Sea off the coast of Indonesia. Since its discovery, the coelacanth has been referred to as a "living fossil," owing its morphological similarities to its fossil ancestors. However, questions have remained as to whether it is indeed evolving slowly, as morphological stasis does not necessarily imply genomic stasis. The modern coelacanth looks remarkably similar to many of its ancient relatives, and its evolutionary proximity to our own fish ancestors provides a glimpse of the fish that first walked on land. The genome of *L. chalumnae* was sequenced in 2013. Through a phylogenomic analysis, Amemiya et al. (2013) conclude that the lungfish, not the coelacanth, is the closest living relative of tetrapods. Coelacanth protein-coding genes are significantly more slowly evolving than those of tetrapods, unlike other genomic features. Analyses of changes in genes and regulatory elements during the vertebrate adaptation to land highlight the genes involved in immunity, nitrogen excretion, and the development of fins, tail, ear, eye, brain, and olfaction. Functional assays of enhancers involved in the fin-to-limb transition and in the emergence of extraembryonic tissues show the importance of the coelacanth genome as a blueprint for understanding the tetrapod evolution.

6.2.1.1.3 *Pacific Bluefin Tuna* (Thunnus orientalis)

Tunas are migratory fishes in offshore habitats and top predators with unique features. The Pacific bluefin tuna is a predatory species found widely in the northern Pacific Ocean, but it is migratory and also recorded as a visitor to the south Pacific. Despite their ecological importance and high market values, the open-ocean lifestyle of tuna, in which effective sensing systems, such as color vision, are required for capture of prey, has been poorly understood. To elucidate the genetic and evolutionary basis of optic adaptation in tunas, Nakamura et al. (2013) determined the genome sequence of the Pacific bluefin tuna using the next-generation sequencing technology. A total of 26,433 protein-coding genes were predicted from 16,802 assembled scaffolds. From these, five common fish visual pigment genes, red-sensitive (middle/long-wavelength sensitive), UV-sensitive (short-wavelength sensitive 1 [SWS1]), blue-sensitive (SWS2), rhodopsin (RH1), and green-sensitive (RH2) opsin genes, were identified. However, sequence comparison revealed that tuna's RH1 gene has an amino acid substitution that causes a short-wave shift in the absorption spectrum (i.e., blue shift). The Pacific bluefin tuna has at least five RH2 paralogs, the most studied among fishes; four of the encoded proteins may be tuned to blue light at the amino acid level. Moreover, phylogenetic analysis suggested that gene conversions have occurred in each of the SWS2 and RH2 loci in a short period. Thus, the Pacific bluefin tuna has undergone evolutionary changes in the three opsin genes (RH1, RH2, and SWS2), which may have contributed to detecting blue-green contrast and measuring the distance to prey in the blue-pelagic ocean. These findings provide basic information on behavioral traits of predatory fish and, thereby, could help to improve the technology to culture such fish in captivity for resource management.

6.2.1.1.4 *Half-Smooth Tongue Sole* (Cynoglossus semilaevis)

The half-smooth tongue sole is found widely in the coastal waters of China. It is a commercially valuable fish of considerable interest as an aquaculture species. In contrast to many other teleost species, classical karyotype analysis and artificial gynogenesis tests have revealed that this species employs a female heterogametic sex-determination system (ZW♀/ZZ♂) and has clear sexual dimorphism, with females growing much faster and reaching final body sizes that are 2–4 times those of males. Genetic sex determination by W and Z chromosomes has developed independently in different groups of organisms. To better understand the evolution of sex chromosomes and the plasticity of sex-determination mechanisms, Chen et al. (2014) sequenced the whole genomes of a male (ZZ) and a female (ZW) half-smooth tongue sole. In addition to insights into adaptation to a benthic lifestyle, the sex chromosomes of these fish are derived from the same ancestral vertebrate protochromosome as the avian W and Z chromosomes were found. Notably, the same gene on the Z chromosome, dmrt1, which is the male-determining gene in birds, showed convergent evolution of features that are compatible with a similar function in the tongue sole. Comparison of the relatively young tongue sole sex chromosomes with those of mammals and birds identified events that occurred during the early phase of sex-chromosome evolution. Pertinent to the current debate about heterogametic sex-chromosome decay, massive gene loss occurred in the wake of sex-chromosome "birth" (Chen et al., 2014). Comparative analysis of the gonadal DNA methylomes of pseudomale, female, and normal male fishes revealed that genes in the sex-determination pathways are the major targets of substantial methylation modification during sexual reversal (Shao et al., 2014). Methylation modification in pseudomales is globally inherited in their ZW offspring, which can naturally develop into pseudomales without temperature incubation. Transcriptome analysis revealed that dosage compensation occurs in a restricted, methylated cytosine-enriched Z chromosomal region in pseudomale testes, achieving equal expression level in normal male testes. In contrast, female-specific W chromosomal genes are suppressed in pseudomales by methylation regulation. It is suggested that epigenetic regulation plays multiple crucial roles in sexual reversal of tongue sole fish. A systemic analysis on epigenetic regulation offers the first clue on the mechanisms behind gene dosage balancing in an organism that undergoes sexual reversal, which is confirmed. Finally, a causal link between the bias sex-chromosome assortment in offspring of pseudomale family and the transgenerational epigenetic inheritance of sexual reversal in tongue sole fish is also confirmed.

6.2.1.1.5 *Rainbow Trout* (Oncorhynchus mykiss)

Rainbow trout is a member of the Salmonid family that is of major ecological interest worldwide. It is one of the most studied fish species and is extensively used for research in many fields such as carcinogenesis, toxicology, immunology, ecology, physiology, and nutrition. It is also an important aquaculture species of major economic importance raised in both hemispheres and on all continents. Due to relatively recent whole genome duplication (WGD), the rainbow trout thus provides a unique opportunity to better understand the early steps of gene fractionation. The genome of rainbow trout was published in 2014 and shows that despite 100 million years of evolution, the two ancestral subgenomes have remained extremely collinear (Berthelot et al., 2014). Only half of the protein-coding genes have been retained as duplicated copies and genes have been lost mostly through pseudogenization. In striking contrast with protein-coding genes is the fate of miRNA genes that have almost all been retained as duplicated copies. Together, the analysis of the rainbow trout genome reveals a slow and stepwise rediploidization process after the salmonid-specific 4th WGD that challenges the current hypothesis that WGD is followed by massive and rapid genomic reorganizations and gene deletions.

6.2.1.1.6 *Mudskippers*

Mudskippers (family Gobiidae; subfamily Oxudercinae) are the largest group of amphibious teleost fishes that are uniquely adapted to live on mud flats. They include four main genera, that is, *Boleophthalmus*, *Periophthalmodon*, *Periophthalmus*, and *Scartelaos* comprising diverse species that represent a continuum of adaptations toward terrestrial life with some being more terrestrial than the others. Thus, mudskippers are a useful group for gaining insights into the genetic changes underlying the terrestrial adaptations of amphibious fishes. Whole genome sequencing of four representative species of mudskippers, including *Boleophthalmus pectinirostris* (BP or blue-spotted mudskipper), *Scartelaos histophorus* (SH or blue mudskipper), *Periophthalmodon schlosseri* (PS or giant mudskipper), and *Periophthalmus magnuspinnatus* (PM or giant-fin mudskipper), have been finished (You et al., 2014). BP and SH are predominantly aquatic and spend less time out of water, whereas PS and PM are primarily terrestrial and spend extended periods of time on land. The analysis of four mudskipper genomes has provided insights into a variety of molecular mechanisms associated with land adaptation of these amphibious fishes. An expansion in innate immune-system-associated genes such as the TLR13 family was identified, which may offer defense against novel pathogens encountered by mudskippers on land. Some core genes in the ammonia excretion pathway, such as *Rhcg1* and *Nhe*, have undergone differential positive selection and structural changes, which may facilitate efficient ammonia excretion from the gills. Mudskippers have experienced loss or divergence of some vision-related genes that might have facilitated aerial vision, which is crucial for courtship and avoiding predators on land and navigating back to water. Mudskippers lack olfactory receptors belonging to the alpha and gamma group required for detecting airborne chemicals and therefore seem to be using certain vomeronasal receptors for detecting airborne odorants on land. Transcriptomic analyses of mudskippers exposed to air provided further insights into the molecular adaptations of mudskippers for survival during long durations on land. In a word, the genomic and transcriptomic data developed in this study provide a useful resource for gaining further insights into genetic changes associated with water-to-land transition of vertebrates.

6.2.1.2 Cartilaginous Fishes

Cartilaginous fishes include sharks, rays, skates, and elephant sharks. The last group belongs to a different subclass (Holocephali) from the others (Elasmobranchii).

6.2.1.2.1 *Elephant Shark* (Callorhinchus milii)

The emergence of jawed vertebrates (gnathostomes) from jawless vertebrates was accompanied by major morphological and physiological innovations, such as hinged jaws, paired fins, and immunoglobulin-based adaptive immunity. Gnathostomes subsequently diverged into two groups: the cartilaginous fishes and the bony vertebrates. The whole genome analysis of the elephant shark was reported in 2014. Its genome size is estimated to be 970 Mb (Venkatesh et al., 2014), which is the smallest among all the cartilaginous fishes and around one-third of the human genome. The *C. milii* genome is the slowest evolving of all known vertebrates, including the "living fossil" coelacanth, and features extensive synteny conservation with tetrapod genomes, making it a good model for comparative analyses of gnathostome genomes. The functional studies suggest that the lack of genes encoding secreted calcium-binding phosphoproteins in cartilaginous fishes explains the absence of bone in their endoskeleton. Furthermore, the adaptive immune system of cartilaginous fishes is unusual: it lacks the canonical CD4 coreceptor and most transcription factors, cytokines, and cytokine receptors related to the CD4 lineage, despite the presence of polymorphic MHC class II molecules (Venkatesh et al., 2014). It thus presents a new model for understanding the origin of adaptive immunity.

6.2.2 Marine Reptiles

Marine reptiles have become secondarily adapted for an aquatic or semiaquatic life in a marine environment. Some marine reptiles rarely ventured onto land and gave birth in the water. Others, such as sea turtles and saltwater crocodiles, return to shore to lay their eggs. After the mass extinction at the end of the Cretaceous period, marine reptiles are less numerous while are still declining.

6.2.2.1 *Green Sea Turtle* (Chelonia mydas)

The green sea turtle is a large sea turtle of the family Cheloniidae. It is the only species in the genus *Chelonia*. Its range extends throughout tropical and subtropical seas around the world, with two distinct populations in the Atlantic and Pacific Oceans. The common name comes from the usually green fat found beneath its carapace. This species is endangered with extinction due to a variety of causes that include consumption of their eggs by humans and other animals, loss of beach habitat, and drowning in fishing nets. These turtles are known to migrate hundreds or thousands of kilometers from their foraging grounds to mating sites and nesting beaches and have been found in waters off the coast of Alaska and Great Britain. The genome of *C. mydas* was sequenced in 2013 to examine the development and evolution of the turtle body plan (Wang et al., 2013). The results indicated the close relationship of the turtles to the bird–crocodilian lineage, from which they split ~267.9–248.3 million years ago (Upper Permian to Triassic). Olfactory receptor genes are expanded extensively in the turtle genome. Embryonic gene expression analysis identified an hourglass-like divergence of turtle and chicken embryogenesis, with maximal conservation around the vertebrate phylotypic period, rather than at later stages that show the amniote-common pattern. Wnt5a expression was found in the growth zone of the dorsal shell, supporting the possible co-option of limb-associated Wnt signaling in the acquisition of this turtle-specific novelty. Turtle evolution was accompanied by an unexpectedly conservative vertebrate phylotypic period, followed by turtle-specific repatterning of development to yield the novel structure of the shell.

6.2.3 Seabirds

Seabirds have adapted to life within the marine environment. Most species nest in colonies. Many species are famous for undertaking long annual migration across the equator or circumnavigating the earth in some cases. As examples of the tripolar animals, penguins were picked out for whole genome sequencing in the BGI.

6.2.3.1 *Penguins*

Penguins (order Sphenisciformes, family Spheniscidae) are a group of aquatic, flightless birds living almost exclusively in the Southern Hemisphere, especially in Antarctica. Most penguins feed on krill, fish, squid, and other forms of sea life caught while swimming underwater. They spend about half of their lives on land and half in the oceans. Penguins have undergone multiple morphological adaptations associated with underwater life, including scalelike feathers, reduced distal wing musculature, and improved visual sensitivity in water. The Antarctic penguins are an excellent model for studying how animals adapt to harsh environments and how climate changes affect the population dynamics. The draft genome sequences of two Antarctic dwelling penguin species, Adélie penguin (*Pygoscelis adeliae*) and emperor penguin (*Aptenodytes forsteri*), have been completed (Li et al., 2014). Phylogenetic dating suggests that early penguins arose ~60 million years ago (mya) and links between these biological patterns and global climate change. Analysis of effective population sizes reveals that the two penguin species experienced population expansions from ~1 mya to ~100 kya, but responded differently to the climate cooling of the last glacial period. Comparative genomic analyses identified molecular changes in genes related to epidermal changes, taste transduction, phototransduction, lipid metabolism, and forelimb morphology. These molecular changes reflect both shared and diverse adaptations of the two penguin species to the marine and cold environment in Antarctica. The genomic resources lay the foundation for further genomic and molecular studies of penguins and will likely facilitate related researches such as avian evolution, polar biology, and climate changes.

6.2.4 Marine Mammals

Marine mammals, including dolphins, whales, seals, otters, and walruses, form a diverse group of over 140 species that rely on the oceans for their existence. They have a number of physiological and anatomical features to overcome the unique challenges associated with aquatic living. The fossil record demonstrates that mammals reentered the marine realm, suggesting the evolution of marine mammals back to the sea after 300 million years (Uhen, 2007).

6.2.4.1 *Common Minke Whale* (Balaenoptera acutorostrata)

The common minke whale or northern minke whale is an interesting whale species within the suborder of baleen whales. It is the smallest member of the rorquals and the second smallest species of baleen whales. Although first ignored by whalers due to its small-size and low-oil yield, it began to be exploited by various countries beginning in the early twentieth century. As other species declined, larger numbers of common minke whales were caught largely for their meat. It is now one of the primary targets of the whaling industry. The shift from terrestrial to aquatic life by whales was a substantial evolutionary event. The whole genome sequencing and de novo assembly of the minke whale genome, as well as the whole genome sequences of a fin whale, a bottlenose dolphin, and a finless porpoise were reported in 2013 (Yin et al., 2014). The comparative genomic analysis identified an expansion in the whale lineage of gene families associated with stress-responsive proteins and anaerobic metabolism, whereas gene families related to body hair and sensory receptors were contracted. The analysis also identified whale-specific mutations in genes encoding antioxidants and enzymes controlling blood pressure and salt concentration. Overall the whale genome sequences exhibited distinct features that are associated with the physiological and morphological changes needed for life in an aquatic environment, marked by resistance to physiological stresses caused by a lack of oxygen, increased amounts of reactive oxygen species, and high salt levels.

6.2.4.2 *Polar Bears*

The polar bear (*Ursus maritimus*) is a carnivorous bear whose native range lies largely within the Arctic Circle, encompassing the Arctic Ocean, its surrounding seas, and surrounding land masses. Although it is the sister species of the brown bear (*Ursus arctos*), it has evolved to occupy a narrower ecological niche, with many body characteristics adapted for cold temperatures; for moving across snow, ice, and open water; and for hunting the seals, which make up most of its diet. Establishing a reliable time frame for when the polar bear emerged as a species is essential for our understanding of what evolutionary processes drove speciation and how fast novel adaptations to extreme environments can arise in a large mammal. A deep-sequenced and de novo assembled polar bear reference genome at a depth of 101× and resequenced at 3.5×–22× coverage 79 polar bears and 10 brown bears were reported in 2014 (Liu et al., 2014). Eighty-nine complete genomes of polar bear and brown bear using population genomic modeling were analyzed, and the results show that the species diverged only 479–343 thousand years ago (kya). Genes in the polar bear lineage have been under stronger positive selection than in brown bears; nine of the top 16 genes under strong positive selection are associated with cardiomyopathy and vascular disease, implying important reorganization of the cardiovascular system. One of the genes showing the strongest evidence of selection, *APOB*, encodes the primary lipoprotein component of low-density lipoprotein (LDL); functional mutations in *APOB* may explain how polar bears are able to cope with life-long elevated LDL levels that are associated with high risk of heart disease in humans. This polar bear genomics study reveals the strength of using a population genomic approach to resolve the evolutionary history of a nonmodel organism in terms of divergence time, demographic history, selection, and adaptation. It is remarkable that a majority of the top genes under positive selection in polar bears have functions related to the cardiovascular system and most of them to cardiomyopathy, in particular when considering their divergence from brown bears no more than 479–343 kya. Such a drastic genetic response to chronically elevated levels of fat and cholesterol in the diet has not previously been reported (Liu et al., 2014). It certainly encourages a move beyond the standard model organisms in polar bear genome search for the underlying genetic causes of human cardiovascular diseases.

6.2.5 Others

Others refer to agnatha and cephalochordate. The former has two groups (~120 species in total), that is, the lampreys and the hagfish, which are still surviving today. The latter is represented by the lancelets, an important family of model organisms. Among the 32 species in total, the first lancelet was reported with whole genome sequences in 2008, and more species have been under genome sequencing for academic purposes.

6.2.5.1 Sea Lamprey (Petromyzon marinus)

Lampreys are an order of jawless fish, the adult of which is characterized by a toothed, funnel-like sucking mouth. The lampreys are a very ancient lineage of vertebrates that diverged from our own ~500 million years ago, though their exact relationship to hagfishes and jawed vertebrates is still a matter of dispute. The sea lamprey is a parasitic animal found in the northern Atlantic Ocean along shores of Europe and North America, in the western Mediterranean Sea, and in the Great Lakes. The genome of sea lamprey was sequenced in 2013. Because of its early divergence from other living vertebrates, the sea lamprey genome is uniquely poised to provide insights into the ancestry of vertebrate genomes and the underlying principles of vertebrate biology (Smith et al., 2013). This sequencing effort revealed that the lamprey has unusual GC content and amino acid usage patterns compared to other vertebrates. Analyses of the assembly indicate that two WGDs likely occurred before the divergence of ancestral lamprey and gnathostome lineages. Moreover, the results help define key evolutionary events within vertebrate lineages, including the origin of myelin-associated proteins and the development of appendages. Therefore, the lamprey genome provides an important resource for reconstructing vertebrate origins and the evolutionary events that have shaped the genomes of extant organisms.

6.2.5.2 Lancelets

The lancelets, also known as amphioxus, are small wormlike marine animals with a global distribution in shallow temperate (as far north as Scotland) and tropical seas, usually found half buried in sand. They are the modern representatives of the subphylum Cephalochordata, formerly thought to be the sister group of the craniates. The structure and gene content of the highly polymorphic ~520 Mb genome of the Florida lancelet (*Branchiostoma floridae*) were described, and the context of chordate evolution was analyzed (Putnam et al., 2008). Extensive conserved synteny between the genomes of amphioxus and various vertebrates lends unprecedented clarity and coherence to the history of genome-scale events in vertebrate evolution. The human and other jawed vertebrate genomes show widespread quadruple-conserved synteny relative to the amphioxus sequence, which extends earlier regional studies and provides a unified explanation for paralogous chromosomal regions in vertebrates. The result provides conclusive evidence for two rounds of complete genome duplication on the jawed vertebrate stem. The amphioxus genome contains a basic set of chordate genes involved in development and cell signaling, including a 15th Hox gene. This set includes many genes that were co-opted in vertebrates for new roles in neural crest development and adaptive immunity. However, where amphioxus has a single gene, vertebrates often have two, three, or four paralogs derived from two WGD events. In addition, several transcriptional enhancers are conserved between amphioxus and vertebrates—a very wide phylogenetic distance. In contrast, urochordate genomes have lost many genes, including a diversity of homeobox families and genes involved in steroid hormone function. The amphioxus genome also exhibits derived features, including duplications of opsins and genes proposed to function in innate immunity and endocrine systems. The amphioxus genome hence is elemental to an understanding of the biology and evolution of nonchordate deuterostomes, invertebrate chordates, and vertebrates (Holland et al., 2008).

6.3 Wide-Area Applications of Marine Genomics

Genomic approaches are now being applied to a diverse catalogue of questions in marine biology. Genome-wide studies of the marine organisms have great impact for comparative genomics: it will provide a deep understanding of the recent half billion years of evolution in vertebrates and of more recent era that led to an extreme diversification of particular subgroups of the Teleostei (Spaink et al., 2014). It will also provide enormous opportunities for data mining and will provide the possibility to trace back the origins of genes from the organisms closest to the earliest evolutionary branches to its origins within invertebrates. For this purpose, it is fortunate that many invertebrate species such as the tunicates are also increasingly being analyzed with genomics technologies. That this can lead to unexpected findings is nicely illustrated by the recent discovery of a completely novel fluorescent protein in the Japanese eel (Kumagai et al., 2013). Furthermore, it can lead to new insights into the origin of individual genes, for instance, the interesting example of horizontal gene transfer of a transposon between lamprey species and their hosts indicates that transfer of genetic material between species mediated by parasite–host interactions could be very frequent (Kuraku et al., 2012).

Although marker-assisted selection (MAS) and genomic selection have not been widely used in aquaculture, their application in breeding programmers is expected to be a fertile area of research (Yue, 2014). Genome-wide DNA markers in linkage disequilibrium with the causative mutations can be used to capture the effect of all these loci (Meuwissen et al., 2001). The equation that predicts breeding value from SNP genotypes must be estimated from a sample of animals, known as the reference population, that have been measured for the traits and genotyped for the SNPs. This prediction equation can then be used to predict breeding values for selection candidates based on their genotypes alone. The candidates are ranked on these estimated breeding values, and the best ones are selected to breed the next generation (Hayes et al., 2013). The advantages of MAS are obvious as compared with the conventional selective breeding. MAS is especially useful for traits that are difficult to measure, exhibit low heritability, and/or are expressed late in development. Implementation of MAS requires DNA markers that are tightly linked to quantitative trait loci (QTL) for traits of interest based on QTL mapping or association studies. Ideally, the DNA markers should be the causative mutation underlying the phenotypic variation. QTL studies in aquaculture species covered a wide range of traits including growth, meat quality, egg production, disease resistance, stress resistance, and reproduction. The results of these studies provide a good starting point to search for QTL within breeding populations. Of the QTL from experimental crosses, only a small number of them have been followed up by confirmation and fine mapping. The responsible mutations in genes have not been described for detected QTL. However, there are already a few applications of MAS in commercial breeding programs in aquaculture species (Ozaki et al., 2012).

In addition to molecular breeding research, there will also be important applied aspects, for instance, in the development of marine medicine (e.g., thrombolytic drugs, conotoxins, antibiotic peptides), nature conservation biology and the impact of ancient climate change on species diversification or extinction processes. The latter research could lead to better prediction models for the effects of current estimated climate changes on biodiversity of the marine organisms and thereby could provide better guidelines for knowledge-based fishery regulations (Schwartz et al., 2007). The state of the art in genomics of the marine organisms has advanced in exploiting the enormous phylogenetic diversity of marine organisms to explore the evolution of developmental processes, characterizing the marine ecosystems that play key roles in global geochemical cycles, searching for novel biomolecules, and understanding ecological interactions within important marine ecosystems. More and more applications are under deep investigation for both academic and commercial purposes.

6.4 Summary

In this chapter, we overviewed the recent advancement and wide-area applications of whole genome sequencing in marine vertebrates, ranging from fishes, reptiles, birds to mammals. In the past 3 years, genome sequencing has been explosively applied into large number of marine species, although only around 10 reports are available. This chapter also deals with few unpublished genomic information of several marine animals, that is, mudskippers (amphibious fishes) and penguins (an attractive example of the seabirds), for a wide range of comparison. By far, marine genomics has been utilized mainly for marker-assisted molecular breeding, development of marine medicine, and conservation of marine species.

References

Amemiya, C.T. et al., 2013. The African coelacanth genome provides insights into tetrapod evolution. *Nature* 496:311–316.

Aparicio, S. et al., 2002. Whole-genome shotgun assembly and analysis of the genome of *Fugu rubripes*. *Science* 297:1301–1310.

Berthelot, C. et al., 2014. The rainbow trout genome provides novel insights into evolution after whole-genome duplication in vertebrates. *Nat. Commun.* 5:1–10.

Chen, S. et al., 2014. Whole-genome sequence of a flatfish provides insights into ZW sex chromosome evolution and adaptation to a benthic lifestyle. *Nat. Genet.* 46:253–260.

Cock, J.M. et al., 2010. *Introduction to Marine Genomics*. New York: Springer.

Hayes, B.J. et al., 2013. The future of livestock breeding: Genomic selection for efficiency, reduced emissions intensity, and adaptation. *Trends Genet.* 29:206–214.

Holland, L.Z. et al., 2008. The amphioxus genome illuminates vertebrate origins and cephalochordate biology. *Genome Res.* 18:1100–1111.

Kumagai, A. et al., 2013. A bilirubin-inducible fluorescent protein from eel muscle. *Cell* 153:1602–1611.

Kuraku, S. et al., 2012. Horizontal transfers of Tc1 elements between teleost fish and their vertebrate parasites, lampreys. *Genome Biol. Evol.* 4:929–936.

Li, C. et al., 2014. Two Antarctic penguins reveal insights into their evolutionary history and molecular changes related to the Antarctic environment. *GigaScience* 3:27.

Liu, S.P. et al., 2014. Population genomics reveal recent speciation and rapid evolutionary adaptation in polar bears. *Cell* 157:785–794.

Meuwissen, T.H. et al., 2001. Prediction of total genetic value using genome-wide dense marker maps. *Genetics* 157:1819–1829.

Nakamura, Y. et al., 2013 Evolutionary changes of multiple visual pigment genes in the complete genome of Pacific bluefin tuna. *Proc. Natl. Acad. Sci. U.S.A.* 110:11061–11066.

Ozaki, A. et al., 2012. Progress of DNA marker-assisted breeding in maricultured finfish. *Bull. Fish. Res. Agency* 35:31–37.

Putnam, N.H. et al., 2008. The amphioxus genome and the evolution of the chordate karyotype. *Nature* 453:1064–1072.

Schwartz, M.K. et al., 2007. Genetic monitoring as a promising tool for conservation and management. *Trends Ecol. Evol.* 22:25–33.

Shao, C. et al., 2014. Epigenetic modification and inheritance in sexual reversal of fish. *Genome Res.* 24:604–615.

Smith, J.J. et al., 2013. Sequencing of the sea lamprey (*Petromyzon marinus*) genome provides insights into vertebrate evolution. *Nat. Genet.* 45:415–421.

Spaink, H.P. et al., 2014. Advances in genomics of bony fish. *Brief. Funct. Genomics* 13:144–156.

Star, B. et al., 2011. The genome sequence of Atlantic cod reveals a unique immune system. *Nature* 477:207–210.

Uhen, M.D., 2007. Evolution of marine mammals: Back to the sea after 300 million years. *Anat. Rec.* 290:514–522.

Venkatesh, B. et al., 2014. Elephant shark genome provides unique insights into gnathostome evolution. *Nature* 505:174–179.

Wang, Z. et al., 2013. The draft genomes of soft-shell turtle and green sea turtle yield insights into the development and evolution of the turtle-specific body plan. *Nat. Genet.* 45:701–706.

Yin, H.S. et al., 2014. Minke whale genome and aquatic adaptation in cetaceans. *Nat. Genet.* 46:88–92.

You, X. et al., 2014. Mudskipper genomes provide insights into terrestrial adaptation of amphibious fishes. *Nat. Commun.* 5:5594.

Yue, G.H., 2014. Recent advances of genome mapping and marker-assisted selection in aquaculture. *Fish Fish.* 15:376–396.

7

Puffer Genome for Tracking Human Genes

S.S. Khora and Pagolu Navya

CONTENTS

7.1 Introduction

Most of the genetic information that governs how humans develop and function is encoded in the human genome sequence. So comparisons between the genomes of different animals will guide future approaches to understanding gene function and regulation. A decade ago, analysis of the compact genome of the pufferfish was proposed as a cost-effective way to illuminate the human sequence through comparative analysis within the vertebrates.

FIGURE 7.1
Pufferfish *Fugu rubripes*.

Pufferfish (Figure 7.1) is a teleost fish belonging to the order Tetraodontiformes (four-toothed puffers) and a member of the gnathostomes (jawed vertebrates). There are over 100 species of pufferfish with diverse saltwater and freshwater habitats. The Tetraodontidae have been estimated to diverge from Diodontidae between 89 and 138 million years ago (Hamilton, 1822). Pufferfish are also known as "balloonfish" or "blowfish," because of their ability to fill their highly elastic stomach with large amount of water or even air to give it a shape of a balloon or ball, which helps to deter predators to swallow it. Puffers are the second most poisonous after golden poison frog and still there is no antidote available (Keiichi and Tyler, 1998). They are typically small to medium in size; even very few species reach a length of 100 cm (Keiichi and Tyler, 1998).

Mostly, pufferfish are short-bodied thickest fishes, and the skin is naked, but is generally studded with small spines. As a rule, these fishes are not edible, and many of the species are known to be highly poisonous. The skin and some internal organs of pufferfish are highly toxic even to humans. Pufferfish poisoning is considered to be the most common cause of fish poisoning along the Asian coast (Chew et al., 1983). Species of puffers contain a deadly poison (tetrodotoxin), which is found to be derived either from the food or by some physiological process peculiar to the puffers (Khora et al., 1997).

First, pufferfish are vertebrate despite their apparent differences; fish have nearly all of the same organ systems and physiology as humans, in contrast to the more distantly related invertebrate animals already sequenced, like flies and worms. Just as important, however, is that the pufferfish genome is unusually small for a vertebrate. Even among fish, puff-erfish is special; most fish genomes are several times longer than puffers. Pufferfish have the smallest known vertebrate genomes, around 350–400 million bases long, or 350–400 megabases (Brenner et al., 1993; Elgar et al., 1999). Puffer has 22 pairs of chromosomes, though these have no direct correspondence with the 23 pairs of human chromosomes. For comparison, the human genome is about three billion bases long. Despite this size difference, however, both pufferfish and humans are expected to have a similar repertoire of genes. This is the main reason pufferfish was chosen for sequencing as a cost-effective, more genes for the buck shortcut to a vertebrate gene set. There is also an evolutionary relationship between pufferfish and humans. Puffers are our very distant cousins, shar-ing a common ancestor with us nearly half a billion years ago. Remarkably, this common ancestry is still recorded in our genes.

In 1968, Hinegardner surveyed the nuclear DNA contents of over 200 teleost fishes. The 10-fold difference in size ranged from 4400 Mb in some of the armored catfish to 400 Mb in some Tetraodontidae species. One of the latter, the Japanese pufferfish (*Fugu rubripes*), with a small genome of 400 Mb (Hinegardner, 1968), was adopted as a model genome by Sydney Brenner as a tool to aid in deciphering the human genomic data. At the time, sequencing technology was such that only the bacteria and invertebrates were within the capabilities of whole-genome sequencing, and their use in exploiting the human genome was limited by their evolutionary distance, simplified physiology, and reduced gene content. Hence, a vertebrate genome was required, ideally with a relatively low nuclear DNA content, to act as an evolutionary bridge between these species and humans. Pufferfish is not an easy experimental animal; there are problems with aquacultural breeding, and it grows up to 1 kg in its first year, so it is not ideal for the average laboratory aquarium. Therefore, research has concentrated on its power to decipher genomic data in organisms with larger genomes such as mammals. Investigations since that time have characterized the pufferfish genome with its high gene density and paucity of repeat sequences (Brenner et al., 1993; Edwards et al., 1998; Elgar et al., 1999) and established it as a comparative model genome.

The main technology associated with the *fugu* project is that of gene identification using sequence scanning (Elgar et al., 1999). This technology is fast, cheap, efficient, and tolerant of sequencing errors, so genes can be identified from basic local alignment search tool (BLAST) similarity searching of single-pass sequencing. In the past, in *fugu*, this methodology has mainly been used to construct short-range linkage maps, but in an era of ever-increasing genome sequencing, it is providing an efficient tool for longer-range mapping and regional identification to pinpoint areas for full sequence analysis. The recent availability of a bacterial artificial chromosome (BAC) library has greatly expanded genome analysis in *fugu*. This library is currently being fingerprinted, and contigs are being constructed of target regions using a combination of sequence scanning and sequence-tagged site (STS) mapping. Gene order can be established within these contigs using STS mapping if there is sufficient clone coverage.

It has been amply demonstrated that many human genes including, for example, "disease genes" like dystrophin, whose mutation causes muscular dystrophy, have close relatives in pufferfish. These related genomic features can be detected computationally by comparing the two genomes and looking for similar sequences. The pufferfish and human genomes are similar by virtue of their shared vertebrate heritage. Of course, humans and pufferfish will have their own unique genes that are special for humanness and fishness that the genome comparisons will also bring to light. But even these fish and human-specific genes are likely to share a common genetic heritage.

Sequencing the pufferfish is better than another mammal like mouse or rat. The pufferfish genome provides a more distant evolutionary comparison (400 million years versus 100 million years for mouse and rat) that permits a more accurate triangulation of genome function than mouse or rat alone. Genomic features that are common to pufferfish and humans will focus attention on the essential core genes that define being a vertebrate.

Tetraodon and *F. rubripes* are considered to be important species to sequence because they are vertebrates and because their genomes are eight times as compact as the human genome, having many of the same genes and regulatory content as humans but with much less "junk" DNA. As a result, scientists say, finding genes and regulatory sequences in the pufferfish genome will be easier. This in turn will help researchers identify analogous genes and DNA regions in the human genome. *Tetraodon* is a freshwater pufferfish that is 20–30 million years distant from the *F. rubripes*, another type of pufferfish whose genome sequence was announced the same day by the department of Energy's Joint Genome Institute. Along with the genome sequences of other species, the sequences of the two pufferfish will provide key tools for gaining insights into the human genome, which will in turn translate into practical knowledge toward developing better therapies in the future.

The draft sequencing and assembly of the *fugu* (pufferfish) genome, announced last October, marked the first publicly released animal genome after the human sequence and the first vertebrate genome publicly sequenced and assembled using the whole-genome shotgun method. The *fugu* genome sequence, along with other information about the project, is available on the World Wide Web at http://www.jgi.doe.gov/fugu and http://www.fugubase.org.

Pufferfish is therefore proposed as a model organism with a reference genome, an archive of densely packed sequence information that will provide, together with data from other model organisms, a translation for the sequence of mammals. Not until the human genome is fully sequenced will the real power of comparative genomics be realized. The aim of this study is to describe how pufferfish genome has been characterized, its general properties, and how these properties may help in comparative genomics in number of different areas.

7.2 Phylogenetic Position

About 450–500 million years ago, the first vertebrates (animals with segmented backbones made of cartilage or bone) appeared in the early oceans. Their descendents split into two main groups: the ray-finned fishes that include pufferfish and most fish familiar to us from the dinner table and the lobe-finned fishes, a more obscure group with fleshy paddle-like appendages in place of the paper-thin fins of the ray-finned fishes. Over millions of years, these lobe-fins evolved into the limbs possessed by all four-limbed creatures (the tetrapods, including reptiles, amphibians, birds, and mammals). So pufferfish are our very distant cousins, sharing a common ancestor with us nearly half a billion years ago. Remarkably, this common ancestry is still recorded in our genes. Table 7.1 summarizes genome sizes of mammals and fishes.

The pufferfish are classified under the order Tetraodontidae, which is nested within the series Percomorpha. The phylogenetic relationship of pufferfish to other members of the order Tetraodontidae is not well established and neither is the relationship of Tetraodontidae to other percomorphs. While some shared morphological characters associate tetraodontids with surgeonfish (family Acanthuridae, order Perciformes), another set of characters suggests that the order Zeiformes, which includes fish such as dories and parazen, is the sister group. This ambiguity might be resolved by phylogenetic analyses of appropriate molecular data. The phylogenetic placement of pufferfish should help in tracing the evolutionary origin of pufferfish and humans and in understanding the mechanism underlying its genome compaction.

TABLE 7.1

Genome Sizes of Mammals and Fishes

	Haploid DNA Content (pg)	Haploid Genome Size (Mb)	No. of Chromosomes (n)
Mammals			
Human (*Homo sapiens*)	3.5	3000	23
Mouse (*Mus musculus*)	3.5	3000	20
Rat (*Rattus norvegicus*)	3.5	3000	21
Fish			
Zebra fish (*Danio rerio*)	1.8	1700	25
Medaka (*Oryzias latipes*)	1.1	1100	24
Pufferfish (*Fugu rubripes*)	0.4	400	22

7.3 Pufferfish Genome

The choice of pufferfish as a model was based on simple criteria. The ideal genome should be compact, with a similar gene repertoire to mammals. In 1968, Ralph Hinegardner assayed the DNA content of cells from a variety of different teleost fish. He found a wide range of haploid genome sizes with the smallest genomes belonging to the Tetraodontidae family, which includes the pufferfish and globefish. The figure of 0.4–0.5 pg for these fish compares with 3 pg in humans and therefore equates to 400 Mb. A detailed sequence analysis confirmed the previous estimates of genome size for this family of teleosts. About 600 sequences, generated randomly from the genome, were analyzed, and the amount of coding sequence with high database homology to mammalian genes was estimated. This represented 0.8% of the total sequence. By calculating the amount of mammalian coding sequence in the database (3×10^6 bp at that time), it was calculated that the same analysis of the human genome would yield 0.1% coding sequence ($3 \times 10^6 / 3 \times 10^9$).

Assuming the same amount of coding sequence in each genome, the genome size of pufferfish was calculated at 400 Mb, in close agreement with the earlier estimates. By single-copy probing of a genomic library of known insert size, a figure of 400 Mb was also reached. It was therefore confirmed that pufferfish has a genome 7.5 times smaller than humans but with the same amount of coding sequence. It was also possible to estimate the frequency as well as the nature of repetitive DNA in the genome. Database searches revealed no homology to intersperse, or minisatellite repeats but microsatellite DNA is abundant. The frequency of (CA)n repeats is quite high—one every 4 kb, which is 7.5 times more frequent than in humans, interestingly the same factor by which the genomes differ in size. Microsatellites comprise about 2% of the genome and clustered minisatellites about 5%. A further 1%–2% of the genome is made up of ribosomal RNA genes and a few very low frequencies of interspersed repeats giving a repetitive fraction of less than 10%. This paucity of repetitive DNA has a significant practical aspect in that both cross-species and pufferfish-specific hybridizations are extremely straightforward and sequencing projects are easy to assemble.

Brenner et al. (1993) also found that intron sizes were small and that intron positions were conserved with those known in mammals. In the human polycystic kidney disease 1 (PKD1) gene on chromosome 16, there is a 3.5 kb exon. In the pufferfish homologue of this gene, this exon is split by number of introns, suggesting that these introns have been lost in humans. Table 7.2 summarizes data from some of these genes. Although the average size of 80% of introns in pufferfish is less than 150 bp (Elgar et al., 1996), all genes sequenced to date in pufferfish have at least one large intron (i.e. greater than 400 bp) and it may be that there is some essential sequence within these larger introns that plays a part in splicing. However, there is no clear homology between introns from equivalent human and pufferfish genes or between different pufferfish genes. Splice junctions conform very closely to the consensus sequences for other vertebrates (Senapathi et al., 1990). Codon usage is biased, like many vertebrates, toward G or C in the third position, with 70% of codons ending in guanine or cytosine (unpublished data). This is particularly useful in the design of degenerate PCR primers for the amplification of fragments of pufferfish genes.

The coding sequence homology between pufferfish and mammalian genes varies greatly, not only between different genes but also within a single gene. For instance, the amino acid homology across exon 6 in the pufferfish p55 gene is greater than 90%, whereas in the first and second exon there is very little similarity at all (Elgar et al., 1996). This marked difference in similarity is probably accounted for by the fact that exon 6 codes for

TABLE 7.2

Data from a Selection of Genes Sequenced in Pufferfish Detailing Gene Size
and Number of Introns and, Where Known, the Comparative Human Data

Gene Name	Size in Human	Size in Pufferfish	Introns in Human	Introns in Pufferfish	References
TSC2	50	17	41	42	Maheshwar et al. (1996)
PKD1	50(36)	36	46	52	(a)
PAH	90(37)	8	12	12	(a)
DLST	17	2.2	12	12	Trower et al. (1996)
HD	170	23	67	67	Baxendale et al. (1995)
β-cyto_A	3.4	4	5	5	Venkatesh et al. (1993)
α-skel_A	2.8	3	6	6	Venkatesh et al. (1993)
Utrophin	~900(39)	40–60	–	>50	(b)
C9	~90(40)	3–5	10	10	(c)

Source: Elgar, G., *Hum. Mol. Gen.*, 5, 1437, 1996.
(a) R. Sandford, unpublished, (b) G. Elgar, unpublished, (c) G. Yeo, unpublished. *TSC2, tuberous sclerosis2; PKD1, autosomal polycystic kidney disease1; PAH, phenylalanine hydroxylase; DLST, dihydrolipoamide succinyltransferase; HD, Huntington's disease gene; V-tRNA-S, valyl-tRNA synthetase; β-cyto_A, beta-cytoplasmic actin; α-skel_A, alpha-skeletal actin; C9, complement component C9; –, data deficiency.

an SH3 domain, a tyrosine kinase motif, whereas exons 1 and 2 may code for a signal peptide that is not present in the mature polypeptide. Data are now emerging from pufferfish gene that allows definition of essential regions of genes by homology, providing a useful reference for mutation screening. It is here that the large evolutionary distance between pufferfish and mammals becomes important as it allows identification of critical domains in genes. There is little evidence of pseudogenes in pufferfish. Venkatesh et al. (1996) used degenerate PCR in order to amplify DNA fragments of the actin family. This approach, which would also amplify pseudogene fragments, of which there are a large number in humans, showed no evidence of any actin pseudogenes in the pufferfish genome.

In 2005, Wataru Kai et al. have constructed the male and female genetic maps for pufferfish (*fugu*) and arranged them in 22 linkage groups in which 94.3% of the markers are linked to at least one other marker, suggesting a high probability that the markers may be linked to genetic traits. Such a map would be useful for the initial low-resolution genetic mapping of traits. Analysis of the *fugu* (pufferfish) genetic map shows that males have an overall reduction in recombination ratio relative to females. While a large number of the common intervals between the male map and female map show expansion in the female map, some intervals also show expansion in the male map. Since the male-specific reduction in the recombination ratio seems to exist in diverse species of teleosts such as trout, zebra fish, and *fugu* (pufferfish), it is likely that this feature is shared by all teleosts.

This finding has practical significance in facilitating efficient experimental design in the genome-wide linkage analysis for species differences among *Takifugu* species (Singer et al., 2002). The decreased rate of recombination in males is an advantage for mapping genetic traits in initial low-resolution analyses, especially when analyzing quantitative trait loci (Glazier et al., 2002). For example, if the phenotype shows complete or partial dominance in F1 hybrids between two *Takifugu* species, it is adequate to use a mapping population produced from a cross between a male hybrid and a female wild species for primary mapping. On the other hand, it would be necessary to use a higher frequency of recombination in females for fine mapping of these loci. Comparisons of the genetic map of *fugu* (pufferfish) with that of other teleosts revealed that, as expected, the synteny between the two puff-erfishes, *Fugu* and *Tetraodon*, is highly conserved. However, there is a highly conserved synteny between *fugu* and medaka, which shared a common ancestor 95 million years ago.

Comparisons of the teleost maps and human and mouse maps showed that syntenic relationships are more conserved in the teleost fish lineage than in the mammalian lineage. For example, even though the divergence period between pufferfish and zebra fish is about threefold longer than that between the human and mouse, a similar degree of conserved synteny was found between the two fish lineages and between the two mammalian lineages. By comparing the complete physical map of human and the "partial physical map" of *Tetraodon*, Jaillon et al. (2004) reconstructed the karyotype of the last common ancestor of the teleost fish and mammals. Their model suggests a lower frequency of major interchromosomal exchanges in the teleost lineage than in the human lineage. Naruse et al. (2004) also inferred the ancestral karyotype using the medaka, zebra fish, and human maps. Their result suggested that karyotype evolution in teleosts occurred mainly by inversion and not by fusion or fission of chromosomes that is frequent in mammalian lineages. It shows the lower frequency of interchromosomal exchanges in the teleost lineage and provides further support to the models proposed by Jaillon et al. (2004) and Naruse et al. (2004). It also shows that the extent of conserved synteny between pufferfish and zebra fish is less than that between pufferfish and medaka. This result is consistent with the phylogenetic positions of the three fish species, that is, zebra fish being basal to both pufferfish and medaka. However, this can also have other explanations. The larger genome size of zebra fish compared to pufferfish and medaka (Venkatesh et al., 2000) implies that the genome of this teleost contains a significant proportion of transposable elements.

7.4 Puffer Genome versus Human Genome

The major goal of the human genome project must be to identify sequence and characterize and assign specific function to all the genes spread through our 3000 Mbp of haploid DNA. Due to the uniformity and simplicity of the DNA code, it is not an easy task to identify genes even after the region in which they lie has been fully sequenced. The average length of a human coding sequence in the DNA databases is approximately 1.2 kbp, and sensible estimates of the total number of genes in the human genome lie between 50 and 100,000. A gene number of 70,000 would give a total coding sequence of 85 megabases, less than 3% of our genome. Herein lies one of the major problems. A 3% return on investment, even when genes are identifiable, is rather poor, especially when sequencing is an expensive business. As if that is not enough, a large percentage of highly reiterated dispersed repeats serve to exacerbate the problem.

Consequently, other more direct approaches are being used, mostly to identify coding sequences within large genomic regions of DNA, and it is only by using a combination of these more elegant strategies that "gene hunters" are able to operate economically. Some of these methods compare human sequences with sequences from other organisms, using the premise that conserved sequences have some function. An extension of this, particularly among mammals but also with chicken, is to identify conserved linkage groups, and this may have particular value in positional cloning projects. Conserved linkage, or conserved synteny, can in fact be used to great advantage in comparative genomics, particularly if a genome is smaller and easier to work with than the human genome.

It was Sydney Brenner who realized the need for a simple genome approach to gene identification and characterization, and he also defined the criteria by which the model should be selected. He reasoned that all vertebrates would have a similar repertoire of genes, due to the way in which genomes have evolved. Thus, a vertebrate with a small genome should be the starting point for investigation. Vertebrates have a similar early developmental pattern and are physiologically similar. Sydney Brenner suggested that it was the way in which genes are regulated rather than different genes that gave rise to vertebrate diversity. It has been known since the late 1960s, when Ralph Hinegardner assayed the haploid DNA content of a large number of teleosts, that some species of fish (pufferfish) have very small genomes.

Pufferfish and humans have similar gene content, but the pufferfish genome is so much smaller than the human genome. In actual fact, only a few percent of the human genome actually represents "coding sequence," the functional parts of genes. The rest of the human sequence is dominated by highly repetitive nongene DNA—for example, regions that read "ACACACAC ..." for hundreds of bases or has longer sequences that are scattered throughout the human genome hundreds of thousands of times. While these repeats make up 40% of the human sequence, the pufferfish genome has much less repetitive content for mysterious reasons that should be illuminated by the genomic sequence now in hand. But it is not only the relative lack of repeats that makes pufferfish special, pufferfish genes themselves are also more compact than human genes and are packed more tightly on the genome.

7.5 *Fugu* Genome Project

The *fugu* genome project was initiated in 1989 in Cambridge by Sydney Brenner and his colleagues Greg Elgar, Sam Aparicio, and Byrappa Venkatesh. With the impending completion of the Human Genome Project, the value of the *fugu* genome for comparative purposes was noted, and in November 2000, the International *Fugu* Genome Consortium was formed, headed by the Joint Genome Institute, the Institute of Molecular and Cell Biology (IMCB) Singapore, and the Human Genome Mapping Project (HGMP) Cambridge. During the following year, the *fugu* genome was sequenced and assembled using the whole-genome shotgun method pioneered by Celera Genomics. *Fugu* is the first vertebrate genome to be publicly sequenced and assembled in this manner and is the first vertebrate genome made publicly available after the human genome. The consortium is currently proceeding to close and selectively finish the *fugu* sequence.

A "draft" sequence of the *fugu* genome was determined by the International *Fugu* Genome Consortium in 2002 using the "whole-genome shotgun" sequencing strategy. The results of the assembly (v2) are reported by Aparicio et al. 2002. *Fugu* is the second vertebrate genome to be sequenced, the first being the human genome. This webpage presents the annotation made on the fifth assembly (v5) by the IMCB team using the Ensembl annotation pipeline.

7.5.1 Assembly Statistics

Second *fugu* genome assembly was updated in May 2002 (v2). The v2 assembly includes 12,403 scaffolds, of length >2 kb, constituting 320 Mb of the 400 Mb genome. A preliminary analysis of the annotated genome is reported in *Science* 297:1301–1310 (2002). The v3 (third *fugu* genome assembly) assembly updated in August 2002 includes 8597 scaffolds, of length >2 kb, constituting 329 Mb of the 400 Mb genome. The v4 (fourth *fugu* genome assembly) assembly was updated in October 2004, based on ~8.7× coverage of the genome, and includes 7213 scaffolds, constituting 393 Mb of the genome. Ninety percent of the genome is represented on 1118 scaffolds. The largest scaffold is 7 Mb, and 74 scaffolds are larger than 1 Mb each. Fifth *fugu* genome assembly wasupdated in January 2010 and it generated by filling some gaps in the v4 assembly and by organizing scaffolds into chromosomes based on a genetic map of *fugu*. V5 assembly comprises 7119 scaffolds covering 393 Mb. About 72% of the assembly (282 Mb) is organized into 22 chromosomes. Another 14% of the assembly (56 Mb) is assigned to chromosomes, but the orientation and order of these scaffolds are not known (Chr_n_un). The remaining 14% of the assembly (55 Mb) is concatenated into a single sequence (Chr_un). Finally, they concluded that *fugu* genome has known protein-coding genes of 1,138, novel protein-coding genes of 18,093, RNA genes of 593, Genscan gene of 27,982, gene transcripts of 19,945, base pairs of 392,376,244, and golden path length of 392,800,674.

The initial analysis of the *fugu* sequence helps illuminate the human genome. By comparing the human and *fugu* sequences, common functional elements such as genes and regulatory sequences can be recognized as having been preserved in the two genomes over the course of the 450 million years since the species diverged from their common ancestor. By contrast, nonfunctional sequences are randomized over this long time period. Over 30,000 *fugu* genes have been identified in our analysis. The great majority of human genes have counterparts in *fugu*, and vice versa, with notable exceptions including genes of the immune system, metabolic regulation, and other physiological systems that differ in fish and mammals. Nearly, 1000 previously unrecognized human genes are predicted by comparing the two genomes. Remarkable stretches of sequence were found containing dozens of genes whose linkage is conserved between humans and *fugu*, shedding light on the large-scale evolutionary processes that shape genomes. The analysis of the *fugu* genome is ongoing and promises further insights into genome structure and evolution.

Many laboratories are now using *fugu* (pufferfish) as a comparative part of larger research projects. This has been made possible largely because of the availability and centralization of *fugu* resources at the UK HGMP resource center. Because of the common origin of all resources, the data generated may be archived through a central database. The *fugu* landmark mapping project is an attempt to coordinate and administer resources and to generate physical data on a global basis. To this end, cosmid clones are being "sample sequenced" whereby a small number of randomly generated one pass sequences are analyzed per cosmid. This low redundancy approach, both rapid and economical, provides about 50% coverage of each cosmid. Clones are then batch searched by cosmid against public DNA and amino acid databases and against each other with relevant database "hits" recorded.

All data from this project, including sequences and searches and other additional information, are available on the World Wide Web (http://fugu.hgmp.mrc.ac.uk; UK HGMP Resource Centre Fugu Group home page) (Elgar et al., 1999) in raw form. Other data are also collected (such as G+C content of each cosmid), which will be used in a series of whole-genome analyses. Analysis will include an assessment of physical linkage within each cosmid compared with data from other organisms such as zebra fish, chicken, and mammals. Ultimately, this not only will provide data for a number of individual regions but also will allow a statistical estimate of the number of breakpoints in the *fugu* genome compared with mammals. Plans are underway to integrate the physical map of *fugu* with the genetic map of the zebra fish.

7.6 Comparison of Pufferfish Genome with Human Genome

A BAC library has greatly expanded analysis in pufferfish. This library is currently being fingerprinted, and contigs are being constructed of target regions using a combination of sequence scanning and STS mapping.

The gene mapping and analysis of regions share synteny with human chromosomal regions 11p, 20q, and 6p21.3 (the major histocompatibility complex [MHC]). Each of these will demonstrate a particular strength of puffer genomics (Melody et al., 2001):

11p → Identification of conserved noncoding regions

20q → Conservation of synteny and potential genome duplications in fish

6p21.3 → Evolution of chromosomal regions

7.6.1 Comparison of the Human WAGR Region on Chromosome 11p13 and Syntenic Region on the Pufferfish Genome

The 2.2 Mb human region is covered by a series of BAC and cosmid clones. The WAGR region (for Wilms tumor, aniridia, genitourinary abnormalities, and retardation [mental] on human chromosome 11p13) contains the Wilms tumor 1 (WT1) and paired box 6 (PAX6) genes and is the site of a series of complex genetic disorders. There are also a number of expressed sequenced tags (ESTs) mapped to the region between cleavage stimulation factor 3 (CSTF3) and GA17 (Francke et al., 1979).

The 200 kb pufferfish region has the same genes in the same relative order and orientation. In addition, a number of other genes have been identified in pufferfish and mapped back to the same positions in the human contig (Miles et al., 1998). The 200 kb contig sequenced in pufferfish equates to 2.5 Mb of human chromosome 11p13, which includes the WAGR region, several upstream genes, and one gene downstream of PAX6. Within this human region are a number of well-characterized genes (such as WT1, PAX6, and CSTF3), well-defined ones with little known about function (such as GA17 and reticulocalbin 1 [RCN1]), and poorly defined genes, only identified from ESTS, partial transcripts, or predictions (such as TR2 and G2).

The pufferfish contig contains all these genes, in the same order and orientation, and by a combination of gene prediction and comparative sequence analysis, it was possible to define all the genes structurally (pufferfish and human genes tend to share the same intron/exon structure). It also has been possible to identify a number of "new" genes by comparing the human and pufferfish data and by coplotting gene prediction with BLAST hits against the pufferfish contig (Melody et al., 2001).

7.6.2 Comparison of Conservation of Synteny and Gene Duplication of Human Chromosome 20q with Pufferfish Genome

Human 20q is the second region that is being comparatively mapped in *fugu*. In humans, deletions of this region have been associated with hematological disorders, in particular myeloid malignancies. The extent of conservation of gene orders between *fugu* and human 20q is substantially less than that between *fugu* and WAGR region on human 11p13.

When a human topoisomerase 1 (*TOP1*) gene probe from human 20q11.2-12 was used to probe a *fugu* cosmid library, two *fugu* TOP 1 paralogues were identified. Each contig (approx. 300 kb) shares synteny with human 20q, but gene orders are not well conserved between these *fugu* regions and human 20q. At least two other single-copy human 20q genes (snail 1 [*SNAI1*] and kreisler [*KRML*]) are found in duplicate in *fugu*, one paralogue of each duplicated gene on each *fugu* contig. In humans, *SNAI1* is separated from these two genes by several megabases of DNA, although *KRML* is found close to *TOP* (Melody et al., 2001).

7.6.3 Comparison of Genome Evolution with Special Reference to the MHC Region on Human Chromosome 6p21.3 with Pufferfish Genome

The MHC region on human chromosome 6p21.3 has been completely sequenced (MHC Sequencing Consortium, 1999) and mapped in a diverse variety of organisms (Namikawa-Yamada et al., 1997; Chardon et al., 1999; Michalova et al., 2000). This is the third region that is being investigated in *fugu*.

The MHC is a 4 Mb region intimately associated with immune function in all vertebrates, containing genes essential to both the adaptive and innate (complement) immune systems. It is the most gene-dense region of the human genome identified so far, and out of the 224 identified gene loci, 128 are predicted to be expressed, with 40% of them having a proposed immune function. The region is divided into three sections (class II, class III, and class I) according to gene content. Specifically, the class I and class II regions have antigen-presenting molecules.

It was hypothesized that the particular organization of the mammalian MHC, with its relatively high concentration of immune-response-related genes and extremely high levels of genic polymorphism, has a functional significance. Certainly, there are particular gene combinations in this human region, such as tubulin (TUBB), tenascin XB (TNXB), pre-B-cell leukemia transcription factor 2 (PBX2), NOTCH4, retinoic acid receptor b (RXRB), and ribosomal protein S18 (RPS18), which are linked in *Drosophila melanogaster* and *Caenorhabditis elegans* (Trachtulec et al., 1997). While the gene complement of the MHC remains relatively stable, rearrangements have occurred in mammals (Chardon et al., 1999; Lewin et al., 1999). In *Xenopus laevis*, genes from the mammalian MHC remain linked, but the precise arrangement comprises class II, class I, and class III, as opposed to class II, class III, and class I. In teleosts, the class I, class II, and class III genes are not linked (Sato et al., 2000) and are thought to represent a derived arrangement of the MHC.

7.7 Tracking the Human Genes Using Pufferfish Genome

Recently, two whole-genome approaches have been taken to identify highly conserved sequences in vertebrate genomes: one using identical mouse, human alignments over at least 200 bp (Bejerano et al., 2004), and the other using alignments between the human and *fugu* genomes. The human–mouse study did not mask coding sequence and, as a result, identifies a mixture of coding and noncoding sequences, some of which associate with transcription factors. The *fugu*–human study is less stringent, as is appropriate with much more evolutionarily divergent vertebrates (BLAST alignments of at least 100 bp in length, including an exact match of 20 bp or more) but masks exons. As a result, only noncoding sequences are identified, almost all of which are associated, at least in terms of genomic location, with genes implicated in developmental regulation.

With a significant number of conserved regions identified between the human and *fugu* genomes (nearly 1400), it is possible to start looking for signatures within the sequences that may help elucidate their function or allow their subgrouping into possible functional units. Yet, remarkably, despite their high degree of conservation throughout vertebrates, these sequences bear little resemblance to each other within a species, and there are no clear patterns within them that allow how regulatory circuits, hardwired into the genomes of vertebrates, control the complex and intricate process of early gene regulation. Nevertheless, the identification of at least some of the sequences involved provides biologists with an invaluable resource, which will greatly accelerate our understanding of development (Elgar, 2004).

7.7.1 Hox Genes

One of the most intensely studied gene families in development are the Hox genes. In mammals, there are four paralogous clusters, each containing a different set of between 9 and 11 genes, co-coordinately regulated both temporally and spatially (Ferrier and Minguillon, 2003). It is, therefore, not surprising that the first example of conserved noncoding sequence between *fugu* and mammals with demonstrated function was found in the Hoxb-1 gene (Marshall et al., 1994). The sequences identified not only were shown to be sensitive to retinoic acid but were also able to recapitulate Hoxb-1 neuroectoderm expression in developing mouse embryos. This work was applied more widely to the Hoxb-4 gene (Aparicio, 1995) where an elegant combination of comparative genomic analysis and transgenic testing in mouse embryos identified further regulatory regions. More recently, Spitz et al. (2003) have made a comprehensive study of the Hoxd cluster region, using targeted enhancer-trap assays to identify a global control region capable of driving the expression of a number of unrelated genes in a tissue-specific manner. After identifying a 40 kb region of high sequence conservation between mouse and human, comparison with pufferfish allowed much smaller core elements to be identified. Remarkably, a much smaller pufferfish sequence was able to reproduce the central nervous system expression in mouse embryos, but not the digit expression that was observed using the 40 kb fragment from mouse. One of the clear benefits from using *fugu* sequence in this analysis is that the signal to noise ratio is much higher, due to the large evolutionary distance between *fugu* and mammals.

7.7.2 PAX6 Gene and Its Neighbors

The PAX6 gene is a transcription regulator, which plays a key role in the development of the eye, brain, and pancreas. Miles et al. (1998) isolated a single 45 kb *fugu* cosmid that contained both the WT1 and PAX6 genes, a section of human chromosome 11p13 known as the WAGR region. By contrast, the two genes are 700 kb apart in the human genome. Nevertheless, gene order and orientation were found to be conserved between the two species, with a third gene, RCN, found between WT1 and PAX6. Early sequencing of the human region allowed genomic comparison with the *fugu* cosmid leading to the identification of a number of conserved noncoding sequences, both 5_ and 3_ to the PAX6 gene, as well as in PAX6 introns (most notably intron 7). A number of these sequences have now been tested functionally and appear to regulate the PAX6 gene in a number of different ways (Kammandel et al., 1999; Griffin et al., 2003; Kleinjan et al., 2004). Interestingly, the 3_ sequences are located in the last intron of another gene (called the "PAXNEB gene") leading to speculation that the location of regulatory sequences in neighboring genes may have consequences for synteny between genomes (Kleinjan et al., 2002).

7.7.3 Wnt1

Sequence comparison between *fugu* and mammalian genomic DNA has identified conserved noncoding sequence on a number of other occasions, and in each case, the genes under study are implicated in developmental control. The Wnt1 gene is an essential component of midbrain–hindbrain patterning. Rowitch et al. (1998) identified a 110 bp downstream enhancer sequence in the murine Wnt1 gene that was 69% identical to a 102 bp region in *fugu*. Interestingly, the *fugu* element was in the opposite orientation to the mouse sequence but was still able to direct expression within the midbrain at neural plate stages (representing correct activation of Wnt1). However, after neural tube closure, the *fugu* sequence was unable to recapitulate the complete profile of the mouse enhancer suggesting that the sequence differences between the two enhancers had functional consequences. From this type of analysis, it is possible to putatively assign different functional roles to different parts of an enhancer element.

7.7.4 Dlx Bigene Clusters

The Dlx bigene clusters are organized as three sets of two genes, with each pair in a tail to tail orientation, a few kilobases apart (Stock et al., 1996). Ghanem et al. (2003) used phylogenetic foot printing with human, mouse, *fugu*, *Sphoeroides* (another pufferfish), and zebra fish DNA to identify highly conserved noncoding elements within the *Dlx1/Dlx2* and *Dlx5/Dlx6* clusters. Two regions were identified in each cluster where sequence was >75% conserved across all five species. These sequences demonstrate enhancer activity in mouse and zebra fish embryos, three in the forebrain and one chiefly in the branchial arches. Despite the overlapping patterns of expression shared between the first three elements, they show little similarity to each other. In addition, four other regions were identified, which were conserved only between the three fish species. This study really highlights the power of multispecies alignment in defining functionally relevant sequences. Alignment between mammalian sequences alone generates too large a number of conserved sequences to consider functional analysis. The use of more distantly related organisms allows a more targeted strategy to be undertaken.

7.7.5 Voltage-Gated Calcium Channel α1-Subunit Genes

Calcium influx through voltage-gated calcium channels (VGCCs) mediates a myriad of cytoplasmic responses, such as neurotransmitter release, excitation–contraction coupling, and secretion and regulation of gene expression. The VGCC α1-subunit gene family in *fugu* shares many similarities with that of human. Studies show that *fugu*, despite being a basal vertebrate, parallels human in terms of heterogeneity in VGCCs. Remarkably, orthologs for all the 10 human VGCC α1-subunit genes responsible for generating different Ca^{2+} currents were found in *fugu*. In addition, multiple copies of the various α1 subtypes were retrieved from the *fugu* genome except for α1C, α1B, and α1I subtypes in which only a single copy of each was found. In total, fugu has 21 copies of α1-subunit genes, 11 more than that in human. Conceivably, the extra copies of α1-subunit genes may contribute toward the diversity of VGCCs in *fugu*, leading to a more extensive VGCC heterogeneity in *fugu* compared to human (Wong et al., 2006).

7.7.6 Recombinase-Activating Gene 1 and 2 Locus

Recombinase-activating gene 1 and 2 (RAG1 and RAG2) are two endonucleases that catalyze the site-specific V(D)J recombination process that occurs during the development of T- and B-lymphocytes. V(D)J recombination assembles the variable (V), diversity (D), and joining (J) germ line segments into a functional antigen receptor gene-encoding immunoglobulins and the T cell receptor. Mice lacking either *RAG1* or *RAG2* are unable to initiate V(D)J recombination and have no mature B- or T-lymphocytes (Lewis, 1994). In the case of B-lymphocytes, both genes are expressed primarily on those immature cells in the bone marrow that undergo receptor editing. However, a continued *RAG* expression has been detected in immature bone marrow and splenic B cells that have already acquired low levels of surface immunoglobulin M (Peixoto et al., 2000). The periodic expression of these two genes indicates that this locus is tightly regulated, and unraveling conserved *cis*-acting elements would shed light on the understanding of the role of *RAG* genes on the development of lymphocytes.

RAG1 and *RAG2* have been studied in many different vertebrate organisms (Oettinger et al., 1990; Ichihara et al., 1992; Miller and Rosenberg, 1997; Willett et al., 1997a). They are closely linked and convergently transcribed. The size of the intergenic region between both genes varies, but the conserved genomic structure indicates that a fraction of the *cis*-acting elements may have been maintained throughout vertebrate evolution. The pufferfish has a haploid genome of 400 Mb, which is one-eighth of that of man (Brenner et al., 1993). With the same number of genes, less intergenic regions, and clustered repetitive DNA, the puffer genome has ideal properties for human genome studies (Peixoto et al., 2000).

7.7.7 Survival Motor Neuron Genes

Spinal muscular atrophy (SMA) is an autosomal recessive disease characterized by degeneration of motor neurons in spinal cord and brain, causing progressive muscle weakness and atrophy that is severely disabling and potentially fatal. It has an incidence of 1 in 6000 live births and a carrier frequency of ~1 in 35 people (Pearn, 1978). The disease is caused by mutations in the telomeric copy (survival motor neuron 1 [SMN1]) of the SMN gene and not in the centromeric copy (SMN2). Both the human SMN (huSMN) genes are located in chromosome 5q11.2-q13.3 as inversely duplicated, almost identical copies (Brzustowicz et al., 1990; Lefebvre et al., 1995). There is only one SMN1 copy, whereas from 0 to 4 copies of SMN2 per chromosome are present (Feldkotter et al., 2002). In contrast, SMN2 differs from SMN1 by 11 nucleotide changes (Lefebvre et al., 1995; Monani et al., 1999). Thus, upregulation of the SMN2 full-length transcript will be a potential therapeutic strategy in SMA. Indeed, clinical trials in SMA are currently underway using drugs to upregulate SMN protein expression. With this strategy in mind, identification of the functional elements of SMN may permit the identification of agents mediating more specific upregulation of the gene.

The identification of functional upregulating elements in the large ~36 kb of huSMN will be difficult. One solution will be to apply comparative functional genome analysis of a significantly smaller SMN homologue of a vertebrate with close structural and functional similarities to huSMN. The pufferfish (*fugu*) genome has a single fsmn that is 13.4 times smaller than huSMN. fsmn and huSMN are highly similar in their genome organization and tissue expression patterns. The functional domains of the pufferfish (*fugu*) SMN and huSMN molecules are also highly conserved. In human MCF-7 cells, expression of fsmn protein resulted in the formation of "gems" in the cytoplasm and nucleus, similar to observations reported for huSMN protein. In these cells, fsmn RNA was also processed correctly and produced alternatively spliced transcripts like huSMN2. It indicates close structural and functional similarities between fsmn and huSMN, suggesting that regulation of the two genes may also be similar and supporting the use of fsmn in comparative genome studies for the identification of functional regulatory elements of huSMN (Paramasivam et al., 2008).

7.7.8 *SOX9* Gene

Regulated gene expression depends to a large extent on the stage- and tissue-specific interactions of transcription factors with regulatory DNA sequences such as enhancers and silencers. These regulatory sequences are usually located in the vicinity of a gene, frequently within a few hundred base pairs of the transcription start site. Occasionally, however, they are found considerable distances from the gene. Examples of such long-range regulatory elements are suspected to exist but have not been identified in the human gene *SOX9*.

SOX9 is a small gene of 5.4 kb located at chromosome 17q24.3-q25.1 that encodes a transcription factor of the SOX family of proteins (Bowles et al., 2000). Heterozygous mutations in and around *SOX9* have been shown to cause the semilethal skeletal malformation syndrome campomelic dysplasia (CD) with associated XY sex reversal (Foster, 1994). Consistent with the defects in skeletal and gonad development in CD patients, *SOX9* has been shown to be a factor essential for chondrocyte differentiation and to be expressed at the stage of testis determination and early testis differentiation in many vertebrate classes (Koopman, 2001). Most CD patients have loss-of-function mutations in the *SOX9* coding region, which indicates that haploinsufficiency for *SOX9* causes CD. A subgroup of CD patients has two intact *SOX9* alleles but shows a chromosomal rearrangement in the vicinity of *SOX9*.

One approach to identify potential regulatory elements is to search for evolutionarily conserved sequences by comparing genomic sequences among species. Such comparative analyses between human and the Japanese pufferfish *F. rubripes* (Elgar et al., 1996). The 430 million years of evolutionary distance between *F. rubripes* and mammals, combined with the compact *F. rubripes* genome, greatly facilitates identification of conserved regulatory sequences, as only the most essential sequence elements required for functions common to all vertebrates are spared from sequence divergence over time. The validity of this approach has been documented in a number of studies (Aparicio, 1995).

7.7.9 AP1 Genes

The activator protein 1 (AP1) transcription factor is known to regulate cell proliferation and differentiation, immune response, and cell death. The prototypical AP1 transcription factor is composed of a Jun and Fos proteins. In mammals, there are three Jun family members (c-Jun, JunB, and JunD) and four Fos family members (c-Fos, FosB, Fra1, and Fra2). The Fos proteins do not form homodimers, but the Jun proteins are capable of forming both homodimers and heterodimers among themselves (Chinenov and Kerppola, 2001). Both the Fos and Jun proteins will dimerize with other transcription factors, for example, the ATF/CREB (Hai and Curran, 1991) or MAFR/NRL (Kerppola and Curran, 1994) protein families. Thus, the repertoire of AP1 dimers is complex, is determined by a variety of extracellular stimuli, and varies among different cell types. This diversity, in mammals, has complicated our understanding of Jun and Fos functions and has resulted in a lack of information about the role of individual AP1 members in cellular processes. The complete sequence of pufferfish (*fugu*) offers a unique opportunity to elucidate the individual roles of the AP1 transcription factor genes, possibly without the complexities associated with this endeavor in mammals (Cottage et al., 2003).

7.8 Computational Tools

Among the most important tools in comparative molecular work are alignment algorithms. An alignment of more than two sequences might subsequently be subjected to phylogenetic analysis, thus visualizing the evolutionary scenario implied by the alignment. The use of bioinformatics tools and applications, coupled with the public availability of whole-genome sequence from a growing number of organisms, massively increases the scope for comparative analyses. Table 7.3 contains information about computer programs for comparing two genomes.

TABLE 7.3

Computer Programs for Comparing Two Genomes

Program	Purpose	References	Website
AGenDa	Gene prediction	Taher et al. (2003)	http://bibiserv.techfak.uni-bielefeld.de/agenda/
ATV	View phylogenies	Zmaskek and Eddy (2001)	http://www.genetics.wustl.edu/eddy/atv/
AVID/MAVID	Genome alignment	Bray et al. (2003)	http://baboon.math.berkeley.edu/mavid/
BLASTZ	Genome alignment	Schwartz et al. (2003)	http://www.bx.psu.edu/miller_lab/
BLAST	Database search	Altschul et al. (1990)	http://www.ncbi.nlm.nih.gov/BLAST/
BLAT	Database search	Kent (2002)	http://www.genomeblat.com/genomeblat/
Clustal W	Multiple sequence alignment	Thompson et al. (1994)	http://bioweb.pasteur.fr/seqanal/interfaces/clustalw.html
Coalator	Drawing coalescent trees	Unpublished	http://adenine.biz.fh-weihenstephan.de/
drawStrees	Drawing suffix trees	Unpublished	http://adenine.biz.fh-weihenstephan.de/
DoubleScan	Gene prediction	Meyer and Durbin (2002)	http://www.sanger.ac.uk/Software/analysis/doublescan/
FASTA	Database search	Pearson and Lipman (1988)	http://fasta.bioch.virginia.edu/
Genscan	Gene prediction	Burge and Karlin (1997)	http://genes.mit.edu/GENSCAN.html
gff2aplot	Plotting sequence comparisons	Abril et al. (2004)	http://genome.imim.es/software/
ms	Coalescent simulations	Hudson (2002)	http://home.uchicago.edu/~rhudson1/source.html
MUMmer	Genome alignment	Delcher et al. (1999)	http://www.tigr.org/software/
PHYLIP	Phylogeny reconstruction	Felsenstein (1993)	http://evolution.genetics.washington.edu/phylip.html
PPH	Haplotype reconstruction	Chung and Gusfield (2003)	http://wwwcsif.cs.ucdavis.edu/~gusfield/pph.html
SGP-2	Gene prediction	Parra et al. (2003)	http://www1.imim.es/software/sgp2/
SLAM	Gene prediction	Pachter et al. (2001)	http://bio.math.berkeley.edu/slam/
TAP	Transcript assembly	Kan et al. (2001)	—
TwinScan	Gene prediction	Korf et al. (2001)	http://genes.cs.wustl.edu/

7.9 Future of Pufferfish Genomics

Fish, as a highly diversified group of the vertebrate family, experience an astonishing range of environmental conditions to which their physiologies, body shapes, and lifestyles have adapted. However, a common denominator of all fish species is their aquatic habitat, meaning that water is in direct contact with several tissues and internal compartments of the animal, potentially inducing a high sensitivity to waterborne parameters such as temperature, oxygen levels, salinity, and sometimes toxic chemicals. This intimate relationship between an organism and a wide range of different environments has recently prompted the view that fish could be used as models for "environmental genomics" (Cossins and Crawford, 2005), in other words the study of the interface between an organism and its environment using genomic approaches.

In addition to counting several member species elevated to the status of genomic models such as zebra fish or pufferfish, fish also represent a major source of food for humans. The European Union thus also recognized the need for improving aquaculture research by funding the AQUAFIRST consortium to identify genes associated with stress and disease resistance in sea bream, sea bass, and rainbow trout in order to provide a physiological and genetic basis for marker-assisted selective breeding. Genomic techniques and sequence comparisons with genomic models are cornerstones of this project, which illustrates how fish genomics may grow in coming years outside of fundamental research labs toward more applied objectives nonetheless essential to human welfare.

Today, a great deal of attention is focused on the part of eukaryote genomes that do not code for proteins but are nevertheless functional. Elements that may carry out specific functions in these regions include noncoding RNA genes and regulatory control regions. One of the most powerful techniques to identify such elements is to align genomic sequences of distantly related organisms and look for regions that have remained similar during evolution, thus suggesting that a functional constraint is acting to preserve the sequence from mutations. The advantage of fish in this context is the long evolutionary distance, approximately 450 million years, since their last ancestor with mammals. Neutral mutations have since saturated the genome to a point where any conserved region between, for instance, human and pufferfish is indicative of a functional constraint. This comparative approach was first applied on a genome scale to identify coding exons (Roest-Crollius et al., 2000) and more recently to identify ultraconserved regions (UCRs) of unknown function (Sandelin et al., 2004; Woolfe et al., 2004). However, an additional assumption that can be deduced from the discovery of UCRs conserved between such distant organisms is their fundamental importance across vertebrates. In line with this, UCRs found in this way lie in clusters around genes involved in the regulation of development.

Several advantages in studying the genome of the pufferfish are that it is much faster to get from one end of a pufferfish gene to the other end and from one gene to the next when determining DNA sequence on continuous stretches of chromosomes. This is because the pufferfish genome is only about an eighth of the size of the human genome—400 million DNA bases. But the pufferfish is not deficient in its total number of genes. Rather, the pufferfish genome contains less of what seems to be irrelevant DNA, sometimes called "junk." This junk DNA separates genes from one another like the space that separates words in a sentence. It also breaks genes into sections like syllables.

The human genome is diluted with so much junk DNA that genes are contained in only three percent of it—compared to fifteen percent in the pufferfish.

According to Elgar and Brenner, the sequence of the pufferfish genome will be more useful in filling any gaps that may remain in the human genome. Elgar believes that "one way to bridge those gaps, if you have conserved regions, is simply to look at them in other organisms" like the pufferfish. Compared to organisms that are closer to humans on the evolutionary scale, such as laboratory mice and primates, the pufferfish genome has the advantage of a compact genome with less room for junk sequences. In addition, the pufferfish is distant enough from humans, compared to mammals, for the junk sequences to become random. When comparing human and pufferfish genomes, "if you find a conserved homology block even quite small conserved blocks they seem to have some significance," says Elgar. It is easier to recognize an important sequence among oceans of irrelevant DNA.

In 1995, Elgar and his colleagues identified and sequenced the pufferfish counterpart of the human Huntington's disease gene, which had already been sequenced. The pufferfish gene turned out to be only 23,000 DNA bases long seven and a half times shorter than the human gene. Although the pufferfish gene has the same 67 interruptions, they are rarely over 1,000 DNA bases compared to interruptions as long as 12,500 DNA bases in the human gene for Huntington's. The actual gene, however, is very similar to the human gene and provides no further information about the protein.

Later Mike Trower in collaboration with Brenner identified and sequenced one of the Alzheimer's disease genes in both human and pufferfish genomes. They knew which human chromosome the gene was on and what other sequenced genes were in the vicinity. The usual way to get to a gene of interest, when the only available information is about the neighboring genes, is to start at the DNA location of one of the neighbors and "walk" away until the telltale signs of the next gene appear. This may be a long hike with the human genome because of the entire junk DNA. Trower devised a more efficient approach. He assumed that genes are in the same order on the human and pufferfish chromosomes. This is a phenomenon known as conserved synteny and is seen between the genomes of more closely related organisms, such as human and mouse. If this assumption was true, Trower would reach the Alzheimer's disease gene much faster by following the lead to the human gene but sequencing the more compact pufferfish genome.

In the near future, it is hoped that *in situ* hybridization with pufferfish chromosomes will be available so that physical mapping of clones may be carried out on a different scale. Other resources are also planned, in particular a BAC library and possibly a radiation hybrid mapping panel. Many of the cosmids in the landmark mapping project are being built up into contigs so that longer-range physical mapping by sequence scanning will be undertaken. As an extension to this, some regions may be completely sequenced, where full genomic sequence is available from the human genome so that a systematic base by base comparison of sequences may be made, so pinpointing any conserved regions not identified by other methods. The pufferfish genome is small and gene dense. It contains the basic vertebrate DNA blueprint and should be a reference genome for many different forms of comparative genomics: from positional cloning projects to the identification of regulatory motifs and from gene characterization to genome structure. For a modest outlay, the evolution of complex genomes may be better understood.

7.10 Summary

Puffer fish is a teleost fish belonging to the order Tetraodontiformes and a member of the gnathostomes. The puffer fish Genome Project is an international program aimed at determining the complete DNA sequence of the genome of the pufferfish, Fugu rubripes. Despite the obvious differences between fish and humans, it is expected that comparisons of the human genome with that of fish will shed light on the common genetic systems shared by these two animals, and help us understand the information encoded in the human genome. Comparing and contrasting them will allow us to discover new human genes and importantly, elements which control or regulate the activity of genes. Sequence of pufferfish is better rather than another mammal like mouse and rat, because it is essential that a broad range of animal genomes be sequenced, to shed light on the underlying similarities and essential differences between species. Pufferfish is just the beginning. The ongoing mouse and rat genome projects are critical for biomedicine and will be particularly powerful tools because these animals are mammals (more closely related to humans) and can be bred and studied more easily than pufferfish. The pufferfish genome provides a more distant evolutionary comparison (400 million years, versus 100 million years for mouse and rat) that permits a more accurate triangulation of genome function than mouse or rat alone. Genomic features that are common to pufferfish, rodents, and humans will focus on the essential core genes that define being a vertebrate.

Although the recently completed draft sequence of the human genome offers an initial look at the human gene content, to fully unravel the important information from the human genome, scientists will have to compare it to the genome sequences of many other species. That is because evolution preserves the most important genetic information across species; if genes and regulatory elements have survived hundreds of millions of years of evolution, they would be functionally important. But other genes may not have survived evolution because they may no longer be important for survival in the new environment. So researchers need comparative sequences that are both closely and distantly related to humans, because different genetic elements in humans would call for comparison with different species.

References

Abril, J. F., R. Guig, and T. Wiehe. 2004. gff2aplot: Plotting sequence comparisons. *Bioinformatics* 19: 2477–2479.

Altschul, S. F., W. Gish, W. Miller, E. W. Myers, and D. J. Lipman. 1990. Basic local alignment search tool. *Journal of Molecular Biology* 215: 403–410.

Aparicio, S. 1995. Detecting conserved regulatory elements with the model genome of the Japanese puffer fish, *Fugu rubripes. Proceedings of National Academy of Sciences of the United States of America* 92: 1684–1688.

Aparicio, S., J. Chapman, E. Stupka, N. Putnam, J. M. Chia, P. Dehal, A. Christoffels, S. Rash, S. Hoon, and A. Smit. 2002. Whole-genome shotgun assembly and analysis of the genome of *Fugu rubripes. Science* 297: 1301–1310.

Baxendale, S., S. Abdulla, G. Elgar, D. Buck, M. Berks, G. Micklem, R. Durbin et al. 1995. Comparative sequence analysis of the human and puffer fish Huntington's disease genes. *Nature of Genetics* 10: 67–76.

Bejerano, G., M. Pheasant, I. Makunin, S. Stephen, and W. J. Kent. 2004. Ultra conserved elements in the human genome. *Science* 304: 1321–1325.

Bowles, J., G. Schepers, and P. Koopman. 2000. Phylogeny of the SOX family of developmental transcription factors based on sequence and structural indicators. *Developmental Biology* 227: 239–255.

Bray, N., I. Dubchak, and L. Pachter. 2003. AVID: A global alignment program. *Genome Research* 13: 97–102.

Brenner, S., G. Elgar, R. Sandford, A. Macrae, B. Venkatesh, and S. Aparicio. 1993. Characterization of the pufferfish (*Fugu*) genome as a compact model vertebrate genome. *Nature* 366: 265–268.

Brzustowicz, L. M., T. Lehner, L. H. Castilla, G. K. Penchaszadeh, K. C. Wilhelmsen, R. Daniels, K. E. Davies, M. Leppert, F. Ziter, and D. Wood. 1990. Genetic mapping of chronic childhood-onset spinal muscular atrophy to chromosome 5q11.2-13.3. *Nature* 344: 540–541.

Burge, C. and S. Karlin. 1997. Prediction of complete gene structures in human genomic DNA. *Journal of Molecular Biology* 268: 78–94.

Chardon, P., C. Renard, and M. Vaiman. 1999. The major histocompatibility Complex in swine. *Immunological Reviews* 167: 179–192.

Chew, S. K., C. H. Goh, K. W. Wang, P. K. Mah, and B.Y. Tan. 1983. Pufferfish (Tetrodotoxin) poisoning: Clinical report and role of anti-cholinesterase drugs in therapy. *Singapore Medical Journal* 24: 168–171.

Chinenov, Y. and T. K. Kerppola. 2001. Close encounters of many kinds: Fos-Jun interactions that mediate transcription regulatory specificity. *Oncogene* 20: 2438–2452.

Chung, H. R. and G. Gusfield. 2003. Perfect phylogeny haplotyper: Haplotype inferral using a tree model. *Bioinformatics* 19: 780–781.

Clark, M. S., S. F. Smith, and G. Elgar. 2001. Use of the Japanese Pufferfish (*Fugu rubripes*) in comparative genomics. *Marine Biotechnology* 3: 130–140.

Cossins, A. R. and D. L. Crawford. 2005. Fish as models for environmental genomics. *Nature Reviews Genetics* 6(4): 324–333.

Cottage, A. J., Y. J. K. Edwards, and G. Elgar. 2003. AP1 genes in *Fugu* indicate a divergent transcriptional control to that of mammals. *Mammalian Genome* 14: 514–525.

Delcher, A. L., S. Kasti, R. D. Fleischmann, J. Peterson, W. White, and S. L. Salzberg. 1999. Alignment of whole genomes. *Nucleic Acids Research* 27: 2369–2376.

Edwards, Y. J. K., G. Elgar, M. S. Clark, and M. J. Bishop. 1998. The identification and characterisation of microsatellites in the compact genome of the Japanese pufferfish, *Fugu rubripes*: Perspectives in functional and comparative genomic analyses. *Journal of Molecular Biology* 278: 843–854.

Elgar, G. 1996. Quality not quantity: The puffer fish genome. *Human Molecular Genetics* 5: 1437–1442.

Elgar, G. 2004. Identification and analysis of *cis*-regulatory elements in development using comparative genomics with the pufferfish, *Fugu rubripes. Seminars in Cell & Developmental Biology* 15: 715–719.

Elgar, G., M. S. Clark, S. E. Meek, S. Smith, S. Warner, Y. J. K. Edwards, N. Bouchireb et al. 1999. Generation and analysis of 25 mega bases of genomic DNA from the pufferfish (*Fugu rubripes*) by sequence scanning. *Genome Research* 9: 960–971.

Elgar G., R. Skndford, S. Aparicio, A. Macrae, B. Venkatesh, and S. Brenne. 1996 Small is beautiful: Comparative genomics with the puffer fish (*Fugu rubripes*). *Trends in Genetics* 12: 145–149.

Feldkotter, M., V. Schwarzer, R. Wirth, T. F. Wienker, and B. Wirth. 2002. Quantitative analyses of SMN1 and SMN2 based on real-time lightCycler PCR: Fast and highly reliable carrier testing and prediction of severity of spinal muscular atrophy. *American Journal of Human Genetics* 70: 358–368.

Felsenstein, J. 1993. PHYLIP (phylogeny inference package) version 3.5. Department of Genetics, University of Washington, Seattle, WA.

Ferrier, D. E. and C. Minguillon. 2003. Evolution of the Hox/ParaHox gene clusters. *The International Journal of Developmental Biology* 47: 605–611.

Foster, J. W. 1994. Campomelic dysplasia and autosomal sex reversal caused by mutations in an *SRY*-related gene. *Nature* 372: 525–530.

Francke, U., L. B. Holmes, L. Atkins, and V. M. Riccardi. 1979. Aniridia-Wilms' tumor association: Evidence for specific deletion of 11p13. *Cytogenetics and Cell Genetics* 24: 185–192.

Ghanem, N., O. Jarinova, A. Amores, Q. Long, and G. Hatch. 2003. Regulatory roles of conserved intergenic domains in vertebrate Dlx Bigene clusters. *Genome Research* 13: 533–543.

Glazier, A. M., J. H. Nadeau, and T. J. Aitman. 2002 Finding genes that underlie complex traits. *Science* 298: 2345–2349.

Griffin, C., D. A. Kleinjan, B. Doe, and V. Van Heyningen. 2003. New 3_ elements control Pax6 expression in the developing pretectum, neural retina and olfactory region. *Mechanisms of Development* 112: 89–100.

Hai, T. and T. Curran. 1991. Cross-family dimerization of transcription factors Fos/Jun and ATF/CREB alters DNA binding specificity. *Proceedings of National Academy of Sciences of the United States of America* 88: 3720–3724.

Hamilton-Buchanan, F. 1822. *An Account of the Fishes of River Ganges and its Branches.* London, U.K.: George Ramsay and Co., 405pp.

Ichihara, Y., M. Hirai, and Y. Kurosawa. 1992. Sequence and chromosome assignment to 11p13-p12 of human RAG genes. *Immunology Letters* 33(3): 277–284.

Hinegardner, R. 1968. Evolution of cellular DNA content in teleost fishes. *American Midland Naturalist Journal* 16: 517–523.

Hudson, R. R. 2002. Generating samples under a Wright-Fisher neutral model of genetic variations. *Bioinformatics* 18: 337–338.

Jaillon, O., J. M. Aury, F. Brunet, J. L. Petit, N. Stange-Thomann, E. Mauceli, L. Bouneau et al. 2004. Genome duplication in the teleost fish *Tetraodon nigroviridis* reveals the early vertebrate proto-karyotype. *Nature* 431: 946–957.

Kammandel, B., K. Chowdhury, A. Stoykova, S. Aparicio, and S. Brenner. 1999. Distinct *cis*-essential modules direct the time-space pattern of the Pax6 gene activity. *Developmental Biology* 205: 79–97.

Kan, Z., E. Rouchka, W. Gish, and D. States. 2001. Gene structure prediction and alternative splicing analysis using genomically aligned ESTs. *Genome Research* 11: 889–900.

Kathirvel, P., W.-P. Yu, B. Venkatesh, C.-C. Lim, P.-S. Lai, and W.-C. Yee. 2008. *Fugu rubripes* and human survival motor neuron genes: Structural and functional similarities in comparative genome studies. *Gene* 424: 108–114.

Keiichi, M. and Tyler, C. J. 1998. Encyclopedia of fishes in: Paxton, J. R. and Eschmeyer, W. N. (eds.). *Encyclopedia of Fishes.* San Diego, CA: Academic Press, pp. 230–231.

Kent, W. J. 2002. BLAT—The BLAST-like alignment tool. *Genome Research* 12: 656–664.

Kerppola, T. K. and T. Curran. 1994. A conserved region adjacent to the basic domain is required for recognition of an extended DNA binding site by Maf/Nrl family proteins. *Oncogene* 9: 3149–3158.

Khora, S. S., K. K. Panda, and B. B. Panda. 1997. Genotoxicity of tetrodotoxin from pufferfish tested in root meristem cells of *Allium cepa* L. *Mutagenesis* 4(12): 265–269.

Kleinjan, D. A., A. Seawright, A. J. Childs, and V. Van Heyningen. 2004. Conserved elements in Pax6 intron 7 involved in (auto) regulation and alternative transcription. *Developmental Biology* 265: 462–477.

Kleinjan, D. J., A. Seawright, G. Elgar, and V. Van Heyningen. 2002. Characterization of a novel gene adjacent to PAX6, revealing synteny conservation with functional significance. *Mammalian Genome* 13: 102–107.

Koopman, P. 2001. The genetics and biology of vertebrate sex determination. *Cell* 105: 843–847.

Korf, I., P. Flicek, D. Duan, and M. R. Brent. 2001. Integrating genomic homology into gene structure prediction. *Bioinformatics* 17: 140–148.

Lefebvre, S., L. Burglen, S. Reboullet, O. Clermont, P. Burlet, L. Viollet, B. Benichou, C. Cruaud, P. Millasseau, and M. Zeviani. 1995. Identification and characterization of a spinal muscular atrophy-determining gene. *Cell* 80: 155–165.

Lewin, H. A., G. C. Russell, and E. J. Glass. 1999. Comparative organization and function of the major histocompatibility complex of domesticated cattle. *Immunological Reviews* 167: 145–158.

Lewis, S. M. 1994. The mechanism of V (D) J joining: Lessons from molecular, immunological and comparative analysis. *Advances in Immunology* 56: 27–150.

Maheshwar, M. M., R. Sandford, M. Nellist, J. P. Cheadle, B. Sgotto, M. Vaudin, and J. R. Sampson. 1996. Comparative-analysis of the tuberous-sclerosis-2 (TSC2) gene in human and pufferfish. *Human Molecular Genetics* 5: 133–137.

Marshall, H., M. Studer, H. Popperl, S. Aparicio, and A. Kuroiwa. 1994. A conserved retinoic acid response element required for early expression of the homeobox gene Hoxb-1. *Nature* 370: 567–571.

Meyer, I. and R. Durbin. 2002. Comparative ab initio prediction of gene structures using pair HMMs. *Bioinformatics* 18: 1309–1318.

MHC Sequencing Consortium. 1999. Complete sequence and gene map of a human major histocompatibility complex. *Nature* 401: 921–923.

Michalova, V., B. W. Murray, H. Sultmann, and J. Klein. 2000. A contig map of the MHC class I genomic region in the zebrafish reveals ancient synteny. *Journal of Immunology* 164(10): 5296–5305.

Miles, C., G. Elgar, E. Coles, D. J. Kleinjan, V. Van Heyningen, and N. Hastie. 1998. Complete sequencing of the *Fugu* WAGR region from WT1 to PAX6: Dramatic compaction and conservation of synteny with human chromosome 11p13. *Proceedings of National Academy of Sciences of the United States of America* 95: 13068–13072.

Miller, R. D. and G. H. Rosenberg. 1997. Recombination activating gene-1 of the opossum *Monodelphis domestica. Immunogenetics* 45: 341–342.

Monani, U. R., C. L. Lorson, D. W. Parsons, T.W. Prior, E. J. Androphy, A. H. Burghes, and J. D. McPherson. 1999. A single nucleotide difference that alters splicing patterns distinguishes the SMA gene SMN1 from the copy gene SMN2. *Human Molecular Genetics* 8: 1177–1183.

Namikawa-Yamada, C., K. Naruse, H. Wada, A. Shima, N. Kuroda, M. Nonaka, and M. Sasaki. 1997. Genetic linkage between the LMP2 and LMP7 genes in the medaka fish, a teleost. *Immunogenetics* 46: 431–433.

Naruse, K., M. Tanaka, K. Mita, A. Shima, J. Postlethwait, and H. Mitani. 2004. A medaka gene map: the trace of ancestral vertebrate proto-chromosomes revealed by comparative gene mapping. *Genome Research* 14: 820–828.

Oettinger, M. A., D. G. Schatz, C. Gorka, and D. Baltimore. 1990. RAG-1 and RAG-2, adjacent genes that synergistically activate V (D) J recombination. *Science* 248: 1517–1523.

Pachter, L., M. Alexandersson, and S. Cawley. 2002. Applications of generalized pair hidden Markov models to alignment and gene finding problems. *Journal of Computational Biology* 9: 389–400.

Parra, G., P. Agarwal, J. Abril, T. Wiehe, J. Fickett, and R. Guig. 2003. Comparative gene prediction in human and mouse. *Genome Research* 13: 108–117.

Pearn, J. 1978. Incidence, prevalence, and gene frequency studies of chronic childhood spinal muscular atrophy. *Journal Medical Genetics* 15: 409–413.

Pearson, W. R. and D. J. Lipman. 1988. Improved tools for biological sequence comparison. *Proceedings of National Academy of Sciences of the United States of America* 85: 2444–2448.

Peixoto, B. R., Y. Mikawa, and S. Brenner. 2000. Characterization of the recombinase activating gene-1 and 2 locus in the Japanese pufferfish, *Fugu rubripes. Gene* 246: 275–283.

Rowitch, D. H., Y. Echelard, P. S. Danielian, K. Gellner, and S. Brenner. 1998. Identification of an evolutionarily conserved 110 base-pair *cis*-acting regulatory sequence that governs Wnt-1 expression in the murine neural plate. *Developmental Biology* 125: 1735–1746.

Roest Crollius, H., O. Jaillon, A. Bernot, C. Dasilva, L. Bouneau, C. Fizames, P. Wincker, Brottier, F. Quetier, and W. Saurin. 2000. Estimate of human gene number provided by genome-wide analysis using *Tetraodon nigroviridis* DNA sequence. *Nature Reviews Genetics* 25: 235–238.

Sato, S., T. Ogata, V. Borja, C. Gonzales, Y. Fukuyo, and M. Kodama. 2000. Frequent occurrence of paralytic shellfish poisoning toxins as dominant toxins in marine puffer from tropical water. *Toxicon* 38: 1101–1109.

Sandelin, A., W. Alkema, P. Engström, W. Wasserman, and B. Lenhard. 2004. JASPAR: An open-access database for eukaryotic transcription factor binding profiles. *Nucleic Acids Research* 32: 91–94.

Schwartz, S., W. J. Kent, A. Smit, Z. Zhang, R. Baertsch, R. C. Hardison, D. Haussler, and W. Miller. 2003. Human–mouse alignments with BLASTZ. *Genome Research* 13: 103–107.

Singer, A., H. Perlman, Y. Yan, C. Walker, G. Corley-Smith, B. Brandhorst, and J. Postlethwait. 2002. Sex-specific recombination rates in zebrafish (*Danio rerio*). *Genetics* 160: 649–657.

Spitz, F., F. Gonzalez, and D. Duboule. 2003. A global control region defines a chromosomal regulatory landscape containing the HoxD cluster. *Cell* 113: 405–417.

Stock, D. W., D. L. Ellies, Z. Zhao, M. Ekker, and F. H. Ruddle. 1996. The evolution of the vertebrate Dlx gene family. *Proceedings of National Academy of Sciences of the United States of America* 93: 10858–10863.

Taher, L., O. Rinner, S. GargS, A. Sczyrba, M. Brudno, S. Batzoglou, and B. Morgenstern. 2003. Agenda: Homology-based gene prediction. *Bioinformatics* 12: 1575–1577.

Thompson, J. D., D. G Higgins, and T. J. Gibson. 1994. CLUSTAL X: Improving the sensitivity of progressive multiple sequence alignment through sequence weighting, position-specific gap penalties and weight matrix choice. *Nucleic Acids Research* 22: 4673–4680.

Trachtulec, Z., R. M. J. Hamvas, J. Forejt, H. R. Lehrach, V. Vincek, and J. Klein. 1997. Linkage of TATA-binding protein and proteasome subunit C5 genes in mice and humans reveals synteny conserved between mammals and invertebrates. *Genomics* 44: 1–7.

Trower, M. K., S. M. Orton, I. J. Purvis, P. Sanseau, J. Riley, C. Chritodoulou, D. Burt et al. 1996. Conservation of synteny between the genome of the pufferfish (*Fugu rubripes*) and the region on human chromosome 14(14q24.3) associated with familial Alzheimer's disease (AD3 locus). *Proceedings of National Academy of Sciences of the United States of America* 93: 1366–1369.

UK HGMP Resource Centre Fugu Group home page. http://fugu.hgmp.mrc.ac.uk.

Venkatesh, B., P. Gilligan, and S. Brenner. 2000. Fugu: A compact vertebrate reference genome. *FEBS Letters* 476: 3–7.

Venkatesh, B., B. H. Tay, G. lgar, and S. Brenner. 1996. Isolation, characterization and evolution of nine pufferfish *Fugu rubripes* actin genes. *Journal of Molecular Biology* 259: 655–665.

Willett, B. J., M. J. Hosie, J. C. Neil, J. D. Turner, and J. A. Hoxie. 1997a. Common mechanism of infection by lentiviruses [letter]. *Nature* 385: 587.

Wong, E., W.-P. Yu, W. H. Yap, B. Venkatesh, and T. W. Soong. 2006. Comparative genomics of the human and Fugu voltage-gated calcium channel α1-subunit gene family reveals greater diversity in Fugu. *Gene* 366(1): 117–127.

Woolfe, A., M. Goodson, D. K. Goode, P. Snell, G. K. McEwen, T. Vavouri, S. F. Smith, P. North, H. Callaway, and K. Kelly. 2004. Highly conserved non-coding sequences are associated with vertebrate development. *PLoS Biology* 3: 7.

Zmaskek, C. M. and S. R. Eddy. 2001. ATV: Display and manipulation of annotated phylogenetic trees. *Bioinformatics* 17: 383–384.

Section III

Marine Metagenomics

8

Coral Holobiont Omics: Microbes and Dinoflagellates

Valliappan Karuppiah, Kumarappan Alagappan, and Zhiyong Li

CONTENTS

8.1 Background of the Coral Holobiont Research

The coral reef affords a structurally and environmentally complex group of habitats, which maintain an extensive microbial diversity and which control both host function and in due course maintain the progression of the ecosystem. Microbial processes and metabolisms robustly control biogeochemical and ecological functions surrounded by the reef environment, such as food webs, organism life cycles, and chemical and nutrient cycling. Microbial functions are also key drivers of the several factors that control the elasticity of the coral reef environment, such as larval enrollment, colonization, and overall species diversity. For example, chemical signals from the benthic microbial population influence the arrangement of larvae of many keystone species such as corals and sea urchins (Webster et al. 2004; Huggett et al. 2008). Also, endosymbiosis flanked by corals and the eukaryotic dinoflagellate genus *Symbiodinium* is important for the evolutionary achievement of stony corals in the shallow tropics and the long-term existence of the coral reef ecosystem (Rohwer et al. 2001; Yellowlees et al. 2008). It is amazing that little research has been intended at understanding the linkages between the symbionts and coral (communities of bacteria, archaea, viruses, fungi, and dinoflagellates) and macroecological shifts on reefs. Only in recent times, the role of microbial diversity and host–microbe interactions in the response of reef ecosystems to environmental change has been studied.

During 1990s, marine microbial diversity has occurred to be more evident than previous studies, and more than 95% of the marine microbes could not be cultured (Fuhrman and Campbell 1998) because most of the marine microbes need complex nutritional and/or physical environment; until now, it could not be simulated in culture conditions. Additionally, microorganisms frequently survive in mutually dependent on particular metabolic units. These associations cannot be provided in the laboratory conditions. Also, microorganisms frequently survive in highly definite associations with their hosts and the lack of key host aspects hinder the culture. However, Rohwer et al. (2001) effectively used the culture-independent techniques to detect the microbial populations associated with the Caribbean coral *Montastraea franksi*. Each coral species has specific and unique microbial communities that are different from the water column (Frias-Lopez et al. 2002; Rohwer et al. 2002).

Metagenomics and transcriptomics approach has been applied to investigate the functional roles of coral symbionts during temperature and bleaching (Littman et al. 2011), nutrient cycling, and antimicrobial protection for the coral holobiont (Ritchie 2006). Shifts in the holobionts community structures have been observed in coral bleaching and disease. The coral environment allows proliferation of potentially harmful bacterial ribotypes with specific antimicrobial activity or competitive advantage in host-derived antimicrobial activity. Alternatively, growth of specific symbionts might be promoted or less favored through the production of secondary metabolites by the coral holobiont, which directly or indirectly lead to a favorable environment for certain species. This ribotype structure and diversity of coral-associated holobionts are assessed using denaturing gradient gel electrophoresis (DGGE) fingerprinting (Kvennefors et al. 2010). The composition of the dominant and potentially important bacteria caused by coral disease has been examined by terminal-restriction fragment length polymorphism (T-RFLP) analysis (Frias-Lopez et al. 2004).

Complete picture of coral holobiont association has been elucidated by next-generation sequencing (NGS), which shows the diversified holobionts in the coral structure and its surroundings. NGS, specifically 454 pyrosequencing approaches to metagenomics and transcriptomics, has greatly expanded our capacity to answer complex environmental questions by enabling analysis of the near-complete collection of genes within identified niches. Pyrosequencing studies so far revealed that coral spatial and species variations in bacteria, viruses, fungi, and symbiotic eukaryotes are associated with corals (Amend et al. 2012; Lee et al. 2012; Šlapeta and Linares 2013). Coral-associated viruslike particles (VLPs) have been also sequenced using Nextera XT MiSeq 250 bp paired-end sequencing (Weynberg et al. 2014).

Gene expression studies have been restricted to small-scale quantitative polymerase chain reaction (PCR) analyses of candidate genes of coral and hybridization on microarrays (DeSalvo et al. 2012). Parallel-sequencing (or NGS) or whole-genome or whole-transcriptome analyses have become a realistic option for genetic nonmodel organisms (Ellegren et al. 2012). Transcriptomic sequences are essential tools for tracing the functional role of the coral holobionts (Amend et al. 2012; DeSalvo et al. 2012; Mehr et al. 2013). For example, transcriptome studies proved that essential amino acids for coral are synthesized by *Symbiodinium* (Shinzato et al. 2014). Hence, in this chapter, we discuss about current knowledge on coral holobionts using metagenomics and transcriptomics.

8.2 Metagenomics Techniques

The classification and functions of microorganisms have been traditionally studied based on their similarities in morphological, developmental, and nutritional characteristics. Biochemical-based identification of microorganisms was changed to molecular-based identification by the pioneering work of Woese (1987), which showed that rRNA genes provide evolutionary chronometers. The early studies were technically challenging, relying on direct sequencing of RNA or sequencing of reverse transcription–generated DNA copies. The next technical step was the development of PCR and primers lead to amplifying the entire gene. This advancement in technology provided a larger microbial diversity that was not distorted by culturing bias.

In coral, 16S rRNA surveys of bacteria have elucidated an astonishing diversity of bacterial ribotypes, many of which are not closely related to cultivated or uncultivated microorganisms identified in previous studies (Rohwer et al. 2002). Metagenomics approach results in identification of culturable and unculturable microbes from the environment. This has significantly improved the potential for sequencing large samples. Technological advances have created new opportunities for large-scale sequencing projects that would have been difficult to imagine several years ago. The next key development is an emerging appreciation for the importance of complex microbial communities in the marine environment. Due to the overwhelming majority of unculturable microbes in the marine environment, metagenomic search results in the identification of unknown genes and proteins.

8.2.1 Isolation of Metagenomic DNA

The coral holobiont includes diverse intracellular and extracellular bacterial communities. Isolating genomic DNA from the coral sample (tissues, mucus, and surrounding environment) is the critical step in metagenomic analysis. Metagenomic approach for extracting DNA directly from environmental samples and constructing libraries by cloning into microorganisms amenable to genetic manipulations requires at least microgram scales of DNA that maintain sufficient DNA length and represent complete microbial diversity.

Several methods are available for extracting DNA from coral samples. Commercial kits are also available for extraction and purification of metagenomic DNA, which are easy to use with significant reproducibility. Isolation of metagenomic DNA from samples can be classified into direct, indirect, and enrichment methods. The ultimate methodology should include cell lysis and DNA extraction. The direct method is based on cell lysis within the sample matrix and subsequent separation of DNA from the matrix and cell debris. Indirect DNA isolation is based on the separation of cells from the matrix, followed by cell lysis and DNA extraction. The enrichment method involves the separation of cells from the matrix, followed by cell enrichment, lysis, and DNA extraction (Yongchang et al. 2010).

However, some specimens like scleractinian coral and limited quantity of coral DNA isolation by previous methods are not sufficient for metagenomic approach. To overcome the limitation of insufficient DNA yield, various methods for whole-genome amplification (WGA) have been developed like PCR-based methods for WGA, such as primer–extension preamplification (Zhang et al. 1992), degenerate oligonucleotide-primed (DOP) PCR (Telenius et al. 1992), generate nonspecific amplification artifacts (Cheung and Nelson 1996), and multiple displacement amplification (MDA) using ϕ29 polymerase (Yokouchi et al. 2006).

8.2.2 Library Construction

Coral metagenomic DNA has the complete story of holobiont that is beneficial or harmful relationship, stress response, and diseases. Metagenomic library construction mainly deals with the identification of microbes and microbial functional genes. Since the genes encoding the biosynthesis are normally clustered, clones in the metagenomic libraries should contain larger DNA. Two types of libraries with respect to average insert size can be generated: small-insert libraries in plasmid vectors (less than 10 kb) and large-insert libraries in cosmid and fosmid vectors (up to 40 kb) or bacterial artificial chromosome (BAC) vectors (more than 40 kb). The selection of a vector system for library construction depends on the quality of the isolated coral DNA, the desired average insert size of the library, the copy number required, the host, and the screening strategy that will be used (Daniel 2005). Small-insert metagenomic libraries are useful for the isolation of single genes or small operons encoding novel biomolecules. Isolation of high-molecular-weight DNA facilitates the cloning of DNA into BACs and allows the characterization of large regions of the genomes (Yongchang et al. 2010).

8.2.3 Terminal-Restriction Fragment Length Polymorphism

T-RFLP analysis is a technique used to study complex microbial communities based on variation in the 16S rRNA gene. T-RFLP analysis can be used to examine microbial community structure and community dynamics in response to changes in different environmental parameters and diseases in corals (Frias-Lopez et al. 2002; Sharp et al. 2012). It is mainly used to identify the most abundant bacteria present in corals. T-RFLP analysis was performed by two steps: first is to amplify the 16S rRNA gene and second is digestion by restriction enzymes. Different restricted bands with fluorescent labels have been performed in Genetic Analyzer. Using the GeneMapper 4.0, the lengths of the fluorescently labeled fragments have determined by comparison with the MapMarker 1000 size standard using the local Southern algorithm and a peak amplitude threshold set at 50 rfu (1% of the maximum peak amplitude) for all dyes. Fragment analysis has been performed between the range of 50 bp and 800 bp, within the linear range of internal size standards. For differential migration of X-Rhodamine-labeled standards and the FAM-labeled fragments, a second-order equation describing the relationship between actual migration distance values of X-Rhodamine-labeled and FAM-labeled size standards of the same size has been calculated and used to transform the fragment migration values. Predicted terminal-restriction fragment lengths of cloned sequences have been obtained using *in silico* digestion on Sequencher (Gene Codes Corporation) and compare with Microbial Community Analysis III. Profiles of environmental samples have been compared with the predicted clone fragment lengths to score the presence of bacterial taxa in specimens. A taxon scored by the fulfillment of two criteria: (1) the predicted fragment length matched the actual fragment length within 2 bp and (2) at least two out of three of the restriction enzyme digests yielded a positive identification of the predicted fragment length of a clone from the sample (Sharp et al. 2012).

8.2.4 Denaturing Gradient Gel Electrophoresis

The coral holobiont includes diverse intracellular and extracellular bacterial communities that differ from those of the overlying water column and display some level of spatial variability from local to geographic scales (Kvennefors et al. 2010). DGGE analysis of 16S rRNA gene segments has been used to profile complex microbial communities and to infer the phylogenetic affiliation of the community members.

Coral tissues, mucus, surrounding sediment, and water column samples were used for microbial community analyses. For 16S rRNA PCR amplification, universal primers 27f and 1492r have been used (Lane 1991), and a nested PCR has been performed using the internal primers GC358f (5′-CGC CCG CCG CGC CCC GCG CCC GTC CCG CCG CCC CCG CCC CCC TAC GGG AGG CAG CAG-3′) and 517r (5′-ATT ACC GCG GCT GCT GG-3′) (Muyzer et al. 1993). DGGE is a particular type of gel electrophoresis in which a constant heat (about 60°C) and increasing concentration of denaturing reagents at which DNA unwinds are said to have melted. This determines the melting domains that are defined as stretches of base pairs with an identical melting temperature. In other words, base pairs formed by nucleotides A (adenine) and T (thymine) and those formed by C (cytosine) and G (guanine) are chemically melted apart. Basically, what happens is that hydrogen bonding between the base pairs is broken by the temperature and the increasing gradient of denaturing chemicals (urea and formamide). Any variation of DNA sequences within these domains will result in different melting temperatures, thus causing different sequences to migrate at different positions in the gel. Sharply featured bands occurring more than once have been excised using sterile scalpels from a minimum of two different samples. Dominant bands that occurred in one sample only have been cut from two different gels to ensure repeatability of results. Ribotypes with high band intensity have been presumed to be more abundant members of the overall bacterial community. DGGE band patterns have been evaluated and bacterial populations have been identified by sequencing individual bands. Moreover, to get a better resolution in DGGE, the forward primer has been modified to incorporate a 40 bp GC clamp (Ferris et al. 1996). Several studies have shown that DGGE of functional genes (e.g., genes involved in sulfur reduction, nitrogen fixation, and ammonium oxidation) can provide information about microbial function and phylogeny simultaneously (Mårtensson et al. 2009; Tabatabaei et al. 2009).

8.3 Transcriptomics Techniques

Transcriptomics constitutes a meaningful resource to develop a large number of popular molecular markers, such as single-nucleotide polymorphisms and microsatellites. Proteomics may not tell about the clear picture about the functions of coral. Measuring the intermediate step between genes and proteins—in other words, transcripts of messenger RNA (mRNA)—bridges the gap between the genetic code and the functional molecules that run the cells. Transcriptomics completely reveals the coral gene expression during environmental stress (DeSalvo et al. 2012), disease, microbial association (Amend et al. 2012; Shinzato et al. 2014), and regulation of metabolic functions.

The genome of a scleractinian coral, *Acropora digitifera*, has been decoded using NGS technology (Shinzato et al. 2011). Transcriptome databases are available for several anthozoan cnidarians including several coral species: *Acropora millepora* (Meyer et al. 2009a; Moya et al. 2012), *Acropora palmata* (Polato et al. 2011), and *Pocillopora damicornis* (Traylor-Knowles et al. 2011). Transcriptomics studies reveal the coral bleaching in *A. palmata* during abiotic stresses, and increasing seawater temperatures are due to involvement of chaperone and antioxidant upregulation, growth arrest, and metabolic modifications (DeSalvo et al. 2012).

mRNA has been isolated from coral samples and converted to double-stranded complementary DNA (cDNA). The libraries of cDNA have been constructed and sequenced using a next-generation sequencer. Coral transcriptome assembly and annotation of sequence data were assembled using CLC Genomics Workbench. All assembled contigs have been compared with the national centre for biotechnology information nonredundant database (NCBI NR) protein database using BLASTX. The resulting output has been imported into MEGAN 3.9 for taxonomic assignment. The BLAST output of putatively coral transcripts (only transcripts that were assigned by MEGAN/LCA to kingdom Fungi) has imported into Blast2Go for gene mapping and annotation. Matches that mapped at 440% similarity and with E-values <10^{-5} to putative proteins have been retained and annotated with general GO-slim categories. The resultant GO-slim annotated transcripts related to metabolic processes: notably enzymes involved in metabolism of complex molecules such as proteins, polysaccharides, carbohydrates and lipids, nitrogen metabolism; enzymes involved in nucleic acid, amine, and cellular nitrogen compound metabolism; and enzymes involved in glutamate and glutamine pathways and nitrate and nitrite reductase.

8.4 Next-Generation Sequencing

NGS refers to non-Sanger-based high-throughput DNA sequencing technologies. Millions or billions of DNA strands can be sequenced in parallel, yielding substantially more throughput and minimizing the need for the fragmented cloning methods that are often used in Sanger sequencing of genomes. NGS includes three platforms, namely, 454 pyrosequencer (Roche Diagnostic Corporation), Illumina MiSeq system (Illumina), and SOLiD instrument (Life Technologies Corporation).

Even though the enormous quantity of phylogenetic data has been used for microbial identification using Sanger sequencing, new sequencing technologies are having a significant application on metagenomics. DNA pyrosequencing was developed to sequence the

DNA with a basically altered method in the mid-1990s (Ronaghi et al. 1996). Sequencing happens by a DNA polymerase driven production of inorganic pyrophosphate with the creation of ATP and ATP-dependent transfer of luciferin to oxyluciferin. The oxyluciferin produces the release of light pulses, and the amplitude of every signal is frankly related to the existence of one or more nucleosides. The major disadvantage of pyrosequencing is its incapability to sequence longer stretches of DNA. Hence, the hypervariable regions within bacterial 16S rRNA genes were amplified by the PCR and subjected to DNA pyrosequencing. DNA pyrosequencing has been efficiently useful in a variety of fields such as genotyping, single-nucleotide polymorphism, and microbial identification (Marsh 2007). Pyrosequencing has been successfully used to categorize the microbes based on the hypervariable regions within the 16S rRNA gene and signature sequence similarity (Jonasson et al. 2002). After DNA sequencing, sequences should be analyzed carefully to facilitate precise bacterial identification.

Illumina executes solid-surface PCR amplification of immobilizing random DNA fragments on a surface. The resultant cluster DNA fragments are sequenced with reversible terminators by synthesis process. The cluster density is massive with hundreds of millions sequence reads for every surface channel. In total, 16 channels are run at the same time in the HiSeq2000 instrument. The sequence read length is about 150 bp and it can be sequenced from both ends. By overlapping both sequences, base pair reads can reach up to 300 bp. Approximately, 60 Gbp can be expected in a single channel. As compared with 454, Illumina needs only a few nanogram of starting material and provides a superior amount of unassembled sequence reads. The limitation of Illumina is the run time; it takes approximately 10 days, but 454 take only 1 day (Thomas et al. 2012).

Apart from this, some other sequencing technologies are also available, but they need to prove themselves for metagenomic application. Among them, the Applied Biosystems SOLiD sequencer has been broadly used in genome resequencing. SOLiD probably provides the smallest error rate compared to other NGS sequencing technologies, but the consistent read length is much afar 50 nucleotides. This will limit its applicability for direct gene annotation of unassembled reads or assembly to large contigs. Ion Torrent is another emerging technology and is based on the principle that nucleotide incorporation can be detected by proton release during DNA polymerization. This system promises read lengths of more than 100 bp.

8.5 Coral Microbiome

8.5.1 Diversity of Coral Microbes

Sequence-based estimation of microbial assemblages, which involve random sampling of bacterial rRNA genes amplified from nucleic acid (Olsen et al. 1986), offers a high taxonomic resolution for environmental samples across huge datasets based on nucleotide heterogeneity. Coral 16S rRNA gene analyses of bacteria revealed an amazing diversity of bacterial ribotypes, several of which are not closely related to cultivated or uncultivated microorganisms. For example, Rohwer et al. (2002) identified the bacterial group of 3 Caribbean species and evaluated the presence of 6000 ribotypes in libraries from 14 coral samples. Further studies on examining bacterial assemblages of multiple coral species and geographic regions were also found similar results (Rohwer et al. 2001; Bourne and

Munn 2005; Klaus et al. 2005; Koren and Rosenberg 2006; Sekar et al. 2006; Kapley et al. 2007; Wegley et al. 2007; Lampert et al. 2008; Hong et al. 2009; Littman et al. 2009b; Reis et al. 2009). Similarly, most microbial assemblages in coral contain diverse clusters of closely related taxa. Microbial assemblages such as plankton communities in corals are dominated by a few different taxonomic units with a long tail of the species–distribution curve (Rohwer et al. 2002), suggesting that much of the diversity within the coral microbiome exists within the "rare" biosphere (Sogin et al. 2006).

The existence of coral–bacterial specificity is extensively accepted, but the spatial and temporal stability of these interactions is debated. In coral, few studies have shown that species-specific bacteria are geographically consistent (Rohwer et al. 2001, 2002). For example, Rohwer et al. (2002) proved that bacteria associated with three coral species in Panama contained similar ribotypes to those of the same coral species in Bermuda. The contrary trend has also been identified, in which bacterial assemblages contained different ribotypes between geographic locations, but similar corals contained similar ribotypes (Klaus et al. 2005; Guppy and Bythell 2006; Littman et al. 2009b). Results observed by sequence library investigations of uncultivated communities are normally consistent with those using fingerprinting approaches, which have a lower taxonomic resolution but provide a greater qualitative assessment of large numbers of samples or assemblages. These divergences could be enlightened, in part, by differences in methods (clone sequencing versus T-RFLP and DGGE), coral taxonomic resolution (comparing coral species within the same genus versus different genera), and the operator-defined taxonomic resolution of sequence analyses (Rohwer et al. 2001, 2002; Klaus et al. 2005; Guppy and Bythell 2006; Littman et al. 2009b). The varied trends over geographic scales and with host species may also reflect different species responses (host and/or microbiota) to site-specific factors (Hong et al. 2009; Littman et al. 2009a).

Determinations of coral–microbial associations are necessary to understand the coral holobiont. Apprill et al. (2009) evaluated the association between microbes and coral, *Pocillopora meandrina*, by comparing bacterial T-RFLP profiles between prespawned oocyte bundles, spawned eggs, and week-old planulae. It revealed that different ribotypes were associated with each stage, but that bacterial cells were not integrated until the planulae were fully developed (Apprill et al. 2009). This suggests that bacteria make associations with *P. meandrina* by horizontal uptake.

There is also proof that coral-associated bacteria vary between adults and juveniles of coral. Nonmetric multidimensional scaling demonstrations of bacterial profiles, evaluated through random sequencing of clone libraries, DGGE, and T-RFLP, were all consistent in indicating that adult *Acropora tenuis* and *A. millepora* showed tight grouping, while there was no evident relationship between profiles of juveniles (Littman et al. 2009a). The bacterial complement of juvenile corals was also more diverse, and while there was some conservation in bacterial ribotypes between adult and juvenile corals, the large numbers of adult-associated bacterial ribotypes were not detected in juveniles. This proposes a successional development whereby associates of adult corals steadily restore the diverse bacterial consortia of juveniles (Littman et al. 2009a).

Microbial taxonomy varied significantly between the different niches. It has been demonstrated that marine microbes broadly fall into two trophic categories, defined as copiotrophic and oligotrophic organisms (Lauro et al. 2009). Oligotrophs are generally highly abundant, but slow-growing microbes that dominate within environments characterized by stable and low-nutrient conditions (Lauro et al. 2009; Tout et al. 2014). Though oligotrophs dominated the noncoral niches, Tout et al. (2014) observed a rise in the numbers of sequences similar to copiotrophic organisms in the coral-associated samples.

This incorporated statistically significant raise in the relative consequence of copiotrophic bacteria, including *Pseudomonas*, *Vibrio*, *Shewanella*, *Pseudoalteromonas*, *Mycobacterium*, and *Alteromonas*. These increases in copiotrophic organisms are reliable with the elevated concentrations of organic material detected near the surfaces of corals (Wild et al. 2004a,b, 2010).

Particularly, several copiotrophic bacteria recognized can symbolize either beneficial bacteria or potential pathogens for the corals. A number of microbial genera enriched in the reef crest–coral niche have been recognized as potential coral pathogens. For example, *Vibrio* species have been concerned with numerous coral diseases. *Vibrio shiloi* and *Vibrio coralliilyticus* have been exposed to be implicated in coral bleaching (Banin et al. 2001; Kimes et al. 2012; Krediet et al. 2013), *Vibrio owensii* has identified as the basis for the *Montipora* white syndrome (Ushijima et al. 2012), and a group of *Vibrio* are believed to be implicated in yellow-band disease (Sato et al. 2009, 2010; Bourne et al. 2011; Ushijima et al. 2012). However, *V. shiloi* and *V. coralliilyticus* involved in bleaching have been identified by several studies (Benin et al. 2001; Kimes et al. 2012); it is still not clear whether other *Vibrio* species found in association with diseased corals (Sato et al. 2009, 2010; Bourne et al. 2011; Ushijima et al. 2012) are the disease-causing agent or opportunistic colonizers (Raina et al. 2009).

In addition to potential pathogens, several of the bacterial genera enriched in the reef crest–coral metagenome may have valuable effect for the coral. Among them, members of the *Pseudomonas* have been revealed to reduce the growth of potential pathogens (Salasia and Lammler 2008), and members of the *Pseudoalteromonas* genus have been revealed to reduce the attachment of the coral pathogen *V. shiloi* (Nissimov et al. 2009).

The most leading archaeal class among all metagenomes belongs to an unclassified *Thaumarchaeota* phylum that was highly distributed in the sandy substrate niche. There is little proof for ecological relations between *Thaumarchaeota* and corals, and its distribution across different coral reef niches is unknown. Tout et al. (2014) showed the significance of *Thaumarchaeota* in the sandy substrate, lagoon–coral, and reef crest–coral niches than within open water, representing that this group may have better ecological significance in a coral reef environment.

Euryarchaeota have been distributed in warm, shallow waters related to coral reefs (Kellogg 2004) and can comprise the important quantity in the coral microbial communities (Wegley et al. 2004; Erwin et al. 2013). Tout et al. (2014) showed the distribution of *Euryarchaeota* around a single reef ecosystem. Halobacteria was distributed in all niches, while the remaining eight archaeal classes are being distributed in different abundances throughout the niches. Notably, Archaeoglobi was the only class distributed in more abundant in the coral niches than the noncoral niches, and Methanopyri was more abundant in the reef crest niche than any other niche. Among the *Crenarchaeota*, Thermoprotei was the only class characterized and revealed to form associations with coral (Siboni et al. 2008). The distribution and abundances of *Crenarchaeota* and *Euryarchaeota* around Heron Island Reef are not uniform, still research is required to know the function of archaea associated with coral reefs.

8.5.2 Roles of Coral Microbes

The associations of coral microbes have long been established (DI Salvo and Gundersen 1971). But little is known about how these microbes contribute their role to coral. There is increasing evidence that coral microbiota are crucially involved in the biogeochemical cycle and pathogen resistance of the host's physiology. Recently, independent techniques have demonstrated that coral holobionts likely play a role in coral reef biogeochemistry (Williams et al. 1987; Szmant et al. 1990; Shashar et al. 1994; Ferrier-Pages et al. 2000; Lesser et al. 2007; Wegley et al. 2007; Chimetto et al. 2008; Olson et al. 2009; Raina et al. 2009; Kimes et al. 2010). For example, bacteria-possessing genes for nitrogen fixation have been identified within multiple coral species from varying geographic regions (Lesser et al. 2004; Wegley et al. 2007; Olson et al. 2009; Kimes et al. 2010). In addition, some studies have found evidence that members of coral-associated microbiota may also be involved in nitrogen-cycling processes, including nitrification, ammonium assimilation, ammoni-fication, and denitrification (Wegley et al. 2007; Kimes et al. 2010). There is also evidence that coral-associated microbial assemblages function in carbon and sulfur cycling (Ferrier-Pages et al. 1998, 2000; Wegley et al. 2007; Raina et al. 2009; Kimes et al. 2010). Genes that regulate carbon fixation, carbon degradation, and methanogenesis have been detected in coral-associated bacteria (Wegley et al. 2007; Kimes et al. 2010), those that regulate the assimilation of organic and inorganic sulfur sources (Wegley et al. 2007; Raina et al. 2009; Kimes et al. 2010).

Tout et al. (2014) found that considerable differences in less abundant, yet more specific and active, metabolic progression also occurred between the different niches. For instance, the two coral seawater metagenomes (lagoon–coral and reef crest–coral) had compara-tively elevated abundances of genes related to chemotaxis and motility, cell signaling and regulation, phages, prophages, transposable elements, and plasmids, whereas genes asso-ciated with photosynthesis have been more common in the open-water niche. This dem-onstrates that microbes reside in coral reefs have a broad stock of core genes, but specific niches support variability in the significance of more definite functional genes.

The microbial populations associated with the corals may have a vital role in the minor pH impact (Meron et al. 2012). Decreased pH resulted in a rise in bacteria related to coral disease and stress, but not in any exterior symptoms of coral disease in a Red Sea coral species (Meron et al. 2011). But in contrast to the laboratory study, there was also no major raise in definite pathogens or bacterial communities coupled with diseased or stressed corals in natural pH gradient of *Balanophyllia europaea* and *Cladocora caespitosa*. For exam-ple, *Rhodobacteraceae* (Alphaproteobacteria) increased in diseased, injured, or stressed marine invertebrates (Sekar et al. 2006; Sunagawa et al. 2009; Meron et al. 2011) in the laboratory experiments, but it did not increase in abundance at lower pH in natural pH gradient (Meron et al. 2012). Similarly, *Alteromonadaceae* and *Vibrionaceae*, often associated with diseased and stressed corals, increased in abundance at lower pH in the laboratory experiments (Meron et al. 2011); these have been also absent or did not change in abun-dance in natural pH gradient. In the case of *Vibrionaceae* associated with *B. europaea*, an increase in abundance has been observed at lower pH, but these bacteria represented only 6% of the total microbial community at pH 7.3, and no homologs of known pathogens were detected. It is interesting to note that a significant correlation has been observed in *B. euro-paea* between *Flavobacteriales* group and reduction in pH (Meron et al. 2012).

8.5.2.1 Bacterial Motility and Chemotaxis

Genes related with bacterial motility and chemotaxis also varied considerably between metagenomes, mainly due to their higher account in the reef crest–coral niche. Whereas oligotrophic bacteria like SAR11 and *Prochlorococcus* are not motile, many other marine bacterioplankton are greatly motile (Grossart et al. 2001; Tout et al. 2014), and the copio-trophic bacteria found in eminent abundance in the reef crest–coral and lagoon–coral niches, including *Pseudomonas*, *Vibrio*, *Shewanella*, *Pseudoalteromonas*, and *Alteromonas*, are all motile. It is possible that motility and chemotaxis are predominantly essential for microbial communities existing near to biotic surfaces on coral reefs, where highest chemical gradients are coupled with benthic organisms. The chemical products liberate from corals, and algal exudates are frequently strong chemoattractants for motile marine bacteria (Garren et al. 2014). Coral mucus and the exudates of *Symbiodinium* are prosperous in numerous organic compounds together with amino acids, sugars, and dimethylsulfoniopropionate (DMSP) (Von Holt and Von Holt 1968; Meikle et al. 1988; Wild et al. 2005; Raina et al. 2009, 2010; Wild et al. 2010), which are recognized as chemoattractants for marine bacteria (Miller et al. 2004; Sharp et al. 2012; Stocker and Seymour 2012), and have newly been revealed to attract important coral pathogens (Garren et al. 2014). Microscale gradients in these compounds in the microenvironment straightaway closer to the coral surface might encourage chemotactic movement of bacterioplankton cells to the coral holobiont (Garren et al. 2014).

8.5.2.2 Regulation and Cell Signaling

Genes related to regulation and cell signaling were also considerably more abundant in the reef crest–coral metagenome (Garren et al. 2014). Not only coral mucus offers a nutrient-loaded environment for surrounded microbes, but it can also include chemical signals implicated in the microbial communities' performance and function (Alagely et al. 2011). The elevated abundance of regulation and cell-signaling genes in the coral seawater niches recommends that the microbial populations related to corals employ in elevated levels of cell–cell communication. Regulation and cell signaling are possible to be involved in vital functions on coral reefs, where bacteria may perhaps use signaling processes, including quorum sensing, to systematize cellular functions to inhabit in corals (Taylor et al. 2004; Tait et al. 2010; Alagely et al. 2011). Chemical signaling also potentially permits bacteria to shield the holobiont from invading pathogens by changing behaviors such as swarming, biofilm formation, and the production of antimicrobial compounds (Ritchie 2006; Mao-Jones et al. 2010; Tait et al. 2010; Alagely et al. 2011; Hunt et al. 2012). Contrastingly, quorum sensing can control virulence in a few bacteria (Mao-Jones et al. 2010), which could facilitate pathogens to more eagerly infect the host and outcompete beneficial microbes (Rice et al. 2005; Hunt et al. 2012). The prominent occurrence of regulation and cell-signaling genes in the coral seawater niches specifies that these processes, which will affect the interactions between microbes and the coral host, are mainly essential in the coral holobiont.

8.5.2.3 Photosynthesis

Genes concerned in photosynthesis were most abundant in the open-water niche with an elevated level of photosynthetic microbes by the genera *Synechococcus* and *Prochlorococcus* (Tout et al. 2014). Dinsdale et al. (2008) detected changes in the genes related with photosystems I and II (PSI and PSII) between coral reef ecosystems. An elevated abundance of genes related to PSI was detected in the open-water niche, while PSII-associated genes were detected to be elevated in the reef crest–coral niche. These variations likely imitate subtle changes in the structure of the phototrophic microbial community (e.g., the shifts in the relative importance of *Synechococcus* and *Prochlorococcus*) across the reef.

Coral hosts photosynthetic symbionts such as *Oceanospirillales* sp., *Roseivirga* sp., *Alteromonas* sp., *Pseudoalteromonas* sp., *Halomonas* sp., *Pseudomonas* sp., and *Flavobacteriaceae* sp. (Bourne et al. 2013). These symbionts are all associated with the species concerned with the metabolism of complex organic molecules such as DMSP and dimethyl sulfide. *Halomonas* sp. have been revealed to be capable of metabolize DMSP and its breakdown product, acrylic acid (Todd et al. 2010). Whereas member of the *Flavobacteriaceae* reacts quickly to high DMSP concentrations in phytoplankton blooms, the genetic pathways for metabolizing this compound is unknown (Howard et al. 2011). Further, sulfur-based organic compounds resulting from photosymbionts control the microbial populations of marine invertebrates by providing nutrient for associated microbes. Even though compounds such as DMSP appear to have a role in structuring microbial populations associated with the corals, there are many other organic exudates derived from photosymbionts that also influence microbial associations. In addition, host animal factors can have an important role in structuring microbial communities. Recent studies emphasize that members of the *Oceanospirillales*, specifically *Endozoicomonas* sp., have been commonly found in marine invertebrates with and without photosymbionts and potentially have significant functional character within their host species (Yang et al. 2010; Nishijima et al. 2012; Speck and Donachie 2012).

8.5.2.4 UV-Damage Protection

Reef-building corals usually reside in shallow and relatively clear tropical waters and therefore continuously exposed to elevated intensity of UV irradiation. While elevated solar radiation occasionally roots for coral bleaching (Gleason and Wellington 1993), one fascinating problem is how corals guard themselves against UV damage. UV-absorbing materials mainly act as photoprotective compounds. These incorporate mycosporine-like amino acids (MAAs), scytonemin, carotenoids, and other unknown chemical constitutions (Shick et al. 1999; Reef et al. 2009). Even though a few photoprotective substances have been isolated from corals (Rastogi et al. 2010), it is frequently uncertain whether symbiotic dinoflagellates and/or bacteria generate the photoprotective compounds, or whether the corals itself can autonomously produce them.

A recent research of the cyanobacterium, *Anabaena variabilis*, showed a four-gene cluster (encoding dehydroquinate synthase (DHQS)-like, O-methyltransferase (O-MT), ATP-grasp, and nonribosomal peptide synthase (NRPS)-like enzymes) that converts pentose–phosphate metabolites into shinorine, one MAAs (Balskus and Walsh 2010). Exploration of cnidarian gene models for components of the shinorine gene cluster showed that this four-gene pathway is present in both *Acropora* and *Nematostella*, but not in *Hydra* (Shinzato et al. 2011). This robustly recommends that both *Acropora* and *Nematostella* can produce shinorine by themselves, which could be a pioneer for photoprotective compounds.

Additionally, the molecular phylogenetic investigation shows that homologous proteins in *Acropora* have more sequence identical to those of bacteria and dinoflagellates (Shinzato et al. 2011). These genes might have been obtained by means of horizontal gene transfer (Starcevic et al. 2008). For instance, through the evolution of cnidarian stinging cells, a subunit of bacterial poly-γ-glutamate (PGA) synthase was transmitted to an animal predecessor via horizontal gene transfer (Denker et al. 2008). It has been anticipated that in marine environments, horizontal gene transfer is vital for adapting to ecological variations (Keeling 2009).

The UV blocker, scytonemin, is detected entirely in cyanobacteria. In *Nostoc punctiforme*, their biosynthesis is regulated by a cluster of 18 genes (Soule et al. 2007). The cluster encompasses one subcluster of genes implicated in aromatic amino acid biosynthesis and a novel subcluster of genes of unknown function (Soule et al. 2009). The former comprise tyrA, dsbA, aroB, trpE, trpC, trpA, tyrP, trpB, trpD, and aroG. The latter comprises source, sub, science, scyD, saucy, and scyF.

The *A. digitifera* genome includes only 6 of the 18 genes: specifically, scyA, scyB, scyF, dsbA, aroB, and tyrP (Shoguchi et al. 2013). This recommends that coral cannot produce scytonemin autonomously. Molecular phylogenetic investigations show that coral scyA and scyB are related to bacterial genes for acetolactate synthase and glutamate dehydrogenase, respectively. This recommends that these enzymes are attached with PGA/amino acid biosynthesis in corals. Additionally, scyA, scyB, and aroB (DHQS-like) are probably derived by horizontal transfer from bacteria.

8.5.2.5 Nitrogen Fixation and Transformation

Most coral reefs are distributed in oligotrophic waters, and the water surrounding coral reefs is commonly very little in dissolved inorganic nitrogen (DIN) where the typical capacity of total DIN is <1 mmol. Despite these very low DIN concentrations, coral reefs support high biomass and biodiversity that can only be possible by higher inputs of nutrients relative to the accumulation of organic matter and/or very efficient recycling of nutrients such as nitrogen. Scleractinian corals are efficient recyclers of nutrients between the coral host and its symbiotic dinoflagellates (zooxanthellae), which contribute to their ecological success in optically clear, low-productivity waters (Lesser 2004). In fact, low in hospite concentrations of DIN have long been presumed to play an important regulatory role in controlling the growth rates of zooxanthellae, thereby maintaining the stability of the symbiosis (Falkowski et al. 1993).

In the coral reef ecosystems, the key resource of new nitrogen is nitrogen fixation, and several important members of the reef population have the capability to fix nitrogen through mutualistic symbiotic associations with nitrogen-fixing cyanobacteria and heterotrophic bacteria (Kneip et al. 2008). In addition to their symbiotic zooxanthellae, corals have a range of bacteria associated with their tissues and endoskeleton (Rohwer et al. 2001, 2002; Wegley et al. 2007; Ainsworth et al. 2009; Thurber et al. 2009), and these associations are extensively dispersed, stable, and nonpathogenic; their function remains unknown. Among the different population of coral bacterial symbionts, various possible nitrogen-fixing bacteria associated with reef-building corals have been identified (Rohwer et al. 2002). Nitrogen fixation has formerly been recognized to bacteria associated with living coral tissue, and current metagenomic studies have revealed that several corals contain cyanobacterial genes and also the genes associated with nitrogen fixation (Wegley et al. 2007; Thurber et al. 2009). Nitrogenase (nifH) gene sequences from the Hawaiian corals *Montipora capitata* and *Montipora flabellata* showed a high diversity of both heterotrophic and cyanobacterial nitrogen-fixing bacteria (Olson et al. 2009) and a direct relationship between nifH transcript abundance and the population density of zooxanthellae. Although different communities of bacteria were identified, a conserved phylogenetic cluster of bacteria in the family *Vibrionaceae* was detected only in association with *M. capitata*, and a less-conserved cluster of g-proteobacteria was associated with *M. flabellata*. This symbiont specificity suggests coevolved, species-specific, symbiotic associations.

In the Caribbean, large numbers of endosymbiotic cyanobacteria described as being related to coccoid cyanobacteria such as *Synechococcus* sp. or *Prochlorococcus* sp. were found to occur in the tissues of a common scleractinian coral, *Montastraea cavernosa* (Lesser et al. 2004). The expression of nitrogenase is shown by Western blots (Lesser et al. 2004), and consequent studies have shown that these corals fix nitrogen, resulting in depleted d15N stable isotope signatures in the zooxanthellae fraction, representing that the products of nitrogen fixation are used by the zooxanthellae (Lesser et al. 2004, 2007).

The distribution of *M. cavernosa* colonies with endosymbiotic cyanobacteria increases significantly with depth and might be related to the amount of time available when microaerophilic conditions exist to support the energy requirements of nitrogen fixation without inhibiting the enzyme nitrogenase (Lesser et al. 2007). Colonies of *M. cavernosa* with cyanobacteria showed higher rates of nitrogen fixation in the early morning and in the evening, when oxygen concentrations in coral host tissue are lower, and it has been hypothesized that these cyanobacteria are utilizing the products of zooxanthellae photosynthesis (e.g., glycerol) as an energy source (Lesser et al. 2007).

In addition to nitrogen fixation, prior studies on corals illustrated an important production of NO_3 with the substrate for this reaction (NH_3) coming from the coral host (Wafar et al. 1990). Both nitrification and denitrification genes of archaeal origin have been detected in association with corals (Beman et al. 2007; Siboni et al. 2008). Whether these prokaryotes are to be found in the mucus or tissues, or both, of corals, they would be exposed to daily patterns of tissue hypoxia/anoxia and hyperoxia (Lesser 2004) that create microenvironments where these two processes could occur and be temporally separated. The latest investigation of the *Montastraea faveolata* holobiont using the GeoChip 2.0 microarray supports previous result that detected functional genes for nitrogen fixation as well as nitrification, denitrification, and anaerobic ammonium oxidation (ANAMMOX) (Kimes et al. 2010). Additionally, several sequences recognized as belonging to a diverse group of microorganisms yielding further evidence of the diverse and complex nature of the coral-associated microbial community.

8.6 Coral Viromes

Genes concerning to phages, prophages, transposable elements, and plasmids have been most rich in the reef crest–coral metagenome and constantly responsible for the most important differences between niches. The improved occurrence of these genes in the reef crest seawater niche reflects the rise in sequences associated with bacteriophages and provides support for the association of virus within the coral holobiont (Vega-Thurber and Correa 2011).

Traditionally, virus discovery required propagation of the virus in cell culture, a proven technique responsible for the identification of the vast majority of viruses known to date. However, many viruses cannot be easily propagated in cell culture, thus limiting our knowledge of viruses. Viral metagenomic analyses of environmental samples suggest that the field of virology has explored less than 1% of the extant viral diversity. Shotgun-based metagenomic sequencing approaches are often required to characterize the viral genome (Mokili et al. 2012).

VLPs from coral samples and seawater were concentrated using different methods like filtration method using 100 kDa tangential flow filter and purified via passage through a 0.22 mm Sterivex, and the resulting viral concentrate has been preserved in molecular biology grade chloroform (2% final concentration) and using cesium chloride density gradient ultracentrifugation, with buoyant densities ranging from 1.2 to 1.7 g mL^{-1} before addition of samples (Vega-Thurber et al. 2009, 2011). DNA has been isolated using an organic extraction protocol (Vega-Thurber et al. 2009, 2011) and amplified using nonspecific MDA. The coral viromes have been barcoded and sequenced using NGS technology for further analysis (Goecks et al. 2010).

The reduced illustration of many viral taxa in public databases limits the capability to understand the results from viral metagenomic sequence data (Weynberg et al. 2014). For instance, a small amount of archaeal viruses of coral viromes might consequence the lack of representative sequences in the public databases. The greater part of dsDNA related to bacteriophages has been detected in corals (Patten et al. 2008). It is noteworthy that the majority of RNA metagenome data of *A. tenuis* matched with MCP gene found in the ssRNA algal virus H. circularis RNA virus (HcRNAV) that infects dinoflagellates (Tomaru et al. 2004). This shows the possible existence of a virus allied with the dinoflagellate endosymbiont, *Symbiodinium* (Correa et al. 2013). As viral genomes include DNA or RNA, double or single stranded (King et al. 2011), coral viromic research should require to isolate both RNA and DNA viruses in an effort to identify all related virus community as possible.

8.7 Coral–Dinoflagellate Association

8.7.1 Diversity of Coral Dinoflagellates

The capability to identify and enumerate the abundance of dinoflagellates associated with corals is necessary for understanding holobiont physiology, susceptibility to stress, and, ultimately, the resilience of corals to environmental change. Quigley et al. (2014) proved that sequencing of the internal transcribed spacer (ITS)-2 region using 454 NGS was capable to identify the existence of cooccurring *Symbiodinium* types D1, C1, C3, and A13 from *Acropora hyacinthus* and *A. digitifera*. Amplicon sequencing of the ITS-1 region for *Symbiodinium* types associated with acroporid corals from Palau also confirmed that this NGS approach could identify haplotype variants of *Symbiodinium microadriaticum* ITS-1 populations when in hospite and distinguish differences in their frequencies among colonies and locations. Quigley et al. (2014) detected and quantified rare, low-abundant haplotype variation within symbiont types.

Coral–dinoflagellate symbioses are defined as mutualistic. Though, Quigley et al. (2014) observed that the symbiosis between a Pacific coral and the *Symbiodinium* clade A lineage as a reduction in the health state of the coral. Clade A *Symbiodinium* are rarely reported in coral hosts; however, this group has been described as fast growing and opportunistic because it is found in corals recovering from bleaching events (Little et al. 2004; Abrego et al. 2008). Quigley et al. (2014) showed that *Acropora cytherea* from Hawaii harboring clade A exhibit suboptimal health status and an increased incidence of disease as compared with corals sampled on the same reef harboring *Symbiodinium* clade C.

Stat et al. (2008) showed that genetic diversity was actually existed within a single lineage of *Symbiodinium*, clade C. Such high levels of genetic divergence within the *Symbiodinium* genus supported that the group contains members with highly diverse functions and physiologies, some of which may provide them with the capacity to form symbiotic interactions with coral.

Examining the zooxanthellae revealed that both *B. europaea* and *C. caespitosa* contained *Symbiodinium* B2, but only *B. europaea* contained *Symbiodinium* clade A (Meron et al. 2012). No change in *Symbiodinium* diversity was observed following the 7-month exposure to reduced pH. It has been proposed that tolerance to various environmental stresses, such as light or temperature, is influenced by *Symbiodinium* type (Baker 2003;

Rodolfo-Metalpa et al. 2006; Jones et al. 2008). Indeed, clade A has been shown to impart resistance to short-term increases in temperature under experimental conditions to its host (Rodolfo-Metalpa et al. 2006).

8.7.2 Roles of Coral Dinoflagellates

8.7.2.1 Amino Acid Biosynthesis Pathways

Generally, 20 amino acids are found in proteins, whereas all animals studied to date either lack the ability to produce one or more of these amino acids or are unable to produce enough quantities to meet their metabolic needs. These amino acids are called "essential" and must be acquired from the diet. For vertebrates, eight or more amino acids are essential; threonine, valine, methionine, leucine, isoleucine, phenylalanine, lysine, and tryptophan are necessary for all vertebrates, while arginine and/or histidine are also important in some cases (Furst and Stehle 2004). The issue of amino acid necessities is complicated by the existence of symbiotic algae (Swanson and Hoegh-Guldberg 1998; Wang and Douglas 1999). A genome-wide assessment of amino acid biosynthetic pathway mechanism in *A. digitifera* revealed that *Acropora* corals may be able to produce 10 nonessential amino acids, but not cysteine (Shinzato et al. 2011). Shinzato et al. (2014) reconstructed the amino acid biosynthetic pathways in the *Porites* holobiont based on kyoto encyclopedia of genes and genomes (KEGG) IDs. Most of the enzymes involved in amino acid biosynthetic pathways were detected in the transcriptomes of *Porites* and *Symbiodinium*, while four enzymes cannot be detected in the lysine anabolic pathway. Interestingly, many enzymes involved in essential amino acid biosynthesis are detected only in *Symbiodinium* contigs. Genetic ontology terms in *Symbiodinium* contigs imply that *Symbiodinium* transport essential amino acids to host cells. In contrast, enzymes for nonessential amino acid pathways have been detected in both *Symbiodinium* and *Porites*. Glutamic acid, glutamine, aspartic acid, and alanine are probably produced in both *Porites* and *Symbiodinium*. According to Shinzato et al. (2014), asparagine is synthesized in *Symbiodinium*, while cysteine is synthesized in *Porites*. *Acropora* corals lack an essential enzyme for cysteine biosynthesis, cystatione β-synthase (Shinzato et al. 2011); however, *Porites* seem to possess it indicates that it does not depend upon *Symbiodinium* for cysteine biosynthesis and might account for its greater resilience to environmental stresses. An interesting example is the methionine biosynthesis pathway. Half of the enzymes reside in the *Porites* and *Symbiodinium* contigs, respectively, suggesting that methionine might be produced by intimate cooperation between host and symbiont in coral holobionts.

8.7.2.2 Histones and Nucleosome-Associated Proteins

Dinoflagellates acquire not only a fundamental nucleosome machinery such as H2A.Z (Zilberman et al. 2008) and H3.3 (Ahmad and Henikoff 2002; Malik and Henikoff 2003; Hake and Allis 2006) but also specific histones that are identified to be involved in transcriptional and epigenetic regulation. In contrast to the histones H2B and H4, the histones H2A and H3 have highly conserved ubiquitously expressed variants with specialized functions. H2A.Z is associated with the promoter region of actively transcribed genes linked to transcriptional competence and is also involved in epigenetic regulation (Zilberman et al. 2008). Dinoflagellates might be able to use the nucleosome machinery for transcriptional regulation through the regulation of the methylation status of specific loci. The variation observed in *Symbiodinium* H3-like proteins is also suggestive of the role of the nucleosome machinery in transcriptional regulation, with possible subfunctionalization of the multiple variants. *Symbiodinium* species contain various genes for the modification of histones including methylation and acetylation as well as orthologs of the histone-specific chaperones ASF1 and CAF1. ASF1 is involved in the modulation of local chromatin structure, whereas CAF1 is mainly associated with processes involving DNA, such as DNA replication and DNA repair (Ramirez-Parra and Gutierrez 2007).

8.7.2.3 Transcription Factors in Symbiodinium

The unusual chromatin structure, low concentration of proteins in the nucleus, and very large genomes of dinoflagellates raise the question whether gene regulation is realized with the same mechanisms as in other eukaryotes. Transcriptional regulation might play a minor role in dinoflagellates as opposed to other mechanisms of regulation. Only a small number of proteins with sequence-specific nucleic acid–binding domains (i.e., putative transcription factors) are present in dinoflagellates. Transcription factors have been shown to scale with genome size (van Nimwegen 2003) and make up 6%–9% of all genes of higher eukaryotic transcriptomes. The percentages found here for dinoflagellates are much lower than those for other protists such as *Plasmodium*, even though *Plasmodium* has a reduced genome due to its parasitic lifestyle. The assemblage of transcription factors in *Symbiodinium* seems to be completely different from other eukaryotes. Common domains such as zinc fingers, helix–loop–helix, AP2, or homeobox domains are rare or absent. This is also true for the other dinoflagellates represented in the set of expressed sequence tag (EST) analyzed here, as the set of transcription factor domains and their abundances are quite similar to those found in *Symbiodinium*. This low abundance of transcription factors appears to be a genomic signature of the dinoflagellate clade.

Dinoflagellates may also contain yet undescribed transcription factor families that would represent part of the "missing" transcriptional regulatory machinery. Thus, it is interesting that more than 60% of the putative transcription factors identified to carry a "cold-shock" domain (CSD). This domain is not very common in other eukaryotes, suggesting a lineage-specific expansion in dinoflagellates. Such lineage-specific expansions of different transcription factor domains have been found in multiple taxa throughout the tree of life (Aravind et al. 2000; Lespinet et al. 2002). Originally identified as a reaction to cold shock in *Escherichia coli*, proteins with CSD domains have now been associated with a wide range of functions. They can act as transcription factors, by binding DNA (i.e., Y-box factors), but many of them interact with RNA rather than DNA. They are involved in regulation of transcription, splicing, and translation and influence mRNA stability as RNA chaperones (Mihailovich et al. 2010). This observation fits with the notions that (1) regulation in dinoflagellates may take place after transcription and (2) that RNA editing is widespread (Lin et al. 2007). Thus, proteins with CSDs may be responsible for much of the transcriptional regulation in dinoflagellates. Considering that *Symbiodinium* undergoes a dramatic change in its environment and lifestyle upon entering invertebrate hosts, a need for efficient regulation of a large number of genes might be advantageous. However, as all data gathered here are based on *Symbiodinium* grown in cultures, it is possible that more or different types of transcription factors are expressed in the symbiotic state. As *Symbiodinium* genomes are currently being sequenced, this question can be conclusively answered in the near future as the genome sequence becomes available.

8.7.2.4 Antioxidative Response

The impact of reactive oxygen species (ROS) on the symbiosis of *Symbiodinium* and its marine invertebrate hosts has possibly influenced the mechanisms to cope with photosynthesis-generated ROS in order to prevent the break of symbiotic relationship. *Symbiodinium* also contains a rich antioxidant gene repertoire but surprisingly appears to lack or transcribe below detection limit the enzyme catalase, one of the central enzymes in eukaryotic cellular redox chemistry. Transcriptome sequences analyzed from cultured zooxanthellae revealed that the catalase gene is only expressed in hospite. One of the main differences between *Symbiodinium* and diatoms or plants is the presence of several prokaryotic Ni-type super oxide dismutase SODs) in both *Symbiodinium* species, which are not present in the plant species and are only represented by a single gene in the diatom *Thalassiosira pseudonana*. The presence of bacterial proteins is probably due to the lateral gene transfer between prokaryotes and eukaryotes, especially protist (Andersson 2005; Keeling and Palmer 2008). Furthermore, several genes of bacterial origin have already been identified in *Symbiodinium* (Leggat et al. 2007). To our surprise, we found that some of the Ni-type SOD genes identified in *Symbiodinium* also encode an additional ubiquitin domain. The ubiquitin domain is a 76 amino acid domain found in eukaryotes, whereas the SOD_Ni domain is evidently of prokaryotic origin (Schmidt et al. 2009), suggesting that these transcripts might represent fusions of prokaryotic and eukaryotic genes. *Symbiodinium* species appear to possess an unexpectedly high number of Trx domain–encoding genes compared to plants and in stark contrast to the substantially smaller number found in diatoms. The Trx super family proteins fulfill diverse cellular functions. These include the maintenance of cell homeostasis and the regulation of the redox state of the cell (Papp et al. 2003; Eckardt 2007). They play key roles in the oxidative stress response (Vieira and Rey 2006; Niwa 2007) and have been shown to be differentially expressed in response to high temperature, salinity, and ultraviolet radiation in corals (Edge et al. 2005; Desalvo et al. 2008; Aranda et al. 2011).

8.7.2.5 Photosynthesis

Recent research designated that the component of the coral holobiont particularly *Symbiodinium* and dinoflagellates controls microbial population structure through the discharge of complex carbon-containing exudates including DMSP (Ikeda and Miyachi 1995; Raina et al. 2009, 2010). DMSP can be degraded to dimethyl sulfide, an inner molecule in the universal sulfur cycle, which diffuses from the ocean into the atmosphere where it influences cloud formation, with consequences for atmospheric chemistry, local climate, and water temperature (Ayers and Gras 1991; Andreae and Crutzen 1997). In the marine environment, DMSP has been the focus of extensive consideration due to its primary role as carbon and sulfur sources for bacteria (Sievert et al. 2007). Coral reefs are one of the main producers of DMSP with the resource thought to be consequent from marine invertebrates harboring symbiotic dinoflagellates (Broadbent et al. 2002; van Alstyne et al. 2006). In fact, the concentrations of DMSP and its breakdown products dimethyl sulfide and acrylate in reef-building corals are the maximum evidence in the marine environment (Broadbent and Jones 2004).

8.7.2.6 Response to Stress of Symbiodinium

The need of a transcriptional response to experimental treatment proposes that *Symbiodinium* cells may react to alteration in the external environment. Alterations in external environmental conditions have been revealed to bring out strong transcriptional responses in corals (DeSalvo et al. 2008; Meyer et al. 2011). Heat stress considerably changed 27-fold of coral gene expression across hundreds of cnidarian genes (Barshis et al. 2013). During the end of heat exposure, the majority of corals demonstrated clear signs of bleaching and fractional mortality. Heat treatment should have caused considerable physiological stress in both the coral and *Symbiodinium*, and although a stress response was apparent in the gene expression of the coral host (Barshis et al. 2013), there was no obvious gene expression response of the symbionts. However, there are reliable and widespread differences between the transcriptional profiles of the two different *Symbiodinium* types, several of which persisted after 12 months of common garden acclimation and are evident within two identical host genetic backgrounds.

Symbiodinium gene expression alters across eight genes considered to be essential in the stress response were less than 2-fold different compared with up to 10-fold differences in similar coral host genes over an 8-day heat exposure (Leggat et al. 2011). *Symbiodinium* in larvae of the host coral *P. damicornis* showed no change in expression of five putative stress genes despite 9 days of elevated temperature and pCO_2 exposure and considerable changes in protein levels of Ribulose-1,5-Bisphosphate Carboxylase (Rubisco) (Putnam et al. 2013). While, Rosic et al. (2010) detected up to 4-fold changes in *Symbiodinium* cytochrome P450 gene expression in response to both a fast and regular heat exposure and, in a later study, up to a 57% (1.57-fold) increase and 89% (1.89-fold) decrease in Hsp70 and Hsp90 in *Symbiodinium* under heat stress, both in culture and when *Symbiodinium* within a coral host (Ross et al. 2011). Additionally, Baumgarten et al. (2013) found that the majority of stresses (salinity, cold, and dark stresses) elicited few changes in gene expression, although many genes responded significantly to heat stress and heat shock. In terms of the stress-related genes, across all 128 putative matches to Hsp70, Hsp90, or CytP450 family genes, only two were significant for heat stress, whereas none were significant for heat stress (Barshis et al. 2014).

One possible clarification for the lack of short-term gene expression response is that the primary mechanisms to regulate the size and composition of the available protein pool in these *Symbiodinium* may act after transcription and/or after translation (Barshis et al. 2014). Bayer et al. (2012) explained a dearth of transcription factors in the transcriptomes of *Symbiodinium*, which could indicate little capability for transcriptional regulation in these taxa. Fagan et al. (1999) detected differences in the concentration of glyceraldehyde-3-phosphate dehydrogenase protein concentration and mRNA levels in circadian. Additionally, Baumgarten et al. (2013) detected the existence of several small RNAs (smRNA) that interrelated in large quantity with approximately 3500 transcript targets, suggesting that smRNA posttranscriptional regulation (i.e., gene silencing) could act on a range of cellular processes in *Symbiodinium*.

On the other hand, the transcriptional stability of *Symbiodinium* observed that experimental treatments could indicate a consequent stability in the *Symbiodinium* protein composition. This could effect from a few types of in hospite buffering or protection of *Symbiodinium* cells existing within host tissues (Barshis et al. 2014). The deficiency of the transcriptional response to the stress of the *Symbiodinium* indicates some sort of host buffering of the intracellular environment of in hospite symbionts, basically protecting them from physiological stress. Rosic et al. (2011) found comparable patterns of Hsp70 and Hsp90 expression in *Symbiodinium* both in symbiosis with a coral and in culture with significant changes during heat stress in both situations, suggesting little effect of residing within host tissues on the expression levels of these genes.

8.8 Future Perspectives

The diversity of coral-associated bacteria is much clearer than before. However, the behaviors of archaeal communities are yet to be studied. Further, the role of in situ microbial activity and interactions is still incomplete. It has turned out to be clear that studying the tropical reef environment needs some new technology from traditional microbial ecology to precisely calculate vital parameters, and these technologies could help to reveal the bacterial abundance, growth rates, and production rates. In future, it is essential to assess and identify the multifaceted functions of coral holobionts that preside over ecological change in these ecosystems. Although few assumptions of coral holobionts reported in the literature have been analyzed, in order to address the overall roles of microbes and host–microbe interactions, the research on coral holobionts should be focused toward the following:

1. Association and interaction of coral and symbionts throughout their life cycle.
2. Influence of environmental changes in the community structure of coral holobionts.
3. Did the community structure of coral holobionts are species specific?
4. Effect of coral disease, recovery, and death during the alteration in the community structure of coral holobionts.
5. The application of metagenomics and transcriptomics should help to define the relationship between microbial communities and environmentally driven macroecological change on reefs. Finally, this microbial perspective will advance our attitude toward the corals and its holobionts to conserve these important ecosystems.

Acknowledgment

Financial support from the National Major Scientific Research Program of China (2013CB956103) is greatly acknowledged.

References

Abrego, D., K. E. Ulstrup, B. L. Willis, and M. J. H. van Oppen. 2008. Species–specific interactions between algal endosymbionts and coral hosts define their bleaching response to heat and light stress. *Proceedings of the Royal Society B: Biological Sciences* 275:2273–2282.

Ahmad, K. and S. Henikoff. 2002. Histone H3 variants specify modes of chromatin assembly. *Proceedings of the National Academy of Sciences of the United States of America* 99:16477–16484.

Ainsworth, T. D., R. V. Thurber, and R. D. Gates. 2009. The future of coral reefs: A microbial perspective. *Trends in Ecology and Evolution* 25:233–240.

Alagely, A., C. J. Krediest, K. B. Ritchie, and M. Teplitski. 2011. Signaling-mediated cross-talk modulates swarming and biofilm formation in a coral pathogen *Serratia marcescens*. *The ISME Journal* 5:1609–1620.

Amend, A. S., D. J. Barshis, and T. A. Oliver. 2012. Coral-associated marine fungi form novel lineages and heterogeneous assemblages. *The ISME Journal* 6:1291–1301.

Andersson, J. O. 2005. Lateral gene transfer in eukaryotes. *Cellular and Molecular Life Sciences* 62:1182–1197.

Andreae, M. O. and P. J. Crutzen. 1997. Atmospheric aerosols: Biogeochemical sources and role in atmospheric chemistry. *Science* 276:1052–1058.

Apprill, A., H. Q. Marlow, M. Q. Martindale, and M. S. Rappe. 2009. The onset of microbial associations in the coral *Pocillopora meandrina*. *The ISME Journal* 3:685–699.

Aranda, M., A. T. Banaszak, T. Bayer, J. R. Luyten, M. Medina, and C. R. Voolstra. 2011. Differential sensitivity of coral larvae to natural levels of ultraviolet radiation during the onset of larval competence. *Molecular Ecology* 20:2955–2972.

Aravind, L., H. Watanabe, D. J. Lipman, and E. V. Koonin. 2000. Lineage-specific loss and divergence of functionally linked genes in eukaryotes. *Proceedings of the National Academy of Sciences of the United States of America* 97:11319–11324.

Ayers, G. P. and J. L. Gras. 1991. Seasonal relationship between cloud condensation nuclei and aerosol methanesulphonate in marine air. *Nature* 353:834–835.

Baker, A. C. 2003. Flexibility and specificity in coral-algal symbiosis: Diversity, ecology and biogeography of *Symbiodinium*. *Annual Review of Ecology and Systematics* 34:661–689.

Balskus, E. P. and C. T. Walsh. 2010. The genetic and molecular basis for sunscreen biosynthesis in cyanobacteria. *Science* 329:1653–1656.

Banin, E., T. Israely, M. Fine, Y. Loya, and E. Rosenberg. 2001. Role of endosymbiotic zooxanthellae and coral mucus in the adhesion of the coral-bleaching pathogen *Vibrio shiloi* to its host. *FEMS Microbiology Letters* 199:33–37.

Barshis, D. J., J. T. Ladner, T. A. Oliver, and S. R. Palumbi. 2014. Lineage-specific transcriptional profiles of *Symbiodinium* spp. unaltered by heat stress in a coral host. *Molecular Biology and Evolution* 31:1343–1352.

Barshis, D., J. Ladner, T. A. Oliver, F. Seneca, N. Traylor-Knowles, and S. R. Palumbi. 2013. Genomic basis for coral resilience to climate change. *Proceedings of the National Academy of Sciences of the United States of America* 110:1387–1392.

Baumgarten, S., T. Bayer, M. Aranda, Y. J. Liew, A. Carr, G. Micklem, and C. R. Voolstra. 2013. Integrating microRNA and mRNA expression profiling in *Symbiodinium* microadriaticum, a dinoflagellate symbiont of reef-building corals. *BMC Genomics* 14:704.

Bayer, T., M. Aranda, S. Sunagawa, L. Yum, M. DeSalvo, E. Lindquist, M. A. Coffroth, C. R. Voolstra, and M. Medina. 2012. *Symbiodinium* transcriptomes: Genome insights into the dinoflagellate symbionts of reef-building corals. *PLoS ONE* 7:e35269.

Beman, M. J., K. J. Roberts, L. Wegley, F. Rohwer, and C. A. Francis, 2007. Distribution and diversity of archaeal ammonia monooxygenase genes associated with corals. *Applied and Environmental Microbiology* 73:5642–5647.

Bourne, D. G., P. G. Dennis, S. Uthicke, R. M. Soo, G. W. Tyson, and N. Webster. 2013. Coral reef invertebrate microbiomes correlate with the presence of photosymbionts. *The ISME Journal* 7:1452–1458.

Bourne, D. G., A. Muirhead, and Y. Sato. 2011. Changes in sulfate-reducing bacterial populations during the onset of black band disease. *The ISME Journal* 5:559–564.

Bourne, D. G. and C. B. Munn. 2005. Diversity of bacteria associated with the coral *Pocillopora damicornis* from the Great Barrier Reef. *Environmental Microbiology* 7:1162–1174.

Broadbent, A. D. and G. B. Jones. 2004. DMS and DMSP in mucus ropes, coral mucus, surface films and sediment pore waters from coral reefs in the Great Barrier Reef. *Marine and Freshwater Research* 55:849–855.

Broadbent, A. D., G. B. Jones, and R. J. Jones. 2002. DMSP in corals and benthic algae from the Great Barrier Reef. *Estuarine, Coastal and Shelf Science* 55:547–555.

Cheung, V. G. and S. F. Nelson. 1996. Whole genome amplification using a degenerate oligonucleotide primer allows hundreds of genotypes to be performed on less than one nanogram of genomic DNA. *Proceedings of the National Academy of Sciences of the United States of America* 93:14676–14679.

Chimetto, L. A., M. Brocchi, C. C. Thompson, R. C. R. Martins, H. R. Ramos, and F. L. Thompson. 2008. *Vibrios* dominate as culturable nitrogen-fixing bacteria of the Brazilian coral *Mussismilia hispida*. *Systematic and Applied Microbiology* 31:312–319.

Correa, A. M. S., R. M. Welsh, and R. L. V. Thurber. 2013. Unique nucleocytoplasmic dsDNA and +ssRNA viruses are associated with the dinoflagellate endosymbionts of corals. *The ISME Journal* 7:13–27.

Daniel, R. 2005. The metagenomics of soil. *Nature Review Microbiology* 3:470–478.

Denker, E., E. Bapteste, H. Le Guyader, M. Manuel, and N. Rabet. 2008. Horizontal gene transfer and the evolution of cnidarian stinging cells. *Current Biology* 18:R858–R859.

DeSalvo, M. K., A. Estrada, S. Sunagawa, and M. Medina. 2012. Transcriptomic responses to darkness stress point to common coral bleaching mechanisms. *Coral Reefs* 31:215–228.

DeSalvo, M. K., C. R. Voolstra, S. Sunagawa, J. A. Schwarz, J. H. Stillman, M. A. Coffroth, A. M. Szmant, and M. Medina. 2008. Differential gene expression during thermal stress and bleaching in the Caribbean coral *Montastraea faveolata*. *Molecular Ecology* 17:3952–3971.

Di Salvo, L. and K. Gundersen. 1971. Regenerative functions and microbial ecology of coral reefs Part 1: Assay for microbial population. *Canadian Journal of Microbiology* 17:1081–1089.

Dinsdale, E. A., O. Pantos, S. Smriga, R. A. Edwards, F. Angly, L. Wegley et al. 2008. Microbial ecology of four coral atolls in the Northern Line Islands. *PLoS ONE* 3:1–7.

Eckardt, N. A. 2007. Oxidation pathways and plant development: Crosstalk between thioredoxin and glutaredoxin pathways. *The Plant Cell Online* 19:1719–1721.

Edge, S. E., M. B. Morgan, D. F. Gleason, and T. W. Snell. 2005. Development of a coral cDNA array to examine gene expression profiles in *Montastraea faveolata* exposed to environmental stress. *Marine Pollution Bulletin* 51:507–523.

Ellegren, H., L. Smeds, R. Burri, P. I. Olason, N. Backström, T. Kawakami et al. 2012. The genomic landscape of species divergence in *Ficedula flycatchers*. *Nature* 491:756–760.

Erwin P. M., M. C. Pineda, N. S. Webster, X. Turon, and S. Lopez-Legentil. 2013. Down under the tunic: Bacterial biodiversity hotspots and widespread ammonia-oxidizing archaea in coral reef ascidians. *The ISME Journal* 8:575–588.

Fagan, T., D. Morse, and J. W. Hastings. 1999. Circadian synthesis of a nuclear-encoded chloroplast glyceraldehyde-3-phosphate dehydrogenase in the dinoflagellate *Gonyaulax polyedra* is translationally controlled. *Biochemistry* 38:7689–7695.

Falkowski, P. G., Z. Dubinsky, L. Muscatine, and L. McCloskey. 1993. Population control in symbiotic corals. *BioScience* 43:606–611.

Ferrier-Pages, C., J. P. Gattuso, G. Cauwet, J. Jaubert, and D. Allemand. 1998. Release of dissolved organic carbon and nitrogen by the zooxanthellate coral *Galaxea fascicularis*. *Marine Ecology Progress Series* 172:265–274.

Ferrier-Pages, C., N. Leclercq, J. Jaubert, and S. P. Pelegri. 2000. Enhancement of pico- and nano-plankton growth by coral exudates. *Aquatic Microbial Ecology* 21:203–209.

Ferris, M. J., G. Muyzer, and D. M. Ward. 1996. Denaturing gradient gel electrophoresis profiles of 16S rRNA-defined populations inhabiting a hot spring microbial mat community. *Applied Environmental Microbiology* 62:340–346.

Frias-Lopez, J., J. S. Klaus, G. T. Bonheyo, and B. W. Fouke. 2004. Bacterial community associated with black band disease in corals. *Applied Environmental Microbiology* 70:5955–5962.

Frias-Lopez, J., A. L. Zerkle, G. T. Bonheyo, and B. W. Fouke. 2002. Partitioning of bacterial communities between seawater and healthy, black band diseased, and dead coral surfaces. *Applied and Environmental Microbiology* 68:2214–2228.

Fuhrman, J. A. and L. Campbell. 1998. Microbial microdiversity. *Nature* 393:410–411.

Furst, P. and P. Stehle. 2004. What are the essential elements needed for the determination of amino acid requirements in humans? *Journal of Nutrition* 134:1558S–1565S.

Garren, M., K. Son, J. B. Raina, R. Rusconi, F. Menolascina, O. H. Shapiro, J. Tout, D. G. Bourne, J. R. Seymour, and R. Stocker. 2014. A bacterial pathogen uses dimethylsulfoniopropionate as a cue to target heat stressed corals. *The ISME Journal* 8:999–1007.

Gleason, D. F. and G. M. Wellington. 1993. Ultraviolet radiation and coral bleaching. *Nature* 36:5836–5838.

Goecks, J., A. Nekrutenko, J. Taylor, and Galaxy Team. 2010. Galaxy: A comprehensive approach for supporting accessible, reproducible, and transparent computational research in the life sciences. *Genome Biology* 11:R86.

Grossart, H. P., L. Riemann, and F. Azam. 2001. Bacterial motility in the seas and its ecological impli-cations. *Aquatic Microbial Ecology* 25:247–258.

Guppy, R. and J. C. Bythell. 2006. Environmental effects on bacterial diversity in the surface mucus layer of the reef coral *Montastraea faveolata*. *Marine Ecology Progress Series* 328:133–142.

Hake, S. B. and C. D. Allis. 2006. Histone H3 variants and their potential role in indexing mamma-lian genomes: The "H3 barcode hypothesis". *Proceedings of the National Academy of Sciences of the United States of America* 103:6428–6435.

Hong, M. J., Y. T. Yu, C. A. Chen, P. W. Chiang, and S. L. Tang. 2009. Influence of species specificity and other factors on bacteria associated with the coral *Stylophora pistillata* in Taiwan. *Applied Environmental Microbiology* 75:7797–7806.

Howard, E. C., S. Sun, C. R. Reisch, D. A. del Valle, H. Burgmann, R. P. Kiene et al. 2011. Changes in dimethylsulfoniopropionate demethylase gene assemblages in response to an induced phyto-plankton bloom. *Applied Environmental Microbiology* 77:534–531.

Huggett, M. J., G. R. Crocetti, S. Kjelleberg, and P. D. Steinberg. 2008. Recruitment of the sea urchin *Heliocidaris erythrogramma* and the distribution and abundance of inducing bacteria in the field. *Aquatic Microbial Ecology* 53:161–171.

Hunt, L. A., S. A. Smith, K. R. Downum, and L. A. Mydlarz. 2012. Microbial regulation in gorgonian corals. *Marine Drugs* 10:1225–1243.

Ikeda, Y. and S. Miyachi. 1995. Carbon dioxide fixation by photosynthesis and calcification for a solitary coral, *Fungia* sp. *Bulletin de l'Institut Oceanographique* 14:61–67.

Jonasson, J., M. Olofsson, and H. J. Monstein. 2002. Classification, identification and subtyping of bacteria based on pyrosequencing and signature matching of 16S rDNA fragments. *Acta Pathologica, Microbiologica, et Immunologica Scandinavica* 110:263–272.

Jones, A. M., R. Berkelmans, M. J. H. van Oppen, J. C. Mieog, and W. Sinclair. 2008. A commu-nity change in the algal endosymbionts of a scleractinian coral following a natural bleaching event: Field evidence of acclimatization. *Proceedings of the Royal Society B: Biological Sciences* 275:1359–1365.

Kapley, A., S. Siddiqui, K. Misra, S. M. Ahmad, and H. J. Purohit. 2007. Preliminary analysis of bacterial diversity associated with the *Porites* coral from the Arabian Sea. *World Journal of Microbiology and Biotechnology* 23:923–930.

Keeling, P. J. 2009. Functional and ecological impacts of horizontal gene transfer in eukaryotes. *Current Opinion in Genetics and Development* 19:613–619.

Keeling, P. J. and J. D. Palmer. 2008. Horizontal gene transfer in eukaryotic evolution. *Nature Review Genetics* 9:605–618.

Kellogg, C. 2004. Tropical Archaea: Diversity associated with the surface microlayers of corals. *Marine Ecology Progress Series* 273:81–88.

Kimes, N. E., C. J. Grim, W. R. Johnson, N. A. Hasan, B. D. Tall, M. H. Kothary et al. 2012. Temperature regulation of virulence factors in the pathogen *Vibrio coralliilyticus*. *The ISME Journal* 6:835–846.

Kimes, N. E., J. D. Van Nostrand, E. Weil, J. Z. Zhou, and P. J. Morris. 2010. Microbial functional structure of *Montastraea faveolata*, an important Caribbean reef-building coral, differs between healthy and yellow-band diseased colonies. *Environmental Microbiology* 12:541–556.

King, A. M. Q., M. J. Adams, E. B. Carstens, and E. J. Lefkowitz. 2011. *Virus Taxonomy: Ninth Report of the International Committee on Taxonomy of Viruses*. San Diego, CA: Elsevier Academic Press.

Klaus, J. S., J. Frias-Lopez, G. T. Bonheyo, J. M. Heikoop, and B. W. Fouke. 2005. Bacterial commu-nities inhabiting the healthy tissues of two Caribbean reef corals: Interspecific and spatial variation. *Coral Reefs* 24:129–137.

Kneip, C., P. Lockhart, C. Voß, and U. Maier. 2008. Nitrogen fixation in eukaryotes—New models for symbiosis. *BMC Evolutionary Biology* 7:55.

Koren, O. and E. Rosenberg. 2006. Bacteria associated with mucus and tissues of the coral *Oculina patagonica* in summer and winter. *Applied Environmental Microbiology* 72:5254–5259.

Krediet, C. J., K. B. Ritchie, A. Alagely, and M. Teplitski. 2013. Members of native coral microbiota inhibit glycosidases and thwart colonization of coral mucus. *The ISME Journal* 7:980–990.

Kvennefors, E. C. E., E. Sampayo, T. Ridgway, A. C. Barnes, and O. Hoegh-Guldberg. 2010. Bacterial communities of two ubiquitous Great Barrier Reef corals reveals both site- and species-specificity of common bacterial associates. *PLoS ONE* 5:e10401.

Lampert, Y., D. Kelman, Y. Nitzan, Z. Dubinsky, A. Behar, and R. T. Hill. 2008. Phylogenetic diversity of bacteria associated with the mucus of Red Sea corals. *FEMS Microbiology and Ecology* 64:187–198.

Lane, D. J. 1991. 16S/23S rRNA sequencing. In *Nucleic Acid Techniques in Bacterial Systematic*, E. Stackebrandt and M. Goodfellow (eds.), pp. 115–147. New York: John Wiley & Sons.

Lauro, F. M., D. McDougald, T. J. Williams, S. Egan, S. Rice, M. Z. DeMaere et al. 2009. A tale of two lifestyles: The genomic basis of trophic strategy in bacteria. *Proceedings of the National Academy of Sciences of the United States of America* 106:15527–15533.

Lee, O. O., J. Yang, S. Bougouffa, Y. Wang, Z. Batang, R. Tian, A. Al-Suwailem, and P. Qian. 2012. Pyrosequencing reveals spatial and species variations in bacterial communities associated with corals from the Red Sea. *Applied Environmental Microbiology* 78:7173–7184.

Leggat, W., O. Hoegh-Guldberg, S. Dove, and D. Yellowlees. 2007. Analysis of an EST library from the dinoflagellate (*Symbiodinium* sp.) symbiont of reef-building corals. *Journal of Phycology* 43:1010–1021.

Leggat, W., F. Seneca, K. Wasmund, L. Ukani, D. Yellowlees, and T. Ainsworth. 2011. Differential responses of the coral host and their algal symbiont to thermal stress. *PLoS ONE* 6:e26687.

Lespinet, O., Y. I. Wolf, E. V. Koonin, and L. Aravind. 2002. The Role of lineage-specific gene family expansion in the evolution of eukaryotes. *Genome Research* 12:1048–1059.

Lesser, M. P. 2004. Experimental biology of coral reef ecosystems. *Journal of Experimental Marine Biology and Ecology* 300:217–252.

Lesser, M. P., L. I. Falcon, A. Rodriguez-Roman, S. Enriquez, O. Hoegh-Guldberg, and R. Iglesias-Prieto. 2007. Nitrogen fixation by symbiotic cyanobacteria provides a source of nitrogen for the scleractinian coral *Montastraea cavernosa*. *Marine Ecology Progress Series* 346:143–152.

Lesser, M. P., C. H. Mazel, M. Y. Gorbunov, and P. G. Falkowski. 2004. Discovery of symbiotic nitrogen-fixing cyanobacteria in corals. *Science* 305:997–1000.

Lin, S., H. Zhang, and M. W. Gray. 2007. RNA editing in dinoflagellates and its implications for the evolutionary history of the editing machinery. In *RNA and DNA EDITING: Molecular Mechanisms and Their Integration into Biological Systems*, H. Smith (ed.), pp. 280–309. John Wiley & Sons, Inc. New York.

Little A. F., M. J. H. van Oppen, and B. L. Willis. 2004. Flexibility in algal endosymbioses shapes growth in reef corals. *Science* 304:1492–1494.

Littman, R. A., B. L. Willis, and D. G. Bourne. 2009a. Bacterial communities of juvenile corals infected with different *Symbiodinium* (dinoflagellate) clades. *Marine Ecology Progress Series* 389:45–59.

Littman, R., B. L. Willis, and D. G. Bourne. 2011. Metagenomic analysis of the coral holobiont during a natural bleaching event on the Great Barrier Reef. *Environmental Microbiology Reports* 3:651–660.

Littman, R. A., B. L. Willis, C. Pfeffer, and D. G. Bourne. 2009b. Diversities of coral-associated bacteria differ with location, but not species, for three acroporid corals on the Great Barrier Reef. *FEMS Microbiology and Ecology* 68:152–163.

Malik, H. S. and S. Henikoff. 2003. Phylogenomics of the nucleosome. *Nature Structural and Molecular Biology* 10:882–891.

Mao-Jones, J., K. B. Ritchie, L. E. Jones, and S. P. Ellner. 2010. How microbial community composition regulates coral disease development. *PLOS Biology* 8:1–16.

Marsh, S. 2007. Pyrosequencing applications. *Methods in Molecular Biology* 373:15–24.

Mårtensson, L., D. Beatriz, W. Ingvild, Z. Weiwen, E. Rehab, and R. Ulla. 2009. Diazotrophic diversity, nifH gene expression and nitrogenase activity in a rice paddy field in Fujian, China. *Plant Soil* 325:207–218.

Mehr, S. F. P., R. DeSalle, H. T. Kao, A. Narechania, Z. Han, D. Tchernov, V. Pieribone, and D. F Gruber. 2013. Transcriptome deep-sequencing and clustering of expressed isoforms from Favia corals. *BMC Genomics* 14:546.

Meikle, P., G. N. Richards, and D. Yellowlees. 1988. Structural investigations on the mucus from six species of coral. *Marine Biology* 99:187–193.

Meron, D., E. Atias, L. Iasur Kruh, H. Elifantz, D. Minz, M. Fine et al. 2011. The impact of reduced pH on the microbial community of the coral *Acropora eurystoma*. *The ISME Journal* 5:51–60.

Meron, D., R. Rodolfo-Metalpa, R. Cunning, A. C. Baker, M. Fine, and E. Banin. 2012. Changes in coral microbial communities in response to a natural pH gradient. *The ISME Journal* 6:1775–1785.

Meyer, E., G. V. Aglyamova, and M. V. Matz. 2011. Profiling gene expression responses of coral larvae (*Acropora millepora*) to elevated temperature and settlement inducers using a novel RNA-Seq procedure. *Molecular Ecology* 20:3599–3616.

Meyer, E., G. V. Aglyamova, S. Wang, J. Buchanan-Carter, D. Abrego, J. K. Colbourne et al. 2009a. Sequencing and de novo analysis of a coral larval transcriptome using 454 GSFlx. *BMC Genomics* 10:219.

Meyer, E., S. Davies, S. Wang, B. L. Willis, D. Abrego, T. E. Juenger, and M. V. Matz. 2009b. Genetic variation in responses to a settlement cue and elevated temperature in the reef-building coral *Acropora millepora*. *Marine Ecology Progress Series* 392:81–92.

Mihailovich, M., C. Militti, T. Gabaldon, and F. Gebauer. 2010 Eukaryotic cold shock domain proteins: Highly versatile regulators of gene expression. *BioEssays* 32:109–118.

Miller, T. R., K. Hnilicka, A. Dziedzic, P. Desplats, and R. Belas. 2004. Chemotaxis of *Silicibacter* sp. strain TM1040 toward dinoflagellate products. *Applied Environmental Microbiology* 70:4692–4701.

Mokili, J. L., F. Rohwer, and B. E. Dutilh. 2012. Metagenomics and future perspectives in virus discovery. *Current Opinion in Virology* 2:63–77.

Moya, A., L. Huisman, E. E. Ball, D. C. Hayward, L. C. Grasso, C. M. Chua et al. 2012. Whole transcriptome analysis of the coral *Acropora millepora* reveals complex responses to CO(2)-driven acidification during the initiation of calcification. *Molecular Ecology* 21:2440–2454.

Muyzer, G., E. C. De Waal, and A. G. Uitterlinden. 1993. Profiling of complex microbial populations by denaturing gradient gel electrophoresis analysis of polymerase chain reaction-amplified genes coding for 16S rRNA. *Applied Environmental Microbiology* 59:695–700.

Nishijima, M., K. Adachi, A. Katsuta, Y. Shizuri, and K. Yamasato. 2012. *Endozoicomonas numazuensis* sp. nov., a gammaproteobacterium isolated from marine sponges, and emended description of the genus *Endozoicomonas Kurahashi* and Yokota 2007. *International Journal of Systematic and Evolutionary Microbiology* 63:709–714.

Nissimov, J., E. Rosenberg, and C. B. Munn. 2009. Antimicrobial properties of resident coral mucus bacteria of *Oculina patagonica*. *FEMS Microbiology Letters* 292:210–215.

Niwa, T. 2007. Protein glutathionylation and oxidative stress. *Journal of Chromatography B* 855:59–65.

Olsen, G. J., D. J. Lane, S. J. Giovannoni, N. R. Pace, and D. A. Stahl. 1986. Microbial ecology and evolution—A ribosomal RNA approach. *Annual Review of Microbiology* 40:337–365.

Olson, N. D., T. D. Ainsworth, R. D. Gates, and M. Takabayashi. 2009. Diazotrophic bacteria associated with Hawaiian Montipora corals: Diversity and abundance in correlation with symbiotic dinoflagellates. *Journal of Experimental Marine Biology and Ecology* 371:140–146.

Papp, E., G. Nardai, C. Soti, and P. Csermely. 2003. Molecular chaperones, stress proteins and redox homeostasis. *BioFactors* 17:249–257.

Patten, N. L., P. L. Harrison, and J. G. Mitchell. 2008. Prevalence of virus-like particles within a staghorn scleractinian coral (*Acropora muricata*) from the Great Barrier Reef. *Coral Reefs* 27:569–580.

Polato, N. R., J. C. Vera, and I. B. Baums. 2011. Gene discovery in the threatened elkhorn coral: 454 sequencing of the *Acropora palmata* transcriptome. *PLoS ONE* 6:e28634.

Putnam, H. M., A. B. Mayfield, T. Y. Fan, C. S. Chen, and R. D. Gates. 2013. The physiological and molecular responses of larvae from the reef-building coral *Pocillopora damicornis* exposed to near-future increases in temperature and pCO$_2$. *Marine Biology* 160:2157–2173.

Quigley, K. M., S. W. Davies, C. D. Kenkel, B. L. Willis, M. V. Matz, and L. K. Bay. 2014. Deep-sequencing method for quantifying background abundances of *Symbiodinium* types: Exploring the rare *Symbiodinium* biosphere in reef-building corals. *PLoS ONE* 9:e94297.

Raina, J. B., E. Dinsdale, B. L. Willis, and D. G. Bourne. 2010. Do organic sulphur compounds DMSP and DMS drive coral microbial associations? *Trends in Microbiology* 18:101–108.

Raina, J. B., D. Tapiolas, B. L. Willis, and D. G. Bourne. 2009. Coral-associated bacteria and their role in the biogeochemical cycling of sulfur. *Applied and Environmental Microbiology* 75:3492–3501.

Ramirez-Parra, E. and C. Gutierrez. 2007. The many faces of chromatin assembly factor 1. *Trends in Plant Science* 12:570–576.

Rastogi, R. P., Richa, R. P. Sinha, S. P. Singh, and D. P. Hader. 2010. Photoprotective compounds from marine organisms. *Journal of Industrial Microbiology and Biotechnology* 37:537–558.

Reef, R., S. Dunn, O. Levy, S. Dove, E. Shemesh, I. Brickner et al. 2009. Photoreactivation is the main repair pathway for UV-induced DNA damage in coral planulae. *Journal of Experimental Biology* 212:2760–2766.

Reis, A. M. M., S. D. Araujo, R. L. Moura, R. B. Francini, G. Pappas, A. M. A. Coelho, R. H. Kruger, and F. L. Thompson. 2009. Bacterial diversity associated with the Brazilian endemic reef coral *Mussismilia braziliensis. Journal of Applied Microbiology* 106:1378–1387.

Rice, S. A., K. S. Koh, S. Y. Queck, M. Labbate, K. W. Lam, and S. Kjelleberg. 2005. Biofilm formation and sloughing in *Serratia marcescens* are controlled by quorum sensing and nutrient cues. *Journal of Bacteriology* 187:3477–3485.

Ritchie, K. B. 2006. Regulation of microbial populations by coral surface mucus and mucus-associated bacteria. *Marine Ecology Progress Series* 322:1–14.

Rodolfo-Metalpa, R., C. Richard, D. Allemand, C. N. Bianchi, and C. Morri. 2006. Response of zooxanthellae in symbiosis with the Mediterranean corals *Cladocora caespitosa* and *Oculina patagonica* to elevated temperatures. *Marine Biology* 150:45–55.

Rohwer F, M. Breitbart, J. Jara, F. Azam, and N. Knowlton. 2001. Diversity of bacteria associated with the Caribbean coral *Montastraea franksi. Coral Reefs* 20:85–91.

Rohwer, F., V. Seguritan, F. Azam, and N. Knowlton. 2002. Diversity and distribution of coral-associated bacteria. *Marine Ecology Progress Series* 243:1–10.

Ronaghi, M., S. Karamohamed, B. Pettersson, M. Uhlen, and P. Nyren. 1996. Real-time DNA sequencing using detection of pyrophosphate release. *Analytical Biochemistry* 242:84–89.

Rosic, N., M. Pernice, S. Dove, S. Dunn, and O. Hoegh-Guldberg. 2011. Gene expression profiles of cytosolic heat shock proteins Hsp70 and Hsp90 from symbiotic dinoflagellates in response to thermal stress: Possible implications for coral bleaching. *Cell Stress Chaperones* 16:69–80.

Rosic, N., M. Pernice, S. Dunn, S. Dove, and O. Hoegh-Guldberg. 2010. Differential regulation by heat stress of novel cytochrome P450 genes from the dinoflagellate symbionts of reef-building corals. *Applied Environmental Microbiology* 76:2823–2829.

Salasia, S. I. O. and C. Lammler. 2008. Antibacterial property of marine bacterium *Pseudomonas* sp. associated with a soft coral against pathogenic *Streptococcus equi* subsp. *zooepidemicus. Journal of Coastal Development* 11:113–120.

Sato, Y., D. G. Bourne, and B. L. Willis. 2009. Dynamics of seasonal out-breaks of black band disease in an assemblage of *Montipora* species at Pelorus Island (Great Barrier Reef, Australia). *Proceedings of the Royal Society B: Biological Sciences* 276:2795–2803.

Sato, Y., B. L. Willis, and D. G. Bourne. 2010. Successional changes in bacterial communities during the development of black band disease on the reef coral, *Montipora hispida. The ISME Journal* 4:203–214.

Schmidt, A., M. Gube, A. Schmidt, and E. Kothe. 2009. In silico analysis of nickel containing superoxide dismutase evolution and regulation. *Journal of Basic Microbiology* 49:109–118.

Sekar, R., D. K. Mills, E. R. Remily, J. D. Voss, and L. L. Richardson. 2006. Microbial communities in the surface mucopolysaccharide layer and the black band microbial mat of black band-diseased *Siderastrea siderea. Applied Environmental Microbiology* 72:5963–5973.

Sharp, K. H., D. Distal, and V. J. Paul. 2012. Diversity and dynamics of bacterial communities in early life stages of the Caribbean coral *Porites astreoides. The ISME Journal* 6:790–801.

Shashar, N., Y. Cohen, Y. Loya, and N. Sar. 1994. Nitrogen fixation (acetylene reduction) in stony corals: Evidence for coral-bacteria interactions. *Marine Ecology Progress Series* 111:259–264.

Shick, J. M., S. Romaine-Lioud, C. Ferrier-Pages, and J. P. Gattuso. 1999. Ultraviolet-B radiation stimulates shikimate pathway-dependent accumulation of mycosporine-like amino acids in the coral *Stylophora pistillata* despite decreases in its population of symbiotic dinoflagellates. *Limnology and Oceanography* 44:1667–1682.

Shinzato, C., M. Inoue, and M. Kusakabe. 2014. A Snapshot of a coral "Holobiont": A transcriptome assembly of the scleractinian coral, *Porites*, captures a wide variety of genes from both the host and symbiotic zooxanthellae. *PLoS ONE* 9:e85182.

Shinzato, C., E. Shoguchi, T. Kawashima, M. Hamada, K. Hisata, M. Tanaka et al. 2011. Using the *Acropora digitifera* genome to understand coral responses to environmental change. *Nature* 476:320–323.

Shoguchi, E., M. Tanaka, T. Takeuchi, C. Shinzato, and N. Satoh. 2013. Probing a coral genome for components of the photoprotective scytonemin biosynthetic pathway and the 2-aminoethylphosphonate pathway. *Marine Drugs* 11:559–570.

Siboni, N., E. Ben-Dov, A. Sivan, and A. Kushmaro. 2008. Global distribution and diversity of coral-associated Archaea and their possible role in the coral holobiont nitrogen cycle. *Environmental Microbiology* 10:2979–2990.

Sievert, S. M., R. P. Kiene, and H. N. Schulz-Vogt. 2007. The sulfur cycle. *Oceanography* 20:117–123.

Šlapeta, J. and M. C. Linares. 2013. Combined amplicon pyrosequencing assays reveal presence of the apicomplexan "type-N" (cf. *Gemmocystis cylindrus*) and *Chromera velia* on the Great Barrier Reef, Australia. *PLoS ONE* 8:e76095.

Sogin, M. L., H. G. Morrison, J. A. Huber, D. Mark Welch, S. M. Huse, P. R. Neal, J. M. Arrieta, and G. J. Herndl. 2006. Microbial diversity in the deep sea and the underexplored "rare biosphere". *Proceedings of the National Academy of Sciences of the United States of America* 103:12115–12120.

Soule, T., K. Palmer, Q. Gao, R. M. Potrafka, V. Stout, and F. Garcia-Pichel. 2009. A comparative genomics approach to understanding the biosynthesis of the sunscreen scytonemin in cyanobacteria. *BMC Genomics* 10:336.

Soule, T., V. Stout, W. D. Swingley, J. C. Meeks, and F. Garcia-Pichel. 2007. Molecular genetics and genomic analysis of scytonemin biosynthesis in *Nostoc punctiforme* ATCC 29133. *Journal of Bacteriology* 189:4465–4472.

Speck, M. D. and S. P. Donachie. 2012. Widespread oceanospirillaceae bacteria in *Porites* spp. *Journal of Marine Biology* 2012: 746720.

Starcevic, A., S. Akthar, W. C. Dunlap, J. M. Shick, D. Hranueli, J. Cullum et al. 2008. Enzymes of the shikimic acid pathway encoded in the genome of a basal metazoan, *Nematostella vectensis*, have microbial origins. *Proceedings of the National Academy of Sciences of the United States of America* 105:2533–2537.

Stat, M., E. Morris, and R. D. Gates. 2008. Functional diversity in coral–dinoflagellate symbiosis. *Proceedings of the National Academy of Sciences of the United States of America* 105:9256–9261.

Stocker, R. and J. R. Seymour. 2012. Ecology and physics of bacterial chemotaxis in the ocean. *Microbiology and Molecular Biology Reviews* 76:792–812.

Sunagawa, S., T. Z. DeSantis, Y. M. Piceno, E. L. Brodie, M. K. DeSalvo, C. R. Voolstra et al. 2009. Bacterial diversity and White Plague Disease-associated community changes in the Caribbean coral *Montastraea faveolata*. *The ISME Journal* 3:512–521.

Swanson, R. and O. Hoegh-Guldberg. 1998. Amino acid synthesis in the symbiotic sea anemone *Aiptasia pulchella*. *Marine Biology* 131:83–93.

Szmant, A. M., L. M. Ferrer, and L. M. FitzGerald. 1990. Nitrogen excretion and O:N ratios in reef corals: Evidence for conservation of nitrogen. *Marine Biology* 104:119–127.

Tabatabaei, M., M. R. Zakaria, R. A. Rahim, A. G. Wright, Y. Shirai, N. Abdullah et al. 2009. PCR-based DGGE and FISH analysis of methanogens in anaerobic closed digester tank treating palm oil mill effluent. *Electronic Journal of Biotechnology* 12:1–12.

Tait, K., Z. Hutchison, F. L. Thompson, and C. B. Munn. 2010. Quorum sensing signal production and inhibition by coral-associated *vibrios*. *Environmental Microbiology Reports* 2:145–150.

Taylor, M. W., P. J. Schupp, H. J. Baillie, T. S. Charlton, R. de Nys, S. Kjelleberg, and P. D. Steinberg. 2004. Evidence for acyl homoserine lactone signal production in bacteria associated with marine sponges. *Applied Environmental Microbiology* 70:4387–4389.

Telenius, H., N. P. Carter, C. E. Bebb, M. Nordenskjold, B. A. Ponder, and A. Tunnacliffe. 1992. Degenerate oligonucleotide-primed PCR: General amplification of target DNA by a single degenerate primer. *Genomics* 13:718–725.

Thomas, T., J. Gilbert, and F. Meyer. 2012. Metagenomics—A guide from sampling to data analysis. *Microbial Informatics and Experimentation* 2:1–12.

Thurber, R., D. Willner Hall, B. Rodriguez Mueller, C. Desnues, R. Edwards, F. Angly et al. 2009. Metagenomic analysis of stressed coral holobionts. *Environmental Microbiology* 11:2148–2163.

Todd, J. D., A. R. Curson, N. Nikolaidou-Katsaraidou, C. A. Brearley, N. J. Watmough, Y. Chan et al. 2010. Molecular dissection of bacterial acrylate catabolism—Unexpected links with dimethyl-sulfoniopropionate catabolism and dimethyl sulfide production. *Environmental Microbiology* 12:327–343.

Tomaru, Y., N. Katanozaka, K. Nishida, Y. Shirai, K. Tarutani, M. Yamaguchi, and K. Nagasaki. 2004. Isolation and characterization of two distinct types of HcRNAV, a single-stranded RNA virus infecting the bivalve-killing microalga *Heterocapsa circularisquama*. *AME* 34:207–218.

Tout, J., T. C. Jeffries, N. S. Webster, R. Stocker, P. J. Ralph, and J. R. Seymour. 2014. Variability in microbial community composition and function between different niches within a coral reef. *Microbial Ecology* 67:540–552.

Traylor-Knowles, N., B. R. Granger, T. J. Lubinski, J. R. Parikh, S. Garamszegi, Y. Xia et al. 2011. Production of a reference transcriptome and transcriptomic database (PocilloporaBase) for the cauliflower coral, Pocillopora damicornis. *BMC Genomics* 12:585.

Ushijima, B., A. Smith, G. S. Aeby, and S. M. Callahan. 2012. *Vibrio owensii* induces the tissue loss disease *Montipora* white syndrome in the Hawaiian reef coral *Montipora capitata*. *PLoS ONE* 7:e46717.

van Alstyne, K., P. Schupp, and M. Slattery. 2006. The distribution of dimethylsulfoniopropionate in tropical Pacific coral reef invertebrates. *Coral Reefs* 25:321–327.

van Nimwegen, E. 2003. Scaling laws in the functional content of genomes. *Trends in Genetics* 19:479–484.

Vega-Thurber, R. L. and A. M. S. Correa. 2011. Viruses of reef-building scleractinian corals. *Journal of Experimental Marine Biology and Ecology* 408:102–113.

Vega-Thurber, R. L., D. Willner-hall, B. Rodriguez-mueller, C. Desnues, R. A. Edwards, F. Angly et al. 2009. Metagenomic analysis of stressed coral holobionts. *Environmental Microbiology* 11:2148–2163.

Vieira, D. S. C. and P. Rey. 2006. Plant thioredoxins are key actors in the oxidative stress response. *Trends in Plant Science* 11:329–334.

Von Holt, C. and M. Von Holt. 1968. The secretion of organic compounds by zooxanthellae isolated from various types of *Zoanthus*. *Comparative Biochemistry and Physiology* 24:83–92.

Wafar, M., S. Wafar, and J. J. David. 1990. Nitrification in reef corals. *Limnology and Oceanography* 35:725–730.

Wang, J. T. and A. E. Douglas. 1999. Essential amino acid synthesis and nitrogen recycling in an alga-invertebrate symbiosis. *Marine Biology* 135:219–222.

Webster, N. S., L. D. Smith, A. J. Heyward, J. E. Watts, R. I. Webb, L. L. Blackall, and A. P. Negri. 2004. Metamorphosis of a Scleractinian coral in response to microbial biofilms. *Applied Environmental Microbiology* 70:1213–1221.

Wegley, L., R. Edwards, B. Rodriguez-Brito, H. Liu, and F. Rohwer. 2007. Metagenomic analysis of the microbial community associated with the coral *Porites astreoides*. *Environmental Microbiology* 9:2707–2719.

Wegley, L., Y. Yu, M. Breitbart, V. Casas, D. I. Kline, and F. Rohwer. 2004. Coral-associated Archaea. *Marine Ecology Progress Series* 273:89–96.

Weynberg, K. D., E. M. Wood-Charlson, C. A. Suttle and M. J. van Oppen. 2014. Generating viral metagenomes from the coral holobiont. *Frontiers in Microbiology* 5:1–11.

Wild, C., W. Holger, and M. Huettel. 2005. Influence of coral mucus on nutrient fluxes in carbonate sands. *Marine Ecology Progress Series* 287:87–98.

Wild, C., M. Huettel, A. Klueter, S. G. Kremb, M. Y. M. Rasheed, and B. Jorgensen. 2004a. Coral mucus functions as an energy carrier and particle trap in the reef ecosystem. *Nature* 428:66–70.

Wild, C., M. Naumann, W. Niggl, and A. Haas. 2010. Carbohydrate composition of mucus released by scleractinian warm and cold water reef corals. *Aquatic Biology* 10:41–45.

Wild, C., M. Rasheed, U. Werner, U. Franke, R. Johnstone, and M. Huettel. 2004b. Degradation and mineralization of coral mucus in reef environments. *Marine Ecology Progress Series* 267:159–171.

Williams, W. M., A. B. Viner, and W. J. Broughton. 1987. Nitrogen fixation (acetylene reduction) associated with the living coral *Acropora variabilis*. *Marine Biology* 94:531–535.

Woese, C. R. 1987. Bacterial evolution. *Microbiological Reviews* 51:221–271.

Yang, C. S., M. H. Chen, A. B. Arun, C. A. Chen, J. T. Wang, and W. M. Chen. 2010. *Endozoicomonas montiporae* sp. nov., isolated from encrusting pore coral *Montipora aequituberculata*. *International Journal of Systematic and Evolutionary Microbiology* 60:1158–1162.

Yellowlees, D., T. A. Rees, and W. Leggat. 2008. Metabolic interactions between algal symboints and invertebrate hosts. *Plant Cell and Environment* 31:679–694.

Yokouchi, H., Y. Fukuoka, D. Mukoyama, R. Calugay, H. Takeyama, and T. Matsunaga. 2006. Whole-metagenome amplification of a microbial community associated with scleractinian coral by multiple displacement amplification using φ29 polymerase. *Environmental Microbiology* 8:1155–1163.

Yongchang, O., S. Dai, L. Xie, M. S. Ravi Kumar, W. Sun, H. Sun, D. Tang, and X. Li. 2010. Isolation of high molecular weight DNA from marine sponge bacteria for BAC library construction. *Marine Biotechnology* 12:318–325.

Zhang, L., X. Cui, K. Schmitt, R. Hubert, W. Navidi, and N. Arnheim. 1992. Whole genome amplification from a single cell: Implications for genetic analysis. *Proceedings of the National Academy of Sciences of the United States of America* 89:5847–5851.

Zilberman, D., D. Coleman-Derr, T. Ballinger, and S. Henikoff. 2008. Histone H2A.Z and DNA methylation are mutually antagonistic chromatin marks. *Nature* 456:125–129.

9

Metagenomics of Marine Actinomycetes: From Functional Gene Diversity to Biodiscovery

Sergey B. Zotchev, Olga N. Sekurova, and D. İpek Kurtböke

CONTENTS

9.1 Introduction

Since the golden era of antibiotics, natural products played an important role in the development of drugs (Hopwood, 2007a; Bérdy, 2005, 2012; Zotchev, 2012; Kurtböke et al., 2014; Müller and Wink, 2014). So far 63% of new drugs have been classified as naturally derived, including unmodified or modified natural products, or synthetic compounds deriving from the pharmacophore scaffolds of natural products (Imhoff et al., 2011). Continued mining of these known scaffolds has been an effective approach in delivering new drugs. Examples include exploitation of bioactive synthetic scaffolds like the oxazolidinones and diaminopyrimidines leading to the delivery of new compounds and proceeding into late-stage clinical trials (Imhoff et al., 2011). Eight quinolone and fluoroquinolone derivatives were also introduced to the market through such approaches (Butler and Cooper, 2011; Wright, 2012), and 68% of antibiotics against bacterial, fungal, parasitic, and viral infections and 63% of drugs for cancer treatment introduced between 1981 and 2008 are derived from natural sources (Cragg et al., 2009; Imhoff et al., 2011). Interest in the discovery of unique pharmacophores expanded to marine environments when the rediscovery of the known compounds originating mostly from terrestrial microorganisms became an obstacle (Hopwood, 2007b; Zotchev, 2012). Stemming from the urgent current need for novel and potent therapeutic agents for the treatment of existing and emerging diseases and the increasing rate of recent marine-derived biodiscoveries supported the intensive search for new substances from marine organisms, including marine actinomycetes (Molinski et al., 2009; Demain and Vaishnav, 2011; Imhoff et al., 2011; Schumacher et al., 2011; Zotchev, 2012; Kurtböke et al., 2014). Moreover, rapid progress in metagenomics has revealed the true extent of marine microbial, including actinobacterial, diversity via direct access to the genomes of numerous uncultivable microorganisms (Li and Qin, 2005).

Actinomycetes are ubiquitous gram-positive bacteria known for their unprecedented capacity to synthesize structurally diverse secondary metabolites with various biological activities. More than 50% of all antibiotics used today in medical practice are derived from these organisms (Hopwood, 2007a; Bérdy, 2012). For many years, actinomycetes have been mostly isolated from terrestrial sources, and their screening for bioactive substances led to the discoveries of such antibiotics as streptomycin, amphotericin, erythromycin, vancomycin, and many other clinically used ones (Hopwood, 2007a; Bérdy, 2012; Newman and Cragg, 2012). However, with passing time, the rediscovery rate became too high, forcing pharmaceutical companies and research organizations to search for alternative sources for actinomycetes (Hopwood, 2007b; Bérdy, 2012). Marine environment turned out to be a very promising source in this respect, yielding many new actinomycetes and novel bioactive compounds (Newman and Cragg, 2004; Fiedler et al., 2005; Newman and Hill, 2006; Bull and Stach, 2007; Gulder and Moore, 2009; Lane and Moore, 2011; Gerwick and Moore, 2012; Zotchev, 2012). At the same time, genome sequencing of marine actinomycetes has revealed vast biosynthetic potential that remains unexploited (Fenical and Jensen, 2006; Udwary et al., 2007; Baltz, 2008). The main problem of biodiscovery from marine actinomycetes remains their often suboptimal growth under laboratory conditions. This obstacle detrimentally impacts production yields of bioactive compounds as well as hampering the efforts for proper characterization of the isolates (Fenical, 1993; Kurtböke et al., 2014). Metagenomics, over the recent years, has been an appealing approach used for natural product discovery providing information on the genetic information needed to govern microbial secondary metabolite biosynthesis (Banik and Brady, 2010). Metagenomics as a culture-independent technology might therefore provide a viable alternative for biodiscovery from noncultivated microorganisms via detection of biosynthetic genes and expression of these marine actinomycete-derived pathways in alternative microbial hosts (Li and Qin, 2005; Brady et al., 2009).

This chapter provides an overview of the recent discoveries and future perspectives based on the metagenomics advances unraveling biotechnological potential of marine actinomycetes.

9.2 Marine Environments and Biodiscovery

The need for novel drugs for the treatment of severe human diseases such as cancer, microbial infections, and inflammatory processes, combined with the recognition that marine organisms provide a rich potential source of such compounds, supports the intensive search for new substances from marine organisms (Hopwood, 2007b; Molinski et al., 2009; Mayer et al., 2010). Concept of "Drugs from the Sea" was actually introduced over 50 years ago, while true discoveries from marine environments were reported to be eventuated much later (Imhoff et al., 2011). The high hit rates from marine-derived compounds in screening for drug leads thus made the search in marine organisms quite attractive (Gulder and Moore, 2009; Williams, 2009).

By 1974, two marine-derived natural products cytarabine and vidarabine were part of the pharmacopeia used to treat cancer and viral infections, respectively (Mayer et al., 2010). Over 30 years passed for the medicinal use of two other marine-derived drugs, ziconotide (Prialt®) and Yondelis® (Mayer et al., 2010), and marine sources continue to offer novel chemical classes not found in terrestrial ones (Imhoff et al., 2011; Cragg and Newman, 2013). Among the most promising candidates for anticancer drug development were psymberin from a *Psammocinia* species of sponge and simultaneously described irciniastatin A from another sponge *Ircinia ramosa*. Both of these compounds displayed significant selective toxicity toward solid tumor cells (Piel et al., 2005). Dermatotoxicity has been suggested as a natural biological role of these pederin-type compounds, which might act as defensive agents against predators or epibionts of the sponges (Piel et al., 2005). Most of the marine-derived bioactive compounds were later confirmed to be produced by marine microorganisms in symbiotic association with their hosts (Fenical, 1993; Proksch et al., 2002; Haefner, 2003; Cragg and Newman, 2006; Piel, 2006; Bull and Stach, 2007; Cragg and Newman, 2013). Consequently, modern marine biotechnology has moved its focus from marine macroorganisms to microorganisms for the discovery of new pharmaceuticals (Fiedler et al., 2005; Newman and Hill, 2006; Hopwood, 2007b; Molinski et al., 2009; Mayer et al., 2010; Lane and Moore, 2011). In particular, bacterial and fungal symbionts of marine macroorganisms such as sponges, corals, and algae were found to produce potent bioactive compounds with selective activities against many different targets, including those in tumor cells (Fenical, 1993; Bull and Stach, 2007; Cragg et al., 2009; Wiese et al., 2009; Gulder and Moore, 2010).

Most marine sessile eukaryotic hosts provide a unique surface for microbial colonization (Egan et al., 2008). Biofilm communities are now also gaining importance for the discovery of microbially mediated compound secretions during the formation of biofilms. As stated by Egan et al. (2008), chemically mediated interactions between the host and the colonizers as well as the interactions between microorganisms in the biofilm community and surface-specific physical and chemical conditions can impact differently on the diversity and function of surface-associated microbial assemblages compared with those in suspended systems. As a result, such environments can provide different clues on the bioactive functions of these microorganisms (Penesyan et al, 2009). Thus, understanding the diversity and ecology of surface-associated microbial communities can greatly contribute to the discovery of next-generation bioactive compounds, such as the toxins, signalling molecules, and other secondary metabolites, as effective competition and defense molecules subsequently generating information for biodiscovery efforts (Egan et al., 2008).

Another novel approach has been the recent investigations into the fish microbiome to define vertebrate-derived bacteria as an environmental niche for the discovery of unique marine natural products (Sanchez et al., 2012).

9.2.1 Marine Actinomycetes

Fenical (1993) provided an in-depth review on the history of marine actinomycete-based biodiscovery, including the efforts of Japanese scientists to deliver the early discoveries. Examples included the aplasmomycins A–C from shallow mud samples taken from Sagami Bay. The compounds were only produced by a marine *Streptomyces griseus* species (SS-20) when cultured with Kobu Cha (dried and pulverized powder produced in Japan from the brown seaweed *Laminaria*) containing media or when grown under conditions (27°C and very low nutrients) that relate to the natural environment of Sagami Bay (Okami, 1986).

In the early days of the marine actinomycetes studies, those belonging to the genus *Streptomyces* predominated. However, other nonstreptomycete actinomycetes were also found to produce rare chemical class of compound such as the one reported from the Maduramycete group of actinomycetes. Maduralide was a member of a rare 24-membered ring lactone group represented at that time only by rectilavendomycin (Fenical, 1993). More recently, following the successful culturing of *Salinispora* species, sediment-derived actinomycetes also gained attention (Jensen et al., 2005b, 2007; Jensen and Mafnas, 2006) and delivered an impressive range of bioactive compounds (Eustáquio et al., 2011), including bioactive compounds against the malaria parasite (Prudhomme et al., 2008). Other examples include the recent screening of marine actinomycetes by Gallagher et al. (2010) that has led to the discovery of a new lineage of biologically active hybrid isoprenoid (HI) secondary metabolite producers. Their studies also provided clues on the horizontal gene transfer among members of the distantly related families of actinomycetes. Engelhardt et al. (2010) have also recently reported the production of a new thiopeptide antibiotic, TP-1161, by a marine *Nocardiopsis* species. Other examples include prenylated depsipeptide cyclomarin A(5), which was originally reported from a marine-derived *Streptomyces* sp. (Renner et al., 1999) and subsequently isolated from the marine actinomycete *Salinispora arenicola* (Schultz et al., 2008; Gallagher et al., 2010). These studies indicate that marine actinomycetes can be a promising resource for natural product discovery. Moreover, they can also reveal the evolutionary history and ecological functions of unusual group of secondary metabolites such as the biologically active HIs (Gallagher et al., 2010).

Generation of increased knowledge on the existence, ecological roles, and functional diversity of actinomycetes in marine ecosystems (Penn and Jensen, 2012) as well as the genetic basis for natural product biosynthetic diversity (Nett et al., 2009) will provide a stronger platform for their further utilization as sources of unique bioactive compounds (Jensen et al., 2005; Fenical and Jensen, 2006). Past experiences and wealth of knowledge generated over the last 50 years on marine biodiscovery will also provide a strong platform to the future investigations into marine-derived bioactive natural products (Gerwick and Moore, 2012).

9.2.1.1 Actinomycetes as Symbionts of Marine Macrofauna

Microbial, in particular actinomycete, origin of many marine metabolites has been demonstrated (König et al., 2006; Egan et al., 2008). Thiocoraline is a thiodepsipeptide antitumor compound that was reported to be produced by *Micromonosporae* isolated from two marine invertebrates (a soft coral and a mollusc) of the Indian Ocean coast of Mozambique (Lombó et al., 2006). Screening of mucus of marine macroorganisms has also revealed the associated presence of actinomycetes (Lampert et al., 2006; Nithyanand et al., 2011).

Diverse actinomycete genera were also found to be associated with marine sponges (Webster et al., 2001; Kim et al., 2005), and sponge-associated bioactive compounds of microbial origin have been often detected (Hill, 2004; Taylor et al., 2007; Schneemann et al., 2010). Examples include sponge-associated actinomycetes as new sources of rifamycins (Kim et al., 2006), as well as isoprenoids (Izumikawa et al., 2010) and antiparasitic compounds, in particular those active against reemerging pathogen *Leishmania* (Pimentel-Elardo et al., 2010). New tetromycin derivatives with antitrypanosomal and protease inhibitory activities were also reported to be produced by a *Streptomyces* species associated with the Mediterranean sponge *Axinella polypoides* (Pimentel-Elardo et al., 2011). Manzamine class of antimalarial compounds was also recently reported to be produced by a *Micromonospora* species associated with the Indo-Pacific sponge *Acanthostrongylophora ingens* (Waters et al., 2014).

Woodhouse et al. (2013) with the application of tag-encoded FLX amplicon pyrosequencing targeting nonribosomal peptide synthetase (NRPS) and polyketide synthase (PKS) genes identified diverse natural product biosynthesis genes within the sponge microbiome. Such approaches will thus bring new insights into the biological processes that influence the production of secondary metabolites in sponges and other marine environments.

9.2.1.2 Actinomycetes in Marine Sediments

Both nearshore and deep-sea marine sediments have also been a prolific source for the discovery of novel bioactive compounds where actinomycetes were detected in abundance (Jensen et al., 2005b; Maldonado et al., 2005; Pathom-aree et al., 2006), in particular the members of the family Micromonosporaceae (Magarvey et al., 2004; Bredholdt et al., 2007; Bredholt et al., 2008; Eccleston et al., 2008; Prieto-Davó et al., 2008; Zhang et al., 2012a). Novel phylogenetic diversity (Gontang et al., 2007) and detection of new taxa have also been reported from marine sediments such as the proposed new genus *Sciscionella* from northern South China Sea sediments by Tian et al. (2009). *Verrucosispora maris*, a novel species isolated from a deep-sea marine sediment, was also reported to produce abyssomicins (Goodfellow et al., 2012) and formed a new genus again within the family Micromonosporaceae. Proximicins with strong cytostatic effect against various human tumor cell lines were also detected from this species (Fiedler et al., 2008). Another novel species of this genus *Verrucosispora sediminis* was isolated from deep-sea marine sediment and reported to produce cyclodipeptide (Dai et al., 2010).

Further examples include recent discovery of novel anticancer and antifungal phenazine derivative from a marine actinomycete BM-17 (Gao et al., 2012). Production of new bisindole alkaloids (spiroindimicins A–D and indimicins A–E) from *Streptomyces* species isolated from a sediment sample collected in the Bengal Bay of the Indian Ocean was also reported (Zhang et al., 2012a, 2014a). These compounds were also found to be satisfactory in their interaction against drug target enzymes when *in silico* molecular docking was used for their preclinical evaluation by Saurav et al. (2014). The same species was also reported to produce polyketide macrolactams (heronamides D–F) indicating the production of broad range of metabolites (Zhang et al., 2014b). New diketopiperazine derivatives from a deep-sea sediment-derived *Nocardiopsis alba* was also recently been reported (Zhang et al., 2012b) as well as the fluostatins I–K from *Micromonospora rosaria* isolated from the South China Sea–derived sediment (Zhang et al., 2012c).

So far one of the most investigated bioactive marine actinomycete genera has been the *Salinispora* with widespread distribution in marine sediments of tropical and subtropical marine environments (Mincer et al., 2005; Jensen and Mafnas, 2006; Freel, 2012; Ahmed et al., 2013). It has also been reported to be presented both in nearshore and deep-sea marine sediments (Prieto-Davó et al., 2013). Comparative genomics studies on this genus also revealed evidence on the functional marine adaptation (Penn et al., 2009; Penn and Jensen, 2012). The genus is currently composed of three closely related species, which are a rich source of secondary metabolites that are produced in species-specific patterns (Jensen et al., 2007, Gontang et al., 2010). The rich and diverse natural product chemistry produced by the genus members has been extensively reported by the Scripps Institution of Oceanography (San Diego, CA) (Oh et al., 2008; Schultz et al. 2008; Fenical et al., 2009; Asolkar et al., 2010; Lane et al., 2013) since the detection of the new taxon by

Mincer et al. (2002). Among the genome sequenced species of this genus, complex secondary metabolome for *Salinispora tropica* has been reported (Udwary et al., 2007). In addition, domain phylogeny investigations on *Salinispora pacifica* by Freel et al. (2011) indicated vertical inheritance from a common ancestor shared with *S. tropica*, which produces related compounds in the salinosporamide series (Feling et al., 2003). Information generated on the distributions and phylogenies of the biosynthetic genes of this genus members is now providing insight into the complex processes driving the evolution of secondary metabolism among closely related species of this genus (Freel et al., 2011).

9.3 Metagenomics

Molecular biology advances starting from the 1980s led to the development of robust methods for isolation of total genomic DNA from complex environmental samples (Handelsman et al., 1998). High-throughput sequencing of such environmental DNA (eDNA) greatly expanded our understanding of the true diversity of the bacterial communities (Brady et al., 2009). Microorganisms in any environmental sample thus were placed into two different categories: (1) "culturable" and (2) "unculturable" (Miao and Davies, 2009). Shotgun cloning method of examining microbial communities without necessarily culturing them in the laboratory environments was developed, and the term metagenomics was coined (Handelsman et al., 1998; Miao and Davies, 2009). Subsequently, vast amount of information has been generated on the existing biosynthetic pathways of "unculturable" microorganisms in the natural environments (Miao and Davies, 2009). To reveal the site-to-site variation of the microbial diversity, two different approaches were also adapted (1) "functional metagenomics" and (2) "heavy-duty sequencing of DNA fragments obtained from environmental bacteria" (Johnston et al., 2005).

In the "functional metagenomics" approach, the metagenomic libraries were constructed from pooled bacterial genomic DNAs and subsequently screened for clones exhibiting a particular phenotype, which might exist in the culturable libraries (Johnston et al., 2005). The second approach involved high-throughput sequencing of DNA fragments obtained from environmental samples, including those with potentially limited biodiversity, such as the extremely acidic mine drainage sites. This approach provided snapshots of microorganisms present in the environment in this particular time and space (Johnston et al., 2005). In yet another approach, the combination of DNA stable-isotope probing (SIP) method and metagenomics facilitated the detection of rare low-abundance species from metagenomic libraries, thus allowing the detection of novel enzymes and bioactive compounds (Chen and Murrell, 2010). Another successful trend has been the use of homology-guided metagenomic screening targeting the discovery of new members of traditionally rare, biomedically relevant natural product families (Chang and Brady, 2013).

9.3.1 Discovery of Novel Compounds via Metagenomical Approaches

Metagenomics of diverse environmental niches, including marine, has greatly contributed to the discovery of new genes and proteins potentially involved in the biosynthesis of bioactive compounds. Subsequently, screening of metagenomic libraries has contributed toward the discovery of an array of novel bioactive compounds and generated strains, which can produce novel natural products or enzymes (Knietsch et al., 2003;

Johnston et al., 2005; Piel et al., 2005; Zotchev et al., 2012). In the drug discovery field, since its introduction, metagenomics has identified a significant number of the genes of interest (GOIs) such as those for type I and type II PKS and NRPS, which are key genes involved in the synthesis of many bioactive compounds by actinomycete bacteria (Streit and Schmitz, 2004; Nikolouli and Mossialos, 2012).

In general, the process of metagenome-based biodiscovery can follow two different strategies, which converge at the stage of expression of metagenome-derived genes in various heterologous hosts (Figure 9.1). The first, conventional strategy, is based on digestion and cloning of environmental DNA into plasmid vectors, usually in *Escherichia coli*, followed by functional screening that may be preceded by transfer of the metagenomic library into various bacterial hosts (alternative hosts for functional (meta)genome analysis (Liebl et al., 2014). The latter is very important for the discovery of novel secondary metabolites with biological activity, since functional expression of such genes, especially if they originate from GC-rich actinomycete genomes, requires regulatory machinery, tRNA, and precursor pools absent or deficient in *E. coli*. Until recently, the latter strategy has been utilized mostly for the discovery of various enzymes encoded by single genes (Zhang et al., 2011; Zhu et al., 2013) and only rarely yielded bioactive secondary metabolites (He et al., 2012; Chang and Brady, 2013). In order to circumvent the latter problem, which is often due to the lack of functional expression of biosynthetic gene clusters in certain bacterial hosts, a homology-guided approach can be used. This approach, mostly utilizing PCR amplicons

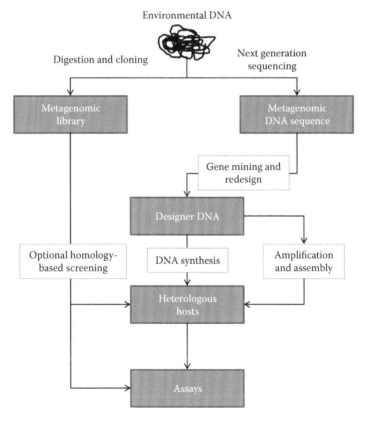

FIGURE 9.1
Alternative strategies for metagenomics-based discovery of bioactive compounds.

for conserved genes as probes, is useful in the discovery of new members of the known classes of compounds, for which core biosynthetic genes are known (Chang and Brady, 2011; Owen et al., 2013).

Mapping gene clusters within arrayed metagenomic libraries can be used to expand the structural diversity of biomedically relevant natural products. Still, this rather conventional strategy most likely fails to capture the true diversity of natural products encoded by metagenomes due to the fragmentation of large biosynthetic gene clusters in the process of original library construction and lack of functional expression in heterologous hosts. Only a small number of compounds apparently originating from actinomycete bacteria have been discovered using this strategy that was amended by reassembly and expression of the discovered core biosynthetic genes in actinomycete host (Wilson and Piel, 2013). In particular, amino acid sequences for ketosynthases (KSs) involved in the biosynthesis of aromatic polyketides eracidin and fluostatin discovered through the latter strategy (Figure 9.2) display high degree of identity (90% and 77%, respectively) to their counterparts from *Streptomyces* bacteria. Other molecules, fasamycin and UT-X26 (Figure 9.2), were discovered using similar approach (Feng et al., 2012). Although the DNA sequences have not been released, successful functional expression of metagenomics clones in *Streptomyces albus* strongly suggests the actinomycete origin of respective biosynthetic gene clusters.

The second, the most advanced strategy, becomes possible due to the rapid development of next-generation DNA sequencing and synthesis technologies (Figure 9.1). Here, the environmental DNA is first deep-sequenced to generate a collection of metagenomic DNA contigs, which can be mined for specific GOIs, for example, hydrolytic enzymes and antibiotic biosynthesis pathways (Do et al., 2014; Wakimoto et al., 2014).

Recently developed online tool antiSMASH can be extremely useful for mining of metagenome sequences for biosynthetic gene clusters (Blin et al., 2013). Once the GOIs are identified, there exist at least two routes to achieving their functional expression. The first

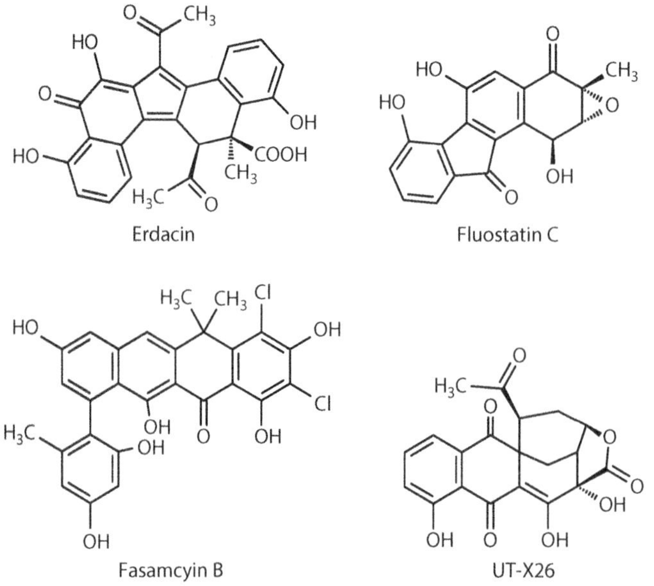

FIGURE 9.2
Novel aromatic polyketides of apparently actinomycete origin discovered via homology-based screening of metagenomic libraries followed by gene reassembly and heterologous expression in *Streptomyces*.

one, providing the highest degree of flexibility, is based on redesign and *de novo* synthesis of the GOIs that are in this way adapted for expression in a particular host. The redesign (sometimes also called refactoring) may include codon optimization and randomization, replacement of regulatory regions affecting gene expression such as promoters, operators, terminators, and ribosome binding sites (Smanski et al., 2014). The success of this route is highly dependent on the sound design rules that may vary greatly depending on the kind of genes/gene clusters targeted for expression, as well as heterologous host. Here, computational tools will play a decisive role, providing guidelines for gene design and synthesis (Medema et al., 2012, 2014). The second route in this strategy relies on PCR amplification of parts or entire biosynthetic gene clusters from metagenomic DNA followed by their assembly is suitable vectors. The success of this approach mainly depends on three parameters: (1) high-fidelity amplification of relatively large DNA fragments, (2) employment of appropriate DNA assembly methods, and (3) selection of host for heterologous expression.

Recent advances in engineering of high-fidelity DNA polymerases such as Phusion and Advantage are helpful in reducing the number of mutations introduced during PCR. However, error-free amplification of GC-rich DNA fragments over 3 kb originating from actinomycete genomes remains a challenge. Successful amplification of long GC-rich DNA fragments depends on the addition of enhancer mixtures that contain chemicals such as betaine and DMSO that reduce the fidelity of DNA synthesis even by most accurate DNA polymerases.

The development and implementation of DNA assembly methods have been expanding rapidly over the last few years. Some of these methods that are applicable for the assembly of the entire biosynthetic gene clusters are listed in Table 9.1. The general strategy here is based on PCR amplification of either overlapping DNA fragments

TABLE 9.1

Methods of Seamless DNA Assembly Applicable for Direct Cloning of Biosynthetic Gene Clusters from Metagenomes

Method	Principle	References
Gibson assembly	Isothermal one-pot in vitro assembly of multiple DNA fragments with >25 bp homologous flanking sequences.	Gibson et al. (2009)
Methylation-assisted tailorable ends rational ligation	Seamless assembly of long complicated DNA constructs. Parts are PCR amplified to contain specifically recognizable by endonuclease *Msp*JI nonmethylated sites.	Chen et al. (2013)
Circular polymerase extension cloning	Single PCR step catalyzed by Phusion DNA polymerase. It allows to clone multiple inserts into any vector, which is useful for cloning of complex combinatorial libraries and pathways.	Quan and Tian (2009)
Seamless ligation cloning extract	Fast, one-tube, homology-mediated DNA assembly using RecA-deficient bacterial cell extract. It can be used as a convenient cloning method for creating recombinant plasmids.	Zhang et al. (2012)
Transformation-assisted recombination)	PCR amplified or obtained by restriction digestion DNA fragments with overlapping ends are transformed into *Saccharomyces cerevisiae* and assembled together with vector via homologous recombination. This method makes it possible to assemble the entire synthetic microbial genomes and pathways.	Shao et al. (2009)
Domino	Assembly of large DNA fragments in *Bacillus subtilis* using special integrative vectors. Long homologous ends are required.	Itaya et al. (2008)
DNA assembly in *E. coli*	Cloning of entire biosynthetic pathways via homologous recombination in *E. coli*, ensured by overexpression of Rac prophage proteins RecE and RecT.	Fu et al. (2012)

encompassing a gene cluster, amplification of entire clusters if they are relatively small (<10 kb), or direct excision of an entire cluster from genomic DNA. In certain cases, it can also be possible to include gene cluster refactoring step into the scheme, for example, for replacement of gene regulatory regions. However, complex gene rearrangements (e.g., to arrange certain genes into distinct controllable operons) that are easy to implement via the *de novo* DNA synthesis are too labor intensive and inefficient when using assembly methods with GC-rich DNA.

9.4 Hosts for the Heterologous Expression of Actinomycete-Derived Biosynthetic Gene Clusters

One of the critical issues in producing a potentially novel compound based on metagenomics DNA is heterologous expression of biosynthetic gene clusters in a suitable host. The ability to express the biosynthetic GOIs by a new host can be a main limitation in this approach. Among a number of different microbial hosts used for heterologous expression, *E. coli, Saccharomyces cerevisiae,* and *Streptomyces* spp. have been shown to express large secondary metabolite gene clusters (for review see Baltz, 2010).

E. coli, being a "working horse" for molecular biologists for many years, could be a good choice in terms of availability of variety of genetic tools. However, in many cases, it is not suitable as its transcription machinery often does not recognize actinomycete promoters. In addition, it cannot provide the whole pool of precursors for the biosynthesis of secondary metabolites, such as methylmalonyl-CoA, or ensure the proper folding of, for example, PKS type I proteins typical for actinomycetes (Betancor et al., 2008). The differences in codon usage, absence of regulatory elements, or toxicity of synthesized product can also lessen the chance to successfully express the actinomycete-derived GOIs in *E. coli* (Ekkers et al., 2012; Schmieder and Edwards, 2012). *Streptomyces,* as natural producers of vast majority of known secondary metabolites, have been used as heterologous hosts for many years, and there are a considerable number of success stories proving them to be suitable host organisms. They are probably an optimal choice for expressing genes from other actinomycetes, if we take into consideration the similarities in metabolism in closely related species. The original, nonengineered *S. albus* J1074 has been used successfully as a heterologous host for production of various compounds (Feng et al., 2011, 2012). Currently, new engineered *Streptomyces* strains are being used more frequently, since they provide certain advantages.

Streptomyces coelicolor is genetically the most studied *Streptomycete* strain. Its genome is completely sequenced and characterized, and various genetic tools are available to manipulate the genes. Some *S. coelicolor* strains modified by deletion of several well-known antibiotic biosynthesis gene clusters are also available. Deletion or silencing of gene clusters limits the competition for precursors and energy in a host to a benefit for production of heterologous compounds. Among a number of heterologous compounds successfully produced in these strains are, for example, lantibiotics and chloramphenicol (for review see Gomez-Escribano and Bibb, 2014) as well as indolocarbazoles staurosporine and streptocarbazoles both originating from marine-derived *Streptomyces sanyensis* (Li et al., 2013). Moreover, heterologous expression of cluster for polyketide-terpenoid merochlorins from marine streptomycete in *S. coelicolor* led to the discovery of novel enzymatic reactions (Kaysser et al., 2012).

Mutant strains of *Streptomyces avermitilis* were engineered by deleting more than 1.4 Mb from the 9.03 Mb linear chromosome and successfully used for heterologous expression of more than 20 different clusters for biosynthesis of secondary metabolites. This number includes also those that could not be produced by the parental *S. avermitilis* strain as, for example, novobiocin and chloramphenicol. Production titers for some compounds were higher than in the native hosts, and clusters cryptic in the original hosts became active in engineered *S. avermitilis* (Komatsu et al., 2013). *Streptomyces venezuelae* was also recently engineered as a heterologous host for expression of exogenous terpene synthases, and successful production of bisabolene was demonstrated. Bisabolene synthases are not bacterial enzymes, and gene for bisabolene cyclase from plant was codon optimized, synthesized, cloned, and expressed in *S. venezuelae*. Further metabolic engineering led to a 5-fold increase in bisabolene titer over the base production strain (Phelan et al., 2014). Despite the availability of all these hosts, the success in heterologous expression of a novel, uncharacterized biosynthetic gene cluster from metagenome is by no means guaranteed. Optimal strategy here is to try as many various hosts as possible for each gene cluster.

9.5 Metagenomics-Based Insights into Biosynthetic Potential of Marine Actinomycetes

Modern metagenomics based on NGS and high-throughput DNA sequence analysis potentially offers unprecedented insights into biosynthetic potential of marine actinomycetes that remain uncultivated. Although there are no reports so far on targeted analysis of actinomycete DNA in metagenomes coding for specific enzymes or biosynthetic pathways, it seems plausible that such bioinformatics methods may be developed based on high GC content. Keeping in mind that such genes are preferentially expressed in actinomycetes hosts, such as engineered *Streptomyces* (see Section 9.2.1), this approach may streamline the procedure of functional screening via assembly/synthesis of such genes followed by introduction into well-characterized *Streptomyces* hosts. The probability of success in such an approach is supported by the relatively recent genome sequencing and analyses of actinomycetes cultivated from marine sediments, in particular those belonging to the genus *Salinispora*. Analysis of genomes from 75 *Salinispora* strains for genes involved in polyketide and nonribosomal peptide biosynthesis identified 124 and predicted further 229 pathways (Ziemert et al., 2014). Also sponge-associated *Streptomyces* were shown to harbor unique secondary metabolite biosynthesis gene clusters not found in the genomes of their terrestrial counterparts belonging to the same species (Ian et al., 2014). These studies suggest that exchange of the biosynthetic genes between actinomycetes in the marine environment occurs relatively frequently, probably due to the significant role secondary metabolites play in adaptation to a particular ecological niche. Moreover, Edlund et al. (2010) following their extensive search for bioactive isolates in distant geographical locations noted that locations can influence secondary metabolism as subpopulations from specific locations maintain distinct sets of biosynthetic genes.

Metagenomics approach targeted at specific enzymes or pathways has recently provided insights into the biosynthetic potential of as yet uncultivated actinomycetes associated with marine sponges. PCR amplification and sequencing of the gene fragment encoding adenylation domain of the NRPSs from the metagenome of marine sponge *Aplysina aerophoba* yielded domain with predicted specificity for hydrohyphenylglycine

(Pimentel-Elardo et al., 2012). Mining the same metagenome with PCR-amplified probe specific for halogenases yielded a novel, apparently actinomycete-derived secondary metabolite biosynthesis gene cluster (Bayer et al., 2013). Trindade-Silva et al. (2013) explored microbiome of the marine sponge *Arenosclera brasiliensis*, targeting genes encoding various types of PKSs. Phylogenetic analysis of the KS domains yielded a diverse group of apparently actinomycete-derived sequences (ca 20% of all KS sequences), strongly suggesting that this sponge hosts many actinomycetes capable of synthesizing various types of polyketides. Metagenomic analysis of the microorganisms associated with marine sponge *Discodermia dissoluta* revealed the presence of diverse PKS genes (Schirmer et al., 2005). Functional gene guided approach by Sun et al. (2012) has also recently revealed the presence of type II PKSs in the genomes of coral-associated actinomycetes.

9.6 Future Prospects

Metagenomics can greatly facilitate the studies on structure and function of microbial communities in complex environments, thus making *in situ* monitoring of the microfloral responses and associated activities on an ecosystem level feasible (Simon and Daniel, 2009). Generation of metagenomic sequencing data for marine environmental samples (e.g., Sargasso Sea near Bermuda (Venter et al., 2004) or marine sponges (Kim and Fuerst, 2006; Pimentel-Elardo et al., 2012; Bayer et al., 2013), together with increasing knowledge on the metabolism and functional diversity of unculturable microorganisms is changing the direction of marine biodiscovery. Only through generation of in-depth understanding of the true genetic diversity of marine microorganisms with respect to the biosynthesis of bioactive compounds will full potential of marine environments as sources of novel drugs be revealed (Li and Qin, 2005; Subramani and Aalbersberg, 2012; Manivasagan et al., 2013).

The lack of in-depth knowledge of the nutritional requirements of marine microorganisms (Fenical, 1993) is still hampering the efforts on sustainable cultivation, thus precluding their effective development as sources of new drug leads (Joint et al., 2010). As stated by Hoover et al. (2007), combining knowledge on metabolite biosynthesis, their structure, and function with genomic data for their producers will lead to a better understanding of many marine ecological interactions (e.g., predator–prey interactions, competition for space and nutrients). A better understanding of biosynthesis may also assist in the production of large quantities of biomedically valuable compounds with a minimal impact on natural populations of marine organisms, which are used as the original source of the compound of interest (Hoover et al., 2007). Differences in the growth requirements of marine microorganisms should be well documented especially for the interphase organisms that adapt to both marine and terrestrial conditions (Kurtböke et al., 2014). Using objective strategies such as cultivation with and without seawater can simply provide a hint on whether a particular environmental sample is rich in genuinely marine actinomycetes (Figure 9.3). The latter will make such sample attractive for using metagenomics approaches to biodiscovery.

Bacteriophages can also indicate the presence of closely related actinobacterial taxa in different marine environments and can be utilized as indicators of the presence of the strains most useful for biodiscovery (Kurtböke, 2005, 2009, 2011, 2012). They can also be used to treat environmental samples prior to isolation of eDNA in order to enrich the metagenomics libraries DNA from particular genera.

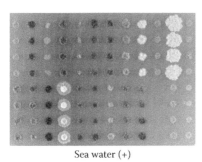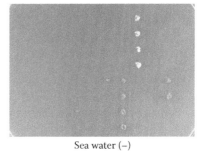

Sea water (+)   Sea water (−)

FIGURE 9.3
Differences in the seawater-dependent growth of Norwegian fjord sediment-derived actinomycete isolates. (Photo courtesy of H. Bredholt.)

Metagenomics targeted to actinomycete bacteria now provides excellent opportunities to access until now practically untapped reservoir of genes from uncultivated actinomycetes, which govern biosynthesis of bioactive secondary metabolites. Detailed analysis of such genes, combined with heterologous expression, will lead to new exciting discoveries not only of bioactive compounds, but also enzymes capable of performing novel chemical reactions. The latter knowledge can be applied to modify already known compounds or create hybrid molecules with beneficial drug-like properties. In order to achieve these goals, more efforts must be directed toward the development of robust and well-characterized hosts for heterologous expression of biosynthetic gene clusters. Moreover, the rapidly advancing field of synthetic biology is bound to contribute greatly to the biodiscoveries through metagenomics. Such principles of synthetic biology as abstraction, standardization, and modelling coupled to the design of novel regulatory circuits will be directed at solving problems of heterologous expression through gene cluster refactoring.

Advances in DNA synthesis along with the DNA assembly technologies will allow robust and inexpensive *de novo* synthesis of redesigned biosynthetic gene clusters. The latter will be much better fitted for heterologous expression, as they will be tailored to specific host and regulatory circuits. Once efficient heterologous expression has been achieved and new compound produced, its analogues with improved pharmacological properties can be generated using both biosynthetic engineering and synthetic chemistry. Thus, combination of technical advances in preparation and sequencing of eDNA, construction of metagenomics libraries, high-throughput bioinformatics, high-throughput screening and analytical methods, metabolic engineering, and synthetic biology will facilitate future marine biodiscovery (Baltz, 2008; Bologa et al., 2013; Hopwood, 2013; Charlop-Powers et al., 2014).

Acknowledgments

Sergey B. Zotchev and Olga N. Sekurova acknowledge in-part support by the Research Council of Norway for the research findings presented in this chapter.

References

Ahmed, L., Jensen, P.R., Freel, K.C., Brown, R., Jones, A.L., Kim, B.Y., and Goodfellow, M. (2013) *Salinispora pacifica* sp. nov., an actinomycete from marine sediments. *Antonie Van Leeuwenhoek* 103(5), 1069–1078.

Asolkar, R.N., Kirkland, T.N., Jensen, P.R., and Fenical, W. (2010) Arenimycin, an antibiotic effective against rifampin- and methicillin-resistant *Staphylococcus aureus* from the marine actinomycete *Salinispora arenicola*. *The Journal of Antibiotics (Tokyo)* 63(1), 37–39.

Baltz, R.H. (2008) Renaissance in antibacterial discovery from actinomycetes. *Current Opinion in Pharmacology* 8(5), 557–563.

Baltz, R.H. (2010) *Streptomyces* and *Saccharopolyspora* hosts for heterologous expression of secondary metabolite gene clusters. *Journal of Industrial Microbiology and Biotechnology* 37, 722–759.

Banik, J.J. and Brady, S.F. (2010) Recent application of metagenomic approaches towards the discovery of antimicrobials and other bioactive small molecules. *Current Opinion in Microbiology* 13(5), 603–609.

Bayer, K., Scheuermayer, M., Fieseler, L., and Hentschel, U. (2013) Genomic mining for novel FADH$_2$-dependent halogenases in marine sponge-associated microbial consortia. *Marine Biotechnology* 15(1), 63–72.

Betancor, L., Fernández, M.J., Weissman, K.J., and Leadlay, P.F. (2008) Improved catalytic activity of a purified multienzyme from a modular polyketide synthase after coexpression with *Streptomyces* chaperonins in *Escherichia coli*. *ChemBioChem* 9, 2962–2966.

Bérdy, J. (2005) Bioactive microbial metabolites. *The Journal of Antibiotics* 58(1), 1–26.

Bérdy, J. (2012) Thoughts and facts about antibiotics: Where we are now and where we are heading. *The Journal of Antibiotics* 65, 385–395.

Blin, K., Medema, M.H., Kazempour, D., Fischbach, M.A., Breitling, R., Takano, E., and Weber, T. (2013) antiSMASH 2.0—A versatile platform for genome mining of secondary metabolite producers. *Nucleic Acids Research* 41, W204–W212.

Bologa, C.G., Ursu, O., Oprea, T.I., Melançon III, C.E., and Tegos, G.P. (2013) Emerging trends in the discovery of natural product antibacterials. *Current Opinion in Pharmacology* 13, 678–687.

Brady, S.F., Simmons, L., Kim, J.H., and Schmidt, E.W. (2009) Metagenomic approaches to natural products from free-living and symbiotic organisms. *Natural Product Reports* 26(11), 1488–1503.

Bredholt, H., Fjarvik, E., Johnsen, G., and Zotchev, S.B. (2008) Actinomycetes from sediments in the Trondheim Fjord, Norway: Diversity and biological activity. *Marine Drugs* 6(1), 12–24.

Bredholdt, H., Galatenko, O.A., Engelhardt, E., Fjærvik, E., Terekhova, L.P., and Zotchev, S.B. (2007) Rare actinomycete bacteria from the shallow water sediments of the Trondheim fjord, Norway: Isolation, diversity and biological activity. *Environmental Microbiology* 9(11), 2756–2764.

Bull, A.T. and Stach, J.E.M. (2007) Marine actinobacteria: New opportunities for natural product search and discovery. *Trends in Microbiology* 15, 491–499.

Butler, M.S. and Cooper, M.A. (2011) Antibiotics in the clinical pipeline in 2011. *The Journal of Antibiotics* 64, 413–425.

Chang, F.Y. and Brady, S.F. (2011) Cloning and characterization of an environmental DNA-derived gene cluster that encodes the biosynthesis of the antitumor substance BE-54017. *Journal of the American Chemical Society* 133(26), 9996–9999.

Chang, F.Y. and Brady, S.F. (2013) Discovery of indolotryptoline antiproliferative agents by homology-guided metagenomic screening. *Proceedings of the National Academy of Sciences United States of America* 110(7), 2478–2483.

Charlop-Powers, Z., Milshteyn, A., and Brady, S.F. (2014) Metagenomic small molecule discovery methods. *Current Opinion in Microbiology* 19, 70–75.

Chen, W.H., Qin, Z.J., Wang, J., and Zhao, G.P. (2013) The MASTER (methylation-assisted tailorable ends rational) ligation method for seamless DNA assembly. *Nucleic Acids Research* 41(8), e93.

Chen, Y. and Murrell, J.C. (2010) When metagenomics meets stable-isotope probing: Progress and perspectives. *Trends in Microbiology* 18(4), 157–163.

Cragg, G.M., Grothaus, P.G., and Newman, D.J. (2009) Impact of natural products on developing new anti-cancer agents. *Chemical Reviews* 109, 3012–3043.

Cragg, GM. and Newman, D.J. (2006) Natural product sources of drugs: Plants, microbes, marine organisms and animals. In: P.D. Kennewell, D. Triggle, and J. Taylor (Eds.), *Comprehensive Medicinal Chemistry* ii. Vol. 1. Elsevier, Oxford, U.K., pp. 355–403.

Cragg, G.M. and Newman, D.J. (2013) Natural products: A continuing source of novel drug leads. *Biochimica et Biophysica Acta (BBA)—General Subjects* 1830(6), 3670–3695.

Dai, H.-Q., Wang, J., Xin, Y.-H., Pei, G., Tang, S.-K., Ren, B., Ward, A., Ruan, J.-S., Li, W.-J., and Zhang, L.-X. (2010) *Verrucosispora sediminis* sp. nov., a cyclodipeptide-producing actinomycete from deep-sea sediment. *International Journal of Systematic and Evolutionary Microbiology* 60, 1807–1812.

Demain, A.L. and Vaishnav, P. (2011) Natural products for cancer chemotherapy. *Microbial Biotechnology* 4(6), 687–699.

Do, T.H., Nguyen, T.T., Nguyen, T.N., Le, Q.G., Nguyen, C., Kimura, K., and Truong, N.H. (2014) Mining biomass-degrading genes through Illumina-based *de novo* sequencing and metagenomic analysis of free-living bacteria in the gut of the lower termite *Coptotermes gestroi* harvested in Vietnam. *Journal of Bioscience and Bioengineering* 118(6), 665–671.

Eccleston, G.P., Brooks, P.R., and Kurtböke, D.I (2008) The occurrence of bioactive Micromonosporae in aquatic habitats of the Sunshine Coast in Australia. *Marine Drugs* 6, 243–261.

Edlund, A., Loesgen, S., Fenical, W., and Jensen, P.R. (2010) Geographic distribution of secondary metabolite genes in the marine actinomycete *Salinispora arenicola*. *Applied and Environmental Microbiology* 77(17), 5916–5925.

Egan, S., Thomas, T., and Kjelleberg, S. (2008) Unlocking the diversity and biotechnological potential of marine surface associated microbial communities. *Current Opinion in Microbiology* 11, 219–225.

Ekkers, D.M., Cretoiu, M.S., Kielak, A.M., and Elsas, J.D. (2012) The great screen anomaly—A new frontier in product discovery through functional metagenomics. *Applied Microbiology and Biotechnology* 93(3), 1005–1020.

Engelhardt, K., Degnes, K.F., Kemmler, M., Bredholt, H., Fjærvik, E., Klinkenberg, G., Sletta, H., Ellingsen, T.E., and Zotchev, S.B. (2010) Production of a new thiopeptide antibiotic, TP-1161, by a marine *Nocardiopsis* species. *Applied and Environmental Microbiology* 76(15), 4969–4976.

Eustáquio, A.S., Nam, S.J., Penn, K., Lechner, A., Wilson, M.C., Fenical, W., Jensen, P.R., and Moore, B.S. (2011) The discovery of salinosporamide K from the marine bacterium "*Salinispora pacifica*" by genome mining gives insight into pathway evolution. *ChemBioChem* 12(1), 61–64.

Feling, R.H., Buchanan, G.O., Mincer, T.J., Kauffman, C.A., Jensen, P.R., and Fenical, W. (2003) Salinosporamide A: A highly cytotoxic proteasome inhibitor from a novel microbial source, a marine bacterium of the new genus *Salinospora*. *Angewandte Chemie International Edition* 42(3), 355–357.

Feng, Z., Chakraborty, D., Dewell, S.B., Reddy, B.V., and Brady, S.F. (2012) Environmental DNA-encoded antibiotics fasamycins A and B inhibit FabF in type II fatty acid biosynthesis. *Journal of American Chemical Society* 134(6), 2981–2987.

Feng, Z., Kallifidas, D., and Brady, S.F. (2011) Functional analysis of environmental DNA-derived type II polyketide synthases reveals structurally diverse secondary metabolites. *Proceedings of the National Academy of Sciences United States of America* 108(31), 12629–12634.

Fenical, W. (1993) Chemical studies of marine bacteria: Developing a new resource. *Chemical Reviews* 93(5), 1673–1683.

Fenical, W. and Jensen, P.R. (2006) Developing a new resource for drug discovery: Marine actinomycete bacteria. *Nature Chemical Biology* 2(12), 666–673.

Fenical, W., Jensen, P.R., Palladino, M.A., Lam, K.S., Lloyd, G.K., and Potts, B.C. (2009) Discovery and development of the anticancer agent Salinosporamide A (NPI-0052). *Bioorganic and Medicinal Chemistry* 17(6), 2175.

Fiedler, H.P., Bruntner, C., Bull, A.T., Ward, A.C., Goodfellow, M., Potterat, O., Puder, C., and Mihm, G. (2005) Marine actinomycetes as a source of novel secondary metabolites. *Antonie van Leeuwenhoek* 87(1), 37–42.

Fiedler, H.-P., Bruntner, C., Riedlinger, J., Bull, A.T., Knutsen, G., Goodfellow, M., Jones, A. et al. (2008) Proximicin A, B and C, Novel aminofuran antibiotic and anticancer compounds isolated from marine strains of the actinomycete *Verrucosispora*. *The Journal of Antibiotics* 61(3), 158–163.

Freel, C.K., Edlund, A., and Jensen, P.P. (2012) Microdiversity and evidence for high dispersal rates in the marine actinomycete genus *Salinispora*. *Environmental Microbiology* 14, 480–493.

Freel, K.C., Nam, S.J., Fenical, W., and Jensen, P.R. (2011) Evolution of secondary metabolite genes in three closely related marine actinomycete species. *Applied and Environmental Microbiology* 77(20), 7261–7270.

Fu, J., Bian, X., Hu, S., Wang, H., Huang, F., Seibert, P.M., Plaza, A. et al. (2012) Full-length RecE enhances linear–linear homologous recombination and facilitates direct cloning for bioprospecting. *Nature Biotechnology* 30(5), 440–446.

Gallagher, K.A., Fenical, W., and Jensen, P.J. (2010) Hybrid isoprenoid secondary metabolite production in terrestrial and marine actinomycetes. *Current Opinion in Biotechnology* 21, 794–800.

Gao, X., Lu, Y., Xing, Y., Ma, Y., Lu, J., Bao, W., Wang, Y., and Xi, T. (2012) A novel anticancer and antifungus phenazine derivative from a marine actinomycete BM-17. *Microbiological Research* 167(10), 616–622.

Gerwick, W.H. and Moore, B.S. (2012) Lessons from the past and charting the future of marine natural products drug discovery and chemical biology. *Chemistry and Biology* 19, 85–98.

Gibson, D.G., Young, L., Chuang, R.Y., Venter, J.C., Hutchison, C.A., and Smith, H.O. (2009) Enzymatic assembly of DNA molecules up to several hundred kilobases. *Nature Methods* 6(5), 343–345.

Gomez-Escribano, J.P. and Bibb, M.J. (2014) Heterologous expression of natural product biosynthetic gene clusters in *Streptomyces coelicolor*: From genome mining to manipulation of biosynthetic pathways. *Journal of Industrial Microbiology and Biotechnology* 41, 425–431.

Gontang, E.A., Fenical, W., and Jensen, P.R. (2007) Phylogenetic diversity of Gram-positive bacteria cultured from marine sediment. *Applied and Environmental Microbiology* 73(10), 3272–3282.

Gontang, E.A., Gaudêncio, S.P., Fenical, W., and Jensen, P.R. (2010) Sequence-based analysis of secondary-metabolite biosynthesis in marine actinobacteria. *Applied and Environmental Microbiology* 76(8), 2487–2499.

Goodfellow, M., Stach, J.E.M, Brown, R., Bonda, A.N.V., Jones, A.L., Mexson, J., Fiedler, H.-P., Zucchi, T.D., and Bull, A.T. (2012) *Verrucosispora maris* sp. nov., a novel deep-sea actinomycete isolated from a marine sediment which produces abyssomicins. *Antonie Van Leeuwenhoek* 101(1), 185–193.

Gulder, T.A. and Moore, B.S. (2009) Chasing the treasures of the sea-bacterial marine natural products. *Current opinion in Microbiology* 12(3), 252–260.

Gulder, T.A. and Moore, B.S. (2010) Salinosporamide natural products: Potent 20 S proteasome inhibitors as promising cancer chemotherapeutics. *Angewandte Chemie International Edition*, 49(49), 9346–9367.

Haefner, B. (2003) Drugs from the deep: Marine natural products as drug candidates. *Drug Discovery Today* 8(12), 536–544.

Handelsman, J., Rondon, M.R., Brady, S.F., Clardy, J., and Goodman, R.M. (1998) Molecular biological access to the chemistry of unknown soil microbes: A new frontier for natural products. *Chemistry and Biology* 5, R245–R249.

He, R., Wakimoto, T., Egami, Y., Kenmoku, H., Ito, T., Asakawa, Y., and Abe, I. (2012) Heterologously expressed β-hydroxyl fatty acids from a metagenomic library of a marine sponge. *Bioorganic and Medicinal Chemistry Letters* 22(24), 7322–7325.

Hill, R.T. (2004) Microbes from marine sponges: A treasure trove of biodiversity for natural products discovery. In: A.T. Bull (Ed.), *Microbial Diversity and Bioprospecting.* ASM Press, Washington, DC, pp. 177–190.

Hoover, C.A., Slattery, M., and Marsh, A.G. (2007) A functional approach to transcriptome profiling: Linking gene expression patterns to metabolites that matter. *Marine Biotechnology* 9, 411–419.

Hopwood, D.A. (2007a) (Ed.) *Streptomycetes in Nature and Medicine: The Antibiotic Makers.* Oxford University Press, New York.

Hopwood, D.A. (2007b) Therapeutic treasures from the deep. *Nature Chemical Biology* 3(8), 457–458.

Hopwood, D.A. (2013) Imaging mass spectrometry reveals highly specific interactions between actinomycetes to activate specialized metabolic gene clusters. *mBio* 4, e00612–e00613.

Ian, E., Malko, D.B., Sekurova, O.N., Bredholt, H., Rückert, C., Borisova, M.E., Albersmeier, A., Kalinowski, J., Gelfand, M.S., and Zotchev, S.B. (2014) Genomics of sponge-associated *Streptomyces* spp. closely related to *Streptomyces albus* J1074: Insights into marine adaptation and secondary metabolite biosynthesis potential. *PLoS ONE* 9(5), e96719.

Imhoff, J.F., Labes, A., and Wiese, J. (2011) Bio-mining the microbial treasures of the ocean: New natural products. *Biotechnology Advances* 29, 468–482.

Itaya, M., Fujita, K., Kuroki, A., and Tsuge, K. (2008) Bottom-up genome assembly using the *Bacillus subtilis* genome vector. *Nature Methods* 5(1), 41–43.

Izumikawa, M., Khan, S.T., Takagi, M., and Shin-ya, K. (2010) Sponge-derived *Streptomyces* producing isoprenoids via the mevalonate pathway. *Journal of Natural Products* 73, 208–212.

Jensen, P.R., Gontang, E., Mafnas, C., Mincer, T.J., and Fenical, W. (2005b) Culturable marine actinomycete diversity from tropical Pacific Ocean sediments. *Environmental Microbiology* 7, 1039–1048.

Jensen, P.R. and Mafnas, C. (2006) Biogeography of the marine actinomycete *Salinispora*. *Environmental Microbiology* 8(11), 1881–1888.

Jensen, P.R., Mincer, T.J., Williams, P.G., and Fenical, W. (2005a) Marine actinomycete diversity and natural product discovery. *Antonie van Leeuwenhoek* 87, 43–48.

Jensen, P.R., Williams, P.G., Oh, D.C., Zeigler, L., and Fenical, W. (2007) Species-specific secondary metabolite production in marine actinomycetes of the genus *Salinispora*. *Applied and Environmental Microbiology* 73(4), 1146–1152.

Johnston, A.W.B., Li, Y., and Ogilvie, L. (2005) Metagenomic marine nitrogen fixation–feast or famine? *Trends in Microbiology* 13(9), 416–420.

Joint, I., Mühling, M., and Querellou, J. (2010) Culturing marine bacteria—an essential prerequisite for biodiscovery. *Microbial Biotechnology* 3(5), 564–575.

Kaysser, L., Bernhardt, P., Nam, S.-J., Loesgen, S., Ruby, J.G., Skewes-Cox, P., Jensen, P.R., Fenical, W., and Moore, B.S. (2012) Merochlorins A–D, cyclic meroterpenoid antibiotics biosynthesized in divergent pathways with vanadium-dependent chloroperoxidases. *Journal of American Chemical Society* 134(29), 11988–11991.

Kim, T.K. and Fuerst, J.A. (2006) Diversity of polyketide synthase genes from bacteria associated with the marine sponge *Pseudoceratina clavata*: Culture-dependent and culture-independent approaches. *Environmental Microbiology* 8, 1460–1470.

Kim, T.K., Garson, M.J., and Fuerst, J.A. (2005) Marine actinomycetes related to the 'Salinospora' group from the Great Barrier Reef sponge *Pseudoceratina clavata*. *Environmental Microbiology* 7, 509–518.

Kim, T.K., Hewavitharana, A.K., Shaw, P.N., and Fuerst, J.A. (2006) Discovery of a new source of Rifamycin antibiotics in marine sponge actinobacteria by phylogenetic prediction. *Applied and Environmental Microbiology* 72(3), 2118–2125.

Knietsch, A., Waschkowitz, T., Bowien, S., Henne, A., and Daniel, R. (2003) Metagenomes of complex microbial consortia derived from different soils as sources for novel genes conferring formation of carbonyls from short-chain polyols on *Escherichia coli*. *Journal of Molecular Microbiology and Biotechnology* 5(1), 46–56.

Komatsu, M., Komatsu, K., Koiwai, H., Yamada, Y., Kozone, I., Izumikawa, M., Hashimoto, J. et al. (2013) Engineered *Streptomyces avermitilis* host for heterologous expression of biosynthetic gene cluster for secondary metabolites. *ACS Synthetic Biology* 2(7), 384–96.

König, G.M., Kehraus, S., Seibert, S.F., Lateff, A.A., and Müller, D. (2006) Natural products from marine organisms and their associated microbes. *ChemBioChem* 7, 229–238.

Kurtböke, D.I. (2005) Actinophages as indicators of actinomycete taxa in marine environments. *Antonie van Leeuwenhoek* 87, 19–28.

Kurtböke, D.I. (2009) Use of phage-battery to isolate industrially important rare actinomycetes. In: H.T. Adam (Ed.), *Contemporary Trends in Bacteriophage Research*. NOVA Science Publishers, New York, pp. 119–149.

Kurtböke, D.I. (2011) Exploitation of phage battery in the search for bioactive actinomycetes. *Applied Microbiology and Biotechnology* 89(4), 931–937.

Kurtböke, D.I. (2012) Biodiscovery from rare actinomycetes: An eco-taxonomical perspective. *Applied Microbiology and Biotechnology* 93(5), 1843–1852.

Kurtböke, D.I., Grkovic, T., and Quinn, R.J. (2014) Marine actinomycetes in biodiscovery: Current trends and future prospects. *Springer Handbook of Marine Biotechnology*, in press.

Lampert, Y., Kelman, D., Dubinsky, Z., Nitzan, Y., and Hill, R.T. (2006) Diversity of culturable bacteria in the mucus of the Red Sea coral *Fungia scutaria*. *FEMS Microbiology Ecology* 58, 99–108.

Lane, A.L. and Moore, B.S. (2011) A sea of biosynthesis: Marine natural products meet the molecular age. *Natural Product Reports* 28(2), 411–428.

Lane, A.L., Nam, S.J., Fukuda, T., Yamanaka, K., Kauffman, C.A., Jensen, P.R., Fenical, W., and Moore, B.S. (2013) Structures and comparative characterization of biosynthetic gene clusters for cyanosporasides, enediyne-derived natural products from marine actinomycetes. *Journal of the American Chemical Society* 135(11), 4171–4174.

Li, T., Du, Y., Cui, Q., Zhang, J., Zhu, W., Hong, K., and Li, W. (2013) Cloning, characterization and heterologous expression of the indolocarbazole biosynthetic gene cluster from marine-derived *Streptomyces sanyensis* FMA. *Marine Drugs* 11, 466–488.

Li, X. and Qin, L. (2005) Metagenomics-based drug discovery and marine microbial diversity. *Trends in Biotechnology* 23(11), 539–543.

Liebl, W., Angelov, A., Juergensen, J., Chow, J., Loeschcke, A., Drepper, T., Classen, T. et al. (2014) Alternative hosts for functional (meta) genome analysis. *Applied Microbiology and Biotechnology* 98(19), 8099–8109.

Lombó, F., Velasco, A., Castro, A., de la Calle, F., Braña, A.F., Sánchez-Puelles, J.M., Méndez, C., and Salas, J.A. (2006) Deciphering the biosynthesis pathway of the antitumor thiocoraline from a marine actinomycete and its expression in two *Streptomyces* species. *ChemBioChem* 7, 366–376.

Magarvey, N.A., Keller, J.M., Bernan, V., Dworkin, M., and Sherman, D.H. (2004) Isolation and characterization of novel marine-derived actinomycete taxa rich in bioactive metabolites. *Applied and Environmental Microbiology* 70(12), 7520–7529.

Maldonado, L.A., Stach, J.E.M., Pathom-aree, W., Ward, A.C., Bull, A.T., and Goodfellow, M. (2005) Diversity of cultivable actinobacteria in geographically widespread marine sediments. *Antonie van Leeuwenhoek* 87, 11–18.

Manivasagan, P., Venkatesan, J., Sivakumar, K., and Kim, S.K. (2013) Marine actinobacterial metabolites: Current status and future perspectives. *Microbiological Research* 168(6), 311–332.

Mayer, A.M.S., Glaser, K.B., Cuevas, C., Jacobs, R.S., Kem, W., Little, R.D., McIntosh, J.M., Newman, D.J., Potts, B.C., and Shuster, D.E. (2010) The odyssey of marine pharmaceuticals: A current pipeline perspective. *Trends in Pharmacological Sciences* 31, 255–265.

Medema, M.H., Cimermancic, P., Sali, A., Takano, E., and Fischbach, M.A. (2014) A systematic computational analysis of biosynthetic gene cluster evolution: Lessons for engineering biosynthesis. *PLoS Computational Biology* 10(12), e1004016.

Medema, M.H., van Raaphorst, R., Takano, E., and Breitling, R. (2012) Computational tools for the synthetic design of biochemical pathways. *Nature Reviews Microbiology* 10(3), 191–202.

Miao, V. and Davies, J. (2009) Metagenomics and antibiotic discovery from uncultivated bacteria. In: S.S. Epstein (Ed.), *Uncultivated Microorganisms*. Springer, Berlin, Germany, pp. 217–236.

Mincer, T.J., Fenical, W., and Jensen, P.R. (2005) Culture-dependent and culture-independent diversity within the obligate marine actinomycete genus *Salinispora*. *Applied and Environmental Microbiology* 71(11), 7019–7028.

Mincer, T.J., Jensen, P.R., Kauffman, C.A., and Fenical, W. (2002) Widespread and persistent popula-tions of a major new marine actinomycete taxon in ocean sediments. *Applied and Environmental Microbiology* 68(10), 5005–5011.

Molinski, T.F., Dalisay, D.S., Lievens, S.L., and Saludes, J.P. (2009) Drug development from marine natural products. *Nature Reviews Drug Discovery* 8, 69–85.

Müller, R. and Wink, J. (2014) Future potential for anti-infectives from bacteria-How to exploit biodi-versity and genomic potential. *International Journal of Medical Microbiology* 304, 3–13.

Nett, M., Ikeda, H., and Moore, B.S. (2009) Genomic basis for natural product biosynthetic diversity in the actinomycetes. *Natural Product Reports* 26(11), 1362–1384.

Newman, D.J. and Cragg, G.M. (2004) Marine natural products and related compounds in clinical and advanced preclinical trials. *Journal of Natural Products* 67, 12160–1238.

Newman, D.J. and Cragg, G.M. (2012) Natural products as sources of new drugs over the 30 years from 1981 to 2010. *Journal of Natural Products* 75, 311–338.

Newman, D.J. and Hill, R.T. (2006) New drugs from marine microbes: The tide is turning. *Journal of Industrial Microbiology and Biotechnology* 33(7), 539–544.

Nikolouli, K. and Mossialos, D. (2012) Bioactive compounds synthesized by non-ribosomal peptide synthetases and type-I polyketide synthases discovered through genome-mining and metage-nomics. *Biotechnology Letters* 34(8), 1393–1403.

Nithyanand, P., Manju, S., and Pandian, S.K. (2011) Phylogenetic characterization of culturable acti-nomycetes associated with the mucus of the coral *Acropora digitifera* from Gulf of Mannar. *FEMS Microbiology Letters* 314, 112–118.

Oh, D.-C., Gontang, E.A., Kauffman, C.A., Jensen, P.R., and Fenical, W. (2008) Salinipyrones and Pacificanones, mixed-precursor polyketides from the marine actinomycete *Salinispora pacifica*. *Journal of Natural Products* 71(4), 570.

Okami, Y. (1986) Marine microorganisms as a source of bioactive agents. *Microbial Ecology* 12(1), 65–78.

Owen, J.G., Reddy, B.V., Ternei, M.A., Charlop-Powers, Z., Calle, P.Y., Kim, J.H., and Brady, S.F. (2013) Mapping gene clusters within arrayed metagenomic libraries to expand the structural diver-sity of biomedically relevant natural products. *Proceedings of the National Academy of Sciences United States of America* 110(29), 11797–802.

Pathom-aree, W., Stach, J.E.M., Ward, A.C., Horikoshi, K., Bull, A.T., and Goodfellow, M. (2006) Diversity of actinomycetes isolated from Challenger Deep sediment (10,898 m) from the Mariana Trench. *Extremophiles* 10, 181–189.

Penesyan, A., Marshall-Jones, Z., Holmstrom, C., Kjelleberg, S., and Egan, S. (2009) Antimicrobial activity observed among cultured marine epiphytic bacteria reflects their potential as a source of new drugs. *FEMS Microbiology Ecology* 69, 1130–124.

Penn, K., Jenkins, C., Nett, M., Udwary, D.W., Gontang, E.A., McGlinchey, R.P., Foster, B. et al. (2009) Genomic islands link secondary metabolism to functional adaptation in marine Actinobacteria. *The ISME Journal* 3, 1193–1203.

Penn, K. and Jensen, P.R (2012) Comparative genomics reveals evidence of marine adaptation in *Salinispora* species. *BMC Genomics* 13, 86.

Phelan, R.M., Sekurova, O.N., Keasling, J.D., and Zotchev, S.B. (2014) engineering terpene biosynthe-sis in *Streptomyces* for production of the advanced biofuel precursor bisabolene. *ACS Synthetic Biology* 2014 Jul 9 [Epub ahead of print; doi: 10.1021/sb5002517].

Piel, J. (2006) Bacterial symbionts: Prospects for the sustainable production of invertebrate-derived pharmaceuticals. *Current Medicinal Chemistry* 13, 39–50.

Piel, J., Butzke, D., Fusetani, N., Hui, D., Platzer, M., Wen, G., and Matsunaga, S. (2005) Exploring the chemistry of uncultivated bacterial symbionts: Antitumor polyketides of the pederin family. *Journal of Natural Products* 68, 472–479.

Pimentel-Elardo, S.M., Buback, V., Gulder, T.A.M., Bugni, T.S., Reppart, J., Bringmann, G., Ireland, C.M., Schirmeister, T., and Hentschel, U. (2011) New Tetromycin derivatives with anti Trypanosomal and protease inhibitory activities. *Marine Drugs* 9, 1682–1697.

Pimentel-Elardo, S.M., Grozdanov, L., Proksch, S., and Hentschel, U. (2012) Diversity of nonribosomal peptide synthetase genes in the microbial metagenomes of marine sponges. *Marine Drugs* 10(6), 1192–1202.

Pimentel-Elardo, S.M., Kozytska, S., Bugni, T.S., Ireland, C.M., Moll, H., and Hentschel, U. (2010) Anti-parasitic compounds from *Streptomyces* sp. strains isolated from Mediterranean sponges. *Marine Drugs* 8, 373–380.

Prieto-Davó, A., Fenical, W., and Jensen, P.R. (2008) Comparative actinomycete diversity in marine sediments. *Aquatic Microbial Ecology* 52, 1–11.

Prieto-Davó, A., Villarreal-Gómez, L.J., Forschner-Dancause, S., Bull, A.T., Stach, J.E.M., Smith, D.C., Rowley, D.C., and Jensen, P.R. (2013) Targeted search for actinomycetes from near-shore and deep sea marine sediments. *FEMS Microbiology Ecology* 84(3), 510–518.

Proksch, P., Edrada, R., and Ebel, R. (2002) Drugs from the seas–current status and microbiological implications. *Applied Microbiology and Biotechnology* 59(2–3), 125–134.

Prudhomme, J., McDaniel, E., Ponts, N., Bertani, S., Fenical, W., Jensen, P., and Le Roch, K. (2008) Marine actinomycetes: A new source of compounds against the human malaria parasite. *PLoS ONE* 3(6), e2335.

Quan, J. and Tian, J. (2009) Circular polymerase extension cloning of complex gene libraries and pathways. *PLoS ONE* 4, e6441.

Saurav, K., Zhang, W., Saha, S., Zhang, H., Li, S., Zhang, Q., Wu, Z., Zhang, G., Zhu, Y., and Verma, G. (2014) *In silico* molecular docking, preclinical evaluation of spiroindimicins AD, lynamicin A and D isolated from deep marine sea derived *Streptomyces* sp. SCSIO 03032. *Interdisciplinary Sciences: Computational Life Sciences* 6(3), 187–196.

Subramani, R. and Aalbersberg, W. (2012) Marine actinomycetes: An ongoing source of novel bioactive metabolites. *Microbiological Research* 167, 571–580.

Sun, W., Peng, C., Zhao, Y., and Li, Z. (2012) Functional gene-guided discovery of type ii polyketides from culturable actinomycetes associated with soft coral *Scleronephthya* sp. *PLoS ONE* 7(8), e42847.

Renner, M.K., Shen, Y.C., Cheng, X.C., Jensen, P.R., Frankmoelle, W., Kauffman, C.A., Fenical, W., Lobkovsky, E., and Clardy, J. (1999) Cyclomarins A–C, new antiinflammatory cyclic peptides produced by a marine bacterium (*Streptomyces* sp.). *Journal of the American Chemical Society* 121, 11273–11276.

Sanchez, L.M., Wong, W.R., Riener, R.M., Schulze, C.J., and Linington, R.G. (2012) Examining the fish microbiome: Vertebrate-derived bacteria as an environmental niche for the discovery of unique marine natural products. *PLoS ONE* 7(5), e35398.

Schneemann, I., Nagel, K., Kajahn, I., Labes, A., Wiese, J., and Imhoff, J.F. (2010) Comprehensive investigation of marine *actinobacteria* associated with the sponge *Halichondria panicea*. *Applied and Environmental Microbiology* 76(11), 3702–3714.

Schmieder, R. and Edwards, R. (2012) Insights into antibiotic resistance through metagenomic approaches. *Future Microbiology* 7(1), 73–89.

Schirmer, A., Gadkari, R., Reeves, C.D., Ibrahim, F., DeLong, E.F., and Hutchinson, C.R. (2005) Metagenomic analysis reveals diverse polyketide synthase gene clusters in microorganisms associated with the marine sponge *Discodermia dissolute*. *Applied and Environmental Microbiology* 71(8), 4840–4849.

Schultz, A.W., Oh, D.C., Carney, J.R., Williamson, R.T., Udwary, D.W., Jensen, P.R., Gould, S.J., Fenical, W., and Moore, B.S. (2008) Biosynthesis and structures of cyclomarins and cyclomarazines, prenylated cyclic peptides of marine actinobacterial origin. *Journal of the American Chemical Society* 130, 4507–4516.

Schumacher, M., Kelkel, M., Dicato, M., and Diederich, M. (2011) Gold from the sea: Marine compounds as inhibitors of the hallmarks of cancer. *Biotechnology Advances* 29, 531–547.

Shao, Z.Y., Zhao, H., and Zhao, H.M. (2009) DNA assembler, an *in vivo* genetic method for rapid construction of biochemical pathways. *Nucleic Acids Research* 37, e16.

Simon, C. and Daniel, R. (2009) Achievements and new knowledge unravelled by metagenomic approaches. *Applied Microbiology and Biotechnology* 85, 265–276.

Smanski, M.J., Bhatia, S., Zhao, D., Park, Y., Woodruff, L.B., Giannoukos, G., Ciulla, D. et al. (2014) Functional optimization of gene clusters by combinatorial design and assembly. *Nature Biotechnology* 32(12), 1241–1249.

Streit, W.R. and Schmitz, R.A. (2004) Metagenomics—The key to the uncultured microbes. *Current Opinion in Microbiology* 7, 492–498.

Taylor, M.W., Radax, R., Steger, D., and Wagner, M. (2007) Sponge-associated microorganisms: Evolution, ecology, and biotechnological potential. *Microbiology and Molecular Biology Reviews* 71, 295–347.

Tian, X.-P., Zhi, X.-Y., Qiu, Y.-Q., Zhang, Y.-Q., Tang, S.-K., Xu, L.-H., Zhang, S., and Li, W.-J. (2009) *Sciscionella marina* gen. nov., sp. nov., a marine actinomycete isolated from a sediment in the northern South China Sea. *International Journal of Systematic and Evolutionary Microbiology* 59, 222–228.

Trindade-Silva, A.E., Rua, C.P., Andrade, B.G., Vicente, A.C., Silva, G.G., Berlinck, R.G., and Thompson, F.L. (2013) Polyketide synthase gene diversity within the microbiome of the sponge *Arenosclera brasiliensis*, endemic to the Southern Atlantic Ocean. *Applied and Environmental Microbiology* 79(5), 1598–605.

Udwary, D.W., Zeigler, L., Asolkar, R.N., Singan, V., Lapidus, A., Fenical, W., Jensen, P.R., and Moore, B.S. (2007) Genome sequencing reveals complex secondary metabolome in the marine actinomycete *Salinispora tropica*. *Proceedings of the National Academy of Sciences United States of America* 104(25), 10376–10381.

Venter, J.C., Remington, K., Heidelberg, J.F., Halpern, A.L., Rusch, D., Eisen, J.A., Wu, D. et al. (2004) Environmental genome shotgun sequencing of the Sargasso Sea. *Science* 304, 66–74.

Wakimoto, T., Egami, Y., Nakashima, Y., Wakimoto, Y., Mori, T., Awakawa, T., Ito, T. et al. (2014) Calyculin biogenesis from a pyrophosphate protoxin produced by a sponge symbiont. *Nature Chemical Biology* 10, 648–655.

Waters, A.L., Peraud, O., Kasanah, N., Sims, J.W., Kothalawala, N., Anderson, M.A., Abbas, S.H. et al. (2014) An analysis of the sponge *Acanthostrongylophora ingens'* microbiome yields an actinomycete that produces the natural product manzamine A. *Frontiers in Marine Science* 1, article 54.

Webster, N.S., Wilson, K.J., Blackall, L.L., and Hill, R.T. (2001) Phylogenetic diversity of bacteria associated with the marine sponge *Rhopaloeides odorabile*. *Applied and Environmental Microbiology* 67(1), 434–444.

Wiese, J., Thiel, V., Nagel, K., Staufenberger, T., and Imhoff, J.F. (2009) Diversity of antibiotic-active bacteria associated with the brown alga *Laminaria saccharina* from the Baltic Sea. *Marine Biotechnology* 11, 287–300.

Williams, P.G. (2009) Panning for chemical gold: Marine bacteria as a source of new therapeutics. *Trends in Biotechnology* 27(1), 45–52.

Wilson, M.C. and Piel, J. (2013) Metagenomic approaches for exploiting uncultivated bacteria as a resource for novel biosynthetic enzymology. *Chemistry and Biology* 20(5), 636–647.

Woodhouse, J.N., Fan, L., Brown, M.V., Thomas, T., and Neilan, B.A. (2013) Deep sequencing of non-ribosomal peptide synthetases and polyketide synthases from the microbiomes of Australian marine sponges. *The ISME Journal* 7, 1842–1851.

Wright, G.D. (2012) Antibiotics: A new hope. *Chemistry and Biology* 19, 3–10.

Zhang, Q., Li, S., Chen, Y., Tian, X., Zhang, H., Zhang, G., Zhu, Y., Zhang, S., Zhang, W., and Zhang, C. (2012b) New diketopiperazine derivatives from a deep-sea-derived *Nocardiopsis alba* SCSIO 03039. *The Journal of Antibiotics* 66(1), 31–36.

Zhang, Q., Li, S., Chen, Y., Tian, X., Zhang, H., Zhang, G., Zhu, Y., Zhang, S., Zhang, W., and Zhang, C. (2012c) Fluostatins I–K from the South China Sea-derived *Micromonospora rosaria* SCSIO N160. *Journal of Natural Products* 75(11), 1937–1943.

Zhang, W., Li, S., Zhu, Y., Chen, Y., Chen, Y., Zhang, H., Zhang, G. et al. (2014b) Heronamides D–F, Polyketide macrolactams from the deep-sea-derived *Streptomyces* sp. SCSIO 03032. *Journal of Natural Products* 77(2), 388–391.

Zhang, W., Liu, Z., Li, S., Yang, T., Zhang, Q., Ma, L., Tuan, X. et al. (2012a) Spiroindimicins A–D: New bisindole alkaloids from a deep-sea-derived actinomycete. *Organic Letters* 14(13), 3364–3367.

Zhang, W., Ma, L., Li, S., Liu, Z., Chen, Y., Zhang, H., Zhang, G. et al. (2014a) Indimicins A–E, Bisindole Alkaloids from the deep-sea-derived *Streptomyces* sp. SCSIO 03032. *Journal of Natural Products* 77(8), 1887–1892.

Zhang, Y., Werling, U., and Edelmann, W. (2012) SLiCE: A novel bacterial cell extract-based DNA cloning method. *Nucleic Acids Research* 40, e55.

Zhang, Y., Zhao, J., and Zeng, R. (2011) Expression and characterization of a novel mesophilic protease from metagenomic library derived from Antarctic coastal sediment. *Extremophiles* 15(1), 23–29.

Zhu, Y., Li, J., Cai, H., Ni, H., Xiao, A., and Hou, L. (2013) Characterization of a new and thermostable esterase from a metagenomic library. *Microbiological Research* 168(9), 589–597.

Ziemert, N., Lechner, A., Wietz, M., Millán-Aguiñaga, N., Chavarria, K.L., and Jensen, P.R. (2014) Diversity and evolution of secondary metabolism in the marine actinomycete genus *Salinispora*. *Proceedings of the National Academy of Sciences United States of America* 111(12), E1130–E1139.

Zotchev, S.B. (2012) Marine actinomycetes as an emerging resource for the drug development pipelines. *Journal of Biotechnology* 158, 168–175.

Zotchev, S.B., Sekurova, O.N., and Katz, L. (2012) Genome-based bioprospecting of microbes for new therapeutics. *Current Opinion in Biotechnology* 23, 941–947.

10

Marine Microbial Diversity:
A Pyrosequencing Perspective

**Balamurugan Sadaiappan, Sivakumar Kannan,
Sivasankar Palaniappan, and Poongodi Subramaniam**

CONTENTS

10.1 Introduction

Marine environments are packed with copious microorganisms that are responsible for the biogeochemical processes in the oceans (Atlas and Bartha, 1981). These organisms are studied continuously, and the microbial marginal count of the water surface is >10^5 cells (Hobbie et al., 1977), which indicates that the microbial cells in the massive ocean embrace various communities such as bacteria, fungi, archaea, cyanobacteria, and protists (Whitman et al., 1998) and are alone accountable for >98% of the oceanic biomass. As per the International Census of Marine Microbes (http://www.coml.org/projects/international-census-marine-microbes-icomm), the estimated microbial diversity of the ocean is 3.6 × 10^{30}, which forms the base for the customary oceanic food cycle.

Influence of marine microorganisms in the ocean is inexplicable since they pay off more than 95% of the ocean respiration (Del Giorgio and Duarte, 2002). Nevertheless, diversity of the microbes in the marine environment is huge, but the recovery and culture of these organisms in the laboratory is merely 0.1% (Ferguson et al., 1984) for the reason that all the marine microbes are not cultivable under laboratory conditions. However, this lacuna between the microorganisms and the laboratory can be overcome with the advent of advanced nucleic acid technologies; vast phylogenetic diversity is integrated with orthologous gene sequences, and the resulting microbial diversity is at least a minimum of 100 times greater than the previously estimated culture-reliant assessments (Pace, 1997). The 16S rRNA gene-based sequences are used to estimate the phylogenetic diversity of microorganisms (Giovannoni et al., 1990; Rappe and Giovannoni, 2003), without culturing them.

In recent times, culture-independent analysis of environmentally derived samples has revolutionized our understanding of unculturable microbial diversity (Hugenholtz and Pace, 1996; Tseng and Tang, 2014). Technological advances, such as pyrosequencing (Ronaghi, 2001), enable prompt categorization of microbial population communities more rapidly with greater sequence complexity (Sogin et al., 2006), and the development of pyrosequencing techniques has brought unprecedented opportunities for environmental microbiological studies (Wang et al., 2013).

10.2 Pyrosequencing of Marine Microbial Diversity

10.2.1 Seawater

Qian et al. (2011) explored the microbial communities of the least-explored ecosystems, the Atlantis II Deep and Discovery Deep in the Red Sea. Taxonomic classification of 16S rRNA gene amplicons showed vertical stratification of microbial diversity from the surface water to 1500 m below the surface. Significant differences in both the bacterial and archaeal diversity were observed in the upper (2 and 50 m) and deeper layers (200 and 1500 m). The dominant bacterial community in the upper layer was *Cyanobacteria*, whereas the deeper layer harbored a large proportion of *Proteobacteria*. Among *Archaea*, *Euryarchaeota*, especially *Halobacteriales*, was found to be dominant in the upper layers but diminished drastically in the deeper layers where *Desulfurococcales* belonging to *Crenarchaeota* became the dominant group. Thus, this study has indicated that the microbial communities sampled from the water columns were different from that of the other parts of the world.

Ottesen et al. (2011) studied the planktonic microbial community diversity using tag sequence method. The study employed the environmental sample processor (ESP) platform and 16S rRNA where the ESP preserved the active RNAs in the environment. An average of 100,000 sequences was obtained. The analysis revealed that all the four samples showing the oxygenic photoautotrophs were predominantly eukaryotic, while the bacterial community was dominated by *Polaribacter*-like *Flavobacteria* and *Rhodobacterales*, sharing high similarity with *Rhodobacterales* sp. HTCC2255. However, each time point was associated with distinct species abundance and gene transcript profiles. These laboratory and field tests confirmed that autonomous collection and preservation are a feasible and useful approach for characterizing the expressed genes and environmental responses of marine microbial communities.

The pyrosequence approach by Quaiser et al. (2011) from the Sea of Marmara assessed microbial community and diversity. The main focus was on 454 pyrosequencing analysis on deep plankton to relate the metagenome data to small subunit (SSU) rRNA gene-based diversity of archaea, bacteria, and eukaryotes from the same samples. An average of 100,000 reads was used for the analyses. Taxonomic investigation showed *Gamma-* and *Alphaproteobacteria*, followed by *Bacteroidetes*, which dominated the bacterial fraction in Marmara deep-sea plankton, whereas *Planctomycetes* and *Delta-* and *Gammaproteobacteria* were the most abundant groups in high bacterial-diversity sediments. Group I Crenarchaeota/Thaumarchaeota dominated the archaeal plankton fraction, although group II and III *Euryarchaeota* were also present. Eukaryotes were highly diverse in SSU rRNA gene libraries, with group I (*Duboscquellida*) and II (*Syndiniales*) alveolates and Radiozoa dominating the plankton

Opisthokonta and Alveolates, in the sediments. *Pelagibacter ubique* and *Nitrosopumilus maritimus* were highly represented in 1000 m deep plankton.

Dramatic decreases in the extent of Arctic multiyear ice (MYI) and its impact on the polar microbial community were studied by Bowman et al. (2012). MYI community appeared less diverse in both Shannon's index and Simpson's index. MYI had lower richness than the underlying surface water and the diversity using Simpson' and Shannon's indices was t = 0.65, P = 0.56 for Simpson, and t = 0.25, P = 0.84 for Shannon. *Cyanobacteria*, comprising 6.8% of reads obtained from MYI, were observed for the first time in Arctic sea ice. In addition, several low-abundance clades not previously reported in sea ice were present, including the phylum TM7 and the classes Spartobacteria and Opitutae. Members of *Coraliomargarita*, a recently described genus of the class Opitutae, were present in sufficient numbers, suggesting the niche occupation within MYI; this pyrosequence helped undertake a detailed census of the microbial community in MYI though it has been characterized by previous studies.

Exploration of total and active bacterial community structure in a gradient, covering the surface waters, from the Mackenzie River to the coastal Beaufort was reported by Ortega-Retuerta et al. (2013). Pyrosequencing analyses yielded a total of 60,544 reads form six samples with the maximum in the coast than the fresh waters and open sea. Also, it indicated that richness and diversity were higher in the coastal water and river than the open sea. Taxonomic classification showed that the particle-attached (PA) bacterial communities were different only 15.7% from the free-living (FL) ones in the open-sea sample. Furthermore, PA samples generally showed higher diversity (Shannon, Simpson, and Chao indices) than the FL samples. At the class level, Opitutae was most abundant in the PA fraction of the sea sample, followed by *Flavobacteria* and *Gammaproteobacteria*, while the FL sea sample was dominant with *Alphaproteobacteria*. In the coast and river samples, in both PA and FL fractions, *Betaproteobacteria*, *Alphaproteobacteria*, and *Actinobacteria* were dominant. The result showed that the influence of river discharges and particle loads have important roles in determining the community diversity.

10.2.2 Deep Sea

Nankai methane deep-sea sediments were assessed by Nunoura et al. (2012) for the microbial community composition. About 8709 and 7690 SSU rRNA gene tag sequences were obtained from methane seep sediments at 5 and 25 cm, respectively, and 679 and 718 archaeal reads were obtained by tag pyrosequence method. The *Proteobacteria* were dominant followed by *Acidobacteria*, *Bacteroidetes*, *Chloroflexi*, and *Planctomycetes*; in the deep-sea hydrothermal vent, euryarchaeotic group 6, marine group I (MGI), and deep-sea archaeal group were found dominant. Some taxa varied significantly between the samples.

Briggs et al. (2012) investigated the microbial abundance in the methane hydrates of Andaman Sea with 16S rRNA pyrosequencing tool. A total of 72,523 primer-tagged pyrosequencing reads were obtained, and after the quality check, qualified reads revealed *Firmicutes* as the dominant phylum with 52%, 53%, and 74% of reads followed by phyla *Actinobacteria*, *Bacteroidetes*, and *Proteobacteria*. In the gas hydrate occurrence zone (GHOZ), the classes of *Gamma-* and *Betaproteobacteria* were predominant, whereas *Bacteroidetes* abundance showed decrease in frequency to ~2%. *Actinobacteria* and *Alphaproteobacteria* had the highest frequency above the GHOZ. Deep subseafloor of Andaman habitat showed a unique microbial community with low cell abundance when compared to other methane hydrates and seafloor.

One among the highly diverse microbial ecosystems is subseafloor sediments that accumulate large amounts of organic and inorganic materials. Zhu et al. (2013) have investigated the microbial community and diversity in such region in the South China Sea. Over 265,000 amplicons falling on the V3 hypervariable region of the 16S ribosomal RNA (rRNA) gene from 16 sediment samples were collected from multiple locations in the northern region of South China Sea from depths ranging from 35 to 4000 m. A total of 9726 unique operational taxonomic units (OTUs) at the 97% threshold from the 16 samples and maximum of 2819 OTUs at 0.03% cluster distance in sample 19 were reported. Principal component analysis showed that the deep-sea sediments showed a higher degree of notable variation from shallow sea. Taxonomic classification revealed that *Proteobacteria, Firmicutes, Planctomycetes, Acidobacteria, Actinobacteria, Chloroflexi, Bacteroidetes, Gemmatimonadetes,* and *Nitrospirae* are the dominant phyla, whereas *Proteobacteria* was the abundant phylum in all the samples and represented 37%–80% of all bacterial amplicons, where more than 54.7% of the sequences belong to the candidate phylum OP9; in a 5000 m depth benthic sample from the Indian Ocean, the class *Alphaproteobacteria* reached 93.4% in total sequences. Overall, the study of Zhu et al. (2013) showed that the geographical locations have a strong impact on microbial community composition.

Vertical stratification of the bacterial populations in two hydrothermal and hypersaline deep-sea pools in the Red Sea rift was studied using 454 sequencing technology by Bougouffa et al. (2013); 454 sequences yielded 190,000 qualified reads of which 18,867 were *Archaea*; 162,870, *Bacteria*; and the remaining 10,600, unclassified. Averages of 16,000 quality reads were analyzed and the bacterial richness and the diversity were high in DIS-LCL with OTUs of 2,297 having the maximum Shannon index (8.53), among the eight samples. In the case of *Archaea*, maximum OTUs were in ATL-BWI. UniFrac UPGMA and PCoA analyses showed that DIS-BWI sample was exceptionally unique from the rest of the samples. Taxonomic approach revealed that *Proteobacteria* was evidently abundant except the brine-water interface at 2028 m. Abundance change in depth, that is, *Alphaproteobacteria* was the dominant class in the deep-sea water, while *Gammaproteobacteria* (*Alteromonadales* and *Oceanospirillales*) were predominant in the lower convective layers. *Deferribacteres* were found to be dismissed toward the brine pool form the deep seawater. In *Archaea*, *Crenarchaeota* and *Euryarchaeota* were the only two phyla found in both sites, and among them, MGI was abundant in the brine-overlying deep water sample and brine-water interface samples, but it was dramatically lower in the brine pools. Multivariate analysis indicated that temperature and salinity were the major influencing factors in shaping the communities.

10.2.3 Hydrothermal Chimneys

Sylvan et al. (2012) studied the microbial diversity and ecology of the hydrothermal chimneys that represent a globally dispersed habitat on the seafloor associated with midocean ridge spreading centers. The use of tag pyrosequencing of the V6 region of the 16S rRNA and full-length 16S rRNA sequencing on inactive hydrothermal sulfide chimney samples, collected from 9°N on the East Pacific Rise, revealed the bacterial composition, metabolic potential, and succession from venting to nonventing (inactive) regimes. Alpha-, beta-, delta-, and gammaproteobacteria and members of the phylum *Bacteroidetes* dominated the inactive sulfides. Greater than 26% of the V6 tags obtained were closely related to lineages involved in sulfur, nitrogen, iron, and methane cycling. *Epsilonproteobacteria* represented <4% of the V6 tags recovered from inactive sulfides and 15% of the full-length clones, despite their high abundance in active chimneys. Members of the phylum *Aquificae* were common in the active vents. In both the analyses, a proportion of *Alphaproteobacteria*, *Betaproteobacteria*, and members of the phylum *Bacteroidetes* was greater than those found in the active hydrothermal sulfides.

10.2.4 Marine Sediments

Kouridaki et al. (2010) were the first to compare the bacterial communities living in the deep-sea ecosystems, using the comparative analysis of five 16S rRNA gene clone libraries from deep-sea sediments (water column depth: 4000 m) of the Northeastern Pacific Ocean and Eastern Mediterranean Sea. The estimated chlorophyll *a*, organic carbon, and C/N ratio provided with the evidence of significant differences in the trophic state of the sediments between the Northeastern Pacific Ocean and the much warmer Eastern Mediterranean Sea. A diverse range of 16S rRNA gene phylotypes was found in the sediments of both the regions that were represented by 11 different taxonomic groups, with *Gammaproteobacteria* predominating in the Northeastern Pacific Ocean sediments and *Acidobacteria*, in the Eastern Mediterranean microbial community.

Zinger et al. (2011) studied the global ocean microbial diversity using pyrosequence, and about 9 million reads were obtained from 509 samples around the global ocean's surface to the deep-sea floor. A total of 44,493 OTUs were reported from the pelagic region of which coastal and benthic coastal regions showed higher number of OTUs, that is, 20,364 and 59,051, respectively. Taxonomic classification showed that *Alphaproteobacteria* were dominant in the pelagic and *Gammaproteobacteria* in the benthic sediments. Benthic vents showed a maximum of 30% abundance of *Epsilonproteobacteria*. The study (2011) also examined the bacterial community composition of the samples, using OTUs. Bacterial community in the pelagic and benthic ecosystems was found to share an average of 7.1 % ± 0.01% of OTUs at 3%. In contrast, comparison between different ecosystems showed that higher percentage of shared OTUs was at 3% distance. Maximum shared OTUs were observed between the different ecosystems, excluding vents and anoxic sediments. Also, differences in the physical mixing might play a fundamental role in the distribution patterns of marine bacteria, as benthic communities showed a higher dissimilarity with increasing distance than the pelagic communities. Thus, this study showed interesting perspectives for the definition of biogeographical biomes for bacteria of ocean waters and the seabed.

Sundarakrishnan et al. (2012) studied the microbial diversity of 1000 m deep pelagic sediments off the coast of the Andaman Sea by a culture-independent technique. They obtained 19,721 reads by amplifying the hypervariable region of the SSU rRNA gene covering V6–V9, from the metagenomic DNA. A total of 305 OTUs were obtained corresponding to the members of *Firmicutes*, *Proteobacteria*, *Planctomycetes*, *Actinobacteria*, *Chloroflexi*, *Bacteroidetes*, and *Verrucomicrobium*. *Firmicutes* was the predominant phylum, which was largely represented by the family, *Bacillaceae*. More than 44% of the sequence reads could not be classified up to the species level and more than 14% of the reads could not be assigned to any genus. Thus, the data indicated the possibility for the presence of uncultivable or unidentified novel bacterial species. In addition, the community structure identified in this study significantly differed with the other reports from the marine sediments.

Jeffries et al. (2012) studied the relative shift in abundance of microbial community composition in response to salinity in the Coorong lagoon using tag pyrosequence. The study revealed rare taxa that could not be accessed by traditional clone libraries. A total of 392,483 DNA sequences obtained from four sediment samples were used to compare the genomic characteristics along the gradient. Taxonomic comparison showed that *Proteobacteria*, *Roseobacter*, and *Roseovarius* peaked in abundance at intermediate salinities and the cyanobacterial genus *Euhalothece* dominated the community at the most saline and nutrient-rich site. Goodall and UniFrac indices showed that the Coorong lagoon is distinct from other habitats, implying that the nature of the particular habitat would have a definite role in shaping the microbial community structure.

Mangroves are complex ecosystems that regulate nutrient and sediment fluxes to the open sea. Importance of bacteria and fungi in regulating the nutrient cycles is well known but the diversity of *Archaea* and their role is not yet studied clearly. Pires et al. (2012) have used tag pyrosequence to reveal the unprecedented archaeal diversity in the mangrove sediments and rhizosphere samples. Ribosomal database project (RDP) classifier showed the archaeal phyla, *Euryarchaeota* and Crenarchaeota, representing five classes, namely, Halobacteria, Methanobacteria, Methanomicrobia, Thermoplasmata, and Thermoprotei. Halobacteria, Methanobacteria, unclassified Thermoprotei, and unclassified *Archaea* were found dominant in the rhizosphere region, whereas Thermoplasmata and Thermoprotei were dominant in the bulk sediment sample. The rhizosphere microhabitats of *Rhizophora mangle* and *Laguncularia racemosa* hosted distinct archaeal assemblages and also various rhizosphere associated groups of *Archaea*, having possibly some important roles in sediment nutrient cycling. This study has thus revealed the important but unprecedented archaeal diversity and composition in the mangrove environment.

Wang et al. (2013) studied the bacterial community of the intertidal zones of Bohai Bay, China, adopting pyrosequencing-based approach to analyze the 16S rRNA gene of bacteria in the sediments collected from two typical intertidal zones—Qikou (Qi) and Gaoshaling (Ga). Results showed that, at a 0.03 distance, the sequences from the Qi sediments had 3252 OTUs belonging to 34 phyla, 69 classes, and 119 genera, and the 3740 OTUs from the Ga sediments had 33 phyla, 66 classes, and 146 genera. While comparing the bacterial communities of the two intertidal sediments, significant difference was observed in the dominant composition and distribution at phylum, class, and genus levels. Canonical correspondence analysis showed that the median grain size and DO were the most important factors, regulating the bacterial abundance and diversity.

Zhu et al. (2013) studied the bacterial community composition of marine sediments using pyrosequencing-based 16S rRNA analysis in the South China Sea. They collected 16 sediment samples from varying depths of 35 to 4000 m and got 9726 OTUs belonging to 40 bacterial phyla containing 22 officially described phyla and 18 candidate phyla, with most diverse *Proteobacteria, Firmicutes, Planctomycetes, Actinobacteria, and Chloroflexi. Gammaproteobacteria* was the most plentiful phylotype, scoring 42.6%, which showed its total preponderance in all the samples. More interestingly, among the 18 candidate phyla observed, 12 were the first-time reports from the South China Sea.

Carr et al. (2013) analyzed the microbial abundance and composition in the marine sediments beneath the Ross Ice Shelf, Antarctica, using pyrosequence. A total of 27,962 sequences represented 15 phyla included *Actinobacteria, Bacteroidetes, Chloroflexi,* and *Firmicutes,* and the subphyla b-, d-, and c-Proteobacteria were dominant. The top 5 cm showed 1130 unique sequences having maximum Shannon index value of 5.7 and the lowest of 3.1 and 2.1 in 120–125 and 150–153 cm, respectively. SSU rRNA gene sequences revealed a more comprehensive understanding of the microbial communities and carbon cycling in the marine sediments, including those of this unique ice shelf environment. From 190,000 qualified reads, 18,867 were *Archaea* and 162,870 were *Bacteria*, and the remaining 10,600, unclassified.

10.2.5 Reefs

Gaidos et al. (2011) showed that the bacterial phyla *Proteobacteria, Firmicutes,* and *Actinobacteria,* the archaeal order *Nitrosopumilales,* and the uncultivated divisions marine group III (*Euryarchaeota*) and marine benthic group C (*Crenarchaeota*) constituted the dominant microbial community in the coral reef sediments of Hawaii, based on ribosomal tag libraries. The four dominant bacterial OTUs were the members of the genera that include pathogens and facultative anaerobes, dissimilatory nitrate reducers, and denitrifiers.

Pyrosequencing of bacterial symbionts associated with the marine sponge, *Axinella corrugata*, was done by White et al. (2012). They collected seven individuals of the sponge from two locations of Florida reef area and analyzed for 16S rRNA amplicons using multiplex 454 pyrosequencing. As a whole, 265,512 sequences were generated and 9,444 distinct OTUs were identified. As approximately 24% sequences did not match with the bacteria at the class level, they could represent novel taxa. Interestingly, 330 *Albinaria corrugata*–specific OTUs were identified related to *Ectothiorhodospiraceae* with more than 34.5% amplicon core.

Coral-associated microbial diversity and relationships with environmental factors and host species remain unclear. To find the sequence diversity and relationship, Lee et al. (2012) have used pyrosequence technique in three stony *Scleractinia* and two soft *Octocorallia* corals, in three locations of the Red Sea; a total of 43,579 quality reads were obtained with an average number of 4,842 reads per sample. Diversity and richness were higher in *Sarcophyton* sp. 2 from site 2 and *Pocillopora verrucosa* from site 1 with the lowest diversity of *Astreopora myriophthalma* from sites 1 and 3. QIIME pipeline and RDP pipeline analyses of taxonomic classification showed the presence of *Proteobacteria, Actinobacteria, Bacteroidetes, Cyanobacteria, Chloroflexi, Deinococcus–Thermus*, and *Firmicutes*. In the class, *Gammaproteobacteria* was the most dominant group, followed by *Chloroflexi* and *Chlamydiae*. Most interestingly, the candidate phylum WS3 was reported in corals for the first time. The location of the corals and environmental conditions greatly influenced the associated bacterial communities, which varied significantly with the same coral species in different locations. The same corals from disturbed areas appeared to share more similar bacterial communities, but larger variations in the community structures were observed between different coral species from pristine waters. Salinity and depth had influenced on the abundance of *Vibrio, Pseudomonas, Serratia, Stenotrophomonas, Pseudoalteromonas*, and *Achromobacter* in the corals. Some bacteria were found to be more sensitive to the coral species such as *Chloracidobacterium* and *Endozoicomonas*, suggesting that the host can also influence the associated bacterial community along with the environmental factors.

10.2.6 Marine Sponges

Jackson et al. (2012) have elucidated the bacterial community profiles associated with the marine sponges, *Raspailia ramosa* and *Stelligera stuposa*, sampled from a single geographical location in Irish waters and with ambient seawater. Four bacterial phyla (*Actinobacteria, Bacteroidetes, Firmicutes*, and *Proteobacteria*) represented among ~200 isolates as compared to 10 phyla found, while using pyrosequencing. Bacterial OTUs (2109) at 95% sequence similarity, from 10 bacterial phyla, were recovered from *R. ramosa*, 349 OTUs from *S. stuposa* representing 8 phyla, and 533 OTUs from 6 phyla were found in the surrounding seawater. Bacterial communities differed significantly between the sponge species and the seawater. Analysis of the data revealed that 2.8% of the reads, classified from the sponge *R. ramose*, can be defined as sponge specific, while 26% of *S. stuposa* sequences as sponge-specific bacteria. Novel sponge-specific clusters were identified, whereas the majority of the previously reported sponge-specific clusters (e.g., *Poribacteria*) were absent in these sponge species.

Lee et al. (2011) have studied the species-specific microbial communities associated with Red Sea sponges such as *Hyrtios erectus, Stylissa carteri*, and *Xestospongia testudinaria* using the pyrosequencing technology. The study revealed 1000 diverse microbial OTUs and the estimated species richness was up to 2000, in 26 bacterial phyla. Among them, 11 phyla have not been previously reported from the surrounding waters and 4 phyla were new records in sponges. A single sponge revealed archaeal species of 100 OTUs with estimated richness up to 300 species. This is the highest record for the archaeal diversity from the sponges. Thus, the sponge-associated microbial communities are consistently divergent than the surrounding waters, and they are highly sponge species specific and are unaffected by environmental disturbances (Lee et al., 2011).

10.3 Summary

These days, sequencing of microbial nucleic acids offers ample information that can be used in many areas of biological sciences, and the study of the unculturable microbes, in this regard, could greatly help us obtain useful information about the ecosystems, their microbial community structure, and functions in the different niches of the natural marine environment. Such microbes may hold potential genes that can act as metabolic gene factories when explored. However, studies on the marine microbial populations and their community structure using pyrosequencing technology are still needed, and in-depth studies are essential in all the marine ecosystems to explore the novel and uncommon microbial diversity of the vast oceans, for deriving a lot of benefits for humans. In this context, a present review on the culture-independent diversity of the marine microorganisms of various habitats, based on the data of pyrosequencing, would help in the future pursuit (Figure 10.1).

FIGURE 10.1
A graphical illustration representing the pyrosequencing of different marine habitats.

Acknowledgment

Authors are thankful to Prof. L. Kannan for critically going through the manuscript and offering valuable suggestions and comments.

References

Atlas, R. M. and R. Bartha. 1981. *Microbial Ecology: Fundamentals and Applications*. Addison-Wesley Publishing Company, Massachusetts.

Bolhuis, H. and L. J. Stal. 2011. Analysis of bacterial and archaeal diversity in coastal microbial mats using massive parallel 16S rRNA gene tag sequencing. *The ISME Journal* 5: 1701–1712.

Bougouffa, S., J. K. Yang, O. O. Lee, Y. Wang, Z. Batang, A. Al-Suwailem, and P. Qian. 2013. Distinctive microbial community structure in highly stratified deep-sea brine water columns. *Applied and Environmental Microbiology* 79(11): 3425–3437.

Bowman, J. S., S. Rasmussen, N. Blom, J. W. Deming, S. Rysgaard, and T. Sicheritz-Ponten. 2012. Microbial community structure of Arctic multiyear sea ice and surface seawater by 454 sequencing of the 16S RNA gene. *The ISME Journal* 6: 11–20.

Briggs, B. R., F. Inagaki, Y. Morono, T. Futagami, C. Huguet, A. Rosell-Mele, T. D. Lorenson, and F. S. Colwell. 2012. Bacterial dominance in subseafloor sediments characterized by methane hydrates. *FEMS Microbial Ecology* 81: 88–98.

Carr, S. A., S. W. Vogel, R. B. Dunbar, J. Brandes, J. R. Spear, R. Levy, T. R. Naish, R. D. Powell, S. G. Wakeham, and K. W. Mandernack. 2013. Bacterial abundance and composition in marine sediments beneath the Ross Ice Shelf, Antarctica. *Geobiology* 11(4): 377–395.

Del Giorgio, P. A. and C. M. Duarte. 2002. Respiration in the ocean. *Nature* 420: 379–384.

Ferguson, R. L., E. N. Buckley, and A. V. Palumbo. 1984. Response of marine bacterioplankton to different filtration and confinement. *Applied and Environmental Microbiology* 47(1): 49–55.

Gaidos, E., A. Rusch, and M. Ilardo. 2011. Ribosomal tag pyrosequencing if DNA and RNA from benthic coral reef microbiota: Community spatial structure, rare members and nitrogen-cycling guilds. *Environmental Microbiology* 13(5): 1138–1152.

Giovannoni, S. J., T. B. Britschgi, C. L. Moyer, and K. G. Field. 1990. Genetic diversity in Sargasso Sea bacterioplankton. *Nature* 345: 60–63.

Hobbie, J. E., R. J. Daley, and S. Jasper. 1977. Use of nuclepore filters for counting bacteria by fluorescence microscopy. *Applied and Environmental Microbiology* 33(5): 1225–1228.

Hugenholtz, P. and N. R. Pace. 1996. Identifying microbial diversity in the natural environment: A molecular phylogenetic approach. *Trends in Biotechnology* 14(6): 190–197.

Jackson, S. A., J. Kennedy, J. P. Morrissey, F. O'Gara, and A. D. Dobson. 2012. Pyrosequencing reveals diverse and distinct sponge-specific microbial communities in sponges from a single geographical location in Irish waters. *Microbial Ecology* 64(1):105–116.

Jeffries, T. C., J. R. Seymour, K. Newton, R. J. Smith, L. Seuront, and J. G. Mitchell. 2012. Increase in the abundance of microbial genes encoding halotolerance and photosynthesis along a sediment salinity gradient. *Biogeosciences* 9: 815–825.

Kouridaki, I., P. N. Polymenakou, A. Tselepides, M. Mandalakis, K. L. Jr. Smith. 2010. Phylogenetic diversity of sediment bacteria from the deep Northeastern Pacific Ocean: A comparison with the deep Eastern Mediterranean Sea. *International Microbiology* 13(3): 143–150.

Lee, O. O., Y. Wang, J. Yang, F. F. Lafi, A. Al-Suwailem, and P. Y. Qian. 2011. Pyrosequencing reveals highly diverse and species-specific microbial communities in sponges from the Red Sea. *The ISME Journal* 5: 650–664.

Lee, O. O., J. Yang, S. Bougouffa, Y. Wang, Z. Batang, R. Tian, A. Al-Suwailem, and P. Y. Qian. 2012. Spatial and species variations in bacterial communities associated with corals from the Red Sea as revealed by pyrosequencing. *Applied and Environmental Microbiology* 78(20): 7173–7184.

Nunoura, T., Y. Takaki, H. Kazama, M. Hirai, J. Ashi, H. Imachi, and K. Takai. 2012. Microbial diversity in deep-sea methane seep sediments presented by SSU rRNA gene tag sequencing. *Microbes and Environments* 27(4): 382–390.

Ortega-Retuerta, E., F. Joux, W. H. Jefferey, and J. F. Ghiglione. 2013. Spatial variability of particle-attached and free-living bacterial diversity in surface waters from the Mackenzie River to the Beaufort Sea (Canadian Arctic). *Biogeosciences* 10: 2747–2759.

Ottesen, E. A., R. M. Ill, C. M. Preston, C. R. Young, J. P. Ryan, C. A. Scholin, and E. F. DeLong. 2011. Metatranscriptomic analysis of autonomously collected and preserved marine bacterioplankton. *The ISME Journal* 5(12): 1881–1895.

Pace, N. R. 1997. A molecular view of microbial diversity and the biosphere. *Science* 276: 734–740.

Pires, A. C. C., D. F. R. Cleary, A. Almeida, A. Cunha, S. Dealtry, L. C. S. Mendonca-Hagler, K. Smalla, and N. C. Gomes. 2012. Denaturing gradient gel electrophoresis and barcoded pyrosequencing reveal unprecedented archaeal diversity in mangrove sediment and rhizosphere samples. *Applied and Environmental Microbiology* 78(16): 5520–5528.

Qian, P. Y., Y. Wang, O. O. Lee, S. C. Lau, J. Yang, F. F. Lafi, A. Al-Suwailem, and T. Y. Wong. 2011. Vertical stratification of microbial communities in the Red Sea revealed by 16S rDNA pyrosequencing. *The ISME Journal* 5(3): 507–518.

Quaiser, A., Y. Zivanovic, D. Moreira, and P. Lopez-Garcia. 2011. Comparative metagenomics of bathypelagic plankton and bottom sediment from the Sea of Marmara. *The ISME Journal* 5: 285–304.

Rappe, M. S. and S. J. Giovannoni. 2003. The uncultured microbial majority. *Annual Review of Microbiology* 57: 369–394.

Ronaghi, M. 2001. Pyrosequencing sheds light on DNA sequencing. *Genome Research* 11: 3–11.

Sogin, M. L., H. G. Morrison, J. A. Huber, D. M. Welch, S. M. Huse, P. R. Neal, J. M. Arrieta, and G. J. Herndl. 2006. Microbial diversity in the deep sea and the underexplored "rare biosphere". *Proceedings of the National Academy of Sciences of the United States of America* 103(32): 12115–12120.

Sundarakrishnan, B., M. Pushpanathan, S. Jayashree, J. Rajendhran, N. Sakthivel, S. Jayachandran, and P. Gunasekaran. 2012. Assessment of microbial richness in pelagic sediment of Andaman Sea by bacterial tag encoded FLX titanium amplicon pyrosequencing (bTEFAP). *Indian Journal of Microbiology* 52(4): 544–550.

Sylvan, J. B., B. M. Toner, and K. J. Edwards. 2012. Life and death of deep-sea vents: Bacterial diversity and ecosystem succession on inactive hydrothermal sulfides. *MBio* 3(1): e00279-11.

Tseng, C. H. and S. L. Tang. 2014. Marine microbial metagenomics: From individual to the environment. *International Journal of Molecular Sciences* 15: 8878–8892.

Wang, L., L. Liu, B. Zheng, Y. Zhu, and X. Wang. 2013. Analysis of the bacterial community in the two typical intertidal sediments of Bohai Bay, China by pyrosequencing. *Marine Pollution Bulletin* 72(1): 181–187.

White, J. R., J. Patel, A. Ottesen, G. Arce, P. Blackwelder, and J. V. Lopez. 2012. Pyrosequencing of bacterial symbionts within *Axinella corrugata* sponges: Diversity and seasonal variability. *PLoS ONE* 7(6): e38204.

Whitman, W. B., D. C. Coleman, and W. J. Wiebe. 1998. Prokaryotes: The unseen majority. *Proceedings of the National Academy of Sciences of the United States of America* 95(12): 6578–6583.

Zhu, D., S. H. Tanabe, C. Yang, W. Zhang, and J. Sun. 2013. Bacterial community composition of South China Sea sediments through pyrosequencing-based analysis of 16S rRNA genes. *PLoS ONE* 8(10): e78501.

Zinger, L., L. A. Amaral-Zettler, J. A. Fuhrman, M. C. Horner-Devine, S. M. Huse, D. B. M. Welch, J. B. H. Martiny, M. Sogin, A. Boetius, and A. Ramette. 2011. Global patterns of bacterial beta-diversity in seafloor and seawater ecosystems. *PLoS ONE* 6(9): e24570.

11

Gene-Targeted Metagenomics for the Study of Biogeochemical Cycling from Coastal-Saline Ecosystems

Basit Yousuf, Avinash Mishra, and Bhavanath Jha

CONTENTS

11.1 Introduction

The metagenomics of marine ecosystems is one of intriguing areas of marine ecological studies. The marine environment is extremely diverse and constitutes the largest habitat, as it covers more than 70% of the Earth's surface and dominated by the microbial community members, namely, archaea, bacteria, and eukarya, in both abundance and diversity (Azam and Malfatti, 2007). The entire ocean is known to consist of millions of species and a single microliter of it is estimated to harbor thousands of different bacteria and archaea (Curtis et al., 2002). Bacteria represent a major portion of the Earth's biota with a number of microbial cells, estimated to be more than 10^{30} (Whitman et al., 1998). These bacteria operate as highly interdependent communities, which is fundamental to ecosystem functioning and sustainability. The bacterial communes act as a driver in different processes, such as primary production, organic matter remineralization, pollution remediation, and global biogeochemical cycling of the key elements including carbon, nitrogen, and sulfur (Falkowski et al., 2008). Microbes are the foundation of the biosphere, as they play a crucial role in biotechnology (Curtis et al., 2003), agriculture (Kennedy and Smith, 1995), environment (Xu et al., 2010), and human health (Ley et al., 2006).

The coastal-saline ecosystems encompass considerable portion of the biosphere as it covers vast land of a number of countries, consisting of marine coastal soil and sediments. These coastal ecosystems are expected to be rich in taxonomically and functionally important extremophiles, as these sites always remain influxed by tidal seawater (Keshri et al., 2013a). The barren coastal-soil ecosystem, predominantly devoid of vegetation, is generally regarded as a source of CO_2. Plant litter and mineralization of organic matter through microbes are the main source of CO_2 flux toward the atmosphere (Raich and Schlesinger, 1992; Raich and Potter, 1995). However, photo- and chemolithoautotrophic CO_2-fixing bacteria are known to sequester carbon, significantly, from the atmosphere (Yuan et al., 2012a; Salek et al., 2013). These two processes, mineralization of organic matter and CO_2 assimilation, occur concomitantly and the net effect depends on the dominant microbial groups. The role of microbes in CO_2 sequestration can be used as an avenue to mitigate climate changes, particularly global warming. The bacterial community structure and metabolic activity in barren saline ecosystems are less studied compared to other ecosystems. One gram of soil or sediment has been reported to contain up to 10 billion microorganisms and thousands of different microbial species (Rosello and Amann, 2001). The coastal-saline ecosystem has the potential to be explored for novel genes and gene products that play a key role in biogeochemical cycling of essential elements and has various biotechnological applications. Most of these microbes ($\approx$99%) are difficult to culture or have not been cultured yet (Amann et al., 1995; Rappe and Giovannoni, 2003). Unravelling the hidden diversity of coastal-saline ecosystem requires applications of a powerful molecular tool, such as metagenomics.

The term "metagenome of the soil" was for the first time used by Handelsman et al. (1998) for the mixture of microbial genomes, extracted directly from the soil. Metagenomics is also known by other popular names, such as environmental DNA cloning (Stein et al., 1996), multigenomic cloning (Cowan, 2000), environmental genomics (Beja et al., 2000b), eDNA cloning (Brady and Clardy, 2000), soil DNA cloning (MacNeil et al., 2001), recombinant environmental cloning (Courtois et al., 2003), and community genome analysis (Tyson et al., 2004). The culture-independent metagenomic methods have emerged to rescue the hurdles, which occur in bacterial identification and isolation, and it can complement or even replace the culture-based approaches (Simon and Daniel, 2011). Further, the comparative metagenomics approach can be helpful for determining the significant variation in sequence composition (Foerstner et al., 2005), genome size (Raes et al., 2007), evolutionary rates (von Mering et al., 2007), and metabolic capabilities (Tringe et al., 2005; Dinsdale et al., 2008), among physically dissimilar environment. The metagenomics is relatively a young research area, driven by different sequencing technologies, and can be achieved by three different methodologies: (1) sequence-based gene-targeted (GT) metagenomics, which is crucial for understanding the role and distribution of well-conserved phylogenetic or functional marker genes of interest; (2) function-based metagenomics, which provides information about the enzyme of interest, expressed in some host and usually done with BAC or Fosmid libraries; and (3) next-generation sequencing technology, the more recent approach of which includes Roche 454 sequencing (Margulies et al., 2005), the SOLiD system of Applied Biosystems (Bentley, 2006), and the Genome Analyzer of Illumina (Bentley, 2006) platform, in which the whole metagenome is investigated for a diverse collection of functional and phylogenetic marker genes, without constructing a metagenomic library (Shendure and Ji, 2008; Harismendy et al., 2009). The GT sequence-based metagenomics can also be performed in conjunction with quantitative real-time polymerase chain reaction (qRT-PCR)-based methods. The GT metagenomics approach is generally utilized for vast array of genes, which possess a sufficiently conserved region for designing specific or degenerate primer pairs. The technique is also commonly used to unravel the genes, directly responsible for important ecosystem functions or ecological processes, including biogeochemical cycles and biodegradation (Iwai et al., 2010). The GT approach provides rapid detailed information about bacterial diversity, community structure (gene/species richness and distribution), and interrelationship with the environment in a broader perspective (Figure 11.1). In addition, this approach also provides useful insight into the functional (metabolic), potential, ecological, and evolutionary pattern of important gene(s), found in microorganisms of different environmental niches (Rondon et al., 2000; Schloss and Handelsman, 2003; Iwai et al., 2010).

One of the important research quests is to understand the relationship between gene diversity and function of an important environmental process. There is enormous genetic diversity in nature, among which certain crucial genes, encoding for the major redox reactions, are indispensable for biogeochemical cycles (Falkowski et al., 2008). There are six major geochemical pathways that involve in the cycling of major elements, such as H, C,

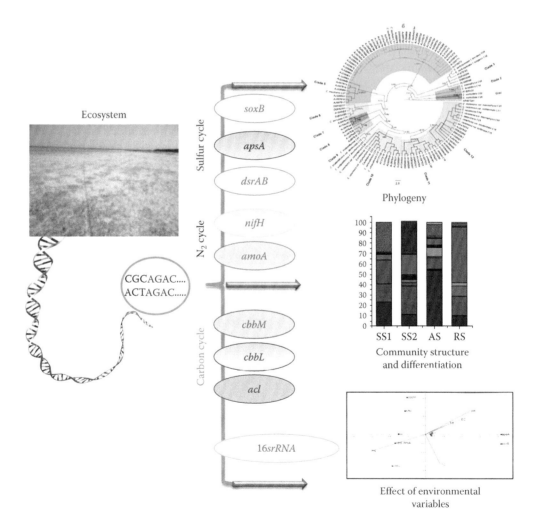

FIGURE 11.1
Gene-targeted-based metagenomics approach to study an ecosystem.

N, O, S, and P (Schlesinger, 1997). The cycling processes are largely driven by microbial-catalyzed reactions. Thus, microbial communities act as biogeochemical engineers as they drive major biogeochemical cycles having significant impact on climatic changes and are indispensable for the health of the Earth (Falkowski et al., 2000, 2008). Metagenomics-based studies of bacterial genome provide the role of bacteria in biogeochemical engineering and elemental cycling in a defined ecosystem (Falkowski et al., 2008).

11.2 Study of Microbes Using the Gene-Targeted Approach

The metagenome or total genome content of the environment/ecosystem can be accessed using specific primers/probes, for example, the primer for universal genes such as 16S rRNA for analyzing the plethora of microbial communities (Figure 11.2). The use of the 16S rRNA gene, as a marker, is well established and has the advantage of being a phylogenetically and taxonomically relevant target for the estimation of microbial diversity (Weisburg et al., 1991). From the past two decades, there are numerous studies, based on molecular fingerprinting techniques, such as denaturing-temperature gradient electrophoresis, single-strand conformation polymorphism, (terminal) restriction fragment length polymorphism, and Sanger sequencing of 16S rRNA gene, to investigate the biogeography and diversity of specific bacterial groups at different environmental niches (Giovannoni et al., 1996; Muyzer and Smalla, 1998; Cottrell et al., 2005; Sogin, et al., 2006; Keshri et al., 2013a,b). The remarkable key studies, based on short-gun sequencing (Venter et al., 2004; Rusch et al., 2007; Yooseph et al., 2007), have discovered numerous new genes and proteins

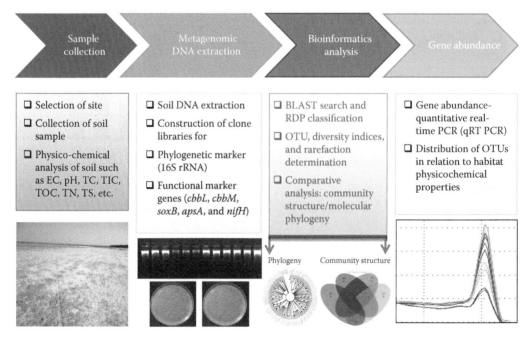

FIGURE 11.2
Steps involved in the study of microbes using the gene-targeted approach.

and provided glimpses about the prevalence of functional genes at different habitats. High-throughput sequencing approaches are advantageous, than 16S rRNA gene amplicon sequencing, because of high resolution for studying microbial communities. But the pyrosequencing data generate about 250–500 bp length sequences, which cannot be used for the assignment of function and reconstruction of the metabolic pathways (Morozova and Marra, 2008; Kennedy et al., 2010). The GT approach, implying small-insert libraries (Fosmid), based on Sanger dideoxy sequencing that provides read of more than 900 bp, is commonly used to infer phylogeny and also identify the putative function of genes. Additionally, this approach can also pave the way for reconstructing metabolic pathways of individual members within a particular microbial consortium.

Because of the widespread phylogenetic distribution of different elemental cycling pathways, it is impractical to infer the functional diversity solely from 16S rRNA gene data. In addition, detection of organisms, which are involved in elemental cycling but do not belong to the most abundant community members, are thus difficult to analyze, solely on the basis of 16S rRNA gene approaches. To circumvent these limitations, functional gene approaches have emerged as a powerful tool to assess the diversity, abundance, and activity of organisms, possessing specific functions. The functional gene(s) encodes key enzyme(s) that is involved in rate-limiting steps of important biogeochemical cycles. The use of functional marker genes was started by the pioneering work of Paul and Pichard (1996) for determining the community structure of specific functional guilds. Their study was based on the *cbbL* gene, encoding a large subunit of the most abundant enzyme ribulose-1,5-bisphosphate carboxylase/oxygenase (RuBisCO) to detect organisms that utilize the Calvin–Benson–Bassham (CBB) cycle for carbon fixation in the environment. The key enzymes, encoded by gene(s), are critical in the transformation of carbon, nitrogen, and sulfur compounds, which in turn lead to successful accomplishment of nutrient cycling that is required for the sustainability of the Earth (Falkowski et al., 2008; Bonavita et al., 2011). Thus, bacteria, harboring the gene(s) that encodes enzyme(s) involved in important metabolic pathways of nutrient cycling, can be targeted with specialized functional marker genes, such as *cbbL*, *cbbM*, *aclB*, *apsA*, *dsrAB*, *soxB*, *amoA*, and *nifH*. These functional molecular marker genes are considered as a robust phylogenetic marker because of conserved gene sequences, which make them a valuable tool to establish the evolutionary relationship and unravel the metabolic potential of microbial communities. The use of molecular approaches to study the terrestrial/aquatic ecosystems resulted in the discovery of many unexpected evolutionary lineages (Beja et al., 2000a). Most of these lineages are distantly related to the known organisms, and numerous of such lineages are sufficiently abundant to have a quintessential impact on the chemistry of biosphere.

11.3 Carbon Metabolism

The biological oxidation of the Earth is driven by photosynthesis (Falkowski and Godfrey, 2008), in which the energy of light oxidizes an electron donor (H_2O in oxygenic photosynthesis, whereas HS^-, H_2, or Fe^{2+} in anoxygenic photosynthesis). Electrons and protons, generated in the process, are utilized to reduce inorganic carbon to organic matter. Presently, six mechanisms of the CO_2 fixation are known that maintain environment, of which three mechanisms are reported exclusively in *Archaea* (Berg et al., 2010).

The most crucial and predominant biogeochemical process, occurring on the surface of the Earth, is the carbon cycle. The sequestration of CO_2, by autotrophic bacteria, is considered to be one of the key processes of this biogeochemical cycle (Figure 11.3). The predominant route for CO_2 fixation is the CBB reductive pentose phosphate pathway that is ubiquitously prevalent among the aerobic members of the *Alpha-*, *Beta-*, and *Gammaproteobacteria* (Tourova et al., 2010). The microbial (oxygenic and anoxygenic) photosynthesis is maintained by different crucial genes, among which key and prevalently occurring genes are *cbbL* and *cbbM*. These genes have highly conserved large subunit, which is being used extensively to explore the phylogenetic diversity and distribution of putative photo- and chemolithoautotrophic bacteria in aquatic systems, hypersaline habitats, and volcanic deposits (Nanba et al., 2004; van der Wielen, 2006; Hügler et al., 2010; Tourova et al., 2010; Alfreider et al., 2011; Kovaleva et al., 2011; Kato et al., 2012). Additionally, molecular ecological studies have also been performed on different agricultural/rhizosphere soil ecosystems, using these two marker genes (Selesi et al., 2005; Tolli and King, 2005; Yousuf et al., 2012a; Yuan et al., 2012b; Xiao et al., 2013), but only limited studies are devoted toward extreme terrestrial habitats (Freeman et al., 2009; Videmšek et al., 2009; Yousuf et al., 2012b). These reports have suggested that the *cbbL* form IC gene sequences are exclusively dominated in various terrestrial ecosystems (agroecosystem, pine forest, paddy soils) and also indicated the less diversity of the form IA *cbbL* compared to the form IC. The form II RuBisCO is known to function in low O_2 and high CO_2 environment (Tabita, 1999; Tabita et al., 2007). The occurrence of *cbbM* gene has been exclusively investigated for chemolithoautotrophy from aquatic habitats like hydrothermal vents (Kato et al., 2012), hypersaline habitats (Tourova et al., 2010), soda lake sediments (Kovaleva et al., 2011), thermal spring (Hall et al., 2008), and Movile Cave (Chen et al., 2009) with only one study from terrestrial ecosystem (Xiao et al., 2013).

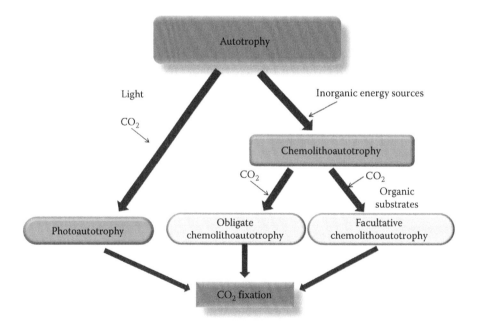

FIGURE 11.3
Schematic diagram of bacterial autotrophy.

The second most important pathway, involved in carbon fixation is the TCA cycle, in which ATP citrate lyase is the key enzyme encoded by the *acl* gene (Buchanan and Arnon, 1990). The reductive (reverse) tricarboxylic acid (rTCA)-based autotrophy has been demonstrated in deep-sea hydrothermal vents, using primers that target the key enzyme ATP citrate lyase (Campbell et al., 2003; Campbell and Cary, 2004; Voordeckers et al., 2008). However, to the best of our literature search, there is no report of rTCA-based autotrophic bacterial distribution from terrestrial soil ecosystem so far.

The other four pathways, such as reductive acetyl CoA pathway, 3-hydroxypropionate bicycle, 3-hydroxypropionate/4-hydroxybutyrate pathway, and dicarboxylate/4-hydroxybutyrate cycle, are also involved in carbon fixation, but the key enzymes of these pathways have not been addressed till now in terrestrial ecosystems.

11.4 Nitrogen Metabolism

The only biological process that makes atmospheric N_2 accessible for the synthesis of proteins and nucleic acids is nitrogen fixation, which is a reductive process that converts nitrogen gas (N_2) into ammonia (NH_3). The recycling of nitrogen contributes substantially in nutrient fluxing and sustainable soil fertility in the terrestrial ecosystem (Hsu and Buckley, 2009; Cavalcante et al., 2012). Around half of the annual nitrogen is fluxed into the biosphere (Vitousek et al., 1997), natural ecosystems (Cleveland et al., 1999), and agricultural systems (Peoples and Crasswell, 1992) by biological nitrogen fixation. This complex process is catalyzed by the nitrogenase reductase enzyme (sensitive to oxygen) and mediated by various groups of bacteria and a few members of *Archaea* in symbiotic, in associative, or under free-living conditions (Chien et al., 2000; Zehr et al., 2003). The enzyme consists of two components: metalloproteins, the iron (Fe) protein encoded by *nifH*, and the molybdenum-iron (MoFe) protein encoded by *nifD* and *nifK* (Zehr et al., 2003). Microbes catalyze both the oxidative (nitrification) and reductive (denitrification, nitrate ammonification, and nitrogen fixation) processes of N_2 fixation (Herbert, 1999). The oxidative nitrification process involves conversion of ammonia to nitrate, carried out by ammonia-oxidizing bacteria, such as *Nitrosomonas* sp. and *Nitrobacter*. The key enzyme of the pathway is ammonia monooxygenase, encoded by *amoA* gene that has been widespread in both bacteria and *Archaea* (Flood et al., 2015). The enzyme performs the rate-limiting step in nitrification and, hence, makes this process an important component of the global biogeochemical nitrogen cycle. Additionally, almost all of the ammonia-oxidizing microorganisms are autotrophic and thus, also known as potential biological sinks for carbon dioxide (Flood et al., 2015). The denitrification, mediated by certain bacterial groups, such as *Pseudomonas* and *Clostridium* in anaerobic conditions, is also extremely important as it involves the reduction of nitrates back into the largely inert nitrogen gas (N_2) to complete the nitrogen cycle.

Nitrogen-fixing diazotrophs are primarily characterized by the *nifH* gene that is one of the oldest existing functional gene in the history of gene evolution. The phylogenetic relationships among bacteria, based on the sequence divergences of the *nifH* gene, have been reported to be in agreement with the phylogeny that inferred from 16S rRNA gene sequences (Ueda et al., 1995; Borneman et al., 1996). Out of three structural genes (*nifH*, *nifD*, and *nifK*), the best phylogenetic resolution has been illustrated by *nifH* (Hirsch et al., 1995). The *nifH* gene is highly conserved among closely related microorganisms, therefore can be used to study relationship among diazotrophic bacteria. A few metagenomic, culture-independent studies have been performed on nitrogen-fixing microbial communities and their functional diversities in moderate and/or extreme terrestrial, bulk, and rhizosphere soil environment, such as rice, forest, grasses, soybean, and sediments (Hirano et al., 2001; Poly et al., 2001; Chowdhury et al., 2009; Teng et al., 2009; Xiao et al., 2010; Orr et al., 2011, 2012; Cavalcante et al., 2012; Desai et al., 2013), but has not been fairly addressed from saline soil ecosystems (Keshri et al., 2013b).

11.5 Sulfur Metabolism

The sulfur-oxidizing and sulfate-reducing bacterial communities are critical knobs for the operation of sulfur cycle and generating metabolic energy (Muyzer and Stams, 2008). Microbial sulfur oxidation is one of the vital processes for biogeochemical sulfur cycle in soil as well as marine ecosystems and is closely linked to the cycling of other essential elements, such as carbon, nitrogen, and oxygen (Murillo et al., 2014; Tourna et al., 2014). Sulfur-oxidizing prokaryotes are phylogenetically and metabolically highly diverse (Lane et al., 1992; Brock et al., 2006) and occur predominantly in extreme environments, such as volcanic hot springs and deep-sea hydrothermal vents. Thiosulfate is the central compound that plays a key role in sulfur cycling. The oxidation of different sulfur compounds, an essential component of sulfur and carbon cycling, by photo- and chemoautotrophs is considered as an ancient bacterial metabolism (Zopfi et al., 2004; Brocks et al., 2005). The photo- and chemoautotrophic microorganisms gain energy through the oxidation process, which is used for the fixation of inorganic carbon. The sulfur oxidation is well established in anoxygenic–phototrophic purple and green sulfur bacterial groups. Additionally, it is also known to be found in some purple nonsulfur bacteria, chemolithotrophic bacteria, and *Cyanobacteria* (Pfennig, 1989). Sulfur-based assimilation of inorganic carbon occurs through the complex sulfur oxidation mechanism, known as the Sox pathway, that involved complete oxidation of reduced sulfur compounds to sulfate (Kelly et al., 1997; Friedrich et al., 2001). Another assimilation mechanism, known as adenosine-5-phosphosulfate (APS) pathway, implemented through APS that is synthesized as an intermediate.

The component of the periplasmic thiosulfate-oxidizing Sox enzyme complex is encoded by the *soxB* gene, which is known to be pervasive among the several phylogenetic groups of sulfur-oxidizing bacteria (Meyer et al., 2007). The use of *soxB* gene as a phylogenetic marker has been demonstrated in a number of environments, including coastal aquaculture, marine sediments, hydrothermal vents, and soda lakes (Hügler et al., 2010; Krishnani et al., 2010; Luo et al., 2011; Lenk et al., 2012; Tourova et al., 2012; Akerman et al., 2013; Headd and Engel, 2013). The *soxB* gene has also been investigated in many reference strains of the *Chlorobi, Alpha-, Beta-, Gamma-,* and *Epsilonproteobacteria* (Petri et al., 2001; Meyer et al., 2007; Nakagawa et al., 2007; Sievert et al., 2008), which revealed that the Sox pathway is operating in a wide range of photo- and chemoautotrophic bacteria that oxidize sulfide and thiosulfate to sulfate (Tourova et al., 2012). Sulfide is a ubiquitous electron donor and commonly generated by sulfate-reducing prokaryotes. There are several enzymatic systems in the Sox pathway that have been associated with the oxidation of reduced-sulfur compounds (Ghosh and Dam, 2009) and widely distributed among photosynthetic microorganisms (Friedrich et al., 2001; Ghosh and Dam, 2009).

The *apsA* gene, encodes APS reductase a key enzyme of APS pathway, ubiquitously occurs in the sulfate-reducing prokaryotes. Moreover, homologues of the *apsA* gene also exist in several sulfur-oxidizing photo- and chemoautotrophic bacteria (Trüper and Fischer, 1982; Dahl and Trüper, 1994; Hipp et al., 1997). Primers, specific to the *apsA* or *soxB* genes, can reveal the occurrence of both sulfate-reducing and sulfur-oxidizing prokaryotes (Blazejak et al., 2006; Meyer and Kuever, 2007; Meyer et al., 2007).

The *dsrAB* gene that encodes dissimilatory sulfite reductase, a key enzyme of sulfate-reducing and some sulfur-oxidizing prokaryotes, has also been demonstrated as a useful phylogenetic marker gene for sulfur cycling. Sulfate-reducing bacteria are known to be involved in anaerobic carbon cycling, decomposition, and sulfur elemental cycling and thus act as a biogeochemical driver, especially in marine ecosystem (Colin et al., 2013; Jackson et al., 2014). Sulfate-reducing bacteria are abundant microbial community, residing in coastal ecosystems (Colin et al., 2013). The *dsrAB* gene diversity has not been addressed adequately from soil ecosystems, especially coastal-saline soil ecosystem. The *Desulfobacteraceae* and *Desulfobulbaceae* are dominant sulfate-reducing bacteria and expected to play a key role in the biogeochemical transformations of carbon and sulfur in various coastal ecosystems (Leloup et al., 2005, 2007; Mussmann et al., 2005; Gittel et al., 2008; Jiang et al., 2009). The occurrence of these three functional genes (*soxB*, *apsA*, and *dsrAB*) has been investigated to determine the distribution of microbial community implicated in sulfur cycle in different environments, such as saline-alkaline soil, hydrothermal vents, and salinity gradient (Ben-Dov et al., 2007; Hügler et al., 2010; Keshri et al., 2013a; Yang et al., 2013). Based on the detection of *dsrB* gene, the diversity and abundance of sulfur-reducing bacteria was evaluated in a salt marsh, indicating the prevalence of *Desulfovibrionaceae* in heavy metal-contaminated surface sediments (Quillet et al., 2012). Similarly, but based on the use of two genes (*apsA* and *dsrAB*), the sulfur cycling communities were assessed in hydrothermal vents (Frank et al., 2013), terrestrial mud volcanoes (Green-Saxena et al., 2012), ponds used for wastewater stabilization (Belila et al., 2013), and mangrove sediments (Varon-Lopez et al., 2014).

The sulfur oxidation pathway predominantly belongs to chemolithotrophic *Alphaproteobacteria* (e.g., *Paracoccus*), *Betaproteobacteria*, (e.g., *Thiobacillus*), and *Gammaproteobacteria* (e.g., *Thiomicrospira*) but is also used by photolithoautotrophs, such as *Rhodovulum* of *Alphaproteobacteria*. Comprehension about the earliest metabolic lineages could be understood from the physiology and biochemistry of sulfur oxidation, which has been extensively reviewed in phototrophs (Trüper and Fischer, 1982; Brune, 1989; 1995) and chemotrophs (Kelly, 1982; Kelly et al., 1997; Friedrich et al., 2005). The metabolic pathways in biogeochemical cycles are not necessarily directly related and sometimes are catalyzed by diverse, multispecies microbial interactions. Biogeochemistry of sulfur, which is vital for life but considered as a minor element, affects the productivity of an ecosystem. For example, the addition of sulfur fertilizer in naturally sulfur-deficient soils dramatically increases the primary production (Bunemann and Condron, 2007) and also affects the rate of coupled biogeochemical cycles such as C, N, and P cycling (McGill and Cole, 1981; Chapman, 1997; Kirkby et al., 2011). Thus, the sulfur cycle seems one of the vital biogeochemical drivers in the ecosystem interlinking three important biochemical pathways like carbon, sulfur, and nitrogen; still little is known regarding the sulfur bacteria communities in these vast coastal soil ecosystems.

11.6 Bacterial Community Abundance by Quantitative Real-Time PCR

The quantification of microbial guilds, based on qPCR, is highly sophisticated, sensitive, and very effective technique to detect abundance of a specific gene in an environmental sample (D'haene et al., 2010). The qPCR has emerged as a promising tool for studying soil microbial communities (Stubner, 2002; Kabir et al., 2003; Okano et al., 2004). The quantitative approach has been used for the copy number determination of 16S rRNA, *cbbL*, *cbbM*, *apsA*, and *nifH* from various environmental samples (Ben-Dov et al., 2007; He et al., 2007; Selesi et al., 2007; Hall et al., 2008; Levy-Booth and Winder, 2010; Kong et al., 2012; Yousuf et al., 2014). For the qPCR, generally a standard curve is made using recombinant plasmids containing a copy of the targeted gene. Experimental qPCR was performed, the efficiency of qPCR was calculated, data were analyzed by comparative C_T method, and the copy number of the targeted gene was determined (Videmšek et al., 2009; Yousuf et al., 2012a). The real-time PCR efficiency (*e*) and copy number (*c*) were calculated using the following equations:

$$e = \left[10^{-\left(\frac{1}{slope} \right)} \right] - 1 \quad e = [10^{-(1/\text{slope})}] - 1$$

$$c = Antilog \left[\frac{C_T - constant}{m} \right] \quad c = \text{Antilog} \left[\frac{C_T - constant}{m} \right]$$

Slope (*m*) and constant were determined by the regression equation $y = mx + const$. Calculated on the graph and plotted between the C_T value and copy number of the standard (targeted genes).

11.7 A Case Study

11.7.1 Functional Metagenomics of a Coastal-Saline Soil

The study of microbial diversity is crucial for the understanding of structure, function, and evolution of biological communities. In the case study, we have unraveled the occurrence and possible function of microbial communities in coastal-saline soil ecosystems by analyzing key protein-encoding functional genes, such as *cbbL*, *cbbM*, *apsA*, *soxB*, and *nifH*, using the GT metagenomics approach (Figure 11.4). These metabolic marker genes were targeted to get an insight of microbial organisms that are involved in carbon, sulfur, and nitrogen metabolisms of selected coastal-saline soil ecosystems. Selected coastal-saline soil had salinity of about 7.0 dS m^{-1} and pH 8.0, and targeted key enzymes are a large subunit of RuBisCO (genes *cbbL* and *cbbM*), adenosine phosphosulfate reductase (gene *apsA*), sulfate thiohydrolase (gene *soxB*), and nitrogenase reductase (gene *nifH*). Additionally, abundance of active functional community members of selected coastal-saline sites was also assessed by quantifying copy numbers of targeted functional genes, *cbbL*, *cbbM*, *apsA*, *soxB*, and *nifH*, using real-time qPCR.

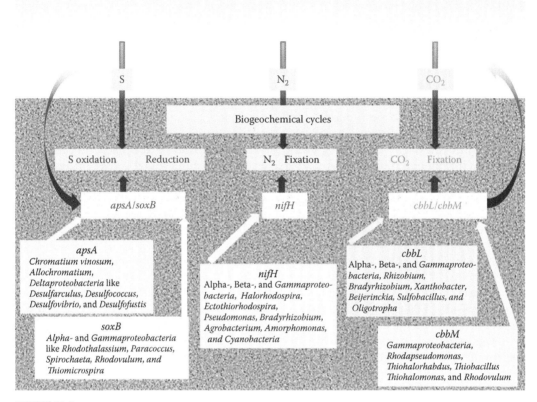

FIGURE 11.4
Study of biogeochemical cycle of coastal-saline soil using functional metagenomics. (Adapted from Yousuf, B. et al., *PLoS ONE*, 2014, 9(9):e107025.)

The clone library was constructed for each gene, sequenced and analyzed using bioinformatics. Results of environmental clone libraries revealed that *cbbL* clones were assigned to some recognized autotrophic bacteria, such as *Nitrosospira, Paracoccus, Rhodobacter,* and *Salinisphaera* that are known to have potential for carbon fixation and a number of uncultured clones. The *cbbM* clone library suggested the predominance of phylotypes related to the *Gammaproteobacteria* clones, *Rhodopseudomonas palustris, Thiohalorhabdus denitrificans, Thiohalomonas denitrificans,* and *Thiohalomonas nitratireducens*. About 75% clone sequences did not cluster with cultured representatives and could be considered as novel, which will sufficiently add to the genetic pool of these genes at these habitats as well as to the databases. The RuBisCO gene library demonstrated the presence of highly diversified and partially unique *cbbL* and *cbbM* gene sequences, which could belong to possibly yet unknown potent CO_2-fixing bacteria.

The *soxB* gene library indicated a high unprecedented novel diversity. Numerous phylotypes showed a very clear connection to potential sulfur oxidizers, belonging to *Alpha-* and *Gammaproteobacteria*, such as *Thiomicrospira crunogena, Rhodothalassium, Paracoccus,* and *Rhodovulum adriaticum*. The *apsA* clone library revealed oxidative and reductive mode of sulfur metabolism. Results envisaged that 40% *apsA* and *soxB* clones did not cluster with the gene sequences from any cultured representatives and more than 70% clones had less than 90% nucleotide identity, therefore could be considered as novel lineages. Most of these bacterial groups are reported to be associated with biogeochemical transformations in the extreme environments.

Majority of phylotypes from the *nifH* gene library were related to *Gammaproteobacteria* and *Alphaproteobacteria*. The phylotypes related to *Gammaproteobacterial* genera, such as *Halorhodospira*, *Ectothiorhodospira*, and *Pseudomonas* (photolithoautotrophic sulfur oxidizers), were prominent. Majority of the *nifH* gene sequences showed close affiliation with sequences from uncultured organisms and revealed lower homology of less than 90% nucleotide identity with the reported clones and genera.

The qPCR analysis showed heterogeneous distribution of functional genes among the sampling sites. The data generated from the clone library approach revealed that saline environmental niches harbor diverse and variable microbial communities. After clone library analysis, the next approach was to investigate the abundance of genes implied in various ecologically important pathways and infer their relative distribution using quantitative real-time PCR (Yousuf et al., 2014). The quantification of microbial guilds, based on qPCR, is a highly sophisticated, sensitive, and very effective technique to detect specific genes in the environmental sample. The qPCR results showed that the DNA-targeted RuBisCO form I (*cbbL*) gene copy numbers significantly outnumbered RuBisCO form II (*cbbM*). The gene copy numbers of *cbbL* at selected coastal-saline soil were observed in the range of $2.33 \times 10^7 – 5.0 \times 10^7$, whereas *cbbM* gene copy numbers were $6.67 \times 10^5 – 4.00 \times 10^6$. To the best of our knowledge, there is only one report of *apsA* gene abundance in soil ecosystems (Keshri et al., 2013a) where 3.23×10^5 copies of *apsA* gene was observed in saline-alkaline soil. The copy number of *soxB* genes was found to be $6.6 \times 10^6 – 7.07 \times 10^7$, whereas average values of the *nifH* gene copies per gram soil was $2.0 \times 10^7 – 1.66 \times 10^8$. Such difference in the copy number may occur because of the variation in the bioavailability of organic carbon and the concentration of sulfur and nitrogen (Andrade et al., 2012; Quillet et al., 2012). Overall, from the higher abundance/diversity of *nifH* and *cbbL* genes, it can be deciphered that nitrogen and carbon-fixing microbes could be significant contributors to the nutrient cycling in the coastal-saline ecosystem (Yousuf et al., 2014).

11.8 Conclusion

The GT metagenomics can provide fundamental insights into the structure, diversity, and abundance of chemolithoautotrophic, photoautotrophic, and diazotrophic microbial communities, involved in carbon, sulfur, and nitrogen cycling in various environmental niches. These biogeochemical pathways are indispensable for environment and thus need to be assessed particularly at unexplored marine-coastal soil ecosystems that are expected to be rich in taxonomically and functionally important repertoire of microbial flora, involved in various metabolic pathways. From the case study, we hypothesized that novel *Gammaproteobacteria* may play a prominent role in the primary production and could be strongly involved in the cycling of carbon, nitrogen, and sulfur compounds principally at selected coastal-saline soil sites. This finding certainly merits further studies, as it puts us on the track of potentially novel autotrophic bacterial communities, including sulfur oxidizers and nitrogen fixers that could have putative ecological role in the habitat. This GT approach will be efficient for detecting a pattern in the distribution of functional microbial guilds and has proven to be a highly efficient, accurate, rapid, and sensitive method to analyze metabolic functioning of hitherto untapped microbial flora involved in the elemental cycling.

11.9 Future Prospects

To gain a better understanding of the functional groups involved in life-sustaining biogeochemical cycles, functional metagenomic studies should be combined with meta-transcriptomic/metaproteomic approaches, imaging, and stable-isotope probing that will reveal the functional dynamics of a given community and provide useful insight of novel physiology and structure–function relationship of resident microbial communities (Figure 11.5). A better understanding of the resident bacterial communities and their functionalities in the coastal soils should shed light on the role of coastal-saline soil as a possible CO_2 sink. This approach can also provide in-depth knowledge for the designing of culture media for ecological understanding and ascertaining autotrophic physiology at these sites.

Additionally, there is a need for rapid development in different sequencing technologies that are involved in metagenomic studies. The resolution of these high-throughput technologies needs to be improved so that it can be utilized for reconstructing the metabolic pathways of individual members, generating a large amount of useful data, and unraveling of previously unrecognized level of diversity and their taxonomic assignment up to the species level. This will allow us to develop a fundamental understanding of the biogeochemical processes occurring in different ecosystems that could lead to discover novel metabolic pathways, optimize existing processes, and explore microbial community for high-value products.

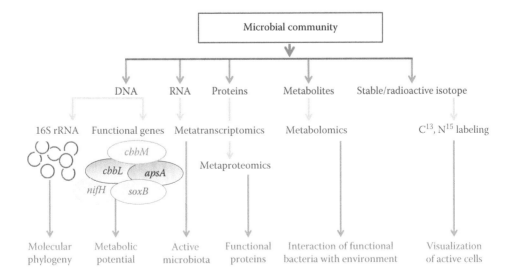

FIGURE 11.5
Different approaches to be integrated for the study of microbial community and their interaction with the environment.

Acknowledgment

CSIR-CSMCRI Communication No. PRIS-141/2014. CSIR Network Project (CSC0102-TapCoal) is thankfully acknowledged for research work conducted in the authors' laboratory.

References

Akerman, N. H., Butterfield, D. A., Huber, J. A., and Paul, J. B. 2013. Phylogenetic diversity and functional gene patterns of sulfur-oxidizing subseafloor Epsilonproteobacteria in diffuse hydrothermal vent fluids. *Frontiers in Microbiology* 4:1–14.

Alfreider, A., Schirmer, M., and Vogt, C. 2011. Diversity and expression of different forms of RuBisCO genes in polluted groundwater under different redox conditions. *FEMS Microbiology Ecology* 79:649–660.

Amann, R. I., Ludwig, W., and Schleifer, K. H. 1995. Phylogenetic identification and in situ detection of individual cells without cultivation. *Microbiological Reviews* 59:143–169.

Andrade, L. L., Leite, D., Ferreira, E. et al. 2012. Microbial diversity and anaerobic hydrocarbon degradation potential in an oil-contaminated mangrove sediment. *BMC Microbiology* 12:186.

Azam, F. and Malfatti, F. 2007. Microbial structuring of marine ecosystems. *Nature Review Microbiology* 5:782–791.

Beja, O., Aravind, L., Koonin, E. V. et al. 2000a. Bacterial rhodopsin: Evidence for a new type of phototrophy in the sea. *Science* 289:1902–1906.

Beja, O., Suzuki, M. T., Koonin, E. V. et al. 2000b. Construction and analysis of bacterial artificial chromosome libraries from a marine microbial assemblage. *Environmental Microbiology* 2:516–529.

Belila, A., Abbas, B., Fazaa, I. et al. 2013. Sulfur bacteria in wastewater stabilization ponds periodically affected by the 'red-water' phenomenon. *Applied Microbiology and Biotechnology* 97:379–394.

Ben-dov, E., Brenner, A., and Kushmaro, A. 2007. Quantification of sulfate-reducing bacteria in industrial wastewater, by real-time polymerase chain reaction (PCR) using dsrA and apsA genes. *Microbial Ecology* 54:439–451.

Bentley, D. R. 2006. Whole-genome re-sequencing. *Current Opinion in Genetics and Development* 16:545–552.

Berg, I. A., Ramos-Vera, W. H., Petri, A., Huber, H., and Fuchs, G. 2010. Study of the distribution of autotrophic CO_2 fixation cycles in Crenarchaeota. *Microbiology* 156:256–269.

Blazejak, A., Kuever, J., Erséus, C., Amann, R., and Dubilier, N. 2006. Phylogeny of 16S rRNA, ribulose-1,5-bisphosphate carboxylase/oxygenase, and adenosine-5′-phosphosulfate reductase genes from gamma- and alphaproteobacterial symbionts in gutless marine worms (Oligochaeta) from Bermuda and the Bahamas. *Applied and Environmental Microbiology* 72:5527–5536.

Bonavita, M. C., Godfroy, A., and Sarradin, P. 2011. Diversity and function in microbial mats from the Lucky Strike hydrothermal vent field. *FEMS Microbiology Ecology* 76: 524–540.

Borneman, J., Skroch, P. W., O'Sullivan, K. M. et al. 1996. Molecular microbial diversity of an agricultural soil in Wisconsin. *Applied and Environmental Microbiology* 62:1935–1943.

Brady, S. F. and Clardy, J. 2000. Long-chain N-acyl amino acid antibiotics isolated from heterologously expressed environmental DNA. *Journal of American Chemical Society* 122:12903–12904.

Brock, T. D., Madigan, M. T., and Martinko, J. M. 2006. *Biology of Microorganisms.* Upper Saddle River, NJ: Prentice Hall.

Brocks, J. J., Love, G. D., Summons, R. E., Knoll, A. H., Logan, G. A., and Bowden, S. A. 2005. Biomarker evidence for green and purple sulfur bacteria in a stratified Palaeoproterozoic sea. *Nature* 437:866–870.

Brune, D. C. 1989. Sulfur oxidation by phototrophic bacteria. *Biochimica Biophysica Acta Journal* 975:189–221.

Brune, D. C. 1995. Sulfur compounds as photosynthetic electron donors. In *Anoxygenic Photosynthetic Bacteria*, R. E. Blankenship, M. T. Madigan, and C. E. Bauer (eds.), pp. 847–870. Dordrecht, the Netherlands: Kluwer Academic Publishers.

Buchanan, B. B. and Arnon, D. I. 1990. A reverse KREBS cycle in photosynthesis: Consensus at last. *Photosynthesis Research* 24:47–53.

Bunemann, E. K. and Condron, L. M. 2007. Phosphorus and sulfur cycling in terrestrial ecosystems. In *Nutrient Cycling in Terrestrial Ecosystems*, P. Marschner and Z. Rengel (eds.), pp. 65–94. New York: Springer-Verlag.

Campbell, B. J. and Cary, S. C. 2004. Abundance of reverse tricarboxylic acid cycle genes in free-living microorganisms at deep-sea hydrothermal vents. *Applied and Environmental Microbiology* 70:6282–6289.

Campbell, B. J., Stein, J. L., and Cary, S. C. 2003. Evidence of chemolithoautotrophy in the bacterial community associated with *Alvinella pompejana*, a hydrothermal vent polychaete. *Applied and Environmental Microbiology* 69:5070–5078.

Cavalcante, A., Dias, F., Cassia, M. et al. 2012. Abundance and genetic diversity of *nifH* gene sequences in anthropogenically affected Brazilian mangrove sediments. *Applied and Environmental Microbiology* 78:7960–7967.

Chapman, S. J. 1997. Carbon substrate mineralization and sulfur limitation. *Soil Biology and Biochemistry* 29:115–122.

Chen, Y., Wu, L., Boden, R. et al. 2009. Life without light: Microbial diversity and evidence of sulfur- and ammonium-based chemolithotrophy in Movile Cave. *International Society for Microbial Ecology Journal* 3:1093–1104.

Chien, Y. T., Auerbuch, V., Brabban, A. D., and Zinder, S. H. 2000. Analysis of genes encoding an alternative nitrogenase in the archaeon methanosarcina barkeri 227. *Journal of Bacteriology* 182:3247–3253.

Chowdhury, S. P., Schmid, M., Hartmann, A., and Tripathi, A. K. 2009. Diversity of 16S-rRNA and *nifH* genes derived from rhizosphere soil and roots of an endemic drought tolerant grass, *Lasiurus sindicus*. *European Journal of Soil Biology* 45:114–122.

Cleveland, C. C., Townsend, A. R., Schimel, D. S. et al. 1999. Global patterns of terrestrial biological nitrogen (N_2) fixation in natural ecosystems. *Global Biogeochemical Cycles* 13:623–645.

Colin, Y., Goñi-Urriza, M., Caumette, P., and Guyoneaud, R. 2013. Combination of high throughput cultivation and *dsrA* sequencing for assessment of sulfate-reducing bacteria diversity in sediments. *FEMS Microbiology Ecology* 83:26–37.

Cottrell, M. T., Waidner, L. A., Yu, L., and Kirchman, D. L. 2005. Bacterial diversity of metagenomic and PCR libraries from the Delaware River. *Environmental Microbiology* 7:1883–1895.

Courtois, S., Cappellano, C. M., Ball, M. et al. 2003. Recombinant environmental libraries provide access to microbial diversity for drug discovery from natural products. *Applied and Environmental Microbiology* 69:49–55.

Cowan, D. A. 2000. Microbial genomes—The untapped resource. *Trends in Biotechnology* 18:14–16.

Curtis, T. P., Head, I. M., and Graham, D. W. 2003. Theoretical ecology for engineering biology. *Environmental Science and Technology* 37:64A–70A.

Curtis, T. P., Sloan, W. T., and Scannell, J. W. 2002. Estimating prokaryotic diversity and its limits. *Proceedings of the National Academy of Sciences United States of America* 99:10494–10499.

D'haene, B., Vandesompele, J., and Hellemans, J. 2010. Accurate and objective copy number profiling using real-time quantitative PCR. *Methods* 50:262–270.

Dahl, C. and H. G. Trüper. 1994. Enzymes of dissimilatory sulfide oxidation in phototrophic sulfur bacteria. *Method in Enzymology* 243:400–421.

Desai, M. S., Assig, K., and Dattagupta, S. 2013. Nitrogen fixation in distinct microbial niches within a chemoautotrophy-driven cave ecosystem. *International Society for Microbial Ecology Journal* 7:2411–2423.

Dinsdale, E. A., Edwards, R. A., Hall, D. et al. 2008. Functional metagenomic profiling of nine biomes. *Nature* 452:629–632.

Falkowski, P., Scholes, R. J., Boyle, E. et al. 2000. The global carbon cycle: A test of our knowledge of earth as a system. *Science* 290:291–296.

Falkowski, P. G., Fenchel, T., and E. F. DeLong. 2008. The microbial engines that drive Earth's biogeochemical cycles. *Science* 320:1034–1039.

Falkowski, P. G. and Godfrey, L. V. 2008. Electrons, life and the evolution of Earth's oxygen cycle. *Philosophical Transactions of the Royal Society B: Biological Sciences* 363:2705–2716.

Flood, M., Frabutt, D., Floyd, D. et al. 2015. Ammonia-oxidizing bacteria and archaea in sediments of the Gulf of Mexico. *Environmental Technology* 36(1):124–135.

Foerstner, K. U., von Mering, C., Hooper, S. D., and Bork, P. 2005. Environments shape the nucleotide composition of genomes. *European Molecular Biology Organisation Reports* 6:1208–1213.

Frank, K. L., Rogers, D. R., Olins, H. C., Vidoudez, C., and Girguis, P. R. 2013. Characterizing the distribution and rates of microbial sulfate reduction at Middle Valley hydro-thermal vents. *International Society for Microbial Ecology Journal* 7:1–11.

Freeman, K. R., Pescador, M. Y., Reed, S. C., Costello, E. K., Robeson, M. S., and Schmidt, S. K. 2009. Soil CO_2 flux and photoautotrophic community composition in high-elevation, "barren" soil. *Environmental Microbiology* 11:674–686.

Friedrich, C. G., Bardischewsky, F., Rother, D., Quentmeier, A., and Fischer, J. 2005. Prokaryotic sulfur oxidation. *Current Opinion in Microbiology* 8:253–259.

Friedrich, C. G., Rother, D., Bardischewsky, F., and Quentmeier, A. 2001. Oxidation of reduced inorganic sulfur compounds by bacteria: Emergence of a common mechanism. *Applied and Environmental Microbiology* 67:2873–2882.

Ghosh, W. and B. Dam. 2009. Biochemistry and molecular biology of lithotrophic sulfur oxidation by taxonomically and ecologically diverse bacteria and archaea. *FEMS Microbiology Reviews* 33:999–1043.

Giovannoni, S. J., M. S. Rappé, D. Gordon, E. Urbach, M. Suzuki, and K. G. Field. 1996. Ribosomal RNA and the evolution of bacterial diversity. In *Evolution of Microbial Life*, M. D. Roberts, P. Sharp, G. Alderson, and M. Collins (eds.), pp. 63–85. Society for General Microbiology Symposium 54. Cambridge, U.K.: Cambridge University Press.

Gittel, A., Mußmann, M., Sass, H., Cypionka, H., and Könneke, M. 2008. Identity and abundance of active sulfate-reducing bacteria in deep tidal flat sediments determined by directed cultivation and CARD-FISH analysis. *Environmental Microbiology* 10:2645–2658.

Green-Saxena, A., Feyzullayev, A., Hubert, C. R. et al. 2012. Active sulfur cycling by diverse mesophilic and thermophilic microorganisms in terrestrial mud volcanoes of Azerbaijan. *Environmental Microbiology* 14:3271–3286.

Hall, J. R., Mitchell, K. R., Jackson-Weaver, O. et al. 2008. Molecular characterization of the diversity and distribution of a thermal spring microbial community by using rRNA and metabolic genes. *Applied and Environmental Microbiology* 74:4910–4922.

Handelsman, J., Rondon, M. R., and Brady, S. F. 1998. Molecular biological access to the chemistry of unknown soil microbes: A new frontier for natural products. *Chemical Biology* 5:R245–R249.

Harismendy, O., Ng, P., Strausberg, R. et al. 2009. Evaluation of next generation sequencing platforms for population targeted sequencing studies. *Genome Biology* 10:R32.

He, J. Z., Shen, J. P., Zhang, L. M. et al. 2007. Quantitative analyses of the abundance and composition of ammonia-oxidizing bacteria and ammonia-oxidizing archaea of a Chinese upland red soil under long-term fertilization practices. *Environmental Microbiology* 9:2364–2374.

Headd, B. and A. S. Engel. 2013. Evidence for niche partitioning revealed by the distribution of sulfur oxidation genes collected from areas of a terrestrial sulfidic spring with differing geochemical conditions. *Applied and Environmental Microbiology* 79:1171–1182.

Herbert, R. A. 1999. Nitrogen cycling in coastal marine ecosystems. *FEMS Microbiology Reviews* 23:563–590.

Hipp, W. M., Pott, A. S., Thum-Schmitz, N., Faath, I., Dahl, C., and Trüper, H. G. 1997. Towards the phylogeny of APS reductases and sirohaem sulfite reductases in sulfate-reducing and sulfur-oxidizing prokaryotes. *Microbiology* 143:2891–2902.

Hirano, K., Hayatsu, M., Nioh, I., and Nakai, H. 2001. Comparison of nitrogen-fixing bacterial flora of rice rhizosphere in the fields treated long-term with agrochemicals and non-agrochemicals. *Microbes Environment* 16:155–160.

Hirsch, P., Eckhardt, F. E. W., and Palmer, Jr., R. J. 1995. Fungi active in weathering of rock and stone monuments. *Canadian Journal of Botany* 73:S1384–S1390.

Hsu, S. F. and D. H. Buckley. 2009. Evidence for the functional significance of diazotroph community structure in soil. *International Society for Microbial Ecology Journal* 3:124–136.

Hügler, M., Gärtner, A., and Imhoff, J. F. 2010. Functional genes as markers for sulfur cycling and CO_2 fixation in microbial communities of hydrothermal vents of the Logatchev field. *FEMS Microbiology Ecology* 73:526–537.

Iwai, S., Chai, B., Sul, W. J., Cole, J. R., Hashsham, S. A., and Tiedje, J. M. 2010. Gene-targeted-metagenomics reveals extensive diversity of aromatic dioxygenase genes in the environment. *International Society for Microbial Ecology Journal* 4:279–285.

Jackson, K. L., Whitcraft, C. R., and Dillon, J. G. 2014. Diversity of Desulfobacteriaceae and overall activity of sulfate-reducing microorganisms in and around a salt pan in a southern California Coastal Wetland. *Wetlands* 34(5):969–977.

Jiang, L., Zheng, Y., Peng, X., Zhou, H., Zhang, C., Xiao, X., and Wang, F. 2009. Vertical distribution and diversity of sulfate-reducing prokaryotes in the Pearl River estuarine sediments, Southern China. *FEMS Microbiology Ecology* 70:249–262.

Kabir, S., Rajendran, N., Amemiya, T., and Itoh, K. 2003. Real-time quantitative PCR assay on bacterial DNA: In a model soil system and environmental samples. *Journal of General and Applied Microbiology* 49:101–109.

Kato, S., Nakawake, M., Ohkuma, M., and Yamagishi, A. 2012. Distribution and phylogenetic diversity of cbbM genes encoding RuBisCO form II in a deep-sea hydrothermal field revealed by newly designed PCR primers. *Extremophiles* 16:277–283.

Kelly, D. P. 1982. Biochemistry of the chemolithotrophic oxidation of inorganic sulfur. *Philosophical Transactions of the Royal Society of London. Series B: Biological Sciences* 298:499–528.

Kelly, D. P., Shergill, J. K., Lu, W. P., and Wood, A. P. 1997. Oxidative metabolism of inorganic sulfur compounds by bacteria. *Antonie Van Leeuwenhoek* 71:95–107.

Kennedy, A. C. and Smith, K. L. 1995. Soil microbial diversity and the sustainability of agricultural soils. *Plant Soil* 170:75–86.

Kennedy, J., Flemer, B., Jackson, S. A. et al. 2010. Marine metagenomics: New tools for the study and exploitation of marine microbial metabolism. *Marine Drugs* 8(3):608–628.

Keshri, J., Mishra, A., and Jha, B. 2013a. Microbial population index and community structure in saline-alkaline soil using gene targeted metagenomics. *Microbiology Research* 168:165–173.

Keshri, J., Mody, K., and Jha, B. 2013b. Bacterial community structure in a semi-arid haloalkaline soil using culture independent method. *Geomicrobiology Journal* 30:517–529.

Kirkby, C. A., Kirkegaard, J. A., Richardson, A. E., Wade, L. J., Blanchard, C., and Batten, G. 2011. Stable soil organic matter: A comparison of C:N:P:S ratios in Australian and other world soils. *Geoderma* 163:197–208.

Kong, W., Dolhi, J. M., Chiuchiolo, A., Priscu, J., and Morgan-Kiss, R. M. 2012. Evidence of form II RubisCO (*cbbM*) in a perennially ice-covered Antarctic lake. *FEMS Microbiology Ecology* 82:491–500.

Kovaleva, O. L., Tourova, T. P., Muyzer, G., Kolganova, T. V., and Sorokin, D. Y. 2011. Diversity of RuBisCO and ATP citrate lyase genes in soda lake sediments. *FEMS Microbiology Ecology* 75:37–47.

Krishnani, K. K., Gopikrishna, G., Pillai, S. M., and Gupta, B. P. 2010. Abundance of sulfur-oxidizing bacteria in coastal aquaculture using *soxB* gene analyses. *Aquatic Research* 41:1290–1301.

Lane, D. J., Harrison, A. P. J., and Stahl, D. 1992. Evolutionary relationships among sulfur and iron-oxidizing eubacteria. *Journal of Bacteriology* 174:269–278.

Leloup, J., Loy, A., Knab, N. J., Borowski, C., Wagner, M., and Jørgensen, B. B. 2007. Diversity and abundance of sulfate-reducing microorganisms in the sulfate and methane zones of a marine sediment, Black Sea. *Environmental Microbiology* 9:131–142.

Leloup, J., Petit, F., Boust, D. et al. 2005. Dynamics of sulfate-reducing microorganisms (dsrAB genes) in two contrasting mudflats of the Seine estuary (France). *Microbial Ecology* 50:307–314.

Lenk, S., Moraru, C., Hahnke, S. et al. 2012. Roseobacter clade bacteria are abundant in coastal sediments and encode a novel combination of sulfur oxidation genes. *International Society for Microbial Ecology Journal* 6:2178–2187.

Levy-Booth, D. J. and Winder, R. S. 2010. Quantification of nitrogen reductase and nitrite reductase genes in soil of thinned and clear-cut Douglas-fir stands by using real-time PCR. *Applied and Environmental Microbiology* 76:7116–7125.

Ley, R. E., Turnbaugh, P. J., Klein, S., and Gordon, J. I. 2006. Microbial ecology: Human gut microbes associated with obesity. *Nature* 444:1022–1023.

Luo, J. F., Lin, W. T., and Guo, Y. 2011. Functional genes based analysis of sulfur-oxidizing bacteria community in sulfide removing bioreactor. *Applied Microbiology and Biotechnology* 90:769–778.

MacNeil, I. A., Tiong, C. L., Minor, C. et al. 2001. Expression and isolation of antimicrobial small molecules from soil DNA libraries. *Journal of Molecular Microbiology and Biotechnology* 3:301–308.

Margulies, M., Egholm, M., Altman, W. E. et al. 2005. Genome sequencing in microfabricated high-density picolitre reactors. *Nature* 437:376–380.

McGill, W. B. and Cole, C. V. 1981. Comparative aspects of cycling of organic C, N, S and P through soil organic matter. *Geoderma* 26:267–286.

Meyer, B., Imhoff, J. F., and Kuever, J. 2007. Molecular analysis of the distribution and phylogeny of the *soxB* gene among sulfur-oxidizing bacteria-evolution of the Sox sulfur oxidation enzyme system. *Environmental Microbiology* 9:2957–2977.

Meyer, B. and Kuever, J. 2007. Molecular analysis of the distribution and phylogeny of dissimilatory adenosine-5-phosphosulfate reductase-encoding genes (aprBA) among sulfur-oxidizing prokaryotes. *Microbiology* 153:3478–3498.

Morozova, O. and Marra, M. A. 2008. Applications of next-generation sequencing technologies in functional genomics. *Genomics* 92:255–264.

Murillo, A. A., Ramírez-Flandes, S., DeLong, E. F., and Ulloa, O. 2014. Enhanced metabolic versatility of planktonic sulfur-oxidizing γ-proteobacteria in an oxygen-deficient coastal ecosystem. *Frontiers in Marine Science* 1:18.

Mussmann, M., Ishii, K., Rabus, R., and Amann, R. 2005. Diversity and vertical distribution of cultured and uncultured Deltaproteobacteria in an intertidal mud flat of the Wadden Sea. *Environmental Microbiology* 7:405–418.

Muyzer, G. and Smalla, K. 1998. Application of denaturing gradient gel electrophoresis (DGGE) and temperature gradient gel electrophoresis (TGGE) in microbial ecology. *Antonie Van Leeuwenhoek* 73:127–141.

Muyzer, G. and Stams, A. J. 2008. The ecology and biotechnology of sulfate-reducing bacteria. *Nature Reviews Microbiology* 6:441–454.

Nakagawa, S., Takaki, Y., Shimamura, S., Reysenbach, A. L., Takai, K., and Horikoshi, K. 2007. Deep-sea vent epsilon-proteobacterial genomes provide insights into emergence of pathogens. *Proceedings of National Academy of Sciences United States of America* 104:12146–12150.

Nanba, K., King, G. M., and Dunfield, K. 2004. Analysis of facultative lithotroph distribution and diversity on volcanic deposits by use of the large subunit of ribulose 1,5-bisphosphate carboxylase/oxygenase. *Applied and Environmental Microbiology* 70:2245–2253.

Okano, Y., Hristova, K. R., Leutenegger, C. M. et al. 2004. Application of real-time PCR to study effects of ammonium on population size of ammonia oxidizing bacteria in soil. *Applied and Environmental Microbiology* 70:1008–1016.

Orr, C. H., James, A., Leifert, C., Cooper, J. M., and Cummings, S. P. 2011. Diversity and activity of free-living nitrogen-fixing bacteria and total bacteria in organic and conventionally managed soils. *Applied and Environmental Microbiology* 77:911–919.

Orr, C. H., Leifert, C., Cummings, S. P., and Cooper, J. M. 2012. Impacts of organic and conventional crop management on diversity and activity of free-living nitrogen fixing bacteria and total bacteria are subsidiary to temporal effects. *PLoS ONE* 7(12):e52891.

Paul, J. H. and L. Pichard. 1996. Molecular approaches to studying natural communities of autotrophs. In *Proceedings of the Eighth International Symposium on C-1 Compozlnds*, M. E. Lidstrom and F. R. Tabita (eds.), pp. 301–309. San Diego, CA: Kluwer.

Peoples, M. B. and Crasswell, E. T. 1992. Biological nitrogen fixation: Investments, expectations and actual contributions to agriculture. *Plant Soil* 141:P13–P39.

Petri, R., Podgorsek, L., and Imhoff, J. F. 2001. Phylogeny and distribution of the soxB gene among thiosulfate-oxidizing bacteria. *FEMS Microbiology Letters* 197:171–178.

Pfennig, N. 1989. Ecology of phototrophic purple and green sulfur bacteria. In: *Autotrophic Bacteria*, H. G. Schlegel and B. Bowien (eds.), pp. 81–96. New York: Springer-Verlag.

Poly, F., Ranjard, L., Nazaret, S., Gourbiere, F., and Monrozier, L. J. 2001. Comparison of nifH gene pools in soils and soil microenvironments with contrasting properties. *Applied and Environmental Microbiology* 67:2255–2262.

Quillet, L., Besaury, L., Popova, M., Paisse, S., Deloffre, J., and Ouddane, B. 2012. Abundance, diversity and activity of sulfate-reducing prokaryotes in heavy metal-contaminated sediment from a salt marsh in the Medway Estuary. *Marine Biotechnology* 14:363–381.

Raes, J., Foerstner, K. U., and Bork, P. 2007. Get the most out of your metagenome: Computational analysis of environmental sequence data. *Current Opinion in Microbiology* 10:490–498.

Raich, J. W. and Potter, C. S. 1995. Global patterns of carbon dioxide emissions from soils. *Global Biogeochemical Cycles* 9:23–36.

Raich, J. W. and Schlesinger, W. H. 1992. The global carbon dioxide flux in soil respiration and its relationship to vegetation and climate. *Tellus B* 44:81–99.

Rappe, M. S. and Giovannoni, S. J. 2003. The uncultured microbial majority. *Annual Reviews of Microbiology* 57:369–394.

Rondon, M. R., August, P. R., Bettermann, A. D. et al. 2000. Cloning the soil metagenome: A strategy for accessing the genetic and functional diversity of uncultured microorganisms. *Applied and Environmental Microbiology* 66:2541–2547.

Rosello, M. R. and Amann, R. 2001. The species concept for prokaryotes. *FEMS Microbiology Reviews* 25:39–67.

Rusch, D. B., Halpern, A. L., Sutton, G. et al. 2007. The Sorcerer II Global Ocean sampling expedition: Northwest Atlantic through eastern tropical Pacific. *PLoS Biology* 5:e77.

Salek, S. S., Kleerebezem, R., Jonkers, H. M., Witkamp, G. J., and Van Loosdrecht, M. C. 2013. Mineral CO_2 sequestration by environmental biotechnological processes. *Trends in Biotechnology* 31:141–148.

Schlesinger, W. H. 1997. *Biogeochemistry: An Analysis of Global Change*. San Diego, CA: Academic Press.

Schloss, P. D. and Handelsman, J. 2003. Biotechnological prospects from metagenomics. *Current Opinion in Biotechnology* 14:303–310.

Selesi, D., Pattis, I., Schmid, M., Kandeler, E., and Hartmann, A. 2007. Quantification of bacterial RuBisCO genes in soils by *cbbL* targeted real-time PCR. *Journal of Microbiological Methods* 69:497–503.

Selesi, D., Schmid, M., and Hartmann, A. 2005. Diversity of green-like and red-like genes (*cbbl*) in differently managed agricultural soils. *Applied and Environmental Microbiology* 71:175–184.

Shendure, J. and Ji, H. 2008. Next-generation DNA sequencing. *Nature Biotechnology* 26:1135–1145.

Sievert, S. M., Scott, K. M., Klotz, M. G. et al. 2008. Genome of the epsilonproteobacterial chemolithoautotroph Sulfurimonas denitrificans. *Applied and Environmental Microbiology* 74: 1145–1156.

Simon, C. and Daniel, R. 2011. Metagenomic analyses: Past and future trends. *Applied and Environmental Microbiology* 77:1153–1163.

Sogin, M. L., Morrison, H. G., Huber, J. A. et al. 2006. Microbial diversity in the deep sea and the underexplored "rare biosphere". *Proceedings of the National Academy of Sciences United States of America* 103(32):12115–12120.

Stein, J. L., Marsh, T. L., Wu, K. Y., Shizuya, H., and DeLong, E. F. 1996. Characterization of uncultivated prokaryotes: Isolation and analysis of a 40-kilobase-pair genome fragment from a planktonic marine archaeon. *Journal of Bacteriology* 178:591–599.

Stubner, S. 2002. Enumeration of 16S rDNA of *Desulfotomaculum* lineage 1 in rice field soil by real-time PCR with Sybr Green detection. *Journal of Microbiological Methods* 50:155–164.

Tabita, F. R. 1999. Microbial ribulose-1,5-bisphosphate carboxylase/oxygenase: A different perspective. *Photosynthesis Research* 60:1–28.

Tabita, F. R., Hanson, T. E., Li, H., Satagopan, S., Singh, J., and Chan, S. 2007. Function, structure, and evolution of the RuBisCO-like proteins and their RuBisCO homologs. *Microbiology and Molecular Biology Reviews* 71:576–599.

Teng, Q., Sun, B., Fu, X., Li, S., Cui, Z., and H. Cao. 2009. Analysis of nifH gene diversity in red soil amended with manure in Jiangxi, South China. *Journal of Microbiology (Seoul, Korea)* 47:135–141.

Tolli, J. and King, G. M. 2005. Diversity and structure of bacterial chemolithotrophic communities in pine forest and agroecosystem soils. *Applied and Environmental Microbiology* 71:8411–8418.

Tourna, M., Maclean, P., Condron, L., O'Callaghan, M., and Wakelin, S. A. 2014. Links between sulfur oxidation and sulfur-oxidising bacteria abundance and diversity in soil microcosms based on *soxB* functional gene analysis. *FEMS Microbiology Ecology* 88:538–549.

Tourova, T. P., Kovaleva, O. L., Sorokin, D. Y., and Muyzer, G. 2010. Ribulose-1,5-bisphosphate carboxylase/oxygenase genes as a functional marker for chemolithoautotrophic halophilic sulfur-oxidizing bacteria in hypersaline habitats. *Microbiology* 156:2016–2025.

Tourova, T. P., Slobodova, N. V., Bumazhkin, B. K., Kolganova, T. V., Muyzer, G., and Sorokin, D. Y. 2012. Analysis of community composition of sulfur-oxidizing bacteria in hypersaline and soda lakes using soxB as a functional molecular marker. *FEMS Microbiology Ecology* 84:280–289.

Tringe, S. G., Mering, C. V., Kobayashi, A. et al. 2005. Comparative metagenomics of microbial communities. *Science* 308:554–557.

Trüper, H. G. and U. Fischer. 1982. Anaerobic oxidation of sulfur-compounds as electron-donors for bacterial photosynthesis. *Philosophical Transactions of the Royal Society of London. Series B: Biological Sciences* 298:529–542.

Tyson, G. W., Chapman, J., Hugenholtz, P. et al. 2004. Community structure and metabolism through reconstruction of microbial genomes from the environment. *Nature* 428:37–43.

Ueda, T., Suga, Y., Yashiro, N., and Matsuguchi, T. 1995. Remarkable N_2 fixing bacterial diversity detected in rice roots by molecular evolutionary analysis of *nifH* gene sequences. *Journal of Bacteriology* 177:1414–1417.

Van der Wielen, P. W. 2006. Diversity of ribulose-1,5-bisphosphate carboxylase/oxygenase large-subunit genes in the $MgCl_2$-dominated deep hypersaline anoxic basin discovery. *FEMS Microbiology Letters* 259:326–331.

Varon-Lopez, M., Dias, A. C. F., Fasanella, C. C., Durrer, A., Melo, I. S., Kuramae, E. E., and Andreote, F. D. 2014. Sulfur-oxidizing and sulfate-reducing communities in Brazilian mangrove sediments. *Environmental Microbiology* 16:845–855.

Venter, J. C., Remington, K., Heidelberg, J. F. et al. 2004. Environmental genome shotgun sequencing of the Sargasso Sea. *Science* 304:66–74.

Videmšek, U., Hagn, A., Suhadolc, M. et al. 2009. Abundance and diversity of CO_2-fixing bacteria in grassland soils close to natural carbon dioxide springs. *Microbial Ecology* 58:1–9.

Vitousek, P. M., Mooney, H. A., Lubchenco, J., and Melillo, J. M. 1997. Human domination of Earth's ecosystems. *Science* 277:494–499.

Von Mering, C., Hugenholtz, P., Raes, J. et al. 2007. Quantitative phylogenetic assessment of microbial communities in diverse environments. *Science* 315:1126–1130.

Voordeckers, J. W., Do, M. H., Hügler, M., Ko, V., Sievert, M., and Vetriani, C. 2008. Culture dependent and independent analyses of 16S rRNA and ATP citrate lyase genes: A comparison of microbial communities from different black smoker chimneys on the Mid-Atlantic Ridge. *Extremophiles* 12:627–640.

Weisburg, W. G., Barns, S. M., Pelletier, D. A., and Lane, D. J. 1991. 16S ribosomal DNA amplification for phylogenetic study. *Journal of Bacteriology* 173:697–703.

Whitman, W. B., Coleman, D. C., and Wiebe, W. J. 1998. Prokaryotes: The unseen majority. *Proceedings of National Academy of Science United States of America* 95:6578–6583.

Xiao, C. H., Tang, H., Pu, L. J., Sun, D. M., Ma, J. Z., Yu, M., and Duan, R. S. 2010. Diversity of nitrogenase (*nifH*) genes pool in soybean field soil after continuous and rotational cropping. *Journal of Basic Microbiology* 50:373–379.

Xiao, K. Q., Bao, P., Bao, Q. L., Jia, Y., Huang, F. Y., Su, J. Q., and Zhu, Y. G. 2013. Quantitative analyses of ribulose-1,5-bisphosphate carboxylase/oxygenase (RuBisCO) large-subunit genes (*cbbL*) in typical paddy soils. *FEMS Microbiology Ecology* 87:89–101.

Xu, M., Wu, W. M., Wu, L. et al. 2010. Responses of microbial community functional structures to pilot-scale uranium in situ bioremediation. *International Society for Microbial Ecology Journal* 4:1060–1070.

Yang, J., Jiang, H., Dong, H., Wu, G., Hou, W., Zhao, W., Sun, Y., and Lai, Z. 2013. Abundance and diversity of sulfur-oxidizing bacteria along a salinity gradient in four qinghai-tibetan lakes, china. *Geomicrobiology Journal* 30:851–860.

Yooseph, S., Sutton, G., Rusch, D. B. et al. 2007. The Sorcerer II global ocean sampling expedition: Expanding the universe of known protein families. *PLoS Biology* 5:e16.

Yousuf, B., Keshri, J., Mishra, A., and Jha, B. 2012a. Application of targeted metagenomics to explore abundance and diversity of CO_2-fixing bacterial community using *cbbL* gene from the rhizosphere of *Arachis hypogaea*. *Gene* 1:18–24.

Yousuf, B., Kumar, R., Mishra, A., and Jha, B. 2014. Unravelling the carbon and sulfur metabolism in coastal soil ecosystems using comparative cultivation-independent genome-level characterisation of microbial communities. *PLoS ONE* 9(9):e107025.

Yousuf, B., Sanadhya, P., Keshri, J., and B. Jha. 2012b. Comparative molecular analysis of chemolithoautotrophic bacterial diversity and community structure from coastal saline soils, Gujarat, India. *BMC Microbiology* 12:150.

Yuan, H., Ge, T., Chen, C., O'Donnell, A. G., and Wu, J. 2012a. Significant role for microbial autotrophy in the sequestration of soil carbon. *Applied and Environmental Microbiology* 78:2328–2336.

Yuan, H., Ge, T., Wu, X. et al. 2012b. Long-term field fertilization alters the diversity of autotrophic bacteria based on the ribulose-1,5-biphosphate carboxylase/oxygenase (RuBisCO) large-subunit genes in paddy soil. *Applied Microbiology and Biotechnology* 95:1061–1071.

Zehr, J. P., Jenkins, B. D., Short, S. M., and Steward, G. F. 2003. Nitrogenase gene diversity and microbial community structure: A cross-system comparison. *Environmental Microbiology* 5:539–554.

Zopfi, J., Ferdelman, T. G., and Fossing, H. 2004. Distribution and fate of sulfur intermediates—Sulfite, tetrathionate, thiosulfate, and elemental sulfur—In marine sediments. In *Sulfur Biogeochemistry—Past and Present*, J. P. Amend, K. J. Edwards, and T. W. Lyons (eds.), pp. 97–116. Boulder, CO: Geological Society of America.

Section IV

Marine Glycomics

12

Analysis of Carrageenan Molecular Composition by NMR

Leonel Pereira

CONTENTS

12.1 Introduction

Seaweeds have been utilized by mankind for several hundreds of years, directly for food, for medicinal purposes, and for agriculture fertilizers. Today, seaweed is used in many countries for very different purposes: directly as food; phycocolloids extraction; extraction of compounds with antiviral, antibacterial, or antitumor activity; and biofertilizers (Rudolph 2000, Pereira 2010a,b, 2011).

About 4 million tonnes of seaweed are harvested annually worldwide. The major producers are the Chinese and Japanese, followed by Americans and Norwegians. France used to import Japanese seaweed in the 1970s but 10 years later, it went on to produce algae for food and for biological product users. Contrary to what happens in East Asia, the West is more interested in thickeners and gelling properties of some polysaccharides extracted from seaweeds. The use of seaweeds for fertilizer and soil improvement was also well known in Europe, and both large brown algae and calcified red algae have been collected for this purpose (Guiry and Bluden 1991, Pereira 2010b).

More than its vegetable nature, algae is due to the conjunction of its variety of colors (and shapes) and the blue ocean, a great selling point, both for food and for cosmetics, especially after certain substances of animal origin have become suspicious of transmitting the bovine spongiform encephalopathy, commonly known as mad cow disease (Pereira 2004, 2010a,b, 2011).

The marine algae are a rich mine of health—vitamins and trace elements—and also offer a dizzying variety of flavors, fragrances, and textures. In fact, the algae, as well as being a vitamin and mineral treasure, are low in fat, an essential feature in weight loss diets. In addition, the algae are rich in dietary fiber, which may facilitate intestinal transit, lowering the rate of blood cholesterol and reducing certain diseases such as colon cancer (Pereira 2010a,b, 2011).

With regard to the structural analysis of polysaccharides from seaweed, in the 1980s, carbon 13 nuclear magnetic resonance (^{13}C-NMR) began to be used on a regular basis for identifying the fine structure of polysaccharides, particularly those of carrageenan and agars. This spectroscopic technique offers a high reliability in the detection of the different fractions, as well as the nature and position of substituent's (Zinoun et al. 1993). The early work of Yarotsky et al. (1977) and Bhattacharjee et al. (1978) allowed to assign the peaks observed in the spectra corresponding to the carbons skeleton of the polysaccharide units. The classification of the main types of phycocolloids was established by the same authors, based primarily on the identification of anomeric peaks. Later, other authors had this work their basis in order to give a definitive assignment of these chemical shifts. Since then, the identification of carrageenan is done by simple comparison of the obtained peaks with those reported in the literature (Bhattacharjee et al. 1978, Usov et al. 1980, Bellion 1982, Greer and Yaphe 1984a,b, Usov 1984, Rochas and Lahaye 1989).

12.2 Phycocolloids

Phycocolloids refer to those polysaccharides extracted from both freshwater and marine algae. Until now, only the polysaccharides extracted from marine red and brown algae, such as agar, carrageenan, and algin, are of economic and commercial significance, since these polysaccharides exhibit high molecular weights, high viscosity, and excellent gelling, stabilizing, and emulsifying properties. They are also extracted in fairly high amount from the algae. All these polysaccharides are water soluble and could be extracted with hot water or alkaline solution (Minghou 1990).

12.2.1 Commercial Hydrocolloids

Colloids are compounds that form colloidal solutions, an intermediate state between a solution and a suspension, and are used as thickeners, gelling agents, and stabilizers for suspensions and emulsions. Hydrocolloids are carbohydrates that when dissolved in water form viscous solutions. The phycocolloids are hydrocolloids extracted from algae and represent a growing industry, with more than 1 million tons of seaweeds extracted annually for hydrocolloid production (Ioannou and Roussis 2009, Pereira et al. 2009a, Pereira 2010b).

Many seaweeds produce hydrocolloids, associated with the cell wall and intercellular spaces. Members of the red algae (Rhodophyta) produce sulfated galactans (e.g., carrageenans and agars), and the brown algae (Heterokontophyta, Phaeophyceae) produce uronates (alginates) (Pereira 2010b, Bixler and Porse 2011, Jiao et al. 2011, Pereira and van de Velde 2011).

The different phycocolloids used in food industry as natural additives are as follows (European codes of phycocolloids):

Alginic acid—E400

Sodium alginate—E401

Potassium alginate—E402

Ammonium alginate—E403

Calcium alginate—E404

Propylene glycol alginate—E405

Agar—E406

Carrageenan—E407

Semirefined carrageenan or processed eucheuma seaweed—E407A

12.3 Sulfated Galactans

The most common and the most abundant cell wall constituents yet encountered in the Rhodophyta are families of galactans bearing the trivial names agar and carrageenan. The pioneering studies conducted in Japan and Canada established that the backbone structure of both agars and carrageenans were based on repeating galactose and 3,6-anhydrogalactose residues linked β-(1 → 4) and α-(1 → 3), respectively. The principal feature distinguishing the highly sulfated carrageenans from the less sulfated agars was the presence of D-galactose and anhydro-D-galactose in the former and D-galactose, L-galactose, or anhydro-L-galactose in the latter. Classification of these polysaccharides based on their solubility and gelling properties proved unsatisfactory, so attention was focused on the common underlying structural patterns (Anderson et al. 1965, Craigie 1990). The seminal concept of the masked repeating structure first reported for the agar-like porphyran (Anderson and Rees 1965) is now widely accepted for both agars and carrageenan (Craigie 1990).

12.3.1 Agar

Agar is the phycocolloid of most ancient origin. In Japan, agar is considered to have been discovered by Minoya Tarozaemon in 1658, and a monument in Shimizu-mura commemorates the first time it was manufactured. Originally, and even in the present times, it was made and sold as an extract in solution (hot) or in gel form (cold), to be used promptly in areas near the factories; the product was then known as "tokoroten." Its industrialization as a dry and stable product started at the beginning of the eighteenth century and it has since been called "kanten." The word "agar–agar," however, has a Malayan origin, and agar is the most commonly accepted term, although in French- and Portuguese-speaking countries it is also called "gelosa" (Armisen and Galatas 1987, Minghou 1990).

Agar is a phycocolloid, the name of which comes from Malaysia and means "red alga" in general, and has traditionally been applied to what we now know taxonomically as *Eucheuma*. Ironically, we now know this to be the commercial source of iota-carrageenan. Agar is composed of two polysaccharides: namely, agarose and agaropectin. The first is responsible for gelling, while the latter has thickening properties (Armisen and Galatas 2000, Pereira 2011).

12.3.2 Carrageenans

The first formal recognition of the gelling properties of boiled *Fucus crispus* was discovered by Turner in 1809, and this mucilaginous matter was named "carrageenin" by Pereira (1840). The gelatinous, hot-water-soluble mucilage of *Chondrus crispus* was first isolated by Schmidt (1844) (Cardoso et al. 2014). Coincidentally, in the same year, Forchhammer reported on the high sulfur content of the ash from *C. crispus* (Buggeln and Carige 1973). The term "carrageenin" was later changed to "carrageenan" so as to comply with the "-an" suffix of terminology as applied to polysaccharides (Pereira et al. 2009a, Pereira 2011).

The modern carrageenan industry dates from the 1940s, receiving its impetus from the dairy applications, where carrageenan was found to be the ideal stabilizer for the suspension of cocoa in milk chocolate (van de Velde and de Ruiter 2002, van de Velde et al. 2004, Pereira et al. 2009a).

The commercial carrageenans are normally divided into three main types: kappa (κ)-, iota (ι)-, and lambda (λ)-carrageenan. The idealized disaccharide repeating units of these carrageenans are given in Figure 12.1. Generally, seaweeds do not produce these idealized and pure carrageenans, but more likely a range of hybrid structures (Table 12.4) and/or precursors (see Figure 12.1). Several other carrageenan-repeating units exist: for example, xi (ξ), theta (θ), beta (β), mu (μ), and nu (ν) (Figure 12.1). The precursors (mu and nu), when exposed to alkali conditions, are modified into kappa and iota, respectively, through formation of the 3,6-anhydrogalactose bridge (Myslabodski 1990, Rudolph 2000, Pereira et al. 2009a, Pereira and van de Velde 2011). This is a feature used extensively in extraction and industrial modification.

Carrageenans are the third most important hydrocolloid in the food industry, after gelatin (animal origin) and starch (plant origin) (van de Velde and de Ruiter 2002). The most commonly used, commercial carrageenans are extracted from *Kappaphycus alvarezii* and *Eucheuma denticulatum* (Gigartinales, Rhodophyta) (McHugh 2003, Pereira 2011).

Primarily, wild-harvested genera such as *Chondrus, Furcellaria, Gigartina, Chondracanthus, Sarcothalia, Mazzaella, Mastocarpus,* and *Tichocarpus* (Gigartinales, Rhodophyta) are also mainly cultivated as carrageenan raw materials and producing countries include Argentina, Canada, Chile, Denmark, France, Japan, Mexico, Morocco, Portugal, North Korea, South Korea, Spain, Russia, and the United States (Pereira et al. 2009c, Bixler and Porse 2011).

The original source of carrageenans was from the red seaweed *C. crispus*, which continues to be used, but in limited quantities. *Betaphycus gelatinum* is used for the extraction of beta (β)-carrageenan. Some South American red algae used previously only in minor quantities have, more recently, received attention from carrageenan producers, as they seek to increase diversification of raw materials in order to provide for the extraction of new carrageenan types with different physical functionalities and therefore increased product development, which in turn stimulates demand (McHugh 2003). *Gigartina skottsbergii, Sarcothalia crispata,* and *Mazzaella laminarioides* are currently the most valuable

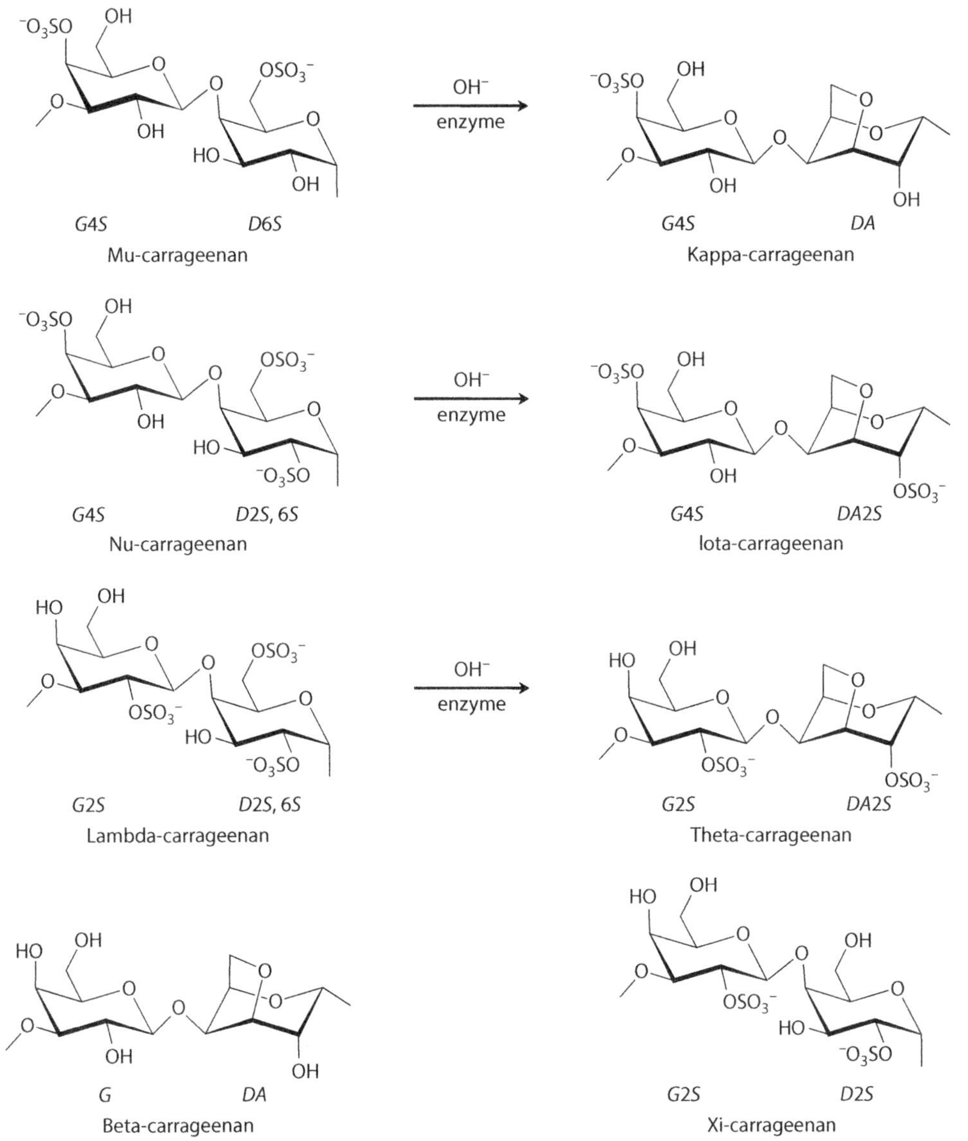

FIGURE 12.1
Idealized units of the main types of carrageenan. (Adapted from Pereira, L. et al., *Food Hydrocoll.*, 23, 1903, 2009a; Pereira, L. et al., *Biodevices*, 131, 2009b; Pereira, L. et al., *J. Appl. Phycol.*, 21, 599, 2009c; Pereira, L. and van de Velde, F., *Carbohydr. Polym.*, 84, 614, 2011.)

species and all are harvested from natural populations in Chile and Peru. We cannot let to mention the recent earthquake in Chile (February 27, 2010), which caused the elevation of intertidal areas and the consequent large losses of harvestable biomass (Pereira 2011). Small quantities of *Gigartina canaliculata* are harvested in Mexico and *Hypnea musciformis* has been used in Brazil (Furtado 1999). The use of high-value carrageenophytes as a dissolved organic nutrient sink to boost economic viability of integrated multitrophic aquaculture operations has been considered (Pereira 2004, Chopin 2006, Sousa-Pinto and Abreu 2011).

Large carrageenan processors have fueled the development of *K. alvarezii* (which goes by the name "cottonii" to the trade) and *E. denticulatum* (commonly referred to as "spinosum" in the trade) farming in several countries including the Philippines, Indonesia, Malaysia, Tanzania, Kiribati, Fiji, Kenya, and Madagascar (McHugh 2003). Indonesia has recently overtaken the Philippines as the world's largest producer of dried carrageenophytes biomass (Pereira 2011).

Shortages of carrageenan-producing seaweeds suddenly appeared in the mid-2007, resulting in doubling of the price of carrageenan; some of this price increase was due to increased fuel costs and a weak U.S. dollar (most seaweed polysaccharides are traded in U.S. dollars). The reasons for shortages of the raw materials for processing are less certain: perhaps it is a combination of environmental factors, sudden increases in demand, particularly from China, and some market manipulation by farmers and traders. Most hydrocolloids are experiencing severe price movements.

However, the monocultures of some carrageenophytes (viz., *K. alvarezii*) have several problems due to environmental change and also diseases. The problems with ice-ice and epiphytes have resulted in large-scale crop losses (Hurtado et al. 2006, 2014, Vairappan et al. 2008, Hayashi et al. 2010).

12.4 Methods and Techniques for Analyzing Phycocolloids

12.4.1 Phycocolloid Extraction

Before phycocolloid extraction, the ground dry material was rehydrated and pretreated in acetone followed by ethanol to eliminate the organosoluble fraction.

For extraction of the "native" phycocolloid, the samples were placed in distilled water (50 mL/g), pH 7 at 85°C for 3 h. For extraction of the native phycocolloid, the seaweed samples were placed in distilled water (50 mL/g), pH 7 at 85°C for 3 h. For an alkaline extraction (resembling the industrial method), the samples were placed in a solution (150 mL/g) of NaOH (1 M) at 80°C–85°C for 3–4 h according to Pereira and Mesquita (2004) and neutralized to pH 6–8 with HCl (0.3 M).

The solutions were hot filtered, twice, under vacuum, through cloth and glass fiber filter. The extract was evaporated under vacuum to one-third of the initial volume. The carrageenan was precipitated by adding the warm solution to twice its volume of ethanol (96%), according to the method described by Pereira (2006) and Pereira et al. (2009b).

12.4.2 Spectroscopic Analysis Methods (Vibrational Spectroscopy)

In order to identify the seaweed carrageenan, agar, and alginate nature, vibrational spectroscopy can reveal detailed information concerning the properties and structure of materials at a molecular level. Until now, this type of analysis required the extraction of colloids, through lengthy and complicated procedures. With the development of Fourier transform infrared (FTIR) diffuse-reflectance spectroscopy, it became possible to directly analyze ground, dried seaweed material (Chopin and Whalen 1993).

Pereira et al. (2003, 2009a,b,c) developed an analysis technique based on FTIR–attenuated total reflectance (FTIR-ATR) and FT-Raman spectroscopy, which allowed for the accurate identification of diverse phycocolloids.

12.4.2.1 Fourier Transform Infrared Spectroscopy

Infrared (IR) spectroscopy was, until recently, the most frequently used vibrational technique for the study of the chemical composition of phycocolloids. This technique presents two main advantages: it requires minute amounts of sample (milligrams), and it is non-aggressive method with reliable accuracy (Pereira et al. 2003). However, conventional IR spectroscopy requires laborious procedures to obtain spectra with a good signal/noise ratio (Chopin and Whalen 1993). This limitation was overcome with the development of interferometric IR techniques (associated with the Fourier transform algorithm), known as FTIR spectroscopy. More recently, Pereira and collaborators had used a technique of analysis on the basis of FTIR-ATR (from attenuated total reflectance) spectroscopy, allowing for the determination of the composition of the different phycocolloids from dried ground seaweed, without having to prepare tablets of KBr (Pereira and Mesquita 2004, Pereira 2006, Pereira et al. 2009a).

12.4.2.2 Raman Spectroscopy

In contrast with FTIR, the application of traditional Raman spectroscopy was limited until recently, due to the laser-induced fluorescence (strong background signal that is detected when some samples, such as biochemical compounds, are excited with visible lasers) and risk of sample destruction by light energy. The use of Nd:YAG lasers operating at 1064 nm has been generalized to decrease the fluorescence level. Optoelectronic devices have progressed dramatically in the past decade as a consequence of major achievements in solid-state technology. As a result, compact, efficient, and reliable diode lasers are now available from the visible to the IR that have been demonstrated to work properly in Raman instruments in combination with suitable filter sets (Pereira 2006, Pereira et al. 2009a,b).

Raman spectroscopy comprises the family of spectral measurements made on molecular media based on inelastic scattering of monochromatic radiation. During this process, energy is exchanged between the photon and the molecule such that the scattered photon is of higher or lower energy than the incident photon. The difference in energy is made up by a change in the rotational and vibrational energy of the molecule and gives information on its energy levels. The modern FT-Raman spectrometers have been used to produce good quality Raman spectra from seaweed samples (Matsuhiro 1996, Pereira et al. 2003, Dyrby et al. 2004, Pereira 2006).

12.4.3 Nuclear Magnetic Resonance Spectroscopy

Since natural carrageenans are mixtures of different sulfated polysaccharides, their composition differs from batch to batch. Therefore, the quantitative analysis of carrageenan batches is of greatest importance for both ingredient suppliers and food industries to ensure ingredient quality. From the pioneering work of Usov and coworkers (Yarotsky et al. 1977, Usov 1984), NMR spectroscopy is nowadays one of the preferred techniques to determine and quantify the composition of carrageenan batches (van de Velde et al. 2002). Starting from the early work of Usov and coworkers, chemical shifts of carrageenan resonances are generally converted to values relative to tetramethylsilane (TMS) via an internal dimethyl sulfoxide (DMSO) or methanol (MeOH) standard (Usov et al. 1980, Usov 1984, Knutsen et al. 1994).

The use of DMSO or MeOH as internal standard resulted in a generally accepted set of chemical shifts for different types of carrageenans as summarized by van de

Velde et al. (2002). However, to convert chemical shifts from aqueous internal DMSO or MeOH to values relative to TMS is not obvious as TMS is only sparingly soluble in highly polar solvents, such as water or D_2O. Therefore, the International Union of Pure and Applied Chemistry (IUPAC) commission for molecular structure and spectroscopy recently recommended the use of 2,2-dimethyl-2-silapentane-3,3,4,4,5,5-d_6-5-sulfonate sodium salt (DSS) as the primary reference for both ^1H and ^{13}C-NMR spectroscopy in highly polar solvents. For most purposes, the difference between DSS and TMS when dissolved in the same solvent is negligible, and, therefore, the data from DSS and TMS scales may be validly compared without correction (see Table 12.1; Harris et al. 2001, Pereira et al. 2003, van de Velde et al. 2004).

However, application of the IUPAC recommendations in NMR spectroscopy of carrageenans results in changed chemical shifts for all common carrageenan types. Pereira et al. (2003) reported chemical shift values relative to DSS for κ- and ι-carrageenan that are 2.5 ppm larger than those reported by Usov (1984). Therefore, detailed application of DSS as an internal standard for NMR spectroscopy of carrageenans is published by van de Velde et al. (2004) and Pereira (2004). The chemical shifts of all carbon atoms and the anomeric protons (see Table 12.1) are reported relative to DSS internal standard ($\delta = 0.000$ ppm) according to IUPAC recommendations (Harris et al. 2001).

Carrageenan from 18 different carrageenophytes (Gigartinales, Rhodophyta), belonging to 16 genera and 4 families, were analyzed by NMR spectroscopy: *Agardhiella* sp. (Solieriaceae); *B. gelatinum* (Esper) Doty ex P. C. Silva (Solieriaceae); *E. denticulatum* (N. L. Burman) F. S. Collins & Hervey (Solieriaceae); *K. alvarezii* (Doty) Doty ex P. C. Silva (Solieriaceae); *Sarconema scinaioides* Børgesen (Solieriaceae); *C. crispus* Stackhouse (Gigartinaceae); *G. skottsbergii* Setchell & N. L. Gardner (Gigartinaceae); *Gigartina pistillata* (S. G. Gmelin) Stackhouse (Gigartinaceae); *Chondracanthus teedei* (Mertens ex Roth) Kützing and *C. teedei* var. *lusitanicus* (J. E. De Mesquita Rodrigues) Bárbara & Cremades; *S. crispata* (Bory de Saint-Vincent) Leister (Gigartinaceae); *Chondracanthus acicularis* (Roth) Fredericq (Gigartinaceae); *M. laminarioides* (Bory de Saint-Vincent) Fredericq (Gigartinaceae); *Mastocarpus stellatus* (Stackhouse) Guiry (Phyllophoraceae); *Ahnfeltiopsis devoniensis* (Greville) P. C. Silva & DeCew (Phyllophoraceae); *Gymnogongrus crenulatus* (Turner) J. Agardh (Phyllophoraceae); *Calliblepharis jubata* (Goodenough & Woodward) Kützing; and *H. musciformis* (Wulfen) J. V. Lamouroux (Cystocloniaceae).

TABLE 12.1

Chemical Shifts for Common Internal Standards for Nuclear Magnetic Resonance Spectroscopy in Aqueous Systems

Compound Abbreviation	Chemical Shift (ppm)		Compound Chemical Name
	^{13}C	^1H	
DSS	0.000	0.000	2,2-Dimethyl-2-silapentane-3,3,4,4,5,5-d_6-5-sulfonate sodium salt
TSP	−0.18	−0.017	3-(Trimethylsilyl)propionic-2,2,3,3-d_4 acid sodium salt
MeOH	51.43	3.337	Methanol
DMSO	41.53	2.696	Dimethyl sulfoxide
Acetone	32.69	2.208	Acetone

Sources: Adapted from van de Velde, F. et al., *Carbohydr. Res.*, 339, 2309, 2004; Pereira, L. and van de Velde, F., *Carbohydr. Polym.*, 84, 614, 2011.
Recorded in D_2O containing Na_2HPO_4 (20 mM) at 65°C.

12.4.3.1 *^{13}C-NMR Molecular Analysis of Carrageenan*

^{13}C-NMR analysis was carried out on samples of commercial and carrageenan extracted from the following species: *B. gelatinum*, *C. teedei* var. *lusitanicus*, *E. denticulatum*, *G. skottsbergii*, *K. alvarezii*, and *S. scinaioides*. Assignment of the ^{13}C-NMR spectra as based on the chemical shift data is summarized in the literature (van de Velde et al. 2002, Pereira et al. 2003). The data for the chemical shift spectra are summarized in Table 12.2.

12.4.3.2 *^{1}H-NMR Molecular Analysis of Carrageenan*

^{1}H-NMR shift data of representative carrageenan samples containing the most common repeating units (see Figure 12.1), for example, kappa-, iota-, lambda-, mu-, nu-, theta-, beta-, and xi-carrageenan, are presented in Table 12.3.

Additionally, pyruvate modifications were recognized as 5.49 ppm, and the presence of floridean starch was indicated at 5.35 ppm.

TABLE 12.2

^{13}C-NMR Chemical Shifts for the Most Common Carrageenan Structural Units[a]

Carrageenan	Unit[b]	Chemical Shift (ppm) Relative to DSS as Internal Standard					
		C-1	C-2	C-3	C-4	C-5	C-6
β (beta)	G	104.81	71.72	82.58	68.56	77.55	63.49
	DA	96.81	72.40	81.64	80.33	79.26	71.72
ι (iota)	G4S	104.43	71.53	79.06	74.34	77.04	63.52
	DA2S	94.29	77.15	80.04	80.55	79.29	72.02
κ (kappa)	G4S	104.70	71.72	80.98	76.25	77.00	63.49
	DA	97.34	72.11	81.41	80.54	79.07	71.72
λ (lambda)	G2S	105.61	79.61	77.99	66.35	76.51	63.45
	D2S,6S	93.85	77.04	71.76	82.61	70.89	70.25
μ (mu)[c]	G4S	107.00	72.69	80.54	76.25	77.1	63.48
	D6S	100.26	70.7	72.8	81.40	70.5	69.89
ν (nu)	G4S	106.96	72.40	82.42	73.36[d]	77.15	63.58
	D2S,6S	100.53	78.58	70.37	82.13	70.37	70.02
θ (theta)[e]	G2S	102.57	79.8	79.4	69.97	77.05	63.38
	DA2S	97.81	77.05	79.6	81.75	79.2	72.35
ξ (xi)[f]	G2S	105.44			66.92		
	D2S	94.94					

Sources: Adapted from Pereira, L., Estudos em macroalgas carragenófitas (Gigartinales, Rhodophyceae) da costa portuguesa—aspectos ecológicos, bioquímicos e citológicos, PhD thesis, Departamento de Botânica, FCTUC, Universidade de Coimbra, Coimbra, Portugal, 2004; Pereira, L. and van de Velde, F., *Carbohydr. Polym.*, 84, 614, 2011.

[a] Carrageenan (30 mg/mL), DSS (10 mM), and Na_2HPO_4 (20 mM) in D_2O recorded at 65°C.

[b] Codes refer to the nomenclature developed by Knutsen et al. (1994).

[c] Chemical shifts given with one decimal place are obtained from literature and corrected with the offset calculated for the chemical shifts of the anomeric carbon atoms (Ciancia et al. 1993).

[d] Chemical shift differs from the value given in literature (Ciancia et al. 1993).

[e] Chemical shifts given with one decimal place are recalculated from literature (Falshaw and Furneaux 1994); moreover, they may be interchanged. TSP is used as the internal standard; however, chemical shifts are given relative to DSS.

[f] Only a few resonances of the ξ-carrageenan spectrum are assigned in literature (Falshaw and Furneaux 1995). TSP is used as the internal standard; however, chemical shifts are given relative to DSS.

TABLE 12.3

Chemical Shifts (ppm) of the α-Anomeric Protons of Carrageenans Referred to DSS as Internal Standard at 0 ppm[a]

Carrageenan	Monosaccharide[b]	Chemical Shift (ppm)
β (beta)	DA	5.074
ι (iota)	DA2S	5.292
κ (kappa)	DA	5.093
λ (lambda)	D2S,6S	5.548
ν (nu)	D2S,6S	5.501
μ (mu)	D6S	5.238
θ (theta)	DA2S	5.30
ξ (xi)	D2S	5.49

Sources: Adapted from van de Velde, F. et al., *Carbohydr. Res.*, 339, 2309, 2004; Pereira, L., Estudos em macroalgas carragenófitas (Gigartinales, Rhodophyceae) da costa portuguesa—aspectos ecológicos, bioquímicos e citológicos, PhD thesis, Departamento de Botânica, FCTUC, Universidade de Coimbra, Coimbra, Portugal, 2004.

[a] Carrageenan (30 mg/mL), DSS (10 mM), and Na_2HPO_4 (20 mM) in D_2O recorded at 65°C.

[b] Codes refer to the nomenclature developed by Knutsen et al. (1994).

¹H-NMR spectra of alkali-extracted carrageenan from the gametophytes of *A. devoniensis* (Figure 12.2), in the region of anomeric protons, showed two main signals at 5.29 and 5.09 ppm. These signals corresponded to the anomeric proton of iota-carrageenan (DA2S) and kappa-carrageenan (DA), respectively. A minor component detected in the spectrum revealed in a weak signal at 5.50 ppm that corresponded to the presence of nu-carrageenan (D2S, 6S). The intensities of the resonances earlier were used to quantify each component in the carrageenan extracted from *A. devoniensis* (Table 12.4).

The spectrum of *C. jubata* (Figure 12.3) carrageenan showed three main signals: 5.50 ppm (nu-carrageenan), 5.29 ppm (iota-carrageenan), and 5.09 ppm (kappa-carrageenan) (Table 12.4).

The spectrum of *C. acicularis* (female gametophytes) alkali-extracted carrageenan (Figure 12.4) presented two prominent signals at 5.29 ppm (iota-carrageenan) and 5.09 ppm (kappa-carrageenan) and two weaker signals at 5.49 ppm (pyruvate) and 5.35 ppm (floridean starch). The water-extracted carrageenan spectrum (from female gametophytes) showed the same signals, only missing the 5.35 ppm peak referring to floridean starch. The spectrum of *C. acicularis* (tetrasporophytes) displayed signals at 5.49 ppm (xi), 5.30 ppm (theta), and 5.09 ppm (kappa) (see Table 12.4).

C. teedei var. *lusitanicus* (nonfructified and female gametophyte thalli) carrageenan spectra (Figure 12.5) showed two prominent signals: that is, 5.29 and 5.09 ppm. These signals corresponded to the anomeric proton of iota-carrageenan (DA2S) and kappa-carrageenan (DA), respectively. The minor component detected in the spectra, showed a slight rise at 5.35 ppm, corresponded to anomeric protons of floridean starch, a natural contaminant of some samples of carrageenan (see Table 12.4).

Other minor components found in the native carrageenan spectra produced signals at 5.50 ppm (i.e., mu-carrageenan) and 5.24 ppm (i.e., nu-carrageenan) and corresponded to biological precursors of iota- and kappa-carrageenan, respectively. ¹H-NMR spectra showed signals at 5.49 ppm (i.e., xi-carrageenan) and 5.30 ppm (i.e., theta-carrageenan) in all tetrasporophyte samples of *C. teedei* var. *lusitanicus* (Figure 12.6 and Table 12.4).

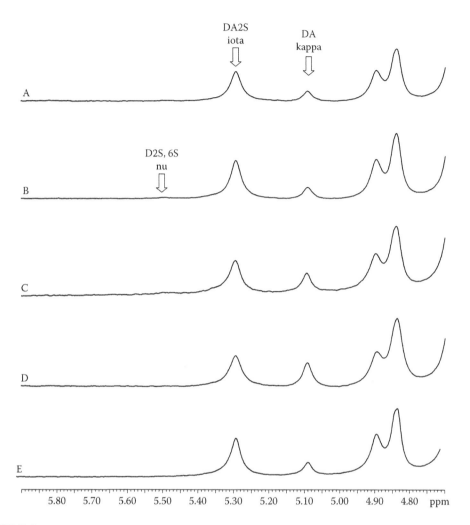

FIGURE 12.2

¹H-NMR spectra of *Ahnfeltiopsis devoniensis* carrageenans. The letter codes refer to the nomenclature of Knutsen et al. (1994).

¹H-NMR spectra of alkali-extracted carrageenans of *M. stellatus* (Figure 12.7) and *G. crenulatus* (Figure 12.8) showed two main signals: that is, 5.09 ppm (kappa-carrageenan) and 5.29 ppm (iota-carrageenan); two additional signals were present at 5.35 ppm (i.e., floridean starch) and 5.49 ppm (i.e., pyruvate) (see Table 12.4).

¹H-NMR spectra of alkali-extracted carrageenans from *C. crispus* and *G. pistillata* (female gametophytes and tetrasporophytes) showed six distinct signals: 5.09 ppm (kappa), 5.29 ppm (iota), 5.55 ppm (lambda), 5.49 ppm (xi), 5.30 ppm (pyruvate), and 5.35 ppm (floridean starch) (see Table 12.4).

The biological precursors (mu and nu) of the gelling carrageenans (kappa and iota) contain a sulfate ester group at position 6 of α-D-galactose 4-connected (see Figure 12.1). This type of structure reduces the ability of carrageenan to form a gel, due to the interruption of sequences of the repeating units responsible for the formation of the double helix structure typical of gel phase. Most of the 6-sulfated units are converted into the

TABLE 12.4

Molecular Composition of Carrageenan Determined by ^{1}H-NMR

Code	Species	Extraction Method	Carrageenan Composition (mol%)							Other Components		
			Kappa	Iota	Mu	Nu	Lambda	Theta	Xi	Pyruvate (mol%)	n. i.	Starch (%)
A	*Ahnfeltiopsis devoniensis* (G)	Alkaline	19.8	80.2	—	—	—	—	—	—	—	—
B	*A. devoniensis* (G)	Soft alkaline	16.7	81.1	—	2.2	—	—	—	—	—	—
C	*A. devoniensis* (G)	Alkaline	29.4	70.6	—	—	—	—	—	—	—	—
D	*A. devoniensis* (G)	Alkaline	22.3	77.7	—	—	—	—	—	—	—	—
E	*A. devoniensis* (G)	Alkaline	18.5	81.5	—	—	—	—	—	—	—	—
F	*Calliblepharis jubata* (NF)	Soft alkaline	2.0	89.3	—	8.7	—	—	—	—	—	25
G	*C. jubata* (NF)	Alkaline	—	100.0	—	—	—	—	—	—	—	23.8
H	*C. jubata* (FG)	Soft alkaline	—	79.7	—	20.1	—	—	—	—	—	—
I	*C. jubata* (T)	Soft alkaline	1.7	79.8	—	18.5	—	—	—	—	—	—
J	*Chondracanthus acicularis* (FG)	Alkaline	60.4	34.4	—	—	—	—	—	5.2	—	6.4
L	*C. acicularis* (FG)	Aqueous	58.8	34.5	—	—	—	—	—	6.8	—	—
M	*C. acicularis* (G + T)	Alkaline	22.8	—	—	—	—	31.4	45.9	—	—	—
N	*Chondracanthus teedei* var. *lusitanicus* (NF)	Alkaline	49.8	50.2	—	—	—	—	—	—	—	—
O	*C. teedei* var. *lusitanicus* (NF)	Alkaline	55.8	44.2	—	—	—	—	—	—	—	—
P	*C. teedei* var. *lusitanicus* (FG)	Alkaline	58.1	41.9	—	—	—	—	—	—	—	5.0
Q	*C. teedei* var. *lusitanicus* (FG)	Alkaline	55.7	44.3	—	—	—	—	—	—	—	11.4
R	*C. teedei* var. *lusitanicus* (NF)	Aqueous	47.2	45.8	0.7	6.3	—	—	—	—	—	—
S	*C. teedei* var. *lusitanicus* (FG)	Aqueous	52.7	37.5	4.6	5.2	—	—	—	—	—	—
T	*C. teedei* var. *lusitanicus* (T)	Alkaline	—	—	—	—	—	33.0	67.0	—	—	—
U	*C. teedei* var. *lusitanicus* (T)	Aqueous	—	—	—	—	—	33.0	67.0	—	—	—
V	*C. teedei* var. *lusitanicus* (T)	Alkaline	—	—	—	—	—	33.0	67.0	—	—	—
X	*Chondrus crispus* (T)	Alkaline	—	—	—	—	100.0	—	—	—	—	—
*	*C. crispus*	Alkaline	75.0	25.0	—	—	—	—	—	—	—	—
**	*C. crispus*	Soft alkaline	70.0	28.0	2.0	—	—	—	—	—	—	2.0
**	*C. crispus*	Alkaline	64.0	36.0	—	—	—	—	—	—	—	—

(Continued)

TABLE 12.4 (*Continued*)

Molecular Composition of Carrageenan Determined by ¹H-NMR

Code	Species	Extraction Method	Carrageenan Composition (mol%)							Other Components		
			Kappa	Iota	Mu	Nu	Lambda	Theta	Xi	Pyruvate (mol%)	n. i.	Starch (%)
Z	*G. crenulatus* (TB)	Soft alkaline	60.0	28.9	4.0	—	—	—	—	7.1	—	4.9
W	*G. crenulatus* (TB)	Alkaline	64.1	30.8	—	—	—	—	—	5.1	—	11.2
K	*Mastocarpus stellatus* (G)	Alkaline	61.8	35.9	—	—	—	—	—	2.3	—	7.4
A₁	*M. stellatus* (G)	Soft alkaline	59.0	4.0	—	—	—	—	—	—	—	—
*	*Agardhiella* sp.	Alkaline	1.0	97.0	—	—	—	—	—	2.0	—	11.0
*	*Eucheuma denticulatum*	Alkaline	4.0	96.0	—	—	—	—	—	—	1	—
**	*E. denticulatum*	Alkaline	4.0	96.0	—	—	—	—	—	4.0	—	12.0
*	*Gigartina skottsbergii*	Alkaline	59.0	41.0	—	—	—	—	—	—	—	1
*	*Hypnea musciformis*	Alkaline	98.0	2.0	—	—	—	—	—	—	—	1
*	*Kappaphycus alvarezii*	Alkaline	93.0	7.0	—	—	—	—	—	—	—	—
**	*K. alvarezii*	Alkaline	90.0	10.0	—	—	—	—	—	—	—	1
*	*Mazzaella laminarioides*	Soft alkaline	55.0	43.0	2.0	—	—	—	—	—	—	—
*	*Sarconema scinaioides*	Soft alkaline	4.0	68.0	—	—	2.0	—	—	26.0	14	15.0
*	*Sarcothalia crispata*	Alkaline	57.0	42.0	1	—	—	—	—	—	—	—

Sources: Adapted from Pereira, L., Estudos em macroalgas carragenófitas (Gigartinales, Rhodophyceae) da costa portuguesa—aspectos ecológicos, bioquímicos e citológicos, PhD thesis, Departamento de Botânica, FCTUC, Universidade de Coimbra, Coimbra, Portugal, 2004; van de Velde, F. et al., *Carbohydr. Res.,* 340(6), 1113, 2005; Pereira, L. and van de Velde, F., *Carbohydr. Polym.,* 84, 614, 2011.

Code A to W (see Figures 12.3 through 12.9); T, tetrasporophytes; FG, female gametophytes; G, gametophytes; NF, nonfructified thalli; TB, tetrasporoblastic thalli; A₁, laboratory-cultivated carrageenophyte.

* Commercial carrageenan from CP Kelco.

** Commercial carrageenan from Degussa (currently Cargill).

n. i., not identified.

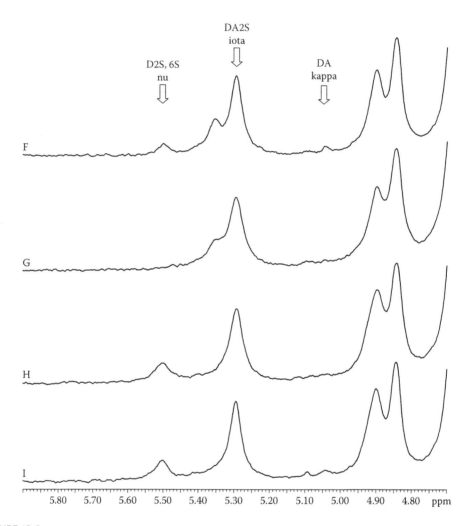

FIGURE 12.3
^{1}H-NMR spectra of *Calliblepharis jubata* carrageenans. The letter codes refer to the nomenclature of Knutsen et al. (1994).

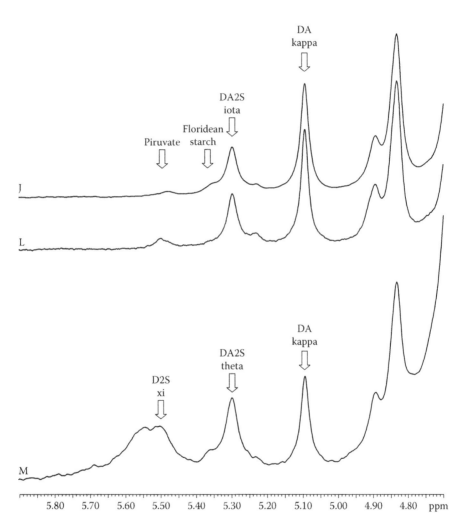

FIGURE 12.4

^{1}H-NMR spectra of *Chondracanthus acicularis* carrageenans. The letter codes refer to the nomenclature of Knutsen et al. (1994).

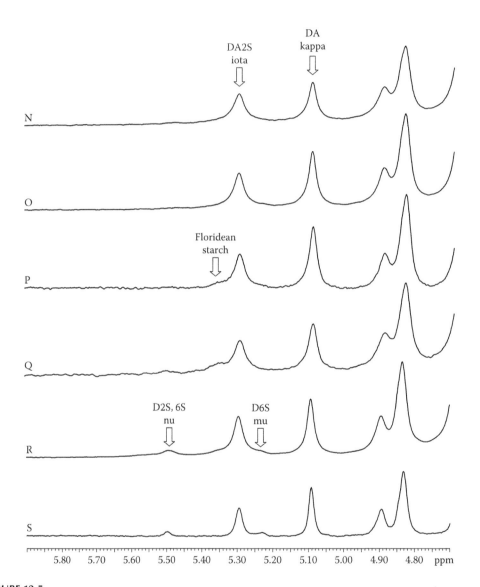

FIGURE 12.5

^{1}H-NMR spectra of *Chondracanthus teedei* var. *lusitanicus* carrageenans (Gametophytes). The letter codes refer to the nomenclature of Knutsen et al. (1994).

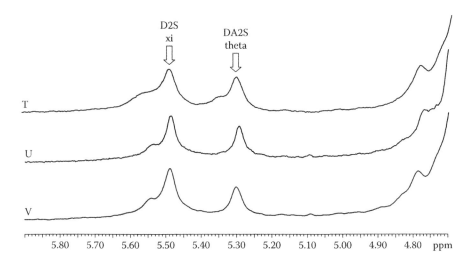

FIGURE 12.6
^{1}H-NMR spectra of *Chondracanthus teedei* var. *lusitanicus* carrageenans (Tetrasporophytes). The letter codes refer to the nomenclature of Knutsen et al. (1994).

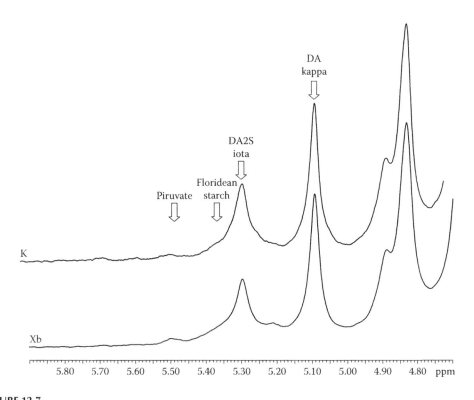

FIGURE 12.7
^{1}H-NMR spectra of *Mastocarpus stellatus* carrageenans. The letter codes refer to the nomenclature of Knutsen et al. (1994).

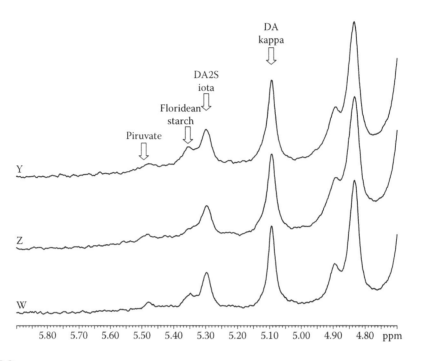

FIGURE 12.8
¹H-NMR spectra of *Gymnogongrus crenulatus* carrageenans. The letter codes refer to the nomenclature of Knutsen et al. (1994).

corresponding 3,6-anhydro form, during a long, highly alkaline (usually 0.1 M Ca[OH]$_2$) extraction process. At the industrial level, the extraction processes are intended to enhance profitability and increase the gelling power of the carrageenan obtained (Pereira 2004, Pereira and van de Velde 2011).

12.5 Conclusion

The ¹H-NMR spectroscopy is useful in identifying and quantifying the different constituent fractions of carrageenan samples. The proton resonance spectroscopy can be used in rapid qualitative analysis of small samples (5 mg) and the quantitative analysis of different samples of hybrid constituents or mixtures of carrageenan. Proton magnetic resonance allows to overcome the limitations of carbon 13 spectroscopy, since the latter is unable to detect carrageenan units present in quantities less than 5 mol%.

12.6 Summary

Carrageenan from 14 different carrageenophytes (Gigartinales, Rhodophyta), belonging to 11 genera and 4 families, were analyzed by NMR spectroscopy. In order to characterize the different carrageenan types, ¹H-NMR spectroscopy was used to identify and quantify the different carrageenan fractions in the extracted phycocolloids (both water and alkali extractions). Thereby, detailed information concerning the properties and structure of these polysaccharides at molecular level was revealed.

Acknowledgments

The author acknowledges the financial support from the Portuguese Foundation for Science and Technology of the IMAR-CMA (Institute of Marine Research).

References

Anderson, N.S., T.C.S. Dolan, and D.A. Rees. 1965. Evidence for a common structural pattern in polysaccharide sulphates of Rhodophyceae. *Nature* 205(4976):1060–1068.

Anderson, N.S. and D.A. Rees. 1965. Porphyran: A polysaccharide with a masked repeating structure. *J. Chem. Soc.* 5880–5887.

Armisen, R. and F. Galatas. 1987. Production, properties and uses of agar—Chapter 1. In *Production and Utilization of Products from Commercial Seaweeds*, ed. D.J. McHugh. Rome, Italy: FAO Fisheries Technical Paper 288, 189pp.

Armisen, R. and F. Galatas. 2000. Agar. In *Handbook of Hydrocolloids*, eds. G. Phillips and P. Williams. Boca Raton, FL: CRC Press, pp. 21–40.

Bellion, C. 1982. Utilisation d'enzymes et the spectroscopie de RMN du carbone 13 pour la caractérisation des carraghénanes. PhD thesis, Tolouse, France: Université Paul Sebatier.

Bhattacharjee, S.S., W. Yaphe, and G.K. Hamer. 1978. C-13-NMR spectroscopic analysis of agar, kappa-carrageenan and lambda-carrageenan. *Carbohydr. Res.* 60(1):C1–C3.

Bixler, H.J. and H. Porse. 2011. A decade of change in the seaweed hydrocolloids industry. *J. Appl. Phycol.* 23(3):321–335.

Buggeln, J.S. and R.G. Carige. 1973. The physiology and biochemistry of *Chondrus crispus* Stackhouse. In *Chondrus crispus*, eds. M.J. Harvey and J. MacLachlan. Halifax, Nova Scotia, Canada: The Nova Scotian Institute of Science (NSIS), pp. 81–102.

Cardoso, S.M., L.G. Carvalho, P.J. Silva, M.S. Rodrigues, O.R. Pereira, and L. Pereira. 2014. Bioproducts from macroalgae: A review with focus on the Iberian Peninsula. *Curr. Org. Chem.* 18(7):896–917.

Chopin, T. 2006. Integrated multi-trophic aquaculture. What it is, and why you should care... and don't confuse it with polyculture. *North. Aquacult.* 12:4.

Chopin, T. and E. Whalen. 1993. A new and rapid method for carrageenan identification by FT-IR diffuse-reflectance spectroscopy directly on dried, ground algal material. *Carbohydr. Res.* 246:51–59.

Ciancia, M., M.C. Matulewicz, P. Finch, and A.S. Cerezo. 1993. Determination of the structures of cystocarpic carrageenans from *Gigartina skottsbergii* by methylation analysis and NMR-spectroscopy. *Carbohydr. Res.* 238:241–248.

Craigie, J.S. 1990. Cell walls. In *Biology of the Red Algae*, eds. K.M. Cole and R.G. Sheath. Cambridge, U.K.: Cambridge University Press, pp. 221–257.

Dyrby, M., R.V. Petersen, B. Larsen, B. Rudolph, L. Norgaard, and S.B. Engelsen. 2004. Towards on-line monitoring of the composition of commercial carrageenan powders. *Carbohydr. Polym.* 57:337–348.

Falshaw, R. and R.H. Furneaux. 1994. Carrageenan from the tetrasporic stage of *Gigartina decipiens* (Gigartinaceae, Rhodophyta). *Carbohydr. Res.* 252:171–182.

Falshaw, R. and R.H. Furneaux. 1995. Carrageenans from the tetrasporic stages of *Gigartina glavifera* and *Gigartina alveata* (Gigartinaceae, Rhodophyta). *Carbohydr. Res.* 276(1):155–165.

Furtado, M.R. 1999. Alta lucratividade atrai investidores em hidrocolóides. *Química e Derivados* 377:20–29.

Greer, C.W. and W. Yaphe. 1984a. Characterization of hybrid (beta-kappa-gamma) carrageenan from *Eucheuma gelatinae* J. Agardh (Rhodophyta, Solieriaceae) using carrageenases, infrared and C-13-NMR spectroscopy. *Botan. Mar.* 27(10):473–478.

Greer, C.W. and W. Yaphe. 1984b. Hybrid (iota-nu-kappa) carrageenan from *Eucheuma nudum* (Rhodophyta, Solieriaceae), identified using iota-carrageenases and kappa-carrageenases and C-13-NMR spectroscopy. *Bot. Mar.* 27(10):479–484.

Guiry, M.D. and G. Blunden. 1991. *Seaweed Resources in Europe: Uses and Potential.* Chichester, U.K.: John Wiley & Sons.

Harris, R.K., E.D. Becker, S.M. Cabral-de-Menezes, R. Goodfellow, and P. Granger. 2001. NMR nomenclature. Nuclear spin properties and conventions for chemical shifts (IUPAC Recommendations 2001). *Pure Appl. Chem.* 73(11):1795–1818.

Hayashi, L., A.G. Hurtado, F.E. Msuya, G. Bleicher-Lhonneur, and A.T. Critchley. 2010. A review of *Kappaphycus* farming: Prospects and constraints. In *Seaweeds and Their Role in Globally Changing Environments* (Cellular Origin, Life in Extreme Habitats and Astrobiology), ed. J. Seckbach. Dordrecht, the Netherlands: Springer, pp. 251–283.

Hurtado, A.Q., B.L. Genevieve, and A.T. Critchley. 2006. *Kappaphycus* "cottonii" farming. In *World Seaweed Resources—An Authoritative Reference System V1.0*, eds. A.T. Critchley, M. Ohno, and D.B. Largo. the Netherlands: ETI Information Services Ltd. Hybrid Windows and Mac DVD-ROM.

Hurtado, A.Q., R.P. Reis, R.R. Loureiro, and A.T. Critchley. 2014. *Kappaphycus* (Rhodophyta) cultivation: Problems and the impacts of Acadian marine plant extract powder. In *Marine Algae— Biodiversity, Taxonomy, Environmental Assessment, and Biotechnology*, eds. L. Pereira and J.M. Neto. New York: CRC Press.

Ioannou, E. and V. Roussis. 2009. Natural products from seaweeds. In *Plant-Derived Natural Products: Synthesis, Function, and Application*, A.E. Osbourn and V. Lanzotti (eds.), Springer Science & Business Media, LLC 2009, pp. 51–81.

Jiao, G., G. Yu, J. Zhang, and H.S. Ewart. 2011. Chemical structures and bioactivities of sulfated polysaccharides from marine algae. *Mar. Drugs* 9:196–223.

Knutsen, S.H., D.E. Myslabodski, B. Larsen, and A.I. Usov. 1994. A modified system of nomenclature for red algal galactans. *Bot. Mar.* 37(2):163–169.

Matsuhiro, B. 1996. Vibrational spectroscopy of seaweed galactans. *Hydrobiologia* 327:481–489.

McHugh, D.J. 2003. A guide to the seaweed industry. Rome, Italy: FAO, Fisheries Technical Paper 441, pp. 73–90.

Minghou, J. 1990. Processing and extraction of phycocolloids. In *Regional Workshop on the Culture and Utilization of Seaweeds*, Vol. II. Cebu City, Philippines: FAO Technical Resource Papers.

Myslabodski, D.E. 1990. Red-algae galactans: Isolation and recovery procedures—Effects on the structure and rheology. PhD thesis, Trondheim, Norway: Norwegian Institute of Technology.

Pereira, L. 2004. Estudos em macroalgas carragenófitas (Gigartinales, Rhodophyceae) da costa portuguesa—aspectos ecológicos, bioquímicos e citológicos. PhD thesis, Departamento de Botânica, FCTUC, Coimbra, Portugal: Universidade de Coimbra.

Pereira, L. 2006. Identification of phycocolloids by vibrational spectroscopy. In *World Seaweed Resources—An Authoritative Reference System V1.0*, eds. A.T. Critchley, M. Ohno, and D.B. Largo. the Netherlands: ETI Information Services Ltd. Hybrid Windows and Mac DVD-ROM.

Pereira, L. 2010a. Seaweed: An unsuspected gastronomic treasury. *Chaîne de Rôtisseurs Magazine* 2:50.

Pereira, L. 2010b. *Littoral of Viana do Castelo—Algae. Uses in Agriculture, Gastronomy and Food Industry* (Bilingual). Viana do Castelo, Portugal: Município de Viana do Castelo.

Pereira, L. 2011. A review of the nutrient composition of selected edible seaweeds. In *Seaweed: Ecology, Nutrient Composition and Medicinal Uses*, ed. V.H. Pomin. New York: Nova Science Publishers Inc., pp. 15–47.

Pereira, L., A.M. Amado, A.T. Critchley, F. van de Velde, and P.J.A. Ribeiro-Claro. 2009a. Identification of selected seaweed polysaccharides (phycocolloids) by vibrational spectroscopy (FTIR-ATR and FT-Raman). *Food Hydrocoll.* 23:1903–1909.

Pereira, L., A.M. Amado, P.J.A. Ribeiro-Claro, and F. van de Velde. 2009b. Vibrational spectroscopy (FTIR-ATR and FT-RAMAN)—A rapid and useful tool for phycocolloid analysis. *Biodevices* 131–136.

Pereira, L., A.T. Critchley, A.M. Amado, and P.J.A. Ribeiro-Claro. 2009c. A comparative analysis of phycocolloids produced by underutilized *versus* industrially utilized carrageenophytes (Gigartinales, Rhodophyta). *J. Appl. Phycol.* 21:599–605.

Pereira, L. and J.F. Mesquita. 2004. Population studies and carrageenan properties of *Chondracanthus teedei* var. *lusitanicus* (Gigartinaceae, Rhodophyta). *J. Appl. Phycol.* 16(5):369–383.

Pereira, L., A. Sousa, H. Coelho, A.M. Amado, and P.J.A. Ribeiro-Claro. 2003. Use of FTIR, FT-Raman and ^{13}C-NMR spectroscopy for identification of some seaweed phycocolloids. *Biomol. Eng.* 20:223–228.

Pereira, L. and F. van de Velde. 2011. Portuguese carrageenophytes: Carrageenan composition and geographic distribution of eight species (Gigartinales, Rhodophyta). *Carbohydr. Polym.* 84:614–623.

Rochas, C. and M. Lahaye. 1989. Solid-state C-13-NMR spectroscopy of red seaweeds, agars and carrageenans. *Carbohydr. Polym.* 10(3):189–204.

Rudolph, B. 2000. Seaweed products: Red algae of economic significance. In *Marine and Freshwater Products Handbook*, eds. R.E. Martin, E.P. Carter, L.M. Davis, and G.J. Flich. Lancaster, PA: Technomic Publishing Company Inc., pp. 515–529.

Sousa-Pinto, I. and H. Abreu. 2011. Aquacultura multitrófica integrada—o que é? In *Macroalgas na aquacultura multitrófica integrada peninsular*, eds. U.V. Ferreiro, M.I. Filgueira, R.F. Otero, and J.M. Leal. Iberomare, Portugal: Valorização da Biomassa, CETMAR, pp. 10–27.

Usov, A.I. 1984. NMR-spectroscopy of red seaweed polysaccharides—Agars, carrageenans, and xylans. *Bot. Mar.* 27(5):189–202.

Usov, A.I., S.V. Yarotsky, and A.S. Shashkov. 1980. ^{13}C-NMR spectroscopy of red algal galactans. *Biopolymers* 19(5):977–990.

Vairappan, C., C. Chung, A. Hurtado, F. Soya, G. Lhonneur, and A. Critchley. 2008. Distribution and symptoms of epiphyte infection in major carrageenophyte-producing farms. *J. Appl. Phycol.* 20:477–483.

van de Velde, F., A.S. Antipova, H.S. Rollema, T.V. Burova, N.V. Grinberg, L. Pereira et al. 2005. The structure of kappa/iota-hybrid carrageenans II. Coil-helix transition as a function of chain composition. *Carbohydr. Res.* 340(6):1113–1129.

van de Velde, F. and G.A. de Ruiter. 2002. Carrageenan. In *Biopolymers V6, Polysaccharides II, Polysaccharides from Eukaryotes*, eds. E.J. Vandamme, S.D. Baets, and A. Steinbèuchel. Weinheim, Germany: Wiley, pp. 245–274.

van de Velde, F., S.H. Knutsen, A.I. Usov, H.S. Rollema, and A.S. Cerezo. 2002. H-1 and C-13 high resolution NMR spectroscopy of carrageenans: Application in research and industry. *Trends Food Sci. Technol.* 13(3):73–92.

van de Velde, F., L. Pereira, and H.S. Rollema. 2004. The revised NMR chemical shift data of carrageenans. *Carbohydr. Res.* 339:2309–2313.

Yarotsky, S.V., A.S. Shashkov, and A.I. Usov. 1977. Analysis of C-13-NMR spectra of some red seaweed galactans. *Bioorg. Khim.* 3(8):1135–1137.

Zinoun, M., J. Cosson, and E. Deslandes. 1993. Physicochemical characterization of carrageenan from *Gigartina teedii* (Rooth) Lamouroux (Gigartinales, Rhodophyta). *J. Appl. Phycol.* 5(1):23–28.

13

Marine Oligosaccharides:
The Most Important Aspects in the Recent Literature

Antonio Trincone

CONTENTS

13.1 Introduction

It is not over the top to say that science about polysaccharides is of paramount importance for the life of humankind on our planet. Topics include not only food and health issues, related to nutritional domain, but many other aspects are present being polysaccharides such as cellulose, starch, chitin, and many others constituting the most part of the biomass. Eventually, many disciplines are involved in the study of polysaccharides, and interdisciplinary cross-fertilization is seen as very beneficial. Polysaccharides constitute a variety of biological polymers with diverse composition, structures, physical characteristics, and biological activity. They have been used in the food industry as thickeners and protective coatings; in cosmetics, textiles, and medicines, their uses are based on rheological, emulsifying, and stabilizing properties. More recently, in the field of smart materials, interesting technological applications are emerging for polysaccharides like cellulose; in particular, the possible intelligent behaviors in reaction to environmental conditions (Qiu and Hu 2013) and very recently ecomaterials based on polysaccharides and food proteins, as possible replacement of synthetic polymers with specific applications, have been reviewed (Yihu and Qiang 2014).

The potential in bioactivity of marine polysaccharides is still considered underexploited; however, these molecules, including the derived oligosaccharides, are an extraordinary source of chemical diversity. As far as the relation between chemical structure and activity/function is concerned, it can be said that it has not been an area of in-depth study even though carbohydrates have a long history among topical scientific subjects. In our current -omic age, the attention to both bioactivity and the relation to structure activity/function of marine polysaccharides increased, as reflected, for example, in a recent proposal of disseminating fucanomics and galactanomics related to interesting sulfated fucans and sulfated galactans of marine origin, recently advanced (Pomin 2012). A foresight for an expected increase of research in projects of such nature has also been prospected (Trincone 2014).

Polysaccharides from the marine environment can provide a useful alternative to glycosaminoglycans (GAGs) (Senni et al. 2011). This optimistic view is justified by the successes in the chemical biology (biosynthesis, structure, and function) of complex glycans of mammalian origin showing the important role of this class of molecules in disease modulation. It has to be said that this animal origin increases the risk for the presence of infectious agents and makes the availability of this material a troublesome topic. Indeed, as far as the quality control of pharmaceutical preparations based on such complex molecules is concerned, different challenging aspects emerge (due to inherent complexity of structural characteristics and also to a certain unfavorable batch-to-batch variability). Active research is welcome in this particular field for widening current limitations for the use of these molecules as ingredients in food supplements and cosmetics. In this recent study (Luhn et al. 2014), the evaluation of a microplate assay for quality screening of marine sulfated polysaccharide is reported. *Delesseria sanguinea* sulfated polysaccharides and several other algae-originating compounds were investigated with a technique, previously devised by the authors, based on the increase of fluorescence intensity of a sensor molecule linked to the glycan.

In this chapter, after a general view of marine polysaccharides, the literature up to middle 2014 on marine oligosaccharides is analyzed. Excellent review articles can be found in literature on single aspects of marine polysaccharides, and to avoid repetition, a subsequent general excursus on sources of these molecules, dealing with extraction issues, production, purification, and structural determination, is presented, avoiding detailed descriptions that can be found in specialized articles. This is useful in preparing the ground for a better detailed analysis of more recent works about marine oligosaccharides reported in the literature of 2014.

13.2 General View on Marine Polysaccharides and Oligosaccharides

The attracting interest in developing potential drugs for chronic diseases has been evidenced in a recent review on biofunction of marine oligosaccharides (Ji et al. 2011). The authors listed many fields for application of these molecules including food industry, cosmetics, biomedicine, agriculture, environmental protection, and wastewater management. However, they concluded that it is hard to explain how exactly these molecules exert their activity and future research should be directed toward the understanding of details of molecular level of activity.

As bioactive molecules with defined structures, low viscosity, and generally better solubility with respect to high-MW progenitors, marine oligosaccharides will be of focused interest for increasing the technology readiness level of related research, easily bridging the gap from basic discovery to pharmacological market application.

In a recent white paper by GlycoMar, examples of carbohydrate-based drugs in use or in development were tabulated (GlycoMar 2012), with heparin manufactured from animal sources as notable historical example. For marine-derived cases, which are increasingly recognized and investigated, some are already in use as nutraceuticals and others tested in preclinical trials were listed. Oral delivery of these compounds seemed to be suitable with no reported problems of immunogenicity at Marinova (Australia). At this Tasmanian biotech company, they are involved in the development of a galactofucan sulfate by modulation of sulfation and acetylation (Fitton 2011). Others, in a public research institute (IFREMER, France), are investigating also different applications (cardiovascular, infection, oncology, and inflammation), while important industrial partners (Sanofi-Aventis) are interested in the preclinical state for the application of sea squirt heparin in arterial thrombosis.

Oligosaccharides are relatively short molecules; hence, it is relatively easy to establish a defined structure that can be precisely related to bioactivity, mechanism, regulation, etc. rendering them properly suitable for high standards required by pharmaceutical industry. Marine oligosaccharides originate by hydrolysis of marine polysaccharides, and in view of the huge list of the effects they play in cell events, sustainable ways based on enzymes to access these molecules are particularly important.

Numerous polysaccharidases are already reviewed (Giordano et al. 2006; Trincone 2013a). Those of marine origin are in the spotlight because these catalysts are the natural tools of polysaccharide metabolism in the living ecosystems. Interestingly, it has been recently published a revised understanding of the ecology and biochemistry of chitin degradation (Beier and Bertilsson 2013).

Modern techniques for purification and for the investigation of chemical structures are also important with mass spectrometry techniques competing efficiently with NMR-based tools (Lang et al. 2014).

Marine organisms representing good sources of polysaccharides are numerous and include seaweed, microalgae, bacteria, and animals (fish, shellfish, mollusks, etc.). Scientific knowledge present in literature for each of these sources greatly differs, from the well-known seaweed (Rinaudo 2007) to relatively new sources of these compounds such as extremophiles and microalgae (de Jesus et al. 2013).

Alginates are polysaccharides obtained from brown algae especially from Northern seas for food, pharmaceutical, and other technical application. As established in the 1960s, these are linear copolymers of homopolymeric blocks of different structures: (1–4)-linked α-D-mannuronate (M) or its C-5 epimer α-L-guluronate (G) residues, respectively. The blocks are covalently linked together in different sequences (Figure 13.1).

Carrageenan is a sulfated polysaccharide obtained by extraction with water or alkaline water of certain *Rhodophyceae* (red seaweed). It consists mainly of sulfate esters salts (potassium, sodium, magnesium, and calcium) of galactose and 3,6-anhydrogalactose copolymers (Figure 13.2).

In addition to alginates, other polyanionic polysaccharides include fucoidans formerly seen as by-products of alginate industry, although these sulfated polysaccharides may be responsible for the bioactivity of these algae (Kusaykin et al. 2008). Sulfated fucans were isolated also from some marine animals. In addition, a recent review from the same

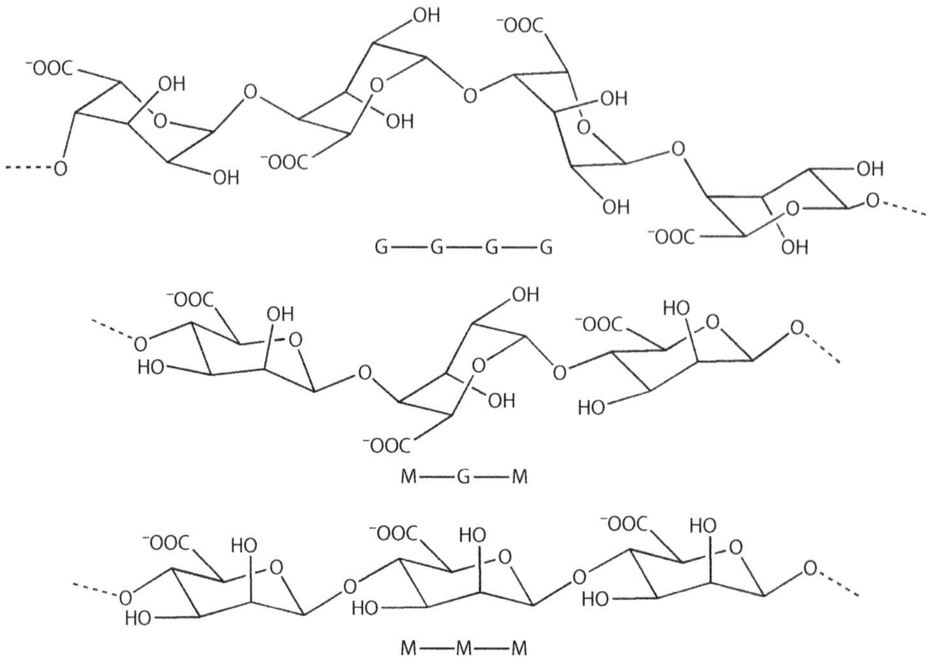

FIGURE 13.1
Alginate types.

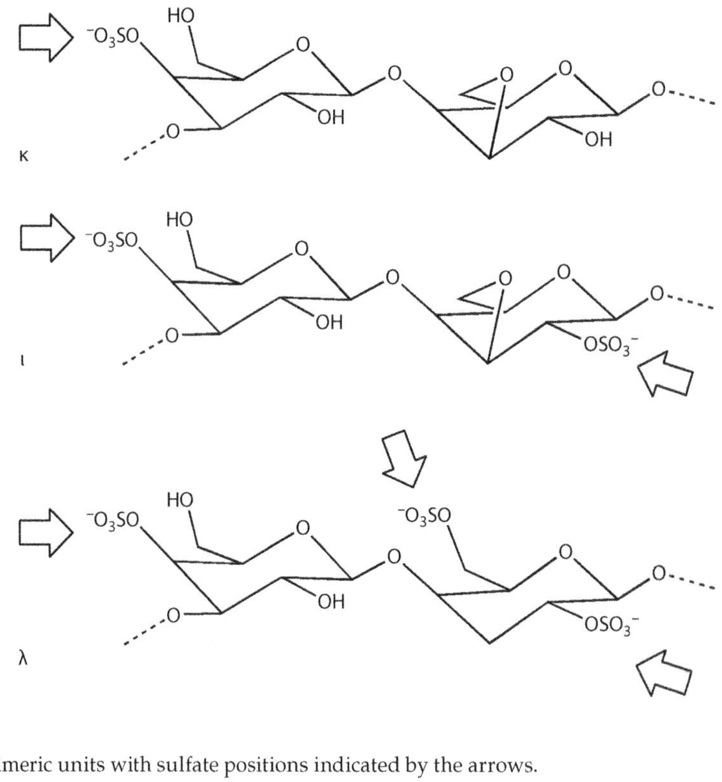

FIGURE 13.2
Carrageenan dimeric units with sulfate positions indicated by the arrows.

group of authors stated that many polysaccharides and/or their depolymerized and semisynthetic derivatives obtained by chemical modifications demonstrate anticancer and cancer preventive properties (Fedorov et al. 2013). These molecules have a complex and heterogeneous structure mostly composed by 2- and/or 3-sulfated α-L-fucopyranosyl units with 3- and 4-linkages (alternate) and a possible presence of nonsulfated and branched units.

Cultivable or uncultivable marine microorganisms with particular emphasis on extremophiles are worth the mention. Structural characteristics, rheological properties, and absence of pathogenicity of exopolysaccharides (EPSs) from extremophiles make these compounds considerable for biotechnological applications in food, pharmaceutical, and cosmetics industries (Nicolaus et al. 2010; Satpute et al. 2010). Although EPS biosynthesis is still poorly understood, manipulation of genes responsible could enable the modulation of chemical–physical characteristics of polysaccharides produced, hence the properties and potential applications. The structures of EPSs present in microbial cells vary greatly in their sugar composition. Neutral carbohydrates are present, but the majority of EPSs are polyanionic for the presence of D-glucuronic, D-galacturonic, and D-mannuronic acids or ketal-linked pyruvate. Sulfate with other inorganic residues such as phosphate contributes to polyanionic status.

Among sources of marine polysaccharides, the waste from fishery products should also be listed (Venugopal 2011). Chitin is composed by units of *N*-acetyl glucosamine covalently linked by β-1,4 linkages. The case of chitin from shell waste or from Antarctic krill is known. It is convincing idea that with the tools of marine biotechnology, marine bioprocess industry can convert seafood wastes to valuable polysaccharides.

The underexploited chemical diversity present in marine polysaccharides from new sources obviously is of great interest for the development of new marine oligosaccharides. Especially, polysaccharides from microorganism present a real potential, being possible accessing to them by fermentation, thus increasing their appeal for application (Finore et al. 2014). Production by fermentation in laboratory conditions enables for a stable supply not influenced by climatic, seasonal, and physiological variations.

13.3 Literature

In the period 1995–2014, the hits (scientific journals, books, reference material) retrieved from Science Direct database using two search entries "marine oligosaccharides" and "marine polysaccharides" are plotted in Figure 13.3. Numbers are significantly high although retrieved material could contain spurious results. However, a close inspection of titles and abstracts revealed a low number of false-positives. Further, a comparison with data retrieved from PubMed resulted in the same ratio of articles for the two databases confirming the validity of the analysis.

Inspection of the graph obtained in Figure 13.3 shows at least two different segments. The trends of both searches are parallel at two different values from 1995 to 2004. The second segment is from the end of this period up to the first 3 months of the current year in which there is an increasing difference value each year for "marine polysaccharides." A third segment could also be individuated from 2009 to 2010 in which a sharp rise is visible for "marine polysaccharides." This graph is a good starting point for this chapter.

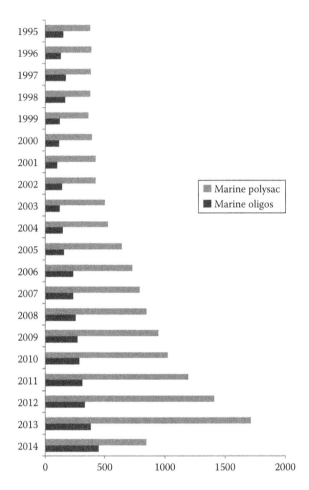

FIGURE 13.3
Number of total hits searching for "marine oligosaccharides" and "marine polysaccharides" using Advanced Search feature of Science Direct, a commercial database, offering ca. 2,500 journals and 20,000 books. Searches are conducted in all fields without double quotation marks (AND Boolean operator retrieves documents that contain both words even far apart from each other). Values for 2014 include only first 3 months of the year. Retrieval date, March 2014.

The analysis lead to a foresight for the next years related to an expected increase of studies about marine oligosaccharides (efforts for sustainable access by specific and high-yield enzymatic hydrolysis, new enabling methodologies for straightforward purification techniques, structural determination, activity assay, and mechanism of action).

13.4 Sources

In this section, the most common sources of marine polysaccharides are shortly reviewed referring to the amount of literature published on this topic. Obviously, this quick mention that should not be considered exhaustive is useful in dealing with marine oligosaccharides derived from polysaccharides by selective hydrolysis.

Seaweed, microalgae, bacteria, and animals (fish, shellfish, mollusks, etc.) are the marine organisms representing good sources of polysaccharides. Scientific knowledge present in literature for each of these sources is not uniform going from the well-known seaweed to sources of polysaccharides such as extremophiles and microalgae (de Jesus Raposo et al. 2013) that are relatively new. Algae are among the oldest known living organisms and are the sources of many important polysaccharides from the point of view of applications (Rinaudo 2007). Marine benthic algae are multicellular and have a large body (macrothallic), and generally they are included in the category of "plants." Most are green (1200 species), brown (2000 species), or red (6000 species). The structural diversity of polysaccharides depends on taxonomy and on other important aspects such as life stage, habitat, and geographical location. As for microalgae, the high biodiversity of polysaccharides and how they can serve the biotechnological field have been recently pointed out (Barra et al. 2014; Stengel et al. 2011). As for cultivable or uncultivable marine microorganisms, particular emphasis on extremophiles is important. Structural characteristics, rheological properties, and absence of pathogenicity of EPSs from them make these compounds considerable for food, pharmaceutical, and cosmetics industries (Finore et al. 2014; Nicolaus et al. 2010; Satpute et al. 2010). The underexploited chemical diversity present in marine polysaccharides from new sources obviously is of great interest for the development of new marine oligosaccharides.

13.4.1 Animals

Illustrative examples of sulfated marine polysaccharides have been recently tabulated; some of their well-defined repetitive oligomeric structures (tetrasaccharides, trisaccharides, disaccharides) make these data extremely interesting for bioprospecting useful enzymes to access small structures by selective hydrolysis (Pomin 2012). Examples are reported from ascidians (sea squirts or tunicates, *Urochordata*, *Ascidiacea*), sea cucumber (*Echinodermata*, Holothuroidea), and sea urchins (*Echinodermata*, *Echinoidea*). Different physiological functions are known for these polysaccharides found in animals: assembling of body walls, building jelly coat surrounding gametes, etc.

13.4.2 Seaweed

Seaweeds are commonly included in the general category of "plants." Thousands of species are known for the green (Chlorophyta), brown (Phaeophyta), and red algae (Rhodophyta). They are used for human food such as the species *Porphyra* (nori for sushi), *Laminaria* (for kombu), and *Undaria* (for wakame). As far as the content of major polysaccharides is concerned, alginates, fucans/fucoidans, and laminarins characterize brown algae. Red seaweed contains mainly agar/agarose, and carrageenans as most abundant sulfated polysaccharide.

The general structure of laminarin is presented in Figure 13.4. It is made up of β-(1–3)-glucan with branches in β-(1–6)-positions. Ramification depends on the origin and generally is 3:1 (β-1–3-/β-1–6-).

Fucoidans were historically considered as 1,2-α-L-fucans. However, recently, α–1,3-backbones or repeating disaccharide units of α–1,3- and α–1,4-linked fucose residues are identified with branching at C2 positions. Sulfation position is possible at C4, C2, or both C2 and C4 positions of fucose units. Acetyl groups can be also found in some fucoidans from algae. After this general structural description, the reader is referred to specific reviews on the subject (Kusaykin et al. 2008; Pomin 2011).

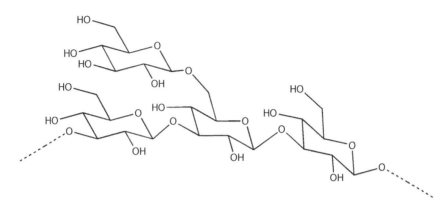

FIGURE 13.4
General structure of laminarin.

Many other minor sulfated polysaccharides are known with their names derived from the algae source: hypnean, porphyran, furcellaran, funoran, etc. (Rinaudo 2007).

13.4.3 Microalgae and Exopolysaccharides from Extremophiles

In recent years, the use of microalgae for biotechnological applications (bioremediation, nutraceutical and pharmaceutical fields, bioenergy production) has been increasing. A recent review (Dismukes et al. 2008) on the use of biofuels derived from cyanobacteria, algae, and diatoms is of great interest. Physiological aspects of these organisms, with polysaccharides typically constituting 80%–95% of their total carbohydrate content, are particularly attractive for integration into renewable biofuel applications. Some species excrete polysaccharides (EPSs) into the culture medium with potential applications such as adhesives, gelling, flocculant, and thickening agents. Working with microalgae is considered advantageous for the ease of growing, culturing, and for the harvest not related to climate or season; however, high content of polysaccharides is known to characterize the stationary phase of growth (Haug and Myklestad 1976). Microalgae are photosynthetic microorganisms growing under controlled conditions and can produce polysaccharides with stable structures. However, other components (lipids, proteins, and other carbohydrates) are produced in large amounts, and these bioprocesses are increasingly becoming of interest in biofuel industry.

EPSs from extremophiles are of considerable importance for food, pharmaceutical, and cosmetics industries (Satpute et al. 2010). With their novel and unique physical characteristics, they have found extensive applications. The EPSs are important molecules in microbial interaction and emulsification of various hydrophobic substrates. Reports related to *Bacillus, Halomonas, Planococcus, Enterobacter, Alteromonas, Pseudoalteromonas, Vibrio, Rhodococcus, Zoogloea,* and also to Archaea such as *Haloferax* and *Thermococcus* have been published describing these species as producers and studying possible hyperproduction of biopolymers in different fermentation conditions (Finore et al. 2014).

Many polysaccharides and sulfated EPSs have been tabulated (de Jesus Raposo et al. 2013) showing to be heteropolymers constituted mainly of xylose, galactose, and glucose in different proportions. However, only a few of these polymers have been well studied in relation to fine structure and biochemical composition.

13.5 Extraction

Numerous methodologies are in use for the extraction of polysaccharides from natural biomass. The topic of the efficiency of extraction of polysaccharides is manifested and a current one in modern literature. Efficiency of procedures is dependent by a number of factors such as pH, solvent type, size of material, temperature, and time, and this chapter cannot cover it exhaustively. Common methods currently in use for marine examples are summarized in Table 13.1 (Rinaudo 2007). Impurities contaminating the extracts can be generally eliminated by precipitation using organic solvents and/or detergents. Selective fractionation with increasing volumes of precipitants is also possible. Organic solvents (ethanol, isopropanol, acetone) or detergents (CPC, Triton, sodium deoxycholate, etc.) can be used.

In the case of ulvan built by disaccharides, repeating sequences composed of sulfated rhamnose and glucuronic acid, iduronic acid, or xylose, a study of the influence of extraction conditions has been carried out showing that pH was the most significant factor affecting not only the yield but also monosaccharide composition and macromolecular characteristics of the ulvan being extracted (Yaich et al. 2013). Small-MW components are present at low pH of extraction originating from the degradation of high-MW components.

TABLE 13.1

Extraction Methods

Polysaccharides	Extraction
Alginates	The material (brown algae) is washed, macerated, and then extracted in alkaline conditions (sodium carbonate): thus, insoluble alginic acid is converted into sodium alginate, which is soluble into the aqueous phase. The extract is filtrated and alginate salt can be transformed to alginic acid by treatment with dilute HCl. After a series of purifications, the alginate is dried and powdered. It can be isolated in different ionic forms and used especially for foods as acid (E 400), sodium (E 401), potassium (E 402), ammonium (E 403), or calcium form (E 404).
Fucoidans	Fucoidans partly consisting of proteoglycans are extracted from algae in dilute acid conditions. To remove pigments and other low-MW compounds, preliminary treatment of the algal biomass with a $CHCl_3$–MeOH–H_2O solutions is possible. Water-soluble polysaccharides are extracted from defatted biomass at 85°C, and acid polysaccharides precipitated from the extract by the action of a mixture of different quaternary ammonium salts and transformed into water-soluble sodium salts. Crude fucoidan is purified and fractionated by ion-exchange chromatography on DEAE-Sephacel.
Laminarin	HPLC-grade water and 1% (w/v) $CaCl_2$ were used for the extraction at 85°C. After centrifugation, unsoluble polymers (crude galactofucan) can be precipitated using EtOH–H_2O (95:5, v/v). The supernatant represent crude laminarin (Rioux et al. 2010).
Agar	Agar is soluble in boiling water but insoluble in cold water. The extraction of agar is performed from boiled seaweed; the extract is then frozen and thawed. During this last step, water separates from agar, carrying with it soluble impurities. Washing, bleaching, sterilization, and drying are performed depending on the applications. Treatment in the presence of alkali increases the gel strength by formation of the 3,6-anhydrogalactose.

13.6 Enzymatic Production of Marine Oligosaccharides: Hydrolysis and Synthesis

Hydrolysis together with reliable synthesis of oligosaccharides is the method to achieve sufficient yields of pure material for biological studies. Selective hydrolysis of polysaccharides to form oligosaccharides to establish a precise chemical identity of these molecules is a well-established common practice in structural studies. As indicated in an important report about the chemical synthesis of a sulfated disaccharide involved in the aggregation process of the marine sponge *Microciona prolifera*, synthetic research and development programs date back to the beginning of new century (Henricus et al. 2000) or even before. Chemoenzymatic synthesis of GAGs has been also developed (De Angelis et al. 2013).

A widespread group of enzymes named glycoside hydrolases (glycosidases) are involved in the carbohydrate metabolism being responsible for the hydrolysis of glycosidic linkages; they can act as exo- or endoglycosidases and are involved in a series of important biological events such as energy uptake, in processes inherent cell wall metabolism, in glycan processing during in vivo glycoprotein synthesis, and in many physiopathological events. Applications of commercial mixtures of glycosyl hydrolases have long since been reported for the structural determination of many natural products and are still in use for shortening carbohydrate sequences attached to steroidal skeletons (De Marino et al. 2000). Actually, the higher the selectivity of hydrolysis and the purity of the biocatalyst, the better the performance and fidelity of information that can be obtained by the use of such enzymes. Selectivity was also achieved in a mild-acid hydrolysis of sulfated fucans even though this method is generally considered nonspecific for depolymerization of polysaccharides (Pomin et al. 2005). Random enzymatic hydrolysis is also of interest for shortening polysaccharide length thus increasing solubility and other rheological characteristics for specialized applications that are not compatible with presence of harsh chemicals.

Marine originating enzymes for the hydrolysis of polysaccharides are in the spotlight being the natural tools for the metabolism. The case of chitinases is illustrative; chitin, as the major component of crustacean, represents a nutrient source for the ecosystem (Beier and Bertilsson 2013), and recently, a revised understanding of the ecology and biochemistry of this process has been achieved. It is useful to recall the importance of these enzymes in the bioconversion of shrimp and crab shells especially for the efforts to isolate chitinase-producing strain by using shrimp shells or squid pens as the sole C/N sources (Wang et al. 2011). Different polysaccharidases were already reviewed (Giordano et al. 2006; Trincone 2013a), and only the latest few are now given in a nonexhaustive list shown in Table 13.2. Interesting examples of these enzymes are reported in the list of application in the following text, and many others will be reported in the paragraph dedicated to the literature of 2014.

Protein engineering, metagenomics, and reaction engineering have led to the discovery of an expanding number of novel enzymes and to the setting up of new biobased processes for the formation of glycosidic linkage. In recent years, glycoside hydrolases used for enzymatic hydrolysis have been used also in synthesis. Recent scientific literature related to examples of uniqueness of the biocatalyst in the synthesis of glycosides (yield, substrate specificity, regioselectivity, resistance to a particular reaction condition, etc.) has been reviewed (Trincone 2013b). The invention of glycosynthases is just one of

TABLE 13.2

Polysaccharidases from Marine Environment

Enzyme Class	References
Agarases	Gupta et al. (2013); Yang et al. (2014); Seo et al. (2014); Liu et al. (2014)
Xylanases	Chen et al. (2013)
Chitinases	Halder et al. (2014)
κ-, ι-, and λ-carrageenases	Liu et al. (2011); Anastyuk et al. (2011)
Laminarinases	Wu (2014); Kumagai et al. (2014); Menshova et al. (2014); Kim et al. (2013); Kumagai and Ojima (2010); Cota et al. (2011)
Fucoidan-hydrolyzing enzymes	Silchenko et al. (2013, 2014); Yu et al. (2013, 2014); Descamps et al. (2006); Bilan et al. (2005); Rodriguez-Jasso et al. (2010)
Mannanases	Zahura et al. (2012)
Hyaluronidases	Madokoro et al. (2011); Kiriake et al. (2014)

the aspects that has thrust forward the research in this field. Glycosynthases are mutant glycosyl hydrolases with no hydrolytic activity that synthesize glycans in near quantitative yields. However, if in 2004, 11 different glycoside hydrolases from bacteria, eukarya, and archaea were listed as modified as efficient glycosynthases (Perugino et al. 2004), a quadrupled value was found in two successive independent reports, one from the same laboratory in 2012 (Cobucci-Ponzano and Moracci 2012) and another of 2011 from Wong group (Schmaltz et al. 2011). The glycosynthase approach is certainly of great help while keeping molecular diversity offered by different natural glycoside hydrolases, including features such as resistance to temperature and organic solvent. However, the obvious success is due to high yields that can be reached in transfer reactions. Using a thermophilic glycosynthase, the synthesis of target branched oligosaccharides was already investigated long time ago (Trincone et al. 2000). A list of natural carbohydrates prepared by glycosynthase approach has been published (Cobucci-Ponzano and Moracci 2012) with particular emphasis on products and possible applications. The cases of endo-β-glycosynthases are of particular interest. As for the mutant produced from an endo-β-glucanase (Fairweather et al. 2003; Hrmova et al. 2002) from barley, the biocatalyst was used to synthesize a series of polymeric glucans, by self-condensation of α-laminaribiosyl fluoride donor including branched examples by modulating the donor structures and using suitable acceptors. Other cases of endoenzymes are similarly of interest; they account for half the number of cases published. Preparative autocondensation of donor substrates (2- and 4-nitrophenyl β-D-glucopyranosides) and a series of transglycosylation reactions using a the thermophilic glycosynthase from *Sulfolobus solfataricus* were also studied in depth (Trincone et al. 2005). In the first case, branched oligoglucosides of up to four sugar units were formed with high productivity. In the transglycosylation reactions, 2-nitrophenyl β-D-glucopyranoside was used as the donor and different aryl and alkyl mono- and disaccharide substrates containing α- and β-glycosidic linkages were used as acceptors. Some interesting effects governed by the anomeric configuration and lipophilicity of hetero acceptors on the regioselectivity and yield of the reactions were found for the first time in glycosynthase-catalyzed reactions. More recently (Pozzo et al. 2014), three mutants of the enzyme β-glucosidase 1A from *Thermotoga neapolitana* (TnBgl1A E349G, TnBgl1A E349A, and TnBgl1A E349S), substituted at the catalytic nucleophile, were produced to evaluate their glycosynthase activity in the synthesis of oligosaccharides. Two known approaches were used for the syntheses, both utilizing 4-nitrophenyl β-D-glucopyranoside as an acceptor molecule: the classical one using an α-glucosyl fluoride donor at a low temperature (35°C) in a classical glycosynthase reaction or the one devised for *Sulfolobus* enzyme (Trincone et al. 2000) using the donor 4-nitrophenyl β-D-glucopyranoside in the presence of formate at a high temperature (70°C). β-1,3-Linkages were preferentially formed and flavonoid glycosylation was also possible.

13.7 Structures and Biological Activity

Nature of sugars, interglycosidic linkages connecting them, and substitutions of sugar hydroxyl functionalities with other groups (amine, ester, carboxylate, phosphate, or sulfate groups) constitute the structural bases for the variability of polysaccharides with a consequent unrivalled number of possible different combinations and chemical information that can be delivered in this manner (Turnbull and Field 2007). Most polysaccharides produced by marine microorganisms are heteropolysaccharides composed of repeating unit of several types of monosaccharides linked by glycosidic bonds. The monomers most commonly found in marine polysaccharides include hexose (glucose, galactose, and mannose), pentose (arabinose, xylose, and ribose), and deoxysugar (fucose and rhamnose).

As far as bioactivity of marine oligosaccharides is concerned, it is evident that more and more articles are appearing in literature in recent years. Few years ago, some authors summarized the bioactivity of oligosaccharides derived from seaweed in stimulating defense responses and protection against pathogens in plants. In particular, seaweed oligosaccharides were able to mediate to a series of actions, such as (1) expression of genes for antifungal and antibacterial protein production, (2) activation of other defense enzymes, and (3) production of active natural products (Vera et al. 2011), all converging toward enhanced protection against pathogens.

A concise summing-up of principal actions of marine oligosaccharides is reported in Table 13.3 (Ji et al. 2011). From the many ways of hypothesized molecular level of action, it is hard to explain how exactly these molecules exert their activity.

Different specific examples of low-MW marine oligosaccharides obtained by chemical or enzymatic methods can be found in literature. Their structural chemistry (agarans, carrageenans, ulvans, fucans) and their biological activities (anticoagulant/antithrombotic, antiviral, immunoinflammatory, antilipidemic, and antioxidant activities) are summarized in Table 13.4.

The general structural features of sulfated galactans from different marine organisms have been interestingly related to phylogenetic occurrence. As shown (Pomin 2011) in a hypothetical cladogram, Pomin reported that the 3-β-D-Galp-1 units are preserved among species of specific phyla in marine environment for green algae, red seaweeds, and marine sea grass (*Angiospermae*, *Spermatophyta*) and in invertebrates (sea urchins, clams, and tunicates) and vertebrates such as fish, which express keratan sulfate (basic repeating disaccharide unit is -3-Gal-β-1–4-GlcNAc-β-1-). However, as far as the sulfation sites of this 3-β-D-Galp-1 units are concerned, 2-sulfation for animals, 4-sulfation for plants (algae and marine angiosperms), and 6-sulfation for the dispersive distribution can be generalized. Conserved galactosyltransferases but variation in the activity of sulfotransferases result in the situation practically found in which basic backbones of polysaccharides from different

TABLE 13.3

Summing-Up of Principal Actions of Marine Oligosaccharides

Effect	Way of Action	Structures
Immuno-enhancing	Humoral immunity	Chitosan oligosaccharides, carrageenan oligosaccharides
	Cellular immunity	Laminarin oligosaccharides, carrageenan oligosaccharides, alginate oligosaccharides
Antitumor	Killing tumor cells	Carrageenan sulfated oligosaccharide fraction
	Inhibiting the growth of vessel	Chitosan oligosaccharides (DP2–7) and other marine-derived oligosaccharides, λ-carrageenan oligosaccharides, sulfated disaccharides, and neoagarotetraose
Anti-inflammatory	Suppression of inflammatory cytokines	Alginate oligosaccharides
Antioxidant	Radical scavenging activity	Chitosan oligosaccharides, alginate oligosaccharides, fucoidan oligosaccharides
	Regulation of enzyme level	Chitosan oligosaccharides, carrageenan oligosaccharides
Anticoagulation		Depolymerized marine dermatan sulfate, low-MW sulfated fucan, sulfated chitosan oligosaccharides
Antimicrobial and anti-infection	Antibacterial	Chitosan oligosaccharides
	Antiviral	Chitosan oligosaccharides, depolymerized *O*-acylated carrageenan
	Antifungal	Chitosan oligosaccharides

species are the same with great difference in the position of the sulfates. In an attempt to obtain low-MW derivatives from 112 sulfated fucooligosaccharides, results based on mild-acid hydrolysis were also reported (Pomin et al. 2005). Authors were able to prepare oligosaccharides with well-defined molecular size as shown by narrow bands in poly-acrylamide gel electrophoresis.

Among body fluids, the one rich in oligosaccharides is the milk. In an interesting article related to evolution of lactation in marine mammals, it has been stated that pro-found evolutionary divergence occurred among mammalian taxa resulting in a diversity of lactation patterns, milk composition, yield, duration of lactation, etc. As far as milk composition is concerned, it is of interest the notation about production of low levels of lactose and other more complex sugars in marine milks. The significance of oligosaccharides found is not clear, although by analogy to human milk oligosaccharides, antimicrobial functions have been hypothesized. Oligosaccharides found are up to octa- and nonasaccharides with lactose at the reducing as well as varying combinations of galactose, *N*-acetyl galactosamine, *N*-acetyl glucosamine, fucose, and sialic acid. Some marine mammals appear to produce milks devoid of lactose (Oftedal and Olav 2011; Oftedal et al. 2014).

Triterpene glycosides found in marine sponges are interesting compounds as far as the glycosidic part is concerned. Carbohydrates attached to the steroid aglycon are highly important for the biological activities as demonstrated by the recent attention of synthetic chemists in the preparation of the trisaccharides of the steroid glycoside Sokodoside B (Dasgupta et al. 2007). Glycosylation occurs at a late stage of the biosynthetic pathway but the roles of characteristic glycosyl transferases and/or glycosidases involved are not clear.

TABLE 13.4

Marine Oligosaccharides, Structures, and Activities

Biological Activity	Structural Chemistry	Organism	References
Anticoagulant	Oligosaccharides from fucoidan from *Nemacystus decipiens* MW of approx. 50,000	Purified fucoidanase from *Patinopecten yessoensis*	Kitamura et al. (1992)
Antithrombotic activity	Low-MW fucoidan of 8 kDa obtained by chemical degradation of a high-MW fraction	Fucoidan from *Ascophyllum nodosum*	Colliec-Jouault et al. (2003)
Antiviral	Highly acidic oligosaccharides composed by mono- and disulfated fucose units alternatively bound by α-1,4 and α-1,3 glycosidic linkages, with a mass below 10,000 obtained by a fucan-degrading hydrolase isolated from a marine bacterium	Fucoidan from *Pelvetia canaliculata*	Klarzynski et al. (2003)
Antitumor	Low-MW carrageenans (MW of 1,726) obtained with a carrageenase isolated from the cell-free medium of a culture of marine *Cytophaga* sp. MCA-2 sulfonated with formamide-chlorosulfonic acid (κ-neocarratetraose sulfate and κ-neocarrahexaose sulfate nature, respectively, by MALDI-TOF-MS and ^{13}C-NMR spectroscopy	Commercial carrageenans or polysaccharides extracted from marine red alga *Chondrus ocellatus*	Haijin et al. (2003)
Antivenom activity	Hydrolyzed low-MW fragments of 5 kDa (Boc-5) and 10 kDa (Boc-10) of sulfated galactans	Sulfated galactans from *Botryocladia occidentalis*	Toyama et al. (2010)
Immunostimulation	Low-MW fractions (below 20 kDa) obtained by free-radical depolymerization and mild-acid hydrolysis of a carrageenan, based mainly on 3,6-anhydrogalactose (DA unit), a (DA2S-G4S)-type structure	*Solieria chordalis*	Bondu et al. (2010)
Immunostimulation	Carrageenan-like oligosaccharides with an average MW of 428 kDa generated by acid hydrolysis of a polysaccharide: 1,4-linked 3,6-anhydrogalactose (40%); 1,3-linked 4-sulfated-galactose (30%); 1,3-linked galactose (20%); 1,4-linked galactose (8%); and 1,4-linked 3-*O*-methyl-galactose (2%)	*Furcellaria lumbricalis*	Yang et al. (2011)
Inhibition of phospholipase A2 enzymatic activity from *Crotalus durissus cascavella*	Low-MW sulfated galactan fragments of 5 or 10 kDa	Red algae *Botryocladia occidentalis*	Toyama et al. (2010)
Antitumor effects by promoting the immune system	Carrageenan oligosaccharides MW of 681 and 798 Da	*Kappaphycus striatum*	Hu et al. (2006)

In a recent excellent review, literature content about glycosides from sponges (Porifera, *Demospongiae*) is reviewed (Kalinin et al. 2012). Biological roles listed for these compounds include defensive role against predatory fish, involvement in the attachment by biofilm forming bacteria, fouling by invertebrates, and other chemical interactions among organisms in field and laboratory assays.

The antioxidant activities of three types of marine oligosaccharides (alginate, chitosan, and fucoidan) were investigated using several antioxidant assays with mixtures of compounds of average molecular weight (MW) of ca. 5000 Da corresponding to 20–25 monosaccharide units, which were obtained by enzymatic hydrolysis of corresponding polymers. Intriguingly, the results showed that these oligosaccharides exhibited different activities in various assays (Wang et al. 2007) in relation to their structures.

Attenuation of neurotoxicity induced by amyloid protein and hydrogen superoxide in human neuroblastoma cells has been reported for acidic oligosaccharide obtained from brown algae *Ecklonia kurome* by depolymerization. Amino and hydroxyl groups attached to free positions of the pyranose rings can react with unstable free radicals to form stable macromolecule chelating metal ions. Alginate-derived oligosaccharides of 373–571 Da and chito-oligosaccharides of 855–1671 Da obtained by enzymolysis with alginate lyase and chitosanase, respectively, were investigated for cell regulation, erythrocytes hemolysis inhibition, and antioxidant capacity (Hu et al. 2004).

Others stressed on the anti-UV radiation potential of both alginate-derived oligosaccharides and chito-oligosaccharides and discussed the potential for development of UV radiation protector agent in the area of functional foods (He et al. 2013).

Particular attention has been reserved to fucoidans for complex structures and for bioactivity they revealed. Studies to improve polymer extraction efficiency are of current interest (Rodriguez-Jasso et al. 2011) as well as those related to enzymes (Table 13.2) as the fucoidanase reported from *Lambis* possessing endo-1,4 cleaving-type action (Silchenko et al. 2014) that will be of interest as a tool for the structure determination and for the access to specific products.

The case of carrageenans is interesting with their action on the immune system, as antiviral and as antiproliferative. They are considered structural analogs of animal sulfated polysaccharides found in extracellular matrix (heparin, chondroitin, keratan, dermatan), hence exhibiting bioactivity related to some functions exerted by these GAGs (Carlucci et al. 2012). Their potent HSV-1 and HSV-2 inhibition is related to their analogy to heparan sulfate, which is the primary receptor for adsorption of HSV in the cells. Blocking the interaction virus-HS is the principal mode of action of carrageenans.

13.8 Landscape of the Literature in 2014

In addition to the earlier examples in Table 13.4, in this paragraph, a selection of the most important articles appeared in 2014 on the subject of this chapter is reported. After a brief consideration about the topical importance of safety issue related to the use of these molecules, a list of articles is commented without grouping them under any category; however, molecular interactions, new enzymes, and other applications are the selected areas of interest.

An important aspect of the use of oligosaccharides is their safety level; to use them as a biofunctional ingredient, a complete safety assessment is important for establishing the level of no observed adverse effect. This has been recently addressed in the case of low-MW chitosans obtained by enzymatic hydrolysis with a chitosanase, studying the effect on reverting renal lesions in mice. The oligosaccharides are considered to be safe at 1.0 g/kg in rats that is equivalent to a daily dose of 0.16 g in humans (Chang et al. 2014).

Oligosaccharides or low-MW fragments obtained by means of degradation can be used more advantageously for a number of reasons including increasing water solubility, better absorption, and increased bioactivity. In a recent case, antioxidant activities of low-MW fragments originated by degradation of polysaccharides from marine microalgae *Pavlova viridis* and *Sarcinochrysis marina* were studied. Polysaccharide from *P. viridis* was composed of rhamnose, D-fructose, glucose, and mannose. However, sulfate content and monosaccharide composition changed according to different chemical degradation systems. Interestingly, the results indicated that free-radical scavenging, inhibition of lipid peroxidation, and hemolysis of red blood cells confirmed effective antioxidant activity of both native polysaccharide and low-MW fragments, the latter with increased activity. In specific cases, low-MW fragments hold a promise as potential antioxidant additives for the food or pharmaceutical industry (Sun et al. 2014).

With a similar rationale, oligosaccharides derived from *Laminaria japonica* were prepared by hydrolysis with H_2O_2. The resulting product contained 94.82% sugar, of which the average degree of polymerization was approximately 8, and showed high hydroxyl radical scavenging activity (91.31%) at the concentration of 100 µg/mL (Wu 2014). The topic is of interest, in fact related enzymes with endo-β-1,3-glucanases (laminarinase, EC 3.2.1.6) from marine environment are currently investigated. An enzyme found in the digestive fluid of a marine gastropod *Aplysia kurodai* has been recently reported for the complete hydrolysis of laminarin to glucose. Interestingly, high transglycosylation activity has been evidenced; thus, it can produce a series of laminarioligosaccharides larger than laminaritetraose from the donor substrate (laminaribiose) and the acceptor (laminaritriose) (Kumagai et al. 2014).

It is known that a combination of antibiotics treatments with the use of marine oligosaccharides leads to an increase in health-promoting benefits. Both alginate-derived oligosaccharides and chito-oligosaccharides are utilized in food. As related to their safety, chitosan is "generally recognized as safe" in many countries (FDA Code of Federal Regulation) and sodium or potassium alginate salts are recognized as safe products. In a recent paper, it has been shown that alginate-derived oligosaccharides and chito-oligosaccharides in conjunction with antibiotics can effectively inhibit the growth of wild-type and resistant *Pseudomonas aeruginosa*. In this case, the preparation of oligosaccharides was based on the use of alginate lyase and chitosanase as described by the same authors (He et al. 2013). The exact mechanism of the antimicrobial activity remains to be elucidated although a role in interfering with bacterial quorum sensing systems seems to be evident. Moreover, a functional role was assigned to charged functional groups in the oligosaccharides for the interaction with the positively/negatively charged cell membrane components; this can lead to the damage of membrane protection with leakage of intracellular constituents (He et al. 2014). Indeed, in a recent excellent review article, the group of Prof. Pomin, which is investigating many sulfated fucans, galactans, and GAGs during the last 25 years, concluded that the electronegative-charge density (sulfation) is not the sole structural determinant for anticoagulant effect (Pomin and Mourão 2014). Sulfate position and interglycosidic linkage are combined elements that seem to be correlated to specific activity of these molecules. Furthermore, the structural effects on the anticoagulant activity and cytotoxicity of two seaweed polysaccharides (agarose and carrageenan) and their sulfated degradation products were very recently investigated confirming that the substitution position of sulfate groups, rather than the substitution degree, showed the biggest impact on bioactivities (Liang et al. 2014). The secondary structures of glycans hence play a key role in biological activities, and these molecules could be seriously considered in biomedical applications after carefully tailoring the sulfate groups.

As known, the fucoidans from algae are better studied than fucan from sea cucumber whose structure was studied only in few species.

After producing low-MW preparations by enzymatic degradation, authors reported the structure from sea cucumber *Acaudina molpadioides* as composed by a tetrafucose repeating unit with 1–3-linkages and a distinctive sulfation pattern as indicated in Figure 13.5. The oligosaccharides were prepared by employing a partially purified enzyme of marine bacterial strain *Flavobacteriaceae* CZ1127. The presence of a sulfatase activity cleaving the sulfate esters in the partially purified preparation rendered difficult to deduce the structure of the originating polymer from that of the oligosaccharides (Yu et al. 2014).

As for other cases of unique structures, the report of an extracellular polysaccharide obtained from the fermented broth of the fungus *Aspergillus ochraceus* derived from coral *Dichotella gemmacea* seems interesting. Marine originating galactomannans generally consist of 1,2-linked α-mannopyranoses as main chain with differences in the side chains of galactofuranoses. The structure of this new polysaccharide was investigated by chemical and spectroscopic methods, including methylation analysis, 1D and 2D nuclear magnetic resonance, and electrospray mass spectrometry with collision-induced dissociation spectroscopy. The results showed a usual 1,2-linked α-mannopyranoses as main chain with two positions on the mannoses substituted by an additional α-mannopyranose unit (at C-6), and a 1,5-linked galactofuranose oligosaccharide of various length of which this fungus is seen as a convenient source (Guo et al. 2014). A very

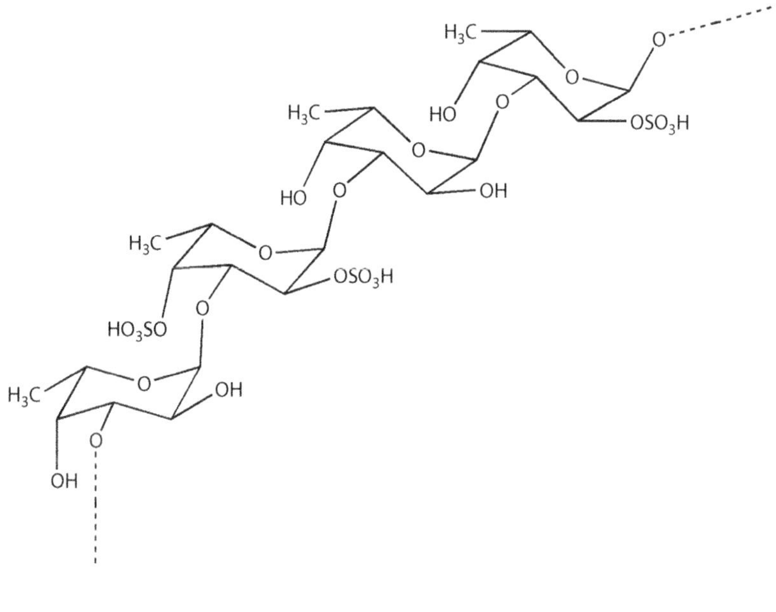

FIGURE 13.5
Tetrafucose repeating unit in fucoidan from sea cucumber *Acaudina molpadioides*.

interesting complete assessment on recent knowledge related to structure, synthesis, and applications of hexofuranosides recently appeared (Peltier et al. 2008). As regarding the structural determination by modern mass spectroscopy techniques, a recent review article providing an overview of these approaches specific for marine oligosaccharides is found in the literature of 2014. Technical challenges for marine oligosaccharides are discussed and examples include agaran, carrageenan, alginate, sulfated fucan, chitosan, GAG, and GAG-like polysaccharides (Lang et al. 2014).

Another outstanding example of preparation and detailed NMR analyses for a series of oligo-carrageenans recently appeared. Tetra- and hexasaccharides or di- and tetrasaccharides as end products were prepared by the enzymatic action of two different ι-carrageenases, and desulfation was possible using 4S-carrageenan sulfatase. Enzymatic production of a wide range of standard and hybrid oligo-carrageenans with a controlled distribution of sulfate ester groups was possible, and full NMR data were acquired for these molecules other than envisioning potential applications and studying of biological activity (Préchoux and Helbert 2014).

Hyaluronidases of marine origin are interesting biocatalysts. This enzyme class is capable of the hydrolysis of hyaluronic acid (HA), a biopolymer abundant in extracellular matrices and widely used in human health and cosmetics (produced from *Streptococcus equi*). The study of hydrolysis-transfer reactions of hyaluronidases is of interest for the identification of new selectivities in biocatalytic steps as a tool for the manipulation of HA structure. Bovine enzyme is the commercial representative of hyaluronidases, but new examples are found in other nature environments. The venoms of two classes of fish, the freshwater stingray (members of the genus *Potamotrygon*) and the stonefish (members of the genus *Synanceia*), contain hyaluronidases that are considered spreading factors facilitating the diffusion of toxins in the tissues, by degrading hyaluronan. Owing to their quick action, these biocatalysts can possess very interesting features for biocatalytic manipulation of carbohydrate-related molecules (Kiriake et al. 2014; Madokoro et al. 2011) also in view of bioactivity expressed by HA oligosaccharides (Ariyoshi et al. 2012). As regards this interest for HA and related enzymes, it is important to mention the characterization of a fish originating HA (Charalampos et al. 2014). Heparin used in pharmaceutical field is derived from mammals (mucosal tissues); for chondroitin sulfate, this is true as well, although semisynthetic processes are being of interest for its commercialization. However, bacterial fermentation technology for producing HA is also of current interest. The GAGs present in the dorsal hump of the female fish *Cyclopterus lumpus* were isolated and purified, and their structure was determined by NMR. Characterization using biochemical and biological cell-based assays is also reported. The presence of galactosamine sulfated in four and six positions was confirmed by analyzing the disaccharides formed by digestion of the sample using either chondroitinase ABC lyase or chondroitinase B or heparinase II (Grampian enzymes) under recommended buffer conditions.

It is known that cyanobacterial EPSs have a complex structure with a long list of various monosaccharides and a different range of linkage types. Although investigated in detail in the past (De Philippis and Vincenzini 1998), interest for new structures is still high and recently EPSs of *Synechocystis aquatilis* were shown to be sulfated arabinofucans containing *N*-acetyl-fucosamine. The special characteristics of these novel structures are numerous and include high sulfation, presence of rare *N*-acetyl-fucosamine, and low glucose content. The NMR analyses of the polymer from *S. aquatilis* led to a proposed linear structure with side chains of single sugar units. It is of interest the similarity of structures with fucoidan for accordance in linearity, linkage types of fucose residues and sulfation pattern, and expectations of useful biological activities (Flamm and Blaschek 2014).

13.9 Summary

Apart the structural role of complex carbohydrates in the cells, involvement of such molecules in physiological and/or pathological events is undeniable. However, technical problems still hinder a complete multidisciplinary biomolecular approach in the study of these molecules. With a chemical diversity strongly underpinning their molecular action, the lack of a sequencing tool such as in proteomics or genomics, for example, is the most important shortage in this field. Nonetheless glycobiology continues to be a major research field allowing more and more the comprehension of molecular level of human diseases and the discovery of novel therapeutic compounds, in these years. The big wave of these successes is present in the analysis of literature as found in Figure 13.3, also reflecting the marine domain. With these data, not only a foresight is possible for the next years, related to an expected increase of studies about marine oligosaccharides, but also—as detailed in this chapter—the situation is multifaceted including different elements such as sustainable access by specific and high-yield enzymatic hydrolysis, new enabling methodologies for straightforward purification techniques, and fine structural detail information including secondary structures, activity assay, and mechanism of action of these molecules. The structural inherent complexity of carbohydrate molecules greater than proteins and nucleic acids adds challenges for the control of pharmaceutical preparations based on such complex molecules, and most works can be expected for widening current limitations for the use of these molecules as ingredients in food supplements and cosmetics. As far as polysaccharides from marine microorganisms are concerned, the production by fermentation in laboratory conditions will make a stable supply, possibly avoiding influences by climatic, seasonal, and/or physiological variations.

Another important aspect resulting from the earlier analysis is the worldwide interest in this research field. In Western countries, and especially in Europe, the research funding Horizon 2020 programs with its starting enthusiasm focused on marine domain will make possible big research efforts to unlock potential of the sea and oceans. In the Eastern countries, traditional interests for these biomasses in food technology are in a renewing era, with the discovery of therapeutic potential of constituent carbohydrate molecules; thus, flourishing of molecular technologies suitable for interdisciplinary study of these molecules will be also possible.

Acknowledgment

The support for bibliographic search facilities is provided by Consiglio Nazionale delle Ricerche, Italy.

References

Anastyuk, S.D., A.O. Barabanova, G. Correc, E.L. Nazarenko, V.N. Davydova, W. Helbert, P.S. Dmitrenok, and I.M. Yermaka 2011 Analysis of structural heterogeneity of κ/β-carrageenan oligosaccharides from *Tichocarpus crinitus* by negative-ion ESI and tandem MALDI mass spectrometry. *Carbohydr. Polym.* 86:546–554.

Ariyoshi, W., N. Takahashi, H.D. Knudson, and W. Knudson 2012 Mechanisms involved in enhancement of the expression and function of aggrecanases by hyaluronan oligosaccharides. *Arthritis Rheum.* 64:187–197.

Barra, L.R., R. Chandrasekaran, F. Corato, and C. Brunet 2014 The challenge of ecophysiological biodiversity for biotechnological applications of marine microalgae. *Mar. Drugs* 12:1641–1675.

Beier, S. and S. Bertilsson 2013 Bacterial chitin degradation-mechanisms and ecophysiological strategies. *Front. Microbiol.* 4:149.

Bilan, M.I., M.I. Kusaykin, A.A. Grachev, E.A. Tsvetkova, T.N. Zvyagintseva, N.E. Nifantiev, and A.I. Usov 2005 Effect of enzyme preparation from the marine mollusk *Littorina kurila* on fucoidan from the brown alga *Fucus distichus*. *Biochemistry* (*Moscow*) 70:1321–1326.

Bondu, S., E. Deslandes, M.S. Fabre, C. Berthou, and G. Yu 2010 Carrageenan from *Soliera chordalis* (Gigartinales): Structural analysis and immunological activities of the low molecular weight fractions. *Carbohydr. Polym.* 81:448–460.

Carlucci, M.J., C.G. Mateu, M.C. Artuso, and L.A. Scolaro 2012 Polysaccharides from red algae: Genesis of a renaissance. In: *The Complex World of Polysaccharides*, Dr. D.N. Karunaratne (ed.). ISBN: 978-953-51-0819-1, InTech, doi: 10.5772/48107. 535–554. Available from: http://www.intechopen.com/books/the-complex-world-of-polysaccharides/polysaccharides-from-red-algae-genesis-of-a-renaissance.

Chang, Y.M., Y.J. Lee, J.W. Liao, J.K. Jhan, C.T. Chang, and Y.C. Chung 2014 In vitro and in vivo safety evaluation of low molecular weight chitosans prepared by hydrolyzing crab shell chitosans with bamboo shoots chitosanase. *Food Chem. Toxicol.* 71:10–16.

Charalampos, G., C.G. Panagos, D. Thomson, C. Moss, C.D. Bavington, H.G. Ólafsson, and D. Uhrín 2014 Characterisation of hyaluronic acid and chondroitin/dermatan sulfate from the lump suckerfish, *C. lumpus. Carbohydr. Polym.* 106:25–33.

Chen, S., M.G. Kaufman, K.L. Miazgowicz, M. Bagdasarian, and E.D. Walker 2013 Molecular characterization of a cold-active recombinant xylanase from *Flavobacterium johnsoniae* and its applicability in xylan hydrolysis. *Bioresour. Technol.* 128:145–155.

Cobucci-Ponzano, B. and M. Moracci 2012 Glycosynthases as tools for the production of glycan analogs of natural products. *Nat. Prod. Rep.* 29:697–709.

Colliec-Jouault, S., J. Millet, D. Helley, C. Sinquin, and A.M. Fischer 2003 Effect of low-molecular-weight fucoidan on experimental arterial thrombosis in the rabbit and rat. *J. Thromb. Haemost.* 1:1114–1115.

Cota, J., T.M. Alvarez, A.P. Citadini, C.R. Santos, M. de Oliveira Neto, R.R. Oliveira, G.M. Pastore et al. 2011 Mode of operation and low-resolution structure of a multidomain and hyperthermophilic endo-β-1,3-glucanase from *Thermotoga petrophila*. *Biochem. Biophys. Res. Commun.* 406:590–594.

Dasgupta, S., K. Pramanik, and B. Mukhopadhyay 2007 Oligosaccharides through reactivity tuning: Convergent synthesis of the trisaccharides of the steroid glycoside Sokodoside B isolated from marine sponge *Erylus placenta. Tetrahedron* 63:12310–12316.

De Angelis, P.L., J. Liu, and R.J. Linhardt 2013 Chemoenzymatic synthesis of glycosaminoglycans: Re-creating, re-modeling and re-designing nature's longest or most complex carbohydrate chains. *Glycobiology* 23:764–777.

de Jesus Raposo, M.F., R.M.S.C. de Morais, and A.M.M.B. de Morais 2013 Bioactivity and applications of sulphated polysaccharides from marine microalgae. *Mar. Drugs* 11:233–252.

De Marino, S., M. Iorizzi, F. Zollo, C.D. Amsler, S.P. Greer, and J.B. McClintock 2000 Starfish saponins LVI—Three new asterosaponins from the starfish *Goniopecten demonstrans. Eur. J. Org. Chem.* 24: 4093–4098.

De Philippis, R. and M. Vincenzini 1998 Exocellular polysaccharides from cyanobacteria and their possible application. *FEMS Microbiol. Rev.* 22:151–175.

Descamps, V., S. Colin, M. Lahaye, M. Jam, C. Richard, P. Potin, T. Barbeyron, J.C. Yvin, and B. Kloareg 2006 Isolation and culture of a marine bacterium degrading the sulfated fucans from marine brown algae. *Mar. Biotechnol.* 8:27–39.

Dismukes, G.C., D. Carrieri, N. Bennette, G.M. Ananyev, and M.C. Posewitz 2008 Aquatic phototrophs: Efficient alternatives to land-based crops for biofuels. *Curr. Opin. Biotechnol.* 19:235–240.

Fairweather, J.K., M. Hrmova, S.J. Rutten, G.B. Fincher, and H. Driguez 2003 Synthesis of complex oligosaccharides by using a mutated (1,3)-β-D-glucan endohydrolase from barley. *Chem. Eur. J.* 9:2603–2610.

Fedorov, S.N., S.P. Ermakova, T.N. Zvyagintseva, and V.A. Stonik 2013 Anticancer and cancer preventive properties of marine polysaccharides: Some results and prospects. *Mar. Drugs* 11:4876–4901.

Finore, I., P. Di Donato, V. Mastascusa, B. Nicolaus, and A. Poli 2014 Fermentation technologies for the optimization of marine microbial exopolysaccharide production. *Mar. Drugs* 12:3005–3024.

Fitton, J.H. 2011 Therapies from fucoidan; multifunctional marine polymers. *Mar. Drugs* 9(10):1731–1760.

Flamm, D. and W. Blaschek 2014 Exopolysaccharides of *Synechocystis aquatilis* are sulfated arabino-fucans containing N-acetyl-fucosamine. *Carbohydr. Polym.* 101:301–306.

Giordano, A., G. Andreotti, A. Tramice, and A. Trincone 2006 Marine glycosyl hydrolases in the hydrolysis and synthesis of oligosaccharides. *Biotechnol. J.* 1:511–530.

GlycoMar Limited 2012 Oligosaccharides in drug discovery Argyll Scotland. www.glycomar. om. http://www.drugdevelopment-technology.com/contractors/clinical_development/glyco-mar/. (Accessed March 2016).

Guo, S., W. Mao, M. Yana, C. Zhao, N. Li, J. Shan, C. Lin, X. Liu, T. Guo, T. Guo, and S. Wang 2014 Galactomannan with novel structure produced by the coral endophytic fungus *Aspergillus ochraceus*. *Carbohydr. Polym.* 105:325–333.

Gupta, V., N. Trivedi, M. Kumar, C.R.K. Reddy, and B. Jha 2013 Purification and characterization of exo-β-agarase from an endophytic marine bacterium and its catalytic potential in bioconversion of red algal cell wall polysaccharides into galactans. *Biomass Bioenerg.* 49:290–298.

Haijin, M., J. Xiaolu, and G. Huashi 2003 A κ-carrageenan derived oligosaccharide prepared by enzymatic degradation containing anti-tumor activity. *J. Appl. Phycol.* 15:297–303.

Halder, S.K., C. Maity, A. Jana, K. Ghosh, A. Das, T. Paul, P.K. Das Mohapatra, B.R. Pati, and K.C. Mondal 2014 Chitinases biosynthesis by immobilized *Aeromonas hydrophila* SBK1 by prawn shells valorization and application of enzyme cocktail for fungal protoplast preparation. *J. Biosci. Bioeng.* 117:170–177.

Haug, A. and S. Myklestad 1976 Polysaccharides of marine diatoms with special reference to *Chaetoceros* species. *Mar. Biol.* 34:217–222.

He, X., H. Hwang, W.G. Aker, P. Wang, Y. Lin, X. Jiang, and X. He 2014 Synergistic combination of marine oligosaccharides and azithromycin against *Pseudomonas aeruginosa*. *Microbiol. Res.* 169(9–10):759–767.

He, X., R. Li, G. Huang, H. Hwang, and X. Jiang 2013 Influence of marine oligosaccharides on the response of various biological systems to UV irradiation. *J. Funct. Foods* 5:858–868.

Henricus, J.V., J.P. Kamerling, and J.F.G. Vliegenthart 2000 Synthesis and conjugation of a sulfated disaccharide involved in the aggregation process of the marine sponge *Microciona prolifera*. *Tetrahedron Asymm.* 11:539–547.

Hrmova, M., T. Imai, S.J. Rutten, J.K. Fairweather, L. Pelosi, V. Bulone, H. Driguez, and G.B. Fincher 2002 Mutated barley (1,3)-β-D-glucan endohydrolases synthesize crystalline (1,3)-β-D-glucans. *J. Biol. Chem.* 277:30102–30111.

Hu, J., M. Geng, J. Li, X. Xin, J. Wang, M. Tang, J. Zhang, X. Zhang, and J. Ding 2004 Acidic oligosaccharide sugar chain, a marine-derived acidic oligosaccharide, inhibits the cytotoxicity and aggregation of amyloid beta protein. *J. Pharmacol. Sci.* 95:248–255.

Hu, X., X. Jiang, E. Aubree, P. Boulenguer, and A.T. Critchley 2006 Preparation and in vivo antitumor activity of kappa-carrageenan oligosaccharides. *Pharm. Biol.* 44:646–650.

Ji, J., L.C. Wang, H. Wu, and H. Luan 2011 Bio-function summary of marine oligosaccharides. *Int. J. Biol.* 3(1):74–86.

Kalinin, V.I., N.V. Ivanchina, V.B. Krasokhin, T.N. Makarieva, and V.A. Stonik 2012 Glycosides from marine sponges (Porifera, Demospongiae): Structures, taxonomical distribution, biological activities and biological roles. *Mar. Drugs* 10:1671–1710.

Kim, E.J., A. Fathoni, G.T. Jeong, H.D. Jeong, T.J. Nam, I.S. Kong, and J.K. Kim 2013 *Microbacterium oxydans*, a novel alginate-and laminarin-degrading bacterium for the reutilization of brown-seaweed waste. *J. Environ. Manage.* 130:153–159.

Kiriake, A., M. Madokoro, and K. Shiomi 2014 Enzymatic properties and primary structures of hyaluronidases from two species of lionfish (*Pterois antennata* and *Pterois volitans*). *Fish Physiol. Biochem.* doi 10.1007/s10695-013-9904-5. 40(4):1043–1053.

Kitamura, K., M. Matsuo, and T. Tsuneo 1992 Enzymic degradation of fucoidan by fucoidanase from the hepatopancreas of *Patinopecten yessoensis*. *Biosci. Biotechnol. Biochem.* 56(3):490–494.

Klarzynski, O., V. Descamps, B. Plesse, J.C. Yivin, B. Kloareg, and B. Fritig 2003 Sulfated fucan oligosaccharides elicit defense responses in tobacco and local and systemic resistance against tobacco mosaic virus. *Mol. Plant Microbe Interact.* 16:115–122.

Kumagai, Y., T. Satoh, A. Inoue, and T. Ojima 2014 A laminaribiose-hydrolyzing enzyme, AkLab, from the common sea hare *Aplysia kurodai* and its transglycosylation activity. *Comp. Biochem. Physiol. Part B* 167:1–7.

Kumagai, Y. and T. Ojima 2010 Isolation and characterization of two types of β-1,3-glucanases from the common sea hare *Aplysia kurodai*. *Comp. Biochem. Physiol. B Biochem. Mol. Biol.* 155:138–144.

Kusaykin, M., I. Bakunina, V. Sova, S. Ermakova, T. Kuznetsova, N. Besednova, T. Zaporozhets, and T. Zvyagintseva 2008 Structure, biological activity, and enzymatic transformation of fucoidans from the brown seaweeds. *Biotechnol. J.* 3:904–915.

Lang, Y., X. Zhao, L. Liu, and G. Yu 2014 Applications of mass spectrometry to structural analysis of marine oligosaccharides. *Mar. Drugs* 12:4005–4030.

Liang, W., X. Mao, X. Peng, and S. Tang 2014 Effects of sulfate group in red seaweed polysaccharides on anticoagulant activity and cytotoxicity. *Carbohydr. Polym.* 101:776–785.

Liu, G., Y. Li, Z. Chi, and Z.M. Chi 2011 Purification and characterization of κ-carrageenase from the marine bacterium *Pseudoalteromonas porphyrae* for hydrolysis of κ-carrageenan. *Proc. Biochem.* 46:265–271.

Liu, N., X. Mao, Z. Du, B. Mu, and D. Wei 2014 Cloning and characterization of a novel neoagarotetraose-forming β-agarase, AgWH50A from *Agarivorans gilvus* WH0801. *Carbohydr. Res.* 388:147–151.

Lühn, S., J.C. Grimm, and S. Alban 2014 Simple and rapid quality control of sulfated glycans by a fluorescence sensor assay exemplarily developed for the sulfated polysaccharides from red algae *Delesseria sanguine*. *Mar. Drugs* 12:2205–2227.

Madokoro, M., A. Ueda, A. Kiriake, and K. Shiomi 2011 Properties and cDNA cloning of a hyaluronidase from the stonefish *Synanceia verrucosa* venom. *Toxicon* 58:285–292.

Menshova, R.V., S.P. Ermakova, S.D. Anastyuk, V.V. Isakov, Y.V. Dubrovskaya, M.I. Kusaykin, B.H. Um, and T.N. Zvyagintseva 2014 Structure, enzymatic transformation and anticancer activity of branched high molecular weight laminaran from brown alga *Eisenia bicyclis*. *Carbohydr. Polym.* 99:101–109.

Nicolaus, B., M. Kambourova, and E.T. Oner 2010 Exopolysaccharides from extremophiles: From fundamentals to biotechnology. *Environ. Technol.* 31(10):1145–1158.

Oftedal, O.T. 2011 Milk of marine mammals. In: *Encyclopedia of Dairy Science*, Fuquay, J.W., Fox, P.F. and McSweeney, P.L.H. (eds.), pp. 563–580. San Diego, CA: Academic Press.

Oftedal, O.T., R. Eisert, and G.K. Barrell 2014 Comparison of analytical and predictive methods for water, protein, fat, sugar, and gross energy in marine mammal milk. *J. Dairy Sci.* 97:4713–4732.

Peltier, P., R. Euzen, R. Daniellou, C. Nugier-Chauvin, and V. Ferrieres 2008 Recent knowledge and innovations related to hexofuranosides: Structure, synthesis and applications. *Carbohydr. Res.* 343:1897–1923.

Perugino, G., A. Trincone, M. Rossi, and M. Moracci 2004 Oligosaccharide synthesis by glycosynthases. *Trends Biotechnol.* 22:31–37.

Pomin, V.H. 2011 Structure and use of algal sulfated fucans and galactans. In: *Handbook of Marine Macroalgae: Biotechnology and Applied Phycology*, S.-K. Kim (ed.), pp. 229–261. Chichester, U.K.: John Wiley & Sons, Ltd.

Pomin, V.H. 2012 Fucanomics and galactanomics: Marine distribution, medicinal impact, conceptions, and challenges. *Mar. Drugs* 10:793–811.

Pomin, V.H. and P.A.S. Mourão 2014 Specific sulfation and glycosylation a structural combination for the anticoagulation of marine carbohydrates. *Front. Cell. Infect. Microbiol.* 6(4):33.

Pomin, V.H., A.P. Valente, M.S. Pereira, and P.A.S. Mourão 2005 Mild acid hydrolysis of sulfated fucans: A selective 2-desulfation reaction and an alternative approach for preparing tailored sulfated oligosaccharides. *Glycobiology* 15:1376–1385.

Pozzo, T., M. Plaza, J. Romero-García, M. Faijesc, E.N. Karlsson, and A. Planas 2014 Glycosynthases from *Thermotoga neapolitana* β-glucosidase 1A: A comparison of α-glucosyl fluoride and in situ-generated α-glycosyl formate donors. *J. Mol. Cat. B Enzym.* 107:132–139.

Préchoux, A. and W. Helbert 2014 Preparation and detailed NMR analyses of a series of oligo-α-carrageenans *Carbohydr. Polym.* 101:864–887.

Qiu, X. and S. Hu 2013 "Smart" materials based on cellulose: A review of the preparations, properties, and applications. *Materials* 6(3):738–781.

Rinaudo, M. 2007 Seaweed polysaccharides. In: *Comprehensive Glycoscience from Chemistry to Systems Biology Vol. 2: Analysis of Glycans; Polysaccharide Functional Properties*, H. Kamerling (ed.), pp. 691–735. Oxford, U.K.: Elsevier.

Rioux, L.E., S.L. Turgeon, and M. Beaulieu 2010 Structural characterization of laminaran and galactofucan extracted from the brown seaweed *Saccharina longicruris*. *Phytochemistry* 71:1586–1595.

Rodriguez-Jasso, R.M., S.I. Mussatto, L. Pastrana, C.N. Aguilar, and J.A. Teixeira 2010 Fucoidan-degrading fungal strains: Screening, morphometric evaluation, and influence of medium composition. *Appl. Biochem. Biotechnol.* 162:2177–2188.

Rodriguez-Jasso, R.M., S.I. Mussatto, L. Pastrana, C.N. Aguilar, and J.A. Teixeira 2011 Microwave-assisted extraction of sulfated polysaccharides (fucoidan) from brown seaweed. *Carbohydr. Polym.* 86(3):1137–1144.

Satpute, S.K., I.M. Banat, P.K. Dhakephalkar, A.G. Banpurkar, A. Balu, and B.A. Chopade 2010 Biosurfactants, bioemulsifiers and exopolysaccharides from marine microorganisms. *Biotechnol. Adv.* 28:436–450.

Schmaltz, R.M., S.R. Hanson, and C.-H. Wong 2011 Enzymes in the synthesis of glycoconjugates. *Chem. Rev.* 111:4259–4307.

Senni, K., J. Pereira, F. Gueniche, C. Delbarre-Ladrat, C. Sinquin, J. Ratiskol, G. Godeau, A.M. Fischer, D. Helley, and S. Colliec-Jouault 2011 Marine polysaccharides: A source of bioactive molecules for cell therapy and tissue engineering *Mar. Drugs* 9:1664–1681.

Seo, Y.B., Y. Lu, W.J. Chi, H.R. Park, K.J. Jeong, S.K. Hong, and Y.K. Chang 2014 Heterologous expression of a newly screened β-agarase from *Alteromonas* sp. GNUM1 in *Escherichia coli* and its application for agarose degradation. *Proc. Biochem.* 49:430–436.

Silchenko, A.S., M.I. Kusaykin, V.V. Kurilenko, A.M. Zakharenko, V.V. Isakov, T.S. Zaporozhets, A.K. Gazha, and T.N. Zvyagintseva 2013 Hydrolysis of fucoidan by fucoidanase isolated from the marine bacterium *Formosa algae*. *Mar. Drugs* 11:2413–2430.

Silchenko, A.S., M.I. Kusaykin, A.M. Zakharenko, R.V. Menshova, H.H.N. Khanh, P.S. Dmitrenok, V.V. Isakov, and T.N. Zvyagintseva 2014 Endo-1,4-fucoidanase from Vietnamese marine mollusk *Lambis* sp. which producing sulphated fucooligosaccharides *J. Mol. Cat. B Enzym.* 102:154–160.

Stengel, D.B., S. Connan, and Z.A. Popper 2011 Algal chemodiversity and bioactivity: Sources of natural variability and implications for commercial application. *Biotechnol. Adv.* 29:483–501.

Sun, L., L. Wang, J. Li, and H. Liu 2014 Characterization and antioxidant activities of degraded poly-saccharides from two marine *Chrysophyta*. *Food Chem*. 160:1–7.

Toyama, M.H., D.O. Toyama, V.M. Torres, G.C. Pontes, W.R.I. Farias, F.R. Melo, S.C.B. Oliveira, F.H.R. Fagundes, E.B.S. Diz Filho, and B.S. Cavada 2010 Effects of low molecular weight sulfated galactan fragments from *Botryocladia occidentalis* on the pharmacological and enzymatic activity of Spla2 *Crotalus durissus cascavella*. *Protein J*. 29(8):567–571.

Trincone, A. (ed.) 2013a *Marine Enzymes for Biocatalysis: Sources, Biocatalytic Characteristics and Bioprocesses of Marine Enzymes*. Biomedicine, Series No. 38. Cambridge, U.K.: Woodhead Publishing. http://www.woodhead publishing.com/en/book. aspx?bookID=2822.

Trincone, A. 2013b Angling for uniqueness in enzymatic preparation of glycosides. *Biomolecules* 3:334–350.

Trincone, A. 2014 Molecular fishing: Marine oligosaccharides. *Front. Mar. Sci*. 1:1–5.

Trincone, A., A. Giordano, G. Perugino, M. Rossi, and M. Moracci 2005 Highly productive auto-condensation and transglycosylation reactions with *Sulfolobus solfataricus* glycosynthase. *ChemBioChem* 6:1431–1437.

Trincone, A., G. Perugino, M. Rossi, and M. Moracci 2000 A novel thermophilic glycosynthase that effects branching glycosylation. *Bioorg. Med. Chem. Lett*. 10:365–368.

Turnbull, J.E. and R.A. Field 2007 Emerging glycomics technologies. *Nat. Chem. Biol*. 3:74–77.

Venugopal, V. 2011 *Marine Polysaccharides Food Applications*. Boca Raton, FL: Taylor & Francis Group.

Vera, J., J. Castro, A. Gonzalez, and A. Moenne 2011 Seaweed polysaccharides and derived oligosac-charides stimulate defense responses and protection against pathogens in plants. *Mar. Drugs* 9:2514–2525.

Wang, P., X. Jiang, Y. Jiang, X. Hu, H. Mou, M. Li, and H. Guan 2007 In vitro antioxidative activities of three marine oligosaccharides. *Nat. Prod. Res*. 21:646–654.

Wang, S.L., T.W. Liang, and Y.H. Yen 2011 Bioconversion of chitin-containing wastes for the produc-tion of enzymes and bioactive materials. *Carbohydr. Polym*. 84:732–742.

Wu, S.J. 2014 Preparation and antioxidant activity of the oligosaccharides derived from *Laminaria japonica*. *Carbohydr. Polym*. 106:22–24.

Yaich, H., H. Garna, S. Besbes, M. Paquot, C. Blecker, and H. Attia 2013 Effect of extraction conditions on the yield and purity of ulvan extracted from *Ulva lactuca*. *Food Hydrocoll*. 31:375–382.

Yang, B., G. Yu, X. Zhao, W. Ren, G. Jiao, L. Fang, Y. Wang et al. 2011 Structural characterization of bioactivities of hybrid carrageenan-like sulphated galactan from red alga *Furcellaria lumbrica-lis*. *Food Chem*. 124:50–57.

Yang, M., X. Mao, N. Liu, Y. Qiu, and C. Xue 2014 Purification and characterization of two agarases from *Agarivorans albus* OAY02. *Proc. Biochem*. 49:905–912.

Yihu, S. and Z. Qiang 2014 Ecomaterials based on food proteins and polysaccharides. *Polym. Rev*. 54:514–571.

Yu, L., L. Ge, C. Xue, Y. Chang, C. Zhang, X. Xu, and Y. Wang 2014 Structural study of fucoidan from sea cucumber *Acaudina molpadioides*: A fucoidan containing novel tetrafucose repeating unit. *Food Chem*. 142:197–200.

Yu, L., X. Xu, C. Xue, Y. Chang, L. Ge, Y. Wang, C. Zhang, G. Liu, and C. He 2013 Enzymatic prepara-tion and structural determination of oligosaccharides derived from sea cucumber (*Acaudina molpadioides*) fucoidan. *Food Chem*. 139:702–709.

Zahura, R.M.M., A. Inoue, and T. Ojima 2012 Characterization of a β-D-mannosidase from a marine gastropod, *Aplysia kurodai*. *Comp. Biochem. Physiol. B Biochem. Mol. Biol*. 162:24–33.

14

Study of Indian Marine Mollusks: A Glycomic Approach

Ramachandran Karthik and Ramachandran Saravanan

CONTENTS

14.1 Introduction

The marine milieu has proven to be an unexploited reservoir of great assorted assortments of bioactivates (bioactive metabolites) from marine flora and fauna with immense potential for pharmaceutical/biotechnological and other applications (Saravanan, 2014). Biphasic life cycle systems are present in the marine invertebrate that includes a pelagic larval stage that is morphologically distinctive from the adult form. Their larvae naturally metamorphose into juveniles or adults concurrently with, or unswervingly following completion out of the water column, they will consequently colonize and structure the benthic communities. Generally, the common features of marine invertebrates consist of remodeling of tissue and discrimination to be underneath hormonal or transcriptional control and are interceded by disparity gene and protein expressions to the cell (Chia and Rice, 1978; Okazaki and Shizuri, 2000a; Snelgrove et al., 2001; Fuentes et al., 2002; Fusetani and Clare, 2006). Figure 14.1 represents the taxonomy of marine invertebrates.

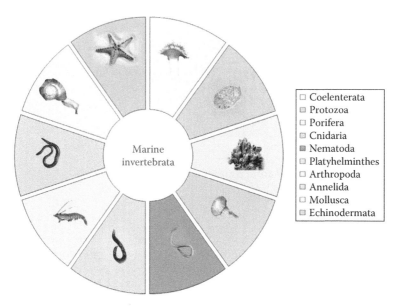

FIGURE 14.1
Taxonomy of marine invertebrates.

14.2 Marine Invertebrates

The current marine invertebrate taxonomy has been carefully correlated, examined, and named based on Aristotle's detailed descriptions. Mainly, marine invertebrates are classified into 10 phyla, which are evidently pointed out with suitable examples in Figure 14.2. This information of the taxonomy of marine invertebrate has been written in Greek language. Moreover, based on Aristotle's classification, the Latin scientific names of animals are sometimes a little bit confusing. In addition, marine invertebrates have different assemblages of animals together with species not very well known, for example, other groups like mammals or birds (Koukouras et al., 1992). Currently, there is a serious threat to the biodiversity of marine mollusks especially in the southeast cost of India, as crisis arises with the desertion of various taxa day by day as a consequence of uncontrolled anthropogenic activities highlighting marine pollution and global warming. Their acquaintance wanted an expertise in the marine field to classify the molluskan phyla based on molecular biology/bioinformatics tool. Consequently, the special familiarity of authors on the marine invertebrates such as Indian marine mollusk and their database of literature are not much cited in scientific literature. There is a long ritual to discover marine molluskan diversity in an attempt to find out new biomolecules and categorize the bioactive compounds, since a marine mollusk generates ample collection of small molecules to large molecules, which is not found in land.

14.2.1 Marine Mollusks

The phylum marine mollusks are the second largest group of soft-bodied heterogeneous animal next to fishes, which are broadly classified into bivalve, gastropod, and cephalopods (Figure 14.2). There are 3370 numbers of the mollusks found in India, which belong

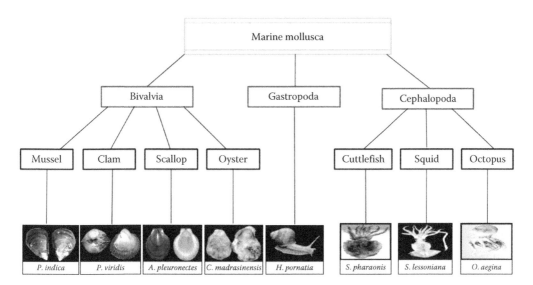

FIGURE 14.2
Major classification of Indian marine mollusks.

to 220 families and 591 genera, of which 1900 are gastropods, 1100 are bivalves, and 210 are cephalopods. The majority of mollusks inhabit marine biotopes, and they occur from the backwater zone, mangroves, intertidal, shelf, and down to deeper waters (Venkatraman and Venkatraman, 2012).

14.2.1.1 Bivalves

Bivalves (first appeared in the late Cambrian about 400 million years ago) have very fascinating shells that can rally round exemplified to researchers' dissimilar livelihood territory (Venkatraman and Venkatraman, 2012). Again, the bivalves are subclassified into mussels, clams, scallops, and oysters. They are sieve feeders gaining nourishment by filtering miniature organisms and palatable debris from the water. Some are inactive, attaching themselves to a substrate (oysters, mussels), some burrow and move around on the bottom (clams), and a few can swim (scallops).

14.2.1.2 Gastropods

Gastropoda (known as univalves) are a major part of the phylum Mollusca and are the most highly diversified class in the phylum, with 60,000–80,000 living snail and slug species. The structure, behavior, nourishing, and reproductive adjustments of gastropods diverge significantly from one group to another. Consequently, it is complicated to state numerous generalities for all gastropods.

14.2.1.3 Cephalopods

The cephalopods (head in foot) are the most extremely advanced group of mollusks, the most intelligent group among the invertebrates, and the ancient group that appeared in the late Cambrian period several million years before the first primitive fish began swimming in the ocean. Again, these cephalopods are subclassified into squid, cuttlefish, and octopus.

Squid have excellent eyesight and active swimming lifestyle, and it seems to be more analogous to fish than to other mollusks. Unlike cuttlefish that are mostly lonely, squid frequently move about in shoals. The chalky bone (cuttlebone) is absent in squid, having a thin membrane called the pen for support as a replacement for the cuttlebone. Under certain circumstances, squid spurt a cloud of black ink into the surrounding water to help mask their escape. Cuttlefish are well known for stunning color and skin surface changes, which can indicate their mood. Like other associates of their family, cuttlefish have a comparatively short lifetime of approximately 18 months. Cuttlefish have eight arms and two tentacles. When nourishing on crustaceans and fish, two tentacles swiftly grasp the prey that is strained toward the beak-like mouth beneath the arms. The cuttlebone is a porous internal structure of cuttlefish to control its resilience. After being dried, the so-called cuttlebone is sometimes given to caged birds as a source of calcium, essential salts, and macro- and microminerals.

Octopus uses jet driving force and may cloud the seawater with ink to escape from predators. Although majority of octopi do not present a menace, the blue-ringed octopus is lethal to humans. Several species occur in seawaters, mainly of which do not surpass 20 cm in length. These octopi are usually dingy in manifestation, for instance, blue-ringed octopi on their senses being watchful upon threats such as an approach of a larger animal. Similar to the entire octopi, the mouth of the blue-ringed octopus has a parrotlike beak. When the beak penetrates the skin, venom is injected to the prey/predators. The venom of a blue-ringed octopus is the same venom found in a puffer fish, which is known as tetrodotoxin that causes paralysis to its victims. There is as yet no antidote.

14.3 Marine Bioinformatics

Bioinformatic techniques are being applied to study many land-based organisms and humans and also medical-based systematics, phylogenetics, genomics, and proteomics. This meadow is growing astoundingly during the last decade due to the advancement of life sciences and medical sciences combined with the development of computational skills (computational biology) and the network efforts by different countries in sharing information. As the volume of database for biomolecules, protein, and gene of various land and marine organisms is growing astonishingly, analyzing the data and getting required results for marine-based studies from the common database of land and marine organisms is increasingly becoming complicated.

Glycomics indeed has finally emerged, in spite of being in the backdrop of the post-genomics and proteomics era, and has caught the attention of scientists and researchers around the globe. Scientific explorers have started analyzing the structural and functional aspects of living systems from the glycomic point of view, sealing the lacunae leftover by genomics and proteomics. With over 68 building blocks, attached peptide and lipid linkages, coupled with linear and transverse branching, confer one of the most complex levels of organization to the glycome of an organism. Extremely higher levels of fidelity in glycomic structure and function are thus an indispensable trait in the secular functioning of living systems (Shriver et al., 2004a).

The marine glycome with a higher population percentage constitutes the major portion of the glycome of the life on earth. Though largely remaining unexplored, marine glycome offers immense potentials of unearthing carbohydrate molecules of scientific interest.

TABLE 14.1

Current Status of Glycan-Based Drugs in Market

S. No	Drug	Disease	Clinical Status	Manufacturer
1	Lovenox	Thrombosis	Market	Aventis
2	Fragmin	Thrombosis	Market	Pfizer
3	Aranesp	Anemia	Market	Amgen
4	Sepragel	Antiadhesive	Market	Genzyme
5	Healon	Cataracts	Market	Pfizer
6	Arixtra	Thrombosis	Market	Sanofi
7	Cerezyme	Gaucher's disease	Market	Genzyme
8	Aldurazyme	Mucopolysaccharidoses type I	Market	Genzyme
9	PI-88	Cancer	Phase II	Progen

Source: Shriver, Z. et al., *Nat. Rev. Drug Discov.*, 3, 863, 2004b.

The marine invertebrate glycomics, constituting the largest proportion of the marine biological reservoir, is thus emerging to be an important pit stop in studying human health and diseases (Table 14.1). Hence, there is a need for separate storage, retrieval, and analysis of marine database, which will reduce much time of the students and scientists doing marine-based research and data mining. Marine bioinformatics is the branch of marine sciences that refers to the use of computer module operands to meet, store up, incorporate, analyze, and broadcast data on the allotment of marine organisms, their portrayal, orderly categorization, phylogeny, structure, and sequence data on biomolecules including protein (marine proteomics) and gene (marine genomics). The key aspiration of marine bioinformatics is to switch over information, that is, database, through network access. Glycomics offer a specialized analytical knowledge device to achieve the goal of finding natural bioactive compounds, especially glycosaminoglycans (GAGs) isolated from marine organisms as potential drugs

The idea of a general scientific investigation of living things has been created by Aristotle. According to him, the modern biology includes the science of biology and the philosophy of biology (Lennox, 2001). His biological characters, comprising larger than 25% of the existing Aristotelian corpus, have contentedly been the theme of an escalating amount of concentration recently, ever since both theorists and natural scientists accept that they might assist in the indulgent of additional significant subjects of his philosophy (Gotthelf and Lennox, 1987), and so they introduced a new invention by natural scientists for the prosperity of Aristotle's biological examinations and the problems that encouraged the lives of the scientists (Tipton, 2006).

14.3.1 Glycans

Glycans (GLYs) are having complex structures that are fastened toward the protein and lipid coats for all types of cells and also originate in the extracellular gap between adjacent cells. Based on their backbone chemical structure, GLY can be divided into two broad categories: the straight chain of sugars or polysaccharides, which incorporates reiterating pyranose monosaccharide rings, and the branched sugars, which are saccharide compositions that are based on various associations sandwiched between monosaccharide rings. Branched GLYs are presented by *N*-linked and *O*-linked glycosylation on proteoglycons (PGs) or on glycolipids (Shriver et al., 2004a). Most of the straight-chain sugars are GAGs, which consist of a long polymer of sulfated disaccharide units that are *O*-linked to core protein forming a PG aggregate, which is presented in Figure 14.3 (Perrimon and Bernfield, 2000). Disaccharide units include either of the two amended sugars, *N*-acetylgalactosamine or *N*-acetylglucosamine, and a uronic acid such as glucuronate or iduronate.

Glycome includes two blocks, namely, glycoproteins and glycolipids, which with the action of glycosyltransferases (GTs) convert the formation of glycome from genome, proteome, and lipids. GTs are found enormously in the cell and catalyze the functions from structure and storage to signaling. Exactly, GTs transfer a sugar molecule from an activated

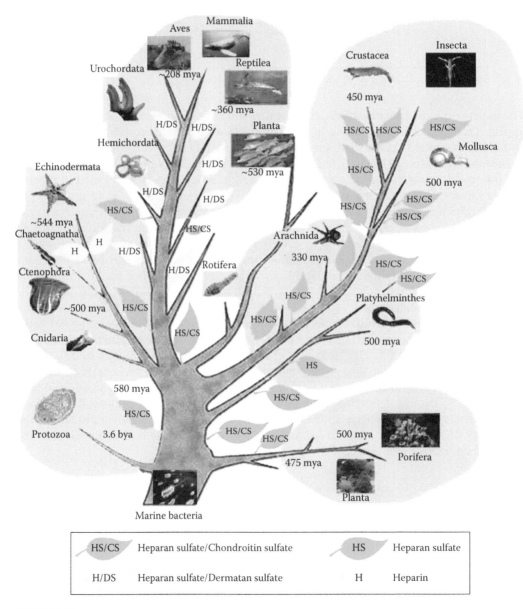

FIGURE 14.3
Origin of tree of marine glycans.

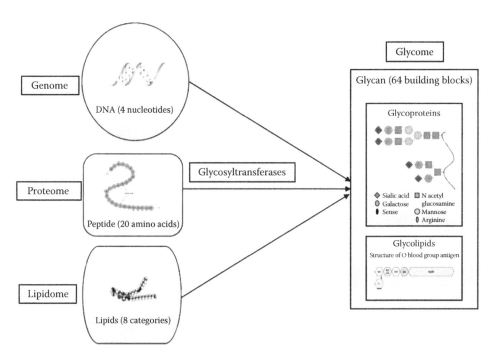

FIGURE 14.4
Pathway of formation of glycome.

donor of sugar nucleotide, which leads to the formation of glycosidic linkages in the GLY (Breton et al., 2006). The biosynthesis of a GLY (which has 64 building blocks) refers to glycosylation, which is an enzymatic process that fastens different sugars collectively to constitute the GLY (Figure 14.4). The major GLYs are glycoproteins (covalent attachment of GLY to proteins to generate glycoproteins), glycopeptides or PGs (peptidoglycans), glycolipids, and lipopolysaccharides. They are classified according to the fastened biomolecule as follows: glycolipids and di-/oligosaccharides attached via hexose sugar such as glucose or galactose to the terminal primary hydroxyl group of lipid moiety. The phospholipids, especially phosphatidylinositol or phosphoethanolamine, strongly bind with GLY to form glycophospholipid and again link the amide and carboxyl terminus of a protein in the cell.

14.3.2 MALDI-TOF MS

Matrix-assisted laser desorption/ionization time-of-flight mass spectrometry (MALDI-TOF MS) was pioneered by Tanaka and Hillenkamp in the 1980s. It has been considered as a fitting technique to investigate the natural macromolecules like proteins, oligosaccharides, and synthetic macromolecules, as a result of its high speed, high throughput, high constancy, etc. Nevertheless, it could not be used to study the small molecules (<500 Da) professionally with the use of traditional matrices, for instance, α-cyano-4-hydroxycinnmaic acid and 2,5-dihydroxybenzoic acid, thanks to the strong matrix-related backdrop noise produced in the low-mass region, which obstructs with the investigation of the target analytes (Liu et al., 2012).

The general aspect of biomolecules (metabolites) is carried out by metabolomics, a relatively latest phase in the meadow of omics techniques, particularly when balanced with move toward exploited to sequence genomes and to RNA profile and proteins. Another stereoscopic technique like nuclear magnetic resonance (NMR), combined with mass spectrometry (MS)-based methods, is one of the most well-known technologies accessible to execute medium- to high-throughput metabolite profiling and has previously been used broadly, for example, for phenotyping of several standard biological replicas or for finding communications of living organisms with their marine environment (marine environmental metabolomics) (Miller, 2007; Viant et al., 2009; Prince et al., 2010).

Gas chromatography (GC) and liquid chromatography (LC) are common techniques to separate biomolecules from marine before identification, even though new techniques have been developed. Successful isolation and purification techniques, in combination with high-resolution MS, are suitable to array with more number of samples that require to be switched concurrently and in undeviating time, and the combined technique of GC-MS has produced extended separations of metabolites profiles, both for qualitative and quantitative determination. The significant features in lieu of a demanding matter for metabolomic analysis of marine sources are the high number of compounds with unidentified structures. In one direction, to trounce this inadequacy is to combine NMR and MS (Weckwerth, 2003). Interestingly, many metabolites isolated from the natural source involved in primary metabolism, in relation with transcripts or proteins, are not so creature explicit; as a consequence, when bioanalytical protocols are effectively useful for their dimension, these protocols can be relocated for the estimation of the same metabolites in other creature (Hollywood et al., 2006). The preface of high-tech glycomic advances to marine mollusks has kindled groundbreaking scientifically and hypothetical progresses, as a result of the development of major new avenues of research.

Recently, Connor and Gracey (2012) reported the high-resolution analysis of metabolic cycles in the intertidal blue mussel *Mytilus californianus* and reviewed both hydrophilic and lipophilic metabolites in a large molecular weight range. But in India, insufficient information available about GLY sequences of marine mollusks is evident; its present attempt aimed to isolate, purify, and structurally characterized PGs from the salivary glands of selected cephalopod (*S. prashadii*) brings local economic interest in the southeast cost of India.

14.3.3 Isolation and Purification of PG from Cuttlefish

The cuttlefishes are obtained from Kasimedu Fish Landing Center, Chennai, Tamil Nadu, in 2012–2013. The specimens have been transported to the laboratory at cold condition, washed with water, and temporarily stored at −20°C until used for further study. The PGs are dissected out from the salivary glands of cuttlefishes and homogenized with 0.01 M phosphate buffer (pH 7.0) in 0.5 M NaCl. The salivary gland extract has then centrifuged and the supernatant is obtained as the crude PG (Ueda et al., 2008). The crude sample has been subjected for further study.

The fractionation and purification of PG from cuttlefish is depicted in Figure 14.5. The crude PG is purified on a 2 × 80 cm column of Sephadex G-100 (Sigma) adopting the method of Ueda et al. (2008). The elution buffer is 0.01 M phosphate buffer saline (pH 7.2) and the volumetric flow rate is 0.33 mL/min. Each 5 mL of the effluent from gel filtration is collected as a fraction and studied at 280 nm in a UV spectrophotometer

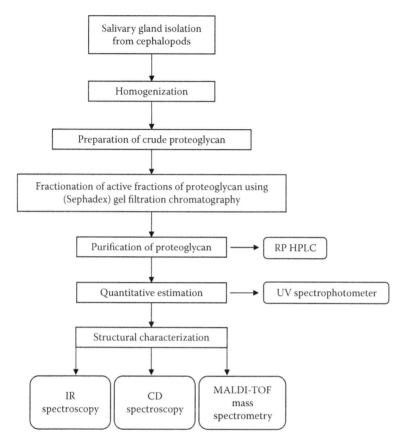

FIGURE 14.5
Flowchart of isolation and purification of proteoglycans from cuttlefish.

(Shimadzu, 1800, Japan). Fractions with maximum absorbance are considered active and are pooled, dialyzed against the same buffer at 4°C, freeze-dried, and subjected to further purification.

The fractionated PG of *Sepia pharaonis* is further purified by reversed-phase high-performance liquid chromatography (RP-HPLC) (Ebada et al., 2008). Acetonitrile/TFA gradient is used as the mobile phase. Acetonitrile/TFA gradient is used as the mobile phase in a Jupiter 5 μ C-18 300A column (Phenomenex 250 × 2.00 mm²) at a flow rate of 1 mL/min. The HPLC system (Agilent Technologies 1200 series) is connected to the UV detector and an HPLC Technology chart recorder and is set to read at 280 and 216 nm wavelengths. Elutants are freeze-dried after each run, kept at −20°C, and used for biochemical analysis (Key et al., 2002). The net yields of total protein and neutral sugar concentrations are recorded as 1.60 mg/g and 69.52 μg/g, respectively.

14.3.4 FT-IR and CD Spectrum of PGs from Cuttlefish

The Fourier transform infrared (FT-IR) spectrum shows the functional groups such as alcohol N–H stretch of secondary amines (3324.91 cm⁻¹), amide or conjugated C≡C (2464.18–2318.62 cm⁻¹), and N–H primary amines bend (1634.47 cm⁻¹) attached to the PG

(Figure 14.6) (Hsu, 1997). Circular dichroism (CD) spectrum is used to analyze the secondary structural features of PG such as α-helix, β-sheet, and random coil. CD spectra can be deconvoluted using a set of basis spectra, but the analysis is difficult and contains bias dependent on the choice of reference spectra. CD is used for PG because of the chiral nature of the structural features of peptides with sugar moieties. The CD spectrum of the PG showed predominant composition of α-helical pattern, evident from the negative peaks obtained within the range of 200–220 nm (Figure 14.7). The α-helical pattern of the PG

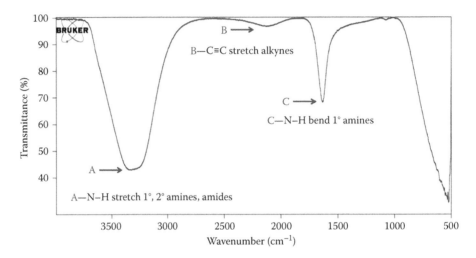

FIGURE 14.6
FT-IR spectrum of proteoglycans from cuttlefish.

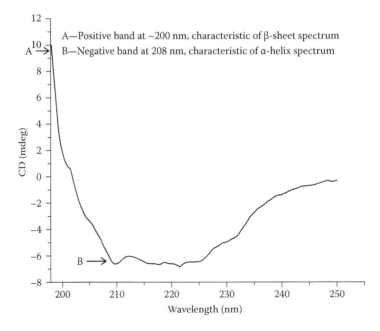

FIGURE 14.7
CD spectrum of proteoglycans from cuttlefish.

is prominent, constituting 59.78% of the structure. β-Sheets accounted for 4.38% and the remaining 35.84% of the protein is randomly coiled, as predicted by the K2D2 tool (Iratxeta and Navarro, 2008). The purified PG has a positive band (β-sheet) at 200 nm and a negative band (α-helix) at 220–225 nm. The percentage composition of α-helix and β-sheet of PG may differ from species to species and genus to genus.

14.3.5 MALDI-TOF MS of PG from Cuttlefish

Nowadays, the application of MALDI has been extensively exploited as a soft ionization MS technique, which provides spectra of biomolecules isolated from marine sources, which are primarily integral quasi-molecular ions (protonated molecule or an ion formed from a molecular ion by loss of a hydrogen atom) with tiny disintegration. Use of MALDI combined with a time of flight (TOF) is an authoritative technique for the characterization of polymers obtained by either synthetic or natural resources (Behrens et al., 2003; Ishida et al., 2005). MALDI-TOF MS can be employed to know the intact compositions of compacted PG isolated from cuttlefish. Together with this method, the disintegration of the object analyte biomolecule upon laser irradiation can be significantly diminished by coprecipitating it with a matrix that attracts the laser energy (Behrens et al., 2003). The MALDI-TOF MS has four advantages that are as follows:

1. High-molecular-weight compounds can be detected.
2. The sensitivity is high in the MALDI-TOF MS technique across a wide range of molecular weight permits to detect the oligomeric compounds from the sample.
3. Only one type of quasi-molecular ion is produced from each parent molecule.
4. The isotope interpretation patterns agree to the discovery of oligomers with small variations in the molecular weights.

Structural analyses of the PG are carried out using an UltrafleXtreme MALDI-TOF MS (Bruker Daltonics, Bremen, Germany) in reflectron (positive ion) mode operated at 25 kV accelerating voltage in the positive ion mode with a time delay of 90 ns. The spectrometer emits at 337 nm using a 50 Hz pulsed nitrogen laser. A set of turbomolecular pumps are operated to maintain the ion source and the flight tube at a pressure of about 7×10^{-7} mbar. A 1:1 (v/v) saturated solution of matrix (α-cyano-4-hydroxycinnamic acid) in the purified PG served as the sample.

Concentrated PGs from cuttlefish, which are irradiated with the MALDI-TOF laser in anticipation of the spectrum, had an intensity of above 500 absorbance units. Depending on the ease of ionization, some of the spectral intensity is observed as high as 822 absorbance units. The synchronized smoothing function is set to high, and the mass spectrometer mass window has set for low-molecular-weight range molecules (500–5000 Da). Matrix hush-up is triggered when production is more (i.e., for highly polydisperse samples).

In contrast, MALDI-TOF MS study of cuttlefish PG fraction produced a very high-quality spectrum (Figure 14.8) with resolution peaks between 520 and 2000 units. The raise in concentration of the baseline after the matrix suppression cutoff at 1966 Da and PG from cuttlefish is a polydispersity. This happens because the baseline intensity after the matrix hush-up cutoff is studied to reduce with decreased polydispersity (McEwen et al., 1997).

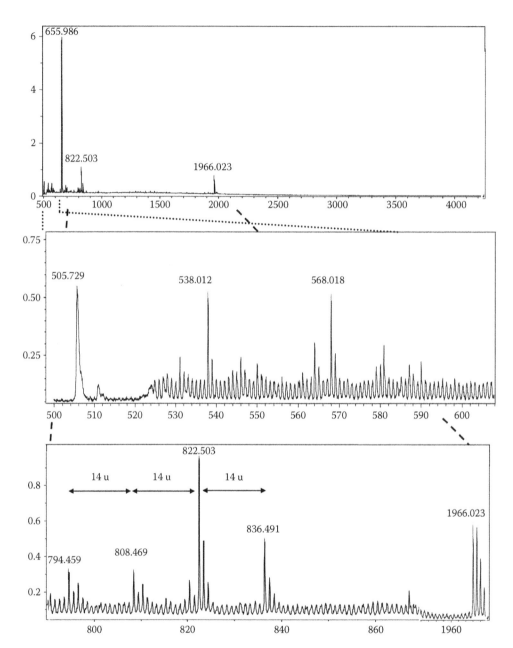

FIGURE 14.8
MALDI-TOF MS of proteoglycans from cuttlefish.

14.4 Conclusions

Investigation into the use of PG from Indian cuttlefish is a new approach. The fractionation of PG is done by using Sephadex-50 gel chromatography and purified by RP-HPLC. The amount of total protein and neutral sugar are observed as 1.60 mg/g and 69.52 µg/g, correspondingly. Majority of the structure is composed of the α-helical pattern of the PG, that is, 59.78%, whereas the β-sheet has been reported to take 4.38% of it and the remaining protein has been aimlessly coiled for 35.84%. The molecular mass of isolated PG from cuttlefish has been recorded as >2000 Da. The biological activity of PG of this cuttlefish has to be carried out and could be considered as an alternative drug in future.

Acknowledgments

The first author acknowledges the Chettinad Academy of Research and Education (CARE) for providing the necessary facility and fellowship in the form of JRF. The authors gratefully acknowledge Professor A. Shanmugam, CAS in Marine Biology, Faculty of Marine Sciences, Annamalai University, Parangipettai, for his valuable guidance about the taxonomy of mollusks. The authors also acknowledged Dr. R. Srinivasan, Bio-organic Department, CLRI; Prof. P. Gautam, Department of Biotechnology, Anna University, Guindy, Chennai; and Dr. P.R. Rajesh, Molecular Biophysics Unit, IISC, Bangalore, India, for providing structural study support.

References

Behrens, A., Maie, N., Knicker, H., and Kogel-Knabner, I. (2003). MALDI-TOF mass spectrometry and PSD fragmentation as means of the analysis of condensed tannins in plant leaves and needles. *Phytochemistry, 62*, 1159–1170.

Breton, C., Snajdrova, L., Jeanneau, C., Koca, J., and Imberty, A. (2006). Structures and mechanisms of glycosyltransferases. *Glycobiology, 16*(2), 29–37.

Chia, F. S. and Rice, M. E. (eds.) (1978). Settlement and metamorphosis of marine invertebrate larvae. *Proceedings of the Symposium on Settlement and Metamorphosis of Marine Invertebrate Larvae, American Zoological Society Meeting,* Toronto, Onatario, Canada, December 27–28, Elsevier, New York.

Connor, K. M. and Gracey, A. Y. (2012). High-resolution analysis of metabolic cycles in the intertidal mussel *Mytilus californianus. The American Journal of Physiology—Regulative, Integrative and Comparative Physiology, 302*, 103–111.

Ebada, S. S., Edrada, R. A., Lin, W., and Proksch, P. (2008). Methods for isolation, purification and structural elucidation of bioactive secondary metabolites from marine invertebrates. *Nature Protocols, 3*(12), 1820–1831.

Fuentes, J., Lopez, J. L., Mosquera, E., Vazquez, J., Villalba, A., and Alvarez, G. (2002). Growth, mortality, pathological conditions and protein expression of *Mytilus edulis* an *M. galloprovincialis* cosses cultured in the Ria d Arousa (NW of Spain). *Aquaculture, 213*, 233–251.

Fusetani, N. and Clare, A. S., and Muller, W. E. G. (Ed) (2006). Antifouling compounds. In *Progress in Molecular and Subcellular Biology* (Subseries Marine Molecular Biotechnology). Berlin, Germany: Springer-Verlag Vol. 42.

Gotthelf, A. and Lennox, J. G. (eds.) (1987). *Philosophical Issues in Aristotle's Biology.* Cambridge, U.K.: Cambridge University Press.

Hollywood, K., Brison, D. R., and Goodacre, R. (2006). Metabolomics: Current technologies and future trends. *Proteomics, 6*, 4716–4723.

Hsu, C. P. Sherman. (Ed). (1997). Infrared spectroscopy. In *Handbook of Instrumental Techniques for Analytical Chemistry,* Vol. 1, pp. 247–263.

Iratxeta, C. P. and Navarro, M. A. A. (2008). K2D2: Estimation of protein secondary structure from circular dichroism spectra. *BMC Structural Biology*, 8, 25.

Ishida, Y., Kitagawa, K., Goto, K., and Ohtani, H., (2005). Solid sampling technique for direct detection of condensed tannins in bark by matrix-assisted laser desorption/ionization mass spectrometry. *Rapid Communications in Mass Spectrometry*, 19, 706–710.

Key, L. N., Boyle, P. R., and Jaspars, M. (2002). Novel activities of saliva from the octopus *Eledone cirrhosa* (Mollusca; Cephalopoda). *Toxicon*, 40(6), 677–683.

Koukouras, I. A., Russoz, A., Koukoura, E. V., Dounas, C., and Chintiroglou, C. (1992). Relationship of sponge macrofauna with the morphology of their Hosts in the North Aegean Sea. *International Review of Hydrobiology*, 77(4), 604–619.

Lennox, J. G. (2001). *Aristotle's Philosophy of Biology*. Cambridge, U.K.: Cambridge University Press.

Liu, J. A., Xiong, L., Zhang, S., Wei, J. C., and Xiong, S. X. (2012). C60 Fluorine derivative as novel matrix for small molecule analysis by MALDI-TOF MS. *Metabolomics*, 1, 1–5.

McEwen, C. N., Jackson, C., and Larsen, B. S. (1997). Instrumental effects in the analysis of polymers of wide polydispersity by MALDI mass spectrometry. *International Journal of Mass Spectrometry and Ion Processes*, 160, 387–394.

Miller, M. G. (2007). Environmental metabolomics: A SWOT analysis (strengths, weaknesses, opportunities, and threats). *Journal of Proteome Research*, 6, 540–545.

Okazaki, Y. and Shizuri, Y. (2000a). Structures of six cDNA expressed specifically at cyprid larvae of barnacles *Balanus amphitrite*. *Gene*, 250, 127–135.

Okazaki, Y. and Shizuri, Y. (2000b). Effect of inducers and inhibitors on the expression of bcs genes involved in cypris larval attachment and metamorphosis of the barnacles *Balanus amphitrite*. *The International Journal of Developmental Biology*, 44, 451–456.

Perrimon, N. and Bernfield, M. (2000). Specificities of heparan sulphate proteoglycans in developmental processes. *Nature*, 404, 725–728.

Prince, E. K., Poulson, K. L., Myers, T. L., Sieg, R. D., and Kubanek, J. (2010). Characterization of allelopathic compounds from the red tide dinoflagellate *Karenia brevis*. *Harmful Algae*, 10, 39–48.

Reed, J. D., Krueger, C. G., and Vestling, M. M. (2005). Review: MALDI-TOF mass spectrometry of oligomeric food polyphenols. *Phytochemistry*, 66, 2248–2263.

Ricardo, A., Frye, F., Carrigan, M. A., Tipton, J. D., Powell, D. H., and Benner, S. A. (2006). 2-Hydroxymethylboronate as a reagent to detect carbohydrates: Application to the analysis of the formose reaction. *Journal of Organic Chemistry*, 71(25), 9503–9505.

Saravanan, R. (2014). Isolation of low molecular weight heparin/heparan sulfate from marine sources. In *Food and Nutrition Research*, Se-Kwon Kim (ed.), (ISSN No. 1043-4526) Academic Press, Elsevier Publications, Vol. 72, pp. 45–60.

Shriver, M. D., Kennedy, G. C., Parra, E. J. et al. (2004a). The genomic distribution of human population substructure in four populations using 8525 SNPs. *Human Genomics*, 1, 274–286.

Shriver, Z., Raguram, Z., and Sasisekharan, R. (2004b). Glycomics: A pathway to a class of new and improved therapeutics. *Nature Reviews Drug Discovery*, 3, 863–872.

Snelgrove, P. V. R., Grassle, J. F., Grassle, J. P., Petrecca, R. F., and Stocks, K. I. (2001). The role of colonization in establishing patterns community composition and diversity in shallow-water sedimentary communities. *Journal of Marine Research*, 59, 813–831.

Ueda, A., Nagai, H., Ishida, M., Nagashima, Y., and Shiomi, K. (2008). Purification and molecular cloning of SE-cephalotoxin, a novel proteinaceous toxin from the posterior salivary gland of cuttlefish *Sepia esculenta*. *Toxicon*, 52, 574–581.

Venkatraman, C. and Venkataraman, K. (2012). Diversity of molluscan fauna along the Chennai coast. *International Day for Biological Diversity, Marine Biodiversity*, UP, India. May 12, 2012, pp. 29–35.

Viant, M. R., Bearden, D. W., Bundy, J. G. et al. (2009). International NMR-based environmental metabolomics intercomparison exercise. *Environmental Science & Technology*, 43, 219–225.

Weckwerth, W. (2003). Metabolomics in systems biology. *Annual Review of Plant Biology*, 54, 669–689.

Section V

Marine Transcriptomics

15

De Novo Transcriptomics in Marine Microalgae: An Advanced Genetic Engineering Approach for Next-Generation Renewable Biofuels

Kashif M. Shaikh, Asha A. Nesamma, and Pavan P. Jutur

CONTENTS

15.1 Introduction

15.1.1 Marine Microalgae

Marine microalgae has gained worldwide attention due to its important role in CO_2 recycling through photosynthesis, thus addressing a critical issue of producing sustainable energy without massive CO_2 release (Matsunaga et al. 2015; Tanaka et al. 2015) and promising source for high-value-added renewables along with sustainable biofuel production (Asha et al. 2015; Jutur and Asha 2015a). These photosynthetic organisms originated as much as 1500 million years ago, evolving as major players in global energy/biomass production and biogeochemical recycling, and are well known for its distribution in all aquatic habitats including freshwater, brackish environments, and marine and fully saline environments throughout the warm temperate and tropical regions of the world (Grossman 2005; Zhang et al. 2012). Perhaps these diverse group of unicellular eukaryotic photosynthetic organisms are responsible for about 40%–50% of marine photosynthesis that occurs on Earth despite of the known fact that photosynthetic biomass represents only 0.2% of that on land (Falkowski et al. 1998; Parker et al. 2008). Marine microalgae may offer a potential feedstock for renewable biofuels capable of converting atmospheric CO_2 to substantial biomass and valuable biofuels, which is of great importance for the food, feed, and energy industries (Jutur and Asha 2015b). In general, microalgal triacylglycerols (TAGs), hydrocarbons, and polysaccharides are potential biofuel precursors, an alternative source of bioenergy, with their higher rates of CO_2 fixation and biomass yields and the fact that they do not compete with food crops for resources (Smith et al. 2010). Microalgae consists of a wide range of biofuel precursor molecules ranging from C9 to C17 alkanes and short fatty acid methyl esters (FAMEs), an alternative for jet fuels and the other main constituents of biodiesel, C9–C23 medium FAMEs (Matsunaga et al. 2015). For sustainability and economic viability of next-generation renewable biofuels from marine microalgae, it is essential to understand the basic to molecular biology of these organisms, especially the effects of environmental stresses on cellular metabolism and the regulation of biosynthetic pathways of fatty acids and TAGs, as well as carbon fixation and allocation (Jutur and Asha 2015a; Singh et al. 2015).

15.1.2 *De Novo* Transcriptomics

Understanding the molecular insights of marine microalgae is important for establishing the foundation for innovative strategies and for the development of ultimate fuel surrogates (Fang 2014). Current advanced technologies like next-generation sequencing (NGS) have gained enormous attention (due to its data quality, robustness, and low noise compared to conventional methods) covering range of applications in understanding the mechanisms involved in biological processes, thus behaving as a molecular microscope that could provide insights on metabolic profiling of these photosynthetic organisms (Buermans and den Dunnen 2014).

Transcriptome is defined as the complete complementation of mRNA molecules generated by a cell or population of cells (McGettigan 2013). Studies involved in global analysis of gene expression or genome-wide expression profiling, as one of the major tools for measuring gene expression levels and uncovering differential expression of whole set of all mRNA/miRNA molecules, or transcripts, produced in a single cell or population of cells, are termed as transcriptomics (Zhang et al. 2010). Thus, understanding whole genome or transcriptome sequencing among microalgae would elucidate the group's unique genetic resources or exploit the genetic basis for their physiological characteristics based on global gene expression (Rudd 2003; Bouck and Vision 2007; Nagaraj et al. 2007). Henceforth, the ultimate application of newly developed high-throughput sequencing technologies has demonstrated that hundreds of thousands of high-quality sequence reads could be produced *de novo* from whole genome or transcriptome templates, thus enriching and enabling immediate inroads to genetic studies of organisms for which little or no sequence data exist (Riggins et al. 2010). For nonmodel organisms, the drawback that still remains challenging is the sequencing and assembling of genomes and their high costs, even with the availability of these developing high-throughput sequencing technologies (Pop and Salzberg 2008). As an alternative, sequencing transcriptomes is less complex and provides faster and cost-effective access to the gene expression profile of an organism (Bouck and Vision 2007; Stinchcombe and Hoekstra 2008). Thus, transcriptomes provide a valuable resource for improving genome annotation (Seki et al. 2002), elucidating phylogenetic relationships (Nishiyama et al. 2003), accelerating gene discovery by expanding gene families (Bourdon et al. 2002; Cheung et al. 2008), facilitating strain improvements by SSR and SNP markers (Gonzalo et al. 2005; Parchman et al. 2010), and allowing large-scale expression analysis (Fei et al. 2004; Eveland et al. 2008) and rapid identification of transcripts involved in specific biological processes (Rismani-Yazdi et al. 2011). Recent studies have demonstrated the success of 454 *de novo* sequencing, assembly, and analysis of transcriptomes in nonmodel organisms with no prior genomic resources (Novaes et al. 2008; Vera et al. 2008; Kristiansson et al. 2009; Meyer et al. 2009; Guarnieri et al. 2011). The generation of such large-scale sequencing data will enable functional analyses that were previously limited to model organisms and their rapid application in ecologically important taxa (Wheat 2010).

Unlike genome sequencing and comparative genomics technologies that mainly focus on DNA, it seems to be static information for any given species and normally does not change significantly in response to short-term external environmental changes. Transcriptomics has enabled quantitative measurements of the dynamic expression of mRNA/miRNA molecules and their variation at the genome scale between different states, thus reflecting the genes that are being actively expressed at any given time, with the exception of mRNA degradation phenomena (Ye et al. 2001; Horak and Snyder 2002). Such studies play pivotal role in the identification of individual transcriptional units, the expression of unique transcripts restricted to cell conditions or cell types, the determination of the level of expression for each gene expressed, the precise assessment of transcript diversity at each transcriptional locus (i.e., splice isoforms, alternative transcription start sites, and polyadenylation sites), and the protein factors that control the transcriptional cassettes of a cell and their mechanisms of action (Hemaiswarya et al. 2013; Jutur and Asha 2015b). Initially, transcriptome analysis was dependent on information obtained from reference genome through gene predictions and expressed sequence tag (EST) evidences that were either partially annotated or biased. But with the recent advancements in NGS or RNA sequencing (RNA-seq), it has become reality to analyze whole transcriptome rapidly and accurately (Wang et al. 2009; Wilhelm and Landry 2009; Marguerat and Bahler 2010; Ozsolak and Milos 2011a,b,c) that has enormous sequencing depth (100–1000 reads per base pair of transcript) with the detection of even the rare regulatory transcripts, revealing almost the complete scenario of a transcriptome.

De novo transcriptome assembly is a strategy that does not require a reference genome for studying the transcriptome; rather, it utilizes the redundancy of short-read sequencing to find overlaps between the reads and then assembles these reads into transcripts (Martin and Wang 2011), thus seems to have advantage over the reference-based transcriptome analysis. Even when the genome sequence is available, performing *de novo* transcriptome assembly can reveal transcripts from that segment of genes that are missing from the genome assembly. Another advantage of *de novo* transcriptome assembly is that it is independent of the requirement for correct alignment of reads to known splice sites and/or prediction of novel splice sites, as well as it is free from long introns (Burset et al. 2000). There are two phases in the *de novo* transcriptomics approach; first is data generation followed by data analysis (Martin and Wang 2011). A successful transcriptome approach requires expertise at both levels, wet lab as well as bioinformatics (Buermans and den Dunnen 2014). In the first phase, total RNA is extracted by standard procedures and assessed for quality with RNA integrity number value of >8, followed by construction of cDNA libraries. These cDNA libraries are subjected to sequencing using NGS platforms such as Illumina, SOLiD, and Roche 454 to produce millions of short reads (Metzker 2010). These short reads are then preprocessed to remove sequencing errors and other artifacts in the second phase, that is, data analysis. Assembly of these preprocessed reads could be done using various *de novo* transcript assemblers such as Rnnotator (Martin et al. 2010), Multiple-k (Surget-Groba

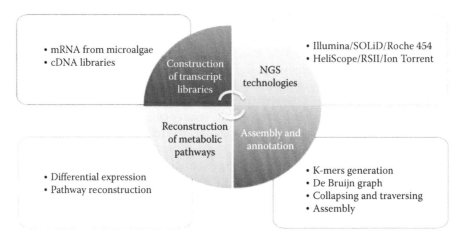

FIGURE 15.1
Schematic representation of *de novo* transcriptomics in marine microalgae.

and Montoya-Burgos 2010), Trans-ABySS (Simpson et al. 2009; Robertson et al. 2010), ALLPATHS (Butler et al. 2008), Trinity (Garber et al. 2011), and Velvet/Oases (Zerbino and Birney 2008; Schulz et al. 2012), and their differential expression is carried out by using edgeR (Robinson et al. 2010) and DESeq (Anders and Huber 2010) (Figure 15.1).

15.2 *De Novo* Transcriptomic Profiling

Currently, technologies are separated into approaches that primarily assess levels of expression and provide precise transcript boundaries and other transcript-focused characteristics, thus glance for existence of the transcription factor–DNA interactions (Jutur and Asha 2015b). Application of transcriptomics have been demonstrated in various microalgal systems to explore genome-wide transcriptional activity (Miller et al. 2010; Rismani-Yazdi et al. 2011, 2012; Dong et al. 2013; Liang et al. 2013; Lv et al. 2013; Yang et al. 2013; Corteggiani Carpinelli et al. 2014; Tanaka et al. 2015). The accuracy of gene expression analysis depends upon the quality of cDNA libraries and sequencing platforms used and how the data are preprocessed before data analysis (Levin et al. 2010).

15.2.1 Construction of Transcript (cDNA) Libraries

All the currently available NGS platforms require preprocessing of RNA into cDNA libraries for sequencing. For increasing the quantity of transcripts, especially the ones that are very low in amount, ribosomal RNA (rRNA) and abundant transcripts are removed during the first steps of library construction (Buermans and den Dunnen 2014). Poly(A) tail–based selection is mostly applicable to enrich mRNAs followed by rRNA removal by hybridization-based depletion methods (He et al. 2010; Chen and Duan 2011). A typical problem occurring during this process is biasness in quantification while using these depletion methods; hence, the use of nondepleted libraries is recommended, if quantification is our primary goal. PCR amplification is a required step in sequencing platforms like Illumina, SOLiD, and Roche 454 to generate polonies (Mitra and Church 1999), so as to increase the signal-to-noise ratio as these systems are less sensitive. As a matter of fact, amplification step for transcripts with high GC content should be avoided as it results in low-sequencing coverage leading to gaps in the assembled transcripts (Kozarewa et al. 2009). For this purpose, techniques such as the single-molecule sequencing technologies from Helicos and Pacific Biosciences, which do not require PCR amplification before sequencing, are employed (Kozarewa et al. 2009; Mamanova et al. 2010; Sam et al. 2011).

15.2.2 Next-Generation Sequencing Technologies

Strand-specific RNA sequences derived from opposite strands of genome are more effective, which tend to overlap with the generated transcripts during assembly and quantification. The choice of sequencing platforms, read length, and whether to use paired-end and/or other methods is the main factor that is mainly focused upon before sequencing a sample (Martin and Wang 2011). Although all NGS sequencing platforms can successfully assemble transcriptomes (Passalacqua et al. 2009; Chen et al. 2010; Schwartz et al. 2010), different sequencing platforms have different strategies to convert sequence libraries into suitable templates and to detect the signal and read the RNA sequence (Buermans and den Dunnen 2014).

Identifying mRNAs that differ in their expression patterns under different experimental conditions and further defining the identity of those respective genes by differential display or producing the snapshot of mRNAs that correspond to fragments of those transcripts are the process also known as serial analysis of gene expression (SAGE) (Velculescu et al. 1995; Harbers and Carninci 2005; van Ruissen and Baas 2007; Hogh and Nielsen 2008; Matsumura et al. 2012). LongSAGE analyses in *Emiliania huxleyi* revealed many new differentially regulated gene sequences and also assigned regulation data to EST sequences with no database homology and unknown function, which highlighted uncharacterized aspects of *E. huxleyi* N and P physiology (Dyhrman et al. 2006). Alternatively assessing changes in the expression pattern of previously defined genes using cDNA or oligonucleotide microarrays (Kagnoff and Eckmann 2001) or chip-based nanoliter-volume reverse transcription polymerase chain reaction (RT-PCR), which measures gene expression for several thousands of genes simultaneously at higher sensitivity and accuracy than microarrays (Stedtfeld et al. 2008).

ESTs are small pieces of the DNA sequence (usually ~200–500 nucleotides long) that are generated by sequencing either one or both ends of an expressed gene. The EST studies (Scala et al. 2002; Maheswari et al. 2005, 2009), together with the first whole genome sequences from diatoms, *Thalassiosira pseudonana* (Armbrust et al. 2004) and *Phaeodactylum tricornutum* (Bowler et al. 2008), revealed that less than 50% of diatom genes can be assigned a putative function using homology-based methods. The diatom EST database (Maheswari et al. 2005, 2009) enabled comparative studies of eukaryotic algal genomes and revealed some interesting differences in the genes involved in basic cell metabolism (Montsant et al. 2005; Herve et al. 2006), along with the key signaling and regulatory pathways (Montsant et al. 2007), carbohydrate metabolism (Maheswari et al. 2005; Kroth et al. 2008), silica metabolism (Lopez et al. 2005; Montsant et al. 2005), and nitrogen metabolism (Allen et al. 2006). Patterns in transcript (ESTs) and protein abundance revealed the fact that *T. pseudonana* was evolved with sophisticated response to phosphorous (P) deficiency that involves multiple biochemical strategies that are essential for its ability to respond to variations in environmental P availability (Dyhrman et al. 2012).

Recent studies in microalgae have demonstrated the use of next-generation sequencers such as Roche 454 sequencer, Applied Biosystems SOLiD, and Illumina systems for direct sequencing of the cDNA converted from whole transcriptomes (Miller et al. 2010; Rismani-Yazdi et al. 2011, 2012; Liang et al. 2013; Lv et al. 2013; Yang et al. 2013; Corteggiani Carpinelli et al. 2014; Tanaka et al. 2015). Some of the salient features in these technologies have been highlighted here.

15.2.2.1 Illumina Technology

The Illumina technology takes place in a flow cell where all the enzymatic processes and imaging steps use bridge amplification method for polony generation and a sequence by synthesis approach for sequencing the templates (Buermans and den Dunnen 2014), generating a high number of cluster reads nearly around ~180 million per lane. The run time for HiSeq and MiSeq are 12 days and 65 h with around 3 billion and 25 million reads per run, respectively. Over many cycles, errors in reads tend to accumulate leading to decrease in quality toward the end of reads. Studies on the transcriptome and lipid biosynthetic pathways in marine *Nannochloropsis* species using Illumina HiSeq 2000 showed a total of 29,203 unigenes that were differentially expressed under conditions of low- and high-lipid-producing phases, wherein among those 195 unigenes were involved in lipid metabolism and 315 unigenes were putatively transcription factors (Zheng et al. 2013).

15.2.2.2 SOLiD Sequencing

The SOLiD systems use an intricate sequencing-by-ligation scheme to read the DNA (Shendure et al. 2005). Similar to Illumina sequencing, SOLiD systems can also generate paired-end reads, but the max read length is limited to ~75 bases due to ligation scheme. The main advantage of this technique is that each base is checked by two octomer ligations, which results in a significant increase in read accuracy (Buermans and den Dunnen 2014). Chromosome scale genome assembly and transcriptome profiling of marine *Nannochloropsis gaditana* under nitrogen depletion revealed traits of genes involved in DNA recombination, RNA silencing, and cell wall synthesis (Corteggiani Carpinelli et al. 2014). This study also showed that the content of lipids increased drastically, but without detectable major changes in expression of the genes involved in their biosynthesis and at the same time significant downregulation of mitochondrial gene expression, suggesting that the acetyl-CoA and NAD(P)H, normally oxidized through the mitochondrial respiration, are available for fatty acid synthesis, thus increasing the flux through lipid biosynthetic pathway (Corteggiani Carpinelli et al. 2014).

15.2.2.3 Roche 454

Based on similar technology, Roche has produced two platforms, the FLX+ and the benchtop GS junior system. These utilize pyrosequencing approach (Margulies et al. 2005), where the sequencing is detected by light generated by luciferase-mediated conversion of luciferin to oxyluciferin upon primer extension. The main advantage of this system is the generation of long reads (~500–800 bp) (Buermans and den Dunnen 2014). Next-generation DNA pyrosequencing technology analyses in *Dunaliella tertiolecta* identified that the majority of lipid and starch biosynthesis and catabolism pathways through assembled transcriptome provide a direct metabolic engineering methodology to enhance the quantity and quality of microalgae-based biofuel feedstock (Rismani-Yazdi et al. 2011).

The oleaginous eukaryotic algae (the B race of *Botryococcus braunii* is unique in producing large amounts of liquid hydrocarbons of terpenoid origin and *de novo* assembly of 1,334,609 next-generation pyrosequencing reads) yielded transcriptomic database of 46,422 contigs with an average length of ~756 bp and reconstruction of pathway highlighted that these pathways divert photosynthetic carbon into tetraterpenoid carotenoids, its isoforms and chlorophyll (Molnar et al. 2012).

Oleaginous microalga *Neochloris oleoabundans* is a unicellular green microalga belonging to chlorophyta is reported to produce total lipids 35%–55% dry cell weight and 10% TAGs in response to the nitrogen stress condition (Li et al. 2008; Griffiths and Harrison 2009). Metabolic measurements revealed by genes and pathway expression under the nitrogen limitation in *N. oleoabundans* showed that carbon is portioned toward TAG production and the overexpression of fatty acid synthesis pathway was bolstered by repression of the β-oxidation pathway along with upregulation of genes encoding for the pyruvate dehydrogenase complex that funnels acetyl-CoA to lipid biosynthesis (Rismani-Yazdi et al. 2012). Studies on the integration of phenotypic, genomic, and transcriptomic data across a *Nannochloropsis* phylogeny provided new insights into the molecular mechanisms driving the diversity and evolution of these oleaginous microalgae (Wang et al. 2014).

15.2.2.4 Ion Torrent Technology

The sequence templates for the proton sequence platforms are produced on a bead or sphere through emulsion PCR (emPCR) method (Dressman et al. 2003; Nakano et al. 2003). Creating oil–water emulsions generates small reaction vesicles, each sphere containing a single library molecule and other reagents needed for amplification. emPCR amplifies each individual library molecule bound to beads and ultimate signal is detected. The Ion Torrent chip contains a flow compartment and solid-state pH sensor microarrayed wells. The detection of incorporated bases depends on release of hydrogen ions (H^+) released with extension of each nucleotide, which leads to pH change within sensor wells.

15.2.2.5 RSII Technology

The RSII system from Pacific Biosciences utilizes the principle of single-molecule real-time (SMRT) sequencing technology that is quite different from previously discussed techniques. This technology works with single-molecule detection, that is, incorporation of single fluorescently labeled nucleotide. Hence, template preparation does not require any amplification step. It also does not rely on interrupted cycles of extension like previously discussed techniques; instead, the signals are recorded in real time (Lundquist et al. 2008). Despite low output of the system per SMRT, this system gives advantage of long-read data, absence of GC bias, and real-time analyses of polymerase during reaction (Buermans and den Dunnen 2014).

15.2.2.6 HeliScope

An alternative advanced technology that was employed in microbial systems not yet reported in microalgal systems is direct RNA-seq technology, which allows massively parallel sequencing of RNA molecules directly without prior synthesis of cDNA or the need for ligation/amplification steps developed by Helicos Genetic Analysis System (Ozsolak et al. 2009; Ozsolak and Milos 2011c), where both the abundance and the identity of mRNA molecules can be determined in one analytical process. The major advantage of NGS over the traditional sequencing method is the dramatically increased degree of parallelism, which can be represented by the number of reads (i.e., the number of DNA templates that can be sequenced simultaneously) in a single sequencing run and the number of sequenced bases per day (Zhang et al. 2010; Jutur and Asha 2015b). HeliScope from Helicos is also based on single-molecule synthesis technology like RSII and offers the same advantages as the RSII system, with reads per run being 500 million and output of around 15 Gb data, quite high as compared to RSII systems but with a longer run time of around 10 days.

15.2.3 Assembly and Annotation

A number of *de novo* transcriptome assemblers such as Rnnotator, Multiple-k, ALLPATHS, and Trans-ABySS have been used to assemble the data sets using De Bruijn graph–based approach (Pevzner et al. 2001; Butler et al. 2008; Simpson et al. 2009; Martin et al. 2010; Robertson et al. 2010; Surget-Groba and Montoya-Burgos 2010). These assemblers reconstruct the transcripts from a broad range of expression levels and merge contigs after removing redundancy. In contrast, other *de novo* assemblers such as Trinity and Velvet/ Oases directly traverse the De Bruijn graph for assembly of each isoform (Zerbino and Birney 2008; Garber et al. 2011; Schulz et al. 2012). *De novo* assembly of higher eukaryotic transcriptomes is a time-consuming task as the data sets are quite large and may require hundreds of gigabytes of memory and can run for days to weeks. Longer reads, such as Roche 454 reads, can also be integrated into these assemblers, which might help in improving their ability to resolve alternative isoforms (Martin and Wang 2011). The assembled sequences are later subjected to BLASTx (Altschul et al. 1990) similarity searches and annotated via Gene Ontology (GO) (Ashburner et al. 2000) and Kyoto Encyclopedia of Genes and Genomes (KEGG) Orthology (KO) (Ogata et al. 1999) identifiers for gene identification.

15.2.4 Reconstruction of Metabolic Pathways

The first *de novo* transcriptome sequencing and annotation were done in an unsequenced microalgae *D. tertiolecta* using Roche 454 pyrosequencing (Rismani-Yazdi et al. 2011). The assembled sequences lead to identification of majority of lipid and starch biosynthesis as well as catabolic pathways in *D. tertiolecta*. This work has been a great advance in the pursuit of high-level systems biology approaches, demonstrating the ability of *de novo* transcriptomics to unravel the pathways and targets of interest. Later, a number of *de novo* transcriptomic studies have been performed in microalgae. The first *de novo* transcriptomic study, the utility of *de novo* transcriptomics as a search model for proteomic analysis, was demonstrated in *Chlorella vulgaris* (Guarnieri et al. 2011). This study revealed the upregulation of fatty acid and TAG biosynthesis in *C. vulgaris* under stress condition (nitrogen depletion). In another study on *Chlamydomonas moewusii*, RNA-seq investigation provided useful insights on the expression profile of contigs associated with starch biosynthesis, glycolysis, anaerobic fermentation, and H_2 photoproduction (Yang et al. 2013). *De novo* transcript profiling of *Botryosphaerella sudeticus* revealed downregulation of ESTs encoding photosynthetic function under nitrogen deprived conditions (Sun et al. 2013). Comparative analysis of three commercially valuable red algae *Betaphycus gelatinus*, *Kappaphycus alvarezii*, and *Eucheuma denticulatum* through *de novo* transcriptomic analysis (Song et al. 2014) revealed many differentially expressed genes involved in glycolysis, carbon fixation, and galactose, fructose, and mannose metabolism as well as genes involved in sulfur metabolism related to carrageenan biosynthesis, an important compound in food, pharmaceutical, cosmetics, printing, and textile industries. Genes involved in carbon fixation and carotenoid biosynthesis have been described in *Trentepohlia jolithus* through *de novo* transcriptomics (Li et al. 2014) demonstrated how this microalgae survives in extreme conditions of cold and high altitude.

15.3 Conclusions

In summary, advances in both reference-based and *de novo* transcriptomics have expanded RNA-seq applications to practically any organism. However, over the past decade, there is a significant advancement in alternative NGS technologies for developing rapid, simplified library preparation methods and high-throughput *de novo* sequencing approaches. Meanwhile, experimental RNA-seq and sequencing protocols are continually improving and should greatly reduce the informatics burdens and challenges (Martin and Wang 2011). Following on the footsteps of the sequencing revolution brought by Sanger sequencing, NGS has gained profound attention in addressing fundamental questions and providing new insights among photosynthetic eukaryotic algae (Kim et al. 2014). Henceforth, a simplified breakthrough approach in *de novo* transcriptomics will be a great asset for genetic engineering of marine microalgae with a special focus on next-generation renewable biofuels.

References

Allen, A. E., A. Vardi, and C. Bowler. 2006. An ecological and evolutionary context for integrated nitrogen metabolism and related signaling pathways in marine diatoms. *Curr Opin Plant Biol* 9(3):264–273.

Altschul, S. F., W. Gish, W. Miller, E. W. Myers, and D. J. Lipman. 1990. Basic local alignment search tool. *J Mol Biol* 215(3):403–410.

Anders, S. and W. Huber. 2010. Differential expression analysis for sequence count data. *Genome Biol* 11(10):R106.

Armbrust, E. V., J. A. Berges, C. Bowler et al. 2004. The genome of the diatom *Thalassiosira pseudonana*: Ecology, evolution, and metabolism. *Science* 306(5693):79–86.

Asha, A. N., M. S. Kashif, and P. P. Jutur. 2015. Chapter 26—Genetic engineering of microalgae for production of value-added ingredients. In *Handbook of Marine Microalgae: Biotechnology Advances*, S. K. Kim (ed.). Amsterdam, the Netherlands: Elsevier Science.

Ashburner, M., C. A. Ball, J. A. Blake et al. 2000. Gene ontology: Tool for the unification of biology. The Gene Ontology Consortium. *Nat Genet* 25(1):25–29.

Bouck, A. and T. Vision. 2007. The molecular ecologist's guide to expressed sequence tags. *Mol Ecol* 16(5):907–924.

Bourdon, V., F. Naef, P. H. Rao et al. 2002. Genomic and expression analysis of the 12p11-p12 amplicon using EST arrays identifies two novel amplified and overexpressed genes. *Cancer Res* 62(21):6218–6223.

Bowler, C., A. E. Allen, J. H. Badger et al. 2008. The *Phaeodactylum* genome reveals the evolutionary history of diatom genomes. *Nature* 456(7219):239–244.

Buermans, H. P. J. and J. T. den Dunnen. 2014. Next generation sequencing technology: Advances and applications. *BBA – Mol Basis Dis* 1842(10):1932–1941.

Burset, M., I. A. Seledtsov, and V. V. Solovyev. 2000. Analysis of canonical and non-canonical splice sites in mammalian genomes. *Nucleic Acids Res* 28(21):4364–4375.

Butler, J., I. MacCallum, M. Kleber et al. 2008. ALLPATHS: *De novo* assembly of whole-genome shotgun microreads. *Genome Res* 18(5):810–820.

Chen, S., P. Yang, F. Jiang, Y. Wei, Z. Ma, and L. Kang. 2010. *De novo* analysis of transcriptome dynamics in the migratory locust during the development of phase traits. *PLoS ONE* 5(12):e15633.

Chen, Z. and X. Duan. 2011. Ribosomal RNA depletion for massively parallel bacterial RNA-sequencing applications. *Methods Mol Biol* 733:93–103.

Cheung, F., J. Win, J. M. Lang et al. 2008. Analysis of the *Pythium ultimum* transcriptome using Sanger and pyrosequencing approaches. *BMC Genomics* 9:542.

Corteggiani Carpinelli, E., A. Telatin, N. Vitulo et al. 2014. Chromosome scale genome assembly and transcriptome profiling of *Nannochloropsis gaditana* in nitrogen depletion. *Mol Plant* 7(2):323–335.

Dong, H. P., E. Williams, D. Z. Wang et al. 2013. Responses of *Nannochloropsis oceanica* IMET1 to long-term nitrogen starvation and recovery. *Plant Physiol* 162(2):1110–1126.

Dressman, D., H. Yan, G. Traverso, K. W. Kinzler, and B. Vogelstein. 2003. Transforming single DNA molecules into fluorescent magnetic particles for detection and enumeration of genetic variations. *Proc Natl Acad Sci USA* 100(15):8817–8822.

Dyhrman, S. T., S. T. Haley, S. R. Birkeland, L. L. Wurch, M. J. Cipriano, and A. G. McArthur. 2006. Long serial analysis of gene expression for gene discovery and transcriptome profiling in the widespread marine coccolithophore *Emiliania huxleyi*. *Appl Environ Microbiol* 72(1):252–260.

Dyhrman, S. T., B. D. Jenkins, T. A. Rynearson et al. 2012. The transcriptome and proteome of the diatom *Thalassiosira pseudonana* reveal a diverse phosphorus stress response. *PLoS ONE* 7(3):e33768.

Eveland, A. L., D. R. McCarty, and K. E. Koch. 2008. Transcript profiling by 3′-untranslated region sequencing resolves expression of gene families. *Plant Physiol* 146(1):32–44.

Falkowski, P. G., R. T. Barber, and V. V. Smetacek. 1998. Biogeochemical controls and feedbacks on ocean primary production. *Science* 281(5374):200–207.

Fang, S.-C. 2014. Chapter 3—Metabolic engineering and molecular biotechnology of microalgae for fuel production. In *Biofuels from Algae*, A. Pandey, D.-J. Lee, Y. Chisti, and C. R. Soccol (eds.). Amsterdam, the Netherlands: Elsevier, pp. 47–65.

Fei, Z., X. Tang, R. M. Alba et al. 2004. Comprehensive EST analysis of tomato and comparative genomics of fruit ripening. *Plant J* 40(1):47–59.

Garber, M., M. G. Grabherr, M. Guttman, and C. Trapnell. 2011. Computational methods for transcriptome annotation and quantification using RNA-seq. *Nat Methods* 8(6):469–477.

Gonzalo, M. J., M. Oliver, J. Garcia-Mas et al. 2005. Simple-sequence repeat markers used in merging linkage maps of melon (*Cucumis melo* L.). *Theor Appl Genet* 110(5):802–811.

Griffiths, M. J. and S. T. L. Harrison. 2009. Lipid productivity as a key characteristic for choosing algal species for biodiesel production. *J Appl Phycol* 21(5):493–507.

Grossman, A. R. 2005. Paths toward algal genomics. *Plant Physiol* 137(2):410–427.

Guarnieri, M. T., A. Nag, S. L. Smolinski, A. Darzins, M. Seibert, and P. T. Pienkos. 2011. Examination of triacylglycerol biosynthetic pathways via *de novo* transcriptomic and proteomic analyses in an unsequenced microalga. *PLoS ONE* 6(10):e25851.

Harbers, M. and P. Carninci. 2005. Tag-based approaches for transcriptome research and genome annotation. *Nat Methods* 2(7):495–502.

He, S., O. Wurtzel, K. Singh et al. 2010. Validation of two ribosomal RNA removal methods for microbial metatranscriptomics. *Nat Methods* 7(10):807–812.

Hemaiswarya, S., R. Raja, R. Ravikumar, A. Y. Kumar, and I. S. Carvalho. 2013. Chapter 18—Microalgal omics and their applications. In *OMICS: Applications in Biomedical, Agricultural, and Environmental Sciences*, D. Barh, V. Zambare, and V. Azevedo (eds.). Boca Raton, FL: CRC Press, pp. 439–450.

Herve, C., T. Tonon, J. Collen, E. Corre, and C. Boyen. 2006. NADPH oxidases in Eukaryotes: Red algae provide new hints! *Curr Genet* 49(3):190–204.

Hogh, A. L. and K. L. Nielsen. 2008. SAGE and LongSAGE. *Methods Mol Biol* 387:3–24.

Horak, C. E. and M. Snyder. 2002. Global analysis of gene expression in yeast. *Funct Integr Genomics* 2(4–5):171–180.

Jutur, P. P. and A. N. Asha. 2015a. Chapter 8—Marine microalgae: Exploring the systems through an omics approach for biofuel production. In *Marine Bioenergy: Trends and Developments*, S. K. Kim and C. G. Lee (eds.). Boca Raton, FL: CRC Press, pp. 149–162.

Jutur, P. P. and A. N. Asha. 2015b. Chapter 24—Genetic engineering of marine microalgae to optimize bioenergy production. In *Handbook of Marine Microalgae: Biotechnology Advances*, S. K. Kim (ed.). Amsterdam, the Netherlands: Elsevier Science.

Kagnoff, M. F. and L. Eckmann. 2001. Analysis of host responses to microbial infection using gene expression profiling. *Curr Opin Microbiol* 4(3):246–250.

Kim, K. M., J. H. Park, D. Bhattacharya, and H. S. Yoon. 2014. Applications of next-generation sequencing to unravelling the evolutionary history of algae. *Int J Syst Evol Microbiol* 64(Pt 2):333–345.

Kozarewa, I., Z. Ning, M. A. Quail, M. J. Sanders, M. Berriman, and D. J. Turner. 2009. Amplification-free Illumina sequencing-library preparation facilitates improved mapping and assembly of (G+C)-biased genomes. *Nat Methods* 6(4):291–295.

Kristiansson, E., N. Asker, L. Forlin, and D. G. Larsson. 2009. Characterization of the *Zoarces viviparus* liver transcriptome using massively parallel pyrosequencing. *BMC Genomics* 10:345.

Kroth, P. G., A. Chiovitti, A. Gruber et al. 2008. A model for carbohydrate metabolism in the diatom *Phaeodactylum tricornutum* deduced from comparative whole genome analysis. *PLoS ONE* 3(1):e1426.

Levin, J. Z., M. Yassour, X. Adiconis et al. 2010. Comprehensive comparative analysis of strand-specific RNA sequencing methods. *Nat Methods* 7(9):709–715.

Li, Q., J. Liu, L. Zhang, and Q. Liu. 2014. *De novo* transcriptome analysis of an aerial microalga *Trentepohlia jolithus*: Pathway description and gene discovery for carbon fixation and carotenoid biosynthesis. *PLoS ONE* 9(9):e108488.

Li, Y., M. Horsman, B. Wang, N. Wu, and C. Q. Lan. 2008. Effects of nitrogen sources on cell growth and lipid accumulation of green alga *Neochloris oleoabundans*. *Appl Microbiol Biotechnol* 81(4):629–636.

Liang, C., S. Cao, X. Zhang et al. 2013. *De novo* sequencing and global transcriptome analysis of *Nannochloropsis* sp. (Eustigmatophyceae) following nitrogen starvation. *BioEner Res* 6(2):494–505.

Lopez, P. J., J. Descles, A. E. Allen, and C. Bowler. 2005. Prospects in diatom research. *Curr Opin Biotechnol* 16(2):180–186.

Lundquist, P. M., C. F. Zhong, P. Zhao et al. 2008. Parallel confocal detection of single molecules in real time. *Opt Lett* 33(9):1026–1028.

Lv, H., G. Qu, X. Qi, L. Lu, C. Tian, and Y. Ma. 2013. Transcriptome analysis of *Chlamydomonas reinhardtii* during the process of lipid accumulation. *Genomics* 101(4):229–237.

Maheswari, U., T. Mock, E. V. Armbrust, and C. Bowler. 2009. Update of the diatom EST database: A new tool for digital transcriptomics. *Nucleic Acids Res* 37(Database issue):D1001–D1005.

Maheswari, U., A. Montsant, J. Goll et al. 2005. The diatom EST database. *Nucleic Acids Res* 33(Database issue):D344–D347.

Mamanova, L., A. J. Coffey, C. E. Scott et al. 2010. Target-enrichment strategies for next-generation sequencing. *Nat Methods* 7(2):111–118.

Marguerat, S. and J. Bahler. 2010. RNA-seq: From technology to biology. *Cell Mol Life Sci* 67(4):569–579.

Margulies, M., M. Egholm, W. E. Altman et al. 2005. Genome sequencing in microfabricated high-density picolitre reactors. *Nature* 437(7057):376–380.

Martin, J., V. M. Bruno, Z. Fang et al. 2010. Rnnotator: An automated *de novo* transcriptome assembly pipeline from stranded RNA-Seq reads. *BMC Genomics* 11:663.

Martin, J. A. and Z. Wang. 2011. Next-generation transcriptome assembly. *Nat Rev Genet* 12(10):671–682.

Matsumura, H., N. Urasaki, K. Yoshida, D. H. Kruger, G. Kahl, and R. Terauchi. 2012. SuperSAGE: Powerful serial analysis of gene expression. *Methods Mol Biol* 883:1–17.

Matsunaga, T., T. Yoshino, Y. Liang, M. Muto, and T. Tanaka. 2015. Chapter 5–Marine Microalgae. In *Springer Handbook of Marine Biotechnology*, S.-K. Kim (ed.). Berlin, Germany: Springer, pp. 51–63.

McGettigan, P. A. 2013. Transcriptomics in the RNA-seq era. *Curr Opin Chem Biol* 17(1):4–11.

Metzker, M. L. 2010. Sequencing technologies—The next generation. *Nat Rev Genet* 11(1):31–46.

Meyer, E., G. V. Aglyamova, S. Wang et al. 2009. Sequencing and *de novo* analysis of a coral larval transcriptome using 454 GSFlx. *BMC Genomics* 10:219.

Miller, R., G. Wu, R. R. Deshpande et al. 2010. Changes in transcript abundance in *Chlamydomonas reinhardtii* following nitrogen deprivation predict diversion of metabolism. *Plant Physiol* 154(4):1737–1752.

Mitra, R. D. and G. M. Church. 1999. *In situ* localized amplification and contact replication of many individual DNA molecules. *Nucleic Acids Res* 27(24):e34.

Molnar, I., D. Lopez, J. H. Wisecaver et al. 2012. Bio-crude transcriptomics: Gene discovery and metabolic network reconstruction for the biosynthesis of the terpenome of the hydrocarbon oil-producing green alga. *Botryococcus braunii* race B (Showa). *BMC Genomics* 13:576.

Montsant, A., A. E. Allen, S. Coesel et al. 2007. Identification and comparative genomic analysis of signaling and regulatory components in the diatom *Thalassiosira pseudonana*. *J Phycol* 43(3):585–604.

Montsant, A., K. Jabbari, U. Maheswari, and C. Bowler. 2005. Comparative genomics of the pennate diatom *Phaeodactylum tricornutum*. *Plant Physiol* 137(2):500–513.

Nagaraj, S. H., R. B. Gasser, and S. Ranganathan. 2007. A hitchhiker's guide to expressed sequence tag (EST) analysis. *Brief Bioinform* 8(1):6–21.

Nakano, M., J. Komatsu, S. Matsuura, K. Takashima, S. Katsura, and A. Mizuno. 2003. Single-molecule PCR using water-in-oil emulsion. *J Biotechnol* 102(2):117–124.

Nishiyama, T., T. Fujita, I. T. Shin et al. 2003. Comparative genomics of *Physcomitrella patens* gametophytic transcriptome and *Arabidopsis thaliana*: Implication for land plant evolution. *Proc Natl Acad Sci USA* 100(13):8007–8012.

Novaes, E., D. R. Drost, W. G. Farmerie et al. 2008. High-throughput gene and SNP discovery in *Eucalyptus grandis*, an uncharacterized genome. *BMC Genomics* 9:312.

Ogata, H., S. Goto, K. Sato, W. Fujibuchi, H. Bono, and M. Kanehisa. 1999. KEGG: Kyoto encyclopedia of genes and genomes. *Nucleic Acids Res* 27(1):29–34.

Ozsolak, F. and P. M. Milos. 2011a. RNA sequencing: Advances, challenges and opportunities. *Nat Rev Genet* 12(2):87–98.

Ozsolak, F. and P. M. Milos. 2011b. Single-molecule direct RNA sequencing without cDNA synthesis. *Wiley Interdiscip Rev RNA* 2(4):565–570.

Ozsolak, F. and P. M. Milos. 2011c. Transcriptome profiling using single-molecule direct RNA sequencing. *Methods Mol Biol* 733:51–61.

Ozsolak, F., A. R. Platt, D. R. Jones et al. 2009. Direct RNA sequencing. *Nature* 461(7265):814–818.

Parchman, T. L., K. S. Geist, J. A. Grahnen, C. W. Benkman, and C. A. Buerkle. 2010. Transcriptome sequencing in an ecologically important tree species: Assembly, annotation, and marker discovery. *BMC Genomics* 11:180.

Parker, M. S., T. Mock, and E. V. Armbrust. 2008. Genomic insights into marine microalgae. *Annu Rev Genet* 42:619–645.

Passalacqua, K. D., A. Varadarajan, B. D. Ondov, D. T. Okou, M. E. Zwick, and N. H. Bergman. 2009. Structure and complexity of a bacterial transcriptome. *J Bacteriol* 191(10):3203–3211.

Pevzner, P. A., H. Tang, and M. S. Waterman. 2001. An Eulerian path approach to DNA fragment assembly. *Proc Natl Acad Sci USA* 98(17):9748–9753.

Pop, M. and S. L. Salzberg. 2008. Bioinformatics challenges of new sequencing technology. *Trends Genet* 24(3):142–149.

Riggins, C. W., Y. Peng, C. N. Stewart Jr., and P. J. Tranel. 2010. Characterization of de novo transcriptome for waterhemp (*Amaranthus tuberculatus*) using GS-FLX 454 pyrosequencing and its application for studies of herbicide target-site genes. *Pest Manage Sci* 66(10):1042–1052.

Rismani-Yazdi, H., B. Z. Haznedaroglu, K. Bibby, and J. Peccia. 2011. Transcriptome sequencing and annotation of the microalgae *Dunaliella tertiolecta*: Pathway description and gene discovery for production of next-generation biofuels. *BMC Genomics* 12:148.

Rismani-Yazdi, H., B. Z. Haznedaroglu, C. Hsin, and J. Peccia. 2012. Transcriptomic analysis of the oleaginous microalga *Neochloris oleoabundans* reveals metabolic insights into triacylglyceride accumulation. *Biotechnol Biofuels* 5(1):74.

Robertson, G., J. Schein, R. Chiu et al. 2010. *De novo* assembly and analysis of RNA-seq data. *Nat Methods* 7(11):909–912.

Robinson, M. D., D. J. McCarthy, and G. K. Smyth. 2010. edgeR: A bioconductor package for differential expression analysis of digital gene expression data. *Bioinformatics* 26(1):139–140.

Rudd, S. 2003. Expressed sequence tags: Alternative or complement to whole genome sequences? *Trends Plant Sci* 8(7):321–329.

Sam, L. T., D. Lipson, T. Raz et al. 2011. A comparison of single molecule and amplification based sequencing of cancer transcriptomes. *PLoS ONE* 6(3):e17305.

Scala, S., N. Carels, A. Falciatore, M. L. Chiusano, and C. Bowler. 2002. Genome properties of the diatom *Phaeodactylum tricornutum*. *Plant Physiol* 129(3):993–1002.

Schulz, M. H., D. R. Zerbino, M. Vingron, and E. Birney. 2012. Oases: Robust de novo RNA-seq assembly across the dynamic range of expression levels. *Bioinformatics* 28(8):1086–1092.

Schwartz, T. S., H. Tae, Y. Yang et al. 2010. A garter snake transcriptome: Pyrosequencing, *de novo* assembly, and sex-specific differences. *BMC Genomics* 11:694.

Seki, M., M. Narusaka, A. Kamiya et al. 2002. Functional annotation of a full-length *Arabidopsis* cDNA collection. *Science* 296(5565):141–145.

Shendure, J., G. J. Porreca, N. B. Reppas et al. 2005. Accurate multiplex polony sequencing of an evolved bacterial genome. *Science* 309(5741):1728–1732.

Simpson, J. T., K. Wong, S. D. Jackman, J. E. Schein, S. J. Jones, and I. Birol. 2009. ABySS: A parallel assembler for short read sequence data. *Genome Res* 19(6):1117–1123.

Singh, R., A. J. Mattam, P. P. Jutur, and S. S. Yazdani. 2015. Chapter 22—Synthetic biology in biofuels production. In *Encyclopedia of Molecular Cell Biology and Molecular Medicine: Synthetic Biology*, Vol. II (PART VI: Chemicals Production), R. A. Meyers (ed.). Verlag GmbH & Co, Wiley-VCH.

Smith, V. H., B. S. Sturm, F. J. Denoyelles, and S. A. Billings. 2010. The ecology of algal biodiesel production. *Trends Ecol Evol* 25(5):301–309.

Song, L., S. Wu, J. Sun et al. 2014. *De novo* sequencing and comparative analysis of three red algal species of Family Solieriaceae to discover putative genes associated with carrageenan biosynthesis. *Acta Oceanol Sin* 33(2):45–53.

Stedtfeld, R. D., S. W. Baushke, D. M. Tourlousse et al. 2008. Development and experimental validation of a predictive threshold cycle equation for quantification of virulence and marker genes by high-throughput nanoliter-volume PCR on the OpenArray platform. *Appl Environ Microbiol* 74(12):3831–3838.

Stinchcombe, J. R. and H. E. Hoekstra. 2008. Combining population genomics and quantitative genetics: Finding the genes underlying ecologically important traits. *Heredity (Edinb)* 100(2):158–170.

Sun, D., J. Zhu, L. Fang, X. Zhang, Y. Chow, and J. Liu. 2013. *De novo* transcriptome profiling uncovers a drastic downregulation of photosynthesis upon nitrogen deprivation in the non-model green alga *Botryosphaerella sudeticus*. *BMC Genomics* 14(1):1–18.

Surget-Groba, Y. and J. I. Montoya-Burgos. 2010. Optimization of *de novo* transcriptome assembly from next-generation sequencing data. *Genome Res* 20(10):1432–1440.

Tanaka, T., Y. Maeda, A. Veluchamy et al. 2015. Oil accumulation by the oleaginous diatom *Fistulifera solaris* as revealed by the genome and transcriptome. *Plant Cell* 27(1):162–176.

van Ruissen, F. and F. Baas. 2007. Serial analysis of gene expression (SAGE). *Methods Mol Biol* 383:41–66.

Velculescu, V. E., L. Zhang, B. Vogelstein, and K. W. Kinzler. 1995. Serial analysis of gene expression. *Science* 270(5235):484–487.

Vera, J. C., C. W. Wheat, H. W. Fescemyer et al. 2008. Rapid transcriptome characterization for a non-model organism using 454 pyrosequencing. *Mol Ecol* 17(7):1636–1647.

Wang, D., K. Ning, J. Li et al. 2014. *Nannochloropsis* genomes reveal evolution of microalgal oleaginous traits. *PLoS Genet* 10(1):e1004094.

Wang, Z., M. Gerstein, and M. Snyder. 2009. RNA-Seq: A revolutionary tool for transcriptomics. *Nat Rev Genet* 10(1):57–63.

Wheat, C. W. 2010. Rapidly developing functional genomics in ecological model systems via 454 transcriptome sequencing. *Genetica* 138(4):433–451.

Wilhelm, B. T. and J. R. Landry. 2009. RNA-Seq-quantitative measurement of expression through massively parallel RNA-sequencing. *Methods* 48(3):249–257.

Yang, S., M. T. Guarnieri, S. Smolinski, M. Ghirardi, and P. T. Pienkos. 2013. *De novo* transcriptomic analysis of hydrogen production in the green alga *Chlamydomonas moewusii* through RNA-Seq. *Biotechnol Biofuels* 6(1):118.

Ye, R. W., T. Wang, L. Bedzyk, and K. M. Croker. 2001. Applications of DNA microarrays in microbial systems. *J Microbiol Methods* 47(3):257–272.

Zerbino, D. R. and E. Birney. 2008. Velvet: Algorithms for *de novo* short read assembly using de Bruijn graphs. *Genome Res* 18(5):821–829.

Zhang, W., F. Li, and L. Nie. 2010. Integrating multiple 'omics' analysis for microbial biology: Application and methodologies. *Microbiology* 156(Pt 2):287–301.

Zhang, X., N. Ye, C. Liang et al. 2012. *De novo* sequencing and analysis of the *Ulva linza* transcriptome to discover putative mechanisms associated with its successful colonization of coastal ecosystems. *BMC Genomics* 13(1):565.

Zheng, M., J. Tian, G. Yang et al. 2013. Transcriptome sequencing, annotation and expression analysis of *Nannochloropsis* sp. at different growth phases. *Gene* 523(2):117–121.

16

Studies on Nervous Necrosis Virus Pathogenesis by Transcriptomic Approaches

Ming-Wei Lu, Chun-Hsi Tso, and Thirunavukkarasu Periyasamy

CONTENTS

16.1 Introduction

Nervous necrosis virus (NNV) belongs to the Nodaviridae family, which typically causes viral encephalopathy and retinopathy (VER). VER was first reported to be present in barramundi (*Lates calcarifer*) farmed in Australia (Glazebrook et al. 1990). There are over 30 species of culture marine fish that could be infected by NNV worldwide (Munday et al. 2002). An acute infected fish typically shows loss of appetite, darker skin color, and abnormal spiral swimming behavior. In histology, this acute infection usually shows serious VER symptoms, such as vacuolating necrosis in the brain, retina, and spinal cord. This acute infection could finally cause such high mortality of up to 100% in larval and juvenile fish (Tanaka et al. 2004).

Virion of NNV is a non-enveloped, icosahedron-shaped virus containing two single-stranded, positive-sense RNA genomes. RNA1 (3.1 kb) and RNA2 (1.4 kb) encode the viral RNA-dependent RNA polymerase (RdRp) and viral capsid protein, respectively (Tan et al. 2001). Moreover, RNA3 (0.38 kb), as a subgenome from RNA1, could be translated to proteins B1 and B2. It is known that B2 can be an antagonist to the RNAi activity of the host cell (Ou et al. 2007). Furthermore, NNV may cause persistent infection in older fish and cause vertical infection to their offspring (Johansen et al. 2002).

Several studies showed that NNV infection induced various gene expressions in both innate and acquired immune responses. Interferon, one of the major antiviral response systems, plays an important role in the suppression of NNV replication in the host cell. Previous studies showed that interferon is involved in the persistence of NNV infection (Lu et al. 2008; Wu and Chi 2006). Interferon also activates the Mx gene to resist the NNV infection by interfering the viral RdRp activity (Wu et al. 2010). The signal transducer and activator of transcription (STAT1) had also been reported to play an important role in the innate immune response against NNV infection (Tso et al. 2013). The expression level of inflammation gene, tumor necrosis factor alpha, and interleukin-1 beta (IL-1β) was also upregulated while NNV was starting to replicate in its target organ (Poisa-Beiro et al. 2008). In a cellular immune response, gene coding for the

IgM light chain showed potent upregulation in its expression after NNV infection (Scapigliati et al. 2010). Moreover, cluster of differentiation 8 alpha, an indispensable membrane glycoprotein on T lymphocytes quickly increased from NNV-infected fish, revealed that NNV also induced specific immune response (Chang et al. 2011). Most of these expression profiles studied earlier depended on quantitative PCR (qPCR) technique. Even though qPCR could provide accurate quantitative result of specific gene response against infection, the throughput of such approaches is still limited at a time. Moreover, the qPCR method requires identified transcript sequences as the template for designing a primer. This limit greatly increases the difficulty of gene profiling on nonmodel studies. For reasons outlined earlier, the overall molecular pathology research of NNV infection in the host was quite fragmented in these studies. This topic needs advanced technique to answer. Over the last decade, DNA microarray and next-generation sequencing were regarded as suitable tools to answer the molecular pathology question of NNV infection. In this chapter, we will review these two methods in order.

16.2 DNA Microarray–Based Gene Profiling

To generally understand that how does the host cell react to the NNV infection, several high-throughput techniques were applied. High-throughput technique could be a powerful platform to analyze the difference between the transcription and expression profile from several infection periods on a transcriptome-wide scale. DNA microarray is a well-established classic high-throughput method. The development of DNA microarrays takes over the place of single-gene approach by analyzing thousands of known or putative transcripts at the same time. This tool presents a robust ability for researchers to analyze the gene profile and detect differential gene expression relative to experimental treatment. Due to its cost-effectiveness, the DNA microarray technique is one of the common methods for researchers to perform transcriptome research in the world.

The first transcriptomic study about immune response to nodavirus infection was approached by suppression subtractive hybridization and microarray in the brain tissue of infected sea bream (Dios et al. 2007). The result showed that ubiquitin-conjugating enzyme 7 interacting protein, interferon-induced protein with helicase C domain 1 (mda-5), and heat shock proteins 70 (Hsp70) were all observed an induction after infection by virus injection *in vivo*. Since ubiquitin-conjugating enzyme and mda-5 were described to play important roles in apoptosis, the upregulated result suggested that apoptosis is involved in the nodavirus infection. This result could provide an explanation that the vacuolation observed in sea bream after nodavirus infection in the brain, spinal cord, and retina may be the result of an apoptotic effect.

A research on turbot produced three expressed sequence tag (EST) libraries from the kidneys, livers, and gills of infected fish with nodavirus stimulated with polyriboinosinic polyribocytidylic acid (poly I:C). These EST clones were constructed into a cDNA microarray to analyze the differential gene expression profile in response to the infection. In the 3-, 6-, 24-, and 72-hour postinfection (hpi) expression profile results, 94 genes were identified to be significantly upregulated after infection. Most of these genes (89 genes) were upregulated at 24 hpi. A large amount of genes assigned to stress-related response and immune defense response, including interferon-inducible protein 35 (IFI35), saxitoxin-binding protein (BiP) 1, serum lectin isoform 4, serum-inducible protein kinase, ceruloplasmin, haptoglobin, and Mx, were found upregulated at 24 hpi. Mx and IFI35 were the two important genes that induced interferon pathway and significantly upregulated with 16.9- and 2.5-fold increase, respectively. On the other hand, 12 genes were downregulated, especially at 6 hpi. These genes included F-box only protein 25, 5-aminolevulinate synthase, phosphatidylino-sitol 4-kinase, and some unidentified genes. There were no significant pathways indicated from these downregulated genes after infection (Park et al. 2009).

To understand the differential ability of NNV resistance in organisms, the differential gene expression of survived orange-spotted grouper larvae from betanodavirus outbreak was also compared to 9600-clone-containing grouper larvae cDNA microarray (Wu et al. 2012). After statistical analysis, 79 clones were found active in healthy larvae and 351 clones were found active in infected larvae. After sequencing and identification, 45 genes and 19 considered novel identifications were submitted to national center for biotechnology information (NCBI) database. These viral-repressed and viral-induced genes were speculated to cause the variation of resistance against NNV in grouper larvae. Various muscle contraction genes, including myosin BiP, myosin light chain, troponin T, adenylate kinase, and parvalbumin, showed significant induction in the healthy larvae than diseased larvae. This result provides a reasonable explanation of classical clinical signs, such as abnormal swimming of larvae in NNV infection. The unnatural swimming behavior of infected larvae might be caused by movement-related gene deficiency.

A challenge experiment with NNV was also carried out in Atlantic cod juveniles and analyzed by microarray (Krasnov et al. 2013). The expression profile showed that the innate immune response reached peak levels at the fifth day postinfection. Various sensors for pathogen recognition, including lymphocyte cells, Toll-like receptors, and nucleotide-binding oligomerization domain (NOD)-like receptors were induced by NNV infection. The Jak-Stat, mitogen-activated protein kinase (MAPK) and nuclear factor of kappa light polypeptide gene enhancer in B cells (NF-κB)-kappa B cascades were found to play an important role in mediating acute inflammation. Several chemokines, cytokines, and their receptors were upregulated. This suggested that the communication between cells and the enrollment of immune cells were activated due to the presence of NNV. The induced downstream immune response includes both cellular (NK-mediated cytotoxicity, antibody-mediated phagocytosis and phagosomes) and humoral (lysozymes, scavengers, heparin binding, and blood coagulation) factors. A noticeable result is that this analysis predicted many previous unidentified genes and improved coverage of several functional groups and pathways in Atlantic cod.

From the earlier microarray study, many antiviral responses of host against nodavirus infection were identified. From these studies, researchers could reveal how the antiviral genes react to infection. However, there is an obvious disadvantage of microarray method on gene expression profiling. The construction of microarray needs prior knowledge on the genome sequence of the query experimental material. Lack of complete genome information could greatly reduce the functionality of microarray. The biggest disadvantage of microarray is that it provides limited novelty information. Moreover, the requirement of genome sequence usually becomes an obstacle for the study of gene expression profiling in a nonmodel organism.

16.3 NGS-Based Gene Profiling by RNA-Seq

Since the development of next-generation sequencing (NGS) is going rapidly, the NGS technologies have become inexpensive and easier to use. The NGS tools are matured in processing large-scale sequencing data in a short period. The NGS also brings a revolution to the revolutionary tool of transcriptomics. The RNA-Seq, also called whole transcriptome shotgun sequencing, is a recently developed approach based on NGS. This sequencing-based approach allows researchers to quantify the RNA expression by sequencing read count with great confidence. This sequencing-based method could be performed without prior genome information. Nevertheless, RNA-Seq could also provide whole transcriptomic information. Hence, the NGS also provides notable advantage in nonmodel animal research in contrast to microarray methods. Therefore, the NGS became a suitable and powerful tool to analyze how the cell reacts to the particular treatment. This technique could provide researchers answer to several biological questions especially in a nonmodel animal. For example, the infection route of NNV in the host is not fully clarified and the cellular response of the host against NNV remains unclear. NNV is known to cause apoptotic death of the infected cells. Some recent studies have shown that virus-induced endoplasmic reticulum (ER) stress and the apoptotic death have close relation (Benali-Furet et al. 2005; He 2006). The virus-induced ER stress even could decide whether infected cells will survive or die. Severe virus-induced ER stress will drive cells to apoptosis. In the infected cells, accumulation of unfolded viral protein caused ER stress and induced immunoglobulin heavy-chain BiP to assist the cells to relieve the stress. The unfolding protein response (UPR) pathway will be activated by BiP. The UPR includes the following three mediators: PKR-like ER kinase (PERK), activating transcription factor 6 (ATF6), and inositol-requiring enzyme 1 (IRE1). These mediators relieve ER stress by assisting protein folding, reducing translocation of newly synthesized protein into ER, and facilitating protein degradation in ER lumen (He 2006).

At the present time, the research of NNV-induced ER stress by NNV infection is incomplete and needs more investigation. In a recent study, the ER stress induced by NNV infection was described by NGS transcriptomic assay on grouper cell line (Lu et al. 2012). The grouper kidney (GK) cell line was used to produce the transcriptome sequence database by NGS. The workflow is shown in Figure 16.1.

After removing the NGS reads with adaptors, reads with unknown nucleotides larger than 5%, and low-quality reads, 51,198,090 paired reads from 4,607,828,100 nucleotides were generated by using the SOAP *de novo* transcriptome assembly program. By overlapping reads into longer fragments, 204,517 contigs with a mean length of 248 bp were combined. The contigs were assembled into a total number of 66,582 unigenes with a mean length of 603 bp. The unigenes were annotated according to two functional annotation conventions, gene ontology (GO) and cluster of orthologous groups (COG). As shown in Table 16.1, the unigenes were classified into three categories of GO, namely, molecular function, cellular component, and biological process, and 51 subcategories (Figure 16.2). Among the 66,582 unigenes, 76,975 unigenes were annotated because multiple GO functions were assigned to a single unigene in some cases. The biological process was the dominant category, which is made up of 49.57%, followed by cellular component (35.45%) and molecular function (14.99%). A total of 19,261 unigenes were grouped into 25 COG categories. The largest group in COG is the "general function prediction only" (3673, 19.07%), followed by group "transcription" (1708, 8.87%) and group "replication, recombination, and repair" (1680, 8.72%). "Defense mechanism" (73, 0.38%), "extracellular structure" (15, 0.08%), and "nuclear structure" (8, 0.04%) represented the smallest groups (Figure 16.3). After these unigenes were mapped into Kyoto Encyclopedia of Genes and Genomes, a total of 555 unigenes were identified to be related with the protein processing in ER. To characterize the gene expression profile in GK cells against NNV infection, the unigene expression was estimated by the reads per kilobase per million read (RPKM) method. The RPKM method could be directly used to compare the differential gene expression from samples.

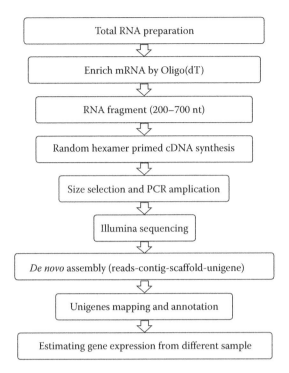

FIGURE 16.1
Workflow of RNA-Seq by using illumina sequencing.

TABLE 16.1

Summary of Gene Ontology Classification for Transcriptome

GO Category	No. of Unigene	Percentage (%)
Biological process	38,153	49.57
Cellular component	27,287	35.45
Molecular function	11,535	14.99
Total	76,975	100

The RPKM value was compared between NNV-infected and mock-infected GK cells at 6 and 33 hpi. A total of 117 unigenes among the 555 ER-related unigenes were affected in NNV-infected GK cells.

Among these 117 genes, 17 important factors (referred from 26 unigenes) in the UPR pathway were upregulated, including BiP, PERK, ATF6, IRE1, GADD34, C/EBP homologous protein (CHOP), and XBP-1. These genes have been included in the NCBI database and summarized with the assigned accession numbers. Further, the quantitative RT-PCR analysis was performed and the results consist of transcriptomic result. The upregulated response of CHOP and GADD34 is the highest and earliest at 6 hpi. Afterwards, the PERK and BiP reached the highest expression level at 33 hpi. Since BiP is the key regulator of UPR pathway, the abundant induction of BiP suggested that there may be an interaction between BiP and viral protein in the NNV-infected GK cell line. From the *de novo* sequence assembly database, the sequence of grouper BiP was identified and jointed from

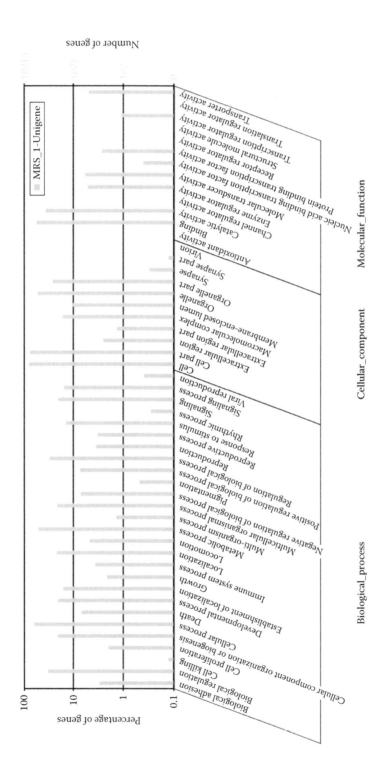

FIGURE 16.2
Histogram of gene ontology classification.

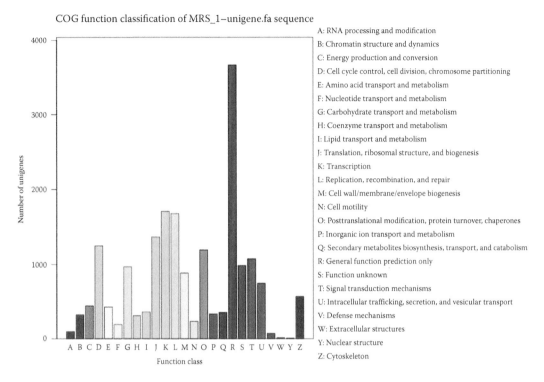

COG function classification of MRS_1−unigene.fa sequence

A: RNA processing and modification
B: Chromatin structure and dynamics
C: Energy production and conversion
D: Cell cycle control, cell division, chromosome partitioning
E: Amino acid transport and metabolism
F: Nucleotide transport and metabolism
G: Carbohydrate transport and metabolism
H: Coenzyme transport and metabolism
I: Lipid transport and metabolism
J: Translation, ribosomal structure, and biogenesis
K: Transcription
L: Replication, recombination, and repair
M: Cell wall/membrane/envelope biogenesis
N: Cell motility
O: Posttranslational modification, protein turnover, chaperones
P: Inorganic ion transport and metabolism
Q: Secondary metabolites biosynthesis, transport, and catabolism
R: General function prediction only
S: Function unknown
T: Signal transduction mechanisms
U: Intracellular trafficking, secretion, and vesicular transport
V: Defense mechanisms
W: Extracellular structures
Y: Nuclear structure
Z: Cytoskeleton

FIGURE 16.3
Histogram of clusters of orthologous groups functional classification.

two overlapped unigenes. A series of protein functional assays of BiP were performed in NNV-infected grouper fin cell line. The interaction between BiP and NNV capsid protein was further confirmed by a coimmunoprecipitation assay. The BiP and capsid protein showed coprecipitation and suggested that capsid protein may have a potential role in the ER stress response. Moreover, there was a recently published study, which shows that BiP also could interact with the RdRp. This interaction may induce the mitochondria-mediated cell death (Su et al. 2011). Hence, these evidence indicated that the BiP may interact with at least NNV capsid protein and RdRp in the NNV-infected cell. In addition to the ER stress study, Lu's transcriptome analysis also revealed the overview of gene regulation among early and later stages of NNV infection at 6 and 33 dpi, respectively (data not published). Several immune-related genes were upregulated after infection, including TLR3, MDA5, NF-κB, and IL-1β. These three genes are members of Toll-like and RIG-I-like signaling pathway. This result suggested that Toll-like and RIG-I-like signaling may play important roles in NNV infection process. Some other immune-related genes such as MDA5 and IRF3 showed relatively high expression level at 6 hpi. TLR3, MDA5, STAT1, IRF3, and IRF7 showed relatively high expression level at 33 hpi. This analysis also detects downregulated genes such as MyD88 and AP1. Although they need further experiment to confirm the interaction between NNV and immune response, the RNA-Seq analysis still provides a great quantity of information at a transcriptome-wide scale. This gene profiling database could bring researchers a greater help. The researchers will be able to discover the details of NNV pathogenesis easily by analyzing the gene profiling database.

16.4 Summary

Every year, NNV could cause devastative loss in various susceptible economical fish, especially in the larvae and juvenile stages. The NNV carrier due to persistent infection on brood fish results in spreading the virus continually and bringing out disease outbreak over and over again. However, the understanding of NNV's pathogenicity is still limited. In this chapter, the transcriptomic approaches of NNV pathogenesis have been generally reviewed. There are two major transcriptomic study tools mentioned, microarray and RNA-Seq. In recent years, the cost reduction of NGS greatly encouraged the researchers to choose the RNA-Seq as the tool for gene profiling. For this reason, the cost-effective advantage of microarray declines a lot. In addition to the cost consideration, the NGS indeed could provide extra advantage for the nonmodel organism transcriptomic study. Sequencing-based approach allows researchers to study any organism without prior genomic sequence by using bioinformatics analysis. This revolution will greatly accelerate the development of transcriptome study and the discovery of gene function. Even though the microarray is still an essential transcriptomic or genomic approach so far, the NGS technology will certainly become essentially important in the near future.

References

Benali-Furet NL, Chami M, Houel L, De Giorgi F, Vernejoul F, Lagorce D, Buscail L, Bartenschlager R, Ichas F, and Rizzuto R. 2005. Hepatitis C virus core triggers apoptosis in liver cells by inducing ER stress and ER calcium depletion. *Oncogene* 24(31):4921–4933.

Chang YT, Kai YH, Chi SC, and Song YL. 2011. Cytotoxic CD8alpha+ leucocytes have heterogeneous features in antigen recognition and class I MHC restriction in grouper. *Fish & Shellfish Immunology* 30(6):1283–1293.

Dios S, Poisa-Beiro L, Figueras A, and Novoa B. 2007. Suppression subtraction hybridization (SSH) and macroarray techniques reveal differential gene expression profiles in brain of sea bream infected with nodavirus. *Molecular Immunology* 44(9):2195–2204.

Glazebrook J, Heasman M, and Beer S. 1990. Picorna-like viral particles associated with mass mortalities in larval barramundi, *Lates calcarifer* Bloch. *Journal of Fish Diseases* 13(3):245–249.

He B. 2006. Viruses, endoplasmic reticulum stress, and interferon responses. *Cell Death & Differentiation* 13(3):393–403.

Johansen R, Ranheim T, Hansen M, Taksdal T, and Totland G. 2002. Pathological changes in juvenile Atlantic halibut *Hippoglossus hippoglossus* persistently infected with nodavirus. *Diseases of Aquatic Organisms* 50(3):161–169.

Krasnov A, Kileng O, Skugor S, Jorgensen SM, Afanasyev S, Timmerhaus G, Sommer AI, and Jensen I. 2013. Genomic analysis of the host response to nervous necrosis virus in Atlantic cod (*Gadus morhua*) brain. *Molecular Immunology* 54(3–4):443–452.

Lu MW, Chao YM, Guo TC, Santi N, Evensen O, Kasani SK, Hong JR, and Wu JL. 2008. The interferon response is involved in nervous necrosis virus acute and persistent infection in zebrafish infection model. *Molecular Immunology* 45(4):1146–1152.

Lu MW, Ngou FH, Chao YM, Lai YS, Chen NY, Lee FY, and Chiou PP. 2012. Transcriptome characterization and gene expression of *Epinephelus* spp in endoplasmic reticulum stress-related pathway during betanodavirus infection in vitro. *BMC Genomics* 13:651.

Munday B, Kwang J, and Moody N. 2002. Betanodavirus infections of teleost fish: A review. *Journal of Fish Diseases* 25(3):127–142.

Ou MC, Chen YM, Jeng MF, Chu CJ, Yang HL, and Chen TY. 2007. Identification of critical residues in nervous necrosis virus B2 for dsRNA-binding and RNAi-inhibiting activity through by bioinformatic analysis and mutagenesis. *Biochemical and Biophysical Research Communications* 361(3):634–640.

Park KC, Osborne JA, Montes A, Dios S, Nerland AH, Novoa B, Figueras A, Brown LL, and Johnson SC. 2009. Immunological responses of turbot (*Psetta maxima*) to nodavirus infection or poly-riboinosinic polyribocytidylic acid (pIC) stimulation, using expressed sequence tags (ESTs) analysis and cDNA microarrays. *Fish & Shellfish Immunology* 26(1):91–108.

Poisa-Beiro L, Dios S, Montes A, Aranguren R, Figueras A, and Novoa B. 2008. Nodavirus increases the expression of Mx and inflammatory cytokines in fish brain. *Molecular Immunology* 45(1):218–225.

Scapigliati G, Buonocore F, Randelli E, Casani D, Meloni S, Zarletti G, Tiberi M et al. 2010. Cellular and molecular immune responses of the sea bass (*Dicentrarchus labrax*) experimentally infected with betanodavirus. *Fish & Shellfish Immunology* 28(2):303–311.

Su YC, Wu JL, and Hong JR. 2011. Betanodavirus up-regulates chaperone GRP78 via ER stress: Roles of GRP78 in viral replication and host mitochondria-mediated cell death. *Apoptosis* 16(3):272–287.

Tan C, Huang B, Chang SF, Ngoh GH, Munday B, Chen SC, and Kwang J. 2001. Determination of the complete nucleotide sequences of RNA1 and RNA2 from greasy grouper (*Epinephelus tauvina*) nervous necrosis virus, Singapore strain. *Journal of General Virology* 82(Pt 3):647–653.

Tanaka S, Takagi M, and Miyazaki T. 2004. Histopathological studies on viral nervous necrosis of sevenband grouper, *Epinephelus septemfasciatus* Thunberg, at the grow-out stage. *Journal of Fish Diseases* 27(7):385–399.

Tso CH, Hung YF, Tan SP, and Lu MW. 2013. Identification of the STAT1 gene and the characterisation of its immune response to immunostimulants, including nervous necrosis virus (NNV) infection, in Malabar grouper (*Epinephelus malabaricus*). *Fish & Shellfish Immunology* 35(5):1339–1348.

Wu MS, Chen CW, Lin CH, Tzeng CS, and Chang CY. 2012. Differential expression profiling of orange-spotted grouper larvae, *Epinephelus coioides* (Hamilton), that survived a betanodavirus outbreak. *Journal of Fish Diseases* 35(3):215–225.

Wu YC and Chi SC. 2006. Persistence of betanodavirus in Barramundi brain (BB) cell line involves the induction of Interferon response. *Fish & Shellfish Immunology* 21(5):540–547.

Wu YC, Lu YF, and Chi SC. 2010. Anti-viral mechanism of barramundi Mx against betanodavirus involves the inhibition of viral RNA synthesis through the interference of RdRp. *Fish & Shellfish Immunology* 28(3):467–475.

Section VI

Marine Metabolomics

17

Untargeted Metabolomics of Halophytes

Manish Kumar Patel, Avinash Mishra, and Bhavanath Jha

CONTENTS

17.1 Introduction

Plants frequently encounter a wide range of environmental stresses, including abiotic stresses during their life cycle. Abiotic stresses, commonly caused by salinity (NaCl), temperature (chilling or freezing), water (drought or flood), light (high and low intensity), and radiations (IR, visible, UV, or ionizing rays such as x-ray and γ-ray), influence physiological responses and biochemical pathways of plants at various levels, individually or in combinations.

Among these abiotic stresses, salinity ingression in the soil is a major constraint throughout the world for plant growth and productivity. Halophytes play a unique role in saline environment and provide an opportunity to understand the mechanistic basis of adaptation to different saline ecological niches. Halophytes account for about 1% of the world flora and flourish naturally in extreme saline environments, such as salt marshes, maritime estuaries, and salt lakes in arid zones (Flowers and Colmer 2008). Halophytes are generally succulent, nonsucculent, or shrubby, grow luxuriantly in high saline soil, and able to complete their life cycle in substantial amount of salt concentration that can be toxic to most of the domesticated crops (Flowers et al. 1977). Halophytes are distributed in a wide range of higher plant families, such as Chenopodiaceae, Poaceae, and Plumbaginaceae;

among these the largest number of halophytic species belongs to the Chenopodiaceae family (Rozema and Flowers 2008). Despite their polyphylogenetic origin, physiological and biochemical studies reveal the basic mechanism of growth and survival of halophytes under saline conditions, which relies on two different mechanisms: salt tolerance and salt avoidance. Different physiological, morphological, and genetical changes in halophytes are the remarkable adaptation for their survival in high salt concentration. Physiologically, plants maintain their growth and survival by ion compartmentalization in the cell vacuoles, accumulation of compatible organic solutes, activation of reactive oxygen species (ROS), and antioxidant defense mechanism, and salt-secreting glands and bladders under abiotic stress conditions, especially salinity (Flowers et al. 1986; Colmer et al. 2005; Flowers and Colmer 2008; Shabala and Mackay 2011). Despite these physiological studies, secondary metabolites also play a key role in stress adaptation. Functional genomics, proteomics, and metabolomics approaches further explained the pathways and role of metabolites that are involved in different mechanisms or pathways. Furthermore, the success in metabolic pathways engineering for compatible solutes, such as glycine betaine, sorbitol, mannitol, trehalose, and proline, which display increased tolerance to various abiotic stresses, has been reported (Chen and Murata 2002). Halophytes are explored worldwide for the metabolites that are regarded as the end products of a cellular process.

Metabolomics is the newest high-throughput technology for qualitative and quantitative analysis of metabolites. The plant metabolic profiling has become an invaluable tool for understanding stress tolerance and response mechanism in concerned plant under abiotic stress (Schauer and Fernie 2006). This would allow the identification of novel metabolic pathways and metabolites involved in different responses toward the specific stress. Abiotic stresses are the key players in differential regulation of metabolic pathways leading to hyper- or hypoaccumulation of metabolites (Obata and Fernie 2012). The comprehensive analyses of metabolites under environmental stresses exhibit new responses and provide biochemical status of a plant (Hasegawa et al. 2000; Chaves et al. 2009; Obata and Fernie 2012). A large number of metabolomic data of plants were reported (Table 17.1); however, there are limited reports on the experimental proof of relationship between pathways and metabolites that function in abiotic stress conditions (Obata and Fernie 2012). An omic study including metabolomics, in combination with and proteomics approach, (Figure 17.1) can be preferred to elucidate the molecular mechanism underlying the abiotic stress tolerance (Urano et al. 2010; Cramer et al. 2011; Arbona et al. 2013; Deshmukh et al. 2014). Different environmental stresses such as salt, drought, and heat may lead to hyperaccumulation of a wide variety of some useful metabolites in plants. Since last few decades, the integration of transcriptome with metabolic profiling has opened new avenues for identification of the secondary metabolites and related genes in plants (Guterman et al. 2002). The intermediary or final end product of biological metabolism is called metabolites, and the complete set of metabolites of an organism is known as *metabolomics* (Fiehn 2002). Identification and quantification of the metabolites is known as "metabolic profiling," which participates in the general metabolic function of the cell (Dunn and Ellis 2005). More than 200,000 metabolites have been estimated in the plant kingdom (Pichersky and Gang 2000; Fiehn 2002). Furthermore, the combination of genome-wide transcriptomics, proteomics, and metabolite analyses would help to elucidate the function and pathways involved in the production of pharmacologically active compounds from plants (Rischer et al. 2006).

In the past decade, many analytical tools, such as gas chromatography mass spectrometry (GC-MS), liquid chromatography mass spectrometry (LC-MS), matrix-assisted laser desorption ionization time-of-flight mass spectrometry (MALDI-TOF-MS), and nuclear magnetic resonance (NMR), are widely used for the identification of metabolites using a

TABLE 17.1

Database and Software to Be Explored for the Analyses of Spectral Peaks Obtained from Modern Analytical Tools

Database/Software	URL Sources
AnalyzerPro	http://www.spectralworks.com/analyzerpro.html
BinBase	http://fiehnlab.ucdavis.edu/projects/binbase_setupx/
BioMap	http://www.maldi-msi.org
Golm metabolome	http://gmd.mpimp-golm.mpg.de/
HMDB	http://www.hmdb.ca/
KEGG	http://www.genome.jp
MASSBANK	http://www.massbank.jp/index-e.html
MetaboAnalyst	http://www.metaboanalyst.ca/MetaboAnalyst/faces/Home.jsp
MetaboSearch	http://omics.georgetown.edu/metabosearch.html
MetAlign	https://www.wageningenur.nl/en/show/MetAlign-1.htm
METLIN	http://metlin.scripps.edu/
MetNetGE	http://metnetonline.org/
MSFACTs	http://www.noble.org/plantbio/ms/
MZmine	http://mzmine.sourceforge.net/index.shtml
NIST	http://www.nist.gov/srd/
omeSOM	http://sourcesinc.sourceforge.net/omesom/
PRIMe	http://prime.psc.riken.jp/
SimMet	http://www.premierbiosoft.com/metabolite/
XCMS	https://xcmsonline.scripps.edu

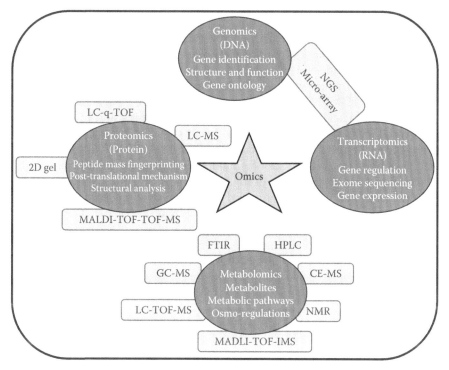

FIGURE 17.1
Integration of omics.

range of database (Table 17.1) under salt stress in different plants for analyzing the response mechanisms and pathways. The application of modern analytical tools and integration of metabolomics with a system biology and biotechnology will elucidate the physiology and metabolic pathways of targeted plant under abiotic stress conditions.

17.2 Metabolomics under Abiotic Stress Conditions

Plants, particularly halophytes, are having an efficient mechanism to adapt in varying environmental conditions. During the onset of adverse condition, primary and secondary metabolites are synthesized as the metabolic response (Atkinson and Urwin 2012). Metabolomics is nowadays used to dissect the metabolic pathways that are involved in the synthesis of various biochemical molecules to sustain the homeostasis within the cell and to ameliorate the stresses caused by environmental factors. Major abiotic stresses and metabolomics are described next.

17.2.1 Salinity

Plants are classified as glycophytes and halophytes based on their physiological behavior and tolerance efficiency to the salinity. Glycophytes can tolerate a low concentration of soil salinity (approximately 50 mM NaCl), whereas halophytes have the ability to sustain more than 200 mM of NaCl concentration. Some halophytes have been reported to withstand salinity about twice of the concentration of seawater (Flowers et al. 1977, 1986).

Halophytes have developed dynamic and efficient mechanism to tolerate salinity during salt stress conditions (Gong et al. 2005). Metabolomic studies of halophytes, such as *Populus euphratica*, *Limonium latifolium*, and *Thellungiella halophila*, have revealed that salinity tolerance can induce a wide range of metabolic pathways, which are involved in primary and secondary metabolism for the growth and development (Gong et al. 2005; Gagneul et al. 2007). Transcriptomics and GC-MS-based metabolomic study of tree *P. euphratica* revealed that the concentration of amino acids, such as proline, valine, β-alanine and myo-inositol, glyceric acid, and glycerol, increases concomitantly with salt stress, whereas concentration of some sugars, such as fructose and mannitol, decreased (Brosché et al. 2005). The comparative study of metabolomics and transcriptomics of a halophyte *T. halophila* and a model glycophyte plant *Arabidopsis* showed that the level of sugars (hexoses), organic acids (malate and citrate), and proline increases in *T. halophila* compared to *Arabidopsis* (Gong et al. 2005). A compatible solute, chiro-inositol, has been reported from a halophyte *L. latifolium* (Gagneul et al. 2007), whereas metabolites, phytoalexins, and phytoanticipins were identified from *Thellungiella salsuginea* ecotype in the response to biotic and abiotic stresses (Pedras and Zheng 2010). Apart from these, a number of polar metabolites were isolated from the leaves of *T. salsuginea* that are analogous to those identified from crucifers, except glucosalsuginin (Pedras and Zheng 2010). It was observed that the syntheses of several metabolites are affected in *Atriplex halimus* under salinity or osmotic stress treatments (Alla et al. 2012). Syntheses of a number of amino acids, such as proline, valine, isoleucine, methionine, and some sugars were upregulated. These compatible metabolites are involved in osmotic adjustment and act as osmoprotectants. Inhibition of growth and downregulation of some metabolites, such as amino acids, lactate, 4-aminobutyrate, malate, choline, and phosphocholine, were observed in root tissue of *Suaeda salsa* by NMR-based metabolomics, whereas elevated quantity of betaine, sucrose, and allantoin was detected under salt stress condition (Wu et al. 2012). About 10 metabolites (flavonoids, saponins, and sterols) were identified from an edible halophyte *Zygophyllum album* by LC-ESI-TOF-MS, among which 5 metabolites, 3-*O*-β-ᴅ-quinovopyranosyl-quinovic acid, 28-β-ᴅ-glucopyranosyl ester, 3-*O*-[β-ᴅ-2-*O*-sulfonylquinovopyranosyl]-quinovic acid-27-*O*-[β-ᴅ-glucopyranosyl] ester, Malvidin 3-rhamnoside, and quercetin 3-sulfate were reported first (Ksouri et al. 2013).

17.2.2 Drought

Drought is defined as a water deficit condition, in which a plant is intended to close the stomata to avoid water loss, occurring through transpiration (Rizhsky et al. 2002; Mahajan and Tuteja 2005; Monneveux et al. 2006). Plants subjected to water deficit conditions undergo a wide range of changes at physiological, morphological, and biochemical level (Verslues et al. 2006; Yamaguchi-Shinozaki and Shinozaki 2006). It was observed that abscisic acid (ABA) plays a crucial role in a water deficit condition by activating the dehydration defense mechanism mediated through primary and secondary metabolites (Boudsocq and Lauriere 2005). Commonly, transcripts and metabolites that are involved in a number of metabolic pathways are upregulated under drought stress condition (Sicher and Barnaby 2012; Silvente et al. 2012).

To cope up with water deficit conditions, a number of metabolic processes are triggered inside the plant cell that lead to the synthesis of low-molecular-weight compatible solutes (Mahajan and Tuteja 2005). These compounds maintain the cellular turgor pressure and osmotic equilibrium and facilitate the movement of water inside the cell by creating low osmotic potential (Mahajan and Tuteja 2005). The osmoprotectants like proline, glycine betaine, sorbitol, glutamate, polyols, and sugar alcohols, such as mannitol, trehalose, and sucrose, are also involved in osmoregulation, but they do not affect the normal function of the cell (Hoekstra et al. 2001; Ramanjulu and Bartels 2002).

Halophytes grow in salt marshes, a physiologically dry condition, and encounter salt-induced osmotic stress because of limited water absorption and creation of low water potential zone in the soil. Study on the osmotic stress tolerance mechanism of a halophyte *Spartina alterniflora* reveals that halophytes undergo active accumulation of various osmotica, which prevents loss of water content from the tissues of the plant (Touchette et al. 2009). It was also observed that the level of different compatible osmoprotectants, such as proline, sugars, polyphenols, and free amino acids, increases under drought stress conditions in an extreme halophyte *Salicornia brachiata* (Mishra et al. 2015).

Traditionally, the analysis of metabolic responses of a plant under drought stress conditions was solely based on physiological responses, such as synthesis of one or two class of compounds that are involved in the development of tolerance. Unfortunately, no emphasis was given on the metabolic network changes that lead to the production of various kinds of untargeted metabolites (Budak et al. 2013). Network analysis of different metabolic pathways among different cultivars and under varying stress conditions can be explored for the identification of various biologically relevant metabolites. Metabolic profiling based on GC-MS and LC-MS showed the reduction of organic acid, changes in the phenylpropanoid pathway, and increase in amino acids, such as Pro, Val, Leu, Thr, and Trp in grapevine (Hochberg et al. 2013). It was observed that scavenging of ROS plays a key role in stress tolerance and accumulation of osmolytes, which prevents photoinhibition, generated in response to the drought environment (Hochberg et al. 2013). A high-throughput NMR-based analysis of metabolites provides useful insight about the shifting of metabolic pathways under drought and other environmental stress conditions. An integrated study of ^{1}H NMR–based metabolic profiling and genotype fingerprinting of soybean combined the physiology to the ability of drought tolerance and elucidated six metabolites from leaves (aspartate, 2-oxoglutaric acid, myo-inositol, pinitol, sucrose, allantoin), whereas two from nodule (2-oxoglutaric acid and pinitol) that were synthesized as osmoprotectant in the physiological response of water stress condition (Silvente et al. 2012). Application of non-targeted metabolomics approach not only provides an unbiased perspective of metabolic profiles of a response but also leads to the discovery of novel metabolic phenotypes.

17.2.3 Temperature

Thermostress conditions are commonly of three types: chilling, freezing, and heat. Chilling stress temperature ranges from 0°C to 15°C and lacks ice crystals formation in plant tissues. Temperature below 0°C leads to freezing stress and causes ice formation in extracellular spaces of the plant tissues (Park and Chen 2006). Low temperature induces various morphological changes and hampers the normal physiology and biochemical response of the plant (Stitt and Hurry 2002). Various symptoms of chilling stress include reduced leaf development, necrosis, wilting, water soaked, and accelerated senescence (Jiang et al. 2002). Various factors that enhance freeze tolerance include change in membrane phospholipids, elevated level of sugars in extracellular space, and expression of cold-responsive genes and metabolites (Levitt 1980). Cold stress also increases the level of certain hormones such as ABA and ethylene. Therefore, it is necessary to understand the every aspect of metabolites as well as plant niche for the development of temperature acclimation (Guy et al. 2008). By exploring the temperature–stress metabolome of *Arabidopsis*, Kaplan et al. (2004) had identified 143 and 311 metabolites under heat and cold stress, respectively, using GC-MS. Afterward, many studies have been done in different ecotypes of *Arabidopsis* for the identification of metabolite that are involved in thermoregulation (Cook et al. 2004; Hannah et al. 2006). It was observed that lipid composition of a plant changes in response to low temperature (Wang et al. 2006) and role of lipids were investigated under freezing stress conditions (Welti et al. 2002). The membrane lipid profile, done by ESI-MS/MS approach, revealed the increase and decrease of membrane lipids, such as phosphatidic acid, lysophospholipids, phosphatidylcholine, phosphatidylethanolamine, and phosphatidylglycerol, under cold and freezing stress conditions (Welti et al. 2002). Heat stress causes oxidative damage in the plant that leads to the imbalance of photosystem II and respiration (Fitter and Hay 2001). There are very few reports on nontargeting metabolomics under heat stress conditions, which revealed that some soluble sugars are synthesized as osmolytes (Diamant et al. 2001) and the level of amino acids increases because of the breakdown of the protein under stress conditions (Guy et al. 2008).

17.3 Metabolite Profiling: An Experimental Approach

17.3.1 Metabolite Extraction: Selection and Optimization of Methods

Commonly, there are two components for sampling: harvesting and drying. Harvesting time and date of samples are very important as metabolic responses vary with growing age of the plant. Similarly, there are different methods of drying such as freeze-drying, oven-drying, and air-drying, which are crucial for the processing of metabolomics (Harbourne et al. 2009).

Optimization of metabolites extraction method for a biological sample is always very important and plays a major role in the isolation and identification of novel bioactive compounds (Broadhurst and Kell 2006). In the past decades, numerous studies have been done regarding the optimization and detection for a wide range of metabolites. The in-depth knowledge of metabolic pathways, mechanisms, and their networking within plant under normal and stress conditions is still at infant stage. For an efficient metabolomics studies, the method must be robust and able to detect all possible metabolites with minimum variable(s) and error (Khakimov et al. 2014). Furthermore, solvent system, extraction time, and temperature are determining factor and play crucial role in the profiling of metabolites.

Preparation of samples and extraction procedures for metabolite profiling generally vary with the techniques to be employed. In general, GC-MS, LC-MS, and MALDI-TOF-IMS techniques are highly sensitive and suitable for the metabolomics study of plants (Figure 17.2), including halophytes for their nutritional and pharmaceutical properties.

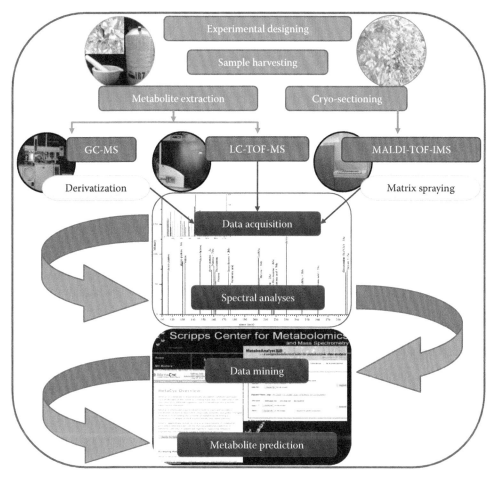

FIGURE 17.2
Metabolomics: an experimental approach.

17.3.2 Analytical Techniques: Commonly Used for Untargeted Metabolomics

17.3.2.1 GC-MS

GC-MS is a robust technique, which is widely used for the quantification of metabolites, and it is a single analytical tool, so far, which can cover the whole metabolome of a plant system (Sumner et al. 2003). It is one of the important tool commonly used for analyzing nontargeted metabolites and different types of primary and secondary metabolites, such as fatty acid, sugars, amino acids, alkaloids, terpenoids, and flavonoids (Sumner et al. 2003).

17.3.2.1.1 Sample Preparation

Isolation of metabolites is a most critical step in GC-MS analysis. Plant samples are first homogenized in liquid N_2 using a mortar and pestle and subsequently mixed with aqueous methanol for the isolation of polar metabolites followed by centrifugation. Chloroform is used for the extraction of nonpolar metabolites like lipids. The extracted mixture is commonly dried in vacuum concentrator and stored at −80°C for further use after gentle purging with nitrogen to deoxygenate the sample (Fiehn et al. 2000). For further analysis, derivatization is required because GC-MS detects only volatile and thermally stable compounds. Some of the metabolites do not have these properties; therefore, their derivatives are needed, which can be done chemically by different derivatizing agents. In the process of derivatization, functional groups are changed, and trimethylsilyl moiety, $-Si(CH_3)_3$, which is a well-known silylation agent, is used to derivatize the functional group of metabolites (Little 1999; Halket et al. 2005).

17.3.2.1.2 Data Acquisition and Analyses

A small volume of derivative samples (depending on the concentration of extracted metabolites) are separated in GC-MS column, and a complex chromatograph having peaks is generated as a result, known as total ion chromatogram. Metabolites are generally identified by comparative spectral analysis using NIST mass spectrum library, and quantification is done by comparing peak area of metabolites present in samples and library. A number of softwares have been developed for the GC-MS, which are available online for the analysis of metabolite spectral data, such as AnalyzerPro (http:// www.spectralworks.com), BinBase (http://fiehnlab.ucdavis.edu), TagFinder (Luedemann et al. 2008), MetaboAnalyst (Xia et al. 2009), MetAlign 3.0 (Lommen and Kools 2012, http://www.metalign.nl), NIST library (Lemmon et al. 2013), and Golm metabolome database (Kopka et al. 2005).

17.3.2.2 Liquid Chromatography/Time-of-Flight/Mass Spectrometry (LC-TOF-MS)

In the past decades, LC-TOF-MS is routinely used for the analysis of untargeted and targeted metabolomics Lu et al. 2008; Wishart et al. 2008; Wikoff et al. 2009; Kaddurah-Daouk et al. 2010; Yanes et al. 2010; Johnson et al. 2011; Trupp et al. 2012). One of the advantages of LC-MS is that derivatization is not required and thermolabile samples can be used for the identification of metabolites. LC-MS is one of the important approach, which covers high-polar plant metabolites, such as amino acids, organic acids, sugars, and secondary metabolites, such as alkaloids, flavonoids, terpenes, polyamines, and phenolics (Huhman and Sumner 2002; Tolstikov et al. 2003; Moco et al. 2006; Rischer et al. 2006). In LC-MS, metabolites (such as semipolar and secondary metabolites) are first ionized by soft ionization technique, such as electrospray ionization (ESI) or atmospheric pressure chemical ionization (APCI), followed by mass spectroscopy. ESI- or APCI-based identification of metabolites largely depends on protonated (positive mode) and deprotonated (negative mode) molecular masses.

17.3.2.2.1 *Experimental Analysis*

For LC-MS analysis, the plant sample is grinded in liquid N_2 using mortar and pestle; thereafter, total metabolites are extracted by adding ice-cold extraction buffer (99.875% v/v methanol solution acidified with 0.125% v/v formic acid) followed by vortexing (De Vos et al. 2007). The mixture is generally kept in ultrasonic water bath for 1 h at a frequency of 40 kHz. Supernatants are collected after centrifugation (20,000 g at 25°C for 10 min), filtered (0.2 μm membrane), and subjected to LC-MS analysis. For peak integration, the background of each spectrum is generally subtracted, the data are made smooth and centered, and peaks are integrated using instrument-operated software, such as MassLynx software version 4.1 of Micromass, Waters. Metabolites are commonly analyzed by LC coupled with TOF-MS and raw data can be analyzed by different online tools and softwares, such as METLIN database (Zhu et al. 2013, http://metlin.scripps. edu/xcms/), MetAlign (Bino et al. 2005; Vorst et al. 2005), XCMS (Nordstrom et al. 2006; Smith et al. 2006, https://xcmsonline.scripps.edu), MZmine (Katajamaa and Oresic 2005), and MetaboAnalyst (Xia et al. 2009).

17.3.2.3 *MALDI-TOF-IMS*

MALDI-TOF-IMS is an emerging powerful technology that enables the metabolic profiling at single cell level to avoid the heterogeneity of plant tissue. This technique is independent and avoids various procedures, such as extraction, derivatization, purification, and separation of samples, which are prerequisite for most of the mass spectrometric techniques as mentioned earlier. The technique (MALDI-TOF-IMS) can detect a wide range of molecules, from a small mass to large mass, with little amount of plant samples (Obel et al. 2009; Westphal et al. 2010; Günl et al. 2011; Mishra et al. 2011; Ropartz et al. 2011). MALDI-TOF-MS is also extensively used in different areas, such as proteomics (Jha et al. 2012), metabolomics (Burrell et al. 2007), oligosaccharide mass profiling (OLIMP) (Mishra et al. 2013), lipidomics (Jackson et al. 2005; Goto-Inoue et al. 2008; Murphy et al. 2009), and glycomics (Ageta et al. 2009; Goto-Inoue et al. 2009). The MALDI-IMS has opened a new prospective in various fields such as agriculture, pharmacology, and unknown biomarkers (Meistermann et al. 2006; Lemaire et al. 2007).

17.3.2.3.1 *Experimental Procedure*

Sample preparation is the most crucial step for the identification of analytes. It is more important for MALDI-IMS, in which the morphology of plant section should be fresh or frozen and plant tissues chemically unmodified (Jehl et al. 1981). The embedding with optimal cutting temperature can contaminate the plant sample, and it may interfere during the time of ionization (Schwartz et al. 2003; McDonnell and Heeren 2007). Frozen and fragile tissue sections must be embedded using carboxymethylcellulose with sodium salt, gelatin, or agarose (Kruse and Sweedler 2003; Altelaar et al. 2005; Stoeckli et al. 2006; Zaima et al. 2010). Washing is the next vital step after cutting, as it improves the performance of MALDI-IMS by removing the matrix. The choice of an organic solvent for washing is very important for the target molecule to remove salt and other contaminates from the tissue. Similarly, selection of a matrix is very difficult for the sample preparation and further analysis of metabolites to obtain a clear spectrum. There are a number of reports on the application of different matrices, such as α-cyano-4-hydroxycinnamic acid (α-CHCA), 2,5-dihydroxybenzoic acid (2,5-DHB), and 9-aminoacridine for the analysis of metabolites. Combination of α-CHCA with 9-aminoacridine allows the identification of amino acids, sugars, and phosphorylated metabolites (Burrell et al. 2007). Xylene and aniline have been used for the lipid profiling (Lemaire et al. 2006; Meriaux et al. 2010). Recently, α-CHCA has been used as the matrix for OLIMP of an extreme halophyte *S. brachiata* using MALDI-TOF-TOF-MS (Mishra et al. 2013). MALDI-TOF-IMS analysis is generally performed with an intensity of 7900 Nd-YAG (neodymium-doped yttrium aluminum garnet) laser, operated with accelerated voltage. Spectral data are processed and can be analyzed by free online available softwares BioMap (http://www.maldi-msi.org), METLIN (http://metlin.scripps.edu/), and KEGG (http://www.genome.ad.jp/kegg/ligand.html).

17.4 Limitations of Analytical MS Techniques

In the global metabolome, all the analytical techniques have some limitations ranging from sample preparation to identification of metabolites. There is no standard or a commercial library available for the analysis of data; therefore, results vary from laboratory to laboratory, instrument to instrument, and person to person (Vinayavekhin and Saghatelian 2010). Each method also has boundaries with concern for identification, sensitivity, quantification, and data processing of metabolites. There are some limitations for GC-MS, such as a sample should be volatile and thermally stable. Many metabolites are not volatile because they have polar groups, such as thiol (–SH), amide (–CONHR), amine (–NH$_2$), and alcoholic (–OH). Further, the sample must be chemically derivatized by processes such as silylation, acylation, or alkylation (Knapp 1979; Evershed et al. 1993). Selection of derivatizing agent generally depends on the metabolites, such as halogens are derivatized by fluorine, and it can be detected in negative chemical ionization only (Peters et al. 2002), which is a time-consuming procedure. Another drawback of GC-MS is that fragmented peaks can be detected in +EI (positive electron ionization) mode only, and identification can be done by comparing compounds available in library.

In LC-TOF-MS, metabolites are identified on the basis of mass only (Xu et al. 2007; Evans et al. 2009; Vinayavekhin and Saghatelian 2010). Additionally, a soft ionization (ESI) is a prerequisite for the LC-MS in both modes (+ve and −ve), which can be done with certain metabolites only. The majority of metabolites require APCI, which is mainly applicable to polar compounds with a molecular weight of less than 1500 Da (Robb et al. 2000); therefore, application of LC-MS is limited for higher-molecular-weight metabolites.

MALDI-TOF-IMS is a very sensitive technique that requires dedicated sample preparation with spatial resolution, organic solvent for washing, and matrix (Chughtai and Heeren 2010; Kaiser et al. 2011). Washing is an important step, and many small metabolites, such as lipids, may diffuse from samples (Lee et al. 2012). Plant tissue section is another key step, and samples used should be of very thin surface layer (1 μm) (Lee et al. 2012). Application of the matrix, lack of mass accuracy, and mass-resolving capacity of the TOF-MS limit its application for the study of metabolites in imaging mode (Heeren et al. 2009; Ernst et al. 2014). There is no single mass spectrometry technique available so far, which is capable of identifying all metabolites present in the plant system.

17.5 Conclusion

Untargeted metabolomics mainly focus on the identification of metabolites that are synthesized in plants under normal or stress conditions. Metabolomics of halophytes are commonly performed to investigate the plant responses toward environmental changes, including abiotic stresses. Over the last decades, advancement in genomics, transcriptomics, and metabolomics has opened a new way to understand the cross talk between various abiotic stresses, which can identify a number of metabolites that are involved in osmoregulation of cell and provide stress endurance by altering physiobiochemicology. After the discovery of a number of metabolites, halophytes are now considered as medicinal plants and a potential source of bioactive compounds, such as alkaloids, flavonoids, triterpenoid, and sterols. An edible halophyte *Z. album* is now explored for anticancer and anti-inflammatory bioactive compounds. A number of analytical tools and techniques are available nowadays to elucidate the metabolic pathways and identify the novel metabolites and comprehensive metabolomics.

17.6 Future Prospective

Metabolome analysis in halophyte has become an invaluable tool for the study of metabolic changes that occur during abiotic stresses. Tools, such as mass spectrometry, allow the identification of metabolites and pathways. Integration of "omics" data like genomics, transcriptomics, and proteomics with metabolomics reveals the biochemical pathways from gene to metabolites. These types of studies are necessary in order to fill the gap area of cellular processes for metabolic regulatory network. The database of metabolites from halophytes is expected to integrate with transcriptomics and proteomics in the future use for the discovery of novel metabolic pathways.

Acknowledgments

CSIR-CSMCRI Communication No. PRIS-155/2014 and CSIR Network Project (BSC0106–BioprosPR) are thankfully acknowledged for the research work conducted in the authors' laboratory.

References

Ageta, H., Asai, S., Sugiura, Y., Goto-Inoue, N., Zaima, N., and Setou, M. 2009. Layer-specific sulfatide localization in rat hippocampus middle molecular layer is revealed by nanoparticle-assisted laser desorption/ionization imaging mass spectrometry. *Medical Molecular Morphology* 42:16–23.

Alla, M. M. N., Khedr, A. H. A., Serag, M. M., Abu-Alnaga, A. Z., and Nada, R. M. 2012. Regulation of metabolomics in *Atriplex halimus* growth under salt and drought stress. *Plant Growth Regulation* 67:281–304.

Altelaar, A. F., Van Minnen, J., Jimenez, C. R., Heeren, R. M., and Piersma, S. R. 2005. Direct molecular imaging of *Lymnaea stagnalis* nervous tissue at subcellular spatial resolution by mass spectrometry. *Analytical Chemistry* 77:735–741.

Arbona, V., Manzi, M., Ollas, C. D., and Gómez-Cadenas, A. 2013. Metabolomics as a tool to investigate abiotic stress tolerance in plants. *International Journal of Molecular Sciences* 14:4885–4911.

Atkinson, N. J. and Urwin, P. E. 2012. The interaction of plant biotic and abiotic stresses: From genes to the field. *Journal of Experimental Botany* 63:3523–3543.

Bino, R. J., De Vos, C. H., Lieberman, M. et al. 2005. The light-hyper responsive high pigment-2dg mutation of tomato: Alterations in the fruit metabolome. *New Phytologist* 166:427–438.

Boudsocq, M. and Laurière, C. 2005. Osmotic signaling in plants: Multiple pathways mediated by emerging kinase families. *Plant Physiology* 138:1185–1194.

Broadhurst, D. I. and Kell, D. B. 2006. Statistical strategies for avoiding false discoveries in metabolomics and related experiments. *Metabolomics* 2:171–196.

Brosché, M., Vinocur, B., Alatalo, E. R. et al. 2005. Gene expression and metabolite profiling of *Populus euphratica* growing in the Negev desert. *Genome Biology* 6:R101.

Budak, H., Kantar, M., and Yucebilgili Kurtoglu, K. 2013. Drought tolerance in modern and wild wheat. *The Scientific World Journal* 2013, Article ID: 548246.

Burrell, M. M., Earnshaw, C. J., and Clench, M. R. 2007. Imaging matrix assisted laser desorption ionization mass spectrometry: A technique to map plant metabolites within tissues at high spatial resolution. *Journal of Experimental Botany* 58:757–763.

Chaves, M. M., Flexas, J., and Pinheiro, C. 2009. Photosynthesis under drought and salt stress: Regulation mechanisms from whole plant to cell. *Annals of Botany* 103:551–560.

Chen, T. H. and Murata, N. 2002. Enhancement of tolerance of abiotic stress by metabolic engineering of betaines and other compatible solutes. *Current Opinion in Plant Biology* 5:250–257.

Chughtai, K. and Heeren, R. M. A. 2010. Mass spectrometric imaging for biomedical tissue analysis. *Chemical Reviews* 110:3237–3277.

Colmer, T., Munns, R., and Flowers, T. 2005. Improving salt tolerance of wheat and barley: Future prospects. *Australian Journal of Experimental Agriculture* 45:1425–1443.

Cook, D., Fowler, S., Fiehn, O., and Thomashow, M. F. 2004. A prominent role for the CBF cold response pathway in configuring the low-temperature metabolome of *Arabidopsis*. *Proceedings of the National Academy of Sciences of the United States of America* 101:15243–15248.

Cramer, G. R., Urano, K., Delrot, S., Pezzotti, M., and Shinozaki, K. 2011. Effects of abiotic stress on plants: A systems biology perspective. *BMC Plant Biology* 11:163–177.

De Vos, R. C., Moco, S., Lommen, A., Keurentjes, J. J., Bino, R. J., and Hall, R. D. 2007. Untargeted large-scale plant metabolomics using liquid chromatography coupled to mass spectrometry. *Nature Protocol* 2:778–791.

Deshmukh, R. K., Sonah, H., Patil, G. et al. 2014. Integrating omic approaches for abiotic stress tolerance in soybean. *Plant Genetics and Genomics* 5:244.

Diamant, S., Eliahu, N., Rosenthal, D., and Goloubinoff, P. 2001. Chemical chaperones regulate molecular chaperones in vitro and in cells under combined salt and heat stresses. *Journal of Biological Chemistry* 276:39586–39591.

Dunn, W. B. and Ellis, D. I. 2005. Metabolomics: Current analytical platforms and methodologies. *TrAC Trends in Analytical Chemistry* 24:285–294.

Ernst, M., Silva, D. B., Silva, R. R., Vêncio, R. Z., and Lopes, N. P. 2014. Mass spectrometry in plant metabolomics strategies: From analytical platforms to data acquisition and processing. *Natural Product Reports* 31:784–806.

Evans, A. M., Dehaven, C. D., Barrett, T., Mitchell, M., and Milgram, E. 2009. Integrated, nontargeted ultra high performance liquid chromatography/electrospray ionization tandem mass spectrometry platform for the identification and relative quantification of the small-molecule complement of biological systems. *Analytical Chemistry* 81:6656–6667.

Evershed, R. P., Blau, K., and Halket, J. M. 1993. *Handbook of Analytical Derivatives for Chromatography*, 2nd ed. New York: John Wiley & Sons.

Fiehn, O. 2002. Metabolomics—The link between genotypes and phenotypes. *Plant Molecular Biology* 48:155–171.

Fiehn, O., Kopka, J., Dörmann, P., Altmann, T., Trethewey, R. N., and Willmitzer, L. 2000. Metabolite profiling for plant functional genomics. *Nature Biotechnology* 18:1157–1161.

Fitter, A. H. and Hay, R. K. M. 2001. *Environmental Physiology of Plants*, 3rd ed. London, U.K.: Academic Press.

Flowers, T. J. and Colmer, T. 2008. Salinity tolerance in halophytes. *New Phytologist* 179:945–963.

Flowers, T. J., Hajibagheri, M. A., and Clipson, N. J. W. 1986. Halophytes. *The Quarterly Review of Biology* 61:313–337.

Flowers, T. J., Troke, P. F., and Yeo, A. R. 1977. The mechanism of salt tolerance in halophytes. *Annual Review of Plant Physiology* 28:89–121.

Gagneul, D., Ainouche, A., Duhaze, C., Lugan, R., Larher, F. R., and Bouchereau, A. 2007. A reassessment of the function of the so-called compatible solutes in the halophytic Plumbaginaceae *Limonium latifolium*. *Plant Physiology* 144:1598–1611.

Gong, Q., Li, P., Ma, S., Indu Rupassara, S., and Bohnert, H. J. 2005. Salinity stress adaptation competence in the extremophile *Thellungiella halophila* in comparison with its relative *Arabidopsis thaliana*. *The Plant Journal* 44:826–839.

Goto-Inoue, N., Hayasaka, T., Sugiura, Y. et al. 2008. High sensitivity analysis of glycosphingolipids by matrix-assisted laser desorption/ionization quadrupole ion trap time-of-flight imaging mass spectrometry on transfer membranes. *Journal of Chromatography B* 870:74–83.

Goto-Inoue, N., Hayasaka, T., Zaima, N., and Setou, M. 2009. The specific localization of seminolipid molecular species on mouse testis during testicular maturation revealed by imaging mass spectrometry. *Glycobiology* 19:950–957.

Günl, M., Kraemer, F. J., and Pauly, M. 2011. Oligosaccharide mass profiling (OLIMP) of cell wall polysaccharides by MALDI-TOF/MS. In *The Plant Cell Wall: Methods and Protocols*, Methods in Molecular Biology, ed. Z. A. Popper, pp. 43–54. New York: Humana Press.

Guterman, I., Shalit, M., Menda, N. et al. 2002. Rose scent genomics approach to discovering novel floral fragrance–related genes. *The Plant Cell* 14:2325–2338.

Guy, C., Kaplan, F., Kopka, J., Selbig, J., and Hincha, D. K. 2008. Metabolomics of temperature stress. *Physiologia Plantarum* 132:220–235.

Halket, J. M., Waterman, D., Przyborowska, A. M., Patel, R. K., Fraser, P. D., and Bramley, P. M. 2005. Chemical derivatization and mass spectral libraries in metabolic profiling by GC/MS and LC/MS/MS. *Journal of Experimental Botany* 56:219–243.

Hannah, M. A., Wiese, D., Freund, S., Fiehn, O., Heyer, A. G., and Hincha, D. K. 2006. Natural genetic variation of freezing tolerance in *Arabidopsis*. *Plant Physiology* 142:98–112.

Harbourne, N., Marete, E., Jacquier, J. C., and O'Riordan, D. 2009. Effect of drying methods on the phenolic constituents of meadowsweet (*Filipendula ulmaria*) and willow (*Salix alba*). *LWT Food Science and Technology* 42:1468–1473.

Hasegawa, P. M., Bressan, R. A., Zhu, J. K., and Bohnert, H. J. 2000. Plant cellular and molecular responses to high salinity. *Annual Review of Plant Biology* 51:463–499.

Heeren, R., Smith, D. F., Stauber, J., Kükrer-Kaletas, B., and MacAleese, L. 2009. Imaging mass spectrometry: Hype or hope? *Journal of the American Society for Mass Spectrometry* 20:1006–1014.

Hochberg, U., Degu, A., Toubiana, D. et al. 2013. Metabolite profiling and network analysis reveal coordinated changes in grapevine water stress response. *BMC Plant Biology* 13:184.

Hoekstra, F. A., Golovina, E. A., and Buitink, J. 2001. Mechanisms of plant desiccation tolerance. *Trends in Plant Science* 6:431–438.

Huhman, D. V. and Sumner, L. W. 2002. Metabolic profiling of saponins in *Medicago sativa* and *Medicago truncatula* using HPLC coupled to an electrospray ion-trap mass spectrometer. *Phytochemistry* 59:347–360.

Jackson, S. N., Wang, H. Y. J., and Woods, A. S. 2005. Direct profiling of lipid distribution in brain tissue using MALDI-TOFMS. *Analytical Chemistry* 77:4523–4527.

Jehl, B., Bauer, R., Dorge, A., and Rick, R. 1981. The use of propane/isopentane mixtures for rapid freezing of biological specimens. *Journal of Microscopy* 123:307–309.

Jha, B., Singh, N. P., and Mishra, A. 2012. Proteome profiling of seed storage proteins reveals the nutritional potential of *Salicornia brachiata* Roxb., an extreme halophyte. *Journal of Agricultural and Food Chemistry* 60:4320–4326.

Jiang, Q. W., Kiyoharu, O., and Ryozo, I. 2002. Two novel mitogen-activated protein signaling components, *OsMEK1* and *OsMAP1*, are involved in a moderate low-temperature signaling pathway in Rice. *Plant Physiology* 1291:1880–1891.

Johnson, C. H., Patterson, A. D., Krausz, K. W. et al. 2011. Radiation metabolomics. 4. UPLC-ESI-QTOFMS-Based metabolomics for urinary biomarker discovery in gamma-irradiated rats. *Radiation Research* 175:473–484.

Kaddurah-Daouk, R., Baillie, R. A., Zhu, H. et al. 2010. Lipidomic analysis of variation in response to simvastatin in the cholesterol and pharmacogenetics study. *Metabolomics* 6:191–201.

Kaiser, N., Quinn, J., Blakney, G., Hendrickson, C. L., and Marshall, A. G. 2011. A novel 9.4 tesla FTICR mass spectrometer with improved sensitivity, mass resolution, and mass range. *Journal of the American Society for Mass Spectrometry* 22:1343–1351.

Kaplan, F., Kopka, J., Haskell, D. W. et al. 2004. Exploring the temperature-stress metabolome of *Arabidopsis*. *Plant Physiology* 136:4159–4168.

Katajamaa, M. and Oresic, M. 2005. Processing software for differential analysis of LC/MS profile data. *BMC Bioinformatics* 6:179.1–179.12.

Khakimov, B., Bak, S., and Engelsen, S. B. 2014. High-throughput cereal metabolomics: Current analytical technologies, challenges and perspectives. *Journal of Cereal Science* 59:393–418.

Knapp, D. R. 1979. *Handbook of Analytical Derivatization Reactions*. New York: Wiley Inter Science.

Kopka, J., Schauer, N., Krueger, S. et al. 2005. GMD@ CSB. DB: The Golm metabolome database. *Bioinformatics* 21:1635–1638.

Kruse, R. and Sweedler, J. V. 2003. Spatial profiling invertebrate ganglia using MALDI MS. *Journal of the American Society for Mass Spectrometry* 14:752–759.

Ksouri, W. M., Medini, F., Mkadmini, K. et al. 2013. LC–ESI-TOF–MS identification of bioactive secondary metabolites involved in the antioxidant, anti-inflammatory and anticancer activities of the edible halophyte *Zygophyllum album* Desf. *Food Chemistry* 139:1073–1080.

Lee, Y. J., Perdian, D. C., Song, Z., Yeung, E. S., and Nikolau, B. J. 2012. Use of mass spectrometry for imaging metabolites in plants. *The Plant Journal* 70:81–95.

Lemaire, R., Menguellet, S. A., Stauber, J. et al. 2007. Specific MALDI imaging and profiling for biomarker hunting and validation: Fragment of the 11s proteasome activator complex, reg alpha fragment, is a new potential ovary cancer biomarker. *Journal of Proteome Research* 6:4127–4134.

Lemaire, R., Wisztorski, M., Desmons, A. et al. 2006. MALDI-MS direct tissue analysis of proteins: Improving signal sensitivity using organic treatments. *Analytical Chemistry* 78:7145–7153.

Levitt, J. 1980. Chilling, freezing, and high temperature stress in responses of plants to environmental stresses.

Little, J. L. 1999. Artifacts in trimethylsilyl derivatization reactions and ways to avoid them. *Journal of Chromatography A* 844:1–22.

Lemmon, E., McLinden, M., and Huber, M. 2013. NIST reference fluid thermodynamic and transport properties. NIST Standard Reference Database 23, version 9.1. Gaithersburg, MD: National Institutes of Standards and Technology (NIST).

Lommen, A. and Kools, H. J. 2012. MetAlign 3.0: Performance enhancement by efficient use of advances in computer hardware. *Metabolomics* 8:719–726.

Lu, W., Bennett, B. D., and Rabinowitz, J. D. 2008. Analytical strategies for LC–MS-based targeted metabolomics. *Journal of Chromatography B* 871:236–242.

Luedemann, A., Strassburg, K., Erban, A., and Kopka, J. 2008. TagFinder for the quantitative analysis of gas chromatography—Mass spectrometry (GC-MS)-based metabolite profiling experiments. *Bioinformatics* 24:732–737.

Mahajan, S. and Tuteja, N. 2005. Cold, salinity and drought stresses, an overview. *Archives of Biochemistry and Biophysics* 444:139–158.

McDonnell, L. A. and Heeren, R. M. 2007. Imaging mass spectrometry. *Mass Spectrometry Reviews* 26:606–643.

Meistermann, H., Norris, J. L., Aerni, H. R. et al. 2006. Biomarker discovery by imaging mass spectrometry: Transthyretin is a biomarker for gentamicin-induced nephrotoxicity in rat. *Molecular and Cellular Proteomics* 5:1876–1886.

Meriaux, C., Franck, J., Wisztorski, M., Salzet, M., and Fournier, I. 2010. Liquid ionic matrixes for MALDI mass spectrometry imaging of lipids. *Journal of Proteomics* 73:1204–1218.

Mishra, A., Joshi, M., and Jha, B. 2013. Oligosaccharide mass profiling of nutritionally important *Salicornia brachiata*, an extreme halophyte. *Carbohydrate Polymers* 92:1942–1945.

Mishra, A., Kavita, K., and Jha, B. 2011. Characterization of extracellular polymeric substances produced by micro-algae *Dunaliella salina*. *Carbohydrate Polymers* 83:852–857.

Mishra, A., Patel, M. K., and Jha, B. 2015. Non-targeted metabolomics and scavenging activity of reactive oxygen species reveal the potential of *Salicornia brachiata* as a functional food. *Journal of Functional Foods* 13:21–31.

Moco, S., Bino, R. J., Vorst, O. et al. 2006. A liquid chromatography-mass spectrometry-based metabolome database for tomato. *Plant Physiology* 141:1205–1218.

Monneveux, P., Rekika, D., Acevedo, E., and Merah, O. 2006. Effect of drought on leaf gas exchange, carbon isotope discrimination, transpiration efficiency and productivity in field grown durum wheat genotypes. *Plant Science* 170:867–872.

Murphy, R. C., Hankin, J. A., and Barkley, R. M. 2009. Imaging of lipid species by MALDI mass spectrometry. *Journal of Lipid Research* 50:317–322.

NIST standard reference database 1 A2011. National Institute of Standard and Technology.

Nordstrom, A., O'Maille, G., Qin, C., and Siuzdak, G. 2006. Nonlinear data alignment for UPLC-MS and HPLC-MS based metabolomics: Quantitative analysis of endogenous and exogenous metabolites in human serum. *Analytical Chemistry* 78:3289–3295.

Obata, T. and Fernie, A. R. 2012. The use of metabolomics to dissect plant responses to abiotic stresses. *Cellular and Molecular Life Sciences* 69:3225–3243.

Obel, N., Erben, V., Schwarz, T., Kühnel, S., Fodor, A., and Pauly, M. 2009. Microanalysis of plant cell wall polysaccharides. *Molecular Plant* 2:922–932.

Park, E. J. and Chen, T. H. 2006. Improvement of cold tolerance in horticultural crops by genetic engineering. *Journal of Crop Improvement* 17:69–120.

Pedras, M. S. C. and Zheng, Q. A. 2010. Metabolic responses of *Thellungiella halophila/salsuginea* to biotic and abiotic stresses: Metabolite profiles and quantitative analyses. *Phytochemistry* 71:581–589.

Peters, F. T., Kraemer, T., and Maurer, H. H. 2002. Drug testing in blood: Validated negative-ion chemical ionization gas chromatographic–mass spectrometric assay for determination of amphetamine and methamphetamine enantiomers and its application to toxicology cases. *Clinical Chemistry* 48:1472–1485.

Pichersky, E. and Gang, D. R. 2000. Genetics and biochemistry of secondary metabolites in plant an evolutionary perspective. *Trends in Plant Science* 5:439–445.

Ramanjulu, S. and Bartels, D. 2002. Drought- and desiccation-induced modulation of gene expression in plants. *Plant Cell Environment* 25:141–151.

Rischer, H., Oresic, M., Seppanen-Laakso, T. et al. 2006. Gene-to-metabolite networks for terpenoid indole alkaloid biosynthesis in *Catharanthus roseus* cells. *Proceedings of the National Academy of Sciences of the United States America* 103:5614–5619.

Rizhsky, L., Liang, H., and Mittler, R. 2002. The combined effect of drought stress and heat shock on gene expression in tobacco. *Plant Physiology* 130:1143–1151.

Robb, D. B., Covey, T. R., and Bruins, A. P. 2000. Atmospheric pressure photoionization: An ionization method for liquid chromatography-mass spectrometry. *Analytical Chemistry* 72:3653–3659.

Ropartz, D., Bodet, P. E., Przybylski, C. et al. 2011. Performance evaluation on a wide set of matrix-assisted laser desorption ionization matrices for the detection of oligosaccharides in a high-throughput mass spectrometric screening of carbohydrate depolymerizing enzymes. *Rapid Communications in Mass Spectrometry* 25:2059–2070.

Rozema, J. and Flowers, T. 2008. Crops for a salinized world. *Science* 322:1478–1480.

Schauer, N. and Fernie, A. R. 2006. Plant metabolomics: Towards biological function and mechanism. *Trends in Plant Science* 11:508–516.

Schwartz, S. A., Reyzer, M. L., and Caprioli, R. M. 2003. Direct tissue analysis using matrix-assisted laser desorption/ionization mass spectrometry: Practical aspects of sample preparation. *Journal of Mass Spectrometry* 38:699–708.

Shabala, S. and Mackay, A. 2011. Ion transport in halophytes. *Advances in Botanical Research* 57:151–199.

Sicher, R. C. and Barnaby, J. Y. 2012. Impact of carbon dioxide enrichment on the responses of maize leaf transcripts and metabolites to water stress. *Physiologia Plantarum* 144:238–253.

Silvente, S., Sobolev, A. P., and Lara, M. 2012. Metabolite adjustments in drought tolerant and sensitive soybean genotypes in response to water stress. *PLoS ONE* 7:e38554.

Smith, C. A., Want, E. J., O'Maille, G., Abagyan, R., and Siuzdak, G. 2006. XCMS: Processing mass spectrometry data for metabolite profiling using nonlinear peak alignment, matching, and identification. *Analytical Chemistry* 78:779–787.

Stitt, M. and Hurry, V. 2002. A plant for all seasons: Alterations in photosynthetic carbon metabolism during cold acclimation in *Arabidopsis*. *Current Opinion in Plant Biology* 5:199–206.

Stoeckli, M., Staab, D., and Schweitzer, A. 2006. Compound and metabolite distribution measured by MALDI mass spectrometric imaging in whole-body tissue sections. *International Journal of Mass Spectrometry* 260:195–202.

Sumner, L. W., Mendes, P., and Dixon, R. A. 2003. Plant metabolomics: Large-scale phytochemistry in the functional genomics era. *Phytochemistry* 62:817–836.

Tolstikov, V. V., Lommen, A., Nakanishi, K., Tanaka, N., and Fiehn, O. 2003. Monolithic silica-based capillary reversed-phase liquid chromatography/electrospray mass spectrometry for plant metabolomics. *Analytical Chemistry* 75:6737–6740.

Touchette, B. W., Smith, G. A., Rhodes, K. L., and Poole, M. 2009. Tolerance and avoidance: Two contrasting physiological responses to salt stress in mature marsh halophytes *Juncus roemerianus* Scheele and *Spartina alterniflora* Loisel. *Journal of Experimental Marine Biology and Ecology* 380:106–112.

Trupp, M., Zhu, H., Wikoff, W. R. et al. 2012. Metabolomics reveals amino acids contribute to variation in response to simvastatin treatment. *PLoS ONE* 7:e38386.

Urano, K., Kurihara, Y., Seki, M., and Shinozaki, K. 2010. 'Omics' analyses of regulatory networks in plant abiotic stress responses. *Current Opinion in Plant Biology* 13:132–138.

Verslues, P. E., Agarwal, M., Katiyar-Agarwal, S., Zhu, J., and Zhu, J. K. 2006. Methods and concepts in quantifying resistance to drought, salt and freezing, abiotic stresses that affect plant water status. *Plant Journal* 45:523–539.

Vinayavekhin, N. and Saghatelian, A. 2010. *Untargeted Metabolomics. Current Protocols in Molecular Biology*. New York: John Wiley & Sons.

Vorst, O., De Vos, C. H. R., Lommen, A. et al. 2005. A non-directed approach to the differential analysis of multiple LC–MS-derived metabolic profiles. *Metabolomics* 1:169–180.

Wang, X., Li, W., Li, M., and Welti, R. 2006. Profiling lipid changes in plant response to low temperatures. *Physiologia Plantarum* 126:90–96.

Welti, R., Li, W., Li, M. et al. 2002. Profiling membrane lipids in plant stress responses role of phospholipase Dα in freezing-induced lipid change in *Arabidopsis*. *Journal of Biological Chemistry* 277:31994–32002.

Westphal, Y., Schols, H. A., Voragen, A. G. J., and Gruppen, H. 2010. MALDI-TOF-MS and CE-LIF fingerprinting of plant cell wall polysaccharide digests as a screening tool for *Arabidopsis* cell wall mutants. *Journal of Agricultural and Food Chemistry* 58:4644–4652.

Wikoff, W. R., Kalisak, E., Trauger, S., Manchester, M., and Siuzdak, G. 2009. Response and recovery in the plasma metabolome tracks the acute LCMV-induced immune response. *Journal of Proteome Research* 8:3578–3587.

Wishart, D. S., Lewis, M. J., Morrissey, J. A. et al. 2008. The human cerebrospinal fluid metabolome. *Journal of Chromatography B* 871:164–173.

Wu, H., Liu, X., You, L. et al. 2012. Salinity-induced effects in the halophyte *Suaeda salsa* using NMR-based metabolomics. *Plant Molecular Biology Reporter* 30:590–598.

Xia, J., Psychogios, N., Young, N., and Wishart, D. S. 2009. MetaboAnalyst: A web server for metabolomic data analysis and interpretation. *Nucleic Acids Research* 37:652–660.

Xu, R. N., Fan, L., Rieser, M. J., and El-Shourbagy, T. A. 2007. Recent advances in high-throughput quantitative bioanalysis by LC–MS/MS. *Journal of Pharmaceutical and Biomedical Analysis* 44:342–355.

Yamaguchi-Shinozaki, K. and Shinozaki, K. 2006. Transcriptional regulatory networks in cellular responses and tolerance to dehydration and cold stresses. *Annual Review of Plant Biology* 57:781–803.

Yanes, O., Clark, J., Wong, D. M. et al. 2010. Metabolic oxidation regulates embryonic stem cell differentiation. *Nature Chemical Biology* 6:411–417.

Zaima, N., Goto-Inoue, N., and Setou, M. 2010. Application of imaging mass spectrometry for analysis of rice *Oryza Sativa*. *Rapid Communications in Mass Spectrometry* 24:2723–2927.

Zhu, Z. J., Schultz, A. W., Wang, J. et al. 2013. Liquid chromatography quadrupole time-of-flight mass spectrometry characterization of metabolites guided by the METLIN database. *Nature Protocol* 8:451–460.

Section VII

Marine Nutrigenomics

18

Marine Nutraceuticals

Katarzyna Chojnacka and Izabela Michalak

CONTENTS

18.1 Introduction

The awareness that health comes from the diet goes back to the ancient times of Hippocrates (Kadam and Prabhasankar 2010). Marine organisms have been always present in the human diet (especially in the Orient) and their beneficial role has been underlined, such as the source of vitamins, minerals, polyunsaturated fatty acids (PUFAs), amino acids, dietary fibers, and other bioactive chemicals (Ganesan et al. 2010). Many compounds isolated from marine organisms (microorganisms, algae, invertebrates, vertebrates) are beneficial to health, considering nutritional and pharmacological values (Kim 2013). The approach that uses the marine resource as the raw material in the production of specialized products is gaining an increasing interest. Marine organisms are able to produce chemicals with health benefits because of their method of adaptation to extreme environmental conditions (Mayakrishnan et al. 2013).

The term "nutraceutical" originates from a combination of two words "nutrition" and "pharmaceutical" and signifies the delivery of active substances with pharmaceutical properties to human by food in order to prevent or treat diseases (Vidanarachchi et al. 2012), including chronic diseases (diabetes, Alzheimer's, colon cancer) (Kuppusamya et al. 2014). Nutraceuticals have biomedical applications. Marine biomolecules are currently in the stage of clinical trials for the treatment of asthma, Alzheimer's disease, and cancer (Libes 2009).

Nutraceuticals are biodegradable, soluble in water, and functionally active (Mayakrishnan et al. 2013). A nutraceutical may be a single natural nutrient in the form of powder, tablets, capsules, or liquid. It is not necessarily a complete food but equally not a drug (Hardy 2000). They are added to food to provide functional properties. Functional food possesses proven physiological health benefits (Shahidi 2009). Both functional foods and nutraceuticals promote health, reduce the risk of disease through prevention with the short-term goal to improve the quality of life, and enhance health and long-term goal to increase life span in good health (Libes 2009; Shahidi 2009; Pandey et al. 2010).

Marine organisms are becoming an important source of nutraceuticals. They constitute half of the global biodiversity (Zhang et al. 2012) and live in oceans that cover 70% of the globe, being a source of a vast array of functional chemicals (Mayakrishnan et al. 2013). This biomass is at present an underutilized resource (Chojnacka 2012). Undergoing permanent biotic and abiotic stress and living in extreme environmental conditions (photodynamic stress, UV radiation, free radicals, herbivores) made these organisms developed protective mechanisms by the synthesis of biologically active compounds (Li et al. 2011). Biomolecules of marine origin found an application not only as nutraceuticals but also as cosmeceuticals, biomedicines, pesticides, and antifouling agents (Libes 2009).

This chapter is devoted to marine organisms that produce bioactive compounds beneficial to human. Special attention was paid to raw biomass that is naturally rich in nutraceuticals and also the categories of nutraceuticals and their functions. At the same time, the difficulties associated with the implementation of new formulations to the market are presented. Finally, the examples of commercially available products are listed, and the future of nutraceuticals is predicted.

18.2 Nutraceuticals

The history of nutraceuticals is dated back to ancient times and related with Chinese and Indian medicine. In Europe, for the first time, probably Hippocrates used juices of mollusks as a laxative agent (Libes 2009). Nowadays, nutraceuticals are a large sector in which marine biotechnology can make a contribution (drugs for problems associated with the heart, joints, osteoporosis; calcium products and other trace elements; antioxidants, astaxanthin, and carotenoids; and marine organisms, such as probiotics) (Pandey et al. 2010).

Generally, nutraceuticals are obtained from natural materials: marine organisms, medicinal plants, fruits, and vegetables. Nutraceuticals are distinguished into the following categories: microorganisms (e.g., *Lactobacillus*), medicinal herbs, plant, secondary metabolites (e.g., alkaloids), and dietary supplements (e.g., olive oil). Nutraceuticals are used mainly in the form of extracts (Kuppusamya et al. 2014).

So far, all therapeutic areas such as antiarthritic, pain killers, cold and cough, sleeping disorders, digestion and prevention of certain cancers, osteoporosis, blood pressure, hypertension, cholesterol, depression, and diabetes have been covered by nutraceuticals (Pandey et al. 2010). They are useful in treatment of inflammations, infections, and chronic diseases (Kim 2013). Therapeutic benefits of nutraceuticals include antidiabetic, antiallergic, anticoagulant, anticancer, anti-inflammatory, antioxidant, immunoprotective, cardiovascular, and immune-enhancing properties (Kim 2013; Kuppusamya et al. 2014). Additionally, compounds isolated from marine resources have a wide array of properties: protect from radiation, bacteria, and viruses (Kim 2013). It should be taken into account that the recommended dosage of nutraceuticals is very important and clearly related to benefit the patients (Kuppusamya et al. 2014). The summary of the source, characteristics, and function of marine nutraceuticals is presented in Table 18.1.

TABLE 18.1

Nutraceuticals: The Sources, Characteristics, and Functions

Source	Characteristics	Function	References
Fish	Fish protein hydrolysate—from muscle flesh	Antihypertensive peptides—angiotensin I–inhibiting ACE; hydrolyzed chemically or biologically to edible protein products; used as milk replacement; food binding and gelling	Kadam and Prabhasankar (2010)
Fish bone and shark cartilage	Source of calcium	Calcium fortification; softened by products from large fish processing; fish bone powder—fortification of fish food (e.g., surimi)	Kadam and Prabhasankar (2010)
Seaweeds	Dietary fibers (polysaccharides) and phytochemicals; antioxidants and carotenoids, phycobilins, fatty acids, vitamins, sterols, tocopherol, phycocyanins, fucoidan, and fucoxanthin; and halogenated compounds	Hydrocolloids important for digestive tract (solubility, viscosity, hydration, ion-exchange capacity); antioxidant (carotenoids), antibacterial and antiviral (polysaccharides—carrageenan), platelet aggregation, antitumour, and hyaluronidase; fucoxanthin—xanthophyll carotenoid (antiangiogenic)—prevention from cancer	Kadam and Prabhasankar (2010); Ganesan et al. (2010); Libes (2009)
Microalgae	Enhance nutritional value of food and feed; pigments (chlorophyll, carotenoids), phycobiliproteins	Antimicrobial, antiviral, antifungal properties; neuroprotective products; therapeutic proteins; drugs	Spolaore et al. (2006)
Crustaceans	Chitin and chitosan	Approved as food additive/dietary supplement; functional food ingredient; immune enhancement, disease recovery, and dietary fiber (weight loss, antilipemic)	Kadam and Prabhasankar (2010); Libes (2009)
Sponges	Collagen, high levels of iodine, brominated cyclohexadienes, polyhydroxybrominated phenols	Therapeutics in clinical and preclinical trials; anti-inflammation, antioxidant, pigments; iodine—treats tumors, goiter, dysentery, and diarrhea	Kim and Dewapriya (2012); Libes (2009)
All marine organisms	Omega-3 fatty acids	Protection from cardiovascular disease, reduction in plasma lipids, and modulation of tumor cell growth	Kadam and Prabhasankar (2010)

18.3 Marine Organisms as a Source of Nutraceuticals

Marine organisms can serve as a valuable source of compounds that are beneficial to human; for example, fish oils and marine bacteria are excellent sources of omega-3 fatty acids, while crustaceans and seaweeds contain powerful antioxidants such as carotenoids and phenolic compounds (Lordan et al. 2011). The characteristics of the most important marine organisms being the source of nutraceuticals are as follows.

18.3.1 Algae

Algae are a valuable raw material used for the production of chemicals, pharmaceuticals, and nutraceuticals. This is the issue realized by a discipline called "algal biotechnology," whereby algae are harvested or cultivated for commercial applications (Chen 2008). In the development of nutraceuticals, either whole cells or extracted functional components are used (Chen 2008). Algal nutraceutical products have the form of extracts or powders (Gellenbeck 2012). Nutraceuticals from marine algae include polyphenols, PUFA, vitamins, minerals, dietary fibers, and antioxidants (Mayakrishnan et al. 2013). It should be underlined that there are some chemicals that are found only in algae, for example, agar-agar, carrageenan, algin, or sulfated polysaccharides (fucoidan, laminarin), and are applied in food, dairy, paper, pharmaceutical, and textile industry for their gelling, thickening, and stabilizing properties. However, there is little scientific evidence that confirm their therapeutic role (Mayakrishnan et al. 2013). Algae refer to various groups of organisms that include microalgae and, among them, cyanobacteria (called "blue-green algae") and macroalgae (called also "seaweed").

18.3.1.1 Microalgae

Eukaryotic microalgae and prokaryotic cyanobacteria can be cultivated photoautotrophically in open or closed ponds (algae require light to grow and produce new biomass) or via heterotrophic methods (algae grown without light and use carbon source, such as sugars, to generate new biomass) (Andersen 2005). Different types of microalgae could become prevalent in food supplements and nutraceuticals. This results from the ability to produce necessary vitamins, essential amino acids, and unsaturated fatty acids and to concentrate essential elements. The main advantage of microalgae over other marine organisms is the high protein content and their amino acid pattern; therefore, they are considered as an alternative protein source (Spolaore et al. 2006; Bishop and Zubeck 2012).

18.3.1.2 Seaweeds

The second group of algae is seaweeds (macroalgae). They are classified into *Chlorophyta* (green algae), *Phaeophyta* (brown algae), and *Rhodophyta* (red algae) by their pigments. Macroalgae can be cultivated in open offshore or in coastal areas (Andersen 2005). The biomass is obtained by wild harvest (*Ascophyllum, Fucus, Laminaria, Macrocystis*) or by controlled planting on ropes (*Porphyra, Ulva, Chondrus*) (Gellenbeck 2012). About 220 species are commercially harvested, half of which are used for food and the remaining for the production of phycocolloids (Libes 2009). Especially, nutraceuticals from seaweeds have been served as a rich source of health-promoting components. Among them, marine brown algae are a rich source of natural bioactive compounds, mainly phlorotannins, and this fact implies their potential as a functional ingredient in the food products, pharmaceuticals, and cosmeceuticals (Li et al. 2011).

18.3.2 Marine Animals

Fish are the most popular marine animals. They are used as raw materials in functional foods and as sources of nutraceuticals. Marine fish is a major source of high-quality peptides, proteins, lipids, and a wide variety of vitamins and minerals (Khora 2013). However, not only the fish meat is a valuable resource of bioactivities but also their by-products. When the fish/shellfish is gutted, headed, and further processed either in onboard fishing vessels or in processing plants on shore, by-products are formed. In the work of Ferraro et al. (2010), it was indicated that fish heads, viscera, skin, tails, offal and blood, and seafood shells possess several high-value-added compounds that are suitable for human health. In this detailed review, the main bioactive compounds with their functions were described (PUFAs, taurine, creatine, chitin, chitosan and their oligomers, collagen and gelatin, hydroxyapatite, antifreeze proteins, enzymes, astaxanthin). Rasmussen and Morrissey (2007) also underlined that some of the most prevalent marine proteins used in foods (collagen, gelatin, and albumin) can be extracted from fish and seafood by-products. In the case of marine animals, a special attention should be paid to omega-3 oils—although originating from marine algae, omega-3 oils are predominant in marine fish and mammals. Lipids from the body of fatty fish (e.g., mackerel and herring), the liver of white lean fish (e.g., cod and halibut), and the blubber of marine mammals (e.g., seals and whales) are rich in long-chain fatty acids (Shahidi 2009). The global production of fish oil is around 1 million tonnes. Fish oil can be used directly in a purified form (nutraceuticals) in a wide range of foods (Pike and Jackson 2010). These authors assume that the use of fish oil in nutraceuticals increases at around 15% per annum.

Production of marine-based food ingredients from these by-products is a growing area of interest as it could help to reduce processing waste and it could result in the development of valuable nutraceutical or functional food formulations (Rasmussen and Morrissey 2007). Additionally, very often, parts of marine organisms that are commonly discarded have unique features, and surprisingly, the highest concentration of minerals, lipids, amino acids, polysaccharides, and proteins is often found in these by-products (Ferraro et al. 2010). The bioactive compounds can be extracted from by-products, purified, and utilized for biotechnological and pharmaceutical applications (Kim and Mendis 2006; Ferraro et al. 2010).

The majority of bioactive marine molecules can be also isolated from benthic species such as ascidians (e.g., *Aplidium albicans*, *Trididemnum inarmatum*), sponges (e.g., *Agelas oroides*, *Axinella damicornis*, *Aplysina cavernicola*), opisthobranchs (a group of soft-bodied mollusks), cnidarians, gorgonian (e.g., *Eunicella cavolini*), coral (e.g., *Dendrophyllia cornigera*), bryozoans (e.g., *Myriapora truncate*), sea cucumber (e.g., *Holothuria polii*), echinoderms, polychaetes, and mollusks (Lloret 2010).

18.4 Functionality of Food Ingredients

Food is functional, in that it was proved that a food beneficially affects one or more functions in the body, besides its nutritional value. Functional food should improve health and reduce the risk of disease (Kadam and Prabhasankar 2010). For example, a functional food could be a lutein-rich food such as chicken, spinach, tomatoes, oranges, or the omega-3 fatty acids found in fish oil (Blas-Valdivia et al. 2012). Marine natural products are secondary metabolites, which means they do not play a direct role in the growth, development, or reproduction of an organism. These compounds are synthesized to control ecological relationships (Libes 2009). Examples of these active compounds include peptides, polysaccharides (sulfated, chitooligo), chitosan, carotenoids, phlorotannins, fucoxanthins (Kim 2013), collagen, gelatin, sterols, lectins (Zhang et al. 2012), PUFAs, omega-3 oils (Kadam and Prabhasankar 2010), acetylapoaranotin, astaxanthin, and siphonaxanthin (Kuppusamya et al. 2014). The maximum reactivity of these compounds is dependent on environmental conditions and is 5°C–12°C, not 30°C–35°C (Libes 2009). Therefore, marine organisms offer a wide array of bioactive ingredients that have a significant promise in the promotion of human health and disease prevention. Potential marine nutraceuticals by class of compounds are presented in Table 18.2.

TABLE 18.2

Potential Marine Nutraceuticals by Class of Compounds

Compound	Isolated From	Application in Industry	References
Lectins	Prokaryotes, algae, marine invertebrates, and vertebrates	Used as prevention from calcium deficiency and in antibiotics in marine vibrios and also antitumor, immunomodulatory, and hemostasis drugs	Zhang et al. (2012)
Bioactive peptides	Produced by enzymatic hydrolysis of marine proteins; by-products of marine processing	Utilized to reduce the risk of cardiovascular diseases; as natural antioxidants in food; for antihypertensive, ACE inhibition, anticoagulant, and antimicrobial drugs; and as support in treating of obesity, stress, hypertension, and migraine	Zhang et al. (2012); Cudennec et al. (2012)
Chitooligosaccharide (COS) derivatives	Crustaceans; COS is obtained by chemical/enzymatic hydrolysis of chitosan	Used in the inhibition of ACE, in antioxidant, antimicrobial, anticancer, antidiabetic, hypocholesterolemic, hypoglycemic, anti-Alzheimer's, and anticoagulant drugs, and in the inhibition of adipogenesis	Zhang et al. (2012)
Phlorotannins	Marine brown and red algae (phloroglucinol-based polyphenols)	Used in anti-HIV, antiproliferative, anti-inflammatory, radioprotective, antidiabetic, anti-Alzheimer's, antimicrobial, and antihypertensive drugs and also radical-scavenging antioxidants in the food industry	Zhang et al. (2012); Li et al. (2011)
Sulfated polysaccharides	Seaweeds: green (ulvan), red (carrageenan), and brown (fucoidan, laminaran)	Applied in anticoagulant, antiviral, and immune inflammatory drugs, cosmetic and pharmacology, food additives (emulsifiers, stabilizers, thickeners), and hypolipidemic drugs by influencing sequestration of bile acid in the intestinal lumen	Zhang et al. (2012); Mayakrishnan et al. (2013)
Polyphenols	Seaweeds	Utilized in the reduction of arterial diseases	Mayakrishnan et al. (2013)
Long-chain omega-3 fatty acids	Single-cell organisms *Crypthecodinium cohnii*, sharks	Eicosapentaenoic acid (EPA), docosahexaenoic acid (DHA) applied for anticardiovascular diseases; added to functional foods, in infant formulas; used to lower blood lipids (cholesterol) and in antithrombotic and anti-inflammatory drugs	Barrow (2010); Libes (2009)

The beneficial effects of marine nutraceuticals and functional foods have been attributed to their components, such as pigments, polyphenols, polysaccharides, PUFAs, vitamins, and minerals. Description of main marine bioactive compounds with their functions is presented next. In addition, the literature with their detailed description is indicated.

18.4.1 Algal Pigments

The main pigments found in the biomass of algae are as follows: chlorophylls, phycobilins (phycocyanin, phycoerythrin), carotenoids, carotenes (α-carotene, β-carotene, lycopene), and xanthophylls (astaxanthin, fucoxanthin, zeaxanthin, lutein). Among the algal proteins, it is worth noting the occurrence of protein–pigment complexes called phycobiliproteins, some of which are currently used as fluorescent markers in the fields of clinical diagnosis and biotechnological applications (Burtin 2003). From over 400 known carotenoids, only very few are used commercially: β-carotene, astaxanthin, and, of lesser importance, lutein, zeaxanthin, and lycopene (Spolaore et al. 2006). Among microalgae, few species are an interesting source of nutraceuticals, for example, *Dunaliella* cultivated in open raceway ponds is a source of β-carotene (a nontoxic precursor of vitamin A). The microalga *Haematococcus* produces carotenoid astaxanthin, which is used in fish aquaculture (red pigment of salmon) and as anti-inflammatory and antioxidant factor in food (Gellenbeck 2012). Astaxanthin represents also between 74% and 98% of the total pigments in shellfish (Ferraro et al. 2010). Marine-derived bioactive compounds are of current interest to cure several ailments including colon cancers, for example, astaxanthin (from microalga *Haematococcus pluvialis*, crab, and marine animals) and siphonaxanthin (from the marine green alga *Codium fragile*) (Kuppusamya et al. 2014). Nutraceutical applications of carotenoids are described in papers of Blas-Valdivia et al. (2012) and Lordan et al. (2011).

18.4.2 Polyphenols

The properties of marine polyphenols—particularly flavonoids (flavones, catechin or flavanols, anthocyanins, isoflavones)—were discussed in several papers (Kuppusamya et al. 2014). It was confirmed that algal polyphenols have different nutraceutical properties, such as antioxidant, anti-inflammatory, anticancer, antibacterial, antiatherogenic, and antiangiogenic (Ngo et al. 2011; Blas-Valdivia et al. 2012; Zhang et al. 2012). Polyphenols are natural phytochemicals used to treat various viral and fungal diseases. In humans, polyphenols—secondary metabolites—can be used as a nutraceutical for colon cancer therapeutics (Kuppusamya et al. 2014).

18.4.3 Polysaccharides

Algae are a rich source of polysaccharides (dietary fiber), some of which are not found in other plants and are believed to possess specific functions (Ruperez et al. 2002). Brown algae contain mainly alginate, fucoidan, and laminarin; red algae contain agar, carrageenan, and porphyran; green algae contain cellulose, xylan, and ulvan. Polysaccharides not only function as dietary fiber but also contribute to the antioxidant activity of marine algae. Antioxidant activity of algal polysaccharides has been reported in the literature (Ruperez et al. 2002; Zhang et al. 2003). It was also shown that polysaccharides derived from algae possess the following properties: anticholesterol, anti–free radical, cardioprotective, antiviral, anticoagulant, immunomodulating, and anticancer (Lordan et al. 2011; Ngo et al. 2011; Zhang et al. 2012). Additionally, bioactive sulfated polysaccharides are a soluble dietary fiber with medicinal benefits—exhibiting cardioprotective properties (Mayakrishnan et al. 2013). Nutraceutical application of sulfated polysaccharides (fucoidans) was described thoroughly in the review papers of Zhang et al. (2012) and Ngo et al. (2011).

In marine polysaccharides, chitin is also ubiquitous (it is one of the major structural components of crustacean shells and shellfish wastes) (Lordan et al. 2011). Chitin and its deacetylated form—chitosan—have a variety of nutraceutical applications. The dietary fiber causes weight loss and possesses antilipemic properties. Chitin is also a source of glucosamine used as a precursor in drug synthesis and as a nutraceutical (Libes 2009).

18.4.4 Polyunsaturated Fatty Acids

PUFAs include omega-3 fatty acids (e.g., eicosapentaenoic acid [EPA] and docosahexaenoic acid [DHA]) and omega-6 fatty acid (e.g., γ-linolenic acid [GLA] and arachidonic acid [AA]). The most common long-chain PUFAs are EPA and DHA, both of which are currently marketed as nutraceuticals (Libes 2009). For example, microalga *Nannochloropsis* is a source of omega-3 fatty acids that are used in cosmeceuticals and skincare products and in heart health protection (Gellenbeck 2012). PUFAs regulate a wide range of functions in the body including blood pressure, blood clotting, and correct development and functioning of the brain and nervous systems (Wall et al. 2010). Their functions were widely described in the available literature (Spolaore et al. 2006; Lordan et al. 2011; Blas-Valdivia et al. 2012).

Marine animals are also a rich source of nutraceuticals (fish oils, shark cartilage and liver oil, fatty acids) (Okada and Morrissey 2007). They are thought to relieve arthritic symptoms, cardiovascular problems, and development of nervous system in infants, and shark cartilage is used in cancer therapy as antiangiogenesis agent (Molyneaux and Lee 1998).

18.4.5 Proteins and Peptides

Proteins from marine sources show promise as functional ingredients in foods because they possess numerous important and unique properties such as film and foaming capacity, gel-forming ability, and antimicrobial activity (Rasmussen and Morrissey 2007). The discovery of the bioregulatory role of different endogenous peptides in the organism, as well as the understanding of the molecular mechanisms of action of some new bioactive peptides obtained from natural sources on specific cellular targets, contributed to consider peptides also as promising lead drug candidates (Aneiros and Garateix 2004). The authors provided detailed description of novel peptides from sponges, ascidians, mollusks, sea anemones, and seaweeds together with their pharmacological properties and production methods. In the literature, it was found that marine bioactive peptides possess antihypertensive, antioxidant, anticoagulant, and antimicrobial activities (Kim and Wijesekara 2010; Lordan et al. 2011; Ngo et al. 2011; Zhang et al. 2012). Bioactive peptides isolated from fish protein hydrolysates have shown numerous properties such as antihypertensive, antithrombotic, anticoagulant, immunomodulatory, and antioxidative (Je et al. 2004; Jun et al. 2004; Rajapakse et al. 2005; Khora 2013). Marine organisms can also serve as a source of taurine, a unique nonessential amino acid that is beneficial to human, for example, algae (Samarakoon and Jeon 2012), marine crustaceans (Hamdi 2011), and fish (Gormley et al. 2007). It was found that oyster *Ostrea gigas* contains highest taurine in all water products, thus bringing obvious and fascinating applicable values (Hamdi 2011). Taurine is known to have antioxidant activity—protecting kidney from oxidative injury. Moreover, it also has inflammatory properties, lowers blood sugar, protects liver, prevents from arrhythmia, and treats rheumatoid arthritis and eye inflammation (Hamdi 2011; Samarakoon and Jeon 2012).

18.4.6 Vitamins and Minerals

Different types of algae, specifically microalgae, could become more prevalent in food supplements and nutraceuticals due to the capability of producing necessary vitamins including A (retinol), B_1 (thiamine), B_2 (riboflavin), B_3 (niacin), B_6 (pyridoxine), B_9 (folic acid), B_{12} (cobalamin), C (L-ascorbic acid), D, E (tocopherol), and H (biotin) (Bishop and Zubeck 2012). On the other hand, fish bones, which are separated after the removal of muscle proteins on the frame, are a valuable source of calcium, which is an essential element for human health (Martínez-Valverde et al. 2000).

18.4.7 Antioxidants

Many of the marine bioactive compounds mentioned earlier are characterized by their antioxidant activity, for example, vitamins and vitamin precursors including α-tocopherol, β-carotene, niacin, thiamin, and ascorbic acid, pigments (chlorophylls and carotenoids), peptides, polyphenols, sulfated polysaccharides, and chitooligosaccharide derivatives (Shahidi 2009; Ngo et al. 2011). Especially interesting are brown algal polyphenols that are called phlorotannins (fucol, phlorethol, fucophlorethol, fuhalol, isofuhalol, eckol) (Li et al. 2011). Their biological activities (antioxidant, bactericidal, anti-HIV, anticancer activity), enzyme inhibitory effect (inhibition of α-glucosidase, α-amylase, acetylcholinesterase and butyrylcholinesterase, angiotensin-I-converting enzyme, matrix metalloproteinases, hyaluronidase, tyrosinase), radioprotective and antiallergic effects, and potential health benefits were described in details in a review of Li et al. (2011), Zhang et al. (2012), and Ngo et al. (2011). Moreover, Shahidi (2009) showed that a number of studies in which the antioxidant activity of marine algae was evaluated revealed high antioxidant efficacy of their extracts, which is equal to or better than that of commercial antioxidants such as butylated hydroxyanisole, butylated hydroxytoluene, and α-tocopherol, and has suggested the use of algal antioxidants in food formulations.

18.5 Technology

Nutraceuticals are isolated from marine biomass by different bioprocessing technologies (Zhang et al. 2012). The choice of suitable extraction method should depend on the raw material and the nature of the extracted compound. Biofunctional components are closed within cells of marine organisms. It is essential to elaborate a technology that would isolate the compounds from the intact cells in the way that the compounds would not be deteriorated (Ryu et al. 2014). The second crucial element is to select optimal process conditions (influence of solvent, pH, temperature, pressure) in order to avoid degradation of extracted compounds (Ibañez et al. 2012; Kadam et al. 2013).

New extraction techniques are proposed, which obtain extracts from the concentrate of biologically active compounds in a solvent-free environment. The obtained products are safe to human and can be used in the food and pharmaceutical industries. In the literature, it is indicated that the following methods are suitable for the extraction of nutraceuticals: supercritical fluid extraction (SFE), microwave-assisted extraction (MAE), ultrasound-assisted extraction (UAE), enzyme-assisted extraction (EAE), and pressurized liquid extraction (PLE) (also known as pressurized fluid extraction, enhanced solvent extraction, high-pressure solvent extraction, or accelerated solvent extraction techniques) (Ibañez et al. 2012; Wijesinghe and Jeon 2012; Kadam et al. 2013).

The majority of studies regarding acquisition of biologically active compounds (e.g., antioxidants) from marine organisms concern the crude extracts from a mixture of non-fractionated compounds. There is a little knowledge on purification and characterization of particular compounds. More research is needed to fully characterize nutraceuticals to add them to the food in the form of pure, isolated compounds (Ngo et al. 2011). Also important is the elaboration of novel methods of delivery and application of nutraceuticals, for example, stabilization by microencapsulation (Barrow 2010).

In the review paper of Michalak and Chojnacka (2014), it was shown that novel extraction techniques can be used for the isolation of a vast array of nutraceuticals. For example, SFE was used for the extraction of pigments such as chlorophylls, carotenoids (e.g., β-carotene, canthaxanthin, astaxanthin, fucoxanthin), lipids such as n-3 (e.g., eicosapentaenoic, DHA) and n-6 (e.g., linoleic acid, GLA, AA), fatty acids, polyphenols such as flavonoids (e.g., isoflavones), and vitamin E from the biomass of algae. The use of MAE enabled the extraction of n-3 fatty acids (e.g., DHA), polysaccharides (e.g., fucoidan), pigments (e.g., fucoxanthin), elements (iodine, bromine), phenols, phytosterols, and phytol from algae. UAE was applied to obtain appropriate biofunctional components from the algal biomass: pigments such as chlorophyll a and carotenoids (e.g., lutein), n-3 fatty acids (e.g., DHA), and major, minor, and trace elements (Ca, K, Na, Mg, Cd, Cr, Cu, Mn, Ni, Pb, and Zn). Antioxidants, fucoxanthin, lipids containing PUFAs, and polysaccharides were extracted from algae using EAE. PLE was used for the extraction of antioxidants, carotenoids (fucoxanthin), phenols, and fatty acids such as phytol, fucosterol, neophytadiene, palmitic, palmitoleic, and oleic acids.

In the case of nutraceuticals isolated from by-products derived from marine animals, three methods can be applied: solvent extraction, enzymatic hydrolysis, or microbial fermentation of marine proteins. However, particularly in food and pharmaceutical industries, the enzymatic hydrolysis of marine proteins is preferred on the account of lack of residual organic solvents or toxic chemicals in the products (Kim and Wijesekara 2010; Lordan et al. 2011). For example, Marealis (a company from Norway) developed a dietary supplement called "Tensiotin," the role of which is to reduce blood pressure. The product is obtained by enzymatic hydrolysis of shrimp shells to peptides (Young 2014). Lipase-catalyzed hydrolysis was demonstrated to be a feasible method for the extraction of n-3 PUFAs from sardine oil for the use in nutraceuticals and in other products (Okada and Morrissey 2007). It should be kept in mind that the process of preparation of fish protein hydrolyzates consists of proteolytic digestion of fish and fish waste at optimal conditions of pH and temperature required by the enzymes. The enzymes can be isolated from plant (papain and ficin), animal (trypsin and pancreatin), or microbial (pronase and alcalase) sources (Loffler 1986). For example, for the hydrolysis of tuna dark muscle, the following enzymes (under optimal conditions) can be used: alcalase, α-chymotrypsin, papain, pepsin, neutrase, and trypsin (Kim and Wijesekara 2010). Ryu et al. (2014) proposed the hydrolysis of peptides at high temperature and pressure conditions and sometimes assisted by ultrawave lysis. Under these conditions, polypeptides are broken down to low-molecular-weight molecules, which improve their bioavailability. Tahergorabi et al. (2014) suggested another method for the recovery of proteins and n-3 fatty acids from fish-processed by-products that are usually discarded. Isoelectric solubilization/precipitation allows efficient recovery of fish protein isolates that can be used in the development of nutraceutical products destined for human consumption.

18.6 Marine Nutraceuticals: Problems and Prospects

Marine nutraceuticals contain compounds that pose a well-defined physiological effect and do not change organoleptic properties of food (Kadam and Prabhasankar 2010). However, the introduction of new products to the market involves many difficulties and limitations. In the last 30 years, over 3000 active compounds from marine organisms were isolated and identified; however, only few have been successfully brought to the market (Libes 2009). Several key factors should be thoroughly considered. The first problem is to manufacture standardized products, with chemically defined composition. This is difficult because the level of active constituents in the biomass of, for example, algae, differs between location, time of harvest, and species, which determines structural and pharmacological differences in characteristics (Mayakrishnan et al. 2013). Other problems related with marine nutraceuticals are possible sensory and physicochemical changes of food and possible contamination with toxic elements (Kadam and Prabhasankar 2010). The next key problem is the lack of appropriate regulations. There are few laws that regulate the production and sale of such products. Because the products are not submitted for standardized toxicology testing, sometimes they may be toxic for human consumption. There are no specific regulations in any country to control nutraceuticals, and they need to be established and should be considered under the same laws that regulate pharmaceuticals and food (Bernal et al. 2011). Health professionals, nutritionists, and regulatory toxicologists should strategically work together to plan an appropriate regulation to provide the ultimate health and therapeutic benefit (Pandey et al. 2010).

Therefore, in the development of new marine nutraceuticals, assessment of efficacy, sourcing, quality assurance, regulatory requirements, and procurement/purchasing are the important factors (Gellenbeck 2012). In the implementation of a new natural product based on living organisms, major quality categories need to be considered: manufacturing flowcharts, solvents (no residues in the final product), quantitative ingredient list, pesticide/chemical residues, contaminants (e.g., from naturally derived organisms, e.g., insects and from processing, e.g., metal flakes from machinery), microbiology, allergens, animal derivations, physical characteristics (e.g., moisture content), stability of the product, GMO status, certifications (e.g., ISO), and safety (e.g., clinical testing) (Gellenbeck 2012).

Nevertheless, the prospects for nutraceuticals are promising. This is due to increased public awareness, taking care of health, and healthy eating. Nowadays, consumers are understandably more interested in the potential benefits of nutritional support for disease control or prevention. Risk of toxicity or adverse effect of drugs led us to consider safer nutraceutical and functional food-based approaches for the health management. This resulted in a worldwide nutraceutical revolution, which will lead into a new era of medicine and health, in which the food industry will become research oriented, similar to the pharmaceutical industry (Pandey et al. 2010). Moreover, according to Brännback and Wiklund (2001), biotechnology is turning a traditionally low-tech industry (food) into a high-tech industry (functional food/nutraceuticals). Innovation processes in the functional food industry should be intensified, including the role of R&D and collaboration.

18.7 Marketing

Marketing strategies have exploited the words "functional food" and "nutraceuticals" in their advertisements. Many entrepreneurs seek to introduce different natural products into the health and nutritional market. The expanding nutraceutical market indicates that final users are seeking for minimally processed food with extra nutritional benefits and organoleptic value (Pandey et al. 2010). At present, the market of nutraceuticals is dynamically expanding. It is estimated that the market size will reach 250 billion U.S. dollars by the year 2018 (Young 2014). The marketed nutraceuticals are produced mainly from the organisms that live in shallow waters because of the ease of harvesting (Libes 2009). The most popular commercial marine nutraceuticals are glucosamine and long-chain omega-3 lipids, which are added to foods and drinks (Barrow 2010).

On implementing the new products in the market, it is important to consider technological aspects (purification, chemistry, functional interactions), as well as utilitarian values and applications (Kim 2013). Also important is the form of supplementation. Nutraceuticals can be delivered in, for example, dairy products, because of the minimum changes in physicochemical properties (Vidanarachchi et al. 2012).

Summarizing, due to the abundant production of beneficial compounds and nutritive contents of marine organisms, the market production for nutraceuticals is imminent.

18.8 Commercial Preparations Available in the Market

There is a need to utilize marine materials for nutraceutical and cosmeceutical applications. Nowadays, algal-based products are manufactured by many producers all over the world. Both whole cells and their ingredients can be applied as nutraceuticals. Different types of algae, specifically microalgae, could become more prevalent in food supplements and nutraceuticals because of their abundant production of beneficial compounds and nutritive contents (Bishop and Zubeck 2012). *Spirulina* is a source of single-cell protein and has valuable nutraceutical properties (Anupama 2000). Many manufacturers produce *Spirulina*; for example, *Parry Organic Spirulina* is produced by Parry Nutraceuticals (India) and is considered as the green food (with over 60% protein content) (www.parrynutraceuticals.com). Another commercial product is *Spirulina Pacifica®* produced by Cyanotech (Hawaii, www.cyanotech.com).

Astaxanthin, β-carotene, and omega-3 fatty acid (DHA and EPA) are also the high-value algae-derived nutraceuticals. Algaetech International (Malaysia) offers *Astaxanthin-60*, which supports the body against oxidative stress, enhances postexercise recovery, supports joint health and cardiovascular health, and helps to fight signs of premature aging (http://algaetech.com.my). Other companies offering algae-derived astaxanthin are Alga Technologies, *AstaPure®* (Israel, www.algatech.com); Cyanotech, *BioAstin®* (Hawaii, www. cyanotech.com); Jingzhou Natural Astaxanthin Inc. (China); Mera Pharmaceuticals Inc., *AstaFactor®* (Hawaii, www.merapharma.com); Valensa International, *Pur-Blue™ SpiruZan®* (Florida, www.valensa.com); and Parry Nutraceuticals, Parry Natural Astaxanthin (India, www.parrynutraceuticals.com).

The main strains of algae for β-carotene production are *Dunaliella, Spirulina platensis, Chlorella,* and *Caulerpa taxifolia.* Some producers of this bioactive compound are Nikken Sohonsha Corporation, *Dunaliella Hard Capsule* (Japan, www.nikken-miho.com), and Parry Nutraceuticals, *Natural Mixed Carotenoids* (India, www.parrynutraceuticals.com).

Important producers of omega-3 fatty (EPA, DHA) acids from algae are Aurora Algae, *A2 EPA Pure™* (United States, www.aurorainc.com); Lonza, *DHAid™* (Switzerland, www. lonza.com); and Photonz, *PNZ0901* (New Zealand, www.photonzcorp.com).

Nutraceuticals derived from marine animals are also commercially available, for example, a nutraceutical called "Lyprinol®." It is a patented and stabilized natural marine lipid extract that comprises a rare combination of lipid groups and unique omega-3 PUFAs. It is derived from the New Zealand green-lipped mussel (*Perna canaliculus*). Lyprinol is a natural anti-inflammatory product for the pain relief from arthritis, asthma, and a range of other inflammatory disorders (Hong Kong, www.lyprinol.com). *Isolutrol®* is an extract of shark bile, and it is used exclusively as an excipient in cosmetics. Isolutrol reduces the oiliness of skin when applied topically (McFarlane Pty. Ltd., Australia, www. nicnas.gov.au). Marealis AS (Norway) produces *Tensiotin®*, which shows a very potent angiotensin-converting enzyme (ACE)-inhibiting effect of peptides from the Arctic cold water shrimp, and *Systolite®*, which is derived by controlled enzymatic hydrolysis of the shell fraction from the cold water shrimp. The hydrolysate is then further refined into a peptide concentrate (www.marealis.com). FMC BioPolymer AS (Sandvika, Norway) produces *NovaMatrix®*—a biocompatible and bioabsorbable biopolymer for use in the pharmaceutical, biotechnology, and biomedical industries, including applications such as in drug delivery, tissue engineering, cell encapsulation, and medical devices (www. novamatrix.biz). They also offer alginate (PRONOVA™ sodium alginates) and chitosan (PROTASAN™ ultrapure chitosan).

The company Setalg (France) offers a wide range of nutraceuticals (www.setalg.com). A product called "Marine Calcium" contains calcium and phosphorus in the form of hydroxyapatite. The hydroxyapatite is an identical crystal formed from the human bone, and it contains a specific ratio of calcium/phosphorus. This product has a particularly high calcium bioavailability. Marine Calcium is obtained by enzymatic hydrolysis of edges of fish. In the offer of this company, there is also *Chitosan*—a molecule that is naturally present in the shell of crustaceans. This polysaccharide is composed almost entirely of repeating units of D-glucosamine and *N*-acetyl-D-glucosamine. Chitosan is obtained by deacetylation of chitin from lobsters, crabs, and shrimps. All the resources used are only from renewable sources. Proteins, minerals, and other remaining insoluble materials have been removed, which gives the chitosan a remarkable purity. Setalg is also a *Fish Cartilage* producer of two types of cartilage, skate cartilage and shark cartilage. It provides natural constituents such as collagen, glucosamine, and chondroitin, which can also be extracted. Another product—*Fish Protein Extract*, is obtained from lean fish such as a cod, saithe, and haddock. To obtain short polypeptide chains with low molecular weight, this extract is obtained by enzymatic hydrolysis in mild conditions of pH and temperature.

18.9 Conclusions

Due to enormous biodiversity, marine organisms (microalgae, seaweeds, fish) are becoming an important source of nutraceuticals. Recent studies provided evidence that marine-derived functional ingredients play a vital role in human health and nutrition. Natural bioactive compounds extracted from marine organisms (pigments, polyphenols, polysaccharides, PUFAs, proteins and peptides, vitamins and minerals) have a wide range of therapeutic properties, including antimicrobial, antioxidant, antihypertensive, anticoagulant, anticancer, and anti-inflammatory. Therefore, these are the valuable raw materials for the preparation of various functional foods and nutraceutical products. However, in order to fully commercialize the marine organisms as nutraceuticals, the performances of research studies using human models or clinical trials are necessary in the future.

Acknowledgments

This work was supported by the grant entitled "Biologically active compounds in extracts from Baltic seaweeds" (2012/05/D/ST5/03379) attributed by the National Science Centre and the grant entitled "Innovative technology of seaweed extracts—components of fertilizers, feed, and cosmetics" (PBS/1/A1/2/2012) attributed by the National Centre for Research and Development in Poland.

References

Andersen, R. 2005. *Algal Culturing Techniques*, 1st edn. Elsevier Academic Press. Burlington, Massachusetts.

Aneiros, A. and A. Garateix. 2004. Review. Bioactive peptides from marine sources: Pharmacological properties and isolation procedures. *Journal of Chromatography B* 803:41–53.

Anupama, P. R. 2000. Value-added food single cell protein. *Biotechnology Advances* 18:459–479.

Barrow, C. J. 2010. Marine nutraceuticals: Glucosamine and omega-3-fatty acids. New trends for established ingredients. *AgroFOOD* 2:20–23.

Bernal, J., J. A. Mendiola, E. Ibanez, and A. Cifuentes. 2011. Advanced analysis of nutraceuticals. *Journal of Pharmaceutical and Biomedical Analysis* 55:758–774.

Bishop, W. M. and H. M. Zubeck. 2012. Evaluation of microalgae for use as nutraceuticals and nutritional supplements. *Journal of Nutrition & Food Sciences* 2:147.

Blas-Valdivia, V., R. Ortiz-Butron, R. Rodriguez-Sanchez, P. Torres-Manzo, A. Hernandez-Garcia, and E. Cano-Europa. 2012. *Microalgae of the Chlorophyceae Class: Potential Nutraceuticals Reducing Oxidative Stress Intensity and Cellular Damage, Oxidative Stress and Diseases*, ed. V. Lushchak, pp. 581–610. InTech, Rijeka, Croatia. Available from: http://www.intechopen.com/books/oxidative-stress-and-diseases/the-microalgae-of-the-chlorophyceae-class-have-potential-as-nutraceuticals-because-their-ingestion-r. Accessed May 26, 2014.

Brännback, M. and P. Wiklund. 2001. A new dominant logic and its implications for knowledge management: A study of the Finnish food industry. *Knowledge & Process Management* 8:197–206.

Burtin, P. 2003. Nutritional value of seaweeds. *EJEAFChe* 2:498–503.

Chen, F. 2008. Microalgae and their biotechnological potential. *Journal of Biotechnology* 136S:519–526.

Chojnacka, K. 2012. Algal extracts. Biological concentrate of the future. *Przemysł Chemiczny* 91:710–712.

Cudennec, B., T. Caradec, L. Catiau, and R. Ravallec. 2012. Upgrading of sea by-products: Potential nutraceutical applications. *Advances in Food and Nutrition Research* 65:479–494.

Ferraro, V., I. B. Cruz, R. F. Jorge, F. X. Malcata, M. E. Pintado, and P. M. L. Castro. 2010. Valorisation of natural extracts from marine source focused on marine by-products: A review. *Food Research International* 43:2221–2233.

Ganesan, P., K. Matsubara, T. Ohkubo, Y. Tanaka, K. Noda, T. Sugawara, and T. Hirata. 2010. Anti-angiogenic effect of siphonaxanthin from green alga, *Codium fragile*. *Phytomedicine* 17:1140–1144.

Gellenbeck, K. W. 2012. Utilization of algal materials for nutraceutical and cosmeceutical applications—What do manufacturers need to know? *Journal of Applied Phycology* 24:309–313.

Gormley, T. R., T. Neumann, J. D. Fagan, and N. P. Brunton. 2007. Taurine content of raw and processed fish fillets/portions. *European Food Research and Technology* 225:837–842.

Hamdi, S. A. H. 2011. Muscle and exoskeleton extracts analysis of both fresh and marine crustaceans *Procambarus clarkii* and *Erugosquilla massavensis*. *African Journal of Pharmacy and Pharmacology* 5:1589–1597.

Hardy, G. 2000. Nutraceuticals and functional foods: Introduction and meaning. *Nutrition* 16:688–689.

Ibañez, E., M. Herrero, J. A. Mendiola, and M. Castro-Puyana. 2012. Extraction and characterization of bioactive compounds with health benefits from marine resources: Macro and micro algae, cyanobacteria, and invertebrates. In: *Marine Bioactive Compounds: Sources, Characterization and Applications*, ed. M. Hayes, pp. 55–98. Springer Science+Business Media, LLC, New York.

Je, J.-Y., P.-J. Park, J. Y. Kwon, and S.-K. Kim. 2004. A novel angiotensin I converting enzyme inhibitory peptide from alaska pollack (*Theragra chalcogramma*) frame protein hydrolysate. *Journal of Agricultural and Food Chemistry* 52:7842–7845.

Jun, S.-Y., P.-J. Park, W.-K. Jung, and S.-K. Kim. 2004. Purification and characterization of an antioxidative peptide from enzymatic hydrolysate of yellowfin sole (*Limanda aspera*) frame protein. *European Food Research and Technology* 219:20–26.

Kadam, S. U. and P. Prabhasankar. 2010. Marine foods as functional ingredients in bakery and pasta products. *Food Research International* 43:1975–1980.

Kadam, S. U., B. K. Tiwari, and C. P. O'Donnell. 2013. Application of novel extraction technologies for bioactives from marine algae. *Journal of Agricultural and Food Chemistry* 61:4667–4675.

Khora, S. S. 2013. Marine fish-derived bioactive peptides and proteins for human therapeutics. *International Journal of Pharmacy and Pharmaceutical Sciences* 5:31–37.

Kim, S. K. 2013. *Marine Nutraceuticals: Prospects and Perspectives*. CRC Press, Boca Raton, FL.

Kim, S. K. and P. Dewapriya. 2012. Bioactive compounds from marine sponges and their symbiotic microbes: A potential source of nutraceuticals. *Advances in Food and Nutrition Research* 65:137–151.

Kim, S. K. and E. Mendis. 2006. Bioactive compounds from marine processing byproducts—A review. *Food Research International* 39:383–393.

Kim, S. K. and I. Wijesekara. 2010. Development and biological activities of marine-derived bioactive peptides: A review. *Journal of Functional Foods* 2:1–9.

Kuppusamya, P., M. M. Yusoffa, G. P. Maniama, S. J. A. Ichwanb, I. Soundharrajanc, and N. Govindana. 2014. Nutraceuticals as potential therapeutic agents for colon cancer: A review. *Acta Pharmaceutica Sinica B* 4:173–181.

Li, Y. X., I. Wijesekara, Y. Li, and S. K. Kim. 2011. Phlorotannins as bioactive agents from brown algae. *Process Biochemistry* 46:2219–2224.

Libes, S. M. 2009. Organic products from the sea: Pharmaceuticals, nutraceuticals, food additives and cosmoceuticals. In: *Introduction to Marine Biogeochemistry*, 2nd edn., ed. S. M. Libes. Elsevier Academic Press. Burlington, Massachusetts. Available at: http://elsevierdirect.com/companions/9780120885305 (July 8, 2014).

Lloret, J. 2010. Human health benefits supplied by Mediterranean marine biodiversity. *Marine Pollution Bulletin* 60:1640–1646.

Loffler, A. 1986. Proteolytic enzymes: Sources and applications. *Food Technology* 40:63–68.

Lordan, S., R. P. Ross, and C. Stanton. 2011. Marine bioactives as functional food ingredients: Potential to reduce the incidence of chronic diseases. *Marine Drugs* 9:1056–1100.

Martínez-Valverde, I., M. Jesús Periago, M. Santaella, and G. Ros. 2000. The content and nutritional significance of minerals on fish flesh in the presence and absence of bone. *Food Chemistry* 71:503–509.

Mayakrishnan, V., P. Kannappan, N. Abdullah, and A. B. A. Ahmed. 2013. Cardioprotective activity of polysaccharides derived from marine algae: An overview. *Trends in Food Science & Technology* 30:98–104.

Michalak, I. and K. Chojnacka. 2014. Review. Algal extracts: Technology and advances. *Engineering in Life Sciences* 2014. 14:581–591.

Molyneaux, M. and C. M. Lee. 1998. The US market for marine nutraceuticals products. *Food Technology* 52:56–57.

Ngo, D. H., I. Wijesekara, T. S. Vo, Q. V. Ta, and S. K. Kim. 2011. Marine food-derived functional ingredients as potential antioxidants in the food industry: An overview. *Food Research International* 44:523–529.

Okada, T. and M. T. Morrissey. 2007. Production of n-3 polyunsaturated fatty acid concentrate from sardine oil by lipase-catalyzed hydrolysis. *Food Chemistry* 103:1411–1419.

Pandey, M., R. K. Verma, and S. A. Saraf. 2010. Nutraceuticals: New era of medicine and health. *Asian Journal of Pharmaceutical and Clinical Research* 3:11–15.

Pike, I. H. and A. Jackson. 2010. Fish oil: Production and use now and in the future. *Lipid Technology* 22:59–61.

Rajapakse, N., W.-K. Jung, E. Mendis, S.-H. Moon, and S.-K. Kim. 2005. A novel anticoagulant purified from fish protein hydrolysate inhibits factor XIIa and platelet aggregation. *Life Sciences* 76:2607–2619.

Rasmussen, R. S. and M. T. Morrissey. 2007. Marine biotechnology for production of food ingredients. *Advances in Food and Nutrition Research* 52:237–292.

Ruperez, P., O. Ahrazem, and J. A. Leal. 2002. Potential antioxidant capacity of sulfated polysaccharides from the edible marine brown seaweed *Fucus vesiculosus*. *Journal of Agricultural and Food Chemistry* 50:840–845.

Ryu, B. M., M. J. Kim, S. W. A. Himaya, K. H. Kang, and S. K. Kim. 2014. Statistical optimization of high temperature/pressure and ultra-wave assisted lysis of *Urechis unicinctus* for the isolation of active peptide which enhance the erectile function *in vitro*. *Process Biochemistry* 49:148–153.

Samarakoon, K. and Y.-J. Jeon. 2012. Bio-functionalities of proteins derived from marine algae—A review. *Food Research International* 48:948–960.

Shahidi, F. 2009. Nutraceuticals and functional foods: Whole versus processed foods. *Trends in Food Science & Technology* 20:376–387.

Spolaore, P., C. Joannis-Cassan, E. Duran, and A. Isambert. 2006. Commercial applications of microalgae. *Journal of Bioscience and Bioengineering* 101:87–96.

Tahergorabi, R., K. E. Matak, and J. Jaczynski. 2014. Fish protein isolate: Development of functional foods with nutraceutical ingredients. *Journal of Functional Foods* 2015. 18:746–756.

Vidanarachchi, J. K., M. S. Kurukulasuriya, A. Malshani Samaraweera, and K. F. Silva. 2012. Applications of marine nutraceuticals in dairy products. *Advances in Food and Nutrition Research* 65:457–478.

Young, L. 2014. *Marine-Derived Nutraceuticals and Cosmetics*. Strategic Business Insights. February. Available from: www.strategicbusinessinsights.com/about/featured/2014/2014-02-marine-nutraceuticals.shtml#.Vv4ND_mLTIU. Accessed on July 8, 2014.

Wall, R., R. P. Ross, G. F. Fitzgerald, and C. Stanton. 2010. Fatty acids from fish: The anti-inflammatory potential of long-chain omega-3 fatty acids. *Nutrition Reviews* 68:280–289.

Wijesinghe, W. A. J. P. and Y. J. Jeon. 2012. Enzyme-Assistant Extraction (EAE) of bioactive components: A useful approach for recovery of industrially important metabolites from seaweeds: A review. *Fitoterapia* 83:6–12.

Zhang, C., X. Li, and S. K. Kim. 2012. Application of marine biomaterials for nutraceuticals and functional foods. *Food Science and Biotechnology* 21:625–631.

Zhang, Q., P. Yu, Z. Li, H. Zhang, Z. Xu, and P. Li. 2003. Antioxidant activities of sulfated polysaccharide fractions from *Porphyra haitanesis*. *Journal of Applied Phycology* 15:305–310.

Section VIII

Marine Pharmacogenomics

19

Unexpected Anticancer Activity of the E-Peptide of Rainbow Trout Pro-IGF-I

Thomas T. Chen, Maria J. Chen, Jay H. Lo, and Chun-Mean Lin

CONTENTS

19.1 Introduction

In all vertebrates, insulin-like growth factors (IGF)-I and IGF-II, members of the insulin family proteins, play important roles in regulating growth, development, and metabolism (de Pablo et al., 1990, 1993; Stewart and Rotwein, 1996). Like all peptide hormones, IGFs are initially synthesized as prepropeptides containing an amino-terminal signal peptide, followed by the mature peptide of B, C, A, and D domains and a carboxyl terminal E-peptide domain. Through intracellular posttranslational processing, the signal peptide and the E-peptide are proteolytically cleaved off from the prepropeptide, and both the mature IGFs and E-peptides are secreted into the bloodstream (Rotein et al., 1986; Duguay, 1999). Molecular cloning and characterization of the IGF-I gene and its cDNA revealed the presence of multiple isoforms of IGF-I mRNA from fish to mammals. In human, three different species of IGF-I mRNAs encoding three isoforms of pro-IGF-I (i.e., pro-IGF-Ia, pro-IGF-Ib, and pro-IGF-Ic) were identified (Rotwein et al., 1986; Duguay, 1999) (Figure 19.1). These three isoforms of pro-IGF-I contain an identical mature IGF-I with 70 amino acid (aa) residues but different E-peptides of 30 aa (Ea), 77 aa (Eb), and 40 aa (Ec), respectively. The a-type E-domains are highly conserved among all vertebrates studied,

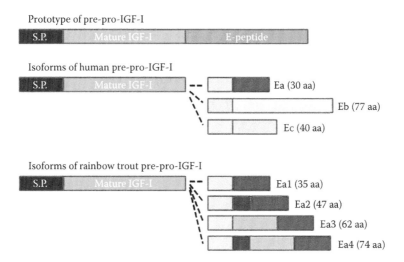

FIGURE 19.1
Schematic representation of isoforms of human and rainbow trout proinsulin-like growth factor-I E-peptides.

but the b-type and c-type E-peptides are conserved in the first 15 aa residues and different thereafter (variable region) among human, rats, and mice. In fish, multiple isoforms of IGF-I mRNA have also been identified, and these different isoforms of IGF-I mRNA encode an identical mature IGF-I but different E-peptides. In rainbow trout (*Oncorhynchus mykiss*), for instance, Shamblott and Chen (1992, 1993) reported the presence of four different IGF-I transcripts encoding four different isoforms of pro-IGF-I (Figure 19.1). Similar to the isoforms of human pro-IGF-I, four different isoforms of the rainbow trout pro-IGF-I contain an identical mature IGF-I (70 aa) and four different lengths of E-peptides (i.e., rtEa1, 35 aa; rtEa2, 47 aa; rtEa3, 62 aa; rtEa4, 74 aa). The first 15 aa residues of these four E-peptides are identical among themselves and the a-type E-peptides of mammals, and the 20 aa residues at the C-termini share 70% identity with their human counterparts. The rtEa1-peptide is composed of the first 15 and the last 20 aa residues. Insertion of either 12 or 27 aa residues between the first and the last segments of rtEa1-peptide results in rtEa2- or rtEa3-peptide, whereas insertion of both segments into rtEa1-peptide gives rise to rtEa4-peptide. Similar to rainbow trout, four different forms of Ea-peptides of pro-IGF-I have also been identified in Chinook salmon, Coho salmon (Duguay et al., 1992; Wallis and Devlin, 1993), and red drum (Faulk et al., 2010). However, not all teleosts possess four different E-peptides. While gilthead sea bream possesses three different E-peptides (viz., Ea1, Ea2, and Ea4), zebrafish and grouper each has only one Ea-peptide: Ea2-peptide in zebrafish and Ea4-peptide in grouper (Chen et al., 2001; Shi et al., 2002; Tiago et al., 2008).

While the biological activities of the mature IGF-I have been extensively studied, the biological activities of E-peptides have been overlooked. It was generally believed that the E-peptide of pro-IGF-I may be biologically inert; however, the following lines of evidence suggest that different E-peptide may possess biological activity. First, many peptide hormones are initially synthesized as complex prohormone molecules and posttranslationally processed into multiple peptides with distinct or similar biological activities. Examples are proopiomelanocortin (Civelli et al., 1986), proglucagon (Bell et al., 1983), and proinsulin (Ido et al., 1997), just to name a few. In a similar manner, generation of E-peptides from pro-IGF-I may suggest a pluripotential role for these peptides. Second, the E-domains of pro-IGF-I are evolutionarily conserved. Third, different isoforms of pro-IGF-I transcripts are expressed in a tissue-specific and developmental stage-specific manner and exhibiting differential responses to growth hormone (Shamblott and Chen, 1993; Duguay et al., 1994; Yang and Goldspink, 2002). Siegfried et al. (1992) reported the first line of evidence documenting that hEb-peptide contained biological activity. In their studies, Siegfried and colleagues showed that a synthetic peptide amide with 23 aa residues (Y-23-R-NH2) derived from hEb-peptide (aa 103–124) exerted mitogenic activity in normal and malignant human bronchial epithelial cells at 2–20 nM. They further demonstrated that Y-23-R-NH2 bound to a specific high-affinity receptor ($K_d = 2.8 \pm 1.4 \times 10^{-11}$ M) is present at 1–2×10^4 binding sites per cell, and the ligand binding was not inhibited by recombinant insulin or IGF-I. Several investigators have also reported recently that the Eb-peptide of rodent pro-IGF-I peptide (same sequence as hEc-peptide) possessed an activity in promoting proliferation of rodent myoblasts, whereas Ea-peptide stimulated differentiation of mature myoblasts (Yang and Goldspink, 2002; Matheny et al., 2010). Besides exerting mitogenic activity in rodent myoblasts by rodent Eb-peptide, murine Ea-peptide of pro-IGF-I also modulates the entry of mature IGF-I protein into C2C12 cells, a murine skeletal muscle cell line (Pfeffer et al., 2009).

The first line of evidence documenting the mitogenic activity of E-peptides of rainbow trout pro-IGF-I came from studies conducted by Tian et al. (1999). Tian et al. produced recombinant rtEa2-, rtEa3-, and rtEa4-peptides by expressing cDNAs of rtEa2, rtEa3, and rtEa4 in *Escherichia coli* cells, and the semipurified recombinant proteins were tested for their mitogenic activity. They showed that recombinant rtEa2-, rtEa3-, and rtEa4-peptides exhibited a dose-dependent mitogenic activity in nononcogenic-transformed NIH 3T3 cells and primary caprine mammary gland epithelial cells. However, the mitogenic activity of these peptides is not observed in oncogenic-transformed cells. Recently, Mark Chen (personal communication) in Taiwan also demonstrated that Ea2-peptide of zebrafish possessed a stimulatory effect on the incorporation of ^{35}S-sulfate into zebrafish gill bronchial cartilage in a sulfation assay. These results are consistent with those reported in mammals (Siegfried et al., 1992; Yang and Goldspink, 2002; Matheny et al., 2010). Further *in vitro* studies conducted in our laboratory showed that both recombinant rtEa4-peptide and synthetic hEb-peptide exerted unexpected anticancer cell activities in established human cancer cell lines such as MDA-MB-231, HT-29, HepG2, SK-N-F1, SKOV-3A, PC-3, and OVCAR-3B (Chen et al., 2002, 2007; Kuo and Chen, 2002; Siri et al., 2006a,b). These anticancer activities include (1) induction of morphological differentiation and inhibition of anchorage-independent cell growth, (2) inhibition of invasion and metastasis, (3) suppression of cancer-induced angiogenesis, and (4) induction of apoptosis. In this chapter, we will review the detail experimental evidence leading to demonstrating the anticancer activities of rtEa4- and hEb-peptide of the pro-IGF-I.

19.2 Anticancer Activities of rtEa4- and hEb-Peptide

As described by Hanahan and Weinberg (2000, 2011) in their review articles, the hallmarks of cancers of all types comprise six biological capabilities acquired during tumorigenesis. These hallmarks include (1) sustaining proliferative signaling, (2) evading growth suppressors, (3) resisting cell death, (4) enabling replicative immortality, (5) inducing angiogenesis, and (6) activating invasion and metastasis. A compound that possesses the activity to control the hallmark activities of cancer but is less toxic or nontoxic to noncancerous cells will meet the requirement as an ideal drug candidate for cancer therapy.

19.2.1 Inhibition of Anchorage-Independent Cell Growth

To investigate whether rtEa4-peptide possesses activity other than the mitogenic activity observed by Tian et al. (1999) in noncancerous cells, Chen et al. (2002) and Kuo and Chen (2002) independently plated single-cell clone of human breast cancer cells (MDA-MB-231) and human neuroblastoma cells (SK-N-F1) in a serum-free medium supplemented with various concentrations of rtEa4-peptide. Unlike rtEa4-peptide untreated cells that showed a rounded-up morphology and attached to the culture chamber loosely, the rtEa4-peptide treated cells attached to the culture chamber tightly and developed several pseudopodia-like structures (multiple processes) (Figure 19.2a). In the case of SK-N-F1 cells, enhanced formation of neurite-like structure was also observed following treatment with rtEa4-peptide (Figure 19.2b). To rule out the possibility that the previously observed morphology was caused by the presence of minor contaminating molecules present in the recombinant rtEa4-peptide preparation, Chen et al. (2002) and Kuo and Chen (2002) introduced a recombinant vector expressing secreted form of rtEa4-peptide and green fluorescence protein (GFP) into MDA-MB-231 and SK-N-F1 cells, respectively, and stable transformed cell clones were isolated. Both types of transformed cells exhibited morphological characteristics similar to those observed in untransfected cells treated with exogenous rtEa4-peptide (Figure 19.3).

In addition to the morphological changes in cancer cells caused by treatment with rtEa4-peptide, does rtEa4- or hEb-peptide cause any other behavior change in cancer cells? Anchorage-independent cell growth, the ability of cells to grow into colonies in the absence of anchorage to extracellular substratum, is among the unique features of cells with tumorigenicity. This behavior change can be easily assessed by an *in vitro* colony formation assay in a soft agar medium. To determine whether rtEa4-peptide may cause change in the anchorage-independent growth behavior of cancer cells, established cancer cells derived from human breast cancer (MDA-MB-231), human colon cancer (HT-29), and neuroblastoma (SK-N-F1) were subjected to colony formation assay in a soft agar medium with or without the presence of various concentrations of rtEa4-peptide (Chen et al., 2002; Kuo and Chen, 2002). As shown in Figure 19.4, a dose-dependent inhibition of colony formation by rtEa4-peptide on MDA-MB-231, HT-29, and SK-N-F1 cells were observed. Similar results were observed when shEb-peptide was tested against these cancer cells. Furthermore, we have also transfected a bicistronic expression vector expressing secreted form of hEb-peptide and EGFP into MDA-MB-231 cells, and several stable transfectants were isolated. Results of colony formation assay of these transfectants on semisolid agar medium showed that much lower colony formation activities were observed in stable transfectants compared to untransfected MDA-MB-231 cells (Figure 19.5). These results suggest that both rtEa4- and shEb-peptides possess inhibitory activity against anchorage-independent cancer cell growth.

Since *in vitro* studies showed that rtEa4- and shEb-peptides inhibited anchorage-independent growth of established human cancer cells such as MDA-MB-231, HT-29, SK-N-F1, HepG-2, and rainbow trout hepatoma cells (RTH) (Chen et al., 2002, 2004; Kuo and Chen, 2002), it would be of great interest to verify whether rtEa4- and shEb-peptides also exert inhibition on cancer cell growth *in vivo*. A step toward answering this question, 10 athymic mice per group each was xenografted with 1.3×10^7 cells of human breast cancer cells (ZR-75-1) mixed with equal volume of matrigel. One group of the mice also received 70 µg of rtEa4-peptide and the other group with the same amount of serum albumin as control. At 1 month postimplantation, the weights of cancer mass in the control animals were twice of the rtEa4-peptide-treated animals (Figure 19.6).

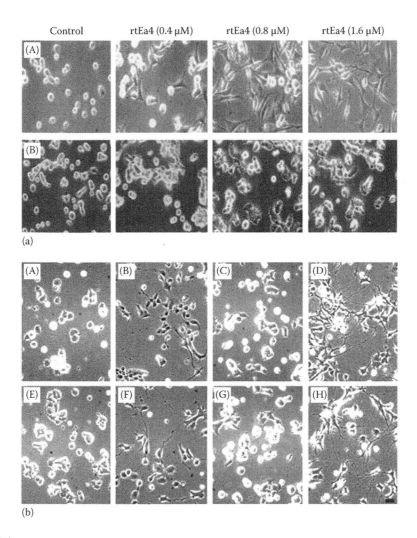

FIGURE 19.2
Effects of rtEa4-peptide on morphological differentiation of cancer cells. (a) $1–2 \times 10^5$ of breast cancer cells (a, MDA-MB-231; b, ZR-75-1) suspended in F12/Dulbecco's modified eagle medium (DMEM) medium supplemented with fetal bovine serum or different concentrations of rtEa4-peptide were plated in a six-well culture chamber and incubated at 37°C in a humidified chamber equilibrated with 5% CO_2. The cells were observed under an Olympus inverted microscope with phase-contrast objectives (200×, original magnification) 24 h after initiation of rtEa4-peptide treatment. (b) Neuroblastoma cells (SK-N-F1) were treated without any growth factor or protein (A), insulin-like growth factor (IGF)-I (5 nM) alone (E), 1.2 μM rtEa4-peptide (B), 1.2 μM rtEa4-peptide and 5 nM IGF-I (F), 3.2 μM synthetic hEa-peptide (C), 3.2 μM synthetic hEa-peptide and 5 nM IGF-I (G), 1.6 μM synthetic hEb-peptide (D), and 1.6 μM synthetic hEb-peptide and 5 nM IGF-I (H). Cell images were taken by random sampling using a MicroMAX CCD camera (Princeton Instrument) after 24 h incubation, Bar, 20 μm. (From Chen, M.J. et al., *Gen. Comp. Endocrinol.*, 126, 342, 2002; Kuo, Y.-H. and Chen, T.T., *Exp. Cell Res.*, 280, 75, 2002. With permission.)

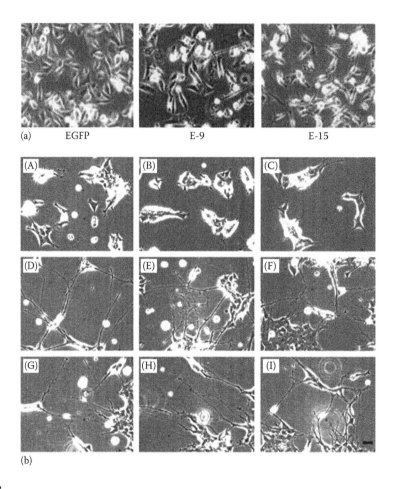

FIGURE 19.3
Morphology of breast cancer cell clones and neuroblastoma cell clones transfected with a transgene construct expressing rtEa4/EGFP or EGFP driven by a cytomegalovirus (CMV) promoter. (a) MDA-MB-231 cell clones and (b) SK-N-F1 cell clones (A–C, cell clones expressing EGFP alone; D–I, cell clones expressing rtEa4-peptide/ EGFP). (From Chen, M.J. et al., *Gen. Comp. Endocrinol.*, 126, 342, 2002; Kuo, Y.-H. and Chen, T.T., *Exp. Cell Res.*, 280, 75, 2002. With permission.)

Chicken embryos are immune deficient during embryonic development, and the chorioallantoic membrane (CAM) of the developing embryos has been used as a convenient model for evaluating many different parameters of tumor growth as well as in antineo-plastic drug screening by many investigators (Chamber et al., 1982; 1992; Brooks et al., 1994; Stan et al., 1999). In addition, the CAM model is also considered as an ideal alternative model system for cancer research because it can conveniently and inexpensively reproduce many tumor characteristics such as tumor mass formation, tumor-induced angiogenesis, and infiltrative growth and metastasis *in vivo* (Chamber et al., 1982, 1992; Brooks et al., 1994; Stan et al., 1999). To test whether rtEa4- or shEb-peptide exhibited inhibitory effect on can-cer cell growth *in vivo*, Chen et al. (2007) adapted the CAM model to access the growth inhibition of MDA-MB-231 cells on the CAM by rtEa4- or shEb-peptide treatment. In this study, they inoculated 1×10^7 cells of MDA-MB-231 cells labeled with GFP protein on 5-day-old embryos. From day 9 to day 13, three doses of shEb-peptide (600 μg/embryo/dose) were applied in a 2-day interval, and the embryos were evaluated on day 15. As shown

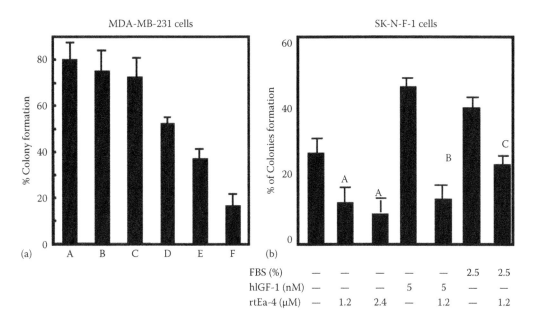

FIGURE 19.4

Effect of rtEa4-peptide on colony formation activity of MDA-MB-231 and SK-N-F-1 cells in a semisolid medium. About 2×10^4 cells were plated in F12/DMEM containing 0.4% agar, 1.25% FBS, and various concentrations of rtEa4-peptide or control proteins, in six-well culture chambers. After the medium was solidified, each well was overlaid with 1 mL of F12/DMEM supplemented with the same concentrations of rtEa4-peptide or control proteins. The plates were incubated at 37°C in a humidified incubator with 5% CO_2 and examined daily under an inverted microscope for 2 weeks. (a) MDA-MB-231 cells: A, F12/DMEM with 1.25% FBS; B, F12/DMEM with 10% FBS; C, F12/DMEM with control protein; D, F12/DMEM with 0.8 mM hEb; E, F12/DMEM with 1.6 mM hEb; and F, F12/DMEM with 2.4 mM hEb. (b) SK-N-F1: A, $P \leq 0.05$; B, $P \leq 0.001$; and C, $P \leq 0.01$. (From Chen, M.J. et al., *Gen. Comp. Endocrinol.*, 126, 342, 2002; Kuo, Y.-H. and Chen, T.T., *Exp. Cell Res.*, 280, 75, 2002. With permission.)

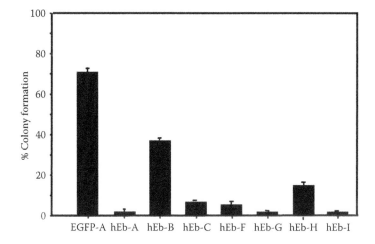

FIGURE 19.5

Colony formation activities of MDA-MB-231 cell colonies transfected with a transgene expressing hEb-peptide/ EGFP. Colony formation activities of MDA-MB-231 transfected clones were determined as described in Figure 19.4.

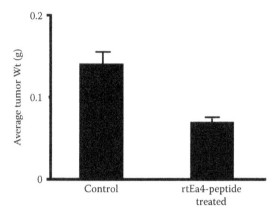

FIGURE 19.6
Effect of rtEa4-peptide on growth of breast cancer cells (ZR-75-1) xenografted in athymic mice. Breast cancer cells (ZR-75-1), 1.3×10^7 cells each, was xenografted into a total of 10 nude mice in each group. Each mouse in the experimental group was treated with 70 mg of rtEa4-peptide or injected with control protein. The weight of tumor mass was measured at 1 month postinjection.

in Figure 19.7 and Table 19.1, while several cancer masses were observed on the CAM of E-peptide untreated embryos, no visible tumor mass was observed in embryos treated with multiple doses of shEb-peptide. Results of *in vitro* and *in vivo* studies clearly confirm that rtEa4- or shEb-peptide inhibits the growth of cancer cells.

19.2.2 Inhibition of Invasion and Metastasis

Metastasis is another hallmark of cancer of all types. Metastasis is a complex series of steps in which cancer cells leave the original tumor site and migrate to other parts of the body via the bloodstream, the lymphatic system, or by direct extension (Chiang and Massague, 2008; Klein, 2008). To do so, malignant cells need to break away from the site of primary tumor, attach to and degrade proteins that make up the surrounding extracellular matrix (ECM), and migrate to the second site for tumorigenesis. In addition to inhibiting anchorage-independent cell growth, rtEa4- and shEb-peptides may also possess activity to inhibit cancer cell invasion and metastasis. Chen et al. (2002) and Siri et al. (2006a) conducted studies to determine the inhibitory activity of rtEa4-peptide on the invasion of HT-1080 and MDA-MB-231 cells through matrigel membrane by an *in vitro* matrigel invasion assay. A dose-dependent inhibition of HT-1080 and MDA-MB-231 cells invading through the matrigel membrane was observed (Figure 19.8). In a separate study with MDA-MB-231 cells cultured on chicken CAM, Chen et al. (2007) observed that in the absence of rtEa4-peptide, not only MDA-MB-231 cells grew into large cancer cell masses, but also some of the MDA-MB-231 cells invaded through CAM and migrated inward into the embryo (Figure 19.9a). In the presence of rtEa4-peptide, none of the MDA-MB-231 cells invaded through the CAM (Figure 19.9b).

19.2.3 Inhibition of Tumor-Induced Angiogenesis

Large-scale growth of a tumor requires a blood supply, which comes from the tumor-induced sprouting of blood vessels (neovascularization) (Folkman, 2002). In this process, endothelial cells lining the existing blood vesicles in the proximity of the tumor are

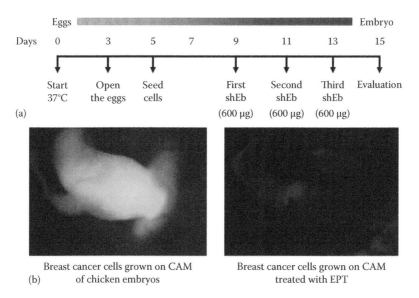

FIGURE 19.7
Suppression of MDA-MB-231 cell growths on chorioallantoic membrane (CAM) of developing chicken embryos by synthetic human Eb-peptide. MDA-MB-231 cells (1×10^7 cells) were seeded on CAM of 5-day-old embryos. From day 9 to day 13, three doses of synthetic hEb-peptide or bovine serum albumin (BSA) (0.6 mg/dose/ embryo) were added to each embryo. On day 15, the CAM was dissected and observed under an inverted microscope (Olympus 1X50 with an appropriate filter to give excitation wavelength at 490 nm and emission wavelength at 520 nm) at 40× magnification. (a) Treatment scheme and (b) breast cancer cell mass on CAM. (From Chen, M.J. et al., *J. Cell. Biochem.*, 101, 1316, 2007. With permission.)

TABLE 19.1

Inhibition of Tumor Growth on Chicken CAM by shEb- or rtEa4-Peptide

Treatment	# Embryos Treated	# Embryos with Tumor Mass	Note
Control	20	20	Multiple tumor masses per embryo
shEb treated	20	2	Single small tumor mass per embryo
rtEa4 treated	20	1	Single small tumor mass per embryo

Source: Chen, M.J. et al., *J. Cell. Biochem.*, 101, 1316, 2007. With permission.
MDA-MB-231 cells (1.5×10^7) were seeded on CAM of chicken embryos and treated with three doses (600 mg/ dose) of control protein, sEb-, or rtEa4-peptide in a 2-day interval.

stimulated by vasculature endothelial growth factor (VEGF) and basic fibroblast growth factor released by tumor cells to differentiate into new capillary sprouts (Gupta and Qin, 2003; Cao, 2005; Weis and Cheresh, 2011). It is generally believed that starving a tumor from its blood supply may be a rational approach to suppress cancer progression (Folkman, 2002). By culturing MDA-MB-231 cells on the chicken CAM model in the presence or absence of rtEa4-peptide, Chen et al. (2007) observed numerous blood vessels radiating from the MDA-MB-231 cell mass on the chicken CAM (Figure 19.10a and b). On the other hand, when MDA-MB-231 cells on the CAM were treated with rtEa4-peptide, no visible cancer cell mass and blood vessels associated with cancer cell mass, like those observed in Figure 19.10a and b, were observed. These results suggest that rtEa4-peptide may possess an anti-angiogenic activity.

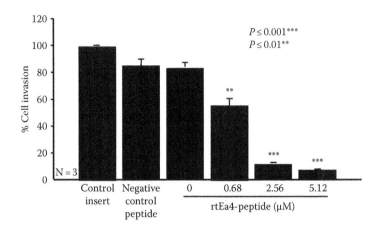

FIGURE 19.8

Effect of rtEa4-peptide on the invasion of MDA-MB-231 cells through matrigel membrane in an invasion chamber. Invasion assay was carried in invasion chambers (Becton–Dickinson) following the protocol provided by the supplier. Percentage of MDA-MB-231 cells, which invaded through the matrigel membrane, were calculated, and the average results from three matrigel inserts were calculated and similar results were obtained. (From Siri, S. et al., *J. Cell. Biochem.*, 99, 1361, 2006a. With permission.)

Among several methods available for assessing the anti-angiogenic activity of a compound, the chicken embryo CAM assay is considered as the simplest, inexpensive, and most reproducible classical assay. Chen et al. (2007) adopted this assay to verify the anti-angiogenic activity of rtEa4- or shEb-peptide. Results presented in Figure 19.11 showed that both rtEa4- and shEb-peptides exerted inhibition of angiogenesis on chicken CAM. At the doses of 2–16 nmoles/egg of rtEa4- or shEb-peptide, a dose-dependent inhibition on angiogenesis was observed (Figure 19.12). Furthermore, the doses of E-peptide that exhibited anti-angiogenic activity on CAM are in the similar order of magnitudes of doses that inhibited cancer cell growth in the chick embryos. These results confirm that rtEa4- and shEb-peptides possess activity to suppress cancer-induced angiogenesis.

19.2.4 Induction of Apoptosis

Programmed cell death, that is, apoptosis, is a crucial process of eliminating unwanted cells in animal life, and it is vital for embryonic development, homeostasis, and immune defense (Elmore, 2007). Apoptosis is characterized by typical morphological and biochemical hallmarks, such as cell shrinkage, nuclear DNA fragmentation, and plasma membrane blebbing (Hengartner, 2000). From extensive studies, it is evident that two pathways, extrinsic pathway (death receptor–triggered pathway) and intrinsic pathway (mitochondrial pathway), are involved in apoptosis (Hengartner, 2000; Fulda and Debatin, 2006). Furthermore, there is ample evidence indicating that these two pathways are tightly linked, and molecules from one pathway can influence the other pathway.

One striking feature of cancer cells is that they evade from apoptosis by reducing expression or mutation of proapoptotic genes such as caspase genes (Ghavami et al., 2009) while increasing the expression of antiapoptotic genes such as Bcl-2 family genes (Youle and Strasser, 2008). Certain anticancer drugs or agents, which have been identified as potential effective cancer therapy, can restore normal apoptotic pathways (Fulda and Debatin, 2006;

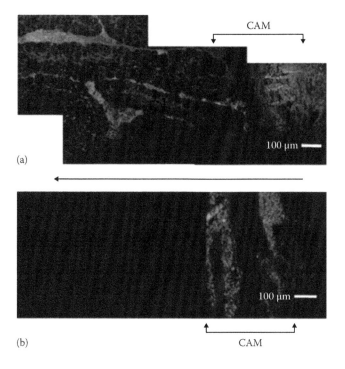

FIGURE 19.9
Effect of rtEa4-peptide on invasion of MDA-MB-231 cells in the developing chicken embryos. Five-day-old chicken embryos were seeded with MDA-MB-231 cells expressing EGFP and treated with rtEa4-peptide or control protein. The chorioallantoic membrane (CAM) with MDA-MB-231 cells was dissected, fixed sequentially in 50% ethanol and 3.7% formaldehyde, and embedded in paraffin with CAM facing up. Serial sections of 5 μm were prepared and stained with anti-enhanced green fluorescence protein (EGFP) IgG and fluorescein isothiocyanate (FITC)-labeled goat antirabbit IgG sequentially. The stained sections were observed under a fluorescence microscope (Olympus 1X50 with an appropriate filter set to give excitation wavelength at 490 nm and emission wavelength at 520 nm) at 200× magnification. (a) Sections of CAM with MDA-MB-231 cells treated with control protein; and (b) Sections of CAM with MDA-MB-231 cells treated with rtEa4-peptide. Arrow indicates the orientation of CAM and direction of sectioning. (From Chen, M.J. et al., *J. Cell. Biochem.*, 101, 1316, 2007. With permission.)

Yang et al., 2008). Since rtEa4- or hEb-peptide has been shown to possess anticancer activities in a variety of human cancer cells (Kuo and Chen, 2002; Chen et al., 2007), it would be of great interest to determine if rtEa4- or hEb-peptide may possess activity to induce apoptosis in cancer cells.

By treating ovarian cancer cells (SKOV-3A) or breast cancer cells (MDA-MB-231C) with recombinant rtEa4-peptide (1.4 and 2.8 μM) for 2 h, Chen et al. (2012) observed that many cells exhibited distinct morphology of membrane blebbing, cell shrinkage, and chromatin condensation and disintegration. While these morphological changes in the rtEa4-peptide-treated cells became more pronounced with time, no such morphological changes were observed in untreated cells or cells treated with control protein (e.g., serum albumin) for 48 h. These morphological characteristics are consistent with those reported for apoptosis by others (Hacker, 2000). To verify that rtEa4-peptide-treated cancer cells exhibited apoptosis, MDA-MB-231C cells were treated with 3.0 μM of recombinant rtEa4-peptide for 48 h, and the treated cells were subjected to TUNEL assay or immunocytochemical staining with a monospecific antibody specific for activated caspase 3. As shown in Figure 19.13, after incubation for 48 h in a serum-free medium containing 3.0 μM of recombinant rtEa4-peptide, most of

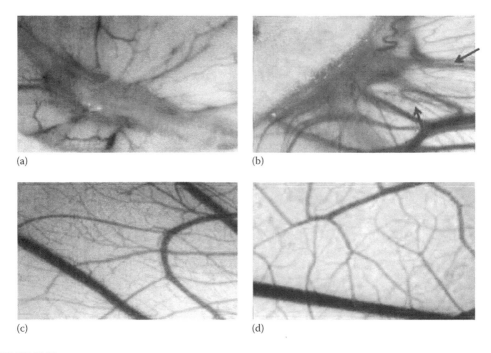

FIGURE 19.10

Effect of rtEa4-peptide on growth of MDA-MB-231 cells on the chorioallantoic membrane (CAM) of chicken embryos. Five-day-old chicken embryos were seeded with MDA-MB-231 cells (1×10^7 cells) and treated with one dose of rtEa4-peptide (600 µg) or the same amount of control proteins. After incubation for 8 days, embryos were observed under a dissecting microscope and photographed with a digital camera. (a) and (b) CAM seeded with MDA-MB-231 cells and 600 µg/embryo of control proteins; (c) CAM seeded with MDA-MB-231 cells mixed with 600 µg/embryo of rtEa4-peptide; and (d) CAM without any treatment. Arrows indicate cancer-induced vessels. (From Chen, M.J. et al., *J. Cell. Biochem.*, 101, 1316, 2007. With permission.)

the cells showed positive in TUNEL assay as well as in immunocytochemical staining for the activated caspase 3. Furthermore, apoptosis was also observed in ovarian cancer cells, SKOV-3A, treated with 1.5 µM of recombinant rtEa4-peptide. These results confirmed that rtEa4-peptide induced apoptosis in human breast cancer cells and ovarian cancer cells.

Does rtEa4-peptide or shEb-peptide induce apoptosis in cancer cells via both extrinsic and intrinsic pathways? It has been reported by many investigators that if apoptosis is induced via extrinsic pathway, upregulation of expression of *TRAIL-RI, FADD*, and *pro-Casp-8* genes will be observed. If apoptosis is induced via intrinsic pathway, downregulation of expression of *Bcl-2, Bcl-XL*, and *Mcl-1* genes and upregulation of expression of *Cyt-C* gene will be observed (Fulda and Debatin, 2006; O'Brien and Kirby, 2008) (Figure 19.14). To address the question of whether rtEa4- or shEb-peptide induces apoptosis in cancer cells via both extrinsic and intrinsic pathways, human breast cancer cells (MDA-MB-231), ovarian cancer cells (OVCAR-3A, SKOV 3B), neuroblastoma cells (SK-N-SH and SK-N-F1), and noncancerous foreskin cells (CCD-1112SK) were treated with 2.0 µM of rtEa4-peptide or 0.5 µM of shEb-peptide for 6 h, and RNA samples were extracted from these cells for determination of levels of *Bcl-2, Bcl-XL, Mcl-1, Casp-3, Casp-8, Casp-9, TRAIL-R1, FADD, Cyt-C, p53*, and *PTEN* mRNAs by quantitative relative real-time RT-PCR analysis. As shown in Table 19.2, while the levels of *Bcl-2, Bcl-XL*, and *Mcl-1* mRNAs in rtEa4-peptide or hEb-peptide treated cancer cells were downregulated significantly, those of *Casp-3, Casp-8, Casp-9, TRAIL-R1, FADD, Cyt-C, p53*, and *PTEN* mRNAs were upregulated. These results provide strong

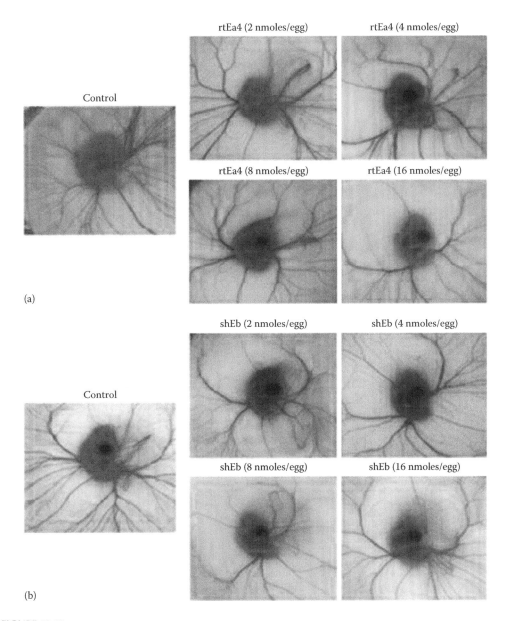

FIGURE 19.11
Anti-angiogenic effect of rtEa4-peptide and shEb-peptide on the chorioallantoic membrane of chicken embryos. Three-day-old chicken embryos were treated with various doses (2–16 nmole) of rtEa4-peptide and/or shEb-peptide, and the embryos were photographed at day 5. (From Chen, M.J. et al., *J. Cell. Biochem.*, 101, 1316, 2007. With permission.)

evidence supporting the hypothesis that rtEa4- or hEb-peptide induces apoptosis in cancer cells via both extrinsic and intrinsic pathways. It is very interesting to note that treatment of noncancerous foreskin cells (CCD-1112SK) with the same concentration of recombinant rtEa4-peptide or shEb-peptide did not result in significant changes in the levels of *Bcl-2, Bcl-XL, Mcl-1, Casp-3, Casp-8, Casp-9, TRAIL-R1, FADD, Cyt-C, p53,* and *PTEN* mRNAs when compared to untreated control (Chen et al., 2012). The resistance of cancer cells to apoptosis

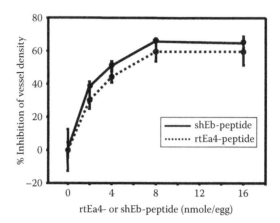

FIGURE 19.12

Dose-dependent response of anti-angiogenic effect of rtEa4-peptide and shEb-peptide on chorioallantoic membrane (CAM) of chicken embryos. Various amounts of rtEa4-peptide or shEb-peptide were applied to the CAM of day 3 chicken embryos and the data were scored at day 5. Percent of inhibition of vessel density is ([vessel density of control – vessel of treated embryos]/vessel density of control) × 100%. Each data point is the average of vessel density from 10 embryos ± standard error. (From Chen, M.J. et al., *J. Cell. Biochem.*, 101, 1316, 2007. With permission.)

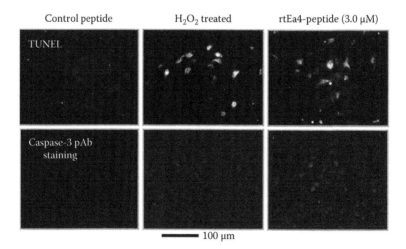

FIGURE 19.13

Apoptotic cells identified by TUNEL assay and immunostaining with monospecific antibodies to activated capase-3. MDA-MB-231 cells in serum-free medium were treated with rtEa4-peptide (3,0 µM) or the same concentration of control protein for 48 h. One sample was treated with H_2O_2 (0.5 mM) for 16 h to induce apoptosis as positive control. TUNEL was conducted following the protocol provided by the manufacturer. Monospecific polyclonal rabbit antiactivated capase-3 immunoglobulin was labeled with Rhodamine Red and used to stain the rtEa4-peptide or H_2O_2-treated cells. The stained cells were observed under an inverted microscope (Olympus IX50) with epifluorescence attachment. (From Chen, M.J. et al., Induction of apoptosis in human cancer cells by human Eb- or rainbow trout Ea4-peptide of pro-insulin like growth factor-I (Pro-IGF-I), *Targeting New Pathways and Cell Death in Breast Cancer*, in: R.L. Aft, ed., InTech, Croatia, 2012, pp. 45–56. With permission.)

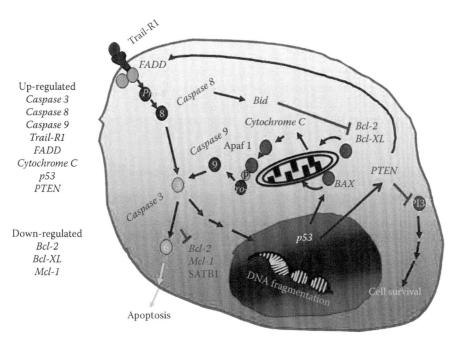

FIGURE 19.14

A schematic presentation of extrinsic and intrinsic pathways of apoptosis. (From Chen, M.J. et al., Induction of apoptosis in human cancer cells by human Eb- or rainbow trout Ea4-peptide of pro-insulin like growth factor-I (Pro-IGF-I) *Targeting New Pathways and Cell Death in Breast Cancer*, in: R.L. Aft, ed., InTech, Croatia, 2012, pp. 45–56. With permission.)

TABLE 19.2

Relative Expression Levels of Apoptosis Genes by E-Peptide of pro-IGF-I

	Cell Lines (Relative Expression Levels)						
	rtEa4						hEb
Genes	CCD-1112SK[a]	MDA-MB-231C	OVCAR-3B	SKOV-3A	SKNSH	SKNF1	MDA-MB-231C
Bcl-2	1.08 ± 0.55	0.19 ± 0.03	0.66 ± 0.10	0.20 ± 0.05	0.79 ± 0.34	0.08 ± 0.02	0.04 ± 0.00
Bcl-XL	0.96 ± 0.04	0.33 ± 0.00	0.49 ± 0.01	0.21 ± 0.01	0.50 ± 0.04	0.18 ± 0.00	0.17 ± 0.01
Mcl-1	0.92 ± 0.04	0.60 ± 0.00	0.29 ± 0.01	0.15 ± 0.02	0.38 ± 0.08	0.07 ± 0.01	0.09 ± 0.00
Casp-3	0.96 ± 0.12	3.48 ± 0.27	26.07 ± 2.50	12.18 ± 2.61	3.32 ± 0.10	13.91 ± 1.81	7.33 ± 0.79
Casp-8	0.91 ± 0.03	3.74 ± 0.28	2.44 ± 0.04	2.67 ± 0.06	5.49 ± 0.56	7.96 ± 0.01	4.15 ± 1.40
Casp-9	1.08 ± 0.01	2.18 ± 0.10	3.13 ± 0.021	1.98 ± 0.02	6.16 ± 0.18	8.70 ± 1.12	15.85 ± 2.86
TRAIL-R1	1.01 ± 0.11	2.04 ± 0.16	2.24 ± 0.01	1.37 ± 0.06	6.50 ± 0.16	7.47 ± 0.78	8.71 ± 0.85
FADD	1.04 ± 0.11	3.61 ± 0.09	1.42 ± 0.03	2.67 ± 0.03	3.31 ± 0.05	8.33 ± 1.14	11.18 ± 1.69
Cyt-C	1.02 ± 0.05	2.54 ± 0.18	1.40 ± 0.00	1.72 ± 0.01	4.31 ± 1.02	10.59 ± 8.20	8.94 ± 0.00
$_p$53	1.08 ± 0.06	2.37 ± 0.05	3.09 ± 0.29	1.03 ± 0.51	3.99 ± 0.47	6.93 ± 1.75	4.68 ± 0.55
PTEN	—	8.30 ± 0.73	—	4.61 ± 0.11	4.09 ± 0.20	30.80 ± 0.15	—

Source: Chen, M.J. et al., *J. Cell. Biochem.*, 101, 1316, 2007. With permission.

Relative expression level = $2^{-(S\Delta CT - C\Delta CT)}$ and standard deviation ($n = 3$), where $S\Delta CT$ is the difference between the CT number of the sample (cancer cells treated with E-peptide) and the housekeeping gene (GAPDH), and $C\Delta CT$ is the difference between the CT of cancer cells without E-peptide treatment and the CT of GAPDH. $2^{-(S\Delta CT-C\Delta CT)} > 1$, upregulation; $2^{-(S\Delta CT - C\Delta CT)} < 1$, downregulation.

[a] CCD-1112SK: nonimmortalized human fibroblast cells.

is one of the major concerns in cancer therapy. Thus, in searching for effective chemotherapeutic drugs, the effectiveness of the drugs to induce apoptosis in a wide variety of cancer types will be the top choice. Although there are numerous chemotherapeutic drugs available on the market to date that have been shown to induce apoptosis in cancer cells, many of these drugs also induce apoptosis in noncancerous cells. Since rtEa4- and shEb-peptides induce apoptosis in a variety of human cancer cells but not in noncancerous cells in the same concentration range, we believe that rtEa4- and hEb-peptides could serve as ideal candidates for development of therapeutic agents for treating human cancers.

19.3 Regulation of Gene Expression by rtEa4-Peptide

As discussed in previous sections, rtEa4- and shEb-peptides possess the following activities against cancer cells: (1) inhibition of anchorage-independent cancer cell growth, (2) inhibition of invasion and metastasis of cancer cells, (3) suppression of cancer-induced angiogenesis, and (4) induction of apoptosis of cancer cells. Is there any molecular evidence supporting these anticancer activities?

19.3.1 Up- and Downexpression of Genes Regulated by rtEa4-Peptide

The introduction of expression vectors directing synthesis of secreted form of rtEa4-peptide or hEb-peptide into human breast cancer cells (MDA-MB-231) (Chen et al., 2002) and human neuroblastoma cells (SK-N-F1) (Kuo and Chen, 2002) resulted in isolation of many stable transfectants. Studies conducted on these stable transfectants revealed that these transfectants lost malignant characteristics such as colony formation in semisolid agar medium and invasion through matrigel membrane. These results suggest that rtEa4- or hEb-peptide may regulate the expression of genes responsible for the malignant activities of cancer cells. To confirm if rtEa4-peptide or hEb-peptide inhibits the growth, metastasis, and angiogenesis and induces apoptosis in MDA-MB-231 cells via up- and/or downregulation of genes related to malignant activities, Chen et al. (2007) conducted DNA microarray analysis to compare the gene expression profiles of MDA-MB-231 cells with and without transfection with rtEa4-peptide expression vector. In this study, labeled cDNA were prepared from RNA of MDA-MB-231 cells transfected with an expression vector that directs the synthesis of secreted form of rtEa4-peptide and control cells and hybridized to a human gene chip containing 9500 unique cDNA clones. A partial list of genes that are up- or downregulated by rtEas4-peptide, as determined by DNA microarray analysis, is presented in Table 19.3. The mRNA levels of some of the genes listed in Table 19.3 were further confirmed by quantitative relative real-time RT-PCR analysis (Table 19.4). Results presented in Table 19.4 are in good agreement with those presented in Table 19.3. Among those genes that their expression were modulated by rtEa4-peptide, *cysteine-rich angiogenic inducer 61* and *VEGF* genes may relate to tumor-induced angiogenesis, genes of *urokinase* (*uPA*) and *plasminogen activator inhibitor* (*PAI1*) relate to invasion and metastasis, genes of fibronectin 1 and *lamin receptor* relate to growth and anchorage-dependent cell growth, and genes of *caspase* 3 and *BCL2* relate to apoptosis.

TABLE 19.3

Partial List of Genes Up- or Downregulated by rtEa4-Peptide in MDA-MB-231 Cells

Name of Genes	Ratio[a]	Accession Number
Upregulated genes		
TYPO3 protein tyrosine kinase	1.3	AA564121
Fibronectin 1	1.2	R42091
Cytochrome C-1	1.0	AA037369
DEAD/H (Asdp-Glu-Ala-Asp/His) box polypeptide 9 (RNA helicase)	1.1	AA028972
DEAD/H (Asdp-Glu-Ala-Asp/His) box polypeptide 5 (RNA helicase 68 kDa)	0.7	H18448
Tumor rejection antigen (gp96) 1	0.6	AA027981
Tissue inhibitor metalloproteinase 1 (erythroid-potentiating activity)	0.6	AA059307
Heat shock 90 kDa protein 1, beta	0.6	AA055974
Heat shock 90 kDa protein 1, alpha	0.6	N20012
Heat shock 70 kDa protein 10	0.6	N26743
Phospholipase C, gamma 2 (phosphatidylinositol specific)	0.6	H57180
Laminin receptor 1 (67 kDa)	0.4	W46382
RAN, member of RAS oncogene family	0.4	r60931
Mitogen-activated kinase 2	0.4	N71990
Downregulated genes		
Tumor-associated calcium signal transducer 2	−1.6	AA029700
EGF-containing fibulin-like ECM protein	−1.5	R26714
Cysteine-rich angiogenic inducer 61	−1.4	W48667
BCL2	−1.3	N38908
Transglutaminase 2 (C polypeptide)	−1.2	H11775
Prion protein (p27–31)	−1.0	H15255
Serum-inducible kinase	−0.9	H52648
Plasminogen activator, urokinase	−0.9	R74194
PAI, type 1	−0.9	R21222
Gelsolin (amyloidosis, finnish type)	−0.9	H06525
Thrombospondin 1	−0.8	AA187185
Keratin 7	−0.8	H02522
Oncogene TC21	−0.8	N23355
BRB7, member of RAS oncogene family	−0.7	AA084368
BCL-like 1	−0.4	H40035
Oncogene TC21	−0.7	R79785
Vav 1 oncogene	−0.4	T65770
Insulin-like growth factor–binding protein 3	−0.3	AA135554
Insulin-like growth factor–binding protein 7	−1.0	N92373

Source: Chen, M.J. et al., *J. Cell. Biochem.*, 101, 1316, 2007. With permission.

[a] Ratio = log (color signal intensity of rtEa4-peptide treated cells/color signal intensity of control cells). When ratio is "+," upregulation; when ratio is "−," downregulation.

TABLE 19.4

Relative Real-Time RT-PCR Analysis of Genes Up- or Downregulated by rtEa4-Peptide

Names of Genes	Relative Expression Levels[a]
uPA	0.52 ± 0.10
PAI 1	0.42 ± 0.04
BCL 2	0.28 ± 0.04
Cysteine-rich angiogenesis inducer	0.86 ± 0.12
Tumor-associated Ca^{++} signal inducer	0.44 ± 0.12
TYPO3 protein tyrosine kinase	3.02 ± 0.80
Tumor rejection antigen (Gp96)	3.46 ± 0.96
Heat shock protein 90 kDa protein 1a	3.40 ± 0.35
Heat shock 70 kDa protein 10	2.92 ± 0.39
Capase 3	4.58 ± 0.35
Fibronectin 1	2.32 ± 0.31
Laminin receptor 1	1.41 ± 0.11

Source: Chen, M.J. et al., *J. Cell. Biochem.*, 101, 1316, 2007. With permission.

[a] Relative expression level = $2^{-(S\Delta CT - C\Delta CT)}$, where $S\Delta CT$ is the difference between the CT of the sample (MDA-MB 231 cells transfected with rtEa4-peptide gene) and the housekeeping gene (β-actin) and $C\Delta CT$ is the difference between CT of the control (MDA-MB 231 cells) and the housekeeping gene (β-actin). $2^{-(S\Delta CT - C\Delta CT)} > 1$ upregulated gene; $2^{-(S\Delta CT - C\Delta CT)} < 1$, downregulated gene.

19.3.2 Effects of Signal Transduction Inhibitors on Gene Expression

In their studies on the effect of rtEa4-peptide on the expression of genes related to the malignant activity of MDA-MB-231 cells, Siri et al. (2006a,b) showed that while the expression of *fibronectin 1* and *laminin receptor* genes is induced (upregulated) by rtEa4-peptide, the expression of *uPA, tPA, PAI1*, and *TIMP1* genes is suppressed (downregulated) by rtEa4-peptide. What are the molecular mechanisms that underlie the regulation of these genes by rtEa4-peptide? To shed some light on this question, Siri et al. (2006a,b) investigated the expression of *fibronectin 1* and *uPA* genes in MDA-MB-231 cells modulated by rtEa4-peptide in the presence of several signal transduction pathway inhibitors such as PD98059, SB202190, SP600125, tamoxifen, and wortmannin. As shown in Figures 19.15 and 19.16, these inhibitors of signal transduction molecules exerted a dose-dependent inhibition on the rtEa4-peptide-modulated expression of *fibronectin 1* and *uPA* genes. Since PD98059 is known as an inhibitor of mitogen-activated protein kinase (MAPK)/ERK kinase (Hotokezaka et al., 2002), SB202190 an inhibitor of p38 MAPK (Manthey et al., 1998), SP600125 an inhibitor of Jun N-terminal kinase (JNK) (Bennett et al., 2001), tamoxifen an inhibitor of protein kinase C (PKC) (Horgan et al., 1986), and wortmannin an inhibitor of phosphatidylinositol 3-kinase (PI3K), the expression of *fibronectin 1* and *uPA* genes may be modulated by rtEa4-peptide via focal adhesion kinase (FAK)/MAPK, JNK1/2, Merk1/2, p38 MAPK, PI3K, and PKC signal transduction cascades.

19.3.3 Identification of the Receptor of rtEa4-Peptide

Identification and characterization of receptor molecules residing on the plasma membrane of cancer cells is the first step toward understanding the molecular mechanism of the antitumor activity of rtEa4- or hEb-peptide. As a step toward this front, Kuo and Chen (2003) initiated

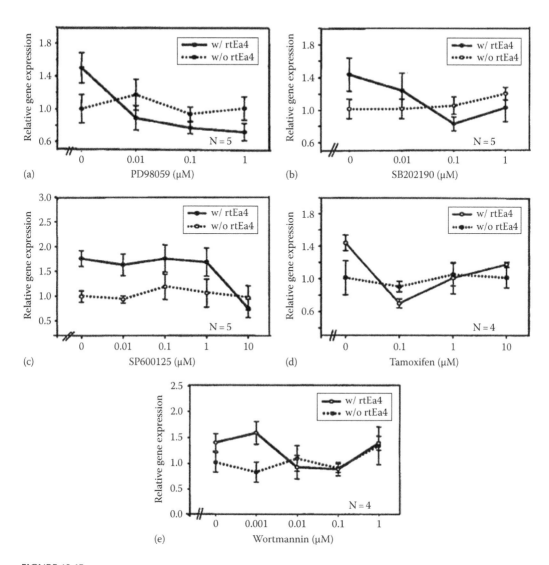

FIGURE 19.15

Effects of focal adhesion kinase (FAK)/mitogen-activated protein kinase (MAPK)-signaling pathway inhibitors on the expression of fibronectin 1 gene in MDA-MB-231 cells treated with rtEa4-peptide. MDA-MB-231 cells were treated with 0–10 µM of inhibitors for the signaling molecules of the FAK/MAPK pathways for 2 h prior to treatment of 2.56 µM of rtEa4-peptide or control proteins for an additional 1.5 h. Total RNA was extracted from treated cells and used for determination of levels of fibronectin 1 mRNA by relative quantitative real-time RT-PCR analysis. (From Siri, S. et al., *J. Cell. Biochem.*, 99, 1524, 2006b. With permission.) (a) PD 98059 (µM), (b) SB 202190 (µM), (c) SP600125 (µM), (d) Tamoxifen (µM), (e) Wortmannin (µM).

studies to identify receptors of rtEa4- or hEb-peptide on the cell membrane of human neuroblastoma cells (SK-N-F1) by the receptor-binding kinetics analysis. High specific activity ^{35}S-labeled rtEa4- and hEb-peptides were prepared by coupled *in vitro* transcription/translation reactions, and the resulting radiolabeled E-peptides were used as ligands in a saturation equilibrium-binding assay with intact cells and membrane fraction of SK-N-F1 cells. At the concentration range of 0.1–50 pM of both ligands, a specific binding was observed. In these saturation equilibrium-binding experiments, full saturation was not reached due to the

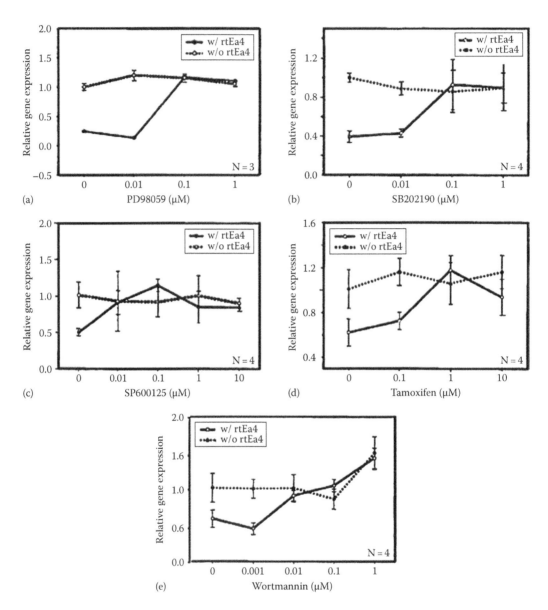

FIGURE 19.16

Effects of inhibitors of signaling molecules in the focal adhesion kinase/mitogen-activated protein kinase pathway on the expression of urokinase-type plasminogen activator (*uPA*) gene in MDA-MB-231 cells treated with rtEa4-peptide. MDA-MB-231 cells (5×10^5 cells) in DMEM/F12 medium were treated with various concentrations of inhibitors of signaling molecules for 2 h. After 2 h of treatment with inhibitors, cells were treated with or without 2.56 mM rtEa4-peptide for 1.5 h. RNA was extracted from treated and control cells and used for determination of levels of *uPA* mRNA by relative quantitative real-time RT-PCR analysis. (From Siri, S. et al., *J. Cell. Biochem.*, 99, 1361, 2006a. With permission.) (a) PD 98059 (μM), (b) SB 202190 (μM), (c) SP600125 (μM), (d) Tamoxifen (μM), (e) Wortmannin (μM).

TABLE 19.5

Binding Characteristics of ^{35}S-rtEa4-Peptide and ^{35}S-hEb-Peptide

Ligand	^{35}S-hEb-Peptide		^{35}S-rtEa4-Peptide	
Saturation Equilibrium Binding	K_d (M)	B_{max} (sites/cell or fmol/mg)	K_d (M)	B_{max} (sites/cell or fmol/mg)
Whole cell	$3.2 \pm 1.9 \times 10^{-11}$	$4.8 \pm 1.7 \times 10^4$	$2.9 \pm 1.8 \times 10^{-11}$	$4.0 \pm 2.6 \times 10^4$
Membrane	$8.0 \pm 4.3 \times 10^{-11}$	294.9 ± 89.1	$9.5 \pm 7.8 \times 10^{-11}$	183.3 ± 50.2
Competition Binding Assay	^{35}S-hEb-Peptide		^{35}S-rtEa4-Peptide	
	IC_{50}-1 (M)	IC_{50}-2 (M)	IC_{50}-1 (M)	IC_{50}-2 (M)
hEb-peptide	$5.8 \pm 4.4 \times 10^{-12}$	$4.8 \pm 2.6 \times 10^{-6}$	$7.3 \pm 1.1 \times 10^{-13}$	$3.9 \pm 2.1 \times 10^{-6}$
rtEa4-peptide	$2.5 \pm 1.9 \times 10^{-12}$	$2.7 \pm 1.2 \times 10^{-6}$	$1.4 \pm 1.1 \times 10^{-12}$	$2.1 \pm 0.6 \times 10^{-6}$

Source: Kuo, Y.-H. and Chen, T.T., *Gen. Comp. Endocrinol.*, 132, 231, 2003. With permission.

K_d, B_{max} and IC_{50} values are shown as mean $\pm$ SD (M) obtained from six independent experiments.

difficulty in producing sufficient quantities of labeled peptides. Scatchard analysis of the binding data resulted in a curvilinear plot, suggesting a negative cooperativity or the existence of multiple independent binding sites (or interconvertible affinity states). In intact cell-binding assays, a high-affinity binding site with a dissociation constant (K_d) of $3.2 \pm 1.9 \times 10^{-11}$ M and a concentration of $4.8 \pm 1.7 \times 10^4$ binding sites per cell was observed (Table 19.5) when radiolabeled hEb-peptide was used as a ligand. Plasma membrane preparation binding assays showed similar results with a dissociation constant (K_d) of $8.0 \pm 4.3 \times 10^{-11}$ M and a concentration of 294.9 ± 89.1 fmol/mg of membrane protein (Table 19.5). Similar to hEb-peptide, specific binding of ^{35}S-labeled rtEa4-peptides (0.1–50 pM) to intact cells and membrane preparation of SK-N-F1 cells was detected. Due to the same reason as that of hEb, full saturation was not achieved in these saturation equilibrium-binding experiments. However, like hEb, Scatchard analysis of the binding data resulted in a curvilinear plot, suggesting a negative cooperativity or the existence of multiple independent binding sites (or interconvertible affinity states). A high-affinity binding site with a dissociation constant (K_d) of $2.9 \pm 1.8 \times 10^{-11}$ M and a concentration of $4.0 \pm 2.6 \times 10^4$ binding sites per cell was observed in intact cell-binding assays. Plasma membrane preparation binding assays showed similar results with a dissociation constant (K_d) of $9.5 \pm 7.8 \times 10^{-11}$ M and a concentration of 183.3 ± 50.2 fmol/mg of membrane protein (Table 19.5).

Due to the limitation of the concentration of radiolabeled E-peptide, the saturation equilibrium-binding assays described earlier were limited within two orders of magnitude; the overall precision of binding parameter estimation and the ability to distinguish multiple binding sites were very limited. Therefore, homologous competitive binding assays were carried out to complement this limitation. The binding of 10 pM ^{35}S-hEb-peptide or ^{35}S-rtEa4-peptide was competed with its respective unlabeled peptide over 12 orders of magnitude (10^{-15} to 10^{-4} M) in both intact SK-N-F1 cells and plasma membrane preparation. A biphasic competition curve revealing two very different binding affinities was observed with either hEb-peptide or rtEa4-peptide. As shown in Table 19.5, the IC_{50} of the high-affinity binding site (IC_{50}-1) fell into the range between 10^{-11} and 10^{-12} M, similar to the dissociation constant (K_d) obtained in saturation equilibrium-binding assays. The second binding site, however, showed a much lower binding affinity with an IC_{50} of $4.8 \pm 2.6 \times 10^{-6}$ M for hEb-peptide and $2.1 \pm 0.6 \times 10^{-6}$ M for rtEa4-peptide. The lower-affinity binding site accounts for about 75% of the specific binding. These results suggest that hEb-peptide and rtEa4-peptide bind to two independent binding sites or one single binding site with two distinct affinity states. To determine whether the same binding site on SK-N-F1 cells binds to both hEb-peptide and rtEa4-peptide, cross-competitive binding assays were conducted on intact cells and membrane preparation of SK-N-F1 cells. As shown in Table 19.5, the values of IC_{50} from the cross-competitive binding assay are comparable with those of the homologous competitive binding assays. These results suggest that hEb-peptide and rtEa4-peptide bind to the same binding site on the cell membrane of SK-N-F1cells. Furthermore, it is interesting to note that the effective concentrations of rtEa4-peptide or hEb-peptide on the *in vitro* anticancer cell studies share in the same concentration range as the low-affinity binding site.

To demonstrate the presence of a specific binding component for hEb-peptide residing on the cell membrane of MDA-MB-231 cells, purified plasma membrane fraction of MDA-MB-231 cells was solubilized in nonionic detergent, and the resulting protein solution was incubated with infrared fluorescent dye (IRDye 800CW) labeled hEb-peptide (IR-hEb) until equilibrium. The resulting mixture was chromatographed on a Sephacryl S-300 HR column (1 × 50 cm). While the free IR-hEb was eluted from the column at 25.5 mL (Figure 19.17), a shift of the IR-hEb to a high molecular peak was

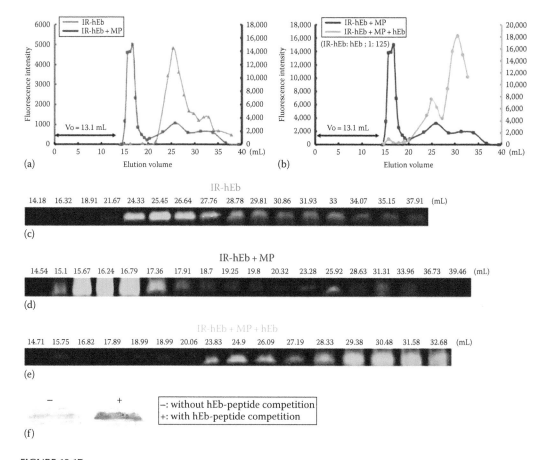

FIGURE 19.17
Demonstration for the presence of specific hEb-binding proteins solubilized from membrane fraction of MDA-MB-231 cells by column chromatography on Sephacryl S-300 gel. (a) and (b), Elution profiles of IR-hEb alone, IR-hEb incubated with solubilized membrane proteins, and IR-hEb incubated with solubilized membrane proteins and unlabeled hEb-peptide; (c–e) detection of IR-hEb in each fraction from (a) and (b); and (f) immunoblot analysis of the identified hEb-peptide binding component with an antibody to glucose-regulated protein.

observed at 16 mL when it was incubated with membrane preparation of MDA-MB-231 cells. This peak disappeared when the binding was carried out in the presence of excess amount of unlabeled hEb, suggesting the presence of a specific binding component(s) for hEb-peptide on basis of its reversible binding characteristic. To further characterize the hEb-specific binding component(s) on the plasma membrane of MDA-MB-231 cells, an affinity purification strategy was developed. In this strategy, double-tagged (his tag and strep tag) recombinant hEb-peptide was immobilized on Strep-Tactin Sepharose beads, and the resulting beads were used for purification of the hEb-peptide-specific binding component(s) from the cell membrane fraction of MDA-MB-231 cells by affinity column chromatography. Membrane proteins were solubilized by resuspending the membrane preparation in phosphate buffered saline (PBS) buffer containing 10% Triton X-100 (a nonionic detergent), and the resulting solution was subsequently diluted with PBS to give a final concentration of 0.2% Triton X-100. The membrane proteins were incubated with Strep-Tactin Sepharose beads until equilibrium, and the resulting slurry was packed in a column and washed with excess PBS containing 0.2% Triton X-100. The column was then eluted with PBS buffer containing hEb-peptide. The eluted fraction was concentrated and resolved into several protein spots by 2D electrophoresis. Among these protein spots, protein X is a predominant spot, and it was picked for determination of its chemical identity by matrix-assisted lazor desorption/ionization tendem mass spectrometry (MALDI-MS/MS) analysis. In comparison with the existing protein database, the result of MALDI-MS/MS analysis revealed the identity of the protein X as glucose-regulated protein (GRP78), a member of the heat-shock protein 70 family. Immunoblot analysis further revealed that this protein shows immunoreactivity with an antibody raised against human GRP78 protein (Figure 19.17f). GRP78 is an endoplasmic reticulum chaperon protein belonging to the heart-shock protein family known to play an essential role in protein biosynthesis (Dudek et al., 2009). Although GRP78 generally resides inside the lumen of endoplasmic reticulum, it is also found at the cell membrane of a wide variety of cancer cells, including neuroblastoma, lung carcinoma, ovarian cancer, colon adenocarcinoma, and prostate cancer (Delie et al., 2012; Mintz et al., 2003; Shin et al., 2003). Since GRP78 was isolated from the cell membrane of MDA-MB-231 cells through its reversible binding to hEb-peptide, it may be a candidate of hEb-peptide receptor. However, its true identity as the receptor of hEb-peptide remains to be confirmed by *in vitro* binding kinetic studies and *in vivo* functional studies.

19.4 Enhancement of Wound Closure by rtEa4-Peptide

Healing of a wound involves multiple cellular events including migration of various cell types into the wound, cell proliferation, angiogenesis, and synthesis and degradation of ECM (Declair, 1999; Best and Hunter, 2000; Li et al., 2001). To identify factors that may promote wound healing, an *in vitro* scratch-wound closure assay is routinely employed (Declair, 1999; Liang et al., 2007; Ranzato et al., 2009a). This assay involves creation of a new artificial gap (scratch wound) on a confluent cell monolayer and observing the cells on the newly created edge to migrate toward the opening of the gap until the gap is closed and the new cell–cell contacts are established again. There are several advantages associated with this simple *in vitro* scratch assay: (1) it mimics to some extent migration of cells *in vivo*; (2) the assay is particularly suitable for studying the regulation of cell migration by cell interaction with ECM and cell–cell interaction; (3) it is compatible with microscopy including live-cell imaging and allowing analysis of intracellular signaling events during cell migration (Fenteany et al., 2000; Ranzato et al., 2009b; Buonomo et al., 2012); and (4) it is the simplest method to study cell migration *in vitro* only using the common and inexpensive supplies in most laboratories capable of culturing cell. To our surprise, when recombinant rtEa4- or shEb-peptide of pro-IGF-I was added to the culture medium, the closure of a scratch wound created on the monolayer cells of CCD-112SK was significantly enhanced. In here, we report the detail characterization of the scratch-wound closure activity induced by rtEa4- or shEb-peptide on monolayer cells of CCD-112SK.

19.4.1 *In Vitro* Scratch-Wound Closure Assay

As mentioned earlier, *in vitro* scratch-wound closure assay is well suited for studying single-cell or cell-sheet migration from the edge of the wound during wound healing (Liang et al., 2007). This assay not only has been used to identify factors that promote migration of population of cells during wound closure by combining with other techniques such as microinjection or gene transfections but also has been used to assess the effect of expression of exogenous genes on migration of individual cells (Etienne-Manneville and Hall, 2001; Abbi et al., 2002). In a preliminary experiment, we observed that addition of various concentrations of rtEa4- or shEb-peptide to a scratch wound created on the monolayer of human breast cancer cells (MDA-MB-231) resulted in widening the gap of the wound in 24 h, but addition of the same peptide to a wound created on monolayer of noncancerous CCD-112SK cells resulted in closure of the wound in 24 h. These results suggested that rtEa4- or shEb-peptide may promote the closure of scratch wound created on noncancerous cells. To confirm this observation, we conducted studies to determine the dose-dependent effect of rtEa4- or shEb-peptide on closure of a scratch wound created on the monolayer cells of CCD-112SK. As shown in Figure 19.18, a dose-dependent wound closure activity was observed for both rtEa4- and shEb-peptide.

19.4.2 Effect of Signal Transduction Inhibitors

Results of studies from various systems showed that the expression of *c-Jun* (Li et al., 2003; Zenz et al., 2003; Katiyar et al., 2007; Gurtner et al., 2008) and phosphatase and tensin homolog gene (PTEN) (Lai et al., 2007; Zhao, 2007; Cao et al., 2011; Mihai et al., 2012) genes is critical for wound closure. Since we have shown that rtEa4-peptide and shEb-peptide enhance the closure of scratch wound created on the monolayer of CCD-112SK cells, these peptides may modulate the expression of *c-Jun* and *PTEN* genes. To address this question, scratch wound was created on the monolayer of CCD-112SK cells and treated with various concentrations of rtEa4- (0.4–47.6 µM) or synthetic hEb-peptide (4.8–237.5 µM), and RNA samples were isolated from E-peptide treated and untreated cells for determination of levels of *c-Jun* and *PTEN* mRNAs by comparative real-time RT-PCR analysis. While a dose-dependent increase of *c-Jun* mRNA levels was observed in CCD-112SK scratch wound after treatment with rtEa4- and shEb-peptides (Figure 19.19), a suppression of PTEN mRNA levels was also observed (data not shown). These results are consistent with those reported by many investigators in other systems (Li et al., 2003; Zenz et al., 2003; Katiyar et al., 2007; Lai et al., 2007; Zhao, 2007; Gurtner et al., 2008; Cao et al., 2011; Mihai et al., 2012).

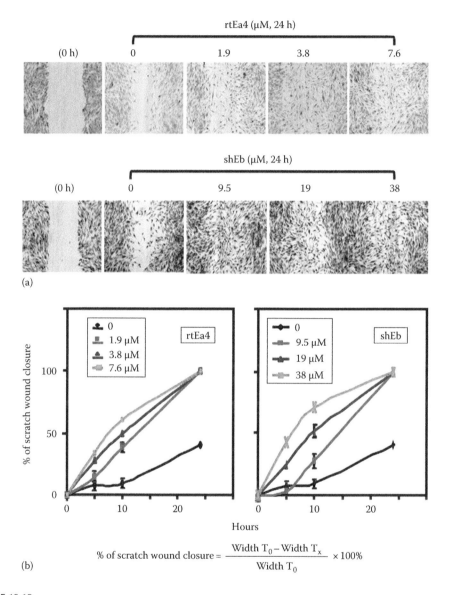

FIGURE 19.18
Dose-dependent scratch-wound closure activity of rtEa4-peptide and shEb-peptide on CCD-1112SK cells. (a) Microscopic images of wound closure and (b) dose-dependent wound closure activity. Scratch wounds were created in confluent monolayer cells of CCD-1112SK cells with a sterile 1 mL pipette tip. After washing off the suspended cells, cultures were refed with serum-free medium supplemented with various concentrations of rtEa4-peptide or shEb-peptide. After incubation at 37°C for 24 h, cells were fixed in 100% methanol and stained by Giemsa staining. Wound-closure space was measured at 0 and 24 h after wounding under an inverted microscope (IX50, Olympus) equipped with a micrometer in the objective.

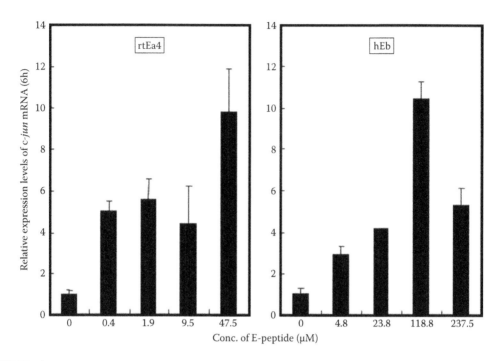

FIGURE 19.19

Relative expression levels of *c-Jun* mRNA in scratch wounds of CCD-1112SK cells treated with rtEa4- or shEb-peptide. Scratch wounds of CCD-1112SK cells were treated with various concentrations of rtEa4- or shEb-peptide for 6 h. Total RNA was extracted from treated cells, and the levels of *c-Jun* mRNA were determined by relative quantitative real-time RT-PCR analysis.

What signal transduction pathway may be involved in the enhancement of scratch-wound closure created on CCD-1112SK cells by rtEa4- or shEb-peptide? Ideally, this question should be initiated from identifying and characterizing the receptor molecules for rtEa4- or shEb-peptide in CCD-1112SK cells. In the event of lacking information on receptor molecules, we tested the effect of known signal transduction inhibitors, namely, U0126, PD098059, SP600125, and SB203580, on rtEa4- or shEb-peptide-induced scratch-wound closure created on the monolayer of CCD-1112SK cells. As shown in Figure 19.20, all of these inhibitors inhibited the rtEa4- or shEb-peptide-induced scratch-wound closure created on the monolayer of CCD-1112SK cells. Furthermore, the induction of *c-Jun* gene expression and the suppression of *PTEN* gene expression in CCD-1112SK cells treated with rtEa4- or shEb-peptide were also abolished by these inhibitors (Figure 19.21). Since U0126 and PD098059 are known as inhibitors of MAPK (Simon et al., 1996; Favata et al., 1998; Newton et al., 2000), SB203580 as an inhibitor of stress-activated protein kinase-2/p38 (Mizuno et al., 2006), and SP600125 as an inhibitor of JNK the rtEa4- or shEb-peptide enhanced the closure of scratch wound on CCD-1112SK cells may be mediated through MAPK/JNK/p38 signal transduction cascades.

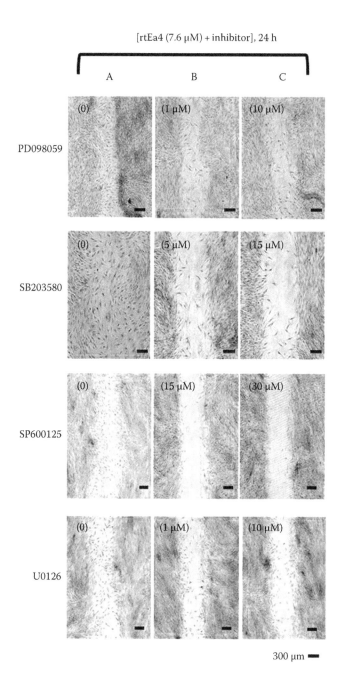

FIGURE 19.20

Effect of signal transduction inhibitors on rtEa4-peptide induced scratch-wound closure of CCD-1112SK cells. Scratch wounds were created on confluent monolayer of CCD-1112SK cells with a sterile 1 mL pipette tip. After washing off the suspended cells, cultures were refed with serum-free medium supplemented with various concentrations of signal transduction inhibitors (PD098059, SB203580, SP600125, or U0126) and rtEa4-peptide (7.6 μM). After incubation at 37°C for 24 h, cells were fixed in 100% methanol and stained by Giemsa staining. Wound-closure space was measured at 0 and 24 h after wounding under an inverted microscope (Olympus IX50) equipped with a micrometer in the objective. A. no inhibitor added; B. inhibitor dose one added; C. inhibitor dose two added.

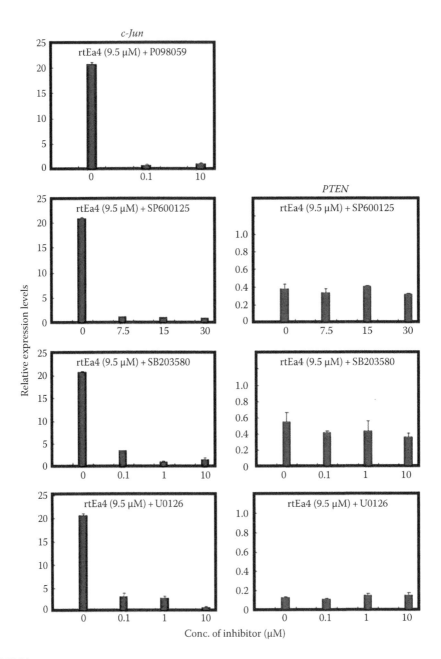

FIGURE 19.21

Effect of signal transduction inhibitors on relative expression levels of *c-Jun* and *PTEN* mRNAs in scratch wounds of CCD-1112SK cells treated with rtEa4-peptide. Scratch wounds of CCD-1112SK cells were treated with rtEa4-peptide (9.4 µM) and various concentrations of signal transduction inhibitors (PD098059, SB203580, SP600125, or U0126) for 24 h. Total RNA was extracted from treated cells, and the levels of *c-Jun* and *PTEN* mRNA were determined by relative quantitative real-time RT-PCR analysis. Relative expression levels were calculated by comparing to the expression level of cells without rtEa4-peptide treatment.

19.5 Summary

In this chapter, the anticancer activity of rainbow trout Ea4-peptide and human Eb-peptide has been reviewed. The anticancer activity of rtEa4-peptide and hEb-peptide includes inhibition of anchorage-independent cell growth, inhibition of invasion and metastasis, suppression of cancer-induced angiogenesis, and induction of cancer cell apoptosis. Interestingly, rtEa4-peptide and hEb-peptide can also stimulate the closure of scratch wounds created on noncancerous cells. Analysis of gene expression profiles revealed that rtEa4-peptide or hEb-peptide modulates the expression of genes related to cancerous activity via FAK/MAPK, JNK1/2, Merk1/2, p38 MAPK, PI3K, and PKC signal transduction cascades.

References

Abbi, S., H. Ueda, C. Zheng, L.A. Cooper, J. Zhao, R. Christopher, and J.L. Guan. 2002. Regulation of focal adhesion kinase by a novel protein inhibitor FIP200. *Molecular Biology of the Cell* 13:3178–3191.

Bell, G.I., R. Sanchez-Pescador, P.J. Laybourn, and R.C. Najarian. 1983. Exon duplication and divergence in the human preproglucagon gene. *Nature* 304:716–718.

Bennett, S.L., D.T. Sasaki, B.W. Murray, E.C. O'Leary, S.T. Sakata, W. Xu, J.C. Leiswten et al. 2001. SP600125, an anthrapyrazolone inhibitor of Jun N-terminal kinase. *Proceedings of the National Academy of Sciences of the United States of America* 98:13681–13686.

Best, T.M. and K.D. Hunter. 2000. Muscle injury and repair. *Physical Medicine and Rehabilitation Clinics of North America* 11:252–266.

Brooks, P.C., A.M. Montgomery, M. Rosenfeld, R.A. Reisfeld, T. Hu, G. Klier, and D.A. Cheresh. 1994. Integrin alpha v beta 3 antagonists promote tumor regression by inducing apoptosis of angiogenic blood vessels. *Cell* 79:1157–1164.

Buonomo, R., F. Giacco, A. Vasaturo, S. Caseerta, S. Guido, V. Pagliara, C. Garbi et al. 2012. PED/PEA-15 controls fibroblast motility and wound closure by ERK1/2-dependent mechanisms. *Journal of Cellular Physiology* 227:2106–2116.

Cao, L., E.O. Graue-Hernandez, V. tran, B. Reid, J. Pu, M.J. Mannis, and M. Zbao. 2011. Downregulation of PTEN at corneal wound sites accelerates wound healing through increased cell migration. *Investigative Ophthalmology & Visual Science* 52:2272–2278.

Cao, Y. 2005. Tumor angiogenesis and therapy. *Medicine and Pharmacotherapy* 59:S340–S343.

Chambers, A.F., R. Shafir, and V. Ling. 1982. A model system for studying metastasis using the embryonic chick. *Cancer Research* 42:4018–4025.

Chambers, A.F., E.E. Schmid, I.C. MacDonald, V.L. Morris, and A.C. Groom. 1992. Early steps in hematogenous metastasis of B16F1 melanoma cells in chick embryos studied by high resolution intravital videomicroscopy. *Journal of the National Cancer Institute* 84:797–803.

Chen, M.H.C., G.H. Lin, H.Y. Gong, C.F. Weng, C.Y. Chang, and J.H. Wu. 2001. The characterization of prepro-insulin-like growth factor-I Ea-2 expression and insulin-like growth factor-I genes(devoid81 bp) in the zebrafish (*Danio rerio*). *Gene* 268:67–75.

Chen, M.J., P. Chiou, B.-Y. Yang, H.-C. Lo, J.-K. Son, J. Hendricks, G. Bailey, and T.T. Chen. 2004. Development of rainbow trout hepatoma cell lines: Effect of pro-IGF Ea4-peptide on morphological changes and anchorage-independent growth. *In Vitro Cellular Developmental Biology—Animal* 40:118–128.

Chen, M.J., P.P. Chiou, P. Lin, C.-M. Lin, S. Siri, K. Peck, and T.T. Chen. 2007. Suppression of growth and cancer-induced angiogenesis of aggressive human breast cancer cells (MDAMB-231) on the chorioallantoic membrane of developing chicken embryos by E-peptide of pro-IGF-I. *Journal of Cellular Biochemistry* 101:1316–1327.

Chen, M.J., Y.H. Kuo, X.C. Tian, and T.T. Chen. 2002. Anti-tumor activities of pro-IGF-I E-peptides: Studies on morphological changes, anchorage-dependent cell division, and invasiveness in tumor cells. *General and Comparative Endocrinology* 126:342–351.

Chen, M.J., C.-M. Lin, and T.T. Chen. 2012. Induction of apoptosis in human cancer cells by rainbow trout Ea4- or human Eb-peptide of pro-insulin-like growth factor-I. In *Targeting New Pathways and Cell Death in Breast Cancer*, ed. R.L. Aft, pp. 45–56, InTech. Croatia.

Chiang, A.C. and J. Massague. 2008. Molecular basis of metastasis. *New England Journal of Medicine* 359:2814–2823.

Civelli, O., J. Douglass, H. Rosen, G. Martnes, and E. Herbert. 1986. Biosynthesis of opioid peptides. In *Opioid Peptides: Molecular Pharmacology, Biosynthesis, and Analysis*, eds. R.S. Rapaka and R.L. Hawaks, NDA Research Monograph 70, U.S. Department of Public Health. Washington D.C. pp. 21–34.

Declair, V. 1999. The importance of growth factors in wound healing. *Ostomy Wound Manage* 45:64–68.

Delie, F., P. Petignat, and M. Cohen. 2012. GRP78 protein expression in ovarian cancer patients and perspectives for a drug-targeting approach. *Journal of Oncology* 2012. pp. 1–5 doi: 10:1155/2012/468615.

de Pablo, F., B. Perez-Villami, J. Serna, P.R. Gonzalez-Guerrero, A. Lopez-Carranza, E.J. de la Rosa, J. Alemany, and T. Caldes. 1993. IGF-I and the IGF-I receptor in development of nonmammalian vertebrates. *Molecular Reproduction and Development* 35:427–433.

de Pablo, F., L.A. Scott, and J. Roth. 1990. Insulin and insulin-like growth factor I in early development: Peptides, receptors and biological events. *Endocrine Review* 11:558–577.

Dudek, J., J. Benedix, S. Cappet, M. Greiner, C. Jalat, L. Muller, and R. Zimmermann. 2009. Functions and pathologies of BiP and its interaction partners. *Cellular and Molecular Life Sciences* 66:1556–1569.

Duguay, S.J. 1999. Post-translational processing of insulin-like growth factors. *Hormone Metabolism Research* 31:43–49.

Duguay, S.J., L.K. Park, M. Samadpour, and W.W. Dickhff. 1992. Nucleotide sequences and tissue distribution of three different insulin-like growth factor-I prohormone in salmon. *Molecular Endocrinology* 6:1202–1210.

Duguay, S.J., P. Swanson, and W.W. Dickhoff. 1994. Differential expression and hormonal regulation of alternatively spliced IGF-I mRNA transcripts in salmon. *Journal of Molecular Endocrinology* 12:25–37.

Elmore, S. 2007. Apoptosis: A review of programmed cell death. *Toxicologic Pathology* 35:495–516.

Etienne-Manneville, S. and A. Hall. 2001. Integrin-mediated activation of Cdc42 controls cell polarity in migrating astrocytes through PKCζ. *Cell* 106:489–498.

Faulk, C.K., R. Perez-Dominguez, K.A. Jr. Webb, and G.J. Holt. 2010. The novel finding of four distinct pre-pro-IGF-I E domains in a perciform fish, *Sciaenops ocellatus*, during ontogeny. *General and Comparative Endocrinology* 169:75–81.

Favata, M.F., K.Y. Horiuchi, E.J. Manos, A.J. Daulerio, D.A. Stradley, W.S. Feeser, D.E. Van Dyk et al. 1998. Identification of a novel inhibitor of mitogen-activated protein kinase kinase. *The Journal of Biological Chemistry* 273:18623–18632.

Fenteany, G., P.A. Janmey, and T.P. Stossel. 2000. Signaling pathways and cell mechanics in wound closure by epithelial cell sheets. *Current Biology* 10:831–838.

Folkman, J. 2002. Role of angiogenesis in tumor growth and metastasis. *Seminars in Oncology* 29:15–18.

Fulda, S. and K.-M. Debatin. 2006. Extrinsic verse intrinsic apoptosis pathways in anticancer chemotherapy. *Oncogene* 25:4798–4811.

Ghavami, S., M. Hashemi, S.R. Ande, B. Yeganeh, W. Xiao, M. Eshraghi, C.J. Bus et al. 2009. Apoptosis and cancer: Mutation with caspase genes. *Journal of Medical Genetics* 46:497–510.

Gupta, M.K. and R.-Y. Qin. 2003. Mechanism and its regulation of tumor-induced angiogenesis. *World Journal of Gastroenterology* 9:1144–1155.

Gurtner, G.C., S. Werner, Y. Barrandon, and M.T. Longaker. 2008. Wound repair and regeneration. *Nature* 453:314–321.

Hacker, G. 2000. The morphology of apoptosis. *Cell and Tissue Research* 301:5–17.

Hanahan, D. and R.A. Weinberg. 2000. The hallmarks of cancer. *Cell* 100:57–70.

Hanahan, D. and R.A. Weinberg. 2011. Hallmarks of cancer: The next generation. *Cell* 144:646–674.

Hengartner, M.O. 2000. The biochemistry of apoptosis. *Nature* 407:770–776.

Horgan, K., E. Cooke, M.B. Hallett, and R.E. Mansel. 1986. Inhibition protein kinase C mediated signal induction by tamoxifen: Important for antitumour activity. *Biochemical Pharmacology* 35:4463–4465.

Hotokezaka, H., E. Sakai, K. Kanaoka, K. Saito, K.-I. Matsuo, H. Kitaura, N. Yoshida, and K. Nakayama. 2002. U0126 and PD98059, specific inhibitors of MEK, accelerate differentiation of RAW264.7 cells into osteoclast-like cells. *The Journal of Biological Chemistry* 277:47366–47372.

Ido, Y., A. Vindigni, K. Chang, L. Stramn, R. Chance, W.F. Heath, R.D. DiMarchi, E. di Cera, and J.R. Williamson. 1997. Prevention of vascular and neural dysfunction in diabetic rats by C-peptide. *Science* 277:563–566.

Katiyar, S., X. Jiao, E. Wagner, M.P. Lisanti, and R.G. Pestell. 2007. Somatic excision demonstrates that cJun induces cellular migration and invasion through induction of stem cell factor. *Molecular and Cellular Biology* 27:1356–1369.

Klein, C.A. 2008. The metastasis cascade. *Science* 321:1785–1787.

Kuo, Y.-H. and T.T. Chen. 2002. Novel activities of pro-IGF-I E-peptides: Regulation of morphological differentiation and anchorage-independent growth in human neuroblastoma cells. *Experimental Cell Research* 280:75–89.

Kuo, Y.-H. and T.T. Chen. 2003. Specific cell surface binding sites shared by human pro-IGF-I Eb-peptide and rainbow trout pro-IGF-I Ea4peptide. *General and Comparative Endocrinology* 132:231–240.

Lai, J.-P., J.T. Dalton, and D.L. Knoell. 2007. Phosphatase and tensin homologue deleted on chromosome ten (PTEN) as a molecular target in lung epithelial wound repair. *British Journal of Pharmacology* 152:1172–1184.

Li, G., C. Gustafson-Brown, S.K. Hanks, K. Nason, J.M. Arbeit, K. Pogliano, R.M. Wisdom, and R.S. Johnson. 2003. C-Jun is essential for organization of the epidermal leading edge. *Development Cell* 4:865–877.

Li, Y., J. Cummins, and J. Huard. 2001. Muscle injury and repair. *Current Opinion in Orthopedics* 12:409–415.

Liang, C.-C., A.Y. Park, and J.-L. Guan. 2007. In vitro scratch assay: A convenient and inexpensive method for analysis of cell migration in vitro. *Nature Protocols* 2:329–331.

Manthey, C.L., S.-W. Wang, S.D. Kinney, and Z. Yao. 1998. SB202190, a selective inhibitor of p38 mitogen-activated protein kinase, is a powerful regulator of LSP-induced mRNAs. *Journal of Leukocyte Biology* 64:409–417.

Matheny, R.W. Jr, B.C. Nindl, and M.L. Adamo. 2010. Minireview: Mechano factor: A putative product of IGF-I gene expression involved in tissue repair and regeneration. *Endocrinology* 151:865–875.

Mihai, C., S. Bao, J.-P. Lai, S.N. Ghadiall, and D.L. Knoell. 2012. PTEN inhibition improves wound healing in lung epithelial through changes in cellular mechanics that enhance migration. *American Journal of Physiology—Lung Cellular and Molecular Physiology* 302:L287–L299.

Mintz, P.J., J. Kim, K.-A. Do, X. Wang, R.G. Zinner, M. Cristofanilli, M.A. Arap et al. 2003. Fingerprinting the circulating repertoire of antibodies from cancer patients. *Nature Biotechnology* 21:57–63.

Mizuno, H., Y.-Y. Cho, W.-Y. Ma, A.M. Bode, and Z. Dong. 2006. Effects of MAP kinase inhibitors on epidermal growth factor-induced neoplastic transformation of human keratinocytes. *Molecular Carcinogenesis* 45:1–9.

Newton, R., L. Cambridge, L.A. Stevens, M.A. Lindsay, and P.J. Barnes. 2000. The MAP kinase inhibitors, PD098059, UO126 and SB203580, inhibit IL-1 β-dependent PGE₂ release via mechanistically distinct processes. *British Journal of Pharmacology* 130:1353–1361.

O'Brien, M.A. and R. Kirby. 2008. Apoptosis: A review of pro-apoptotic and anti-apoptotic pathways and dysregulation in disease. *Journal of Veterinary Emergency and Critical Care* 18:572–586.

Pfeffer, L.A., B.K. Brisson, H. Lei, and E.R. Barton. 2009. The insulin-like growth factor (IGF)-I E-peptides modulate cell entry of the mature IGF-I protein. *Molecular Biology of the Cell* 20:3810–3817.

Ranzato, E., V. Balbo, F. Boccafoschi, L. Mazucco, and B. Burlando. 2009a. Scratch wound closure of C2C12 myoblasts is enhanced by human platelet lysate. *Cell Biology International* 33:911–917.

Ranzato, E., L. Mazzucco, M. Patrone, and B. Burlando. 2009b. Platelet lysate promotes in vitro wound scratch closure of human dermal fibroblasts: Different roles of cell calcium, P38, ERK and PI3K/AKT. *Journal of Cellular and Molecular Medicine* 13:2030–2038.

Rotein, P., K.M. Pollock, D.K. Didier, and G.G. Krivi. 1986. Organization and sequence of the human insulin-like growth factor I gene: Alternative RNA processing produces two insulin-like growth factor I precursor peptides. *The Journal of Biological Chemistry* 261:4828–4832.

Shamblott, M.J. and T.T. Chen. 1992. Identification of a second insulin-like growth factor in a fish species. *Proceedings of the National Academy of Sciences of the United States of America* 89:8913–8917.

Shamblott, M.J. and T.T. Chen. 1993. Age-related and tissue-specific levels of five forms of insulin-like growth factor mRNA in a teleost. *Molecular Marine Biology and Biotechnology* 2:351–361.

Shi, F.T., W.S. Li, C.H. Bai, and H.R. Lin. 2002. IGF-I of orange spotted grouper *Epinephelus coioides*: cDNA cloning, sequencing and expression in *Escherichia coli*. *Fish Physiology and Biochemistry* 27:147–156.

Shin, B.K., H. Wang, A.M. Yim, F. le Naour, F. Briechory, J.H. Jang, R. Zhao et al. 2003. Global profiling of the cell surface proteome of cancer cells uncovers an abundance of proteins with chaperone function. *The Journal of Biological Chemistry* 278:7607–7616.

Siegfried, J.M., P.G. Kasprzyk, A.M. Treston, J.L. Mulshine, K.A. Quinn, and F. Cuttitta. 1992. A mitogenic peptide amide encoded within the E-peptide domain of the insulin-like growth factor IB prohormone. *Proceedings of the National Academy of Sciences of the United States of America* 89:8107–8111.

Simon, C., J. Juarez, G.L. Nicolson, and D. Boyd. 1996. Effect of PD 098059, a specific inhibitor of mitogen-activated protein kinase kinase, on urokinase expression and in vitro invasion. *Cancer Research* 56:5369–5374.

Siri, S., M. Chen, and T.T. Chen. 2006a. Inhibition of human breast cancer cell (MDA-MB 231) invasion by the Ea4-peptide of rainbow trout pro-IGF-I. *Journal of Cellular Biochemistry* 99:1361–1373.

Siri, S., M. Chen, and T.T. Chen. 2006b. Biological activity of rainbow trout Ea4-peptide of the pro-insulin-like growth factor (pro-IGF)-I on promoting attachment of breast cancer cells (MDA-MB 231) via α2- and β1integrin. *Journal of Cellular Biochemistry* 99:1524–1535.

Stan, A.C., D.L. Radu, S. Casares, C.A. Bona, and T.D. Brumeanu. 1999. Antineoplastic efficacy of doxorubicin enzymatically assembled on galactose residues of a monoclonal antibody specific for the carcinoembryonic antigen. *Cancer Research* 59:115–121.

Stewart, C.E. and P. Rotwein. 1996. Growth, differentiation, and survival: Multiple physiological functions for insulin-like growth factors. *Physiological Review* 76:1005–1026.

Tiago, D.M., V. Laize, and M.L. Cancela. 2008. Alternatively spliced transcripts of *Sparus aurata* insulin-like growth factor-I are differentially expressed in adult tissues and during early development. *General and Comparative Endocrinology* 157:107–115.

Tian, X.C., M. Chen, A.G. Pantschenko, T.J. Yang, and T.T. Chen. 1999. Recombinant E-peptides of pro-IGF-I have mitogenic activity. *Endocrinology* 140:3387–3390.

Wallis, A.E. and R.H. Devlin. 1993. Duplicate insulin-like growth factor-I genes in salmon display alternative splicing pathways. *Molecular Endocrinology* 7:409–422.

Weis, S.M. and D.A. Cheresh. 2011. Tumor angiogenesis: Molecular pathways and therapeutic targets. *Nature Medicine* 17:1359–1370.

Yang, S.Y. and G. Goldspink. 2002. Different roles of the IGF-I Ec peptide (MGF) and mature IGF-I in myoblast proliferation and differentiation. *FEBS Letters* 522:156–160.

Yang, S.Y., Sales, K.M., Fuller, B.J., Seifalian, A.M., and Winslet, M.C. 2008. Inducing apoptosis of human colon cancer by an IGF-I D domain analogue peptide. *Molecular Cancer* 7:17–28.

Youle, R.J. and A. Strasser. 2008. The BCL-2 protein family: Opposing activities that mediate cell death. *Nature Reviews/Molecular Cell Biology* 9:47–59.

Zenz, R., H. Scheuch, P. Martin, C. Frank, R. Eferl, L. Kenner, M. Sibilia, and E.F. Wagner. 2003. *c-Jun* regulates eyelid closure and skin tumor development through EGFR signaling. *Developmental Cell* 4:879–889.

Zhao, M. 2007. PTEN: A promising pharmacological target to enhance a epithelial wound healing. *British Journal of Pharmacology* 152:1141–1144.

20

Bioactive Natural Products of Endophytic Fungi from Marine Algae

Vasuki Subramanian, P. Anantharaman, and K. Kathiresan

CONTENTS

20.1 Introduction

In recent years, marine endophytic fungi are recognized as a valuable source for novel secondary metabolites that have the potential to lead to innovations in drug therapy. The literature clearly shows that natural products have been a rich source of compounds that have found many applications in the fields of medicine, pharmacy, and biology. The marine fungi particularly those associated with marine algae, sponge, invertebrates, and sediments appear to be a rich source for secondary metabolites (Belofsky et al. 1998). Endophyte research has yielded potential drug lead compounds with antibacterial, antiviral, antioxidant, insulin mimetic, and antineurodegenerative, and immunosuppressant properties. Several reviewers addressed the issue (e.g., Tan and Zou 2001, Strobel and Daisy 2003, Owen and Hundley 2004, Schulz and Boyle 2005, Gunatilaka 2006, Ravikumar et al. 2010), shedding light into selected aspects of the nature of endophytes. In this chapter, preference is given to recently discovered natural products, as well as to compounds that are particularly promising for future development as drugs in pharmacological studies.

20.2 Seaweeds and Endophytes: In General

Marine macroalgae, commonly known as seaweeds, are benthic organisms and constituents of highly productive ecosystems. They play a major role in nutrient cycling and stabilizing the habitat (Mann 1973). Seaweeds are studied for their role in supporting coastal marine life and nutrient cycling as well as for their bioactive metabolites. They produce diverse structural metabolites such as polyketides, alkaloids, peptides, proteins, lipids, shikimates, glycosides, isoprenoids, and hybrids of these metabolites (Rateb and Ebel 2011; Blunt et al. 2013). The widely accepted definition of endophytes is "microorganisms that spend the whole or part of their life cycle, colonizing tissues of their host plants without causing apparent symptoms of disease" (Bacon and White 2000). They also play a vital role in ecosystem processes such as decomposition and nutrient cycling and have beneficial symbiotic relationships with roots of many plants. Endophytes have an unclear existence and one of their main roles in the ecosystem is their function as decomposers, as they are among the primary colonizers of dead plant tissues (Kumaresan and Suryanarayanan 2002). In particular, the ability to produce a large number of chemically different secondary metabolites is associated mainly with the filamentous fungi (Donadio et al. 2002).

20.3 Host and Endophyte: The Relationship

A hypothesis has been proposed and extensively documented by Schulz and coworkers, postulating the relationship to be a "balanced antagonism" (Schulz and Boyle 2005). The interaction, however, can vary from mutualism to parasitism, based on a fine-tuned balance between the demands of the invader and the plant response (Kogel et al. 2006). Endophytes can be transmitted from one generation to the next through the tissue of host seeds or vegetative propagules (Carroll 1988). An interrelationship exists between host and microbes, such that the host plant provides nutrients to the endophyte, and in return, the endophyte produces unique secondary metabolites to enhance the growth and competitiveness of the host in its natural habitat. For many living organisms, this chemical diversity reflects the impact of evolution in the selection and conservation of self-defense mechanisms that represent the strategies, employed to repel or destroy the predators.

Cladosporium cladosporioides is found usually in wide range of environments and in different types of associations like endophytic (Schulz and Boyle 2005), pathogenic (Briceno and Latorre 2008), and saprophytic (Sadaka and Ponge 2003) with terrestrial plants. *C. cladosporioides* has been long associated with marine algae (Miller and Whitney 1981; Ding et al. 2008). Cooley et al. (2011) revealed a novel algal/fungal relationship *in situ* and in the laboratory. *C. cladosporioides* can be isolated from the declining cultures *in vitro*, can be readily cultured, induced decline when introduced to fungal-free cultures of *Pseudendoclonium submarinum*, and can then be reisolated.

Seaweeds are known to produce secondary metabolites that deter microbial colonization; these metabolites determine the abundance and composition of the microbial community associated with the seaweeds as they select those microbes that can tolerate or detoxify the chemicals. These isolated metabolites originate from different biosynthetic pathways.

Interaction between the algae and colonizing bacteria is suggested by Saha et al. (2011). The high surface concentration of fucoxanthin on *Fucus vesiculosus* is an ecologically relevant antimicrobial agent and significant drivers of the interaction between the alga and colonizing bacteria. As such, there are no such studies in seaweed endophytes (Lachnit et al. 2013).

Such ecological research on algal-derived metabolites with antimicrobial activity has recently received increased attention and is no longer only aimed at identifying novel natural compounds with potential use in applied perspectives. For instance, the red alga *Bonnemaisonia asparagoides* produces surface-bound antibacterial compounds as a chemical defense with a significant impact on the abundance and composition of the associated bacterial community (Nylund et al. 2010). The observation that the spores of *Acremonium fuci* associated with *Fucus serratus* can germinate only in the presence of tissues of the alga and not in seawater alone points to host specificity among seaweed endophytes and adds credibility to the view that marine-derived fungi are not just migrants from land habitats (Zuccaro et al. 2004). In another study, the planktonic bacteria and fungi are selectively eliminated by exposure to marine macroalgae in close proximity. The red algae *Ceramium rubrum* and *Mastocarpus stellatus* and brown alga *Laminaria digitata* demonstrate defenses against epiphytic microorganisms *Cytophaga*, *Flavobacterium*, and *Bacteroides*, Alphaproteobacteria and *Roseobacter* groups, and the marine fungi (Lam et al. 2008). The interrelatedness of endophytes with their respective hosts is vastly underinvestigated. Thus, more background information on a given species and its microbial biology would be remarkably helpful in directing the search for bioactive products.

20.4 Seaweed Endophytes and Biodiversity

The nature of association and factors that determine the diversity and distribution of endophytes with their algal hosts is not clear. Schulz et al. (2008) noticed that the dominant endophyte species isolated from terrestrial plants and fungi associated with marine algae differs. However, the algae support a sizeable and diverse assemblage of endophytes. According to Suryanarayanan (2012), the endosymbiont assemblage of an alga can consist of different ecological groups of fungi such as latent pathogens, true endosymbionts, or casual residents. Sometimes, the endophyte assemblage of the algae comprises a few dominant species and many times less frequently come across.

Some fungal endophyte species have a wide host range. Marine-derived fungi such as *Aspergillus* spp., apart from dominating the endophyte assemblage of seaweeds (Suryanarayanan et al. 2010b), are not casual residents of the seas but have coevolved with the seaweeds (Zuccaro et al. 2004). It would seem that many factors like seasons, age, environment, and location (Kubanek et al. 2003; Keonig et al. 2006; Lam et al. 2008; Harvey and Goff 2010; Mohamed and Martiny 2010) are some of the reasons given for the constant presence of certain marine-derived fungal genera such as *Aspergillus*, *Cladosporium*, and *Penicillium* as endophytes in different seaweeds (Keonig et al. 2006; Zuccaro et al. 2008).

The limited studies so far indicate that genera such as *Aspergillus*, *Cladosporium*, and *Penicillium* are common endophytes isolated from marine algae (Suryanarayanan 2012; Flewelling et al. 2013a). The isolation frequencies of endophytic fungi vary by host, with the highest isolation frequency obtained from the red alga *Palmaria palmata* (87%) and the lowest isolation frequency in *Chondrus crispus* at a rate of 0.08% (Flewelling et al. 2013a). The endophytes isolated from algae, collected from Mandapam and Pondicherry coast, show both higher and lower frequencies in green algae. The endophytes isolated from marine algae collected from the Shetland Islands, United Kingdom, show the highest diversity of isolable fungi from the green algae, whereas fungal endophyte assemblages of brown algae are more diverse than those of the green algae (Flewelling et al. 2013b). None of the fungi isolated from the brown alga *Ascophyllum nodosum* has been identified as *Mycosphaerella ascophylli*, despite it is a well-documented endophyte of that host (Xu et al. 2008). Even genetic variations exist in a single morphological species of seaweed fungal endophytes (Harvey 2002).

Though the role of production of secondary metabolite is the same whether it belongs to terrestrial or marine organisms, a direct comparison of seaweeds with terrestrial counterpart is not ideal. Endophytes such as *Aspergillus terreus* from different marine sources, for instance, in the same species of seaweeds, collected from diverse geographical location or between genus or different parts of seaweeds with complex structures (Suryanarayanan et al. 2010b; Suja et al. 2013), show variations in their secondary metabolite profile. This indicates that screening of fungal species once will not capture the entire metabolite spectrum of the endophyte (Seymour et al. 2004). So, as suggested by Suryanarayanan and Thennarasan (2004), it is equally important to screen the seaweeds during different seasons as the endophyte assemblage of plants is altered by seasons.

At present, no one is quite certain of the role of endophytes in nature and their relationship to various host plant species. We have very limited information on the diversity and nature of association of fungal endophytes of seaweeds. Conventional culture methods along with valid molecular techniques to obtain a more complete picture of the diversity of the endophytes in seaweeds are necessary, as it has been done for endophytes of terrestrial plants (Arnold and Lutzoni 2007; Unterseher and Schnittler 2010; Selim et al. 2012).

20.5 Factors Affecting Metabolite Production

The bioactive compounds are chemical compounds derived from plants, animals, and microorganisms. These compounds may be derived from primary and secondary metabolisms of the organisms (Berdy 2005). The biosynthesis of natural products is usually associated with cell differentiation and development in fungi. The G-protein-mediated growth pathway in *Aspergillus nidulans* regulates both asexual sporulation and natural product biosynthesis (Hicks et al. 1997). Several other molecules and pathways also link the chemical and morphological differentiation processes in fungi (Tag et al. 2000; Calvo et al. 2002). Some of the primary metabolism serves as a branching point of biosynthetic pathways, which lead to the end product of primary and secondary metabolisms (Calvo et al. 2002).

Growth media and incubation conditions strongly influence the secondary metabolite production in many fungi. Such that the yield of bioactive compounds can be significantly increased by the optimization of physical factors such as temperature, salinity, pH, and light (Jain and Pundir 2011) and chemical factors such as media composition or addition of certain precursors and inhibitors to the medium (Llorens et al. 2004). In order to increase the number of secondary metabolites obtainable from one microbial strain, a simple approach called "one strain–many compounds" has been used by Bode et al. (2002).

They were able to isolate up to 20 different metabolites that yield up to 2.6 g L^{-1} from a single organism. This clearly shows that in many microorganisms, genetic possibilities for numerous biosynthetic pathways in one strain exist, which is used for the production of secondary metabolites depending on the environment.

The production of secondary metabolites may be highly dependent on the culture conditions and origin of the strains. To produce these metabolites and to maximize the potential chemical diversity, they need to be grown in various nutrient media. The optimal conditions for achieving maximal mycelial growth and bioactivity of the marine-derived fungus are different as the antibiotic activity of a fungus increases with salinity of the growth medium (Miao et al. 2006). It is suggested that cultivation of marine fungi in NaCl medium can result in the production of metabolites not synthesized or only synthesized in undetectable levels in normal growth medium (Bugni and Ireland 2004). In *Aspergillus terreus* isolated from green seaweed *Codium decorticatum*, maximum growth and bioactive compound is achieved in potato dextrose medium supplemented with sucrose as a carbon source and yeast extract as a nitrogen source *in vitro*. Further process parameters like incubation temperature (25°C), pH (5.5), and NaCl (5 g L^{-1}) have been found to be optimum for the maximal mycelial growth and bioactive metabolite production (Suja et al. 2013). Several *Aspergillus* spp. (Bhattacharyya and Jha 2011; Jain and Pundir 2011) encountered similar chances under different environmental parameters. Many species of the genus *Aspergillus* are halotolerant (Gunde-Cimerman et al. 2009). Most of the species have evolved unique metabolism to cope with the salinity changes. An improved production of bioactive metabolite in a novel endophytic fungus *Fusarium* sp. DF2, isolated from *Taxus wallichiana*, has been reported. The active compound with UV λ-max 270 nm in ethyl acetate has got the lowest minimum inhibitory concentration (MIC) against *Bacillus subtilis*, *Staphylococcus aureus*, and *Escherichia coli* and highest against *Pseudomonas aeruginosa* (Gogoi et al. 2008).

Pesatlone, an antibiotic synthesized by *Pestalotia* sp. (isolated from algae), is found to be effective against methicillin-resistant *S. aureus* when cultured with a marine bacterium (Cueto et al. 2001). Another marine-derived fungus *Phomopsis asparagi* produces novel chaetoglobosins when the growth medium is modified with the sponge metabolite jasplakinolide (Christian et al. 2005). The addition of small molecule elicitors to the fermentation medium precisely influences the synthesis of bioactive molecules by fungi (Pettit 2010). Thus, the synthetic ability of seaweed endophytes can be enhanced by altering the culture conditions.

The biosynthesis of secondary metabolite is complex and a number of factors influence the production. The genes required for biosynthesis of some of natural products are clustered, and it contains all or most of the genes required for natural product biosynthesis. Moreover, natural product biosynthetic gene clusters are conserved between organisms, for example, the sterigmatocystin–aflatoxin biosynthetic gene cluster in several *Aspergillus* spp. The cocultivation of marine-derived fungi with microbe habitats activates the silent gene clusters of the fungi, enabling them to synthesize novel metabolites (Brakhage and Schroeckh 2011; Brakhage 2013).

Secondary metabolites make their hosts more resistant to abiotic and biotic stress (Wagner and Lewis 2000; Arnold et al. 2003; Rodriguez et al. 2008). Over the last few decades, evidence has led to the hypothesis that many of the compounds isolated from marine macroorganisms in fact have a microbial origin (Bewley and Faulkner 1998). Though the seaweed is well known for production of wide array of bioactive metabolites and strong antioxidants, it would be worthwhile to determine the role of endophytes in stress tolerance and also how far their synthetic ability is affected by the presence of fungal endophytes.

A great deal of uncertainty also exists between what endophyte produces in culture and what it may produce in nature. As such, there is no information regarding the negative effect of seaweed endophytes, such as susceptibility to diseases, inducing diseases under suitable environmental conditions or when the host's defense is weakened, and bleaching and reduction in photosynthetic efficiency.

20.6 Chlorophyta

20.6.1 Bioactive Compounds from Green Algae

The phylum Chlorophyta represents only a small percentage (13%) of the marine environment, and thus, they are the least studied macroalgae division for chemistry of marine natural products. The main compounds synthesized by this group are isoprenoid derivatives. Acetogenins, amino acid derivatives, carbohydrates, and shikimate derivatives have also been isolated from these algae (McClintock and Baker 2001).

Activated defense appears to be common among tropical macroalgae. It is clearly seen in *Halimeda* sp., which produces diterpenes, halimedatetraacetate, and halimedatrial metabolites, and these act as feeding deterrents and allow the alga to persist in areas of intense herbivory. It has been further revealed that upon damage, the levels of halimedatetraacetate decreases while the concentration of halimedatrial increases, that is, the alga quickly converts halimedatetraacetate into the more deterrent halimedatrial via a putative enzyme–mediated pathway (Paul and Fenical 1984; Paul and Van Alstyne 1992). Another example of such trigged macroalgal chemical defenses is caulerpenyne. It is isolated from the invasive green alga *Caulerpa taxifolia* and from caulerpenyne, it rapidly transforms into oxytoxin 2 upon wounding (Jung and Pohnert 2001). An acetylene sesquiterpenoid ester is isolated from *Caulerpa prolifera* and it inhibits fouling in the same way as the biocide *bis*(tributyltin) oxide (Smyrniotopoulos et al. 2003). Two novel triterpene sulfate esters, capisterones A and B, have been isolated from *Penicillus capitatus*; it suppresses the growth of the marine fungal pathogen *Lindra thallasiae* (Puglisi et al. 2004). Macroalgae that are generally unfouled in the field are strong leads for the discovery of novel antifoulants, especially if found in habitats where other organisms are highly fouled. So the macroalgal terpenes have the potentials to be utilized commercially as antifoulants.

Many natural peptides are synthesized by a sequence of enzyme-controlled processes carried out by a multifunctional enzyme of modular arrangement, similar to some polyketide synthases. Depsipeptides are a class of nonribosomal peptides cyclized via an ester bond and often contain nonprotein amino acids. Several bioactive depsipeptides, ranging from a C31 tripeptide to a C75 tridecapeptide, collectively known as kahalalides, display a range of biological activities. The most remarkable in terms of bioactive metabolites isolated from marine green algae is the cyclic depsipeptide kahalalide F. The source of this compound is the *Bryopsis* sp., kahalalide F, developed by the Spanish biopharmaceutical company PharmaMar, which is a novel antitumor drug currently in phase II clinical trials and causes oncosis in cancer cells by lysosomal induction and cell membrane permeabilization (Folmer et al. 2010).

Nigricanoside A and B is a new class of ether-linked glycoglycerolipids isolated from *Avrainvillea nigricans* (Portsmouth, Dominica). Nigricanoside A dimethyl ester is found to arrest human breast cancer cells in mitosis with $IC_{50} = 3$ nmol L^{-1}, stimulating polymerization of tubulin. The ability of the potent nigricanosides to promote tubulin polymerization makes these glycoglycerolipids an exciting anticancer drug lead (Williams et al. 2007). Capisterones A and B are isolated from the green algae *Penicillus capitatus* collected from Sweetings Cay, Bahamas. It enhances fluconazole activity in *Saccharomyces cerevisiae* but did not show any inherent antifungal activity when tested against several opportunistic pathogens or cytotoxicity to several cancer and noncancerous cell lines (up to 35 µM). These compounds may have a potential for combination therapy of fungal infections caused by clinically relevant azole-resistant strains (Li et al. 2006). Other metabolites isolated from green algae, like the diterpene halimedatrial from *Udotea flabellum* and *Halimeda* sp. besides the bromophenolic compound isorawsonol from *Avrainvillea rawsonii*, have demonstrated interesting anticancer potential (Folmer et al. 2010). A new sterol, 24-R-stigmasta-4, 25-diene-3β, 6β-diol, among another known compounds, is isolated from *Codium divaricatum*, a traditional Chinese medicine used, as an anticancer agent since the ancient time (He et al. 2010).

20.6.2 Bioactive Compounds from Endophytes of Green Algae

The bioactive metabolites derived from endophytes of marine green algae present bicyclical structures with some oxygenation or even aromatic moieties. Sclerosporin is a rare antifungal and sporogenic cadinane-type sesquiterpene. Sclerosporin and four new hydroxylated sclerosporin derivatives, namely, 15-hydroxysclerosporin, 12-hydroxysclerosporin, 11-hydroxysclerosporin, and 8-hydroxysclerosporin, have been derived from *Cadophora malorum* isolated from *Enteromorpha* sp. The compound 8-hydroxysclerosporin shows a weak fat-accumulation inhibitory activity against 3T3-L1 murine adipocytes (Almeida et al. 2010). The fungus *Coniothyrium cereale* is isolated from the Baltic Sea alga *Enteromorpha* sp., which produces the isoindole conioimide and the polyketide cereoanhydride. Both compounds possess unique carbon skeletons. Conioimide is an enzyme involved in many inflammatory diseases and shows selective inhibitory activity toward the protease human leukocyte elastase (HLE). It shows an inhibitory activity with an IC_{50} of 0.2 µg mL^{-1}, and hence, it is considered as a lead for anti-inflammatory drugs (Elsebai et al. 2012). The genus *Drechslera* is responsible for production of various secondary metabolites. A new naturally seco-sative-type sesquiterpene helminthosporic acid along with three known sesquiterpenes, helminthosporol, drechslerine A, and (+) secolongifolenediol, has been obtained from *Drechslera* sp., isolated from *Ulva* sp. The compounds tested toward antioxidant, antimicrobial, and antifouling effects are of without any considerable activities (Abdel-Lateff et al. 2013).

A new chromone derivative, 2-(hydroxymethyl)-8-methoxy-3-methyl-4H-chromen-4-one (chromanone A) is derived from *Penicillium* sp., isolated from *Ulva* sp. It acts as an active tumor anti-initiating via modulation of carcinogen-metabolizing enzymes and protects from DNA damage (Gamal-Eldeen et al. 2009). The strain *Monodictys putredinis* isolated from the inner tissue of a marine green alga produces four new monomeric xanthones and a benzophenone. The compounds have been examined for their cancer chemopreventive potential (Krick et al. 2007). A new pyrrolidine derivative and eight known steroids have been isolated from the cultures of *Gibberella zeae*, an endophytic fungus isolated from *Codium fragile*. These derivatives show cytotoxicity against A-549 and BEL-7402 cell lines (Liu et al. 2011). Cytoglobosins A–G and seven new cytochalasin derivatives, along with two structurally related known compounds, isochaetoglobosin D and chaetoglobosin F, have been isolated from the cultures of *Chaetomium globosum* QEN-14, an endophytic fungus derived from *Ulva pertusa*. Both cytoglobosin C and D display cytotoxic activity against the A-549 tumor cell line (Cui et al. 2010). The compound derived from ethyl acetate extract of *Aspergillus terreus* isolated from *C. decorticatum* shows excellent cytotoxicity against the HepG2 cancerous cell line. The compound is further identified as galactofuranose, an exopolysaccharide (Suja 2013).

20.7 Phaeophyta

20.7.1 Bioactive Compounds from Brown Algae

The phylum Phaeophyta is almost exclusively marine and is known for producing major metabolites derived from isoprene (McClintock and Baker 2001). Besides the complex diterpenoids, synthesized by brown algae, volatile compounds, fucoidans, phlorotannins, and fucoxanthins exhibit antioxidant, antibiotic, antifungal, antiviral, and anticancer activities (Folmer et al. 2010).

Cystoseira sp., from the Mediterranean Sea, yields two novel meroditerpenes, cystoseirone diacetate and amentol chromane diacetate. Extensive literature is available on antiherbivory of macroalgae. More specific antiherbivore diterpenoid metabolites, acutilol A, acutilol A acetate, and acutilol B have been isolated from *Dictyota acutiloba*, and it is found to be potent feeding deterrents against both temperate and tropical herbivorous fishes and sea urchins (Hardt et al. 1996). Dolabellane is isolated from *Dictyota dichotoma* collected from Spain. It exhibits mild activity in *in vitro* cytotoxicity assays against P-388 mouse lymphoma, A-549 human lung carcinoma, HT-29 human colon cancer carcinoma, and MEL-28 human melanoma tumor cell lines (Duran et al. 1997, 1999). The chemical defensive action against the sea urchin and the generalist fishes was found due to the diterpenoid 10,18-diacetoxy-8-hydroxy-2,6-dolabelladiene found as the major natural product in *Dictyota pfaffii* (Barbosa et al. 2004). Although dolabellane diterpenes exhibit a range of pharmacologically relevant activities, the metabolite has not been pursued so far for further pharmaceutical development.

Lobophorolide, a polyketide, is isolated from the brown alga *Lobophora variegata*. It shows strong antifungal activity against the marine pathogen *Lindra thalassiae*. This compound shows significant antifungal effects against the human pathogen *Candida albicans* (IC_{50} = 1.3 mg mL^{-1}) and against human colon cancer cells (IC_{50} = 0.03 mg mL^{-1}) (Kubanek et al. 2003). Six new tetraprenyltoluquinol derivatives, two new triprenyltoluquinol derivatives, and two new tetraprenyltoluquinone derivatives have been isolated from brown alga *Cystoseira crinita* together with the four known tetraprenyltoluquinol derivatives. Each compound has been evaluated for its antioxidative properties. These hydroquinones are found to have powerful antioxidant activity (Fisch et al. 2003).

Diterpene dictyotadimer A is isolated from the genus *Dictyota* (Viano et al. 2011), while antiviral diterpenes are found in *Dictyota menstrualis* (Cavalcanti et al. 2011). Ayyad et al. (2011) describe the new diterpene amijiol acetate from *Dictyota dichotoma*. It shows potent cytotoxicity against several different cell lines and also antioxidant activities. New diterpenes containing the 2,6-cyclo-xenicane skeleton are isolated from *Dilophus fasciola* and *Dilophus spiralis* (Ioannou et al. 2009). New brominated selinane and cadinane sesquiterpenes have been reported from the genus *Dictyopteris* (Ji et al. 2009; Qiao et al. 2009; Wen et al. 2009). New bisprenylatedquinol is isolated from *Sporochnus comosus* and it shows cytotoxic activity (Oveden et al. 2011). In addition, two dolastane terpenes, isolated from the Brazilian brown alga *Canistrocarpus cervicornis*, are described as promising antiviral compounds (Valliam et al. 2010).

20.7.2 Bioactive Compounds from Endophytes of Brown Algae

The endophytes associated with brown algae highlight a greater structural and bioactivity assortment: naphtho-gamma-pyrone derivatives, macrolides, isobenzofuranone derivative, bicyclic lactones, benzodiazepine, ergosterolide derivatives, etc.

Penicillium brevicompactum is isolated from the marine alga *Pterocladia* sp. and it produces 11 active compounds. Structural elucidation of the pure compounds leads to the identification of di(2-ethyl hexyl) phthalate and fungisterol or one of its isomers, respectively. The pure compounds assessed for cytotoxicity against six different types of tumor cell lines, and the fungisterol reveals anticancer activities (Mabrouk et al. 2011). *Aspergillus versicolor*, an endophytic fungus, is isolated from *Sargassum thunbergii* and the bioactive compounds obtained exhibit antibacterial activities against *E. coli* and *S. aureus*. These also showed lethality against brine shrimp (*Artemia salina*) with an LC_{50} value of 0.5 μg mL^{-1} (Miao et al. 2012).

Epicoccum sp. is isolated from *F. vesiculosus* and it produces a new antioxidant isobenzofuranone derivatives. It is found to be potently active, showing 95% DPPH radical scavenging effects at 25 μg mL^{-1}, and also inhibits the peroxidation of linolenic acid (62% inhibition at 37 μg mL^{-1}) (Abdel-Lateff et al. 2003a). Asperamides A and B, a sphingolipid, and their corresponding glycosphingolipid possess a previously unreported 9-methyl-C20-sphingosine moiety, characterized from the culture extract of *Aspergillus niger* EN-13, an endophytic fungus, isolated from *Colpomenia sinuosa*. Asperamide A displays moderate antifungal activity against *Candida albicans* (Zhang et al. 2007a). Three new naphtho-gamma-pyrones, nigerasperone A, B, and C, together with nine related known compounds have been characterized from *Aspergillus niger* EN-13. Though these compounds do not show any remarkable inhibitory effects against A-549 and SMMC-7721 tumor cell lines, some of the known compounds show weak antifungal activity against *Candida albicans* and moderate activity on DPPH scavenging (Zhang et al. 2007b). Two new phenethyl-α-pyrone derivatives, including isopyrophen and aspergillusol, have been characterized from the culture extract of *Aspergillus niger* EN-13. In addition, four known compounds, including a phenethyl-α-pyrone derivative (pyrophen and three cyclodipeptides), have also been isolated (Zhang et al. 2010).

Brassicaterol, ergosterol, ergosterol peroxide, and 7,22-ergostadiene-β,5α,6β-triol are isolated from a brown alga endophytic fungus NO.ZZF36 from the South China Sea (Yang et al. 2006a). The antimicrobial activities of lasiodiplodins and the 13-acetyl and 12,14-dibromo derivatives of lasiodiplodin have been tested for the first time (Yang et al. 2006). 7-Nor-ergosterolide, a rare 7-norsteroid with an unusual pentalactone B-ring system, is derived from the culture extract of *Aspergillus ochraceus* EN-31, an endophytic fungus isolated from *Sargassum kjellmanianum*. This naturally occurring steroid displays cytotoxicity against NCI-H460, SMMC-7721, and SW1990 cell lines with IC_{50} values of 5.0, 7.0, and 28.0 μg mL^{-1}, respectively. Another compound also presents cytotoxicity against the SMMC-7721 cell line with an IC_{50} value of 28.0 μg mL^{-1} (Cui et al. 2010). Asperolides A–C and tetranorlabdane diterpenoids are produced by *Aspergillus wentii* EN-48, the marine alga–derived endophytic fungus. These compounds have been tested for cytotoxic and antibacterial activities (Sun et al. 2012).

Guignardia sp., an endophytic fungus, is isolated from the leaves of *Undaria pinnatifida*. This fungus produces bioactive metabolites that show potent cytotoxic and antifungal activity (Wang 2012). Two new terpeptin analogs—JBIR-81 and JBIR-82—have been isolated from seaweed-derived fungus *Aspergillus* sp. (Izumikawa et al. 2010). A new chlorinated benzophenone antibiotic, pestalone, is produced by a cultured marine fungus, only when a unicellular marine bacterium, strain CNJ-328, is cocultured in the fungal fermentation. This fungus, an undescribed member of the genus *Pestalotia*, is isolated from the surface of the brown alga *Rosenvingea* sp. collected in the Bahamas Islands. Pestalone exhibits moderate *in vitro* cytotoxicity for 60 human tumor cell lines of the National Cancer Institute (mean GI(50) = 6.0 μM). More significantly, pestalone shows potent antibiotic activity against methicillin-resistant *S. aureus* (MIC = 37 ng mL^{-1}) and vancomycin-resistant *Enterococcus faecium* (MIC = 78 ng mL^{-1}), and it indicates that pestalone should be further evaluated in advanced models of infectious diseases (Cueto et al. 2001).

20.8 Rhodophyta

20.8.1 Bioactive Compounds from Red Algae

The phylum Rhodophyta comprises 98% of the marine environment and the metabolites from these algae present some interesting chemical structure of biological activities such as peptides, polyketides, indoles, terpenes, acetogenins, and phenols to volatile halogenated hydrocarbons. (Cabrita et al. 2010; Fujii et al. 2011). In addition, red algae synthesize large amounts of polysaccharides (agar and carrageenan) as cell wall constituents (Guven et al. 2010; Wijisekara et al. 2011). A wide variety of biological activities is associated with marine red algae metabolites for antibacterial, antiviral, anti-inflammatory, antioxidant, antifouling, cytotoxic, ichthyotoxic, and insecticidal properties.

Marine terpenoids are frequently found with halogenated functionalities and one or more rings, which can have important implications for their biological activities. Isoprenoid metabolites are derived via the classical mevalonate pathway or the more recently discovered deoxyxylulose phosphate pathway. In macroalgae, isoprenoids are the dominant class of secondary metabolites, representing 59% of the metabolites, isolated to date from green algae, 46% from red algae, and 68% from brown algae (Harper et al. 2001).

Laurencia, belonging to the Rhodomelaceae family, is the most studied genus of phylum Rhodophyta and has been intensively investigated. *Laurencia* sp. is having intriguing structures of labdane-based diterpenes, known to possess a board spectrum of biological activities, but so far no biological tests have been reported for these metabolites obtained from the red algae. Only four marine species are so far reported to utilize the cyclization pattern (C-6 and C-11) in secondary metabolism. Two dactylomelane metabolites are isolated from sea hares: *Aplysia dactylomela* (dactylomelol) and *Aplysia punctata* (puctatene acetate). Similar metabolites are also found from two red algal species, *Sphaerococcus coronopifolius* (sphaerolabdadiene-3,14-diol) and *Laurencia* sp. Another *Laurencia* sp. collected from Tenerife, Canary Islands, produced a number of novel and relatively unstable hydroperoxide metabolites, such as dactylohydroperoxide C. But here, also no biological activities are reported for these metabolites.

The polyketides (acetogenin) and fatty acid–based metabolites constitute the second most abundant class found in macroalgae, accounting for 19% of green algal metabolites and 38% of red algal metabolites. Novel derivatives, with mono-, di-, tri-, and tetracyclic structures possessing enzyme and bromoallene functionalities, are frequently reported. Scanlonenyne, chinzallene, okamurallene, laurendecumallenes (A and B), and laurendecumenynes (A and B) are acetogenins commonly isolated from *Laurencia* spp. Though structurally diverse, these known macroalgal acetogenins do not possess any exciting pharmacological activities. Moreover, the ecological role of the compounds has not been well documented and the large database of acetogenins, isolated from various species of *Laurencia*, is used for chemotaxonomic identification of species. Another interesting bioactive halogenated compound is the bicyclic diterpene laurenditerpenol, isolated from the lipid extract of *Laurencia intricate*. Laurenditerpenol is the first marine natural product that inhibits the hypoxia-inducible transcription factor ($IC_{50} = 0.4$ μM), which has recently emerged as a key tumor-selective molecular target for anticancer drug development (Chittiboyina et al. 2007; Jung and Im 2008; Nagle and Zhou 2009).

In *Laurencia* sp., a new brominated diterpene, 10-acetoxyangasiol, is discovered, which exhibits potent antibacterial activities against the clinical bacteria *S. aureus*, *Staphylococcus* sp., and *Vibrio cholera* (Vairappan et al. 2010). One more natural product, (5S)-5-acetoxycaespitol, is isolated among seven new halogenated metabolites from the Brazilian red algae *Laurencia catarinensis*, and it demonstrates cytotoxic activity in different tumor cell lines (Lhullier et al. 2010). From the same genus, *Laurencia*, several other compounds with fascinating structural diversity have been published: cytotoxic oxasqualenoid (Cen-Pacheco et al. 2011); neorogioltriol, a diterpenoid with *in vitro* and *in vivo* anti-inflammatory activity (Chatter et al. 2011); laurefurenynes, a new cyclic ether acetogenin (Abdel-Mageed et al. 2010); nonterpenoid C 15-acetogenins (Gutierrez-Cepeda et al. 2011); and laurenidificin, a brominated C 15-acetogenin (Liu et al. 2010).

Although there are no drugs derived from red algae, it is clear that this phylum represents a potential source of bioactive molecules that should be explored more thoroughly.

Novel bioactive metabolites such as bromophycolides (A and B), callophycoic acid C, and callophycol A have been isolated from the red alga *Callophycus serratus* collected from Fiji Islands. The most cytotoxic metabolite is bromophycolide A, with moderate *in vitro* cytotoxicity against a broad range of tumor types (mean $IC_{50} = 6.7$ mmol L^{-1}). The G1 phase of the cell cycle is arrested when human ovarian cells are exposed to metabolite, suggesting that apoptosis stemmed from cells arrested in G1. Several bromophycolides exhibit activities in the low micromolar range against the human malaria parasite *Plasmodium falciparum*. This represents the first discovery of natural product incorporating a diterpene and benzoate skeleton into a macrolide system (Kubanek et al. 2005, 2006; Lane et al. 2007, 2009).

Halogenated monoterpenes have been known from red algae for a long time, but it is in the early 1990s that a metabolite with pharmaceutical potential moved into preclinical drug development. The pentahalogenated monoterpene halomon is isolated from the extracts of the red alga *Portieria hornemannii*. This halomon is associated with a novel cytotoxicity profile against brain, renal, and colon tumor types in the primary screening. But unfortunately, development of this as an anticancer agent is limited due to its failure to exhibit significant *in vivo* effects (Fuller et al. 1992; Jha and Zi-rong 2004). Manauealides A–C, aplysiatoxin, and debromoaplysiatoxin are the only macrocyclic polyketide metabolites reported from macroalgae isolated from the Hawaiian alga *Gracilaria coronopifolia*. Both aplysiatoxin and debromoaplysiatoxin are causative agents known for food poisonings (Nagai et al. 1997). Peyssonenynes A and B are isolated from the red alga *Peyssonnelia caulifera*, collected from Yanuca Islands, Fiji, and lipid acetylenes, from the red alga *Liagora farinosa*, which is rare for macroalgae. Both show similar activity in a DNA methyltransferase (DNMT-1) enzyme inhibition assay, with an $IC_{50} = 16$ and 9 mmol L^{-1}, respectively (Paul and Fenical 1980; McPhail et al. 2004).

The alga *Delisea pulchra* is well known for producing a series of structurally related brominated furanones. These furanones act as a defense against multiple threats, such as biofouling, antimicrobial colonization, and herbivory (De Nys et al. 1993; Steinberg and

De Nys 2002). Shikimic acid is the biosynthetic precursor to an array of aromatic compounds, including benzoic and cinnamic acids. Red algae are known to be a prolific source of halogenated phenolic metabolites, derived from shikimic acid, comprising approximately 5% of known algal metabolites (Harper et al. 2001). A number of bromophenols have been isolated from the red algae. One of the metabolites, isolated from *Rhodomela confervoides* (Qingdao, China), exhibits moderate activity against *Staphylococcus epidermidis*, with an MIC of 35 mg mL^{-1} (Xu et al. 2003), whereas bromophenol from *Polysiphonia urceolata* displays significant radical scavenging activity, when compared to known antioxidants, with IC$_{50}$ values of 6–36 mmol L^{-1}. Thus, making these metabolites potential leads as antioxidant drugs (Li et al. 2007). A new bromophenol with antioxidant activity is found in *Rhodomela confervoides* (Li et al. 2011b), while lithothamnin A, a new and unique bastadin-like structure, is isolated from *Lithothamnion fragilissimum* (Wyk et al. 2011).

The genus *Chondria* is well known for the production of cyclic polysulfides and terpenoids. Novel indolic metabolites, chondriamides A and B, with antiviral, antifungal, and cytotoxic activities, are isolated from *Chondria* sp. collected from Buenos Aires, Argentina. Chondriamide A exhibits antiviral activity against HSV II with IC$_{50}$ = 1 mg mL^{-1}, while chondriamide B displays antifungal activity against *Aspergillus oryzae* and *Trichophyton mentagrophytes*. Both the metabolites present moderate cytotoxicity against KB cell lines, with IC$_{50}$ values of 0.5 and <1 mg mL^{-1}, respectively. Chondriamide C is another indolic metabolite, isolated from *Chondria atropurpurea*, which exhibits *in vitro* anthelmintic activity against *Nippostrongylus brasiliensis* with an EC$_{80}$ of 90 mmol L^{-1} (Palermo et al. 1992; Davyt et al. 1998). Almazoles (A and B) and another metabolite bearing an oxazolone ring, almazolone, are isolated from *Haraldiophyllum* (Dakar, Senegal). But no biological activity is reported so far for these metabolites (N'Diaye et al. 1994; Guella et al. 2006).

Gross et al. (2006) have isolated two novel 2,7-naphthyridine metabolites, lophocladines A and B, from *Lophocladia*, collected from Savusavu, Fiji Islands. Lophocladine B exhibits moderate activity against lung and breast cancer cells, with IC$_{50}$ values of 64.6 and 3.1 mg mL^{-1}, respectively. NCI-H460 lung cancer cells show morphologic changes when treated with this metabolite, depolymerizing 85% of the microtubules at 45 mmol L^{-1}, but the potency is rather moderate when compared to other tubulin-depolymerizing compounds.

Macroalgae are known to produce biologically active glycosides. Metabolites possessing unusual cyclopropane-containing alkyl chains gracilarioside and gracilamides are isolated from *Gracilaria asiatica*, obtained from Indonesia. This metabolites exhibit weak activity against melanoma cells, with 18.2% cell death at 20 mg mL^{-1} for gracilarioside and 11.7% cell death at 30 mg mL^{-1} for gracilamides (Sun et al. 2006). Glycosidic macrolides, polycavernoside, are isolated from the red alga *Polycavernosa tsudai* previously called *Gracilaria edulis* (Tanguisson Beach, Guam). Human intoxication resulted after ingestion of this red alga, and polycavernoside A has been reported as the illness-causing agent (Yotsu-Yamashita et al. 1993, 2007).

20.8.2 Bioactive Compounds from Endophytes of Red Algae

The metabolites from the endophytes of red algae present some fascinating chemical structures of biological activities such as macrolides, several classes of terpenes and indoloterpenes, sesquiterpenoids, tetracyclic diterpenes and monoterpene, polyoxygenated compounds, steroids with tetrahydroxy and C-16-acetoxy groups, polyketides, aromatic pentaketides, and benzaldehyde derivatives.

Indoloditerpenes are known as potent tremorgenic mycotoxins, which arise tremor in animals by affecting neurotransmission. Two new indoloditerpene derivatives asporyzin A and asporyzin B, one new indoloditerpene asporyzin C, and three known related indoloditerpenes JBIR-03, emindole SB, and emeniveol, have been purified from an endophytic fungus *Aspergillus oryzae*, isolated from the *Heterosiphonia japonica*. These compounds are evaluated preliminarily for insecticidal and antimicrobial activities in order to probe into their chemical defensive function. Indoloditerpene JBIR-03 is more active against brine shrimp (*A. salina*), and indoloditerpene asporyzin C possesses potent activity against *E. coli* (Qiao et al. 2010b). Endophytic fungus *Exophiala oligosperma* (EN-21) is isolated from the inner tissue of *Laurencia similis*. It produces seven compounds (1–7). The 2-phenoxynaphthalene compound is a new natural product from this fungus. This type of the structure is used widely as a substrate in the synthesis of diaryl ethers (Thornton et al. 1989; Li et al. 2011a). *Xylaria psidii* (KT30) and *Mycelium sterillium* (KT31), isolated from red algae, exhibit considerable antibacterial and cytotoxic activities, but only a minor antifungal activity. In some marine fungal strains, freshwater-based medium can be used to enhance the production of bioactive secondary metabolites. Depending on the strains, there are differences in the strength of the bioactivity between freshwater and seawater cultures. A 2-carboxy-8-methoxy-naphthalene-1-ol is isolated as a new natural compound but is found to be inactive in antibacterial test probably due to its high polarity (Tarman et al. 2011).

Penicillium sp. (strain CNL-338) is isolated from *Laurencia* sp. collected in the Bahamas Islands. A novel lumazine peptide, penilumamide, is isolated from the fermentation broth of a marine-derived fungal strain. This has attracted attention because the culture extract shows low micromolar cancer cell cytotoxicity (Meyer et al. 2010). The anthelmintic amino acid kainic acid (KA) is a neuroactive compound, isolated from *Digenea simplex* (Hopkins et al. 2000; Sakai et al. 2005). The structure of KA is strictly related to other neuronal agonist amino acids, such as domoic acid, isolated from the red alga *Chondria armata*, and is also an anthelmintic compound (Sakai et al. 2005). Now, kainoids are used in neurobiological research as a standard reagent, playing an important role in studies of neurophysiological disorders such as Alzheimer, Parkinson, and epilepsy (Smit 2004).

Penicillium chrysogenum QEN-24S is isolated from an unidentified species of the genus *Laurencia*. Penicimonoterpene is a natural inhibitor of pathogenic fungus *Alternaria brassicae* and penicitide A displays moderate activity against *A. brassicae*, and in addition to that, it also exhibits moderate selective cytotoxic activity against HepG2 tumor cell line (Gao et al. 2011a).

Alternaria alternata, a marine endophytic fungus, is derived from an unidentified species of the genus *Laurencia*. It produces two new perylene derivatives, 7-epi-8-hydroxyaltertoxin I and 6-epi-stemphytriol, along with two known compounds, stemphyperylenol and altertoxin I (Gao et al. 2009). Chaetopyranin, a benzaldehyde derivative, exhibits moderate to weak cytotoxic activity toward several tumor cell lines. This compound is isolated from fungus derived from the alga *Polysiphonia urceolata* (Wang et al. 2006). Eleven new botryane metabolites and four known cytochalasins have been derived from the mitosporic fungus *Geniculosporium* sp., isolated from *Polysiphonia* sp. These new natural products show herbicidal, antifungal, and antibacterial activities (Krohn et al. 2005).

P. chrysogenum QEN-24S, an endophytic fungus, is isolated from the genus *Laurencia*. It produces two new polyoxygenated steroids, namely, penicisteroids A and B. In addition to that, seven known steroids have been identified. Penicisteroid A is structurally unique steroid having tetrahydroxy and C-16-acetoxy groups, and it displays potent antifungal and cytotoxic activity in the preliminary bioassays (Gao et al. 2011b). A new steroid with an E-double bond between C-17 and C-20 named as asporyergosterol is identified from the culture extracts of *Aspergillus oryzae*, an endophytic fungus, isolated from *H. japonica*. Along with these, four known steroids have also been identified. These isolates exhibit low activity to modulate acetylcholinesterase (AChE) (Qiao et al. 2010b).

A new oxylipin, a new steroid, 3β,4α-dihydroxy-26-methoxyergosta-7,24(28)-dien-6-one, and four known steroids, episterol, (22E,24R)-ergosta-7,22-dien-3β,5α,6α-triol, (22E,24R)-ergosta-5,22-dien-3β-ol, and (22E,24R)-ergosta-4,6,8(14),22-tetraen-3-one, have been isolated from the cultures of *Aspergillus flavus*. This endophytic fungus is isolated from *Corallina officinalis*, which exhibits low activity to inhibit AChe and no activity against plant pathogenic fungi *Colletotrichum lagenarium* and *Fusarium oxysporum* (Qiao et al. 2011). *Phaeosphaeria spartinae* is an endophyte of the marine alga *Ceramium* sp. Spartinoxide is isolated, which is the enantiomer of the known compound A82775C. Additionally, the known metabolites 4-hydroxy-3-prenyl-benzoic acid and anofinic acid have been obtained. These compounds assayed against the enzymes HLE, trypsin, AChE, and cholesterol esterase. Spartinoxide and anofinic acid show potent inhibition of HLE with IC_{50} values of 1.7 ± 0.30 μg mL^{-1} (6.5 μM) and 1.67 ± 0.32 μg mL^{-1} (8.1 μM), respectively (Elsebai et al. 2010).

20.9 Bioactive Compounds from Unidentified Algae and Its Endophytes

Xanthones, hydroquinone derivatives, prenylated polyketide, and benzophenone derivatives are some of the metabolites from unidentified marine alga endophytes.

Xanthones are described as "privileged structures," because this large group of natural products reveals a broad spectrum of pharmacological activities. They influence several inflammatory mediators such as the arachidonic acid cascade, enzymes such as various kinases and proteases, and receptors such as monoamine oxidase B. They also exhibit antimicrobial activity against a large number of human pathogenic organisms. The uncommon fungus *Monodictys putredinis* is isolated from an unidentified green alga collected in Tenerife, Spain. It produces two novel dimeric chromanones that consist of two uniquely modified xanthone-derived units. Both monodictyochromes A and B exhibit cancer chemopreventive potential and inhibit cytochrome P450 enzymes such as CYP1A activity (involved in carcinogen activation) with IC_{50} values of 5.3 and 7.5 μM, respectively. In addition, both compounds moderately induce quinone reductase (QR) activity as an indication of enhanced carcinogen detoxification (Pontius et al. 2008b).

An algicolous marine fungus *Wardomyces anomalus* is found to produce two new xanthone derivatives, 2,3,6,8-tetrahydroxy-1-methylxanthone and 2,3,4,6,8-pentahydroxy-1-methylxanthone. In addition to that, known xanthone derivative 3,6,8-trihydroxy-1-methylxanthone and the known fungal metabolite 5-(hydroxymethyl)-2-furanocarboxylic acid are inhibitors of p56 (lck) tyrosine kinase (Abdel-Lateff et al. 2003). Synthesis of antioxidants may be a strategy among endophytes to counter the defense reactions of their algal hosts (Dring 2005).

New antioxidant hydroquinone derivatives obtained from a marine fungal isolate, identified as *Acremonium* sp., are 7-isopropenylbicyclo[4.2.0]octa-1,3,5-triene-2,5-diol and 7-isopropenylbicyclo[4.2.0]octa-1,3,5-triene-2,5-diol-5-beta-D-glucopyranoside. Both the compounds possess the most unusual ring system showing significant antioxidant properties (Abdel-Lateff et al. 2002).

Nodulisporium sp. is an endophyte of a Mediterranean alga and produces the novel polyketide noduliprevenone. It consists of two uniquely modified xanthone-derived units. This compound is identified as a competitive inhibitor of cytochrome P450 1A activity with an IC_{50} value of 6.5 ± 1.6 mm and induces at the same time twofold NAD(P) H:QR activity in Hepa 1c1c7 mouse culture cells with a concentration of 5.3 ± 1.1 μm (Pontius et al. 2008a).

Talaroflavone and 1-deoxyrubralactone are natural compounds, isolated from fungal strain derived from unidentified sea algae. These compounds selectively inhibit the activities of families X and Y of eukaryotic DNA polymerases (pols). This is the first report about the selective inhibitors of families X and Y of eukaryotic pols (Naganuma et al. 2008). A novel chiral dipyrrolobenzoquinone derivative, terreusinone, is isolated as a potent UV-A protectant from the marine algicolous fungus *Aspergillus terreus*. The compound exhibits a UV-A absorbing activity with ED_{50} value of 70 μg mL^{-1} (Lee et al. 2003). Another novel benzonaphthyridinedione derivative, chaetominedione, is isolated from marine fungal isolate *Chaetomium* sp. The bioactive compounds 2-furancarboxylic acid and 5-(hydroxymethyl)-2-furancarboxylic acid are the known fungal metabolites. The total extract and chaetominedione has significant inhibitory activity toward p56^{lck} tyrosine kinase (18.7% and 93.6% enzyme inhibition at 200 μg mL^{-1}, respectively) (Abdel-Lateff 2008). Fungal isolate F01V25 is obtained from the alga *Dictyospaeria versluyii*, and from the fungal isolates, two new polyketides, dictyosphaeric acids A and B, along with the known anthraquinone carviolin, have been derived (Bugni et al. 2004).

20.10 Summary

Scientific reports on new bioactive compounds from marine macroalgae have steadily increased since the 1980s (Stout and Kubanek 2010), and at the same time from microorganisms, especially endophytic fungi and bacteria, an increased number of reviews, patents, and original research articles are published every year in the modern field of drug discovery (Tejesvi and Pirttila 2011). Extensive data are available on new marine compounds, yet to be processed. Therefore, advances in microbial cell fermentation technology and metagenomic approach should be a valuable alternative (Schloss and Handelsman 2005; Monciardini et al. 2014) to harness the biotechnological potential of the marine organisms. The other way is the use of metabolomics by way of nuclear magnetic resonance and liquid chromatography tandem mass spectrometry that can help to resolve extracts with the same or similar compounds (Wishart 2008; Berrue et al. 2011; Hou et al. 2012), facilitating rapid progress in research of secondary metabolites.

It is obvious from the opinions of many diverse leaders in the field of natural compounds (Glaser and Mayer 2009) that the need for sources of diverse and pharmacologically active leads is growing constantly and the natural products from marine still maintain the potential to provide the diversity. Hence, identification of biologically active compounds should lead to the development of potential therapeutic drugs against human disease.

References

Abdel-Lateff, A. 2008. Chaetominedione, a new tyrosine kinase inhibitor isolated from the algicolous marine fungus *Chaetomium* sp. *Tetrahedron Lett.* 49: 6398–6400.

Abdel-Lateff, A., K. M. Fisch, A. D. Wright, and G. M. Konig. 2003a. A new antioxidant isobenzofuranone derivative from the algicolous marine fungus *Epicoccum* sp. *Planta Med.* 69: 831–834.

Abdel-Lateff, A., C. Klemke, G. M. Konig, and A. D. Wright. 2003b. Two new xanthone derivatives from the algicolous marine fungus *Wardomyces anomalus*. *J. Nat. Prod.* 66(5): 706–708.

Abdel-Lateff, A., G. M. Konig, K. M. Fisch, U. Holler, P. G. Jones, and A. D. Wright. 2002. New antioxidant hydroquinone derivatives from the algicolous marine fungus *Acremonium* sp. *J. Nat. Prod.* 65: 1605–1611.

Abdel-Lateff, A., T. Okino, W. M. Alarif, and S. S. Al-Lihaibi. 2013. Sesquiterpenes from the marine algicolous fungus *Drechslera* sp. *J. Saudi Chem. Soc.* 17: 161–165.

Abdel-Mageed, W. M., R. Ebel, F. A. Valeriote, and M. J. Jaspars. 2010. Laurefurenynes A-F, new cyclic ether acetogenins from a marine red alga, *Laurencia* sp. *Tetrahedron* 66: 2855–2862.

Almeida, C., E. Eguereva, S. Kehraus, C. Siering, and G. M. Konig. 2010. Hydroxylated sclerosporin derivatives from the marine-derived fungus *Cadophora malorum*. *J. Nat. Prod.* 73: 476–478.

Arnold, A. E. and F. Lutzoni. 2007. Diversity and host range of foliar fungal endophytes: Are tropical leaves biodiversity hotspots? *Ecology* 88: 541–549.

Arnold, A. E., L. C. Mejia, D. Kyllo et al. 2003. Fungal endophytes limit pathogen damage in a tropical tree. *Proc. Natl. Acad. Sci. U.S.A.* 100: 15649–15654.

Ayyad, S. N., M. S. Makki, N. S. Al-kayal et al. 2011. Cytotoxic and protective DNA damage of three new diterpenoids from the brown alga *Dictyota dichotoma*. *Eur. J. Med. Chem.* 46: 165–172.

Bacon, C. W. and J. F. J. White. 2000. Physiological adaptations in the evolution of endophytism in the Clavicipitaceae. In C. W. Bacon and J. F. J. White (eds.), *Microbial Endophytes*. New York: Marcel Dekker Inc., pp. 237–263.

Barbosa, J. P., L. V. L. Teixeira, and R. C. Pereira. 2004. A dolabellane diterpene from the brown alga *Dictyota pfaffii* as chemical defense against herbivores. *Bot. Mar.* 47: 147–151.

Belofsky, G. N., P. R. Jensen, M. K. Renner, and W. Fenical. 1998. New cytotoxic sesquiterpenoid nitrobenzoyl esters from a marine isolate of the fungus *Aspergillus versicolor*. *Tetrahedron* 54: 1715–1724.

Berdy, J. 2005. Bioactive microbial metabolites. *J. Antibiot.* 58(1): 1–6.

Berrue, F., S. T. Withers, B. Haltli, J. Withers, and R. G. Kerr. 2011. Chemical screening method for the rapid identification of microbial sources of marine invertebrate associated metabolites. *Mar. Drugs* 9: 369–381.

Bewley, C. A. and D. J. Faulkner. 1998. Lithistid sponges: Star performers or hosts to the stars. *Angew. Chem. Int. Ed.* 37: 2162–2178.

Bhattacharyya, P. N. and D. K. Jha. 2011. Optimization of cultural conditions affecting growth and improved bioactive metabolite production by a subsurface *Aspergillus* strain TSF 146. *Int. J. Appl. Biol. Pharm. Technol.* 2(4): 133–143.

Blunt, J. W., B. R. Copp, R. A. Keyzers, M. H. G. Munro, and M. R. Prinsep. 2013. Marine natural products. *Nat. Prod. Rep.* 30: 237–323.

Bode, H. B., B. Bethe, R. Heofs, and A. Zeeck. 2002. Big effects from small changes: Possible ways to explore nature's chemical diversity. *Chembiochem.* 3: 619–627.

Brakhage, A. A. 2013. Regulation of fungal secondary metabolism. *Nat. Rev. Microbiol.* 11: 21–32.

Brakhage, A. A. and V. Schroeckh. 2011. Fungal secondary metabolites—Strategies to activate silent gene clusters. *Fungal Genet. Biol.* 48: 15–22.

Briceno, E. X. and B. A. Latorre. 2008. Characterization of *Cladosporium* rot in grapevines, a problem of growing importance in Chile. *Plant Dis.* 92: 1635–1642.

Bugni, T. S. and C. M. Ireland. 2004. Marine-derived fungi: A chemically and biologically diverse group of microorganisms. *Nat. Prod. Rep.* 21: 143–163.

Bugni, T. S., J. E. Janso, R. T. Williamson et al. 2004. Dictyosphaeric acids A and B: New decalactones from an undescribed *Penicillium* sp. obtained from the alga *Dictyosphaeria versluyii*. *J. Nat. Prod.* 67: 1396–1399.

Cabrita, M. T., C. Vale, and A. P. Rauter. 2010. Halogenated compounds from marine algae. *Mar. Drugs* 8: 2301–2317.

Calvo Ana, M., R. A. Wilson, J. Woo Bok, and N. P. Keller. 2002. Relationship between secondary metabolism and fungal development. *Microbiol. Mol. Biol. Rev.* 66(3): 447–459.

Carroll, G. C. 1988. Fungal endophytes in stem and leaves: From latent pathogen to mutualistic symbiont. *Ecology* 69: 2–9.

Cavalcanti, D. N., M. R. Oliveira, J. C. De-Paula et al. 2011. Variability of a diterpene with potential anti-HIV activity isolated from the Brazilian brown alga *Dictyota menstrualis. J. Appl. Phycol.* 23: 873–876.

Cen-Pacheco, F., J. A. Villa-Pulgarin, F. Mollinedo, M. Norte, A. H. Daranas, and J. J. Fernandez. 2011. Cytotoxic oxasqualenoids from the red alga *Laurencia viridis. Eur. J. Med. Chem.* 46: 3302–3308.

Chatter, R., R. B. Othman, S. Rabhi et al. 2011. *In vivo* and *in vitro* anti-inflammatory activity of neoogioltriol, a new diterpene extracted from the red algae *Laurencia glandulifera. Mar. Drugs* 9: 1293–1306.

Chittiboyina, A. G., M. K. Gundluru, and P. Carvalho. 2007. Total synthesis and absolute configuration of laurenditerpenol: A HIF-1 activation inhibitor. *J. Med. Chem.* 50: 6299–6302.

Christian, O. E., J. Compton, K. R. Christian, S. L. Mooberry, F. A. Valeriote, and P. Crews. 2005. Using Jasplakinolide to turn on pathways that enable the isolation of new chaetoglobosins from *Phomospis asparagi. J. Nat. Prod.* 68: 1592–1597.

Cooley, D. R., R. F. Mullins, P. M. Bradley, and R. T. Wilce. 2011. Culturing the upper littoral zone marine alga *Pseudendoclonium submarinum* induces pathogenic interaction with the fungus *Cladosporium cladosporioides. Phycologia* 50: 541–547.

Cueto, M., P. R. Jensen, C. Kauffman, W. Fenical, E. Lobkovsky, and J. Clardy. 2001. Pestalone, a new antibiotic produced by a marine fungus in response to bacterial challenge. *J. Nat. Prod.* 64: 1444–1446.

Cui, C. M., X. M. Li, L. Meng, C. S. Li, C. G. Huang, and B. G. Wang. 2010. 7-Nor-ergosterolide, a pentalactone-containing norsteriod and related steroids from the marine derived endophytic *Aspergillus ochraceus EN-31. J. Nat. Prod.* 73: 1780–1784.

Davyt, D., W. Entz, R. Fernandez, R. Mariezcurrena, A. W. Mombru, J. Saldana, L. Dominguez, J. Coll, and E. Manta. 1998. A new indole derivative from the red alga *Chondria atropurpurea* isolation, structure determination and anthelmintic activity1. *J. Nat. Prod.* 61(12): 1560–1563.

De Nys, R., A. Wright, G. M. Konig, and O. Sticher. 1993. New halogenated furanones from the marine alga *Delisea pulchra* (cf. *fimbriata*). *Tetrahedron* 49(48): 11213–11220.

Ding, L., S. Qin, F. C. Li, X. Y. Chi, and H. Laatsch. 2008. Isolation, antimicrobial activity, and metabolites of fungus *Cladosporium* sp. associated with red alga *Porphyra yezoensis. Curr. Microbiol.* 56: 229–235.

Donadio, S., P. Monciardini, R. Aluina, P. Mazza, C. Chiocchini, L. Cavaletti, M. Sosio, and A. Maria Puglia. 2002. Microbial technologies for the discovery of novel bioactive metabolites. *J. Biotechnol.* 99: 187–198.

Dring, M. J. 2005. Stress resistance and disease resistance in seaweeds: The role of reactive oxygen metabolism. *Adv. Bot. Res.* 43: 175–207.

Duran, R., L. Garrido, M. J. Ortega, A. Rueda, J. Salva, and E. Zubia. 1999. Bioactive natural compounds of algae and invertebrates from the littoral of Cadiz. *Bol. Inst. Esp. Oceanogr.* 15(1–4): 357–361.

Duran, R., E. Zubia, M. J. Ortega, and J. Salva. 1997. New diterpenoids from the alga *Dictyota dichotoma. Tetrahedron* 53: 8675–8688.

Elsebai, M. F., S. Kehraus, M. Gutschow, and G. M. Konig. 2010. Spartinoxide, a new enantiomer of A82775C with inhibitory activity toward HLE from the marine-derived fungus *Phaeosphaeria spartinae. Nat. Prod. Commun.* 5(7): 1071–1076.

Elsebai, M. F., M. Nazir, L. Marcourt et al. 2012. Novel polyketide skeletons from the marine alga-derived fungus *Coniothyrium cereale. Eur. J. Org. Chem.* 31: 6197–6203.

Fisch, K. M., V. Bohm, A. D. Wright, and G. M. Konig. 2003. Antioxidant meroterpenoids from the brown alga *Cystoseira crinita. J. Nat. Prod.* 66(7): 968–975.

Flewelling, A. J., K. T. Ellsworth, J. Sanford, E. Forward, J. A. Johnson, and C. A. Gray. 2013a. Macroalgal endophytes from the Atlantic coast of Canada: A potential source of antibiotic natural products? *Microorganisms* 1: 175–187.

Flewelling, A. J., J. A. Johnson, and C. A. Gray. 2013b. Isolation and bioassay screening of fungal endophytes from North Atlantic marine macroalgae. *Bot. Mar.* 56: 287–297.

Folmer, F., M. Jaspars, M. Dicato, and M. Diederich. 2010. Photosynthetic marine organisms as a source of anticancer compounds. *Phytochem. Rev.* 9: 557–579.

Fujii, M. T., M. Cassano, E. M. Stein, and L. R. Carvalho. 2011. Overview of the taxonomy and of the major secondary metabolites and their biological activities related to human health of the *Laurencia* complex (Ceramiales, Rhodophyta) from Brazil. *Rev. Bras. Farmacogn.* 21: 268–282.

Fuller, R. W., J. H. Cardellina, Y. Kato et al. 1992. A pentahalogenated monoterpene from the red alga *Portieria hornemannii* produces a novel cytotoxicity profile against a diverse panel of human tumor cell lines. *J. Med. Chem.* 35: 3007–3011.

Gamal-Eldeen, A. M., A. Abdel-Lateff, and T. Okino. 2009. Modulation of carcinogen metabolizing enzymes by chromanone A; a new chromone derivative from algicolous marine fungus *Penicillium* sp. *Environ. Toxicol. Pharmacol.* 28: 317–322.

Gao, S. S., X. M. Li, F. Y. Du, C. S. Li, P. Proksch, and B. G. Wang. 2011a. Secondary metabolites from a marine-derived endophytic fungus *Penicillium chrysogenum* QEN-24S. *Mar. Drugs* 9: 59–70.

Gao, S. S., X. M. Li, C. S. Li, P. Proksch, and B. G. Wang. 2011b. Penicisteroids A and B, antifungal and cytotoxic polyoxygenated steroids from the marine alga-derived endophytic fungus *Penicillium chrysogenum* QEN-24S. *Bioorg. Med. Chem. Lett.* 21: 2894–2897.

Gao, S. S., X. M. Li, and B. G. Wang. 2009. Perylene derivatives produced by *Alternaria alternate*, an endophytic fungus isolated from *Laurencia* species. *Nat. Prod. Commun.* 4: 1477–1480.

Glaser, K. B. and A. M. S. Mayer. 2009. A renaissance in marine pharmacology: From preclinical curiosity to clinical reality. *Biochem. Pharmacol.* 78: 440–448.

Gogoi, D. K., H. P. Deka Boruah, R. Saikia, and T. C. Bora. 2008. Optimization of process parameters for improved production of bioactive metabolite by a novel endophytic fungus *Fusarium* sp. DF2 isolated from *Taxus wallichiana* of North East India. *World J. Microbiol. Biotechnol.* 24(1): 79–87.

Gross, H., D. E. Goeger, P. Hills et al. 2006. Lophocladines, bioactive alkaloids from the red alga *Lophocladia* sp. *J. Nat. Prod.* 69(4): 640–644.

Guella, G., I. N'Diaye, M. Fofana, and I. Mancini. 2006. Isolation synthesis and photochemical properties of almazolone a new indole alkaloids from a red alga of Senegal. *Tetrahedron* 62(6): 1165–1170.

Gunatilaka, A. A. L. 2006. Natural products from plant-associated microorganisms: Distribution, structural diversity, bioactivity, and implications of their occurrence. *J. Nat. Prod.* 69: 509–526.

Gunde-Cimerman, N., J. Ramos, and A. Plemenitas. 2009. Halotolerant and halophilic fungi. *Mycol. Res.* 113: 1231–1241.

Gutierrez-Cepeda, A., J. J. Fernandez, L. V. Gil, M. Opez-Rodriguez, M. Norte, and M. L. Souto. 2011. Nonterpenoid C15 acetogenins from *Laurencia marilzae*. *J. Nat. Prod.* 74: 441–448.

Guven, K. C., A. Percot, and E. Sezik. 2010. Alkaloids in marine algae. *Mar. Drugs* 8: 269–284.

Hardt, I. H., W. Fenical, G. Cronin, and M. E. Hay. 1996. Acutilols, potent herbivore feeding deterrents from the tropical brown alga, *Dictyota acutiloba*. *Phytochemistry* 43: 71–73.

Harper, M. K., T. S. Bugni, B. R. Copp et al. 2001. Introduction to the chemical ecology of marine natural products. In J. B. McClintock and B. J. Basker (eds.), *Marine Chemical Ecology*, Marine Science Series. Washington, DC: CRC Press, pp. 8–10.

Harvey, J. B. J. 2002. Intraspecific variation in *H. irritans*, a fungal endosymbiont of marine brown algae of the North American Pacific. *J. Phycol.* 38: 16.

Harvey, J. B. J. and L. J. Goff. 2010. Genetic co-variation of the marine fungal symbiont *Haloguignardia irritans* (Ascomycota, Pezizomycotina) with its algal hosts *Cystoseira* and *Halidrys* (Phaeophyceae, Fucales) along the west coast of North America. *Fungal Biol.* 114: 82–95.

He, Z., A. Zhang, L. Ding, X. Lei, J. Sun, and L. Zhang. 2010. Chemical composition of the green alga *Codium divaricatum*, Holmes. *Fitoterapia* 81: 1125–1128.

Hicks, J., J. H. Yu, N. Keller, and T. H. Adams. 1997. *Aspergillus* sporulation and mycotoxin production both require inactivation of the FadA G alpha protein-dependent signaling pathway. *EMBO J.* 16: 4916–4923.

Hopkins, K. J., G. J. Wang, and L. C. Schmued. 2000. Temporal progression of kainic acid induced neural and myelin degeneration in the rat forebrain. *Brain Res.* 864: 69–80.

Hou, Y., D. R. Braub, C. R. Michel et al. 2012. Microbial strain prioritization using metabolomics tools for the discovery of natural products. *Anal. Chem.* 84: 4277–4283.

Ioannou, E., M. Zervou, A. Ismail, L. Ktari, C. Vagias, and V. Roussis. 2009. 2, 6-Cyclo-xenicanes from the brown algae *Dilophus fasciola* and *Dilophus spiralis*. *Tetrahedron* 65: 10565–10572.

Izumikawa, I., J. Hashimoto, M. Takagi, and K. Shin-ya. 2010. Isolation of two new terpeptin analogs-JBIR-81 and JBIR-82-from a seaweed-derived fungus, *Aspergillus* sp. *J. Antibiot.* 63: 389–391.

Jain, P. and R. K. Pundir. 2011. Effect of fermentation medium, pH and temperature variations on antibacterial soil fungal metabolite production. *J. Agric. Technol.* 7(2): 247–269.

Jha, R. K. and X. Zi-rong. 2004. Biomedical compounds from marine organisms. *Mar. Drugs* 2: 123–146.

Ji, N. Y., W. Wen, X. M. Li, Q. Z. Xue, H. L. Xiao, and B. G. Wang. 2009. Brominated selinane sesquiterpenes from the marine brown alga *Dictyopteris divaricata*. *Mar. Drugs* 7: 355–360.

Jung, M. E. and G. Y. J. Im. 2008. Convergent total synthesis of the racemic HIF-1 inhibitor laurenditerpenol. *Tetrahedron Lett.* 49: 4962–4964.

Jung, V. and G. Pohnert. 2001. Rapid wound-activated transformation of the green algal defensive metabolite caulerpenyne. *Tetrahedron* 57(33): 7169–7172.

Keonig, G. M., S. Kehraus, S. F. Seibert, A. Abdel-Lateff, and D. Muller. 2006. Natural products from marine organisms and their associated microbes. *Chembiochem.* 7: 229–238.

Kogel, K. H., P. Franken, and R. Huckelhoven. 2006. Endophyte or parasite—What decides? *Curr. Opin. Plant Biol.* 9: 358–363.

Krick, A., S. Kehraus, C. Gerhauser et al. 2007. Potential cancer chemopreventive *in vitro* activities of monomeric xanthone derivatives from the marine algicolous fungus *Monodictys putredinis*. *J. Nat. Prod.* 70: 353–360.

Krohn, K. I., J. Dai, U. Florke, H. J. Aust, S. Drager, and B. Schulz. 2005. Botryane metabolites from the fungus *Geniculosporium* sp. isolated from the marine red alga *Polysiphonia*. *J. Nat. Prod.* 68(3): 400–405.

Kubanek, J., C. P. Anne, W. S. Terry et al. 2006. Bromophycolides C-I from the fijian red alga *Callophycus serratus*. *J. Nat. Prod.* 69(5): 731–735.

Kubanek, J., P. R. Jensen, P. A. Keifer, M. Cameron Sullards, D. O. Collins, and W. Fenical. 2003. Seaweed resistance to microbial attack: A targeted chemical defense against marine fungi. *Proc. Natl. Acad. Sci. U.S.A.* 100(12): 6916–6921.

Kubanek, J., A. C. Prusak, T. W. Snell et al. 2005. Antineoplastic diterpene-benzoate macrolides from the fijian red alga *Callophycus serratus*. *Org. Lett.* 7(23): 5261–5264.

Kumaresan, V. and T. S. Suryanarayanan. 2002. Endophyte assemblages in young, mature and senescent leaves of *Rhizophora apiculata*: Evidence for the role of endophytes in mangrove litter degradation. *Fungal Divers.* 9: 81–91.

Lachnit, T., M. Fischer, S. Keunzel, J. F. Baines, and T. Harder. 2013. Compounds associated with algal surfaces mediate epiphytic colonization of the marine macroalga *Fucus vesiculosus*. *FEMS Microbiol. Ecol.* 84: 411–420.

Lam, C., A. Stang, and T. Harder. 2008. Planktonic bacteria and fungi are selectively eliminated by exposure to marine macroalga in close proximity. *FEMS Microbiol. Ecol.* 63: 283–291.

Lane, A. L., P. S. Elizabeth, E. H. Mark et al. 2007. Callophycoic acids and callophycols from the fijian red alga *Callophycus serratus*. *Org. Chem.* 72(19): 7343–7351.

Lane, A. L., E. P. Stout, L. An-Shen et al. 2009. Antimalarial bromophycolides J-Q from the Fijian red alga *Callophycus serratus*. *J. Org. Chem.* 74(7): 2736–2742.

Lee, S. M., X. F. Li, H. Jiang et al. 2003. Terreusinone, a novel UV-A protecting dipyrroloquinone from the marine algicolous fungus *Aspergillus terreus*. *Tetrahedron Lett.* 44: 7707–7710.

Lhullier, C., M. Falkenberg, E. Ioannou et al. 2010. Cytotoxic halogenated metabolites from the Brazilian red algae *Laurencia catarinensis*. *J. Nat. Prod.* 73: 27–32.

Li, F., K. Li, X. Li, and B. Wang. 2011a. Chemical constituents of marine agal-derived endophytic fungus *Exophiala oligosperma* EN-21. *Chin. J. Oceanol. Limnol.* 29: 63–67.

Li, K., X. M. Li, J. B. Gloer, and B. G. Wang. 2011b. Isolation characterization and antioxidant activity of bromophenols of the marine red alga *Rhodomela confervoides*. *J. Agric. Food. Chem.* 59: 9916–9921.

Li, K., X. M. Li, N. Y. Ji, and B. G. Wzng. 2007. Bromophenols from the marine red alga *Polysiphonia urceolata* with DPPH radical scavenging activity. *J. Nat. Prod.* 71(1): 28–30.

Li, X. C., M. R. Jacob, Y. Ding et al. 2006. Capisterones A and B from the marine green alga *Penicillus capitatus* that enhance fluconazole activity in *Saccharomyces cerevisiae*. *J. Nat. Prod.* 69(4): 542–546.

Liu, X., X. M. Li, C. S. Li, N. Y. Ji, and B. G. Wang. 2010. Laurenidificin, a new brominated C 15-acetogenin from the marine red alga *Laurencia nidifica*. *Chin. Chem. Lett.* 21: 1213–1215.

Liu, X. H., X. Z. Tang, F. P. Miao, and N. Y. Ji. 2011. A new pyrrolidine derivative and steroids from an algicolous *Gibberella zeae* strain. *Nat. Prod. Commun.* 6(9): 1243–1246.

Llorens, A., R. Matco, M. J. Hinojo, A. Logrieco, and M. Jimenez. 2004. Influence of the interactions among ecological variables in the characterization of a Zearalenone producing isolates of *Fusarium* spp. *Syst. Appl. Microbiol.* 27: 253–260.

Mabrouk, A. M., Z. H. Kheiralla, E. R. Hamed, A. A. Youssry, and A. A. Abd El Aty. 2011. Physiological studies on some biologically active secondary metabolites from marine-derived fungus *Penicillium brevicompactum*. 1(1): 1–15. www.Gate2Biotech.com.

Mann, K. L. 1973. Seaweeds: Their productivity and strategy for growth. *Science* 182: 975–981.

McClintock, J. B. and B. J. Baker. 2001. *Marine Chemical Ecology*. New York: CRC Press.

McPhail, K. L., D. France, S. Cornell-Kennon, and W. H. Gerwick. 2004. Peyssonenynes A and B novel Enediyne oxylipins with DNA methyl transferase inhibitory activity from the red marine alga *Peyssonnelia caulifera*. *J. Nat. Prod.* 67(6): 1010–1013.

Meyer, S. W., F. M. Thorsten, L. Choonghwan, R. J. Paul, W. Fenical, and K. Matthias. 2010. Penilumamide, a novel lumazine peptide isolated from the marine-derived fungus, *Penicillium* sp. CNL-338. *Org. Biomol. Chem.* 8(9): 2158–2163.

Miao, F. P., L. Xiao-Dong, L. Xiang-Hong, R. H. Cichewicz, and J. Nai-Yun. 2012. Secondary metabolites from an algicolous *Aspergillus versicolor* strain. *Mar. Drugs* 10: 131–139.

Miao, L., F. N. Theresa, and K. P. Y. Qian. 2006. Effect of culture conditions on mycelial growth, antibacterial activity, and metabolite profiles of the marine derived fungus *Arthrinium* c.f. *saccharicola*. *Appl. Microbiol. Biotechnol.* 72: 1063–1073.

Miller, J. D. and N. J. Whitney. 1981. Fungi from the Bay of Fundy: Observations on fungi from living and cast seaweeds. *Bot. Mar.* 24: 405–411.

Mohamed, D. J. and J. B. H. Martiny. 2010. Patterns of fungal diversity and composition along a salinity gradient. *ISME J.* 5(3): 379–388.

Monciardini, P., M. Ioria, S. Maffioli, M. Sosio, and S. Donadio. 2014. Discovering new bioactive molecules from microbial sources. *Microb. Biotechnol.* 7(3): 209–220.

Nagai, H. I., T. Yasumoto, and Y. M. Hokama. 1997. Some of the causative agents of a red alga *Gracilaria coronopifolia* poisoning in Hawaii. *J. Nat. Prod.* 60(9): 925–928.

Naganuma, M., M. Nishida, K. Kuramochi, F. Sugawara, H. Yoshida, and Y. Mizushina. 2008. 1-Deoxyrubralactone, a novel specific inhibitor of families X and Y of eukaryotic DNA polymerases from a fungal strain derived from sea algae. *Bioorg. Med. Chem.* 16: 2939–2944.

Nagle, D. G. and Y. Zhou. 2009. Marine natural products as inhibitors of hypoxic signaling in tumors. *Phytochem. Rev.* 8: 415–429.

N'Diaye, I., G. Guella, G. Chiasera, I. Mancini, and F. Pietra. 1994. Almazole A and almazole B unusual marine alkaloids of an unidentified red seaweed of the family Delesseriaceae from the coasts of Senegal. *Tetrahedron Lett.* 35(27): 4827–4830.

Nylund, G. M., F. Persson, M. Lindegarth, G. Cervin, M. Hermansson, and H. Pavia. 2010. The red alga *Bonnemaisonia asparagoides* regulates epiphytic bacterial abundance and community composition by chemical defence. *FEMS Microbiol. Ecol.* 71: 84–93.

Ovenden, S. P. B., J. L. Nielson, C. H. Liptrot et al. 2011. Comosusols A-D and comosone A: Cytotoxic compounds from the brown alga *Sporochnus comosus*. *J. Nat. Prod.* 74: 739–743.

Owen, N. L. and N. Hundley. 2004. Endophytes—The chemical synthesizers inside plants. *Sci. Prog.* 87: 79–99.

Palermo, J. A., P. B. Flower, and A. M. Seldes. 1992. Chondriamides A and B new indolic metabolites from the red alga *Chondria* sp. *Tetrahedron Lett.* 33(22): 3097–3100.

Paul, V. J. and W. Fenical. 1980. Toxic acetylene-containing lipids from the red marine alga *Lamouroux*. *Tetrahedron Lett.* 21(35): 3327–3330.

Paul, V. J. and W. Fenical. 1984. Bioactive diterpenoids from tropical marine algae of genus *Halimeda* (chlorophyta). *Tetrahedron* 40: 3053–3062.

Paul, V. J. and K. L. Van Alstyne. 1992. Activation of chemical defenses in the tropical green algae *Halimeda* spp. *J. Exp. Mar. Biol. Ecol.* 160: 191–203.

Pettit, R. K. 2010. Small-molecule elicitation of microbial secondary metabolites. *Microb. Biotechnol.* 4(4): 471–478.

Pontius, A., A. Krick, S. Kehraus et al. 2008a. Noduliprevenone: A novel heterodimeric chromanone with cancer chemopreventive potential. *Chem. Eur. J.* 14: 9860–9863.

Pontius, A., A. Krick, R. Mersy et al. 2008b. Monodictyochromes A and B dimeric xanthone derivatives from the marine algicolous fungus *Monodictys putredinis*. *J. Nat. Prod.* 71: 1793–1799.

Puglisi, M. P., L. T. Tan, P. R. Jensen, and W. Fenical. 2004. Capisterones A and B from the tropical green alga *Penicillus capitatus*: Unexpected anti-fungal defenses targeting the marine pathogen *Lindra thallasiae*. *Tetrahedron* 60: 7035–7039.

Qiao, M. F., N. Y. Ji, X. H. Liu, F. Li, and Q. Z. Xue. 2010a. Asporyergosterol, a new steroid from an algicolous isolate of *Aspergillus oryzae*. *Nat. Prod. Commun.* 5(10): 1575–1578.

Qiao, M. F., N. Y. Ji, X. H. Liu, K. Li, Q. M. Zhu, and Q. Z. Xue. 2010b. Indoloditerpenes from an algicolous isolates of *Aspergillus oryzae*. *Bioorg. Med. Chem. Lett.* 20: 5677–5680.

Qiao, M. F., N. Y. Ji, F. P. Miao, and X. L. Yin. 2011. Steroids and an oxylipin from an algicolous isolate of *Aspergillus flavus*. *Magn. Reson. Chem.* 49: 366–369.

Qiao, Y. Y., N. Y. Ji, W. Wen, X. L. Yin, and Q. Z. Xue. 2009. A new epoxy-cadinane sesquiterpene from the marine brown alga *Dictyopteris divaricata*. *Mar. Drugs* 7: 600–604.

Rateb, M. E. and R. Ebel. 2011. Secondary metabolites of fungi from marine habitats. *Nat. Prod. Rep.* 28: 290–344.

Ravikumar, S., K. Kathiresan, R. Sengottuvel, and A. Kalaiarasi. 2010. Natural products from endophytes. In S. Ravikumar and K. Kathiresan (eds.), *Marine Pharmacology*, Vol. 1. Karaikudi, Tamil Nadu, India: Alagappa University, pp. 164–181.

Rodriguez, R. J., J. Henson, E. Van et al. 2008. Stress tolerance in plants via habitat-adapted symbiosis. *Int. Soc. Microb. Ecol.* 2: 404–416.

Sadaka, N. and J. F. Ponge. 2003. Fungal colonization of phyllosphere and litter of *Quercus rotundifolia* Lam in a holm oak forest (High Atlas, Morocco). *Biol. Fertil. Soils* 39: 30–36.

Saha, M., M. Rempt, K. Grosser, G. Pohnert, and F. Weinberger. 2011. Surface-associated fucoxanthin mediates settlement of bacterial epiphytes on the rockweed *Fucus vesiculosus*. *Biofouling* 27: 423–433.

Sakai, R., S. Minato, K. Koike, K. Koike, M. K. Jimbo, and H. Kamiya. 2005. Cellular and subcellular localization of Kainic acid in the marine red alga *Digenea simplex*. *Cell Tissue Res.* 322: 491–502.

Schloss, P. D. and J. Handelsman. 2005. Metagenomics for studying unculturable microorganisms: Cutting the Gordian knot. *Genome Biol.* 6(8): 229.

Schulz, B. and C. Boyle. 2005. The endophytic continuum. *Mycol. Res.* 109: 661–686.

Schulz, B., S. Draeger, T. E. Cruz et al. 2008. Screening strategies for obtaining novel, biologically active, fungal secondary metabolites from marine habitats. *Bot. Mar.* 51: 219–234.

Selim, K. A., A. A. El-Beih, T. M. Abdel-Rahman, and A. I. El-Diwany. 2012. Biology of endophytic fungi. *Curr. Res. Environ. Appl. Mycol.* 2(1): 31–82.

Seymour, F. A., J. E. Cresswell, P. J. Fisher et al. 2004. The influence of genotypic variation on metabolite diversity in populations of two endophytic fungal species. *Fungal Genet. Biol.* 41: 721–744.

Smit, A. J. 2004. Medical and pharmaceutical uses of seaweed natural products: A review. *J. Appl. Phycol.* 16: 245–262.

Smyrniopoulos, V., D. Abatis, L. A. Tziveleka et al. 2003. Acetylene sesquiterpenoid esters from the green alga *Caulerpa prolifera*. *J. Nat. Prod.* 66(1): 21–24.

Steinberg, P. D. and R. De Nys. 2002. Chemical mediation of colonization of seaweed surfaces. *J. Phycol.* 38(4): 621–629.

Stout, E. P. and J. Kubanek. 2010. Marine macroalgal natural products. In L. Mander and W. B. Liu (eds.), *Comprehensive Natural Products II, Vol. 2: 41–65 Natural Products Structural Diversity-II*. Elsevier. Amsterdam.

Strobel, G. A. and B. Daisy. 2003. Bioprospecting for microbial endophytes and their natural products. *Microbiol. Mol. Biol. Rev.* 67: 491–502.

Suja, M. 2013. Bioactive potentials and cytotoxicity effect of endophytic fungi from *Codium decorticatum*. PhD dissertation, Annamalai University, Chidambaram, Tamil Nadu, India.

Suja, M., S. Vasuki, and N. Sajitha. 2013. Optimization and antimicrobial metabolite production from endophytic fungi *Aspergillus terreus* KC 582297. *Eur. J. Exp. Biol.* 3(4): 138–144.

Sun, H. F., X. M. Li, L. Meng et al. 2012. Asperolides A-C, tetranorlabdane diterpenoids from the marine alga-derived endophytic fungus *Aspergillus wentii* EN-48. *J. Nat. Prod.* 75(2): 148–152.

Sun, Y., Y. Xu, K. Liu, H. Hua, H. Zhu, and Y. Pei. 2006. Gracilarioside and Gracilamides from the red alga *Gracilaria asiatica*. *J. Nat. Prod.* 69(10): 1488–1491.

Suryanarayanan, T. S. 2012. Fungal endosymbionts of seaweeds. In C. Raghukumar (ed.), *Biology of Marine Fungi, Progress in Molecular and Subcellular Biology*. Berlin, Germany: Springer-Verlag, pp. 53–69.

Suryanarayanan, T. S. and S. Thennarasan. 2004. Temporal variation in endophyte assemblages of *Plumeria rubra* leaves. *Fungal Divers.* 15: 195–202.

Suryanarayanan, T. S., A. Venkatachalam, N. Thirunavukkarasu, J. P. Ravishankar, M. Doble, and V. Geetha. 2010b. Internal mycobiota of marine macroalgae from the Tamilnadu coast: Distribution, diversity and biotechnological potential. *Bot. Mar.* 53: 457–468.

Tag, A., J. Hicks, G. Garifullina et al. 2000. G-protein signalling mediates differential production of toxic secondary metabolites. *Mol. Microbiol.* 38: 658–665.

Tan, R. X. and W. X. Zou. 2001. Endophytes: A rich source of functional metabolites. *Nat. Prod. Rep.* 18: 448–459.

Tarman, K., U. Lindequist, K. Wende, A. Porzel, N. Arnold, and L. A. Wessjohan. 2011. Isolation of a new natural product and cytotoxic and antimicrobial activities of extracts from fungi of Indonesian marine habitats. *Mar. Drugs* 9: 294–306.

Tejesvi, M. V. and A. M. Pirttila. 2011. Potential of tree endophytes as sources for new drug compounds. In A. M. Pirttila and A. C. Frank (eds.), *Endophytes of Forest Trees: Biology and Applications*. Springer, New York, pp. 295–312.

Thornton, T. A., G. A. Ross, D. Patil et al. 1989. Carbon-oxygen bond-cleavage reactions by electron transfer. 4. Electrochemical and alkali-metal reductions of phenoxynaphthalenes. *J. Am. Chem. Soc.* 111(7): 2434–2440.

Unterseher, M. and M. Schnittler. 2010. Species richness analysis and ITS rDNA phylogeny revealed the majority of cultivable foliar endophytes from beech (*Fagus sylvatica*). *Fungal Ecol.* 4: 366–378.

Vairappan, C. S., T. Ishii, T. K. Lee, M. Suzuki, and Z. Zhaoqi. 2010. Antibacterial activities of a new brominated diterpene from Borneon *Laurencia* spp. *Mar. Drugs* 8: 1743–1749.

Vallim, M. A., J. E. Barbosa, D. N. Cavalcanti et al. 2010. *In vitro* antiviral activity of diterpenes isolated from the Brazilian brown alga *Canistrocarpus cervicornis*. *J. Med. Plant Res.* 22: 2379–2382.

Viano, Y., D. Bonhomme, A. Ortalo-Magne et al. 2011. Dictyotadimer A, a new dissymmetric bis-diterpene from a brown alga of the genus *Dictyota*. *Tetrahedron Lett.* 52: 1031–1035.

Wagner, B. L. and L. C. Lewis. 2000. Colonization corn, *Zea mays*, by the endopathogenic fungus *Beauveria bassiana*. *Appl. Environ. Microbiol.* 66: 3468–3473.

Wang, F. W. 2012. Bioactive metabolites from *Guignardia* sp., an endophytic fungus residing in *Undaria pinnatifida*. *Chin. J. Nat. Med.* 10(1): 72–76.

Wang, S., X. L. Li, F. Teuscher et al. 2006. Chaetopyranin, a benzaldehyde derivative and other related metabolites from *Chaetomium globosum*, an endophytic fungus derived from the marine red alga *Polysiphonia urceolata*. *J. Nat. Prod.* 69: 1622–1625.

Wen, W., F. Li, N. Y. Ji et al. 2009. A new cadinane sesquiterpene from the marine brown alga *Dictyopteris divaricata*. *Molecules* 14: 2273–2277.

Wijesekara, I., R. Pangestuti, and S. Kim. 2011. Biological activities and potential health benefits of sulfated polysaccharides derived from marine algae. *Carbohyd. Polym.* 84: 14–21.

Williams, D. E., C. M. Sturgeon, M. Roberge, and R. J. Andersen. 2007. Nigricanosides A and B, antimitotic glycolipids isolated from the green alga *Avrainvillea nigricans* collected in Dominica. *J. Am. Chem. Soc.* 129(18): 5822–5823.

Wishart, D. S. 2008. Quantitative metabolomics using NMR. *Trend Anal. Chem.* 27: 228–237.

Wyk, A. W. W. V., K. M. Zuck, and T. C. McKee. 2011. Lithothamnin A, the first bastadin like metabolite from the red alga *Lithothamnion fragilissimum. J. Nat. Prod.* 74: 1275–1280.

Xu, H., R. J. Deckert, and D. J. Garbary. 2008. *Ascophyllum* and its symbionts. X. Ultrastructure of the interaction between *A. nodosum* (Phaeophyceae) and *Mycophycias ascophylli* (Ascomycetes). *Botany* 86: 185–193.

Xu, N., X. Fan, X. Yan, X. Li, R. Niu, and C. K. Tseng. 2003. Antibacterial bromophenols from the marine red alga *Rhodomela confervoides. Phytochemistry* 62(8): 1221–1224.

Yang, R. Y., C. Y. Li, Y. C. Lin, G. T. Peng, Z. G. She, and S. N. Zhou. 2006a. Lactones from a brown alga endophytic fungus (No. ZZF36) from the South China Sea and their antimicrobial activities. *Bioorg. Med. Chem. Lett.* 16: 4205–4208.

Yang, R. Y., C. Y. Li, Y. C. Lin, G. T. Peng, and S. N. Zhou. 2006b. Study on the sterols from a brown alga endophytic fungus (No. ZZF36) from the South China Sea. *Zhong Yao Cai* 29: 908–909.

Yotsu-Yamashita, M., K. Abe, T. Seki, K. Fujiwara, and T. Yasumoto. 2007. Polyavernoside C and C2, the new analogs of the human lethal toxin polycavernoside A from the red alga, *Gracilaria edulis. Tetrahedron Lett.* 48(13): 2255–2259.

Yotsu-Yamashita, M., R. L. Haddock, and T. Yasumoto. 1993. Polycavernoside A: A novel glycosidic macrolide from the red alga *Polycavernosa tsudai* (*Gracilaria edulis*). *J. Am. Chem. Soc.* 115(3): 1147–1148.

Zhang, Y., X. M. Li, Y. Feng, and B. G. Wang. 2010. Phenethyl α-pyrone derivatives and cyclodipeptides from a marine algous endophytic fungus *Aspergillus niger* EN-13. *Nat. Prod. Res.* 24(11): 1036–1043.

Zhang, Y., X. M. Li, and B. G. Wang. 2007a. Nigerasperones A-C, new monomeric and dimeric naphto-γ-pyrones from a marine alga-derived endophytic fungus *Aspergillus niger* EN-13. *J. Antibiot.* 60: 204–210.

Zhang, Y., S. Wang, X. M. Li, C. M. Cui, C. Feng, and B. G. Wang. 2007b. New sphingolipids with a previously unreported 9-methyl-C20—Sphingosine moiety from a marine algous endophytic fungus *Aspergillus niger* EN-13. *Lipids* 42: 759–764.

Zuccaro, A., C. L. Schoch, J. W. Spatafora, J. Kohlmeyer, S. Draeger, and J. I. Mitchell. 2008. Detection and identification of fungi intimately associated with the brown seaweed *Fucus serratus. Appl. Environ. Microbiol.* 74: 931–941.

Zuccaro, A., R. C. Summerbell, and W. Gams. 2004. A new *Acremonium* species associated with *Fucus* spp., and its affinity with a phylogenetically distinct marine *Emericellopsis* clade. *Stud. Mycol.* 50: 283–297.

21

SUEL/RBL and Their Biomedical Applications

Yasuhiro Koide, Yuki Fujii, Imtiaj Hasan, Yukiko Ogawa, Sultana Rajia,
Sarkar M.A. Kawsar, Robert A. Kanaly, Shigeki Sugawara, Masahiro Hosono,
Jiharu Hamako, Taei Matsui, and Yasuhiro Ozeki

CONTENTS

21.1 Introduction: First Study of a Sea Urchin Egg Lectin

Biological evolution proceeds through competition of progeny and "survival of the fittest." Maternal gamete cells of organisms that do not use internal fertilization are always subject to predation. Because destruction of gametes may lead to loss of genes, many invertebrates and plants have developed specific toxins that are concentrated in oocytes and seeds (respectively) to deter predators. The action of protein toxins such as ricin, abrin (from plant seeds), Shiga toxin, and cholera toxin (from bacteria) involves their targeting to specific glycan structures on the host cell surface. A toxin typically has discrete functional domains with toxic and carbohydrate-binding activities. Following the attachment of the toxin to the host target cell through its carbohydrate-binding domain, it is incorporated into and subsequently kills the cell.

Characteristic glycan-binding proteins (lectins) with a variety of sugar-binding properties are found in oocytes of many species in the phyla Mollusca, Echinodermata, and Chordata (Wu et al., 2000; Lee et al., 2001; Maehashi et al., 2003; Kawsar et al., 2011; Dreon et al., 2013). Many studies during the 1970s and 1980s focused on the effects of glycoconjugates on fertilization processes (Oikawa et al., 1973; Glabe et al., 1981). In 1981, Hajime Sasaki, a doctoral student of the well-known Japanese/Hawaiian reproductive biologist Dr . Ryuzo Yanagimachi, purified a D-galactoside-binding protein from cytoplasmic fluid of unfertilized eggs of the Japanese purple sea urchin (*Anthocidaris crassispina*). This lectin showed greater affinity to α-galactosides such as methyl α-galactopyranoside and melibiose (Galα1–6Glc) than to β-galactosides such as methyl β-galactopyranoside and lactose (Galβ1–4Glc). Sasaki identified the lectin as a disulfide-linked dimer consisting of two 11.5 kDa polypeptides. Experiments using a specific antibody showed that the lectin was transferred to

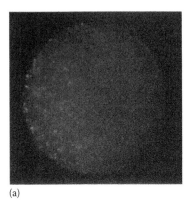
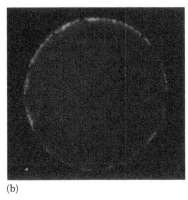

(a)   (b)

FIGURE 21.1
The striking change in distribution of a D-Gal-binding lectin in sea urchin (*Anthocidaris crassispina*) egg before (a) versus after (b) fertilization. (From Sasaki, H. and Aketa, K., *Exp. Cell Res.*, 135, 15, 1981.)

the underside of the vitelline coat from random locations in the cytoplasm within a few minutes after fertilization (Sasaki and Aketa, 1981) (Figure 21.1). This was the first reported case of dynamic relocation of a lectin during a physiological process such as fertilization.

In spite of this exciting observation, 10 more years elapsed before the primary structure of the lectin was elucidated, as described in the next section.

21.2 Structural Properties of SUEL and Other Members of the SUEL/RBL Family

The primary structure of the sea urchin egg lectin (SUEL) discovered by Sasaki was identified in 1991 by two of the present authors (TM, YO) in collaboration with Dr. Koiti Titani, a pioneer in protein chemistry studies in both the United States and Japan. This primary structure consisted of 105 amino acid (aa) residues in the polypeptide (P22031 LEG_HELCR) and was quite novel (distinctive) in comparison with structures of previously known lectins or other proteins (Ozeki et al., 1991).

After the primary structure of SUEL was elucidated in 1991, no similar structures were found in other proteins for several years. Eventually, in 1997, a homologous structure was found in an extracellular domain of the neurotoxin receptor latrophilin (Lelianova et al., 1997), which binds to latrotoxin in widow spiders (genus *Latrodectus*), and belongs to a family of adhesion-class G protein-coupled receptors (Langenhan et al., 2013) that are expressed widely in mammalian brain. In independent studies a short time later, primary structures of rhamnose-binding lectins (RBLs) isolated from fish eggs were shown to be similar to those of SUEL and latrophilin (Tateno et al., 1998; Hosono et al., 1999). RBLs from eggs of various fish species were found to be constructed as dual (STL; Tateno et al., 1998) or triple (SAL [*Silurus asotus* lectin]; Hosono et al., 1999) tandem repeats of SUEL/RBL-like domains. β-galactosidases isolated from various plant species also have SUEL/RBL-like domains (Kotake et al., 2005). Members of the SUEL family can be assigned to at least four categories (proto, dual tandem, triple tandem, chimera). Including sequences of animal and plant lectins, receptors, and galactosidases, 2014 and 197 sequences from protein and translated nucleotide informatics databases (respectively) showing similarity to those of SUEL-type lectin domains are registered in the BLAST database. These findings suggest that proteins having a SUEL-type lectin domain play a wide variety of roles in essential biological phenomena.

The SUEL polypeptide contains nine Cys and no Met, His, or Trp residues (Figure 21.2a). According to the EMBL nomenclature system (www.embl.org), the structure is termed D-galactoside/L-rhamnose-binding SUEL-type lectin or SUEL-type domain (http://www. uniprot.org/uniprot/P22031). The positions of four disulfide bonds (Cys¹⁴–Cys⁴⁴, Cys²³– Cys¹⁰², Cys⁵⁷–Cys⁸⁹, Cys⁷⁰–Cys⁷⁶) and five essential aa's (Glu¹⁵, Asn⁸², Asp⁸⁷, Gly⁹¹, Lys⁹⁴) that interact with α-Gal of Gb3 (Galα1–4Galβ1–4Glc-Cer) and L-rhamnose are conserved in various members of the SUEL/RBL family (Tateno, 2010; Ogawa et al., 2011). Although many L-RBLs have been found in animals, it is debatable whether L-rhamnose sugar is a physiological ligand of these lectins, because carbohydrates are rarely metabolized in animals. More appropriate endogenous ligand sugars should be identified for this lectin family in the future.

A 3D model of SUEL was predicted using a PHYRE2 server (http://www.sbg.bio.ic.ac. uk/phyre2/) based on the primary structure of the protein. Three α-helixes and five β-sheets were predicted to be involved in SUEL polypeptides (Figure 21.2b), and folding of the lectin was quite similar to that of the N-terminal domain of latrophilin (Figure 21.2d). This combination of α-helixes and β-sheets is seen in both SUEL-type and C-type lectin families, although SUEL-type lectins do not require divalent cations.

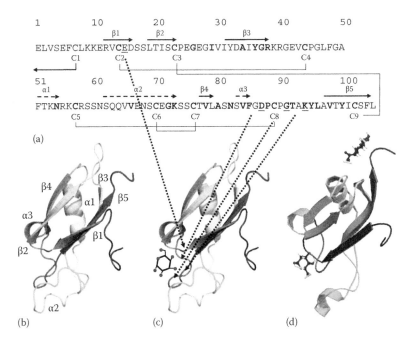

FIGURE 21.2
Structural properties of sea urchin egg lectin (SUEL). (a) Primary structure of SUEL. Secondary structure is indicated by arrows. Disulfide-linked Cys is indicated by connecting lines. (b) Secondary structure indicated in SUEL conformation predicted by PHYRE2 server. (From Kelley, L.A. and Sternberg, M.J.E., *Nat. Protoc.*, 4, 363, 2009; http://www.sbg.bio.ic.ac.uk/phyre2/.) (c) Superimposition of essential amino acids that bind glycans in SUEL conformation. (d) 3D structure of latrophilin for comparison.

21.3 Transient Overexpression of SUEL in Early Development of Sea Urchin Embryo

Immunological studies revealed striking changes of SUEL levels during embryonic development after fertilization. From the two-cell stage to the blastula stage, the arrangement of SUEL molecules in the hyaline layer in the periphery of the fertilized egg was relocated to the extracellular matrix of the epidermal region (Ozeki et al., 1995). In the gastrula stage, SUEL was concentrated in the primitive gut and epidermis. By the pluteus larva stage, SUEL molecules had disappeared from the embryo (Figure 21.3).

Embryonic organs come in contact with pathogens from the external environment (food or seawater). If a lectin in gamete cells acts as a toxin to protect the embryonic body, it would be advantageous to transport the lectin around the surface of the body that is contiguous with the external environment. Some SUEL/RBL have been reported to upregulate various types of cytokines, including interleukins, tumor necrotic factors (Watanabe et al., 2009; Franchi et al., 2011; Peatman et al., 2013; Hosono et al., 2014), colony-stimulating factor (Kawano et al., 2008), and proinflammatory cytokines (López et al., 2011). Members of the SUEL/RBL family evidently function in both body protection and cell growth. Following the disappearance of SUEL in pluteus larvae, stronger and more highly ordered body protection systems for innate immunity begin to function, as the larvae metamorphose and develop very different systems, for example, the water vascular system of the coelom. At this stage, "echinoidin," another divalent cation-dependent C-type lectin found in the same species, functions by aggregating in coelomocytes (Giga et al., 1987). Studies using molecular biology approaches are underway to elucidate the function of SUEL in larvae and the transition to echinoidin in adult sea urchins.

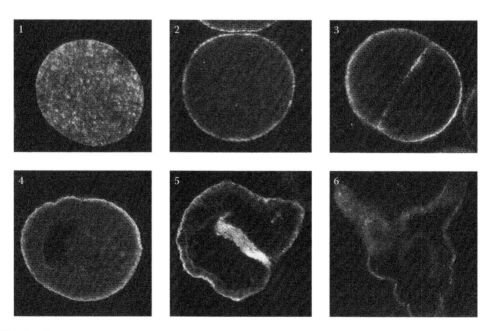

FIGURE 21.3
Localization of sea urchin egg lectin (SUEL) molecules during early development of *Arbacia crassispina*. 1, Before fertilization; 2, after fertilization; 3, two-cell stage; 4, blastula stage; 5, gastrula stage; 6, pluteus larva. Developmental changes of SUEL in eggs and embryos were monitored using fluorescence microscopy.

21.4 New Type of Gene Regulation of Multidrug Resistance Protein-1 by a SUEL/RBL Family, Revealed by Lectin Glycomics

SAL is a member of the SUEL-type lectin family isolated from catfish (*S. asotus*) eggs (Hosono et al., 1999). SAL is a 32 kDa molecule containing three SUEL/RBL domains in the polypeptide arranged as noncovalently associated trimeric polypeptides (Murayama et al., 1997), resulting in a total of nine binding sites in the molecule. A glycan-binding profile based on frontal affinity chromatography (Hirabayashi et al., 2002, 2003) showed that SAL binds specifically to α-galactoside glycosphingolipid Gb3 and has greater affinity for Galα1–4 (Gb3) than for Galα1–3 (Galili-penta and B-tetra) (Hosono et al., 2013) (Figure 21.4a). Gb3 is the target

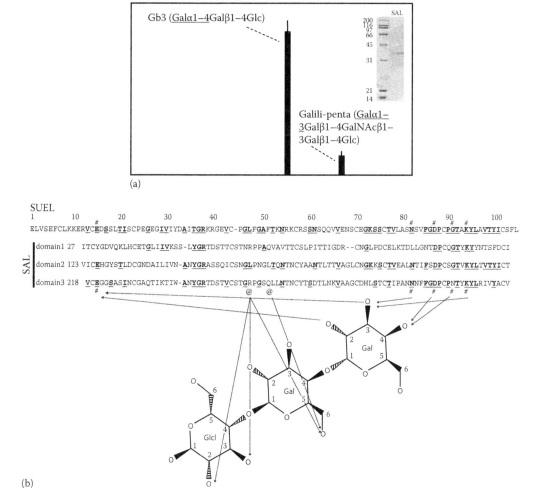

FIGURE 21.4
Binding of *Silurus asotus* lectin (SAL) to Gb3. (a) Gb3-binding pattern of SAL by frontal affinity chromatography. Inset: molecular mass of SAL. (b) Domain construct of SAL and comparison with SUEL. Number at left: N-terminal amino acid (aa) in each SUEL/RBL domain of the polypeptide. Underline: identical aa's in SUEL and SAL. #, @: essential aa's for binding to Gb3 in SAL as predicted by conformation study. (According to Shirai, T. et al., *J. Mol. Biol.*, 391, 390, 2009.)

of Shiga toxin from pathogenic bacterial strain O158 (Lingwood, 2011). In analogy to the specific carbohydrate-binding properties of SUEL and SAL, the lectin domain of latrophilin was found to bind to α-galactoside (melibiose), D-Gal, and L-rhamnose (Vakonakis et al., 2008). It is interesting that the common target ligand of α-Gal/α-galactoside and L-rhamnose is highly conserved; it is recognized by SUEL/RBL family members obtained from organisms ranging from deuterostomes (echinoderms) to vertebrates (fish, mammals).

A crystallographic analysis based on conformation study of CSL3, a member of the SUEL/RBL family, identified seven essential aa's that bind to Gb3 (Shirai et al., 2009). In addition to five essential aa's (designated as # in Figure 21.4b) that bind to D-Gal and L-rhamnose, the study found two aa's (R and Q, designated as @ in Figure 21.4b) that interact with hydroxyl groups at C6 in β-D-Gal and at C2 and C3 in D-glucose at the nonreducing terminal of Gb3. Thus, the Gal-binding SUEL/RBL family displays "micro" diversification in glycan-binding profiles of its members, as well as differences in cell regulatory activities, domain architecture, and multimerization of polypeptides.

SAL had no direct cytotoxic effect on Raji cells. However, the cytotoxic effect of doxorubicin, an anticancer drug that inhibits activity of DNA polymerase and topoisomerase, was enhanced 1.5-fold by cotreatment with SAL (Sugawara et al., 2005). SAL may promote membrane permeability and thereby accelerate the influx of the anticancer drug and water into the cell through binding with Gb3 at the cell surface. Specific cell membrane components are involved in excretion of chemicals for promotion of malignant cell survival, and transporter-channel proteins in cancer cells actively excrete anticancer drugs (Mishra et al., 2008; Schumacher et al., 2012). Multidrug-resistance proteins (MRPs) and multidrug-resistance associated proteins (MRAPs) are ATP-binding cassette (ABC) transporter-channel proteins (Loo and Clarke, 2008). The expression level of these proteins in cells helps maintain normal membrane permeability (Morin et al., 1995), and the proteins may support the resistance of malignant cells against anticancer drugs (Bikadi et al., 2011). We propose that the expression of transporter proteins on Raji cells increases membrane permeability through SAL/Gb3 binding. Detailed studies of these molecular mechanisms will clarify the role of lectins in control of gene expression.

Along this line, we investigated the effect of SAL/Gb3 interaction on the cell surface expression of ABC-transporter proteins. Vincristine, etoposide, and cisplatin applied once after 48 h incubation killed Raji cells at concentrations of 1, 1, and 10 μg/mL, respectively. The cells were also sensitive to vincristine at lower concentrations (0.1–1.0 μg/mL). SAL alone did not kill the cells, even at 25 μg/mL (Figure 21.5a, diagonal striped column). The cells were killed by coadministration of vincristine at less-than-lethal concentration 10 ng/mL and SAL at 6–25 μg/mL (Figure 21.5a). SAL-dependent cytotoxicity was effective against etoposide to the same degree as against vincristine (Figure 21.5b). Cisplatin, which is tolerated by Raji cells even at high doses (1–10 μg/mL), did not have its effect enhanced by cotreatment with SAL to a similar degree (Figure 21.5c). SAL-dependent cytotoxicity in combination with low-concentration anticancer drugs was eliminated by cotreatment with the α-galactoside melibiose (Mel; Galα1–6Glc) but not with the β-glucoside chitobiose (Chito; GlcNAcβ1–4GlcNAc) (Figure 21.5). Taken together, these findings indicate that SAL enhances the effect of anticancer drugs at low concentrations through binding with Gb3 (Fujii et al., 2012).

Whereas SAL specifically agglutinated Raji cells, it did not agglutinate erythroleukemia K562 cells, which do not express Gb3. Real-time polymerase chain reaction (RT-PCR) analyses indicated that Raji cells express mRNA encoding an MRAP-termed MRP1, a type of ABC transporter-channel protein. Quantitative RT-PCR showed that mRNA of MRP1 decreased in a time-dependent manner following SAL treatment, and the decrease was inhibited

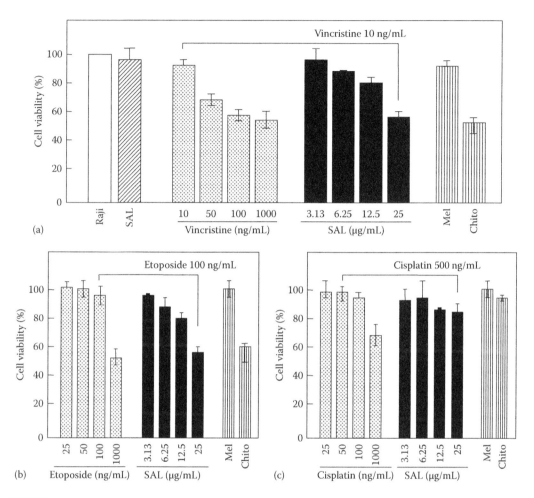

FIGURE 21.5

Enhancement of anticancer drug effect against Burkitt's lymphoma Raji cells by *Silurus asotus* lectin (SAL). Cell viability (%) (vertical axis) was quantified by trypan blue exclusion assay. Raji (white columns): untreated cells (no lectin or anticancer drug). SAL (diagonal stripes): cells treated with SAL (25 μg/mL). Anticancer drugs vincristine (0.01–1 μg/mL) (a), etoposide (0.025–1 μg/mL) (b), or cisplatin (0.25–10 μg/mL) (c) were administered to cultured cells for 48 h at various concentrations (dotted). Following 24 h incubation with various concentration of SAL (3.13–25 μg/mL), cells were treated for 48 h with 10 ng/mL vincristine, 100 ng/mL etoposide, or 500 ng/mL cisplatin (black). The disaccharides melibiose (Mel; 20 mM) and chitobiose (Chito; 20 mM) were coadministered with SAL in Raji cells for 24 h, and then each of the aforementioned drugs was administered for 48 h (horizontal stripes). Solid lines compare the enhancement of cytotoxicity by SAL in SAL-treated vs. -untreated cells.

specifically by the copresence of melibiose. We confirmed the reduction of MRP1 on the cell surface after SAL treatment by an immunological procedure using anti-MRP1 mAb (Figure 21.6a). Cell surface expression of mRNA encoding MRP1 transporter was reduced by SAL treatment in a dose-, time-, and glycan-dependent manner. The role of MRP1 in controlling the effect of vincristine was confirmed by small interfering RNA technique: sensitivity to vincristine was significantly enhanced in MRP1 knockdown cells (Figure 21.6b).

These findings, taken together, indicate that SAL *per se* is not cytotoxic, but downregulates both gene and protein expression levels of the ABC-transporter protein MRP1 on cells through binding with Gb3, a component of glycosphingolipid-enriched microdomains

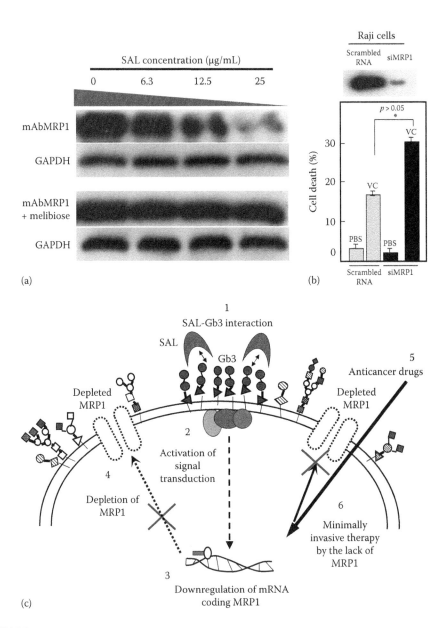

FIGURE 21.6

Dose-dependent reduction of MRP1 in Raji cells by *Silurus asotus* lectin (SAL) treatment. (a) Western blotting. Cells incubated at various SAL concentrations (0–25 µg/mL) were treated with (mAbMRP1 + melibiose) or without (mAbMRP1) melibiose for 24 h. Protein from cell lysates (10 µg/mL) was subjected to SDS-PAGE and transferred onto PVDF membrane with GAPDH as loading control. (b) Enhanced anticancer effect of vincristine resulting from siRNA knockdown of MRP1 gene. The knockdown of MRP1 was confirmed by Western blotting with anti-MRP1 mAb (top). Vincristine (10 ng/mL) (VC) and control PBS were administered to transfected cells (scrambled RNA vs. siMRP1) for 48 h. Cell viability (%) (vertical axis) was quantified by trypan blue exclusion assay. Bars: standard error. * means $p < 0.05$. (c) Proposed mechanism for the enhanced effect of low-concentration anticancer drugs against Raji cells, resulting from downregulation of MRP1 gene expression through a lectin/glycosphingolipid pathway.

(Steelant et al., 2002). We propose that the enhanced effect of low-concentration anticancer drugs against Raji cells resulted from downregulation of MRP1 gene expression through a lectin/glycosphingolipid pathway, leading to increased membrane permeability and incorporation of anticancer drugs. This novel pathway has obvious application for development of new oncotherapeutic strategies or drugs.

21.5 Conclusions and Perspectives for the Future

The "omics" approach has the potential to elucidate the properties of a biological molecule with high-throughput comparison using any standardized scale and provides a useful infrastructure for achieving optimization of life sciences, including biomedical sciences. A recent study by Wu et al. (2014) showed that a SUEL/RBL with triple tandem repeating domains isolated from American purple sea urchin (*Strongylocentrotus purpuratus*) and a fucose-binding lectin isolated from European sea bass (*Dicentrarchus labrax*) both cause apoptosis of cancer cells through protein methyltransferase 5 and transcription factor E2F. These lectins, exogenously induced through a replication-deficient adenovirus vector engineered to carry their respective genes, significantly suppressed levels of antiapoptosis factors such as Bcl-2 and X-linked inhibitor of apoptosis protein. However, the lectins did not enhance caspase activation.

The use of a similar adenovirus vector may allow adaptation of SUEL/RBL family as anticancer drugs. Our study demonstrating SAL-induced depletion of MRP1 in Raji cells (Figure 21.6) did not clarify the blocking mechanism; it may have occurred at the transcription or translation level following SAL/Gb3 binding.

The discovery that SUEL/RBL family members recognize both α-D-Gal and L-rhamnose is an important starting point. More precise future investigations of differences in glycan-binding profiles among prototype, dual tandem–type, triple tandem–type, and chimera-type species within the family will provide the information necessary for their practical biomedical application.

Acknowledgment

The authors are grateful to Dr. Steve Anderson for English editing of the manuscript.

References

Bikadi, Z., I. Hazai, D. Malik, K. Jemnitz, Z. Veres, P. Hari, Z. Ni et al. 2011. Predicting P-glycoprotein-mediated drug transport based on support vector machine and three-dimensional crystal structure of P-glycoprotein. *PLOS ONE* 6(10):e25815.

Dreon, M. S., M. V. Frassa, M. Ceolín, S. Ituarte, J. W. Qiu, J. Sun, P. E. Fernández, and H. Heras. 2013. Novel animal defenses against predation: A snail egg neurotoxin combining lectin and pore-forming chains that resembles plant defense and bacteria attack toxins. *PLOS ONE* 30(8):e63782.

Franchi, N., F. Schiavon, M. Carletto, F. Gasparini, G. Bertoloni, S. C. Tosatto, and L. Ballarin. 2011. Immune roles of a rhamnose-binding lectin in the colonial ascidian *Botryllus schlosseri*. *Immunobiology* 216:725–736.

Fujii, Y., S. Sugawara, D. Araki, T. Kawano, T. Tatsuta, K. Takahashi, S. M. Kawsar et al. 2012. MRP1 expressed on Burkitt's lymphoma cells was depleted by catfish egg lectin through Gb3-glycosphingolipid and enhanced cytotoxic effect of drugs. *The Protein Journal* 31:15–26.

Giga, Y., A. Ikai, and K. Takahashi. 1987. The complete amino acid sequence of echinoidin, a lectin from the coelomic fluid of the sea urchin *Anthocidaris crassispina*. Homologies with mammalian and insect lectins. *Journal of Biological Chemistry* 262:6197–6203.

Glabe, C. G., L. B. Grabel, V. D. Vacquier, and S. D. Rosen. 1981. Carbohydrate specificity of sea urchin sperm bindin: A cell surface lectin mediating sperm-egg adhesion. *Journal of Cell Biology* 94:123–128.

Hirabayashi, J., Y. Arata, and K. Kasai. 2003. Frontal affinity chromatography as a tool for elucidation of sugar recognition properties of lectins. *Methods in Enzymology* 362:353–368.

Hirabayashi, J., T. Hashidate, Y. Arata, N. Nishi, T. Nakamura, M. Hirashima, T. Urashima et al. 2002. Oligosaccharide specificity of galectins: A search by frontal affinity chromatography. *Biochimica et Biophysica Acta* 1572:232–254.

Hosono, M., K. Ishikawa, R. Mineki, K. Murayama, C. Numata, Y. Ogawa, Y. Takayanagi, and K. Nitta. 1999. Tandem repeat structure of rhamnose-binding lectin from catfish (*Silurus asotus*) eggs. *Biochimica et Biophysica Acta* 1472:668–675.

Hosono, M., S. Sugawara, A. Matsuda, T. Tatsuta, Y. Koide, I. Hasan, Y. Ozeki, and K. Nitta. 2014. Binding profiles and cytokine-inducing effects of fish rhamnose-binding lectins on Burkitt's lymphoma Raji cells. *Fish & Physiological Biochemistry*. 40:1559–1572. doi:10.1007/s10695-014-9948-1.

Hosono, M., S. Sugawara, T. Tatsuta, T. Hikita, J. Kominami, S. Nakamura-Tsuruta, J. Hirabayashi et al. 2013. Domain composition of rhamnose-binding lectin from shishamo smelt eggs and its carbohydrate-binding profiles. *Fish & Physiological Biochemistry* 39:1619–1630.

Kawano, T., S. Sugawara, M. Hosono, T. Tatsuta, and K. Nitta. 2008. Alteration of gene expression induced by *Silurus asotus* lectin in Burkitt's lymphoma cells. *Biological and Pharmaceutical Bulletin* 31:998–1002.

Kawsar, S. M., R. Matsumoto, Y. Fujii, H. Matsuoka, N. Masuda, I. Chihiro, H. Yasumitsu et al. 2011. Cytotoxicity and glycan-binding profile of a d-galactose-binding lectin from the eggs of a Japanese sea hare *(Aplysia kurodai)*. *Protein Journal* 30:509–519.

Kelley, L. A. and M. J. E. Sternberg. 2009. Protein structure prediction on the web: A case study using the Phyre server. *Nature Protocols* 4:363–371.

Kotake, T., S. Dina, T. Konishi, S. Kaneko, K. Igarashi, M. Samejima, Y. Watanabe, K. Kimura, and Y. Tsumuraya. 2005. Molecular cloning of a β-galactosidase from radish that specifically hydrolyzes β(1–3)- and β(1–6)-galactosyl residues of Arabinogalactan protein. *Plant Physiology* 138:1563–1576.

Langenhan, T., G. Aust, and J. Hamann. 2013. Sticky signaling-adhesion class G protein-coupled receptors take the stage. *Science Signaling* 6(276):re3.

Lee, J. K., J. Schnee, M. Pang, M. Wolfert, L. G. Baum, K. W. Moremen, and M. K. W. Pierce. 2001. Human homologs of the *Xenopus* oocyte cortical granule lectin XL35. *Glycobiology* 11:65–73.

Lelianova, V. G., B. A. Davletov, A. Sterling, M. A. Rahman, E. V. Grishin, N. F. Totty, and Y. A. Ushkaryov. 1997. α-Latrotoxin receptor, latrophilin, is a novel member of the secretin family of G protein-coupled receptors. *Journal of Biological Chemistry* 272:21504–21508.

Lingwood, C. A. 2011. Glycosphingolipid functions. *Cold Spring Harbor Perspectives in Biology* 3:a004788.

Loo, T. W. and D. M. Clarke. 2008. Mutational analysis of ABC proteins. *Archives of Biochemistry and Biophysics* 476:51–64.

López, J. A., M. G. Fain, and L. F. Cadavid. 2011. The evolution of the immune-type gene family Rhamnospondin in cnidarians. *Gene* 473:119–124.

Maehashi, E., C. Sato, K. Ohta, Y. Harada, T. Matsuda, N. Hirohashi, W. J. Lennarz, and K. Kitajima. 2003. Identification of the sea urchin 350-kDa sperm-binding protein as a new sialic acid-binding lectin that belongs to the heat shock protein 110 family: Implication of its binding to gangliosides in sperm lipid rafts in fertilization. *Journal of Biological Chemistry* 278:42050–42057.

Mishra, J., Q. Zhang, J. L. Rosson, J. Moran, J. M. Dopp, and B. L. Neudeck. 2008. Lipopolysaccharide increases cell surface P-glycoprotein that exhibits diminished activity in intestinal epithelial cells. *Drug Metabolism and Disposition* 36:2145–2149.

Morin, X. K., T. D. Bond, T. W. Loo, D. M. Clarke, and C. E. Bear. 1995. Failure of P-glycoprotein (MDR1) expressed in *Xenopus* oocytes to produce swelling-activated chloride channel activity. *Journal of Physiology* 486(Part3):707–714.

Murayama, K., H. Taka, N. Kaga, T. Fujimura, R. Mineki, N. Shindo, M. Morita, M. Hosono, and K. Nitta. 1997. The structure of *Silurus asotus* (catfish) roe lectin (SAL): Identification of a noncovalent trimer by mass spectrometry and analytical ultracentrifugation. *Analytical Biochemistry* 247:319–326.

Ogawa, T., M. Watanabe, T. Naganuma, and K. Muramoto. 2011. Diversified carbohydrate-binding lectins from marine resources. *Journal of Amino Acids* 2011:ID 838914.

Oikawa, T., R. Yanagimachi, and G. L. Nicolson. 1973. Wheat germ agglutinin blocks mammalian fertilization. *Nature* 241:256–259.

Ozeki, Y., T. Matsui, M. Suzuki, and K. Titani. 1991. Amino acid sequence and molecular characterization of a D-galactose-specific lectin purified from sea urchin (*Anthocidaris crassispina*) eggs. *Biochemistry* 30:2391–2394.

Ozeki, Y., Y. Yokota, K. H. Kato, K. Titani, and T. Matsui. 1995. Developmental expression of D-galactose-binding lectin in sea urchin (*Anthocidaris crassispina*) eggs. *Experimental Cell Research* 216:318–324.

Peatman, E., C. Li, B. C. Peterson, D. L. Straus, B. D. Farmer, and B. H. Beck. 2013. Basal polarization of the mucosal compartment in *Flavobacterium columnare* susceptible and resistant channel catfish (*Ictalurus punctatus*). *Molecular Immunology* 56:317–327.

Sasaki, H. and K. Aketa. 1981. Purification and distribution of a lectin in sea urchin (*Anthocidaris crassispina*) egg before and after fertilization. *Experimental Cell Research* 135:15–19.

Schumacher, U., N. Nehmann, E. Adam, D. Mukthar, I. N. Slotki, H. P. Horny, M. J. Flens, B. Schlegelberger, and D. Steinemann. 2012. MDR-1-overexpression in HT 29 colon cancer cells grown in SCID mice. *Acta Histochemistry* 114:594–602.

Shirai, T., Y. Watanabe, M. S. Lee, T. Ogawa, and K. Muramoto. 2009. Structure of rhamnose-binding lectin CSL3: Unique pseudo-tetrameric architecture of a pattern recognition protein. *Journal of Molecular Biology* 391:390–403.

Steelant, W. F., Y. Kawakami, A. Ito, K. Handa, E. A. Bruyneel, M. Mareel, and S. Hakomori. 2002. Monosialyl-Gb5 organized with cSrc and FAK in GEM of human breast carcinoma MCF-7 cells defines their invasive properties. *FEBS Letters* 531:93–98.

Sugawara, S., M. Hosono, Y. Ogawa, M. Takayanagi, and K. Nitta. 2005. Catfish egg lectin causes rapid activation of multidrug resistance 1 P-glycoprotein as a lipid translocase. *Biological and Pharmaceutical Bulletin* 28:434–441.

Tateno, H. 2010. SUEL-related lectins, a lectin family widely distributed through organisms. *Bioscience, Biotechnology, and Biochemistry* 74:1141–1144.

Tateno, H., A. Saneyoshi, T. Ogawa, K. Muramoto, H. Kamiya, and M. Saneyoshi. 1998. Isolation and characterization of rhamnose-binding lectins from eggs of steelhead trout (*Oncorhynchus mykiss*) homologous to low density lipoprotein receptor superfamily. *Journal of Biological Chemistry* 273:19190–19197.

Vakonakis, I., T. Langenhan, S. Prömel, A. Russ, and I. D. Campbell. 2008. Solution structure and sugar-binding mechanism of mouse latrophilin-1 RBL: A 7TM receptor-attached lectin-like domain. *Structure* 16:944–953.

Watanabe, Y., H. Tateno, S. Nakamura-Tsuruta, J. Kominami, J. Hirabayashi, O. Nakamura, T. Watanabe et al. 2009. The function of rhamnose-binding lectin in innate immunity by restricted binding to Gb3. *Developmental & Comparative Immunology* 33:187–197.

Wu, A. M., S. C. Song, Y. Y. Chen, and N. Gilboa-Garber. 2000. Defining the carbohydrate specificities of aplysia gonad lectin exhibiting a peculiar D-galacturonic acid affinity. *Journal of Biological Chemistry* 275:14017–14024.

Wu, L., X. Yang, X. Duan, L. Cui, and G. Li. 2014. Exogenous expression of marine lectins DlFBL and SpRBL induces cancer cell apoptosis possibly through PRMT5-E2F-1 pathway. *Scientific Reports* 4:4505.

Section IX

Marine Bioinformatics

22

Marine Informatics: A New Area of Research in Biology

V. Alexandar, R. Saravanan, and Shiek S.S.J. Ahmed

CONTENTS

22.1 Introduction

Current developments in technologies and high-throughput research generate an unprecedented amount of biological data (Luscombe et al. 2001). This large digital information from biology causes data overload, which could not be processed by the human brain (Rhee et al. 2006). As a result, computers have become essential to biological research to handle large quantities of data in order to extract meaningful biological information. This leads to the emergence of a new field of science called "bioinformatics," which integrates biology and information technology.

Bioinformatics is a field to manage, organize, and understand the information associated with biological macromolecules using the concepts in computer science, engineering, and statistics. It has grown enormously during the last decade, with increased advancement in biotechnology, computational skills, and information sharing effort developed by various countries. It is further enriched by grouping and compiling of huge data from several genome projects by modernized Internet concepts using the supremacy of algorithms. Even to the current technological advancement, the biological systems remain complex and challenging than any other man-made objects. Thus, bioinformatics aims to solve the multifaceted biological problems associated to biomolecules, including DNA, RNA, proteins, and metabolites to fill the vital gaps between molecules to a physiological outcome via exchange of information and databases.

22.1.1 Bioinformatics Resources

Several bioinformatics resources are available to dissect complex biological mechanisms through the analysis of large experimental biological data in the form of sequences and structures. Though few of these resources are commercialized, most of them are publicly available for research application (Ouzounis 2012). The biological information of DNA, RNA, and proteins is available as sequences in the database like National Center for Biotechnology Information (NCBI) (http://www.ncbi.nlm.nih.gov/), European Molecular Biology Laboratory (EMBL) (http://www.embl.org/), and DDBJ (http://www.ddbj.nig.ac.jp/), while 3D structures are available in PDB (http://www.rcsb.org/pdb/home/home.do) and ExPASy (http://www.expasy.org/). These databases are predominantly used in various bioinformatics analyses. Overall, the current view in biology is changing from classification of organisms to genomics and proteomics with the availability of several high-throughput tools and databases. New insights are made in several fields of biology with the development of recent technologies to explore molecular biology of microbes, plants, animals, and marine organisms.

22.1.2 Marine Biology

The marine ecosystem is tremendously diverse that it plays vital roles in global climatic changes. Though every part of the land around the world has been explored and represented, not much is known about the marine environment that covers about 70% of the earth's surface (Vermeulen 2013). Especially, life in the deep oceans and microorganisms found in very high and low extreme conditions, including saline, temperature, pressure, and methanogenic environment, still remain a mystery. Furthermore, organisms living in such extreme conditions produce a wide spectrum of biomolecules, which reveal structural and chemical properties that are not found in terrestrial biomolecules. Recently, the importance of marine organisms, compounds, and drugs has become an extensive focus of research.

22.1.3 Marine Bioinformatics

Marine bioinformatics uses computational algorithms and databases to analyze the digital information from marine resources to determine organism's distribution, taxonomy, phylogeny, and biomolecular structures and sequences (McKillen et al. 2005). In addition, several bioinformatics tools are available to identify natural bioactive compounds from marine organisms as potential drugs, antifouling compounds, biomaterials, etc.

Tremendous advancement in molecular biotechnology and computational approach increased the data volume after the Human Genome Project (Edwards and Batley 2004). The successful completion of the project widened the understanding about molecular mechanism, evolution, and genetic disease association. This leads to the genome sequencing of several organisms from terrestrial, aerial, and aquatic environments, which resulted in the accumulation of overflooding of biological data. In particular, storing, retrieving, and analyzing the data of marine organisms are becomingly complex with increased volume of data from marine genomics, transcriptomics, proteomics, and natural products. Data from these multiple marine platforms are made available in common biological databases, which further increase the complicity of data interpretation. Hence, it is essential to create separate storage and analytical bioinformatics tools for researchers working on marine organisms.

22.2 Genome

The ocean is filled with remarkable biodiversity: algae, cnidarians, mangroves, microorganisms, molluscs, phytoplankton, sponges, tunicates, plants, and animals. These organisms play the vital role in food and energy resources. Also, most of the marine natural products are used in human medicines (Selig et al. 2013). In addition, several marine species are used as model organisms for studying biological systems to understand molecular mechanism. This leads to the sequencing of several marine organisms to elucidate the biodiversity and their adaptation to environmental changes.

22.2.1 Marine Genomics

Remarkable advances in high-throughput genomic techniques such as polymerase chain reaction (PCR) and next-generation sequencing led to the exploration of various genomes. The sequencing of free-living organisms was initiated more than 10 years ago. Initially, the modest microbes were analyzed followed by the complex eukaryotic genomes and multicellular plants and animals. Several plant species belonging to various families and genera were sequenced (Schatz et al. 2012). The first-generation capillary sequencing method was implicated to explore the genome of *Arabidopsis thaliana* (The Arabidopsis Genome Initiative 2000), *Oryza sativa* (Matsumoto et al. 2005), and *Zea mays* (Schnable et al. 2009). Though, various plant genomes were sequenced with automated first-generation sequencers with the speed of thousand base pairs per day, whereas the current high-throughput second-generation instruments are capable of sequencing billions of base pairs a day (Zhou et al. 2010). These instruments with high accuracy reduce the cost of sequencing and more importantly time. The sequence of deep-sea bacteria led toward the development of DNA primers and genomic technologies like PCR. PCR is an innovative method developed by Kary Mullis during 1980s, which is a simple and inexpensive method to amplify a segment of DNA into billion copies (Rahman et al. 2013). PCR uses the capability of DNA polymerase to synthesize new DNA strand complementary to the parent strand. The application of PCR extends from DNA cloning, phylogeny, diagnosis, genetic fingerprinting to the detection of microbial infection and disorders. Currently, these technologies are employed to explore several other genomes across the tree of life, including marine organisms for enriching gene annotation, providing evidence to compare genomes to understand mechanism and evolution. So far, more than 200 marine microbes were sequenced, and their whole genome sequence data are available in databases like CAMERA (Wu and Smyth 2012). These data sets are applied to understand structural and functional characteristics of marine genomes.

Structural genomics characterizes the structure of the genome to delineate the protein folds, by providing 3D representations of proteins in an organism to understand protein functions (Goldsmith-Fischman and Honig 2003). Several marine organisms are sequenced and the data is available in databases (Table 22.1) that are essential to determine

TABLE 22.1

Marine Genome Sequenced

Organism	Investigator	Year
Loktanella vestfoldensis SKA53	ÅkeHagström	2005
Idiomarina baltica OS145	Ingrid Brettar, Manfred Höfle	2005
Roseovarius sp. 217	J. Colin Murrell, Hendrik Schaefer	2005
Nitrobacter sp. Nb-311A	John Waterbury	2005
Vibrio splendidus 12B01	Martin Polz	2005
Sulfitobacter sp. EE-36	Mary Ann Moran	2005
Roseovarius nubinhibens ISM	Mary Ann Moran	2005
Erythrobacter sp. NAP1	Paul Falkowski	2005
Prochlorococcus marinus str. MIT 9211	Penny Chisholm	2005
Janibacter sp. HTCC2649	Stephen J. Giovannoni, Jang-Cheon Cho	2005
Oceanicola batsensis HTCC2597	Stephen J. Giovannoni, Jang-Cheon Cho	2005
Croceibacter atlanticus HTCC2559	Stephen J. Giovannoni, Jang-Cheon Cho	2005
Oceanicaulis alexandrii HTCC2633	Stephen J. Giovannoni, Jang-Cheon Cho	2005
Rhodobacterales bacterium HTCC2654	Stephen J. Giovannoni, Jang-Cheon Cho	2005
Parvularcula bermudensis HTCC2503	Stephen J. Giovannoni, Jang-Cheon Cho	2005
Bacillus sp. NRRL B-14911	Janet Siefert	2005
Dokdonia donghaensis strain MED134	JaronePinhassi, Carlos Pedrós-Alió	2005
Polaribacter dokdonensis strain MED152	JaronePinhassi, Carlos Pedrós-Alió	2005
Roseobacter sp. MED193	JaronePinhassi, Carlos Pedrós-Alió	2005
Leeuwenhoekiella blandensis strain MED217	JaronePinhassi, Carlos Pedrós-Alió	2005
Vibrio sp. MED222	JaronePinhassi, Carlos Pedrós-Alió	2005
Marinomonas sp. MED121	JaronePinhassi, Carlos Pedrós-Alió	2005
Sulfitobacter sp. NAS-14.1	Mary Ann Moran	2005
Erythrobacter litoralis HTCC2594	Stephen J. Giovannoni, Jang-Cheon Cho	2005
Sphingomonas sp. SKA58	ÅkeHagström	2006
Marine actinobacterium PHSC20C1	Alison Murray	2006
Polaribacter irgensii 23-P	Alison Murray	2006
Psychromonas sp. CNPT3	Art Yayanos	2006
Aurantimonas sp. SI85-9A1	Bradley Tebo	2006
Synechococcus sp. WH 5701	David Scanlan	2006
Synechococcus sp. RS9917	David Scanlan	2006
Synechococcus sp. WH 7805	David Scanlan	2006
Photobacterium profundum 3TCK	Doug Bartlett	2006
Flavobacteria bacterium BBFL7	Farooq Azam	2006
Alteromonas macleodii Deep ecotype	Francisco Rodriguez-Valera	2006
Stappia aggregata IAM 12614	Gary King	2006
Reinekea sp. MED297	JaronePinhassi, Carlos Pedrós-Alió	2006
Oceanobacter sp. RED65	JaronePinhassi, Carlos Pedrós-Alió	2006
Nitrococcus mobilis Nb-231	John Waterbury	2006
Pseudoalteromonas tunicata D2	Mary Ann Moran, StaffanKjelleberg	2006
Vibrio angustum S14	Rick Cavicchioli, StaffanKjelleberg	2006
Blastopirellula marina DSM 3645	Rudolf Amann	2006
Congregibacter litoralis KT71	Rudolf Amann	2006
Flavobacteriales bacterium HTCC2170	Stephen J. Giovannoni, Jang-Cheon Cho	2006

(Continued)

TABLE 22.1 (*Continued*)

Marine Genome Sequenced

Organism	Investigator	Year
Robiginitalea biformata HTCC2501	Stephen J. Giovannoni, Jang-Cheon Cho	2006
Marine gammaproteobacterium HTCC2143	Stephen J. Giovannoni, Jang-Cheon Cho	2006
Marine gammaproteobacterium HTCC2080	Stephen J. Giovannoni, Kevin Vergin	2006
Photobacterium sp. SKA34	ÅkeHagström	2006
Mariprofundus ferrooxydans PV-1	Dave Emerson	2006
Synechococcus sp. BL107	David Scanlan	2006
Synechococcus sp. RS9916	David Scanlan	2006
Alteromonadales TW-7	Farooq Azam	2006
Oceanospirillum sp. MED92	JaronePinhassi, Carlos Pedrós-Alió	2006
Psychroflexus torquis ATCC 700755	John P. Bowman	2006
Lyngbya aestuarii CCY9616	Lucas Stal	2006
Nodularia spumigena CCY9414	Lucas Stal	2006
Vibrio alginolyticus 12G01	Martin Polz	2006
Oceanicola granulosus HTCC2516	Stephen J. Giovannoni, Jang-Cheon Cho	2006
Marine gammaproteobacterium HTCC2207	Stephen J. Giovannoni, Jang-Cheon Cho	2006
Candidatus Pelagibacter ubique HTCC1002	Stephen J. Giovannoni, Jang-Cheon Cho	2006
Fulvimarina pelagi HTCC2506	Stephen J. Giovannoni, Jang-Cheon Cho	2006
Roseovarius sp. HTCC2601	Stephen J. Giovannoni, Jang-Cheon Cho	2006
Alphaproteobacterium HTCC2255	Stephen J. Giovannoni, Jang-Cheon Cho	2006
Pedobacter sp. BAL39	ÅkeHagström	2007
Unidentified. eubacteri.um SCB49	ÅkeHagström	2007
Vibrio campbellii AND4	ÅkeHagström	2007
Alphaproteobacterium BAL199	ÅkeHagström	2007
Brevundimonas sp. BAL3	ÅkeHagström	2007
Hydrogenivirga sp. 128–5-R1–1	Anna-Louise Reysenbach	2007
Aciduliprofundum boonei T469	Anna-Louise Reysenbach	2007
Shewanella benthica KT99	Art Yayanos	2007
Bacillus sp. SG-1	Bradley Tebo	2007
Erythrobacter sp. SD-21	Bradley Tebo	2007
Alcanivorax sp. DG881	Christopher Bolch	2007
Caminibacter mediatlanticus TB-2	Costa Vetriani	2007
Marinobacter sp. DG893	David Green	2007
Carnobacterium sp. AT7	Doug Bartlett	2007
Vibrio fischeri MJ11	Eric Stabb, Ned Ruby	2007
Vibrio shiloi AK1	Eugene Rosenberg	2007
Vibrionales bacterium SWAT-3	Farooq Azam	2007
Flavobacteriales bacterium ALC-1	Farooq Azam	2007
Methylophaga sp. DMS010	Hendrik Schaefer	2007
Hoefleaphototrophica DFL-43	Irene Wagner-Dobler	2007
Oceanibulbus indolifex HEL-45	Irene Wagner-Dobler	2007
Stappia alexandrii DFL-11	Irene Wagner-Dobler	2007
Verrucomicrobiales bacterium DG1235	Mark Hart	2007
Nitrosococcus oceani AFC-27	Martin Klotz	2007
Synechococcus sp. PCC 7335	Nicole Tandeau de Marsac	2007

(*Continued*)

TABLE 22.1 (*Continued*)

Marine Genome Sequenced

Organism	Investigator	Year
Microcoleus chthonoplastes PCC 7420	Nicole Tandeau de Marsac	2007
Roseovarius sp. TM1035	Robert Belas	2007
Planctomyces maris DSM 8797T	Rudolf Amann	2007
Kordia algicida OT-1	Sang-Jin Kim	2007
Phaeobacter gallaeciensis 2.10	StaffanKjelleberg	2007
Lentisphaera araneosa HTCC2155	Stephen J. Giovannoni, Jang-Cheon Cho	2007
Phaeobacter gallaeciensis BS107	Thorsten Brinkhoff, Boris Wawrik, Meinhard	2007
Roseobacter litoralis Och 149	Thorsten Brinkhoff, Boris Wawrik, Meinhard	2007
Octadecabacter antarcticus str. 307	Thorsten Brinkhoff, Boris Wawrik, Meinhard	2007
Thermococcus MP	Viggo Thór Marteinsson	2007
Flavobacteria bacterium BAL38	ÅkeHagström	2007
Moritella sp. PE36	Art Yayanos	2007
Marinobacter sp. ELB17	Bess Ward	2007
Roseobacter sp. SK209-2-6	Bess Ward	2007
Roseobacter sp. AzwK-3b	Chris Francis	2007
Limnobacter sp. MED105	JaronePinhassi, Carlos Pedrós-Alió	2007
Plesiocystis pacifica SIR-1	Larry Shimkets	2007
Cyanothece sp. CCY0110	Lucas Stal	2007
Microscilla marina ATCC 23134	Margo Haygood	2007
Sagittula stellata E-37	Mary Ann Moran	2007
Roseobacter sp. CCS2	Mary Ann Moran	2007
Algoriphagus sp. PR1	Nicole King	2007
Prochlorococcus marinus str. AS9601	Penny Chisholm	2007
Prochlorococcus marinus str. MIT 9515	Penny Chisholm	2007
Prochlorococcus marinus str. NATL1A	Penny Chisholm	2007
Prochlorococcus marinus str. MIT 9303	Penny Chisholm	2007
Prochlorococcus marinus str. MIT 9301	Penny Chisholm	2007
Bacillus sp. B14905	Rob Edwards	2007
Methylophilales bacterium HTCC2181	Stephen J. Giovannoni, Kevin Vergin	2007
Rhodobacterales bacterium HTCC2150	Stephen J. Giovannoni, Kevin Vergin	2007
Marinitoga piezophila KA3	Doug Bartlett, ElizavetaBonch-Osmolovskaya	2008
Thermococcus sp. AM4	Doug Bartlett, ElizavetaBonch-Osmolovskaya	2008
Rhodobacterales bacterium Y4I	Mary Ann Moran, Alison Buchan	2008
Campylobacterales bacterium GD 1	Matthias Labrenz, Klaus Jürgens	2008
Betaproteobacterium KB13	Michael Rappe	2008
Roseobacter sp. GAI101	Rob Edwards	2008
Pseudovibrio sp. JE062	Russell Hill	2008
Ruegeria sp. R11	StaffanKjelleberg, Rebecca Case	2008
Rhodobacterales bacterium HTCC2083	Stephen J. Giovannoni, Jang-Cheon Cho	2008
Stenotrophomonas sp. SKA14	ÅkeHagström	2008
Cyanobium sp. PCC 7001	Emily Lilly	2008
Carboxydibrachium pacificum str. DSM 12653	Juan Gonzalez, Tatyana Sokolova	2008
Prochlorococcus marinus str. MIT 9202	Penny Chisholm	2008
Vibrio parahaemolyticus 16	Rob Edwards	2008
Gammaproteobacterium NOR5-3	Rudolf Amann, Bernhard Fuchs	2008

(*Continued*)

TABLE 22.1 (*Continued*)

Marine Genome Sequenced

Organism	Investigator	Year
Gammaproteobacterium NOR51-B	Rudolf Amann, Bernhard Fuchs	2008
Marine gammaproteobacterium HTCC2148	Rudolf Amann, Bernhard Fuchs, Stephen J. Giovannoni	2008
Rhodobacteraceae bacterium KLH11	Russell Hill	2008
Pelagibacter HTCC7211	Stephen J. Giovannoni, Torsten Thomas	2008
Gammaproteobacterium HTCC5015	Stephen J. Giovannoni, Kevin Vergin	2008
Octadecabacter antarcticus str. 238	Thorsten Brinkhoff, Boris Wawrik, Meinhard	2008
Citreicella sp. SE45	Alison Buchan	2009
Silicibacter lacuscaerulensis ITI-1157	Erik Zinser, Alison Buchan	2009
Thalassiobium sp. R2A62	Marcelino Suzuki	2009
Alphaproteobacterium HIMB114	Michael Rappé	2009
Synechococcus sp. WH8109	Debbie Lindell	2009
Silicibacter sp. TrichCH4B	Kathy Barbeau, Elizabeth Mann	2009
Loktanella sp. SE62	Alison Buchan	2010
Sulfurospirillum sp. Am-N	Barbara Campbell	2010
Thauera aromatica 3CB2	Bongkeun Song	2010
Pseudoalteromonas luteoviolacea 2ta16	Eric Allen	2010
Pseudoalteromonas flavipulchra 2ta6	Eric Allen	2010
Alteromonas macleodii EZ55	Erik Zinser	2010
Alteromonas BB2AT2	Kay Bidle	2010
Pseudoalteromonas Tw2	Kay Bidle	2010
Ahrensia R2A130	Marcelino Suzuki	2010
OM60 HIMB55	Michael Rappé	2010
ND anaconda	Victor Gallardo	2010
Beggiatoa sp. *Orange Guaymas*	Andreas Teske	2010
ND BG20	Chris Francis	2010
ND BD31	Chris Francis	2010
Gloeothece sp. PCC6909/1	Dilara Sharif, Geertje Van Keulen	2010
Thalassospira TrichSKD10	Elizabeth Mann, Kathy Barbeau	2010
Roseibium TrichSKD4	Elizabeth Mann, Kathy Barbeau	2010
Synechococcus CB0205	Feng Chen, Eric Wommack	2010
Synechococcus CB0101	Feng Chen, Eric Wommack	2010
Desulfonemalimicola Jadebusen (DSM 2076)	Gerard Muyzer	2010
Desulfosarcina variabilis Montpellier	Gerard Muyzer	2010
Desulfonema magnum Montpellier (DSM 2077)	Gerard Muyzer	2010
Dehalobium chlorocoercia DF-1	Kevin Sowers, Harold May	2010
SAR11 HIMB5	Michael Rappé	2010
SAR11 HIMB59	Michael Rappé	2010
Methanogenium frigidum Ace-2	Rick Caviccholi, Kevin Sowers	2010
Thioploca araucae Tha-CCL	Victor Gallardo	2010
Roseobacter R2A57	Marcelino Suzuki	2010
OM252 HIMB30	Michael Rappé	2010
Prochlorococcus sp. UH18301	Erik Zinser	2010
Calothrix sp. SC01	Jonathan Zehr	2010
Acaryochloris sp. CCMEE 5410	Scott Miller	2010

the function of gene and its products (Skolnick et al. 2000). In this context, there has been a need on 3D structural information on macromolecules, which are emphasized by structural genomics initiatives worldwide (Westbrook 2003).

In functional genomics, recent technologies are employed to monitor the expression of thousands of genes simultaneously to provide clear understanding of gene interactions (Searls 2000). Current technologies are excessively applied in human genomics as well as several terrestrial organisms. Now, these technologies are stepping into marine genomics. The ultimate goal of marine functional genomics is to identify gene function corresponding to the marine environment. Information gathered in databases from such studies would help in understanding of adaptation and evolution and more significantly to identify novel compounds or molecules for human welfare.

22.2.2 Database and Tools

Most of the nucleic acid–associated data sets are available in public domains (Table 22.2). Nucleic Acids Research online molecular biology database consists of more than 1170 collections, in which the majority is terrestrial and very few of them are marine associated

TABLE 22.2

Genomic Database Tools

Name	Type	Link
BioSample	Sequence	http://www.ebi.ac.uk/biosamples
DDBJ[a]	Sequence	http://www.ddbj.nig.ac.jp
EBI patent sequences	Sequence	http://www.ebi.ac.uk/patentdata/nr
European Nucleotide Archive	Sequence	http://www.ebi.ac.uk/ena
GenBank[a]	Sequence	http://www.ncbi.nlm.nih.gov
NCBI BioSample	Sequence	http://www.ncbi.nlm.nih.gov/biosample
neXtProt	Sequence	http://www.nextprot.org
The Sequence Read Archive	Sequence	http://www.ncbi.nlm.nih.gov/sra
Animal Genome Size	General	http://www.genomesize.com
CAMERA[a]	General	http://camera.calit2.net
EBI Genomes[a]	General	http://www.ebi.ac.uk/genomes
Entrez Gene[a]	General	http://www.ncbi.nlm.nih.gov/sites/entrez?db=gene
GeneNest	General	http://genenest.molgen.mpg.de
Genome Project[a]	General	http://www.ncbi.nlm.nih.gov/entrez/genomeprj
GOLD	General	http://www.genomesonline.org
KaryotypeDB	General	http://www.nenno.it/karyotypedb
KEGG[a]	General	http://www.genome.jp/kegg
MBGD	General	http://mbgd.genome.ad.jp
Microbial Resource[a]	General	http://cmr.jcvi.org
TMBETA-GENOME[a]	General	http://tmbeta-genome.cbrc.jp/annotation
BacMap	General	http://wishart.biology.ualberta.ca/BacMap
diArk	General	http://www.diark.org
Greglist	Structure	http://tubic.tju.edu.cn/greglist
GRSDB	Structure	http://bioinformatics.ramapo.edu/GRSDB2
MINAS	Structure	http://www.minas.uzh.ch
NDB	Structure	http://ndbserver.rutgers.edu
NNDB	Structure	http://rna.urmc.rochester.edu/NNDB

[a] Database contains marine organism–associated information.

(Galperin and Cochrane 2009). Though databases like GeneNest (http://genenest.molgen. mpg.de/) and ZFIN (http://zfin.org) are present, the marine-based databases and resources are highly required. Presently, the marine-based genomic data sets are largely generated from scientists across the globe, and it is essential to store, manage, and retrieve data using bioinformatics concepts. In general, several genomic data repositories are available for analyzing and extracting the biological information, for instance, the genomic databases such as EMBL, NCBI, and DDBJ. These databases collaborate and exchange their biological data systematically to maintain a common platform that makes information compatibility shared among them. Though these databases are flooded with human genomic data, an increasing number of sequences from other species are noticed, such as mouse, rat, fish, frog, and microbes that have been added to these genomic databases (Bernardi 2008; Rast and Messier-solek 2008; Heidelberg et al. 2010; Overview and Illumina). Increased volume of genomic data leads to the development of a new method called "metagenomics." This technique uses genetic sequences to characterize biodiversity of organisms in various environments. A few metagenomic databases such as UniMES (Patient and Martin 2011) and MetaBioME (Sharma et al. 2009) are currently used to determine the species diversity and gene prediction based on DNA sequence.

The sequence information available within the database is easily accessible to the entire scientific community through web-based tools called genome browsers. The UCSC genome browser (Karolchik et al. 2011) and the EBI/Ensembl browser (Flicek et al. 2013) are mainly used to provide essential genomic information such as sequence, structure, and function. In addition, the comparison between sequence data is achieved using an alignment algorithm such as basic local alignment search tool (BLAST) (Johnson et al. 2008) and FASTA (Akram et al. 2011). These alignment tools are publicly available through web servers, www.ncbi.nlm.nih.gov/BLAST and www.ebi.ac.uk, respectively. These programs are potentially used for contig assembly, SNP/mutation detection, and phylogenetic analysis.

22.3 Transcriptome

Transcriptome is referred to as gene expression profile that provides the quantitative level of mRNAs in a cell, tissue, or organism. The study of transcriptome is defined as transcriptomics, which is considered as a subfield of functional genomics and also acts as the precursor of proteome, expressed by a genome. In addition, transcriptome is considered as an intermediate phenotype that directly relates genotype that determines the final phenotype of an organism. The analysis of transcriptome provides the information regarding the expression level of each gene in a genome over a condition that includes disease, treatment, and environmental and toxicological effects.

Large-scale, high-throughput methods such as quantitative PCR (qPCR), expressed sequence tag (EST), serial analysis of gene expression (SAGE), microarray, and RNA-Seq have been developed for transcriptomic analysis to examine and compare the gene expression profile of an organism (Zhou et al. 2010; Martin and Wang 2011; Mutz et al. 2013). Among these methods, microarray and RNA-Seq are widely used because microarray gives an absolute transcript abundance and RNA-Seq produces direct quantification of gene expression more efficiently than large-scale EST. The microarray is a hybridization technique that hybridizes the cDNA from the sample to a very large set of oligonucleotide probes spotted on to the microarray chip to provide the relative expression between the samples. The RNA-Seq follows a similar method like EST sequencing. However, RNA-Seq uses the pool of cDNA for next-generation sequencing, which generates millions of short EST sequences (reads). These reads are then mapped in a genome to quantify mRNA expression. Overall, both the techniques generate a large amount of data that requires standard algorithm, tools, and software for the analysis and data interpretation.

22.3.1 Marine Transcriptomics

The transcriptome profiling of marine organism is rapidly growing with the increase in the advancement in technology (Table 22.3). Most marine species living in extreme environmental conditions has been attributed to show an environmental response through gene expression. Understanding the transcriptome of such organisms will provide information on environmental associated functional processes of marine communities. The transcriptome of several marine organisms is determined using microarray and next-generation sequencing. For instance, *Ochromonas*, *Prymensiales*, *Striatellales*, *Corethrales*, and *Stylonematales* spp. were profiled using RNA-Seq technology, while *Fucus vesiculosus*,

TABLE 22.3

Marine Transcriptome

Author	Organisms	Technique
Christine M. Smith	*Calanus finmarchicus*	EST
Jonathon H. Stillman	*Petrolisthes cinctipes*	EST
Carla Flöthe	*Fucus vesiculosus*	Expression array
Paulo A. Zaini	*Gracilaria tenuistipitata*	Expression array
Cock JM	*Ectocarpus siliculosus*	Expression array
O'Donnell	*Lytechinus pictus*	Expression array
Robert J. Calin-Jageman	*Aplysia californica*	qPCR
David Caron	*Uronema* sp.	RNA-Seq
Boris Wawrik	*Karenia brevis*	RNA-Seq
John Archibald	*Oxyrrhis marina*	RNA-Seq
Sonya Dyhrman	*Prorocentrum minimum*	RNA-Seq
John Archibald	*Cryptomonas paramecium*	RNA-Seq
John Archibald	*Hemiselmis andersenii*	RNA-Seq
David Caron	*Prymnesium parvum*	RNA-Seq
John Archibald	*Lotharella oceanica*	RNA-Seq
Ginger Armbrust	*Rhodosorus marinus*	RNA-Seq
David Caron	*Ochromonas* sp.	RNA-Seq
Ginger Armbrust	*Grammatophora oceanica*	RNA-Seq
Thomas Mock	*Fragilariopsis kerguelensis*	RNA-Seq
Sonya Dyhrman	*Chaetoceros affinis*	RNA-Seq
Sonya Dyhrman	*Chaetoceros affinis*	RNA-Seq
Alexandra Worden	*Dolichomastix tenuilepis*	RNA-Seq
Alexandra Worden	*Nephroselmis pyriformis*	RNA-Seq
Aurora Nedelcu	*Polytomella parva*	RNA-Seq
Aurora Nedelcu	*Chlamydomonas euryale*	RNA-Seq
Dyhrman ST	*Emiliania huxleyi*	SAGE

Gracilaria tenuistipitata, *Lytechinus pictus*, and *Ectocarpus siliculosus* were done using micro-array expression. However, understanding and retrieving meaningful information from the transcriptome data require the bioinformatics tools and softwares to develop database resources for marine transcriptomics.

22.3.2 Database and Tools

Databases are the major data recourses for bioinformatics that uses algorithms to solve and to retrieve meaningful biological information from the transcriptomic gene expression data (Garber et al. 2011). The transcriptome profile data obtained from microarray and RNA-Seq were deposited in the public repositories such as GEO at NCBI, ArrayExpress at EBI, CIBEX at DDB, and CAMERA at DDC. All the data deposited in these repositories follow a standard, common data exchange format to make the maximum data usage for research. However, analysis on these data requires high-level mathematical and statistical knowledge, which is made easy with the help of user-friendly software tools. GeneSpring (Agilent Technology), GeneChip Array (Affymetrix), and GenomeStudio (Illumina) are widely used softwares to manage and analyze the transcriptomic data. In addition, several other tools are available to solve and to make a challenging decision on transcriptome analysis (Table 22.4).

TABLE 22.4

Transcriptomics Database Tools

Name	Type	Link
RNA Bricks	Structure	http://iimcb.genesilico.pl/rnabricks
RNA CoSSMos	Structure	http://cossmos.slu.edu
RNA FRABASE	Structure	http://rnafrabase.ibch.poznan.pl/
RNA SSTRAND	Structure	http://www.rnasoft.ca/sstrand/
RNAJunction	Structure	http://rnajunction.abcc.ncifcrf.gov/
SARS-CoV RNA SSS	Structure	http://www.liuweibo.com/sarsdb/
16S and 23S Ribosomal RNA Mutation Database	Sequence	http://ribosome.fandm.edu/
Database for Bacterial Group II Introns	Sequence	http://www.fp.ucalgary.ca/group2introns/
deepBase	Sequence	http://deepbase.sysu.edu.cn/
European rRNA database	Sequence	http://www.psb.ugent.be/rRNA/
miR2Disease	Sequence	http://www.mir2disease.org/
MODOMICS	Sequence	http://genesilico.pl/modomics/
Rfam	Sequence	http://rfam.sanger.ac.uk/
RNA Modification Database	Sequence	http://rna-mdb.cas.albany.edu/RNAmods/
sRNAMap	Sequence	http://srnamap.mbc.nctu.edu.tw/
tRNAdb	Sequence	http://trnadb.bioinf.uni-leipzig.de/
ArrayExpress	Expression	http://www.ebi.ac.uk/arrayexpress/
Stanford Microarray Database[a]	Expression	http://smd.stanford.edu/
Gene Expression Omnibus[a]	Expression	http://www.ncbi.nlm.nih.gov/geo/
Expression Atlas[a]	Expression	http://www.ebi.ac.uk/gxa
CAMERA[a]	Sequence	http://camera.calit2.net

[a] Database contains marine organism–associated information.

22.4 Proteome

Proteome was introduced during the year 1990 to represent the protein, a translated product of gene. Proteomics is the study of all expressed proteins in a biological sample—cell, tissue, and organism (Graves and Haystead 2002). It emerged as a significant field of focus in the "omics" era. The knowledge gained from proteomics significantly enhances the understanding of genome, transcriptome, and metabolome of an organism. Proteome reflects the changes in gene regulation, mRNA expression, stability, and posttranslational modifications (PTMs) in response to ecological changes. Large-scale analysis of proteome bridges the gaps between genotypic and phenotypic variations by determining the expression, structure, and function. Remarkable progress has been made in the past decades in generating enormous data from high-throughput technologies such 2D gel (2D gel electrophoresis), liquid chromatography/mass spectrometry (LC–MS), electrospray ionization, matrix-assisted laser desorption/ionization–time of flight, protein array, selected reaction monitoring, and isobaric tags for relative and absolute quantification (Gevaert 2000; Mehta and Steven 2014). These technological advances provide information on protein sequences, PTMs, protein–protein interactions, and differential protein expression in various conditions. In addition, improvements in managing and organizing these large data sets are initiated for proteomics to accomplish its potential (Palagi et al. 2006). However, proteomic approaches are slowly employed within the field of marine microbiology, biotechnology, and toxicology because of difficulties in sample processing, optimization, data analysis, and interpretation (Slattery et al. 2012; Wang et al. 2014). Hence, the proteomic information on marine organisms remains highly unexplored.

22.4.1 Marine Proteomics

The proteomic studies generally use 2D gel electrophoresis coupled with mass spectroscopy instruments to study the proteome of an organism. So far, this technique is applied to explore the proteome of several organisms, including *Homo sapiens*. Current advancements in efficient protein separation methods such as affinity and ion-exchange chromatography are promising than 2D-gel-based techniques. Also, improvements in MS techniques provide more reliable data with high sensitivity and specificity. To date, proteomic research is focused on marine organisms for which the genome sequences are available, thus providing access to structure and functions of the proteins. Several marine vertebrates and invertebrates are sequenced to understand the physiology, genetic inconsistency, and adaptation to environmental changes. For instance, the marine microbes such as *Vibrio cholerae*, *Vibrio vulnificus*, *V. parahaemolyticus*, and *Vibrio salmonicida* serve as model organisms, which provide insights into microbial taxonomy, physiology, and biofilm formation. Accumulation of the vast amount of marine data produced from these techniques requires storage repositories and efficient bioinformatics tools and softwares to analyze and retrieve meaningful information (Table 22.5).

22.4.2 Structural Proteomics

Structural proteomics is one of the core areas of proteomics that describes the 3D structure that provides the functional clues of a protein. It uses nuclear magnetic resonance (NMR) and x-ray crystallography to experimentally determine the 3D structure of proteins (Christendat et al. 2000). However, these experimental studies are expensive and

TABLE 22.5

Proteomic Database Tools

Name of the Database	Type	Link
FusionDB	Functional	http://igs-server.cnrs-mrs.fr/FusionDB
InterDom	Functional	http://interdom.i2r.a-star.edu.sg/
iPfam	Functional	http://ipfam.sanger.ac.uk/
OMA	Functional	http://www.omabrowser.org
PANTHER[a]	Functional	http://www.pantherdb.org/
Pfam	Functional	http://pfam.sanger.ac.uk/
PIRSF	Functional	http://pir.georgetown.edu/pirsf/
Protein Clusters	Functional	http://www.ncbi.nlm.nih.gov/sites/ entrez?db=proteinclusters
ProtoNet	Functional	http://www.protonet.cs.huji.ac.il/
TIGRFAMs	Functional	http://www.jcvi.org/cgi-bin/tigrfams/index.cgi
COMBREX	Sequence	http://combrex.bu.edu
NCBI Protein database[a]	Sequence	http://www.ncbi.nlm.nih.gov/protein
PIR[a]	Sequence	http://pir.georgetown.edu/
PRF	Sequence	http://www.prf.or.jp/
TCDB	Sequence	http://www.tcdb.org/
UniProt[a]	Sequence	http://www.uniprot.org/
CharProtDB	Sequence	http://www.jcvi.org/charprotdb/
EXProt	Sequence	http://www.cmbi.kun.nl/EXProt/
MIPS resources	Sequence	http://mips.gsf.de/
DisProt	Structure	http://www.dabi.temple.edu/disprot/
DSMM	Structure	http://projects.villa-bosch.de/dbase/dsmm/
EMDataBank[a]	Structure	http://emdatabank.org/
Genomes To Protein[a]	Structure	http://spock.genes.nig.ac.jp/~genome/gtop.html
IEDB-3D	Structure	http://www.immuneepitope.org/bb_structure.php
IMGT/3Dstructure-DB	Structure	http://www.imgt.org/
MolMovDB	Structure	http://bioinfo.mbb.yale.edu/MolMovDB/
PDB[a]	Structure	http://www.rcsb.org/pdb/
PDBj	Structure	http://pdbj.org/
Pocketome	Structure	http://www.pocketome.org
3DSwap	Sequence	http://caps.ncbs.res.in/3DSwap

[a] Database contains marine organism–associated information.

time-consuming, which makes computational methods for predicting protein structure promising. The computational methods use sequence homology and structural information to model structures of unknown proteins. It principally applies an approach of selecting a protein homolog (template) to map/model the unknown protein using bioinformatics concepts, which addresses the structure, function, and phylogenetic distribution.

22.4.3 Functional Proteomics

Functional proteomics is an emerging area of research, which focuses to understand the biological and cellular mechanisms of proteins at the molecular level (Monti et al. 2005). With the increase in proteome projects, there is a tremendous growth in the accumulation of protein sequences whose function is unidentified. Functional proteomics allows a selected group of proteins to study, using affinity chromatography, yeast two-hybrid systems, and MS, and to understand the protein function (Godovac-Zimmermann and Brown 2001). Determining the protein functions reveals the protein–protein interactions, biological functions, molecular mechanism, pathways, and disease pathology.

22.4.4 Expression Proteomics

Expression proteomics provides large-scale information on protein expression of a cell, tissue or organism. Recent technologies such as 2D gel electrophoresis, chromatography, and MS are widely used in expression proteomics (Aebersold and Cravatt 2002; Savino et al. 2012). These techniques help in identification of differentially expressed protein by relative analysis of samples. In addition, it reveals information and post translational modification which together provide clues on biomarkers discovery and drug targets.

22.4.5 Databases and Tools

The discovery of Sanger sequencing method and successive improvements in proteomic techniques initiated the creation of biological databases. In general, proteomic databases are classified into sequence, structure, and functional databases. Most of the sequence databases are attributed for storage and retrieval of amino acid sequences, whereas the structural and functional databases provide information on 3D structures of proteins, biological process, molecular function, and cellular components (Rameshwari 2011).

Several protein sequence databases are freely available such as Protein Information Resource (PIR) and Universal Protein Resource (UniProt). These databases are mainly utilized to retrieve and compare the newly determined sequences with existing sequences using alignment tools. Furthermore, they act as a source for predicting protein structure, function, and hypothesis testing for evolution. PIR is an integrated and nonredundant public domain protein sequence database that contains sequence information, covering the entire taxonomic range (Wu et al. 2003). UniProt knowledge base provides a comprehensive, high-quality and freely accessible resource of protein sequences. The UniProt data are derived from genome sequencing projects and research publications. Currently, this database contains more than 5,00,000 entries from ~2,50,000 references to support genomic, transcriptomic, proteomic and systems biological research (Bairoch et al. 2005). Though these databases provide information on sequences, the structure of corresponding sequence is necessary to understand the biological function.

In this context, the illustration of protein structure is generated using techniques such as NMR and crystallography. The data derived from these techniques are stored in structure databases. Protein Data Bank (PDB) is a major structural database that stores structural data of macromolecules, molecular weight ranging from 1000 Da to 1000 kDa (Berman et al. 2000). This database also favors depositing the predicted protein structure from sequence using bioinformatics tools, CASP (http://www.predictioncenter.org), HHpred (http://toolkit.tuebingen.mpg.de/hhpred), and Ps2 (http://ps2.life.nctu.edu.tw) (Kryshtafovych et al. 2005; Söding et al. 2005). In addition, several other databases such as PDBe (http://www.ebi.ac.uk/pdbe), CATH (http://www.cathdb.info), and SBKB (http://sbkb.org) provide structural information on protein molecules (Orengo et al. 2003; Gabanyi et al. 2011; Gutmanas et al. 2014).

The protein functional databases help to determine the molecular and biological function of proteins. Understanding the function is essential to elucidate their molecular mechanism to identify drug targets. The functional databases are available in public domains such as PANTHER (http://www.pantherdb.org), KEGG pathway (http://www.genome.jp/kegg/pathway.html), STRING (http://string-db.org) and TMFunction (http://tmbeta-genome.cbrc.jp/TMFunction) (Von Mering et al. 2003; Kanehisa et al. 2006; Gromiha et al. 2009; Mi et al. 2009). Also, ProtFun (http://www.cbs.dtu.dk/services/ProtFun), PredictProtein (https://www.predictprotein.org), and RaptorX (http://raptorx.uchicago.edu) are the databases that predict the protein function from sequence/structure.

22.5 Natural Products

Nature provides resources for the treatment of several diseases by means of natural products and their derivatives. Several drugs such as vinblastine (*Catharanthus roseus*), camptothecin (*Camptotheca acuminata*), and ipomeanol (*Ipomoea batatas*) are of natural origin (Sithranga Boopathy and Kathiresan 2010). Though natural products fill the earth, only a few of them are well studied and our understanding of the metabolites remains unexplored. Metabolome represents complete metabolites of a cell, tissue, or organism, which is the outcome of gene, transcriptome and protein regulation (Rochfort 2005). The major goal of metabolomics is not only to identify metabolites in a sample but also to estimate the concentration and fluxes at a given spatial and temporal condition (Bundy et al. 2009). Advancement in chromatography, mass spectroscopy, and other separation techniques led to the identification and quantification of metabolites with high sensitivity and specificity (Goulitquer et al. 2012). Analytical instruments such as NMR, chemical shift imaging, infrared (Fourier transform infrared spectroscopy), and LC–MS are extensively employed to study metabolomics to determine the potential compounds toward the therapeutic application. However, analyzing huge data from these techniques requires potential contribution of bioinformatics databases, software, and tools.

22.5.1 Marine Metabolomics

Marine system is a highly diverse and renowned source of a wide variety of metabolites/natural products. NMR and MS techniques are highly used technologies for high-throughput metabolite profiling, which facilitate understanding biological process, phenotypic variations, and biodiversity of marine organisms. Metabolomic profiling of several marine organisms such as algae, bacteria, and animals (vertebrates and invertebrates) is currently explored. However, the metabolomics of micro- and macroalgae was extensively studied for pigment profiling to distinguish the taxa of algae, including the red algae, the green algae, Rhizaria, dinoflagellates, and Stramenopiles. Also, the algae like *Cocconeis scutellum, Cylindrotheca closterium, Seminavis robusta, Gracilaria vermiculophylla, Palmaria palmata, and Ulva conglobata* have been profiled to determine the molecules such as nucleotides, carbohydrates, fatty acids, amino acids, secondary metabolites, and natural products. In bacteria, the metabolome of *Actinobacteria, Roseobacter clade, Saccharophagus degradans, Firmicutes*, and *Alpha-* and *Gammaproteobacteria* were studied, which provides the understanding of metabolites and chemical ecology. In addition, few metabolomic studies were conducted on marine vertebrates and invertebrates such as *Engraulis, Mytilus californianus*, and *Dysidea herbacea* to determine lipids and other metabolites (Goulitquer et al. 2012). Overall, these profiling enhance the understanding of marine chemical diversity that provides an opportunity toward therapeutic applications.

22.5.2 Therapeutic Applications of Marine Compounds

Currently, the marine compounds are the major focus for novel drug discovery. Studies suggest that marine species are viable in producing biological active compounds in the treatment of microbial infections, cancer, inflammation, and other diseases. On this basis, several bioactive compounds were derived from marine organisms, for instance, eleutherobin (*Eleutherobia* sp.), diterpenoids (*Sarcodictyon roseum*) and dolastatins (*Dolabella auricularia*), halichondrin B (*Halichondria okadai*), laulimalide (*Cacospongia mycofijiensis*), peloruside A (*Mycale* sp.), hemiasterlins (*Auletta* sp.), vitilevuamide (*Didemnum cuculiferum*), dehydrodidemnin B (*Aplidium albicans*), and Kahalalide F (*Bryopsis* sp.) (Haefner 2003; Molinski et al. 2009). In addition to these, few marine molecules are FDA approved, while others are in various stages of clinical trials.

22.5.3 Databases and Tools

For the past two decades, the identification of novel compounds and drugs from natural origin has increased periodically (Table 22.6). Initially, the terrestrial plants and animals were the major source of natural products. Very recently, the importance of marine eco- system has been acknowledged as it contains high species diversity. The advancement in analytical techniques resulted in the vast amount of data on natural products and drugs. Hence, it is essential to store and retrieve these information for various applications including novel drug identification and modification of existing drugs to amplify their effectiveness. Currently, databases for drugs and natural products are developed but the marine-associated databases are limited (Füllbeck et al. 2006).

The SuperNatural Database (http://bioinformatics.charite) contains ~50,000 natural compounds collected from 15 different databases and suppliers (Dunkel et al. 2006). It provides information associated with compound substructure matching, possible medi- cal application, similar drugs, and natural products. The marine compound database (http://www.progenebio.in/mcdb/index.html) encompasses chemical entities and marine natural products collected from literature evidence (Babu et al. 2008). It contains 182 com- pounds that possess bioactivity against various human diseases. In addition, the Seaweed Metabolite Database is a literature-centered database that provides information on known compounds from seaweed and their corresponding biological activity (Davis and Vasanthi 2011). It provides literature evidence on geographical origin, extraction procedure, and chemical descriptors for cheminformatics analysis. In addition, databases such as NaPDoS,

TABLE 22.6

Compound Databases

Database	Link
ChEBI	www.ebi.ac.uk/chebi
ChemBank	chembank.broad.harvard.edu
ChemID	chem.sis.nlm.nih.gov/chemidplus
NCI	cactus.nci.nih.gov/ncidb3/download_ncidb3.html
PubChem	pubchem.ncbi.nlm.nih.gov
AntiBase	http://eu.wiley.com/WileyCDA
Chapman & Hall	www.crcpress.com/
Chemical Abstracts	www.cas.org
ROMPP Online	www.roempp.com/thieme-chemistry/np/info
The Merck Index	www.merckbooks.com

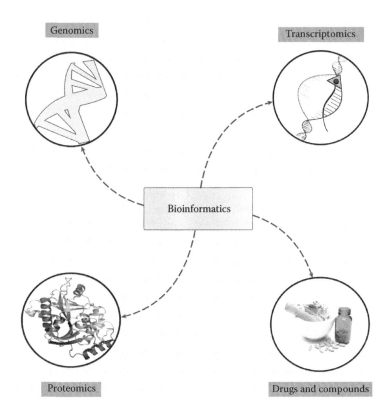

FIGURE 22.1
Application of bioinformatics in various domains of molecular biology.

UCSD, and Environmental Surveyor of Natural Product Diversity (eSNaPD) provide viable information about marine compounds. The NaPDoS (http://www.biokepler.org/use_cases/napdos) is a secondary metabolite database that provides the phylogeny information on the secondary metabolite (Ziemert et al. 2012). The UCSD Marine Natural Products Database is a repository that contains publicly available natural product data, mostly from marine microorganisms and exosymbionts. Also, eSNaPD is a bioinformatics web tool aid to discover genes encoding natural compound from genomic data (Reddy et al. 2014). It identifies gene clusters using reference genomic data to elucidate biosynthetic diversity. Overall, the application of bioinformatics on marine natural products is gradually progressing; in the near future, it may aid in early diagnosis and predict drug effectiveness and disease susceptibility (Chang 2013; Figure 22.1).

References

Aebersold, R. and B.F. Cravatt. 2002. Proteomics-advances, applications and the challenges that remain. *Trends in Biotechnology* 20 (12 Suppl): S1–S2.

Akram, M., M. Saim Jamil, Z. Mehmood, M. Akram, M.K. Waqas, Z. Iqbal, A.A. Khan, and H.M. Asif. 2011. Fast alignment (FASTA): A review article. *Journal of Medicinal Plant Research* 5: 6931–6933.

Babu, P.A., S.S. Puppala, S.L. Aswini, and M.R. Vani. 2008. A database of natural products and chemical entities from marine habitat. *Bioinformation* 3: 142–143.

Bairoch, A., R. Apweiler, C.H. Wu, W.C. Barker, B. Boeckmann, S. Ferro, E. Gasteiger et al. 2005. The universal protein resource (UniProt). *Nucleic Acids Research* 33: D154–D159.

Berman, H.M., J. Westbrook, Z. Feng, G. Gilliland, T.N. Bhat, H. Weissig, I.N. Shindyalov, and P.E. Bourne. 2000. The protein data bank. *Nucleic Acids Research* 28: 235–242.

Bernardi, G. 2008. Fish genomics: A mini-review on some structural and evolutionary issues. *Marine Genomics* 1 (1): 3–7.

Bundy, J., M. Davey, and M. Viant. 2009. Environmental metabolomics: A critical review and future perspectives. *Metabolomics* 5: 3–21.

Chang, H.-w. 2013. Drug discovery and bioinformatics of marine natural products. *Journal of Marine Science: Research & Development* 4 (1): 1–2.

Christendat, D., A. Yee, A. Dharamsi, Y. Kluger, M. Gerstein, C.H. Arrowsmith, and A.M. Edwards. 2000. Structural proteomics: Prospects for high throughput sample preparation. *Progress in Biophysics and Molecular Biology* 73 (5): 339–345.

Davis, G.D.J. and A.H.R. Vasanthi. 2011. Seaweed metabolite database (SWMD): A database of natural compounds from marine algae. *Bioinformation* 5 (8): 361–364.

Dunkel, M., M. Fullbeck, S. Neumann, and R. Preissner. 2006. SuperNatural: A searchable database of available natural compounds. *Nucleic Acids Research* 34 (Database issue): D678–D683.

Edwards, D. and J. Batley. 2004. Plant bioinformatics: From genome to phenome. *Trends in Biotechnology* 22 (5): 232–237.

Flicek, P., I. Ahmed, M. Ridwan Amode, D. Barrell, K. Beal, S. Brent, D. Carvalho-Silva et al. 2013. Ensembl 2013. *Nucleic Acids Research* 41: D48–D55.

Füllbeck, M., E. Michalsky, M. Dunkel, and R. Preissner. 2006. Natural products: Sources and databases. *Natural Product Reports* 23 (3): 347–356.

Gabanyi, M.J., P.D. Adams, K. Arnold, L. Bordoli, L.G. Carter, J. Flippen-Andersen, L. Gifford et al. 2011. The structural biology knowledgebase: A portal to protein structures, sequences, functions, and methods. *Journal of Structural and Functional Genomics* 12: 45–54.

Galperin, M.Y. and R.Gy. Cochrane. 2009. The molecular biology database collection: 2005 update. *Nucleic Acids Research* 37: D1–D4.

Garber, M., M.G. Grabherr, M. Guttman, and C. Trapnell. 2011. Computational methods for transcriptome annotation and quantification using RNA-Seq. *Nature Methods* 8: 469–477.

Gevaert, K. 2000. Protein identification, methods in proteomics. *Electrophoresis* 21: 1145–1154.

Godovac-Zimmermann, J. and L.R. Brown. 2001. Perspectives for mass spectrometry and functional proteomics. *Mass Spectrometry Reviews* 20 (1): 1–57.

Goldsmith-Fischman, S. and B. Honig. 2003. Structural genomics: Computational methods for structure analysis. *Protein Science: A Publication of the Protein Society* 12: 1813–1821.

Goulitquer, S., P. Potin, and T. Tonon. 2012. Mass spectrometry-based metabolomics to elucidate functions in marine organisms and ecosystems. *Marine Drugs* 10: 849–880.

Graves, P.R. and T.A.J. Haystead. 2002. Molecular biologist's guide to proteomics *Microbiology and Molecular Biology Reviews* 66 (1): 39–63.

Gromiha, M.M., Y. Yabuki, M. Xavier Suresh, A. Mary Thangakani, M. Suwa, and K. Fukui. 2009. TMFunction: Database for functional residues in membrane proteins. *Nucleic Acids Research* 37: D201–D204.

Gutmanas, A., Y. Alhroub, G.M. Battle, J.M. Berrisford, E. Bochet, M.J. Conroy, J.M. Dana et al. 2014. PDBe: Protein data bank in Europe. *Nucleic Acids Research* 42: D285–D291.

Haefner, B. 2003. Drugs from the deep. *Marine Natural* 8 (12): 536–544.

Heidelberg, K.B., J.A. Gilbert, and I. Joint. 2010. Marine genomics: At the interface of marine microbial ecology and biodiscovery. *Microbial Biotechnology* 3 (5): 531–543.

Johnson, M., I. Zaretskaya, Y. Raytselis, Y. Merezhuk, S. McGinnis, and T.L. Madden. 2008. NCBI BLAST: A better web interface. *Nucleic Acids Research* 36: W5–W9.

Kanehisa, M., S. Goto, M. Hattori, K.F. Aoki-Kinoshita, M. Itoh, S. Kawashima, T. Katayama, M. Araki, and M. Hirakawa. 2006. From genomics to chemical genomics: New developments in KEGG. *Nucleic Acids Research* 34: D354–D357.

Karolchik, D., A.S. Hinrichs, and W.J. Kent. 2011. The UCSC genome browser. *Current Protocols in Human Genetics* 71: 18.6.1–18.6.33.

Kryshtafovych, A., Č. Venclovas, K. Fidelis, and J. Moult. 2005. Progress over the first decade of CASP experiments. *Proteins: Structure, Function and Genetics* 61: 225–236.

Luscombe, N.M., D. Greenbaum, and M. Gerstein. 2001. What is bioinformatics? *Gene Expression* 40: 83–100.

Martin, J.A. and Z. Wang. 2011. Next-generation transcriptome assembly. *Nature Reviews Genetics* 12: 671–682.

Matsumoto, T., J.Z. Wu, H. Kanamori, Y. Katayose, M. Fujisawa, N. Namiki, H. Mizuno et al. 2005. The map-based sequence of the rice genome. *Nature* 436: 793–800.

McKillen, D.J., Y.A. Chen, C. Chen, M.J. Jenny, H.F. Trent, J. Robalino, D.C. McLean et al. 2005. Marine genomics: A clearing-house for genomic and transcriptomic data of marine organisms. *BMC Genomics* 6: 34.

Mehta, M. and B. Steven. 2014. Recent updates in the treatment of glioblastoma: Introduction. *Seminars in Oncology* 41: S1–S3.

Mi, H., Q. Dong, A. Muruganujan, P. Gaudet, S. Lewis, and P.D. Thomas. 2009. PANTHER Version 7: Improved phylogenetic trees, orthologs and collaboration with the gene ontology consortium. *Nucleic Acids Research* 38: D204–D210.

Molinski, T.F., D.S. Dalisay, S.L. Lievens, and J.P. Saludes. 2009. Drug development from marine natural products. *Nature Reviews Drug Discovery* 8 (1): 69–85.

Monti, M., S. Orrù, D. Pagnozzi, and P. Pucci. 2005. Functional proteomics. *Clinica Chimica Acta; International Journal of Clinical Chemistry* 357: 140–150.

Mutz, K.-O., A. Heilkenbrinker, M. Lönne, J.-G. Walter, and F. Stahl. 2013. Transcriptome analysis using next-generation sequencing. *Current Opinion in Biotechnology* 24: 22–30.

Orengo, C.A., F.M.G. Pearl, and J.M. Thornton. 2003. The CATH domain structure database. *Methods of Biochemical Analysis* 44: 249–271.

Ouzounis, C.A. 2012. Rise and demise of bioinformatics? Promise and progress. *PLOS Computational Biology* 8(4): e1002487.

Palagi, P.M., P. Hernandez, D. Walther, and R.D. Appel. 2006. Proteome informatics I: Bioinformatics tools for processing experimental data. *Proteomics* 20: 5435–5444.

Patient, S. and M. Martin. 2011. Annotating UniProt metagenomic and environmental sequences in UniMES. *Bioinformatics 2011—Proceedings of the International Conference on Bioinformatics Models, Methods and Algorithms*, pp. 367–368.

Rahman, M.T., M.S. Uddin, R. Sultana, A. Moue, and M. Setu. 2013. Polymerase chain reaction (PCR): A short review. *Anwer Khan Modern Medical College Journal* 4: 30–36.

Rameshwari, R. 2011. Systematic and integrative analysis of proteomic data using bioinformatics tools. *International Journal of Advanced Computer Science and Applications* 2: 29–35.

Rast, J.P. and C. Messier-solek. 2008. Marine invertebrate genome sequences and our evolving understanding of animal immunity. *Biological Bulletin* 214: 274–283.

Reddy, B., A. Milshteyn, Z. Charlop-Powers, and S. Brady. 2014. ESNaPD: A versatile, web-based bio-informatics platform for surveying and mining natural product biosynthetic diversity from metagenomes. *Chemistry and Biology* 21: 1023–1033.

Rhee, S.Y., J. Dickerson, and D. Xu. 2006. Bioinformatics and its applications in plant biology. *Annual Review of Plant Biology* 57: 335–360.

Rochfort, S. 2005. Metabolomics reviewed: A new 'Omics' platform technology for systems biology and implications for natural products research. *Journal of Natural Products* 68: 1813–1820.

Savino, R., S. Paduano, M. Preianò, and R. Terracciano. 2012. The proteomics big challenge for biomarkers and new drug-targets discovery. *International Journal of Molecular Sciences* 13: 13926–13948.

Schatz, M.C., J. Witkowski, and W. Richard McCombie. 2012. Current challenges in de Novo plant genome sequencing and assembly. *Genome Biology* 13(4):243.

Schnable, P.S., D. Ware, R.S. Fulton, J.C. Stein, F. Wei, S. Pasternak, C. Liang et al. 2009. The B73 maize genome: Complexity, diversity, and dynamics. *Science* 326: 1112–1115.

Searls, D.B. 2000. Bioinformatics tools for whole genomes. *Annual Review of Genomics and Human Genetics* 1: 251–279.

Selig, E.R., C. Longo, B.S. Halpern, B.D. Best, D. Hardy, C.T. Elfes, C. Scarborough, K.M. Kleisner, and S.K. Katona. 2013. Assessing global marine biodiversity status within a coupled socioecological perspective. *PLOS ONE* 8(4): e60284.

Sharma, V.K., N. Kumar, T. Prakash, and T.D. Taylor. 2009. MetaBioME: A database to explore commercially useful enzymes in metagenomic datasets. *Nucleic Acids Research* 38: D468–D472.

Sithranga Boopathy, N. and K. Kathiresan. 2010. Anticancer drugs from marine flora: An overview. *Journal of Oncology* 2010 (January): 214186.

Skolnick, J., J.S. Fetrow, and A. Kolinski. 2000. Structural genomics and its importance for gene function analysis. *Nature Biotechnology* 18: 283–287.

Slattery, M., S. Ankisetty, J. Corrales, K. Erica Marsh-Hunkin, D.J. Gochfeld, K.L. Willett, and J.M. Rimoldi. 2012. Marine proteomics: A critical assessment of an emerging technology. *Journal of Natural Products* 75 (10): 1833–1877.

Söding, J., A. Biegert, and A.N. Lupas. 2005. The HHpred interactive server for protein homology detection and structure prediction. *Nucleic Acids Research* 33: W244–W248.

Teufel, A., M. Krupp, A. Weinmann, and P.R. Galle. 2006. Current bioinformatics tools in genomic biomedical research (review). *International Journal of Molecular Medicine* 17: 967–973.

The Arabidopsis Genome Initiative. 2000. Analysis of the genome sequence of the flowering plant *Arabidopsis thaliana*. *Nature* 408: 796–815.

Vermeulen, N. 2013. From Darwin to the census of marine life: Marine biology as big science. *PLOS ONE* 8(1): e54284.

Von Mering, C., M. Huynen, D. Jaeggi, S. Schmidt, P. Bork, and B. Snel. 2003. STRING: A database of predicted functional associations between proteins. *Nucleic Acids Research* 31(1): 258–261.

Wang, D.-Z., Z.-X. Xie, and S.-F. Zhang. 2014. Marine metaproteomics: Current status and future directions. *Journal of Proteomics* 97: 27–35.

Westbrook, J. 2003. The protein data bank and structural genomics. *Nucleic Acids Research* 31: 489–491.

Wu, C.H., L.S. Yeh, H. Huang, L. Arminski, J. Castro-Alvear, Y. Chen, Z. Hu et al. 2003. The protein information resource. *Nucleic Acids Research* 31: 345–347.

Wu, D. and G.K. Smyth. 2012. Camera: A competitive gene set test accounting for inter-gene correlation. *Nucleic Acids Research* 40: e133.

Zhou, X., L. Ren, Q. Meng, Y. Li, Yu. Yude, and J. Yu. 2010. The next-generation sequencing technology and application. *Protein & Cell* 1: 520–536.

Ziemert, N., S. Podell, K. Penn, J.H. Badger, E. Allen, and P.R. Jensen. 2012. The natural product domain seeker NaPDoS: A phylogeny based bioinformatic tool to classify secondary metabolite gene diversity. *PLOS ONE* 7(3): e34064. doi:10.1371/journal.pone.0034064.

Section X

Marine Omics and Its Application in Nanotechnology

23

Omics in Marine Nanotechnology

Nabeel M. Alikunhi and Kandasamy Saravanakumar

CONTENTS

Preface

Marine environments enthrall human beings due to its incredible species diversity. Species thrives in marine ecosystems are unique in nature and complex in association due to its extreme physicochemical characters. Nanotechnology, a technology with billionth of a meter, is the convergence of several scientific divisions. Let us now ponder about merging of marine science into nanotechnology, the marine nanotechnology. Marine environments exhibit remarkable biomimetic nanostructures in its cell wall, shells, pearls, and skeletons that motivated researchers of the world. As an emerging science in its infancy, marine nanotechnology promises the nanoscale manufacture of materials using marine organisms or influenced from marine environment. Integration of marine nanotechnology with molecular biology, particularly omics, a buzzword of the modern biologists, would be an amazing prospect that is believed to reach the hitherto unreachable and unimagined stages in future decades. This review summarizes the influence of marine organisms on nanoscience and its future prospects through integration with omics.

23.1 Marine Nanotechnology

Nanoscience is an exciting field of research that deals with fundamentals of tiny molecules and structures approximately between 1 and 100 nm. By virtue of smaller size, nanoparticles have a large fraction of surface atoms that increases the surface energy compared to their bulk counterparts. High surface/volume ratio together with size effects (quantum effects) gives nanoparticles distinctive chemical, electronic, optical, magnetic, and mechanical properties than bulk material with indispensable societal impacts. Recent development in nanotechnology has influenced various fields of science and technology with a wide spectrum of potential applications in many areas of human activity. Many commercial products that are currently marketed and are used in routine human life such as sunscreens, paints, semiconductors, and cosmetics have nanoparticles (PEN, 2012). The global markets for nanoparticle products are expected to reach over $1 billion by 2015 (DEFRA, 2007).

Bionanotechnology in general and marine bionanotechnology in particular are more dynamic due to various reasons. Marine environment occupies 71% of the earth encompassing most productive, biologically diverse, and exceedingly valuable ecosystems with a wide variety of flora and fauna. Many marine organisms produce remarkable biomimetic nanostructures in their cell wall, shells, pearls, and skeletons with potential application in nanotechnology. Many commercially available nanomaterial-based products are built by the motivation from nature. For example, sharks and whales are known to have smooth skin that is not affected by fouling organisms (Hui et al., 2006). The smooth skin is actually not so, when observed under scanning electron microscope. The skin surface of those animals is rippled due to the presence of nanoridges that enclose a pore size of 0.2 μm^2, which is below the size of marine fouling organisms. Hence, there is no attachment of biofouling organisms. Based on this, new nanoparticle coating is synthesized to prevent biofouling of ship hulls, which cause increased fuel consumption and thereby economic loss to the shipping industry. Other classic biomimetic nanostructures in marine environment are by diatoms and sponges that are constructed with nanostructured cover of silica, arranged in remarkable architectures (van der Woude et al., 1995). Many organisms in marine environments are still under exploration due to its rich abundance and diversity. For instance, a liter of seawater constitutes more than a billion individual microbes that contribute about 98% of oceanic biomass. The biodiversity-rich marine environment provides great prospects of marine bionanotechnology. Besides being biologically diverse, marine environment comprises rich hot brine pools and cold seeps with distinctive ecological features and extreme physicochemical parameters, creating the most complex environments on earth. The organism surviving in such extreme environments might produce unique metabolites with exclusive chemical properties that are currently being explored for nanoparticle synthesis.

Interaction of biotechnology with nanoscience integrated biology and material science, which resulted in several positive outcomes beneficial for mankind. Integration of biotechnology, particularly genomics and proteomics, in nanotechnology is an emerging field of research with greater potential. Genomics, proteomics, and other omics tools provide essential information about the gene or protein expression in any biological species during nanoparticle synthesis. This helps in understanding the underlying mechanism of nanoparticle synthesis as well as elucidating the pathway of responses induced by nanoparticles in other biological organisms. Nanotechnology with omics, together known as nano-omics, has recently emerged as a new and efficient approach

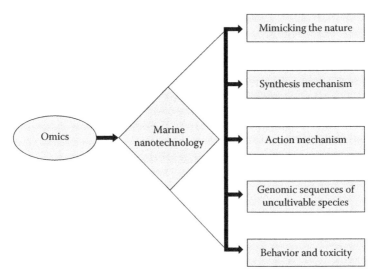

FIGURE 23.1
The conceptual framework of the chapter: integration of molecular biology and marine nanotechnology.

for medical diagnostics and therapy. Involvement of omics in nanoscience is a research at infancy but is beckoning as a promising prospect in various aspects of human life. Omics is also a tool for investigating fate, behaviors, and toxicity of involuntarily released synthetic nanoparticle in the ecosystems and its living components. This chapter deals with involvement of omics in marine organism–derived nanoparticles. The conceptual framework of the chapter is given in Figure 23.1. Possible outcomes expected through the integration of omics with marine nanotechnology are (1) synthesis of nanomaterial by mimicking marine species, particularly the surface structure, (2) understanding the molecular aspect of synthesis mechanism for controlled production of nanoparticles of desired characteristic, (3) delineating the action mechanism of nanoparticles during its interaction with other biological species, (4) identifying the genetic sequences of uncultivable marine species with potential nanoparticle synthesis efficiency, and (5) last but not least, investigating the fate and behavior of nanoparticles on sensitive marine environment and organisms.

23.2 Why Marine Nanoparticles?

Synthesis of nanoparticles is still challenging in material science. Several manufacturing techniques are being employed in synthesis of nanoparticles of superior characteristics. The use of chemical and physical methods for the synthesis of nanoparticles is capital intensive due to inefficient material and energy usage. Further, such methods involve the use of hazardous chemicals or produce toxic chemicals absorbed on the surface that may have adverse effects during application of the synthesized nanoparticles. Hence, there is a growing need to develop environmentally benign synthesis procedures. Consequently, many researchers turned to biological systems for eco-friendly and sustainable synthesis of nanoparticles with different sizes, shapes, and controlled dispersion for advanced biological applications.

It is well known that many biological organisms are capable of producing inorganic materials through intracellular and extracellular methods (Simkiss and Wilbur, 1989). This property is exploited in the biological synthesis of nanoparticles that is now termed as green chemistry approach. In this regard, several marine organisms are recently being experimented for synthesis of various kinds of nanoparticles (Table 23.1). Biological genomic groups from marine environment involved in nanoparticle synthesis are discussed here.

23.2.1 Bacterial Genomes

Many bacterial species are reported to synthesize inorganic nanoparticles through extracellular and intracellular process with properties similar to chemically synthesized materials. Majority of bacterial species reported for synthesis of nanomaterials are confined to metals, metal sulfide, and very low oxides. Nanoparticles synthesized by marine bacterial genomes are with unique properties such as structure, stability, and aggregation than its terrestrial counterpart due to the fact that marine bacteria thrive in extreme environmental conditions. Many of them exist in the sea bottom over millions of years, naturally capable of reducing vast inorganic elements in the deep sea. Deep-sea strain of *Pseudomonas aeruginosa* was capable of synthesizing silver nanoparticles of size between 13 and 76 nm with antimicrobial, antibiofilm, and cytotoxic activity (Ramalingam et al., 2013). Another deep-sea bacterial species *Marinobacter pelagius* has been reported for the synthesis of gold nanoparticle with the size lesser than 10 nm (Sharma et al., 2012). Apart from deep-sea bacteria, shallow-water bacterial genomes isolated from mangrove environments are also widely proven candidates for nanoparticle synthesis. Marine strains of *Escherichia coli* AUCAS 112 isolated from mangrove sediments produced extracellular spherical silver nanoparticles of 5–20 nm size with antimicrobial activity against human bacterial and fungal pathogens (Kathiresan et al., 2010). Mangrove *Streptomyces* sp. BDUKAS10 synthesized bactericidal silver nanoparticles (Sivalingam et al., 2012). Similar to deep-sea bacteria, mangrove-associated bacterial species also survives in extreme environmental conditions. However, instead of stress due to pressure at deep-sea bottom environments, mangrove-associated bacterial species has to overcome highly anaerobic and hypersaline conditions, which makes them comparatively more potent than its terrestrial counterparts.

Many other bacterial species of marine environment have been tested for synthesis of nanoparticles with many beneficial applications. A novel strain of *Pseudomonas* sp. 591786 was reported to produce intracellular silver nanoparticles, polydispersed with different size groups ranging from 20 to 100 nm (Muthukannan and Karuppiah, 2011), while *Vibrio alginolyticus* produced silver nanoparticles with size ranges from 50 to 100 nm (Rajeshkumar et al., 2013). The marine-derived silver-resistant bacteria isolate MER1 is known to produce the nanoparticles with an antimicrobial effect against gram-positive and gram-negative bacteria, yeasts, and fungus (Youssef and Abdel-Aziz, 2013). A majority of these studies were on silver nanoparticle synthesis using bacterial species, while very meager studies focused on other nanoparticle or bacterial groups. Halotolerant *Bacillus megaterium* produced the selenite nanoparticles (Mishra et al., 2011). Marine cyanobacterium, *Oscillatoria willei* NTDM01 produced silver nanoparticles (MubarakAli et al., 2011), while *Phormidium tenue* NTDM05 produced cadmium sulfide nanoparticles of 5 nm (MubarakAli et al., 2012). Marine actinomycetes, *Streptomyces* spp. (VITSTK7 and LK-3), produced silver nanoparticles with antifungal activity (Thenmozhi et al., 2013) and gold nanoparticles with antimalarial activity (Karthik et al., 2014). The iron nanoparticle synthesized by marine actinomycetes *Nocardiopsis* sp. MSA13A has a significant influence on growth and glycolipid biosurfactant production (Kiran et al., 2014). Few researchers

TABLE 23.1

Nanoparticles Synthesized from Different Marine Organisms

Marine Organism	Name of the Species	Type	Size (nm)	Biological Application	References
Marine plants	*Sesuvium portulacastrum*	Silver	50–90	Antimicrobial activity	Nabikhan et al. (2010)
	Rhizophora mucronata	Silver	60–95	Larvicidal	Gnanadesigan et al. (2011)
	Citrullus colocynthis	Silver	85–100	Anticancer	Satyavani et al. (2011)
	Xylocarpus mekongensis	Silver	5–20	Antimicrobial activity	Kathiresan et al. (2012)
	Prosopis chilensis	Silver	5–25	Antibacterial activity	Kathiresan et al. (2012)
	Suaeda monoica	Gold	12.96	Free radical scavenging	Rajathi et al. (2014)
Marine bacteria	*Escherichia coli*	Silver	5–20	Antimicrobial activity	Kathiresan et al. (2010)
	Brevibacterium casei	Silver	—	—	Kiran et al. (2011)
	Bacillus megaterium	Selenite	—	—	Mishra et al. (2011)
	Oscillatoria willei	Silver	—	—	MubarakAli et al. (2011)
	Marinobacter pelagius	Gold	10	—	Sharma et al. (2012)
	Pseudomonas aeruginosa	Silver	10.4	Antimicrobial activity	Boopathi et al. (2012)
	Streptomyces sp.	Silver	21–48	Antibacterial activity	Sivalingam et al. (2012)
	Phormidium tenue	Cadmium	5	Biolabel application	MubarakAli et al. (2012)
	Vibrio alginolyticus	Silver	50–100	—	Rajeshkumar et al. (2013)
	Marine bacteria MER1	Silver	—	Antimicrobial activity	Youssef and Abdel-Aziz (2013)
	Pseudomonas aeruginosa	Silver	—	—	Shivakrishna et al. (2013)
	Streptomyces sp.	Silver	35.2	Antifungal activity	Thenmozhi et al. (2013)
	Streptomyces sp.	Gold	5–50	Antimalarial activity	Karthik et al. (2014)
	Bacillus subtilis MSBN17	Silver	60	—	Sathiyanarayanan et al. (2013)
	Bacillus megaterium	Gold	5–20	Antibacterial activity	Sathiyanarayanan et al. (2014)
	Penicillium fellutanum	Silver	5–25	—	Kathiresan et al. (2009)

(Continued)

TABLE 23.1 (*Continued*)

Nanoparticles Synthesized from Different Marine Organisms

Marine Organism	Name of the Species	Type	Size (nm)	Biological Application	References
Marine fungi	*Aspergillus niger*	Silver	35	Antimicrobial activity	Kathiresan et al. (2010)
	Aspergillus terreus	Lead	—	—	Jacob and Chakravarthy (2014)
	Aspergillus sydowii	Gold	8.7–15.6	—	Vala et al. (2014)
Marine algae	*C. calcitrans, C. salina, I. galbana, and T. gracilis*	Silver	53.2–71.9	Antimicrobial activity	Merin et al. (2010)
	Porphyra vietnamensis	Silver	13 ± 3	Antibacterial activity	Venkatpurwar and Pokharkar (2011)
	Stoechospermum marginatum	Gold	18.7–93.7	Antimicrobial activity	Rajathi et al. (2012)
	Galaxaura elongata	Gold	3.85–77.13	Antibacterial activity	Abdel-Raouf et al. (2013)
	Pterocladia capillacea, Jania rubins, Ulva fasciata, and Colpomenia sinuosa	Silver	7, 7, 12, and 20	Antibacterial activity	El-Rafie et al. (2013)
	Codium capitatum	Silver	—	—	Kannan et al. (2013a)
	Chaetomorpha linum	Silver	30	—	Kannan et al. (2013b)
	Caulerpa peltata, Hypnea valencia, and Sargassum myriocystum	Zinc	36	Antimicrobial activity	Nagarajan and Kuppusamy (2013)
	Turbinaria conoides	Silver and gold	Ag, 2–17; Au, 2–19	Antimicrofouling	Vijayan et al. (2014)
	Sargassum swartzii	Gold	35	Cytotoxic activity	Dhas et al. (2014b)
	Sargassum plagiophyllum	Silver	21–48	Antibacterial activity	Dhas et al. (2014a)
	Sargassum muticum	Zinc	30–57	—	Azizi et al. (2014)
	Amphora sp.	Silver	20–25	Antimicrobial activity	Jena et al. (2012)
Marine sponge	*Acanthella elongata*	Gold	7–20	—	Inbakandan et al. (2010)
	Acanthella elongata	Silver	—	Antimicrofouling	Inbakandan et al. (2013)

used bacterial derivatives for nanoparticle synthesis. For instance, glycolipid derived from marine bacteria *Brevibacterium casei* MSA19 produced stable silver nanoparticles (Kiran et al., 2011). Polysaccharide bioflocculant of marine sponge–associated *Bacillus subtilis* MSBN17 produced silver nanoparticles (Sathiyanarayanan et al., 2013). Recently, exopolysaccharide derived from marine bacterium *B. megaterium* was used as the reducing and stabilizing agent for the synthesis of the gold nanoparticles with significant antibacterial activity against clinical human pathogens (Sathiyanarayanan et al., 2014).

23.2.2 Fungal Genomes

The fungal community is ubiquitous in marine environment, involved in the nutrient cycles and reduction of the metal pollution. Marine fungus received remarkable attention in bioprospecting since they secrete large amounts of enzymes of unique property and are simpler to deal in the laboratory. They received due attention in nanotechnology because of the same reasons. The use of fungi is potentially exciting and several researchers reported marine fungal species capable of nanoparticle synthesis with significant applications. *Penicillium fellutanum* isolated from mangrove environments was the first report on synthesis of the silver nanoparticles by marine fungus (Kathiresan et al., 2009). Later, marine strains of *Aspergillus niger* AUCAS 237 isolated from mangrove sediments produced extracellular silver nanoparticles of less than 35 nm (Kathiresan et al., 2010) with antimicrobial properties against human pathogens. Further, few other researchers reported synthesis of various nanoparticles using different marine fungal species, dominated by *Aspergillus* species. Marine strain of *A. flavus* produced silver, *A. sydowii* synthesized gold (Vala et al., 2014), and *A. terreus* was used for lead nanoparticle synthesis (Jacob and Chakravarthy, 2014). Besides fungi, marine yeast species were reported for nanoparticle synthesis. Intracellular synthesis of gold nanoparticles of 15 nm size by the tropical marine yeast *Yarrowia lipolytica* NCIM 3589 was the first reported dimorphic marine fungus (Agnihotri et al., 2009). Of several marine yeast species screened, the culture filtrate of mangrove-derived yeast *Pichia capsulata* exhibited the most efficient production of silver nanoparticles, which occurred within minutes (Subramanian et al., 2010). Marine yeast, *Rhodosporidium diobovatum* was reported to be capable of intracellular synthesis of stable lead sulfide nanoparticles (Seshadri et al., 2011).

23.2.3 Plant Genomes

Many coastal plants have been investigated for synthesis of nanoparticles, which was termed as green nanoscience. Plant-based methods can be easily conducted at room temperature and pressure, without any sophisticated technical requirements. Furthermore, plant-based nanoparticle synthesis approaches are easy to conduct due to easy accessibility and environment friendliness. Plants, particularly those of coastal origin, serve as excellent reducing agents, which include phenolic compounds, alkaloids, and sterols. These coastal plant metabolite materials are rich in natural capping agents and devoid of toxicants. Several marine floral species of salt marsh, mangrove, seaweeds, and microalgae have reported for nanoparticle syntheses. Synthesis of antimicrobial silver nanoparticles by leaf callus of *Sesuvium portulacastrum* was proved better than leaf extract of the same species (Nabikhan et al., 2010). Coastal plant, *Citrullus colocynthis*, extracts produced silver nanoparticles with anticancer properties (Satyavani et al., 2011). Halophyte plant species *Suaeda monoica* produced gold nanoparticles with higher free radical scavenging property (Rajathi et al., 2014). Mangrove plant species were also tested for nanoparticle synthesis. *Rhizophora mucronata* and *Avicennia marina* extracts produced silver nanoparticles with larvicidal and antimicrobial activity, respectively (Gnanadesigan et al., 2011, 2012). Kathiresan et al. (2012) screened 13 plants of coastal origin for silver nanoparticle synthesis. They observed mangrove species, *Xylocarpus mekongensis*, followed by salt marsh plant *Suaeda maritima* as the most potential candidates of silver nanoparticle synthesis. *Prosopis chilensis* produced antibacterial silver nanoparticles to control vibriosis in *Penaeus monodon* culture (Kandasamy et al., 2013). Marine algae are also proven sources for synthesis of various nanoparticles with potential features. Marine algae, such as *Chaetoceros calcitrans*, *Chlorella salina, Isochrysis galbana,* and *Tetraselmis gracilis*, produced antimicrobial silver nanoparticles of 53.2–71.9 nm size (Merin et al., 2010). Marine diatom *Amphora* sp. synthesized antimicrobial silver nanoparticles (Jena et al., 2012). Marine seaweed species were tested for nanoparticle synthesis. *Sargassum myriocystum* produced zinc oxide nanoparticles of 36 nm with significant antimicrobial activity against the human pathogens (Nagarajan and Kuppusamy, 2013). Similarly, the polysaccharide aqueous extract of *Sargassum muticum* showed significant synthesis of zinc oxide nanoparticles of size 5–15 nm, which was smaller than the earlier mentioned (Azizi et al., 2014). Other *Sargassum* species, *S. swartzii* produced silver nanoparticles of 35 nm size with higher cytotoxic activity against the human cervical carcinoma cell line (Dhas et al., 2014b), while *S. plagiophyllum* synthesized antibacterial silver nanoparticle (Dhas et al., 2014a). Brown alga, *Stoechospermum marginatum*, showed remarkable synthesis of gold nanoparticles with higher antimicrobial activity (Rajathi et al., 2012). Silver and gold nanoparticles synthesized by *Turbinaria conoides* exhibited antibiofilm activity against marine biofilm forming bacteria (Vijayan et al., 2014). Yet another seaweed species, *Galaxaura elongata*, synthesized gold nanoparticles with higher antibacterial activity (Abdel-Raouf et al., 2013).

23.2.4 Marine Faunal Species

Few marine faunal species were tested for nanoparticle production. Inbakandan et al. (2010) reported the potential of marine sponge *Acanthella elongata* on the synthesis of gold nanoparticles 7–20 nm in size. The same sponge species was later proved to be an excellent source of silver nanoparticles with antimicrofouling effect (Inbakandan et al., 2013). Biosynthesis of antimicrobial silver nanoparticles of 40–90 nm size using marine polychaete extract at room temperature was reported (Singh et al., 2014).

23.3 Synthesis Mechanisms

Biological approaches for synthesis of nanoparticles have been attempted with several marine organisms. Both unicellular and multicellular marine species have been investigated for production of various nanoparticles. The basic procedure in biosynthesis is the incorporation of the growth media with desired metal salts to prepare its own metal nanoparticles. A number of synthesis mechanisms have been postulated for marine organism–mediated nanoparticle production.

Nanoparticles are synthesized both at extracellular and intracellular location of marine organisms. Many marine bacterial species deposit crude metal inside their bodies at the metallurgical sites. The enzymes reduce these metals from trivalent to zerovalent state and to deposit it in the intracellular region of the cell in the form of nanoparticles, termed as intracellular nanoparticles. Extracellular synthesis of nanoparticles is based on electrostatic forces of attraction between negatively charged phospholipids occupying the membranes and positively charged metal ions. It is widely believed that nanoparticles synthesized at intracellular locations are with superior quality due to their comparatively smaller size than that of extracellular nanoparticles. Size and shape of the nanoparticles has large implication for any particular application process. However, there exist certain complications in retrieving the nanoparticles synthesized at intracellular locations, which is mainly due to the additional reduction and isolation steps. These processes are usually time consuming and expensive due to high energy requirements. Hence, the downstream processing of intracellular nanoparticles is difficult, while nanoparticles synthesized at extracellular regions of marine organisms could be immediately tapped for intended applications.

Marine bacteria have been explored for nanoparticle synthesis of potential benefits. Among marine bacterial species, prokaryotic group have been extensively researched for synthesis of metallic nanoparticles. Both extracellular and intracellular driven synthesis of nanoparticles could be achieved using marine bacteria. One of the mainly believed mechanisms of metal nanoparticle biosynthesis by microorganisms is bioreduction in which myriads of proteins, carbohydrates, and biomembranes are involved. In this method, nanoparticles form on cell wall surfaces by trapping the metal ions, probably occurring due to the electrostatic interaction between the metal ions and positively charged groups in enzymes present at the cell wall. This is followed by enzymatic reduction of the metal ions, leading to their aggregation and the formation of nanoparticles. The microbial cell reduces metal ions by the use of specific reducing enzymes like NADH-dependent reductase or nitrate-dependent reductase. Marine bacteria possess a rich diversity enzymes usually produced in order to survive in extreme environmental conditions. In one of our earlier studies on silver nanoparticle synthesis using *E. coli* isolated from anaerobic muddy mangrove environment, the culture filtrate with silver nitrate was detected with protein of molecular weight estimated as 50 kDa (Kathiresan et al., 2010). This indicates secretion of some protein components into the medium by the bacterial biomass when added with silver nitrate. This protein was hypothesized as nitrate reductase enzyme that might have played an important role in the reduction of the metal ions in the form of nanoparticles. Similarly during the synthesis of gold nanoparticles using *M. pelagius*, the analysis of the spectra indicated the presence of proteins in the sample that may have been used for the reduction and/or capping and stabilization of the nanoparticles. Protein can bind to the nanoparticles either by interaction through its free amine group or by cysteine residues or through the amine groups (Sharma et al., 2012). The marine cyanobacterium, *O. willei* NTDM01, secreted protein that supported the reduction of silver ions and stabilization of silver nanoparticles (MubarakAli et al., 2011). The possible mechanism assumed during nanoparticle synthesis by *P. aeruginosa* is also of high protein and secondary metabolite content (Ramalingam et al., 2013). Hence, identification and characterization of various proteins responsible for nanoparticle synthesis are imperative for a better understanding of the possible mechanism. Studies on enzyme structure and the genes that code these enzymes may help improve our understanding of metal nanoparticle synthesis.

Marine fungal species generally are advantageous in nanoparticle synthesis than its bacterial counterparts particularly due to high extracellular enzyme-secreting efficiency. They release very vital enzymes and proteins in sufficient concentrations enabling easier bioreduction of corresponding metal salts to form the biochemically reduced metallic ions as zerovalent nanoparticles. However, a significant drawback is the genetic manipulation of eukaryotic organisms through overexpressing specific enzymes that are identified to be involved in synthesis of metallic nanoparticles is relatively much more difficult than that in prokaryotes. In one of our studies on extracellular synthesis of silver nanoparticles by a marine fungus, *P. fellutanum*, a single prominent protein band with molecular weight of

70 kDa was detected in the culture filtrate (Kathiresan et al., 2009). Similarly during silver nanoparticle synthesis using *A. niger*, the purification of protein in culture filtrate with silver nitrate was 1.73-fold and the molecular weight of the protein was estimated as 50 kDa (Kathiresan et al., 2010). These studies indicate the secretion of some enzymes such as nitrate reductase components into the medium by the fungal biomass, which might play an important role in the reduction of the metal ions in the form of nanoparticles. The reductase gains electrons from NADH and oxidizes it to NAD+. The enzyme is then oxidized by the simultaneous reduction of metal ions. Possible mechanisms for silver nanoparticle biosynthesis by marine yeasts may be more or less the same. The molecular weight of the ammonium sulfate–precipitated protein sample in the culture filtrate of marine yeast *P. capsulata* revealed a single protein band, with a molecular weight of 55 kDa, which is similar to that of NADH-dependent nitrate reductase (Subramanian et al., 2010). Along with this, the membrane-bound (as well as cytosolic) oxidoreductases and quinones might have played an important role in this process. The inherent presence of reductases or proteases may play a vital role in the reduction of the gold salt into nanoparticles (Agnihotri et al., 2009). However, in the case of *R. diobovatum*, a sulfur-rich peptide acted as a capping agent for the synthesis of lead sulfide nanoparticles (Seshadri et al., 2011). Probably multiple factors together determine the nanoparticle synthesis by any marine organism. However, the exact reaction mechanism leading to the formation of nanoparticles by marine organisms is yet to be elucidated.

Plant-mediated nanoparticle synthesis received significant attention due to the fact that this route of nanoparticle synthesis involves lesser technical hurdles and nontoxic substrates, but synthesized products could be exploited for multiple applications. The plant materials are easy and inexpensive for their culture as compared to those of microorganisms. Plant extracts, particularly those of coastal origin, are believed to be rich in reducing agents and stabilizing agents that play a key role in nanoparticle synthesis. The presence of several bioactive compounds in the leaf extracts such as alkaloids, amino acids, alcoholic compounds, and several other chelating proteins are also responsible for the reduction of nanoparticles. Alcoholic intermediates such as those of quinol and chlorophyll pigments reduce silver ions to the zerovalent forms. During one of our previous studies (Nabikhan et al., 2010) on silver nanoparticle synthesis by callus and leaf extract of salt marsh plant *S. portulacastrum*, Fourier transform infrared spectroscopy (FTIR) measurements revealed prominent peaks corresponding to amide I, II, and III, aromatic rings, geminal methyls, and ether linkages. The observed peaks suggest the presence of various compounds such as flavanones and terpenoids. A similar observation was noticed in a subsequent study conducted for the screening of mangroves and other coastal plants for antimicrobial silver nanoparticle synthesis (Kathiresan et al., 2012). Mangrove plant extract, *X. mekongensis*, showed prominent peaks corresponding to amide I, II, and III, aromatic rings, geminal methyls, and ether linkages. The adsorption on the surface of metal nanoparticles is a characteristic of flavanones and terpenoids (Shankar et al., 2004), which may be able possibly by interaction through carbonyl groups or π-electrons in the absence of other strong ligating agents in sufficient concentration (Shankar et al., 2004). The presence of reducing sugars could be another reason responsible for the reduction of metal ions and formation of silver nanoparticles. It is also possible that the terpenoids play a role in reduction of metal ions by oxidation of aldehydic groups in the molecules to carboxylic acids (Shankar et al., 2004). The presence of the peaks in the amide I and II regions characteristic of proteins/enzymes has been found to be responsible for the reduction of metal ions for the synthesis of metal nanoparticles (Ahmad et al., 2003; Mukherjee et al., 2001a,b). It is well known that proteins can bind to silver nanoparticle through either free amine groups or cysteine residues in the proteins (Gole et al., 2001). The coastal plants are generally rich in polyphenols like tannic acids and these plant-derived compounds are efficient reducing agent in the synthesis of silver nanoparticles

(Sivaraman et al., 2009). Phenolic substances have a considerable role in nanoparticle synthesis using seaweed *S. myriocystum* (Nagarajan and Kuppusamy, 2013). During synthesis of silver nanoparticles using *S. swartzii*, proteins were detected in FTIR (Dhas et al., 2014b). Thus, as in the case of microbial species, multiple factors may determine nanoparticle synthesis of coastal plants. However, the exact mechanism is yet to be elucidated.

The biosynthesis of nanoparticles by biological species particularly those of coastal origin depends on culture conditions. It is necessary to mimic the natural conditions at marine environment to optimize the synthesis. Standardization of the culture conditions for synthesis of nanoparticles is highly imperative. Nanoparticle synthesis depends critically on the operational parameters such as pH and temperature. Variations in the physical conditions often lead to different size ranges of the synthesized particles. The size of the synthesized nanoparticles is a very vital parameter for specific application because the novelty in terms of major physicochemical properties will be more pronounced at smaller sizes. Further, the biochemical and molecular mechanism of these processes needs to be elucidated for the advancement in the synthesis of nanoparticles. The molecular-level investigation of the carbohydrates, lipids, DNA, and proteins involved in the synthesis of nanoparticles is very much essential. Genetic modification could generate the improvements in size, shape, and other characteristics of the nanoparticles. Hence, it is important to study the synthesis mechanism of nanoparticles and to elucidate biochemical pathways involved in metal ion reduction by the different marine organisms.

23.4 Omics in Nanotechnology

Nanobiotechnology in general is the integration of biotechnology with nanotechnology for developing biosynthetic and eco-friendly technology for synthesis of nanomaterials with remarkable application beneficial for human beings. This interdisciplinary research has amplified the prospects of material research, drawing new initiative from biological systems. Marine organism itself exhibits remarkable merging of the biotechnology and nanotechnology. Corals and sponges synthesize inorganic nanoparticles in nature either inside or outside cells. Coral synthesis of calcium and sponges does possess a silicon dioxide structure called spicules. Mechanisms that control nanofabrication and properties of nanoparticles and composite structures such as seashells, pearls, bone, and silica are yet to be understood. The ability to design synthetic materials at the same level of complexity of nature is unachieved but a vital challenge in the field of nanoscience. Recently, many researchers have initiated technology development for biological self-assembly processes using molecular methods. Such kind of studies includes unveiling the complexity of naturally occurring nanostructures, particularly those in the biologically diverse and relatively unexplored marine environment. A possible route to the development of bioinspired and biomimetic systems is the understanding and exploitation of this self-assembly through molecular tools, mainly the omics technology. The omics comprising genomics, transcriptomics, proteomics, and metabolomics are advanced tools in understanding biological system processes. There is an accelerated progression in convergence of omics into nanotechnology for understanding different aspects of the complexity of nature's design.

Nanoparticle synthesis using marine organisms is a complex process. Although many researchers postulate the protein, enzymes, and other biochemical materials, a convincing evidence on the mechanism on nanoparticle production is yet to be revealed. Certainly, nanomaterial productions involve several mechanisms related to omics. If the gene or a particular protein involved in the reduction of metal nanoparticles is revealed, an efficient synthesis of nanoparticles could be achieved. Hence, special focus on nano-omics, particularly on marine biological synthesis of nanoparticles, would be a remarkable achievement for the production of more potent nanoparticles. Another major challenge is the inaccessibility of genomes of unculturable marine organisms including many microbial species at deep-sea environment. Marine organisms are usually very complex in their association and challenging for isolation. Consequently, many marine organisms could not be nurtured in normal laboratory conditions. Although very abundant in marine environment, only a very meager percentage of marine microbial species could be cultured under laboratory conditions and only few of them are taxonomically identified (Allen et al., 2004). Efficient methods for accessing high-quality samples of DNA from extreme marine environment such as hot brine pools and cool seeps have enabled through metagenomic sequencing. Several researchers have developed new enzymes from these unique genomes. Such kind of technological advancements is still awaited in marine nanoparticle synthesis. Access to unculturable marine organisms using next-generation sequencing could help us to explore seemingly inaccessible wealth of gene clusters for nanoparticle synthesis.

The integration of nanotechnology with omics serves as a platform for the development of nontoxic and environmentally friendly procedures for the synthesis and assembly of metal nanoparticles. To effectively utilize these nanoparticles in different scientific fields, it is imperative to have proper understanding of the biochemical and molecular mechanisms of their synthesis. Information on genomics and proteomics involved behind the synthesis would help in obtaining nanoparticles with better chemical composition, shape, size, and monodispersity, which are critical characteristics in application point of view. Control of these properties through molecular mechanisms particularly genomics and proteomics would be an incredible achievement in nanoscience.

Applications of nanoparticles are increasing abruptly due to its characteristic properties. Many of synthetic nanoparticles interact directly with sensitive marine ecosystem and the inhabited organisms. Nanoparticle-based products that are commercialized for human usage are prone to reach marine environments through direct and indirect pathways. Nanoparticle-bounded products such as antifouling paints are even directly applied in marine fields. Toxicology studies through omic-based tools are highly warranted for better understanding of the side effects of the synthetic nanoparticles. Omics tools provide useful platform for studying molecular expression of nanoparticles in biological species. This provides better understanding of the toxicity mechanisms of the nanoparticle interaction with marine biological species. Microarray and real-time PCR techniques are widely employed to elucidate the pathway-specific responses induced due to nanoparticle interaction. The genomic data provide essential information about the nanoparticle response of an organism, while proteomic data provide information on differential protein expression of the organism when exposed to the nanoparticles.

23.5 Nano-Omics in Drug Discovery

Convergences between omics and nanotechnology have brought significant discoveries in modern science particularly in human health and integrative biological systems. The integration of nanotechnology with omics has influenced nanomedicine research that focuses mainly on medical proteomics and genomics on disease diagnosis. The combination of these research arenas has developed highly sensitive diagnosis and therapy tools of biomedical applications. Nanoscale innovations in chip technology, detection capabilities, and molecular imaging provide increased sensitivity, resolution, and detection limits of the available methodologies. The proteins and genes that are involved in cancer and other fatal diseases might be available in very minute quantity especially in their early stages. The integration of the nanotechnology and proteomics, the nano-omics, provides extensive platform for the sensitive detection of low abundant proteins that are biomarkers in many dreadful human diseases (Shrivastava et al., 2008). Basically nanotechnology-integrated proteomics and genomics research are focused not only on disease pathogenesis but also on other clinical applications such as identification of the candidate marker for the diagnosis and early detection of the disease, elucidation of the mechanism of drug action, and identification of the next drug targets. The treatment of any human disease basically depends on the identification of a target and delivery of a therapeutic agent that either causes the function to be restored, switches off inappropriate activity, or in the case of cancer, destroys the cell. However, many pharmaceuticals are limited in their development or application because of poor solubility, poor stability, or side effects in inappropriate tissues. Manipulating the composition of a drug formulation at the nanoscale can resolve some of these issues. Nanosized particles provide the opportunity to deliver drugs, heat, light, or other substances to specific cells in the human body. The nanoparticles can be taken up by specific cells where the nanoparticles dissolve to release the protein or drug. This technology has already been used to effectively treat and eliminate tumors in animal models. By this way, it allows treatment of diseases or injuries within the targeted cells, thereby minimizing the damage to the existing healthy cells in the human body. Hence, the targeted therapeutic drugs are among the largest nanomedicine applications that directly act at disease sites, increasing effectiveness while reducing side effects. However, usage of the nanoparticles in pharmaceutical requires proper understanding on nanoparticle toxicity.

Marine influence on nanomedicine is heavily focused on the synthesis of antimicrobial nanoparticles. For instance, nanoparticles produced using marine organisms have shown remarkable antimicrobial activity against human microbial pathogens. Studies have demonstrated that specially formulated metal oxide nanoparticles have good antibacterial activity. The antimicrobial formulations that comprise nanoparticles could be effective bactericidal materials (Fresta et al., 1995; Hamouda et al., 1999). Because of the developing resistance of bacteria against bactericides and antibodies and the irritant and toxic nature of some antimicrobial agents, biomaterial scientists have focused their research on nanosized metal particles particularly on silver due to naturally recognized antimicrobial activity. Silver, in its many oxidation states (Ag^0, Ag^+, Ag^{2+}, and Ag^{3+}), has been recognized as an element with strong biocidal action against many bacterial strains and microorganisms (Lansdown, 2002). Silver nanoparticles synthesized using marine organisms are among the highly potential antimicrobial agent (Kathiresan et al., 2010). However, the mechanisms of antimicrobial action of these nanoparticles are still unknown. Silver works in a number of ways to disrupt critical functions in a microorganism and several hypotheses have, however, been published. Silver cation binds strongly to electron donor groups containing sulfur, oxygen, or nitrogen. Biological molecules generally contain these components in the form of negatively charged side groups such as thio(sulfhydryl), aminoimidazole, imidazole, carboxylate, and phosphate groups or other charged groups distributed throughout microbial cells. This binding reaction alters the molecular structure of the macromolecule. Silver simultaneously attacks multiple sites within the cell to inactivate critical physiological functions such as cell wall synthesis and translation, protein folding and function, and electron transport, which is important in generating energy for the cell. The binding of silver to bacterial DNA may inhibit a number of important transport processes such as phosphate and succinate uptake and can interact with cellular oxidation processes as well as the respiratory chain (Vermieren et al., 2002). However, exact mechanisms of antimicrobial action from silver nanoparticles are yet to be understood. As in the case of silver nanoparticles, the bioactivity produced by nanoparticles from marine organisms is not thoroughly investigated at molecular level. Nano-omics provides an avenue for this kind of research, which would help us to enhance the potential benefits of these tiny particles.

23.6 Other Applications of Marine Nanoparticles

Metal nanoparticles have been used in diverse applications due to their novel and improved physical, chemical, and biological properties and function by virtue of their nanoscale size. The synthesis of metallic nanoparticles particularly using marine organisms is an active area of application-oriented research that has elicited interest of various researchers. Apart from significant medical application of nanoparticles, it has numerous other applications in diverse areas such as electronics, agriculture, environmental, cosmetics, coatings, packaging, etc. For example, nanoparticles can be induced to merge into a solid at relatively lower temperatures, often without melting, leading to improved and easily manufactured coatings for electronics applications (e.g., capacitors). Nanoparticles are used in enhancing agricultural yield and food security (Mohammed Fayaz et al., 2010). Nanoporous zeolites can be used for slow release of water and fertilizers and nanosensors for monitoring soil quality and plant health. Silver nanoparticles are used in bioremediation process (Bystrzejewska-Piotrowska et al., 2009). They are capable of degrading pesticides and killing human pathogenic bacteria. Nanoparticles are also used as the antifouling agent in the civil contraction materials. The bactericidal effect of the silver nanoparticles increased the interest to use in the food preservation and prevent the bacterial spoilage in the consumer goods of market products (Bradford et al., 2009; Fabrega et al., 2011; Mueller and Nowack, 2008). With increasing usage of various nanoparticles in diverse field, it is necessary to develop nanoparticles with novel characteristics. Marine environment would be an excellent source for such type of exploration studies because many of the marine species survive in extreme environmental conditions by producing novel metabolites. Marine-based nanoparticles with intervention of omics could be a breakthrough in many research arenas.

23.7 Toxicology of Nanoparticles

Many metal nanoparticles are directly or indirectly discarded to sensitive marine environments without proper treatment procedures. Like its bulk counterparts, metal nanoparticles mainly reach to marine ecosystems through aerial deposition, effluents dumping, and river runoff. Nanoparticles are also widely found in nature in the form of ash, dusts, aerosols, and metal oxide particles of the earth's crust. Many of the commercially important nanoparticles including silver are considered as contaminants in marine environment particularly due to their toxic properties on biological organisms when available above the threshold. The toxicology of such nanoparticles that are inevitably released to marine ecosystems is of complex concern. The disposed nanoparticles will have significant negative impacts on the sensitive marine ecosystems like coral reef and associated species. Several researchers studied the biological and toxicity of the metal nanoparticles on the aquatic biota due to increased usage in recent years. The dissolution of nanoparticles such as Ag, Cu, or Zn in the environment induces the oxidative stresses and sublethal effects to many organisms, and this could result in chronic health impacts (Doiron et al., 2012). Nanoparticles in aquatic systems promote rapid sedimentation rate, distress in benthic organisms, iron transports and respiratory system of aquatic animals, and alteration in feeding mechanism of faunal species by clogging of gills, particularly in filter-feeding invertebrates (Baker et al., 2014). Embryos of many marine species settle in shallow coastal waters, putting them at a greater risk from point source of nanoparticle discharge outlets. Nanoparticles released to the ecosystem may affect the natural marine microbial communities (Fabrega et al., 2011), due to the fact that many nanoparticles are with immense antimicrobial potential (e.g., silver) (Kathiresan et al., 2012). This leads to significant loss of the bacterial community in the environment (Cauerhff and Castro, 2013), which affects the environmental process such as biogeochemical cycling (Fabrega et al., 2011). Nanoparticles also adversely affect other primary producers in the marine environment. TiO_2 nanoparticles are toxic to phytoplankton when illuminated by sunlight (Miller et al., 2012). Adverse effects of nanoparticles on primary producers in marine environment like phytoplankton will have extensive influence on nutrient production, nutrient cycling, and food availability in marine trophic webs. Biomagnification during trophic transfer that finally affects the human health of the end consumer is yet another drawback of nanoparticle bioaccumulation in marine environment.

Although the toxicity of nanoparticles has adverse effect on marine environment and organisms, the positive perspective of this could be utilized through several means. Many nanoparticles, particularly silver, are of superior antimicrobial potential. Many *in vivo* and *in vitro* studies of silver nanoparticles against bacterial and fungal pathogens exhibited positive outcome (Kathiresan et al., 2010, 2012; Ramalingam et al., 2013). The antimicrobial potential of the nanoparticles could be used an alternative solution to antibiotic application in artificial culture systems. We tested the role of silver nanoparticles against vibriosis (caused due to gram-negative *Vibrio* bacteria) in tiger prawn *P. monodon* culture (Kandasamy et al., 2013). The exposure to *Vibrio*-affected tiger prawn exhibited higher survival levels when treated with silver nanoparticles similar to those of unexposed prawns. The exposure of nanoparticles maintained or increased hemolymph phenol oxidase activity, as well as hemocyte counts, which suggests that silver nanoparticles were taken up by the prawns and may have been responsible for reduced ATP production in the pathogens, as well as cell membrane damage through interacting with membrane-bound proteins. These results are encouraging as vibriosis is major setback in commercial shrimp farming, which is a million-dollar industry. However, the long-term effects of nanoparticles on the prawns and the environment, including the potential chances of bioaccumulation, must be investigated. Similarly, exposure of mussel larvae to iron nanoparticles has shown protective effect against the toxicity (Kadar et al., 2010). The authors hypothesize that aggregations of both forms of iron-scavenged compounds may cause toxicity at lower pH or may buffer pH changes around the cell. Such kind of protection has significant implications in modern days considering long-term ocean acidification or acute acidic runoff (Baker et al., 2014).

The earlier studies explain the negative and positive applications of nanoparticle toxicity. Hence, a better understanding about the long-term consequences of these particles in the marine environment is highly warranted. Assessment studies for understanding proper balance between positive and negative effects of nanoparticle toxicity should be established for their sustainable usage.

23.8 Anticipated Prospects

The use of nanoparticle-based products continues to grow rapidly on a consistent basis. A critical need in the field of nanotechnology is the development of a reliable and eco-friendly process for synthesis of metallic nanoparticles. Marine ecosystem has provided numerous sources for synthesis of nanoparticles with superior quality. The combination of marine biotechnology, particularly omics with nanotechnology, is imperative to fill the existing knowledge gaps in the synthesis mechanism. The integration of nanotechnology with omics, termed as nano-omics, has drawn the attention of scientists because of their extensive application particularly in the field of medicine and biotechnology. The following are possible research areas with the integration of marine biotechnology and nanotechnology:

1. Many marine species exhibit remarkable natural nanotechnology. Mimicking those structures and characteristics would help us in the production of novel products as witnessed in case of antifouling paints motivated from the skin structure of sharks and whales. In order to imitate some of these natural wonders, it is imperative to understand the molecular mechanism behind this.

2. Identification and production of naturally available marine genomic sequences with excellent ability in synthesis of nanoparticles with superior quality. There is a need to delineate the genomic sequences of the uncultivable organism that could be potentially utilized for this process. This will also help in developing large-scale production methodology with less destruction of the critical marine species.

3. For efficient utilization of nanoparticles in different scientific fields, understanding the underlying biochemical and molecular mechanisms of synthesis is required. This enables to obtain nanoparticles with superior qualities such as better chemical composition, shape, size, and monodispersity. This could be achieved by understanding the genomics and proteomics of organisms during the nanoparticle synthesis. Further, fate and behavior of nanoparticles in biological organisms should be investigated using these molecular tools for enhancing the efficiency of synthetic nanoparticles.

4. The uses of nanoparticles are increasing abruptly due to its numerous applications in diverse field. Many of synthetic nanoparticles interact directly with sensitive marine ecosystem and the inhabited marine organisms. Currently, marine environments around the globe are vulnerable to anthropogenic risks. Many nanoparticle-incorporated products are commercialized for human daily uses and hence are prone to reach these critical environments. Nanoparticle-bounded products such as antifouling paints are even worse as they are directly applied in marine field. Toxicology studies using omic-based tools are highly warranted for better understanding of the side effects of the synthetic nanoparticles to the marine environment and organisms.

Hence, the integration of marine nanotechnology and omics is imperative to mimic the marine environment as nanotechnology notions, to delineate the molecular mechanism of the nanoparticle synthesis, to develop novel products beneficial to mankind, and to understand its toxicity in marine ecosystem and organism.

References

Abdel-Raouf, N., I.B.M. Ibraheem, and N.M. Al-Enazi. 2013. Green biosynthesis of gold nanoparticles using *Galaxaura elongata* and characterization of their antibacterial activity. *Arabian Journal of Chemistry*. doi:10.1016/j.arabjc.2013.11.044.

Agnihotri, M., S. Joshi, A. R. Kumar, S. Zinjarde, and S. Kulkarni. 2009. Biosynthesis of gold nanoparticles by the tropical marine yeast *Yarrowia lipolytica* NCIM 3589. *Materials Letter* 63: 1231–1234.

Ahmad, A., P. Mukherjee, D. Mandal, S. Senapati, M. I. Khan, R. Kumar, and M. Sastry. 2003. Extracellular biosynthesis of silver nanoparticles using the fungus, *Fusarium oxysporum*. *Colloids and Surfaces B: Biointerfaces* 28: 313–318.

Allen, M. J., S. C. Edbergb, and D. J. Reasoner. 2004. Heterotrophic plate count bacteria—What is their significance in drinking water? *International Journal of Food Microbiology* 92: 265–274.

Azizi, S., F. Namvar, M. Mahdavi, M.B. Ahmad, and R. Mohamad. 2013. Biosynthesis of silver nanoparticles using brown marine macroalga, *Sargassum muticum*. *Aqueous Extract Materials* 6: 5942–5950.

Baker, E. E. G., A. J. Taylor, B. C. Sayers, E. A. Thompson, and J. C. Bonner. 2014. Nickel nanoparticles cause exaggerated lung and airway remodeling in mice lacking the T-box transcription factor, TBX21 (T-bet). *Particle and Fibre Toxicology* 11: 7.

Boopathi, S., S. Gopinath, T. Boopathi, V. Balamurugan, R. Rajeshkumar, and M. Sundararaman. 2012. Characterization and antimicrobial properties of silver and silver oxide nanoparticles synthesized by cell-free extract of a Mangrove-associated *Pseudomonas aeruginosa* M6 using two different thermal treatments. *Industrial and Engineering Chemistry Research* 51: 5976–5985.

Bradford, A., R. D. Handy, J. W. Readman, A. Atfield, and M. Mühling. 2009. Impact of silver nanoparticle contamination on the genetic diversity of natural bacterial assemblages in estuarine sediments. *Environmental Science and Technology* 43: 4530–4536.

Bystrzejewska-Piotrowska, G., J. Golimowski, and P. L. Urban. 2009. Nanoparticles: Their potential toxicity, waste and environmental management. *Waste Management* 29: 2587–2595.

Cauerhff, A. and G. R. Castro. 2013. Bionanoparticles, a green nanochemistry approach. *Electronic Journal of Biotechnology*. 16 doi: 10.2225/vol16-issue3-fulltext-3.

DEFRA. 2007. Characterizing the potential risks posed by engineered nanoparticles. Report, Department for Environment, Food and Rural Affairs, London, U.K.

Dhas, T. S., V. G. Kumar, V. Karthick, K. J. Angel, and K. Govindaraju. 2014a. Facile synthesis of silver chloride nanoparticles using marine alga and its antibacterial efficacy. *Spectrochimica Acta Part A: Molecular and Biomolecular Spectroscopy* 120: 416–420.

Dhas, T. S., V. G. Kumar, V. Karthick, K. Govindaraju, and T. Shankara Narayana. 2014b. Biosynthesis of gold nanoparticles using *Sargassum swartzii* and its cytotoxicity effect on HeLa cells. *Spectrochimica Acta Part A: Molecular and Biomolecular Spectroscopy* 133: 102–106.

Doiron, K., E. Pelletier, and K. Lemarchand. 2012. Impact of polymer-coated silver nanoparticles on marine microbial communities: A microcosm study. *Aquatic Toxicology* 24: 22.

El-Rafie, H.M., M.H. El-Rafie, and M.K. Zahran. 2013. Green synthesis of silver nanoparticles using polysaccharides extracted from marine macro algae. *Carbohydrate Polymers* 96: 403–410.

Fabrega, J., R. Zhang, J. C. Renshaw, W. T. Liu, and J. R. Lead. 2011. Impact of silver nanoparticles on natural marine biofilm bacteria. *Chemosphere* 85: 961–966.

Fresta, M., G. Puglisi, G. Giammona, G. Cavallaro, N. Micali, and P. M. Furneri. 1995. Pefloxacine mesi-
late-loaded and ofloxacin-loaded polyethylcyanoacrylate nanoparticles—Characterization of
the colloidal drug carrier formulation. *Journal of Pharmacological Science* 84: 895–902.

Gnanadesigan, M., M. Anand, S. Ravikumar, M. Maruthupandy, M. Syed Ali, V. Vijayakumar, and
A. K. Kumaragu. 2012. Antibacterial potential of biosynthesised silver nanoparticles using
Avicennia marina mangrove plant. *Applied Nanoscience* 2: 143–147.

Gnanadesigan, M., M. Anand, S. Ravikumar, M. Maruthupandy, V. Vijayakumar, S. Selvam,
M. Dhineshkumar, and A. K. Kumaraguru. 2011. Biosynthesis of silver nanoparticles by using
mangrove plant extract and their potential mosquito larvicidal property. *Asian Pacific Journal
of Tropical Medicine* 4: 799–803.

Gole, A., C. Dash, V. Ramakrishnaan, S. R. Sainkar, A. B. Mandal, M. Rao, and M. Sastry. 2001.
Pepsin-gold colloid conjugates: Preparation, characterization and enzymatic activity. *Langmuir*
17: 1674–1679.

Hamouda, T., M. Hayes, Z. Cao, R. Tonda, K. Johnson, W. Craig, J. Brisker, and J. Baker. 1999. A novel
surfactant nanoemulsion with broad-spectrum sporicidal activity against *Bacillus* species.
Journal of Infectious Disease 180: 1939–1949.

Hui, J., H. Yan-Bing, T. Ai-Wei, M. Xian-Guo, and T. Feng. 2006. Photoconductive properties of
MEH-PPV/CuS-nanoparticle composites. *Chinese Physics Letters* 23: 693.

Inbakandan, D., C. Kumar, L. S. Abraham, R. Kirubagaran, R. Venkatesan, and S. A. Khan. 2013.
Silver nanoparticles with anti microfouling effect: A study against marine biofilm forming
bacteria. *Colloids and Surfaces B: Biointerfaces* 111: 636–643.

Inbakandan, D., R. Venkatesan, and S. A. Khan. 2010. Biosynthesis of gold nanoparticles utilizing
marine sponge *Acanthella elongate* (Dendy 1905). *Colloids and Surfaces B: Biointerfaces* 81: 634–639.

Jacob, A. and K. Chakravarthy. 2014. Engineering magnetic nanoparticles for thermo-ablation and
drug delivery in neurological cancers. *Cureus* 6(4): e170.

Jena, P., S. Mohanty, R. Mallick, B. Jacob, and A. Sanawane. 2012. Toxicity and antibacterial assessment of chitosan-coated silver nanoparticles on human pathogens and macrophage cells.
International Journal of Nanomedicine 7: 1805–1818.

Kadar, E., F. Simmance, O. Martin, N. Voulvoulis, S. Widdicombe, S. Mitov, J. R. Lead, and J. W.
Readman. 2010. The influence of engineered Fe(2)O(3) nanoparticles and soluble (FeCl(3)) iron
on the developmental toxicity caused by CO(2)-induced seawater acidification. *Environmental
Pollution* 158: 3490–3497.

Kandasamy, K., N. M. Alikunhi, G. Manickaswami, A. Nabikhan, and G. Ayyavu. 2013. Synthesis
of silver nanoparticles by coastal plant *Prosopis chilensis* (L.) and their efficacy in controlling
vibriosis in shrimp *Penaeus monodon*. *Applied Nanoscience* 3: 65–73.

Kannan, R.R.R., W.A. Stirk, and J. Van Staden. 2013a. Synthesis of silver nanoparticles using the
seaweed *Codium capitatum* P.C. Silva (Chlorophyceae). *South African Journal of Botany* 86: 1–4.

Kannan, R.R.R., Arumugam, R., Ramya, D., Manivannan, K., and Anantharaman, P., 2013b. Green
synthesis of silver nanoparticles using marine macroalga *Chaetomorpha linum*. *Applied
Nanoscience* 3: 229–233.

Karthik, L., G. Kumar, A. V. Kirthi, A. A. Rahuman, and K. V. Bhaskara Rao. 2014. *Streptomyces* sp.
LK$_3$ mediated synthesis of silver nanoparticles and its biomedical application. *Bioprocess and
Biosystems Engineering* 37: 261–267.

Kathiresan, K., S. Manivannan, M. A. Nabeel, and B. Dhivya. 2009. Studies on silver nanoparticles synthesized by a marine fungus, *Penicillium fellutanum* isolated from coastal mangrove
sediment. *Colloids and Surfaces B: Biointerfaces* 7: 133–137.

Kathiresan, K., M. A. Nabeel, P. SriMahibala, N. Asmathunisha, and K. Saravanakumar. 2010.
Analysis of antimicrobial silver nanoparticles synthesized by coastal strains of *Escherichia coli*
and *Aspergillus niger*. *Canadian Journal of Microbiology* 56: 1050–1059.

Kiran, G. S., L. A. Nishanth, S. Priyadharshini, K. Anitha, and J. Selvin. 2014. Effect of Fe nanoparticle on growth and glycolipid biosurfactant production under solid state culture by marine
Nocardiopsis sp. MSA13A. *BMC Biotechnology* 14: 48.

Kiran, G. S., J. Selvin, A. Manilal, and S. Sujith. 2011. Biosurfactants as green stabilizer for the biological synthesis of nanoparticles. *Critical Review in Biotechnology* 31: 354–364.

Lansdown, A. B. 2002. Silver: Its antibacterial properties and mechanism of action. *Journal of Wound Care* 11: 125.

Merin, D. D., S. Prakash, and B. V. Bhimba. 2010. Antibacterial screening of silver nanoparticles synthesized by marine micro algae. *Asian Pacific Journal of Tropical Medicine* 10: 797–799.

Miller, R. J., S. Bennett, A. A. Keller, S. Pease, and H. S. Lenihan. 2012. TiO_2 nanoparticles are phototoxic to marine phytoplankton. *PLOS ONE* 7(1): e30321.

Mishra, A., S. K. Tripathy, R. Wahab, S. H. Jeong, I. Hwang, Y. B. Yang, Y. S. Kim, H. S. Shin, and S. I. Yun. 2011. Microbial synthesis of gold nanoparticles using the fungus *Penicillium brevicompactum* and their cytotoxic effects against mouse mayo blast cancer C 2 C 12 cells. *Applied Microbiology and Biotechnology* 92: 617–630.

Mohammed Fayaz, A., K. Balaji, M. Girilal, R. Yadav, P. T. Kalaichelvan, and R. Venkatesan. 2010. Biogenic synthesis of silver nanoparticles and their synergistic effect with antibiotics: A study against Gram-positive and Gram-negative bacteria. *Nanomedicine* 6: 103–109.

MubarakAli, D., V. Gopinath, N. Rameshbabu, and N. Thajuddin. 2012. Synthesis and characterization of CdS nanoparticles using C-phycoerythrin from the marine cyanobacteria. *Materials Letters* 74: 8–11.

MubarakAli, D., N. Thajuddina, K. Jeganathanb, and M. Gunasekaran. 2011. Plant extract mediated synthesis of silver and gold nanoparticles and its antibacterial activity against clinically isolated pathogens. *Colloids and Surfaces B: Biointerfaces* 85: 360–365.

Mueller, N. C. and B. Nowack, 2008. Exposure modeling of engineered nanoparticles in the environment. *Environmental Science and Technology* 42: 4447–4453.

Mukherjee, P., A. Ahmad, D. Mandal, S. Senapati, S. R. Sainkar, M. I. Khan, R. Parischa et al. 2001a. Fungus mediated synthesis of silver nanoparticles and their immobilization in the mycelial matrix: A novel biological approach to nanoparticle synthesis. *Nano Letters* 1: 515–519.

Mukherjee, P., A. Ahmad, D. Mandal, S. Senapati, S. R. Sainkar, M. I. Khan, R. Ramani et al. 2001b. Bioreduction of $AuCl_4$-ions by the fungus, *Verticillium* sp. and surface trapping of the gold nanoparticles formed. *Angewandte Chemie International Edition* 40: 3585–3588.

Muthukannan, R. and B. Karuppiah. 2011. Rapid synthesis and characterization of silver nanoparticles by novel *Pseudomonas* sp. "ram bt-1." *Journal of Ecobiotechnology* 3: 24.

Nabikhan, A., K. Kandasamy, A. Raj, and N. M. Alikunhi. 2010. Synthesis of antimicrobial silver nanoparticles by callus and leaf extracts from saltmarsh plant, *Sesuvium portulacastrum*. *Colloids and Surfaces B: Biointerfaces* 79: 488–493.

Nagarajan, S. and A. Kuppusamy. 2013. Extracellular synthesis of zinc oxide nanoparticle using seaweeds of gulf of Mannar, India. *Journal of Nanobiotechnology* 11: 39.

PEN. 2012. Woodrow Wilson International Centre for Scholars. Project on emerging nanotechnologies. Consumer Products Inventory of Nanotechnology Products. Accessed December 2012. http://www.nanotechproject.org/inventories/consumer/.

PEN. 2012. Silver nanotechnology: A database of silver nanotechnology in commercial products.

Rajathi, F. A. A., R. Arumugam, S. Saravanan, and P. Anantharaman. 2014. Phytofabrication of gold nanoparticles assisted by leaves of *Suaeda monoica* and its free radical scavenging property. *Journal of Photochemistry and Photobiology B: Biology* 13: 75–80.

Rajathi, F. A. A., C. Parthiban, V. Ganesh Kumar, and P. Anatharaman. 2012. Biosynthesis of antibacterial gold nanoparticles using brown alga, *Stoechospermum marginatum* (Kutzing). *Spectrochimica Acta Part A: Molecular and Biomolecular Spectroscopy* 99: 166–173.

Rajeshkumar, S., C. Malarkodi, K. Paulkumar, M. Vanaja, G. Gnanajobitha, and G. Annadurai. 2013. Intracellular and extracellular biosynthesis of silver nanoparticles by using marine bacteria *Vibrio alginolyticus*. *Nanoscience and Nanotechnology* 3: 21–25.

Ramalingam, B., S. Mukherjee, C. J. Mathai, K. Gangopadhyay, and S. Gangopadhyay. 2013. Sub-2 nm size and density tunable platinum nanoparticles using room temperature tilted-target sputtering. *Nanotechnology*. 24 doi: 10.1088/0957-4484/24/20/205602.

Sathiyanarayanan, G., G. S. Kiran, and J. Selvin. 2013. Synthesis of silver nanoparticles by polysaccharide bioflocculant produced from marine *Bacillus subtilis* MSBN17. *Colloids and Surfaces B: Biointerfaces* 102: 13–20.

Sathiyanarayanan, G., V. Vignesh, G. Saibaba, A. Vinothkanna, K. Dineshkumar, M. B. Viswanathana, and J. Selvin. 2014. Synthesis of carbohydrate polymer encrusted gold nanoparticles using bacterial exopolysaccharide: A novel and greener approach. *RSC Advances* 4: 22817–22827.

Satyavani, K., T. Ramanathan, and S. Gurudeeban. 2011. Plant mediated synthesis of biomedical silver nanoparticles by using leaf extract of *Citrullus colocynthis*. *Research Journal of Nanoscience and Nanotechnology* 1: 95–101.

Seshadri, S., K. Saranya, and M. Kowshik. 2011. Green synthesis of lead sulfide nanoparticles by the lead resistant marine yeast, *Rhodosporidium diobovatum*. *Biotechnology Progress* 27: 1464–1469.

Shankar, S. S., A. Rai, A. Ahmad, and M. Sastry. 2004. Rapid synthesis of Au, Ag, and bimetallic Au core-Ag shell nanoparticles using neem (*Azadirachta indica*) leaf broth. *Journal of Colloid and Interface Science* 275: 496–502.

Sharma, N., A. K. Pinnaka, M. Raje, A. Fnu, M. S. Bhattacharyya, and A. R. Choudhury. 2012. Exploitation of marine bacteria for production of gold nanoparticles. *Microbial Cell Factories* 11: 86.

Shivakrishna, P., M.R.P.G. Krishna, and M.A.S. Charya. 2013. Synthesis of silver nanoparticles from marine bacteria *Pseudomonas aerogenosa*. *Octa Journal of Biosciences* 1: 108–114.

Shrivastava, S., T. Bera, A. Roy, G. Singh, P. Ramachandrarao, and D. Dash. 2008. Characterization of enhanced antibacterial effects of novel silver nanoparticles. *Nanotechnology* 18: 103–112.

Simkiss, K. and K. M. Wilbur. 1989. *Biomineralization*. Academic Press, New York.

Singh, K., M. Panghal, S. Kadyan, U. Chaudhary, and J. P. Yadav. 2014. Antibacterial activity of synthesized silver nanoparticles from *Tinospora cordifolia* against multi drug resistant strains of *Pseudomonas aeruginosa* isolated from burn patients. *Journal of Nanomedicine and Nanotechnology* 5: 192.

Sivalingam, P., J. J. Antony, D. Siva, S. Achiraman, and K. Anbarasu. 2012. Mangrove *Streptomyces* sp. BDUKAS10 as nanofactory for fabrication of bactericidal silver nanoparticles. *Colloids and Surfaces A* 98: 12–17.

Sivaraman, S. K., I. Elango, S. Kumar, and V. Santhanam. 2009. A green protocol for room temperature synthesis of silver nanoparticles in seconds. *Current Science* 97: 1055.

Subramanian, M., N. M. Alikunhi, and K. Kandasamy. 2010. In vitro synthesis of silver nanoparticles by marine yeasts from coastal mangrove sediment. *Advanced Science Letters* 3: 428–433.

Thenmozhi, S., P. Rajeswari, M. Kalpana, M. Haemalatha, and P. Vijayalakshmi. 2013. Antibactericital activity of silver nitrate on biofilm forming *Aeromonas* spp. isolated from drinking water. *International Journal of Pure and Applied Bioscience* 1: 117–125.

Vala, A. K., S. Shah, and R. Patel. 2014. Biogenesis of silver nanoparticles by marine-derived fungus *Aspergillus flavus* from Bhavnagar Coast, Gulf of Khambhat, India. *Journal of Marine Biology and Oceanography* 3: 1.

van der Woude, I., H. W. Visser, M. B. ter Beest, A. Wagenaar, M. H. Ruiters, J. B. Engberts, and D. Hoekstra. 1995. Parameters influencing the introduction of plasmid DNA into cells by the use of synthetic amphiphiles as a carrier system. *Biochimica et Biophysica Acta* 1240: 34–40.

Venkatpurwar, V. and V. Pokharkar. 2011. Green synthesis of silver nanoparticles using marine polysaccharide: Study of in-vitro antibacterial activity. *Materials Letters* 65: 999–1002.

Vermieren, L., F. Devlieghere, and J. Debevere. 2002. Effectiveness of some recent antimicrobial packaging concepts. *Food Additives and Contaminants* 19: 163–171.

Vijayan, S. R., P. Santhiyagu, M. Singamuthu, N. K. Ahila, R. Jayaraman, and K. Ethiraj. 2014. Synthesis and characterization of silver and gold nanoparticles using aqueous extract of seaweed, *Turbinaria conoides*, and their antimicrofouling activity. *The Scientific World* 10. 2014. Article ID:938272.

Youssef, A. M. and M. S. Abdel-Aziz. 2013. Preparation of polystyrene nanocomposites based on silver nanoparticles using marine bacterium for packaging. *Polymer-Plastics Technology and Engineering* 52: 607–613.

24

Advances in Nanobiotechnology in Enhancing Microalgal Lipid Production, Harvesting, and Biodiesel Production

Avinesh R. Reddy and Munish Puri

CONTENTS

24.1 Introduction

Biodiesel is a renewable fuel of commercial interest, driven in part by the rising cost of fossils and negative environmental impacts associated with conventional fuel sources (Sadeghinezhad et al., 2014). First-generation biodiesel was produced from biomass, vegetable oils, animal fats, and waste oils by transesterification (van Eijck et al., 2014). The use of edible oils for first-generation biodiesel has resulted in the undesired effect of increasing global competition for food crops (Kumar and Sharma, 2015; Pinzi et al., 2014). Second-generation biofuels utilize noncompeting crops such as jatropha, karanja, jojoba, and mahua as well as the use of waste cooking oil, grease, and animal fats. While the use of "inedible" crops reduces competition with food crops, second-generation biofuels still compete for arable land, which ultimately negatively impacts the world food supply and can contribute to the destruction of soil resources, and these feedstock cannot meet the current energy demands (Nautiyal et al., 2014; Zhu et al., 2014). Sustainable alternatives that can provide the large fuel volumes required must be found, while not compromising the use of agricultural land for food production. Microalgae have attracted attention as a means of addressing rapidly growing demand for feedstock in biodiesel production (Baganz et al., 2014; Bellou et al., 2014; Chen et al., 2015b). Microalgae are easy to cultivate and can be grown on nonarable land, as well as accumulate more lipids as a percentage of their biomass than plants (Challagulla et al., 2015).

Several studies have been reported on different autotrophic and heterotrophic microalgae for biodiesel production (Li et al., 2014; Ogbonna and McHenry, 2015). Autotrophic microalgae utilize carbon dioxide as carbon source and solar energy to synthesize cellular components that include lipids for their growth. Most of the autotrophic algae cultivation for biodiesel production is reported in indoor photobioreactors. The autotrophic algae cultivation mainly depends on climatic conditions during outdoor production, whereas indoor photobioreactors require energy for illumination (Yadala and Cremaschi, 2014). The solar energy availability influences the biomass productivity and that ultimately leads to the low level of lipid accumulation. And the other important variable is the cultivation temperature, which can also alter the productivity of the final product (Ramos Tercero et al., 2014). In comparison, heterotrophic microalgae are easier to cultivate in controlled environment, thus producing a large amount of lipids (Shen et al., 2010). It has been reported that the lipid accumulation from heterotrophic microalgae was more than three-folds compared to autotrophic algae (Miao and Wu, 2004).

Nanobiotechnology (synergy between biotechnology and nanotechnology) encompasses the application and development of biological process using nanomaterials on the nanometer scale (López-Serrano et al., 2014). Nanosized materials offer a large surface area and unique physiochemical properties such as reactivity, tenacity, elasticity, and strength (Zhang et al., 2013). Among nanomaterials, nanoparticles (NPs) have been used in a range of applications including crop production, cosmetics, drug delivery, photonic crystals, analysis, food, coatings, paints, bioremediation, catalysis, and materials science (Husen and Siddiqi, 2014). Recently, the application of nanotechnology in microalgae biodiesel production has been explored to solve the challenges faced during the production process (Pugh et al., 2011; Verma, 2015). This chapter focuses on the developments in terms of microalgae biodiesel production using nanotechnology approaches, which includes microalgae cultivation, harvesting, lipid extraction, and transesterification steps using nanomaterial-immobilized biocatalyst. The outlay of the process and steps involved are presented in Figure 24.1.

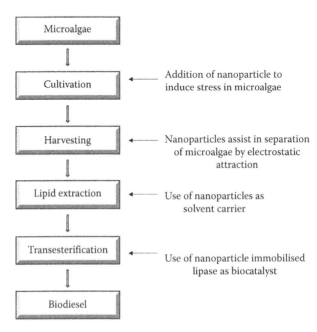

FIGURE 24.1
Application of nanobiotechnology at different stages of algal biodiesel production.

24.2 Application of Nanoparticles in Lipid Production

Synthesis of lipids from heterotrophic microalgae for biodiesel production requires the growth of microbes on expensive nutrients, particularly glucose as a carbon source (Slade and Bauen, 2013). Glucose can constitute 80% of the total medium cost and up to a third of total fermentation production costs (Li et al., 2007). To reduce process costs and assist in the commercialization of biodiesel production from heterotrophic microalgae, sustainable low-cost carbon sources are required. One approach is to couple microalgae cultivation with the use of inexpensive carbon sources such as carbon-rich wastewater, lignocellulose hydrolysate, or use of glycerol derived from the biodiesel industry (Di Caprio et al., 2015; Gupta et al., 2015; Xin et al., 2010). The important issues to be addressed in this step are the control of the unwanted microorganism such as fungi and bacteria, which compete with the growth of microalgae.

Screening of high lipid-producing microalgae is a key step toward the success of large-scale cultivation and biodiesel production from microalgae. These microalgae are specific to particular environments and certain climate conditions (Brennan and Owende, 2010; Williams and Laurens, 2010). These organisms occur in various environments such as freshwater, lacustrine lakes, brackish water, marine water, maturation ponds, and other hypersaline zones (Mutanda et al., 2011). During the selection of suitable microalgae for lipid production, organisms that have high biomass productivity and lipid accumulation are preferred for mass cultivation than those that have high lipid storage with low productivity (Amaro et al., 2011; Mata et al., 2010). The accumulation of oil in microalgae cell has been enhanced by various approaches such as nitrogen deprivation (Farooq et al., 2013), cocultivation of antimicrobial agents (Kim et al., 2013), salt stress, CO_2 influence, temperature, salinity, and heavy metal stress (Sibi et al., 2015). NPs have been reported to enhance the microbial activities; however, addition of NPs in cultivation medium could enhance the lipid accumulation. For example, the addition of minerals such as iron and magnesium in nanoform improved the photosynthetic process in autotrophic microalgae and led to the enhanced lipid accumulation in *Chlorella vulgaris* (Liu et al., 2008). A recent study compared the effect of soluble iron and synthetic nanoscale zerovalent iron NPs on lipid accumulation on three different microalgae, where the presence of nanoscale iron resulted in increased lipid production (Kadar et al., 2012). This further explained the generation of reactive oxygen species via Fenton-type reaction, which causes the oxidative stress to microbes, which drives the lipid accumulation (Kadar et al., 2012).

Magnesium sulfate ($MgSO_4$) NPs were supplemented in a cultivation medium, which enhanced the photosynthesis process and also the carbon source uptake in *Chlorella*, resulting in an increased lipid production (Sarma et al., 2014). The potential of silica NPs of size 38–190 nm was investigated in *C. vulgaris* to increase the lipid accumulation, and a slight increase in biomass productivity was observed (San et al., 2014). The effect of TiO_2 NP–induced stress under UV environment was investigated and optimized on *Chlorella*, and the optimal conditions were found to be 0.1 g TiO_2/L with 2 days of UV induction, which increased lipid accumulation by 15% (Kang et al., 2014). Some of the studies have indicated the toxic effects of NPs by inhibiting the cell growth. This indicates that sound knowledge is required regarding the toxic effects of NPs based on the particle and dosage as well as microalgae species. Table 24.1 lists the use of various NPs in microalgae for improving lipid accumulation for biodiesel production.

TABLE 24.1

List of Types of Nanoparticles Employed for Microalgae Lipid Production

Microalgae	Nanoparticles	Observations	References
Picochlorum sp.	Magnetite iron oxide (Fe₃O₄) NPs	Increase in algae growth at a particle size of 20 nm	Hazeem et al. (2015)
Green microalgae	Zerovalent iron NPs	Increase in growth and polyunsaturated fatty acids	Pádrová et al. (2015)
Chlamydomonas reinhardtii, Cyanothece 51142	Silver NPs	Increase in cell growth by 30%	Torkamani et al. (2010)
C. vulgaris	Silver NPs, gold nanorods	Increase in accumulation of chlorophyll and carotenoids	Eroglu et al. (2013)
Pavlova lutheri, Isochrysis galbana, and *Tetraselmis suecica*	Nanoscale zerovalent iron	Increase in lipid accumulation by up to 15% in *P. lutheri*	Kadar et al. (2012)
C. vulgaris	MgSO₄ NPs	Increase in lipid yield as compared with control	Sarma et al. (2014)
C. vulgaris	Laser ablation–prepared silica NPs	Increase in cell growth	San et al. (2014)
C. vulgaris UTEX 265	TiO₂ NPs	Increase in lipid accumulation by 15%	Kang et al. (2014)
C. vulgaris	Mg-aminoclay NPs	Increase in cell size and oil accumulation	Lee et al. (2015)

24.3 Application of Nanoparticles in Microalgae Harvesting

In general, microalgae harvesting is one of the challenging steps in biodiesel production because it requires more energy with the existing techniques. The small size, low density, and other culture conditions such as pH negatively influence the harvesting process, resulting in more energy investments (Abdelaziz et al., 2013). Thus, an efficient and cost-effective microalgae harvesting technique is mandatory for commercial production because it consumes approximately 30% of the total process cost. However, several physicochemical methods that include centrifugation, filtration, and air flotation have been employed so far. These methods are energy intensive and are not easy to scale-up.

Flocculent loading is an emerging approach for large-scale microalgal harvesting. Recently, the application of NPs as an organic or inorganic flocculants with their unique properties in microalgae harvesting has been reported as a simple low-cost technique (Vandamme et al., 2013). Generally, microalgae possess negative surface charge, which results in a stable cell suspension. However, the addition of NP in algal harvesting is more efficient, provided it satisfies the requirements for recyclability, low toxicity, and inexpensive cost of nanomaterials (Lee et al., 2013a).

Microalgae harvesting using magnetic NPs is getting more attention because of its mild processing conditions, easy to scale-up, and unique advantages. This process follows (1) attachment of the microalgae to magnetic NPs based on the cationic charge below the isoelectric point or negative above it, which results in (2) generation of electrostatic interaction between negative-zeta-potential microalgae with positive-zeta-potential metal oxides and (3) separation of complex influenced by external magnetic field. Figure 24.2 demonstrates the triazabicyclodecene magnetic

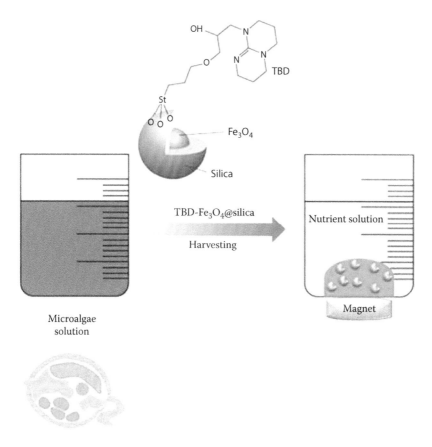

FIGURE 24.2
Proposed use of magnetic nanoparticles for harvesting microalgae. (Reproduced and modified from Chiang et al., 2015.)

NPs as an efficient support for achieving microalgae harvesting. Liu and Guo et al. reported 95% of harvesting efficiency in freshwater (*Botryococcus braunii* and *Chlorella ellipsoidea*) and marine (*Nannochloropsis maritime*) microalgal species using magnetic particles (Fe_3O_4) in less than 5 min (Hu et al., 2013; Xu et al., 2011). A continuous magnetic separator was developed for harvesting *C. ellipsoidea* using magnetic NPs. The microalgae cells and NP aggregates were resulted from the attraction of magnetic force by a permanent magnetic drum. It was observed that the harvesting efficiency was decreased by increasing the culture broth flow rate (Hu et al., 2014b). *Chlorella* sp. was efficiently harvested using bare Fe_3O_4 particle at lower pH because of the high electrostatic attraction between cells and Fe_3O_4 particles (Lee et al., 2014a). In spite of the similar surface charge of Fe_3O_4 and microalgae, there is no change in harvesting efficiency. This is due to the ion exchange between positively charged molecules (glycoproteins) on the surface of microalgae and NPs, which induces the electrostatic interaction (Prochazkova et al., 2013).

Iron oxide, a high-molecular-weight compound, has been compared to conventional chemical flocculants to achieve high harvesting efficiency. The ratio of microalgae to iron oxide NP resulted in high-level aggregation of NP and microalgae cells in sediments that facilitated harvesting (Lee et al., 2014a). Few attempts have been made to recover the

TABLE 24.2

List of Types of Nanoparticles Employed in Microalgae Harvesting

Microalgae	Nanoparticle	Harvesting Efficiency (%)	Time (min)	References
Chlorella sp.	Trifunctionality of Fe_3O_4-embedded carbon microparticles	99	ND	Seo et al. (2015)
Scenedesmus rubescens and *Dunaliella tertiolecta*	Magnetic microparticles	91	5	Vergini et al. (2015)
C. vulgaris 211-11b	Cellulose-based nanocrystals	95	ND	Vandamme et al. (2015)
Chlorella sp. KR-1	Aminoclay-templated nanoscale zerovalent iron	100	3	Lee et al. (2014c)
B. braunii	Iron oxide and cationic polyacrylamide	95	10	Wang et al. (2014)
C. ellipsoidea	Magnetic nanocomposites coated with polyethylenimine	97	2	Hu et al. (2014a)
C. ellipsoidea	Fe_3O_4 NPs	95	0.6	Hu et al. (2014b)
Chlorella sp. KR-1	Aminoclay-conjugated TiO_2 NPs	85	10	Lee et al. (2014b)
N. maritime	Magnetic particles (Fe_3O_4)	95	5	Hu et al. (2013)
B. braunii and *C. ellipsoidea*	Magnetic particles (Fe_3O_4)	95	4	Xu et al. (2011)

magnetic NPs by dissolving them in hydrochloric acid (HCl) and posttreatments such as coprecipitation and filtration (Xu et al., 2011). Another approach of using sulfuric acid (10% v/v H_2SO_4) and ultrasonication treatment at 40°C resulted in 100% recovery of NPs (Prochazkova et al., 2013).

Due to the variation in microalgae surface charge by various factors such as algal species, pH, and medium strength, NPs coated with functional groups are developed to obtain high harvesting efficiencies. Two different shapes of NPs, rod-shaped Fe_3O_4 NPs and NPs (rod and spherical) coated with poly diallyldimethylammonium chloride, were synthesized and investigated for *Chlorella* sp. harvesting. The rod-shaped NPs were observed to be efficient in harvesting (99%) compared to spherical shaped NPs (Lim et al., 2012). Polyethylenimine was successfully utilized as a coating material for NPs, and this resulted in 97% harvesting efficiency from *C. ellipsoidea* within 2 min (Hu et al., 2014a). Chitosan, a polysaccharide, was utilized as a functional coating material on magnetic NPs for achieving 95% microalgae harvesting efficiency (Toh et al., 2014). The application of nanotechnology in microalgae harvesting is relatively new; however, further efforts in developing low-cost and stable NPs will assist in developing cost analysis for scale-up. Table 24.2 summarizes the list of NPs and their coating with functional groups in microalgae harvesting.

24.4 Microalgae Lipid Extraction Using Nanoparticles

A large number of lipid extraction procedures are used for microalgal biomass, and their extraction efficiencies have been documented. All microalgae are single-cell organisms having an individual cell wall. Furthermore, some species of microalgae have high lipid levels, but they are located inside a tough cell wall. The presence of tough cell walls of microalgae increases the lipid extraction cost, since a cell disruption step is required for effective lipid extraction. Cell disruption is an energy-consuming process but an important step in liberating intracellular lipid molecules from microalgae (Byreddy et al., 2015; Lee et al., 2012). A number of cell disruption methods are available for lipid extraction from microalgae. Different microalgae have different disruption propensities, and as a result, no single method of disruption can be universally applied to the various microalgae species (Halim et al., 2012).

Cell disruption methods generally used at the laboratory scale are classified according to the working procedure by which microalgal cells are disrupted, for example, mechanical and nonmechanical (Chisti and Moo-Young, 1986; Günerken et al., 2015; Middelberg, 1995). Mechanical methods include bead mill, press, high-pressure homogenization, ultrasonication, autoclave, lyophilization, and microwave, while nonmechanical methods often involve the lysis of the microalgal cells with acids, alkalis, enzymes, or osmotic shocks (Chisti and Moo-Young, 1986; Günerken et al., 2015).

Recently, the application of NPs in microalgae cell disruption and lipid extraction has been implemented. It was hypothesized that this approach would be a sustainable way of microalgae lipid extraction, since NPs will replace the usage of toxic chemicals (such as solvents) and reduce the energy consumption in solvent–lipid separation during conversion process (Cao et al., 2013). Lee et al. (2013) reported the application of organic nanoclays for lipid extraction from wet microalgal biomass (*Chlorella*). They observed that the lipid extraction efficiency was increased, and the conversion of fatty acids into biodiesel was 100% (Lee et al., 2013c). A positively charged organic nanoclay plays a major role in weakening the algal cell walls and favors the solvent penetration in oil recovery. Microalgal cells were disrupted by hydroxyl (OH) free radical generated by Fenton-like reaction with Fe-based aminoclay in the presence of hydrogen peroxide (H_2O_2) (Lee et al., 2013d). This indicates the use of NP as an alternative to solvents, which breaks the microalgae cells for lipid extraction, which can be further recycled to reduce cost.

24.5 Nanoparticle-Immobilized Lipase for Biodiesel Production

The basic requirements for biodiesel production are the availability of (1) a cost-effective feedstock (oil), (2) an alcohol, and (3) a catalyst (e.g., base/acid or enzyme). The biodiesel production occurs in the following steps: production of free fatty acids from triacylglycerols (TAGs) and transesterification of free fatty acids to methanol, which result in the formation of new chemical compounds called methyl esters. The important process variables during the production of biodiesel are reaction temperature, ratio of alcohol to vegetable oil, amount of catalyst, mixing intensity (RPM), use of raw oils, and catalyst (Marchetti et al., 2007). In the alkali-based production process, generally sodium hydroxide (NaOH) or potassium hydroxide (KOH) is used as the catalyst. During the initial phase of the reaction, the catalyst reacts with the alcohol and forms alkoxy, and this reacts with the TAG to produce biodiesel and glycerol. Glycerol and biodiesel can be separated easily. There may be a chance of soap formation by free acid or water contamination that makes the separation process difficult (Barnwal and Sharma, 2005; Gnanaprakasam et al., 2013). Sulfuric acid is the most preferable catalyst in the acid-based method (Wang et al., 2009). The acid catalyst gives high yield of esters, but the reaction rate for conversion requires a long time. Although the production cost of biodiesel from the conventional methods such as acid and alkali catalysis is cheap, but associated with some difficulties, that is, the yield of methyl esters is normal, recovery of glycerol is difficult, and repeated washing is required to purify methyl esters (Marchetti et al., 2007).

Biodiesel production by an enzymatic method, an alternative method, has been found to be sustainable in recent times. The use of lipases (triacylglycerol hydrolase; EC 3.1.1.3) can solve the problems associated with the recovery of glycerol and methyl esters (Tan et al., 2010). In addition to that, lipase can perform both reactions such as hydrolysis of triglycerides and transesterification of fatty acids (Véras et al., 2011). The limitations in the production of biodiesel using lipases are high cost of enzyme, longer reaction time, organic solvents, and water requirement in the reaction mixture (Zhao et al., 2015). Recently, many researchers have focused on lipase-based biodiesel production, downstream processing, immobilization of lipase, bioreactors development, process optimization, simulation, and technoeconomic evaluation (Zhao et al., 2015). Lipase immobilization can be done by various techniques, which included adsorption, covalent binding, entrapment, encapsulation, and cross-linking, as have been reviewed elsewhere (Verma et al., 2014; Zhang et al., 2012). Immobilizing lipases onto NPs showed encouraging results in terms of its reusability and stability. It was observed that the NPs are too small, thus mandating energy-intensive process to recover from the reaction mixture. This issue was overcome by introducing the magnetic NPs in enzyme immobilization, which facilitated easier separation by using external magnetic field (Puri et al., 2013). It indicates that the immobilization of lipase on magnetic NPs enhances the reusability of the biocatalyst. Some of the NPs used for immobilizing lipases for biodiesel production are presented in Table 24.3.

Royon et al. (2007) reported 97% of the biodiesel yield using *Candida antarctica* lipase immobilized on macroporous resin. In another study, Dizge et al. (2008) developed a lipase immobilization method from hydrophilic polyurethane foams using polyglutaraldehyde. The immobilized lipase at optimized reaction condition showed 90% efficiency conversion of methyl esters and remained 100% stable after 10 batches. Kyu et al. (2013) immobilized lipase source from *Candida rugosa* (Amano AY-30) to polyvinylidene fluoride (PVDF) membrane via adsorption. The transesterification conditions with the immobilized enzyme were also optimized by response surface methodology for biodiesel production. The transesterification efficiency of more than 95% was achieved.

A functionalized magnetic NP was used for covalent immobilization of lipase for 30 min and is applied in biodiesel production using soybean oil in a continuous reaction (Wang et al., 2011). An organic and inorganic nanosupport was synthesized to immobilize lipase via encapsulation method. The immobilized lipase remains active after five cycles of reaction, and the product yield was higher than a free enzyme (Macario et al., 2013). The high content of free fatty acids present in waste grease was used as a source for enzymatic biodiesel production. Lipases from *Thermomyces lanuginosus* and *C. antarctica* lipase B were covalently immobilized onto magnetic NP that achieved 97% product conversion in 12 h (Ngo et al., 2013). Several attempts have been made in immobilizing lipases onto NPs for improving biodiesel production and lowering the production cost.

TABLE 24.3

List of Various Lipase Immobilization Methods for Biodiesel Production Using a Variety of Raw Materials

Raw Material	Name of Enzyme	Lipase Source	Method	Carrier	Solvent Used	Yield/Conversion	References
Sunflower	Novozyme 435	*Mucor miehei*, *Candida antarctica*	Adsorption	Macroporous resin	—	—	Ranganathan et al. (2008)
Cotton seed	Novozyme 435	*Candida antarctica*	Adsorption	Macroporous resin	*t*-Butanol	97%	Royon et al. (2007)
Canola	Lipozyme TL	*Thermomyces lanuginosus*	Covalent binding	Polyurethane foams	Methanol	90%	Dizge and Keskinler (2008)
Soybean	Lipase	*Candida rugosa*	Covalent binding	Magnetic chitosan	Methanol	87%	Xie and Wang (2012)
Soybean	Lipase	*Pseudomonas cepacia*	Covalent binding	Activated magnetic silica nanocomposite particles	Methanol	54%	Kalantari et al. (2013)
Jatropha	Lipase	*Burkholderia cepacia*	Cross-linking	Modified attapulgite (ATP)	Methanol	94%	You et al. (2013)
Simulated waste cooking	Lipase	*Candida antarctica*, *Pseudomonas cepacia*	Adsorption	Ceramic beads	Methanol and *n*-hexane	40% without *n*-hexane	Al-Zuhair et al. (2009)
Sunflower and soybean waste cooking	Lipase	*Thermomyces lanuginosus*	Covalent binding	Microporous polymeric biocatalyst (bead, powder, monolithic)	Methanol	63.8% (sunflower), 55.5% (soybean), 50.9% (waste cooking)	Dizge et al. (2009)
Sapium sebiferum	Lipase	*Pseudomonas cepacia* G63	—	—	Methanol	96.22%	Li and Yan (2010)
Waste grease	Lipase	*Thermomyces lanuginosus* lipase, *Candida antarctica* lipase B	Covalent binding	Iron oxide (magnetic nanobiocatalyst aggregates)	Methanol	>97%	Ngo et al. (2013)

(Continued)

TABLE 24.3 (*Continued*)

List of Various Lipase Immobilization Methods for Biodiesel Production Using a Variety of Raw Materials

Raw Material	Name of Enzyme	Lipase Source	Method	Carrier	Solvent Used	Yield/Conversion	References
Palm	Novozyme 435, Lipozyme TL IM	*Candida antarctica* B, *Thermomyces lanuginosus*	Adsorption, covalent binding	Macroporous acrylic resin, silica gel	Dimethyl carbonate (DMC)	90.5%, 11.6%	Zhang et al. (2010)
Waste, corn	Lipase	*Penicillium expansum*	Adsorption	Resin D4020	Methanol	92.8%	Li et al. (2009)
Soybean	Lipase	*Candida rugosa* (Amano AY-30)	Adsorption	PVDF membrane	Methanol, *n*-hexane	97.2%	Kuo et al. (2013)
Soybean	Lipozyme TL	*Thermomyces lanuginosa*	Covalent binding	Magnetic NPs	Methanol	>90%	Xie and Ma (2010)
Chlorella protothecoides	Lipase	*Candida* sp. 99–125	Adsorption	Macroporous resin	Methanol	98%	Xiong et al. (2008)
Chlorella vulgaris ESP-31	Lipase	*Burkholderia* sp. C20	Covalent binding	Alkyl-grafted Fe_3O_4-SiO_2	Methanol	97.3 wt% oil	Tran et al. (2012)
Chlorella sp. KR-1	Novozyme 435	*Candida antarctica*	Adsorption	Macroporous resin	DMC mixture	>90%	Lee et al. (2013b)
Chlorella pyrenoidosa, *Chlorella vulgaris*, *Botryococcus braunii* (BB763, BB764)	Novozyme 435[a]	*Candida antarctica*,[a] *Penicillium expansum*	Adsorption	Acrylic resins	Ionic liquid [BMIm][PF$_6$], *tert*-butanol	90.7% and 86.2% in ionic liquid; 48.6% and 44.4% in *tert*-butanol	Lai et al. (2012)
Chlorella protothecoides	Lipase	*Candida* sp. 99–125	Adsorption	—	Methanol	98.15%	Li et al. (2007)
Tetraselmis sp.	Lipase	*Candida* sp.		Sodium alginate	Methanol	Sevenfolds increased	Teo et al. (2014)

24.6 Conclusion and Future Directions

Many critical issues need to be resolved before the practical application of NPs in microalgae lipid production, harvesting, and transesterification. With respect to lipid production, manipulating culture conditions such as pH, temperature, and carbon to nitrogen ratio in the presence of NPs needs further investigation. Specifically, the stability of NPs in the growth conditions at minimum dosages while maintaining a higher growth rate must be developed. The separation of NPs from microalgal and NP aggregates should be further examined. The utilization of nanomaterials as an alternative to organic solvents with more efficiency in oil recovery and reusability will lead to better biomass disruption. The major advantages of magnetic NPs are easy separation and reusability. The synthesis of magnetic NPs from waste generated by steel and related industries could be considered.

Lipase-catalyzed transesterification for biodiesel production is more popular due to mild reaction conditions, environmental friendliness, and wide adaptability for feedstock. Lipase immobilization on various NPs improves its operational stability of the enzyme, thus improving the reusability of the enzyme. The enzyme reusability should be increased by enhancing the interaction between enzyme and carrier. Various process parameters have been observed to influence the biodiesel production yield and stability of immobilized enzyme. The optimization of important parameters such as water content, enzyme loading, alcohol to oil ratio, temperature, and reaction media would lead to high biodiesel yield. A detailed technical and economical evaluation of the biodiesel production process is a prerequisite for graduating proposed pathway to commercial-scale production.

Acknowledgment

The authors are thankful to Professor Colin J. Barrow, Director CCB, for providing necessary facility to carry out research work in the area of algal biofuels.

References

Abdelaziz, A.E.M., Leite, G.B., Hallenbeck, P.C. 2013. Addressing the challenges for sustainable production of algal biofuels: II. Harvesting and conversion to biofuels. *Environmental Technology*, **34**(13–14), 1807–1836.

Al-Zuhair, S., Dowaidar, A., Kamal, H. 2009. Dynamic modeling of biodiesel production from simulated waste cooking oil using immobilized lipase. *Biochemical Engineering Journal*, **44**(2–3), 256–262.

Barnwal, B.K., Sharma, M.P. 2005. Prospects of biodiesel production from vegetable oils in India. *Renewable and Sustainable Energy Reviews*, **9**(4), 363–378.

Byreddy, A., Gupta, A., Barrow, C., Puri, M. 2015. Comparison of cell disruption methods for improving lipid extraction from thraustochytrid strains. *Marine Drugs*, **13**(8), 5111.

Cao, H., Zhang, Z., Wu, X., Miao, X. 2013. Direct biodiesel production from wet microalgae biomass of *Chlorella pyrenoidosa* through in situ transesterification. *BioMed Research International*, **2013**, 6.

Chisti, Y., Moo-Young, M. 1986. Disruption of microbial cells for intracellular products. *Enzyme and Microbial Technology*, **8**(4), 194–204.

Dizge, N., Aydiner, C., Imer, D.Y., Bayramoglu, M., Tanriseven, A., Keskinler, B. 2009. Biodiesel production from sunflower, soybean, and waste cooking oils by transesterification using lipase immobilized onto a novel microporous polymer. *Bioresource Technology*, **100**(6), 1983–1991.

Dizge, N., Keskinler, B. 2008. Enzymatic production of biodiesel from canola oil using immobilized lipase. *Biomass and Bioenergy*, **32**(12), 1274–1278.

Eroglu, E., Eggers, P.K., Winslade, M., Smith, S.M., Raston, C.L. 2013. Enhanced accumulation of microalgal pigments using metal nanoparticle solutions as light filtering devices. *Green Chemistry*, **15**(11), 3155–3159.

Farooq, W., Lee, Y.C., Ryu, B.G., Kim, B.H., Kim, H.S., Choi, Y.E., Yang, J.W. 2013. Two-stage cultivation of two *Chlorella* sp. strains by simultaneous treatment of brewery wastewater and maximizing lipid productivity. *Bioresource Technology*, **132**, 230–238.

Gnanaprakasam, A., Sivakumar, V.M., Surendhar, A., Thirumarimurugan, M., Kannadasan, T. 2013. Recent strategy of biodiesel production from waste cooking oil and process influencing parameters: A review. *Journal of Energy*, **2013**, 10.

Günerken, E., D'Hondt, E., Eppink, M.H.M., Garcia-Gonzalez, L., Elst, K., Wijffels, R.H. 2015. Cell disruption for microalgae biorefineries. *Biotechnology Advances*, **33**(2), 243–260.

Gupta, A., Abraham, R.E., Barrow, C.J., Puri, M. 2015. Omega-3 fatty acid production from enzyme saccharified hemp hydrolysate using a novel marine thraustochytrid strain. *Bioresource Technology*, **184**, 373–378.

Halim, R., Harun, R., Danquah, M.K., Webley, P.A. 2012. Microalgal cell disruption for biofuel development. *Applied Energy*, **91**(1), 116–121.

Hazeem, L., Waheed, F., Rashdan, S., Bououdina, M., Brunet, L., Slomianny, C., Boukherroub, R., Elmeselmani, W. 2015. Effect of magnetic iron oxide (Fe_3O_4) nanoparticles on the growth and photosynthetic pigment content of *Picochlorum* sp. *Environmental Science and Pollution Research*, **22**(15), 11728–11739.

Hu, Y.-R., Guo, C., Wang, F., Wang, S.-K., Pan, F., Liu, C.-Z. 2014a. Improvement of microalgae harvesting by magnetic nanocomposites coated with polyethylenimine. *Chemical Engineering Journal*, **242**, 341–347.

Hu, Y.-R., Guo, C., Xu, L., Wang, F., Wang, S.-K., Hu, Z., Liu, C.-Z. 2014b. A magnetic separator for efficient microalgae harvesting. *Bioresource Technology*, **158**, 388–391.

Hu, Y.-R., Wang, F., Wang, S.K., Liu, C.Z., Guo, C. 2013. Efficient harvesting of marine microalgae *Nannochloropsis maritima* using magnetic nanoparticles. *Bioresource Technology*, **138**, 387–390.

Husen, A., Siddiqi, K. 2014. Phytosynthesis of nanoparticles: Concept, controversy and application. *Nanoscale Research Letters*, **9**(1), 1–24.

Kadar, E., Rooks, P., Lakey, C., White, D.A. 2012. The effect of engineered iron nanoparticles on growth and metabolic status of marine microalgae cultures. *Science of the Total Environment*, **439**, 8–17.

Kalantari, M., Kazemeini, M., Arpanaei, A. 2013. Evaluation of biodiesel production using lipase immobilized on magnetic silica nanocomposite particles of various structures. *Biochemical Engineering Journal*, **79**, 267–273.

Kang, N., Lee, B., Choi, G.-G., Moon, M., Park, M., Lim, J., Yang, J.-W. 2014. Enhancing lipid productivity of *Chlorella vulgaris* using oxidative stress by TiO_2 nanoparticles. *Korean Journal of Chemical Engineering*, **31**(5), 861–867.

Kim, S., Kim, H., Ko, D., Yamaoka, Y., Otsuru, M., Kawai-Yamada, M., Ishikawa, T. et al. 2013. Rapid induction of lipid droplets in *Chlamydomonas reinhardtii* and *Chlorella vulgaris* by Brefeldin A. *PLOS ONE*, **8**(12), e81978.

Kuo, C.H., Peng, L.T., Kan, S.C., Liu, Y.C., Shieh, C.J. 2013. Lipase-immobilized biocatalytic membranes for biodiesel production. *Bioresource Technology*, **145**, 229–232.

Lai, J.-Q., Hu, Z.-L., Wang, P.-W., Yang, Z. 2012. Enzymatic production of microalgal biodiesel in ionic liquid [BMIm][PF6]. *Fuel*, **95**, 329–333.

Lee, A.K., Lewis, D.M., Ashman, P.J. 2012. Disruption of microalgal cells for the extraction of lipids for biofuels: Processes and specific energy requirements. *Biomass and Bioenergy*, **46**, 89–101.

Lee, K., Lee, S.Y., Na, J.G., Jeon, S.G., Praveenkumar, R., Kim, D.M., Chang, W.S., Oh, Y.K. 2013a. Magnetophoretic harvesting of oleaginous *Chlorella* sp. by using biocompatible chitosan/magnetic nanoparticle composites. *Bioresource Technology*, **149**, 575–578.

Lee, K., Lee, S.Y., Praveenkumar, R., Kim, B., Seo, J.Y., Jeon, S.G., Na, J.G., Park, J.Y., Kim, D.M., Oh, Y.K. 2014a. Repeated use of stable magnetic flocculant for efficient harvest of oleaginous *Chlorella* sp. *Bioresource Technology*, **167**, 284–290.

Lee, O.K., Kim, Y.H., Na, J.G., Oh, Y.K., Lee, E.Y. 2013b. Highly efficient extraction and lipase-catalyzed transesterification of triglycerides from *Chlorella* sp. KR-1 for production of biodiesel. *Bioresource Technology*, **147**, 240–245.

Lee, Y.-C., Huh, Y.S., Farooq, W., Chung, J., Han, J.-I., Shin, H.-J., Jeong, S.H., Lee, J.-S., Oh, Y.-K., Park, J.-Y. 2013c. Lipid extractions from docosahexaenoic acid (DHA)-rich and oleaginous *Chlorella* sp. biomasses by organic-nanoclays. *Bioresource Technology*, **137**, 74–81.

Lee, Y.-C., Huh, Y.S., Farooq, W., Han, J.I., Oh, Y.K., Park, J.Y. 2013d. Oil extraction by aminoparticle-based H_2O_2 activation via wet microalgae harvesting. *RSC Advances*, **3**(31), 12802–12809.

Lee, Y.-C., Lee, H.U., Lee, K., Kim, B., Lee, S.Y., Choi, M.-H., Farooq, W. et al. 2014b. Aminoclay-conjugated TiO_2 synthesis for simultaneous harvesting and wet-disruption of oleaginous *Chlorella* sp. *Chemical Engineering Journal*, **245**, 143–149.

Lee, Y.-C., Lee, K., Hwang, Y., Andersen, H.R., Kim, B., Lee, S.Y., Choi, M.-H. et al. 2014c. Aminoclay-templated nanoscale zero-valent iron (nZVI) synthesis for efficient harvesting of oleaginous microalga, *Chlorella* sp. KR-1. *RSC Advances*, **4**(8), 4122–4127.

Lee, Y.-C., Lee, K., Oh, Y.-K. 2015. Recent nanoparticle engineering advances in microalgal cultivation and harvesting processes of biodiesel production: A review. *Bioresource Technology*, **184**, 63–72.

Li, N.W., Zong, M.H., Wu, H. 2009. Highly efficient transformation of waste oil to biodiesel by immobilized lipase from *Penicillium expansum*. *Process Biochemistry*, **44**(6), 685–688.

Li, Q., Yan, Y. 2010. Production of biodiesel catalyzed by immobilized *Pseudomonas cepacia* lipase from *Sapium sebiferum* oil in micro-aqueous phase. *Applied Energy*, **87**(10), 3148–3154.

Li, X., Xu, H., Wu, Q. 2007. Large-scale biodiesel production from microalga *Chlorella protothecoides* through heterotrophic cultivation in bioreactors. *Biotechnology and Bioengineering*, **98**(4), 764–771.

Li, Y.-R., Tsai, W.-T., Hsu, Y.-C., Xie, M.-Z., Chen, J.-J. 2014. Comparison of autotrophic and mixotrophic cultivation of green microalgal for biodiesel production. *Energy Procedia*, **52**, 371–376.

Lim, J.K., Chieh, D.C.J., Jalak, S.A., Toh, P.Y., Yasin, N.H.M., Ng, B.W., Ahmad, A.L. 2012. Rapid magnetophoretic separation of microalgae. *Small*, **8**(11), 1683–1692.

Liu, Z.-Y., Wang, G.-C., Zhou, B.-C. 2008. Effect of iron on growth and lipid accumulation in *Chlorella vulgaris*. *Bioresource Technology*, **99**(11), 4717–4722.

López-Serrano, A., Olivas, R.M., Landaluze, J.S., Cámara, C. 2014. Nanoparticles: A global vision. Characterization, separation, and quantification methods. Potential environmental and health impact. *Analytical Methods*, **6**(1), 38–56.

Macario, A., Verri, F., Diaz, U., Corma, A., Giordano, G. 2013. Pure silica nanoparticles for liposome/lipase system encapsulation: Application in biodiesel production. *Catalysis Today*, **204**, 148–155.

Marchetti, J.M., Miguel, V.U., Errazu, A.F. 2007. Possible methods for biodiesel production. *Renewable and Sustainable Energy Reviews*, **11**(6), 1300–1311.

Miao, X., Wu, Q. 2004. High yield bio-oil production from fast pyrolysis by metabolic controlling of *Chlorella protothecoides*. *Journal of Biotechnology*, **110**(1), 85–93.

Middelberg, A.P.J. 1995. Process-scale disruption of microorganisms. *Biotechnology Advances*, **13**(3), 491–551.

Ngo, T.P.N., Li, A., Tiew, K.W., Li, Z. 2013. Efficient transformation of grease to biodiesel using highly active and easily recyclable magnetic nanobiocatalyst aggregates. *Bioresource Technology*, **145**, 233–239.

Ogbonna, J., McHenry, M. 2015. Culture systems incorporating heterotrophic metabolism for biodiesel oil production by microalgae. In: *Biomass and Biofuels from Microalgae*, N.R. Moheimani, M.P. McHenry, K. de Boer, P.A. Bahri (eds.), Vol. 2, Springer International Publishing, pp. 63–74.

Pádrová, K., Lukavský, J., Nedbalová, L., Čejková, A., Cajthaml, T., Sigler, K., Vítová, M., Řezanka, T. 2015. Trace concentrations of iron nanoparticles cause overproduction of biomass and lipids during cultivation of cyanobacteria and microalgae. *Journal of Applied Phycology*, **27**(4), 1443–1451.

Prochazkova, G., Safarik, I., Branyik, T. 2013. Harvesting microalgae with microwave synthesized magnetic microparticles. *Bioresource Technology*, **130**, 472–477.

Pugh, S., McKenna, R., Moolick, R., Nielsen, D.R. 2011. Advances and opportunities at the interface between microbial bioenergy and nanotechnology. *Canadian Journal of Chemical Engineering*, **89**(1), 2–12.

Puri, M., Verma, M.L., Barrow, C.J. 2013. Enzyme immobilization on nanomaterials for biofuel production. *Trends in Biotechnology*, **31**(4), 215–216.

Ramos Tercero, E.A., Domenicali, G., Bertucco, A. 2014. Autotrophic production of biodiesel from microalgae: An updated process and economic analysis. *Energy*, **76**, 807–815.

Ranganathan, S.V., Narasimhan, S.L., Muthukumar, K. 2008. An overview of enzymatic production of biodiesel. *Bioresource Technology*, **99**(10), 3975–3981.

Royon, D., Daz, M., Ellenrieder, G., Locatelli, S. 2007. Enzymatic production of biodiesel from cotton seed oil using t-butanol as a solvent. *Bioresource Technology*, **98**(3), 648–653.

San, N.O., Kurşungöz, C., Tümtaş, Y., Yaşa, Ö., Ortaç, B., Tekinay, T. 2014. Novel one-step synthesis of silica nanoparticles from sugarbeet bagasse by laser ablation and their effects on the growth of freshwater algae culture. *Particuology*, **17**, 29–35.

Sarma, S.J., Das, R.K., Brar, S.K., Le Bihan, Y., Buelna, G., Verma, M., Soccol, C.R. 2014. Application of magnesium sulfate and its nanoparticles for enhanced lipid production by mixotrophic cultivation of algae using biodiesel waste. *Energy*, **78**, 16–22.

Seo, J.Y., Lee, K., Praveenkumar, R., Kim, B., Lee, S.Y., Oh, Y.-K., Park, S.B. 2015. Tri-functionality of Fe_3O_4-embedded carbon microparticles in microalgae harvesting. *Chemical Engineering Journal*, **280**, 206–214.

Shen, Y., Yuan, W., Pei, Z., Mao, E. 2010. Heterotrophic culture of *Chlorella protothecoides* in various nitrogen sources for lipid production. *Applied Biochemistry and Biotechnology*, **160**(6), 1674–1684.

Sibi, G., Shetty, V., Mokashi, K. 2015. Enhanced lipid productivity approaches in microalgae as an alternate for fossil fuels—A review. *Journal of the Energy Institute*.

Tan, T., Lu, J., Nie, K., Deng, L., Wang, F. 2010. Biodiesel production with immobilized lipase: A review. *Biotechnology Advances*, **28**(5), 628–634.

Teo, C.L., Jamaluddin, H., Zain, N.A.M., Idris, A. 2014. Biodiesel production via lipase catalysed transesterification of microalgae lipids from *Tetraselmis* sp. *Renewable Energy*, **68**, 1–5.

Toh, P.Y., Ng, B.W., Chong, C.H., Ahmad, A.L., Yang, J.W., Chieh Derek, C.J., Lim, J. 2014. Magnetophoretic separation of microalgae: The role of nanoparticles and polymer binder in harvesting biofuel. *RSC Advances*, **4**(8), 4114–4121.

Torkamani, S., Wani, S.N., Tang, Y.J., Sureshkumar, R. 2010. Plasmon-enhanced microalgal growth in miniphotobioreactors. *Applied Physics Letters*, **97**(4), 043703.

Tran, D.T., Yeh, K.L., Chen, C.L., Chang, J.S. 2012. Enzymatic transesterification of microalgal oil from *Chlorella vulgaris* ESP-31 for biodiesel synthesis using immobilized *Burkholderia* lipase. *Bioresource Technology*, **108**, 119–127.

Vandamme, D., Eyley, S., Van den Mooter, G., Muylaert, K., Thielemans, W. 2015. Highly charged cellulose-based nanocrystals as flocculants for harvesting *Chlorella vulgaris*. *Bioresource Technology*, **194**, 270–275.

Vandamme, D., Foubert, I., Muylaert, K. 2013. Flocculation as a low-cost method for harvesting microalgae for bulk biomass production. *Trends in Biotechnology*, **31**(4), 233–239.

Véras, I.C., Silva, F.A.L., Ferrão-Gonzales, A.D., Moreau, V.H. 2011. One-step enzymatic production of fatty acid ethyl ester from high-acidity waste feedstocks in solvent-free media. *Bioresource Technology*, **102**(20), 9653–9658.

Vergini, S., Aravantinou, A., Manariotis, I. 2015. Harvesting of freshwater and marine microalgae by common flocculants and magnetic microparticles. *Journal of Applied Phycology*, 1–9.

Verma, M.L., Chaudhary, R., Tsuzuki, T., Barrow, C.J., Puri, M. 2013. Immobilization of β-glucosidase on a magnetic nanoparticle improves thermostability: Application in cellobiose hydrolysis. *Bioresource Technology*, **135**, 2–6.

Verma, M.L., Puri, M., Barrow, C.J. 2015. Recent trends in nanomaterials immobilised enzymes for biofuel production. *Critical Reviews in Biotechnology*, 1–12.

Wang, S.-K., Wang, F., Hu, Y.-R., Stiles, A.R., Guo, C., Liu, C.-Z. 2014. Magnetic flocculant for high efficiency harvesting of microalgal cells. *ACS Applied Materials and Interfaces*, **6**(1), 109–115.

Wang, X., Liu, X., Zhao, C., Ding, Y., Xu, P. 2011. Biodiesel production in packed-bed reactors using lipase–nanoparticle biocomposite. *Bioresource Technology*, **102**(10), 6352–6355.

Wang, Z., Wu, C., Yuan, Z., Lee, J., Park, S. 2009. Repeated use of methanol and sulfuric acid to pretreat jatropha oil for biodiesel production. In: *Proceedings of ISES World Congress 2007* (Vols. I–V), D.Y. Goswami, Y. Zhao (eds.). Springer, Berlin, Germany, pp. 2413–2417.

Xie, W., Ma, N. 2010. Enzymatic transesterification of soybean oil by using immobilized lipase on magnetic nano-particles. *Biomass and Bioenergy*, **34**(6), 890–896.

Xie, W., Wang, J. 2012. Immobilized lipase on magnetic chitosan microspheres for transesterification of soybean oil. *Biomass and Bioenergy*, **36**, 373–380.

Xiong, W., Li, X., Xiang, J., Wu, Q. 2008. High-density fermentation of microalga *Chlorella protothecoides* in bioreactor for microbio-diesel production. *Applied Microbiology and Biotechnology*, **78**(1), 29–36.

Xu, L., Guo, C., Wang, F., Zheng, S., Liu, C.Z. 2011. A simple and rapid harvesting method for microalgae by in situ magnetic separation. *Bioresource Technology*, **102**(21), 10047–10051.

Yadala, S., Cremaschi, S. 2014. Design and optimization of artificial cultivation units for algae production. *Energy*, **78**, 23–39.

You, Q., Yin, X., Zhao, Y., Zhang, Y. 2013. Biodiesel production from jatropha oil catalyzed by immobilized *Burkholderia cepacia* lipase on modified attapulgite. *Bioresource Technology*, **148**, 202–207.

Zhang, B., Weng, Y., Xu, H., Mao, Z. 2012. Enzyme immobilization for biodiesel production. *Applied Microbiology and Biotechnology*, **93**(1), 61–70.

Zhang, L., Sun, S., Xin, Z., Sheng, B., Liu, Q. 2010. Synthesis and component confirmation of biodiesel from palm oil and dimethyl carbonate catalyzed by immobilized-lipase in solvent-free system. *Fuel*, **89**(12), 3960–3965.

Zhang, X.L., Yan, S., Tyagi, R.D., Surampalli, R.Y. 2013. Biodiesel production from heterotrophic microalgae through transesterification and nanotechnology application in the production. *Renewable and Sustainable Energy Reviews*, **26**, 216–223.

Zhao, X., Qi, F., Yuan, C., Du, W., Liu, D. 2015. Lipase-catalyzed process for biodiesel production: Enzyme immobilization, process simulation and optimization. *Renewable and Sustainable Energy Reviews*, **44**, 182–197.

25

Marine Nanobiomaterials: Their Biomedical and Drug Delivery Applications

Sougata Jana and Subrata Jana

CONTENTS

25.1 Introduction

The nanotechnology in medical sciences has created new opportunities in applications of nanomaterials for drug delivery, molecular imaging, and diagnosis. It can serve faster diagnosis, improved imaging, enhanced therapeutic efficacy, and reduced adverse side effects compared to the conventional methods (Hu and Zhang 2009; Jee et al. 2012). The fabrication of biomaterials has been studied extensively as a new drug delivery vehicle for the past few decades. The research scientists all over the world are attempting to develop marine biodegradable polymeric nanoparticles (NPs) for drug delivery and biomedical application, such as gene delivery, tissue engineering, and imaging. This chapter is focused in view of their applications in controlling the release of drugs for prolonged period, stabilizing heat labile molecules (e.g., proteins, peptides, or DNA) from degradation and site-specific drug targeting (Singh and Lillard 2009).

A variety of marine biomaterials can be extracted from marine plants and animal organisms, of which alginate (AL), carrageenan, and chitosan (CH) are the more popular polymeric biomaterials in controlled drug delivery for prolonged periods. Seaweeds are the most abundant source of polysaccharides, such as ALs, agar, and agarose, as well as carrageenans. Chitin and CH are derived from the exoskeleton of marine crustaceans (Laurienzo 2010).

25.2 Alginate

AL is an anionic polymer, produced by brown algae (Figure 25.1) and bacteria. AL is a polyanionic copolymer of 1–4 linked β-ᴅ-mannuronic acid and α-ʟ-glucuronic acid (Murano 1998; Smidsrød and Skjåk-Braek 1990). AL is a nontoxic, biodegradable, cost-effective, and readily available natural polymer. Besides these, it has been found to be a mucoadhesive, biocompatible, and nonimmunogenic substance. AL can also be chemically modified to alter its properties (Pawar and Edgar 2012; Yang et al. 2011a). At neutral pH, carboxyl groups of AL are deprotonated to make the polymer highly negatively charged. Addition of divalent cations, such as Ca^{2+}, can induce cross-linking of the polymer, which subsequently leads to the formation of gel (Yang et al. 2011b). Applying these principles, different types of AL-based biomaterials that are used for biomedical and drug delivery applications are produced, which are summarized in Table 25.1.

25.2.1 Applications

25.2.1.1 Colorectal-Specific Delivery

A high-performance protocol for the photodynamic detection of colorectal cancer was developed by Yang et al. (2011). In their work, AL is complexed with folic acid–modified CH to form NPs with improved drug release efficiency in the cellular lysosome, containing 5-aminolevulinic acid (5-ALA). Folic acid is a folate receptor ligand used for targeting the cell membrane and increasing NP endocytosis. The average diameter of the prepared NPs was 115 nm. The loading efficiency of 5-ALA is 27%. The loaded 5-ALA was released in the lysosome, and this was promoted by the reduced force of attraction between CH

Brown seaweed

FIGURE 25.1
Source and structure of alginate.

TABLE 25.1

Modified Alginate and Their Applications

Modified AL	Drugs/Other Loading Materials	Delivery/Targeted Region	Reference
AL–folic acid–CH	5-ALA	Colorectal-specific delivery for fluorescent endoscopic detection	Yang et al. (2011)
Thiolated AL–albumin	TMX	Delivery of anticancer drugs	Martínez et al. (2011)
AL–cysteine	TMX	Treatment of breast cancer	Martínez et al. (2012)
Montmorillonite–AL	Irinotecan	Drug delivery system in chemotherapy	Irina et al. (2014)
AgNP–AL composite	Silver	Antibacterial application	Zahran et al. (2014)
Silica/AL	Magnetic iron oxide colloids and fluorescent carboxyfluorescein	Hybrid magnetic carriers	Boissière et al. (2007)
AL/CH	Insulin	Oral delivery system	Zhang et al. (2010)
Inhalable AL NPs	INH, RIF, and PZA	Antitubercular drug delivery	Zahoor et al. (2005)
CH-coated sodium AL–CNPs	5-FU	Ocular delivery	Nagarwal et al. (2012)

and 5-ALA via the deprotonated AL, resulting in a higher intracellular protoporphyrin IX accumulation for the photodynamic detection. These studies revealed that the AL and folic acid–incorporated complex with CH NPs (CNPs) was an excellent carrier for colorectal-specific delivery of 5-ALA for fluorescent endoscopic detection.

25.2.1.2 Delivery of Anticancer Drugs

The major problem for the treatment of cancer chemotherapy is to supply the required concentration of the drug at the tumor site for a prolonged period. To overcome this type of problem, Martínez et al. (2011) prepared tamoxifen (TMX)-loaded NPs by coacervation method stabilized by disulfide bond formation, which is based on thiolated AL (ALG-CYS) and disulfide bond reduced albumin (BSA-SH). The study showed the cubic-shaped NPs with partial spherical tendency, and the particle size varied in the range of 42–388 nm. The NPs were characterized by Fourier transform infrared (FTIR) spectroscopy and Thermogravimetric analysis (TGA). The maximum TMX released (23%–61% of loaded TMX) between 7 and 75 h and the amount of released TMX can be influenced by the percentage of ALG-CYS in the particle.

The *in vivo* antitumor activity of two of these formulations in xenograft nude mice was compared with TMX-free doses. Martínez et al. (2012) also reported TMX-loaded NPs for the treatment of metastatic estrogen-dependent breast cancer. For the treatment of estrogen-dependent advanced or metastatic breast cancer, TMX has been used first among the family of drugs known as "selective estrogen receptor modulators." *In vivo* studies were also performed using TMX-loaded NPs in an MCF-7 nude mice xenograft model. These systems showed an enhancement of the TMX led antitumor activity by lowering tumor evolutions and tumor growth rates. Histological and immune histochemical studies revealed that treatments with TMX-loaded NPs showed the most regressive and less proliferative tumor tissues. TMX biodistribution studies revealed that TMX-loaded NPs caused more accumulation of the drug into the tumor site with undetectable levels of TMX in plasma, reducing the possibility of delivering TMX to other nontargeted organs and consequently developing possible side effects. TMX nanoparticulate systems will provide a novel approach to the treatment of breast cancer in the future.

Iliescu et al. (2014) fabricated and evaluated irinotecan (I) nanocomposite beads based on montmorillonite (Mt) and sodium AL as drug carriers. I-Mt-AL nanocomposite beads were prepared by ionotropic gelation technique. The structure and surface morphology of the hybrid and composite materials were established by means of x-ray diffraction (XRD), IR spectroscopy (FTIR), thermal analysis (TG-DTA), and scanning electron microscopy (SEM). Irinotecan incorporation efficiency in Mt and AL beads was determined both by UV–vis spectroscopy and thermal analysis and was found to be high. The *in vitro* study in simulated intestinal fluid (SIF) (pH 7.4 at 37°C) is conducted to establish that if upon administering the beads at the site of a targeted colorectal tumor, the delivery of the drug is sustained and can represent an alternative to the existing systemic chemotherapy. Thus, nanocomposite beads are a promising drug delivery system in chemotherapy.

25.2.1.3 Antibacterial Application

The current development of biomaterial containing substances with antimicrobial activity has drawn attention into a variety of biomedical applications. Zahran et al. (2014) investigated antimicrobial activities by applying silver NP (AgNP)–AL composite on cotton fabric, using a simple one-step rapid synthetic route by reduction of silver nitrate using alkali-hydrolyzed AL solution, which acts as both a reducing and a capping agent. Physical deposition of AgNP–AL composite on the fabric was determined by FTIR spectra, color coordinates, silver content, % of silver release, and SEM analyses. The NP fabrics showed excellent antibacterial activity against the tested bacteria, *Escherichia coli*, *Staphylococcus aureus*, and *Pseudomonas aeruginosa*.

25.2.1.4 Hybrid Magnetic Carriers

The silica-based NPs with tailored formulation have several advantages such as enhanced and controllable mechanical and chemical stability. Boissière et al. (2007) reported and formulated silica/AL NPs incorporating magnetic iron oxide colloids and fluorescent carboxyfluorescein. These nanocomposites were characterized by electron microscopy, XRD, and magnetic measurements. The release of fluorophore was investigated *in vitro* and was demonstrated to occur in 3T3 fibroblast cells. Further grafting of organic moieties on particle surface was also described. These data suggested that hybrid NPs are flexible platforms for the development of multifunctional biosystems.

25.2.1.5 Oral Delivery System

Polymeric polyelectrolyte complexes are formed when oppositely charged polyelectrolytes are mixed together and interact through electrostatic interaction. Zhang et al. (2010) developed insulin-loaded AL/CNP system. Polymeric NPs have shown potential for use as controlled drug delivery systems. Insulin was protected by forming complexes with cationic β-cyclodextrin polymers (CPβCDs), which were synthesized by polycondensation from β-cyclodextrin (β-CD), epichlorohydrin, and choline chloride. The electrostatic attraction between insulin and CPβCDs, as well as the assistance of its polymeric chains, could effectively protect insulin under simulated gastrointestinal conditions. The NPs have size lower than 350 nm and association efficiency up to 87%. The cumulative percentage of insulin release in SIF was significantly higher (40%) than that without CPβCDs (18%). It is observed that insulin was mainly retained in the core of the NPs and well protected against degradation in simulated gastric fluid.

25.2.1.6 Antitubercular Drug Carriers

The pharmacokinetic and chemotherapeutic studies of aerosolized AL-encapsulated NPs, containing isoniazid (INH), rifampicin (RIF), and pyrazinamide (PZA), were evaluated by Zahoor et al. (2005). The NPs were prepared by cation-induced gelification of AL. The average size of the particle was 235 nm, and its drug encapsulation efficiency was 70%–90% for INH and PZA and 80%–90% for RIF. The relative bioavailability of all drugs encapsulated in AL NPs was significantly higher compared with oral-free drugs. All drugs detected in the organs (lungs, liver, and spleen) were above the minimum inhibitory concentration until 15 days postnebulization, while free drugs stayed for 1 day only. The chemotherapeutic efficacy of three doses of drug-loaded AL NPs nebulized 15 days apart was comparable with 45 daily doses of oral-free drugs. The inhalable AL NPs act as an ideal carrier for the controlled release of the antitubercular drug.

25.2.1.7 Ocular Delivery

Nagarwal et al. (2012) investigated CH-coated sodium AL–CH (SA–CH) NPs, that is, CH-SA–CH NPs, loaded with 5-flurouracil (5-FU) for ophthalmic delivery. The drug-loaded NPs (DNPs) were prepared by ionic gelation technique using SA and CH and then suspended in CH solution. The mean size of NPs and their morphology were characterized by dynamic light scattering, SEM, atomic force microscopy, and zeta potential. The *in vitro* release was studied by dialysis membrane technique. The DNPs showed sustained release of 5-FU compared to the 5-FU solution with high burst effect. *In vivo* study in rabbit eye showed significantly greater level of 5-FU in aqueous humor compared to 5-FU solution. The enhanced mucoadhesiveness of CH-SA–CH DNPs results in higher bioavailability as compared to the uncoated NPs.

25.3 Carrageenans

Carrageenan is an important marine biomaterial, which fulfills the criteria of polysaccharide; it is a natural carbohydrate (polysaccharide) obtained from edible red seaweeds. Carrageenan is a naturally occurring anionic sulfated linear polysaccharides extracted from certain red seaweed (Kirk and Othmer 1992) of the Rhodophyceae family (Chen et al. 2002) particularly from *Chondrus crispus,* commonly known as carrageen moss or Irish moss, *Eucheuma, Gigartina stellata, Iridaea, Hypnea, Solieria, Agardhiella,* and *Sarconema* (Chiovitti et al. 1988; McCandles et al. 1982; Mollet et al. 1988; Mollion 1983; Murano et al. 1977; Parekh et al. 1979).

25.3.1 Types and Structure

There are several different carrageenans with slightly varied chemical structures and properties (McHugh 2003). Carrageenan is formed by alternate units of D-galactose and 3,6-anhydrogalactose joined by α-1,3 and β-1,4-glycosidic linkage. According to the literature survey, carrageenan can be classified into different types based on the amount and position of sulfate groups, their family, and its properties. Carrageenan is classified into λ (lambda), κ (kappa), ι (iota), υ (nu), μ (mu), θ (theta), and ξ (Ksi) depending on the amount and position of the SO_3^- groups, all containing about 22%–35% of sulfate groups (Figure 25.2) (Stanley 1987). Carrageenan is also classified mainly into three types based on the family (Anderson et al. 1969; Greer and Yaphe 1984; Hoffmann et al. 1996).

25.3.2 Applications

The biological applicability of carrageenan as a natural occurring polysaccharide has been increasing widely for biomedical applications and creates a strong position in the related research field. This natural source can be used in the different applications, varying from tissue engineering to the preparation of drug vehicles for controlled release due to their different chemical structures and physical properties. Different types of biomedical and drug delivery applications of modified carrageenan are summarized in Table 25.2.

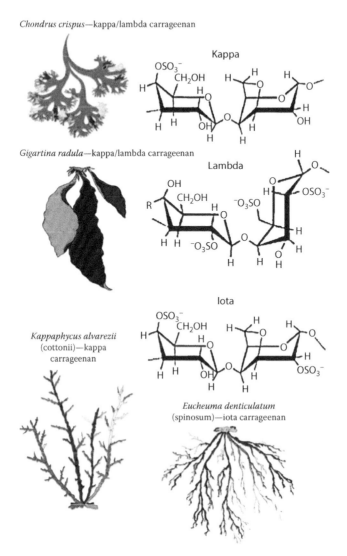

FIGURE 25.2
Source and structure of carrageenans.

25.3.2.1 Antimicrobial Activity

The nanobiotechnology of polymeric hydrogel is widely used in the biomedical field such as tissue engineering and drug delivery with focusing the treatment of antibacterial disease. Jayaramudu et al. (2013) developed biodegradable silver nanocomposite hydrogels for bacterial inactivation applications. Biodegradable silver nanocomposite hydrogels were prepared by a green process using acrylamide with I-carrageenan. The AgNPs were prepared as silver colloid by reducing $AgNO_3$ with leaf extracts of *Azadirachta indica* (neem leaf) that (Ag^0) formed the hydrogel network. The formation of biodegradable AgNPs in the hydrogels was characterized using UV–vis spectroscopy, TGA, XRD, SEM, and TEM studies. The biodegradable hydrogels of AgNP composite were tested for antibacterial activities. The antibacterial activity of the silver nanocomposite hydrogels was

TABLE 25.2

Modified Carrageenan and Their Applications

Modified Carrageenan	Drugs/Other Loading Materials	Delivery/Targeted Region	Reference
Iota-carrageenan-based biodegradable Ag^0 nanocomposite	*A. indica* (neem leaf)	Antimicrobial activity	Jayaramudu et al. (2013)
Clay-modified κ-carrageenan and AgNPs	Clay mineral	Antimicrobial activity against both gram-positive and gram-negative bacteria	Rhim and Wang (2014)
CH/carageenan	Nisin	Antimicrobial activity	Chopra et al. (2014)
κ-Carrageenan-coated magnetic iron oxide NPs	Iron oxide	Dye removal (MB)	Salgueiro et al. (2013)
κ-Carrageenan nanogels	MB	Controlled drug delivery	Daniel-da-Silva et al. (2011)
κ-Carrageenan hydrogel nanocomposites	MB	Development of drug delivery systems	Daniel-da-Silva et al. (2012)
κ-Carrageenan nanocomposite hydrogels	MB	GIT-controlled drug delivery	Hezaveh and Muhamad (2012)
CH/carrageenan	Cross-linking with TPP	Mucosal delivery	Rodrigues et al. (2012)
Drug/carrageenan	Water-soluble compounds	Drug delivery	Dai et al. (2007)

studied by the inhibition zone method against *Bacillus* and *E. coli*, which suggested that the silver nanocomposite hydrogels were effective and potential candidates for antimicrobial applications.

Rhim and Wang (2014) prepared bionanocomposite films with κ-carrageenan and AgNPs and chemically modified clay mineral (Cloisite® 30B). The tensile strength of the nanocomposite films increased by 14%–26% and water vapor permeability decreased by 12%–27% compared to those of the neat carrageenan film. The nanocomposite films exhibited characteristic antimicrobial activity against gram-positive (*Listeria monocytogenes*) and gram-negative (*E. coli* O157:H7) pathogenic bacteria. While AgNP-included nanocomposite film showed strong antimicrobial activity against gram-negative bacteria, the clay-included nanocomposite films showed strong antimicrobial activity against gram-positive bacteria.

The activity of the preservatives generally decreases in the direct application where it can bind to the food matrix. The loss of activity can also happen due to the interaction with food components. The food scientists developed and explored naturally occurring food antimicrobials for the production of safe and effective healthy food from contamination and spoilage of fruits, vegetables, grains, and food by various microbes and their toxic metabolites. Chopra et al. (2014) prepared nisin-loaded CH/carrageenan nanocapsules using an ionic complex forming method and evaluated for antibacterial activity. The concentration of polymers and surfactant affected particle size and encapsulation efficiency. *In vitro* release study over a period of 2 weeks indicated slow and sustained release from the formulation. Furthermore, *in vitro* evaluation of antibacterial activity against *Micrococcus luteus* microbial type culture collection (MTCC) 1809, *P. aeruginosa* MTCC 424, *Salmonella enterica* MTCC 1253, and *Enterobacter aerogenes* MTCC2823 in tomato juice for 6 months indicated that encapsulated nanocapsules showed better antibacterial effect on microbes for prolonged period as compared to the components evaluated separately.

25.3.2.2 Dye Removal

Methylene blue (MB) is a cationic dye commonly present in the discharged aqueous waste solution from the textile and related industries. Salgueiro et al. (2013) proposed and synthesized κ-carrageenan-coated magnetic iron oxide NPs, which was tested as adsorbents for the magnetically assisted removal of MB. Thus, κ-carrageenan-coated magnetic NPs are very promising ecofriendly materials for removing MB from wastewater using magnetic separation.

25.3.2.3 Controlled Drug Delivery

Daniel-da-Silva et al. (2011) reported cross-linked κ-carrageenan hydrogel NPs with an average size smaller than 100 nm and prepared, using reverse microemulsions, the method combined with thermally induced gelation. The size of the nanogels varied with biopolymer concentration at a constant water/surfactant concentration ratio. The nanogels were found to be thermosensitive in a temperature range acceptable for living cells (37°C–45°C) undergoing reversible volume transitions in response to thermal stimuli. These formulations are used to explore the application of nanogels in smart therapeutics such as thermosensitive drug carriers. Similarly, the sustained release of MB from the nanogels was evaluated under *in vitro* conditions as proof of concept experiments, and the release rate was found to be controlled with temperature.

The incorporation of magnetic NPs in polysaccharide hydrogels is currently being explored as a strategy to confer to the hydrogels' novel functionalities, valuable for specific bioapplications. The hydrogel is hydrophilic, with high water-holding capacity, soft, biocompatible, and biodegradable with three-dimensional networks of water-soluble polymers showed unique physicochemical properties that could be beneficial for controlled delivery of drug. So exploiting the properties of soft material, Daniel-da-Silva et al. (2012) used an important class of biomaterials derived from marine natural polysaccharides such as κ-carrageenan for drug delivery. Within this context, κ-carrageenan magnetic hydrogel nanocomposites have been prepared, and the effect of magnetic (Fe_3O_4) nanofillers in the swelling of the hydrogels and in the release kinetics and mechanism of a model drug (MB) has been investigated. *In vitro* release studies demonstrated the applicability of the composites in sustained drug release. The mechanism of controlling the release seems to be determined by the strength of the gel network and the extent of gel swelling, both being affected by the incorporation of nanofillers.

Hezaveh and Muhamad (2012) synthesized a gastrointestinal tract (GIT)-controlled drug delivery tool in modified κ-carrageenan NP hydrogels using *in situ* method. The effect of metallic NPs in GIT release of a model drug (MB) has been investigated. The effect of NPs loading and genipin cross-linking on GIT release of nanocomposite is also studied to finally provide the most suitable drug carrier system. *In vitro* release studies revealed that using metallic nanocomposite hydrogels in GIT could improve the drug release in the intestine and minimize it in the stomach. It was found that cross-linking and nanofiller loading can significantly improve the targeted release.

25.3.2.4 Mucosal Delivery

Rodrigues et al. (2012) developed CH/carrageenan/tripolyphosphate NPs by polyelectrolyte complexation/ionic gelation. The NP matrix was characterized by FTIR, x-ray photoelectron spectroscopy (XPS), and TOF-SIMS. The size and zeta potential are decreased from 450–500 to 150–300 nm and from +75 to +85 mV to +50 to +60 mV, respectively, whereas the production yield and stability are increased from 15%–20% to 25%–35% and from 1 week to up to 9 months, respectively. Also, a correlation between positive and negative charge ratios in the formulations and the aforementioned characteristics was established. The small size and high positive surface charge make the developed CH/carrageenan/tripolyphosphate NPs potential tools for mucosal delivery.

25.3.2.5 Drug Delivery

Dai et al. (2007) fabricated drug/polymeric nanocomplex to increase both solubility and dissolution rate of poor water-soluble compounds. A hydrophilic polymer, lambda-carrageenan, was first complexed with a poorly water-soluble model compound to increase the compounds' aqueous solubility. The compound/carrageenan complex was further nanosized by wet-milling to enhance the dissolution rate after lyophilization. This approach of nanosizing a drug/carrageenan complex increased the aqueous solubility of the compound from 1 to 39 μg/mL.

25.4 Chitosan

CH was derived from naturally occurring sources, like exoskeleton of insects, crustaceans, and fungi that have been shown to be biocompatible and biodegradable (Figure 25.3). CH polymers are semisynthetically derived aminopolysaccharides that have unique structures, with multidimensional properties and highly sophisticated functionality. They are very compatible and effective biomaterials and have a wide range of applications in biomedical and other industrial areas (Dash et al. 2011).

25.4.1 Structure

CH is an aminopolysaccharide composed of randomly distributed β-1,4-linked glucosamine and *N*-acetyl-ᴅ-glucosamine units. It is obtained from deacetylation of chitin, a natural polysaccharide found in the outer skeleton of insects, crabs, shrimp, and lobsters along with the internal structure of other invertebrates (Najafabadi et al. 2014). It is an FDA GRAS (generally recognized as safe)–approved material and has been widely utilized in many applications including drug delivery, tissue engineering, and food technology. CH has been used in the development of various pharmaceutical formulations. Although CH is a very promising biopolymer for use as a carrier material in drug delivery systems, it has a limited capacity for controlling drug release from oral dosage forms due to its fast dissolution in the stomach. To overcome this disadvantage, chemical modification of CH has been carried out in the development of drug delivery systems (Jana et al. 2013b).

Chitosan obtained from the shells of
shrimp and other sea crustaceans

Chitosan

FIGURE 25.3
Source and structure of chitosan.

25.4.2 Applications

Various types of biomedical and drug delivery application of modified CH are summarized in Table 25.3.

25.4.2.1 Targeted Drug Delivery and Cellular Imaging

Shen et al. (2012) prepared a novel folate-conjugated carboxymethyl CH ferrosoferric oxide–doped cadmium telluride quantum dot NPs (CFLMNPs). The resulting CFLMNPs possessed intense super paramagnetic effect and photoluminescence property at room temperature. The size range of CFLMNPs was from 170 to 190 nm under simulated physiological environment. The anticancer drug adriamycin was used for the human liver cancer treatment. The multifunctional CFLMNPs showed high drug-loading efficiency, low cytotoxicity, and favorable cell compatibility, which make this material a very promising candidate for carboxymethyl CH–based targeted drug delivery and cellular imaging.

TABLE 25.3

Modified Chitosan and Their Applications

Modified CH	Drugs/Loading Materials	Target Organ/Delivery/ Biomedical Applications	Reference
Carboxymethyl CH–based folate/ Fe_3O_4/CdTe NP	Adriamycin (ADM)	Targeted drug delivery and cellular imaging	Shen et al. (2012)
Cetuximab-conjugated O-carboxymethyl CH	Paclitaxel	Targeting EGFR-overexpressing cancer cells	Maya et al. (2013)
CH/PEG blended	Paclitaxel	Tumor drug delivery	Parveen and Sahoo (2011)
CH–clay nanocomposite	Doxorubicin (DOX)	Tissue engineering and controlled drug delivery to cancer cells	Yuan et al. (2010)
O-carboxymethyl CH	Curcumin		Anitha et al. (2011)
CH–GO	Fluorescein sodium (FL)	Transdermal drug delivery system	Justin and Chen (2014)
Carbopol gel containing CH–egg albumin NPs	Aceclofenac	Transdermal delivery	Jana et al. (2014)
CNPs	Dopamine	Brain delivery	Trapani et al. (2011)
CNPs	Bovine serum albumin	Protein delivery carrier	Gan and Wang (2007)
CNPs	EGF and FGF	Growth factor delivery system for tissue engineering applications	Rajam et al. (2011)
CH-FITC	PEG 600, PEG 1000	Pulmonary drug delivery	Sharma et al. (2012)
LMW CH	Erythrocytes	Vascular drug delivery	Fana et al. (2012)
CH-based NPs	Alprazolam	Oral drug delivery	Jana et al. (2013)

25.4.2.2 Targeting Cancer Cells

Maya et al. (2013) fabricated O-carboxymethyl CH (O-CMC) surface-conjugated NPs, with cetuximab (Cet) from targeted delivery of paclitaxel (PTXL) to epidermal growth factor receptor (EGFR)-overexpressing cancer cells. NPs were prepared through simple ionic gelation technique and size around 180 nm. The alamarBlue assay revealed superior anticancer activity compared to nontargeted NPs. The nanoformulation triggered enhanced cell death (confirmed by flow cytometry) due to its higher cellular uptake. The selective uptake of Cet-PTXL-O-CMC NPs by EGFR-positive cancer cells (A549, A431, and SKBR3) compared to EGFR-negative MIAPaCa-2 cells confirmed the active targeting and delivery of PTXL via the targeted nanomedicine. Cet-PTXL-O-CMC functioned as a promising candidate for the targeted therapy of EGFR-overexpressing cancers.

Parveen and Sahoo (2011) formulated PTXL-loaded CH with polyethylene glycol–coated PLGA (PLGA–CS–PEG) NPs, characterized to efficiently encapsulate hydrophobic drugs, enhancing the bioavailability of the drug. This formulation may contain long-circulating drug delivery carriers for tumor drug delivery.

Yuan et al. (2010) developed CH–aluminosilicate nanocomposite carrier that was characterized by controlled and extended release of drug. Drug release from the nanocomposite particle carrier occurred by degradation of the carrier to its individual components or nanostructures with a different composition. In both the layers of aluminosilicate-based mineral and CH–aluminosilicate nanocomposite carriers, the positively charged chemotherapeutic drug strongly bound to the negatively charged aluminosilicate and release of the drug was slow. Furthermore, the pattern of drug release from the CH–aluminosilicate nanocomposite carrier was affected by pH and the CH/aluminosilicate ratio. The study points to the potential application of this hybrid nanocomposite carrier in biomedical applications, including tissue engineering and controlled drug delivery.

Anitha et al. (2011) investigated curcumin-loaded *O*-carboxymethyl CNP derivatives for biomedical applications because of their nontoxic and biodegradable properties and increases in the oral bioavailability of curcumin. The prepared NPs were characterized by DLS, AFM, SEM, FTIR, XRD, and TG/DTA. Particle size analysis studies revealed the spherical particles with a diameter of about 150 nm. NPs entrapment efficiency of curcumin was 87%. *In vitro* drug release profile was studied at 37°C under different pHs (7.4 and 4.5) with and without lysozyme. MTT assay indicated that curcumin-*O*-CMC NPs were toxic to cancer and nontoxic to normal cells. Cellular uptake of the curcumin-*O*-CMC NPs was analyzed by fluorescence microscopy and FACS. Overall, these studies indicated that *O*-CMC was a promising nanomatrix carrier for cancer cell–targeted applications.

25.4.2.3 Transdermal Drug Delivery System

Justin and Chen (2014) developed biodegradable CH–graphene oxide (GO) nanocomposites with pH-sensitive and controlled transdermal drug delivery system. CH nanocomposites containing varying GO contents and drug-loading ratios were investigated. The nanocomposites with 2 wt% GO provided the optimal combination of mechanical properties and drug-loading capacity. The drug delivery profiles of the nanocomposite were dependent on the drug-loading ratio, that is, drug to GO was 0.84:1 for quick and efficient release of the loaded drug. The pH-sensitive nanocomposite was releasing 48% less drug in an acidic condition than in a neutral environment.

Jana et al. (2014) prepared aceclofenac-loaded CH–egg albumin NPs by heat coagulation method. These NPs were characterized by FE-SEM, FTIR, DSC, and P-XRD analyses. The *in vitro* drug release from NPs showed sustained drug release over 8 h. The DNPs showed highest drug entrapment 96.32%, average particle diameter 352.90 nm, and zeta potential –22.10 mV, which was used for further preparation of Carbopol 940 gel for transdermal application. The prepared gel exhibited sustained *ex vivo* permeation of aceclofenac over 8 h through excised mouse skin. The *in vivo* anti-inflammatory activity in carrageenan-induced rats demonstrated comparative higher inhibition of the swelling of rat paw edema by the prepared gel compared with that of the marketed aceclofenac gel over 4 h.

25.4.2.4 Brain Delivery

Trapani et al. (2011) prepared CH-based NPs and evaluated their potential application for brain delivery of the neurotransmitter dopamine (DA) to treat Parkinson's disease. DA was adsorbed onto the external surface of such NPs characterized by XPS analysis. The cytotoxic effect of the CNPs and DA/CNPs was assessed using the MTT test, and it was found that the nanovectors are less cytotoxic than the neurotransmitter DA after 3 h of incubation time. Transport studies across MDCKII-MDR1 cell line showed that DA/CNPs (5) give rise to a significant transport-enhancing effect compared with the control and are greater than the corresponding DA/CNPs (1). The measurement of reactive oxygen species suggested a low DA/CNPs neurotoxicity after 3 h. *In vivo* brain microdialysis experiments in rat showed that intraperitoneal acute administration of DA/CNPs (5) (6–12 mg/kg) induced a dose-dependent increase in striatal DA output.

25.4.2.5 Protein Delivery Carrier

Gan and Wang (2007) fabricated CH–BSA–tripolyphosphate (TPP) NPs, where BSA was used as a model protein, which was encapsulated by using the polyanion TPP as the coacervation cross-link agent. The BSA-loaded CH –TPP NPs were characterized for particle size, morphology, zeta potential, BSA encapsulation efficiency, and the release kinetics. The size of BSA-loaded NPs ranges from 200 to 580 nm and exhibits a high positive zeta potential.

25.4.2.6 Growth Factor Delivery System for Tissue Engineering Applications

Rajam et al. (2011) reported about the sustained delivery of growth factors that are essential in cellular signaling for migration, proliferation, differentiation, and maturation. Epidermal growth factor (EGF) and fibroblast growth factor (FGF) loaded CNP, either individually or in combination, which could ultimately be impregnated into engineered tissue construct. CNPs were characterized by FTIR spectroscopy, zeta sizer, and high-resolution transmission electron microscope. The particles were in the size range of 50–100 nm. The incorporated EGF and FGF into CNP showed the sustained manner and were free from toxicity against fibroblasts up to 4 mg/mL of culture medium. The *in vitro* results showed that the delivery of growth factors from CNP for cellular signaling and western blotting results also revealed that poor inflammatory response showing less expression of proinflammatory cytokines such as IL-6 and TNF in the macrophage cell line J774 A-1.

25.4.2.7 Pulmonary Drug Delivery

Sharma et al. (2012) prepared pressurized metered dose inhalers (pMDIs) for deep lung delivery of cross-linked CNPs, by ionic gelation method. The formulations were prepared from cross-linked CH alone and addition of PEG 600, PEG 1000, and PEG 5000 for dispersion in aerosol propellant, hydrofluoroalkane 227. NPs were spherical, smooth-surfaced, cationic particles of mean particle size less than 230 nm. Following actuation from pMDIs, the fine particle fraction (FPF) for cross-linked CH–PEG 1000 NPs, determined using a next-generation impactor, was 34.0% ± 1.4% with a mass median aerodynamic diameter of 4.92 ± 0.3 nm. The FPFs of cross-linked CH, CH–PEG 600, and CH–PEG 5000 NPs were 5.7% ± 0.9%, 11.8% ± 2.7%, and 17.0% ± 2.1%, respectively. These results indicate that cross-linked CH–PEG 1000-based NPs are promising candidates for delivering therapeutic agents, particularly in lung delivery using pMDIs.

25.4.2.8 Vascular Drug Delivery

Fana et al. (2012) investigated CNPs for intravascular drug delivery. In this preparation, monodisperse, low-molecular-weight (LMW) CNPs were prepared by ionic gelation technique, and these NPs were investigated with regard to their erythrocyte compatibility. Confocal microscopy was used for any interactions between erythrocytes and fluorescence-labeled LMW CNPs. These results revealed that erythrocytes-loaded LMW CNPs were used as carriers of potential vascular drug delivery system.

25.4.2.9 Oral Drug Delivery

Jana et al. (2013a) developed oral sustained drug delivery of lipophilic interpolymeric complexation of cationic CH and anionic egg albumin stabilized with PEG 400 by heat coagulation method at pH 5.4 and 80°C. The drug entrapment efficiency of these NPs was in the range of 68%–99%. These NPs were characterized by FTIR, DSC, P-XRD, and FE-SEM analyses. The mean particle diameter, polydispersity index, and zeta potential of these NPs were found to be 259.60 nm, 0.501, and –9.00 mV, respectively. The *in vitro* drug release from these alprazolam-loaded NPs showed sustained drug release over a period of 24 h. So these novel formulations of CH–egg albumin–PEG NPs were used as a promising vehicle for sustained release of lipophilic drugs.

25.5 Conclusion

The polysaccharide-based biomaterials having carbohydrates with repeating monosaccharide unit are of tremendous interest due to their enormous application in biomedical research. Natural biopolymers and their biomedical application, particularly drug delivery point of view, are preferred because they are inert, safe, nontoxic, biocompatible, biodegradable, affordable, ecofriendly, and abundantly available in nature. Biomaterials can be obtained from a number of marine sources such as seaweeds, plants, bacteria, fungi, insects, crustaceans, and animals. Thus, the large varieties of biomedical applications as well as its steadily increasing number of research have engaged the researchers to discover new avenue for the applications of AL, carrageenan, and CH. It makes these biopolymers versatile and gains more significant interest in future.

References

Anderson, N. S., J. W. Campbell, M. M. Harding, D. A. Rees, and J. W. B. Samuel. 1969. X-ray diffraction studies of polysaccharide sulphates: Double helix models for κ and ι carrageenans. *Journal of Molecular Biology* 45: 85–97.

Anitha, A., S. Maya, N. Deepa, K. P. Chennazhi, S. V. Nair, H. Tamura, and R. Jayakumar. 2011. Efficient water soluble *O*-carboxymethyl chitosan nanocarrier for the delivery of curcumin to cancer cells. *Carbohydrate Polymers* 83: 452–461.

Boissière, M., J. Allouche, C. Chanéac, R. B. J. J. Livage, and T. Coradin. 2007. Potentialities of silica/alginate nanoparticles as hybrid magnetic carriers. *International Journal of Pharmaceutics* 344: 128–134.

Chen, Y., M. L. Liao, and D. E. Dustan. 2002. The rheology of carrageenan as a weak gel. *Carbohydrate Polymers* 50: 109–116.

Chiovitti, A., A. Bacic, D. J. Craik, G. T. Kraft, M. L. Liao, and R. Falshaw. 1988. A pyruvated carrageenan from Australian specimens of the red alga *Sarconema filiforme*. *Carbohydrate Research* 310: 77–83.

Chopra, M., P. Kaur, M. Bernela, and R. Thakur. 2014. Surfactant assisted nisin loaded chitosan-carrageenan nanocapsule synthesis for controlling food pathogens. *Food Control* 37: 158–164.

Dai, W., L. C. Dong, and Y. Song. 2007. Nanosizing of a drug/carrageenan complex to increase solubility and dissolution rate. *International Journal of Pharmaceutics* 342: 201–207.

Daniel-da-Silva, A. L., L. Ferreira, A. M. Gil, and T. Trindade. 2011. Synthesis and swelling behavior of temperature responsive κ-carrageenan nanogels. *Journal of Colloid and Interface Science* 355: 512–517.

Daniel-da-Silva, A. L., J. Moreira, R. Neto, A. C. Estrada, A. M. Gil, and T. Trindade. 2012. Impact of magnetic nanofillers in the swelling and release properties of κ-carrageenan hydrogel nanocomposites. *Carbohydrate Polymers* 87: 328–335.

Dash, M., F. Chiellini, R. M. Ottenbrite, and E. Chiellini. 2011. Chitosan—A versatile semi-synthetic polymer in biomedical applications. *Progress in Polymer Science* 36: 981–1014.

Fana, W., W. Yanb, Z. Xu, and H. Ni. 2012. Erythrocytes load of low molecular weight chitosan nanoparticles as a potential vascular drug delivery system. *Colloids and Surfaces B: Biointerfaces* 95: 258–265.

Gan, Q. and T. Wang. 2007. Chitosan nanoparticle as protein delivery carrier—Systematic examination of fabrication conditions for efficient loading and release. *Colloids and Surfaces B: Biointerfaces* 59: 24–34.

Greer, C. W. and W. Yaphe. 1984. Characterization of hybrid (beta-kappa-gamma)carrageenan from *Eucheuma gelatinae* J. Agardh (Rhodanhyta, Solieriaceae) using carrageenases, infrared and 13C-nuclear magnetic resonance spectroscopy. *Botinica Marina* 27: 473–478.

Hezaveh, H. and I. I. Muhamad. 2012. The effect of nanoparticles on gastrointestinal release from modified κ-carrageenan nanocomposite hydrogels. *Carbohydrate Polymers* 89: 138–145.

Hoffmann, R. A., A. R. Russell, and M. J. Gidley. 1996. Molecular weight distribution of carrageenans. In G. O. Phillips, P. J. Williams, and D. J. Wedlock (eds.), *Gums and Stabilisers for the Food Industry*, Vol. 8. Oxford, U.K.: IRL Press at the Oxford University Press, pp. 137–148.

Hu, C. M. and L. Zhang. 2009. Therapeutic nanoparticles to combat cancer drug resistance. *Current Drug Metabolism* 10: 836–841.

Iliescu, R. I., E. Andronescu, C. D. Ghitulica, G. Voicu, A. Ficai, and M. Hoteteu. 2014. Montmorillonite–alginate nanocomposite as a drug delivery system—Incorporation and in vitro release of irinotecan. *International Journal of Pharmaceutics* 463: 184–192.

Jana, S., N. Maji, A. K. Nayak, K. K. Sen, and S. K. Basu. 2013a. Development of chitosan-based nanoparticles through inter-polymeric complexation for oral drug delivery. *Carbohydrate Polymers* 98: 870–876.

Jana, S., S. Manna, A. K. Nayak, K. K. Sen, and S. K. Basu. 2014. Carbopol gel containing chitosan-egg albumin nanoparticles for transdermal aceclofenac delivery. *Colloids and Surfaces B: Biointerfaces* 114: 36–44.

Jana, S., A. Saha, A. K. Nayak, K. K. Sen, and S. K. Basu. 2013b. Aceclofenac-loaded chitosan-tamarind seed polysaccharide interpenetrating polymeric network microparticles. *Colloids and Surfaces B: Biointerfaces* 105: 303–309.

Jayaramudu, T., G. M. Raghavendra, K. Varaprasad, R. Sadiku, and K. Ramam. 2013. Iota-Carrageenan-based biodegradable Ag0 nanocomposite hydrogels for the inactivation of bacteria. *Carbohydrate Polymers* 95: 188–194.

Jee, J., J. Na, S. Lee, S. H. Kim, K. Choi, Y. Yeo, and I. C. Kwon. 2012. Cancer targeting strategies in nanomedicine: Design and application of chitosan nanoparticles. *Current Opinion in Solid State and Materials Science* 16: 333–342.

Justin, R. and B. Chen. 2014. Characterisation and drug release performance of biodegradable chitosan–graphene oxide nanocomposites. *Carbohydrate Polymers* 103: 70–80.

Kirk, R. E. and D. F. Othmer. 1992. In J. I. Kroschwitz and M. Howe-Grant (eds.), *Encyclopedia of Chemical Technology*, Vol. 4. New York: John Wiley & Sons, p. 942.

Laurienzo, P. 2010. Marine polysaccharides in pharmaceutical applications: An overview. *Marine Drugs* 8: 2435–2465.

Martínez, A., I. Iglesias, R. Lozano, J. M. Teijón, and M. D. Blanco. 2011. Synthesis and characterization of thiolated alginate-albumin nanoparticles stabilized by disulfide bonds. Evaluation as drug delivery systems. *Carbohydrate Polymers* 83: 1311–1321.

Martínez, A., E. Muñiz, I. Iglesias, J. M. Teijón, and M. D. Blancoc. 2012. Enhanced preclinical efficacy of tamoxifen developed as alginate–cysteine/disulfide bond reduced albumin nanoparticles. *International Journal of Pharmaceutics* 436: 574–581.

Maya, S., L. G. Kumar, B. Sarmento, N. S. Rejinold, D. Menon, S. V. Nair, and R. Jayakumar. 2013. Cetuximab conjugated *O*-carboxymethyl chitosan nanoparticles for targeting EGFR overexpressing cancer cells. *Carbohydrate Polymers* 93: 661–669.

McCandles, E. L., J. A. West, and M. D. Guiry. 1982. Carrageenan patterns in the Phyllophoraceae. *Biochemical Systematics and Ecology* 10: 275–284.

McHugh, D. J. 2003. Chapter 7: Carrageenan. In A guide to the seaweed industry: FAO fisheries technical paper 441. Rome, Italy: Food and Agriculture Organization of the United Nations.

Mollet, J. C., A. Rahaoui, and Y. Lemoine. 1988. Yield, chemical composition and gel strength of agarocolloids of *Gracilaria gracilis, Gracilariopsis longissima* and the newly reported *Gracilaria vermiculophylla* from Roscoff, France. *Journal of Applied Phycology* 10: 59–66.

Mollion, J. 1983. Exploitation of algal carrageenans: Global situation and perspectives in Senegal. In J. M. Kornprobst (ed.), *Marine Chemistry for Development*, pp. 84–95.

Murano, E. 1998. Use of natural polysaccharides in the microencapsulation techniques. *Journal of Applied Ichthyology* 14: 245–249.

Murano, E., R. Toffanin, E. Cecere, R. Rizzo, and S. H. Knutsen. 1977. Investigation of the carrageenans extracted from *Solieria filiformis* and *Agardhiella subulata* from Mar Piccolo, Taranto. *Marine Chemistry* 58: 319–325.

Nagarwal, R. C., R. Kumar, and J. K. Pandit. 2012. Chitosan coated sodium alginate–chitosan nanoparticles loaded with 5-FU for ocular delivery: In vitro characterization and in vivo study in rabbit eye. *European Journal of Pharmaceutical Sciences* 47: 678–685.

Najafabadi, A. H., M. Abdouss, and S. Faghihi. 2014. Synthesis and evaluation of PEG-*O*-chitosan nanoparticles for delivery of poor water soluble drugs: Ibuprofen. *Materials Science and Engineering C* 41: 91–99.

Parekh, R. G., S. K. Garg, B. R. Mehta, and D. J. Mehta. 1979. Studies on extraction of carrageenan from hypnea and its characterization. In *International Symposium on Marine Algae of the Indian Ocean Region*, Bhavnagar, India, p. 42.

Parveen, S. and S. K. Sahoo. 2011. Long circulating chitosan/PEG blended PLGA nanoparticle for tumor drug delivery. *European Journal of Pharmacology* 670: 372–383.

Pawar, S. N. and K. J. Edgar. 2012. Alginate derivatization: A review of chemistry, properties and applications. *Biomaterials* 33: 3279–3305.

Rajam, M., S. Pulavendran, C. Rose, and A. B. Mandal. 2011. Chitosan nanoparticles as a dual growth factor delivery system for tissue engineering applications. *International Journal of Pharmaceutics* 410: 145–152.

Rhim, J. and L. Wang. 2014. Preparation and characterization of carrageenan-based nanocomposite films reinforced with clay mineral and silver nanoparticles. *Applied Clay Science* 97–98: 174–181.

Rodrigues, S., A. M. Rosa da Costa, and A. Grenha. 2012. Chitosan/carrageenan nanoparticles: Effect of cross-linking with tripolyphosphate and charge ratios. *Carbohydrate Polymers* 89: 282–289.

Salgueiro, A. M., A. L. Daniel-da-Silva, A. V. Girão, P. C. Pinheiro, and T. Trindade. 2013. Unusual dye adsorption behavior of κ-carrageenan coated superparamagnetic nanoparticles. *Chemical Engineering Journal* 229: 276–284.

Sharma, K., S. Somavarapu, A. Colombani, N. Govind, and K. M. G. Taylor. 2012. Crosslinked chitosan nanoparticle formulations for delivery from pressurized metered dose inhalers. *European Journal of Pharmaceutics and Biopharmaceutics* 81: 74–81.

Shen, J., W. Tang, X. Zhang, T. Chen, and H. Zhang. 2012. A novel carboxymethyl chitosan-based folate/Fe_3O_4/CdTe nanoparticle for targeted drug delivery and cell imaging. *Carbohydrate Polymers* 88: 239–249.

Singh, R. and J. W. Lillard Jr. 2009. Nanoparticle-based targeted drug delivery. *Experimental and Molecular Pathology* 86: 215–223.

Smidsrød, O. and G. Skjåk-Braek. 1990. Alginate as immobilization matrix for cells. *Trends in Biotechnology* 8: 71–78.

Stanley, N. F. 1987. Carrageenans. In D. J. McHugh (ed.), *Production and Utilization of Products from Commercial Seaweeds*. FAO Fisheries Technical Papers, Campbell, Australian Capital Territory, Australia.

Trapani, A., E. D. Giglio, D. Cafagna, N. Denora, G. Agrimi, T. Cassano, S. Gaetani, V. Cuomo, and G. Trapani. 2011. Characterization and evaluation of chitosan nanoparticles for dopamine brain delivery. *International Journal of Pharmaceutics* 419: 296–307.

Yang, J. S., Y. J. Xie, and W. He. 2011a. Research progress on chemical modification of alginate: A review. *Carbohydrate Polymers* 84: 33–39.

Yang, S., F. Lin, H. Tsai, C. Lin, H. Chin, J. Wong, and M. Shieh. 2011b. Alginate-folic acid-modified chitosan nanoparticles for photodynamic detection of intestinal neoplasms. *Biomaterials* 32: 2174–2182.

Yuan, Q., J. Shah, S. Hein, and R. D. K. Misra. 2010. Controlled and extended drug release behavior of chitosan-based nanoparticle carrier. *Acta Biomaterialia* 6: 1140–1148.

Zahoor, A., S. Sharma, and G. K. Khuller. 2005. Inhalable alginate nanoparticles as antitubercular drug carriers against experimental tuberculosis. *International Journal of Antimicrobial Agents* 26: 298–303.

Zahran, M. K., H. B. Ahmeda, and M. H. El-Rafie. 2014. Surface modification of cotton fabrics for antibacterial application by coating with AgNPs–alginate composite. *Carbohydrate Polymers* 108: 145–152.

Zhang, N., J. Li, W. Jiang, C. Ren, J. Li, J. Xin, and K. Li. 2010. Effective protection and controlled release of insulin by cationic β-cyclodextrin polymers from alginate/chitosan nanoparticles. *International Journal of Pharmaceutics* 393: 212–218.

26

Omics and Its Application in Clinical Nanotechnology and Nanodiagnostics

Renesha Srivastava, Sushrut Sharma, Ananya Srivastava,
Pankaj Suman, and Pranjal Chandra

CONTENTS

26.1 Introduction

Omics refers to the study of various relationships of biological molecules, their actions and roles, and the collective technologies that are used to explore them. It helps to study the processes involved in a particular disease and the mode of action of compounds. It is divided in to three categories—genomics, proteomics, and metabolomics. Large number of DNA sets can be sequenced to analyze the structure and function of genomes, profiling whole mRNA genome, miRNA expression (transcriptomics), and epigenetic modifications on the genome of mRNA (epigenomics). These are classified under genomics. It is, nowadays, no more a tedious job, courtesy of the sophisticated technologies available now that facilitate analysis of millions of sequences at once. The genetic approach toward a disease is an important aspect to know the underlying mechanism of the disease [1,2]. Computer technologies such as MRI and CT can help in the imaging of the tissue anatomy more effectively, provide molecular information, and help in the effective study of omics [3] (Figure 26.1).

Proteomics is concerned with the cell product, that is, proteins and their expression. Proteins, which are formed as an outcome of a cell's translation machinery, can be analyzed by 2D polyacrylamide gel electrophoresis and mass spectrometry (MS). The identification and quantification of metabolites in a biological system, such as a cell, tissue, or organ, and their profiling in urine and blood is the field concerning metabolomics. It mainly deals with the metabolic response to biological stimuli, by analyzing the complex sample and then quantifying the molecules in such samples. The most commonly used metabolomics techniques are liquid chromatography–mass spectrometry (LC-MS), gas chromatography–mass spectrometry, and nuclear magnetic resonance (NMR). Nutritional genomics, another domain of the omics, helps to find and explore the effects of nutrients on genome, gene expression patterns, and metabolic responses to dietary interventions. Pharmacogenomics and pharmacogenetics can altogether aid in the application of genetics for the development of a drug [4]. The gene expression analysis and the changes in mRNA expression (transcriptomics) can prove opportunistic in foodomics [5]. The profiling of information of the RNA pool, or the transcriptome, and the resulting cellular functions mediated by mRNA is encountered with in transcriptomics [6]. The RNA interference pathway, translation, protein degradation pathways, and signaling pathways can be dealt with next-generation sequencing and oligonucleotide microarrays, which are the two genome-scale analysis tools [7].

The omics technologies help to study the toxicity of many components on mRNA, proteins, metabolites, genes, and mutations. Such applications of toxicogenomics help to

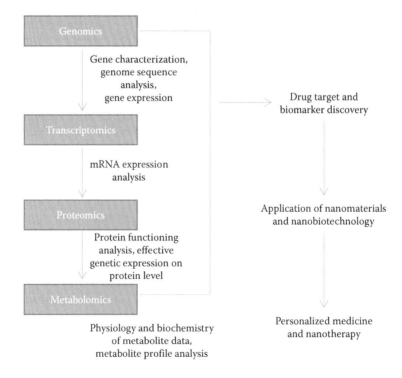

FIGURE 26.1
Various omics in clinical nanotechnology.

give the insight of the mechanisms that the toxic compounds have on our body and the risk that they pose on our health. Every human disease has a connection with the genes. A scrupulous understanding of the genomics can help pave the way for new approaches to diagnose as well as treat the ailment. Its application in the development of the drugs, the targeted drug approach for localized treatment, and the analysis of a single cell using nanobiomics is a promising approach for disease treatment.

26.2 Omics Categories

26.2.1 Genomics

Genomics has helped us to map the genes and retrieve the information about the sequences, which are applicable to basic research, biology, and health. It gives an insight about the structural and functional components of the genome and identifies the gene responsible for the disease and the protein interaction with the cells, allowing us to design a drug for that disease. The Human Genome Project (HGP) has helped to assess the human DNA, providing us the salient features of the human genome. Human genome is the largest, containing 3164.7 million nucleotide bases, out of which 99.9% is identical and only 2% are the coding regions and 50% of the genome being noncoding DNA. HGP can help us to understand the molecular mechanism of the disease.

Genomics can be categorized as structural genomics, comparative genomics, and functional genomics. The structural genomics, example being the HGP itself, relates to the genome analysis and creates the physical and transcript maps of the organism. Functional genomics helps to identify the proteins that the gene codes and thereby understand the function of gene sequences. Bioinformatics predicts the biological function by analyzing the genome, with the help of primary and secondary sequence and structural databases. Comparative genomics marks the differences and similarities in the genome, and what percent of it is matching with the human. For example, there is 98.4% genetic similarity between humans and chimpanzees. The similarity may include a whole genome, rather than an individual gene if two species are being compared, forming the underlying principle of the comparative genomics [8].

26.2.2 Proteomics

Proteomics is the study of the proteins encoded by the genes, which includes their time of expression, extent of posttranslational modification, function of the protein in the cell, and its cellular compartmentalization. The techniques involved basically start with the protein itself and trace back the gene that encoded it. The proteome, that is, the complete set of proteins encoded by a genome, is dependent on the kind of organism and the tools used for investigation, and thus, there is no defined end point for proteomics. The proteins are isolated, separated, and identified, and mainly techniques such as 2D gel electrophoresis (2DGE) and MS are used. The nucleic acid information is transcribed and finally translated to understand functioning of the cells and the disease invasion [9].

Proteomics can be categorized as expression proteomics, which studies the global expression of the proteins and cell map proteomics, for the study of protein–protein interactions. 2DGE is the most widely used tool for proteome profiling and forms the most important experimental step. The next step is the protein purification, that is, extraction of protein from cell lysate. However, the techniques have limitations in terms of the protein representation in the proteome. There are different cells in the lysate and each has a different proteome. For example, hydrophobic proteins are not easily extracted and represented on a 2D gel pattern. The subcellular-level proteome complexity is less if the subcellular components (plasma membrane and organelles) are enriched with the protein. The result of the 2DGE is not enough to identify the protein, since the posttranslational modification can affect the isoelectric point and molecular weight, which poses a challenge to the classical proteomics [8,9]. Differential proteomics, emphasizing on 1DGE/2DGE and protein digestion, followed by mass spectrometric analysis is another category. Shotgun proteomics focuses on multidimensional LC and automated mass spectroscopy. Usage of mesoporous silica for protein digestion can be taken as a new step toward nanobiotechnology [10].

Analysis of serum and tissue lysates has provided with a large amount of information to allow early detection of disease, which makes use of low-resolution MS. Diagnosis, prognosis, and drug therapy are useful applications of DNA microarrays. With the advancement in nanoscience and molecular biology, they can be applied to the central nervous system and hematological malignancies, providing a great scope for proteomic application in clinical care.

26.2.3 Toxicogenomics

Toxicogenomics is the scientific field that tells how a genome responds to environment stressors. The studies of genetics, mRNA expression (transcriptomics), proteomics, metabolite profiling (metabonomics), and bioinformatics along with the conventional toxicology help to understand the interactions of the gene and the environment. This integration has the potential to synergize our understanding of the relationship between toxicological outcomes and molecular genetics. Thousands of genes' expression can be studied at a time with the help of molecular technologies like DNA microarrays and protein chips, and they accelerate the discovery of the toxicant pathways along with specific chemical and drug pathways.

The three foremost goals of toxicogenomics are to accept the established connection linking environmental stress and human disease vulnerability rate, to categorize disease biomarkers and disclosure to toxic substances, and to explicate molecular mechanisms of a study. This field has evolved from early gene expression studies describing the response of a biological system to a toxicant, and that the investigations will integrate several omics domains with toxicology and pathology data. Extensive genome sequencing and annotation efforts, response profiles of toxicologically important species (rat, dog, human, etc.), and research consortia have played a crucial role in the evolution of toxicogenomics. Initiation of toxicogenomics was with development of "toxicology-specific" cDNA microarrays to quantify xenobiotic-metabolizing enzymes and acute phase, for instance, cytochrome P450s. But these were outmoded because commercial platforms were developed for toxicologically vital species, for instance, rats. It is now feasible to use commercial oligonucleotide microarrays to quantify expression response in a diverse range of species of nematodes, frogs, zebrafish, rodents (rats and mice), and nonhuman primates to humans.

The chemical stressors can pose many challenges: the diverse properties of thousands of chemicals and stressors, the time and dosage that relates with the exposure to a chemical and disease, and the genetic diversity of human populations, which act as surrogates to bear the adversities of a toxicant. The profile of a gene can be verified by real-time PCR, Northern and Western blotting. The gene expression can be analyzed by its exposure to a toxicant dose and the time of exposure. The cell types can be identified by immunohistochemistry and *in situ* hybridization [11].

26.2.4 Metabolomics

Specific mRNAs, proteins, or metabolites can act as diagnostic markers and help us to predict a particular disease and give way for its therapy. Metabolomics deals with the investigations of the metabolite network. It is quite related with transcriptomics and proteomics. It can be easily applied to large populations and is of higher throughput [12]. It is the branch of omics that deals with the quantification and the identification of the metabolites, which are a subset of metabolome. It finds its use in proteomics, clinical chemistry, and toxicology, as well as functional genomics. Metabolomics is deeply integrated with both physiology and biochemistry of the metabolite data [13]. Metabolites, which occur in pathways and classes of compounds, have been identified by quantitative methods, called as metabolite profiling. Semiquantitative NMR is based on metabolite fingerprinting, where the peaks detected in NMR indicate the structure of the metabolite. Metabolomics can give a better insight of the organism's phenotype by amplifying proteome changes and can reduce the differences between the understanding of genotype and phenotype [11,14].

26.2.5 Marine Omics

The oceans and all other water bodies are the Earth's prime ecosystem, casing 70% of the planet and providing a mainstream of goods and services to a majority of the world's population. Considering the multifaceted abiotic and biotic developments on the micro- to macroscale is the key to leverage, protect, understand, and withstand the marine ecosystem. Marine microorganisms control the global cycle's energy and organic matter. A multidisciplinary methodology, fetching together research on genomics, oceanography, and biodiversity, is now required to recognize multipart responses of the marine ecosystem. The marine omics approach will bring enhanced indulgence of the intricate interplay of the organisms with their respective environment, but will disclose an affluence of new metabolic progressions and functions, which have extraordinary latent for biotechnological applications [15] (Figure 26.2).

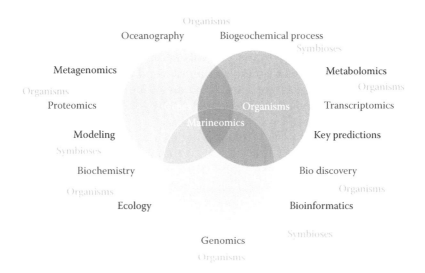

FIGURE 26.2
Marine omics and its ramifications.

Since the past decade, the configuration and deeds of marine and microbial life from varied habitats have been the attention of penetrating research being stimulated largely by developments in molecular biology and genomic technologies, thus its recent entry into the age of the "omics" —metabolomics, (meta) genomics, transcriptomics, and proteomics with possibly others on the growth [16].

26.3 Nanotherapy

Nanocarriers are employed for getting itself delivered into the human body, and are linked with molecules that bind to antigens or receptors that have been overexpressed in the tumor cells. Their specificity can be elevated by the biomarkers present in the body. Nanomedicine and its applications have an immense impact on identifying the biomarkers and the accurate drug delivery.

26.3.1 Nanotherapy in Cancer and Clinical Care

In cancer, the conventional therapeutic approach does not necessarily seem to differentiate between the normal and tumor cells, adversely affecting other tissues and leading to system toxicity, accompanied by cardiomyopathy and neurotoxicity. Reduced water solubility, nonspecific targeting, and low therapeutic index are common problems of the conventional drug delivery systems.

26.3.1.1 Nanoparticles

Nanoparticles can effectively target specific cell populations by directly reaching the metastatic sites. Nanomaterials have high surface to volume ratio, which is responsible for its high reactivity. Their ability to penetrate the cells and their compatibility are few of the deciding factors for their application commercially. Carbon nanotubes, fullerenes, particles of metal oxides, and quantum dots are the common examples of nanomaterials [17,18].

Nanoparticles have been very effective to treat cancer. They can circulate in the bloodstream for a greater time by virtue of their small size and characteristics of their surface, and can actively approach the target if linked to an antibody or ligand. They have high permeability and retention effect. Nanochannel electroporation cites a good example to deliver siRNA to a living cell [17].

Biomarkers help to predict the clinical outcomes and assist to enhance the personalized therapies. Practically, it is difficult to identify biomarkers. Nanoparticles can prove useful to identify these biomarkers in a noninvasive way and help to detect the tumor cells circulating within the body, which can be captured by magnetic nanoparticles. Using quantum dots is a good way for early cancer detection. Nanoporous silica chips aid in the stabilization of low-molecular-weight species in the proteome and help in the peptide and protein stabilization. These low-molecular-weight biomarkers are detected and stabilized effectively, providing a trustworthy fractionation system if these chips are coupled with MS. Biolabels such as antibodies, therapeutic drugs, and imaging labels can be integrated to a single platform, thereby pointing toward the multifunctional nature of nanoparticles and thereby help to deliver drugs and identify the disease biomarkers. This is useful and at par with the concept of personalized medicine. Nanoparticles and nanobiosensors have increased the efficiency of the drug delivery and disease diagnosis. There are, however, some physiological barriers such as extracellular matrix, reticular endothelial cells, high interstitial fluid pressure, and reduced receptor expression on target cells. The design of the nanoparticles and the way of treatment should consider these to accomplish the therapy [17,19]. Nanomaterials show tissue permeability as well as they are electrically reactive, explaining their use in medicines and as a safety biomarker [20]. The recognition element for a biosensor can be a nanoparticle that has a biological origin, such as proteins and immunoagents, as well as enzymes, which articulate with a transducer to produce an electrical signal as a response measured for the biological event [21,22,23,24,25,26].

The core of the disease treatment lies in its diagnostics, and omics can play a great role in it. Genomics had advanced, and the sequencing of the human genome has led to a better understanding of the disease in context of the pathology. MS has improved and has now helped to analyze complex proteomes from the body fluids, wherein the noninvasive methods come into play. The diseases can be now combated at the molecular level, due to advancements in genomic and proteomic techniques [19,27].

It has been discussed earlier that drugs coupled with nanoparticles can be used to target a particular organ or the site of infection. There can be an alternative to this—using oral path, which may enhance the future prospects of the DNA and peptide-based therapeutics. The acidic conditions of the gut would act as barrier in the efficacy of the drug delivery. However, the use of compounds such as eudragit, which is a pH-receptive and fast-melting polymer; polyethyleneglycol–ethylcellulose capsule, which is sensitive to pressure and the capsule breaks open under the pressure of peristalsis; or Carbopol, which holds up the clearance of the intestine and can be used to aid up oral delivery. Eudragit and ethyl cellulose make up the backing layer, and Carbopol conjugated with the drug is added in it, and the whole is cut into a 1 mm patch. Nanoimprinting, hydrodynamic focusing, and rapid expansion of supercritical fluids are some of the encroachments made in the fabrication technologies. The geometry, that is, shape and size of the nanoparticles, can affect the intracellular delivery process. Nonviral synthetic delivery systems, such as liposomes, lipoplexes, and nanoparticles, have gained prominence, with the aim to transfer the genes to a particular cell type. Hepatocyte gene delivery can employ lipids, polycations such as poly-L-lysine and polyethyleneimine, and bionanocapsules. Attaching site-directing molecules on the nanovectors provide stability in the biofluids and the specificity toward receptors [27,28].

26.4 Nanoproteomics

Nanotechnology deals with the components ranging from 1 to 100 nm. There are limitations in currently reaching a target protein *in vivo* due to barriers present biologically. These nanoparticles can be devised and tuned for therapeutic applications, anticancer drugs, and nanosensors, which detect the biomarkers in the blood that are generally present in minute quantities [29]. Clinically, the advent of personalized medicine as a result of the molecular analysis of the biological specimens is promising, and the proteomic analysis can prove helpful in the diagnosis of the disease and molecular medicine.

26.4.1 Nanobiochips and Nanoparticles

Nanobiochips display a cost-effective and an analytical approach for disease diagnosis. The range of size of proteins and peptides are nanometers, and therefore, there is a need to employ particles that are in the nanorange, thereby facilitating the treatment of the disease. The proteomic evaluation should be preceded by removal of abundant proteins such as albumin, since they can interfere with the information that can be extracted from low-abundance proteins. Nanoparticles bioconjugate with the low-abundance-molecule surfaces. They need to be tuned to fractionate the low-molecular-weight serum peptidome, for downstream processing. Tools for nanopore fractionation are used, which can be modified to comprise of quantum dots or fluorescent tags. A surface having an affinity to these low-abundance molecules is created, which can compete with carrier proteins [30,31].

Enzyme stabilization and immobilization can be achieved through nanobiocatalytic approaches, which employ nanostructured materials, that provide a large surface area, uniform size distribution, along with magnetic and conductivity properties. The nanoporous materials can enable improved protein digestion, by adsorbing trypsin on the pores, which are of nanoscale dimensions. Nanoparticles play a role in the recovery of enzyme immobilization system, postprotein digestion. In proteomics, protein digestion is vital, and trypsin is generally used, but the process is slothful. To speed up, nonclassical conditions, like microwave-assisted digestion, ultrasound-assisted digestion, and pressure-assisted digestion, are carried out [32]. Fourier transform ion cyclotron resonance MS (FTICR-MS) has been a useful way to identify the peptide sequences and is highly sensitive with high mass measurement accuracy [33]. Immunoprecipitation experiments done with LC-MS as the proteomic approach is useful for protein identification, which is affinity purified. The antibodies used are gel coupled, by virtue of the porous nature of the gel beads. However, nanosized superparamagnetic beads, monodispersed, speed up the process and separation, but the aggregation of the beads due to magnetic properties hinders the process. Similarly, modified gold nanoparticles have been used for biosensor fabrication and protein detection [34].

FTICR-MS provides high mass measurement accuracy and is a sensitive technique for proteomic analysis. The uniqueness of molecular mass and the LC elution forms the basis of the accurate mass and time tag approach, which can increase the sensitivity of proteomic analysis, using LC-FTICR-MS as the technique. However, there is a need to increase the limit of LC-MS analysis to a greater genome size, which top down proteomics goals upon [35]. Online nanoscale electrostatic repulsion hydrophilic interaction chromatography can be used as a tool to enhance the MS-based shotgun proteomics [33,36].

Nanoproteomics can help in the identification of proteins that play a vital role in cell physiology. Nucleic Acid Programmable Protein Assay (NAPPA) along with synchrotron radiation helps to typify these proteins at an atomic level by structural proteomics. The 3D structure of a protein can be determined by NAPPA in amalgamation with a DNA analyzer along with bioinformatics and nanocrystallography. This protein assay can assist in the study of T lymphocytes [37]. Potential high-throughput applications, low-cost equipments, and a high level of sensitivity of nanotechniques makes their application in proteomics very advantageous. Their ability to reach the subcellular locations and quantitatively less sample requirement adds to the pro [38].

Protein quantification can be accomplished by nanofluidic proteomic immunoassay (NIA) that works on nanofluidic isoelectric focusing, which is capillary based, and chemiluminescence analysis, which is antibody based, analyzing protein in nanogram quantities, and finds its source from tissue lysates [39]. Sample profiling, biomarker identification, and drug discovery are some of the important applications of proteomics. The protein sample is in small amounts, and there is a need to shift from the conventional arrays to a newer one. The fluorescent-labeled proteins can be dispersed by a piezoelectric inkjet printer in nanoliter quantity, unaffecting the sensitivity and specificity, forming the roots of nanoprobe arrays [40].

26.4.2 Identification of Drug Targets

Proteome, genome, transcriptome, metabolome, and epigenome can be analyzed for study of drug targets. Genes and pathways are identified by the multiomics analysis and play a role in the molecular pathogenesis of diseases [41]. *In vitro* drug testing may give off unexpected results, paving a need for *in situ* probing of the protein targets. Soluble nanopolymers act as drug delivery mediators and imaging contrast agents and are a proteomic approach for *in vitro* identification of a drug target. Dendrimers are such branched nanomolecules that are nontoxic to cells and highly soluble. They are linked to the drug, and the polymer is incubated with the cells to complete the delivery process. The cells are then lysed to recover the protein, followed by mass spectrometric analysis.

Poly(amidoamine) generation 4.0 dendrimer, having 64 amine groups with a diameter of 4.5 nm, can conjugate with methotrexate (MTX) and hydroxylamine as the "handling group." Dihydrofolate reductase (DHFR) aids to study the effectiveness of MTX, by the fact that DHFR is a MTX target. Free MTX was added to act as a competitor at different concentrations to cell lysate, and the affinity can be confirmed by Western blotting. Dendrimer–MTX enrichment and quantitative proteomics will lead to the determination of MTX targets. Stable isotope labeling with amino acids in cell culture (SILAC) experiment, which is a metabolic isotope labeling method, introduces amino acids with substituted stable isotope nuclei. However, in human B-cells grown in amino acids with natural isotope abundance are labeled as "light" and $^{13}C_6$-bearing version of lysine and arginine media labeled as "heavy," which is done parallely. The proteins that were enriched differentially in the "light" sample can be said as putative MTX targets. The dendrimer reagent is taken up by the cell lines, namely, human B-cells and hela cells. The dendrimer would form a protein complex that was captured on the aldehyde agarose beads after cell lysis. The MTX-interacting proteins are identified, which confirms that drug interacted with the protein. The ability to have multiple sites of attachment, immobilization of hydrophobic, and negatively charged drugs, and its combination with MS to provide an accurate identification, makes the application of dendrimers in nanomedicine a potential option [42].

Protein corona around a gold nanoparticle can be modulated to act as a therapeutic target in cancer. The protein corona formed as a result of the interaction of the gold nanoparticle with the protein informs about the disease and function of the nanoparticle in the biological system. They incubate both positively and negatively charged gold nanoparticles with normal and malignant ovarian cell lysates. Hepatoma-derived growth factor (HDGF) can act as a potential target of the positively charged gold nanoparticles, which can be inferred from the proteomic analysis. Western blot can lead to the estimation of HDGF level, which is followed by tap down of this factor in A2780 cells by siRNA, resulting in a decrease in level of proliferation in these cells by approximately 70%, thereby verifying HDGF as a potential therapeutic target in ovarian cancer [43].

26.4.3 Biomarker Discovery

Of the total proteome, antibodies and albumin account for around 90%. The low-molecular-weight proteins and metabolites, which account for unit percentage, can act as an immense possible source of biomarkers and potential information. 2DGE is a conventional method to detect protein and lacks the sensitivity for quantification and detection of low-molecular-weight proteins and metabolites. In gastric cancer, annexin A1 can act as an effective biomarker, which is involved in metastasis of tumor and invasion. These low-molecular-weight proteins tend to be linked to the carrier proteins (which are heavy weighed molecules), which are removed in the MS, thereby leading to their loss. Isoelectric focusing, size-exclusion chromatography, and ion-exchange chromatography are used for fractionation. However, fractionation of serum with nanoporous substrates can unravel the problem of loss of desired low-molecular-weight biomarkers, but may consume time and decrease the quantity of product. Therefore, this decreased abundance of biomarkers is beyond the scope of detection by conventional methods. Moreover, the resident proteins (antibodies and albumin) endogeneously and noncovalently mask these biomarkers and decrease their accessibility. Also, the biomarkers degenerate by proteases soon after blood withdrawal from the patient. These physiological barriers hinder the biomarker discovery [44,45].

Nanoparticles act as an alternative. It parts away the required molecules away from highly occurring molecules, like albumin. Affinity chromatography and molecular sieve chromatography seems to be taking place together. The proteases fail to degrade the molecule that has been bound to affinity matrix. There are affinity baits, which trap a particular type of biomarker, ranging from proteins to metabolites to fatty acids. Due to these baits, the proteolytic enzyme is unable to digest the protein. These proteins therefore get segregated from a cluster of other biomolecules and remain intact and uncleaved. Nanoparticles are stored in a lyophilized state and have towering reproducibility. Mesoporous silica films can prove to be a trusted multifractionation system that have been fabricated via triblock copolymer template pathway. Polymer concentration and templates were varied in the precursor solution and were applied for low-molecular-weight protein recovery. The pore structure, size, and surface properties were some of the parameters that were determined [45,46]. The peptide analysis can be done by MALDI-TOF. The flexibility of nanomaterials, reduced sample requirement, and cheap production are the reasons for a successful application of nanotechnology to biomarker discovery and proteomic estimation. ELISA is a label-based technique for biomarker identification and quantification. MS approaches include SILAC, which allows identification of cell surface proteins that are expressed in varying amounts, and SELDI-TOF. Microelectrochemical cantilevers, surface plasmon resonance, microfluidic purification chips, and functionalized nanopipette probes are some of the label-free detection strategies, which assists in study of real-time kinetics of reactions taking place in bimolecular state. Protein arrays, including NAPPA and atomic force microscopy, are used to monitor protein–protein interactions.

Nanoproteomics uses electrochemical biosensors that are based on nanomaterials for the detection of biomolecules. Gold nanoparticles have a great scope in cancer diagnosis and therapy. They are labeled with biomolecules of high molecular weight, and by virtue of their optical and electrical properties, they are used in biosensors and *in vivo* imaging. Carbon nanotubes, owing to its high-signal amplification, less sample requirement, and a feasible detection time, prove to be of immense potent as electrical biosensors, with a varied use as catalase, peroxidase, dehydrogenase, and oxidase biosensors. Quantum dots can label proteins and oligonucleotides and can detect manifold biomarkers from the biological system. They are readily compatible with the biological system, and their property of high quantum yield, yet being low on toxicity, makes it an effective probe for biomarker analysis. Another approach includes composite nanomatrices, citing a common example of the combination of cytochrome P450 with anodic porous alumina (APA). NAPPA has also been used with APA to detect a few macromolecules [38].

26.4.4 Protein Nanoarray

It is the ultraminiaturized way of the microarray to measure the interactions between the molecules up to nanolevel. It is advantageous by the fact that it can be used to screen individual protein against drugs as the target and can be incorporated into the sensors. They need a small amount of protein as the input and need not be removed from the solution like the conventional microarrays. Biomarkers are proteins that are secreted in extremely small amount and thus cannot be amplified like DNA. Nanoscale protein array can also help in studying stem cells. Electron beam lithography and dip pen nanolithography are few of the techniques for fabrication of the protein nanoarray. Phat L. Tran et al. have provided a new approach to the protein nanoarray fabrication wherein the antibody-conjugated fluorescent nanobeads are assembled into the nanowells by size-dependent self-assembly. These wells were created on polymethyl methacrylate by electron beam lithography and it detected a transcriptional protein, Oct4, found in differentiating stem cells, with the help of fluorescence resonance energy transfer. This multicomponent protein nanoarray, with different sized beads filling the respective sized wells, provides a highly sensitive analysis of the transcriptional proteins [47]. Inkjet printing can create replicate arrays and is scaleable to printing of many compounds. Parylene C is a polymer that is compatible biologically to act as the template and can be used for patterning biomolecular arrays. Tan et al., in this print-and-peel, method have conjugated inkjet printing with parylene peel-off, which finds its application in making nanoarrays of antibodies [48]. Multiplexed detection array, which utilizes DNA tiles and their self-assembly, finds its use in the detection of nucleic acids and diagnosing diseases [49]. Nanowell arrays can be formed on a gold film and have great deal of applications and advantages, such as large number of wells per array, and are highly variable ranging for different samples [50]. Prediction of molecular properties, with a small test sample requirement and high sensitivity, can be achieved by nanoarrays of the biomolecules [51].

26.5 Genomics and Nanotechnology

Genomics is the study of all genes of a cell, or a tissue. Nearly every human ailment has a genetic root. Cancer, birth defects, diabetes, or even a carrier, which may express the gene, can be carried forward genetically to other progenies. Through a thorough understanding of the genomics, efforts can be made toward removing the proclivity of these diseases. About 38% of the DNA is noncoding, called as junk DNA, and only 1%–3% code for proteins [52]. Gene expression analysis has conventionally been done by microarray technology. A DNA analyzer has been introduced to improve the imaging from the fluorescent DNA microarray. Nanoporous alumina has found its application in DNA microarray as a substrate [53].

Light microscopy had limitations, thereby calling for a need to improve the imaging techniques to study the DNA. To go up to the level of genes, the cells are lysed and then DNA is extracted and then fractionated. The protein pattern is scanned followed by mapping of the genome. Austin et al. [54] worked on the lac operon model system, with an aim to find the binding pattern of the lac repressor that prevents the transcription upon binding to DNA, with a high resolution. They emphasized on near-field excitation, to excite a molecule locally on the smallest scale possible. The idea laid stress on a slit that was nanofabricated in a film made up of aluminum, and let DNA pass in an evanescent field through the slit. The slit was back illuminated and the fluorescent tags attached to the DNA excite. The high-resolution imaging of a DNA molecule involved nanoslits, for a high spatial resolution, as well as nanochannels to keep DNA in a straight path. However, poor optical resolution and poor DNA stretching posed a problem. One method devised in the paper to reduce the nanochannel grating from 100 nm was controlled sputtering of silicon dioxide SiO_2 at various angles, thereby reducing it to 50 nm. They also explored two ways to nanofabricate the slits. The first being e-beam lithography, the other one being FIB tool, for direct nanomachining. The former was time consuming. However, the second method was way accurate and resolved up to 20 nm. It employed Ga+ ions as the energetic beams to remove away the materials and resolve efficiently. They also created polynorbornene, which is an unzipping polymer, by both e-beam lithography and imprinting technology [54].

In genomics, sequencing of DNA holds the key. The underlying principle of nanopore sequencing is the possibility of driving DNA(single stranded) or RNA by an electrical force via an ion channel, present in a lipid bilayer [55]. Omics advancements, particularly proteogenomics and transcriptomics, have led to genome reannotation accurately. *In silico* predictions, carried on a multiomics platform, sequencing of mRNA, and genome-scale metabolic reconstruction on *Saccharomyces erythraea*, have led to improvement in the genome annotation [56]. Singer et al. [57] performed sequence-specific DNA detection, wherein they hybridized dsDNA with peptide nucleic acid probes (PNA). Nanopores were used as single-molecule sensors, and bis-PNA (two PNA molecules connected by linker) was used as probe. PNA was tagged to a specific fragment of the DNA, which propelled through the nanopore, which was fabricated on a SiN membrane using an electron beam [57]. Nanopore allows single-molecule detection, wherein the signals reflect the order of the nucleobases. It is advantageous by the fact of feasible sample preparation, low priced than the Sanger method of sequencing, and ruling out the need of nucleotides and polymerases during readout. A small concentration of DNA (less than a microgram) can provide sequence coverage six times than the conventional methods. However, the nanopore channels practically are more than 5 nm since no constituent material of the nanopore had the quality to sense one nucleotide at a time. Also, the pore surface interactions causes DNA translocation and fluctuation in translocation kinetics and adds to the challenges in nanopore technology [58]. There is always a need for new single-molecule approaches to analyze the genome due to the shortcomings of the amplification steps involved in the approach. Jo et al. [59] have shown that poly(dimethylsiloxane) can make up a disposable device, which bears nanoslits for analysis of the DNA, and that DNA stretch can be achieved in buffers that have low ionic strength [59]. Bioinformatics and electronic medical records can play a role in the disease treatment and diagnosis, that is, implementing research clinically [60]. Genome mapping can be done using nanochannel arrays [61]. DNA is suitable to make scaffolds that can clutch other molecules in place. The nanoscale folding of DNA, which creates 2D/3D shapes, is DNA origami. There is very low abundance of the protein biomarkers, which makes it difficult for detection. DNA origami can be used for cell analysis for high sensitivity and gives an intracellular understanding of the living cell and an *in vitro* diagnostic tool [62]. The interaction between the genome and external agents, like drugs, is studied under toxicogenomics. The introduction of nanomaterials inside the body can affect the cellular mechanisms, and hence, toxicogenomics can act as a tool to monitor the disturbances in the cell.

26.6 Conclusion

Omics has shown a new way to define and discuss the various aspects of the biotechnology. With the insight of nanotechnology, the application of omics has seen a new era. Nanomaterials, which exhibit a very small size and high versatility, have a varied application in proteomics, high-throughput technologies, and accurate results. NIA, nanobiochips, and mass spectrometric technologies pose a promising growth and response in the future. Pharmaceutical application of genomics, namely, targeted drug therapy, personalized medicine, and nanobiomics have a secured use in medicine. Its application in forensics, which allows small-scale DNA identification and the production of programmable bimolecular devices, has displayed great trust. Nanoparticles, mesoporous silica films, and electrochemical biosensors can detect the low-abundance biomarkers. Dendrimers and gold nanoparticles have shown their worth in delivering the drugs at the accurate site. These nanomaterials have a huge commercial scope for drugs, DNA, cancer treatment, and biosensors for the accurate and sensitive disease diagnosis. The impending metatranscriptomic sequencing technique detailing about the environmental influences and the regulation of microbial activities is governed by amount of community expression captured in sequence libraries, and the sensitivity with which alterations are observed. By demonstrating unabridged community gene wealth and expression in absolute units (per volume or mass of environment), "omics" statistics can be enhanced and leveraged to expand understanding of microbial mediated processes in the ocean. This can help provide remarkable insights into gene expression in a marine microbial community [63], experimental contribution to marine omics [64], and genomics for microalgae as fuel [65], bio discovery, and insights into regulation of biogeochemical processes in coastal ocean.

Acknowledgments

Authors are thankful to Dr. Ashok K. Chauhan, president of Amity University, Uttar Pradesh, India, and Dr. W. Selwamurthy, president of Amity Science, Technology and Innovation Foundation, for the encouragement and research facility.

References

1. R.H. Austin, J.O. Tegenfeldt, H. Cao, S.Y. Chou, and E.C. Cox. 2002. Scanning the controls: Genomics and nanotechnology. *IEEE Transactions on Nanotechnology* 1(1): 12–18.
2. S.M. Bhaktiar, A. Ali, S.M. Baig, D. Barh, A. Miyoshi, and V. Azevedo. 2014. Identifying human disease genes—Advances in molecular genetics and computer advancements. *Genetics and Molecular Research* 13(3): 5073–5087.
3. H.-M. Huang and Y.-Y. Shih. 2014. Pushing CT and MR imaging to the molecular level for studying the "omics"—Current challenges and advancements. *BioMed Research International* 2014: 1–17.
4. R.B. Altman, D.L. Rubin, and T.E. Klein. 2004. An "omics" view of drug development. *Drug Development Research* 62: 81–85.

5. A. Cifuentes. 2012. Food analysis—Present, future and foodomics. *ISRN Analytical Chemistry* 2012: 1–17.

6. E.M. Tsapakis, A. Basu, and K.J. Aitchinson. 2004. Transcriptomics and proteomics—Understanding of psychiatric pharmacogenomics. *Clinical Neuropsychiatry* 1: 117–124.

7. Q. Nguyen, L.K. Nielsen, and S. Reid. 2013. Genome scale transcriptomics of baculovirus-insect interactions. *Viruses* 5: 2721–2747.

8. G.A. Niazi and S. Riaz-ud-Din. 2006. Biotechnology and genomics in medicine—A review. *World Journal of Medical Sciences* 1(2): 72–81.

9. D. Figeys. 2005. Proteomics: The basic overview. In *Industrial Proteomics. Applications for Biotechnology in Pharmaceuticals*, ISBN 978-0-471-45714-5. John Wiley & Sons, Inc., pp. 1–62.

10. R. Savino, F. Casadonte, and R. Terracciano. 2011. In mesopore protein digestion: A new forthcoming strategy in proteomics. *Molecules* 16: 5938–5962.

11. M. Waters et al. 2003. Systems toxicology and the chemical effects in biological systems (CEBS) knowledge base. *Environmental Health Perspectives* 6: 811–824.

12. K. Hollywood, D.R. Brison, and R. Goodacre. 2006. Metabolomics—Current technologies and future trends. *Proteomics* 6: 4716–4723.

13. D.S. Wishart. 2007. Current progress in computational metabolomics. *Briefings in Bioinformatics* 5: 279–293.

14. A.K. Chaudhary, D. Dhakal, and J.K. Sohng. 2013. An insight into the "omics" based engineering of *Streptomycetes* for secondary metabolite overproduction. *BioMed Research International* 2013: 1–15.

15. F.O. Glöckner and I. Joint. 2010. Marine microbial genomics in Europe—Current status and perspectives. *Microbial Biotechnology* 3(5): 523–530.

16. K.B. Heidelberg, J.A. Gilbert, and I. Joint 2010. Marine genomics at the interface of marine microbial ecology and biodiscovery. *Microbial Biotechnology* 3(5): 531–543.

17. S. Liu. 2012. Epigenetics advancing personalized nanomedicine in cancer therapy. *Advanced Drug Delivery Reviews* 64 (13): 1532–1543.

18. A. Poma and M.L. Di Giorgio. 2008. Toxicogenomics to improve comprehension of the mechanisms underlying responses of in vitro and in vivo systems to nanomaterials: A review. *Current Genomics* 9(8): 571–585.

19. J.H. Sakamoto et al. 2010. Enabling individualized therapy through nanotechnology. *Pharmacological Research* 62(2): 57–89.

20. H. Nabeshi, T. Yoshikawa, T. Imazawa, S.-I. Tsunoda, and Y. Tsutsumi. 2010. Safety assessment of nanomaterials using toxicokinetics and toxicoproteome analysis. *Journal of the Pharmaceutical Society of Japan* 130(4): 465–470.

21. P. Chandra, H.-B. Noh, and Y.-B. Shim. 2013. Cancer cell detection based on the interaction between an anticancer drug and cell membrane components. *Chemical Communications* 49(19): 1900–1902.

22. Y. Zhu, P. Chandra, and Y.-B. Shim. 2013. Ultrasensitive and selective electrochemical diagnosis of breast cancer based on a hydrazine–Au nanoparticle–Aptamer bioconjugate. *Analytical Chemistry* 85(2): 1058–1064.

23. S.-Y. Won, P. Chandra, T.S. Hee, and Y.-B. Shim. 2013. Simultaneous detection of antibacterial sulfonamides in a microfluidic device with amperometry. *Biosensors and Bioelectronics* 39(1): 204–209.

24. H.-B. Noh, P. Chandra, J.O. Moon, and Y.-B. Shim. 2012. In vivo detection of glutathione disulfide and oxidative stress monitoring using a biosensor. *Biomaterials* 33(9): 2600–2607.

25. Y. Zhu, P. Chandra, K.-M. Song, C. Ban, and Y.-B. Shim. 2012. Label-free detection of kanamycin based on the aptamer-functionalized conducting polymer/gold nanocomposite. *Biosensors and Bioelectronics* 36(1): 29–34.

26. P. Chandra et al. 2014. Prospects and advancements in C-reactive protein detection. *World Journal of Methodology* 4(1): 1–5.

27. J.S. Zimmer et al. 2006. Advances in proteomics data analysis and display using an accurate mass and time tag approach. *Mass Spectrometry Reviews* 25(3): 450–482.

28. A. Pathak et al. 2008. Nano-vectors for efficient liver specific gene transfer. *International Journal of Nanomedicine* 3(1): 31–49.

29. P. Karimi et al. 2014. Implementation of proteomics for cancer research: Past, present, and future. *Asian Pacific Journal of Cancer Prevention* 15(6): 2433–2438.

30. D.H. Geho et al. 2007. Nanotechnology in clinical proteomics. *Nanomedicine* 2(1): 1–5.

31. J.V. Jokerst and J.T. McDevitt. 2010. Programmable nano-bio-chips: Multifunctional clinical tools for use at the point-of-care. *Nanomedicine* 5(1): 143–155.

32. J. Kim et al. 2010. Nanobiocatalysis for protein digestion in proteomic analysis. *Proteomics* 10(4): 687–699.

33. J.S. Zimmer et al. 2006. Advances in proteomics data analysis and display using an accurate mass and time tag approach. *Mass Spectrometry Reviews* 25(3): 450–482.

34. P.-C. Cheng, H.-K. Chang, and S.-H. Chen. 2009. Quantitative nanoproteomics for protein complexes (QNanoPX) related to estrogen transcriptional action. *Molecular and Cellular Proteomics* 9(2): 209–224.

35. J.D. Tipton et al. 2012. Nano-LC FT-ICR tandem mass spectrometry for top-down proteomics: Routine baseline unit mass resolution of whole cell lysate proteins up to 72 kDa. *Analytical Chemistry* 84(5): 2111–2117.

36. E.P. de Jong and T.J. Griffin. 2012. Online nanoscale ERLIC-MS outperforms RPLC-MS for shotgun proteomics in complex mixtures. *Journal of Proteome Research* 11(10): 5059–5064.

37. C. Nicolini and E. Pechkova. 2010. Nanoproteomics for nanomedicine. *Nanomedicine* 5(5): 677–682.

38. N. Dasilva, P. Díez, S. Matarraz, M. González-González, S. Paradinas, A. Orfao, and M. Fuentes. 2012. Biomarker discovery by novel sensors based on nanoproteomics approaches. *Sensors* 12: 2284–2308.

39. D.A. Drew et al. 2013. Nanoproteomic analysis of extracellular receptor kinase-1/2 post-translational activation in microdissected human hyperplastic colon lesions. *Proteomics* 13(9): 1428–1436.

40. V.J. Nagaraj, S. Eaton, and P. Wiktor. 2011. NanoProbeArrays for the analysis of ultra-low-volume protein samples using piezoelectric liquid dispensing technology. *Journal of Laboratory Automation* 16: 126–133.

41. Y. Kanai and E. Arai. 2014. Multilayer omics analyses of human cancers: Exploration of biomarkers and drug targets based on the activities of the International Human Epigenome Consortium. *Frontiers in Genetics* 5: 1–7.

42. L. Hu et al. 2011. Identification of drug targets in vitro and in living cells by soluble nanopolymer-based proteomics. *Angewandte Chemie International Edition* 50(18): 4133–4136.

43. R.R. Arvizo et al. 2012. Identifying new therapeutic targets via modulation of protein corona formation by engineered nanoparticles. *PLOS ONE* 7(3): 1–8.

44. A. Luchini, C. Fredolini, B.H. Espina, F. Meani, A. Reeder, S. Rucker, E.F. Petricoin III, and L.A. Liotta. 2010. Nanoparticle technology. Addressing the fundamental roadblocks to protein biomarker discovery. *Current Molecular Medicine* 10(2): 133–141.

45. Z.-Q. Zhang, X.-J. Li, G.-T. Liu, Y. Xia, X.-Y. Zhang, and H. Wen. 2013. Identification of Annexin A1 protein expression in human gastric adenocarcinoma using proteomics and tissue microarray. *World Journal of Gastroenterology* 19(43): 7795–7803.

46. J. Fan, J.W. Gallagher, H.-J. Wu, M.G. Landry, J. Sakamoto, M. Ferrari, and Y. Hu. 2012. Low molecular weight protein enrichment on mesoporous silica thin films for biomarker discovery. *Journal of Visualized Experiments* 62: 1–5.

47. P.L. Tran et al. 2010. FRET detection of Octamer-4 on a protein nanoarray made by size-dependent self-assembly. *Analytical and Bioanalytical Chemistry* 398(2): 759–768.

48. C.P. Tan et al. 2010. Nanoscale resolution, multi-component biomolecular arrays generated by aligned printing with parylene peel-off. *Nano Letters* 10(2): 719–725.

49. C. Lin et al. 2007. Self-assembled combinatorial encoding nanoarrays for multiplexed biosensing. *Nano Letters* 7(2): 507–512.

50. N.J. Wittenberg et al. 2011. Facile assembly of micro- and nanoarrays for sensing with natural cell membranes. *ACS Nano* 5(9): 7555–7564.

51. M. Palma et al. 2011. Selective biomolecular nanoarrays for parallel single-molecule investigations. *Journal of the American Chemical Society* 133(20): 7656–7659.

52. J. Politz and A. Pombo. 2002. Genomics meets nanoscience. Probing genes and the cell nucleus at 10^{-9} meters. *Genome Biology* 3(3): 1–3.

53. C. Nicolini. 2006. Nanogenomics for medicine. *Nanomedicine* 1(2): 147–151.

54. R.H. Austin, J.O. Tegenfeldt, H. Cao, S.Y. Chou, and E.C. Cox. 2002. Scanning the controls: Genomics and nanotechnology. *IEEE Transactions on Nanotechnology* 1(1): 12–18.

55. H. Stranneheim and J. Lundeberg. 2012. Stepping stones in DNA sequencing. *Biotechnology Journal* 7: 1063–1073.

56. E. Marcellin et al. 2013. Re-annotation of the *Saccharopolyspora erythraea* genome using a systems biology approach. *BMC Genomics* 14: 1–8.

57. A. Singer et al. 2010. Nanopore-based sequence-specific detection of duplex DNA for genomic profiling. *Nano Letters* 10(2): 738–742.

58. D. Branton et al. 2008. The potential and challenges of nanopore sequencing. *Nature Biotechnology* 26(10): 1146–1153.

59. K. Jo, D.M. Dhingra, T. Odijk, J.J. de Pablo, M.D. Graham, R. Runnheim, D. Forrest, and D.C. Schwartz. 2007. A single-molecule barcoding system using nanoslits for DNA analysis. *Proceedings of the National Academy of Sciences of the United States of America* 8: 2673–2678.

60. I.Y. Choi et al. 2013. Perspectives on clinical informatics: Integrating large-scale clinical, genomic, and health information for clinical care. *Genomics & Informatics* 4: 186–190.

61. A.R. Hastie et al. 2013. Rapid genome mapping in nanochannel arrays for highly complete and accurate de novo sequence assembly of the complex *Aegilops tauschii* genome. *PLOS ONE* 2: 1–10.

62. Q. Mei et al. 2011. Stability of DNA origami nanoarrays in cell Lysate. *Nano Letters* 11(4): 1477–1482.

63. S.M. Gifford, S. Sharma, J.M. Rinta-Kanto, and M.A. Moran. 2011. Quantitative analysis of a deeply sequenced marine microbial metatranscriptome. *The ISME Journal* 5: 461–472.

64. C. Cravo-Laureau and R. Duran. 2014. Marine coastal sediments microbial hydrocarbon degradation processes: Contribution of experimental ecology in the omics'era. *Frontiers in Microbiology* 5: 1–8.

65. R.J.W. Brooijmans and R.J. Siezen. 2010. Genomics of microalgae, fuel for the future? *Microbial Biotechnology* 3(5): 514–522.

27

Application of Nanoparticles in Marine Biofilm Control and Characterization

Dhinakarasamy Inbakandan

CONTENTS

> Biofilms are the most successful form of life on earth and tolerate high concentrations of biocidal substances.
>
> **—Flemming and Ridgway (2009)**

27.1 Introduction to Marine Biofilms

Communities of microbes colonizing the surfaces in a marine or seawater environment are tiny but powerful entities to form the marine biofilms (Peltonen et al. 2007). The ecological succession of a biofilm happens in a sequential way. The initial conditioning layer is formed by the adhesion of organic molecules to the surfaces. The organic nutrient available in this conditioning layer made the planktonic cells to adhere, and further it was influenced by the bacterial attachment, which was initially reversible and then irreversible (Callow and Callow 2002, 2006, Ploux et al. 2007, Inbakandan et al. 2010). The irreversible attachment is followed by recruitment of bacteria and microalgae to form a matured biofilm through quorum sensing signaling (Callow and Callow 2002, Joint et al. 2007). The maturity of the biofilm was successful with bacterial adhesion and characterized by their production of extracellular polymeric substances (EPSs). The detachment and succession of biofilm depend on the nutrients trapped by the EPS from the surrounding environment. These processes were also influenced by the flow of water, shear physical forces, and the abiotic–biotic interfaces (Van Houdt and Michiels 2005, Denkhaus et al. 2007, Flemming and Ridgway 2009).

Biofilm that protects the bacterial population dwelling in it is also a problem for industries where seawater application is in usage. The habitation of the bacterial consortia and planktonic cells in the marine biofilm will survive even in a harsh situation influenced by the concentration of any antifouling (AF) or antimicrobial agents (Lazarova and Manem 1995, Flemming and Ridgway 2009). These biofilms protected bacteria that are roofed and sheltered by the EPS matrix from environmental stress, physical disturbance, biocides, chemicals, and detergents (Castonguay et al. 2006, Kujundzic et al. 2007). Thus, the resistance and the existence of bacteria in biofilm are a big challenge in AF control in marine industries (Chambers et al. 2006).

Microbial adhesion on the surfaces or biofilm formation is the triggering switch for this worldwide marine pest problem (Hayes et al. 2005). Thus, biofoulings are the biggest disadvantages and undesirable efforts faced by the marine industries on their immersed structures. This marine fouling increases the dragging forces of the ships and vessels, which in turn reflects large fuel consumption and raises the cost inventory in ship transportation and logistics (Callow and Callow 2002, Munk et al. 2009, Bai et al. 2013). This biofilm also affects the oil and petroleum pipelines, which leads to biocorrosion and biodegradation of the material surfaces (Rajasekar et al. 2007, Stephenson et al. 2014). Cooling water systems and power plant cooling water circuits boarded with head exchangers face frictional loss in fluid flow due to the presence of biofilms, resulting in altered heat transfer coefficient (Murthy et al. 2005). Data buoys moored with sensors, underwater detectors, and imaging systems receive data with more noises due to inhabitation of marine microbes and larval settlement on the surfaces of the sensors (Dineshram et al. 2009). This results in inaccurate and meaningless data acquisition and deterioration of the sensors and detectors due to accumulation of macrofoulers (Zervakis et al. 2003).

A number of treatments and techniques have been developed through chemical and mechanical methods to solve the problem on the accumulation of marine foulers on the surfaces (Flemming and Ridgway 2009). Chemical methods involve biocidal and biostatic agents like chemicals/paints and detergents (Murthy et al. 2005). Surface modification, self-cleaning/self-release coatings, and electrochemical methods are considered as a mechanical way of control. A surface modification approach progressively depends on the bacterial adhesion, where the material surface topography and roughness influence the microbial sticking. Surface cleaning/self-release coatings are architecture with altered hydrophilic and hydrophobic quality of the surfaces, which are directly correlated with the irreversible and reversible attachment of bacterial colonies of the marine biofilms (Yarbrough et al. 2006, Flemming and Ridgway 2009). Surface roughness and surface energy play a major role in the surface modification approach of a material that altered the bacterial adhesion (Vladkova 2009). Electrochemical method involves the modification of the surfaces based on the electric charges and conductive nature of the coated materials (Wang et al. 2009). Up to 2008, organotin-based AF paints were used to combat the marine biofilm problems (Gipperth 2009, Law et al. 2012). Due to the harmful effects and the sex reversal of some of the marine living resources by the accumulation of tin, environmental protection agencies banned the use of such AF paints legally worldwide (Batley et al. 1992, Berto et al. 2012). After this ban, till date yet the marine industries are searching and researching for a new solution to hit the market.

27.2 Introduction to Nanomaterials

Nanoparticles are insoluble materials and intermediate particles between bulk and atomic structures with the size ranging from 1 to 100 nm. Like bulk materials, nanoparticles are not bearing any constant properties, but they are known for their optical, mechanical, and electrical properties associated with them. The fast and speedy progresses of recent knowhow need a systematic consideration of mechanics of materials, structures, and processes at the micro- and nanoscales. Nanoparticles, nanomaterials, nanotubes, nanowires, nano-composites, nanoscale thin films and coatings, and micro- and nanofabrication processes are all patterns of materials systems, technologies, and processes whose uninterrupted development depends upon elementary perception of material properties at small length scales. Advances in the growth of new materials for multifunctional reasons will entail design, fabrication, and characterization at the nanometer length scale. Materials that can be tailored at nanolevel to achieve greater mechanical properties along with their electrical, optical, thermal, and other functional properties are essential for future applications in many industry sectors (Wang et al. 2009, 2011a and 2011b).

AF materials are commercially important products in marine and seawater-associated industrial application. A recent documentation by research outcomes points out the novel use of nanomaterials in biofouling control (Vishwakarma et al. 2009, Inbakandan et al. 2013, Sahoo et al. 2013). Antimicrobial, antiadhesive, hydrophobic, self-cleaning, and anti-corrosive properties made these materials fitting for the battle against marine biofilms and microfouling (Flemming and Ridgway 2009). Research and development of such materials congregates the scientists in an interdisciplinary way to integrate different disciplines like physics, chemistry, biology, mathematics, and informatics. These integrations address the challenges effectively with a suitable know-how that gives appropriate data regarding the application of nanomaterials. This review attempts to describe the techniques used for the characterization of coated/assembled nanoparticles and methods used for the assessment of their efficacy against marine biofilms/microfouling succession (Table 27.1).

27.3 Relevance of Nanomaterials in Marine Biofouling Control

The possibility of nanostructured silicon oxide–type coatings deposited on glass slides from a hexamethyldisiloxane precursor by plasma-assisted chemical vapor deposition to manage the aquatic biofouling was reported by Akesso et al. (2009). They correlated surface energies with the degree of surface oxidation and hydrocarbon contents. Tapping mode atomic force microscopy (AFM) revealed a range of surface topologies with R_a values and RMS roughness. Settlement of spores of the green alga *Ulva* was significantly less, and detachment under shear was significantly more on the lowest surface energy coatings. Removal of young plants (sporelings) of *Ulva* under shear was positively correlated with reducing the surface energy of the coatings. The most hydrophobic coatings also showed good performance against a freshwater bacterium, *Pseudomonas fluorescens*, significantly reducing initial attachment and biofilm formation and reducing the adhesion strength of the attached bacterial cells under shear. Taken together, these results indicated the potential for further investigation of these coatings for applications such as heat exchangers and optical instruments.

The role of nanoroughness in AF was reported by Scardino et al. (2009). Nanoengineered superhydrophobic surfaces have been investigated for potential fouling resistance properties. Integration of hydrophobic materials with nanoscale roughness generates surfaces with superhydrophobicity that have water contact angles (θ) >150° and concomitant low hysteresis (<10°). Three superhydrophobic coatings (SHCs) differing in their chemical composition and architecture were tested against major fouling species (*Amphora* sp., *Ulva rigida*, *Polysiphonia sphaerocarpa*, *Bugula neritina*, *Amphibalanus amphitrite*) in settlement assays. The SHC that had nanoscale roughness alone deterred the settlement of all the tested fouling

TABLE 27.1

Instrumentation and Know-How to Study Nanomaterials and Marine Biofilm Control

Techniques Used for Nanoparticle Characterizations	Techniques Used to Coat or Deposit Nanoparticles	Techniques Used to Characterize Nanocoatings or Deposition	Staining Techniques Used to Study the Biofilm Formation Coupled with Microscopy	Basic Techniques Used to Study the Biofilm Formation	Molecular Techniques Used to Study the Biofilm Formation
UV–visible spectroscopy	Manual coating	Scanning electron microscopy	Crystal violet	Protein, carbohydrate, lipid, chlorophyll content of the biofilm	Polymerase chain reaction
High-resolution transmission electron microscopy	Electroless coating	Atomic force microscopy	Acridine orange	Mass/dry weight of the biofilm	Real-time quantitative polymerase chain reaction
Wide-angle x-ray diffractometry	Electroplating	Confocal scanning laser microscopy	DAPI (4,6-diamidino-2-phenylindole)	Bacterial cell enumeration	Fluorescence *in situ* hybridization
Particle size analysis	Chemical vapor deposition	Attenuated total reflection—Fourier transform infrared spectroscopy	1,9-Dimethyl-methylene blue	Diffusion test	Denaturing gradient gel electrophoresis
Energy-dispersive x-ray analysis	Sputter deposition/physical vapor deposition	X-ray photoelectron spectroscopy	Fluorescein diacetate (3,6-diacetylfluorescein)	Minimum inhibitory concentration	Next-generation sequencing
Raman spectroscopy	Spin coating	Contact angle goniometer and tensiometer	Live/dead backlight (syto R9 dye and propidium iodide)	Minimum bactericidal concentration	
Inductively coupled plasma mass spectrometry			Resazurin(7-hydroxy-3H-phenoxazin-3-one 10-oxide)	ATP assay	

Sources: Based on Denkhaus, E. et al. *Microchim. Acta*, 158, 1, 2007; Weir, E. et al., *Analyst*, 133, 835, 2008; Briand, J.F. *Biofouling*, 25, 297, 2009; Pantanella, F. et al., *Ann. Ig.*, 25, 31, 2013; Neu, T.R. and Lawrence, J.R., *Trends Microbiol.*, 23, 233, 2015.

organisms, compared to selective settlement on the SHCs with nano- and microscale archi-tectures. The presence of air incursions or nanobubbles at the interface of the SHCs when immersed was characterized using small-angle x-ray scattering, a technique sensitive to local changes in electron density contrast resulting from partial or complete wetting of a rough interface. The coating with broad-spectrum AF properties (SHC) had a notice-ably larger amount of unwetted interface when immersed, likely due to the comparatively high work of adhesion required for creating solid/liquid interface from the solid/vapor interface. This was the first example of a nontoxic, fouling-resistant surface against a broad spectrum of fouling organisms ranging from plant cells and nonmotile spores to complex invertebrate larvae with highly selective sensory mechanisms. The only physical property differentiating the immersed surfaces was the nanoarchitectured roughness that supports long-standing air incursions providing a novel nontoxic broad-spectrum mechanism for the prevention of biofouling.

Marine biofouling field tests, settlement assay, and footprint (FP) micromorphology of cyprid larvae of *Balanus amphitrite* on model surfaces were studied by Phang et al. (2010). AFM, laboratory settlement assays, and field tests were used to correlate cyprid FP mor-phology with the behavior of cyprids on different substrata. AFM imaging under labora-tory conditions revealed more porous and larger FPs on glass exposing a CH_3 surface than on aminosilane functionalized (NH_2-) surfaces. The secreted FP volume was found to be similar on both the substrata. Laboratory settlement assays and marine field tests were performed on three substrata, namely, untreated clean glass, NH_2 glass, and CH_3 glass. The results distinguished settlement preferences for NH_2 glass and untreated glass over CH_3-terminated surfaces, suggesting that cyprids favor settling on hydrophilic over hydrophobic surfaces. On combining observations from different length scales, it was speculated that the confined FP size on NH_2 glass may induce a higher concentration of the settlement inducing protein complex. Settlement may be further facilitated by a stron-ger adherence of FP adhesives to the NH_2-surface via Coulombic interactions.

The skin surface of the pilot whale (*Globicephala melas*) for its nonfouling and self-cleaning abilities was observed and documented by Baum et al. (2002). In their study, they exam-ined the skin surface of the pilot whale. Employing cryoscanning electron microscopic techniques combined with various sample preparations, the skin displayed an average nanorough surface characterized by a pattern of nanoridge-enclosed pores. The average pore size was below the size of most marine biofouling organisms. Further, the implica-tions of this type of surface to the self-cleaning abilities of the skin of pilot whales were discussed, based on reduced availability of space for biofouler attachment, the lack of any particular microniches as shelters for biofoulers, and the challenges of turbulent water flow and liquid–air interfaces during surfacing and jumping of the dolphin.

A preliminary investigation on the fouling behavior of smooth and roughened SHCs was carried out by Zhang et al. (2005). The effect of nanoscale interfacial roughness on the adhesion of single and mixed cultures of microfoulant for periods of up to 6 months was assessed using visual and wettability measurements. Detailed analysis indicated that vir-tually no microorganism got attached to the superhydrophobic surfaces in the first week of immersion. As a result, by comparison with smooth substrates, which exhibited foul-ing within a day, very rough surfaces exhibited high resistance to fouling over a 6-month period. However, after periods exceeding 2 months under ocean conditions, both films showed limited AF properties.

The role of surface energy and water wettability in aminoalkyl/fluorocarbon/hydrocarbon-modified xerogel surfaces in the control of marine biofouling was studied by Bennett et al. (2010). Xerogel films with uniform surface topography, as determined by scanning electron microscopy (SEM), AFM, and time-of-flight secondary ion mass spectrometry, were prepared from aminopropylsilyl-, fluorocarbonsilyl-, and hydrocarbonsilyl-containing precursors. Young's modulus was determined from AFM indentation measurements. The xerogel coatings gave reduced settlement of zoospores of the marine fouling alga *Ulva* compared to a poly(dimethylsiloxane) elastomer (PDMSE) standard. Increased settlement was correlated with decreased water wettability as measured by the static water contact angle, θ_{Ws}, or with decreased polar contribution (γ_P) to the surface free energy (γ_S) as measured by comprehensive contact angle analysis. The strength of attachment of 7-day sporelings (young plants) of *Ulva* on several of the xerogels was similar to that on PDMSE, although no overall correlation was observed with either θ_{Ws} or γ_S. For sporelings attached to the fluorocarbon/hydrocarbon-modified xerogels, the strength of attachment increased with increased water wettability. The aminopropyl-modified xerogels did not follow this trend.

Developments in superhydrophobic surfaces and their relevance to marine fouling were reviewed by Genzer and Efimenko (2006). In their review, a brief synopsis of superhydrophobicity (i.e., extreme nonwettability) and its implications on marine fouling were presented. A short overview of wettability and recent experimental developments aimed at fabricating superhydrophobic surfaces by tailoring their chemical nature and physical appearance (i.e., substratum texture) were presented. The formation of responsive smart surfaces, which adjust their physicochemical properties to variations in some outside physical stimulus, including light, temperature, electric field, or solvent, was also described. Finally, implications of tailoring the surface chemistry, texture, and responsiveness of the surfaces on the design of effective marine fouling coatings were considered and discussed.

The relationship between contact angle analysis, surface dynamics, and biofouling characteristics of cross-linkable, random perfluoropolyether (PFPE)-based graft terpolymers was studied by Yarbrough et al. (2006). Their objective was the rational design of minimally adhesive, mechanically stable, nontoxic fouling release (FR) coatings as responsible and practical alternatives to AF technologies. In their report, they covered the synthesis and characterization of a series of cross-linkable PFPE graft terpolymers containing various alkyl (meth)acrylate monomers with glycidyl methacrylate as the cure-site monomer. These materials were targeted for use as coatings to prevent marine biofouling. A series of terpolymers were prepared through the application of the macromonomer approach, allowing for control of cross-link density, T_g, and modulus. Structure/property relationships were established through compositional variation with regard to the three classes of monomers. The first monomer class was an alkyl (meth)acrylate used to create the continuous phase of the microphase-separated graft terpolymers. Variation between methyl methacrylate and *n*-butyl acrylate provided materials with a low and a high temperature (T_g) for the continuous phase. This was a mean of isolating the effect of modulus and T_g on surface properties, while the basic chemical nature of the monomer remained unchanged. The second monomer class contained a curable functional group. Through incorporation of glycidyl methacrylate in the monomer feed and manipulation of curing conditions, the relative effect of cross-link density on surface dynamics was evaluated. The third monomer class was the PFPE macromonomer itself. The incorporation of this macromonomer was used to enhance the release properties of the resulting materials, which relied on the surface enrichment of the low-surface-energy PFPE component. Dynamic surface properties of these materials have been evaluated through dynamic surface tensiometry. It was demonstrated that contact angle hysteresis can be significantly mitigated by as much as 50° through variation in bulk polymer composition, the chemical nature of monomers, cross-link density, modulus, and environmental conditions at the time of cure. The AF and FR potential of the experimental coatings were also evaluated by laboratory assays employing the green fouling macroalga *Ulva*. The results from these initial studies suggested promising AF properties, especially with regard to spore settlement, which was strongly inhibited on the experimental surfaces. Additionally, those that did settle were only weakly attached with one sample set, exhibiting fairly moderate release of the young *Ulva* plants.

Rosenhahn et al. (2010) explained the advanced nanostructures for the control of biofouling in the FP6 EU Integrated Project AMBIO. The colonization of man-made structures by marine or freshwater organisms or "biofouling" was a problem for maritime and aquaculture industries. Increasing restrictions on the use of toxic coatings that prevent biofouling create a gap in the market that required new approaches to produce novel nonbiocidal alternatives. Their review detailed the systematic strategy adopted by the FP6 EU Integrated Project "AMBIO" to develop fundamental understanding of key surface properties that influence settlement and adhesion of fouling organisms. By this approach, the project contributed to the understanding of fundamental phenomena involved in biofouling and to the development of environmentally benign solutions by coating manufacturers within the consortium.

Differences in the colonization of five marine bacteria on two types of glass surfaces were studied by Mitik-Dineva et al. (2009). The retention patterns of five taxonomically different marine bacteria after attachment on two types of glass surfaces, as-received and chemically etched, were investigated. Contact angle measurements, AFM, SEM, confocal laser scanning microscopy, x-ray fluorescence spectroscopy, and x-ray photoelectron spectrometry were employed to investigate the impact of nanometer-scale surface roughness on bacterial attachment. Chemical modification of glass surfaces resulted in a decrease in the average surface roughness (R_a) and the root-mean-squared roughness (R_q) and a decrease in the surface height and the peak-to-peak ratio (R_{max}) and the 10-point average roughness (R_z). The study revealed amplified bacterial attachment on the chemically etched, nanosmoother glass surfaces. This was a consistent response, notwithstanding the taxonomic affiliation of the selected bacteria. Enhanced bacterial attachment was accompanied by elevated levels of secreted EPSs. An expected correlation between cell surface wettability and the density of the bacterial attachment on both types of glass surfaces was also reported, while no correlation was established between cell surface charge and the bacterial retention pattern.

Structures and AF properties of low-surface-energy nontoxic AF coatings modified by nano-SiO_2 powder were reported by Chen et al. (2008). The low-surface-energy marine coating was an entirely nontoxic alternative, which reduces the adhesion strength of marine organisms, facilitating their hydrodynamic removal at high speeds. Novel low-surface-energy nontoxic marine AF coatings were prepared with modified acrylic resin, nano-SiO_2, and other pigments. The effects of nano-SiO_2 on the surface structure and elastic modulus of coating films were studied, and the seawater test was carried out in the Dalian Bay. Their results showed that micro- and nanolayered structures on the coating films and the lowest surface energy and elastic modulus could be obtained when an appropriate mass ratio of resin, nano-SiO_2, and other pigments in coatings is approached.

Biofouling studies on nanoparticle-based metal oxide coatings on glass coupons exposed to marine environment were conducted by Dineshram et al. (2009). Titania, niobia, and silica coatings, derived from their respective nanoparticle dispersions or sols and fabricated on soda lime glass substrates, were subjected to field testing in marine environment for antimacrofouling applications for marine optical instruments. Settlement and enumeration of macrofouling organisms like barnacles, hydroides, and oysters on these nanoparticle-based metal oxide coatings subjected to different heat treatments up to 400°C were periodically monitored for a period of 15 days. They observed differences in the AF behavior between the coated and uncoated substrates and discussed their results based on the solar ultraviolet light–induced photocatalytic activities as well as hydrophilicities of the coatings in case of titania and niobia coatings and the inherent hydrophilicity in the case of silica coating. The effect of heat treatment on the photocatalytic activity of the coatings was also discussed.

The development of nano cerium oxide–incorporated aluminum alloy sacrificial anode for marine applications was explained by Shibli et al. (2008). Aluminum–zinc alloy sacrificial anodes were extensively used for cathodic protection. The performance of the sacrificial anodes can be significantly improved by incorporation of microalloying elements in the aluminum matrix. In this study, nano cerium oxide particles of different concentrations, ranging from 0 to 1 wt%, were incorporated for activating and improving the performance of the anode. The electrochemical test results revealed the increased efficiency of the anode. The electrochemical impedance spectroscopy revealed the information that the presence of nano cerium oxide in the anode matrix caused effective destruction of the passive alumina film, which facilitated enhancement of galvanic performance of the anode. Moreover, the biocidal activity of cerium oxide prevented the bioaccumulation considerably, which enabled the anodes to be used in aggressive marine conditions.

The challenges in the development of new nontoxic AF solutions were reviewed by Marechal and Hellio (2009). Marine biofouling was of major economic concern to all marine industries. The shipping trade was particularly keen to the development of new AF strategies, especially green AF paint, as international regulations regarding the environmental impact of the compounds actually incorporated into the formulations were becoming more and stricter. It was also recognized that vessels play an extensive role in invasive species propagation as ballast waters transport potentially threatening larvae. It was then crucial to develop new AF solutions combining advances in marine chemistry and topography, in addition to the knowledge on biofoulers, with respect to the marine environment. Their review presented the progress made in the field of new nontoxic AF solutions (new microtexturing of surfaces and FR coatings, with a special emphasis on marine natural antifoulants) as well as the perspectives for future research directions.

Attachment tendencies of *Escherichia coli* K12, *Pseudomonas aeruginosa* ATCC 9027, and *Staphylococcus aureus* CIP 68.5 onto glass surfaces of different degrees of nanometer-scale roughness were studied by Mitik-Dineva et al. (2009). Contact angle and surface charge measurements, AFM, SEM, and confocal laser scanning microscopy were employed to characterize substrata and bacterial surfaces. Modification of the glass surface resulted in nanometer-scale changes in the surface topography, whereas the physicochemical characteristics of the surfaces remained almost constant. AFM analysis indicated that the overall surface roughness parameters were reduced. SEM, CLSM, and AFM analyses clearly demonstrated that although *E. coli*, *P. aeruginosa*, and *S. aureus* presented significantly different patterns of attachment, all of the species exhibited a greater propensity for adhesion to the "nanosmooth" surface. The bacteria responded to the surface modification with a remarkable change in cellular metabolic activity, as shown by the characteristic cell morphologies, production of EPSs, and an increase in the number of bacterial cells undergoing attachment.

The influence of the ultrafine crystallinity of commercial purity grade 2 (as-received) titanium and titanium modified by equal channel angular pressing (modified titanium) on bacterial attachment was studied by Truong et al. (2009). A topographic profile analysis of the surface of the modified titanium revealed a complex morphology of the surface. The undulating surfaces were nanosmooth and with height variations. These surface topography characteristics were distinctly different from those of the as-received samples, where broad valleys were detected, whose inner surfaces exhibited asperities. It was found that each of the three bacteria strains used in this study as adsorbates, namely, *S. aureus* CIP 68.5, *P. aeruginosa* ATCC 9025, and *E. coli* K12, responded differently to the two types of titanium surfaces. Extreme grain refinement resulted in substantially increased number of cells attached to the surface compared to as-received titanium. This enhanced degree of attachment was accompanied with an increased level of EPS production by the bacteria.

The control of marine biofouling on xerogel surfaces with nanometer-scale topography was studied by Gunari et al. (2011). Mixtures of *n*-octadecyltrimethoxysilane (C18, 1–5 mole-%), *n*-octyltriethoxysilane, and tetraethoxysilane gave xerogel surfaces of varying topographies. Segregation of the coating into alkane-rich and alkane-deficient regions in the xerogel was observed by IR microscopy. Immersion in ASW for 48 h gave no statistical difference in surface energy for the xerogel. Settlement of barnacle cyprids and removal of juvenile barnacles, settlement of zoospores of the alga *Ulva linza*, and strength of attachment of 7-day sporelings were compared among the xerogel formulations. Settlement of barnacle cyprids was significantly lower in comparison to glass and polystyrene standards. The xerogels were comparable to PDMSE with respect to the removal of juvenile barnacles and sporeling biomass, respectively.

The progress of marine biofouling and AF technologies was reviewed (Cao et al. 2011). In their review, physical and biochemical developments in the field of marine biofouling, which involve biofilm formation and macroorganism settlement, were discussed. The major AF technologies based on traditional chemical methods, biological methods, and physical methods were presented. The chemical methods included self-polishing types such as tributyltin (TBT) self-polishing copolymer coatings, which despite good performance were banned in 2008 because of serious environmental impact. Therefore, other methods have been encouraged, which include coatings with copper compounds and biocide boosters to replace the TBT coatings. Biological extracts of secreted metabolites and enzymes were anticipated to act as antifoulants. Physical methods such as modification of surface topography, hydrophobic properties, and charge potential have also been considered to prevent biofouling. Their review proposed that AF technologies would be the ultimate AF solution because of their broad-spectrum effectiveness and zero toxicity.

Reduction in biofouling on titanium surface by electroless deposition of antibacterial copper nanofilms was attempted and documented by Therasa et al. (2010). The main objective of their work was to study the antibacterial properties of copper thin nanofilms on titanium surface deposited by electroless plating technique for biofouling-free condenser tube applications. The electroless deposition of copper nanofilms on titanium substrates was done, and Cu films were also postannealed for 1 h at 600°C under vacuum condition to increase the particle size of the films. Surface characteristics of the films were studied using GIXRD, SEM, and AFM. Antibacterial properties of the surface were evaluated by exposure studies in seawater using total viable count and epifluorescence microscopic techniques. Excellent antibacterial activity was exhibited by the electroless-plated copper nanofilm on the titanium surface, showing more than two-order decrease in the bacterial density compared to titanium surface with no copper film.

Recent developments of nanomaterial-doped paints for the minimization of biofouling in submerged structures were reviewed and reported by Rawat et al. (2010). It was necessary to develop antibiofouling paints that were innocuous to both the environment and the structures. One alternative approach might be the incorporation of nanoparticles and prevention of bacterial biofilm formation and the attachment of larger organisms. Various nanoparticles of metal and their oxides had been recognized to possess antibacterial properties. The development of such materials was a challenge to both the chemist and the biologist, where effective choice of methods that provided relevant information regarding application of the metal nanoparticles in antifouling materials became the central objective. Their review covered the area of nanoparticle-doped AF paints. In addition, some experimental studies, which concentrated on biofouling paints with dispersed nanoparticles of zinc oxide (ZnO), CuO, Al_2O_3, MgO, TiO_2, and Co_3O_4, were also pointed out. In addition, the techniques used for the characterization of the nanoparticle-doped materials and the methods for the determination of their efficacy against biofilm formation were also covered in their review.

The efficacy of CeO_2–TiO_2 mixed oxide–incorporated high-performance hot-dip zinc coating was studied by Shibli and Chacko (2011). The greatest limitation in using zinc-coated steel in industrial environments was the biofilm formation and greater dissolution rate of zinc. If initial bacterial adhesion to the coating surface has been inhibited, biofilm formation would be prevented. Effective corrosion resistance and better AF characteristics for the coatings were achieved by the incorporation of CeO_2–TiO_2 mixed oxides. Based on the performance of the coatings during physicochemical and electrochemical characterization, the concentration of the mixed oxide in the bath was optimized. The interior layers of the mixed oxide–incorporated coatings exhibited greater stability toward corrosion. The morphological results from SEM and AFM analyses revealed the structural refinement and better quality of mixed oxide–incorporated coatings. The incorporation of the CeO_2–TiO_2 mixed oxide into the galvanic bath yielded coatings with effective barrier protection, better corrosion resistance, good AF characteristics, and improved surface quality. The coating was nonporous in nature and has potential scope for high industrial utility.

The area of self-assembled ultrafine particulate-based composites (nanocomposites) has been a major thrust in advanced material development. Thus, Schnur et al. (1994) reported on the application of biologically derived, self-assembled cylindrical microstructures to form advanced composite materials for controlled release applications. Those microstructures had many applications in the material sciences. They focused on the potential for rationally controlling the fabrication of submicron microstructures for controlled release applications.

Polysiloxane coatings containing chemically bound (tethered) quaternary ammonium salt (QAS) moieties were investigated for potential application as environment-friendly coatings to control marine biofouling by Majumdar et al. (2008). A combinatorial/high-throughput approach was applied to the investigation to enable multiple variables to be probed simultaneously and efficiently. The variables investigated for the moisture-curable coatings included QAS composition, that is, alkyl chain length, and concentration as well as silanol-terminated polysiloxane molecular weight. A total of 75 compositionally unique coatings were prepared and characterized using surface characterization techniques and biological assays. Biological assays were based on two different marine microorganisms, a bacterium, *Cellulophaga lytica*, and a diatom, *Navicula incerta*, as well as a macrofouling alga, *Ulva*. The results of the study showed that all three variables influenced coating surface properties as well as AF and FR characteristics. The incorporation of QAS moieties into a polysiloxane matrix generally resulted in an increase in coating surface hydrophobicity. Characterization of coating surface morphology revealed a heterogeneous, two-phase morphology for many of the coatings investigated. A correlation was found between water contact angle and coating surface roughness, with the contact angle increasing with increasing surface roughness. Coatings based on the QAS moiety containing the longest alkyl chain (18 carbons) displayed the highest microroughness and, thus, the most hydrophobic surfaces. With regard to AF and FR properties, coatings based on the 18-carbon QAS moieties were very effective at inhibiting biofilm formation by *C. lytica* and enabling easy removal of *Ulva* sporelings (young plants), while coatings based on the 14-carbon QAS moieties were very effective at inhibiting biofilm growth of *N. incerta*.

Carbon nanotubes (CNTs)/polymer hybrid films have been studied (Miao et al. 2011) for the hydrophobic surface in recent years. Multiwalled carbon nanotubes (MWCNTs)/ fluorinated silane containing waterborne polyurethane (FSiPUA) hybrid films based on the excellent hydrophobicity have been prepared for the long-term objective to prevent the marine fouling. MWCNTs/FSiPUA hybrid emulsion was prepared, respectively, by the methods of solution mixing with ultrasonic excitation and *in situ* emulsion polymerization. Chemical and physical properties of the films were investigated by FTIR, contact angle meter, and water absorption. The results showed that lauryl sodium sulfate was a better dispersant than BYK-154 prepared via solution mixing with ultrasonic excitation. The performance of hybrid films approached by solution mixing with ultrasonic excitation was better than that by *in situ* emulsion polymerization when the same amount of MWCNTs were filled. The study indicated that the film surface properties reached best by both technologies when the content of MWCNTs was equal to 0.15 wt%.

An assay was developed by D'Souza et al. (2010) to accurately quantify the growth and release behavior of bacterial biofilms on several test reference materials and coatings, using the marine bacterium *Cobetia marina* as a model organism. This assay could be used to investigate the inhibition of bacterial growth and release properties of many surfaces compared to a reference. The method was based on the staining of attached bacterial cells with the nucleic acid–binding, green-fluorescent SYTO 13 stain. A strong linear correlation exists between the fluorescence of the bacterial suspension measured using a plate reader and the total bacterial count measured with epifluorescence microscopy. This relationship allowed the fluorescent technique to be used for the quantification of bacterial cells attached to surfaces. As the bacteria proliferate on the surface over a period of time, the relative fluorescence unit measured using the plate reader also showed an increase with time. This was observed on all three test surfaces (glass, Epikote, and Silastic® T2) over a period of 4 h of bacterial growth, followed by a release assay, which was carried out by the application of hydrodynamic shear forces using a custom-made rotary device. Different fixed rotor speeds were tested, and based on the release analysis, 12 knots were used to provide the standard shear force. The assay developed was then applied for assessing three different AF coatings of different surface roughness. The novel assay allowed the rapid and sensitive enumeration of attached bacteria directly on the coated surface. This was the first plate reader assay technique that allows estimation of irreversibly attached bacterial cells directly on the coated surface without their removal from the surface or extraction of a stain into solution.

The settling of marine diatoms *Amphora coffeaeformis* onto a photocatalytic titanium dioxide surface was monitored by Kemmitt et al. (2011) using attenuated total reflection infrared (ATR-IR) spectroscopy. Attachment of the diatoms via their mucilaginous secretions was monitored via observation of protein amide I and II linkages, carboxylate linkages from uronic acid polysaccharide, and sulfate ester groups. Exposure to UV-A light resulted in the rapid removal of the groups attached to the photocatalyst surface, demonstrating the potential AF capability of a TiO_2 anatase surface.

The influence of the dispersion quality of MWCNTs in a silicone matrix on the marine FR performance of the resulting nanocomposite coatings was reported (Beigbeder et al. 2010). A first set of coatings filled with different nanofiller contents was prepared by the dilution of a silicone/MWCNTs master batch within a hydrosilylation-cured polydimethylsiloxane (PDMS) resin. The FR properties of the nanocomposite coatings were studied through laboratory assays with the marine alga (seaweed) *Ulva*, a common fouling species. As reported previously, the addition of a small (0.05%) amount of CNTs substantially improved the FR properties of the silicone matrix. Their work showed that the improvement was dependent on the amount of filler, with a maximum obtained with 0.1 wt% of MWCNTs. The method of dispersion of CNTs in the silicone matrix was also shown to significantly influence the FR properties of the coatings. Dispersing 0.1% MWCNTs using the master batch approach yielded coatings with ~40% improved FR properties over those where MWCNTs were dispersed directly into the polymeric matrix. This improvement was directly related to the state of nanofiller dispersion within the cross-linked silicone coating.

Fouling-resistant surface characteristics and their mechanisms of action against settling larvae of barnacles were reviewed by Aldred and Clare (2008). The role of the barnacle in marine fouling was discussed in the context of its life cycle and the behavioral ecology of its cypris larva. The temporary and permanent adhesion mechanisms of cyprids were covered in detail, and an overview of adult barnacle adhesion was presented. Due to the recent legislation, researches focus firmly on environmentally inert marine coatings. Therefore, the actions of traditional biocides on barnacles have been included here. The discussion was restricted to those surface modifications that interfere with settlement-site selection and adhesion of barnacle cypris larvae, specifically, textural engineering of surfaces, development of inert *nonfouling* surfaces, and the use of enzymes in AF.

Bacterial pattern on periodic nanostructure arrays has been studied and reported by Hochbaum and Aizenberg (2010). Surface-associated bacteria typically form self-organizing communities called biofilms. Spatial segregation has been reported to be important for various bacterial processes associated with cellular and community development. They demonstrated bacterial ordering and oriented attachment on the single-cell level induced by nanometer-scale periodic surface features. These surfaces cause spontaneous and distinct patterning phases, depending on their periodicity, which was observed for several strains, both Gram positive and negative. This patterning was a general phenomenon that could control natural biofilm organization on the cellular level.

The synthesis, characterization, and antimicrobial properties of functionalized copper nanoparticles (CuNPs)/polymer composites were covered by Anyaogu et al. (2008). CuNPs were stabilized by surface attachment of the acrylic functionality that could be copolymerized with other acrylic monomers, thus becoming an integral part of the polymer backbone. Biological experiments showed that CuNP/polymer composites exhibited antimicrobial activity similar to that of conventional copper-based biocides. Atomic absorption spectroscopy showed the smallest amount of copper ions leaching from chemically bound acrylated CuNPs compared to the nonfunctionalized biocides. These composites were reported to have a strong potential for use in antibacterial or marine AF coatings.

The adhesion of nonmotile bacteria *Streptococci* consortium and motile *P. fluorescens* was studied by Diaz et al. (2007). Substrates with micro- and nanopatterned topography were used. The influence of surface characteristics on bacterial adhesion was investigated using optical and epifluorescence microscopy, SEM, and AFM. Their results showed an important influence of the nature of the substratum. On microrough surfaces, initial bacterial adhesion was less significant than on smooth surfaces. In contrast, nanopatterned samples showed more bacterial attachment than the smooth control. A remarkable difference in morphology, orientation, and distribution of bacteria between the smooth and the nanostructured substrate was also observed. The results showed the important effect of substratum nature and topography on bacterial adhesion, which depended on the relation between roughness characteristics dimensions and bacterial size.

The silicone-, phosphorous-, and sulfur-containing nanocoatings using diglycidyl ether of bisphenol-A type epoxy resin (DGEBA) as a base material, tris (p-isocyanatophenyl) thiophosphate (DESMODUR) as a modifier, and POSS-NH$_2$ (polyhedral oligomeric silsesquioxane) as nanoreinforcement were developed and characterized by Ananda Kumar and Sasikumar (2010). The nanocoatings were cured by Aradur 140 (polyamidoimidazoline) and XY 54 (polyamidoamine) curatives. The corrosion- and fouling-resistant properties of these coatings were evaluated by potentiodynamic polarization, electrochemical impedance, salt spray tests, and AF tests. It was interesting to observe that the molecular structure of curing agents and the nanoreinforcing effect of POSS-NH$_2$ significantly influenced their corrosion and fouling protection behavior. For example, coating system "1" alone retained an impedance value of 10^9 Ω cm^2 even after 30 days of immersion, indicating no deterioration on the coating. However, the impedance values of other specimens decreased from 10^9 to 10^7 and 10^6 Ω cm^2. This observation clearly indicated the synergistic effect of curing agent and nanoreinforcing effect of POSS-NH$_2$ toward corrosion resistance. Similar observation was made from the AF study, with a marked inhibition of bacterial adhesion on such coated panels.

The synthesis of hydrophobic, hydrophilic, and amphiphilic polyurethane coatings with tethered hydrophilic and/or hydrophobic moieties was reported by Goel et al. (2009). These coatings have been characterized and tested for mechanical properties and surface characteristics using advanced instruments such as the scanning probe microscope (SPM), dynamic contact angle analyzer (DCA), adhesion tester, and nanoindenter. The surfaces with tethered hydrophobic or hydrophilic moieties, when immersed in water, showed remarkable changes in the surface topography, hence their dynamic surface characteristics. The amphiphilic surfaces, containing both hydrophobic and hydrophilic moieties, showed intelligent behavior in response to the external environment. The ability to tailor surfaces with predictable behavior upon exposure to the external environment opens up enormous opportunities for their potential end-use applications.

The current research on the interaction of living cells with both native and nanostructured surfaces and the role that these surface properties play in the different stages of cell attachment were reviewed by Bazaka et al. (2011). Whereas the employment of nanotechnology in electronics and optics engineering has been relatively well established, the use of nanostructured materials in medicine and biology has been undoubtedly novel. Certain nanoscale surface phenomena were being exploited to promote or prevent the attachment of living cells. However, as yet, it has not been possible to develop methods that completely prevent cells from attaching to solid surfaces since the mechanisms by which living cells interact with the nanoscale surface characteristics of these substrates are still poorly understood. Recently, novel and advanced surface characterization techniques have been developed that allow the precise molecular and atomic scale characterization of both living cells and the solid surfaces to which they attached. Based on these additional capabilities, it might be possible to define boundaries, or minimum dimensions, at which a surface feature could exert influence over an attaching living organism.

The real-time quantification of microscale bioadhesion events *in situ* using imaging surface plasmon resonance was reported by Aldred et al. (2011). From macro- to nanoscales, adhesion phenomena were all-pervasive in nature yet remain poorly understood. In recent years, studies of biological adhesion mechanisms, terrestrial and marine, have provided inspiration for biomimetic adhesion strategies and important insights for the development of fouling-resistant materials. Although the focus of most contemporary bioadhesion research was on large organisms such as marine mussels, insects, and geckos, adhesion events on the micro/nanoscale were critical to understand the important underlying mechanisms. Observing and quantifying adhesion at this scale were particularly relevant for the development of biomedical implants and in the prevention of marine biofouling. However, such characterization has so far been restricted by insufficient quantities of material for biochemical analysis and the limitations of contemporary imaging techniques. Thus, they introduced an optical method that allowed precise determination of adhesive deposition by microscale organisms *in situ* and in real time: a capability not before demonstrated. In this extended study, they used the cypris larvae of barnacles and a combination of conventional and imaging surface plasmon resonance techniques to observe and quantify adhesive deposition onto a range of model surfaces (CH_3-, COOH-, NH_3-, and mPEG-terminated SAMs and a PEGMA/HEMA hydrogel). They then correlated this deposition to passive adsorption of a putatively adhesive protein from barnacles. This way, they were able to rank surfaces in order of effectiveness for preventing barnacle cyprid exploration and demonstrate the importance of observing the natural process of adhesion rather than predicting surface effects from a model system. As well as contributing fundamentally to the knowledge on the adhesion and adhesives of barnacle larvae, a potential target for future biomimetic glues, this method also provided a versatile technique for laboratory testing of fouling-resistant chemistries.

The preparation of low-protein and cell-binding multilayer thin films, formed by the alternate deposition of a block copolymer comprising polystyrene sulfonate and poly(poly(ethylene glycol) methyl ether acrylate) (PSS-*b*-PEG), and polyallylamine hydrochloride (PAH) was reported by Cortez et al. (2010). Film buildup was followed by quartz crystal microgravimetry (QCM), which showed linear growth and a high degree of hydration of the PSS-*b*-PEG/PAH films. Protein adsorption studies with bovine serum albumin using QCM demonstrated that multilayer films of PSS/PAH with a terminal layer of PSS-*b*-PEG were up to fivefold more protein resistant than PSS-terminated films. Protein binding was found dependent on the ionic strength at which the terminal layer of PSS-*b*-PEG was adsorbed, as well as the pH of the protein solution. It was also possible to control the protein resistance of the films by coadsorption of the final layer with another component (PSS), which showed an increase in protein resistance as the proportion of PSS-*b*-PEG in the adsorption solution was increased. In addition, protein resistance could also be controlled by the location of a single PSS-*b*-PEG layer within a PSS/PAH film. Finally, the buildup of PSS-*b*-PEG/PAH films on colloidal particles was demonstrated. PSS-*b*-PEG-terminated particles exhibited enhancement in cell-binding resistance compared with PSS-terminated particles. The stability of PSS-*b*-PEG films, combined with their low-protein and cell-binding characteristics, provided opportunities for the use of the films as low fouling coatings in devices and other surfaces requiring limited interaction with biological interfaces.

Biomimetic silica encapsulation of enzymes for the replacement of biocides in AF coatings was reported by Kristensen et al. (2010). Current AF technologies for ship hulls are based on metals such as cuprous oxide and cobiocides like zinc pyrithione. Due to persistent adverse environmental effects of these biocides, enzyme-based AF paints were proposed as a bio-based, nonaccumulating alternative. Thus, a hydrogen peroxide–producing system composed of hexose oxidase (HOX, EC 1.1.3.5), glucoamylase (GA, EC 3.2.1.3), and starch was tested for the chemical and physical functionalities necessary for successful incorporation into a marine coating. The activity and stability of the enzymes in seawater were evaluated at different temperatures, and paint compatibility was assessed by measuring the distribution and activity of the enzymes incorporated into prototype coating formulations. They used a biomimetic encapsulation procedure for HOX through polyethylenimine-templated silica coprecipitation. The coprecipitation and formulation of a powder for mixing into a marine paint was performed in a one-step economical and gentle formulation process, in which silica coprecipitated HOX was combined with GA and starch to form the AF system. Silica coprecipitation significantly improved the stability and performance of the AF system in marine-like conditions. For example, encapsulation of HOX resulted in 46% higher activity at pH 8, and its stability in artificial seawater increased from retaining only 3.5% activity after 2 weeks to retaining 55% activity after 12 weeks. A coating comprising the full enzyme system released hydrogen peroxide at rates exceeding a target of 36 nmol cm^{-2} day^{-1} for 3 months in a laboratory assay and had potential for prolonged action through incorporation in a self-polishing coating.

FR coatings were prepared by Marabotti et al. (2009) from blends of a fluorinated/siloxane copolymer with a PDMS matrix to couple the low modulus character of PDMS with the low surface tension typical for fluorinated polymers. The content of the surface-active copolymer was varied in the blend over a broad range (0.15–10 wt% with respect to PDMS). X-ray photoelectron spectroscopy depth profiling analyses were performed on the coatings to establish the distribution of specific chemical constituents throughout the coatings. Enrichment in fluorine of the outermost layers of the coating surface was bound. Addition of the fluorinated/siloxane copolymer to the PDMS matrix resulted in a concentration-dependent decrease in the settlement of barnacle, *B. amphitrite,* cyprids. The release of young plants of *Ulva*, a soft fouling species, and young barnacles showed that adhesion strength on the fluorinated/siloxane copolymer was significantly lower than the siloxane control. However, differences in adhesion strength were not directly correlated with the concentration of copolymer in the blends.

Surface exploration of *A. amphitrite* cyprids on microtextured surfaces was studied by Chaw et al. (2011). Microtopography was one of the several strategies used by marine organisms to inhibit colonization by fouling organisms. While replicates of natural micro-textures discourage settlement, details of larval interactions with the structured surfaces remain sparse. Close-range microscopy was used to quantify the exploration of cyprids of *A. amphitrite* on cylindrical micropillars with heights of 5 and 30 μm and diameters ranging from 5 to 100 μm. While 5 μm high structures had little impact, 30 μm high pillars significantly influenced cyprid exploration. The observed step length decreased and step duration increased on diameter pillars attributed to the small dimensions of the voids excluding the cyprid's attachment disc and consequently reducing the area of adhesive contact. When exploring larger diameter pillars, cyprids preferred the use of the voids to form temporary attachment points. This might enhance their resistance to flow. No-choice assay settlement patterns mirrored this exploration behavior, albeit in a pattern counter to what was predicted.

In situ ATR-IR spectroscopic and electron microscopic analyses of settlement secretions of kelp (*Undaria pinnatifida*) spores were done by Petrone et al. (2011). Knowledge about the settlement of marine organisms on substrates was important for the development of environmentally benign new methods for the control of marine biofouling. The adhesion to substrates by spores of *U. pinnatifida*, a kelp species invasive to several countries, was studied by SEM/transmission electron microscopy (TEM) as well as by *in situ* ATR-IR spectroscopy. The IR spectra showed that adhesive secretion began approximately 15 min after the initial settlement and that the adhesive bulk material contained protein and anionic polysaccharides. Energy-dispersive x-ray microanalysis of the adhesive identified sulfur and phosphorus as well as calcium and magnesium ions, which facilitated the gelation of the anionic polysaccharides in the seawater. The adhesive might be secreted from the Golgi bodies in the spore, which were imaged by TEM of spore thin sections. Additionally, an *in situ* settlement study on TiO_2 particle film by ATR-IR spectroscopy revealed the presence of phosphorylated moieties directly binding the substrate. The presence of anionic groups dominating the adhesive suggested that inhibition of spore adhesion is favored by negatively charged surfaces.

The effects of formulation variables, such as type of polyol, solvent type and solvent content, and coating application method, on the surface properties of siloxane–polyurethane FR coatings were explored by Bodkhe et al. (2011). FR coatings allowed easy removal of marine organisms from a ship's hull via the application of a shear force to the surface. Application of self-stratified siloxane–polyurethane coatings was a new approach to a tough FR coating system. Combinatorial high-throughput experimentation was employed to formulate and characterize 24 different siloxane–polyurethane coatings applied using drawdown and drop-casting methods. The resulting coatings were tested for surface energy using contact angle measurements. The FR performance of the coatings was tested using a number of diverse marine organisms including bacteria (*Halomonas pacifica* and *Cytophaga lytica*), sporelings (young plants) of the green macroalga (*U. linza*), diatom (microalga *N. incerta*), and barnacle (*A. amphitrite*). The performance of the majority of the coatings was found to be better than the silicone standards, Intersleek® and Silastic T2. An increase in solvent content in the formulations increased the surface roughness of the coatings. Coatings made with polycaprolactone polyol appeared to be somewhat rougher compared to coatings made with the acrylic polyol. The adhesion strength of sporelings of *Ulva* increased with an increase in solvent content and increase in surface roughness. The adhesion strengths of *Ulva* sporelings, *C. lytica*, and *N. incerta* were independent of the application method (cast or drawdown) in contrast to *H. pacifica* adhesion, which was dependent on the application method.

Multiseasonal barnacle (*Balanus improvisus*) protection achieved by trace amounts of a macrocyclic lactone (ivermectin) included in rosin-based coatings was studied by Pinori et al. (2011). Rosin-based coatings loaded with 0.1% (w/v) ivermectin were found to be effective in preventing colonization by barnacles (*B. improvisus*) both on test panels and on yachts for at least two fouling seasons. The leaching rate of ivermectin was determined by mass spectroscopy (LC/MS-MS) to be 0.7 ng cm^{-2} day^{-1}. This low leaching rate, as deduced from the Higuchi model, was a result of the low loading, low water solubility, high affinity to the matrix, and high molar volume of the model biocide. Comparison of ivermectin and control areas of panels immersed in the field showed undisturbed colonization of barnacles after immersion for 35 days. After 73 days, the mean barnacle base plate area on the controls was 13 mm^2, while on the ivermectin coating, it was 3 mm^2. After 388 days, no barnacle was observed on the ivermectin coating, while the barnacles on the control coating had reached a mean of 60 mm^2. In another series of coated panels, ivermectin was dissolved in a cosolvent mixture of propylene glycol and glycerol formal prior to the addition to the paint base. This method further improved the antibarnacle performance of the coatings. An increased release rate (3 ng cm^{-2} day^{-1}) and dispersion of ivermectin, determined by fluorescence microscopy, and decreased hardness of the coatings were the consequences of the cosolvent mixture in the paint. The AF mechanism of macrocyclic lactones, such as avermectins, needs to be clarified through further studies. Besides chronic intoxication, ivermectin was slowly released from the paint film (even contact intoxication occurring inside the coatings), triggered by penetration of the coating by barnacles.

Use of amphiphilic block copolymer/PDMS blends and nanocomposites for improved FR was reported by Martinelli et al. (2011). Amphiphilic diblock copolymers, Sz6 and Sz12, consisting of a PDMS block and a PEGylated-fluoroalkyl-modified polystyrene block were prepared by atom transfer radical polymerization. Coatings were obtained from blends of block copolymer (1–10 wt%) with a PDMS matrix. The coating surface presented a simultaneous hydrophobic and lipophobic character, owing to the strong surface segregation of the lowest surface energy fluoroalkyl chains of the block copolymer. Surface chemical composition and wettability of the films were affected by exposure to water. Block copolymer Sz6 was also blended with PDMS and a 0.1 wt% amount of MWCNT. The excellent FR properties of these new coatings against the macroalga *U. linza* were essentially resulted from the inclusion of the amphiphilic block copolymer, while the addition of CNT did not appear to improve the FR properties.

The design of a photocurable amphiphilic conetwork consisting of PFPE and PEG segments that display outstanding nonfouling characteristics with respect to spores of green fouling alga *Ulva* when cured under high humidity conditions was reported by Wang et al. (2011b). The analysis of contact angle hysteresis revealed that the PEG density at the surface was enhanced when cured under high humidity. The nonfouling behavior of non-biocidal surfaces against marine fouling was rare because such surfaces usually reduce the adhesion of organisms rather than inhibit colonization. They proposed that the resultant surface segregation of these materials induced by high humidity might be a promising strategy for achieving nonfouling materials. Such an approach was more important than simply concentrating on PEG moieties at an interface with low surface energy.

Silicone and fluoropolymer technologies were compared by Dobretsov and Thomason (2011) and studied the development of marine biofilms on two commercial nonbiocidal coatings. The antimicrobial performance of two FR coating systems, Intersleek 700® (IS700; silicone technology) and Intersleek 900® (IS900; fluoropolymer technology), and a tie coat (TC, control surface) was investigated in a short-term (10 days) field experiment conducted at a depth of 0.5 m in the Marina Bandar Rowdha (Muscat, Oman). Microfouling on coated glass slides was analyzed using epifluorescence microscopy and adenosine-5′-triphosphate (ATP) luminometry. All the coatings developed biofilms composed of heterotrophic bacteria, cyanobacteria, seven species of diatoms (two species of *Navicula*, *Cylindrotheca* sp., *Nitzschia* sp., *Amphora* sp., *Diploneis* sp., and *Bacillaria* sp.), and algal spores (*Ulva* sp.). IS900 had significantly thinner biofilms with fewer diatom species, no algal spores, and the least number of bacteria in comparison with IS700 and the TC. The ATP readings did not correspond to the number of bacteria and diatoms in the biofilms. The density of diatoms was negatively correlated with the density of the bacteria in biofilms on the IS900 coating, and, conversely, diatom density was positively correlated with biofilms on the TC. The higher AF efficacy of IS900 over IS700 might lead to lower roughness and thus lower fuel consumption for those vessels that utilize the IS900 FR coating.

Enzymes finding places in AF strategies were reviewed by Cordeiro and Werner (2011). During the past decades, much effort has been made to find efficient alternative solutions to prevent and/or disrupt the adhesion of fouling organisms to surfaces. The use of enzymes emerged as one of the favorite candidate AF technologies due to biodegradability and affordable prices. An overview of different enzymatic AF strategies was presented, highlighting the most promising groups of enzymes, and their utilization upon surface confinement to control biofouling. The main strategies adapted to control marine biofouling included the degradation of secreted adhesives and the production of AF compounds. The main concepts to control pathogenic biofilms were based on cell lysis and degradation of extracellular matrix polymers. However, immobilization could improve enzyme stability, activity, and AF performance. The successful incorporation of enzymes into coatings, yielding surfaces with broad AF spectrum and long-term efficacy, remains a challenge.

Commercially available epoxy resin modified with nonfunctionalized nano–zinc oxide (nZnO) was examined to get information on its AF, and anticorrosive properties were studied by Saravanan et al. (2014). Epoxy nanohybrid coating was synthesized using nZnO and DGEBA type of epoxy resin. The curing behavior of these materials was ascertained from FT-IR spectral studies. The anticorrosive properties of the nanohybrid were investigated using salt spray and electrochemical polarization studies. The surface morphology images were taken by SEM analysis. This study indicated that nZnO particles were dispersed homogenously through the polymer matrix. The nZnO-incorporated coating was found to exhibit enhanced anticorrosive performance. Approximately 50% reduction in fouling attachment was achieved with coatings containing 3 wt% of nZnO. Alternative coatings should be as effective as conventional paints but with lower toxicity.

An epoxy nanocomposite coating was developed by Palanivelu et al. (2014) using amine functionalized nZnO as the dispersed phase and a commercially available epoxy resin as the matrix phase. The structural features of these materials were ascertained by FT-IR spectral studies. The anticorrosive properties of the epoxy/nZnO hybrid coatings in comparison with a virgin coating were investigated by a salt spray test and an electrochemical impedance spectroscopy technique. The surface morphology determined by SEM indicated that nZnO particles were dispersed homogenously through the epoxy polymer matrix. The results showed improved AF and anticorrosive properties for epoxy/nZnO hybrid coatings.

Engineered micro-/nanoscale topographies that significantly impact bacterial surface attachment were reviewed by Graham and Cady (2014). According to them, bacterial surface fouling has been problematic for a wide range of applications and industries, including, but not limited to, medical devices (implants, replacement joints, stents, pacemakers), municipal infrastructure (pipes, wastewater treatment), food production (food-processing surfaces, processing equipment), and transportation (ship hulls, aircraft fuel tanks). One method to combat bacterial biofouling was to modify the topographical structure of the surface in question, thereby limiting the ability of individual cells to attach to the surface, colonize, and form biofilms. Multiple research groups have demonstrated that micro- and nanoscale topographies significantly reduce bacterial biofouling, for both individual cells and bacterial biofilms. AF strategies that utilized engineered topographical surface features with well-defined dimensions and shapes have demonstrated a greater degree of controllable inhibition over initial cell attachment, in comparison to undefined, texturized, or porous surfaces.

The performances of innovative nanomodified surfaces were discussed by Janabi and Malayeri (2014) when subjected to $CaSO_4$ scale deposition during convective heat transfer. Two types of nonstructured and structured nanomodified surfaces were experimentally examined. The experimental results demonstrated that such coatings would increase the induction time before fouling starts and also reduce the subsequent fouling rate, in comparison with untreated stainless steel surfaces. After laboratory runs, the coatings that performed best in laboratory were utilized in a desalination plant in a plate heat exchanger. The field findings resulted in applying thinner thickness of the attempted coatings with no sign of fouling. Considering these promising results, their work continued to assess the requirements for better thermal and mechanical stabilities that such coatings would have to satisfy.

Marine eukaryotic microbial community was analyzed by Sun and Zhang (2014) in the context of its complexity and dynamic process in the early stage of biofouling. Samples were collected from five different surfaces: the control group (coating CG), 0.2% silicium (coating A), 2% carbon nanopowder (coating B), 0.2% graphite (coating C), and 2% silicium plus 2% carbon nanopowder (coating F). These coatings were sea tested for 11 months for their antibiofouling performance. Coatings A and B did not have antibiofouling ability just like coating CG, while coatings C and F have excellent antibiofouling capacity. Single-stranded conformation polymorphisms were employed to analyze diversity level and dynamic changes of marine fouling eukaryotic communities adhering to different coatings during different sampling time (days 2, 4, 6, 9, and 14). The results indicated that composition, distribution, and diversity level of marine fouling eukaryotic microbial community presented obviously differences as sampling time and coating surfaces changed. Coatings with good antibiofouling performance tend to have low diversity level and different adhering dynamics of marine fouling eukaryotic microbial community.

Ultrasound treatment was used by Guo et al. (2014) to reduce the adhesion of newly metamorphosed barnacles up to 2 days old. This was observed in the reduction of adhesion strength of the newly settled barnacles from ultrasound-treated cyprids on silicone substrate compared to the adhesion strength of barnacles metamorphosed from cyprids not exposed to ultrasound. AFM was used to analyze the effect of ultrasound on barnacle cyprid FPs, which were protein adhesives secreted when the larvae explore surfaces. The ultrasound-treated cyprids were found to secrete less FPs, which appeared to spread a larger area than those generated by untreated cyprids. The evidence from this study suggested that ultrasound treatment resulted in a reduced cyprid settlement and FP secretion and might affect the subsequent recruitment of barnacles onto FR surfaces by reducing the ability of early settlement stage of barnacles (up to 2 days old) from firmly adhering to the substrates. Ultrasound therefore could be used in combination with FR coatings to offer a more efficient AF strategy.

Developments in marine AF polymer brushes and coatings that were tethered to material surfaces and did not actively release biocides were reviewed by Yang et al. (2014). Polymer brush coatings have been designed to inhibit molecular fouling, microfouling, and macrofouling through incorporation or inclusion of multiple functionalities. Hydrophilic polymers, such as PEG, hydrogels, zwitterionic polymers, and polysaccharides, resisted attachment of marine organisms effectively due to extensive hydration. FR polymer coatings, based on fluoropolymers and PDMSEs, minimized adhesion between marine organisms and material surfaces, leading to easy removal of biofoulants. Polycationic coatings were effective in reducing marine biofouling partly because of their good bactericidal properties. Recent advances in controlled radical polymerization and click chemistry have also allowed better molecular design and engineering of multifunctional brush coatings for improved AF efficacies.

In laboratory experiments, the AF properties of ZnO nanorod coatings were investigated by Fori et al. (2014) using the marine bacterium *Acinetobacter* sp. AZ4C, larvae of the bryozoan *Bugula neritina*, and the microalga *Tetraselmis* sp. ZnO nanorod coatings that were fabricated on microscope glass substrata by a simple hydrothermal technique using different molar concentrations of zinc precursors. These coatings were tested under artificial sunlight and in the dark (no irradiation). In the presence of light, both the ZnO nanorod coatings significantly reduced the density of *Acinetobacter* sp. AZ4C and *Tetraselmis* sp. in comparison to the control (microscope glass substratum without a ZnO coating). High mortality and low settlement of *B. neritina* larvae were observed on ZnO nanorod coatings subjected to light irradiation. In darkness, neither mortality nor enhanced settlement of larvae was observed. Larvae of *B. neritina* were not affected by Zn^{2+} ions. The AF effect of the ZnO nanorod coatings was thus attributed to the reactive oxygen species produced by photocatalysis. It was concluded that ZnO nanorod coatings effectively prevented marine micro- and macrofouling in static conditions.

Condensation on the hierarchically structured lotus leaf, which could facilitate self-propulsion of water droplets off the surface, was studied and reported by Watson et al. (2014). Droplets on leaves inclined at high angles could be completely removed from the surface by self-propulsion with the assistance of gravity. Due to the small size of mobile droplets, light breezes might also fully remove the propelled droplets, which were typically projected beyond the boundary layer of the leaf cuticle. Moreover, the self-propelled droplets/condensate was able to remove contaminants (e.g., silica particles) from the leaf surface. The biological significance of this process might be associated with maintaining a healthy cuticle surface when the action of rain to clean the surface via the lotus effect was not possible (due to no precipitation). Indeed, the native lotus plants in this study were located in a region with extended time periods (several months) without rain. Thus, dew formation on the leaf might provide an alternative self-cleaning mechanism during times of drought and optimized the functional efficiency of the leaf surface as well as protecting the surface from long-term exposure to pathogens such as bacteria and fungi.

Novel surface materials and its microtopographies were investigated by Bloecher et al. (2013) to deter hydroid *Ectopleura larynx* settlement on salmon aquaculture cages in Norway. The settlement preferences of hydroid larvae for 12 materials with wettabilities ranging from hydrophobic (54°) to hydrophilic (112°) were tested in a no-choice bioassay. Although settlement differed between materials, with the highest average settlement on polytetrafluoroethylene (95%) and the lowest on untreated polyurethane (53%), no trend regarding the tested wettabilities could be found, and none of the tested materials was able to reduce average settlement below 50%. Furthermore, nine high-density polyethylene (100–600 μm microtopographies) and seven PDMS (40–400 μm microtopographies) micro-textured surfaces were tested. There was no systematic effect of microtopography on the settlement of *E. larynx* larvae. However, there was a preference for settlement in channels on PDMS microtopographies between 80 and 300 μm. Similarly, there were no preferences for any of the examined microtopographies in a 12-day field test using PDMS surfaces at a commercial fish farm. The study indicated that neither surface wettability (hydrophilicity–phobicity) nor microtopographies were effective at deterring the settlement of the hydroid *E. larynx*. The high plasticity of the aboral pole and the hydrorhiza of the hydroids might explain settlement even under unfavorable conditions, highlighting the successful colonization traits of these dominant biofouling species.

27.4 Conclusion

Understanding the linkage among marine bacterial adhesion, larval settlement, and surface topography/architecture plays a major role in marine industries to use nanobased technologies in biofouling control. This review discusses the impact of adhering organisms from bacteria to barnacle when exposed to nanomaterials. The need for novel antifoulants comes from the ban on tin-based AF paints, which accelerates the science to control micro- and macrofouling. Consequently, new methods for marine biofouling control are badly needed. Nanotechnology, the use of materials with dimensions on the atomic or molecular scale, has become increasingly utilized for marine industrial applications and is of great interest as an approach to killing or reducing the activity of numerous biofilms forming bacterial attachment. Nanoparticles interact and penetrate into bacteria and provide unique bactericidal mechanisms. Day by day, the number of new researches and reports has been added into the tag "marine biofouling control," and the need of new and effective AF materials is still in demand on the market. Thus, marine industries are in a great need to sort out this universal problem and to save the cost inventories. Nanomaterials may be one of the answers to address the problems faced by marine industries. This review thus summarizes the nanotechnology-based challenges on designing and studying the impact of biofouling control through antibacterial paints, lowering the surface energy, superhydrophobicity, self-cleaning abilities, composites, and coatings.

Acknowledgment

The author is grateful to the chancellor and directors of Sathyabama University, Chennai, for their immense support.

References

Akesso, L., P.D. Navabpour, D. Teer, M.E. Pettitt, M.E. Callow, C. Liu et al. 2009. Deposition parameters to improve the fouling-release properties of thin siloxane coatings prepared by PACVD. *Applied Surface Science* 255: 6508–6514.

Aldred, N. and A.S. Clare. 2008. The adhesive strategies of cyprids and development of barnacle-resistant marine coatings. *Biofouling* 24: 351–363.

Aldred, N., T. Ekblad, O. Andersson, B. Liedberg, and A.S. Clare. 2011. Real-time quantification of microscale bioadhesion events in situ using imaging surface plasmon resonance (iSPR). *ACS Applied Materials & Interfaces* 3: 2085–2091.

Ananda Kumar, S. and A. Sasikumar. 2010. Studies on novel silicone/phosphorus/sulphur containing nano-hybrid epoxy anticorrosive and antifouling coatings. *Progress in Organic Coatings* 68: 189–200.

Anyaogu, K.C., A.V. Fedorov, and D.C. Neckers. 2008. Synthesis, characterization, and antifouling potential of functionalized copper nanoparticles. *Langmuir* 24: 4340–4346.

Bai, X.-C., I.S. Fernandez, G. McMullan, and S.H.W. Scheres. 2013. Ribosome structures to near-atomic resolution from thirty thousand cryo-EM particles. *eLIFE* 2: e00461.

Batley, G.E., M.S. Scammell, and C.I. Brockbank. 1992. The impact of the banning of tributyltin-based antifouling paints on the Sydney rock oyster, *Saccostrea commercialis*. *The Science of the Total Environment* 122: 301–314.

Baum, C., W. Meyer, R. Stelzer, and L.G. Fleischer. 2002. Average nanorough skin surface of the pilot whale (*Globicephala melas*, Delphinidae): Considerations on the self-cleaning abilities based on nanoroughness. *Marine Biology* 140: 653–657.

Bazaka, K., M.V. Jacob, R.J. Crawford, and E.P. Ivanova. 2011. Plasma-assisted surface modification of organic biopolymers to prevent bacterial attachment. *Acta Biomaterialia* 7: 2015–2028.

Beigbeder, A., R. Mincheva, M.E. Pettitt, M.E. Callow, J.A. Callow, M. Claes, and P. Dubois. 2010. Marine fouling release silicone/carbon nanotube nanocomposite coatings: On the importance of the nanotube dispersion state. *Journal of Nanoscience and Nanotechnology* 10: 2972–2978.

Bennett, S.M., J.A. Finlay, N. Gunari, D.D. Wells, A.E. Meyer, G.C. Walker, M.E. Callow, J.A. Callow, F.V. Bright, and M.R. Detty. 2010. The role of surface energy and water wettability in aminoalkyl/fluorocarbon/hydrocarbon-modified xerogel surfaces in the control of marine biofouling. *Biofouling* 26: 235–246.

Berto, D., R.B. Brusa, F. Cacciatore, S. Covelli, F. Rampazzo, O. Giovanardi, and M. Giani. 2012. Tin free antifouling paints as potential contamination source of metals in sediments and gastropods of the southern Venice lagoon. *Continental Shelf Research* 45: 34–41.

Bloecher, N., R. de Nys, A.J. Poole, and J. Guenther. 2013. The fouling hydroid *Ectopleura larynx*: A lack of effect of next generation antifouling technologies. *Biofouling* 29: 237–246.

Bodkhe, R.B., S.E.M. Thompson, C. Yehle, N. Cilz, J. Daniels, S.J. Stafslien, M.E. Callow, J.A. Callow, and W.C. Dean. 2011. The effect of formulation variables on fouling-release performance of stratified siloxane-polyurethane coatings. *Journal of Coatings Technology and Research* 9: 235–249.

Briand, J.F. 2009. Marine antifouling laboratory bioassays: An overview of their diversity. *Biofouling* 25: 297–311.

Callow, M.E. and J.A. Callow. 2002. Marine biofouling: A sticky problem. *Biologist* 49: 1–5.

Cao, S., J.D. Wang, H.S. Chen, and D.R. Chen. 2011. Progress of marine biofouling and antifouling technologies. *Chinese Science Bulletin* 56: 598–612.

Castonguay, M.H., S. van der Schaaf, W. Koester, J. Krooneman, W. van der Meer, H. Harmsen, and P. Landini. 2006. Biofilm formation by *Escherichia coli* is stimulated by synergistic interactions and co-adhesion mechanisms with adherence-proficient bacteria. *Research in Microbiology* 157: 471–478.

Chambers, L., K. Stokes, F. Walsh, and R. Wood. 2006. Modern approaches to marine antifouling coatings. *Surface and Coatings Technology* 201: 3642–3652.

Chaw, K.C., G.H. Dickinson, K. Ang, J. Deng, and W.R. Birch. 2011. Surface exploration of *Amphibalanus amphitrite* cyprids on microtextured surfaces. *Biofouling* 27: 413–422.

Chen, M., Y.Y. Qu, Y. Li, and H. Gao. 2008. Structures and antifouling properties of low surface energy non-toxic antifouling coatings modified by nano-SiO$_2$ powder. *Science in China Series B: Chemistry* 51: 848–852.

Cordeiro, A.L. and C. Werner. 2011. Enzymes for antifouling strategies. *Journal of Adhesion Science and Technology* 25: 2317–2344.

Cortez, C., J.F. Quinn., X. Hao, C.S. Gudipati, M.H. Stenzel, T.P. Davis, and F. Caruso. 2010. Multilayer buildup and biofouling characteristics of PSS-b-PEG containing films. *Langmuir* 26: 9720–9727.

Denkhaus, E., S. Meisen, U. Telgheder, and J. Wingender. 2007. Chemical and physical methods for characterisation of biofilms. *Microchimica Acta* 158: 1–27.

Diaz, C., M.C. Cortizo, P.L. Schilardi, S.G. Saravia, and M.A. de Mele. 2007. Influence of the nano-micro structure of the surface on bacterial adhesion. *Materials Research* 10: 11–14.

Dineshram, R., R. Subasri, K.R.C. Somaraju, K. Jayaraj, L. Vedaprakash, L. Ratnam, S.V. Joshi, and R. Venkatesan. 2009. Biofouling studies on nanoparticle-based metal oxide coatings on glass coupons exposed to marine environment. *Colloids and Surface B: Biointerfaces* 74: 75–83.

Dobretsov, S. and J.C. Thomason. 2011. The development of marine biofilms on two commercial non-biocidal coatings: A comparison between silicone and fluoropolymer technologies. *Biofouling* 27: 869–880.

D'Souza, F., A. Bruin, R. Biersteker, G. Donnelly, J. Klijnstra, C.P. Rentrop, and P. Willemsen. 2010. Bacterial assay for the rapid assessment of antifouling and fouling release properties of coatings and materials. *Journal of Industrial Microbiology & Biotechnology* 37: 363–370.

Fori, M., S. Dobretsov, M.T. Myint, and J. Dutta. 2014. Antifouling properties of zinc oxide nanorod coatings. *Biofouling* 30: 871–882.

Genzer, J. and K. Efimenko. 2006. Recent developments in superhydrophobic surfaces and their relevance to marine fouling: A review. *Biofouling* 22: 339–360.

Gipperth, L. 2009. The legal design of the international and European Union ban on tributyltin antifouling paint: Direct and indirect effects. *Journal of Environmental Management* 90: S86–S95.

Goel, A., R.G. Joshi, and V. Mannari. 2009. Intelligent polymeric surfaces through molecular self-assembly. *Journal of Coatings Technology and Research* 6: 123–133.

Graham, M.V. and N.C. Cady. 2014. Nano and microscale topographies for the prevention of bacterial surface fouling. *Coatings* 4: 37–59.

Gunari, N., L.H. Brewer, S.M. Bennett, A. Sokolova, N.D. Kraut, J.A. Finlay et al. 2011. The control of marine biofouling on xerogel surfaces with nanometer-scale topography. *Biofouling* 27: 137–149.

Guo, S., B.C. Khoo, S.L.M. Teo, S. Zhong, C.T. Lim, and H.P. Lee. 2014. Effect of ultrasound on cyprid footprint and juvenile barnacle adhesion on a fouling release material. *Colloids and Surfaces B: Biointerfaces* 115: 118–124.

Hayes, K.R., R. Cannon, K. Neil, and G. Inglis. 2005. Sensitivity and cost considerations for the detection and eradication of marine pests in ports. *Marine Pollution Bulletin* 50: 823–834.

Hochbaum, A.I. and J. Aizenberg. 2010. Bacteria pattern spontaneously on periodic nanostructure arrays. *Nano Letters* 10: 3717–3721.

Inbakandan, D., C. Kumar, L. Stanley Abraham, R. Kirubagaran, R. Venkatesan, and S. Ajmal Khan. 2013. Silver nanoparticles with anti microfouling effect: A study against marine biofilm forming bacteria. *Colloids and Surfaces B: Biointerfaces* 111: 636–643.

Inbakandan, D., P.S. Murthy, R. Venkatesan, and S.A. Khan. 2010. 16S rDNA sequence analysis of culturable marine biofilm forming bacteria from a ship's hull. *Biofouling* 26: 893–899.

Janabi, A. and M.R. Malayeri. 2014. Nano-coated surfaces to mitigate fouling in thermal water services. *Desalination and Water Treatment*. 55:1–9. doi: 10.1080/19443994.2014.940205.

Joint, I., K. Tait, and G. Wheeler. 2007. Cross-kingdom signalling: Exploitation of bacterial quorum sensing molecules by the green seaweed *Ulva*. *Philosophical Transactions of the Royal Society B: Biological Sciences* 362: 1223–1233.

Kemmitt, T., A.G. Young, and C. Depree. 2011. Observation of diatom settling and photocatalytic antifouling on TiO_2 using ATR-IR spectroscopy. *Materials Science Forum* 700: 227–230.

Kristensen, J.B., R.L. Meyer, C.H. Poulsen, K.M. Kragh, F. Besenbacher, and B.S. Laursen. 2010. Biomimetic silica encapsulation of enzymes for replacement of biocides in antifouling coatings. *Green Chemistry* 12: 387–394.

Kujundzic, E., A.C. Fonseca, E.A. Evans, M. Peterson, A.R. Greenberg, and M. Hernandez. 2007. Ultrasonic monitoring of early stage biofilm growth on polymeric surfaces. *Journal of Microbiological Methods* 68: 458–467.

Lazarova, V. and J. Manem. 1995. Biofilm characterization and activity analysis in water and wastewater treatment. *Water Research* 29: 2227–2245.

Majumdar, P., E. Lee, N. Patel, K. Ward, J. Stafslien, J. Daniels et al. 2008. Combinatorial materials research applied to the development of new surface coatings IX: An investigation of novel antifouling/fouling-release coatings containing quaternary ammonium salt groups. *Biofouling* 24: 185–200.

Marabotti, I., A. Morelli, L.M. Orsini, E. Martinelli, G. Galli, E. Chiellini et al. 2009. Fluorinated/siloxane copolymer blends for fouling release: Chemical characterization and biological evaluation with algae and barnacles. *Biofouling* 25: 481–493.

Marechal, J.P. and C. Hellio. 2009. Challenges for the development of new non-toxic antifouling solutions. *International Journal of Molecular Sciences* 10: 4623–4637.

Martinelli, E., M. Suffredini, G. Galli, A. Glisenti, M.E. Pettitt, M.E. Callow, J.A. Callow, D. Williams, and G. Lyall. 2011. Amphiphilic block copolymer/poly(dimethylsiloxane) (PDMS) blends and nanocomposites for improved fouling-release. *Biofouling* 27: 529–541.

Miao, M., Y.H. Qi, Z.P. Zhang, and Z. Zhang. 2011. Preparation and characterization of MWCNTs/FSiPUA hybrid film. *Materials Science Forum* 688: 233–237.

Mitik-Dineva, N., J. Wang, V.K. Truong, P.R. Stoddart, F. Malherbe, R.J. Crawford, and E.P. Ivanova. 2009. Differences in colonisation of five marine bacteria on two types of glass surfaces. *Biofouling* 25: 621–631.

Munk, P.S., E.M. Staal, N. Butt, K. Isaksen, and A.I. Larsen. 2009. High-intensity interval training may reduce in-stent restenosis following percutaneous coronary intervention with stent implantation: A randomized controlled trial evaluating the relationship to endothelial function and inflammation. *American Heart Journal* 158: 734–741.

Murthy, P.S., R. Venkatesan, K.V.K. Nair, D. Inbakandan, S. Syed Jahan, D. Magesh Peter, and M. Ravindran. 2005. Evaluation of sodium hypochlorite for fouling control in plate heat exchangers for seawater application. *International Biodeterioration and Biodegradation* 55: 161–170.

Neu, T.R. and J.R. Lawrence. 2015. Innovative techniques, sensors, and approaches for imaging biofilms at different scales. *Trends in Microbiology* 23: 233–242.

Palanivelu, S., D. Dhanapal, and A.K. Srinivasan. 2014. Studies on silicon containing nano-hybrid epoxy coatings for the protection of corrosion and bio-fouling on mild steel. *Silicon* 6: 1876–9918. doi: 10.1007/s12633-014-9202-6.

Pantanella, F., P. Valenti, T. Natalizi, D. Passeri, and F. Berlutti. 2013. Analytical techniques to study microbial biofilm on abiotic surfaces: Pros and cons of the main techniques currently in use. *Annali di Igiene, Medicina Preventiva e di Comunita* 25: 31–42.

Petrone, L., R. Easingwood, M.F. Barker, and A.J. McQuillan. 2011. In situ ATR-IR spectroscopic and electron microscopic analyses of settlement secretions of *Undaria pinnatifida* kelp spores. *Journal of the Royal Society Interface* 6: 410–422.

Phang, Y., N. Aldred, Y. Ling, J. Huskens, S. Anthony, G. Clare, and J. Vancso. 2010. Atomic force microscopy of the morphology and mechanical behaviour of barnacle cyprid footprint proteins at the nanoscale. *Journal of the Royal Society Interface* 7: 285–296.

Pinori, E., M. Berglin, L.M. Brive, M. Hulander, M. Dahlstrom, and H. Elwing. 2011. Multi-seasonal barnacle (*Balanus improvisus*) protection achieved by trace amounts of a macrocyclic lactone (ivermectin) included in rosin-based coatings. *Biofouling* 27: 941–953.

Ploux, L., S. Beckendorff, M. Nardin, and S. Neunlist. 2007. Quantitative and morphological analysis of biofilm formation on self-assembled monolayers. *Colloids and Surfaces B: Biointerfaces* 57: 174–181.

Rajasekar, A., T. Ganesh Babu, S. Karutha Pandian, S. Maruthamuthu, N. Palaniswamy, and A. Rajendran. 2007. Biodegradation and corrosion behavior of manganese oxidizer *Bacillus cereus* ACE4 in diesel transporting pipeline. *Corrosion Science* 49: 2694–2710.

Rawat, J., S. Ray, P.V.C. Rao, and V.C. Nettem. 2010. Recent developments of nanomaterial doped paints for the minimization of biofouling in submerged structures. *Materials Science Forum* 657: 75–82.

Rosenhahn, A., S. Schilp, H.J. Kreuzer, and M. Grunze. 2010. The role of "inert" surface chemistry in marine biofouling prevention. *Physical Chemistry Chemical Physics* 12: 4275–4286.

Sahoo, B.N., K. Balasubramanian, and T. Amrutha. 2013. Effect of TiO_2 powder on the surface morphology of micro/nanoporous structured hydrophobic fluoropolymer based composite material. *Journal of Polymers* 2013: Article ID 615045.

Saravanan, P., D. Duraibabu, and S.A. Kumar. 2014. Development of environmentally acceptable nano-hybrid coatings for bio-fouling protection. *Advanced Materials Research* 938: 269–274.

Scardino, A.J., D. Hudleston, Z. Peng, N.A. Paul, and R. de Nys. 2009. Biomimetic characterisation of key surface parameters for the development of fouling resistant materials. *Biofouling* 25: 83–93.

Schnur, J.M., B.R. Ratna, J.V. Selinger, A. Singh, G. Jyothi, and K.R. Easwaran. 1994. Diacetylenic lipid tubules: Experimental evidence for a chiral molecular architecture. *Science* 13: 945–947.

Shibli, S.M.A., S.R. Archana, and P. Muhamed Ashraf. 2008. Development of nano cerium oxide incorporated aluminium alloy sacrificial anode for marine applications. *Corrosion Science* 50: 2232–2238.

Shibli, S.M.A. and F. Chacko. 2011. Development of nano TiO_2-incorporated phosphate coatings on hot dip zinc surface for good paintability and corrosion resistance. *Applied Surface Science* 257: 3111–3117.

Stephenson, S.N., G.G. Roberts, M.J. Hoggard, and A.C. Whittaker. 2014. A cenozoic uplift history of Mexico and its surroundings from longitudinal river profiles. *Geochemistry, Geophysics, Geosystems* 15: 4734–4758.

Sun, Y. and Z. Zhang. 2014. Differential early dynamic process of marine eukaryotic microbial community on anti-biofouling and biofouling-enhancing micro/nano surfaces. *Journal of Chemical and Pharmaceutical Research* 6: 183–188.

Therasa, J.J., V. Vinita, R.P. George, M. Kamruddin, S. Kalavathi, N. Manoharan, A.K. Tyagi, and R.K. Dayal. 2010. Reducing biofouling on titanium surface by electroless deposition of antibacterial copper nano films. *Current Science* 99: 1079–1083.

Truong, V.K., S. Rundell, R. Lapovok, Y. Estrin, J.Y. Wang, C. Christopher et al. 2009. Effect of ultrafine-grained titanium surfaces on adhesion of bacteria. *Applied Microbiology and Biotechnology* 83: 925–937.

Van Houdt, R. and C.W. Michiels. 2005. Role of bacterial cell surface structures in *Escherichia coli* biofilm formation. *Research in Microbiology* 156: 626–633.

Vishwakarma, V., J. Josephine, R.P. George, R. Krishnan, S. Dash, M. Kamruddin, S. Kalavathi, N. Manoharan, A.K. Tyagi, and R.K. Dayal. 2009. Antibacterial copper–nickel bilayers and multilayer coatings by pulsed laser deposition on titanium. *Biofouling* 25: 705–710.

Vladkova, T. 2009. Surface modification approach to control biofouling. *Marine and Industrial Biofouling, Springer Series on Biofilms* 4: 135–163.

Wang, Y., J.A. Finlay, D.E. Betts, T.J. Merkel, J.C. Luft, M.E. Callow et al. 2011a. Amphiphilic co-networks with moisture-induced surface segregation for high-performance nonfouling coatings. *Langmuir* 27: 10365–10369.

Wang, Y., H. Liu, K. Wang, H. Eiji, Y. Wang, and H. Zhou. 2009. Synthesis and electrochemical performance of nano-sized $Li_4Ti_5O_{12}$ with double surface modification of Ti(III) and carbon. *Journal of Materials Chemistry* 19: 6789–6795.

Wang, Y., L.M. Pitet, J.A. Finlay, L.H. Brewer, G. Cone, D.E. Betts et al. 2011b. Investigation of the role of hydrophilic chain length in amphiphilic perfluoropolyether/poly(ethylene glycol) networks: Towards high-performance antifouling coatings. *Biofouling* 27: 1139–1150.

Watson, G.S., M. Gellender, and J.A. Watson. 2014. Self-propulsion of dew drops on lotus leaves: A potential mechanism for self cleaning. *Biofouling* 30: 427–434.

Weir, E., A. Lawlor, A. Whelan, and F. Regan. 2008. The use of nanoparticles in anti-microbial materials and their characterization. *Analyst* 133: 835–845.

Yang, W.J., K.G. Neoh, E.T. Kang, S.L.M. Teo, and D. Rittschof. 2014. Polymer brush coatings for combating marine biofouling. *Progress in Polymer Science* 39: 1017–1042.

Yarbrough, J.C., J.P. Rolland, J.M. DeSimone, M.E. Callow, J.A. Finlay, and J.A. Callow. 2006. Contact angle analysis, surface dynamics, and biofouling characteristics of cross-linkable, random perfluoropolyether-based graft terpolymers. *Macromolecules* 39: 2521–2528.

Zervakis, V., G. Papadoniou, C. Tziavos, and A. Lascaratos. 2003. Seasonal variability and geostrophic circulation in the eastern Mediterranean as revealed through a repeated XBT transect. *Annals of Geophysics* 21: 33–47.

Zhang, H., R. Lamb, and J. Lewis. 2005. Engineering nanoscale roughness on hydrophobic surface—Preliminary assessment of fouling behaviour. *Science and Technology of Advanced Materials* 6: 236–239.

28

Predicting the Impacts of Ocean Acidification: Approaches from an omics Perspective

Dhinakarasamy Inbakandan

CONTENTS

28.1 Introduction to Marine omics

Marine omics is the application of omics technologies, including genomics, transcriptomics, proteomics, and metabolomics, for enhanced understanding of the association of living resources with marine and costal environmental on par with genetic factors, toxicity mechanisms, and modes of action in response to both acute and chronic exposure to environmental changes and, in the long term, development of diseases or disorders caused or influenced by these exposures. Marine omics is still in the early stage of omics data collection and validation of molecular profiles for identifying mechanisms of adaptation, assimilation, toxicity signatures, biomarkers, and pathways after exposure to marine environmental changes. Marine omics research can be roughly divided into three categories (Figure 28.1). The first category focuses mainly on global climate change and environmental monitoring, enabling risk assessment. The second category focuses on health outcomes of living resources and environmental impacts, while the last category focuses on ecological functions and environmental adaptation. The primary goal of these research fields is similar, namely, to identify molecular changes, especially changes at the expression levels of mRNA, proteins, and metabolites, in cells or tissues exposed to global and local environmental changes, and to relate these molecular changes to ecological and health outcomes (Ge et al. 2013).

Additionally, omics approaches have an impact on marine ecological monitoring and sustainability through high throughput and simultaneous measurement of multiple genomic, proteomic, and metabolomic profiles and parameters in a given system under defined marine environmental conditions. Marine omics, at its early stage of omics data collection, is transitioning from "profiling omics" to "functional omics." Beyond providing a list of molecular changes, omics approaches emphasize the biological significances of the identified molecular changes, which are key to the success of marine environmental omics development. There is a future need for marine environmental functional omics approaches that may help to solve global environmental challenges, such as the rise in

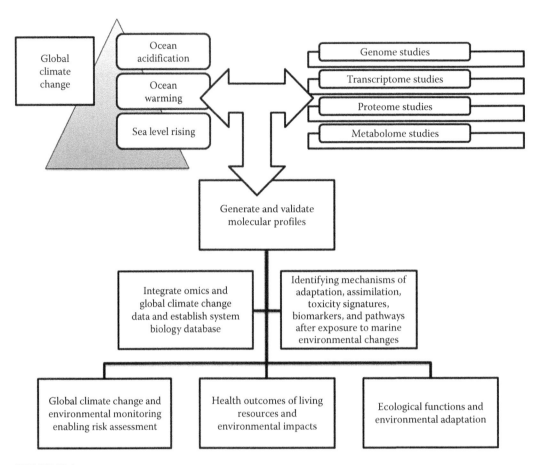

FIGURE 28.1

Diagrammatic representation of marine omics and its application to understand "ocean acidification, ocean warming, and sea level rising" at the global level.

sea level and sea surface temperature, global warming, and ocean acidification. Other needs identified for marine environmental omics are the combined analysis of multiple omics data sets, including genomics, proteomics, and metabolomics, and the integration of omics data with global ocean climate change data to define the links between omics data and the impact of ocean acidification. The integration of omics data with classical ecology end points and biological observation will allow more sensitive and earlier detection of adverse effects of climate change in the marine environment.

Genes, mRNA, proteins, and metabolites are frequently modified in response to alteration of pH or to the acidified ocean waters. It has been widely recognized that knowledge of a gene, mRNA, or protein and its sequence is just a prelude to understanding the role of that molecule and its products within the cell and is only a starting point for a full description of its function. While the modified form of the molecule is essential to the understanding of molecular functions and mechanisms, more detailed studies are expected on the modified forms of the genes and functional molecules at different concentrations of carbon dioxide (CO_2), which were chosen to reflect the CO_2 emission scenarios predicted by the Intergovernmental Panel on Climate Change (2007). Oxidative stress and protein oxidation are examples that stress the importance, challenges, and directions of omics applications for the identification of gene and protein modifications. Several studies have already examined the effects of ocean acidification on calcification for tropical marine organisms (Hoegh-Guldberg et al. 2007), but future studies should include the physiological effects of hypercapnia on cellular pH and various metabolic pathways including ROS production and apoptosis (Abele et al. 2011). Therefore, using proteomic technologies to detect protein changes and modifications in response to ocean acidification-mediated oxidative stresses has been an important research topic. Redox proteomics is probably one of the most extensively studied areas for the characterization of protein modifications under acidification stresses.

Although technologies for the detection of oxidized proteins have been advanced recently and offer some advantages for identifying oxidative stress biomarkers, studies on the identification of oxidized protein biomarkers to determine acidification-induced oxidative stress and injury are still severely lacking. As proteomics studies progress, the goal will not only be to identify proteins in mixtures but also to derive more information about protein modifications from the samples analyzed. The ability to identify oxidized proteins will yield in-depth information on protein structure and function, which, in turn, can facilitate the pace of studies to understand ocean pH–altering processes, pathways, and mechanisms of global climate changes. In addition to protein modifications, RNA and metabolite modifications could also function as mechanistic linkages between altered acidified environment and outcomes and serve as important biomarkers of ocean pH change and effects. However, little is known about RNA and metabolite modifications.

28.2 What Is Ocean Acidification?

CO_2 is being released into the atmosphere of our living planet at an alarming rate due to the industrial revolution and human activities, with records showing a higher level of increase compared with the past years. Coupled with the issues of global warming, the rise in CO_2 level has an adverse effect on the chemical composition of the world's oceans. It is estimated and recorded that 25%–30% of the atmospheric CO_2 has been dissolved in the global oceanic waters. This reaction increases the concentration of hydrogen ions and decreases the concentration of carbonate ions. This major deviation, called ocean acidification, was predicted to cause a reduction in global oceans' pH by 0.1 units from preindustrial times and then expected to decrease by 0.45 units in the next century. But as per today's records, the predicted change in global ocean pH is higher and faster than any experienced occurrence in the last 300 million years.

Sea surface temperature and acidity are changing rapidly at an unprecedented rate. These changes in the ocean environment have a great impact on marine biodiversity. Rise in ocean pH will affect marine living resources that may fail to survive or acclimatize and adapt to the global ocean changes. The fate of marine organisms has been forecasted by several experiments and documented to know the behavior, adaptation, and acclimatization to the predicted acidified environment (Table 28.1). Thus, the challenges for the marine biodiversity in the acidified ocean have been focused on as per the following criteria: (1) changes in the physiological mechanisms to tolerate the acidification stress; (2) interaction of other stressors constrained with the adaptive response to ocean pH change; (3) variations in populations due to acidification and its impact on other populations; (4) change in physiology due to the change in mean environmental pH, which is interlinked with an organism's body pH; and (5) spatiotemporal variation in ocean pH and its impact on gene flow and organismal lifespan.

TABLE 28.1

List of Documentations on Genomics- and Proteomics-Based Approaches to Understand
the Impact of Predicted Ocean Acidification

Experimental Animal Used	Scientific Name	Objectives of the Experiments	References
European eelpout	*Zoarces viviparus*	Role of Na^+/HCO_3^- cotransporter NBC1 and Cl^-/HCO_3^- exchanger AE1 in gills	Deigweiher et al. (2008)
Purple sea urchin	*Strongylocentrotus purpuratus*	DNA microarrays and quantitative PCR of calcification responses	Hofmann et al. (2008)
Sea urchins	*Lytechinus pictus*	Biomineralization, ion regulation, and acid–base balance pathways	O'Donnell et al. (2009)
Red sea urchins	*Strongylocentrotus franciscanus*	Expression of the molecular chaperone hsp70	O'Donnell et al. (2009)
Purple sea urchin	*Strongylocentrotus purpuratus*	Transcriptomic response to biomineralization, cellular stress response, metabolism, and apoptosis	Todgham and Hofmann (2009)
Coral	*Acropora formosa*	Photoprotective role of photorespiration within dinoflagellates	Crawley et al. (2010)
Red abalone	*Haliotis rufescens*	Genes involved in shell formation	Zippay and Hofmann (2010)
Eastern oyster	*Crassostrea virginica*	Gene expression responses to temperature and pH	Chapman et al. (2011)
Baltic sea green crab	*Carcinus maenas*	Gene expression involved in osmoregulation and acid–base regulation and cellular stress response	Fehsenfeld et al. (2011)
Common cuttlefish	*Sepia officinalis*	In situ hybridization and expression patterns of mRNA	Hu et al. (2011)
Diatom	*Thalassiosira pseudonana*	Expression of δ-carbonic anhydrase and small subunit of RuBisCO	Katharine et al. (2011)
Sea urchin	*Paracentrotus lividus*	Proteins involved in development and biomineralization	Martin et al. (2011)
Coccolithophore	*Emiliania huxleyi*	Genes of calcification (alpha- and gamma-CA, Ca_2^+ channel, and Gene description: polysacharide-associated, calcium-binding protein (gpa))	Richier et al. (2011)
Purple sea urchin	*Strongylocentrotus purpuratus*	Expression of metabolism, calcification, and ion regulation genes	Stumpp et al. (2011)
Barnacle	*Balanus amphitrite*	Proteins involved in phosphorylation and glycosylation	Wong et al. (2011)
Oyster larvae	*Crassostrea gigas*	Global protein expression with a decrease or loss of proteins	Dineshram et al. (2012)
Shore crabs	*Carcinus maenas*	1H NMR spectroscopy of hemolymph, gills, and leg muscle	Hammer et al. (2012)
Purple sea urchin	*Strongylocentrotus purpuratus*	Transcript-level responses of spiculogenesis	Hammond and Hofmann (2012)
Sea urchin	*Hemicentrotus pulcherrimus*	Expression of the spicule protein matrix genes SM30 and SM50	Kurihara et al. (2012)
Pearl oyster	*Pinctada fucata*	Gene expression patterns of aspein, calmodulin, nacrein, she-7-F10, and hsp70	Liu et al. (2012)
Corals	*Acropora millepora*	Expression of heat shock proteins Bcl-2 family members	Moya et al. (2012)

(Continued)

TABLE 28.1 (*Continued*)

List of Documentations on Genomics- and Proteomics-Based Approaches to Understand the Impact of Predicted Ocean Acidification

Experimental Animal Used	Scientific Name	Objectives of the Experiments	References
Coccolithophore	*Emiliania huxleyi*	Microarray-based transcriptome profiling of signal transduction and ion transport	Rokitta et al. (2012)
		Acclimation or adaptation to calcification responses	Benner et al. (2013)
Sea urchin	*Echinoplutei*	Arm growth response	Byrne et al. (2013)
Marine metazoan	*Platynereis dumerilii*	Metabolic plasticity and adaptation	Calosi et al. (2013)
Hermatypic coral	*Stylophora pistillata*	Proteomics of biomineralization-related proteins	Drake et al. (2013)
Purple sea urchin	*Strongylocentrotus purpuratus*	Transcriptional response to present day pH regimes	Evans et al. (2013)
Brittle star	*Amphiura filiformis*	Gene expression of acid–base and metabolic genes	Hu et al. (2013)
Blue mussel	*Mytilus edulis*	Mantle gene expression patterns	Hüning et al. (2013)
Purple sea urchin	*Strongylocentrotus purpuratus*	Larval development, skeletal growth, metabolism, and patterns of gene expression	Padilla-Gamino et al. (2013)
		Genetic variation	Pespeni et al. (2013)
Reef coral	*Pocillopora damicornis*	Respiration and RuBisCO protein expression	Putnam et al. (2013)
Japanese rice fish	*Oryzias latipes*	Acid–base regulation and metabolism	Tseng et al. (2013)
Scleractinian coral	*Pocillopora damicornis*	Transcriptomic modifications	Vidal-Dupiol et al. (2013)
Gastropod	*Patella vulgata*	Transcriptomics of mantle tissue	Werner et al. (2013)
Atlantic halibut	*Hippoglossus hippoglossus*	Energy metabolism proteins	De Souza et al. (2014)
Bluegill	*Lepomis macrochirus*	Tissue-specific gene expression	Dennis et al. (2014)
Silver carp	*Hypophthalmichthys molitrix*	Tissue-specific gene expression	Dennis et al. (2014)
Antarctic sea urchin	*Sterechinus neumayeri*	Transcriptome and the mitogenome architecture	Dilly et al. (2015)
Purple sea urchin	*Strongylocentrotus purpuratus*	Biomineralization	Evans and Wynn (2014)
Eastern oyster	*Crassostrea virginica*	Calcification rates, sizes, shell thicknesses, metamorphosis, RNA:DNA ratios, and lipid contents	Gobler and Talmage (2014)
		Metal accumulation, intracellular ATP/ubiquitin-dependent protein degradation, stress response, and energy metabolism	Götze et al. (2014)
Boreal spider crab	*Hyas araneus*	Acid–base, metabolic, and stress-related processes	Harms et al. (2014)
Squid	*Sepioteuthis lessoniana*	Branchial acid–base regulation	Hu et al. (2014)
Hard shell clam	*Mercenaria mercenaria*	Metabolism and immune-related functions	Ivanina et al. (2014)

(Continued)

TABLE 28.1 (*Continued*)

List of Documentations on Genomics- and Proteomics-Based Approaches to Understand the Impact of Predicted Ocean Acidification

Experimental Animal Used	Scientific Name	Objectives of the Experiments	References
Intertidal snail	*Concholepas concholepas*	Phenotypic adaptation	Lardies et al. (2014)
Chinese mitten crab	*Eriocheir sinensis*	Calmodulin (multifunctional calcium sensor protein) expression	Li et al. (2014)
Coccolithophores	*Emiliania huxleyi*	pH regulation, carbon transport, calcification, and photosynthesis	Lohbeck et al. (2014)
Atlantic herring	*Clupea harengus*	Larva's proteome expression pattern	Maneja et al. (2014)
Atlantic cod	*Gadus morhua*	Endocrine and biotransformation systems	Olsen et al. (2014)
Cyanobacteria	*Synechocystis* sp.	Proteomic and metabolomic characterization of acid-response network	Ren et al. (2014)
Orange-spotted groupers	*Epinephelus coioides*	Growth rate and/or growth hormones	Shao et al. (2014)
Cold-water corals	*Desmophyllum dianthus*	Calcification and respiration	Carreiro-Silva et al. (2014)
Marine cyanobacterium	*Trichodesmium erythraeum*	Nitrogen acquisition, glutamine synthetase activity, C-uptake rates, intracellular adenosine triphosphate (ATP) concentration	Spungin et al. (2014)
Calanoid copepod	*Pseudocalanus acuspes*	Transgenerational effects	Thor and Dupont (2015)
Primary producers	*Alexandrium tamarense*	Paralytic shellfish poisoning toxin content	Van de Waal et al. (2014)
Oyster	*Crassostrea gigas*	Integrated proteomic and metabolomic approaches	Wei et al. (2015)
Phaeophyte kelp	*Saccharina japonica*	CO_2 concentration mechanism	Ye et al. (2014)
Abalone	*Haliotis iris*	Growth, metabolism, and biomineralization	Cunningham et al. (2015)
Mediterranean gilt-head sea bream	*Sparus aurata*	Expression patterns of hsp70 and hsp90	Feidantsis et al. (2015)
Two coral species	*Pocillopora damicornis* and *Oculina patagonica*	Ocean acidification induce tissue-specific apoptosis	Kvitt et al. (2015)
Corals	*Acropora millepora*	Heat shock proteins and heat shock factors	Moya et al. (2015)
Sydney rock oysters	*Saccostrea glomerata*	Energy production, cellular stress responses, the cytoskeleton, protein synthesis, and cell signaling	Thompson et al. (2015)
Oyster	*Crassostrea gigas*	Integrated proteomic and metabolomic approaches	Wei et al. (2015)
Cyanobacteria	*Microcystis strains*	Microcystin production and mcyD gene expression	Yu et al. (2015)

28.3 Glimpses on the Genomic and Proteomic Approaches toward Ocean Acidification

To fully understand the significant factors of the effects of ocean acidification on an individual organism's physiology, the scientific research community attempted to predict the effects of ocean acidification on marine ecosystems. Hofmann et al. (2008) explored the use of genomics-based tools that measure gene expression—DNA microarrays and quantitative PCR—and found they could assist in the effort and revealed aspects of how calcifying marine organisms would respond to ocean acidification. They provided a brief overview of biomineralization processes in corals and purple sea urchin larvae (*Strongylocentrotus purpuratus*) and then linked those pathways to ways in which gene expression analysis could reveal physiological responses and mechanisms and, further, could define new testable hypotheses.

Benthic eelpout *Zoarces viviparus* was acclimated to 10,000 ppm CO_2 by Deigweiher et al. (2008) to describe the role of HCO_3^- transporter family (Na^+/HCO_3^- cotransporter NBC1 and Cl^-/HCO_3^- exchanger AE1) in gills of marine fish. Hypercapnia did not affect whole animal oxygen consumption over a period of 4 days. During a time series of 6 weeks, NBC1 mRNA levels first decreased by about 40% (8–24 h) but finally increased threefold over control. mRNA expression of AE1 decreased transiently by 50% at day 4 but recovered to control levels only. Reduced mRNA levels were also found for two Na^+/H^+ exchangers (NHE1A and NHE1B) during the first days (by 50%–60% at days 1 and 2), followed by restoration of control levels. This pattern was mirrored in a slight decrease of NHE1 protein contents and its subsequent recovery. In contrast, Na^+-K^+-ATPase (NKA) mRNA and protein contents, as well as maximum activity, rose steadily from the onset of hypercapnia and reached up to twofold control levels at the end. These results indicated shifting acclimation patterns between short- and long-term CO_2 exposures. Overall, ion gradient–dependent transporter mRNA levels were transiently downregulated in the beginning of the disturbance. Upregulation of NBC1 on long time scales stresses the importance of this transporter in the hypercapnia response of marine teleosts. Long-term rearrangements included NKA at higher densities and capacities, indicating a shift to elevate rates of ion and acid–base regulation under environmental hypercapnia.

Gene expression profiling with a DNA microarray was employed by O'Donnell et al. (2009) to understand the consequences of ocean acidification at the community and ecosystem levels that involve range shifts and population declines using morphometric methods. They explored the effects of elevated CO_2 conditions on *Lytechinus pictus* echinoplutei, which form a calcium carbonate endoskeleton during pelagic development. Larvae were raised from fertilization to the pluteus stage in seawater with elevated CO_2. Morphometric analysis showed the significant effects of enhanced CO_2 on both size and shape of larvae. Also, larvae grown in a high CO_2 environment were smaller and had a more triangular body than those reared in normal CO_2 conditions. Gene expression profiling showed that genes central to energy metabolism and biomineralization were downregulated in the larvae in response to elevated CO_2, whereas only a few genes involved in ion regulation and acid–base balance pathways were upregulated. Taken together, these results suggest that, although the larvae are able to form an endoskeleton, development at elevated CO_2 levels has consequences for larval physiology as shown by changes in the larval transcriptome.

Ocean acidification interferes with the fixation of calcium carbonate to form shells or calcified skeletons, and future ocean chemistry may significantly alter the physiology of calcifying marine organisms. These alterations may manifest themselves directly in the calcification process or have synergistic effects with other environmental factors such as elevated temperatures. To understand these consequences, O'Donnell et al. (2009) raised sea urchins (*Strongylocentrotus franciscanus*) under conditions simulating future atmospheric CO_2 levels of 540 and 970 ppm. When larvae raised under elevated CO_2 conditions were subjected to 1 h acute temperature stress, their ability to mount a physiological response (as measured by the expression of the molecular chaperone hsp70) was reduced relative to those raised under ambient CO_2 conditions. Their results represented the first use of gene expression assays to study the effects of OA on sea urchin development. They highlighted the importance of looking at multiple environmental factors simultaneously as this approach might reveal previously unsuspected biological impacts of atmospheric changes.

Microarray-based transcriptomic analysis was carried out by Todgham and Hofmann (2009) to know the physiological response of larvae of a calcifying marine invertebrate, the purple sea urchin, *S. purpuratus*, to CO_2-driven seawater acidification. In laboratory-based cultures, larvae were raised under conditions approximating current ocean pH conditions (pH 8.01) and at projected, more acidic pH conditions (pH 7.96 and 7.88) in seawater aerated with CO_2 gas. Targeting expression of 1000 genes involved in several biological processes, this study captured changes in gene expression patterns that characterize the transcriptomic response to CO_2-driven seawater acidification of developing sea urchin larvae. In response to both elevated CO_2 scenarios, larvae underwent broad-scale decreases in gene expression in four major cellular processes: biomineralization, cellular stress response, metabolism, and apoptosis. This study underscored that physiological processes beyond calcification were impacted greatly, suggesting that overall physiological capacity and not just a singular focus on biomineralization processes was essential for forecasting the impact of future CO_2 conditions on marine organisms. Conducted on targeted and vulnerable species, genomics-based studies, such as the one highlighted here, have the potential to identify potential "weak links" in physiological function that might ultimately determine an organism's capacity to tolerate future ocean conditions.

Ocean acidification is expected to lower the net accretion of coral reefs; yet, little is known about its effect on coral photophysiology. This study investigated the effect of increasing CO_2 on photosynthetic capacity and photoprotection in *Acropora formosa* by Crawley et al. (2010). Photoprotective role of photorespiration within dinoflagellates (genus *Symbiodinium*) has largely been overlooked due to the focus on the presence of a carbon-concentrating mechanism despite the evolutionary persistence of a form II RuBisCO. The photorespiratory fixation of oxygen produces phosphoglycolate that would otherwise inhibit carbon fixation through the Calvin cycle if it were not converted to glycolate by phosphoglycolate phosphatase (PGPase). Glycolate is then either excreted or dealt with by enzymes in the photorespiratory glycolate and/or glycerate pathways adding to the pool of carbon fixed in photosynthesis. Crawley and team found that CO_2 enrichment led to enhancing photoacclimation (increased chlorophyll a per cell) to subsaturating light levels. Light-enhanced dark respiration per cell and xanthophyll de-epoxidation increased, with resultant decreases in photosynthetic capacity (Pn_{max}) per chlorophyll. The conservative CO_2 emission scenario (A1B; 600–790 ppm) led to a 38% increase in the Pn_{max} per cell, whereas the "business-as-usual" scenario (A1F1; 1160–1500 ppm) led to a 45% reduction in PGPase expression and no change in Pn_{max} per cell. These findings support an important functional role for PGPase in dinoflagellates that were potentially compromised under CO_2 enrichment.

In laboratory experiments designed to mimic seawater chemistry in future oceans, Zippay and Hofmann (2010) examined the effect of pH reduction, driven by CO_2 acidification of seawater, on larvae of the red abalone, *Haliotis rufescens*. Following development under CO_2-acidified conditions, they measured two indicators of physiological response to low pH in four stages of larval development: (1) tolerance of acute thermal challenges and (2) quantitative real-time polymerase chain reaction–determined expression of two genes involved in shell formation (engrailed and ap24). Their results showed that low pH (pH 7.87 vs. 8.05 for control treatments) had a significant effect and decreased larval thermal tolerance in some developmental stages (pretorsion and late veliger) but not in others (posttorsion and premetamorphic veligers). In contrast to the thermal tolerance data, decreased pH did not affect the expression pattern of the two shell formation genes in any of the abalone larval stages. The results indicated that larval stages were differentially sensitive to low pH conditions, and this variability may play into the resilience of individual species to withstand environmental change in the longer term.

To investigate the sensitivity of prefeeding (2 days postfertilization) and feeding (4 and 7 days postfertilization) pluteus larvae, Stumpp et al. (2011) raised *S. purpuratus* embryos in control (pH 8.1 and pCO_2 41 Pa, e.g., 399 microatmospheres [µatm]) and CO_2-acidified seawater at a pH of 7.7 (pCO_2 134 Pa, e.g., 1318 µatm). They observed the growth, calcification, and survival of the embryos. At three time points (day 2, day 4, and day 7 postfertilization), they measured the expression of 26 representative genes important for metabolism, calcification, and ion regulation using quantitative real-time reverse transcription-PCR (RT-qPCR). They found an upregulation of metabolic genes (ATP-synthase, citrate synthase, pyruvate kinase, and thiolase) and downregulation of calcification-related genes (msp130, SM30B, and SM50). Thus, they concluded that a stressor induces an alteration in the speed of development, which was crucial to employ experimental designs with a high time resolution to correct for developmental artifacts.

To understand how ocean acidification would affect the initial life stages of the sea urchin *Paracentrotus lividus*, a common species that is widely distributed in the Mediterranean Sea and the NE Atlantic was studied and observed by Martin et al. (2011). The effects of decreased pH (elevated pCO_2) were investigated through physiological and molecular analyses on both embryonic and larval stages. Eggs and larvae were reared in Mediterranean seawater at six pH levels, that is, pHT 8.1, 7.9, 7.7, 7.5, 7.25, and 7.0. Fertilization success, survival, growth, and calcification rates were monitored over a 3-day period. The expression of genes coding for key proteins involved in development and biomineralization was also monitored. *P. lividus* appeared to be extremely resistant to low pH, with no effect on fertilization success or larval survival. Larval growth slowed down when exposed to low pH but with no direct impact on relative larval morphology or calcification down to pHT 7.25. Consequently, at a given time, larvae exposed to low pH were present at a normal but delayed larval stage. More surprisingly, candidate genes involved in development and biomineralization were upregulated by factors of up to 26 at low pH. Their results revealed plasticity at the gene expression level that allows a normal, but delayed, development under low pH conditions.

To promote the understanding of the responses of green crab acid–base regulatory epithelia to high pCO_2, Baltic Sea green crabs, *Carcinus maenas*, were exposed to a pCO_2 of 400 Pa by Fehsenfeld et al. (2011). Gills were screened for differentially expressed gene transcripts using a 4462-feature microarray and quantitative real-time PCR. Crabs responded mainly through fine-scale adjustment of gene expression to elevate pCO_2. Most of the genes known to code for proteins involved in osmoregulation and acid–base regulation, and cellular stress response, were not impacted by elevated pCO_2. However, after 1 week of exposure, significant changes were detected in a calcium-activated chloride channel, a hyperpolarization-activated nucleotide-gated potassium channel, a tetraspanin, and an integrin. Furthermore, a putative syntaxin-binding protein, a protein of the transmembrane 9 superfamily, and a Cl^-/HCO_3^- exchanger of the SLC 4 family were differentially regulated. These genes were also affected in a previously published hypoosmotic acclimation response study. Their moderate but specific response of *C. maenas* gill gene expression indicated that (1) seawater acidification did not act as a strong stressor on the cellular level in gill epithelia, (2) the response to hypercapnia was to some degree comparable to a hypoosmotic acclimation response, (3) the specialization of each of the posterior gill arches might go beyond what has been demonstrated to date, and (4) a reconfiguration of gill epithelia might occur in response to hypercapnia.

Benthic marine invertebrates, barnacle species, *Balanus amphitrite*, were cultured from nauplius to the cyprid stage in the present (control) and in the projected elevated concentrations of CO_2 for the year 2100 by Kelvin et al. (2011). Cyprid response to ocean acidification was analyzed at the total proteome level and two protein posttranslational modification (phosphorylation and glycosylation) levels using a two-dimensional gel electrophoresis (2-DE)-based proteomic approach. The cyprid proteome showed ocean acidification-driven changes. Proteins that were differentially upregulated or downregulated by ocean acidification come from three major groups, namely, those related to energy metabolism, respiration, and molecular chaperones, illustrating a potential strategy that the barnacle larvae may employ to tolerate ocean acidification stress. The differentially expressed proteins were tentatively identified as responsive to ocean acidification, effectively creating unique protein expression signatures for ocean acidification scenario of 2100. This study showed the promise of using a sentinel and nonmodel species to examine the impact of ocean acidification at the proteome level.

Changes in gene expression of eastern oyster *Crassostrea virginica*, associated with the physicochemical conditions and metal contaminants in their environment, were studied and evaluated by Chapman et al. (2011). The results indicated that transcript signatures could effectively disentangle the complex interactive gene expression responses to the environment and were also capable of disentangling the complex dynamic effects of environmental factors on gene expression. Thus, mapping of environment for gene and gene to environment was reciprocal and mutually reinforcing. In general, the response of transcripts to the environment was driven by major factors known to affect oyster physiology, such as temperature, pH, salinity, and dissolved oxygen, with pollutant levels playing a relatively small role, at least within the range of concentrations found in the oyster habitats studied. Further, two environmental factors that dominate these effects (temperature and pH) interacted in a dynamic and nonlinear fashion to impact gene expression. Transcriptomic data obtained in their study provided insights into the mechanisms of physiological responses to temperature and pH in oysters that were consistent with the known effects of these factors on physiological functions of ectotherms and indicated the important linkages between transcriptomics and physiological outcomes. Thus, these linkages would be grasped in further studies and other organisms and might provide a novel-integrated approach for assessing the impacts of climate change, ocean acidification, and anthropogenic contaminants on aquatic organisms via relatively inexpensive microarray platforms.

To demonstrate that Na^+/K^+-ATPase (soNKA), a V-type H^+-ATPase (soV-HA), and Na^+/HCO_3^- cotransporter (soNBC) were colocalized in NKA-rich cells in the gills of *Sepia officinalis*, in situ hybridization and immunohistochemical methods were used and reported by Hu et al. (2011). Expression patterns of mRNA of these transporters and selected metabolic genes were examined in response to moderately elevated seawater pCO_2 (0.16 and 0.35 kPa) over a time course of 6 weeks in different ontogenetic stages. The applied CO_2 concentrations were relevant for ocean acidification scenarios projected for the coming decades. By their results, they determined strong expression changes in late-stage embryos and hatchlings, with one to three log_2-fold reductions in soNKA, soNBCe, socCAII, and COX. In contrast, no hypercapnia-induced changes in mRNA expression were observed in juveniles during both short- and long-term exposure. However, a transiently increased ion regulatory demand was evident during the initial acclimation reaction to elevated seawater pCO_2. Gill NKA activity and protein concentration were increased by 15% during short-term (2–11 days) but not during long-term (42-days) exposure. Their findings supported the hypothesis that the energy budget of adult cephalopods was not significantly compromised during long-term exposure to moderate environmental hypercapnia. However, the downregulation of ion regulatory and metabolic genes in late-stage embryos, taken together with a significant reduction in somatic growth, indicated that cephalopod early life stages were challenged by elevated seawater pCO_2.

The effect of ocean acidification conditions has been investigated in cultures of the diatom *Thalassiosira pseudonana* CCMP1335 by Katharine et al. (2011). Expected end-of-the-century pCO_2 (aq) concentrations of 760 µatm (equivalent to pH 7.8) were compared with present-day conditions (380 µatm CO_2, pH 8.1). Batch culture pH changed rapidly because of CO_2 (aq) assimilation, and pH targets of 7.8 and 8.1 could not be sustained. Long-term (100 generation) pH-auxostat, continuous cultures could be maintained at target pH when cell density was kept low (<2 × 105 cells mL^{-1}). After 3 months of continuous culture, the C:N ratio slightly decreased under high CO_2 conditions and red fluorescence per cell slightly increased. However, no change was detected in the photosynthetic efficiency (Fv/Fm) or functional cross section of PS II (σPSII). Elevated pCO_2 has been predicted to be beneficial to diatoms due to reduced cost of carbon concentration mechanisms. There was reduced transcription of one putative δ-carbonic anhydrase (CA-4) after 3 months growth at increased CO_2, but three other δ-CAs and the small subunit of RuBisCO showed no change. There was no evidence of adaptation or clade selection of *T. pseudonana* after ~100 generations at elevated CO_2. On the basis of this long-term culture, pH change of this magnitude in future oceans might have little effect on *T. pseudonana* in the absence of genetic adaptation.

Changes in gene expression induced by variations in pH/pCO_2 in the widespread and abundant coccolithophore *Emiliania huxleyi* were reported by Richier et al. (2011). Batch cultures were subjected to increased partial pressure of CO_2 (pCO_2; i.e., decreased pH), and the changes in expression of four functional gene classes directly or indirectly related to calcification were investigated. Increased pCO_2 did not affect the calcification rate and only CA transcripts exhibited a significant downregulation. They observed that elevated pCO_2 induced only limited changes in the transcription of several transporters of calcium and bicarbonate and gave new significant elements to understand cellular mechanisms underlying the early response of *E. huxleyi* to CO_2-driven ocean acidification.

Whole-transcriptome analysis was used to compare the effects of such "acute" exposure to elevated pCO_2 with a longer period of exposure beginning immediately postfertilization in juvenile corals (*Acropora millepora*) by Moya et al. (2012). Far fewer genes were differentially expressed under the 9 days treatment, and although the transcriptome data implied wholesale disruption of metabolism and calcification genes in the acute treatment experiment, expression of most genes was at control levels after prolonged treatment. There was little overlap between the genes responding to the acute and prolonged treatments, but heat shock proteins (hsps) and heat shock factors (HSFs) were overrepresented among the genes responding to both treatments. Among these was an hsp70 gene previously shown to be involved in acclimation to thermal stress in a field population of another acroporid coral. The most obvious feature of its molecular response to a 9-day treatment experiment was the upregulation of five distinct Bcl-2 family members, the majority predicted to be antiapoptotic. They suggested that an important component of the longer term response to elevate CO_2 is the suppression of apoptosis. It therefore appeared that juvenile *A. millepora* have the ability to rapidly acclimate to elevated pCO_2, a process mediated by the upregulation of specific hsps and a suite of Bcl-2 family members.

The synergistic effects of elevated seawater temperature and declined seawater pH on gene expression patterns of aspein, calmodulin, nacrein, she-7-F10, and hsp70 in the pearl oyster *Pinctada fucata* were studied and demonstrated by Liu et al. (2012) under four treatments: (1) ambient pH (8.10) and ambient temperature (27°C) (control condition), (2) ambient pH and elevated temperature (+3°C), (3) declined pH (7.70) and ambient temperature, and (4) declined pH and elevated temperature. Their results revealed that under warming and acidic seawater conditions, expression of aspein and calmodulin showed no significant differences among different time points in condition 8.10 T. But the levels of aspein and calmodulin in conditions 8.10 T + 3, 7.70 T, and 7.70 T + 3 and levels of nacrein and she-7-F10 in all the four treatments changed significantly. Low pH and pH × temperature interaction influenced the expression of aspein and calmodulin significantly after 48 and 96 h. Significant effects of low pH and pH × temperature interaction on the expression of nacrein were observed at 96 h. The expression level of she-7-F10 was affected significantly by pH after 48 and 96 h. The expression of hsp70 was significantly affected by temperature, pH, temperature × pH interaction at 6 h, and by temperature × pH interaction at 24 h. Their study suggested that declined pH and pH × temperature interaction induced the downregulation of calcification-related genes, and the interaction between declined seawater pH and elevated temperature caused upregulation of hsp70 in *P. fucata*. Their results demonstrated that the declined seawater pH and elevated temperature would impact the physiological process and potentially the adaptability of *P. fucata* to future warming and acidified ocean.

The effects of ocean acidification on the calcification process at the molecular level were analyzed by Kurihara et al. (2012), who evaluated the expression of three biomineralization-related genes in the sea urchin *Hemicentrotus pulcherrimus* exposed under control, 1000, and 2000 ppm CO_2 from egg to pluteus larval stage. They found that the expression of the gene msp130, which was proposed to transport Ca^{2+} to the calcification site, was suppressed by increased CO_2 at pluteus larval stage. Meanwhile, the expression of spicule protein matrix genes SM30 and SM50 was apparently not affected. Their results suggested that the combined effects of ocean acidification on the expression of skeletogenesis-related genes and the change in seawater carbonate chemistry affect the biomineralization ability of sea urchins.

The impact of ocean acidification on photosynthesis and calcification in the coccolithophore *E. huxleyi*, a cosmopolitan calcifier that significantly contributes to the regulation of the biological carbon pumps, was studied by Rokitta et al. (2012). Comparative microarray-based transcriptome profiling was used to screen the underlying cellular processes and allowed to follow up interpretations derived from physiological data. In the diplont, the observed increases in biomass production under ocean acidification were likely caused by the stimulated production of glycoconjugates and lipids. The observed lowered calcification under ocean acidification could be attributed to impaired signal transduction and ion transport. The haplont utilized distinct genes and metabolic pathways, reflecting the stage-specific use of certain portions of the genome. With respect to functionality and energy dependence, however, the transcriptomic ocean acidification responses resembled those of the diplont. In both life-cycle stages, ocean acidification affected the cellular redox state as a master regulator and thereby caused a metabolic shift from oxidative toward reductive pathways, which involved a reconstellation of carbon flux networks within and across compartments. Whereas signal transduction and ion homeostasis appeared equally sensitive to ocean acidification under both light intensities, the effects on carbon metabolism and light physiology were clearly modulated by light availability. These interactive effects could be attributed to the influence of ocean acidification and light on the redox equilibria of NAD and NADP, which function as major sensors for energization and stress. This generic mode of action of ocean acidification might therefore provoke similar cell physiological responses in other protists.

Dineshram et al. (2012) used 2-DE coupled with mass spectrophotometer to compare the global protein expression pattern of oyster larvae (*Crassostrea gigas*) exposed to ambient and to high-CO_2. Exposure to ocean acidification resulted in marked reduction of global protein expression with a decrease or loss of 71 proteins (18% of the expressed proteins in control), indicating a widespread depression of metabolic genes expression in larvae reared under ocean acidification. Thus, the team reported the first proteome analysis that provided insights into the link between physiological suppression and protein downregulation under ocean acidification in oyster larvae.

The metabolic response to elevated levels of CO_2 in shore crabs (*Carcinus maenas*) was reported by Hammer et al. (2012). 1H NMR spectroscopy was used to analyze the extracts of hemolymph, gills, and leg muscle from shore crabs, which were exposed to different levels of CO_2-acidified seawater with pH 7.4, 6.6, and 6.3 ($pCO_2 \sim$ 2,600, 16,000, and 30,000 µatm, respectively) for 2 weeks (level-dependent exposure). In addition, the metabolic response was followed for up to 4 weeks of exposure to seawater pH 6.9 ($pCO_2 \sim$ 7600 µatm). A partial least-squares regression analysis of data showed an increased differentiation between metabolic fingerprints of controls and exposed groups for all sample types with increasing CO_2 levels. Difference between controls and animals subjected to time-dependent exposure was observed after 4 weeks in the hemolymph and gills and after 48 h of exposure in the leg muscle. Changes in metabolic profiles were mainly due to a reduced level of important intracellular osmolytes such as amino acids (glycine, proline), while the level of other metabolites varied between the different sample types. Their results were similar to what was observed in animals exposed to hypoosmotic stress and might suggest disturbances in intracellular isoosmotic regulation. Their results might also reflect increased catabolism of amino acids to supply the body fluids with proton-buffering ammonia. Alternatively, their findings might reflect an exhaustive effect of CO_2 exposure.

Transcript-level responses to elevated seawater CO_2 during gastrulation and the initiation of spiculogenesis, two crucial developmental processes in the purple sea urchin, *S. purpuratus*, were studied by Hammond and Hofmann (2012). Embryos were reared at the current, accepted oceanic CO_2 concentration of 380 µatm and by the elevated levels of 1000 and 1350 µatm, simulating predictions for oceans and upwelling regions, respectively. The seven genes of interest comprised a subset of pathways in the primary mesenchymal cell gene regulatory network shown to be necessary for the regulation and execution of gastrulation and spiculogenesis. Of the seven genes, qPCR analysis indicated that elevated CO_2 concentrations only had a significant but subtle effect on two genes: one important for early embryo patterning, Wnt8, and the other an integral component in spiculogenesis and biomineralization, SM30b. The protein levels of another spicule matrix component, SM50, demonstrated significant variable responses to elevated CO_2. These data linked the regulation of crucial early developmental processes with the environment that these embryos would be developing within, situating the study of organismal responses to ocean acidification in a developmental context.

Benthic habitats were observed by Hu et al. (2013) to record information regarding physiological and behavioral responses of benthic marine organisms in seawater acidification. According to their study determination of pO_2 and pCO_2 gradients within the burrows of the brittle star *Amphiura filiformis* during environmental hypercapnia demonstrated that besides hypoxic conditions, increases of environmental pCO_2 were additive to the already high pCO_2 (up to 0.08 kPa) within the burrows. In response to up to 4 weeks exposure to pH 7.3 (0.3 kPa pCO_2) and pH 7.0 (0.6 kPa pCO_2), metabolic rates of *A. filiformis* significantly reduced in pH 7.0 treatments accompanied by increased ammonium excretion rates. Gene expression analyses demonstrated significant reductions of acid–base (NBCe and AQP9) and metabolic (G6PDH, LDH) genes. The determination of extracellular acid–base status indicated an uncompensated acidosis in CO_2-treated animals, which could explain depressed metabolic rates. Metabolic depression was associated with a retraction of filter feeding arms into sediment burrows. Regeneration of lost arm tissues following traumatic amputation was associated with significant increases in the metabolic rate, and hypercapnic conditions (pH 7.0, 0.6 kPa) dramatically reduce the metabolic scope for regeneration reflected in 80% reductions in the regeneration rate. Their results demonstrated that elevated seawater pCO_2 significantly affects the environment and the physiology of infaunal organisms like *A. filiformis*.

The progeny of adult purple sea urchins (*S. purpuratus*) collected from the northeast Pacific region was cultured by Evans et al. (2013) in CO_2-acidified seawater, representing present-day and near-future ocean scenarios, and gene expression was monitored using transcriptomics. Their observations hypothesized that persistent exposure to upwelling during evolutionary history would have selected for increased pH tolerance in sea urchins population and that their transcriptomic response to low pH seawater would provide insight into mechanisms underlying pH tolerance in a calcifying species. Expression patterns revealed that *S. purpuratus* larvae might alter the bioavailability of calcium and adjust skeletogenic pathways to sustain calcification in a low pH ocean, and larvae used different strategies for coping with different magnitudes of pH stress, initiating a robust transcriptional response to present-day pH regimes but a muted response to near-future conditions. Their results concluded that an enhanced capacity to cope with present-day pH variation might not translate into success in future oceans.

The impacts of an 8-week acclimation period to four seawater pCO_2 treatments (39, 113, 243, and 405 Pa/385, 1120, 2400, and 4000 µatm) on mantle gene expression patterns in the blue mussel *Mytilus edulis* from the Baltic Sea were elucidated by Huning et al. (2013). Based on the *M. edulis* mantle tissue transcriptome, the expression of several genes involved in metabolism, calcification, and stress responses was assessed in the outer (marginal and pallial zone) and the inner mantle tissues (central zone) using quantitative real-time PCR. The expression of genes involved in energy and protein metabolism (F-ATPase, hexokinase, and elongation factor alpha) was strongly affected by acclimation to moderately elevated CO_2 partial pressures. Expression of a chitinase, potentially important for the calcification process, was strongly depressed (maximum ninefold), correlating with a linear decrease in shell growth observed in the experimental animals. Interestingly, shell matrix protein candidate genes were less affected by CO_2 in both tissues. A compensatory process toward enhanced shell protection was indicated by a massive increase in the expression of tyrosinase, a gene involved in periostracum formation (maximum 220-fold). Using correlation matrices and a force-directed layout network graph, they were able to uncover possible underlying regulatory networks and the connections between different pathways, thereby providing a molecular basis of observed changes in animal physiology in response to ocean acidification.

To demonstrate the adaptive capacity of species to ocean acidification, Pespeni et al. (2013) cultured purple sea urchins (*S. purpuratus*) under different CO_2 levels, which generated striking patterns of genome-wide selection. They examined genetic change at 19,493 loci in larvae from seven adult populations cultured under realistic future CO_2 levels. Although larval development and morphology showed little response to elevated CO_2, they resulted in substantial allelic change in 40 functional classes of proteins involving hundreds of loci. Pronounced genetic changes, including excess amino acid replacements, were detected in all populations and occurred in genes for biomineralization, lipid metabolism, and ion homeostasis—gene classes that build skeletons and interact in pH regulation. Such genetic change represented a neglected and important impact of ocean acidification that may influence populations that showed few outward signs of response to acidification. Their results demonstrated the capacity for rapid evolution in the face of ocean acidification and showed that standing genetic variations could be a reservoir of resilience to climate change in coastal upwelling ecosystems. However, an effective response to strong natural selection demands large population sizes and might be limited in species impacted by other environmental stressors.

A comprehensive meta-analysis to date by synthesizing the results of 228 studies examines biological responses to ocean acidification. The results revealed decreased survival, calcification, growth, development, and abundance in response to acidification when the broad range of marine organisms is pooled together. However, the magnitude of these responses varied among taxonomic groups, suggesting there was some predictable trait-based variation in sensitivity, despite the investigation of approximately 100 new species in recent research. The results also revealed an enhanced sensitivity of mollusk larvae but suggest that an enhanced sensitivity of early life history stages was not universal across all taxonomic groups. In addition, the variability in species' responses was enhanced when they were exposed to acidification in multispecies assemblages, suggesting that it was important to consider indirect effects and exercise caution when forecasting abundance patterns from single-species laboratory experiments. Furthermore, the results suggested that other factors, such as nutritional status or source population, could cause substantial variations in organisms' responses. Their results also highlighted a trend toward enhanced sensitivity to acidification when taxa were concurrently exposed to elevated seawater temperature.

Over the next century, elevated quantities of atmospheric CO_2 are expected to penetrate into the oceans, causing a reduction in pH (−0.3/−0.4 pH unit in the surface ocean) and the concentration of carbonate ions (so-called ocean acidification). The impacts that this will have on marine and estuarine organisms and ecosystems is of growing concern. Marine shelled mollusks, which colonized a large latitudinal gradient and can be found from intertidal to deep-sea habitats, are economically and ecologically important species providing essential ecosystem services, including habitat structure for benthic organisms, water purification, and a food source for other organisms.

The effects of ocean acidification on the growth and shell production by juvenile and adult shelled mollusks were variable among species and even within the same species, precluding the drawing of a general picture. This was, however, not the case for pteropods, with all species tested so far, being negatively impacted by ocean acidification, as per Gazeau et al. (2013). The blood of shelled mollusks might exhibit lower pH with consequences for several physiological processes (e.g., respiration, excretion, etc.) and, in some cases, increased mortality in the long term. While fertilization might remain unaffected by elevated pCO_2, embryonic and larval development would be highly sensitive to important reductions in size and decreased survival of larvae, increases in the number of abnormal larvae, and an increase in the developmental time. There are big gaps in the current understanding of the biological consequences of an acidifying ocean on shelled mollusks. For instance, the natural variability of pH and the interactions of changes in the carbonate chemistry with changes in other environmental stressors such as increased temperature and changing salinity, the effects of species interactions, and the capacity of the organisms to acclimate and/or adapt to changing environmental conditions are poorly described.

An integrative approach to forecast the impact of future ocean conditions on larval purple sea urchins (*S. purpuratus*) from the northeast Pacific Ocean was taken by Padilla-Gamino et al. (2013). In laboratory experiments that simulated ocean warming and ocean acidification, they examined larval development, skeletal growth, metabolism, and patterns of gene expression using an orthogonal comparison of two temperature (13°C and 18°C) and pCO_2 (400 and 1100 μatm) conditions. Simultaneous exposure to increased temperature and pCO_2 significantly reduced larval metabolism and triggered a widespread downregulation of histone-encoding genes. pCO_2, but not temperature, impaired skeletal growth and reduced the expression of a major spicule matrix protein, suggesting that skeletal growth will not be further inhibited by ocean warming. Importantly, shifts in skeletal growth were not associated with developmental delay. Collectively, their results indicated that global change variables would have additive effects that exceed thresholds for optimized physiological performance in this keystone marine species.

The growth of arm in sea urchin echinoplutei was assessed as a proxy of larval calcification, to increase seawater acidity/pCO_2 and decreased carbonate mineral saturation in a global synthesis of data from 15 species by Byrne et al. (2013). Phylogenetic relatedness did not influence the observed patterns. Regardless of habitat or latitude, ocean acidification impeded larval growth with a negative relationship between arm length and increased acidity/pCO_2 and decreased carbonate mineral saturation. In multiple linear regression models incorporating these highly correlated parameters, pCO_2 exerted the greatest influence on decreased arm growth in the global data set and also in the data subsets for polar and subtidal species. Thus, reduced growth appeared to be largely driven by the organism hypercapnia. For tropical species, decreased carbonate mineral saturation was most important. No single parameter played a dominant role in arm size reduction in the temperate species. For intertidal species, the models were equivocal. Levels of acidification causing a significant reduction in arm growth varied between species. In 13 species, reduction in the length of arms and supporting skeletal rods was evident in larvae reared in near-future (pCO_2 800+ μatm) conditions, whereas greater acidification (pCO_2 1000+ μatm) reduced growth in all species. Although multistressor studies were few, when temperature was added to the stressor mix, near-future warming could reduce the negative effect of acidification on larval growth. Overall, responses of larvae from regions across the world showed similar trends despite disparate phylogeny, environments, and ecology. Larval success might be the bottleneck for species success with flow-on effects for sea urchin populations and marine ecosystems.

The combined effect of pCO_2 and temperature on the coccolithophore *E. huxleyi* for more than 700 generations was studied by Benner et al. (2013). Cells increased inorganic carbon content and calcification rate under warm and acidified conditions compared with ambient conditions, whereas organic carbon content and primary production did not show any change. In contrast to findings from short-term experiments, their results suggested that long-term acclimation or adaptation could change, or even reverse, negative calcification responses in *E. huxleyi* and its feedback to the global carbon cycle. Genome-wide profiles of gene expression using RNA-seq revealed that genes thought to be essential for calcification were not those that were most strongly differentially expressed under long-term exposure to future ocean conditions.

Brooded larvae from the reef coral *Pocillopora damicornis* collected from Nanwan Bay, southern Taiwan, were exposed to ambient or elevated temperature (25°C or 29°C) and pCO_2 (415 or 635 µatm) in a factorial experiment for 9 days by Putnam et al. (2013), and a variety of physiological and molecular parameters were measured. Respiration and RuBisCO protein expression decreased in larvae exposed to elevated temperature, while those incubated at high pCO_2 were larger in size. Collectively, these findings highlighted the complex metabolic and molecular responses of this life history stage and the need to integrate the understanding across multiple levels of biological organization. Their results also suggested that for this pocillopora larval life stage, the impacts of elevated temperature were likely a greater threat under near-future predictions for climate change than ocean acidification.

Little is known of long-term metabolic plasticity and potential for metabolic adaptation in marine ectotherms exposed to elevated pCO_2. Thus, Calosi et al. (2013) carried out a series of in situ transplant experiments using a number of tolerant and sensitive polychaete species living around a natural CO_2 vent system. They observed that a marine metazoan (i.e., *Platynereis dumerilii*) was able to adapt to chronic and elevated levels of pCO_2. The vent population of *P. dumerilii* was physiologically and genetically different from nearby populations that experience low pCO_2, as well as smaller in body size. In contrast, different populations of *Amphiglena mediterranea* showed marked physiological plasticity, indicating that adaptation and acclimatization were both viable strategies for the successful colonization of elevated pCO_2 environments. In addition, sensitive species showed either a reduced or increased metabolism when exposed acutely to elevated pCO_2. Thus it was concluded that the impacts of elevated temperature on pocillopora larval life stage were likely a greater threat under near-future predictions for climate change than ocean acidification.

Gene expression profiling of genes involved in acid–base regulation and metabolism was used to investigate the effects of seawater hypercapnia on developing Japanese rice fish (medaka; *Oryzias latipes*) by Tseng et al. (2013). Their results demonstrated that embryos respond with delayed development during the time window of 2–5 dpf when exposed to a seawater pCO_2 of 0.12 and 0.42 kPa. This developmental delay was associated with strong downregulation of genes from major metabolic pathways including, glycolysis (G6PDH), Krebs cycle (CS), and the electron transport chain (CytC). In a second step, they identified acid–base relevant genes in different ontogenetic stages (embryos, hatchlings, and adults) and tissues (gill and intestine) that were upregulated in response to hypercapnia, including NHE3, NBCa, NBCb, AE1a, AE1b, ATP1a1a.1, ATP1a1b, ATP1b1a, Rhag, Rhbg, and Rhcg. Interestingly, NHE3 and Rhcg expressions increased in response to environmental hypercapnia in all ontogenetic stages and tissues tested, indicating the central role of these proteins in acid–base regulation. Furthermore, the increased expression of genes from amino acid metabolism pathways (ALT1, ALT2, AST1a, AST1b, AST2, and GLUD) suggested that energetic demands of hatchlings were fueled by the breakdown of amino acids. Thus, their study provided a first detailed gene expression analysis throughout the ontogeny of a euryhaline teleost in response to seawater hypercapnia, indicating highest sensitivity in early embryonic stages, when functional ion regulatory epithelia were not yet developed.

Using liquid chromatography–tandem mass spectrometry analysis of proteins extracted from the cell-free skeleton of the hermatypic coral, *Stylophora pistillata*, combined with a draft genome assembly from the cnidarian host cells of the same species, Drake et al. (2013) identified 36 coral skeletal organic matrix proteins. The proteome of the coral skeleton contained an assemblage of adhesion and structural proteins as well as two highly acidic proteins that may constitute a unique coral skeletal organic matrix protein subfamily. They compared 36 skeletal organic matrix protein sequences to genome and transcriptome data from three other corals, three additional invertebrates, one vertebrate, and three single-celled organisms. This work represented a unique extensive proteomic analysis of biomineralization-related proteins in corals from which they identify a biomineralization "toolkit," an organic scaffold on which aragonite crystals could be deposited in specific orientations to form a phenotypically identifiable structure.

The application of illumina sequencing to develop a transcriptome from the adult mantle tissue of gastropod *Patella vulgata* was reported by Werner et al. (2013). They obtained 47,237,104 paired-end reads of 51 bp, trialed de novo assembly methods, and settled on the additive multiple K method followed by redundancy removal as resulting in the most comprehensive assembly. Their results yielded 29,489 contigs of at least 500 bp in length. They then used three methods to search for candidate genes relevant to biomineralization: searches via BLAST and hidden Markov models for homologues of biomineralizing genes from other mollusks, searches for predicted proteins containing tandem repeats, and searches for secreted proteins that lacked a transmembrane domain. From the results of those searches, they selected 15 contigs for verification by RT-PCR, of which 14 were successfully amplified and cloned. These included homologues of Pif-177/BSMP, Perlustrin, SPARC, AP24, Follistatin-like, and CA, as well as three containing extensive G-X-Y repeats as found in nacrein. They selected two for further verification by in situ hybridization, demonstrating expression in the larval shell field. They concluded that de novo assembly of illumina data offers a cheap and rapid route to a predicted transcriptome that could be used as a resource for further biological study.

Global transcriptomic modifications in a scleractinian coral (*P. damicornis*) exposed to pH 7.4 compared to pH 8.1 during a 3-week period were investigated by Vidal-Dupiol et al. (2013). The RNAseq approach showed that 16% of their transcriptome was affected by the treatment with 6% of upregulations and 10% of downregulations. A more detailed analysis suggested that the downregulations were less coordinated than the upregulations and allowed the identification of several biological functions of interest. To better understand the links between these functions and the pH, a transcript abundance of 48 candidate genes was quantified by q-RT-PCR (corals exposed at pH 7.2 and 7.8 for 3 weeks). The combined results of these two approaches suggested that pH $\geq$ 7.4 induced an upregulation of genes coding for proteins involved in calcium and carbonate transport, conversion of CO_2 into HCO_3^-, and organic matrix that might sustain calcification. Concomitantly, genes coding for heterotrophic- and autotrophic-related proteins were upregulated. This reflects that low pH might increase the coral energy requirements, leading to an increase in the energetic metabolism with the mobilization of energy reserves. In addition, uncoordinated downregulations measured could reflect a general trade-off mechanism that might enable energy reallocation.

The immediate physiological response in the coccolithophore *E. huxleyi* to ocean acidification may be partially compensated by evolutionary adaptation, and the underlying molecular mechanisms were studied by Lohbeck et al. (2014). They reported the expression levels of 10 candidate genes putatively relevant to pH regulation, carbon transport, calcification, and photosynthesis in short-term exposure of *E. huxleyi* populations to ocean acidification conditions after acclimation (physiological response) and after 500 genera-tions of high CO_2 adaptation (adaptive response). The physiological response revealed downregulation of candidate genes, well reflecting the concomitant decrease of growth and calcification. In the adaptive response, putative pH regulation and carbon transport genes were upregulated, matching partial restoration of growth and calcification in high CO_2-adapted populations. Adaptation to ocean acidification in *E. huxleyi* likely involved improved cellular pH regulation, presumably indirectly affecting calcification. Adaptive evolution might thus have the potential to partially restore cellular pH regulatory capacity and thereby mitigate adverse effects of ocean acidification.

Global gene expression profiling through RNA sequencing was used by Harms et al. (2014) to study the transcriptional responses to ocean acidification and warming in gills of the boreal spider crab *Hyas araneus* exposed medium term (10 weeks) to intermediate (1120 µatm) and high (1960 µatm) pCO_2 at different temperatures (5°C and 10°C). Their analysis revealed shifts in steady-state gene expression from control to intermediate and from intermediate to high CO_2 exposures. At 5°C, acid–base, energy metabolism, and stress response–related genes were upregulated at intermediate pCO_2, whereas high pCO_2 induced a relative reduction in expression to levels closer to controls. A similar pattern was found at elevated temperature (10°C). There was a strong coordination between acid–base, metabolic, and stress-related processes. Hemolymph parameters at intermediate pCO_2 indicate enhanced capacity in acid–base compensation potentially supported by upregula-tion of a V-ATPase. The likely enhanced energy demand might be met by the upregulation of the electron transport system but might lead to increasing oxidative stress reflected in upregulated antioxidant defense transcripts. These mechanisms were attenuated by high pCO_2, possibly as a result of limited acid–base compensation and metabolic downregula-tion. Their studies indicated a pCO_2-dependent threshold beyond which compensation by acclimation fails progressively. They also indicated a limited ability of this stenoecious crustacean to compensate for the effects of ocean acidification with and without concomi-tant warming.

Meta-analyses were performed by Evans and Wynn (2014) as a means of minimizing experimental discrepancies and resolving broader-scale trends regarding the effects of ocean acidification on gene expression in the purple sea urchin (*S. purpuratus*). Analyses across eight studies and four urchin species largely support prevailing hypotheses about the impact of ocean acidification on marine calcifiers. The predominant expression pattern involved the downregulation of genes within energy-producing pathways, a clear indication of metabolic depression. Genes with functions in ion transport were significantly overrepresented and were most plausibly contributing to intracellular pH regulation. Expression profiles provided extensive evidence for an impact on biominer-alization, epitomized by the downregulation of seven spicule matrix proteins. In con-trast, expression profiles provided limited evidence for CO_2-mediated developmental delay or induction of a cellular stress response. Congruence between studies of gene expression and the ocean acidification literature in general validated the accuracy of gene expression in predicting the consequences of ocean change and justifies its continu-ous use in future studies.

The geographic variation in CO_2 seawater concentrations was examined by Lardies et al. (2014) in the phenotype and in the reaction norm of physiological traits using a laboratory mesocosm approach with short-term acclimation in two contrasting populations (from Antofagasta and Calfuco) of the intertidal snail *Concholepas concholepas*. Their results revealed that elevated pCO_2 conditions increase standard metabolic rates in both popula-tions of the snail juveniles, likely due to the higher energy cost of homeostasis. Juveniles of

C. concholepas in the Calfuco (southern) population showed a lower increment of metabolic rate in high-pCO_2 environments concordant with a lesser gene expression of a hsp with respect to the Antofagasta (northern) population. Grouping these results indicated a negative effect of ocean acidification on whole-organism functioning of *C. concholepas*. Finally, a significant population × pCO_2 level interaction in both studied traits indicated that there was variation between populations in response to high-pCO_2 conditions.

A long-term, 8-month experiment was conducted by Carreiro-Silva et al. (2014) to compare the physiological responses of cold-water corals *Desmophyllum dianthus* to ocean acidification at both the organismal and gene expression levels under two pCO_2/pH treatments: ambient pCO_2 (460 µatm, pHT = 8.01) and elevated pCO_2 (997 µatm, pHT = 7.70). At the organismal level, no significant differences were detected in the calcification and respiration rates of *D. dianthus*. Conversely, significant differences were recorded in gene expression profiles, which showed an upregulation of genes involved in cellular stress (hsp70) and immune defense (mannose-binding c-type lectin). Expression of alpha-CA, a key enzyme involved in the synthesis of the coral skeleton, was also significantly upregulated in corals under elevated pCO_2, indicating that *D. dianthus* was under physiological reconditioning to calcify under these conditions. Thus, gene expression profiles revealed physiological impacts that were not evident at the organismal level. Consequently, understanding the molecular mechanisms behind the physiological processes involved in a coral's response to elevated pCO_2 is critical to assess the ability of cold-water corals to acclimate or adapt to future OA conditions.

The impact of the ocean's CO_2 concentrations on the growth and physiology of larval stages of Eastern oyster, *C. virginica*, was experimented by Gobler and Talmage (2014). *C. virginica* larvae that are grown in present-day pCO_2 concentrations (380 µatm) displayed higher growth and survival than individuals grown at both lower (250 µatm) and higher pCO_2 levels (750 and 1500 µatm). *C. virginica* larvae manifested calcification rates, sizes, shell thicknesses, metamorphosis, RNA:DNA ratios, and lipid contents that paralleled trends in survival, with maximal values for larvae grown at 380 µatm pCO_2 and reduced performance in higher and lower pCO_2 levels. While some physiological differences among oysters could be attributed to CO_2-induced changes in size or calcification rates, the RNA:DNA ratios at ambient pCO_2 levels were elevated, independent of these factors. Likewise, the lipid contents of individuals exposed to high pCO_2 levels were depressed even when differences in calcification rates were considered. These findings revealed the cascading, interdependent impact that high CO_2 can have on oyster physiology. *C. virginica* larvae were significantly more resistant to elevated pCO_2 than other North Atlantic bivalves, such as *Mercenaria mercenaria* and *Argopecten irradians*, a finding that may be related to the biogeography and/or evolutionary history of these species and may have important implications for future bivalve restoration and aquaculture efforts.

Juvenile orange-spotted groupers (*Epinephelus coioides*) were exposed by Shao et al. (2014) to seawater of different levels of acidification: a condition predicted by the Intergovernmental Panel on Climate Change (pH 7.8–8.0), and a more extreme condition (pH 7.4–7.6) that may occur in coastal waters in the near future. After 6 weeks of exposure, the growth rates of fish in pH 7.4–7.6 were less than those raised in control water (pH 8.1–8.3). Furthermore, exposure at pH 7.4–7.6 increased blood pCO_2 and HCO_3^- significantly; exposure at pH 7.8–8.0, meanwhile, did not affect the acid–base chemistry. Moreover, exposure to pH 7.4–7.6 resulted in lower levels of hepatic *igf1* (insulin-like growth factor I) mRNA but did not affect levels of pituitary *gh* (growth hormone) or hypothalamus *psst2* and *psst3* (preprosomatostatin II and III). These results revealed that highly acidified seawater suppresses growth of juvenile grouper, which may be a consequence of reduced levels of IGF-1 but not due to diminished growth hormone release.

The first de novo transcriptome and complete mitochondrial genome of Antarctic sea urchin, *Sterechinus neumayeri*, was sequenced by Dilly et al. (2015). To identify transcripts important to ocean acidification and thermal stress, the transcriptome was created by pooling, and 13 larval samples representing developmental stages on day 11 (late gastrula), 19 (early pluteus), and 30 (mid pluteus) were maintained at three CO_2 levels (421, 652, and 1071 µatm), along with four additional heat-shocked samples. The normalized complementary DNA (cDNA) pool was sequenced using emulsion PCR (pyrosequencing), resulting in 1.34M reads with an average read length of 492 bp. A total of 40,994 isotigs were identified, averaging 1,188 bp with a median coverage of 11×. Additional primer design and gap sequencing were required to complete the mitochondrial genome. The mitogenome of *S. neumayeri* was a circular DNA molecule with a length of 15,684 bp that contains all 37 genes normally found in metazoans. Dilly and coworkers detailed the main features of the transcriptome and the mitogenome architecture and investigated the phylogenetic relationships of *S. neumayeri* within Echinoidea. In addition, they provided comparative analyses of *S. neumayeri* with its closest relative, *S. purpuratus*, including a list of potential OA gene targets. The resources described by them will support a variety of quantitative (genomic, proteomic, multistress, and comparative) studies to interrogate physiological responses to OA and other stressors in this important Antarctic calcifier.

Juvenile Atlantic cod (*Gadus morhua*) was divided into groups and exposed by Olsen et al. (2014) to three different water bath perfluoroalkyl sulfonate (PFOS) exposure regimes (0 [control], 100, and 200 µg L^{-1}) for 5 days at 1 h day^{-1}, followed by three different CO_2 levels (normocapnia, moderate [0.3%], and high [0.9%]). The moderate CO_2 level is the predicted near-future (within the year 2300) level, while 0.9% represents severe hypercapnia. Tissue samples were collected at 3, 6, and 9 days after CO_2 exposure. The effects on the endocrine and biotransformation systems were examined by analyzing levels of sex steroid hormones (*E2, T, 11-KT*) and transcript expression of estrogen responsive (ER) genes (*ERα, Vtg-α, Vtg-β, ZP2,* and *ZP3*). In addition, transcripts for genes encoding xenobiotic metabolizing enzymes (*cyp1a* and *cyp3a*) and hypoxia-inducible factor (*HIF-1α*) were analyzed. Hypercapnia alone produced increased levels of sex steroid hormones (*E2, T, 11-KT*) with concomitant mRNA level increase of ER genes, while PFOS produced weak and time-dependent effects on E2-inducible gene transcription. Combined PFOS and hypercapnia exposure produced increased effects on sex steroid levels as compared to hypercapnia alone, with transcript expression patterns that were indicative of time-dependent interactive effects. Exposure to hypercapnia singly or in combination with PFOS produced modulations of the biotransformation and hypoxic responses that were apparently concentration- and time-dependent. Loading plots of principal component analysis produced a significant grouping of individual scores according to the exposure scenarios at days 6 and 9. Overall, the analysis produced a unique clustering of variables that signifies a positive correlation between exposure to high PFOS concentration and mRNA expression of E2 responsive genes. Notably, this pattern was not evident for individuals exposed to PFOS concentrations in combination with elevated CO_2 scenarios. Thus, their study evaluated such effects using combined exposure to a PFOS and elevated levels of CO_2 saturation, representative of future oceanic climate change, in any fish species or lower vertebrate.

To understand the CO_2 concentration mechanism in the Phaeophyta kelp, *Saccharina japonica*, a full-length cDNA-encoding CA was cloned by Ye et al. (2014) from the gametophytes based on the two screened clones from a suppressive subtracted cDNA library. The cDNA sequence was of 2804 bp in length, including a 166-bp 5′-untranslated region (UTR), a 1765-bp

3'-UTR, and an 873-bp open reading frame. No intron separating this gene was found after comparing its cDNA and DNA sequences. The deduced precursor protein of *S. japonica* CA consisted of 290 amino acids with a typical signal peptide cleavage site between Gly 20 and Val 21 from the N terminus. The mature protein contained three conserved His residues chelated with a zinc ion constituting a catalytic active site. This cloned CA gene from *S. japonica* could be grouped into the α-type as shown by a constructed phylogenetic tree and most identity within the conserved domains characteristic of the α-CA. Quantitative real-time PCR results demonstrated that the diurnal transcription of this CA gene in the gametophytes cultured under the addition of CO_2 or HCO_3^- was not significantly different from those cultured only with filtered air supply at any sampling time, suggesting that this CA might not be periplasmic. After preparation of polyclonal antibody with this recombinant CA in *Escherichia coli*, the gold immunolocalization showed that this α-CA was associated with the chloroplast envelopes and thylakoid membranes, suggesting that this α-CA could provide the chloroplasts with sufficient CO_2 for carbon assimilation.

The impact of elevated pCO_2 on paralytic shellfish poisoning toxin (PST) content and composition in two strains of *Alexandrium tamarense*, Alex5 and Alex2, were investigated by Van de Waal et al. (2014). Experiments were carried out as batches to keep carbonate chemistry unaltered over time. They observed only minor changes with respect to growth and elemental composition in response to elevated pCO_2. For both strains, the cellular PST content, and in particular the associated cellular toxicity, was lower in the high CO_2 treatments. In addition, Alex5 showed a shift in its PST composition from a nonsulfated analogue toward less toxic sulfated analogues with increasing pCO_2. Transcriptomic analyses suggest that the ability of *A. tamarense* to maintain cellular homeostasis is predominantly regulated on the posttranslational level rather than on the transcriptomic level. Furthermore, genes associated with secondary metabolite and amino acid metabolism in Alex5 were downregulated in the high CO_2 treatment, which may explain the lower PST content. Elevated pCO_2 also induced upregulation of a putative sulfotransferase sxtN homologue and a substantial downregulation of several sulfatases. Such changes in sulfur metabolism might explain the shift in PST composition toward more sulfated analogues. Their results indicated that elevated pCO_2 would have minor consequences for growth and elemental composition but might potentially reduce the cellular toxicity of *A. tamarense*.

The effects of short-term (few hours) to medium-term (up to 168 h) seawater acidification on pelagic squids *Sepioteuthis lessoniana* were investigated by Hu et al. (2014). Routine metabolic rates, NH_4^+ excretion, and extracellular acid–base balance were monitored during exposure to control (pH 8.1) and acidified conditions of pH 7.7 and 7.3 along a period of 168 h. Metabolic rates were significantly depressed by 40% after exposure to pH 7.3 conditions for 168 h. Animals fully restored extracellular pH accompanied by an increase in blood HCO_3^- levels within 20 h. This compensation reaction was accompanied by increased transcript abundance of branchial acid–base transporters, including V-type H^+-ATPase (VHA), Rhesus protein (RhP), Na^+/HCO_3^- cotransporter (NBC), and cytosolic carbonic anhydrase. Immunocytochemistry demonstrated the subcellular localization of NKA, VHA in basolateral, and Na^+/H^+-exchanger 3 (NHE3) and RhP in apical membranes of the ion-transporting branchial epithelium. Branchial VHA and RhP responded with increased mRNA and protein levels in response to acidified conditions, indicating the importance of active NH_4^+ transport to mediate acid–base balance in cephalopods. Their results demonstrated that cephalopods have well-developed branchial acid–base regulatory machinery. However, pelagic squids that evolved a lifestyle at the edge of energetic limits were probably more sensitive to prolonged exposure to acidified conditions compared to their more sluggish relatives, including cuttlefish and octopods.

A proteomics approach was used by De Souza et al. (2014) to assess the adverse physiological and biochemical changes that might occur from the exposure to ocean acidification and warming. The team analyzed gills and blood plasma of Atlantic halibut (*Hippoglossus hippoglossus*) exposed to temperatures of 12°C (control) and 18°C (impaired growth) in combination with control (400 µatm) or high-CO_2 water (1000 µatm) for 14 weeks. The proteomic analysis was performed using 2-DE followed by Nanoflow LC-MS/MS using an LTQ-Orbitrap. The high-CO_2 treatment induced the upregulation of immune system–related proteins, as indicated by the upregulation of the plasma proteins complement component C_3 and fibrinogen β chain precursor in both temperature treatments. Changes in gill proteome in the high-CO_2 (18°C) group were mostly related to increased energy metabolism proteins (ATP-synthase, malate dehydrogenase, thermostable malate dehydrogenase, and fructose 1,6-bisphosphate aldolase), possibly coupled with a higher energy demand. Gills from fish exposed to high CO_2 at both temperature treatments showed changes in proteins associated with increased cellular turnover and apoptosis signaling (annexin 5, eukaryotic translation elongation factor 1γ, receptor for protein kinase C, and putative ribosomal protein S27). Their study indicated that moderate CO_2-driven acidification, alone and combined with high temperature, could elicit biochemical changes that might affect fish health.

The interactive effects of partial pressure of carbon dioxide (P_2CO) (pCO_2) and trace metals such as cadmium (Cd) exposure on metal levels, metabolism, and immune-related functions in hemocytes of two ecologically and economically important bivalve species, *M. mercenaria* (hard shell clam) and *C. virginica* (Eastern oyster), were studied by Ivanina et al. (2014). Clams and oysters were exposed to combinations of three P_2CO pCO_2 levels (~400, 800, and 2000 µatm P_2CO pCO_2, corresponding to present-day conditions and projections for years 2100 and 2250, respectively) and two Cd concentrations (0 and 50 µg L^{-1}) in seawater. Following 4 weeks of exposure to Cd, hemolymph of both species contained similar Cd levels (50–70 µg L^{-1}), whereas hemocytes accumulated intracellular Cd burdens up to 15–42 mg L^{-1}, regardless of the exposure to P_2CO pCO_2. Clam hemocytes had considerably lower Cd burdens than those of oysters (0.7–1 ng 10^{-6} cells vs. 4–6 ng 10^{-6} cells, respectively). Cd exposure suppressed hemocyte metabolism and increased the rates of mitochondrial proton leak in normocapnia, indicating partial mitochondrial uncoupling. This Cd-induced mitochondrial uncoupling was alleviated in hypercapnia. Cd exposure suppressed immune-related functions in hemocytes of clams and oysters, and these effects were exacerbated at elevated P_2CO pCO_2. Thus, elevated P_2CO pCO_2 combined with Cd exposure resulted in a decrease in phagocytic activity and adhesion capacity, as well as lower expression of mRNA for lectin and hsp (hsp70) in clam and oyster hemocytes. In oysters, combined exposure to elevated P_2CO pCO_2 and Cd also led to reduced activity of lysozyme in hemocytes and hemolymph. Overall, their study showed that moderately elevated P_2CO pCO_2 (~800–2000 µatm P_2CO pCO_2) potentiates the negative effects of Cd on immunity and thus might sensitize clams and oysters to pathogens and diseases during seasonal hypercapnia and/or ocean acidification in polluted estuaries.

The interactive effects of CO_2 and two common metal pollutants, copper (Cu) and Cd, on metal accumulation, intracellular adenosine triphosphate (ATP)/ubiquitin-dependent protein degradation, stress response, and energy metabolism in two common estuarine bivalves—*C. virginica* (eastern oyster) and *M. mercenaria* (hard shell clam)—were studied by Gotze et al. (2014). Bivalves were exposed for 4–5 weeks to clean seawater (control) and to either 50 µg L^{-1} Cu or 50 µg L^{-1} Cd at one of three partial pressures of CO_2 (P_2CO pCO_2 ~ 395, ~800, and ~1500 µatm) representative of present-day conditions and projections of the Intergovernmental Panel for Climate Change for the years 2100 and 2250, respectively. Clams accumulated lower metal burdens than oysters, and elevated P_2CO pCO_2 enhanced the Cd and Cu accumulation in mantle tissues in both species. Higher Cd and Cu burdens were associated with elevated mRNA expression of metal-binding proteins metallothionein and ferritin. In the absence of added metals, proteasome activities of clams and oysters were robust to elevated P_2CO pCO_2, but P_2CO pCO_2 modulated the proteasome response to metals. Cd exposure stimulated the chymotrypsin-like activity of the oyster proteasome at all CO_2 levels. In contrast, trypsin- and caspase-like activities of the oyster proteasome were slightly inhibited by Cd exposure in normocapnia, but this inhibition was reversed at elevated P_2CO pCO_2. Cu exposure inhibited the chymotrypsin-like activity of the oyster proteasome, regardless of the exposure to P_2CO pCO_2. The effects of metal exposure on proteasome activity were less pronounced in clams, likely due to the lower metal accumulation. However, the general trends (i.e., an increase during Cd exposure, inhibition during exposure to Cu, and overall stimulatory effects of elevated P_2CO pCO_2) were similar to those found in oysters. The levels of mRNA for ubiquitin and tumor suppressor p53 were suppressed by metal exposures in normocapnia in both species, but this effect was alleviated or reversed at elevated P_2CO pCO_2. Cellular energy status of oysters was maintained at all metal and CO_2 exposures, while in clams, the simultaneous exposure to Cu and moderate hypercapnia (~800 µatm P_2CO pCO_2) led to a decline in glycogen, ATP, and ADP levels and an increase in AMP, indicating energy deficiency. Their data suggested that environmental CO_2 levels can modulate accumulation and physiological effects of metals in bivalves in a species-specific manner, which can affect their fitness and survival during global changes in estuaries.

Proteins intimately associated with precipitated calcium carbonate in three metazoan phyla, Cnidaria, Echinodermata, and Mollusca, were compared by Drake et al. (2014). Specifically, the scientific team used a cluster analysis and gene ontology approach to compare ~1500 proteins, from over 100 studies, extracted from calcium carbonates in stony corals, in bivalve and gastropod mollusks, and in adult and larval sea urchins to identify common motifs and differences. Their analysis suggested that there were few sequence similarities across all three phyla, supporting the independent evolution of biomineralization. However, there were core sets of conserved motifs in all three phyla they examined. These motifs included acidic proteins that appeared to be responsible for the nucleation reaction and inhibition, structural and adhesion proteins that determine spatial patterning, and signaling proteins that modify enzymatic activities. Based on this analysis and the fossil record, they proposed that biomineralization was an extremely robust and highly controlled process in metazoans that could withstand extremes in pH predicted for the coming century, similar to their persistence through the Paleocene-Eocene Thermal Maximum.

Calmodulin, a multifunctional calcium sensor protein, cDNA (EsCaM) was identified from Chinese mitten crab *Eriocheir sinensis*, and its mRNA expression patterns in response to ambient (salinity and pH) stress and immune challenges were examined by Li et al. (2014). EsCaM encoded a 149-amino acid protein with a calculated molecular mass of 16.8 kDa and an isoelectric point of 4.09. In unstimulated healthy *E. sinensis*, EsCaM mRNA transcript was detected in all tested tissues with predominant expression in hepatopancreas and the lowest expression in hemocytes. Ambient salinity (15‰ and 30‰ salinities) and pH (pH 6 and 8.5) stress significantly altered EsCaM mRNA expression in gill, hepatopancreas, hemocytes, intestine, and muscle in Chinese mitten crab. In addition, EsCaM gene expression was significantly and rapidly induced as early as 2 h after LPS and poly(I:C) immune stimulations in hemocytes in vitro. Furthermore, EsCaM expression was significantly upregulated in *E. sinensis* hemocytes, gill, hepatopancreas, intestine, and muscle in response to *Edwardsiella tarda* and *Vibrio anguillarum* challenges. Collectively, their findings suggested that EsCaM was an important stress and immune response gene in Chinese mitten crab.

The combined effects of phosphorus (P) limitation and pCO_2, forecast under ocean acidification scenarios, on *Trichodesmium erythraeum* IMS101 cultures were documented by Spungin et al. (2014). They measured nitrogen acquisition, glutamine synthetase activity, uptake rates, intracellular ATP concentration, and the pool sizes of the related key proteins. Their results suggested that cellular energy reallocation enabled the higher growth and N_2 fixation rates detected in *Trichodesmium* cultured under high pCO_2. This was reflected in altered protein abundance and metabolic pools. Also modified were particulate organic carbon and nitrogen production rates, enzymatic activities, and cellular ATP concentrations. They also suggested that adjusting these cellular pathways to change environmental conditions enables *Trichodesmium* to compensate for low P availability and to thrive in acidified oceans. Moreover, elevated pCO_2 could provide *Trichodesmium* with a competitive dominance that would extend its niche, particularly in P-limited regions of the tropical and subtropical oceans.

The molecular response of fishes to acute hypercarbia exposure was studied by Dennis et al. (2014). Bluegill (*Lepomis macrochirus*) and silver carp (*Hypophthalmichthys molitrix*) were exposed to either 30 mg L^{-1} CO_2 ($pCO_2 \approx 15,700$ µatm) or ambient (10 mg L^{-1} CO_2; $pCO_2 \approx$ 920 µatm) conditions for an hour, and the expression of a variety of genes, across three tissues, was compared. Exposure of bluegill and silver carp to 30 mg L^{-1} CO_2 resulted in an increase in c-fos, hif1-α, and gr-2 transcripts, while silver carp alone showed increases in hsp70 and hsc70-2 mRNA. Their study demonstrated that acute hypercarbia exposure impacts gene expression in a species- and tissue-specific manner, which could be useful in identifying potential mechanisms for hypercarbia tolerance between species, and pinpoints specific tissues that were sensitive to hypercarbia exposure.

Maneja et al. (2014) exposed Atlantic herring larvae to ambient (370 µatm) and elevated (1800 µatm) pCO_2 for 1 month to study their proteome expression pattern. The proteome structure of the larvae was examined using 2-DE and mass spectrometry. The length of herring larvae was marginally less in the elevated pCO_2 treatment compared to the control. The proteome structure was also different between the control and treatment. The expression of a small number of proteins was altered by a factor of less than twofold at elevated pCO_2. Their comparative proteome analysis suggested that the proteome of herring larvae was resilient to elevate pCO_2. Their observations also suggested that herring larvae could cope with levels of CO_2 projected for the near future without significant proteome-wide changes.

By screening gene knockout mutants for all 44 putative response regulator (RR)-encoding genes of cyanobacteria, *Synechocystis* sp. PCC 6803 grown under acid stress, Ren et al. (2014) found that a mutant of slr1909 (previously known as rre9), which encoded an orphan RR, grew poorly in BG11 medium at pH 6.2–6.5 when compared with the wild type. Using a quantitative iTRAQ-LC-MS/MS proteomics approach coupled with GC-MS-based metabolomics and RT-qPCR, they further determined the possible acid response network mediated by Slr1909. Their results showed that the signal transduction pathway mediated by Slr1909 might be independent from that mediated by SphS–SphR previously discovered, as none of the proteins and their coding genes regulated by SphS–SphR were differentially regulated in the Δslr1909 mutant grew under acid stress. Only 24 and 10 proteins were upregulated and downregulated in the Δslr1909 mutant when compared with the wild type under acid stress condition, respectively. Notably, three proteins, Slr1259,

Slr1260, and Slr1261, whose encoding genes seem to be located in an operon were down-regulated upon the knockout of the slr1909 gene, suggesting their roles in acid tolerance. In addition, metabolomic analysis allowed identification of a dozen metabolites important for the discrimination of the Δslr1909 mutant and the wild type under acid stress, including several monosaccharide and fatty acids. Their study provided a proteomic and metabolomic characterization of the acid-response network mediated by an orphan regulator Slr1909 in *Synechocystis*.

Little is known about the transgenerational effects of parental environments or natural selection on the capacity of populations to counter detrimental ocean acidification effects. To study the transgenerational effects, six laboratory populations of the calanoid copepod *Pseudocalanus acuspes* were established at three different CO_2 partial pressures (pCO_2 of 400, 900, and 1550 µatm) and grown for two generations at these conditions by Thor and Dupont (2015). Their research output revealed an evidence of alleviation of ocean acidification effects as a result of transgenerational effects in *P. acuspes*. Second-generation adults showed a 29% decrease in fecundity at 900 µatm CO_2 compared to 400 µatm CO_2. This was accompanied by a 10% increase in metabolic rate indicative of metabolic stress. Reciprocal transplant test is an experiment that involves introducing organisms from each of two or more environments into the others. Furthermore, these tests showed that at a pCO_2 exceeding the natural range experienced by *P. acuspes* (1550 µatm), fecundity would have decreased by as much as 67% compared to at 400 µatm CO_2 as a result of this plasticity. However, transgenerational effects partly reduced OA effects so that the loss of fecundity remained at a level comparable to that at 900 µatm CO_2. This also relieved the copepods from metabolic stress, and respiration rates were lower than at 900 µatm CO_2. These results highlight the importance of tests for transgenerational effects to avoid overestimation of the effects of OA.

Incubation of two coral species (*P. damicornis* and *Oculina patagonica*) under reduced pH conditions (pH 7.2) simulating past ocean acidification induced tissue-specific apoptosis that led to the dissociation of polyps from coenosarcs (Kvitt et al. 2015). This in turn led to the breakdown of the coenosarc and, as a consequence, to loss of coloniality. Their data revealed that apoptosis was initiated in the polyps and that once dissociation between polyp and coenosarc terminates, apoptosis subsides. After reexposure of the resulted solitary polyps to normal pH (pH 8.2), both coral species regenerated coenosarc tissues and resumed calcification. These results indicated that the regulation of coloniality was under the control of the polyp, the basic modular unit of the colony. A mechanistic explanation for several key evolutionarily important phenomena that occurred throughout coral evolution was proposed, including mechanisms that permitted species to survive the third tier of mass extinctions.

Reduced seawater pH and elevated pCO_2 are important considerations in tank-based abalone aquaculture, while sea-based farms may be at risk to ocean acidification reductions in pH. Thus juvenile *Haliotis iris* (5–13 and 30–40 mm shell length) were reared in two, 100-day experiments at ambient pH (~8.1, 450 µatm CO_2), pH 7.8 (~1000 µatm CO_2) and pH 7.6 (~1600 µatm CO_2) and studied by Cunningham et al. (2015). Seawater pH was measured and adjusted automatically by bubbling CO_2 into water in replicated flow through tanks. Two separate trials were run, in winter (8.8°C) and summer (16.5°C). Survival and growth were monitored every 30 days, and postexperimental measurements of morphometrics and respiration rate undertaken. Growth of shell length and wet weight were negatively affected by reduced pH, with a two to threefold reduction in the growth of both size classes between ambient and pH 7.6 treatments in the summer experiment. For small juveniles, growth reductions were in conjunction with decreases to shell weight, while large juveniles showed greater resilience in shell production. No changes to respiration rate occurred, suggesting that juveniles might maintain physiological functioning while tolerating dissolution pressure or that they were unable to upregulate metabolism to compensate for pH effects. These data revealed that CO_2-driven reductions in pH could impact growth, metabolism, and biomineralization of abalone and indicated that water quality and ocean acidification were of importance in aquaculture of the species.

In an attempt to elucidate the effects of different CO_2 concentrations (270, 380, and 750 μL L^{-1}) on the competition of microcystin-producing (MC-producing) and non-MC-producing microcystis strains during dense cyanobacteria blooms, an in situ simulation experiment was conducted in the Meiliang Bay of Lake Taihu in the summer of 2012 by Yu et al. (2015). The abundance of total microcystis and MC-producing microcystis genotypes was quantified based on the 16S rDNA and mcyD gene using real-time PCR. The results showed that atmospheric CO_2 elevation would significantly decrease the pH value and increase the dissolved inorganic carbon concentration. Changes in CO_2 concentration did not show significant influence on the abundance of total microcystis population. However, CO_2 concentrations might be an important factor in determining the subpopulation structure of microcystis. The enhancement of CO_2 concentrations could largely increase the competitive ability of non-MC-producing over MC-producing microcystis, resulting in a higher proportion of non-MC-producing subpopulation in treatments using high CO_2 concentrations. Concurrently, MC concentration in water declined when CO_2 concentrations were elevated. Therefore, they concluded that the increase in CO_2 concentrations might decrease potential health risks of MC for human and animals in the future.

The effects of exposure to elevated pCO_2 were characterized in gills and hepatopancreas of *C. gigas* using integrated proteomic and metabolomic approaches by Wei et al. (2015). Metabolic responses indicated that high CO_2 exposure mainly caused disturbances in energy metabolism and osmotic regulation marked by differentially altered ATP, glucose, glycogen, amino acids, and organic osmolytes in oysters, and the depletions of ATP in gills and the accumulations of ATP, glucose, and glycogen in hepatopancreas accounted for the difference in energy distribution between these two tissues. Proteomic responses suggested that ocean acidification could not only affect energy and primary metabolisms, stress responses, and calcium homeostasis in both tissues, but also influence the nucleotide metabolism in gills and cytoskeleton structure in hepatopancreas. Thus, their study demonstrated that the combination of proteomics and metabolomics could provide an insightful view into the effects of ocean acidification on oyster *C. gigas*.

Whole-transcriptome analysis was used to compare the effects of such "acute" (3 days) exposure to elevated pCO_2 with a longer ("prolonged"; 9 days) period of exposure beginning immediately postfertilization in juvenile corals (*A. millepora*) by Moya et al. (2015). Far fewer genes were differentially expressed under the 9-day treatment, and although the transcriptome data implied wholesale disruption of metabolism and calcification genes in the acute treatment experiment, expression of most genes was at control levels after the prolonged treatment. There was little overlap between the genes responding to the acute and prolonged treatments, but hsps and HSFs were overrepresented among the genes responding to both treatments. Among these was an hsp70 gene previously shown to be involved in acclimation to thermal stress in a field population of another acroporid coral. The most obvious feature of the molecular response in the 9-day treatment experiment was the upregulation of five distinct Bcl-2 family members, the majority predicted to be antiapoptotic. Their results suggested that an important component of the longer term response to elevate CO_2 was the suppression of apoptosis. It therefore appeared that juvenile *A. millepora* has the capacity to rapidly acclimate to elevated pCO_2, a process mediated by the upregulation of specific hsps and a suite of Bcl-2 family members.

Proteomics was used to investigate the molecular differences between oyster populations in adult Sydney rock oysters and to identify if those formed the basis for observations seen in larvae by Thompson et al. (2015). Adult oysters from a selective breeding line (B2) and nonselected wild types (WT) were exposed for 4 weeks to elevated pCO_2 (856 µatm) before their proteomes were compared to those of oysters held under ambient conditions (375 µatm pCO_2). Exposure to elevated pCO_2 resulted in substantial changes in the proteomes of oysters from both the selectively bred and wild-type populations. When biological functions were assigned, these differential proteins fell into five broad, potentially interrelated categories of subcellular functions in both oyster populations. These functional categories were energy production, cellular stress responses, the cytoskeletal and molt-related gene expressions, protein synthesis, and cell signaling. In the wild-type population, proteins were predominantly upregulated. However, unexpectedly, these cellular systems were downregulated in the selectively bred oyster population, indicating cellular dysfunction. Their results reflected a trade-off, whereby an adaptive capacity for enhanced mitochondrial energy production in the selectively bred population might help to protect larvae from the effects of elevated CO_2, while being deleterious to adult oysters.

Resilience of the Mediterranean gilt-head sea bream (*Sparus aurata*) to acute warming and water acidification, using cellular indicators of systemic to molecular responses to various temperatures and CO_2 concentrations, was studied and assessed by Feidantsis et al. (2015). Tissue metabolic capacity was derived from enzyme measurements, citrate synthase, 3-hydroxyacyl CoA dehydrogenase (HOAD), and lactate dehydrogenase. Cellular stress and signaling responses were identified from expression patterns of hsp70 and hsp90; the phosphorylation of p38-MAPK, JNKs, and ERKs; from protein ubiquitylation; and finally from the levels of transcription factor Hif-1α as an indicator of systemic hypoxemia. Exposure to elevated CO_2 levels at temperatures higher than 24°C generally caused an increase in fish mortality above the rate caused by warming alone, indicating effects of the two factors and a failure of acclimation and thus the limits of phenotypic plasticity to be reached. As a potential reason, tissue-dependent induction and stabilization of Hif-1α indicated hypoxemic conditions. Their exacerbation by enhanced CO_2 levels was linked to the persistent expression of hsp70 and hsp90, oxidative stress, and activation of MAPK and ubiquitin pathways. Antioxidant defense was enhanced by the expression of catalase and glutathione reductase, however, leaving superoxide dismutase suppressed by elevated CO_2 levels. On longer time scales in specimens surviving warming and CO_2 exposures, various metabolic adjustments initiated a preference to oxidize lipid via HOAD for energy supply. Those processes indicated significant acclimation up to a limit and a time-limited capacity to survive extreme conditions passively by exploiting mechanisms of cellular resilience.

28.4 Conclusion

Overall, genomic and proteomic technologies have great potential to contribute to marine environmental research by generating an unbiased snapshot of the global changes, in particular related to ocean acidification. Integration of these measurements is essential to infer the underlying and complex mechanisms in acid–base regulations. For example, simultaneous analyses of gene and protein expression using a same sample set can provide posttranscriptional regulation for biomineralization pathway with either genomic or proteomic measurements. Once genes and proteins of interest are identified, targeted assays can be developed for decision and policy making for rapid environmental changes. However, erroneous genomic and proteomic signatures that were first claimed to be predictive of global climate changes to ocean acidification or warming cast a shadow on genomic and proteomic analyses. Because of the complexity of these types of data, minor and unintentional processing errors can render many findings nonreproducible. Despite the steep learning curve, members of the genomics community are adopting open-source software and tools, such as R for all analyses. Overall, an omics approach gives us a powerful tool to begin to understand how the physiology of marine calcifying organisms is likely to change in the face of a more acidic ocean. Targeted studies of individual species are significant in that each calcifier's response will vary and thus the ecosystem level impact will be transduced through the physiology of key species. Although gene expression is but one technique (there are other approaches in systems biology, e.g., proteomics or metabolomics), there is great potential to learn about the complexity of the compensatory responses in calcification and other metabolic pathways under ocean acidification conditions. Additionally, transcriptome profiling and its ability to reveal subtle, complex patterns will be a powerful approach to tease apart interacting stressors such as the synergistic effects of ocean acidification and warming.

Acknowledgment

The author is grateful to the chancellor and directors of Sathyabama University, Chennai, for their immense support.

References

Abele, D., J.P. Vázquez-Medina, T. Zenteno-Savín. 2011. Introduction to oxidative stress in aquatic ecosystems. In *Oxidative Stress in Aquatic Ecosystems* (D. Abele, J.P. Vázquez-Medina, and T. Zenteno-Savín, eds.). John Wiley & Sons, Ltd, Chichester, U.K.

Benner, I., R.E. Diner, S.C. Lefebvre, D. LiKomada, E.J. Carpenter, J.H. Stillman. 2013. *Emiliania huxleyi* increases calcification but not expression of calcification-related genes in long-term exposure to elevated temperature and pCO_2. *Philosophical Transactions of the Royal Society B* 368: 1627.

Byrne, M., M. Lamare, D. Winter, S.A. Dworjanyn, S. Uthicke. 2013. The stunting effect of a high CO_2 ocean on calcification and development in sea urchin larvae, a synthesis from the tropics to the poles. *Philosophical Transactions of the Royal Society B* 368: 20120439.

Calosi, P., S.P.S. Rastrick, C. Lombardi, H.J. De Guzman, L. Davidson, M. Jahnke et al. 2013. Adaptation and acclimatization to ocean acidification in marine ectotherms: An in situ transplant experiment with polychaetes at a shallow CO_2 vent system. *Philosophical Transactions of the Royal Society B* 368: 20120444.

Carreiro-Silva, M., T. Cerqueira, A. Godinho, M. Caetano, R.S. Santos, R. Bettencourt. 2014. Molecular mechanisms underlying the physiological responses of the cold-water coral *Desmophyllum dianthus* to ocean acidification. *Coral Reefs* 33: 465–476.

Chapman, R., W. Mancia, A. Beal, M. Veloso, A. Rathburn, C. Blair et al. 2011. The transcriptomic responses of the eastern oyster *Crassostrea virginica*, to environmental conditions. *Molecular Ecology* 20: 1431–1449.

Crawfurd, K.J., J.A. Raven, G.L. Wheeler, E.J. Baxter, I. Joint. 2011. The response of *Thalassiosira pseudonana* to long-term exposure to increased CO_2 and decreased pH. *PLOS ONE* 6: 6695.

Crawley, A., D.I. Kline, S. Dunn, K. Anthony, S. Dove. 2010. The effect of ocean acidification on symbiont photorespiration and productivity in *Acropora formosa*. *Global Change Biology* 16: 851–863.

Cunningham, S.C., A.M. Smith, M.D. Lamare. 2015. The effects of elevated pCO_2 on growth, shell production and metabolism of cultured juvenile abalone, *Haliotis iris*. *Aquaculture Research* doi: 10.1111/are.12684. Early View (Online Version of Record published before inclusion in an issue).

De Souza, K.B., F. Jutfelt, P. Kling, L. Forlin, J. Sturvem. 2014. Effects of Increased CO_2 on fish gill and plasma proteome. *PLOS ONE* 9(7): e102901.

Deigweiher, K., N. Koschnick, H. Portner, M. Lucassen. 2008. Acclimation of ion regulatory capacities in gills of marine fish under environmental hypercapnia. *American Journal of Physiology—Regulatory, Integrative and Comparative Physiology* 295: 1660–1670.

Dennis III, C.E., D.F. Kates, M.R. Noatch, C.D. Suski. 2014. Molecular responses of fishes to elevated carbon dioxide. *Comparative Biochemistry and Physiology Part A: Molecular and Integrative Physiology* 187: 224–231.

Dilly, G.F., J.D. Gaitán-Espitia, G.E. Hofmann. 2015. Characterization of the Antarctic sea urchin (*Sterechinus neumayeri*) transcriptome and mitogenome: A molecular resource for phylogenetics, ecophysiology and global change biology. *Molecular Ecology Resources* 15: 425–436.

Dineshram, R., K.K. Wong, S. Xiao, Z. Yu, P.Y. Qian, V. Thiyagarajan. 2012. Analysis of Pacific oyster larval proteome and its response to high-CO_2. *Marine Pollution Bulletin* 64: 2160–2167.

Drake, J.L., T. Mass, P.G. Falkowski. 2014. The evolution and future of carbonate precipitation in marine invertebrates: Witnessing extinction or documenting resilience in the Anthropocene? *Elementa: Science of the Anthropocene* 2: 000026.

Drake, J.L., T. Mass, L. Haramaty, E. Zelzion, D. Bhattacharya, P.G. Falkowski. 2013. Proteomic analysis of skeletal organic matrix from the stony coral *Stylophora pistillata*. *Proceedings of the National Academy of Sciences of the United States of America* 110: 3788–3793.

Evans, T.G., F. Chan, B.A. Menge, G.E. Hofmann. 2013. Transcriptomic responses to ocean acidification in larval sea urchins from a naturally variable pH environment. *Molecular Ecology* 22: 1609–1625.

Evans, T.G., P.W. Wynn. 2014. Effects of seawater acidification on gene expression: Resolving broader-scale trends in sea urchins. *The Biological Bulletin* 226: 237–254.

Fehsenfeld, S., R. Kiko, Y. Appelhans, D.W. Towle, M. Zimmer, F. Melzner. 2011. Effects of elevated seawater pCO_2 on gene expression patterns in the gills of the green crab, *Carcinus maenas*. *BMC Genomics* 12: 488.

Feidantsis, K., H.O. Portner, E. Antonopoulou, B. Michaelidis. 2015. Synergistic effects of acute warming and low pH on cellular stress responses of the gilthead seabream *Sparus aurata*. *Journal of Comparative Physiology B* 185: 185–205.

Gazeau, F., L.M. Parker, S. Comeau, J.-P. Gattuso, W.A. O'Connor, S. Martin et al. 2013. Impacts of ocean acidification on marine shelled molluscs. *Marine Biology* 160: 2207–2245.

Ge, Y., D. Wang, J. Chiu, S. Cristobal, D. Sheehan, F. Silvestre et al. 2013. Environmental omics: Current status and future directions. *Journal of Integrated omics* 3: 75–87.

Gobler, C.J., S.C. Talmage. 2014. Physiological response and resilience of early life-stage Eastern oysters (*Crassostrea virginica*) to past, present and future ocean acidification. *Conservation Physiology* 2: 1–15.

Gotze, S., O.B. Matoo, B. Elia, R. Saborowski. 2014. Interactive effects of CO_2 and trace metals on the proteasome activity and cellular stress response of marine bivalves *Crassostrea virginica* and *Mercenaria mercenaria*. *Aquatic Toxicology* 149: 65–82.

Hammer, K.M., S.A. Pedersen, T.R. Storseth. 2012. Elevated seawater levels of CO_2 change the metabolic fingerprint of tissues and hemolymph from the green shore crab *Carcinus maenas*. *Comparative Biochemistry and Physiology Part D: Genomics and Proteomics* 7: 292–302.

Harms, L., S. Frickenhaus, M. Schiffer, F.C. Mark, D. Storch, C. Held, Portner, H. Magnus, O. Lucassen. 2014. Gene expression profiling in gills of the great spider crab *Hyas araneus* in response to ocean acidification and warming. *BMC Genomics* 15: 789.

Hoegh-Guldberg, O., P.J. Mumby, A.J. Hooten, R.S. Steneck, P. Greenfield, E. Gomez, C.D. Harvell, P.F. Sale. 2007. Coral reefs under rapid climate change and ocean acidification. *Science* 318: 1737–1742.

Hofmann, G.E., M.J. O'Donnell, A.E. Todgham. 2008. Using functional genomics to explore the effects of ocean acidification on calcifying marine organisms. *Marine Ecology Progress Series* 373: 219–225.

Hu, M.Y., I. Casties, M. Stumpp, O. Ortega-Martinez, S. Dupont. 2014. Energy metabolism and regeneration impaired by seawater acidification in the infaunal brittlestar, *Amphiura filiformis*. *The Journal of Experimental Biology* 217: 2411–2421.

Hu, M.Y., Y.C. Tseng, M. Stumpp, M.A. Gutowska, R. Kiko, M. Lucassen, F. Melzner. 2011. Elevated seawater PCO_2 differentially affects branchial acid-base transporters over the course of development in the cephalopod *Sepia officinalis*. *American Journal of Physiology—Regulatory, Integrative and Comparative Physiology* 300: 1100–1114.

Huning, A.K., F. Melzner, J. Thomsen, M.A. Gutowska, L. Kramer, S. Frickenhaus et al. 2013. Impacts of seawater acidification on mantle gene expression patterns of the Baltic Sea blue mussel: Implications for shell formation and energy metabolism. *Marine Biology* 160: 1845–1861.

IPCC. 2007. Climate change 2007: Synthesis report. Contribution of Working Groups I, II and III to the Fourth Assessment Report of the Intergovernmental Panel on Climate Change [Core Writing Team, Pachauri, R.K and Reisinger, A. (eds.)]. Geneva, Switzerland: IPCC, 104pp.

Ivanina, A.V., C. Hawkins, I.M. Sokolova. 2014. Immunomodulation by the interactive effects of cadmium and hypercapnia in marine bivalves *Crassostrea virginica* and *Mercenaria mercenaria*. *Fish & Shellfish Immunology* 37: 299–312.

Kurihara, H., Y. Takano, D. Kurokawa, K. Akasaka. 2012. Ocean acidification reduces biomineralization related gene expression in the sea urchin, *Hemicentrotus pulcherrimus*. *Marine Biology* 159: 2819–2826.

Kvitt, H., E.K. Winter, K. Maor-Landaw, K. Zandbank, A. Kushmaro, H. Rosenfeld, M. Fine, D. Tchernov. 2015. Breakdown of coral colonial form under reduced pH conditions is initiated in polyps and mediated through apoptosis. *Proceedings of the National Academy of Sciences of the United States of America* 112: 2082–2086.

Lardies, M.A., M.B. Arias, M.J. Poupin, P.H. Manriquez, R. Torres, C.A. Vargas, J.M. Navarro, N.A. Lagos. 2014. Differential response to ocean acidification in physiological traits of *Concholepas concholepas* populations. *Journal of Sea Research* 90: 127–134.

Li, S., Z. Jia, X. Li, X. Geng, J. Sun. 2014. Calmodulin is a stress and immune response gene in Chinese mitten crab *Eriocheir sinensis*. *Fish & Shellfish Immunology* 40: 120–128.

Liu, W., X. Huang, J. Lin, M. He. 2012. Seawater acidification and elevated temperature affect gene expression patterns of the pearl oyster *Pinctada fucata*. *PLOS ONE* 7: 33679.

Lohbeck, K.T., U. Riebesell, T.B.H. Reusch. 2014. Gene expression changes in the coccolithophore *Emiliania huxleyi* after 500 generations of selection to ocean acidification. *Philosophical Transactions of the Royal Society B* 281: 20140003.

Maneja, R.H., R. Dineshram, V. Thiyagarajan, A.B. Skiftesvik, A.Y. Frommel, C. Clemmesen, A.J. Geffen, H.I. Browman. 2014. The proteome of Atlantic herring (*Clupea harengus* L.) larvae is resistant to elevated pCO_2. *Marine Pollution Bulletin* 86: 154–160.

Maor-Landaw, K., S. Karako-Lampert, H.W. Ben-Asher, S. Goffredo, G. Falini, Z. Dubinsky, O. Levy. 2014. Gene expression profiles during short-term heat stress in the red sea coral *Stylophora pistillata*. *Global Change Biology* 20: 3026–3035.

Martin, S., S. Richier, M.L. Pedrotti, S. Dupont, C. Castejon, Y. Gerakis et al. 2011. Early development and molecular plasticity in the Mediterranean sea urchin *Paracentrotus lividus* exposed to CO_2-driven acidification. *Journal of Experimental Biology* 214: 1357–1368.

Moya, A., L. Huisman, E.E. Ball, D.C. Hayward, L.C. Grasso, C.M. Chua, H.N. Woo, J.P. Gattuso, S. Foret, D.J. Miller. 2012. Whole transcriptome analysis of the coral *Acropora millepora* reveals complex responses to CO_2-driven acidification during the initiation of calcification. *Molecular Ecology* 21: 2440–2454.

Moya, A., L. Huisman, S. Foret, J.P. Gattuso, D.C. Hayward, E.E. Ball, D.J. Miller. 2015. Rapid acclimation of juvenile corals to CO_2-mediated acidification by up regulation of heat shock protein and Bcl-2 genes. *Molecular Ecology* 24: 438–452.

O'Donnell, M.J., L.M. Hammond, G.E. Hofmann. 2009. Predicted impact of ocean acidification on a marine invertebrate: Elevated CO_2 alters response to thermal stress in sea urchin larvae. *Marine Biology* 156: 439–446.

Olsen, G.P., M.O. Olufsen, S.A. Pedersen, R.J. Letcher, A. Arukwe. 2014. Effects of elevated dissolved carbon dioxide and perfluorooctane sulfonic acid, given singly and in combination, on steroidogenic and biotransformation pathways of Atlantic cod. *Aquatic Toxicology* 155: 222–235.

Padilla-Gamino, J.L., M.W. Kelly, T.G. Evans, G.E. Hofmann. 2013. Temperature and CO_2 additively regulate physiology, morphology and genomic responses of larval sea urchins, *Strongylocentrotus purpuratus*. *Proceedings: Biological Sciences—The Royal Society* 280(1759): 20130155.

Pespeni, M.H., E. Sanford, B. Gaylord, T.M. Hill, J.D. Hosfelt, H.K. Jarisa et al. 2013. Evolutionary change during experimental ocean acidification. *Proceedings of the National Academy of Sciences of the United States of America* 110: 6937–6942.

Putnam, H.M., A.B. Mayfield, T.Y. Fan, C.S. Chen, R.D. Gates. 2013. The physiological and molecular responses of larvae from the reef-building coral *Pocillopora damicornis* exposed to near-future increases in temperature and pCO_2. *Marine Biology* 160: 2157–2173.

Ren, Q., M. Shi, L. Chen, J. Wang, W. Zhang. 2014. Integrated proteomic and metabolomic characterization of a novel two-component response regulator Slr1909 involved in acid tolerance in *Synechocystis* sp. PCC 6803. *Journal of Proteomics* 109: 76–89.

Richier, S., S. Fiorini, M.E. Kerros, P.V. Dassow, J. Gattuso. 2011. Response of the calcifying coccolithophore *Emiliania huxleyi* to low pH/high pCO_2: From physiology to molecular level. *Marine Biology* 158: 551–560.

Rokitta, S.D., U. John, B. Rost. 2012. Ocean acidification affects redox-balance and ion homeostasis in the life-cycle stages of *Emiliania huxleyi*. *PLOS ONE* 7: 52212.

Shao, Y.T., F.Y. Chang, W.C. Fu, H.Y. Yan. 2014. Acidified seawater suppresses insulin-like growth factor I mRNA expression and reduces growth rate of juvenile orange-spotted groupers, *Epinephelus coioides* (Hamilton, 1822). *Aquaculture Research* 47: 721–731.

Spungin, D., I. Berman-Frank, O. Levitan. 2014. *Trichodesmium*'s strategies to alleviate phosphorus limitation in the future acidified oceans. *Environmental Microbiology* 16: 1935–1947.

Stumpp, M., S. Dupont, M.C. Thorndyke, F. Melzner. 2011. CO_2 induced seawater acidification impacts sea urchin larval development II: Gene expression patterns in pluteus larvae. *Comparative Biochemistry and Physiology Part A: Molecular & Integrative Physiology* 160: 320–330.

Thompson, E.L., W. O'Connor, L. Parker, P. Ross, D.A. Raftos. 2015. Differential proteomic responses of selectively bred and wild-type Sydney rock oyster populations exposed to elevated CO_2. *Molecular Ecology* 24: 1248–1262.

Thor, P., S. Dupont. 2015. Transgenerational effects alleviate severe fecundity loss during ocean acidification in a ubiquitous planktonic copepod. *Global Change Biology* 21: 2261–2271.

Todgham, A.E., G.E. Hofmann. 2009. Transcriptomic response of sea urchin larvae *Strongylocentrotus purpuratus* to CO_2-driven seawater acidification. *The Journal of Experimental Biology* 212: 2579–2594.

Tseng, Y.C., M.Y. Hu, M. Stumpp, L. Lin, Y. Melzner, F. Hwang. 2013. CO_2-driven seawater acidification differentially affects development and molecular plasticity along life history of fish (*Oryzias latipes*). *Comparative Biochemistry and Physiology Part A: Molecular & Integrative Physiology* 165: 119–130.

Van de Waal, D.B., T. Eberlein, U. John, S. Wohlrab, B. Rost. 2014. Impact of elevated pCO_2 on paralytic shellfish poisoning toxin content and composition in *Alexandrium tamarense*. *Toxicon* 78: 58–67.

Vidal-Dupiol, J.V., D. Zoccola, E. Tambutte, C. Grunau, C. Cosseau, K.M. Smith, M. Nolwenn, M.F. Dheilly, D. Allemand, S. Tambutte. 2013. Genes related to ion-transport and energy production are upregulated in response to CO_2-driven pH decrease in corals: New insights from transcriptome analysis. *PLOS ONE* 8: e58652.

Wei, L., Q. Wang, H. Wu, C. Ji, J. Zhao. 2015. Proteomic and metabolomic responses of Pacific oyster *Crassostrea gigas* to elevated pCO$_2$ exposure. *Journal of Proteomics* 112: 83–94.

Werner, G.D.A., P. Gemmell, S. Grosser, R. Hamer, S.M. Shimeld. 2013. Analysis of a deep transcriptome from the mantle tissue of *Patella vulgata* Linnaeus (Mollusca: Gastropoda: Patellidae) reveals candidate biomineralising genes. *Marine Biotechnology* 15: 230–243.

Wong, K.K., A.C. Lane, P.T. Leung, Y. Thiyagarajan. 2011. Response of larval barnacle proteome to CO$_2$-driven seawater acidification. *Comparative Biochemistry and Physiology Part D: Genomics and Proteomics* 6: 310–321.

Ye, R., X. Yu, Z. Shi, W.W. Gao, H.J. Bi, Y.H. Zhou. 2014. Characterization of α-type carbonic anhydrase (CA) gene and subcellular localization of α-CA in the gametophytes of *Saccharina japonica*. *Journal of Applied Phycology* 26: 881–890.

Yu, L., F. Kong, X. Shi, Z. Yang, M. Zhang, Y. Yu. 2015. Effects of elevated CO$_2$ on dynamics of microcystin-producing and non-microcystin-producing strains during *Microcystis* blooms. *Journal of Environmental Sciences* 27: 251–258.

Zippay, M.L., G.E. Hofmann. 2010. Effect of pH on gene expression and thermal tolerance of early life history stages of red abalone (*Haliotis rufescens*). *Journal of Shellfish Research* 29: 429–439.

Section XI

Marine Lipidomics

29

Lipidomic Analysis of Marine Microalgae

Tsuyoshi Tanaka, Yue Liang, and Yoshiaki Maeda

CONTENTS

29.1 Overview

Marine microalgae, which are a diverse group of prokaryotic and eukaryotic photosynthetic microorganisms, are the largest primary producers in the marine system. These microorganisms are diversified in terms of size, morphology, life cycle, pigments, etc. The best known microalgae are Cyanophyta (blue-green algae), Chlorophyta (green algae), Bacillariophyta (diatoms), and Dinophyta (dinoflagellates). Marine microalgae have been used by humans for a long period as food and as natural compounds such as polyunsaturated fatty acids (PUFAs), pigments, antioxidants, vitamins, and so on (Arakaki et al. 2012; Borowitzka 2013; Gao 1998; Lebeau and Robert 2003; Milledge 2011; Priyadarshani and Rath; Yen et al. 2012). The pharmaceutical industry has also benefited from marine microalgae due to their enormous chemically distinct secondary metabolites (Gallardo-Rodríguez et al. 2012; Guedes et al. 2011; Skjånes et al. 2013). Recently, marine microalgae are widely accepted as promising biofuel producers, which are better than higher plants because of their high productivity and no competition with agriculture. Microalgal lipids, hydrocarbons, starch, and cellulose are considered as precursors in the production of alternative petroleum fuel such as biodiesel, bio-jet fuel, and bioethanol. To regulate microalgal biofuel productivity and quality, the understanding of microalgal carbon fixation and the metabolism of biofuel precursors becomes fundamental. To achieve this, studies on microalgal lipids through lipidomic approaches are necessary.

Lipidomics is the large-scale identification and quantification of cellular lipid molecules for various purposes, such as discovering biomarkers or bioactive compounds, understanding cellular functions of different lipids, and the elucidation of the lipid metabolic pathway and its regulations (Wenk 2010). The development of mass spectrometry (MS), nuclear magnetic resonance (NMR) spectroscopy, Raman spectrometry, and various chromatographic techniques has significantly promoted the emergence and evolution of modern lipidomics. Recently, owing to the increasing genome sequencing of microalgae (Sasso et al. 2012; Smith et al. 2012), several systems biology studies, including transcriptomics (Cheng et al. 2013; Guarnieri et al. 2011; Rismani-Yazdi et al. 2011, 2012; Wan et al. 2012; Zheng et al. 2013) and proteomics (Guarnieri et al. 2011, 2013), have been applied in research for understanding microalgal lipid metabolism and regulation. However, due to technological difficulties, lipidomic analyses of microalgae were still infrequent. This chapter presents the fundamental knowledge of microalgal lipids, followed by analytical tools for lipidomic research. Then, the recent discoveries on marine microalgal lipid metabolism are summarized.

29.2 Lipids in Marine Microalgae

In general, lipids are defined as hydrophobic or amphiphilic small molecules, while in the narrow sense, they are defined as molecules that originate entirely or partially from ketoacyl or isoprene groups. Biological lipids can be divided into eight categories: fatty acids, glycerolipids, glycerophospholipids, sphingolipids, saccharolipids, polyketides, sterol lipids, and prenol lipids. The hundreds and thousands of cellular lipid molecules are not only the structural elements of membrane systems but are also functional in energy storage and signaling.

It is still hard to determine the kinds of microalgal lipids that exist because of the extremely large number of microalgal species in nature. However, it is well known that the structures of cyanobacteria are similar to plant chloroplast, and the cell structures of most eukaryotic microalgae are similar to a plant cell. Thus, the major microalgal lipid classes can be considered similar to the major lipid classes of higher plants (Figure 29.1). In general, monogalactosyldiacylglycerol (MGDG), digalactosylglycerol, sulfoquinovosyldiacylglycerol (SQDG), and phosphatidylglycerol, which are referred to as photosynthetic lipids, are the major components of the cyanobacteria membrane system, the chloroplast envelope, and thylakoid membrane of eukaryotic microalgae. Phosphatidylcholine (PC), phosphatidylethanolamine, phosphatidylinositol, and phosphatidylserine are basically found in the extra chloroplast membrane system of eukaryotic microalgae. Many eukaryotic microalgae consist of a species of betaine lipid, diacylglyceryltrimethylhomoserine (DGTS), which is also present in the extra chloroplast membrane system (Vieler et al. 2007). Several studies showed that the DGTS can replace PC under phosphate limitation (Van Mooy et al. 2009). Triacylglycerol (TAG) is the major storage lipid of most eukaryotic microalgae, which is usually assembled into an oil body or a lipid droplet after its synthesis. However, in some microalgal species, free fatty acid (FFA) is another major compound of neutral lipids (Kaul et al. 2012). Other lipid classes, such as sphingolipids and sterol lipids, are also believed to be universal in eukaryotic microalgae, yet lack comprehensive studies. The glycerol lipids, including polar lipids and neutral lipids, which possess a glycerol backbone, are the target lipids of this chapter because of their diverse cellular physiological functions and applications in biofuel production.

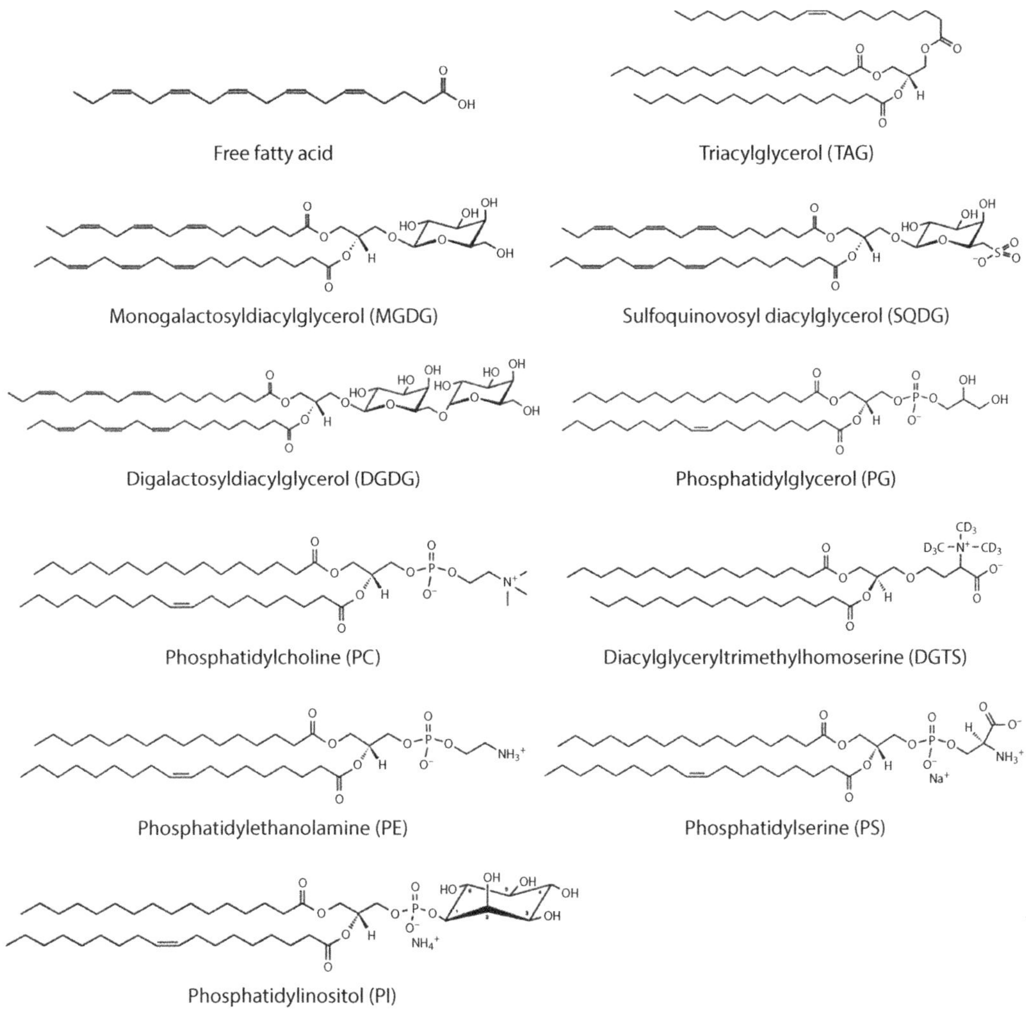

FIGURE 29.1
The structures of major plant lipid classes and one betaine lipid (DGTS) that has been found in several microalgae.

29.2.1 Photosynthetic Lipids

Photosynthetic lipids are not only chloroplast membrane elements but are also involved in photosynthesis and nonphotochemical quenching, in which the MGDG could provide a soluble microenvironment for xanthophyll pigments (Dörmann and Hölzl 2010; Kern et al. 2010; Wada and Mizusawa 2010). Most studies on photosynthetic lipids are undertaken with cyanobacteria because they are almost identical to plant chloroplast and their genomes can be easily manipulated. Yet, the lipid and pigment compositions of cyanobacteria are different from eukaryotic organisms. Thus, green algae, which belong to the "green" lineage, are also commonly studied in photosynthesis. In contrast, the microalgae in the "red" lineage, including red algae and those derived from the secondary endosymbiosis of a red alga (such as heterokonts, haptophyta, cryptomonas, and dinoflagellates), are rarely studied. The structure and molecular composition of their chloroplasts are different from the "green" lineage (McFadden 2001; Stoebe and Maier 2002). For example, the diatom (belonging to heterokonts) chloroplasts are enveloped by four membranes instead of two and they possess a higher content of SQDG and a lower content of MGDG compared to green algae (Lepetit et al. 2012). Further determination of the composition of photosynthetic lipids in various microalgae will not only facilitate the understanding of their physiological roles in photosynthesis but also inspect the understanding of their biosynthesis.

29.2.2 Triacylglycerol

Under stressed conditions, microalgae can not only naturally accumulate large amounts of carbohydrates but also neutral lipids (mainly TAGs) in their cells. The microalgal oil itself or its further processed product fatty acid methyl esters (FAMEs) can be used as biofuel for heating or alternative petroleum diesel, respectively (Lam and Lee 2012). Recent studies revealed that the accumulation of TAGs in microalgae under stressed conditions has more profound physiological significances rather than only for energy stock. For example, as discovered in the green alga *Chlamydomonas reinhardtii*, the accumulated TAGs, which contain three fatty acid chains, may function as the electron sink under stressed conditions to prevent reactive oxygen species (ROS) generation caused by excessive NADPH from photosynthesis and subsequently inhibit the ROS-induced cell death (Li et al. 2012). In addition, microalgal TAGs may also be involved in membrane rearrangement during adaptation to stressed conditions by serving as a depot for certain fatty acids such as PUFA (Solovchenko 2012). Microalgae, especially eukaryotic microalgae, naturally produce large amounts of PUFAs that are usually incorporated into glycerolipids or present in the form of FFAs. Under stressed conditions, PUFA may be released from membrane lipids and incorporated into TAGs during adaptation to stress. Then, the PUFA-enriched TAGs could generate PUFA-containing diacylglycerol backbones for the resynthesis of membrane lipids and the rapid recovery of membrane system when the stress vanishes. However, there is little experimental evidence for this.

29.2.3 Polyunsaturated Fatty Acids

ω-3 long-chain PUFAs such as eicosapentaenoic acid (EPA) and docosahexaenoic acid (DHA) have long been considered as human health supplements. Most red algae and diatoms naturally produce EPA and DHA. On the other hand, high PUFA content will affect biodiesel quality and should be decreased to lower than 1% based on the EN 14214 (a European standard that clarifies the requirements of biodiesel). Although the PUFA content in microalgal cells is sensitive to environmental factors such as temperature (Adarme-Vega et al. 2012; Wen and Chen 2003), metabolic engineering was considered to be more applicable for drastic variations of PUFA content (Li et al. 2009; Lopez et al. 2013; Xue et al. 2013). For most PUFA-producing microalgae species, PUFAs are synthesized from C18:1 fatty acid by a series of desaturation/elongation reactions through the ω-3 and/or ω-6 pathway (Arao et al. 1994; Arao and Yamada 1994; Guschina and Harwood 2006; Lippmeier et al. 2009; Schneider and Roessler 1994; Shiran et al. 1996). A distinct difference between these pathways is the localization of double bonds of the intermediates. The exact biosynthesis pathway for PUFA could be complex and species specific in microalgae. Thus, for the purpose of metabolic engineering, the characterization of fatty acid composition, including the position determination of double bonds and further construction of the PUFA synthesis pathway, is essential.

29.3 Analytical Tools in Lipidomics

As mentioned, the emergence of lipidomics is highly dependent on the rapid development of spectrometric techniques, especially MS. Before the establishment of MS, lipid research relied on thin layer and column chromatography. These chromatographic techniques are still widely applied in lipidomic research. A combination of chromatography and spectrometry techniques in lipidomic analysis can provide detailed quantitative information of hundreds of chemically different lipid species. Lipidomics discoveries are a crucial complement to genomics, transcriptomics, and proteomics.

29.3.1 Mass Spectrometry for Lipid Profiling

MS, which can be regarded as the base of lipidomics, is the mostly applied spectrometric technology in the field of lipid research. In a typical MS test, lipid molecules are ionized to charged molecule ions or fragment ions and their mass-to-charge (m/z) ratios are measured. Through MS analysis, the elemental composition and basic structure of a particular lipid molecule can be determined.

Various ionization techniques have been applied in lipidomic analyses. Different ionization methods have different specificities and limitations, and so the choice of the ionization method should be considered before the test. Electron ionization (EI), electrospray ionization (ESI), and matrix-assisted laser desorption/ionization (MALDI) are usually employed.

EI can produce molecular ions or fragment ions according to the electron energy. Most mass spectrometers use 70 eV of electron energy for EI. Under this electron energy, the molecular ion may be weak or absent, which may be unfavorable for the determination of element composition and molecular weight. Yet, the resulting fragmentation pattern, which can be referred to as the "fingerprint" of a molecule, could provide structural information. EI may provide the most reproducible mass spectra among all the ionization methods, and thus libraries of mass spectra can be searched for EI mass spectral "fingerprint." MALDI can generate molecular ions; however, the fragmentation might be difficult. Thus, MALDI will not provide mass spectra with enough structural information. ESI is a newly developed soft ionization method. Pseudomolecular ions can be obtained via ESI (Han and Gross 2005). By coupling with tandem mass spectrometry (MS/MS), the fragmentation information can also be obtained by ESI-MS/MS. The MS/MS spectra, which are strongly dependent on the instrument parameters, may not be highly reproducible. However, the fragmentation pattern can still provide large structural information of the building block of target lipids, such as the head group and fatty acid chain. Lipid molecules can be divided into several structural moieties, which will give a specific fragmentation pattern in MS/MS analysis (Han and Gross 2005). This discovery significantly facilitated the specific detection of each lipid classes, and thus the ESI tandem MS has now become a powerful tool for the detection and primary identification of lipid structures. Currently, a full structural characterization may only be achieved through NMR.

The generated ions are ready to be separated by a mass analyzer based on their m/z ratios and outputted to the detector where they are detected and converted to digital signals. There are six major mass analyzers: quadrupole mass analyzer, time-of-flight (TOF) mass analyzer, magnetic sector mass analyzer, electrostatic sector mass analyzer, quadrupole ion trap mass analyzer, and ion cyclotron resonance. With high-resolution mass analyzers, the MS resolution ($R = m/\Delta m$, where m is the molecular mass of the target molecule and Δm is full width at half maximum of the peak of interest) could exceed 100,000, which is referred to as high-resolution MS. Soft ionization and high-resolution specificities of ESI-MS can facilitate characterization of element composition and subsequent determination of the molecular formula.

MS is suitable for highly sensitive detection of minor compounds and high-resolution analysis for precise prediction of the molecular composition. The absolute quantification of whole cellular lipids is still problematic due to the lack of ideal quantitative standard samples for the huge number of cellular lipid species. However, targeted and semitargeted absolute quantification can be achieved for specific lipid classes (Ejsing et al. 2009; Merrill et al. 2005; Okazaki et al. 2013; Schuhmann et al. 2011). In addition, the spatial distribution of particular lipid species can also be obtained by imaging MS. For example, MALDI mass spectrometer accompanied by a TOF detector is generally used for the analysis of solid materials, such as tissue samples or living cells. Even single unicellular microalgal cells can be analyzed by MALDI-TOF MS (Amantonico et al. 2010), although this study did not focus on lipids. Besides MALDI MS, the recently developed secondary ion spectrometry and desorption ESI-MS have also been used for the imaging MS; yet, the spatial resolution is still unfavorable for unicellular microalgae.

29.3.2 Chromatography Assisting MS Analysis

In some cases, crude samples are introduced into mass spectrometer through direct infusion. Direct-infusion ESI tandem MS, also known as shotgun lipidomics, may be the simplest way to profile a lipidome based on the mechanism of intrasource separation (Han 2010; Han et al. 2012). Absolute quantification of global or targeted lipidome has been achieved in many organisms, such as yeast (Klose et al. 2012) and mammalian cells (Andreyev et al. 2010; Nie et al. 2010). However, plant lipidome is much more diversified than that of yeast or animal cells, and this incredibly high diversity makes the absolute quantification of plant lipidome through shotgun lipidomics impossible. In addition, the unwanted overlap of isotopic ion peaks, lack of proper standards, and high ion suppression due to the complex sample matrix would not only affect the quantification but also the identification of plant lipidome.

To avoid these difficulties, chromatographic techniques are generally applied prior to the MS analysis. Chromatography, which aims to separate different compounds from mixtures, has long been developed and applied in various research areas. Thin layer chromatography (TLC), solid phase extraction (SPE), gas chromatography (GC), and high-performance liquid chromatography (HPLC) are the mostly applied chromatographic techniques in lipid research.

TLC may be the most basic chromatography for sample pretreatment of lipid analysis. The precoated TLC plate ensured the simple operation and wide application of this technique. By applying TLC, especially high-performance thin layer chromatography (HPTLC), different lipid classes even different lipid species can achieve satisfactory separation. The resulting spots or bands on the TLC plate can be further analyzed by MS. MALDI-TOF MS and ESI-MS are often used for the analysis of the TLC spots or bands. SPE is also commonly used for sample pretreatment. Samples separated from SPE can be induced by simple direct infusion, GC, or HPLC to MS and will significantly improve the signal to noise ratio.

GC is usually coupled with flame ionization detector or EI-MS online for the precise detection of fatty acids and their derivatives such as FAME. Even the fatty acid isomers can be well separated with an appropriate GC program. The relatively simple process compared with HPLC and the high reproducibility make GC-MS the most frequently used technique for lipid researches. With the preseparation of each lipid classes via HPTLC or SPE, and specific lipase treatment, the cellular lipid composition can be roughly determined by GC-MS. In contrast to TLC, SPE, and GC, online HPLC is much more recently applied for lipid analysis owing to the development of MS, which could be linked online with HPLC for the detection of lipids. Currently, HPLC and ultraperformance liquid chromatography (UPLC), which are coupled with ESI-MS online, are widely used for the identification and quantification of global or targeted lipidome. Despite the possible unexpected loss of the lipid molecules during the HPLC or UPLC separation, detailed plant lipidome can be profiled through LC-ESI-MS/MS (Lísa et al. 2011; Okazaki et al. 2013).

Recently, more chromatographic techniques such as solid-phase microextraction, micro-GC/MS, nano-HPLC, supercritical fluid extraction (Solana et al. 2014), and supercritical fluid chromatography (Yamada et al. 2013) have also been developed and applied in lipid analysis, especially for the analysis of samples with small amount and high-throughput detection.

29.4 Marine Microalgae Lipidomics

Hundreds and thousands of cellular lipid molecules play a number of roles such as energy storage (Meyer and Kinney 2010) and photosynthesis (Shimojima et al. 2010) as described earlier. The physiological roles of cellular lipids are strictly dependent on their chemical structures. The profiling and even quantification of the cellular lipids, especially under stressed conditions, will facilitate the understanding of the physiological functions of lipids and their roles in stress tolerance.

Despite the rapid development of lipidomic research in the last decades, the microalgal lipidomic analysis is still in its infancy. One major problem may be the unexpected highly diversified microalgal lipidome. Generally, microalgae tend to produce hundreds of chemically distinct lipid molecules and even various unusual lipid molecules to adapt to environmental conditions. A number of reports regarding the identification of microalgal lipids, especially the characterization of novel lipid species, via traditional technologies like TLC and column chromatography combined with NMR, inferred spectroscopy, etc., were published from the 1960s to 1990s. Currently, the determination and structural characterization of novel bioactive compounds, including lipids, are still conducted worldwide. Meanwhile, studies on microalgal fatty acids profiling and quantification have been largely reported due to the highly prosperous studies on microalgal biodiesel (Roleda et al. 2012; Shekh et al. 2013).

However, just in the last decade, systematical microalgal lipidomic studies have been initiated (Table 29.1). In addition, most of those studies focused on specific lipid classes such as glycolipids mainly for better understanding of photosynthesis or TAG for studying biodiesel. Moreover, most of such researches just reported the establishment of the MS-based method for microalgal lipidomic analyses and lacked thoughtful discussion regarding the lipid metabolism and their regulation relating to environmental changes in microalgae cells.

TABLE 29.1

Summary of Lipidomic Analyses in Microalgae

Year	Microalgal Species	Lipid Classes	Method	References
2007	*Chlamydomonas reinhardtii, Cyclotella meneghiniana*	Total	MALDI-TOF MS and TLC	Vieler et al. (2007)
2010	*Stephanodiscus* sp.	MGDG, digalactosylglycerol (DGDG), SQDG	UPLC-ESI-Q-TOF MS	Xu et al. (2010)
2011	*Phaeodactylum tricornutum, Nannochloropsis salina, Nannochloropsis oculata, Tetraselmis suecica*	TAG	SPE, MALDI-TOF MS, ESI linear ion trap–orbitrap (LTQ Orbitrap) MS, and ^{1}H NMR	Danielewicz et al. (2011)
	Botryococcus braunii, Nannochloropsis gaditana, Neochloris oleoabundans, Phaeodactylum tricornutum, Porphyridium aerugineum, Scenedesmus obliquus	TAG	UHPLC-Orbitrap MS	MacDougall et al. (2011)
	Chattonella subsalsa, Fibrocapsa japonica, Heterosigma akashiwo, Gonyostomum semen	MGDG DGDG	Silica column chromatography, direct-infusion ESI-MS/MS	Roche and Leblond (2011)
	Nannochloropsis oculata	Polar lipids	HPLC-LTQ-FT-ICR MS	He et al. (2011)
2012	*Chlamydomonas nivalis*	Biomarker (polar lipids)	UPLC/Q-TOF-MS and multivariate statistics	Lu et al. (2012)
	Chlorella minutissima	Lipid secretions	LC-QTOF, chip-based direct-infusion MS/MS, GC-MS	Kind et al. (2012)
	Thalassiosira sp.	TAG	ESI-IT-MS	Nurachman et al. (2012)
	Nitzschia closterium	Total lipids	UPLC-ESI-Q-TOF–MS and multivariate statistics	Su et al. (2012)
	Chlorella	Total lipids from crude microalgal lipid extract and hydrocarbon	NP-HPLC with parallel quantitation by an evaporative light scattering detector and MS; NMR, GC-MS	Jones et al. (2012)
	Phaeodactylum tricornutum	TAG	RP-HPLC/MS-APCI	Řezanka et al. (2012)
2013	*Skeletonema marinoi, Phaeodactylum tricornutum, Navicula perminuta_cf, Thalassiosira weissflogii, Haslea ostrearia*	MGDG and DGDG	Direct-infusion ESI-MS/MS	Dodson et al. (2013)
	Not specific	Crude microalgal oil	MSN (mesoporous silica nanoparticles) treatment; ESI, APCI, APPI, MALTI-LTQ-Orbitrap MS; GC-MS	Lee et al. (2013)
	Nannochloropsis salina, Scenedesmus obliquus	Total lipid component	Direct-infusion ESI FT-ICR MS	Holguin and Schaub (2012)
	Dunaliella salina, D. tertiolecta, D. bardawil, D. granulata	TAG and FFA	UPLC-ESI-MS and MS/MS	Samburova et al. (2013)
	Chlamydomonas nivalis	Biomarker (polar lipids)	UPLC-Q-TOF-MS and multivariate statistics	Lu et al. (2013)

(Continued)

TABLE 29.1 (*Continued*)

Summary of Lipidomic Analyses in Microalgae

Year	Microalgal Species	Lipid Classes	Method	References
2014	*Fistulifera solaris*	TAG and FA	GC-MS, SPE, and direct-infusion ESI-Q-Trap MS and MS/MS	Liang et al. (2014a)
	Fistulifera solaris	Polar lipids	SPE and direct-infusion ESI-Q-Trap MS and MS/MS	Liang et al. (2014b)
	Coccomyxa subellipsoidea, Chlamydomonas moewusii, Chlamydomonas minutissima, Chlamydomonas reinhardtii, Chlamydomonas smithii, Chlorella sorokiniana, Chlorella miniata, Chlorella lobophora, Chlorella minutissima, Chlorella saccharophila, Chlorella kessleri, Chlorella protothecoides, Chlorella vulgaris, Gloeococcus minutissima, Volvox tertius, Palmellopsis muralis, Asterococcus superbus, Scenedesmus dimorphus, Scenedesmus parisiensis, Scenedesmus quadricauda, Scenedesmus obliquus, Stigeoclonim pascheri, Dictyochloropsis splendida, Stichococcus bacillaris, Closteriopsis acicularis, Watanabe reniformis	TAG	GC-MS, catalytic saturation and ESI-LC-MS/MS	Allen et al. (2014)
	Synechocystis 6803, Synechococcus 7002, Anabaena 7120	Total lipids (living cell)	Sonic-spray ionization MS (EASI-MS) and multivariate statistics	Liu et al. (2014)
	Haematococcus pluvialis	Total lipids	LC-ESI-MS and MS/MS	Gwak et al. (2014)

However, there were still some reports that revealed novel discoveries on lipid synthesis and regulation in microalgae, even though the techniques used were relatively traditional. For example, a GC-MS analysis combined with TLC preseparation and specific lipase treatments revealed a prokaryotic pathway in the chloroplast for the synthesis of TAG in *Chlamydomonas reinhardtii* (Fan et al. 2011). A further analysis of various green microalgae species suggested an evolutionary divergence in regioisometry of the dominant TAG synthetic pathways by the order *Chlamydomonadales* (Allen et al. 2014). A shotgun lipidomic analysis combined with GC-MS analysis of lipid compounds in marine diatom *Fistulifera solaris* indicated possible pathways for EPA synthesis and its incorporation into glycerolipids (Liang et al. 2014b). Moreover, lipidomic data were usually used as a compensation of transcriptomics data to reveal overall metabolic changes in microalgal cells under stressed conditions (Gwak et al. 2014; Li et al. 2014).

29.5 Summary

In this chapter, general marine microalgal lipid compounds and their functions were first introduced. Then, the concept and development of lipidomics, the common techniques used for lipidomic studies, and the basic application of each technique were reviewed. Several newly developed MS and chromatography for lipid analyses were also briefly introduced. Finally, the recent developments in microalgal lipidomics were summarized. Due to technological difficulties, the profiling and quantification of the overall microalgal lipidome are still under development. However, the current technology can still provide high-quality data to facilitate new discoveries on the metabolism and function of microalgal lipids.

Abbreviations

DGTS   Diacylglyceryltrimethylhomoserine
DHA   Docosahexaenoic acid
EI   Electron ionization
EPA   Eicosapentaenoic acid
ESI   Electrospray ionization
FAME   Fatty acid methyl ester
FFA   Free fatty acid
GC   Gas chromatography
LC   Liquid chromatography
MALDI   Matrix-assisted laser desorption/ionization
MGDG   Monogalactosyldiacylglycerol
MS   Mass spectrometry
NMR   Nuclear magnetic resonance
PC   Phosphatidylcholine
PUFA   Polyunsaturated fatty acid
SPE   Solid phase extraction
SQDG   Sulfoquinovosyldiacylglycerol
TAG   Triacylglycerol
TOF   Time-of-flight

References

Adarme-Vega, TC, DK Lim, M Timmins, F Vernen, Y Li, and PM Schenk. 2012. Microalgal biofactories: A promising approach towards sustainable omega-3 fatty acid production. *Microbial Cell Factories* 11:96.

Allen, JW, CC DiRusso, and PN Black. 2014. Triglyceride quantification by catalytic saturation and LC–MS/MS reveals an evolutionary divergence in regioisometry among green microalgae. *Algal Research* 5:23–31.

Amantonico, A, PL Urban, SR Fagerer, RM Balabin, and R Zenobi. 2010. Single-cell MALDI-MS as an analytical tool for studying intrapopulation metabolic heterogeneity of unicellular organisms. *Analytical Chemistry* 82(17):7394–7400.

Andreyev, AY, E Fahy, Z Guan et al. 2010. Subcellular organelle lipidomics in TLR-4-activated macrophages. *Journal of Lipid Research* 51(9):2785–2797.

Arakaki, A, D Iwama, Y Liang et al. 2013. Glycosylceramides from marine green microalga *Tetraselmis* sp. *Phytochemistry*. 13:107–114.

Arao, T, T Sakaki, and M Yamada. 1994. Biosynthesis of polyunsaturated lipids in the diatom, *Phaeodactylum tricornutum*. *Phytochemistry* 36(3):629–635.

Arao, T and M Yamada. 1994. Biosynthesis of polyunsaturated fatty acids in the marine diatom, *Phaeodactylum tricornutum*. *Phytochemistry* 35(5):1177–1181.

Borowitzka, MA. 2013. High-value products from microalgae—Their development and commercialisation. *Journal of Applied Phycology* 25(3):743–756.

Cheng, R-L, J Feng, B-X Zhang, Y Huang, J Cheng, and C-X Zhang. 2013. Transcriptome and gene expression analysis of an oleaginous diatom under different salinity conditions. *BioEnergy Research* 7(1):192–205.

Danielewicz, MA, LA Anderson, and AK Franz. 2011. Triacylglycerol profiling of marine microalgae by mass spectrometry. *Journal of Lipid Research* 52(11):2101–2108.

Dodson, VJ, JL Dahmen, J-L Mouget, and JD Leblond. 2013. Mono- and digalactosyldiacylglycerol composition of the marennine-producing diatom, *Haslea ostrearia*: Comparison to a selection of pennate and centric diatoms. *Phycological Research* 61:199–207.

Dörmann, P and G Hölzl. 2010. (eds) Hajime Wada and Norio Murata. The role of glycolipids in photosynthesis. In *Lipids in Photosynthesis*. Springer. Berlin, Germany. pp 265–282.

Ejsing, CS, JL Sampaio, V Surendranath et al. 2009. Global analysis of the yeast lipidome by quantitative shotgun mass spectrometry. *Proceedings of the National Academy of Sciences of the United States of America* 106(7):2136–2141.

Fan, J, C Andre, and C Xu. 2011. A chloroplast pathway for the de novo biosynthesis of triacylglycerol in *Chlamydomonas reinhardtii*. *FEBS Letters* 585(12):1985–1991.

Gallardo-Rodríguez, J, A Sánchez-Mirón, F García-Camacho, L López-Rosales, Y Chisti, and E Molina-Grima. 2012. Bioactives from microalgal dinoflagellates. *Biotechnology Advances* 30(6):1673–1684.

Gao, K. 1998. Chinese studies on the edible blue-green alga, *Nostoc flagelliforme*: A review. *Journal of Applied Phycology* 10(1):37–49.

Guarnieri, MT, A Nag, SL Smolinski, A Darzins, M Seibert, and PT Pienkos. 2011. Examination of triacylglycerol biosynthetic pathways via de novo transcriptomic and proteomic analyses in an unsequenced microalga. *PLOS ONE* 6(10):e25851.

Guarnieri, MT, A Nag, S Yang, and PT Pienkos. 2013. Proteomic analysis of *Chlorella vulgaris*: Potential targets for enhanced lipid accumulation. *Journal of Proteomics* 93:245–253.

Guedes, A, HM Amaro, and FX Malcata. 2011. Microalgae as sources of high added-value compounds—A brief review of recent work. *Biotechnology Progress* 27(3):597–613.

Guschina, IA and JL Harwood. 2006. Lipids and lipid metabolism in eukaryotic algae. *Progress in Lipid Research* 45(2):160–186.

Gwak, Y, Y-S Hwang, B Wang et al. 2014. Comparative analyses of lipidomes and transcriptomes reveal a concerted action of multiple defensive systems against photooxidative stress in *Haematococcus pluvialis*. *Journal of Experimental Botany* 65(15):4317–4334.

Han, X. 2010. Multi-dimensional mass spectrometry-based shotgun lipidomics and the altered lipids at the mild cognitive impairment stage of Alzheimer's disease. *Biochimica et Biophysica Acta* 1801(8):774–783.

Han, X and RW Gross. 2005. Shotgun lipidomics: Electrospray ionization mass spectrometric analysis and quantitation of cellular lipidomes directly from crude extracts of biological samples. *Mass Spectrometry Reviews* 24(3):367–412.

Han, X, K Yang, and RW Gross. 2012. Multi-dimensional mass spectrometry-based shotgun lipidomics and novel strategies for lipidomic analyses. *Mass Spectrometry Reviews* 31(1):134–178.

He, H, RP Rodgers, AG Marshall, and CS Hsu. 2011. Algae polar lipids characterized by online liquid chromatography coupled with hybrid linear quadrupole ion trap/Fourier transform ion cyclotron resonance mass spectrometry. *Energy & Fuels* 25(10):4770–4775.

Holguin, FO and T Schaub. 2012. Characterization of microalgal lipid feedstock by direct-infusion FT-ICR mass spectrometry. *Algal Research* 2(1):43–50.

Jones, J, S Manning, M Montoya, K Keller, and M Poenie. 2012. Extraction of algal lipids and their analysis by HPLC and mass spectrometry. *Journal of the American Oil Chemists' Society* 89(8):1371–1381.

Kaul, S, R Jain, S Konathala, D Bangwal, N Atray, and B Kumar. 2012. Evaluation of *Chlorella minutissima* oil for biodiesel production. *Journal of ASTM International* 9(5):1–6.

Kern, J, A Zouni, A Guskov, and N Krauss. 2010. Lipids in the structure of photosystem I, photosystem II and the cytochrome b6f complex. In *Lipids in Photosynthesis*, H. Wada and N. Murata (eds.). Springer Berlin, Germany, pp. 203–242.

Kind, T, JK Meissen, D Yang et al. 2012. Qualitative analysis of algal secretions with multiple mass spectrometric platforms. *Journal of Chromatography A* 1244:139–147.

Klose, C, MA Surma, MJ Gerl, F Meyenhofer, A Shevchenko, and K Simons. 2012. Flexibility of a eukaryotic lipidome—Insights from yeast lipidomics. *PLOS ONE* 7(4):e35063.

Lam, MK and KT Lee. 2012. Microalgae biofuels: A critical review of issues, problems and the way forward. *Biotechnology Advances* 30(3):673–690.

Lebeau, T and J-M Robert. 2003. Diatom cultivation and biotechnologically relevant products. Part II: Current and putative products. *Applied Microbiology and Biotechnology* 60(6):624–632.

Lee, YJ, RC Leverence, EA Smith, JS Valenstein, K Kandel, and BG Trewyn. 2013. High-throughput analysis of algal crude oils using high resolution mass spectrometry. *Lipids* 48(3):297–305.

Lepetit, B, R Goss, T Jakob, and C Wilhelm. 2012. Molecular dynamics of the diatom thylakoid membrane under different light conditions. *Photosynthesis Research* 111(1–2):245–257.

Li, J, D Han, D Wang et al. 2014. Choreography of transcriptomes and lipidomes of Nannochloropsis reveals the mechanisms of oil synthesis in microalgae. *The Plant Cell Online* 26(4):1645–1665.

Li, X, ER Moellering, B Liu et al. 2012. A galactoglycerolipid lipase is required for triacylglycerol accumulation and survival following nitrogen deprivation in *Chlamydomonas reinhardtii*. *The Plant Cell Online* 24(11):4670–4686.

Li, Y-T, M-T Li, C-H Fu, P-P Zhou, J-M Liu, and L-J Yu. 2009. Improvement of arachidonic acid and eicosapentaenoic acid production by increasing the copy number of the genes encoding fatty acid desaturase and elongase into *Pichia pastoris*. *Biotechnology Letters* 31(7):1011–1017.

Liang, Y, Y Maeda, T Yoshino, M Matsumoto, and T Tanaka. 2014a. Profiling of fatty acid methyl esters from the oleaginous diatom *Fistulifera* sp. strain JPCC DA0580 under nutrition-sufficient and -deficient conditions. *Journal of Applied Phycology* 26(6):2295–2302.

Liang, Y, Y Maeda, T Yoshino, M Matsumoto, and T Tanaka. 2014b. Profiling of polar lipids in marine oleaginous diatom *Fistulifera solaris* JPCC DA0580: Prediction of the potential mechanism for eicosapentaenoic acid-incorporation into triacylglycerol. *Marine Drugs* 12(6):3218–3230.

Lippmeier, JC, KS Crawford, CB Owen, AA Rivas, JG Metz, and KE Apt. 2009. Characterization of both polyunsaturated fatty acid biosynthetic pathways in *Schizochytrium* sp. *Lipids* 44(7):621–630.

Lísa, M, E Cífková, and M Holčapek. 2011. Lipidomic profiling of biological tissues using off-line two-dimensional high-performance liquid chromatography–mass spectrometry. *Journal of Chromatography A* 1218(31):5146–5156.

Liu, Y, J Zhang, H Nie et al. 2014. Study on variation of lipids during different growth phases of living cyanobacteria using easy ambient sonic-spray ionization mass spectrometry. *Analytical Chemistry* 86(14):7096–7102.

Lopez, NR, RP Haslam, SL Usher, JA Napier, and O Sayanova. 2013. Reconstitution of EPA and DHA biosynthesis in *Arabidopsis*: Iterative metabolic engineering for the synthesis of n-3 LC-PUFAs in transgenic plants. *Metabolic Engineering* 17:30–41.

Lu, N, D Wei, F Chen, and S-T Yang. 2013. Lipidomic profiling reveals lipid regulation in the snow alga *Chlamydomonas nivalis* in response to nitrate or phosphate deprivation. *Process Biochemistry* 48(4):605–613.

Lu, N, D Wei, X-L Jiang, F Chen, and S-T Yang. 2012. Regulation of lipid metabolism in the snow alga *Chlamydomonas nivalis* in response to NaCl stress: An integrated analysis by cytomic and lipidomic approaches. *Process Biochemistry* 47(7):1163–1170.

MacDougall, KM, J McNichol, PJ McGinn, SJB O'Leary, and JE Melanson. 2011. Triacylglycerol profiling of microalgae strains for biofuel feedstock by liquid chromatography–high-resolution mass spectrometry. *Analytical and Bioanalytical Chemistry* 401(8):2609–2616.

McFadden, GI. 2001. Primary and secondary endosymbiosis and the origin of plastids. *Journal of Phycology* 37(6):951–959.

Merrill Jr AH, MC Sullards, JC Allegood, S Kelly, and E Wang. 2005. Sphingolipidomics: High-throughput, structure-specific, and quantitative analysis of sphingolipids by liquid chromatography tandem mass spectrometry. *Methods* 36(2):207–224.

Meyer, K and AJ Kinney. 2010. (eds) Hajime Wada and Norio Murata. Biosynthesis and biotechnology of seed lipids including sterols, carotenoids and tocochromanols. In *Lipids in Photosynthesis*. Springer. Berlin, Germany. pp 407–444.

Milledge, JJ. 2011. Commercial application of microalgae other than as biofuels: A brief review. *Reviews in Environmental Science and Bio-Technology* 10(1):31–41.

Nie, H, R Liu, Y Yang et al. 2010. Lipid profiling of rat peritoneal surface layers by online normal-and reversed-phase 2D LC QToF-MS. *Journal of Lipid Research* 51(9):2833–2844.

Nurachman, Z, S Anita, EE Anward et al. 2012. Oil productivity of the tropical marine diatom *Thalassiosira* sp. *Bioresource Technology* 108:240–244.

Okazaki, Y, Y Kamide, MY Hirai, and K Saito. 2013. Plant lipidomics based on hydrophilic interaction chromatography coupled to ion trap time-of-flight mass spectrometry. *Metabolomics* 9(1):121–131.

Priyadarshani, I and B Rath. Commercial and industrial applications of micro algae—A review. *Journal of Algal Biomass Utilization*. 2012. 3(4):89–100.

Řezanka, T, J Lukavský, L Nedbalová, I Kolouchová, and K Sigler. 2012. Effect of starvation on the distribution of positional isomers and enantiomers of triacylglycerol in the diatom *Phaeodactylum tricornutum*. *Phytochemistry* 80:17–27.

Rismani-Yazdi, H, B Haznedaroglu, K Bibby, and J Peccia. 2011. Transcriptome sequencing and annotation of the microalgae *Dunaliella tertiolecta*: Pathway description and gene discovery for production of next-generation biofuels. *BMC Genomics* 12(1):148.

Rismani-Yazdi, H, BZ Haznedaroglu, C Hsin, and J Peccia. 2012. Transcriptomic analysis of the oleaginous microalga *Neochloris oleoabundans* reveals metabolic insights into triacylglyceride accumulation. *Biotechnology for Biofuels* 5(1):1–16.

Roche, SA and JD Leblond. 2011. Mono- and digalactosyldiacylglycerol composition of *Raphidophytes* (Raphidophyceae): A modern interpretation using positive-ion electrospray/mass spectrometry/mass spectrometry. *Journal of Phycology* 47(1):106–111.

Roleda, MY, SP Slocombe, RJG Leakey, JG Day, EM Bell, and MS Stanley. 2012. Effects of temperature and nutrient regimes on biomass and lipid production by six oleaginous microalgae in batch culture employing a two-phase cultivation strategy. *Bioresource Technology*. 129:439–449.

Samburova, V, MS Lemos, S Hiibel, SK Hoekman, JC Cushman, and B Zielinska. 2013. Analysis of triacylglycerols and free fatty acids in algae using ultra-performance liquid chromatography mass spectrometry. *Journal of the American Oil Chemists' Society* 90(1):53–64.

Sasso, S, G Pohnert, M Lohr, M Mittag, and C Hertweck. 2012. Microalgae in the postgenomic era: A blooming reservoir for new natural products. *FEMS Microbiology Reviews* 36(4):761–785.

Schneider, JC and P Roessler. 1994. Radiolabeling studies of lipids and fatty acids in *Nannochloropsis* (Eustigmatophyceae). An oleaginous marine alga. *Journal of Phycology* 30(4):594–598.

Schuhmann, K, R Herzog, D Schwudke, W Metelmann-Strupat, SR Bornstein, and A Shevchenko. 2011. Bottom-up shotgun lipidomics by higher energy collisional dissociation on LTQ Orbitrap mass spectrometers. *Analytical Chemistry* 83(14):5480–5487.

Shekh, AY, P Shrivastava, K Krishnamurthi et al. 2013. Stress-induced lipids are unsuitable as a direct biodiesel feedstock: A case study with *Chlorella pyrenoidosa*. *Bioresource Technology*. 138:382–386.

Shimojima, M, H Ohta, and Y Nakamura. 2010. Biosynthesis and function of chloroplast lipids. In *Lipids in Photosynthesis*. Springer. Berlin, Germany. pp. 35–55.

Shiran, D, I Khozin, YM Heimer, and Z Cohen. 1996. Biosynthesis of eicosapentaenoic acid in the microalga *Porphyridium cruentum*. I: The use of externally supplied fatty acids. *Lipids* 31(12):1277–1282.

Skjånes, K, C Rebours, and P Lindblad. 2013. Potential for green microalgae to produce hydrogen, pharmaceuticals and other high value products in a combined process. *Critical Reviews in Biotechnology* 33(2):172–215.

Smith, SR, RM Abbriano, and M Hildebrand. 2012. Comparative analysis of diatom genomes reveals substantial differences in the organization of carbon partitioning pathways. *Algal Research* 1(1):2–16.

Solana, M, CS Rizza, and A Bertucco. 2014. Exploiting microalgae as a source of essential fatty acids by supercritical fluid extraction of lipids: Comparison between *Scenedesmus obliquus*, *Chlorella protothecoides* and *Nannochloropsis salina*. *The Journal of Supercritical Fluids* 92:311–318.

Solovchenko, AE. 2012. Physiological role of neutral lipid accumulation in eukaryotic microalgae under stresses. *Russian Journal of Plant Physiology* 59(2):167–176.

Stoebe, B and U-G Maier. 2002. One, two, three: Nature's tool box for building plastids. *Protoplasma* 219(3–4):123–130.

Su, X, J Xu, X Yan et al. 2013. Lipidomic changes during different growth stages of *Nitzschia closterium f. minutissima*. *Metabolomics* 9(2):300–310.

Van Mooy, BAS, HF Fredricks, BE Pedler et al. 2009. Phytoplankton in the ocean use non-phosphorus lipids in response to phosphorus scarcity. *Nature* 458(7234):69–72.

Vieler, A, C Wilhelm, R Goss, R Süss, and J Schiller. 2007. The lipid composition of the unicellular green alga *Chlamydomonas reinhardtii* and the diatom *Cyclotella meneghiniana* investigated by MALDI-TOF MS and TLC. *Chemistry and Physics of Lipids* 150(2):143–155.

Wada, H and N Mizusawa. 2010. (eds) Hajime Wada and Norio Murata. The role of phosphatidyl-glycerol in photosynthesis. In *Lipids in Photosynthesis*. Springer. Berlin, Germany. pp. 243–263.

Wan, LL, J Han, M Sang et al. 2012. De novo transcriptomic analysis of an oleaginous microalga: Pathway description and gene discovery for production of next-generation biofuels. *PLOS ONE* 7(4):e35142.

Wen, Z-Y and F Chen. 2003. Heterotrophic production of eicosapentaenoic acid by microalgae. *Biotechnology Advances* 21(4):273–294.

Wenk, MR. 2010. Lipidomics: New tools and applications. *Cell* 143(6):888–895.

Xu, J, D Chen, X Yan, J Chen, and C Zhou. 2010. Global characterization of the photosynthetic glycer-olipids from a marine diatom *Stephanodiscus* sp. by ultra performance liquid chromatography coupled with electrospray ionization-quadrupole-time of flight mass spectrometry. *Analytica Chimica Acta* 663(1):60–68.

Xue, Z, PL Sharpe, S-P Hong et al. 2013. Production of omega-3 eicosapentaenoic acid by metabolic engineering of *Yarrowia lipolytica*. *Nature Biotechnology*. 31:734–740.

Yamada, T, T Uchikata, S Sakamoto et al. 2013. Supercritical fluid chromatography/orbitrap mass spectrometry based lipidomics platform coupled with automated lipid identification software for accurate lipid profiling. *Journal of Chromatography A* 1301:237–242.

Yen, H-W, I-C Hu, C-Y Chen, S-H Ho, D-J Lee, and J-S Chang. 2012. Microalgae-based biorefinery–From biofuels to natural products. *Bioresource Technology*. 135:166–174.

Zheng, M, J Tian, G Yang et al. 2013. Transcriptome sequencing, annotation and expression analysis of *Nannochloropsis* sp. at different growth phases. *Gene*. 523(2):117–121.

Section XII

Marine Biocatalysts: Approach and Applications

30

Brazilian Marine-Derived Microorganisms:
Biocatalytic Discoveries and Applications
in Organic Chemistry

Irlon M. Ferreira, Willian G. Birolli, Natália Alvarenga,
Ana M. Mouad, and André L.M. Porto

CONTENTS

30.1 Introduction

The biocatalysis and biotransformation of xenobiotic compounds applied to organic chemistry can be understood as the use of enzymes to modify organic molecules of interest, particularly via enantioselective reactions. The process of biotransformation involves transformation of a substance by complex metabolic pathways from microorganisms.

The principle for the use of biocatalysis in organic transformations of xenobiotic compounds depends upon the great diversity and number of enzymes available in nature. The analysis of the metabolic map shows an extraordinary variety of reaction types catalyzed by enzymes, which range from hydrolysis and oxidoreduction up to reactions that produce carbon–carbon bonds.

The search for efficient biotechnological processes and preservation of the environment has motivated our research group to use the principles of organic chemistry and microbiology for new applications. In this sense, the use of marine microorganisms and their enzymes is the focus of this review, which provides results achieved by the research group over the past 5 years.

30.2 Biocatalysis with Algae and Marine-Derived Microorganisms

In recent years, attention has been dedicated to the syntheses of enantiomeric pure compounds, which are used in the development of modern drugs, agrochemicals, and food industry (Matsuda et al. 2009, Nakamura et al. 2003). Among them, enantiomerically pure secondary alcohols are important chiral compounds for pharmaceuticals and other fine chemicals (Inoue et al. 2005).

Biocatalysis is one of the most important methods for the preparation of chiral alcohols. The asymmetric reduction of prochiral ketones through natural enzymes is particularly noteworthy, as it leads to enantiopure secondary alcohols.

In this way, considerable interest has been directed toward the finding of new enzymes for technological applications through biodiversity (Holsch and Weuster-Botz 2010). For instance, the marine habitat represents a potential source of biocatalysts with special properties for biotechnological applications (Trincone 2010). Fishes, plants, and algae are some examples of biocatalysts sources that have produced special enzymes with properties, such as salt tolerance, barophilicity, and adaptability to extreme temperatures (Debashish et al. 2005). Notably, these enzymes can offer novel chemical and stereochemical properties, which provide chirality to the molecule biosynthesized (Illanes 2008).

Marine algae have been extensively studied in the field of taxonomy and physiology. Similarly, their ability to produce secondary metabolites has been documented (Felício et al. 2010). However, research on the biocatalytic potential of marine algae and their associated microorganisms is scarce in the literature, especially in comparison to filamentous filamentous fungi, yeasts, and bacteria for the reduction of ketones (Utsukihara et al. 2004).

Among seaweeds, the genus *Bostrychia* (Rhodomelaceae, Ceramiales) has been widespread in tropical and warm temperate environments (De Oliveira et al. 2012). Here, we address the use of marine algae *Bostrychia tenella* and *Bostrychia radicans* and their associated microorganisms for the reduction of prochiral ketones, such as halogenated acetophenone derivatives and trifluoromethylketones.

30.2.1 Reduction of Ketones by Marine Red Algae *Bostrychia tenella* and *Bostrychia radicans* and Their Associated Microorganisms

The biocatalytic reduction of acetophenone derivatives was initially exploited with algal biomass from *B. tenella* and *B. radicans* (Table 30.1). The biomass of *B. tenella* was used in the bioreduction of acetophenones with different substituent groups, such as F, Cl, Br, and NO_2 in the *ortho*-position. The reactions were performed under the same experimental conditions, and the results showed that the reduction of *ortho*-fluoroacetophenone yielded (S)-fluoroalcohol with 52% conversion and enantiomeric excess >99%. However, the bulky groups in the *ortho*-position (Cl, Br, and NO_2) led to a decrease in the conversion of the respective (S)-alcohols, but with high enantiomeric excesses (>99% *ee*) (Table 30.1) (Mouad et al. 2011, Mouad 2013).

In studies of algae biomass, the associated microorganisms were isolated from red alga *B. radicans*. Bacteria are opportunistic microorganisms easily found in algae. In this experiment, strain Br62 was isolated from *B. radicans* and identified as *Bacillus*. The associated bacteria *Bacillus* sp. Br62 was employed as biocatalysts in the reduction of *ortho*-acetophenone derivatives (Table 30.1) (Mouad et al. 2011).

The reactions catalyzed by strain Br62 with *ortho*-fluoro and *ortho*-bromoacetophenones led to high conversions and enantiomeric excess (>99%) (Table 30.1), and only *ortho*-nitroacetophenone was obtained with poor conversion and selectivity. The study demonstrated the importance of microorganism screenings, and new results for the production of chiral alcohols have been obtained.

TABLE 30.1

Reduction of *Ortho*-Acetophenones by Algal Biomass of *Bostrychia tenella* and Its Bacteria-Associated *Bacillus* sp. Br62[a]

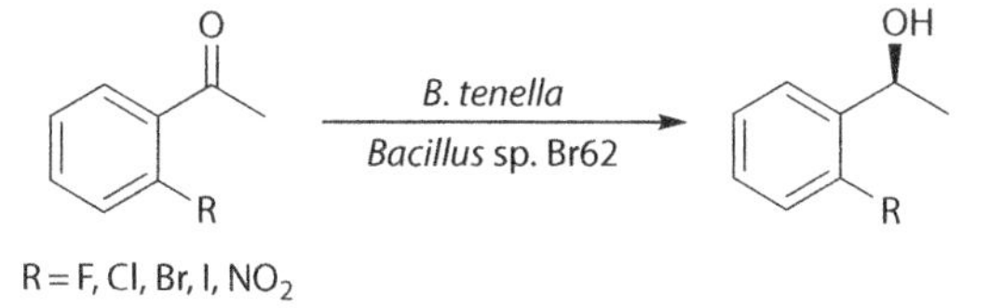

R = F, Cl, Br, I, NO_2

Ortho-Ketones	*c* (%) Ketones	Conv. (%) Alcohols	*ee* (%) (ac) Alcohols
Bostrychia tenella			
F	48	52	99 (S)
Cl	80	20	99 (S)
Br	90	10	99 (S)
NO_2	87	13	99 (S)
Bacillus sp. Br62			
F	27	73	99 (S)
Cl	51	49	99 (S)
Br	5	95	99 (S)
I	34	46	99 (S)
NO_2	93	7	10 (S)

Abbreviations: ac, absolute configuration; *c*, concentration; conv., conversion; *ee*, enantiomeric excess.

[a] Reaction condition (Mouad et al. 2011).

TABLE 30.2

Bioreduction of Iodoacetophenones with Algal Biomass of *B. tenella* and *B. radicans*[a]

Algae	Iodoketones	*c* (%) Ketones	Conv. (%) Alcohols	*ee* (%) (ac)
Bostrychia radicans	*Ortho*	58	42	99 (S)
	Meta	92	8	99 (S)
	Para	83	17	99 (S)
Bostrychia tenella	*Ortho*	60	40	99 (S)
	Meta	95	5	99 (S)
	Para	84	16	99 (S)

Abbreviations: ac, absolute configuration; *c*, concentration; conv., conversion; *ee*, enantiomeric excess.
[a] Reaction condition (Mouad et al. 2011).

A screening of *ortho-*, *meta-*, and *para*-substituted iodoacetophenones by whole fragments of algae from *B. tenella* and *B. radicans* was evaluated in order to produce enantiopure iodoalcohols (Table 30.2) (Mouad et al. 2011, Mouad 2013).

Reactions with *ortho*-iodoacetophenone afforded the best results and provided (S)-alcohol with moderated conversions (40%–42%) by *B. tenella* and *B. radicans* algae (Table 30.2). The reduction of *meta*-iodoacetophenone and *para*-iodoacetophenone with both algae produced lower conversions of the respective (S)-alcohols (Table 30.2).

Although the conversion values were obtained in a low–moderate range, both algae yielded enantiomerically pure (iodophenyl)ethanol with high enantiomeric excesses in all cases (>99% *ee*) (Table 30.2). It is noteworthy that the reduction of acetophenone derivatives containing a substituent group in the *ortho*-position is unusual and indicates the presence of specific enzymes for the production of enantiomerically pure or enriched chiral alcohols (Mouad et al. 2011, Mouad 2013).

In a strategy for finding new biocatalysts, four marine-associated fungi were isolated from red algae *B. radicans*, identified as *Botryosphaeria* sp. CBMAI 1197, *Eutypella* sp. CBMAI 1196, *Hidropisphaera* sp. CBMAI 1194, and *Xylaria* sp. CBMAI 1195 (Mouad 2014). These biocatalysts were employed in reactions with fluorinated ketones, once studies have shown that fluorinated organic compounds are potential drugs (Tables 30.3 through 30.5). The presence of a fluorine atom especially from the trifluoromethyl group confers high stability to the compound and avoids undesirable decompositions (Lam et al. 2009, Nie et al. 2011). The fungus *Botryosphaeria* sp. CBMAI 1197 produced 100% of the respective (S)-alcohol with high enantiomeric excess (>99% *ee*) and was considered an excellent biocatalyst in the reduction of the trifluorinated ketone. The reduction with fungus *Eutypella* sp. CBMAI 1196 also produced the respective (S)-alcohol with excellent enantiomeric excess (>99% *ee*) and conversion of 38% (Table 30.3).

The biocatalytic reactions with *Hidropisphaera* sp. CBMAI 1194 and *Xylaria* sp. CBMAI 1195 produced the (S)-alcohol with >99% *ee*; however, these fungi did not exhibit significant potential in conversions of 2,2,2-trifluoro-1-phenylethanone (Table 30.3). The whole cells of marine fungi catalyzed the bioreduction of the original compound and produced the derivative (S)-alcohols with anti-Prelog selectivity in all cases (Prelog 1964, Yadav et al. 2008).

TABLE 30.3

Reduction of 2,2,2-Triflouro-1-Phenylethanone with Marine Fungi from the Red Marine Alga *Bostrychia radicans*[a]

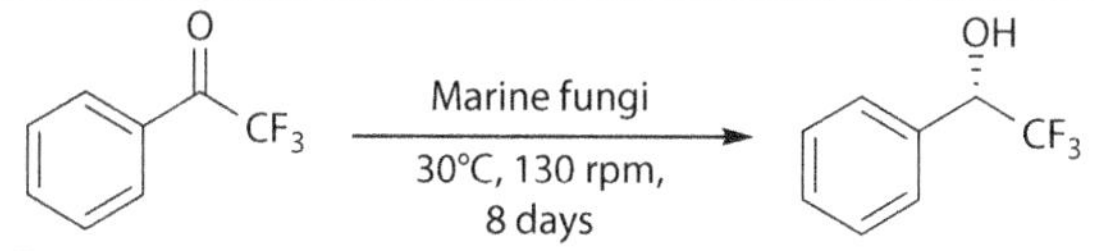

Fungi	*c* (%) Ketone	Conv. (%) Alcohol	*ee* (%) (ac) Alcohol
Botryosphaeria sp. CBMAI 1197	0	100	99 (S)
Eutypella sp. CBMAI 1196	62	38	99 (S)
Hidropisphaera sp. CBMAI 1194	94	6	99 (S)
Xylaria sp. CBMAI 1195	95	5	99 (S)

Abbreviations: ac, absolute configuration; *c*, concentration; conv., conversion; *ee*, enantiomeric excess; (Mouad et al. 2015).

[a] Reaction condition (Mouad 2014).

TABLE 30.4

Reduction of 1-(2-(Trifluoromethyl)phenyl)ethanone with Marine Fungi from the Red Marine Alga *Bostrychia radicans*[a]

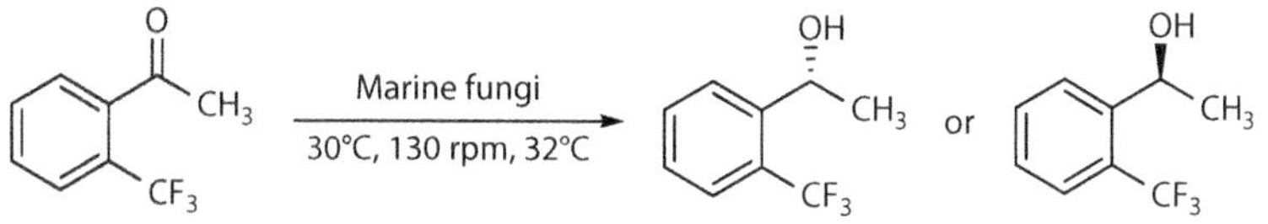

Fungi	*c* (%) Ketone	Conv. (%) Alcohol	*ee* (%) (ac) Alcohol
Botryosphaeria sp. CBMAI 1197	54	46	99 (S)
Eutypella sp. CBMAI 1196	84	16	99 (S)
Hidropisphaera sp. CBMAI 1194	60	40	58 (R)
Xylaria sp. CBMAI 1195	98	2	99 (S)

Abbreviations: ac, absolute configuration; *c*, concentration; conv., conversion; *ee*, enantiomeric excess; (Mouad et al. 2015).

[a] Reaction condition (Mouad 2014).

In the reactions with 1-(2-(trifluoromethyl)phenyl)ethanone, the best biocatalysts were fungi *Botryosphaeria* sp. CBMAI 1197 and *Hidropisphaera* sp. CBMAI 1194, whose active enzymes produced the antipodes of the respective chiral alcohol (*R*- and *S*-enantiomers) (Table 30.4).

The fungus *Botryosphaeria* sp. CBMAI 1197 was selective and it produced a single enantiomer, (*S*)-alcohol with high enantioselectivity (>99% *ee*) and moderate conversion (46%). Enzymes from fungus *Hidropisphaera* sp. CBMAI 1194 formed the antipode, *R*-alcohol, with 40% of conversion and 58% of enantiomeric excess. The reactions with fungi *Eutypella* sp. CBMAI 1196 and *Xylaria* sp. CBMAI 1195 showed low conversions (16% and 2%), respectively (Table 30.4).

The reactions with 1-(2,4,5-trifluorophenyl)ethanone with *Botryosphaeria* sp. CBMAI 1197 produced an excellent result for (*S*)-alcohol with 97% conversion and >99% *ee*. Enzymes from fungus *Hidropisphaera* sp. CBMAI 1194 catalyzed the production of the respective (*S*)-alcohol with 100% conversion. However, a decrease was observed in the enzymatic enantioselectivity (53% *ee*) (Table 30.5).

TABLE 30.5

Reduction of 1-(2,4,5-Trifluorophenyl)ethanone with Marine Fungi from the Red Marine Alga *Bostrychia radicans*[a]

Fungi	c (%) Ketone	Conv. (%) Alcohol	ee (%) (ac) Alcohol
Botryosphaeria sp. CBMAI 1197	3	97	99 (S)
Eutypella sp. CBMAI 1196	100	0	0
Hidropisphaera sp. CBMAI 1194	0	100	53 (S)
Xylaria sp. CBMAI 1195	100	0	0

Abbreviations: ac, absolute configuration; c, concentration; conv., conversion; ee, enantiomeric excess; (Mouad et al. 2015).

[a] Reaction condition (Mouad 2014).

The fungi *Eutypella* sp. CBMAI 1196 and *Xylaria* sp. CBMAI 1195 showed no conversion to alcohols. The reaction conditions did not favor the activity of alcohol dehydrogenases present in the fungi *Eutypella* sp. CBMAI 1196 and *Xylaria* sp. CBMAI 1195 (Table 30.5).

Seaweeds *B. radicans* and *B. tenella* shown be potential biocatalysts for the obtainment of enantiomerically pure alcohols. According to the results, algae and their associated microorganisms can be used as biocatalysts for the reduction of acetophenone derivatives.

30.2.2 Reduction of Ketones by Marine-Derived Fungi Isolated from Marine Sponges

The biocatalytic asymmetric reduction of α-haloketones represents a potentially useful method for the preparation of chiral halohydrins, which are important intermediates in the synthesis of numerous compounds of biological interest, as optically active epoxides (Barbieri et al. 2001).

Chiral (S)-2-chloro-1-phenylethan-1-ol was synthesized by the bioreduction of α-chloroacetophenone by marine-derived fungi (*Trichoderma* sp. Gc1, *Penicillium miczynskii* Gc5, *Aspergillus sydowii* Gc12, *Bionectria* sp. Ce5 [*Bionectria* cf. *ochroleuca*], *A. sydowii* Ce15, *Penicillium raistrickii* Ce16, and *A. sydowii* Ce19) isolated from marine sponges *Geodia corticostylifera* and *Chelonaplysylla erecta* (Table 30.6) (Rocha et al. 2009).

Initially, the microorganisms were grown in an artificial seawater-based medium containing a high concentration of ions Cl⁻ (1.2 M), similar to natural seawater (0.5 M). In some experiments, marine fungi were also grown in the medium with freshwater and the growth was similar to that obtained in a seawater medium. However, the expression of oxidoreductases was directly dependent on the growth conditions similar to those of a marine environment, therefore, the enzymes were active only in seawater-based media (Rocha et al. 2009).

Table 30.6 shows the conversions and selectivities of the (S)-chlorohydrin obtained by the microbial reduction of α-chloroacetophenone in phosphate buffer at pH 7 by whole cells of *Trichoderma* sp. Gc1, *P. miczynskii* Gc5, *A. sydowii* Gc12, *Bionectria* sp. Ce5, *A. sydowii* Ce15, *P. raistrickii* Ce16, and *A. sydowii* Ce19 (Rocha et al. 2009).

TABLE 30.6

Bioreduction of α-Chloroacetophenone to (*S*)-(-)-2-Chloro-1-phenylethanol by Whole Cells of Marine Fungi[a]

Fungi	Time (h)	*c* (%) Haloketone	Conv. Halohydrin	*ee* (%) (ac) Halohydrin
P. miczynskii Gc5	48	5	95	50 (*S*)
Trichoderma sp. Gc1	48	70	30	66 (*S*)
Aspergillus sydowii Gc12	48	77	23	20 (*S*)
A. sydowii Ce19	72	22	78	20 (*S*)
A. sydowii Ce15	72	12	88	35 (*S*)
Bionectria sp. Ce5	72	1	99	22 (*S*)
P. miczynskii Ce16	72	1	99	17 (*S*)

Abbreviation: ac, absolute configuration; *c*, concentration; conv., conversion; *ee*, enantiomeric excess.

[a] Reaction condition (Rocha et al. 2009).

Enantiopure iodoalcohols can be used as versatile building blocks in the synthesis of new chiral compounds by various cross-coupling processes (Ferreira et al. 2014b).

Asymmetric bioreduction of iodoacetophenones was investigated with nine marine fungi (*Aspergillus sclerotiorum* CBMAI 849, *A. sydowii* Ce19, *Beauveria felina* CBMAI 738, *Mucor racemosus* CBMAI 847, *Penicillium citrinum* CBMAI 1186, *P. miczynskii* Ce16, *P. miczynskii* Gc5, *Penicillium oxalicum* CBMAI 1185, and *Trichoderma* sp. Gc1) (Table 30.7) (Rocha et al. 2012b).

The chiral iodinated compounds obtained from marine fungi can be easily converted into interesting chiral blocks by Suzuki reaction, giving enantiopure biphenyl compounds (Scheme 30.1). All the marine fungi produced exclusively (*S*)-*ortho*-iodophenylethanol and (*S*)-*meta*-iodophenylethanol in accordance with the Prelog rule. *Beauveria felina* CBMAI 738, *P. miczynskii* Gc5, *P. oxalicum* CBMAI 1185, and *Trichoderma* sp. Gc1 produced (*R*)-*para*-iodophenylethanol as an anti-Prelog product (Table 30.7) (Prelog and Wieland 1966, Rocha et al. 2012b).

The bioconversion of *p*-iodoacetophenone with whole cells of *P. oxalicum* CBMAI 1185 showed competition between reduction and oxidation reactions. In this case, one of the two redox reactions (Scheme 30.2) was reversible, and this process is closely related to the deracemization reaction (Comasseto et al. 2004). The microbiological reduction was monitored by Gas Chromatography-Mass analysis (GC-MS), and *p*-iodoacetophenone showed a change in the enantiomeric excess for (*R*)-alcohol, initially with *ee* = 85% in 48 h to *ee* > 99% at 168 h, and a decrease in the conversion of the *para*-iodoacetophenone with *ee* = 41% in 48 h, to 16% at 168 h with whole cells of *P. oxalicum* CBMAI 1185. The enantiomer (*S*)-alcohol was frequently oxidized to the *para*-iodoacetophenone, whereas the other enantiomer remained either unchanged or oxidized at a slower rate.

Nine different strains of marine-derived fungi (*A. sydowii* Ce15, *A. sydowii* Ce19, *A. sclerotiorum* CBMAI 849, *Bionectria* sp. Ce5, *B. felina* CBMAI 738, *Cladosporium cladosporioides* CBMAI 857, *M. racemosus* CBMAI 847, *P. citrinum* CBMAI 1186, and *P. miczynskii* Gc5) were used in the asymmetric bioreduction of 1-(4-methoxyphenyl)ethanone to its corresponding 1-(4-methoxyphenyl)ethanol (Scheme 30.3) (Rocha et al. 2012a).

TABLE 30.7

Bioconversion of Iodoacetophenones by Marine-Derived Fungi from Marine Sponges[a]

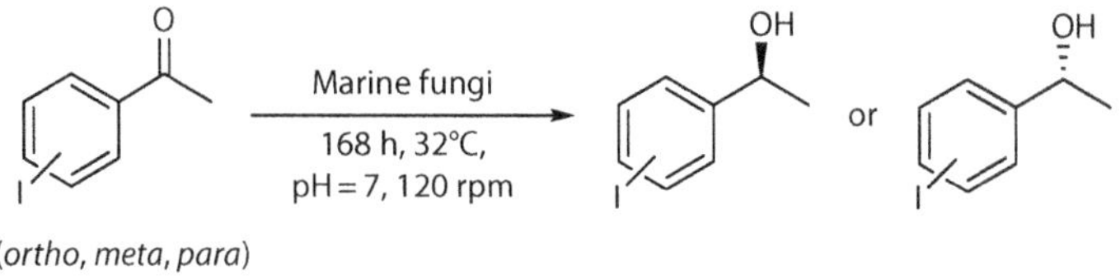

I (ortho, meta, para)

	Ortho-Iodoacetophenone		Meta-Iodoacetophenone		Para-Iodoacetophenone	
Conv. (%) Alcohols		ee (%) (ac)	Conv. (%) Alcohols	ee (%) (ac)	Conv. (%) Alcohols	ee (%) (ac)
Aspergillus sclerotiorum CBMAI 849						
6		74 (S)	31	>99 (S)	37	77 (S)
Aspergillus sydowii Gc19						
—			18	35 (S)	92	>99 (S)
Beauveria felina CBMAI 738						
—		—	—	—	71	>99 (R)
Mucor racemosus CBMAI 847						
23		>99 (S)	29	>99 (S)	84	>99 (S)
Penicillium citrinum CBMAI 1186						
86		>99 (S)	28	>99 (S)	—	—
Penicillium miczynskii Ce16						
4		—	56	94 (S)	72	>99 (S)
Penicillium miczynskii Gc15						
27		>99 (S)	26	62 (S)	25	18 (R)
Penicillium oxalicum CBMAI 1185						
49		>99 (S)	10	>99 (S)	16	>99 (R)
Penicillium miczynskii Gc15						
27		>99 (S)	25	62 (S)	67	32 (R)

Abbreviations: ac, absolute configuration; c, concentration; conv., conversion; ee, enantiomeric excess.

[a] Reaction condition (Rocha et al. 2012b).

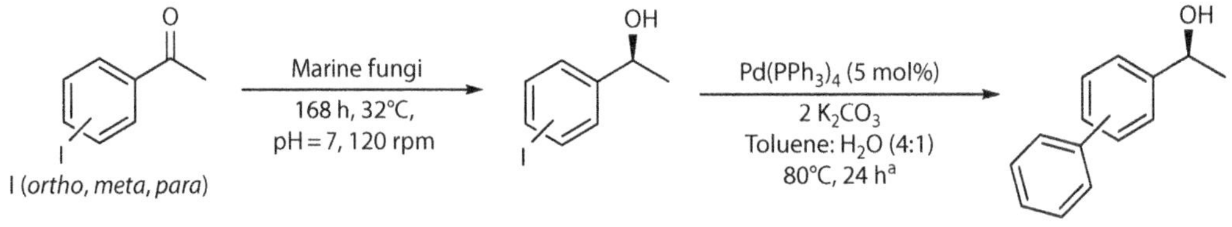

I (ortho, meta, para)

SCHEME 30.1

Bioconversion of iodoacetophenones by marine fungi and synthesis of biphenyl compounds. [a] Rocha et al. (2010b).

Marine fungi *B. felina* CBMAI 738, *A. sclerotiorum* CBMAI 849, and *P. miczynskii* Gc5 catalyzed the bioreduction of 1-(4-methoxyphenyl)ethanone in accordance with the Prelog rule. Fungi *A. sclerotiorum* CBMAI 849 and *B. felina* CBMAI 738 catalyzed the bioreduction of ketone to (S)-alcohol. After purification by silica gel column chromatography, (S)-alcohol was obtained with 90% yield (Scheme 30.3) (Rocha et al. 2012a).

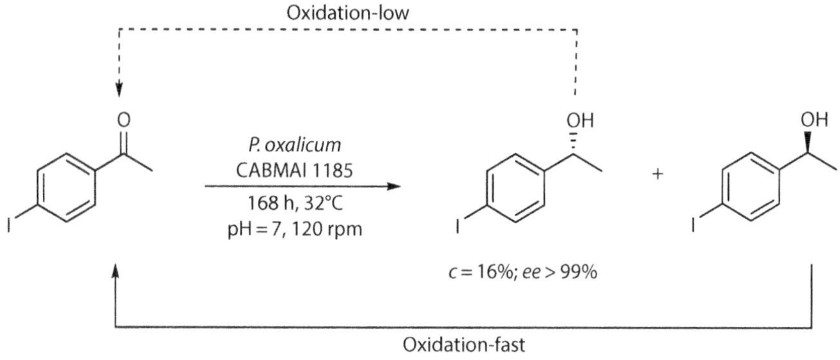

SCHEME 30.2
Competitive oxidation–reduction reactions of *para*-iodophenylethanol with the marine fungus *P. oxalicum* CBMAI 1185.

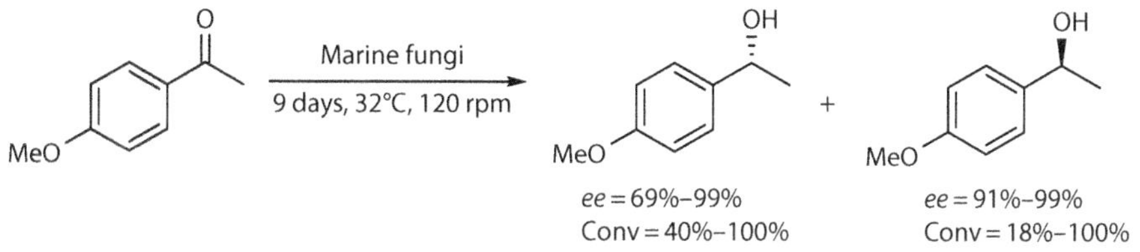

SCHEME 30.3
Bioreduction of 1-(4-methoxyphenyl)ethanone to (S)- or (R)-1-(4-methoxyphenyl)ethanol by marine-derived fungi.

Fungi *A. sclerotiorum* CBMAI 849 and *P. miczynskii* Gc5 catalyzed the bioreduction of 1-(4-methoxyphenyl)ethanone to (S)-alcohol and showed different results (c 100% and $ee >$ 99% and c 18% and ee 91%, respectively). After 9 days of incubation, (S)- and (R)-alcohols were produced with 95% and 15% isolated yield by *A. sclerotiorum* CBMAI 849 and *P. miczynskii* Gc5, respectively. Fungi *M. racemosus* CBMAI 847, *A. sydowii* Ce19, and *C. cladosporioides* CBMAI 857 catalyzed the bioreduction of 1-(4-methoxyphenyl)ethanone at a low conversion rate ($c < 10\%$). The results showed that the use of microbial cells for the bioreduction of 1-(4-methoxyphenyl)ethanone depends on the xenobiotic substrate and the strains used in these experiments (Scheme 30.3) (Rocha et al. 2012a).

Whole cells of marine fungi *A. sydowii* Ce19 were used to catalyze the reduction of 2-bromo-1-phenylethanone to the corresponding (R)-2-bromo-1-phenylethanol in an artificial seawater medium containing a high concentration of chloride ions (1.20 M). The products such as 2-bromo-1-phenylethanol (56%), 2-chloro-1-phenylethanol (9%), 1-phenylethan-1,2-diol (26%), acetophenone (4%), and phenylethanol (5%) were generated spontaneously in the biotransformation reactions by fungi *A. sydowii* Ce19. Marine fungus *A. sydowii* Ce19 showed potential for biodegradation and biotransformation of the α-bromo acetophenone compound (Scheme 30.4) (Rocha et al. 2010a).

The growth curves of marine-derived fungi (*A. sydowii* CBMAI 933, *P. miczynskii* CBMAI 930, and *Trichoderma* sp. CBMAI 932) were obtained over different periods of time. The exponential growth phases were characteristic of each fungal strain growth in the malt extract liquid medium (Melgar et al. 2013).

For *A. sydowii* CBMAI 933, the growth curve showed a log phase between 48 and 144 h, and an accentuated fall in the mycelial mass was observed until 192 h, possibly because the nutrient had been consumed, which led to the death phase (Melgar et al. 2013).

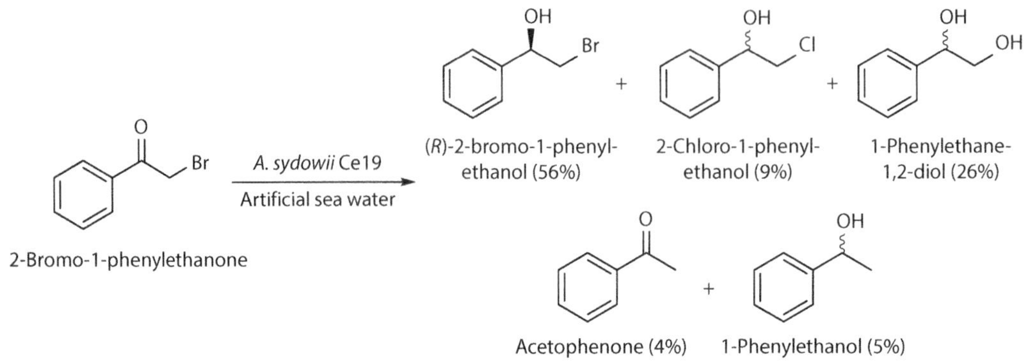

SCHEME 30.4
Biotransformation and biodegradation of 2-bromo-1-phenylethanone by *A. sudowii* Ce19.

For the marine fungus *P. miczynskii* CBMAI 930, the log phase in the growth curve has maximum mass at 192 h, after which the growth curve showed a negative slope at 240 h. *Trichoderma* sp. CBMAI 932 showed 48–144 h log phase growth period, which indicates that the preliminary determination of growth curves of filamentous fungi is important for the optimization of the conditions for the biocatalytic reduction of xenobiotic compounds.

By the result presented in this section, it is possible to conclude that marine fungi (*A. sclerotiorum* CBMAI 849, *A. sydowii* Ce19, *B. felina* CBMAI 738, *M. racemosus* CBMAI 847, *P. citrinum* CBMAI 1186, *P. miczynskii* Ce16, *P. miczynskii* Gc5, *P. oxalicum* CBMAI 1185, and *Trichoderma* sp. Gc1) showed excellent potential as biocatalyst for the enantioselective reduction of acetophenone derivatives. In some cases, the alcohols produced showed a high enantiomeric excess and good conversion. Particularly, this biotransformation by marine fungi is an alternative and attractive method for the production of enantiomerically enriched or pure alcohols, which can be converted into other compounds of pharmacological interest.

30.2.3　Reduction of α,β- and α,β,γ,δ-Unsaturated Ketones by Marine-Derived Fungi

The reduction of carbon–carbon double bond of alkenes or unsaturated ketones catalyzed by enoate reductases (ERs) is undoubtedly one of the more emerging methods in biocatalysis, representing a key step in the synthesis of important substances.

In general, the reduction of carbon–carbon double bond involves chemical methods without chemoselectivity and eco-friendly conditions, with metal salts and complexes, including copper (Nahra et al. 2013), palladium (Sommovigo 1993), and rhodium (Shiomi et al. 2009). So, following the concern over environmental pollution and sustainable development, new chemoselective methods for the reduction of carbon–carbon double bond from enones using enzymes are emerging (Winkler et al. 2012).

Among the most common proteins used in the bioreduction of carbon–carbon double bond, the old yellow enzyme (OYE) has been the center of attention. This family of flavoproteins is based on the OYE, which was first isolated from brewer's bottom yeast. Members of this protein family have been found in other yeasts, bacteria, plants, fungus, and nematodes. Analyses of the three-dimensional structures of those OYE homologs, which have been crystallized, show that they consist of one or more monomers with an α/β-barrel structure and a flavin mononucleotide prosthetic group. In spite of detailed structural knowledge, the physiological role of most of these enzymes remains unknown (Kohli and Massey 1998, Stuermer et al. 2007).

Whole free mycelia of *P. citrinum* CBMAI 1186 were tested for the reduction of the carbon–carbon double bond of chalcone, (*E*)-3-(4-fluorophenyl)-1-phenylprop-2-en-1-one, yielding 98% of the dihydrochalcone 3-(4-fluorophenyl)-1-phenylpropan-1-one after 6 days. Compounds such as (*E*)-3-(4-fluorophenyl)-1-phenylprop-2-en-1-ol and 3-(4-fluorophenyl)-1-phenylpropan-1-ol, which can result from the reduction of the carbonyl group (catalyzed by alcohol dehydrogenase), were not detected (Scheme 30.5). Therefore, the biotransformation of (*E*)-3-(4-fluorophenyl)-1-phenylprop-2-en-1-one by the whole fungus showed excellent chemoselectivity in the reduction of carbon–carbon double bond (Ferreira et al. 2014c).

The reduction of the double bond is directly influenced by substituent groups attached to ring B in chalcones by means of the conjugation of π-bonds on this side (ring B) of the molecule (Silva et al. 2010). Surprisingly, chalcones containing electron-withdrawing or electron-donating groups were chemoselectively reduced to dihydrochalcones in good conversions using whole cells of *P. citrinum* CBMAI 1186 (Scheme 30.6).

Little has been known about the tolerance of marine fungus *P. citrinum* CBMAI 1186 to organic solvents. A preliminary study was conducted in the presence of various organic solvents to evaluate the microbial growth of the fungus. The solvents were chosen according to log *P* values and polarities. The experiment was carried out in duplicate, and the results are summarized in Table 30.8 (Ferreira et al. 2014a).

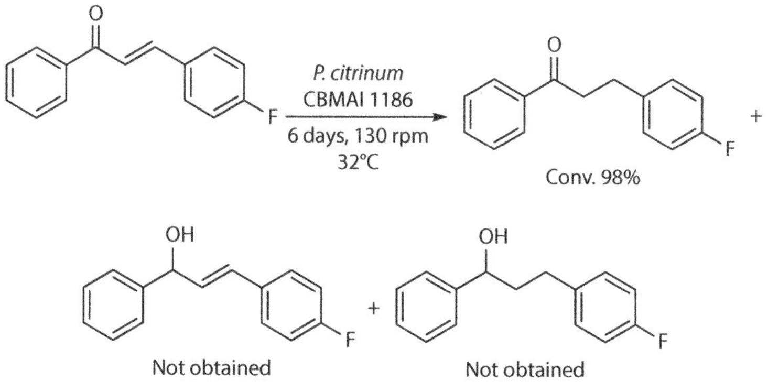

SCHEME 30.5
Reduction of (*E*)-3-(4-fluorophenyl)-1-phenylprop-2-en-1-one by *P. citrinum* CBMAI 1186.

$R_1 = R_2 = H, R_3 = F; R_1 = R_2 = H; R_3 = Br$
$R_1 = R_2 = H, R_3 = OCH_3; R_1 = OCH_3, R_2 = R_3 = H$
$R_1 = R_3 = H, R_2 = NO_2$

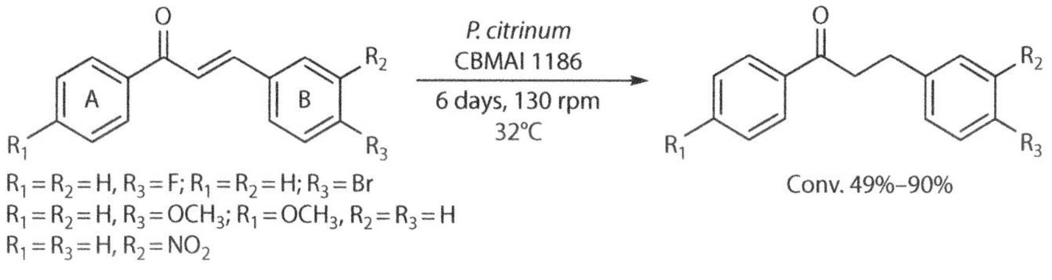

SCHEME 30.6
Chemoselective reduction of chalcones by *P. citrinum* CBMAI 1186.

TABLE 30.8

Growth of Marine Fungus *P. citrinum* CBMAI 1186 in the Presence
of Organic Solvents[a]

Solvents (5%)	log P[b]	Weight Dry (mg)
Artifical seawater	—	783
Acetone	−0.24	193
Ethyl acetate	0.73	16
n-Butanol	0.88	12
Dichloromethane	0.93	12
n-Hexane	3.98	724
Toluene	2.73	32

[a] Reaction condition (Ferreira et al. 2014c).
[b] Logarithm of the partition coefficient (Ferreira et al. 2014a).

The results suggest that the solvent of lowest toxicity to *P. citrinum* CBMAI 1186 was
n-hexane, which is consistent with the empirical rule by which the toxicity of the solvent
is expressed as the logarithm of the partition coefficient (log *P*). A solvent whose log *P* is
greater than 4.0 tends to slowly whittle the toxicity for the microbial cells, while solvents
with low value tend to inactivate the mycelial growth. As log *P* of *n*-hexane is 3.98, it was
considered the best medium for the growth of *P. citrinum* CBMAI 1186 cells.

In another study (Ferreira et al. 2014a), the chemoselective reduction of C=C unsaturated bond was explored with the fungus *P. citrinum* CBMAI 1186 in the hydrogenation of
carbon-carbon bonds in α,β- and α,β,γ,δ-unsaturated ketones employing a biphasic system
of phosphate buffer and *n*-hexane 9:1 (Scheme 30.7).

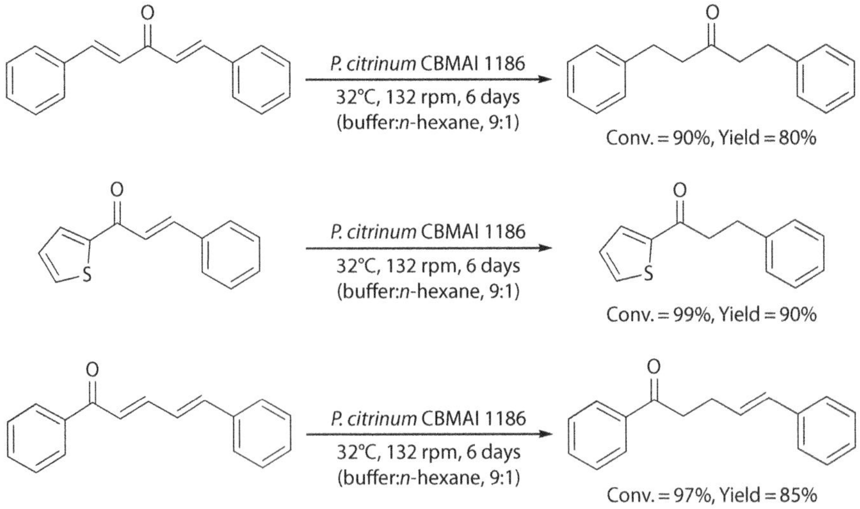

SCHEME 30.7
Bioreduction of carbon–carbon double bond by marine fungus *P. citrinum* in a biphasic system.

The biotransformation of α,β-unsaturated ketones led exclusively to the formation of saturated ketones after 6 days of inoculation with mycelia of *P. citrinum* CBMAI 1186 (32°C and 132 rpm) (Scheme 30.7). The yields obtained in the biotransformation reaction of chalcones to dihydrochalcones were excellent and ranged between 80% and 90%.

In another study, we evaluated the action of the enzymatic broth and whole cells of marine-derived fungi to investigate the bioreduction of carbon–carbon bonds in organic compounds. The enzymatic broth and mycelia of marine-derived fungus *Trichoderma* sp. CBMAI 932 were used in the bioconversion of (1*E*,4*E*)-1,5-bis(4-methoxyphenyl)penta-1,4-dien-3-one to the corresponding products, namely, 1,5-bis(4-methoxyphenyl) pentan-3-one (37%), (*E*)-1,5-bis(4-methoxyphenyl)pent-1-en-3-one (11%), and 1,5-bis(4-methoxyphenyl)pentan-3-ol (9%) (Scheme 30.8); these products resulted from the action of oxidoreductase enzymes like alcohol dehydrogenase and ER. Interestingly, when cells of *Trichoderma* sp. CBMAI 932 were used, the corresponding alcohol 1,5-bis(4-methoxyphenyl)pentan-3-ol was converted as the major product of biotransformation with a yield of 37%; 1,5-bis(4-methoxyphenyl)pentan-3-one, 20%; and (*E*)-1,5-bis(4-methoxyphenyl)pent-1-en-3-one, 4%. The route of the microbial biotransformation of di-α,β-unsaturated ketone initially involved flavin-dependent ERs, which catalyze the asymmetric reduction of double C=C bonds and form the monoreduced compound (Scheme 30.9).

The catalytic mechanism has been solved and studied in detail (Vaz et al. 1995). A hydride derived from the flavin cofactor is transferred stereoselectively onto a carbon atom of the double bond. *Step* A involves the transfer of the hydride by the ER enzyme again, forming the di-reduced compound, which was subsequently reduced to an alcohol group by oxidoreductase enzyme, as shown in *step C*. The formation of the compound

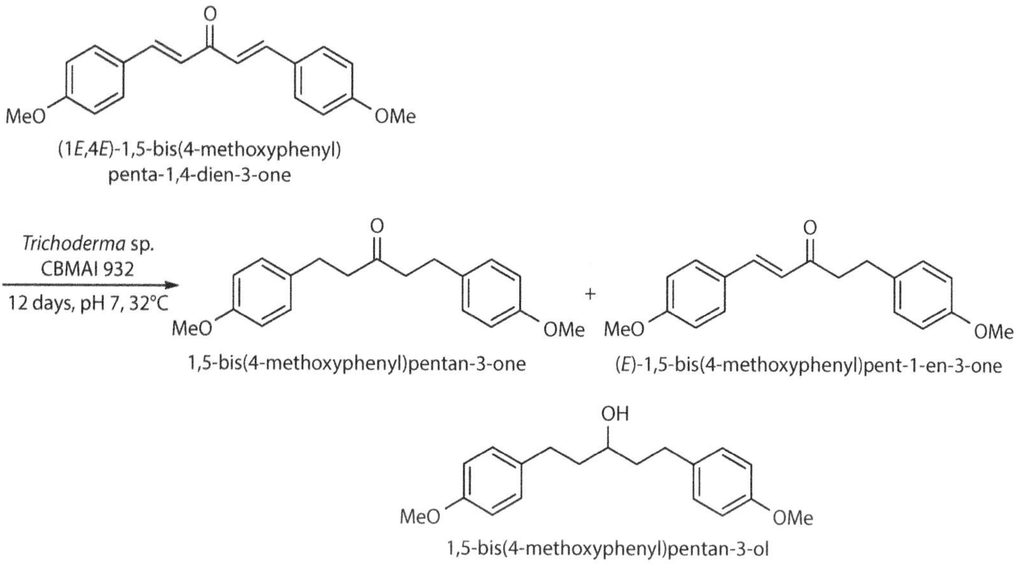

(1*E*,4*E*)-1,5-bis(4-methoxyphenyl)
penta-1,4-dien-3-one

Trichoderma sp.
CBMAI 932

12 days, pH 7, 32°C

1,5-bis(4-methoxyphenyl)pentan-3-one

+

(*E*)-1,5-bis(4-methoxyphenyl)pent-1-en-3-one

1,5-bis(4-methoxyphenyl)pentan-3-ol

SCHEME 30.8
Microbial biotransformation of α,β-unsaturated (1*E*,4*E*)-1,5-bis(4-methoxyphenyl)penta-1,4-dien-3-one.

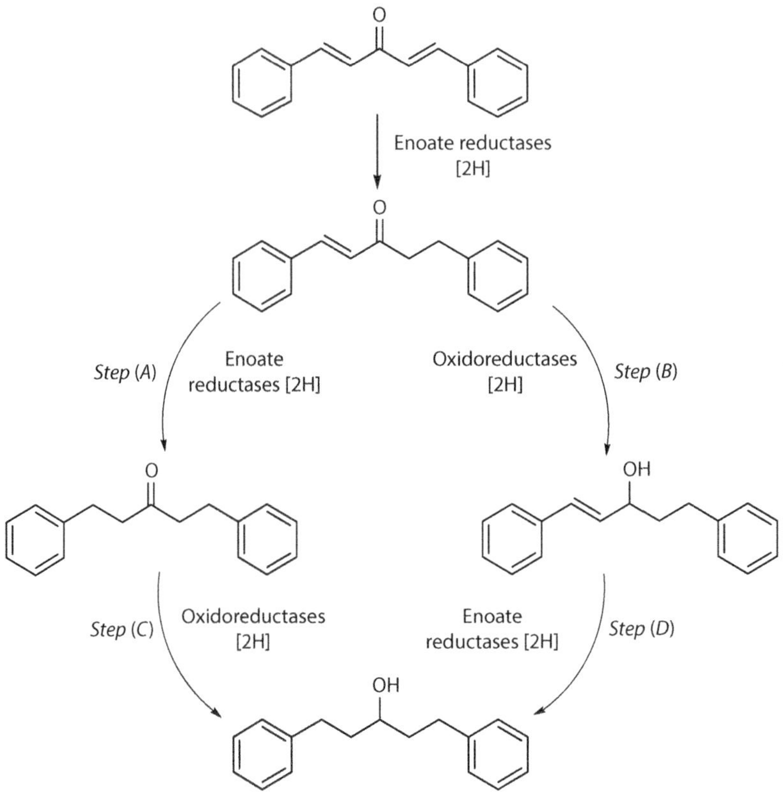

SCHEME 30.9
Biotransformation of α,β-unsaturated (1*E*,4*E*)-1,5-bis(4-methoxyphenyl)penta-1,4-dien-3-one by marine-derived fungi.

enolylic (*E*)-1,5-diphenylpent-1-en-3-ol passes through hydride transfer to the carbonyl compound, generating the allylic alcohol. Then, this subsequently leads to the reduction of the oxidoreductase enzyme, as shown in *step B*, by the action of another ER, yielding the 1,5-diphenylpentan-3-ol product, as shown in *step D* (Scheme 30.9).

The chemoselective and regioselective reduction of the α,β-unsaturated bonds of compounds similar to (2*E*,4*E*)-1,5-diphenylpenta-2,4-dien-1-one in excellent yield is an arduous task for chemists by synthetic methodology, since the reduction does not occur with high selectivity. Catalytic metals, such as InCl$_3$ (Che and Lam 2010), are used in the presence of NaBH$_4$ or TiCl$_4$ with Hantzsch ester (Ojima et al. 1972) or rhodium with trimethylsilane (Brenna et al. 2013). The reduction of (2*E*,4*E*)-1,5-diphenylpenta-2,4-dien-1-one by wet mycelia of the marine-derived microorganism *P. citrinum* CBMAI 1186 is an innovative, simple, and inexpensive method of biocatalysis, as reported by Ferreira et al. (2014a) (Scheme 30.10).

This was the first report of the regioselective and chemoselective bioreduction of α,β- and α,β,γ,δ-unsaturated ketones by marine-derived fungi, and the results suggest that the mycelium of marine-derived fungus *P. citrinum* CBMAI 1186 is a potential biocatalyst for the bioreduction of carbon–carbon bond and eco-friendly reactions.

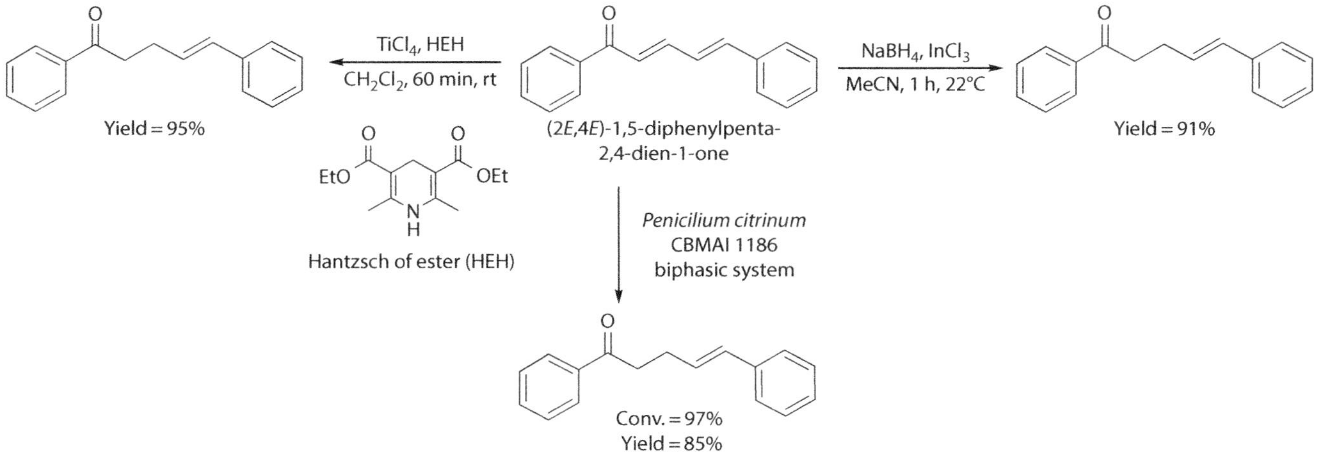

SCHEME 30.10

Chemoselective and regioselective (bio)reduction of α,β,γ,δ-unsaturated (2E,4E)-1,5-diphenylpenta-2,4-dien-1-one by chemical and enzymatic methods.

30.3 Biocatalysis of Nitriles and Epoxides by Marine-Derived Fungi

30.3.1 Hydrolysis of Nitriles by Fungi Isolated from Marine Sponges

Nitriles are important starting materials and intermediates in the synthesis of amines, amides, amidines, carboxylic acids, esters, aldehydes, ketones, and heterocyclic compounds (Banerjee et al. 2002, Jin et al. 2013). In the chemical industry, nitrile compounds are used in the manufacture of a variety of polymers, such as polyacrylonitrile and nylon-6,6 (Banerjee et al. 2002). As a result of their extensive use, industrial nitrile compounds are widespread in the natural environment. Moreover, the nitriles produced by plants such as cyanoglycosides, cyanolipids, ricinine, and phenylacetonitrile are also found (Gong et al. 2012).

Since many nitriles are toxic, mutagenic, and/or carcinogenic because of the cyano group, their biotransformation could be extremely valuable (Banerjee et al. 2002). Various known enzymes are responsible for the metabolism of nitriles by distinct mechanism pathways, for example, by hydrolysis (e.g., nitrile hydratase, amidase, and nitrilase), oxidation (e.g., oxygenase), and reduction (e.g., nitrogenase) (Banerjee et al. 2002). The hydrolysis, catalyzed by nitrilases that directly convert the nitrile compounds to the corresponding acids, is the commonest pathway for the microbial metabolism of nitriles (Banerjee et al. 2002, Jin et al. 2013). This class of enzymes can be divided into three major categories based on the substrate specificity: aromatic nitrilase, aliphatic nitrilase, and arylacetonitrilase (Gong et al. 2012).

The first studies reporting the discovery of nitrilases in organisms were published in the 1960s (Martínková et al. 2009, Winkler et al. 2009). Seeking biotechnological applications, the research on bacterial nitrilases was intensified in the late 1970s. Fungal nitrilases remained unexplored and, the nitrilase from *Fusarium solani* IMI196840 was, for more than a decade, since its discovery, the only characterized fungal nitrilase (Martínková et al. 2009). In fact, fungal nitrilases are still underexplored, and only a few data are available on filamentous fungal conversion of nitriles, making these microorganisms a promising source of new nitrilases (Jin et al. 2013, Petříčková et al. 2012, Šnajdrová et al. 2004).

Nitrilases usually promote the biotransformation of nitriles directly to the corresponding carboxylic acids. Generally, the nitriles are synthetically more accessible than carboxylic acids, and thus they may act as precursors in the synthesis of these compounds. Since the chemical transformation of nitriles into carboxylic acids usually requires the use of strong bases, acids, or heavy metal salts (Kiełbasiński et al. 2008), enzymatic methods emerge as a "greener" alternative once they allow a clean synthesis under mild conditions, showing thermal stability and usually a high yield and selectivity (Jin et al. 2013, Petříčková et al. 2012, Zhu et al. 2008).

The biotransformation of phenylacetonitrile by marine-derived fungi was studied by De Oliveira et al. (2013). First, 12 different strains were screened on solid mineral culture (glucose and agar in artificial seawater) supplemented with phenylacetonitrile (5.0, 10.0, 15.0, and 20.0 μL per plate) and without phenylacetonitrile (control plate) to select strains with biocatalytic potential for nitrile biotransformation. The strains tested were *A. sydowii* Ce15, *A. sydowii* Ce19, *A. sydowii* DR(M3)2, *A. sydowii* Gc12, *Bionectria* sp. Ce5, *Cladosporium* sp. DR(A)2, *P. citrinum* F35, *Penicillium decaturense* DR(F)2, *P. miczynskii* Gc5, *P. oxalicum* F30, *P. raistrickii* Ce16, and *P. raistrickii* DR(B)2. The microorganisms were grown on agar plates containing only glucose and phenylacetonitrile as carbon and nitrogen sources, respectively. In the control plates, where no nitrogen source was present, the marine-derived fungi did not grow on the agar surface. The concentration at which fungal colonies grew best was 15 μL of phenylacetonitrile per plate, whereas at 20.0 μL, the phenylacetonitrile inhibited the growth of the most of the marine fungi (De Oliveira et al. 2013).

Eight strains, which showed better growth, were selected to proceed to experiments with phenylacetonitrile in a liquid medium: *A. sydowii* Ce15, *A. sydowii* Ce19, *A. sydowii* DR(M3)2, *A. sydowii* Gc12, *Bionectria* sp. Ce5, *P. decaturense* DR(F)2, *P. raistrickii* Ce16, and *P. raistrickii* DR(B)2 (Table 30.9).

These experiments were conducted in the malt extract medium (2.0% w/v malt) and mineral medium (1.5% w/v glucose), inoculated with slices of agar-bearing fungi previously induced in the solid medium by phenylacetonitrile. The fungi were cultivated for 5 days in malt extract, after which the phenylacetonitrile (60.0 μL, 0.52 mmol) was added. In mineral medium, since the phenylacetonitrile was the only source of nitrogen for fungal growth, the substrate was added at the time of the fungal inoculation. These experiments showed that the marine fungi catalyzed the biotransformation of phenylacetonitrile to 2-hydroxyphenylacetic acid when cultured in liquid mineral medium, whereas no biotransformation of the nitrile occurred in malt extract medium. Moreover, the marine fungi cultured in liquid mineral medium in the absence of the nitrile showed no nitrile biotransformation. These results show, in agreement with literature data, that nitrilases from filamentous fungi are inducible and not constitutive enzymes, which are similar to most bacterial nitrilases (Martínková et al. 2009). The activity of nitrilases and nitrile hydratases are reported only in the presence of suitable inducers, with the exception of some extremely toxic nitriles, such as mandelonitrile (De Oliveira et al. 2013). However, the mechanism of nitrilase induction in filamentous fungi has not yet been well elucidated (Martínková et al. 2009).

The biotransformation reactions in liquid mineral medium were monitored every 24 h for 8 days. The reactions were analyzed by GC-FID, with a total consumption of phenylacetonitrile at 96 h of reaction for all the tested strains (Table 30.9).

The biotransformation product, 2-hydroxyphenylacetic acid, was isolated and characterized throughout the reaction between the strain *A. sydowii* Ce19 and phenylacetonitrile, as outlined in Figure 30.1 (De Oliveira 2012, De Oliveira et al. 2013).

The biotransformation of nitrile to the corresponding carboxylic acid involves a sequence of enzymatic reactions. In this case, the marine-derived fungi developed an efficient enzymatic pathway to produce the 2-hydroxyphenylacetic acid (Scheme 30.11). The proposed pathway is that the phenylacetonitrile was hydrolyzed to the 2-hydroxyphenylacetic acid and then an enzymatic oxidation occurred at the *ortho*-position of the aromatic ring, leading to 2-hydroxyphenylacetic acid.

An additional experiment was performed in order to determine the sequence of enzymatic reactions. The enzyme pathway for the biotransformation of 4-fluorophenylacetonitrile by *A. sydowii* Ce19 was previously induced by phenylacetonitrile (60 μL, 96 h, 124 rpm), and after that, the substrate was added and the reaction was maintained on an orbital shaker (124 rpm, 32°C) for 24 h, showing that only the first substrate was biotransformed to the corresponding 2-hydroxyphenylacetic acid (100% conversion and 51% yield isolated), while no product of 2-(4-fluorophenyl)acetic acid was obtained (Scheme 30.11). Therefore, it is possible to suggest that the phenylacetonitrile was first hydrolyzed and then oxidized by the marine fungus *A. sydowii* Ce19.

Once it was confirmed in this study that phenylacetonitrile is a good inducer of nitrilase activity, this substrate was employed in the biotransformation of methylphenylacetonitriles by the marine filamentous fungus now registered as *A. sydowii* CBMAI 934 (previously known as *A. sydowii* Ce19) by De Oliveira et al. (2014).

TABLE 30.9

Biotransformation of Phenylacetonitrile to 2-Hydroxyphenylacetic Acid by Whole Cells of Marine Fungi[a]

Marine Fungi	Time (Days)	c (%) Phenylacetonitrile	Conv. (%) 2-Hydroxyphenylacetic Acid
Aspergillus sydowii Ce19	2	100	—
	4	62	38
	6	—	100
	8	—	100
Aspergillus sydowii Ce15	2	92	8
	4	25	75
	6	—	100
	8	—	100[b]
Aspergillus sydowii DR(M3)2	2	100	—
	4	57	43
	6	—	100
	8	—	100
Aspergillus sydowii Gc12	2	100	—
	4	27	73
	6	—	100
	8	—	100
Bionectria sp. Ce5	2	100	—
	4	51	49
	6	—	100
	8	—	100
Penicillium decaturense DR(F)	2	100	—
	4	88	11
	6	13	87
	8	—	100
Penicillium raistrickii Ce16	2	100	—
	4	81	19
	6	10	90
	8	—	100
Penicillium raistrickii DR(B)2	2	100	—
	4	84	16
	6	—	100
	8	—	100

Abbreviations: c, concentration; conv., conversion.

[a] Reaction conditions (De Oliveira et al. 2013).

[b] Isolated yield (51%).

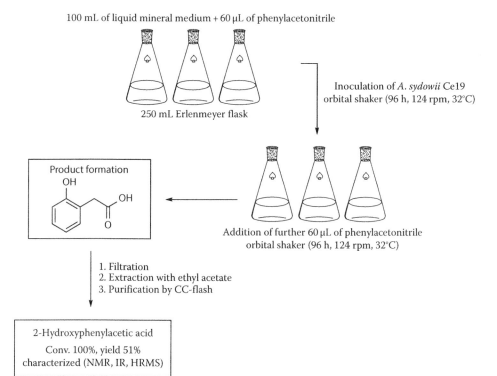

100 mL of liquid mineral medium + 60 μL of phenylacetonitrile
250 mL Erlenmeyer flask
Inoculation of A. sydowii Ce19
orbital shaker (96 h, 124 rpm, 32°C)
Product formation
OH
OH
O
Addition of further 60 μL of phenylacetonitrile
orbital shaker (96 h, 124 rpm, 32°C)
1. Filtration
2. Extraction with ethyl acetate
3. Purification by CC-flash
2-Hydroxyphenylacetic acid
Conv. 100%, yield 51%
characterized (NMR, IR, HRMS)

FIGURE 30.1
Procedures for biotransformation of phenylacetonitrile to 2-hydroxyphenylacetic acid by *A. sydowii* Ce19, with phenylacetonitrile as inducer and substrate (De Oliveira 2012).

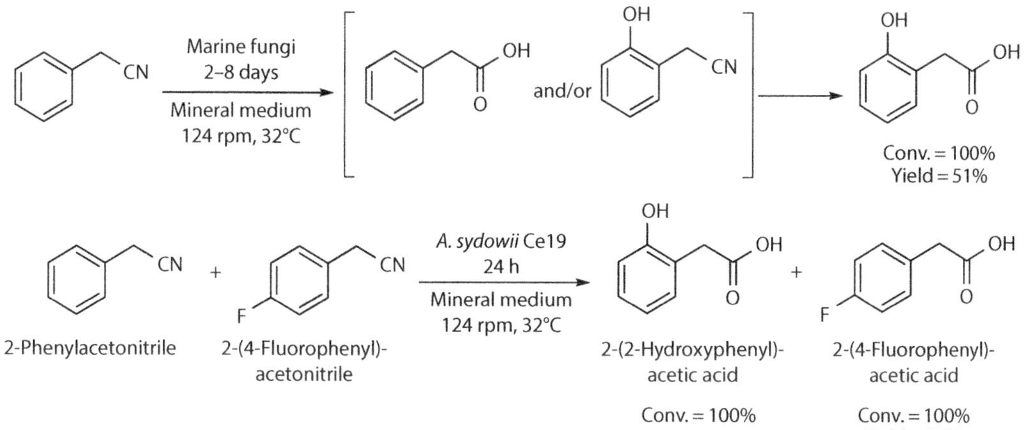

CN
Marine fungi
2–8 days
Mineral medium
124 rpm, 32°C
OH
O
and/or
OH
CN
OH
OH
O
Conv. = 100%
Yield = 51%
CN
+
CN
F
A. sydowii Ce19
24 h
Mineral medium
124 rpm, 32°C
OH
OH
O
+
OH
O
F
2-Phenylacetonitrile
2-(4-Fluorophenyl)-
acetonitrile
2-(2-Hydroxyphenyl)-
acetic acid
2-(4-Fluorophenyl)-
acetic acid
Conv. = 100%
Conv. = 100%
Yield = 51%

SCHEME 30.11
Proposed enzymatic pathway for phenylacetonitrile by *A. sydowii* Ce19 and biotransformation of 4-fluorophenylacetonitrile by *A. sydowii* Ce19 (De Oliveira 2012).

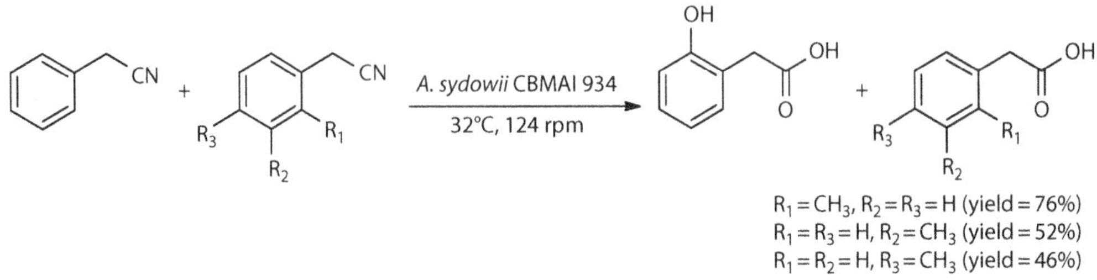

SCHEME 30.12
Biotransformation of phenylacetonitrile and its methylphenylacetonitrile derivatives by marine fungus *A. sydowii* CBMAI 934 (De Oliveira 2012).

The biotransformations of 2-methylphenylacetonitrile, 3-methylphenylacetonitrile, and 4-methylphenylacetonitrile, in which the nitrile group is bonded to the sp^3 carbon atom, produced the corresponding carboxylic acid at high conversion rates with phenylacetonitrile as an inducer. No hydroxylation products were observed for the compounds (De Oliveira et al. 2014).

The whole-cell reaction with methylphenylacetonitriles was conducted at three different substrate concentrations (20.0, 40.0, and 60.0 µL per reaction flask); however, at the higher substrate concentrations (40.0 and 60.0 µL), the nitriles may have inhibited the enzymatic activity of the fungus and the substrates were not converted into the carboxylic acids. Conversely, in reactions with 20.0 µL of substrate concentration, nitriles were quantitatively biotransformed (Scheme 30.12).

The reaction with 2-methylphenylacetonitrile, containing the methyl group in the *ortho*-position of the aromatic ring, showed higher yield than 3-methylphenylacetonitrile and 4-methylphenylacetonitrile, which have the methyl group in the *meta*- and *para*-positions, respectively. Usually, enzymatic reactions in aromatic compounds containing substituents in the *ortho*-position do not occur or are hindered (De Oliveira et al. 2014). According to O'Reilly and Turner (2003), the nitrilases generally prefer *meta*- and *para*-substituted substrates, showing poor activity with *ortho*-substituted substrates (O'Reilly and Turner 2003).

Therefore, the enzymatic system of the marine-derived fungal strain *A. sydowii* CBMAI 934 induced by phenylacetonitrile afforded innovative results in nitrile biotransformation, exhibiting not only aromatic nitrilase activity (phenylacetonitrile biotransformation) (De Oliveira et al. 2013) but also arylaliphatic nitrilase activity (biotransformation of methylphenylacetonitrile), with a preference for *ortho*-substituted substrates (De Oliveira et al. 2014).

30.3.2 Hydrolysis of Epoxides by Fungi Isolated from Marine Sponges

Asymmetric synthesis has become a very important tool and, with the aid of biocatalytic techniques, the synthesis of optically active compounds has been widely studied. Enantiopure compounds have a great potential in the pharmaceutical industry for the synthesis of biologically active drugs (Kotik et al. 2005). In this context, enantiopure epoxides and their corresponding vicinal diols are extensively used as building blocks for the synthesis of various bioactive molecules used as agrochemicals and pharmaceuticals

(Gong and Xu 2005, Simeó and Faber 2006). These compounds are employed as intermediates in the synthesis of chiral compounds with high added value due to their ability to react with a broad variety of nucleophiles (Mischitz et al. 1995).

Glycidyl phenyl ether and its derivatives are important aryloxy epoxides used as intermediates in the synthesis of chiral amino alcohols and β-adrenergic blockers, since (S)-glycidyl phenyl ether is easily converted into the final β-amino alcohol product (Gong and Xu 2005, Martins et al. 2011). One eco-friendly process to produce these chiral compounds under mild conditions is the use of epoxide hydrolases to promote the enantio-specific hydrolysis of racemic epoxides, generating enantiopure epoxides and vicinal diols (Gong and Xu 2005, Kotik et al. 2005). These enzymes were detected in a variety of microorganisms, such as bacteria, yeasts, and filamentous fungi, with a high potential for use in the biotransformation of epoxides (Kotik et al. 2005).

It is inconvenient for the biotransformation of epoxides that most of them may be spontaneously hydrolyzed into vicinal diols without any enantiospecificity in a single aqueous medium. In addition, because of their instability and low solubility in aqueous phase, the biohydrolysis may be limited, causing a decrease in the yield of kinetic resolutions (Gong and Xu 2005). In reactions employing whole cells as biocatalysts, in which there is a multienzymatic system, enzymes may compete and enantioselectivities are often lower and selectivity enhancement is required (Simeó and Faber 2006). For example, excellent results have been obtained with an engineered enzyme derived from *Bacillus subtilis*, which catalyzed the biohydrolysis of (±)-2-methyl glycidyl benzyl ether to afford the (R)-diol and residual (R)-oxirane in high enantiomeric excess (Fujino et al. 2007) (Scheme 30.13).

In our group, Martins et al. (2011) performed the hydrolysis of (±)-2-(benzyloxymethyl) oxirane (benzyl glycidyl ether, BGE) with whole cells of marine-derived fungi. Initially, the spontaneous hydrolysis of BGE was tested in various liquid media: freshwater, freshwater plus malt extract (3% w/v), phosphate buffer, and artificial seawater. Hydrolysis of BGE was observed in artificial seawater, generating the corresponding chlorohydrin (85% yield). Since artificial seawater contains a high concentration of chloride ions (1.20 M), which induced the spontaneous hydrolysis, the reaction with marine-derived fungi and BGE was clearly impossible (Martins et al. 2011).

The strains employed in this study were *A. sydowii* Gc12, *P. raistrickii* Ce16, *P. miczynskii* Gc5, and *Trichoderma* sp. Gc1. In preliminary experiments, strains were cultured in several media for various incubation times to establish the best conditions for BGE hydrolysis. However, when grown in the freshwater-based media, the marine-derived fungi were not capable of catalyzing the hydrolysis of the epoxide. Therefore, the expression of epoxide hydrolases was probably directly dependent on the growth of the fungus under similar condition to its natural marine environment, suggesting that these enzymes may require

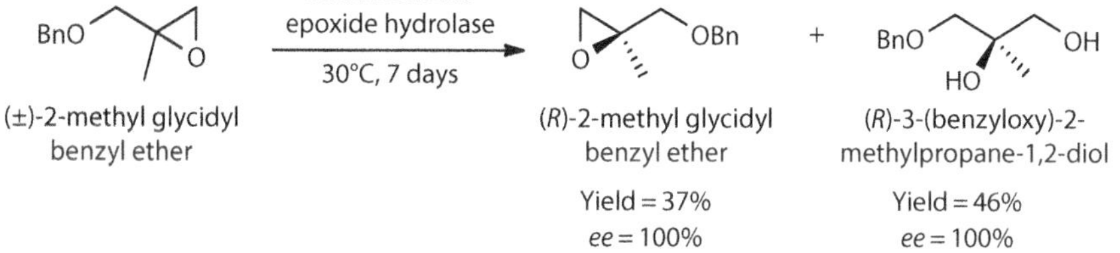

SCHEME 30.13
Biotransformation of (±)-2-methyl glycidyl benzyl ether by *Bacillus subtilis*.

ions for their expression or activity. To solve not only the lack of hydrolase expression in freshwater media but also the spontaneous hydrolysis of BGE in artificial seawater, the reaction conditions were modified by growing the fungi for 5 days in a medium containing malt extract (30 g L^{-1}), soy peptone (3 g L^{-1}), and artificial seawater and then adding increasing the mycelial mass by filtering and resuspending in phosphate buffer (100 mL) containing (R,S)-BGE (100 mg, 0.6 mmol) (Martins et al. 2011).

Mycelia of *P. raistrickii* Ce16 and *P. miczynskii* Gc5 did not catalyze the biotransformation of BGE. However, the strains *A. sydowii* Gc12 and *Trichoderma* sp. Gc1 catalyzed the hydrolysis of racemic BGE in significant amounts (Table 30.10) (Martins et al. 2011).

In reactions between 5.0 g of whole wet *A. sydowii* Gc12 mycelium and the (R,S)-BGE, the diol was formed with a yield of 41% and *ee* of 10%. The residual oxirane was recovered at 59% with an *ee* of 46% for the (R)-enantiomer in 120 h. It is noteworthy that the enantiomeric excess of (R,S)-BGE was enhanced during the reaction time; however, the selectivity was low (E < 10). As mentioned before, this is an inconvenient feature of epoxide reactions with whole cells of microorganisms. Possibly, fungal enzymes with an opposite selectivity to that of the epoxide hydrolase were active. In contrast to *A. sydowii* Gc12, the hydrolases from *Trichoderma* sp. Gc1 exhibited a preference for the (R)-enantiomer of BGE forming the diol (R)-diol with retention of configuration because of the attack occurring at the terminal β-carbon (Scheme 30.14) (Martins et al. 2011).

TABLE 30.10

Enzymatic Hydrolysis of (±)-Benzyl Glycidyl Ether by Marine Fungi[a]

Aspergillus sydowii Gc12 (5.0 g of whole wet mycelium)

Entry	Time (h)	c BGE	ee BGE	ac BGE	Conv. Diol	ee Diol	ac Diol
1	2	100	24	R	—	—	—
2	29	83	27	R	17	—	—
3	48	76	31	R	24	—	—
4	120	59	46	R	41	10	nd

Trichoderma sp. Gc1 (5.0 g of whole wet mycelium)

5	4	95	3	S	5	—	—
6	19	59	45	S	41	—	—
7	25	48	60	S	52	32	R
8	47	12	8	S	88	—	—
9	70	7	—	S	93	—	—

Abbreviations: ac, absolute configuration; c, concentration; conv., conversion; ee, enantiomeric excess; nd, not determined.

[a] Reaction conditions (Martins et al. 2011).

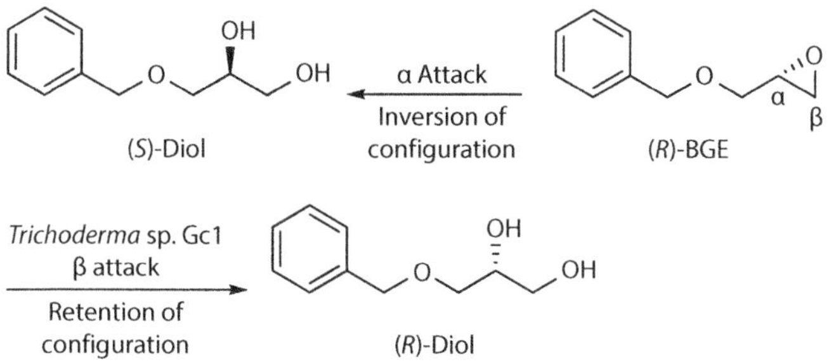

SCHEME 30.14
Hydrolysis of racemic BGE (β attack) by whole cells of *Trichoderma* sp. Gc1.

The formation of diol was relatively fast (Table 30.10, Entries 5–9). After 25 h of incubation, it was possible to obtain the product in 52% yield, the absolute configuration being found by the derivatization of the diol with acetic anhydride and pyridine to obtain the (*R*)-3-benzyloxy)propane-1,2-diyl diacetate in *ee* of 32%. The selectivity of enzymatic hydrolysis with *Trichoderma* sp. Gc1 was also low (E < 10). The hydrolases of the marine-derived fungi exhibited complementary regioselectivity in opening the epoxide ring of BGE, since *A. sydowii* Gc12 showed a preference for (*S*) and *Trichoderma* sp. Gc1 showed a preference for the (*R*) substrate, with retention of the configuration.

In another study, Martins et al. (2012) also performed the enzymatic hydrolysis of (±)-2-((allyloxy)methyl)oxirane (allylglycidyl ether, AGE) with whole cells of *Trichoderma* sp. Gc1 (Scheme 30.15). Reaction conditions were the same as described for the enzymatic hydrolysis of BGE. The enzymatic hydrolysis of AGE produced (*S*)-AGE in a yield of 23% and *ee* of 34% and the corresponding vicinal (*R*)-diol in a yield of 60% after purification. To determine the absolute configuration, the diol was derivatized with acetic anhydride and pyridine, and the corresponding diacetate showed low selectivity (10% *ee*). Thus, as in the reaction with BGE, *Trichoderma* sp. Gc1 exhibited poor enantioselective performance in the hydrolysis of AGE and again showed selectivity with preference for (*R*)-AGE, with retention of configuration (Scheme 30.15) (Martins et al. 2012).

For the results obtained in epoxide biotransformation, it can be concluded that the marine fungal strains are able to catalyze the enzymatic hydrolysis of the epoxides BGE and AGE to produce enantiomerically enriched epoxides and diols, however, with low selectivity.

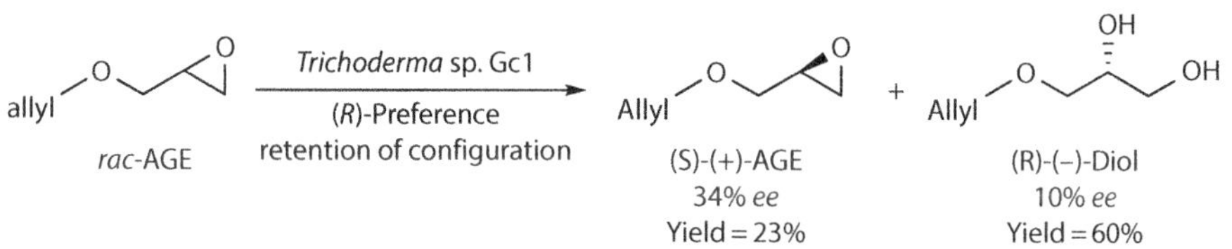

SCHEME 30.15
Enzymatic resolution of oxirane *rac*-AGE by whole cells of *Trichoderma* sp. Gc1.

30.4 Biohydroxylations of Natural Products by Marine-Derived Fungi

Terpenes are an important class of natural products with thousands of identified compounds, which play important roles in physiological and pathological processes (Adam et al. 2002). They occur widely in nature and have attracted commercial attention because of their anticancer, antimicrobial, and insecticidal activities and also act as building blocks for the synthesis of high-value-added compounds (De Carvalho and Da Fonseca 2006). Terpene transformations generate commercially important compounds applied in the production of fragrances, perfumes, and flavors (Monteiro and Veloso 2004). The biotransformation of terpenes is particularly interesting because it enables the production of enantiomerically pure flavors and fragrances under mild reaction conditions (De Carvalho and Da Fonseca 2006).

In a study of the biotransformation of related terpenes, our group investigated the hydroxylation of three important natural products: (–)-Ambrox®, (–)-sclareol, and (+)-sclareolide (Martins 2012, Martins et al. 2015). (–)-Ambrox is produced by oxidative decomposition of Ambergris, a natural product from blue whale with an amber-like odor (Musharraf et al. 2012). (–)-Sclareol is widely distributed in nature and has strong antibacterial and fungistatic activities. It is also used as a fixative in the perfume industry, as flavoring in tobacco, and even as a precursor in the synthesis of (–)-Ambrox (Kouzi and McChesney 1991, Musharraf et al. 2012). (+)-Sclareolide may be isolated from several plant species, as a result of antifungal activity, and is used as a starting material for the synthesis of various bioactive natural products (Cano et al. 2011).

Martins (2012) performed a screening of nine species of marine-derived fungi (*A. sydowii* CBMAI 934, *A. sydowii* CBMAI 935, *Botryosphaeria* sp. CBMAI 1197, *Eutypella* sp. CBMAI 1196, *Hydropisphaera* sp. CBMAI 1194, *P. raistrickii* CBMAI 931, *P. oxalicum* CBMAI 1185, *P. citrinum* CBMAI 1186, and *Xylaria* sp. CBMAI 1195) and selected four strains with oxidoreductase activities to catalyze the biotransformation of the natural products (–)-Ambrox, (–)-sclareol, and (+)-sclareolide (Martins 2012).

The selected strains were *A. sydowii* CBMAI 934, *Xylaria* sp. CBMAI 1195, *Eutypella* sp. CBMAI 1196, and *Botryosphaeria* sp. CBMAI 1197. The marine-derived strain *A. sydowii* CBMAI 934 was able to biotransform only (–)-Ambrox, leading to the hydroxylated derivative (–)-1β-hydroxy-Ambrox (14% yield, 10 days). *Xylaria* sp. CBMAI 1195 catalyzed the biotransformation of only (–)-sclareol, leading to two hydroxylated derivatives, (–)-3β-hydroxy-sclareol and (+)-18-hydroxy-sclareol, isolated in yields of 31% and 10%, respectively, after 3 days of reaction (Scheme 30.16) (Martins 2012, Martins et al. 2015).

The strains *Eutypella* sp. CBMAI 1196 and *Botryosphaeria* sp. CBMAI 1197 catalyzed the biotransformation of all three terpenes, leading to the same monohydroxylated metabolites. Both strains converted (–)-Ambrox into the metabolite (–)-3β-hydroxy-Ambrox (17%, *Botryosphaeria* sp.; 11%, *Eutypella* sp.). Selective hydroxylation of (–)-sclareol led to the formation of 3β-hydroxy-sclareol (69%, *Botryosphaeria* sp.; 55%, *Eutypella* sp.), as in the biotransformation with *Xylaria* sp. CBMAI 1195. Fermentation of (+)-sclareolide with *Eutypella* sp. CBMAI 1196 and *Botryosphaeria* sp. CBMAI 1197 resulted in 3β-hydroxy-sclareolide (7% and 34%, respectively) (Martins 2012, Martins et al. 2015).

In these experiments, it was noteworthy that marine-derived fungal species showed the same selectivity in the hydroxylation reactions, all three substrates being hydroxylated in the 3β-position. Scheme 30.16 shows the biotransformation of (–)-Ambrox, (–)-sclareol, and (+)-sclareolide to yield the respective products of hydroxylation.

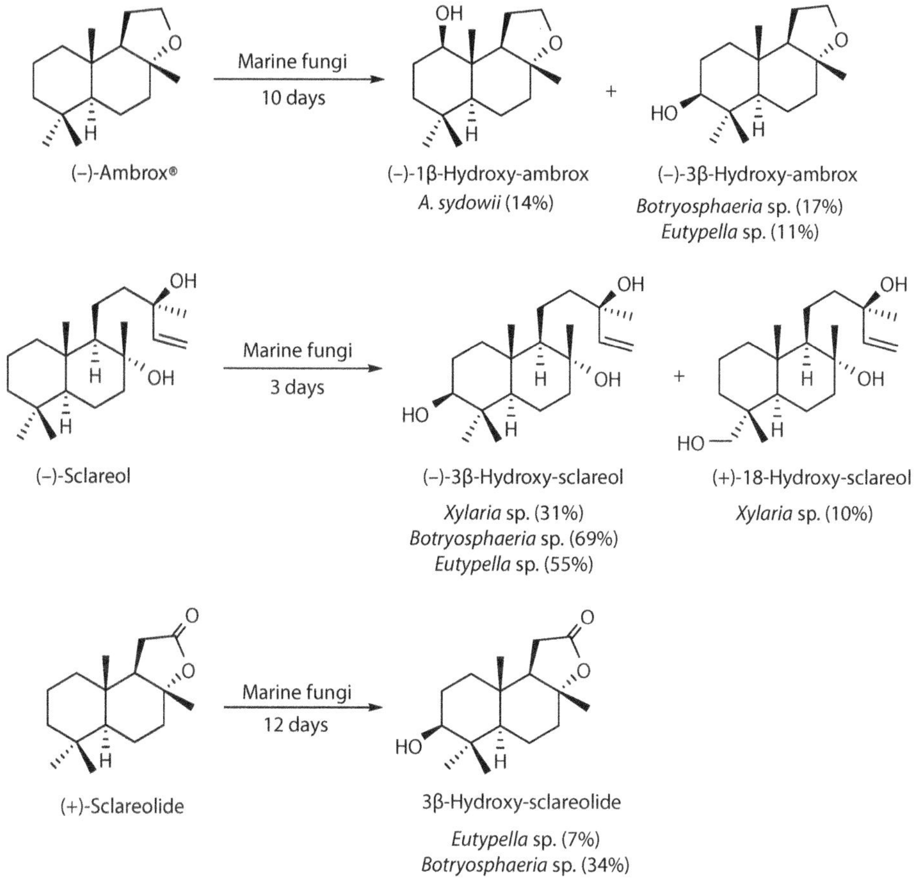

SCHEME 30.16
Biotransformation of (–)-Ambrox, (–)-sclareol, and (+)-sclareolide by marine-derived fungi.

Marine-derived fungi *A. sydowii* CBMAI 934, *Xylaria* sp. CBMAI 1195, *Eutypella* sp. CBMAI 1196, and *Botryosphaeria* sp. CBMAI 1197 are a promising catalyst of natural products under mild conditions (Martins 2012, Martins et al. 2015), yielding regioselective hydroxylated products, which could be of interest for industrial application.

30.5 Immobilization of Whole Mycelium of Marine-Derived Fungi for Biocatalysis

Immobilization is a process by which enzymes/whole cells are confined to a fixed phase, distinct from the one in which the substrates and products are present. This process allows easy removal of the product, reuse of the cells, and enhanced effectiveness due to the reuse of enzyme (Chibata 1978). The reuse of immobilized whole cells is that whole mycelia may be reused several times, reducing the costs of a biocatalyst and thus of the entire process (Chibata 1981).

The technology for immobilizing microbial cells is derived from the process of immobilization of enzymes. There are two types of immobilized living cells. The immobilization of cells after growth in a free culture and immobilization of inoculum followed by incubation for colony growth (Kuek 1986).

Immobilized whole-cell systems have advantages and disadvantages, with respect to free cells (Baklashova et al. 1984, Liu et al. 2013, Surendhiran et al. 2014).

1. Advantages
   a. The physical and chemical interactions that take place between carriers and cells often give rise to increased stability of the entrapped cells, which may, in turn, lead to increased cell productivity.
   b. Immobilized cells undergo repeated or continuous operations in various types of reactors.
   c. Immobilized cells are readily separated from the reaction system, making the separation of product easier.
2. Disadvantages
   a. Yields of products may be lowered by the consumption of substrates such as carbon and energy sources for the maintenance of cell viability or growth.
   b. Unwanted side reactions may occur because metabolic systems other than the desired ones may be active.
   c. Some materials used as supports can inhibit the metabolism of the microorganism.

The use of immobilized filamentous fungi has recently increased because these microorganisms can produce (or transform) many compounds of commercial interest, including organic acids (Vassilev et al. 1992), enzymes (Linko et al. 1996), biodiesel (Koda et al. 2010), and steroids (Peart et al. 2012, 2013). Interest in this technique may be estimated from the growing number of published articles and keyword reviews referring to "immobilized fungus" or "immobilized microorganism" as shown in Figure 30.2.

30.5.1 Immobilization of the Whole Cells of Marine-Derived Fungi on Biopolymers

The techniques of immobilizing whole cells can be divided into two classes:

1. Natural method, which includes formation of biofilms and adhesion/microbial adsorption on natural or artificial media such as chitosan, fibroin, or vegetable loofah (Kourkoutas et al. 2006)
2. Encapsulation in matrices such as calcium alginate, with the use of binders or gelatin (Guo et al. 2012)

Chitosan is one of the most used supports in the immobilization of whole cells. It is the product of chitin deacetylation, which is the second most abundant biopolymer, after cellulose. Chitosan is used for the immobilization of enzymes or whole cells because it is nontoxic, user-friendly and has low protein affinity (Krajewska 2004).

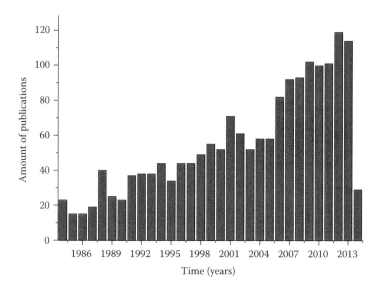

FIGURE 30.2
Profile of the number of scientific publications in indexed international journals, referring to the term "immobilized fungus" in the last 30 years. (Courtesy of SciFinder "online.")

Alginate is also one of the most extensively used polymers for various biomedical applications and in the pharmaceutical industry as a gelling agent (Moreira et al. 2006). Calcium alginate should be a good choice of matrix, given the relatively inert environment within the gel, which avoids damages to the enzyme structure that cause loss of activity. It is also permeable to the passage of substrates and products. Alginate is thus the most suitable matrix for immobilization by encapsulation, owing to the advantages it offers as a polymer: hydrophilic nature, the presence of carboxyl groups, and great stability under the reaction conditions (Moreira et al. 2006).

Silk fibroin is the set of biopolymers derived from the cocoon of the silk worm *Bombyx mori*, a protein of molecular weight between 300 and 420 kDa (Bini et al. 2004). Silk biomaterials are stable to changes in temperature and moisture, as well as being mechanically robust, owing to the extensive network of physical cross-links (β-sheets) formed during the assembly process. Thus, first, as a protein immobilization matrix, this biomaterial offers some important features that suggest its use as a stabilization matrix, such as tensile strength, elasticity, good thermal stability, hygroscopicity, and microbial resistance (Kazemimostaghim et al. 2014).

In a recent study of our research group, whole mycelia of marine fungal strain *P. citrinum* CBMAI 1186, both free and immobilized on cotton (*Gossypium* sp.), silk fibroin (*B. mori*), and a local kapok (*Ceiba speciosa*), catalyzed the chemoselective reduction of chalcone, (*E*)-3-(4-fluorophenyl)-1-phenylprop-2-en-1-one, to dihydrochalcone, 3-(4-fluorophenyl)-1-phenylpropan-1-one (Scheme 30.6) (Ferreira et al. 2014c). When *P. citrinum* CBMAI 1186 hyphal growth was immobilized on cotton fiber, it yielded the dihydrochalcone at 92% conversion (75% isolated yield); on fibroin, 80% conversion (78% isolated yield); and on kapok, 93% conversion (73% isolated yield) (Table 30.11). The same experiment was simultaneously conducted with free whole mycelia, achieving 98% conversion (79% isolated yield) to dihydrochalcone by reduction of the α,β-carbon–carbon double bond. These studies demonstrated that the immobilized whole cells from *P. citrinum* CBMAI 1186 displayed excellent biocatalytic activity on natural supports (Figure 30.3) (Ferreira et al. 2014c).

TABLE 30.11

Biosynthesis of Dihydrochalcone (3-(4-Fluorophenyl)-1-Phenylpropan-1-One) by Whole Cells Free and Immobilized Mycelia of *P. citrinum* CBMAI 1186 (6 Days, 130 rpm, 32°C)[a]

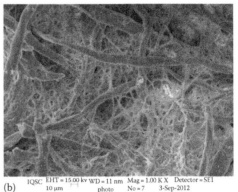

Matrices	Conv. (%)[b]	Conv. (%)[c]
Free mycelia	98	15
Cotton fiber	92	30
Fibroin fiber	80	22
Kapok fiber	93	36

[a] Reaction conditions (Ferreira et al. 2014c).
[b] Chalcone reduced by fresh biocatalyst.
[c] Chalcone reduced after one month biocatalyst storage at 4°C.
Conv., conversion.

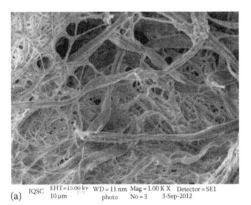

FIGURE 30.3

Scanning electron micrographs: whole hyphae of *P. citrinum* CBMAI 1186 immobilized on (a) cotton fiber, (b) silk fibroin fiber, and (c) kapok fiber.

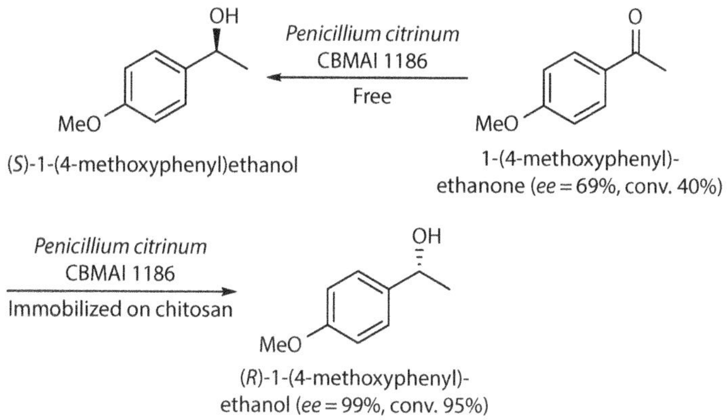

SCHEME 30.17
Biotransformation of ketones by free and immobilized mycelia of *P. citrinum* CBMAI 1186.

To assess the reuse of immobilized whole cells on cotton, fibroin, and kapok, the biore-duction of (*E*)-3-(4-fluorophenyl)-1-phenylprop-2-en-1-one was performed again after the supports had been stored in the refrigerator (4°C) for 30 days. It must be emphasized that after the immobilized fungus was used in the first cycle, the supported mycelium was fil-tered, washed with ethyl acetate, and preserved in the refrigerator for 1 month. A decrease was observed in the formation of the desired product on all supports used (Table 30.11) (Ferreira et al. 2014c). This study shows the potential use of an immobilized filamentous fungus for the chemoselective reduction of chalcones by ERs.

Another interesting approach of application of biopolymers was developed by Rocha et al. (2012a). They demonstrated the use of *P. citrinum* CBMAI 1186 immobi-lized on chitosan in the catalysis of the quantitative reduction of 1-(4-methoxyphenyl)-ethanone to the enantiomer (*S*)-1-(4-methoxyphenyl)-ethanol, with an excellent enantioselectivity (*ee* > 99%, conv. = 95%) (Scheme 30.17). Interestingly, the free myce-lium of *P. citrinum* CBMAI 1156 fungal reduced 1-(4-methoxyphenyl)ethanone with moderate selectivity and conversion to (*R*)-1-(4-methoxyphenyl)ethanol (*ee* = 69%, conv. 40%) (Scheme 30.17).

30.6 Biodegradation of Organopesticides with Whole Cells of Marine-Derived Fungi

Since the green revolution, in the 1950s, traditional agricultural production has changed, with the adoption of new technologies aiming at the intensive production of commodities such as wheat, soy, sugarcane, coffee, and orange. These technologies include the use of syn-thetic pesticides to control pests and plant diseases and, consequently, increase agricultural production. Currently, the massive use of toxic compounds is one of the main threats to the environment, since they often harm nontarget organisms, interfering in the physiology, genetics, behavior, life expectancy, and reproduction of living beings. Moreover, depending on the degree of toxicity and persistence in the environment, such compounds may affect basic ecological processes, such as soil respiration and nutrient cycling (Rebelo et al. 2010).

Several processes can act in pesticide degradation and removal: chemical degradation occurs through various types of reaction (hydrolysis, oxidoreduction, photolysis, etc.), while biodegradation involves the existing biota in soil and water, being more active in the root zone and decreasing as the soil deepens. It is important to note that practically all pesticides are susceptible to such biotransformations, albeit to a greater or lesser extent (Havens 1995, Radosevich et al. 1997, Schnoor 1992).

Scientists are exploring the diverse microbial word in the search for new catalysts because of the capacity of microorganisms to degrade xenobiotics (Patel 2011, Spain 1995). Recently, the biotransformation of pesticides has been investigated by screening micro-organisms with the aid of recombinant DNA technology (Chen and Muchandani 1998, Richins et al. 1997).

Marine-derived microorganisms are naturally exposed and adapted to extreme tem-peratures, acidity, high pressure, and/or salt concentration, which are the hard conditions found in a significant part of the biosphere. Another important characteristic is quick adap-tation to environmental change, since marine currents promote rapid temperature and pH alterations (Dash et al. 2013). In addition, marine microorganisms may show an efficient biodegradation of pesticides because they possess an enzymatic system adapted to highly halogenated and oxygenated marine compounds (Mayer et al. 2013, Saleem et al. 2007), resembling the organochlorine, organophosphorus, and pyrethroid pesticides employed in these studies. Thus, marine-derived fungi might have a great potential for bioremedia-tion applications and thus should be studied.

30.6.1 Biodegradation of Organophosphorus Pesticides

30.6.1.1 *Biodegradation of Organophosphorus Pesticides by Marine-Derived Fungi*

Organophosphate pesticides (OPs) constitute a heterogeneous class of chemical com-pounds. They are employed worldwide in agriculture, public hygiene, disease vector con-trol, and household pest control (Edwards and Tchounwou 2005, Yang et al. 2008, Zheng et al. 2007). OPs of major toxicological and commercial interest are esters or thiols derived from phosphoric, phosphonic, phosphinic, or phosphoramidic acid (Sogorb and Vilanova 2002). The most commonly used derivatives are the phosphates and phosphorothioates (=S and S-substituted) (Bigley and Raushel 2013).

OPs have been used in the replacement of organochlorine pesticides because of their low cost, easy synthesis, high biodegradability, and low accumulation in living species (Santos et al. 2007). However, due to the indiscriminate use of pesticides, the OPs may become a health concern, since they are toxic to nontarget species (Coutinho et al. 2005, Diez 2010). The toxic effects of OPs occur mainly in the central nervous system, where they cause the irreversible phosphorylation of acetylcholinesterase, the enzyme respon-sible for hydrolysis of the neurotransmitter acetylcholine (Ghanem and Raushel 2005, Sogorb and Vilanova 2002). Owing to their high toxicity, several organophosphate for-mulations have had their use restricted by regulatory agencies, as in the case of methyl parathion (MP), which has been restricted by the U.S. Environmental Protection Agency, and profenofos (PF), which is a restricted use pesticide sprayed only on cotton crops in the United States (Da Silva et al. 2013, Diagne et al. 2007, Edwards and Tchounwou 2005).

In the environment, the main processes of pesticide degradation are chemical, microbial, and photodegradation. It is noteworthy that biodegradation plays an impor-tant role in OP decontamination and may determine its fate in the environment as the safest, least disruptive, and most cost-effective treatment method for contaminated soil

(Gavrilescu 2005, Sutherland et al. 2004, Singh and Walker 2016). The main enzyme classes responsible for natural OP decontamination are the phosphotriesterases (PTEs). A wide range of enzymes with PTE activity in OP hydrolysis have been found, including organophosphate hydrolase, MP hydrolase, organophosphorus acid anhydrolase, diisopropylfluorophosphate hydrolase, and paraoxonase 1 (Bigley and Raushel 2013). These enzymes are expressed intracellularly and, therefore, the overall detoxification rate may be limited by the transport of the pesticide across the cell membrane (Chen and Mulchandani 1998).

Biodegradation reactions with marine-derived fungi were conducted with OP pesticides such as PF, MP, and chlorpyrifos (CP). The expected pathway promoted by PTE enzymes is the hydrolysis of the OP, separating the aromatic part from the phosphate or phosphorothioate moiety (Scheme 30.18).

Screenings were performed on solid medium of malt extract (2% w/v in artificial seawater) to select the marine-derived fungi with biocatalytic potential to degrade pesticides. The marine strains evaluated were *A. sydowii* CBMAI 934, *A. sydowii* CBMAI 935, *P. raistrickii* CBMAI 931, *A. sydowii* CBMAI 1241, *P. decaturense* CBMAI 1234, *P. raistrickii* CBMAI 1235, and *Trichoderma* sp. CBMAI 932. All the strains investigated were grown for 10 days in both the presence (5, 10, and 15 μL of pesticide) and the absence of the pesticide (control culture). By measuring the size of the colonies formed on the agar surface, two strains were selected for each pesticide to proceed to quantitative experiments in liquid medium (Figure 30.4).

Alvarenga et al. (2014) studied the biodegradation of MP using *A. sydowii* CBMAI 835 and *P. decaturense* 1234 in a malt extract (2% w/v) liquid medium. The degradation of MP and the formation of *p*-nitrophenol (PNP), the main hydrolysis metabolite of MP, were monitored for 10, 20, and 30 days.

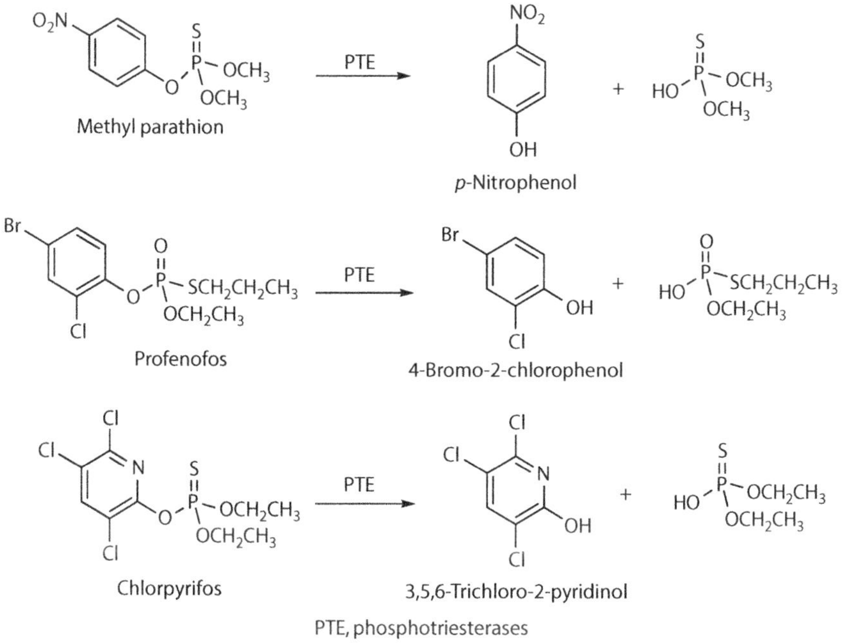

SCHEME 30.18
Hydrolysis of the pesticides methyl parathion, profenofos, and chlorpyrifos by PTE.

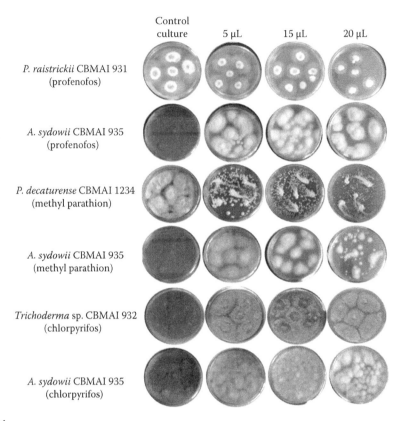

FIGURE 30.4
Marine fungi growing on solid culture medium containing various concentrations of profenofos, methyl parathion, and chlorpyrifos (10 days at 35°C).

The experiments showed that *A. sydowii* CBMAI 935 degraded almost all the pesticide in just 10 days (Table 30.12). The degradation was slower with *P. decaturense* CBMAI 1234, with 87%, 96%, and 99% of biodegradation in 10, 20, and 30 days, respectively (Figure 30.5). Both strains were grown in the absence of pesticide, and the mycelial masses produced after 30 days by *A. sydowii* CBMAI 935 and *P. decaturense* CBMAI 1234 were 0.29 and 0.24 g, respectively. Comparison of the mycelial masses produced by both fungi in the absence and presence of MP indicates that there was a small decrease in the mycelial mass of *A. sydowii* CBMAI 9354 when MP was added. No growth inhibition of *P. decaturense* CBMAI 1234 culture was observed, since its mycelial mass was increased in the presence of MP (Table 30.12) (Alvarenga et al. 2014).

The main products of MP hydrolysis, such as PNP, are persistent organic pollutants in wastewater. These compounds are present in effluents from industries that produce pesticides, explosives, colorants, skin treatment products, and agricultural irrigation runoff water (Oturan et al. 2000). PNP biodegradation experiments were performed to evaluate the efficiency of marine strains in promoting the mineralization of MP. Strains were effective in PNP biodegradation, achieving rates of 51% and 40% for *A. sydowii* CBMAI 935 and *P. decaturense* CBMAI 1234, respectively (Alvarenga et al. 2014). However, since no PNP degradation products were identified, it is supposed that, in addition to biodegradation, the PNP may have suffered conjugation reactions (phase II of metabolic biotransformation) in the liquid medium.

TABLE 30.12

Quantitative Biodegradation of Methyl Parathion by *A. sydowii* CBMAI 935 and *P. decaturense* CBMAI 1234, on the 10th, 20th, and 30th Days of Reaction in a Liquid Medium[a]

Reaction Time (Days)	Fungal Dry Mass (g)	c PNP (mg L^{-1})	c MP (mg L^{-1})	% of MP Degraded
A. sydowii CBMAI 935 (50 mg L^{-1} of methyl parathion)				
10	0.27 ± 0.03	6.6 ± 0.0	1.0 ± 0.9	98
20	0.27 ± 0.01	7.9 ± 3.0	—	100
30	0.24 ± 0.00	5.1 ± 2.1	—	100
P. decaturense CBMAI 1234 (50 mg L^{-1} of methyl parathion)				
10	0.29 ± 0.01	3.4 ± 0.2	6.5 ± 0.2	87
20	0.27 ± 0.00	3.7 ± 0.8	1.9 ± 0.3	96
30	0.26 ± 0.01	2.3 ± 2.0	0.2 ± 0.2	99

Abbreviations: c, concentration; MP, methyl parathion; PNP, *p*-nitrophenol.
[a] Reaction conditions (Alvarenga et al. 2014).

Biodegradation studies were also conducted in liquid mineral medium, in which the marine strains were provided with a high concentration of pesticide (100 mg L^{-1}) as sole source of carbon and KNO$_3$ as the nitrogen source. Both strains showed 80% MP biodegradation, indicating a high biocatalytic potential to degrade MP, even under poor nutrient conditions.

Reactions of MP in the absence of fungi (abiotic control) were performed to evaluate the pesticide stability. In 30 days, only 52% of MP was spontaneously degraded. In comparison with the same period reaction, *A. sydowii* CBMAI 935 and *P. decaturense* CBMAI 1234 were extremely efficient in MP biodegradation, since all the pesticide was degraded.

In further analysis of the reactions between MP and *P. decaturense* CBMAI 1234 by GC-MS and liquid chromatography-mass spectrometry (LC-MS), it was possible to identify the presence of dimethyl hydrogen phosphate, which indicates the fungal conversion of MP to methyl paraoxon, through a desulfurization mechanism promoted by cytochrome enzymes. Based on the metabolites found in the reaction between MP and *P. decaturense*, a possible pathway can be suggested for the biodegradation of MP, involving bioactivation to methyl paraoxon followed by hydrolysis (Scheme 30.19).

Biodegradation of PF by *A. sydowii* CBMAI 935 and *P. raistrickii* CBMAI 931 was studied by Da Silva et al. (2013). The data regarding this reaction conducted in the liquid malt extract (2% w/v) medium are shown in Table 30.13. On day 10, separate analyses were done in the liquid medium and mycelium. The results showed a higher concentration of the pesticide in the mycelium extract than in the liquid medium. In the case of *P. raistrickii* CBMAI 931, it was also clear that most of the metabolite, 4-bromo-2-chlorophenol (BCP) was in the liquid medium (Table 30.13). This fungus may take up PF and, after the reaction, excrete the metabolite into the liquid medium. The higher concentration of PF in the mycelium could be explained by the fact that the PF-degrading enzyme, PTE, is intracellular, as described by Chen and Mulchandani (1998), and that a membrane-based system actively transports the pesticide into the hypha.

Strain *A. sydowii* CBMAI 935 promoted a high rate of biodegradation in 20 days (average 71%). However, the degradation stagnated between 20 and 30 days of reaction. *P. raistrickii* CBMAI 931 promoted excellent biodegradation of PF, with almost complete degradation in 30 days of reaction. The main degradation product of PF, BCP, was formed by enzymatic hydrolysis and reached a higher concentration in the *P. raistrickii* CBMAI 931 reaction than in that of *A. sydowii* CBMAI 935.

A. sydowii CBMAI 935 and *P. raistrickii* CBMAI 931 also promoted an excellent biodegradation and/or bioconjugation of BCP, with average rates of 98% and 97%, respectively.

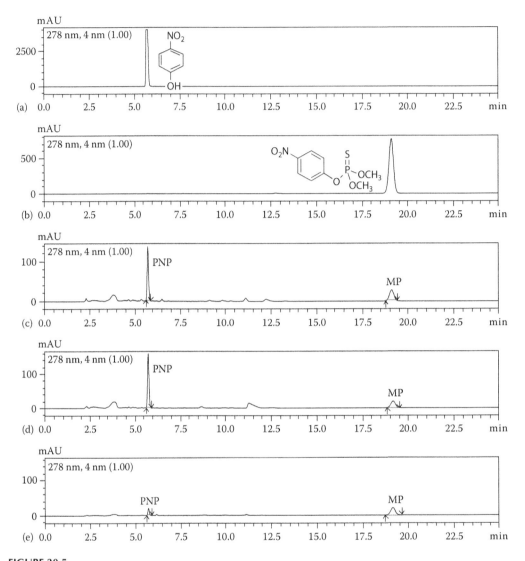

FIGURE 30.5
HPLC chromatogram of methyl parathion biodegradation in the presence of *P. decaturense* CBMAI 1234: (a) standard of PNP; (b) standard of MP; the (c) 10th, (d) 20th, and (e) 30th day of reaction.

SCHEME 30.19
Proposed pathway of methyl parathion biodegradation by *P. decaturense* CBMAI 1234.

TABLE 30.13

Quantitative Biodegradation of Profenofos by *A. sydowii* CBMAI 935 and *P. raistrickii* CBMAI 931, on the 10th, 20th, and 30th Days of Reaction in a Liquid Medium[a]

Reaction Time (Days)	Extraction	c BCP (mg L^{-1})	c PF (mg L^{-1})	% of PF Degraded
A. sydowii CBMAI 935 (50 mg L^{-1} profenofos)				
10	Liquid medium	1.8 ± 0.5	2.5 ± 0.2	
10	Mycelium	8.1 ± 0.6	22.9 ± 3.0	49
20	Liquid medium + mycelium	5.2 ± 0.4	14.7 ± 2.3	71
30	Liquid medium + mycelium	11.5 ± 0.7	14.3 ± 2.5	72
P. raistrickii CBMAI 931 (50 mg L^{-1} profenofos)				
10	Liquid medium	8.9 ± 0.6	2.0 ± 1.3	
10	Mycelium	1.7 ± 0.1	15.0 ± 10.4	66
20	Liquid medium + mycelium	12.4 ± 1.9	9.2 ± 2.5	82
30	Liquid medium + mycelium	19.6 ± 2.5	1.5 ± 1.2	97

Abbreviations: BCP, 4-bromo-2-chlorophenol; *c*, concentration; PF, profenofos.

[a] Reaction condition (Da Silva et al. 2013).

Reactions in the liquid mineral medium containing PF (100 mg L^{-1}) as sole source of carbon and KNO$_3$ as nitrogen source demonstrated, by the high percentage of PF degradation, that both fungi could use the pesticide as a source of carbon. *A. sydowii* CBMAI 935 and *P. raistrickii* CBMAI 931 were able to promote an average biodegradation of 91% and 96%, respectively. In the absence of fungi (abiotic control), in 30 days, 39% of PF was spontaneously degraded so that *A. sydowii* CBMAI 935 and *P. raistrickii* CBMAI 931 demonstrated a high potential for the biocatalytic degradation of PF (Da Silva et al. 2013, Alvarenga et al. 2015).

There are numerous studies on the biodegradation of CP by marine strains *A. sydowii* CBMAI 935 and *Trichoderma* sp. CBMAI 932 (Da Silva 2013). On the 20th day of reaction, extractions of the liquid medium and mycelium were performed separately and showed a higher concentration of CP in the mycelium than in the liquid medium, as in the studies with PF. These results contribute to literature reports that describe the main enzyme in organophosphate degradation, PTE, as intracellular. In 30 days, *A. sydowii* CBMAI 935 and *Trichoderma* sp. CBMAI 932 were able to degrade CP, with an average conversion of 63% and 72%, respectively. Both strains could also reduce the concentration of 3,5,6-trichloro-2-pyridinol, the main metabolite formed in the enzymatic hydrolysis of CP (biodegradation of 25% by *A. sydowii* CBMAI 935 and 39% by *Trichoderma* sp. CBMAI 932). In liquid mineral medium, only *A. sydowii* CBMAI 935 was able to satisfactorily use CP as a sole source of carbon (Da Silva 2013).

From the results obtained by our group in organophosphate biodegradation by marine-derived fungi, it is clear that these strains have a great potential for further application in *in situ* biodegradation of contaminated areas. These data also contribute to the fate elucidation of pesticides in marine environments, since the final destination of these xenobiotics is often the sea and ocean.

30.6.2 Biodegradation of Pyrethroid Pesticides

Synthetic pyrethroids have been developed to improve the specificity and activity of pyrethrin, the natural insecticide produced by the flowers of pyrethrum species (*Chrysanthemum cinerariaefolium* and *C. coccineum*). Pyrethrin is known for its instability in light and air, which limits its effectiveness in crop protection. The synthetic pyrethroids were developed to increase the photostability while retaining the potent and rapid insecticidal activity and relatively low acute mammalian toxicity of pyrethrin. There are about 1000 different structures and some of them are very different from the original pyrethrin insecticide (Soderlund et al. 2002, Sogorb and Vilanova 2002).

Scientists are exploring the microbial world in a search for new microorganisms that degrade xenobiotics, including pyrethroid insecticides. The first step in biodegradation screening experiments is to assess the growth of microorganisms in the presence of the studied xenobiotic. The growth of seven marine-derived fungi (*P. raistrickii* CBMAI 931, *A. sydowii* CBMAI 935, *Cladosporium* sp. CBMAI 1237, *Microsphaeropsis* sp. Dr(A)6, *Acremonium* sp. Dr(F)1, *Westerdykella* sp. Dr(M2)4, and *Cladosporium* sp. Dr(M2)2) in the presence of pyrethroids was assessed by measuring the radial growth of colonies on 3% malt solid medium (Birolli et al. 2014a).

In this study, it was observed that all the strains were able to grow and develop in the presence of the pyrethroid insecticide esfenvalerate (*S,S*-fenvalerate) and its main metabolites (3-phenoxybenzaldehyde [PBAld], 3-phenoxybenzoic acid [PBAc], 3-phenoxybenzyl alcohol [PBALc], and 2-(4-chlorophenyl)-3-methylbutyric acid [CLAc]), showing the biocatalytic potential for esfenvalerate biodegradation of these strains. From observations on the fungal growth, it was proposed that the biodegradation of esfenvalerate in the commercial insecticide (SUMIDAN 150SC) is slower than that of the technical grade active ingredient, since slower biodegradation of esfenvalerate would reduce the accumulation of phenolic compounds and thus the growth inhibition (Birolli et al. 2014a).

In another study, esfenvalerate biodegradation by marine-derived fungi was studied in 3% malt liquid medium. The esfenvalerate (*S,S*-fenvalerate) and its metabolites (PBAld, PBAc, PBALc, and CLAc) were quantitatively analyzed in triplicate experiments by a validated method. All the strains tested (*P. raistrickii* CBMAI 931, *A. sydowii* CBMAI 935, *Cladosporium* sp. CBMAI 1237, *Microsphaeropsis* sp. Dr(A)6, *Acremonium* sp. Dr(F)1, *Westerdykella* sp. Dr(M2)4, and *Cladosporium* sp. Dr(M2)2) were able to degrade esfenvalerate (Birolli 2013).

Quantitative biodegradation experiments were performed with an initial esfenvalerate concentration of 100 mg L^{-1} (Sumidan 150SC) and the strains *Microsphaeropsis* sp. Dr(A)6, *Acremonium* sp. Dr(F)1, and *Westerdykella* sp. Dr(M2)4. The residual esfenvalerate concentration (64.8–95.2 mg L^{-1}) and the formation of FBAc (0.5–7.4 mg L^{-1}), ClAc (0.1–7.5 mg L^{-1}), and PBAlc (0.2 mg L^{-1}) were measured. A biodegradation pathway was proposed, based on liquid chromatography-high resolution mass spectrometry (LC-HRMS) analysis, which includes PBAld, PBAc, PBAlc, ClAc, 3-(hydroxyphenoxy) benzoic acid, and methyl 3-phenoxy benzoate (Birolli 2013).

Initially, esfenvalerate can undergo hydrolysis by carboxylesterases, resulting in CLAc and 2-hydroxy-2-(3-phenoxyphenyl)acetonitrile (Scheme 30.20).

The 2-hydroxy-2-(3-phenoxyphenyl)acetonitrile is biotransformed to PBAld by an oxynitrilase enzyme, which converts cyanohydrins to aldehydes and hydrogen cyanide (Scheme 30.21).

The PBAld might be converted to PBAc by the activity of aldehyde dehydrogenase, which oxidizes aldehydes to carboxylic acids (Scheme 30.22).

The reduction of PBAld to PBAlc aldehyde reductase was also observed. This enzyme catalyzes the reduction of aldehydes to the corresponding alcohol (Scheme 30.23).

The PBAc can be hydroxylated by a monooxygenase enzyme to generate 3-(hydroxyphenoxy)benzoic acid (Scheme 30.24).

The esterification of PBAc to methyl 3-phenoxybenzoate was also observed (Scheme 30.25). This reaction is considered a conjugation reaction, in which abundant *in vivo* groups are added to the xenobiotic molecule, and it may be reversed (Birolli 2013).

A biodegradation comparison was performed between the esfenvalerate in simple suspension and the emulsion-forming commercial insecticide (Birolli et al. 2014a). It was observed

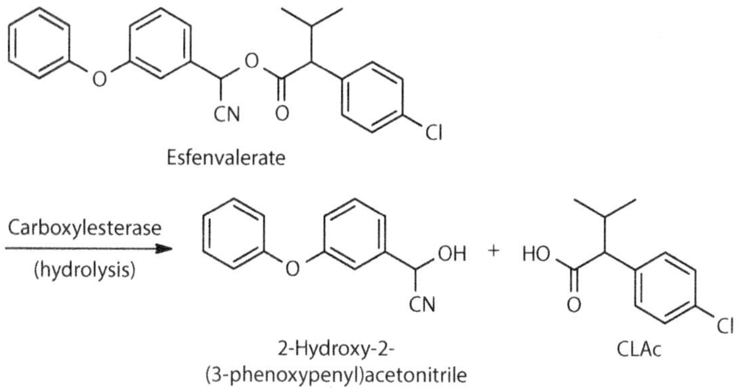

SCHEME 30.20
Esfenvalerate hydrolysis by marine-derived fungi.

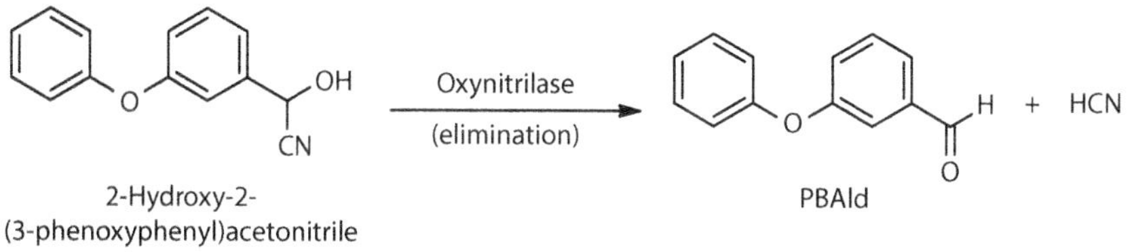

SCHEME 30.21
Biotransformation of 2-hydroxy-2-(3-henoxyphenyl) acetonitrile to PBAld.

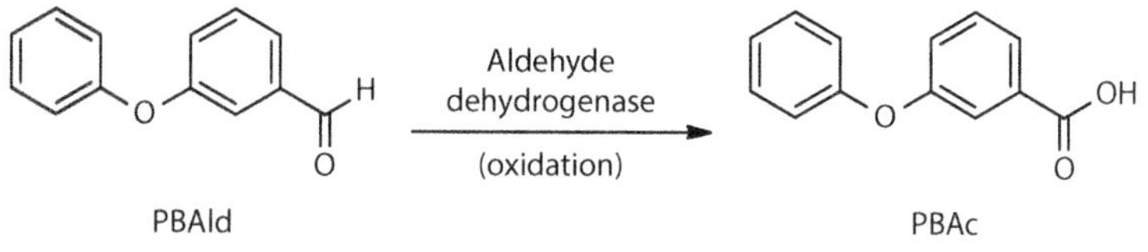

SCHEME 30.22
Oxidation of PBAld to PBAc by marine-derived fungi.

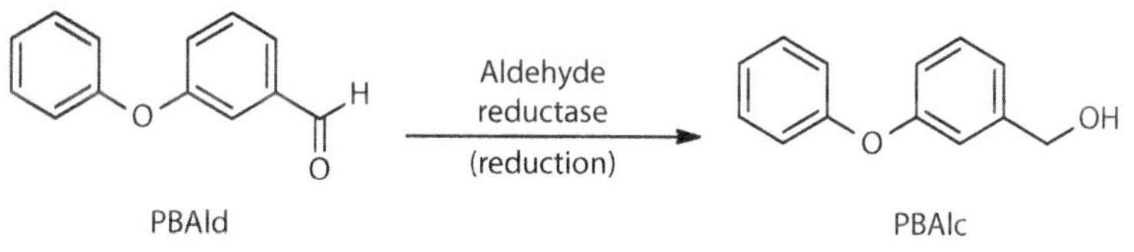

SCHEME 30.23
Reduction of PBAld to PBAlc by marine-derived fungi.

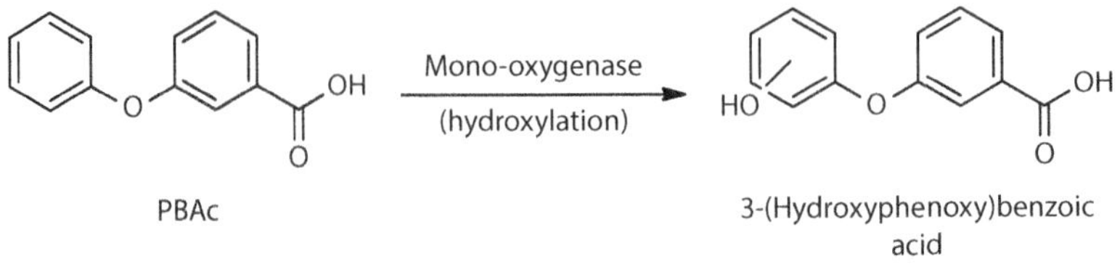

SCHEME 30.24
Hydroxylation of PBAc to 3-(hydroxyphenoxy)benzoic acid by marine-derived fungi.

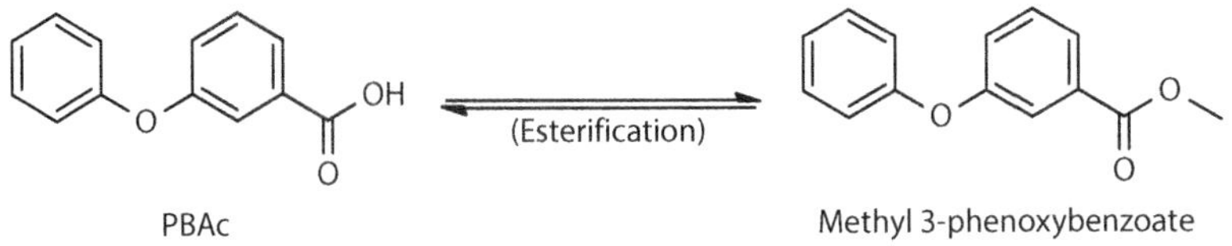

SCHEME 30.25
Esterification of PBAc to methyl 3-phenoxybenzoate by marine-derived fungi.

that the biodegradation of free esfenvalerate (residue of 26.5–41.5 mg L^{-1} after 28 days) was faster than that of the commercial insecticide (residual esfenvalerate of 48.0–52.2 mg L^{-1} after 28 days). The more efficient biodegradation of the technical grade esfenvalerate may be due to the support employed in the emulsifiable insecticide, to which the active ingredient might be adsorbed, hindering the enzymatic degradation by the fungal strains (Birolli 2013).

Concomitantly with the esfenvalerate biodegradation, the formation of PBAc and CLAc was observed with both the technical grade esfenvalerate (the PBAc concentration was 1.9–20.4 mg L^{-1} and the CLAc concentration was 2.9–18.5 mg L^{-1} after 28 days of biodegradation) and commercial insecticide (the PBAc concentration was 2.7–16.6 mg L^{-1} and the CLAc concentration was 6.6–13.4 after 28 days of biodegradation). PBAc and ClAc are considered toxic compounds that can cause health and environmental problems. Thus, for bioremediation purpose, it is more interesting to employ strains that rapidly biotransform these metabolites (Birolli 2013).

It was concluded that marine-derived fungi are able to degrade esfenvalerate and its main metabolites. However, this biodegradation leads to the formation of a significant concentration of FBAc and ClAc, which are toxic compounds. Thus, for bioremediation, it is important to screen for strains that can biodegrade these metabolites. It is also important to carry out future biodegradation and bioremediation studies on the commercial insecticide, since it was observed that there is a significant difference in degradation rate between the technical grade and the commercial product.

30.6.3 Biodegradation of Organochlorine Pesticides by Marine-Derived Fungi

Among all the pesticides used in agriculture, those belonging to the class of organochlorines cause serious concern, which is mainly on account of their high toxicity and persistence in the environment due to their resistance to biotic and abiotic degradation (Singh et al. 2009). Although organochlorine pesticides have been banned for use in agriculture, they are still worth studying because they were widely used on crops in the decades from 1960 to 1980, generating accumulated toxic waste in various ecosystems around the world (Kayser et al. 2001).

The DDD biodegradation was tested with seven marine-derived fungal strains (*A. sydowii* Ce15, *A. sydowii* Ce19, *A. sydowii* Gc12, *Bionectria* sp. Ce5, *P. miczynskii* Gc5, *P. raistrickii* Ce16, and *Trichoderma* sp. Gc1), which were capable of growth in 3% malt solid medium in the presence of DDD. This organochlorine pesticide caused significant growth inhibition in the fungal strains, but, at the chosen sublethal level, this toxic effect was not enough to prevent the biotransformation, since the fungal growth was satisfactory (Ortega et al. 2011).

The strain *Trichoderma* sp. Gc1, which showed the best growth, was employed in the quantitative biodegradation analysis. It was observed that a 100% degradation of DDD was apparently obtained in liquid culture medium when the mycelium was previously cultured for 5 days and supplemented with 5.0 mg of DDD in the presence of hydrogen peroxide. However, the quantitative analysis showed that DDD had accumulated in the mycelium and biodegradation level reached a maximum value of 58% after 14 days, while no intermediate metabolites were observed (Ortega et al. 2011).

The marine-derived fungi *A. sydowii* CBMAI 935, *A. sydowii* CBMAI 933, *P. miczynskii* CBMAI 930, and *Trichoderma* sp. CBMAI 932 isolated from the marine sponges *G. corticostylifera* and *C. erecta* were able to grow in the presence of the pesticide dieldrin (in 3% malt solid medium). *P. miczynskii* CBMAI 930 was selected for quantitative experiments since it showed the highest tolerance to this pesticide (Birolli et al. 2014b).

P. miczynskii CBMAI 930 catalyzed the biotransformation of dieldrin (50 mg L^{-1}) at high conversion rates (90%) after 14 days in 3% malt liquid medium. Kinetic studies showed that 70% of the dieldrin was biotransformed after 4 days of incubation. It is noteworthy that a method validation was carried out in this study, showing satisfactory recovery and standard deviation. The presence of other organochlorine compounds was identified by GC-MS in the reaction medium (endrin [similarity 90], endrin ketone [similarity 89], cyclopentene [similarity 89], aldrin [similarity 91], and chlordene [similarity 88]). However, they were also present in the starting material at similar concentrations (Birolli et al. 2014b).

Although organochlorine pesticides cause some growth inhibition in marine fungal colonies, it was found that these strains were able to grow in the presence of these pesticides and thus have biodegradation potential. Metabolite identification was a serious problem, since no metabolite was observed in the DDD and dieldrin biodegradations probably because of conjugation reactions, in which abundant *in vivo* groups are bonded to the xenobiotic, generating water-soluble products (Alvarenga et al. 2014, Derelanko and Hollinger 2002). Future experiments should focus on different extraction methods, enabling water-soluble metabolites to be identified.

Marine-derived fungi were able to degrade various chemical classes of pesticides and are thus promising strains for biodegradation, bioremediation, and enzymatic studies for the biotransformation of these compounds.

30.7 Collection of Marine-Derived Fungi for Biocatalysis and Biotransformation Reactions

Most microbial work, including that in biocatalysis and biotransformation, requires pure cultures (Tortora 2010), since mixed microorganisms would catalyze several reactions resulting in many products and leading to a decrease in yield, regioselectivity, and

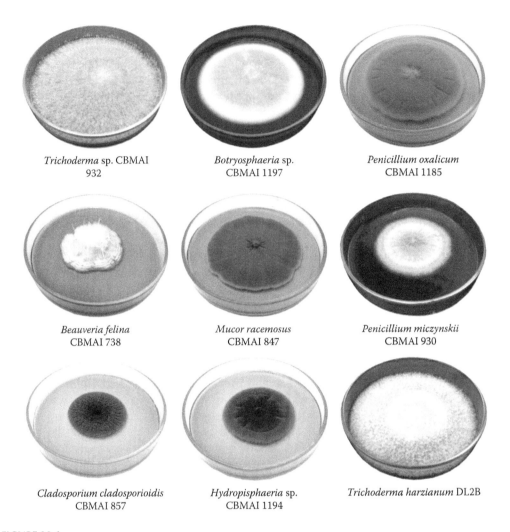

Trichoderma sp. CBMAI 932

Botryosphaeria sp. CBMAI 1197

Penicillium oxalicum CBMAI 1185

Beauveria felina CBMAI 738

Mucor racemosus CBMAI 847

Penicillium miczynskii CBMAI 930

Cladosporium cladosporioides CBMAI 857

Hydropisphaeria sp. CBMAI 1194

Trichoderma harzianum DL2B

FIGURE 30.6
Some marine-derived fungi strains employed in biocatalysis experiments.

stereoselectivity. A visible colony arises from a single spore, a vegetative cell, or a small number of hyphae attached to one another, representing a pure strain. Some marine-derived fungal strains used in biocatalysis are presented in Figure 30.6. Applications of marine-derived fungi described in this chapter are summarized in Tables 30.14 and 30.15.

30.8 Conclusions and Perspectives

Recent advances have been presented by our research group using microorganisms from marine environment in biocatalysis, biotransformation, and biodegradation of organic compounds.

TABLE 30.14

Codes of the Strains Deposited in CBMAI and Described in This Chapter[a]

Strain	Species	CBMAI Code
Beauveria felina	*Beauveria felina*	CBMAI 738
Br09	*Botryosphaeria* sp.	CBMAI 1197
Br23	*Eutypella* sp.	CBMAI 1196
Br27	*Hydropisphaera* sp.	CBMAI 1194
Br61	*Xylaria* sp.	CBMAI 1195
Ce5	*Bionectria* sp.	CBMAI 936
Ce15	*Aspergillus sydowii*	CBMAI 935
Ce16	*Penicillium raistrickii*	CBMAI 931
Ce19	*Aspergillus sydowii*	CBMAI 934
DR(A)2	*Cladosporium* sp.	CBMAI 1237
DR(B)2	*Penicillium raistrickii*	CBMAI 1235
DR(F)2	*Penicillium decaturense*	CBMAI 1234
DR(M3)2	*Aspergillus sydowii*	CBMAI 1241
F30	*Penicillium oxalicum*	CBMAI 1185
F53	*Penicillium citrinum*	CBMAI 1186
GC1	*Trichoderma* sp.	CBMAI 932
GC5	*Penicillium miczynskii*	CBMAI 930
GC12	*Aspergillus sydowii*	CBMAI 933
Penicillium chrysogenum	*Penicillium chrysogenum*	CBMAI 1199
Cs1	*Rhizopus* sp.	CBMAI 1458
CBMAI 847	*Mucor racemosus*	CBMAI 847
CBMAI 849	*Aspergillus sclerotiorum*	CBMAI 849
CBMAI 857	*Cladosporium cladosporioides*	CBMAI 857

Source: Colecao Brasileira de Micro-organismos de Ambiente e Industria, http://webdrm.cpqba. unicamp.br/cbmai/, accessed on March 30, 2016.

[a] CBMAI provides the scientific and industrial community a diverse collection of microbial strains (bacteria, yeasts, and filamentous fungi) for applications in research, testing, and production of enzymes and other metabolites.

Reduction of acetophenone derivatives by oxidoreductases yielded excellent selectivities to obtain enriched or enantiopure alcohols and reduced chalcones by ERs from marine-derived fungi. Other important reactions were presented to hydrolyze epoxides and nitriles from epoxide hydrolases and nitrilases.

Marine-derived fungi catalyzed biodegradation of organochlorine, organophosphate, and pyrethroid pesticides. All reactions were performed with whole cells, showing the potential in bioremediation of organic pollutants.

New advances in our group are involving the cloning and expression of alcohol dehydrogenases, nitrilases, and ERs, as well as immobilization of filamentous fungi on biopolymers, macro- and nanoparticles, and applications of reaction under microwave irradiation using whole cells of marine microorganisms.

Strains of marine-derived fungi are deposited in the Brazilian Collection of Environmental and Industrial Microorganisms used in the biocatalytic reactions presented in this chapter.

Finally, from the presented results, it is clear that marine microorganisms have shown great potential for biotechnological processes.

TABLE 30.15

Applications of Marine-Derived Fungi Described in This Chapter Classified by Strain

Strain	Application	References
Bacillus sp. Br62	Reduction of *ortho*-acetophenones	Mouad (2014)
Botryosphaeria sp. CBMAI 1197	Reduction of 2,2,2-trifluoroacetophenone	Mouad (2014)
	Reduction of 1-(2,4,5-trifluorophenyl)ethanone	Mouad (2014)
	Reduction of 1-(2-(trifluoromethyl)phenyl)ethanone	Mouad (2014)
Eutypella sp. CBMAI 1196	Reduction of 2,2,2-trifluoroacetophenone	Mouad (2014)
	Reduction of 1-(2-(trifluoromethyl)phenyl)ethanone	Mouad (2014)
	Reduction of 1-(2,4,5-trifluorophenyl)ethanone	Mouad (2014)
Hidropisphaera sp. CBMAI 1194	Reduction of 2,2,2-trifluoroacetophenone	Mouad (2014)
	Reduction of 1-(2-(trifluoromethyl)phenyl)ethanone	Mouad (2014)
	Reduction of 1-(2,4,5-trifluorophenyl)ethanone	Mouad (2014)
Xylaria sp. CBMAI 1195	Reduction of 2,2,2-trifluoroacetophenone	Mouad (2014)
	Reduction of 1-(2-(trifluoromethyl)phenyl)ethanone	Mouad (2014)
	Reduction of 1-(2,4,5-trifluorophenyl)ethanone	Mouad (2014)
Penicillium miczynskii CBMAI 930	Reduction of α-chloroacetophenone	Rocha et al. (2009)
	Reduction of iodoacetophenones	Rocha et al. (2012b)
	Reduction of 1-(4-methoxyphenyl) ethanone	Rocha et al. (2012a)
Trichoderma sp. CBMAI 932	Reduction of α-chloroacetophenone	Rocha et al. (2009)
	Hydrolysis of (±)-benzyl glycidyl ether	Martins et al. (2011)
	Hydrolysis of (±)-2-((allyloxy)methyl)oxirane	Martins et al. (2012)
	Biodegradation of DDD	Ortega et al. (2011)
Aspergillus sydowii CBMAI 933	Reduction of α-chloroacetophenone	Rocha et al. (2009)
	Biotransformation of phenylacetonitrile	De Oliveira et al. (2013)
	Hydrolysis of (±)-benzyl glycidyl ether	Martins et al. (2011)
Aspergillus sydowii CBMAI 934	Reduction of α-chloroacetophenone	Rocha et al. (2009)
	Reduction of 1-(4-methoxyphenyl) ethanone	Rocha et al. (2012)
	Biotransformation and biodegradation of 2-bromo-1-phenylethanone	Rocha et al. (2010)
	Biotransformation of phenylacetonitrile	De Oliveira et al. (2013)
Aspergillus sydowii CBMAI 935	Reduction of α-chloroacetophenone	Rocha et al. (2009)
	Reduction of 1-(4-methoxyphenyl) ethanone	Rocha et al. (2012)
	Biotransformation of phenylacetonitrile	De Oliveira et al. (2013)
Bionectria sp. CBMAI 936	Reduction of α-chloroacetophenone	Rocha et al. (2009)
	Reduction of 1-(4-methoxyphenyl) ethanone	Rocha et al. (2012)
	Biotransformation of phenylacetonitrile	De Oliveira et al. (2013)
Penicillium raistrickii CBMAI 931	Reduction of α-chloroacetophenone	Rocha et al. (2009)
	Reduction of iodoacetophenones	Rocha et al. (2012)
	Biotransformation of phenylacetonitrile	De Oliveira et al. (2013)
Aspergillus sclerotiorum CBMAI 849	Reduction of iodoacetophenones	Rocha et al. (2012)
	Reduction of 1-(4-methoxyphenyl) ethanone	Rocha et al. (2012)
Beauveria felina CBMAI 738	Reduction of iodoacetophenones	Rocha et al. (2012)
	Reduction of 1-(4-methoxyphenyl) ethanone	Rocha et al. (2012)

(Continued)

TABLE 30.15 (*Continued*)

Applications of Marine-Derived Fungi Described in This Chapter Classified by Strain

Strain	Application	References
Mucor racemosus CBMAI 847	Reduction of iodoacetophenones	Rocha et al. (2012)
	Reduction of 1-(4-methoxyphenyl) ethanone	Rocha et al. (2012)
Penicillium citrinum CBMAI 1186	Reduction of iodoacetophenones	Rocha et al. (2012)
	Reduction of 1-(4-methoxyphenyl) ethanone	Rocha et al. (2012)
	Reduction of chalcones to dihydrochalcones and biohydrogenation of different α,β-unsaturated ketones	Ferreira et al. (2014c)
Penicillium oxalicum CBMAI 1185	Reduction of iodoacetophenones	Rocha et al. (2012)
Cladosporium cladosporioides CBMAI 857	Reduction of 1-(4-methoxyphenyl) ethanone	Rocha et al. (2012)
Trichoderma sp. CBMAI 932	Biotransformation of di-α,β-unsaturated ketone	Rocha et al. (2014)
Aspergillus sydowii CBMAI 1241	Biotransformation of phenylacetonitrile	De Oliveira et al. (2013)
Penicillium decaturense CBMAI 1234	Biotransformation of phenylacetonitrile	De Oliveira et al. (2013)
Penicillium raistrickii CBMAI 1235	Biotransformation of phenylacetonitrile	De Oliveira et al. (2013)
Aspergillus sydowii CBMAI 934	Hydroxylation of (−)-Ambrox	Martins et al. (2014)
	Hydroxylation of (−)-sclareol	Martins et al. (2014)
Botryosphaeria sp. CBMAI 1197	Hydroxylation of (−)-Ambrox	Martins et al. (2014)
	Hydroxylation of (−)-sclareol	Martins et al. (2014)
	Hydroxylation of (+)-sclareolide	Martins et al. (2014)
Eutypella sp. CBMAI 1196	Hydroxylation of (−)-Ambrox	Martins et al. (2014)
	Hydroxylation of (−)-sclareol	Martins et al. (2014)
	Hydroxylation of (+)-sclareolide	Martins et al. (2014)
Aspergillus sydowii CBMAI 935	Biodegradation of methyl parathion	Alvarenga et al. (2014)
	Biodegradation of profenofos	Da Silva et al. (2013)
	Biodegradation of esfenvalerate	Birolli (2013)
Penicillium decaturense CBMAI 1234	Biodegradation of methyl parathion	Alvarenga et al. (2014)
Penicillium raistrickii CBMAI 931	Biodegradation of profenofos	Da Silva et al. (2013)
Penicillium miczynskii CBMAI 930	Biotransformation of dieldrin	Birolli et al. (2014)
Cladosporium sp. CBMAI 1237	Biodegradation of esfenvalerate	Birolli et al. (2014)
Microsphaeropsis sp. Dr(A)6	Biodegradation of esfenvalerate	Birolli et al. (2014)
Acremonium sp. Dr(F)1	Biodegradation of esfenvalerate	Birolli et al. (2014)
Westerdykella sp. Dr(M2)4	Biodegradation of esfenvalerate	Birolli et al. (2014)
Cladosporium sp. Dr(M2)2	Biodegradation of esfenvalerate	Birolli et al. (2014)

Acknowledgments

The authors acknowledge Conselho Nacional de Desenvolvimento Científico e Tecnológico (CNPq) and Fundação de Amparo a Pesquisa do Estado de São Paulo (FAPESP) for the financial support provided to this research. The authors also acknowledge Prof. Roberto G.S. Berlinck (Instituto de Química de São Carlos—USP) , Prof. Mirna H.R. Seleghim (Universidade Federal de São Carlos—UFSCar), and Prof. Hosana M. Debonsi (Faculdade de Ciências Farmacêuticas de Ribeirão Preto—USP) for providing the marine fungal strains. The English language was reviewed by Timothy Roberts, MSc, a native English speaker, and Prof. Ângela Cristina Pregnolato Giampedro.

References

Adam, P., S. Hecht, W. Eisenreich, J. Kaiser, T. Gräwert, D. Arigoni, A. Bacher, and F. Rohdich. 2002. Biosynthesis of terpenes: Studies on 1-hydroxy-2-methyl-2-(E)-butenyl 4-diphosphate reductase. *Proceedings of the National Academy of Sciences of the United States of America* 99:12108–12113.

Alvarenga, N., W.G. Birolli, M.O.O. Rezende, M.H.R. Seleghim, and A.L.M. Porto. 2015. Biodegradation of chlorpyrifos by whole cells of marine-derived fungi *Aspergillus sydowii* and *Trichoderma* sp. *Microbial & Biochemical Technology* 7:133–139.

Alvarenga, N., W.G. Birolli, M.H.R. Seleghim, and A.L.M. Porto. 2014. Biodegradation of methyl parathion by whole cells of marine-derived fungi *Aspergillus sydowii* and *Penicillium decaturense*. *Chemosphere* 117:47–52.

Baklashova, T., K. Koshcheenko, and G. Skeyabin. 1984. Hydroxylation of indolyl-3-acetic by immobilized mycellium of *Aspergillus niger*. *Applied Microbiology and Biotechnology* 19:217–223.

Banerjee, A., R. Sharma, and U. Banerjee. 2002. The nitrile-degrading enzymes: Current status and future prospects. *Applied Microbiology and Biotechnology* 60:33–44.

Barbieri, C., L. Bossi, P. D'Arrigo, G. Fantoni, and S. Servi. 2001. Bioreduction of aromatic ketones: Preparation of chiral benzyl alcohols in both enantiomeric forms. *Journal of Molecular Catalysis B: Enzymatic* 11:415–421.

Bigley, A.N. and F.M. Raushel. 2013. Catalytic mechanisms for phosphotriesterases. *Biochimica et Biophysica Acta (BBA)—Proteins and Proteomics* 1834:443–453.

Bini, E., D. Knight, and D. Kaplan. 2004. Mapping domain structures in silks from insects and spiders related to protein assembly. *Journal of Molecular Biology* 1:27–40.

Birolli, W.G. 2013. Biodegradação do pesticida esfenvalerato por fungos de ambiente marinho. MSc dissertation, University of São Paulo, Sao Paulo, Brazil.

Birolli, W.G., N. Alvarenga, B. Vacondio, M.H.R. Seleghim, and A.L.M. Porto. 2014a. Growth assessment of marine-derived fungi in the presence of esfenvalerate and its main metabolites. *Journal of Microbial and Biochemical Technology* 6:260–267.

Birolli, W.G., K.Y. Yamamoto, J.R. De Oliveira, M. Nitschke, M.H.R. Seleguim, and A.L.M. Porto. 2014b. Biotransformation of dieldrin by the marine fungus *Penicillium miczynskii* CBMAI 930. *Biocatalysis and Agricultural Biotechnology* 4:39–43.

Brenna, E., F. Gatti, L. Malpezzi, D. Monti, F. Parmeggiani, and A. Sacchetti. 2013. Synthesis of robalzotan, ebalzotan, and rotigotine precursors via the stereoselective multienzymatic cascade reduction of alpha,beta-unsaturated aldehydes. *Journal of Organic Chemistry* 78:4811–4822.

Cano, A., M.T. Ramírez-Apan, and G. Delgado. 2011. Biotransformation of sclareolide by filamentous fungi: Cytotoxic evaluations of the derivatives. *Journal of the Brazilian Chemical Society* 22:1177–1182.

Che, J. and Y. Lam. 2010. Rapid and regioselective hydrogenation of alpha,beta-unsaturated ketones and alkylidene malonic diesters using *Hantzsch ester* catalyzed by titanium tetrachloride. *Synlett* 16:2415–2420.

Chen, W. and A. Mulchandani. 1998. The use of live biocatalysts for pesticide detoxification. *Trends in Biotechnology* 16:71–76.

Chibata, I. and T. Tosa. 1978. Immobilized enzymes and immobilized microbial-cells. *Journal of Synthetic Organic Chemistry Japan* 36:917–930.

Chibata, I. and T. Tosa. 1981. Use of immobilized cells. *Annual Review of Biophysics and Bioengineering* 10:197–216.

Comasseto, J., L. Andrade, A. Omori, L. Assis, and A.L.M. Porto. 2004. Deracemization of aryl ethanols and reduction of acetophenones by whole fungal cells of *Aspergillus terreus* CCT 4083, *A. terreus* CCT 3320 and *Rhizopus oryzae* CCT 4964. *Journal of Molecular Catalysis B: Enzymatic* 29:55–61.

Coutinho, C.F.B., S.T. Tanimoto, A. Galli, G.S. Garbellini, M. Takayama, R.B. Amaral, L.H. Mazo, L.A. Avaca, and S.A.S. Machado. 2005. Pesticidas: Mecanismo de ação, degradação e toxidez. *Pesticidas* 15:65–72.

Da Silva, N.A. 2013. Biodegradação dos pesticidas clorpirifós, metil paratiom and profenofós por fungos de origem marinha. MSc dissertation, University of São Paulo, Sao Paulo, Brazil.

Da Silva, N.A., W.G. Birolli, M.H.R. Seleghim, and A.L.M. Porto. 2013. Biodegradation of the organo-phosphate pesticide profenofos by marine fungi. In *Applied Bioremediation—Active and Passive Approaches*, eds. Y.B. Patil and P. Rao, pp. 149–180. In Tech. Rijek, Croatia.

Dash, H.R., N. Mangwani, J. Chakraborty, S. Kumari, and S. Das. 2013. Marine bacteria: Potential candidates for enhanced bioremediation. *Applied Microbiology and Biotechnology* 2:561–571.

De Carvalho, C.C. and M.M.R. Da Fonseca. 2006. Biotransformation of terpenes. *Biotechnology Advances* 24:134–142.

De Oliveira, A.L.L., D.B. da Silva, N.P. Lopes, and H.M. Debonsi. 2012. Chemical constituents from red algae *Bostrychia radicans* (*Rhodomelaceae*): New amides and phenolic compounds. *Química Nova* 35:2186–2188.

De Oliveira, J.R. 2012. Hidrólise enzimática de nitrilas pelo fungo de origem marinha *Aspergillus sydowii* CBMAI 934. PhD dissertation, University of São Paulo, Sao Paulo, Brazil.

De Oliveira, J.R., C.M. Mizuno, M.H.R. Seleghim, D.C.D. Javaroti, M.O.O. Rezende, M.D. Landgraf, L.D. Sette, and A.L.M. Porto. 2013. Biotransformation of phenylacetonitrile to 2-hydroxyphenylacetic acid by marine fungi. *Marine Biotechnology* 15:97–103.

De Oliveira, J.R., M.H.R. Seleghim, and A.L.M. Porto. 2014. Biotransformation of methylphenyl-acetonitriles by Brazilian marine fungal strain *Aspergillus sydowii* CBMAI 934: Eco-friendly reactions. *Marine Biotechnology* 16:156–160.

Debashish, G., S. Malay, S. Barindra, and M. Joydeep. 2005. Marine enzymes. *Advances in Biochemical and Engineering/Biotechnology* 96:189–218.

Derelanko, M.J. and M.A. Hollinger. 2002. *Handbook of Toxicology*, Vol. 1. NJ: Taylor & Francis Group. Florida.

Diagne, M., N. Oturan, and M.A. Oturan. 2007. Removal of methyl parathion from water by electro-chemically generated Fenton's reagent. *Chemosphere* 66:841–848.

Diez, M. 2010. Biological aspects involved in the degradation of organic pollutants. *Journal of Soil Science and Plant Nutrition* 10:244–267.

Edwards, F.L. and P.B. Tchounwou. 2005. Environmental toxicology and health effects associated with methyl parathion exposure—A scientific review. *International Journal of Environmental Research and Public Health* 2:430–441.

Felício, R., S. Albuquerque, M.C.M. Young, N.S. Yokoya, and H.M. Debonsi. 2010. Trypanocidal, leishmanicidal and antifungal potential from marine red alga *Bostrychia tenella J.* Agardh (*Rhodomelaceae, Ceramiales*). *Journal of Pharmaceutical and Biomedical Analysis* 52:763–769.

Ferreira, I.M., E.B. Meira, I.G. Rosset, and A.L.M. Porto. 2014a. Chemoselective biotransformation of α,β- and $\alpha,\beta,\gamma,\delta$-unsaturated ketones by marine-derived fungus *Penicillium citrinum* CBMAI 1186 in a biphasic system. *Journal of Molecular Catalysis B: Enzymatic* 115:59–65.

Ferreira, I.M., L. Pizzuti, and C. Raminelli. 2014b. Ultrasound-promoted iodination of aromatic com-pounds in the presence of iodine and hydrogen peroxide in water. *Synthetic Communications* 44:2094–2102.

Ferreira, I.M., L.C. Rocha, S.A. Yoshioka, M. Nitschke, A.H. Jeller, L. Pizzuti, and A.L.M. Porto. 2014c. Chemoselective reduction of chalcones by whole hyphae of marine fungus *Penicillium citrinum* CBMAI 1186, free and immobilized on biopolymers. *Biocatalysis and Agricultural Biotechnology* 3:358–364.

Gavrilescu, M. 2005. Fate of pesticides in the environment and its bioremediation. *Engineering in Life Sciences* 5:497–526.

Ghanem, E. and F.M. Raushel. 2005. Detoxification of organophosphate nerve agents by bacterial phosphotriesterase. *Toxicology and Applied Pharmacology* 207:459–470.

Gong, J.S., Z.M. Lu, H. Li, J.S. Shi, Z.M. Zhou, and Z.H. Xu. 2012. Nitrilases in nitrile biocatalysis: Recent progress and forthcoming research. *Microbial Cell Factories* 11:142–160.

Gong, P.F. and J.H. Xu. 2005. Bio-resolution of a chiral epoxide using whole cells of *Bacillus megate-rium* ECU1001 in a biphasic system. *Enzyme and Microbial Technology* 36:252–257.

Havens, P.L. 1995. Fate of herbicides in the environment. In *Handbook of Weed Management Systems*, ed. A. Smith, pp. 245–278. New York: John Wiley & Sons.

Holsch, K. and D. Weuster-Botz. 2010. New oxidoreductases from cyanobacteria: Exploring nature's diversity. *Enzyme and Microbial Technology* 47:228–235.

Illanes, A. 2008. *Enzyme Biocatalysts: Principles and Applications*. Springer Science, Valparaíso, Chile.

Inoue, K., Y. Makino, and N. Itoh. 2005. Production of (R)-chiral alcohols by a hydrogen-transfer bioreduction with NADH-dependent *Leifsonia* alcohol dehydrogenase (LSADH). *Tetrahedron: Asymmetry* 15:2539–2549.

Jin, L.Q., Z.Q Liu, J.M. Xu, and Y.G. Zhen. 2013. Biosynthesis of nicotinic acid from 3-cyanopyridine by a newly isolated *Fusarium proliferatum* ZJB-09150. *World Journal of Microbiology and Biotechnology* 29:431–440.

Kazemimostaghim, M., R. Rajkhowa, K. Patil, T. Tsuzuki, and X. Wang. 2014. Structure and characteristics of milled silk particles. *Powder Technology* 254:488–493.

Kiełbasiński, P., M. Rachwalski, M. Mikołajczyk, and F.P. Rutjes. 2008. Nitrilase-catalysed hydrolysis of cyanomethyl *p*-tolyl sulfoxide: Stereochemistry and mechanism. *Tetrahedron: Asymmetry* 19:562–567.

Koda, R., T. Numata, S. Hama, S. Tamalampudi, K. Nakashima, T. Tanaka, and A. Kondo. 2010. Ethanolysis of rapeseed oil to produce biodiesel fuel catalyzed by *Fusarium heterosporum* lipase-expressing fungus immobilized whole-cell biocatalysts. *Journal of Molecular Catalysis B: Enzymatic* 66:101–104.

Kohli, R. and V. Massey. 1998. The oxidative half-reaction of old yellow enzyme—The role of tyrosine 196. *Journal of Biological Chemistry* 49:32763–32770.

Kotik, M., J. Brichac, and P. Kyslík. 2005. Novel microbial epoxide hydrolases for biohydrolysis of glycidyl derivatives. *Journal of Biotechnology* 120:364–375.

Kourkoutas, Y., M. Kanellaki, A. Koutinas, and C. Tzia. 2006. Effect of storage of immobilized cells at ambient temperature on volatile by-products during wine-making. *Journal of Food Engineering* 74:217–223.

Kouzi, S.A. and J.D. Mcchesney. 1991. Microbial models of mammalian metabolism: Fungal metabolism of the diterpene sclareol by *Cunninghamella* species. *Journal of Natural Products* 54:483–490.

Krajewska, B. 2004. Application of chitin- and chitosan-based materials for enzyme immobilizations: A review. *Enzyme and Microbial Technology* 35:126–139.

Kuek, C. 1986. Immobilized living fungal mycelia for the growth-dissociated synthesis of chemicals. *International Industrial Biotechnology* 6:123–125.

Martínková, L., V. Vejvoda, O. Kaplan, D. Kubáč, A. Malandra, M. Cantarella, K. Bezouška, and V. Křen. 2009. Fungal nitrilases as biocatalysts: Recent developments. *Biotechnology Advances* 27:661–670.

Martins, M.P. 2012. Resolução cinética de haloidrinas racêmicas com a lipase B de *Candida antarctica* e biotransformação de produtos naturais por micro-organismos. PhD dissertation, University of São Paulo, Sao Paulo, Brazil.

Martins, M.P., A.M. Mouad, L. Boschini, M.H.R. Seleghim, L.D. Sette, and A.L.M. Porto. 2011. Marine fungi *Aspergillus sydowii* and *Trichoderma* sp. catalyze the hydrolysis of benzyl glycidyl ether. *Marine Biotechnology* 13:314–320.

Martins, M.P., A.M. Mouad, and A.L.M. Porto. 2012. Hydrolysis of allylglycidyl ether by marine fungus *Trichoderma* sp. Gc1 and the enzymatic resolution of allylchlorohydrin by *Candida antarctica* lipase type B. *Current Topics in Catalysis* 10:27–33.

Martins, M.P., J. Ouazzani, G. Arcile, A.H. Jeller, J.P.F. Lima, M.H.R. Seleghim R., A.L.L. De Oliveira et al. 2015. Biohydroxylation of (–)-Ambrox®, (–)-sclareol, and (+)-sclareolide by whole cells of Brazilian marine-derived fungi. *Marine Biotechnology* 17:211–218.

Matsuda, T., R. Yamanaka, and K. Nakamura. 2009. Recent progress in biocatalysis for asymmetric oxidation and reduction. *Tetrahedron: Asymmetry* 20:513–557.

Mayer, A.M.S., A.D. Rodriguez, O. Taglialatela-Scafati, and N. Fusetani. 2013. Marine pharmacology in 2009–2011: Marine compounds with antibacterial, antidiabetic, antifungal, antiinflammatory, antiprotozoal, antituberculosis, and antiviral activities; affecting the immune and nervous systems, and other miscellaneous mechanisms of action. *Marine Drugs* 11:2510–2573.

Melgar, G.Z., F.V.S. Assis, L.C. Rocha, S.C. Fanti, L.D. Sette, Porto, and A.L.M. 2013. Growth curves of filamentous fungi for utilization in biocatalytic reduction of cyclohexanones. *Global Journal of Science Frontier Research Chemistry* 13:13–19.

Mischitz, M., W. Kroutil, U. Wandel, and K. Faber. 1995. Asymmetric microbial hydrolysis of epoxides. *Tetrahedron: Asymmetry* 6:1261–1272.

Monteiro, J.L.F. and C.O. Veloso. 2004. Catalytic conversion of terpenes into fine chemicals. *Topics in Catalysis* 27:169–180.

Moreira, S., M. Moreira-Santos, U. Guilhermino, and R. Ribeiro. 2006. Immobilization of the marine microalga *Phaeodactylum tricornutum* in alginate for *in situ* experiments: Bead stability and suitability. *Enzyme and Microbial Technology* 38:135–141.

Mouad, A.M. 2013. Biocatálise na produção de moléculas orgânicas: Oxidorredutases de fungos marinhos para a síntese de álcoois quirais e lipase de *Candida antarctica* na produção de amidas fenólicas graxas. PhD dissertation, University of São Paulo, Sao Paulo, Brazil.

Mouad, A.M., A.L.L. De Oliveira, H.M. Debonsi, and A.L.M. Porto. 2015. Bioreduction of fluoroacetophenone derivatives by endophytic fungi isolated from the marine red alga *Bostrychia radicans*. *Biocatalysis* 1:141–147.

Mouad, A.M., M.P. Martins, H.M. Debonsi, A.L.L. De Oliveira, R. De Felicio, N.S. Yokoya, T.M. Fujii, C.B.A. De Menezes, F. Fantinatti-Garboggini, and A.L.M. Porto. 2011. Bioreduction of acetophenone derivatives by red marine algae *Bostrychia radicans*, *B. tenella*, and marine bacteria associated. *Helvetica Chimica Acta* 94:1506–1514.

Musharraf, S.G., S. Naz, A. Najeeb, S. Khan, and M.I. Choudhary. 2012. Biotransformation of perfumery terpenoids, (–)-ambrox® by a fungal culture *Macrophomina phaseolina* and a plant cell suspension culture of *Peganum harmala*. *Chemistry Central Journal* 6:82.

Nahra, F., Y. Macé, D. Lambin, and O. Riant. 2013. Copper/palladium-catalyzed 1,4-reduction and asymmetric allylic alkylation of α,β-unsaturated ketones: Enantioselective dual catalysis. *Angewandte Chemie International Edition* 52:3208–3212.

Nakamura, K., R. Yamanaka, T. Matsuda, and T. Harada. 2003. Recent developments in asymmetric reduction of ketones with biocatalysts. *Tetrahedron: Asymmetry* 14:2659–2681.

Nie, J., G. Hong-Chao, D. Cahard, and M. Jun-An. 2011. Asymmetric construction of stereogenic carbon centers featuring a trifluoromethyl group from prochiral trifluoromethylated substrates. *Chemical Reviews* 32:455–529.

Ojima, I., Y. Nagal, and T. Kogure. 1972. Selective reduction of alfa,beta-unsaturated terpene carbonyl-compounds using hydrosilane-rhodium (I) complex combinations. *Tetrahedron Letters* 49:265–268.

O'reilly, C. and P. Turner. 2003. The nitrilase family of CN hydrolysing enzymes—A comparative study. *Journal of Applied Microbiology* 95:1161–1174.

Ortega, S.N., M. Nitschke, A.M. Mouad, M.D. Landgraf, M.O. Oliveira Rezende, M.H. Regali Seleghim, L.D. Sette, and A.L. Meleiro Porto. 2011. Isolation of Brazilian marine fungi capable of growing on DDD pesticide. *Biodegradation* 22:43–50.

Oturan, M.A., J. Peiroten, P. Chartrin, and A.J. Acher. 2000. Complete destruction of *p*-nitrophenol in aqueous medium by electro-Fenton method. *Environmental Science and Technology* 34:3474–3479.

Patel, R.N. 2011. Biocatalysis: Synthesis of key intermediates for development of pharmaceuticals. *ACS Catalysis* 1:1056–1074.

Peart, P.C., A.R. Chen, W.F. Reynolds, and P.B Reese. 2012. Entrapment of mycelial fragments in calcium alginate: A general technique for the use of immobilized filamentous fungi in biocatalysis. *Steroids* 77:85–90.

Petříčková, A., A.B. Veselá, O. Kaplan, D. Kubáč, B. Uhnáková, A. Malandra, J. Felsberg, A. Rinágelová, P. Weyrauch, and V. Křen. 2012. Purification and characterization of heterologously expressed nitrilases from filamentous fungi. *Applied Microbiology and Biotechnology* 93:1553–1561.

Prelog, V. 1964. Specification of the stereospecificity of some oxido-reductases by diamond lattice sections. *Pure and Applied Chemistry* 9:119–130.

Prelog, V. and P. Wieland. 1966. Über die Bamford-Stevens-Reaktion des cyclodecanon-*p*-toluolsulfonylhydrazons in protischem Lösungsmittel. *Helvetica Chimica Acta* 27:2275–2278.

Radosevich, S.R., J. Holt, and C. Guersa. 1997. *Weed Ecology*. New York: John Wiley & Sons.

Rebelo, R.M., R.A. Vasconcelos, B.D.M.C. Ruys, J.A. Rezende, K.O.C. Moraes, and R.P. Oliveira. 2010. Produtos agrotóxicos e afins comercializados em 2009 no Brasil: Uma abordagem ambiental. Report, IBAMA. http://www.ibama.gov.br/phocadownload/Qualidade_Ambiental/produtos_agrotoxicos_comercializados_brasil_2009.pdf (accessed August 21, 2014).

Richins, R.D., I. Kaneva, A. Mulchandani, and W. Chen. 1997. Biodegradation of organophosphorus pesticides by surface-expressed organophosphorus hydrolase. *Nature Biotechnology* 15:984–987.

Rocha, L.C., H. Ferreira, R. Luiz, L. Sette, and A.L.M. Porto. 2012a. Stereoselective bioreduction of 1-(4-methoxyphenyl)ethanone by whole cells of marine-derived fungi. *Marine Biotechnology* 14:358–362.

Rocha, L.C., H. Ferreira, E. Pimenta, R. Berlinck, M. Rezende, M. Landgraf, M. Seleghim, L. Sette, and A.L.M. Porto. 2010a. Biotransformation of alpha-bromoacetophenones by the marine fungus *Aspergillus sydowii*. *Marine Biotechnology* 12:552–557.

Rocha, L.C., H. Ferreira, E. Pimenta, R. Berlinck, M. Seleghim, D. Javaroti, L. Sette, R. Bonugli, and A.L.M. Porto. 2009. Bioreduction of alpha-chloroacetophenone by whole cells of marine fungi. *Biotechnology Letters* 31:1559–1563.

Rocha, L.C., R.F. Luiz, I.G. Rosset, C. Raminelli, M.H.R. Seleghim, L.D. Sette, and A.L.M. Porto. 2012b. Bioconversion of iodoacetophenones by marine fungi. *Marine Biotechnology* 14:396–401.

Rocha, L.C., I.G. Rosset, R.F. Luiz, C. Raminelli, and A.L.M. Porto. 2010b. Kinetic resolution of iodophenylethanols by *Candida antarctica* lipase and their application for the synthesis of chiral biphenyl compounds. *Tetrahedron: Asymmetry* 21:926–929.

Saleem, M., M.S. Ali, S. Hussain, A. Jabbar, M. Ashraf, and Y.S. Lee. 2007. Marine natural products of fungal origin. *Natural Product Reports* 24:1142–1152.

Santos, V.M.R., C.L. Donnici, J.B.N. Dacosta, and J.M.R. Caixeiro. 2007. Compostos organofosforados pentavalentes: Histórico, métodos sintéticos de preparação e aplicações como inseticidas e agentes antitumorais. *Química Nova* 30:159–170.

Schnoor, J.L. 1992. Fate of pesticides and chemicals in the environment. In *Fate of Pesticides & Chemicals in the Environment*, ed. N.L. Wolfe, pp. 93–104. New York: John Wiley & Sons.

Shiomi, T., T. Adachi, K. Toribatake, L. Zhou, and H. Nishiyama. 2009. Asymmetric β-boration of α,β-unsaturated carbonyl compounds promoted by chiral rhodium–bisoxazolinylphenyl catalysts. *Chemical Communications* 40:5987–5989.

Silva, V., B. Stambuk, and M. Nascimento. 2010. Efficient chemoselective biohydrogenation of 1,3-diaryl-2-propen-1-ones catalyzed by *Saccharomyces cerevisiae* yeasts in biphasic system. *Journal of Molecular Catalysis B: Enzymatic* 63:157–163.

Simeó, Y. and K. Faber. 2006. Selectivity enhancement of enantio- and stereo-complementary epoxide hydrolases and chemo-enzymatic deracemization of (±)-2-methylglycidyl benzyl ether. *Tetrahedron: Asymmetry* 17:402–409.

Singh, B.K. and A. Walker. 2006. Microbial degradation of organophosphorus compounds. *FEMS Microbiology Reviews* 30:428–471.

Šnajdrová, R., V. Kristová-Mylerová, D. Crestia, K. Nikolaou, M. Kuzma, M. Lemaire, E. Gallienne, J. Bolte, K. Bezouška, and V.R. Křen. 2004. Nitrile biotransformation by *Aspergillus niger*. *Journal of Molecular Catalysis B: Enzymatic* 29:227–232.

Soderlund, D.M., J.M. Clark, L.P. Sheets, L.S. Mullin, V.J. Piccirillo, D. Sargent, J.T. Stevens, and M.L. Weiner. 2002. Mechanisms of pyrethroid neurotoxicity: Implications for cumulative risk assessment. *Toxicology* 171(1):3–59.

Sogorb, M.A. and E. Vilanova. 2002. Enzymes involved in the detoxification of organophosphorus, carbamate and pyrethroid insecticides through hydrolysis. *Toxicology Letters* 128:215–228.

Sommovigo, M. and H. Alper. 1993. Mild reduction of α,β-unsaturated ketones and aldehydes with an oxygen-activated palladium catalyst. *Tetrahedron Letters* 34:59–62.

Spain, J.C. 1995. Biodegradation of nitroaromatic compounds. *Annual Review of Microbiology* 49:523–555.

Surendhiran, D., M. Vijay, and A.R. Sirajunnisa. 2014. Biodiesel production from marine micro-alga *Chlorella salina* using whole cell yeast immobilized on sugarcane bagasse. *Journal of Environmental Chemical Engineering* 2:1294–1300.

Sutherland, T., I. Horne, K. Weir, C. Coppin, M. Williams, M. Selleck, R. Russell, and J. Oakeshott. 2004. Enzymatic bioremediation: From enzyme discovery to applications. *Clinical and Experimental Pharmacology and Physiology* 31:817–821.

Toogood, H.S., T. Knaus, and N.S. Scrutton. 2014. Alternative hydride sources for ene-reductases: Current trends. *ChemCatChem* 6:951–954.

Tortora, G.J. 2010. *Microbiology: An Introduction*, 10th edn. Pearson Education, San Francisco.

Trincone, A. 2010. Potential biocatalysts originating from sea environments. *Journal of Molecular Catalysis B: Enzymatic* 66:241–256.

Utsukihara, T., W. Chai, N. Kato, K. Nakamura, and C.A. Horiuchi. 2004. Reduction of (+)- and (–)-camphorquinones by cyanobacteria. *Journal of Molecular Catalysis B: Enzymatic* 31:19–24.

Vassilev, N. and M. Vassileva. 1992. Production of organic acids by immobilized filamentous fungi. *Mycological Research* 96:563–570.

Vaz, A., S. Chakraborty, and V. Massey. 1995. Old yellow enzyme—Aromatization of cyclic enones and the mechanism of a novel dismutation reactions. *Biochemistry* 34:4246–4256.

Winkler, C., G. Tasnadi, D. Clay, M. Hall, and K. Faber. 2012. Asymmetric bioreduction of activated alkenes to industrially relevant optically active compounds. *Journal of Biotechnology* 162:381–389.

Winkler, M., O. Kaplan, V. Vejvoda, N. Klempier, and L. Martínková. 2009. Biocatalytic application of nitrilases from *Fusarium solani* O1 and *Aspergillus niger* K10. *Journal of Molecular Catalysis B: Enzymatic* 59:243–247.

Yadav, J.S., B.V.S. Reddy, C. Sreelaksmi, G.G.K.S.N. Kumar, and A.B. Rao. 2008. Enantioselective reduction of 2-substituted tetrahydropyran-4-ones using *Daucus carota* plant cells. *Tetrahedron Letters* 49:2768–2771.

Yang, C., N. Cai, M. Dong, H. Jiang, J. Li, C. Qiao, A. Mulchandani, and W. Chen. 2008. Surface display of MPH on *Pseudomonas putida* JS444 using ice nucleation protein and its application in detoxification of organophosphates. *Biotechnology and Bioengineering* 99:30–37.

Zheng, Y., W. Lan, C. Qiao, A. Mulchandani, and W. Chen. 2007. Decontamination of vegetables sprayed with organophosphate pesticides by organophosphorus hydrolase and carboxyl-esterase (B1). *Applied Biochemistry and Biotechnology* 136:233–241.

Zhu, D., C. Mukherjee, Y. Yang, B.E. Rios, D.T. Gallagher, N.N. Smith, E.R. Biehl, and L. Hua. 2008. A new nitrilase from *Bradyrhizobium japonicum* USDA 110: Gene cloning, biochemical characterization and substrate specificity. *Journal of Biotechnology* 133:327–333.

Section XIII

Marine Foodomics

31

Polysaccharide-Based Nanoparticles as Drug Delivery Systems

V.L. Sirisha and Jacinta S. D'Souza

CONTENTS

31.1 Introduction

Delivering therapeutic compound to the target site is a major problem in treatment of many diseases. A conventional application of drugs is characterized by limited effectiveness, poor biodistribution, and lack of selectivity (Nevozhay et al. 2007). These limitations and drawbacks can be overcome by controlling drug delivery. In controlled drug delivery systems (DDSs), the drug is transported to the place of action; thus, its influence on vital tissues and undesirable side effects can be minimized. In addition, DDS protects the drug from rapid degradation or clearance and enhances drug concentration in target tissues; therefore, lower doses of drug are required (Nevozhay et al. 2007). This modern form of therapy is especially important when there is a discrepancy between a dose and concentration of a drug and its therapeutic results or toxic effects. Cell-specific targeting can be achieved by attaching drugs to individually designed carriers. Recent developments in nanotechnology have shown that nanoparticles (structures smaller than 100 nm in at least one dimension) have a great potential as drug carriers. Due to their small sizes, the nanostructures exhibit unique physicochemical and biological properties (e.g., an enhanced reactive area as well as an ability to cross cell and tissue barriers) that make them a favorable material for biomedical applications.

31.2 Nanocarriers Used in Drug Delivery System

According to the definition from the National Nanotechnology Initiative, nanoparticles are structures of sizes ranging from 1 to 100 nm in at least one dimension. However, the prefix "nano" is commonly used for particles that are up to several hundred nanometers in size. Nanocarriers with optimized physicochemical and biological properties are taken up by cells more easily than larger molecules, so they can be successfully used as delivery tools for currently available bioactive compounds (Suri et al. 2007). Liposomes, solid lipids nanoparticles, dendrimers, polymers, silicon or carbon materials, and magnetic nanoparticles are the examples of nanocarriers that have been tested as DDSs.

Nanoparticle DDSs have wide range of advantages: (1) they can pass through the smallest capillary vessels because of their ultratiny volume and avoid rapid clearance by phagocytes so that their duration in the blood stream is greatly prolonged; (2) they can penetrate cells and tissue gaps to arrive at target organs such as liver, spleen, lung, spinal cord, and lymph; (3) they could show controlled release properties due to the biodegradability, pH, ion, and/or temperature sensibility of materials; and (4) they can improve the utility of drugs and reduce toxic side effects; etc.

The way of conjugating the drug to the nanocarrier and the strategy of its targeting are very important for a targeted therapy. A drug may be adsorbed or covalently attached to the nanocarrier's surface or else it can be encapsulated into it. When compared to other ways of attaching, covalent linking has the advantage, as it enables to control the number of drug molecules connected to the nanocarrier, that is, an accurate control of the amount of therapeutic compound delivered. Cell-specific targeting with nanocarriers may be accomplished by using active or passive mechanisms.

Once the drug-nanocarrier conjugates reach the diseased tissues, the therapeutic agents are released. A controlled release of drugs from nanocarriers can be achieved through changes in physiological environment such as temperature, pH, osmolarity, or *via* an enzymatic activity.

Presently, the research on nanoparticle DDS is focused on the following:

1. Selection and combination of carrier materials to obtain appropriate drug release speed
2. Modification of nanoparticles surface so as to improve their targeting ability
3. Standardization of nanoparticles preparation, so as to increase their drug delivery capability, and their prospective industrial-scale production
4. Investigation of *in vivo* dynamic process to disclose the interaction of nanoparticles with blood and targeting tissues and organs, etc.

31.3 Polysaccharides as a Source of Nanoparticles

Nanocarriers used for medical applications have to be biocompatible (able to integrate with a biological system without eliciting immune response or any negative effects) and nontoxic (harmless to a given biological system). Many polymeric materials that are biocompatible and biodegradable are used in preparing nanoparticles for drug delivery, which include poly(lactic acid), poly(glycolic acid), polycaprolactone, polysaccharides (particularly chitosan), poly(acrylic acid) family proteins, or polypeptides (such as gelatine). Among these systems, the role of natural polysaccharides in developing prepared nanoparticles has significantly increased (Leonard et al. 2003; Aumelas et al. 2007; Yang et al. 2008a; Zhang et al. 2011).

Polysaccharides are long carbohydrate molecules of repeated monosaccharide units joined together by glycosidic bonds and are often one of the main structural element of plant and animals exoskeleton (cellulose, carrageenan, chitin, chitosan, etc.) or have an important role in plant energy storage (starch, paramylon, etc.; Aminabhavi et al. 1990). In nature, polysaccharides are available from various resources like plant (e.g., pectin, guar gum), animal (chitosan, chondroitin), algal (e.g., alginate), and microbial origin (e.g., dextran, xanthan gum) (Sinha and Kumria 2001). Polysaccharides, depending on their monosaccharide components, can be homopolysaccharides or heteropolysaccharides: if all the monosaccharides are from the same type, the polysaccharide formed is called homopolysaccharide, and when the polysaccharide is composed of more than one type of monosaccharide, it is called heteropolysaccharide. The polysaccharides differ remarkably in their chemical structure, source, positive, or negative charge, among others, which can be observed for homopolysaccharides (Table 31.1) and heteropolysaccharides (Table 31.2). They offer a wide diversity in structure and properties due to their wide range of molecular weight (MW) and chemical composition.

The majority of natural polysaccharides present several hydrophilic groups such as carboxyl, hydroxyl, and amino groups, which endow their solubility in water and the formation of noncovalent bonds with biological tissues and mucosal membranes (Liu et al. 2008). This way, the hydrophilic properties of most of the polysaccharide nanoparticles provide bioadhesion and mucoadhesion characteristics to these biomaterials, as well as

TABLE 31.1

Characteristics as Chemical Structure, Source, and Charge of Some Homopolysaccharides

Polysaccharide	Structure	Source	Charge
Chitosan		Animal	+
Dextran	α-(1–6) + α-(1–4) α-(1–6)	Microbial	−
Pullulan		Microbial	−

giving the possibility of chemical modification of the macromolecules to bind drugs or targeting agents. The hydrophilic nanoparticles also possess the enormous advantage of extended circulation in the blood, which increases the probability of passive targeting of the nanoparticles into the tumor tissues (Mitra et al. 2001).

The most profitable use of polysaccharides as natural biomaterials is their availability in nature and low cost in processing, which makes them quite accessible materials to be used as drug carriers. Moreover, they are highly stable, safe, nontoxic, biodegradable, and hydrophilic (Liu et al. 2008). Thus, they have a large variety of composition and properties that cannot be replicated in a chemical laboratory, and the ease of their production makes various polysaccharides cheaper than synthetic polymers (Coviello et al. 2007). Therefore, the use of polysaccharides as biomaterials is quite promising in terms of biomedical, environmental, and food-related fields and even pharmaceutical applications (Lemarchand et al. 2005; Rinaudo 2008; Park et al. 2010).

Recently, many studies have been using polysaccharides and their derivatives for their potential application as nanoparticles DDSs. The most commonly used ones are chitosan, alginate, hyaluronic acid (HA), pullulan, pectin, cellulose, dextran, and guar gum (Liu et al. 2008; Boddohi et al. 2009; Bhaw-Luximon 2011; Mizrahy and Peer 2012; Saravanakumar et al. 2012). A brief description of their structural features and some of the techniques used to prepare polysaccharide-based nanoparticles is discussed subsequently.

TABLE 31.2

Characteristics as Chemical Structure, Source, and Charge of Some Heteropolysaccharides

Polysaccharide	Structure	Source	Charge
Hyaluronic acid		Human	–
Alginate		Algal	
Gum arabic		Plant	–
Xanthan gum		Microbial	–
Pectin		Plant	–

31.4 Structure and Characteristics of Some of the Polysaccharides

An enormous number of polysaccharides have been in use as DDSs. The chemical structures, features, and applications in various fields of some commonly used polymers are discussed in the following text. This is done emphasizing their role in preparation of DDSs.

31.4.1 Homopolysaccharides

31.4.1.1 Chitosan

Chitosan has appeared the most promising biomaterial for the development of ideal hydrophilic drug vehicles for the controlled drug delivery and therefore has been rigorously investigated over the last two decades (Felt et al. 1998; Janes et al. 2001; Mitra et al. 2001). Chitosan is a linear polysaccharide composed of randomly distributed β-(1–4)-linked D-glucosamine (deacetylated unit) and N-acetyl-D-glucosamine (acetylated unit) (Figure 31.1). It is a hydrophilic biopolymer and is produced commercially by hydrolyzing the aminoacetyl groups of chitin, which is the structural element in the exoskeleton of crustaceans (such as crabs and shrimp) and cell walls of fungi (Tharanathan and Ramesh 2003; Thanou et al. 2005; Zhuangdong 2007) by an alkaline deacetylation treatment (Muzzarelli and Muzzarelli 2005). The amount of deacetylation has been estimated by NMR spectroscopy, and in general, the degree of % deacetylation of commercial chitosan ranges from 60% to 100%. The average MW of industrially produced chitosan is between 3.8 and 20 kDa. Deacetylation of chitin to produce chitosan is performed using the excess of sodium hydroxide as a reagent and water as a solvent. This reaction pathway, when allowed to go to completion (complete deacetylation), yields up to 98% product (Yuan 2007). So, once deacetylation happens, chitosan consists primarily of repeating units of β-(1,4)-2-aminodeoxy-D-glucose (D-glucosamine). Chitosan is positively charged at neutral and alkaline pH. It is a weak base and it is insoluble in water and organic solvents. However, it is soluble in diluted aqueous acidic solution (pH < 6.5), which can convert the glucosamine units into a soluble form with protonated amine groups (Sinha et al. 2004). The solubility of chitosan in water can be increased by removing one or two hydrogen atoms from the amino groups of chitosan and introducing some hydrophilic segments (Srinophakun and Boonmee 2011).

The various biological characteristics of chitosan, such as low or nontoxicity, biocompatibility, biodegradability, low antimicrobial, and immunogenic properties, provide its potential for various applications (Guerrero et al. 2010). Its rare positive charge converts chitosan into a special polysaccharide, since it provides strong electrostatic interaction with negatively charged mucosal surfaces and macromolecules such as DNA and RNA

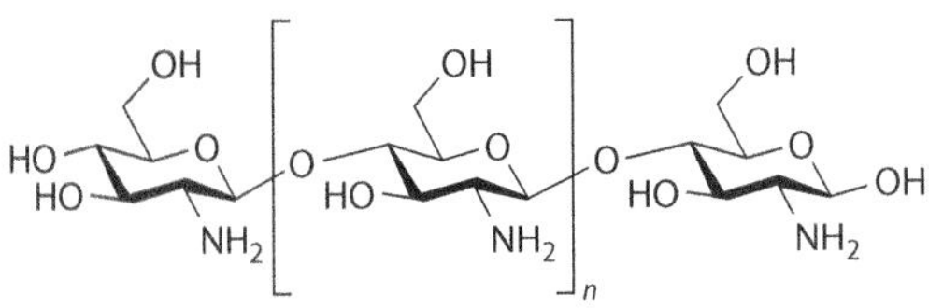

FIGURE 31.1
The chemical structure of chitosan.

(Fang et al. 2006; Morille et al. 2008), which is an attractive feature for the treatment of solid tumors (Li et al. 2009). The polyelectrolyte nature of chitosan can be used as an absorbent of heavy metal ions and textile industry effluents from wastewaters. It has been also used as template for the preparation of mesoporous metal oxides spheres (Braga et al. 2009). However, it is commonly used in various pharmaceutical and biomedical fields due to its biodegradability and biocompatibility. As chitosan can be biodegraded by enzyme action, it has been analyzed as biomaterial for wound healing and as a prosthetic material (Bernardo et al. 2003). It is also reported to find uses as an antimicrobial compound and as a drug in the treatment of hyperbilirubinemia and hypercholesterolemia, and also, it has been prepared and evaluated for its antitumor activity carrying several antineoplastic agents (Blanco et al. 2000).

Chitosan has attracted attention as a matrix for controlled release in the field of nano-medicine due to its reactive functionalities, polycationic nature, easy degradation by enzymes, and nontoxic degradation products. From a long time ago, various natural and synthetic polymers have been used for the preparation of drug-loaded microparticles; out of those, chitosan was found to be very promising and has been extensively investigated (Muzzarelli and Muzzarelli 2005; Davidenko et al. 2009). Because of its bioadhesive prop-erties, chitosan has received significant attention as carrier in novel bioadhesive DDSs that increase the residence time of the drugs at the site of absorption and thereby increase the drug bioavailability (Varum et al. 2008). Hence, some drugs administered via nasal (Learoyd et al. 2008) or gastrointestinal routes have improved their treatment efficacy when they are included into chitosan-based systems (Guerrero et al. 2010).

Considering the various advantages of chitosan, it is found to be a promising matrix for the controlled release of pharmaceutical gents. Both *in vitro* and *in vivo* experiments have clearly demonstrated chitosan as an ideal carrier for a variety of drugs whose effectiveness is increased when they are admitted into these systems.

31.4.1.2 Dextran

It is a polysaccharide consisting of many glucose molecules composed of chains of dif-ferent lengths. It has a significant amount of α-(1–6) glycosidic linkages in its main chain (Figure 31.2) and a variable number of α-(1–2), α-(1–3) and α-(1–4) branched linkages (Misaki et al. 1980). Depending on the type of bacterial strain that synthesizes it, the type of branching will be different. The average MW of dextran is as high as 10^7–10^8 Da (Heinze et al. 2006) but can be reduced by acidic hydrolysis to obtain MW fractions that are of spe-cific interest.

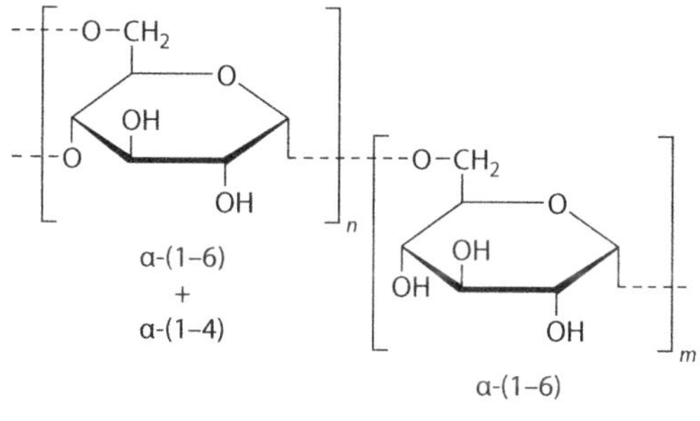

FIGURE 31.2
The chemical structure of dextran.

Dextran is water soluble, neutral, biodegradable, and biocompatible. Its features may vary depending on the MW as well as the distribution, the type of branches, and the degree of branching, which depend on the bacterial synthesis or postsynthesis reactions to form derivatives. It is synthesized by a wide variety of bacterial strains. *Gluconobacter oxydans* produces dextran from maltodextrin, while *Leuconostoc mesenteroides* and *Streptococcus mutans* produce dextran from sucrose (Heinze et al. 2006). It can also be synthesized enzymatically using cell-free extract supernatant (Wang et al. 2011). Dextran can be also produced by chemical synthesis, developing a cationic ring opening polymerization of levoglucosan (Heinze et al. 2006).

Its natural structure can be modified by reacting different hydrophobic molecules with its different hydroxyl groups (Lemarchand et al. 2003). It is reported that by varying the nature of the reacting molecules and the number of grafted, that is, number of hydrophobic groups per 100 glucopyranose units or degree of substitution (DS), various amphiphilic dextran derivatives have been obtained (aromatic rings, aliphatic or cyclic hydrocarbons) (Rotureau et al. 2004).

Dextran has a broad range of applications in varied areas like, clinical, chemical, food, and pharmaceutical industry. It is used as an emulsifier, stabilizer, adjuvant, stabilizer, and thickener in jam and ice cream and mainly as a drug (as blood plasma volume expander). By using matrix of cross-linked dextran gel layer, proteins can be separated and purified by size exclusion chromatography. Dextran and its derivatives (which are produced by structural modifications) are used for the preparation of modified DDSs (Chen et al. 2003; Aumelas et al. 2007; Coviello et al. 2007). In the field of nanomedicine, because of its biocompatibility, good availability, and biodegradability, it is not only used as a nanoparticulate carrier system but also engaged to encapsulate these systems (Gavory et al. 2011).

31.4.1.3 Pullulan

Pullulan is a naturally occurring fungal polysaccharide produced by fermentation of liquefied corn starch by *Aureobasidium pullulans*, a ubiquitous yeastlike fungus. It has a linear structure consisting of predominantly repeating maltotriose units, which are made up of three α-(1–4)-linked glucose molecules (Wallenfels et al. 1965; Catley and Whelan 1971; Carolan et al. 1983), linked by α-(1–6)-glycosidic bonds. The backbone is formed by glycosidic linkages between α-(1–6)-D-glucopyranose and α-(1–4)-D-glucopyranose units in a 1:2 ratio (Figure 31.3). Depending on the growth conditions, the MW of pullulan ranges from 1000 to 2000 kDa (Rekha and Chandra 2007).

Pullulan is stable in aqueous solution over a vast range of pH (pH 3–8). It decomposes upon dry heating and carbonizes at 250°C–280°C. It is soluble in water but insoluble in organic solvents. Aqueous solutions of pullulan are viscous but do not form gels. It forms transparent, water-soluble, odorless, flavorless, fat-resistant, antistatic films.

This polysaccharide has many uses: as an edible, bland, and tasteless polymer, the chief commercial use of pullulan is in the manufacture of edible films that are used in various breath freshener or oral hygiene products such as Listerine Cool Mint PocketPaks. In pharmaceuticals, it is used as a coating agent; in manufacturing and electronics, it is used because of its film- and fiber-forming properties. It is worth noting that pullulan films, formed by drying pullulan solutions, are clear and highly oxygen impermeable and have excellent mechanical properties. It is used as a thickener or as a carrier in the production of capsules for dietary supplements as an alternative to gelatine. It is also used in the production of jams and jellies, confectionery, and some fruit and meat products.

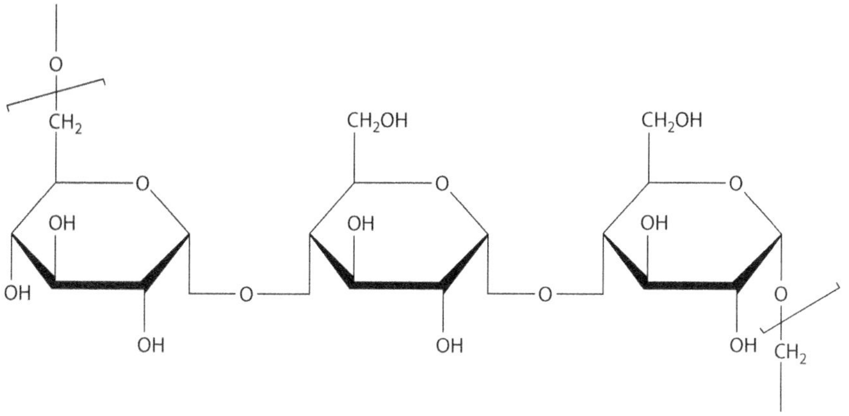

FIGURE 31.3
The chemical structure of pullulan.

It is commonly used as a foaming agent in milk-based desserts and as a texturing agent in chewing gums (Sugimoto 1978; Wiley et al. 1993; Gibbs and Seviour 1996; Madi et al. 1997; Lazarridou et al. 2002).

Because of its noncarcinogenic, nontoxic, nonimmunogenic, and hemocompatible properties (Coviello et al. 2007), the FDA has approved it for various biomedical applications such as drug and gene delivery (Rekha and Chandra 2007), wound healing (Bae et al. 2011), and tissue engineering (Thebaud et al. 2007). Even though pullulan is not a natural gelling polysaccharide, an appropriate chemical extraction of its backbone leads to a polymeric system that is able to form hydrogels. Various papers deal with pullulan hydrogels as DDSs, particularly in the form of micro- and nanogels. Interest in using pullulan nanogels has increased over the last decade due to its related potential applications in the development and implementation of new environmentally responsive or smart materials, artificial muscles, biomimetics, DDSs, biosensors, and chemical separations (Coviello et al. 2007).

To prepare pullulan nanostructures that act as carriers of different drugs, its backbone structure is modified with hydrophobic molecules, results in a molecule of hydrophobized pullulan that self-assembles in water solutions. Vitamin H or hexadecanol cholesterol is some molecules that are attached to the pullulan structure in order to obtain micelles in water solution (Liu et al. 2008).

31.4.2 Heteropolysaccharides

31.4.2.1 Hyaluronic Acid

HA is a carbohydrate, more specifically a mucopolysaccharide, occurring naturally in all living organisms. It is composed of repeating disaccharide units of D-glucuronic acid and N-acetyl-D-glucosamine linked by β-(1–3) and β-(1–4) glycosidic bonds (Figure 31.4) (Cafaggi et al. 2011). Both sugars are spatially related to glucose, which in the beta configuration allows all of its bulky groups (the hydroxyls, the carboxylate moiety, and the anomeric carbon on the adjacent sugar) to be in sterically favorable equatorial positions, while all of the small hydrogen atoms occupy the less sterically favorable axial positions. Thus, the structure of the disaccharide is energetically very stable.

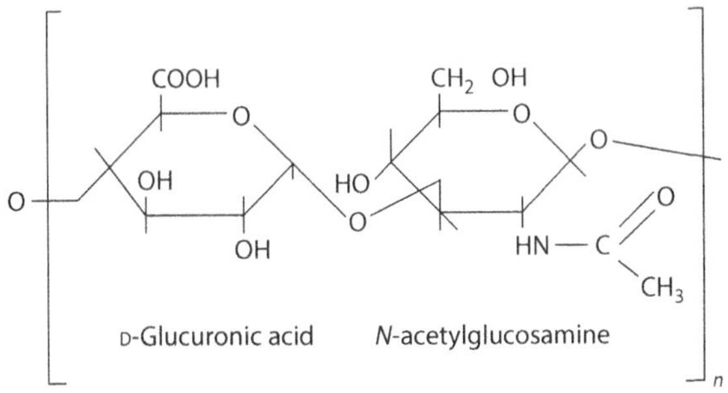

FIGURE 31.4
The chemical structure of hyaluronic acid.

HA can be modified in many ways to alter the properties of the resulting materials, including modifications leading to hydrophobicity and biological activity. The three functional groups that can be modified are the following: the glucuronic acid carboxylic acid, the primary and secondary hydroxyl groups, and the N-acetyl group (Burdick and Prestwich 2011). HA has a MW that can reach as high as 10^7 Da.

HA is the most simple glycosaminoglycan, which is not covalently associated with the core protein and the one that is nonsulfated (Kogan et al. 2007). As hyaluronan is a physiological substance, it is found primarily in the extracellular matrix and pericellular matrix but has also been shown to occur intracellularly. The biological functions of HA include maintenance of the elastoviscosity of liquid connective tissues such as joint synovial and eye vitreous fluid, control of tissue hydration and water transport, supramolecular assembly of proteoglycans in the extracellular matrix, and numerous receptor-mediated roles in cell detachment, mitosis, migration, tumor development and metastasis, and inflammation (Balazs et al. 1986; Toole et al. 2002; Turley et al. 2002; Hascall et al. 2004). Thus, it is a major and important component of cartilage, skin, and synovial fluid.

In several organisms, generally HA is linked to various biopolymers, and this requires several separation procedures such as protease digestion, HA ion-pair precipitation, membrane ultrafiltration, and HA nonsolvent precipitation and/or lyophilization (Mendichi and Soltes 2002) to obtain the pure compound. These methods generate HA with a MW of several thousands to about 2.5 MDa. Nonetheless, some microorganisms such as *Streptococcus zooepidemicus* and *Streptococcus equi* can produce HA with a molar mass in the range of several MDa.

The unique viscoelastic nature of HA along with its biocompatibility and nonimmunogenicity has led to its use in a number of clinical applications, including the supplementation of joint fluid in arthritis (Neo et al. 1997; Barbucci et al. 2002; Uthman et al. 2003; Medina et al. 2006), as a surgical aid in eye surgery, and to facilitate the healing and regeneration of surgical wounds. It has also been associated with several cellular processes, including angiogenesis and the regulation of inflammation (Leach and Schmidt 2004).

Among its various applications, it is widely used as a coating material of various biomaterials for prosthetic cartilage, vascular graft, guided nerve regeneration, and drug delivery (Li et al. 2006). Its function in the body is to bind water and to lubricate movable parts of the body, such as joints and muscles. Its consistency and tissue friendliness allows it to be used in skin-care products as an excellent moisturizer. HA is one of the most hydrophilic molecules in nature and can be described as nature's moisturizer.

Hyaluronan, as other glycosaminoglycans, serves as a targeting vehicle for the delivery of chemotherapeutic agents to cancerous tissues, as many tumors overexpress the hyaluronan CD44 and RHAMM receptors (Yip et al. 2006). Recently, HA has been studied as a drug delivery agent for various administration routes, including ophthalmic, nasal, pulmonary, parenteral, and topical (Brown and Jones 2005). As a drug delivery carrier, HA has several advantages including the negligible nonspecific interaction with serum components due to its polyanionic characteristics (Ito et al. 2006) and the highly efficient targeted specific delivery to the liver tissues with HA receptors (Zhou et al. 2003).

In tissue engineering and regenerative medicine, HA has been identified as an important building block for the creation of new biomaterials (Allison and Grande-Allen 2006; Prestwich 2008). Furthermore, it has been shown that HA binds to cells and effectively promotes new bone formation. As there are wide ranges of applications of HA, it has been used as a successful biomaterial in different field of biomedicine.

31.4.2.2 Alginate

Alginate is a naturally occurring anionic and hydrophilic polysaccharide. It is one of the most abundant biosynthesized materials (Skjak-Braerk et al. 1989; Narayanan et al. 2012) and is derived primarily from brown seaweed and bacteria. Alginate contains blocks of (1–4)-linked β-D-mannuronic acid (M) and α-L-guluronic acid (G) monomers (Figure 31.5). Typically, the blocks are composed of three different forms of polymer segments: consecutive G residues, consecutive M residues, and alternating MG residues. Depending on the source of alginate, it has varied composition and sequence, resulting in an irregular blockwise pattern of GG, MG, and MM blocks. It has a variable MW, depending on the degree of depolymerization caused by its extraction and the enzymatic control during its production. Industrially produced alginates have a molecular weight ranging from 400 to 500 kDa, with an average MW of 200 kDa (Rehm 2009).

The physicochemical properties of alginate have been found to be highly affected by the M/G ratio as well as by the structure of the alternating zones, which can be controlled by enzymatic pathways (Coviello et al. 2007). Smidsrød et al. (1973) reported that the extension of alginate chain was dependent on its composition, with the intrinsic flexibility of the blocks decreasing in the order MG > MM > GG. M blocks show linear and flexible conformation because of the β-(1–4) linkages, while glucuronic acid that produce α-(1–4) linkages serves to produce a steric hindrance around the carboxylic groups and provide

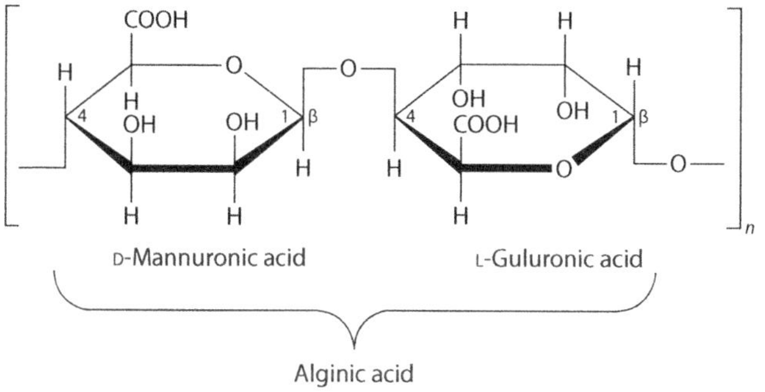

D-Mannuronic acid   L-Guluronic acid

Alginic acid

FIGURE 31.5
The chemical structure of alginate.

folded and rigid structural conformations that are responsible for a pronounced stiffness of the molecular chains (Yang et al. 2011).

Alginate is of specific interest for a wide range of applications as a biomaterial and especially as a supporting matrix or delivery system for tissue regeneration and repair. As it is a biopolymer and a polyelectrolyte considered to be biocompatible, nonimmunogenic, nontoxic, and biodegradable with chelating ability, it has been used in a variety of biomedical applications. A high content of glucuronic acid blocks in alginate is produced in the form of calcium salts, whose structure of the polymer is stabilized by cross-links, thereby producing a rigid gel form. Hence, this property enables alginate solutions to be prepared in the form of beads, films, and sponges (Sujata 2002). However, for cell transplantation and for biohybrid organs, high mannuronic acid alginate capsules are of specific interest because of their less viscosity. Some research groups have identified that alginates with high mannuronic acid cause immunostimulatory activity while immunosuppressive activity caused by alginates with high guluronic acid content. It was also shown that the mannuronic acid oligomers would provoke cytokine release by macrophages by a receptor-mediated mechanism, whereas guluronic oligomers should inhibit this reaction (Orive et al. 2002).

Due to its abundance, low price, and nontoxicity, alginate has been extensively used in different industries. For instance, it has been used as food additive and thickener in salad dressings and ice creams in the alimentary industry (Nair and Laurencin 2007). Moreover, the biocompatibility behavior and the high functionality make alginate a favorable biopolymer material for its use in biomedical applications, such as scaffolds in tissue engineering (Barbosa et al. 2005), immobilization of cells (Lan and Starly 2011), and controlled drug release devices (Pandey and Ahmad 2011).

Alginate exhibits a pH-dependent anionic nature and has the ability to interact with cationic polyelectrolytes and proteoglycans. Therefore, delivery systems for cationic drugs and molecules can be obtained through simple electrostatic interactions. Scaffolds are often used for the delivery of drugs, growth factors, and therapeutically useful cells. As such, scaffolding materials allow protection of biologically active substances or cells from the biological environment. Depending on the site of implantation, the biomaterials are subjected to different pH environments, which affect the degradation properties, mechanical properties, and swelling behavior of the biomaterials. As such, alginate plays an important role in the long-term stability and performance of alginate-based biomaterials *in vitro*. The MW of alginate influences the degradation rate and mechanical properties of alginate-based biomaterials. Basically, higher MW decreases the number of reactive positions available for hydrolysis degradation, which further facilitates a slower degradation rate. In addition, degradation also inherently influences the mechanical properties owing to structural changes both at molecular or macroscopic levels. It is a natural polymer compatible with a wide variety of substances, which does not need multiple and complex drug-encapsulation process. Moreover, it is mucoadhesive and biodegradable, and consequently, it can be used in the preparation of controlled DDSs achieving an enhanced drug bioavailability (Pandey and Ahmad 2011). Therefore, the biocompatibility, availability, and versatility of this polysaccharide make it an important and hopeful tool in the field of nanomedicine, especially in the preparation of nanoparticulate DDSs.

31.4.2.3 Pectin

Pectins are a complex family of heteropolysaccharides that constitute a large proportion of the primary cell walls of dicotyledons and play important roles in growth, development, and senescence (van Buren 1991; Tombs and Harding 1998; Ridley et al. 2001;

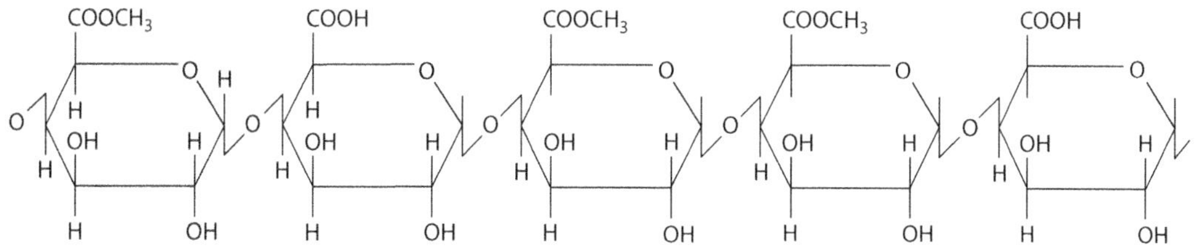

FIGURE 31.6
The chemical structure of pectin.

Willats et al. 2001). The chemical structure of this natural polymer has large amounts of poly(D-galacturonic acid) bonded via α-(1–4)-glycosidic linkage (Figure 31.6). Pectin has a few hundred to about 1000 building blocks per molecule, which corresponds to an average MW of about 50–180 kDa (Sinha and Kumria 2001).

The carboxyl groups are partially in the methyl ester form with different degree of esterification (DE) and amidation, which determine the content of carboxylic acid in pectin chains. The DE and therefore the charge on a pectin molecule are important to the functional properties in the plant cell wall. It also significantly affects their commercial use as gelling and thickening agents (Lapasin and Pricl 1995; Tombs and Harding 1998). High methoxyl (HM) pectins (low charge) form gels at low pH (<4.0) and in the presence of a high amount (>55%) of soluble solids, usually sucrose (Oakenfull 1991). HM pectin gels are stabilized by hydrogen bonding and hydrophobic interactions of individually weak but cumulatively strong junction zones (Figure 31.6) (Morris 1980; Pilnik 1990; Oakenfull 1991; Lopes da Silva and Gonçalves 1994). In contrast, LM pectins (high charge) form electrostatically stabilized gel networks with or without sugar and with divalent metal cations (Morris 1980; Morris et al. 1982; Pilnik 1990; Axelos and Thibault 1991; Oakenfull and Scott 1998), which also depends on the distribution of negative carboxylate groups and structure-breaking rhamnose side chains (Powell et al. 1982; Axelos and Thibault 1991).

Pectins have been used earlier as gelling and thickening agents for a large number of years in food industry, but recently there has been an interest in using pectin gels for pharmaceutical applications (Liu et al. 2003). Interesting uses of pectin in biomedical applications include the specific delivery of unique amino acid sequences, wound healing substances, anti-inflammatory agents, and anticoagulants to specific tissue sites. Moreover, in the physiological environment of stomach and small intestine, pectin remains intact, but the bacterial inhabitants of human colon degrades pectin by secreting pectinases. Because of its controlled drug delivery property (Lui et al. 2003, 2007; Sungthongjeen et al. 2004), long-standing reputation of being nontoxic (generally regarded as safe) (Lui et al. 2003, 2007; Watts and Smith 2009), relatively low production costs (Sungthongjeen et al. 2004), and high availability (Beneke et al. 2009) pectin could be used as a delivery vehicle to assist protein and polypeptide drugs from the mouth to the colon (Sinha and Kumria 2001) orally, nasally, and vaginally (Peppa et al. 2000; Sinha and Kumria 2001; Lui et al. 2003, 2007; Nafee et al. 2004; Valenta 2005; Chelladurai et al. 2008; Thirawong et al. 2008), which are generally well accepted by patients (Lui et al. 2003, 2007; Yadav et al. 2009).

As pectin is not able to safeguard its drug delivery while passing through stomach and small intestine due to its high water solubility (Sinha and Kumria 2001), research is focused on developing water-resistant pectin derivatives. For this purpose, calcium salts that bind noncovalently with the carbohydrate chains of pectin were found out, which can reduce the solubility and are stable in low pH solution while resisting extensive hydration *in vivo* in the gastrointestinal tract. Hence, calcium pectinate is a prospective candidate as a drug carrier for colon-specific delivery in different formulations such as gels or droplets, films, and microspheres (Liu et al. 2003). Another pectin derivative, amidated pectin cross-linked with calcium, can be considered for colonic delivery because of its biodegradability, high tolerance to fluctuations in calcium levels, and different pH (Sinha and Kumria 2001).

The pectin-derived films have shown potential applications in coating, encapsulating, and thickening for food and pharmaceutical uses (Liu et al. 2007). Hoagland and Parris (1996) fabricated chitosan/pectin films with either glycerol or lactic acid as plasticizers giving rise to clear laminated films possessing dynamic mechanical properties similar to those for pectin films alone. To prevent fungal growth on the laminated films, they replaced glycerol with lactic acid without significant change in its dynamic mechanical properties. Recently, pectin and polylactic acid (PLA) film were fabricated for their intended applications in antimicrobial packaging (Liu et al. 2007). They loaded model antimicrobial polypeptide, nisin, into the composite by diffusion method. The nisin-loaded composite suppressed *Lactobacillus plantarum* growth, which was indicated by agar diffusion and liquid phase culture tests.

Pectin in combining with natural polymers or synthetic polymers, various useful novel formulations have been obtained. The combination of pectin and a second polymer into a composite may alter the degree of swelling and change mechanical properties (Liu et al. 2003), improving in the most cases the stability of the drug and controlling the drug release. It has been combined with 4-aminothiophenol (Perera et al. 2010), HA (Pliszczak et al. 2011) chitosan (Fernandez-Hervas and Fell 1998), or poly(lactide-co-glycolide) (Liu et al. 2004), showing good results as controlled drug release devices.

31.4.2.4 Gum Arabic

It is a complex heteropolysaccharide derived from exudates of *Acacia senegal* and *Acacia seyal* trees. The carbohydrate moiety is made up of D-galactose (~40% of residues), L-arabinose (~24%), L-rhamnose (~13%), and two uronic acids, responsible for the polyanionic character of the gum, the D-glucuronic acid (~21%), and the 4-O-methyl-D-glucuronic acid (2%). The main structural feature is a backbone of β-galactopyranose units with 1,3-bonds and side chains of 1,6-linked galactopyranose units terminating in β-D-glucopyranose and 4-O-methyl-β-D-glucopyranose (Street and Andersom 1983). Also, α-L-arabinofuranose and α-L-rhamnopyranose are bound through 1,3- and 1,4-linkages, respectively (Figure 31.7).

Its physicochemical properties display remarkable surface properties. The concomitant presence of hydrophilic sugar residues and hydrophobic amino acids contributes to the gum amphiphilicity that favor its adsorption to air/water (O/W) or O/W interface (Randall et al. 1988; Ray et al. 1995; Dickinson and McClements 1996). Due to the presence of flexible structure that allows molecules to be easily deformed at interfaces (Jayme et al. 1999; Fauconnier et al. 2000), acacia gum is mainly used as an emulsifier/stabilizer. Gum arabic presents high water solubility, low viscosity in aqueous solutions, and good emulsifying abilities, due to the existent protein fraction (Gabas et al. 2007; Kurozawa et al. 2009).

Some reports have shown that at pH above 2.2, GA is negatively charged because the dissociation of its carboxylic groups is suppressed at lower than 2.2 pH values (Ye et al. 2006). However, it provides retention of volatile substances and ensures effective protection of the encapsulated compound or drug against oxidation (Gabas et al. 2007). The cost and limited supply of gum arabic limit its usage in encapsulation.

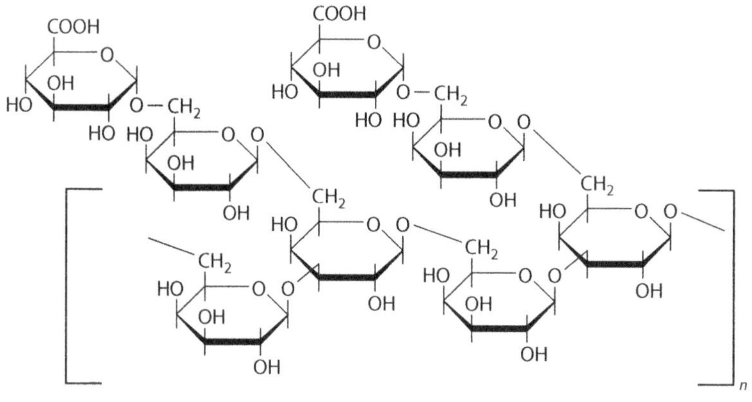

FIGURE 31.7
The chemical structure of gum arabic.

Acacia gum reduces the antibacterial effectiveness of the preservative methyl-*p*-hydroxybenzoate against *Pseudomonas aeruginosa*, presumably by offering physical barrier protection to the microbial cells from the action of the preservative (Kurup et al. 1992). A trypsin inhibitor also has been identified, but the clinical significance of the presence of this enzyme is not known (Duke 1985).

Together maltodextrin and gum arabic–based nanoparticles are being used as catechin delivery systems (Gomes et al. 2010; Peres et al. 2010). To enhance the protein delivery by gum arabic–based nanoparticles, interaction between gum arabic and chitosan is starting to be exploited (Avadi et al. 2010, 2011; Coelho et al. 2011). Very little has been known about the use of gum arabic–based nanoparticles and DDS; it has to be exploited further.

31.5 Preparation of Polysaccharide-Based Nanoparticles

Many studies have demonstrated that nanoparticles have a number of advantages over microparticles (Panyam and Labhasetwar 2003). Alonso et al. (2001) and Prabaharan and Mano (2005) have written excellent reviews that focus on the preparation and application of chitosan nanoparticle carriers. Polysaccharide-based nanoparticles greatly enrich the versatility of nanoparticle carriers in terms of category and function. Polysaccharide-based nanoparticles are basically prepared by four different mechanisms depending on their structural characteristics, namely, covalent cross-linking, ionic cross-linking, polyelectrolyte complexation (PEC), and self-assembly of hydrophobically modified polysaccharides. The select of method depends on a number of factors, such as particle size, particle size distribution, and area of application. Particle size is the most important characteristics of nanoparticles.

31.5.1 Covalently Cross-Linked Polysaccharide Nanoparticles

Preparation of polysaccharide nanoparticles by covalent cross-linking is the earliest method that was adopted. Among various polysaccharides, chitosan is the initial one to be used for nanoparticle preparation. Initially, chitosan-based polysaccharides were cross-linked using glutaraldehyde, a common cross-linker (Zhi et al. 2005; Liu et al. 2007). However, because of the cellular toxicity of glutaraldehyde, its use in drug delivery was limited. Therefore, the use of biocompatible covalent cross-linking agents is quite promising. Presently, various water-soluble condensation agents such as carondiamide and natural di- and tricarboxylic acids, including succinic acid, tartaric acid, malic acid, and citric acid, are being used as intermolecular cross-linkers for chitosan nanoparticles (Bodnar et al. 2005). Hence, biodegradable chitosan nanoparticles were obtained by performing the condensation reaction between carboxylic groups of natural acids and the pendent amino groups of chitosan. This method allows the formation of polycations, polyanions, and polyampholyte nanoparticles. The prepared nanoparticles were stable in aqueous media at low pH, neutral, and mild alkaline conditions. Depending on the pH in the swollen state, the average size of the particles ranged between 270 and 370 nm.

31.5.2 Ionically Cross-Linked Polysaccharide Nanoparticles

Ionic cross-linking has more advantages than covalent cross-linking as the procedure is simple and the preparation conditions are mild. For charged polysaccharides, low MW polyanions and polycations could act as ionic cross-linkers for polycationic and polyanionic polysaccharides, respectively. To date, the most widely used polyanion cross-linker is tripolyphosphate (TPP) as it is nontoxic and has multivalent cations. TPP cross-linked chitosan nanoparticles were first reported in 1997 by Alonso et al. (Calvo et al. 1997). TPP can form a gel by ionic interaction between positively charged amino groups of chitosan and negatively charged counter ions of TPP (Jain et al. 2008). From then on, TPP–chitosan nanoparticles have been widely used to deliver various drugs and macromolecules (Vila et al. 2004; Zhang et al. 2004; Aktas et al. 2005; Luangtana-anan et al. 2005; Qi et al. 2005; Lu et al. 2006; Maestrelli et al. 2006; Gan and Wang 2007; Sun and Wan 2007; de la Fuente et al. 2008; Tsai et al. 2008; Zhang et al. 2008).

Recently, apart from TPP, water-soluble chitosan derivatives were also ionically cross-linked to prepare nanoparticles. Chitosan derivatives can easily dissolve in aqueous media, avoiding the probable toxicity of acids and thereby protecting the bioactivity of loaded biomacromolecules. A water-soluble chitosan derivative, N-(2-hydroxyl)propyl-3-trimethyl ammonium chitosan chloride, was synthesized by Xu et al. (2003) by the reaction between glycidyltrimethylammonium chloride and chitosan. Based on ionic gelation process of the derivative and TPP, nanoparticles of 110–180 nm in size were obtained. In addition, Amidi et al. (2006) prepared N-trimethyl chitosan nanoparticles by ionic cross-linking of N-trimethyl chitosan with TPP and their potential as a carrier system for the nasal delivery of proteins; ovalbumin were evaluated. It is found that the nanoparticles had an average size of about 350 nm and a positive zeta potential. They showed a loading efficiency up to 95% and a loading capacity up to 50% (w/w). The integrity of the entrapped ovalbumin was preserved. Cytotoxicity tests with Calu-3 cells showed no toxic effects of the nanoparticles. The absorption properties of N-trimethyl chitosan/TPP nanoparticles were evaluated by Sandri et al. (2007) using *in vitro* (Caco-2 cells) and *ex vivo* (excised rat jejunum) models. Additionally, carboxymethyl chitosan (CM chitosan) nanoparticles (200–300 nm and in a narrow distribution) were prepared through ionic gelification with calcium ion and evaluated the potential of the nanoparticles as carriers for anticancer drug, doxorubicin (DOX). Effects of DS and MW of CM chitosan on DOX delivery were discussed.

More recently, calcium-cross-linked negatively charged polysaccharide nanoparticles have found efficacy as drug carrier. As some polysaccharides bear carboxylic groups on molecular chains, they can be cross-linked by bivalent calcium ion to form nanoparticles.

By using water-in-oil reverse microemulsion method, You and Peng (2004) prepared Ca-cross-linked alginate nanoparticles. To examine the potency of the nanoparticles for gene delivery, green fluorescent protein-encoding plasmids were encapsulated in the nanoparticles to investigate the degree of endocytosis by NIH 3T3 cells and ensuing transfection rate. Results showed that Ca-alginate nanoparticles with an average size around 80 nm in diameter were very efficient gene carriers. Zahoor et al. (2005) also prepared Ca-alginate nanoparticles (235.50 nm in size) by ion-induced gelification. It is found that the drug encapsulation efficiency in theses nanoparticles were 70%–90% for isoniazid and pyrazinamide and 80%–90% for rifampicin. The bioavailabilies of encapsulated drugs were found to be relatively higher when compared with oral-free drugs. It was observed that all the drugs that are encapsulated were found to be present in the organs (liver, lungs, and spleen) for about 15 days postnebulization, while free drugs stayed up to 1 day. Therefore, these inhalable nanoparticles could serve as carriers for controlled release of drugs. Kim et al. (2006) encapsulated retinol into chitosan nanoparticles by ion complex due to the electrostatic interaction between amine group of chitosan and hydroxyl group of retinol and reconstituted it into aqueous solution for pharmaceutical and cosmetic applications. By encapsulation, it is found that the solubility of retinol is able to increase by more than 1600-fold.

31.5.3 Polysaccharide-Based Nanoparticles by Polyelectrolyte Complexation

Polyelectrolyte polysaccharides can form PEC with oppositely charged polymers by intermolecular electrostatic interaction. PEC nanoparticles of polysaccharides can be obtained by adjusting the MW of polymer components in a specific range. Although theoretically any polyelectrolyte could interact with polysaccharide to form PEC nanoparticles, in practice these polyelectrolytes are restricted to biocompatible and water-soluble polymers in view of safety purpose. Chitosan is the only natural polycationic polysaccharide that fulfils all the needs; however, there are many negative polymers (Figure 31.8) that complex with chitosan to form PEC nanoparticles, peptides, polyacrylic acid family, etc.

31.5.3.1 Negative Polysaccharides Complexed with Chitosan-Based Nanoparticles

Carboxymethyl cellulose (CMC) can complex with chitosan and form stable cationic nanoparticles. These nanoparticles containing plasmid DNA were used as a potential approach to genetic immunization. Cui and Mumper (2001) coated the plasmid DNA on preformed cationic chitosan/CMC nanoparticles. These chitosan-based nanoparticles containing plasmid DNA resulted in both detectable and quantifiable levels of luciferase expression in mouse skin after 24 h topical application, and significant antigen-specific IgG titer expressed β-galactosidase at 28 days. Chen et al. (2003), by employing a simple coacervation process, developed chitosan/dextran sulfate nanoparticles delivery system. The study observed the effect of the weight ratio of the two polymers on particle size, surface charge, entrapment efficiency, and release characteristics of antiangiogenesis peptide. Tiyaboonchai and Limpeanchob (2007) developed a nanoparticulate delivery system for amphotericin B with chitosan and dextran sulfate together with zinc sulfate as a cross-linking and hardening agent. The nanoparticles possessed a mean particle size of 600–800 nm with a polydispersity index of 0.2, indicating a narrow size distribution. Insulin-loaded nanoparticles were prepared by Sarmento et al. (2006) using ionotropic pregelation of alginate with calcium chloride followed by complexation between alginate

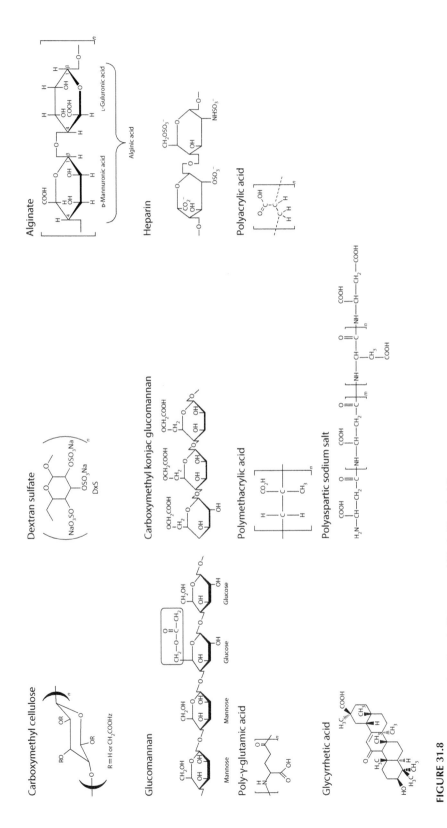

FIGURE 31.8
Negative Polymers Complexed with Chitosan and Their Chemical Structures.

and chitosan. They also studied the structural integrity of insulin after being entrapped into chitosan/alginate nanoparticles (Sarmento et al. 2007). Their results clearly showed that no significant conformational changes of insulin occurred in terms of α-helix and β-sheet content. Alonso-Sande et al. (2006) prepared nanoparticles using two different types of glucomannan (nonphosphorylated and phosphorylated) and two different approaches. The interaction of chitosan and glucomannan in these procedures involved the presence or absence of sodium TPP, which acted as an ionic cross-linking agent for chitosan. The nanoparticles showed a great capacity for the association of insulin and the immunomodulatory protein P1, reaching association efficiency values as high as 89%. Liu et al. (2007), by using PEC, prepared heparin/chitosan nanoparticles. Entrapment studies of the nanoparticles were conducted using bovine serum albumin as a model protein. Their study was focused in detail about the effects of the pH value of the chitosan solution, chitosan concentration, chitosan MW, heparin concentration, and protein concentration on the nanoparticle size, the nanoparticles yield, and the protein entrapment.

31.5.3.2 Negative Peptides Complexed with Chitosan-Based Nanoparticles

By using ionic-gelation method, Lin et al. (2005) prepared poly-γ-glutamic acid/chitosan nanoparticles system. These nanoparticles showed enhanced intestinal paracellular transport in Caco-2 cell monolayers *in vitro*. These nanoparticles, which have chitosan dominated on the surface, could effectively reduce the transepithelial electrical resistance of Caco-2 cell monolayers and opened the tight junctions between Caco-2 cells, thus allowing transport of the nanoparticles *via* the paracellular pathways. Furthermore, these nanoparticles were used for transdermal gene delivery. When compared to chitosan/DNA, chitosan/poly-γ-glutamic acid/DNA improved their penetration depth into the mouse skin and enhanced gene expression. These studies clearly showed that chitosan/poly-γ-glutamic acid/DNA were more compact in their internal structures and had a greater density than their chitosan/DNA counterparts, thus they can penetrate into the skin barrier (Lee et al. 2008) more easily.

31.5.3.3 Polyacrylic Acid Family Complexed with Chitosan-Based Nanoparticles

pH-sensitive polymethacrylic acid/chitosan/polyethylene glycol nanoparticles were produced under mild aqueous conditions (Sajeesh and Sharma 2006). The procedure involves free-radical polymerization of methacrylic acid in the presence of chitosan and polyethylene glycol using a water-soluble initiator. Nanoparticles were obtained instantaneously without adding any organic solvents or surfactants/steric stabilizers. Model proteins like insulin and bovine serum albumin were added into the nanoparticles by diffusion-filling method and at pH 1.2 and 7.4, and their *in vitro* release characteristics were evaluated. The nanoparticles exhibited good protein encapsulation efficiency and pH-responsive release profile under *in vitro* conditions. When polyanion poly(acrylic acid) was dropped into polycation chitosan solution, chitosan/poly(acrylic acid) nanoparticles were formed (Chen et al. 2005). It was reported that plasmid DNA was encapsulated very well in these nanoparticles, giving them great potential in gene delivery. The same group also synthesized hollow polymeric nanospheres by polymerization of acrylic acid monomers inside the chitosan–acrylic acid assemblies. The effects of pH, polymerization concentration, temperature, shell cross-linking, and salt concentration on the size and stability of hollow polymeric nanospheres were studied (Hu et al. 2005).

31.5.3.4 Others

Chitosan/glycyrrhetic acid nanoparticles were prepared by PEC method. The study focused on encapsulation efficiency and *in vitro* release of glycyrrhetic acid (Zheng et al. 2006). Stoilova et al. (1999) prepared PEC nanoparticles between chitosan and poly(2-acryloyl-amido-2-methylpropanesulfonic acid) by mixing aqueous solutions of its components or by free-radical polymerization on chitosan template. The nanoparticles (mean diameter 250 nm and monomodal distribution) were found to be quite stable in acidic and neutral medium, and at pH > 8, they are getting disassociated.

In addition, the nanoparticle formed by complexation between chitosan or its derivative and DNA or RNA is a special type of delivery system, which has been well investigated (Thanou et al. 2002; Kean et al. 2005; Jiang et al. 2007a,b; Kim et al. 2007; Liu et al. 2007; Tan et al. 2007; Yang et al. 2007; Zheng et al. 2007; Xu et al. 2008).

It is noteworthy that the category of this kind of nanoparticles will increase continuously because of the following: (1) chitosan has a lot of water-soluble positively charged derivatives, such as *N*-trimethyl chitosan, glycol chitosan, and *N*-trimethyl chitosan, which can be used as polycation instead of chitosan, and (2) more biocompatible negative polymers will be exploited, in particular, various polyanionic polysaccharides.

31.5.4 Polysaccharide-Based Nanoparticles through Self-Assembly Method

There are several reports that have been carried out to understand the synthesis and potential applications of polysaccharide-based self-assembled nanoparticles as DDSs. When hydrophilic polymeric chains are grafted with hydrophobic segments, amphiphilic copolymers are formed. When these copolymers exposed to aqueous environment, they spontaneously form micelles or micelle-like aggregates by intra- or intermolecular associations between hydrophobic moieties, particularly to minimize the interfacial free energy. Depending on the hydrophilic/hydrophobic constituents, these polymeric micelles display unique characteristics, like thermodynamic stability, unusual rheology feature, and small hydrodynamic radius (less than microsize) with core–shell structure. Polymeric micelles have been recognized as a promising and potential drug carrier as they have a hydrophobic domain, surrounded by a hydrophilic outer shell, which serves as a preservative for various hydrophobic drugs (Letchford and Burt 2007). The hydrophobic molecules used to modify polysaccharides are listed in Table 31.3. In specific, these hydrophobic molecules are divided into linear, cyclic hydrophobic molecules, hydrophobic drug, polyacrylate family, etc.

31.5.4.1 Linear Hydrophobic Molecules

Poly(ε-caprolactone) (PCL) is biodegradable, nontoxic, and biocompatible polyester with excellent mechanical strength. Gref et al. (2002) and Lemerchand et al. (2003) synthesized amphiphilic dextran by coupling between carboxylic function present on preformed PCL monocarboxylic acid and the hydroxyl groups on dextran. Nanoparticles of less than 200 nm were successfully prepared by using the new materials (Rodrigues et al. 2003). Further, in the nanoparticles that were obtained, bovine serum albumin and lectin were incorporated. Lectins found to be adsorbed onto the surface of the nanoparticles. These surface-bound lectins sustained its hemagglutinating activity, suggesting the possible targeted oral administration of such surface-modified nanoparticles. Additionally, in J774 macrophages, it was found that the modification of the surface with dextran significantly

TABLE 31.3

Hydrophobic Molecules and Their Chemical Structures That Are Used to Modify Polysaccharides

Polysaccharides	Hydrophobic Molecules	Chemical Structure of Hydrophobic Molecules
Chitosan	Poly(ethylene) glycol	$HO-CH_2CH_2-O-(CH_2CH_2O)_n-CH_3$
β-Cyclodextrin	Hexanoic acid	$CH_3(CH_2)_4COOH$
	Decanoic acid	$CH_3(CH_2)_8COOH$
Chitosan, amylose	Linoleic acid	$CH_3(CH_2)_4CH=CHCH_2CH=CH(CH_2)_7COOH$
Chitosan	Linolenic acid	$CH_3(CH_2CH=CH)_3(CH_2)_7COOH$
Chitosan	Palmitic acid	$CH_3(CH_2)_{14}COOH$
Chitosan	Stearic acid	$CH_3(CH_2)_{16}COOH$
Chitosan	Oleic acid	$CH_3(CH_2)_7CH=CH(CH_2)_7COOH$
Dextran, chitosan	Poly(ε-caprolactone)	$-[O-(CH_2)_5-CO]_n-$
Heparin, hyaluronic acid	Pluronic	$-(CH_2CH_2O)_n-(CH_2CH(CH_3)O)_m-$
Pullulan	Hexadecanol	$CH_3(CH_2)_{15}OH$
Chitosan, carboxymethyl chitosan, pullulan	Cholesterol	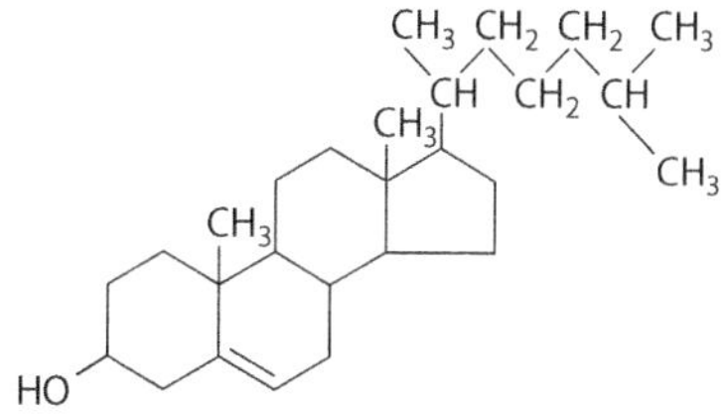
Chitosan, heparin, glycol chitosan	Deoxycholic acid	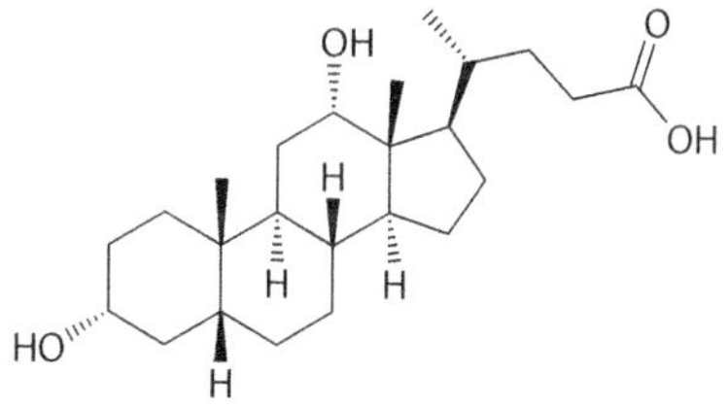
Glycol chitosan	5β-Cholanic acid	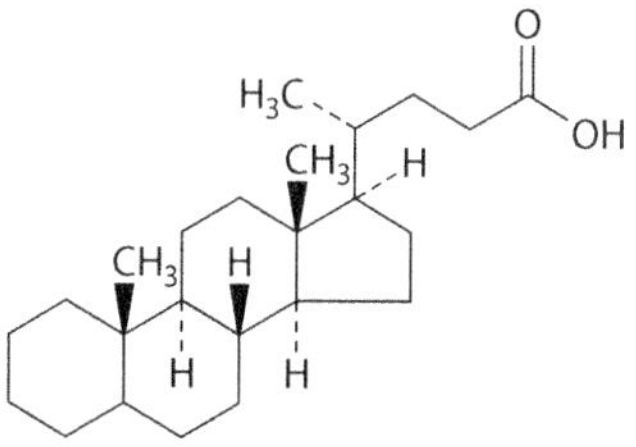
Glycol chitosan	Fluorescein isothiocyanate	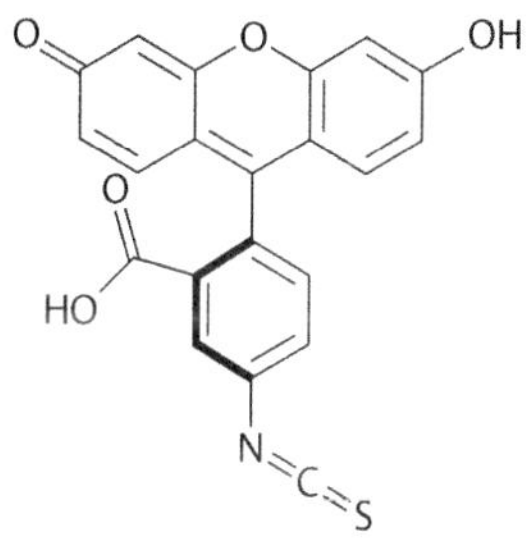

(Continued)

TABLE 31.3 (*Continued*)

Hydrophobic Molecules and Their Chemical Structures That Are Used to Modify Polysaccharides

Polysaccharides	Hydrophobic Molecules	Chemical Structure of Hydrophobic Molecules
Glycol chitosan	Doxorubicin	
Pullulan	Vitamin H	
Glycol chitosan	*N*-acetyl histidine	
Heparin, Dextran	Poly(methyl methacrylate)	
Chitosan, dextran, dextran sulfate, thiolated chitosan, heparin, hyaluronic acid, pectin	Poly(isobutyl cyanoacrylate)	

reduced their cytotoxicity. Biodegradable amphiphilic PCL–graft–chitosan copolymer was synthesized (Jing et al. 2006), and they form spherical or elliptic nanoparticles in water.

Poly(ethylene glycol), another linear hydrophobic molecule, because of its exceptional physicochemical and biological properties including hydrophilic property, nontoxicity, solubility, absence of antigenicity and immunogenicity, and ease of chemical modification, has been used extensively in pharmaceutical and biomedical fields. Therefore, poly(ethylene glycol) has been often used as implantable carriers for DDSs or as surgical repair materials. Many researchers have tried to produce poly(ethylene glycol)-g-derivative chitosan-based nanoparticles (Ouchi et al. 1998; Jung et al. 2006; Opanasopit et al. 2007; Park et al. 2008; Yang et al. 2008b). Yoksan et al. (2004) grafted poly(ethylene glycol) methyl ether onto *N*-phthaloyl chitosan chains. The graft copolymer aggregated to obtain sphere-like nanoparticles. Stable nanospheres could be obtained directly by merely adjusting the hydrophobicity/hydrophilicity of the chitosan chain. Methoxy poly(ethylene glycol) grafted on chitosan was synthesized by Jeong et al. (2006b) to develop polymeric micelles for the drug delivery to brain tumor. The loading efficiency of the micelles was higher than 80% (w/w) for all formulations. It was found that all-trans retinoic acid-incorporated nanoparticles were more effective to inhibit invasion of tumor cells than free all-trans retinoic acid at invasion test using matrigel. Moreover, the CMC of these nanoparticles in water was similar (28 µg/mL). The nanoparticles exhibited a regular spherical shape with core–shell structure with sizes in the range of 100–250 nm. In the inner core of the micelles, camptothecin as a model drug was loaded. Yang et al. (2008) synthesized methoxy poly(ethylene glycol)-grafted-chitosan conjugates by formaldehyde linking method. The critical aggregation concentration (CMC) of the conjugates was 0.07 mg/mL in water. The conjugates formed monodisperse self-aggregated nanoparticles with a spherical shape and a mean diameter of 261.9 nm. Inside the nanoparticles, a poorly water-soluble anticancer drug, methotrexate, was physically entrapped.

Some modifying polysaccharides and some long-chain fatty acids such as linoleic acid, hexanoic acid, decanoic acid, palmitic acid, linolenic acid, oleic acid, and stearic acid have been used in preparing nanoparticles. Choisnard et al. (2006) synthesized decanoate β-cyclodextrin esters (DS, 2–7) and hexanoate β-cyclodextrin esters (DS, 4–8) biocatalyzed by thermolysin from native β-cyclodextrin and vinyl hexanoate or vinyl decanoate used as acyl donors. By a nanoprecipitation technique, both esters self-organized into nanoparticles. Chen et al. (2003) modified chitosan by coupling with linoleic acid through the 1-ethyl-3-(3-dimethylaminopropyl)-carbodiimide-mediated reaction to increase its amphipathicity for improved emulsification. By O/W emulsification with methylene chloride, the micelle formation of linoleic acid–modified chitosan in the 0.1 M acetic acid solution was enhanced in an oil phase, the self-aggregation concentration from 1.0 to 2.0 g/L. Addition of 1 M sodium chloride allowed the self-aggregation of linoleic acid chitosan molecules both with and without emulsification. The micelles produced nanoparticles ranging from 200 to 600 nm. A lipid-soluble model compound, retinyl acetate, was encapsulated in the nanoparticles that showed 50% efficiency. The same team modified chitosan with linolenic acid (the DS 1.8%) using the same reaction. The self-aggregated nanoparticles of linolenic acid–chitosan were also used to immobilize trypsin using glutaraldehyde as cross-linker. The result showed that with increasing concentration of glutaraldehyde up to 0.07% (v/v), the activity of trypsin immobilized onto the nanoparticles increased and then gradually decreased with increasing amount of glutaraldehyde. Additionally, with the increasing concentration of glutaraldehyde (from 0.03% to 0.1% v/v), the particle size was increased from 523 to 1372 nm. The kinetic constant value of trypsin immobilized on nanoparticles (71.9 mg/mL) was higher than that of pure trypsin (50.2 mg/mL). Jiang et al. (2006), by coupling swollen chitosan with palmitic anhydride in dimethyl sulfoxide, prepared water-soluble *N*-palmitoyl chitosan, which could form micelles in water. *N*-palmitoyl chitosan DS was in the range of 1.2%–14.2%, and the CMC of *N*-palmitoyl chitosan micelles was in the range of 2.0×10^{-3} to 37.2×10^{-3} mg/mL. The loading capacity of hydrophobic model drug ibuprofen in the micelles was about 10%. Zhang et al. (2007) developed self-assembled nanoparticles based on oleoyl–chitosan with a mean diameter of 255.3 nm. DOX was efficiently loaded into the nanoparticles with an encapsulation efficiency of 52.6%. The drug was immediately and completely released from the nanoparticles at pH 3.8, whereas at pH 7.4 there was a sustained release after a burst release.

Pluronic triblock copolymers composed of poly(ethylene oxide)–poly(propylene oxide)–poly(ethylene oxide) show lower critical solution temperature behaviors over a broad temperature range depending on the composition and MW. By hydrophobic interaction of the poly(propylene oxide) middle block in the structure, they self-assemble to produce a spherical micellar structure above the lower critical solution temperature. At high concentration above about 25% (v/v), they exhibit sol–gel transition behavior when raising the temperature above the lower critical solution temperature. Choi et al. (2006) prepared pluronic/heparin composite nanocapsules, which when the temperature was cycled between 20°C and 37°C, respectively, exhibited a 1000-fold volume transition (ca. 336 nm at 25°C; ca. 32 nm at 37°C) and a reversible swelling and deswelling behavior. Core–shell nanoparticles with the poly(lactide-co-glycolide) core and the polymeric shell composed of pluronics and HA were prepared, and lysozyme was loaded into the polymeric shell up to 7 wt.% via ionic interaction between HA and lysozyme and the sustained release pattern was observed (Han et al. 2005).

31.5.4.2 Cyclic Hydrophobic Molecules

In animals, cholesterol is a necessary lipid that participates not only in the formation of cell membrane but also acts as a raw material for the synthesis of bile acids, steroid hormones, and vitamin D. When hydrophobic cholesterol conjugates with hydrophilic polysaccharides, amphiphilic copolymers will form, which may further form self-assembly nanoparticles in aqueous solution. Wang et al. (2007) synthesized cholesterol-modified chitosan conjugate with succinyl linkages. In aqueous media, by probe sonication, these conjugates form monodisperse self-aggregated nanoparticles with a roughly spherical shape with a mean diameter of 417.2 nm. By remote-loading technique, epirubicin, a model anticancer drug, was physically trapped inside the nanoparticles. Epirubicin-loaded nanoparticles were almost spherical in shape and their size increased from 338.2 to 472.9 nm with the epirubicin-loading content increasing from 7.97% to 14.0%. Wang et al. (2007) also prepared self-aggregated nanoparticles of cholesterol-modified O-CM chitosan and investigated the interaction between bovine serum albumin and self-aggregated nanoparticles.

It is also reported that cholesterol-bearing pullulans with different MWs of the parent pullulan and DS of the cholesteryl moiety were synthesized (Nishikawa et al. 1996; Akiyoshi et al. 1997). It is observed that all of cholesterol–pullulans provided unimodal and monodisperse self-aggregates in water irrespective of the MW of the parent pullulan and the DS. The size of the self-aggregate decreased with an increase in the DS of the cholesteryl moiety (hydrodynamic radius, 8.4–13.7 nm). But the aggregation number of cholesterol–pullulans in one nanoparticle was almost independent of the DS. However, the polysaccharide density within the self-aggregate (0.13–0.50 g/mL) was affected by both the MW and the DS of cholesterol–pullulans. The characteristic temperature to cause a structural change of the nanoparticles decreased with an increase in the DS and the ionic strength of the medium. Insulin was integrated into the cholesterol–pullulan nanoparticles. It was found that upon complexation onto the nanoparticles, the thermal denaturation and subsequent aggregation of insulin were effectively suppressed. The complexed insulin was extremely protected from enzymatic degradation. Furthermore, by self-assembly of two different hydrophobically modified polymers, thermoresponsive nanoparticles are obtained. They are cholesterol–pullulan and a copolymer of N-isopropylacrylamide and N-[4-(1-pyrenyl)butyl]-N-n-octadecylacrylamide via their hydrophobic moieties (Akiyoshi et al. 2000), as well as hexadecyl group–bearing pullulan self-assembly nanoparticles (Kuroda et al. 2002).

Bile acids, such as deoxycholic acid and 5β-cholanic acid as a result of their amphiphilicity, are known to form micelles in water, and they play an important role in the emulsification, solubilization, and absorption of cholesterol, lipophilic vitamins, and fats in human body. Hence, introduction of deoxycholic acid or 5β-cholanic acid into chitosan would induce self-association to form self-aggregates. Lee et al. (1998a,b), by using carbodiimide-mediated reaction, covalently conjugated deoxycholic acid to chitosan and generated self-aggregated nanoparticles. Inside the self-aggregated nanoparticles, adriamycin was physically entrapped. It is observed that by increasing the loading content of adriamycin, the size of adriamycin-loaded self-aggregates was increased. Adriamycin was slowly released from the self-aggregates in pH 7.2 PBS (Lee et al. 2000). By chemically modifying chitosan oligosaccharides with deoxycholic acid, the deoxycholic acid–chitosan formed self-aggregated nanoparticles in aqueous milieu because of the amphiphilic characters (Chae et al. 2005). The particle size of the nanoparticles ranged between 200 and 240 nm. As efficient gene carriers, the nanoparticles showed superior gene condensation and protection of condensed gene from endonuclease attack than unmodified chitosan. Moreover, deoxycholic acid–chitosan nanoparticles showed great potential for gene carrier with the high level of gene transfection efficiencies, even in the presence of serum. Reports also developed deoxycholic acid–heparin amphiphilic conjugates with different DS of deoxycholic acid (Park et al. 2004), which provided monodispersed self-aggregates in water, with mean diameters (120–200 nm) decreasing with increasing DS. The hydrophobicity of the self-aggregate inner core was enhanced by increasing DS. The mean aggregation number of deoxycholic acid per hydrophobic microdomain indicated that five to nine of deoxycholic acid–heparin chains comprised a hydrophobic domain in the conjugates.

However, for DDSs, chitosan-based self-aggregates were difficult to apply, as these aggregates cannot penetrate into biological solution (pH 7.4) and they get precipitated within few days. To overcome this problem, water-soluble chitosan derivatives have been developed so as to increase their stability in biological solution and to decrease the cytotoxicity induced by acidic solution where chitosan is soluble. Glycol chitosan, a chitosan derivative, was found to be quite promising as a novel drug carrier because of its solubility in water and biocompatibility. A new DDS was adapted by conjugating covalently modified glycol chitosan with deoxycholic acid and investigated in detail the effect of deoxycholic acid attached to glycol chitosan on the formation, physicochemical characteristics, and stability of self-aggregates that are formed in the aqueous media (Kim et al. 2005). They also covalently attached 5β-cholanic acid to glycol chitosan through amide formation using carbodiimide as catalyzer (Kwon et al. 2003; Park et al. 2007). The 5β-cholanic acid–glycol chitosan formed self-aggregates (210–859 nm in diameter) in an aqueous phase by intra- or intermolecular association between hydrophobic 5β-cholanic acids attached to glycol chitosan.

Another hydrophobic molecule, fluorescein isothiocyanate (FITC), is widely used as hydrophobic fluorescein, whose isothiocyanato can readily react with free amine to incorporate fluorescence labeling. DOX, another hydrophobic cyclic molecule, is an antitumor antibiotic, which prevents the synthesis of DNA and RNA and has a therapeutic effect on many tumors. Both FITC and DOX can be conjugated onto hydrophilic polysaccharides and can be conjugated for amphiphilic copolymers. Park et al. (2006) successfully prepared hydrophobically modified glycol chitosans (HGCs) by chemical conjugation of FITC or DOX to the backbone of glycol chitosan. The self-aggregates that are formed were systemically administered on tumor-bearing mice via the tail vein. It is noticed that negligible quantity of self-aggregates were found in the heart and lungs, while small amounts (3.6%–3.8%) were detected in the liver for 3 days after intravenous administration. These results revealed the promising potential of self-aggregates on the basis of glycol chitosan as a carrier for hydrophobic antitumor agents (Park et al. 2006). The same group also chemically conjugated an anthracycline drug, adriamycin (i.e., DOX), onto the backbone of glycol chitosan via an acid-labile *cis*-aconityl linkage (Park et al. 2006).

Polymeric nanoparticles can be delivered to targeted sites either by active targeting or passive targeting. Active targeting was attempted by many researchers since the drug can be delivered directly to the target cells. In order to enhance the intracellular localization into the cancer cell, ligands such as vitamins and sugars have been introduced into the drug carriers. Vitamin H exists in the liver, kidney, and pancreas and in milk. Vitamin H and its derivatives have been used in cancer studies and in the tissue-engineering field. Na et al. (2003) introduced vitamin H to pullulan acetate and prepared corresponding self-assembled nanoparticles (~100 nm) in order to improve their cancer-targeting activity and internalization. Moreover, *N*-succinyl-chitosan was synthesized successfully, which could produce stable nanospheres in distilled water with 50–100 nm in diameter (Zhu et al. 2006). Experimental results indicated that a hydrophobic domain formed within these nanospheres. The assembly mechanisms were believed to be the intermolecular H-bonding of *N*-succinyl-chitosan and hydrophobic interaction among the hydrophobic moieties in *N*-succinyl-chitosan macromolecules. Park et al. (2006) described *N*-acetyl histidine-conjugated glycol chitosan self-assembled nanoparticles as a promising system for intracytoplasmic delivery of drugs.

31.5.4.3 *Polyacrylate Family Molecule-Based Nanoparticles*

Poly(methyl methacrylate) and poly(isobutyl cyanoacrylate) (PIBCA), which belong to polyacrylate family, were widely used for biomaterials. They are hydrophobic molecules containing carboxylic ester in their structure. The significant uptake of injected nanoparticles by cells of the mononuclear phagocyte system restrains the development of long-circulating colloidal drug carriers. The complement system plays an important role in the recognition and opsonization processes of foreign materials. As heparin is an active inhibitor of complement activation, Passirani et al. prepared nanoparticles having heparin covalently bound to poly(methyl methacrylate) and evaluated their interactions with complement was prepared (Passirani et al. 1998a). The particles preserved the complement-inhibiting properties of soluble heparin. Nanoparticles containing covalently linked dextran instead of heparin were found to be weak activators of complement as compared with cross-linked dextran or bare poly(methyl methacrylate) nanoparticles. In both the polysaccharides, apart from the specific activity of bound heparin, the protective effect is found to be due to the presence of a dense brushlike layer on the surface of the particles. Such properties are believed to reduce the uptake by mononuclear phagocyte system *in vivo*. When the *in vivo* blood-circulating time of the two kinds of nanoparticles were evaluated, dextran nanoparticles were eliminated very slowly over 48 h, while bare poly(methyl methacrylate) nanoparticles were found to have a half-life of only 3 min. Both of them were found to be long circulating. The potent opsonization capacity of the poly(methyl methacrylate) core was concealed by the protective effect of either polysaccharide because of the presence of a dense brushlike structure. However, for heparin nanoparticles, the "stealth" effect was probably increased by its inhibiting properties against complement activation (Passirani et al. 1998a).

Yang et al. (2000), by using emulsion polymerization of IBCA in the presence of chitosan as a polymeric stabilizer at low pH, prepared PIBCA-chitosan nanoparticles. Nimodipine as a model drug was successfully integrated into the nanoparticles with mean particle diameter of 31.6 nm and has a positive charge. Furthermore, PIBCA-chitosan, PIBCA-dextran, and PIBCA-dextran sulfate core–shell nanoparticles were prepared by redox radical or anionic polymerization of IBCA in the presence of chitosan, dextran, or dextran sulfate (Bertholon et al. 2006). Complement activation induced by these nanoparticles was studied by investigating the conversion of C3 into C3b in serum incubated with these nanoparticles. It is found that cleavage of C3 increased with size of dextran bound in "loops" configuration, whereas it is decreased when dextran was bound in "brush." Bravo-Osuna et al. (2006, 2007a,b,c) prepared PIBCA-thiolated chitosan nanoparticles by radical emulsion polymerization. They had mean hydrodynamic diameter around 200 nm and positive zeta potential values, demonstrating the presence of the cationic thiolated chitosan at their surface. By radical emulsion polymerization of IBCA in the presence of various polysaccharides (dextran, dextran sulfate, heparin, chitosan, HA, pectin), poly-saccharide-coated nanoparticles were synthesized (Chauvierre et al. 2003). Then, the complement activation induced by different polysaccharide-coated nanoparticles and of the antithrombotic activity of heparin was evaluated. These nanoparticles maintained the heparin antithrombotic properties and inhibited complement activation. This work clearly showed that the hemoglobin loading on nanoparticles surface rather than being encapsu-lated. With a size of 100 nm, these DDSs have found to be suitable tools in the treatment of thrombosis oxygen-deprived pathologies (Chauvierre et al. 2004). For the first time using electronic paramagnetic resonance, they investigated the mobility of dextran chains on the PIBCA nanoparticles. This technique opens an interesting aspect of investigating surface properties of polysaccharide-coated nanoparticles by a new physicochemical approach to further correlate the mobility of the polysaccharide chains with the fate of the nanopar-ticles in biological systems (Vauthier et al. 2004).

31.6 Medical Applications of Polysaccharide-Based Nanoparticles

Polysaccharide-based nanoparticles have received the most promising nanoparticulate DDSs because of their unique properties. They are characterized as particulate dispersions or solid particles with a size ranging from 10 to 1000 nm with various morphologies, like nanocap-sules, nanospheres, nanoliposomes, nanodrugs, and nanomicelles. In this system, the drug is dissolved, encapsulated, entrapped, or attached to the nanoparticle matrix (Kommareddy et al. 2005; Lee and Kim 2005). Polysaccharide-based nanoparticles DDS has wide advan-tages, such as efficient drug protection against enzymatic and chemical degradation, abil-ity to create a controlled release to a specific tissue, high drug encapsulation efficiency, cell internalization, and ability to reverse the multidrug resistance of tumor cells (Soma et al. 1999). Starch-based nanoparticles have received a significant amount of attention because of their biocompatibility, good hydrophobicity, and biodegradability. Hydrophobic-grafted and cross-linked starch nanoparticles were used for drug delivery, and indomethacin was incorporated as the model drug (Abraham and Simi 2007). Hydrophilic amylopectin was modified by grafting hydrophobic PLA chains for the fabrication of polymeric micelles for drug delivery. When these spherical nanoaggregates were used as the drug carrier, it was found that they had a good loading capacity and *in vitro* release properties for hydrophobic indomethacin drug (Brecher et al. 1997; Dufresne et al. 2004).

Nanoparticles of poly(DL-lactide-co-glycolide) (PLGA)-grafted dextran were synthesized for use as oral drug carrier. These nanoparticles have particle size ranging from 50 to 300 nm and were able to form nanoparticles in water by self-aggregating process. Super para-magnetic chitosan–dextran sulfate hydrogels as drug carriers were synthesized. The 5-aminosalicylic acid was incorporated as model drug molecule (Saboktakin et al. 2010).

To overcome the pharmacokinetic problems and to obtain the full benefits of the drug, Anitha et al. (2011) prepared dextran sulfate–chitosan nanoparticles. Self-assembled hydrogel nanoparticles composed of dextran and poly(ethylene glycol) were synthesized and prepared nanoparticles used for drug carrier with hydrophobic model drug *in vitro* (Kim et al. 2000).

Hydrophobized pullulan, specifically cholesterol–pullulan and a copolymer of *N*-isopropylacrylamide and *N*-[4-(1-pyrenyl)butyl]-*N*-*n*-octadecylacrylamide via their hydrophobic moieties, as well as hexadecyl group–bearing pullulan self-assembly nanoparticles (Akiyoshi et al. 1993, 1998; Jung et al. 2004), has been used as DDSs. These hydrophobized pullulan self-associate to form colloidally stable nanoparticles with inner hydrophobic core. This hydrophobic core can encapsulate only the hydrophobic substances like insoluble drugs and proteins (Gupta and Gupta 2004). However, amphiphilic polysaccharides composed of pullulan and PLGA were synthesized to give amphiphilic and biodegradable novel drug carriers. For the controlled release of drugs, PLGA is commonly used because of its biodegradability (Jeong et al. 2006a). *In vivo* studies showed that HGC nanoparticles found to be potential as carriers for anticancer peptides and anticancer drugs because of their biocompatible nature (Kwon et al. 2003; Yoo et al. 2005). Modified chitosan derivatives are emerging as novel carriers of drugs because of their solubility and biocompatibility *in vivo* (Chen et al. 2003a; Sinha et al. 2004; Jiang et al. 2006). Nanoparticles of CM chitosan as carriers for the anticancer drug were prepared by gelification with calcium ions with DOX chosen as a model drug.

31.7 Conclusions and Future Perspectives

The literature enumerated in this review showed that a lot of attention has been aimed in the combination of polysaccharide-based polymers with inorganic nanoparticles, so as to profit from the advantages of both organic and inorganic components. The literature earlier clearly depicted the significant use of polysaccharide-based nanoparticles because of their availability in natural source, renewability, biocompatibility, biodegradability, low cost, and nontoxic nature. Hence, formulations of such bionanocomposites can perform outstanding characteristics, like optical, antimicrobial functionalities, surface coverage, size particles, enzyme degradability, and colloidal stability, and their derivatives for various biotechnological and biomedical applications were explained. The important step of this kind of material depends strongly on the earlier steps of their production and their modification steps, which emphasize the correlation of preparative strategies that rely on their final applications. Until now, these nanoparticles are mostly investigated in terms of their physicochemical properties, *in vitro* toxicity, drug-loading ability, and comparatively simple *in vivo* tests. However, the more critical issues, such as the specific interaction of these nanoparticles with human organs, tissues, cells, or biomolecules, their effect on human's metabolism brought by the nanoparticles, and the wider application of these nanoparticles for drug delivery, need to be focused on in the near future. Furthermore, attempts in finding new methods for the earlier diagnosis of diseases and more efficient therapies to synthesize the new generation of multifunctional nanostructured materials based on polysaccharides, modified polysaccharides, and polysaccharide-based dendrimers are very fast emerging. Hence, in near future, more polysaccharide-based nanoparticles emerge, which greatly enriches the versatility of nanoparticle carriers agents in terms of category and function.

References

Abraham, T.E. and Simi, C.K. 2007. Hydrophobic grafted and cross-linked starch nanoparticles for drug delivery. *Bioprocess and Biosystems Engineering* 30(3): 173–180.

Akiyoshi, K., Deguchi, S., Moriguchi, N., Yamaguchi, S., and Sunamoto, J. 1993. Self-aggregates of hydrophobized polysaccharides in water—Formation and characteristics of nanoparticles. *Macromolecules* 26(12): 3062–3068.

Akiyoshi, K., Deguchi, S., Tajima, H., Nishikawa, T., and Sunamoto, J. 1997. Microscopic structure and thermoresponsiveness of a hydrogel nanoparticle by self-assembly of a hydrophobized polysaccharide. *Macromolecules* 30(4): 857–861.

Akiyoshi, K., Kang, E.C., Kurumada, S., Sunamoto, J., Principi, T., and Winnik, F.M. 2000. Controlled association of amphiphilic polymers in water: Thermosensitive nanoparticles formed by self-assembly of hydrophobically modified pullulans and poly(*N*-isopropylacrylamides). *Macromolecules* 33: 3244–3249.

Akiyoshi, K., Kobayashi, S., Shichibe, S., Mix, D., Baudys, M., Kim, S.W., and Sunamoto, J. 1998. Self-assembled hydrogel nanoparticle of cholesterol-bearing pullulan as a carrier of protein drugs: Complexation and stabilization of insulin. *Journal of Controlled Release* 54: 313–320.

Aktas, Y., Yemisci, M., Andrieux, K., Gursoy, R.N., Alonso, M.J., Fernandez-Megia, E., Novoa-Carballal, R. et al. 2005. Development and brain delivery of chitosan-PEG nanoparticles functionalized with the monoclonal antibody OX26. *Bioconjugate Chemistry* 16: 1503–1511.

Allison, D.D. and Grande-Allen, K.J. 2006. Review. Hyaluronan: A powerful tissue engineering tool. *Tissue Engineering* 12(8): 2131–2140.

Alonso, M.J., Janes, K.A., and Calvo, P. 2001. Polysaccharide colloidal particles as delivery systems for macromolecules. *Advanced Drug Delivery Reviews* 47(1): 83–97.

Alonso-Sande, M., Cuna, M., and Remunan-Lopez, C. 2006. Formation of new glucomannan–chitosan nanoparticles and study of their ability to associate and deliver proteins. *Macromolecules* 39: 4152–4158.

Amidi, M., Romeijn, S.G., Borchard, G., Junginger, H.E., Hennink, W.E., and Jiskoot, W. 2006. Preparation and characterization of protein-loaded *N*-trimethyl chitosan nanoparticles as nasal delivery system. *Journal of Controlled Release* 111: 107–116.

Aminabhavi, T.M., Balundgi, R.H., and Cassidy, P.E. 1990. A review on biodegradable plastics. *Polymer-Plastics Technology and Engineering* 29(3): 235–262.

Anitha, A., Deepagan, V.G., Divya Rani, V.V., Menon, D., Nair, S.V., and Jayakumar, R. 2011. Preparation, characterization, in vitro drug release and biological studies of curcumin loaded dextran sulphate–chitosan nanoparticles. *Carbohydrate Polymers* 84: 1158–1164.

Aumelas, A., Serrero, A., Durand, A., Dellacherie, E., and Leonard, M. 2007. Nanoparticles of hydrophobically modified dextrans as potential drug carrier systems. *Colloids and Surfaces B: Biointerfaces* 59(1): 74–80.

Avadi, M. R., Sadeghi, A. M. M., Mohammadpour, N., Abedin, S., Atyabi, F., Dinarvand, R., and Rafiee-Tehrani, M. 2010. Preparation and characterization of insulin nanoparticles using chitosan and Arabic gum with ionic gelation method. *Nanomedicine: Nanotechnology, Biology and Medicine* 6: 58–63.

Avadi, M., Sadeghi, A., Dounighi, N.M., Dinarvand, R., Atyabi, F., and Rafiee-Tehrani, M. 2011. Ex vivo evaluation of insulin nanoparticles using chitosan and arabic gum. *ISRN Pharmacology* 2011: 860109.

Axelos, M.A.V. and Thibault, J.-F. 1991. The chemistry of low-methoxyl pectin gelation. In: Walter, R.H. (Ed.), *The Chemistry and Technology of Pectin*. San Diego, CA: Academic Press, pp. 109–118.

Bae, H., Ahari, A.F., Shin, H., Nichol, J.W., Hutson, C.B., Masaeli, M., Kim, S.H., Aubin, H., Yamanlar, S., and Khademhosseini, A. 2011. Cell-laden microengineered pullulan methacrylate hydrogels promote cell proliferation and 3D cluster formation. *Soft Matter* 7(5): 1903–1911.

Balazs, E.A., Laurent, T.C., and Jeanloz, R.W. 1986. Nomenclature of hyaluronic acid. *Biochemical Journal* 235(3): 903.

Barbosa, M., Granja, P., Barrias, C., and Amaral, I. 2005. Polysaccharides as scaffolds for bone regeneration. *ITBM-RBM* 26: 212–217.

Barbucci, R., Lamponi, S., Borzacchiello, A., Ambrosio, L., Fini, M., Torricelli, P., and Giardino, R. 2002. Hyaluronic acid hydrogel in the treatment of osteoarthritis. *Biomaterials* 23: 4503–4513.

Beneke, C.M., Viljoen, A.M., and Hamman, J.H. 2009. Polymeric plant-derived excipients in drug delivery. *Molecules* 14: 2602–2620.

Bernardo, M.V., Blanco, M.D., Sastre, R.L., Teijon, C., and Teijon, J.M. 2003. Sustained release of bupivacaine from devices based on chitosan. *Farmaco* 58(11): 1187–1191.

Bertholon, I., Vauthier, C., and Labarre, D. 2006. Complement activation by core–shell poly(isobutyl cyanoacrylate)–polysaccharide nanoparticles: Influences of surface morphology, length, and type of polysaccharide. *Pharmaceutical Research* 23: 1313–1323.

Bhaw-Luximon, A. 2011. Modified natural polysaccharides as nanoparticulate drug delivery devices. In: *Engineered Carbohydrate-Based Materials for Biomedical Applications* (edited by Ravin Narain). Berlin, Germany: John Wiley & Sons, Inc., pp. 355–395.

Blanco, M.D., Gomez, C., Olmo, R., Muniz, E., and Teijon, J.M. 2000. Chitosan microspheres in PLG films as devices for cytarabine release. *International Journal of Pharmaceutics* 202(1–2): 29–39.

Boddohi, S., Moore, N., Johnson, P.A., and Kipper, M.J. 2009. Polysaccharide based polyelectrolyte complex nanoparticles from chitosan, heparin, and hyaluronan. *Biomacromolecules* 10: 1402–1409.

Bodnar, M., Hartmann, J.F., and Borbely, J. 2005. Nanoparticles from chitosan. *Macromolecular Symposia* 227: 321–326.

Braga, T.P., Chagas, E.C., Freitas de Sousa, A., Villarreal, N.L., Longhinotti, N., and Valentini, A. 2009. Synthesis of hybrid mesoporous spheres using the chitosan as template. *Journal of Non-Crystalline Solids* 355: 860–866.

Bravo-Osuna, I., Millotti, G., Vauthier, C., and Ponchel, G. 2007a. *In vitro* evaluation of calcium binding capacity of chitosan and thiolated chitosan poly(isobutyl cyanoacrylate) core–shell nanoparticles. *International Journal of Pharmaceutics* 338(1–2): 284–290.

Bravo-Osuna, I., Ponchel, G., and Vauthier, C. 2007b. Tuning of shell and core characteristics of chitosan-decorated acrylic nanoparticles. *European Journal of Pharmaceutical Sciences* 30(2): 143–154.

Bravo-Osuna, I., Schmitz, T., Bernkop-Schnurch, A., Vauthier, C., and Ponchel, G. 2006. Elaboration and characterization of thiolated chitosan-coated acrylic nanoparticles. *International Journal of Pharmaceutics* 316(1–2): 170–175.

Bravo-Osuna, I., Vauthier, C., Farabollini, A., Palmieri, G.F., and Ponchel, G. 2007c. Mucoadhesion mechanism of chitosan and thiolated chitosan–poly(isobutyl cyanoacrylate) core–shell nanoparticles. *Biomaterials* 28: 2233–2243.

Brecher, M.E., Owen, H.G., and Bandarenko, N. 1997. Alternatives to albumin: Starch replacement for plasma exchange. *Journal of Clinical Apheresis* 12: 146–153.

Brown, M.B. and Jones, S.A. 2005. Hyaluronic acid: A unique topical vehicle for the localized delivery of drugs to the skin. *Journal of European Academy of Dermatology and Venereology* 19: 308–318.

Burdick, J.A. and Prestwich, G.D. 2011. Hyaluronic acid hydrogels for biomedical applications. *Advanced Materials* 23(12): 1521–4095.

Cafaggi, S., Russo, E., Stefani, R., Parodi, B., Caviglioli, G., Sillo, G., Bisio, A., Aiello, C., and Viale, M. 2011. Preparation, characterisation and preliminary antitumour activity evaluation of a novel nanoparticulate system based on a cisplatin-hyaluronate complex and *N*-trimethyl chitosan. *Investigational New Drugs* 29(3): 443–455.

Calvo, P., Remunan-Lopez, C., Vila-Jato, J.L., and Alonso, M.J. 1997. Chitosan and chitosan ethylene oxide propylene oxide block copolymer nanoparticles as novel carriers for proteins and vaccines. *Pharmaceutical Research* 14: 1431–1436.

Carolan, G., Catley, B.J., and McDougal, F.J. 1983. The location of tetrasaccharide units in pullulan. *Carbohydrate Research* 114: 237–243.

Catley, B.J. and Whelan, W.J. 1971. Observations on the structure of pullulan. *Archives of Biochemistry and Biophysics* 143: 138–142.

Chae, S.Y., Son, S., Lee, M., Jang, M.K., and Nah, J.W. 2005. Deoxycholic acid-conjugated chitosan oligosaccharide nanoparticles for efficient gene carrier. *Journal of Controlled Release* 109: 330–344.

Chauvierre, C., Labarre, D., Couvreur, P., and Vauthier, C. 2003. Novel polysaccharide decorated poly(isobutyl cyanoacrylate) nanoparticles. *Pharmaceutical Research* 20(11): 1786–1793.

Chauvierre, C., Marden, M.C., Vauthier, C., Labarre, D., Couvreur, P., and Leclerc, L. 2004. Heparin coated poly(alkylcyanoacrylate) nanoparticles coupled to hemoglobin: A new oxygen carrier. *Biomaterials* 25(15): 3081–3086.

Chelladurai, S., Mishra, M., and Mishra, B. 2008. Design and evaluation of bioadhesive *in-situ* nasal gel of ketorolac tromethamine. *Chemical and Pharmaceutical Bulletin* 56: 1596–1599.

Chen, Q., Hu, Y., Chen, Y., Jiang, X.Q., and Yang, Y.H. 2005. Microstructure formation and property of chitosan–poly(acrylic acid) nanoparticles prepared by macromolecular complex. *Macromolecular Bioscience* 5: 993–1000.

Chen, X.G., Lee, C.M., and Park, H.J. 2003a. O/W emulsification for the self-aggregation and nanoparticle formation of linoleic acid-modified chitosan in the aqueous system. *Journal of Agricultural and Food Chemistry* 51(10): 3135–3139.

Chen, Y., Mohanraj, V.J., and Parkin, J.E. 2003b. Chitosan–dextran sulfate nanoparticles for delivery of an anti-angiogenesis peptide. *Letters in Peptide Science* 10: 621–629.

Choi, S.H., Lee, J.H., Choi, S.M., and Park, T.G. 2006. Thermally reversible pluronic/heparin nanocapsules exhibiting 1000-fold volume transition. *Langmuir* 22: 1758–1762.

Choisnard, L., Geze, A., Putaux, J.L., Wong, Y.S., and Wouessidjewe, D. 2006. Nanoparticles of beta-cyclodextrin esters obtained by self-assembling of biotransesterified betacyclodextrins. *Biomacromolecules* 7: 515–520.

Coelho, S., Moreno-Flores, S., Toca-Herrera, J.L., Coelho, M.A.N., Pereira, M.C., and Rocha, S. 2011. Nanostructure of polysaccharide complexes. *Journal of Colloid and Interface Science* 363(2): 450–455.

Coviello, T., Matricardi, P., Marianecci, C., and Alhaique, F. 2007. Polysaccharide hydrogels for modified release formulations. *Journal of Controlled Release* 119(1): 1873–4995.

Cui, Z.R. and Mumper, R.J. 2001. Chitosan-based nanoparticles for topical genetic immunization. *Journal of Controlled Release* 75: 409–419.

Davidenko, N., Blanco, M.D., Peniche, C., Becherán, L., Guerrero, S., and Teijón, J.M. 2009. Effects of different parameters on characteristics of chitosan–poly(acrylic acid) nanoparticles obtained by the method of coacervation. *Journal of Applied Polymer Science* 111: 2362–2371.

de la Fuente, M., Seijo, B., and Alonso, M.J. 2008. Novel hyaluronan-based nanocarriers for transmucosal delivery of macromolecules. *Macromolecular Biosciences* 8: 441–450.

Dickinson, E. and McClements, D.J. 1996. *Advances in Food Colloids.* Glasgow, Scotland: Chapman & Hall, p. 98.

Dufresne, A., Angellier, H., Choisnard, L., Molina-Boisseau, S., and Ozil, P. 2004. Optimization of the preparation of aqueous suspensions of waxy maize starch nanocrystals using a response surface methodology. *Biomacromolecules* 5(4): 1545–1551.

Duke, J.A. 1985. *Handbook of Medicinal Herbs.* Boca Raton, FL: CRC Press.

Fang, Y., Huang, M.F., Liu, L., Zhang, G.B., and Yuan, G.B. 2006. Preparation of chitosan derivative with polyethylene glycol side chains for porous structure without specific processing technique. *International Journal of Biological Macromolecules* 38(3–5): 191–196.

Fauconnier, M.-L., Blecker, C., Groyne, J., Razafindralambo, H., Vanzeveren, E., Marlier, M., and Paquot, M. 2000. Characterization of two Acacia gums and their fractions using Langmuir film balance. *Journal of Agricultural and Food Chemistry* 48: 2709–2712.

Felt, O., Buri, P., and Gurny, R. 1998. Chitosan: A unique polysaccharide for drug delivery. *Drug Development and Industrial Pharmacy* 24: 979–993.

Fernandez-Hervas, M. and Fell, J. 1998. Pectin/chitosan mixtures as coatings for colon specific drug delivery: An in vitro evaluation. *International Journal of Pharmaceutics* 169: 115–119.

Gabas, A.L., Telis, V.R.N., Sobral, P.J.A., and Telis-Romero, J. 2007. Effect of maltodextrin and arabic gum in water vapor sorption thermodynamic properties of vacuum dried pineapple pulp powder. *Journal of Food Engineering* 82(2): 246–252.

Gan, Q. and Wang, T. 2007. Chitosan nanoparticle as protein delivery carrier—Systematic examination of fabrication conditions for efficient loading and release. *Colloids and Surfaces B: Biointerfaces* 59: 24–34.

Gavory, C., Durand, A., Six, J.L., Nouvel, C., Marie, E., and Leonard, M. 2011. Polysaccharide-covered nanoparticles prepared by nanoprecipitation. *Carbohydrate Polymers* 84: 133–140.

Gibbs, P.A. and Seviour, R.J. 1996. Pullulan. In: Severian, D. (Ed.), *Polysaccharides in Medicinal Applications*. New York: Marcel Dekker, pp. 59–86.

Gomes, J.F.P.S., Rocha, S., Pereira, M.C., Peres, I., Moreno, S., Toca-Herrera, J., and Coelho, M.A.N. 2010. Lipid/particle assemblies based on maltodextrin-gum arabic core as bio-carriers. *Colloids and Surfaces B: Biointerfaces* 76(2): 449–455.

Gref, R., Rodrigues, J., and Couvreur, P. 2002. Polysaccharides grafted with polyesters: Novel amphiphilic copolymers for biomedical applications. *Macromolecules* 35(27): 9861–9867.

Guerrero, S., Teijón, C., Muñiz, E., Teijón, J.M., and Blanco, M.D. 2010. Characterization and in vivo evaluation of ketotifen-loaded chitosan microspheres. *Carbohydrate Polymers* 79: 1006–1013.

Gupta, M. and Gupta, A. 2004. Hydrogel pullulan nanoparticles encapsulating pBUDLacZ plasmid as an efficient gene delivery carrier. *Journal of Controlled Release* 99: 157–166.

Han, S.K., Lee, J.H., Kim, D., Cho, S.H., and Yuk, S.H. 2005. Hydrophilized poly(lactide-coglycolide) nanoparticles with core/shell structure for protein delivery. *Science and Technology of Advanced Materials* 6: 468–474.

Hascall, V.C., Majors, A.K., de la Motte, C.A., Evanko, S.P., Wang, A., Drazba, J.A., Strong, S.A., and Wight, T.N. 2004. Intracellular hyaluronan: A new frontier for inflammation? *Biochimica et Biophysica Acta* 1673: 3–12.

Heinze, T., Liebert, T., Heublein, B., and Hornig, S. 2006. Functional polymers based on dextran. *Advances in Polymer Science* 205: 199–291.

Hoagland, P.D. and Parris, N. 1996. Chitosan/pectin laminated films. *Journal of Agricultural and Food Chemistry* 44: 1915–1919.

Hu, Y., Chen, Y., Chen, Q., Zhang, L.Y., Jiang, X.Q., and Yang, C.Z. 2005. Synthesis and stimuli responsive properties of chitosan/poly(acrylic acid) hollow nanospheres. *Polymer* 46: 12703–12710.

Ito, T., Iida-Tanaka, N., Niidome, T., Kawano, T., Kubo, K., Yoshikawa, K., Sato, T., Yang, Z., and Koyama, Y. 2006. Hyaluronic acid and its derivative as a multi-functional gene expression enhancer: Protection from non-specific interactions, adhesion to targeted cells, and transcriptional activation. *Journal of Controlled Release* 112(3): 382–388.

Jain, D. and Banerjee, R. 2008. Comparison of ciprofloxacin hydrochloride-loaded protein, lipid, and chitosan nanoparticles for drug delivery. *Journal of Biomedical Materials Research* 86B: 105–112.

Janes, K.A., Calvo, P., and Alonso, M.J. 2001. Polysaccharide colloidal particles as delivery systems for macromolecules. *Advanced Drug Delivery Reviews* 47: 83–97.

Jayme, M.L., Dunstan, D.E., and Gee, M.L. 1999. Zeta potential of gum Arabic stabilised oil in emulsions. *Food Hydrocolloids* 13: 459–465.

Jeong, Y., Na, H.S., Oh, J.S., Choi, K.C., Song, C., and Lee, H. 2006a. Adriamycin release from self-assembling nanospheres of poly(DL-lactide-co-glycolide)-grafted pullulan. *International Journal of Pharmaceutics* 322: 154–160.

Jeong, Y.I., Kim, S.H., Jung, T.Y., Kim, I.Y., Kang, S.S., Jin, Y.H., Ryu, H.H. et al. 2006b. Polyion complex micelles composed of all-trans retinoic acid and poly(ethylene glycol)-grafted-chitosan. *Journal of Pharmaceutical Science* 95: 2348–2360.

Jiang, G.B., Quan, D., Liao, K., and Wang, H. 2006. Novel polymer micelles prepared from chitosan grafted hydrophobic palmitoyl groups for drug delivery. *Molecular Pharmaceutics* 3(2): 152–160.

Jiang, H.L., Kim, Y.K., Arote, R., Nah, J.W., Cho, M.H., Choi, Y.J., Akaike, T., and Cho, C.S. 2007a. Chitosan-graft-polyethylenimine as a gene carrier. *Journal of Controlled Release* 117: 273–280.

Jiang, L., Qian, F., He, X.W., Wang, F., Ren, D., He, Y., Li, K., Sun, S., and Yin, C. 2007b. Novel chitosan derivative nanoparticles enhance the immunogenicity of a DNA vaccine encoding hepatitis B virus core antigen in mice. *Journal of Gene Medicine* 9: 253–264.

Jing, X.B., Yu, H.J., Wang, W.S., Chen, X.S., and Deng, C. 2006. Synthesis and characterization of the biodegradable polycaprolactone-graft-chitosan amphiphilic copolymers. *Biopolymers* 83(3): 233–242.

Jung, S.W., Jeong, Y.I., Kim, Y.H., and Kim, S.W. 2004. Self-assembled nanoparticles of poly (ethylene glycol) grafted pullulan acetate as a novel drug carrier. *Archives of Pharmacal Research* 27: 562–569.

Jung, S., Jeong, Y.I., Kim, S.H., Jung, T.Y., Kim, I.Y., Kang, S.S., Jin, Y.H. et al. 2006. Polyion complex micelles composed of all-trans retinoic acid and poly (ethylene glycol)-grafted-citosan. *Journal of Pharmcological Science* 95: 2348–2360.

Kean, T., Roth, S., and Thanou, M. 2005. Trimethylated chitosans as non-viral gene delivery vectors: Cytotoxicity and transfection efficiency. *Journal of Controlled Release* 103: 643–653.

Kim, S.H., Kim, I.S., and Jeong, Y.I. 2000. Self-assembled hydrogel nanoparticles composed of dextran and poly(ethylene glycol) macromer. *International Journal of Pharmaceutics* 205: 109–116.

Kim, D.G., Jeong, Y.I., Choi, C., Roh, S.H., Kang, S.K., Jang, M.K., and Nah, J.W. 2006. Retinolencapsulated low molecular water-soluble chitosan nanoparticles. *International Journal of Pharmaceutics* 319: 130–138.

Kim, K., Kwon, S., Park, J.H., Chung, H., Jeong, S.Y., Kwon, I.C., and Kim, I.S. 2005. Physicochemical characterizations of self-assembled nanoparticles of glycol chitosan–deoxycholic acid conjugates. *Biomacromolecules* 6(2): 1154–1158.

Kim, T.H., Jiang, H.L., Jere, D., Park, I.K., Cho, M.H., Nah, J.W., Choi, Y.J., Akaike, T., and Cho, C.S. 2007. Chemical modification of chitosan as a gene carrier *in vitro* and *in vivo*. *Progress in Polymer Science* 32: 726–753.

Kogan, G., Soltes, L., Stern, R., and Gemeiner, P. 2007. Hyaluronic acid: A natural biopolymer with a broad range of biomedical and industrial applications. *Biotechnology Letters* 29(1): 17–25.

Kommareddy, S., Tiwari, S., and Amiji, M. 2005. Long-circulating polymeric nanovectors for tumor-selective gene delivery. *Technology in Cancer Research Treatment* 4: 615–625.

Kuroda, K., Fujimoto, K., Sunamoto, J., and Akiyoshi, K. 2002. Hierarchical self-assembly of hydrophobically modified pullulan in water: Gelation by networks of nanoparticles. *Langmuir* 18: 3780–3786.

Kurozawa, L.E., Park, K.J., and Hubinger, M.D. 2009. Effect of maltodextrin and gum arabic on water sorption and glass transition temperature of spray dried chicken meat hydrolysate protein. *Journal of Food Engineering* 91(2): 287–296.

Kurup, T.R., Wan, L.S., and Chan, L.W. 1992. Interaction of preservatives with macromolecules: Part I—Natural hydrocolloids. *Pharmaceutica Acta Helvetiae* 67: 301–307.

Kwon, S., Park, J.H., Chung, H., Kwon, I.C., Jeong, S.Y., and Kim, I.S. 2003. Physicochemical characteristics of self-assembled nanoparticles based on glycol chitosan bearing 5 beta-cholanic acid. *Langmuir* 19: 10188–10193.

Lan, S.F. and Starly, B. 2011. Alginate based 3D hydrogels as an in vitro co-culture model platform for the toxicity screening of new chemical entities. *Toxicology and Applied Pharmacology* 256: 62–72.

Lapasin, R. and Pricl, S. 1995. *Rheology of Industrial Polysaccharides: Theory and Applications*. London, U.K.: Blackie.

Lazarridou, A., Roukas, T., Biliaderis, C.G., and Vaikousi, H. 2002. Characterisation of pullulan produced from beet molasses by *Aureobasidium pullulans* in a stirred tank reactor under varying agitation. *Enzyme and Microbial Technology* 31: 122–132.

Leach, J.B. and Schmidt, C.E. 2004. Hyaluronan. In: *Encyclopedia of Biomaterials and Biomedical Engineering* (edited by Gary E. Winek and Gary L. Bowlin). New York: Marcel Dekker, pp. 779–789.

Learoyd, T.P., Burrows, J.L., French, E., and Seville, P.C. 2008. Chitosan-based spray-dried respirable powders for sustained delivery of terbutaline sulfate. *European Journal of Pharmaceutics and Biopharmaceutics* 68(2): 224–234.

Lee, K.Y., Jo, W.H., Kwon, I.C., Kim, Y.H., and Jeong, S.Y. 1998a. Structural determination and interior polarity of self-aggregates prepared from deoxycholic acid-modified chitosan in water. *Macromolecules* 31: 378–383.

Lee, K.Y., Kim, J.H., Kwon, I.C., and Jeong, S.Y. 2000. Self-aggregates of deoxycholic acid modified chitosan as a novel carrier of adriamycin. *Colloid and Polymer Science* 278: 1216–1219.

Lee, K.Y., Kwon, I.C., Kim, Y.H., Jo, W.H., and Jeong, S.Y. 1998b. Preparation of chitosan self aggregates as a gene delivery system. *Journal of Controlled Release* 51: 213–220.

Lee, M. and Kim, S. 2005. Polyethylene glycol-conjugated copolymers for plasmid DNA delivery. *Pharmaceutical Research* 22: 1–10.

Lee, P.W., Peng, S.F., Su, C.J., Mi, F.L., Chen, H.L., Wei, M.C., Lin, H.J., and Sung, H.W. 2008. The use of biodegradable polymeric nanoparticles in combination with a low-pressure gene gun for transdermal DNA delivery. *Biomaterials* 29: 742–751.

Lemarchand, C., Couvreur, P., Vauthier, C., Costantini, D., and Gref, R. 2003. Study of emulsion stabilization by graft copolymers using the optical analyzer Turbiscan. *International Journal of Pharmaceutics* 254(1): 77–82.

Lemarchand, C., Gref, R., Lesieur, S., Hommel, H., Vacher, B., Besheer, A., Maeder, K., and Couvreur, P. 2005. Physico-chemical characterization of polysaccharide-coated nanoparticles. *Journal of Controlled Release* 108(1): 97–111.

Leonard, M., Rouzes, C., Durand, A., and Dellacherie, E. 2003. Influence of polymeric surfactants on the properties of drug-loaded PLA nanospheres. *Colloids and Surfaces B—Biointerfaces* 32(2): 125–135.

Letchford, K. and Burt, H. 2007. A review of the formation and classification of amphiphilic block copolymer nanoparticulate structures: Micelles, nanospheres, nanocapsules and polymersomes. *European Journal of Pharmaceutics and Biopharmaceutics* 65: 259–269.

Li, F., Li, J., Wen, X., Zhou, S., Tong, X., Su, P., Li, H., and Shi, D. 2009. Anti-tumor activity of paclitaxel-loaded chitosan nanoparticles: An *in vitro* study. *Materials Science and Engineering C* 29(8): 2392–2397.

Li, Y., Nagira, T., and Tsuchiya, T. 2006. The effect of hyaluronic acid on insulin secretion in HIT-T15 cells through the enhancement of gap-junctional intercellular communications. *Biomaterials* 27(8): 1437–1443.

Lin, Y.H., Chung, C.K., Chen, C.T., Liang, H.F., Chen, S.C., and Sung, H.W. 2005. Preparation of nanoparticles composed of chitosan/poly-gamma-glutamic acid and evaluation of their permeability through Caco-2 cells. *Biomacromolecules* 6: 1104–1112.

Liu, L., Fishman, M.L., Kost, J., and Hicks, K.B. 2003. Pectin-based systems for colon-specific drug delivery via oral route. *Biomaterials* 24(19): 3333–3343.

Liu, L., Won, Y.J., Cooke, P.H., Coffin, D.R., Fishman, M.L., Hicks, K.B., and Ma, P.X. 2004. Pectin/poly(lactide-co-glycolide) composite matrices for biomedical applications. *Biomaterials* 25(16): 3201–3210.

Liu, L.S., Finkenstadt, V.L., Liu, C.-K., Jin, T., Fishman, M.L., and Hicks, K.B. 2007a. Preparation of poly(lactic acid) and pectin composite films intended for applications in antimicrobial packaging. *Journal of Applied Polymer Science* 106: 801–810.

Liu, X.D., Howard, K.A., Dong, M.D., Andersen, M.O., Rahbek, U.L., Johnsen, M.G., Hansen, O.C., Besenbacher, F., and Kjems, J. 2007b. The influence of polymeric properties on chitosan/siRNA nanoparticle formulation and gene silencing. *Biomaterials* 28: 1280–1288.

Liu, Z., Jiao, Y., Wang, Y., Zhou, C., and Zhang, Z. 2008. Polysaccharides-based nanoparticles as drug delivery systems. *Advanced Drug Delivery Reviews* 60(15): 1650–1662.

Liu, Z.H., Jiao, Y.P., Liu, F., and Zhang, Z.Y. 2007c. Heparin/chitosan nanoparticle carriers prepared by polyelectrolyte complexation. *Journal of Biomedical Material & Research Part A* 83A: 806–812.

Lopes da Silva, J.A. and Gonçalves, M.P. 1994. Rheological study into the ageing process of high methoxyl pectin/sucrose gels. *Carbohydrate Polymers* 24: 235–245.

Lu, B., Xiong, S.B., Yang, H., Yin, X.D., and Zhao, R.B. 2006. Mitoxantrone-loaded BSA nanospheres and chitosan nanospheres for local injection against breast cancer and its lymph node metastases—I: Formulation and *in vitro* characterization. *International Journal of Pharmaceutics* 307: 168–174.

Luangtana-anan, M., Opanasopit, P., Ngawhirunpat, T., Nunthanid, J., Sriamornsak, P., Limmatvapirat, S., and Lim, L.Y. 2005. Effect of chitosan salts and molecular weight on a nanoparticulate carrier for therapeutic protein. *Pharmaceutical Development and Technology* 10: 189–196.

Lui, L., Fishman, M.L., and Hicks, K.B. 2007. Pectin in controlled drug delivery—A review. *Cellulose* 14: 15–24.

Lui, L., Fishman, M.L., Kost, J., and Hicks, K.B. 2003. Pectin-based systems for colon-specific drug delivery via oral route. *Biomaterials* 24: 3333–3343.

Madi, N.S., Harvey, L.M., Mehlert, A., and McNeil, B. 1997. Synthesis of two distinct exopolysaccharide fractions by cultures of the polymorphic fungus *Aureobasidium pullulans*. *Carbohydrate Polymers* 32: 307–314.

Maestrelli, F., Garcia-Fuentes, M., Mura, P., and Alonso, M.J. 2006. A new drug nanocarrier consisting of chitosan and hydoxypropyleyclodextrin. *European Journal of Pharmaceutics and Biopharmaceutics* 63: 79–86.

Medina, J.M., Thomas, A., and Denegar, C.R. 2006. Knee osteoarthritis: Should your patient opt for hyaluronic acid injection? *Journal of Family Practice* 8: 667–675.

Mendichi, R. and Soltes, L. 2002. Hyaluronan molecular weight and polydispersity in some commercial intra-articular injectable preparations and in synovial fluid. *Inflammation Research* 51(3): 115–116.

Misaki, A., Torii, M., Sawai, T., and Goldstein, I.J. 1980. Structure of the dextran of *Leuconostoc mesenteroides* B-1355. *Carbohydrate Research* 84: 273–285.

Mitra, S., Gaur, U., Ghosh, P.C., and Maitra, A.N. 2001. Tumour targeted delivery of encapsulated dextran–doxorubicin conjugate using chitosan nanoparticles as carrier. *Journal of Controlled Release* 74: 317–323.

Mizrahy, S. and Peer, D. 2012. Polysaccharides as building blocks for nanotherapeutics. *Chemical Society Reviews* 41: 2623–2640.

Morille, M., Passirani, C., Vonarbourg, A., Clavreul, A., and Benoit, J.P. 2008. Progress in developing cationic vectors for non-viral systemic gene therapy against cancer. *Biomaterials* 29(24–25): 3477–3496.

Morris, E.R. 1980. Physical probes of polysaccharide conformation and interactions. *Food Chemistry* 6: 15–39.

Morris, E.R., Powell, D.A., Gidley, M.J., and Rees, D.A. 1982. Conformation and interactions of pectins I. Polymorphism between gel and solid states of calcium polygalacturonate. *Journal of Molecular Biology* 155: 507–516.

Muzzarelli, R.A.A. and Muzzarelli, C. 2005. Chitosan chemistry: Relevance to the biomedical sciences. Polysaccharides 1: Structure, characterization and use. *Advances in Polymer Science* 186: 151–209.

Na, K., Lee, T.B., Park, K.H., Shin, E.K., Lee, Y.B., and Cho, H.K. 2003. Self-assembled nanoparticles of hydrophobically-modified polysaccharide bearing vitamin H as a targeted anti-cancer drug delivery system. *European Journal of Pharmaceutical Science* 18: 165–173.

Nafee, N.A., Ismail, F.A., Boraie, N.A., and Mortada, L.M. 2004. Mucoadhesive delivery systems. I. Evaluation of mucoadhesive polymers for buccal tablet formulation. *Drug Development and Industrial Pharmacy* 30: 985–993.

Nair, L.S. and Laurencin, C.T. 2007. Biodegradable polymers as biomaterial. *Progress in Polymer Science* 6: 762–798.

Narayanan, R.P., Melman, G., Letourneau, N.J., Mendelson, N.L., and Melman, A. 2012. Photodegradable iron(III) cross-linked alginate gels. *Biomacromolecules* 13: 2465–2471.

Neo, H., Ishimaru, J.I., Kurita, K., and Goss, A.N. 1997. The effect of hyaluronic acid on experimental temporomandibular joint osteoarthrosis in the sheep. *Journal of Oral and Maxillofacial Surgery* 55: 1114–1119.

Nevozhay, D., Kańska, U., Budzyńska, R., and Boratyński, J. 2007. Current status of research on conjugates and related drug delivery systems in the treatment of cancer and other diseases (Polish). *Postępy Higieny i Medycyny Doświadczalnej* 61: 350–360.

Nishikawa, T., Akiyoshi, K., and Sunamoto, J. 1996. Macromolecular complexation between bovine serum albumin and the self-assembled hydrogel nanoparticle of hydrophobized polysaccharides. *Journal of the American Chemical Society* 118(26): 6110–6115.

Oakenfull, D.G. 1991. The chemistry of high-methoxyl pectins. In: Walter, R.H. (Ed.), *The Chemistry and Technology of Pectin*. San Diego, CA: Academic Press, pp. 87–108.

Oakenfull, D.G. and Scott, A. 1998. Milk gels with low methoxy pectins. In: Phillips, G.O., Williams, P.A., and Wedlock, D.J. (Eds.), *Gums and Stabilisers for the Food Industry*, Vol. 9. Oxford, U.K.: Oxford University Press, pp. 212–221.

Opanasopit, P., Ngawhirunpat, T., Rojanarata, T., Choochottiros, C., and Chirachanchai, S. 2007. Camptothecin-incorporating N-phthaloylchitosan-g-mPEG self-assembly micellar system: Effect of degree of deacetylation. *Colloids and Surface B: Biointerfaces* 60: 117–124.

Orive, G., Ponce, S., Hernandez, R.M., Gascon, A.R., Igartua, M., and Pedraz, J.L. 2002. Biocompatibility of microcapsules for cell immobilization elaborated with different type of alginates. *Biomaterials* 23(18): 3825–3831.

Ouchi, T., Nishizawa, H., and Ohya, Y. 1998. Aggregation phenomenon of PEG-grafted chitosan in aqueous solution. *Polymer* 39: 5171–5175.

Pandey, R. and Ahmad, Z. 2011. Nanomedicine and experimental tuberculosis: facts, flaws, and future. *Nanomedicine* 7: 259–272.

Panyam, J. and Labhasetwar, V. 2003. Biodegradable nanoparticles for drug and gene delivery to cells and tissue. *Advanced Drug Delivery Reviews* 55(3): 329–347.

Park, J.H., Cho, Y.W., Son, Y.J., Kim, K., Chung, H., Jeong, S.Y., Choi, K. et al. 2006. Preparation and characterization of self-assembled nanoparticles based on glycol chitosan bearing adriamycin. *Colloid and Polymer Science* 284: 763–770.

Park, J.H., Kwon, S., Lee, M., Chung, H., Kim, J.H., Kim, Y.S., Park, R.W. et al. 2006. Self-assembled nanoparticles based on glycol chitosan bearing hydrophobic moieties as carriers for doxorubicin: In vivo biodistribution and anti-tumour activity. *Biomaterials* 27: 119–126.

Park, J.H., Saravanakumar, G., Kim, K., and Kwon, I.C. 2010. Targeted delivery of low molecular drugs using chitosan and its derivatives. *Advanced Drug Delivery Reviews* 62(1): 28–41.

Park, J.S., Koh, Y.S., Bang, J.Y., Jeong, Y.I., and Lee, J.J. 2008. Antitumor effect of all-trans retinoic acid-encapsulated nanoparticles of methoxy poly(ethylene glycol)-conjugated chitosan against CT-26 colon carcinoma *in vitro*. *Journal of Pharmaceutical Science* 97(9): 4011–4019.

Park, K., Kim, J.H., Nam, Y.S., Lee, S., Nam, H.Y., Kim, K., Park, J.H. et al. 2007. Effect of polymer molecular weight on the tumour targeting characteristics of self-assembled glycol chitosan nanoparticles. *Journal of Controlled Release* 122: 305–314.

Park, K., Kim, K., Kwon, I.C., Kim, S.K., Lee, S., Lee, D.Y., and Byun, Y. 2004. Preparation and characterization of self-assembled nanoparticles of heparin-deoxycholic acid conjugates. *Langmuir* 20(26): 11726–11731.

Passirani, C., Barratt, G., Devissaguet, J.P., and Labarre, D. 1998a. Interactions of nanoparticles bearing heparin or dextran covalently bound to poly(methyl methacrylate) with the complement system. *Life Sciences* 62(8): 775–785.

Peppa, S.N.A., Bures, P., Leobandung, W., and Ichikawa, H. 2000. Hydrogels in pharmaceutical formulations. *European Journal of Pharmaceutics and Biopharmaceutics* 50: 27–46.

Perera, G., Barthelmes, J., and Bernkop-Schnurch, A. 2010. Novel pectin-4-aminothiophenole conjugate microparticles for colon-specific drug delivery. *Journal of Controlled Release* 145(3): 240–246.

Peres, I., Rocha, S., Pereira, M.C., Coelho, M., Rangel, M., and Ivanova, G. 2010. NMR structural analysis of epigallocatechin gallate loaded polysaccharide nanoparticles. *Carbohydrate Polymers* 82(3): 861–866.

Pilnik, W. 1990. Pectin—A many splendored thing. In: Phillips, G.O., Williams, P.A., and Wedlock, D.J. (Eds.), *Gums and Stabilisers for the Food Industry*, Vol. 5. Oxford, U.K.: IRL Press, pp. 209–222.

Pliszczak, D., Bourgeois, S., Bordes, C., Valour, J.P., Mazoyer, M.A., Orecchioni, A.M., Nakache, E., and Lanteri, P. 2011. Improvement of an encapsulation process for the preparation of pro- and prebiotics-loaded bioadhesive microparticles by using experimental design. *European Journal of Pharmaceutical Science* 44(1–2): 83–92.

Powell, D.A., Morris, E.R., Gidley, M.J., and Rees, D.A. 1982. Conformation and interactions of pectins. II. Influence of residue sequence on chain association in calcium pectate gels. *Journal of Molecular Biology* 155: 317–331.

Prabaharan, M. and Mano, J.F. 2005. Chitosan-based particles as controlled drug delivery systems. *Drug Delivery* 12(1): 41–57.

Prestwich, G.D. 2008. Engineering a clinically-useful matrix for cell therapy. *Organogenesis* 4(1): 42–47.

Qi, L.F., Xu, Z.R., Jiang, X., Li, Y., and Wang, M.Q. 2005. Cytotoxic activities of chitosan nanoparticles and copper-loaded nanoparticles. *Bioorganic and Medicinal Chemistry Letters* 15: 1397–1399.

Randall, R.C., Phillips, G.O., and Williams, P.A. 1988. The role of proteinaceous component on the emulsifying properties of gum Arabic, *Food Hydrocolloids* 2: 131.

Ray, A.K., Bird, P.B., Iacobucci, G.A. and Clark, B.C. 1995. Functionality of gum Arabic fractionation, characterization and evaluation of gum fractions in citrus oil emulsions and model beverages. *Food Hydrocolloids* 9: 123–131.

Rehm, B.H.A. (Ed). 2009. *Alginates: Biology and Applications.* Springer. New York. ISBN: 978-3-540-92678.

Rekha, M.R. and Chandra, P.S. 2007. Pullulan as a promising biomaterial for biomedical applications: A perspective. *Trends in Biomaterials & Artificial Organs* 20: 116–121.

Ridley, B.L., O'Neil, M.A., and Mohnen, D. 2001. Pectins: Structure, biosynthesis and oligogalacturonide-related signalling. *Phytochemistry* 57: 929–967.

Rinaudo, M. 2008. Main properties and current applications of some polysaccharides as biomaterials. *Polymer International* 57(3): 397–430.

Rodrigues, J.S., Santos-Magalhaes, N.S., Coelho, L.C.B.B., Couvreur, P., Ponchel, G., and Gref, R. 2003. Novel core(polyester)–shell(polysaccharide) nanoparticles: Protein loading and surface modification with lectins. *Journal of Controlled Release* 92: 103–112.

Rotureau, E., Leonard, M., Dellacherie, E., and Durand, A. 2004. Amphiphilic derivatives of dextran: Adsorption at air/water and oil/water interfaces. *Journal of Colloid Interface Science* 279(1): 68–77.

Saboktakin, M., Tabatabaie, R., Maharramov, A., and Ramazanov, M. 2010. Synthesis and characterization of superparamagnetic chitosan-dextran sulfate hydrogels as nano carriers for colon-specific drug delivery. *Carbohydrate Polymers* 81: 372–376.

Sajeesh, S. and Sharma, C.P. 2006. Novel pH responsive polymethacrylic acid chitosanpolyethylene glycol nanoparticles for oral peptide delivery. *Journal of Biomedical Materials Research* 76B: 298–305.

Sandri, G., Bonferoni, M.C., Rossi, S., Ferrari, F., Gibin, S., Zambito, Y., Di Colo, G., and Caramella, C. 2007. Nanoparticles based on *N*-trimethylchitosan: Evaluation of absorption properties using in vitro (Caco-2 cells) and ex vivo (excised rat jejunum) models. *European Journal of Pharmaceutics and Biopharmaceutics* 65: 68–77.

Saravanakumar, G., Jo, D.G., and Park, J.H. 2012. Polysaccharide-based nano-particles: A versatile platform for drug delivery and biomedical imaging. *Current Medicinal Chemistry* 19: 3212–3229.

Sarmento, B., Ferreira, D., Veiga, F., and Ribeiro, A. 2006. Characterization of insulin-loaded alginate nanoparticles produced by ionotropic pre-gelation through DSC and FTIR studies. *Carbohydrate Polymers* 66: 1–7.

Sarmento, B., Ferreira, D.C., Jorgensen, L., and van de Weert, M. 2007. Probing insulin's secondary structure after entrapment into alginate/chitosan nanoparticles. *European Journal of Pharmaceutics and Biopharmaceutics* 65: 10–17.

Sinha, V.R. and Kumria, R. 2001. Polysaccharides in colon-specific drug delivery. *International Journal of Pharmaceutics* 224(1–2): 19–38.

Sinha, V.R., Singla, A.K., Wadhawan, S., Kaushik, R., Kumria, R., Bansal, K., and Dhawan, S. 2004. Chitosan microspheres as a potential carrier for drugs. *International Journal of Pharmaceutics* 274(1–2): 1–33.

Skjak-Braerk, G., Grasdalen, H., and Smidsrød, O. 1989. Inhomogeneous polysaccharide ionic gels. *Carbohydrate Polymers* 10: 31–54.

Smidsrød, O., Glover, R.M., and Whittington, S.G. 1973. The relative extension of alginates having different chemical composition. *Carbohydrate Research* 27: 107–118.

Soma, C.E., Dubernet, C., Barratt, G., Nemati, F., Appel, M., Benita, S., and Couvreur, P. 1999. Ability of doxorubicin-loaded nanoparticles to overcome multidrug resistance of tumour cells after their capture by macrophages. *Pharmaceutical Research* 16: 1710–1716.

Srinophakun, T. and Boonmee, J. 2011. Preliminary study of conformation and drug release mechanism of doxorubicin-conjugated glycol chitosan, via cis-aconityl linkage, by molecular modeling. *International Journal of Molecular Science* 12(3): 1672–1683.

Stoilova, O., Koseva, N., Manolova, N., and Rashkov, I. 1999. Polyelectrolyte complex between chitosan and poly(2-acryloylamido-2-methylpropanesulfonic acid). *Polymer Bulletin* 43: 67–73.

Street, C.A. and Andersom, D.M.W. 1983. Refinement of the structures previously proposed for gum Arabic and other Acacia gum exudates. *Talanta* 30: 887–893.

Sugimoto, K. 1978. Pullulan: Production and applications. In: *Fermentation and Industry. Journal of the Fermentation Association, Japan* 36(2): 98–108.

Sujata, V.B. 2002. *Biomaterials*. India: Springer New York. ISBN: 0-7923-7058-9.

Sun, Y. and Wan, A.J. 2007. Preparation of nanoparticles composed of chitosan and its derivatives as delivery systems for macromolecules. *Journal of Applied Polymer Science* 105: 552–561.

Sungthongjeen, S., Sriamornsak, P., Pitaksuteepong, T., Somsiri, A., and Puttipipatkhachorn, S. 2004. Effect of degree of esterification of pectin and calcium amount on drug release from pectin-based matrix tablets. *AAPS PharmSciTech* 5: 1–9.

Suri, S.S., Fenniri, H., and Singh, B. 2007. Nanotechnology-based drug delivery systems. *Journal of Occupational Medicine and Toxicology* 2: 16.

Tan, W.B., Jiang, S., and Zhang, Y. 2007. Quantum-dot based nanoparticles for targeted silencing of HER2/neu gene via RNA interference. *Biomaterials* 28: 1565–1571.

Thanou, M., Florea, B.I., Geldof, M., Junginger, H.E., and Borchard, G. 2002. Quaternized chitosan oligomers as novel gene delivery vectors in epithelial cell lines. *Biomaterials* 23: 153–159.

Thanou, M., Kean, T., and Roth, S. 2005. Trimethylated chitosans as non-viral gene delivery vectors: Cytotoxicity and transfection efficiency. *Journal of Controlled Release* 103(3): 643–653.

Tharanathan, R.N. and Ramesh, H.P. 2003. Carbohydrates—The renewable raw materials of high biotechnological value. *Critical Reviews in Biotechnology* 23(2): 149–173.

Thebaud, N.B., Pierron, D., Bareille, R., Le Visage, C., Letourneur, D., and Bordenave, L. 2007. Human endothelial progenitor cell attachment to polysaccharide-based hydrogels: A pre-requisite for vascular tissue engineering. *Journal of Material Science: Material Medicine* 18: 339–345.

Thirawong, N., Kennedy, R.A., and Sriamornsak, P. 2008. Viscometric study of pectin–mucin interaction and its mucoadhesive bond strength. *Carbohydrate Polymers* 71: 170–179.

Tiyaboonchai, W. and Limpeanchob, N. 2007. Formulation and characterization of amphotericin B–chitosan–dextran sulfate nanoparticles. *International Journal of Pharmaceutics* 329: 142–149.

Tombs, M.P. and Harding, S.E. 1998. *An Introduction to Polysaccharide Biotechnology*. London, U.K.: Taylor & Francis Group.

Toole, B.P., Wight, T.N., and Tammi, M.I. 2002. Hyaluronan–cell interactions in cancer and vascular disease. *Journal of Biological Chemistry* 277: 4593–4596.

Tsai, M.L., Bai, S.W., and Chen, R.H. 2008. Cavitation effects versus stretch effects resulted in different size and polydispersity of ionotropic gelation chitosan-sodium tripolyphosphate nanoparticle. *Carbohydrate Polymers* 71: 448–457.

Turley, E.A., Noble, P.W., and Bourguignon, L.Y. 2002. Signaling properties of hyaluronan receptors. *Journal of Biological Chemistry* 277: 4589–4592.

Uthman, I., Raynauld, J.P., and Haraoui, B. 2003. Intra-articular therapy in osteoarthritis. *Postgraduate Medical Journal* 79: 449–453.

Valenta, C. 2005. The use of mucoadhesive polymers in vaginal delivery. *Advanced Drug Delivery Reviews* 57: 1692–1712.

van Buren, J.P. 1991. Function of pectin in plant tissue structure and firmness. In: Walter, R.H. (Ed.), *The Chemistry and Technology of Pectin*. San Diego, CA: Academic Press, pp. 1–22.

Varum, F.J., McConnell, E.L., Sousa, J.J., Veiga, F., and Basit, A.W. 2008. Mucoadhesion and the gastrointestinal tract. *Critical Reviews in Therapeutic Drug Carrier Systems* 25(3): 207–258.

Vauthier, C., Chauvierre, C., Labarre, D., and Hommel, H. 2004. Evaluation of the surface properties of dextran-coated poly(isobutylcyanoacrylate) nanoparticles by spin-labelling coupled with electron resonance spectroscopy. *Colloid and Polymer Science* 282(9): 1016–1025.

Vila, A., Sanchez, A., Janes, K., Behrens, I., Kissel, T., Jato, J.L.V., and Alonso, M.J. 2004. Low molecular weight chitosan nanoparticles as new carriers for nasal vaccine delivery in mice. *European Journal of Pharmaceutics and Biopharmaceutics* 57: 123–131.

Wallenfels, K., Keilich, G., Bechtler, G., and Freudenberger, D. 1965. Investigations on pullulan. IV. Resolution of structural problems using physical, chemical and enzymatic methods. *Biochemische Zeitschrift* 341: 433–450.

Wang, S., Mao, X., Wang, H., Lin, J., Li, F., and Wei, D. 2011. Characterization of a novel dextran produced by *Gluconobacter oxydans* DSM 2003. *Applied Microbiology and Biotechnology* 91(2): 287–294.

Wang, Y.S., Jiang, Q., Liu, L.R., and Zhang, Q.Q. 2007. The interaction between bovine serum albumin and the self-aggregated nanoparticles of cholesterol-modified *O*-carboxymethyl chitosan. *Polymer* 48: 4135–4142.

Watts, P. and Smith, A. 2009. PecSys: In situ gelling system for optimised nasal drug delivery. *Expert Opinion on Drug Delivery* 6: 543–552.

Wiley, B.J., Ball, D.H., Arcidiacono, S.M., Sousa, S., Mayer, J.M., and Kaplan, D.L. 1993. Control of molecular weight distribution of the biopolymer pullulan produced by *Aureobasidium pullulans*. *Journal of Environmental Polymer Degradation* 1: 3–9.

Willats, W.G.T., McCartney, L., Mackie, W., and Knox, J.P. 2001. Pectin: Cell biology and prospects for functional analysis. *Plant Molecular Biology* 47: 9–27.

Xu, X.M., Capito, R.M., and Spector, M. 2008. Plasmid size influences chitosan nanoparticles mediated gene transfer to chondrocytes. *Journal of Biomedical Materials Research* 84A: 1038–1048.

Xu, Y.M., Du, Y.M., Huang, R.H., and Gao, L.P. 2003. Preparation and modification of *N*-(2-hydroxyl) propyl-3-trimethyl ammonium chitosan chloride nanoparticle as a protein carrier. *Biomaterials* 24: 5015–5022.

Yadav, N., Morris, G.A., Harding, S.E., Ang, S., and Adams, G.G. 2009. Various non-injectable delivery systems for the treatment of diabetes mellitus. *Endocrine, Metabolic & Immune Disorders— Drug Targets* 9: 1–13.

Yang, J.S., Xie, Y.J., and He, W. 2011. Research progress on chemical modification of alginate: A review. *Carbohydrate Polymers* 84: 33–39.

Yang, L.Q., Kuang, J.L., Li, Z.Q., Zhang, B.F., Cai, X., and Zhang, L.M. 2008a. Amphiphilic cholesteryl-bearing carboxymethylcellulose derivatives: Self-assembly and rheological behaviour in aqueous solution. *Cellulose* 15(5): 659–669.

Yang, S.C., Ge, H.X., Hu, Y., Jiang, X.Q., and Yang, C.Z. 2000. Formation of positively charged poly(butyl cyanoacrylate) nanoparticles stabilized with chitosan. *Colloid and Polymer Science* 278: 285–292.

Yang, X.D., Zhang, Q.Q., Wang, Y.S., Chen, H., Zhang, H.Z., Gao, F.P., and Liu, L.R. 2008b. Self-aggregated nanoparticles from methoxy poly(ethylene glycol)-modified chitosan: Synthesis; characterization; aggregation and methotrexate release in vitro. *Colloids and Surface B* 61: 125–131.

Yang, Y., Chen, J.L., Li, H., Wang, Y.Y., Xie, Z., Wu, M., Zhang, H. et al. 2007. Porcine interleukin-2 gene encapsulated in chitosan nanoparticles enhances immune response of mice to piglet paratyphoid vaccine. *Comparative Immunology, Microbiology and Infectious Diseases* 30: 19–32.

Ye, A., Flanagan, J., and Singh, H. 2006. Formation of stable nanoparticles via electrostatic complexation between sodium caseinate and gum arabic. *Biopolymers* 82(2): 121–133.

Yip, G.W., Smollich, M., and Gotte, M. 2006. Therapeutic value of glycosaminoglycans in cancer. *Molecular Cancer Therapy* 5(9): 2139–2148.

Yoksan, R., Matsusaki, M., Akashi, M., and Chirachanchai, S. 2004. Controlled hydrophobic/hydrophilic chitosan: Colloidal phenomena and nanosphere formation. *Colloid and Polymer Science* 282: 337–342.

Yoo, H.S., Lee, J.E., Chung, H., Kwon, I., and Jeong, S.Y. 2005. Self-assembled nanoparticles containing hydrophobically modified glycol chitosan for gene delivery. *Journal of Controlled Release* 103: 235–243.

You, J.O. and Peng, C.A. 2004. Calcium-alginate nanoparticles formed by reverse microemulsion as gene carriers. *Macromolecular Symposia* 219: 147–153.

Yuan, Z. 2007. Study on the synthesis and catalyst oxidation properties of chitosan bound nickel(II) complexes. *Journal of Agricultural and Food Chemistry* 21: 22–24.

Zahoor, A., Sharma, S., and Khuller, G.K. 2005. Inhalable alginate nanoparticles as antitubercular drug carriers against experimental tuberculosis. *International Journal of Antimicrobial Agents* 26: 298–303.

Zhang, H., Oh, M., Allen, C., and Kumacheva, E. 2004. Monodisperse chitosan nanoparticles for mucosal drug delivery. *Biomacromolecules* 5: 2461–2468.

Zhang, J., Chen, X.G., Li, Y.Y., and Liu, C.S. 2007. Self-assembled nanoparticles based on hydrophobically modified chitosan as carriers for doxorubicin. *Nanomedicine:* 3: 258–265.

Zhang, L.M., Lu, H.W., Wang, C., and Chen, R.F. 2011. Preparation and properties of new micellar drug carriers based on hydrophobically modified amylopectin. *Carbohydrate Polymers* 83(4): 1499–1506.

Zhang, Y.Y., Yang, Y., Tang, K., Hu, X., and Zou, G.L. 2008. Physicochemical characterization and antioxidant activity of quercetin-loaded chitosan nanoparticles. *Journal of Applied Polymer Science* 107: 891–897.

Zheng, F., Shi, X.W., Yang, G.F., Gong, L.L., Yuan, H.Y., Cui, Y.J., Wang, Y., Du, Y.M., and Li, Y. 2007. Chitosan nanoparticle as gene therapy vector via gastrointestinal mucosa administration: Results of an in vitro and in vivo study. *Life Sciences* 80: 388–396.

Zheng, Y.L., Wu, Y., Yang, W.L., Wang, C.C., Fu, S.K., and Shen, X.Z. 2006. Preparation, characterization, and drug release in vitro of chitosan-glycyrrhetic acid nanoparticles. *Journal of Pharmaceutical Science* 95: 181–191.

Zhi, J., Wang, Y.J., and Luo, G.S. 2005. Adsorption of diuretic furosemide onto chitosan nanoparticles prepared with a water-in-oil nanoemulsion system. *Reactive and Functional Polymers* 65: 249–257.

Zhou, B., McGary, C.T., Weigel, J.A., Saxena, A., and Weigel, P.H. 2003. Purification and molecular identification of the human hyaluronan receptor for endocytosis. *Glycobiology* 13(5): 339–349.

Zhu, A.P., Chen, T., Yuan, L.H., Wu, H., and Lu, P. 2006. Synthesis and characterization of N-succinyl-chitosan and its self-assembly of nanospheres. *Carbohydrate Polymers* 66(2): 274–279.

Zhuangdong, Y. 2007. Study on the synthesis and catalyst oxidation properties of chitosan bound nickel(II) complexes. *Journal of Agricultural and Food Chemistry* 21(5): 22–24.

Section XIV

Marine Toxicogenomics

32

Current Advances in Biotechnology-Driven Marine Microbial Metagenomics

P.V. Bramhachari and G.P.C.N. Raju

CONTENTS

32.1 Introduction

Marine water constitutes the largest contiguous habitat on the globe, occupying more than 70% of Earth's surface with an average depth of 4 km. Life in this environment is, in contrast to the most terrestrial habitats, dominated by microorganisms, both with regard to metabolism and biomass. These microbes accomplish many unique steps of the biogeochemical cycles, and they also signify a huge and dynamic source of genetic variability (Karl, 2007). Therefore, this environment represents one of the most significant, but still least understood, microbial environments on Earth. In addition, marine microbial communities were among the first microbial communities to be studied using metagenomic approaches (DeLong, 2005), and over the last years, there have been several studies reporting exploration of microorganisms within marine water and sediments (e.g., Venter et al., 2004; Rusch et al., 2007). Unlocking the secrets of marine microbes requires the application of an infinite array of techniques and approaches. Classical microbiology deals with cultivation and characterization of organisms. Merely a tiny portion of the microorganisms we know exist. The vast majority of them (95%–99%) remain "unseen" in pure culture (Amann et al., 1995). For instance, using molecular biology and genomics to establish their community structure and metabolic capabilities (Amann et al., 1995), well-established molecular techniques like fluorescence *in situ* hybridization (FISH) (Amann et al., 1990) have proven crucial in characterizing the community structure of planktonic prokaryotes (Schattenhofer et al., 2009). Metagenomics is a term coined by Handelsman et al. (1998). It describes the sequencing and analysis of whole microbial communities from environmental samples. Community genomics, environmental genomics, and population genomics are often used as synonyms. Direct cloning of DNA from environmental samples was proposed in the late 1980s (Pace et al., 1985). Metagenomics involves DNA isolation from an environmental sample, cloning of the DNA into a vector, and transforming the clones into a host bacterium. Depending on the scientific questions, numerous approaches can be taken from here on. The transformed clones can be screened for phylogenetic markers (e.g., 16S rRNA) of a target organism or clade. Once these are found, the respective clones can be completely sequenced to expose the functional potential of the target clade. Reversely, one could screen for a functional gene of interest first and try to identify the responsible organisms as a next step. Since the sequencing revolution, either random sequencing of clones or high-throughput sequencing of complete DNA libraries offers unprecedented insights into community structure and functional potential on different scales (Handelsman, 2004; Riesenfeld et al., 2004). Furthermore, genome reconstruction from metagenomic samples can narrow down the analysis to specific clades or single organisms in a culture-independent way (Venter et al., 2004; Meyerdierks et al., 2010). Such analysis might reveal key parameters necessary for the successful isolation of yet unculturable organisms. Examples of the unforeseen findings in metagenomic data are plenty. In a keystone work, DeLong and coworkers reported the discovery of an archaeal 16S rRNA gene in a metagenomic library constructed from seawater (Stein et al., 1996). Bacterial rhodopsins were found in an uncultured γ-proteobacterium (Beja et al., 2000). It proved that marine autotrophs possess a light-driven proton pump based on other pigments than chlorophyll. Further metagenomic studies revealed high diversity of bacterial proteorhodopsins (Venter et al., 2004).

This cultivation-independent approach, universally known as metagenomics, provides a quicker way of discovering an unlimited pool of new potential biocatalysts, genes, and biosynthetic pathways from uncultivable marine microorganisms (Daniel, 2005). In recent times, many single genes, such as those coding for cellulolytic enzymes (Handelsman, 2004) and for chitinases (Cottrell et al., 1999), have been successfully collected from environments and expressed by using this metagenomics approach. Developments in metagenomics (Dunlap et al., 2006) have also opened up new opportunities to access the genomes and, hence, metabolic potential of the uncultivable majority of microorganisms. Over the past decade, genome-based studies of marine microorganisms have unveiled the tremendous diversity of the producers of natural products and also contributed to efficiently harness the strain diversity, chemical diversity, and genetic diversity of marine microorganisms for the rapid discovery and generation of new natural products. Marine omics, from metagenomic technologies, which currently offer new and promising strategies for marine biodiscovery, to metabolomics for biofunction exploration, will also be developed in the future, enabling new approaches for the study of chemically mediated interactions (Goulitquer et al., 2012). However, the metagenomics has revolutionized marine microbial ecology. It offers unprecedented insights into the gene pool and diversity of microbial communities, which are inaccessible through culture-dependent genomics. Despite its limitations, metagenomics is one of the most consequential techniques of our time.

Despite many successes have been achieved in marine biotechnology, still many gaps remain to be filled in our basic knowledge on marine science before it could be fully exploited. Still, there remains a variety of problems that needs to be overcome in marine biology, including understanding the role of biodiversity and the impact of global change in maintaining the functionality of ecosystems; understanding relationships between disturbances owing to human behavior and ecosystems; assessing marine ecosystems' health; recovering ecosystem structure and functioning through restoration; conserving, protecting, and managing the seas using the ecosystem approach and spatial planning; and modeling ecosystems for better management (Borja, 2014). The marine ecosystems are still unexploited reservoir of biologically active compounds, which have considerable potential in biotechnology (Kim, 2013). The recent years have clearly shown that these genome studies deciphered important clues on the marine life in terms of symbiosis (Müller et al., 2004), defense mechanisms (Thakur et al., 2005), and biopolymer production (Sogutcu et al., 2012) and expected to accelerate marine biotechnology particularly via improved understanding of the organisms' genetics and metabolism, improvement of cultivation techniques for these marine organisms, discovery of novel pathways for energy and carbon use, development of efficient and profitable production schemes by the use of different hydrocarbons for energy sources, and discovery of novel commercial applications of marine organisms, in near future. In this review, we emphasize the exhilarating prospective that metagenomic-based approaches offer us in gaining access to novel biocatalysts and gene clusters with biotechnological potential from uncultivable marine microorganisms, thereby allowing us to exploit, the potential of this vast and as yet untapped, marine microbial biodiversity resources.

32.2 Metagenomics Technology to Analyze Structure and Function of Marine Microbial Communities

Metagenomics is the study of DNA from a mixed population of organisms and primarily involves the cloning of either total or enriched DNA directly from the environment (eDNA) into a host that can be effortlessly cultivated (Handelsman, 2004). Currently, function-driven analysis and sequence-driven analysis are the two key approaches used for eDNA library screening (Figure 32.1; Kennedy et al., 2010). More recently, advances in next-generation sequencing technologies have allowed isolated eDNA to be sequenced and analyzed directly from environmental samples (Shokralla et al., 2012; Schofield and Sherman, 2013). Nevertheless, this is an effective strategy to access bioactive compounds encoded by the genomes of previously uncultured microbes through introduction of eDNA into a suitable host, bypassing the laborious step of library construction.

32.2.1 Sequence-Based Screening

Improved methods for DNA extraction from environmental sources and polymerase chain reaction (PCR) have made it possible to target a specific type of gene directly from the extracted metagenome. At least four different applications using gene-specific targeting in the study of environmental samples exist. These include (1) the study of microbial diversity, (2) the detection of microorganisms with specific metabolic capabilities, (3) the detection of specific microorganisms, and (4) the discovery and recovery of novel genes/gene clusters. By screening the metagenome of a given environmental sample, a more complete picture of the true microbial biodiversity can be obtained as compared

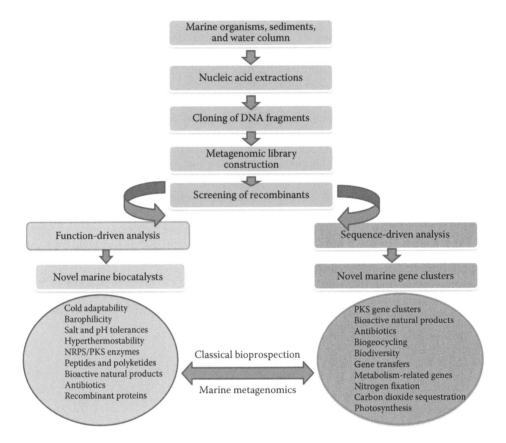

FIGURE 32.1
Scheme of metagenomics for the discovery of novel marine biocatalysts and gene clusters from marine microorganisms.

to culturing alone. The most common approach is based on the amplification of a gene, unique to prokaryotes, the 16S rRNA gene, using a set of universal primers that is specific to a group of organisms, that is, bacteria or archaea. The amplified products are subsequently cloned to generate rRNA gene library and such clones can be sequenced for further phylogenetic analysis.

32.2.2 Function-Based Screening

The function-driven analysis is initiated by identification of clones that express a desired trait, followed by characterization of the active clones by sequence and biochemical analysis. This approach quickly identifies clones that have potential applications in medicine, agriculture, or industry by focusing on natural products or proteins that have useful activities. A powerful yet challenging approach to metagenomic analysis is to identify clones that express a function. Success requires faithful transcription and translation of the gene or genes of interest and secretion of the gene product if the screen or assay requires it to be extracellular. The innovation of new biological motifs will depend on functional analysis of metagenomic clones. Functional screens of marine metagenomic libraries have led to the assignment of functions to numerous "hypothetical proteins" in the databases. Together with this innovation will be required to identify and overcome the barriers to heterologous gene expression and to detect rare clones efficiently in the immense libraries that are required to represent all of the genomes present in complex environments.

32.3 Advances in Marine Microbial Metagenomics of Biocatalysts

Metagenomic studies of marine environments have mainly focused on industrial enzymes (Kennedy et al., 2008; Lee et al., 2010; Zhang and Kim, 2010) or pathways of pharmaceutical interest (Kennedy et al., 2008). Lipases (Hardeman and Sjoling, 2007; Jeon et al., 2009), esterases (Chu et al., 2008; Fu et al., 2011), and monooxygenases (Xu et al., 2008) are the enzymes most often targeted in metagenomic libraries, usually by functional screening. Marine sediments are the environments frequently used for this type of studies, as they are now familiar as deposits of an enormous and mostly unexplored microbial diversity. They were found to be even more phylogenetically diverse than any other environment type including soil (Lozupone and Knight, 2007). In marine biotechnology, the successful expression of enzymes from psychrophilic microorganisms is of particular interest. Due to the misfolding that occurs in thermolabile proteins at higher temperatures, low activities are generally observed at conventional screening temperatures. Decreasing cultivation temperatures of the *Escherichia coli* host can recover folding and activity, but there is a concomitant reduction in growth and heterologous protein synthesis rates (Feller et al., 1998). This problem has been surmounted using an *E. coli* strain carrying chaperone genes from *Oleispira antarctica*, which was able to functionally express a temperature-sensitive esterase (Ferrer et al., 2007). This type of promising system could be tailored for the construction and functional screening of metagenomic libraries from cold environments (Kennedy et al., 2008). In contrast to function-based screening strategies, the application of sequence-driven approaches involves the use of specific primers or probes to screen the library, designed based on conserved regions in the genes of interest. Since they can only be designed based on prior knowledge, variants of known proteins tend to be retrieved by this approach (Song et al., 2005). Nevertheless, this strategy has succeeded in the identification of novel genes, for example, broad-range alkane monooxygenases from deep-sea sediments (Xu et al., 2008). Alternatively, sequences of the genes identified by a functional approach can also be used as a source of *de novo*, unbiased genetic information to maximize the discovery process (Ferrer et al., 2008). In another report, Luo et al. (2009) found that many marine bacterial genomes contain both secreted and cytoplasmic alkaline phosphatases, whereas nearly all genomes contain a suite of genes involved in glycerol phosphate uptake. However, the studies by Quinn et al. (2007) and Martinez et al. (2010) highlighted the widespread distribution of enzymes specific to the degradation of the phosphonates. This group of organic phosphorus was previously thought to be recalcitrant to biological use, hitherto studies like Global Ocean Sampling (GOS) have demonstrated that this phosphonate assimilation is well known in microbial communities and hence represents a significant phosphorus resource for marine microbes. With respect to biocatalysts, several novel hydrolytic enzymes have recently been cloned from Antarctic seawater bacterial metagenomic DNA (Acevedo et al., 2008), while a novel low-temperature-active lipase has also recently been isolated from a metagenomic library of Baltic Sea marine sediment bacteria. This low-temperature-active lipase gene, which displayed 54% amino acid similarity to a *Pseudomonas putida* esterase, was successfully heterologously expressed in *E. coli* and subsequently biochemically characterized (Hardeman and Sjoling, 2007). This highlights the value of employing metagenomics in marine environments to, in this case, identify novel lipases. Lipases have important applications not only in the detergent industry but also in paper processing, as food additives (Jaeger et al., 1999), and in biofuel production through catalyzing the conversion of vegetable oil to methyl alcohol ester (Jaeger and Eggert, 2002). They also have applications in synthetic organic chemistry, due primarily to their enantio-/stereoselectivity coupled with their ability to retain activity in organic solvents. Thus, similar metagenomics approaches in other marine environments may prove fruitful in identifying other novel lipase genes, with biotechnological applications in these areas. Another example where metagenomics has been successfully employed to identify novel enzymes from a marine environment is the recent report of the cloning of two alkane hydroxylase genes from a metagenomic library from deep-sea sediment in the Pacific Ocean. While this is the first report of the genetic characterization of an alkane hydroxylase from a deep-sea environment, it is also interesting to note that these two alkane hydroxylase genes were functionally expressed in a *Pseudomonas fluorescens* strain (Xu et al., 2008). Identification of these novel proteins may help increase the range of alkane hydroxylase biocatalytic applications and again highlights the utility of metagenomics in identifying potential novel biocatalysts.

A clone with conferred lipolytic activity was identified from a metagenomic fosmid library constructed from a South China Sea marine sediment sample. The phylogenetic analysis of amino acid sequence with other lipolytic enzymes revealed that EstF and seven closely related putative lipolytic enzymes comprised a unique clade in the phylogenetic tree (Fu et al., 2011). Yet in another challenging approach to metagenomic analysis, a novel mesophilic protease was characterized from metagenomic library derived from the Antarctic coastal sediment. The amino acid sequence comparison and phylogenetic analysis revealed that it could be classified as a subtilisin-like serine protease, though the highly conserved residue aspargine was replaced by alanine (Zhang et al., 2011). The observations of Jeon et al. (2012) lend support of such view, where the functional metagenomic sequence analysis of the tidal flat sediment of Ganghwa Island in South Korea revealed that the clones contained different open reading frames (ORFs), which showed 50%–57% amino acid identity with putative lipolytic enzyme and formed a distinct and novel subfamily in the family IV of bacterial lipolytic enzymes. A fosmid metagenomic library was constructed using DNA isolated from deepest layer of the extreme environments that displayed high temperature, high salinity, and high concentrations of multiple toxic heavy metals of the Red Sea. Mohamed et al. (2013) demonstrated unusual esterases with optimum temperature of 65°C, halotolerant capability and maintain at least 60% of its activity in the presence of a wide spectrum of heavy metals. Subseafloor sediments are usually inhabited by carbohydrate degrading psychrophilic aerobic bacteria. A variety of carbohydrate-active enzymes were identified by metagenomic analysis of deep-sea sediment bacteria. Klippel et al. (2014) investigated more than 200 ORFs novel genes coding for glycoside hydrolases, lignocellulases, β-glucosidases endomannanase, endoxylanases, and β-xylosidases, which were produced in *E. coli* using recombinant DNA technology. In another study, a metagenomic library was constructed from microorganisms associated with brown alga *Ascophyllum nodosum*. Functional screening of metagenomic library revealed 13 novel putative esterase loci and 2 glycoside hydrolase loci. Additionally, an endo-β-1,4-glucanase enzyme, having less than 50% identity to known cellulases, was also purified and partially characterized, showing activity at low temperature (Martin et al., 2014).

32.4 Advances in Marine Microbial Metagenomics of Gene Clusters

During the past decade, many successful metagenomic studies have been witnessed. Few such examples are, onnamides and theopederins are antitumor polyketides produced by an uncultured *Pseudomonas* sp. symbiont of marine sponge *Theonella swinhoei*. Biosynthesis genes of onnamides and theopederins were isolated from the metagenome of the marine sponge, and bioinformatic analysis of the biosynthetic genes strongly indicated a prokaryotic origin, suggesting that these potent antitumor agents may be produced by the symbiotic bacteria (Piel et al., 2004). By using a function-based metagenomic screening approach, Ren et al. (2004) discovered a halogenated furanone derivatives, (5Z)-4-bromo-5-(bromomethylene)-3-butyl-2(5H)-furanone from a marine macroalgae, which inhibits the QS systems in *E. coli* K-12 (Ren et al., 2004). Hildebrand et al. (2004) identified the bryostatin polyketide synthase (PKS) gene cluster in *Candidatus* sp. bacterial symbiont of the marine bryozoans *Bugula neritina*. PCR of total DNA using degenerated primers led to amplification of 300 bp products of the b-ketoacyl synthase (KSa) that were used as probes in order to screen genomic libraries. In this way, the 65 kb gene cluster of bryostatin was identified, and genetic analysis of the *bryA* gene revealed that it is responsible for the initial steps of bryostatin biosynthesis (Hildebrand et al., 2004). The discovery of bryostatin led eventually to the identification of a vast amount of (PKSs) of biotechnological interest. Successes in the marine environment include identification of the biosynthetic machinery for the cytotoxic peptide patellamide from the cyanobacterial symbiont, *Prochloron*, of a marine didemnid (Zhang and Bryant 2011). In another study, Long et al. (2005) and Schmidt et al. (2005) independently explored the biosynthesis genes for patellamide using a DNA-sequencing approach and functional metagenomics approach. Bacterial genomics of model culturable organisms and metagenomics of uncultured bacterial consortia present in association with marine sponges and soil communities have revealed numerous gene clusters of PKS and nonribosomal peptide-synthetases (NRPS) for which no molecules have been identified (Ginolhac et al., 2004; Piel et al., 2004; Schirmer et al., 2005; Kim and Fuerst, 2006; Donadio et al., 2007). Palenik et al. (2006) studied the genomic variations in *Synechococcus* to understand the genetic differences between coastal and open-ocean cyanobacteria. These findings indicated a near doubling of response regulators in coastal genomes with a significant increase in metalloenzymes, metal storage proteins, and a large decrease in phosphorous uptake genes. Interestingly, Badger et al. (2006) also investigated the genes of the carbon-concentrating mechanism in cyanobacteria. Interestingly, this analysis included the genes involved in inorganic carbon transport and formation of the proteinaceous coat of the carboxysome, Rubisco, and carbonic anhydrases. Kettler et al. (2007) used a comparative genome approach to establish the genetic basis for differences between *Prochlorococcus* and *Synechococcus* and high-light/low-light adaptation. These researchers investigated 1855–3017 genes and studied the patterns and implications of gene gain and loss within the 2 genera; only 33 genes were found to be unique to all *Prochlorococcus* sp.

Functional-based metagenomic screening approaches have also been supplemented with homology-based screens, primarily involving PCR-based approaches targeting novel genes with sequences similar to known genes. This has resulted in the cloning of genes such as PKSs (Seow et al., 1997), beta-xylanases (Sunna and Bergquist, 2003), xylanases (Hayashi et al., 2005), cyclomaltodextrinases (Tang et al., 2006), and alkane hydroxylases (Xu et al., 2008). Recently, novel methods such as preamplification inverse-PCR (Yamada et al., 2008) and metagenomic DNA shuffling (Boubakri et al., 2006) have been employed to isolate new biocatalysts. Genomic information of environmental microbes provides DNA-based evidence of adaptive strategies for different ecological niches. Interestingly, the genome sequence of *Prochlorococcus marinus* contains a low percentage of signal transduction genes, which reflects an abridged need for auto-trophs in the oligotrophic ocean to be able to sense and act in response to extracellu-lar compounds (Dufresne et al., 2003). Coleman et al. (2006) found five genomic islands in two *Prochlorococcus* type strains, MED4 and MIT9312, using comparative genomics (Venter et al., 2004) and are apparently hot spots for genetic recombination in oceanic *Prochlorococcus* that are predisposed to carry important genes for ecological niche dif-ferentiation. A remarkable strategy for avoiding reactive oxygen species generation has developed in the genome of the versatile marine Antarctica bacterium *Pseudoalteromonas haloplanktis* TAC125, which is an alternative to the more typical molybdopterin-dependent metabolism. The elevated percentage of cell envelope genes in the genome reflected a spe-cialized adaptation mechanism that copes with cold temperatures and maintains normal membrane functionality (Medigue et al., 2005).

More recently, metagenomic and metatranscriptomic analysis of ocean surface water has shown the Cyanobacteria (genus *Prochlorococcus*) and *Alphaproteobacteria* (genus *Roseobacter*) to be the best represented taxonomic groups (Frias-Lopez et al., 2008). However, few stud-ies have addressed structural polysaccharide degradation in marine ecosystems and focused on the isolation and identification of bacterial species, including members of the genus *Pseudoalteromonas* (Garsoux et al., 2004; Zeng et al., 2006). Recently, *Saccharophagus degradans* and *Teredinibacter turnerae* have emerged as strong examples of two well-characterized marine bacterial species involved in polysaccharide degradation, and their whole-genome annotations revealed an extensive repertoire of pertinent functional genes (Weiner et al., 2008; Yang et al., 2009). In another study, gene cluster for the biosynthesis of potent antitumor agent psymberin was isolated from the metagenomic library of marine sponge *Psammocinia* aff. *Bulbosa* (Fisch et al., 2009). Edwards et al. (2010) demonstrated the use of an *in situ* cellulose baiting method yielded a marine microbial metagenome consid-erably enriched in functional genes involved in polysaccharide degradation. They have employed an *in silico* approach to investigate the factors randomly sampled data for taxo-nomic markers and functional genes and showed that the community was dominated by *Gammaproteobacteria* and *Bacteroidetes*. The study yielded gene sequences that matched to a custom-made database comprising the members of nine glycoside hydrolase families with extensive repertoire of functional genes predicted to be involved in cellulose utiliza-tion. Based on activity screening of a marine sediment microbial metagenomic library, a total of 19 fosmid clones showing lipolytic activity were identified from the South China Sea. Novel lipolytic genes FLS18C and FLS18D were sequenced and characterized as novel alkaline esterases (Hu et al., 2010). Application of metagenomic approaches in the tropi-cal marine cyanobacterium *Lyngbya bouillonii* resulted in the identification of 58 kb gene cluster encoding a hybrid PKS-NRPS system. Metagenomic DNA libraries were formed using cyanobacterial filaments, while concurrently single cells were isolated and multiple displacement amplification was performed in their whole genome. Combining the two strategies, *apr* gene cluster was detected and proved conscientious for the biosynthesis of the antitumor natural product apratoxin A (Grindberg et al., 2011). In a recent study by Fujita et al. (2011), vibrioferrin biosynthetic gene cluster was cloned and expressed in heter-ologous hosts using marine metagenomic library. Donia et al. (2011) showed the validity of metagenomics also for ribosomal peptides. In their work, they described the isolation and identification of a novel cyanobactin peptide that was called "minimide" from environ-mental DNA extracted from dotting colonies of *Didemnum molle*, ubiquitous ascidian that inhabits diverse tropical marine habitats. Cyclic peptides containing 5-hydroxytryptophan and thiazole moieties were isolated from the marine sponge *Discodermia calyx* collected near Shikine-jima Island, Japan. The analysis of the 16S rDNA sequences obtained from the metagenomic DNA of *D. calyx* revealed the presence of *Candidatus Entotheonella* sp., an unculturable δ-proteobacterium inhabiting the *Theonella* genus and implicated in the biosynthesis of bioactive peptides (Kimura et al., 2012).

Metagenomic natural product discovery in lichen-associated *Nostoc* sp. provided evidence for novel biosynthetic pathways in diverse symbioses. The presence of unique polyketide, nosperin, in such highly dissimilar associations suggests that some bacterial metabolites may be specific to symbioses with eukaryotes and encourages exploration of other symbioses for drug discovery and better understanding of ecological interactions mediated by complex bacterial metabolites (Kampa et al., 2013). Fascinatingly, a novel cold-adapted lipolytic enzyme gene, *est97*, was identified from a high Arctic intertidal zone sediment metagenomic library. Interestingly, the functional and structural studies of this gene primarily related to high number of methionine and glycine residues and flexible loops in the high-resolution structures (Fu et al., 2013). A potential novel fumarate reductase gene designated *frd1A* was isolated by screening a marine metagenomic library through a sequence-based strategy. The gene was subcloned into a pETBlue-2 vector and expressed in *E. coli* Tuner (DE3) pLacI cells. Functional characterization by HPLC technique demonstrated that the recombinant Frd1A protein could catalyze the hydrogenation of fumarate to succinate acid (Jiang et al., 2013). Sulfoquinovosyldiacylglycerols (SQDG) are polar sulfur-containing membrane lipids, whose presence has been associated to a microbial strategy to adapt to phosphate deprivation in marine environments. Diversity and distribution of a key sulfolipid biosynthetic *sqd*B gene coding the uridine 5′-diphosphate-sulfoquinovose synthase involved in SQDG biosynthetic pathway in marine microbial assemblages was explored using metagenomics. The phylogeny of the *sqd*B-coding protein revealed two distinct clusters: one including green algae, higher plants, and cyanobacteria and another one comprising primarily nonphotosynthetic bacteria, as well as other cyanobacteria and algal groups (Villanueva et al., 2014). Very few marine bacterial taxa in the literature are reported to produce cobalamin including selected members of the Cyanobacteria, Alphaproteobacteria, Gammaproteobacteria, and Bacteroidetes (Sanudo-Wilhelmy et al., 2014). However, marine metagenomics has provided strong evidence for specific biogeographical distributions of thaumarchaeotal cobalamin genes. This hypothesis is supported by the recent findings by Doxey et al. (2014) that niche partitioning was evidenced between thaumarchaeotal and cyanobacterial ribosomal and cobalamin synthesis genes across all metagenomic data sets analyzed. The metagenomics-based functional profiling of the photic/aphotic gene sets revealed a superior variety of source organisms in the aphotic zone, although the majority of individual photic and aphotic depth-related clusters of orthologous groups (COGs) are assigned to the same taxa across the different sites. This increase in phylogenetic and functional diversity of the core aphotic related COGs of proteins, most probably reflects selection for the utilization of a broad range of alternate energy sources in the absence of light (Ferreira et al., 2014).

Diverse reductive dehalogenase gene (*rdhA*) homologs in marine subsurface sediment suggested that anaerobic respiration of organohalides is one of the possible energy yielding pathways in the organic-rich sedimentary habitats. However, in a recent study by Kawai and colleagues, a high frequency of phylogenetically diverse reductive dehalogenase-homologous novel gene clusters was explored in deep subseafloor sedimentary metagenomes (Kawai et al., 2014). Interestingly, these researchers also examined the frequency of dissimilatory sulfite reductase genes (*dsrAB*) and observed that the occurrence of *dsrAB* genes was generally smaller *rdhA*-homologous genes. It has become more apparent that photoautotrophic picocyanobacteria can harvest light via phycobilisomes (PBSs) consisting of the pigments phycocyanin and phycoerythrin, encoded by genes in conserved gene clusters. The occurrence and arrangement of these gene clusters give picocyanobacteria characteristic light absorption properties and allow the colonization of specific ecological niches. Larsson et al. (2014) investigated that *Picocyanobacteria* contain a novel pigment gene cluster that dominates the brackish water Baltic Sea. It is, therefore, anticipated that a full understanding of the evolution and distribution of the PBS gene cluster may perhaps contribute to the sequences databases of fresh and brackish water strains. Catalyzed reporter deposition–FISH is a novel and powerful metagenomic approach for the quantification of bacterial taxa. Acinas et al. (2014) have recently demonstrated this quantitative method by using CF319a and CF968 probes that showed significantly different in certain oceanographic regions and that CF968 shows seasonality within marine *Bacteroidetes*, notably large differences between summer and winter that is overlooked by CF319a.

32.5 Advances in Marine Microbial Metagenomics of Biogeocycling

Marine microorganisms, which are extremely diverse and play fundamental roles in global biogeochemical processes, are subjected to fluctuating environments due to changes in the water conditions. Marine bacteria are ubiquitous in marine ecosystems and outstanding in their capacity to persist in oligotrophic waters, an adaptive trait of biological importance and of great interest in marine microbiology (Thomas et al., 2007).

For instance, a marine *Silicibacter pomeroyi* has operons for inorganic carbon monoxide and sulfide oxidation, suggesting that a lithoheterotrophic metabolic strategy is used to supplement their heterotrophy. Encoded motility genes point out the physiological potential to find elevated nutrient niches associated with particles and plankton (Moran et al., 2004). Giovannoni et al. (2005) demonstrated that a marine SAR11 *Pelagibacter ubique* possesses an unpredictably genome by having few redundant DNA sequences and short intergenic lengths with abundant ATP-binding cassette (ABC) transporters genes for nutrient uptake. The study derived few genes that enabled an adaptive strategy for oligotrophic conditions by exploiting the high substrate affinities of ABC transporters to reduce the costs of ATP hydrolysis. A genomics analysis of the Arctic sediment bacterium revealed that *Colwellia psychrerythraea* produced polyhydroxyalkanoates and polyamides, which is a psychrophilic lifestyle relating to the limitation of carbon and nitrogen uptake caused by cold temperatures (Methe et al., 2005). By means of microarray technology, Martiny et al. (2009) experimentally determined the gene clusters from *Prochlorococcus* that are regulated by phosphate availability. In a subsequent comparison of the GOS phase I sites. The same group yet again examined the genomic incorporation of proteins involved in phosphate sensing, uptake, and organophosphate utilization in *Prochlorococcus*. The findings of Hallam et al. (2006) substantiated the global metabolic importance of Crenarchaeota with key steps in the biogeochemical transformation of carbon and nitrogen in marine ecosystems. The fosmid sequences that resulted from uncultivated marine *Cenarchaeum symbiosum* were evidenced to actively contribute in the reconstruction of carbon and energy metabolism. Their genes predicted to encode multiple components of a modified 3-hydroxypropionate cycle of autotrophic carbon assimilation, consistent with utilization of CO_2 as a sole carbon source. Fragment recruitment analysis using metagenomes collected from the GOS expedition (Rusch et al., 2007; Yooseph et al., 2007), which is the largest marine metagenomic library so far, revealed their numerical significance in the ocean. These enriched sequence libraries observed for SAR11 and *Prochlorococcus* ecotypes in the GOS data set implied that genomic locations were genetic hot spots with diverse contents within natural populations (Coleman et al., 2006; Rusch et al., 2007). The authors imply that horizontal gene transfer and mobile genetic elements, specifically plasmids assembled from the metagenomes, could elucidate some of the vast diversity of this genus and assumed to be tightly associated with their adaptation to extreme conditions. The deep-sea bacterium, *Photobacterium profundum*, is a good example (Eloe et al., 2008).

In a groundbreaking study, Mou et al. (2008) demonstrated a coupled immunocapture technique with sequencing to examine microbes actively responding to the presence of dissolved organic carbon substrates, namely DMSP or vanillate. Gilbert et al. (2009) used a combination of fosmid libraries from the western English Channel sampling site, L4, and the metagenomic data demonstrated the ubiquity and importance of phosphonate-degrading bacteria in coastal marine ecosystems. Their study suggested that regardless of an abundance of inorganic phosphate, several microorganisms were vigorously using the organic phosphonate fraction, potentially as a niche diversification strategy to avoid competition. Yet in another study, Sebastian and Ammerman (2009) used the Sargasso Sea data to show that the traditional alkaline phosphatase protein (phoA) used by bacteria to access inorganic and organic phosphate was significantly less abundant in the GOS databases than a more novel phoX phosphatase protein. This was an exceptional example of understanding nutrient cycling, which has been derived from cultured bacteria and may not always be appropriate to the natural ecosystems. Walsh et al. (2009) constructed a fosmid library from an oxygen minimum zone in Saanich Inlet, British Columbia. The metagenomic sequencing of fosmids from the uncultured SUP05 γ-proteobacterial lineage uncovered an extensive metabolic collection, together with the ability to oxidize multiple sulfur compounds and reduce nitrate to fuel autotrophic carbon assimilation via the Calvin–Benson–Bassham cycle. Woyke et al. (2009) also used flow cytometry to isolate a population but focused instead on Flavobacteria. After sorting, multiple displacement amplification was used to amplify the genomes of two individual cells, with subsequent shotgun sequencing and assembly. The two Flavobacteria genomes were considerably dissimilar from cultured *Flavobacteria*, being surprisingly streamlined but far more representative of the *Flavobacteria* found in the GOS data set. Metabolic reconstruction data suggested that these organisms are specialized at the incorporation of organic rather than inorganic carbon, nitrogen, and sulfur, potentially using proteorhodopsin to fuel uptake.

In a most recent report (Morris and Schmidt, 2013), it highlights the presence of high-affinity terminal oxidase genes in sequenced bacterial genomes and shotgun metagenomes. The results indicate that the bacteria with the potential to respire under microoxic conditions are more phylogenetically diverse and environmentally widespread than previously appreciated. A study conducted by Karl (2014) demonstrated that marine microorganisms are primarily responsible for phosphorous (P) assimilation and remineralization, including (P) reduction–oxidation bioenergetic processes in the marine nutrient geocycling. This study highlighted that functional metagenomics unveiled the inextricable connection between the P cycle and cycles of other bioelements and its impacts on nitrogen fixation and carbon dioxide sequestration. Humily et al. (2014) developed a targeted metagenomic approach to study a genomic region involved in light-harvesting complex (PBS) biosynthesis and regulation in marine *Synechococcus*. This study provided novel insights into *Synechococcus* community structure and pigment type diversity at a representative coastal station of the English Channel. Ammonia-oxidizing archaea (AOA) are ubiquitous and contribute significantly to the carbon and nitrogen cycles in the ocean. Park et al. (2014) have recovered the genes of two novel AOA, namely, *Nitrosopumilus koreensis* AR1 and *Nitrosopumilus sediminis*, isolated from the deep marine sediments of Donghae, South Korea, and Svalbard in the Arctic region, by sequencing the enriched metagenomes. Interestingly, the AR1 and AR2 genomes contained few genes pertaining to energy metabolism and carbon fixation as conserved in other AOA but, conversely, had fewer heme-containing proteins and more copper-containing proteins than other AOA. Marine Group I (MGI) Thaumarchaeota are one of the most abundant and cosmopolitan chemoautotrophs within the ocean. Phylogenetic and metagenomic recruitment analysis revealed that MGI single amplified genomes are genetically and biogeographically distinct from existing thaumarchaea cultures obtained from surface waters of subtropical North Pacific and South Atlantic Ocean. Swan et al. (2014) explored the genes encoding proteins for aerobic ammonia oxidation and the hydrolysis of urea, which may be used for energy production, as well as genes involved in 3-hydroxypropionate/4-hydroxybutyrate and oxidative tricarboxylic acid pathways. More recently, Dupont et al. (2014) revealed that the functional trade-offs emphasized a novel salinity-driven divergence mechanism in microbial community composition. Such findings could radically advance our understanding of microbial distributions and stress the need to incorporate salinity in future climate change models.

32.6 Advances in Marine Microbial Metagenomics of Natural Products

Marine ecosystem may be considered a huge container of natural products that could be exploited in medicine. It is progressively more recognized that an enormous number of natural products exists in nonculturable microbes with chemical, biological, and functional activities for potential uses in various industrial and biomedical applications (Handelsman, 2004). Drugs originated from marine microorganisms, such as cephalosporins (antimicrobial), cytarabine (anticancer), and vidarabine (antivirus), are already well known on the market (Haefner, 2003). Many of the bioactive compounds found in marine invertebrates might actually be produced by associated microorganisms. Dolastatin, an anticancer compound produced by the "sea hare" (nudibranch) *Dolabella auricularia*, is produced by its cyanobacterial symbionts. Interestingly, marine cyanobacterial symbionts belonging to *Prochloron* are a rich source of useful secondary metabolites, such as the tubulin-binding compounds *dolastatin* and *curacin* and various actin-binding and neurotoxic compounds (Ramaswamy et al., 2006). Marine macro- and microorganisms, through evolution, acquired the potential to produce secondary metabolites with unique biological activity (Imhoff et al., 2011). In recent times, these compounds have found a wide range of applications as antibacterial (Plaza et al., 2010), antifungal (Nishimura et al., 2010), antimalarial, and antiprotozoa (Santos et al., 2011), as well as being active in diseases related to the cardiovascular, immune, and nervous systems (Sakurada et al., 2010; Mayer et al., 2013). Metagenomics revealed to be a very powerful tool also for the exploitation of bioactive compounds from marine bacterial communities, since it is extremely hard to isolate and cultivate symbiotic bacteria of marine macroorganisms, for example, sponges that have been recently indicated as promising sources of novel compounds, in particular as anticancer (Kennedy et al., 2008). The recent advances of the genomics contributed to provide novel tools for developing of new metagenomic strategies. The single-cell genomics helped to reduce the metagenomic complexity. A single cell can now be isolated from complex microbial mixtures and the genome amplified for sequencing or PCR screening (Kvist et al., 2007). This approach, combined with metagenomic screening, led the isolation *apr* gene cluster that proved to be responsible for the biosynthesis of the antitumor natural product apratoxin A (Grindberg et al., 2011). Polybrominated diphenyl ethers (PBDEs) and polybrominated bipyrroles are natural products that bioaccumulate in the marine food chain. These PBDEs have attracted widespread interest because of their persistence in the environment and potential toxicity to humans. However, the natural origins of PBDE biosynthesis are not known. Recently, Agarwal et al. (2014) reported few biosynthetic motifs from the metagenomic sequence databases to discover unrealized marine bacterial producers of organobromine compounds.

32.7 Advances in Marine Methanobacterial Metagenomics

Methanogenic bacteria are a phylogenetically diverse group of strictly anaerobic Euryarchaeota. They grow with enzymatic formation of CH_4 from H_2 plus CO_2, acetate, or other C1 compounds, and play an important role in the global carbon cycle because of their large contribution to global CH_4 (<600 Tg CH4 year^{-1}) in both terrestrial and marine environments. Few studies identified a diverse functional community and highlighted the potential importance of the Methylococcaceae and the Methylophilaceae species (Kalyuzhnaya et al., 2008; Chistoserdova, 2011a). Metagenome-based metabolic reconstruction of these species has indicated that at least some of them are capable of denitrification, signifying that they may be adaptable to an anaerobic/microaerobic lifestyle (Kalyuzhnaya et al., 2008; 2009). A metagenomic approach by Biddle et al. (2008) also suggested that genes in the methanogenesis pathway accounted for 1.4% of those amplified from Peru Margin sediments (<50 m below the seafloor), whereas uncultivated Crenarchaeota were major groups. In methanotrophs, this reaction is carried out by an enzyme encoded by *mxaFI* genes (Chistoserdova and Lidstrom, 2013). *mxaFI* genes are also essential in methanol oxidation by other groups of methylotrophs (Chistoserdova and Lidstrom, 2013). However, recently methylotrophs capable of methanol oxidation were described not possessing *mxaFI* genes, and in these, either *mdh2* or *xoxF* genes have been implicated in this function (Chistoserdova and Lidstrom, 2013). Previous metagenomic analysis of Lake Washington populations suggested that the abundant Methylophilaceae phylotypes lacked *mxaFI* genes but possessed multiple copies of *xoxF* (Kalyuzhnaya et al., 2008). Genomes of both alpha- and gammaproteobacterial methanotrophs are also known to encode *xoxF* genes (Chistoserdova, 2011b). Aerobic degradation of methylphosphonate (MPn) by marine bacterioplankton has been hypothesized to contribute significantly to the ocean's methane supersaturation, yet little information is available in literature about MPn utilization by marine microbes. Metagenomic, metatranscriptomic, and functional screening of microcosm perturbation experiments revealed that few microbial taxa of the Vibrionales and Rhodobacterales orders and their metabolic functions associated with MPn-driven methane production can utilize MPn aerobically under conditions of P limitation using the C–P lyase pathway and thereby elicit a significant increase in the dissolved methane concentration (Martinez et al., 2013). The metagenomic and metaproteomic data of methane-oxidizing microbial consortia in sulfidic marine sediments of

Oregon and California, USA, revealed to utilize scarce micronutrients in addition to nickel and molybdenum. Additionally, genetic machinery for cobalt-containing vitamin B_{12} biosynthesis was reported in both anaerobic methanotrophic archaea and sulfate-reducing bacteria. Interestingly, several proteins affiliated with the tungsten-containing form of formylmethanofuran dehydrogenase were recombinantly expressed. Glass et al. (2014) found that these consortia used specialized biochemical strategies to overcome the challenges of metal availability in sulfidic environments.

32.8 Concluding Remarks

Microbial communities, in particular those inhabiting marine environments, represent an invaluable source of genetic information that can help crack critical issues for our society. Over the past decade, metagenomics has undoubtedly benefited the scientific world in rapidly analyzing changes in microbial communities in the marine environment. Due to the absolute volume of marine microbial metagenomic data that is constantly being generated, development of novel methods for analysis, data storage, and sharing is warranted. Marine metagenomics, together with *in vitro* evolution and high-throughput screening technologies, provides the industry with an unprecedented chance to bring biomolecules into industrial application. In spite of the potential of finding new biocatalysts from uncultured marine microorganisms, it often appears that they need to be improved at molecular level to meet industrial and biotechnological requirements through genetic engineering and protein engineering. However, the advances in genetic engineering, bioinformatics, and combinatorial chemistry will further facilitate discovery or synthetic modification of potentially therapeutic compounds that are much needed. It is therefore anticipated that this promising environment continues to be explored in the search of new biotechnological applications. In this review, we highlighted the importance of metagenomics for marine bioprospecting, and moreover, this study allows the biotechnological implementation of marine biodiversity proved by the huge number of novel gene clusters, novel biocatalysts, and compounds discovered recently. It is expected that metagenomics will acquire further interest, despite improvements of culturing techniques. It is therefore foreseeable that this promising environment continues to be explored in the search of new biotechnological applications. Microbial communities, in particular those inhabiting marine environments, represent an invaluable source of genetic information that can help solve pressing issues for our society. The bioprospection of these communities, which involves searching for products and activities of interest, is currently experiencing a revolution due to the development of an expanding methodological toolkit that allows the mining of biodiversity with a depth never possible before. This review highlighted the most promising research areas in marine metagenomics and describes advanced methodologies used for the bioprospection of marine microorganisms.

Acknowledgment

The author gratefully acknowledges DST-SERB, Government of India, for financial support under the Grant No: SR/FT/LS-109-2011.

References

Acevedo, J. P., Reyes, F., Parra, L. P., Salazar, O., Andrews, B. A., and Asenjo, J. A. (2008). Cloning of complete genes for novel hydrolytic enzymes from Antarctic sea water bacteria by use of an improved genome walking technique. *Journal of Biotechnology*, 133(3), 277–286.

Acinas, S. G., Ferrera, I., Sarmento, H., Díez-Vives, C., Forn, I., Ruiz-González, C., Cornejo-Castillo, F. M., Salazar, G., and Gasol, J. M. (2014). Validation of a new catalyzed reporter deposition–fluorescence *in situ* hybridization probe for the accurate quantification of marine *Bacteroidetes* populations. *Environmental Microbiology*, 17(10), 3557–3569.

Agarwal, V., El Gamal, A. A., Yamanaka, K., Poth, D., Kersten, R. D., Schorn, M., Allen, E. E., and Moore, B. S. (2014). Biosynthesis of polybrominated aromatic organic compounds by marine bacteria. *Nature Chemical Biology*, 10(8), 640–647.

Amann, R. I., Krumholz, L., and Stahl, D. A. (1990). Fluorescent-oligonucleotide probing of whole cells for determinative, phylogenetic, and environmental studies in microbiology. *Journal of Bacteriology*, 172(2), 762–770.

Amann, R. I., Ludwig, W., and Schleifer, K. H. (1995). Phylogenetic identification and in situ detection of individual microbial cells without cultivation. *Microbiological Reviews*, 59(1), 143–169.

Badger, M. R., Price, G. D., Long, B. M., and Woodger, F. J. (2006). The environmental plasticity and ecological genomics of the cyanobacterial CO_2 concentrating mechanism. *Journal of Experimental Botany*, 57(2), 249–265.

Beja, O., Aravind, L., Koonin, E. V., Suzuki, M. T., Hadd, A., Nguyen, L. P. et al. (2000). Bacterial rhodopsin: Evidence for a new type of phototrophy in the sea. *Science*, 289(5486), 1902–1906.

Biddle, J. F., Fitz-Gibbon, S., Schuster, S. C., Brenchley, J. E., and House, C. H. (2008). Metagenomic signatures of the Peru Margin subsea floor biosphere show a genetically distinct environment. *Proceedings of the National Academy of Sciences of the United States of America*, 105, 10583–10588.

Borja, A. (2014). Grand challenges in marine ecosystems ecology. *Frontiers in Marine Science*, 1, 1.

Boubakri, H., Beuf, M., Simonet, P., and Vogel, T. M. (2006). Development of metagenomic DNA shuffling for the construction of a xenobiotic gene. *Gene*, 375, 87–94.

Chistoserdova, L. (2011a). Methylotrophy in a lake: From metagenomics to single-organism physiology. *Applied and Environmental Microbiology*, 77(14), 4705–4711.

Chistoserdova, L. (2011b). Modularity of methylotrophy, revisited. *Environmental Microbiology*, 13(10), 2603–2622.

Chistoserdova, L. and Lidstrom, M. E. (2013). Aerobic methylotrophic prokaryotes. In *The Prokaryotes*, Springer, Berlin, Germany, (eds. E. Rosenberg, E. F. DeLong, S. Lory, E. Stackebrandt, and F. Thompson), pp. 267–285.

Chu, X., He, H., Guo, C., and Sun, B. (2008). Identification of two novel esterases from a marine metagenomic library derived from South China Sea. *Applied Microbiology and Biotechnology*, 80(4), 615–625.

Coleman, M. L., Sullivan, M. B., Martiny, A. C., Steglich, C., Barry, K., DeLong, E. F., and Chisholm, S. W. (2006). Genomic islands and the ecology and evolution of *Prochlorococcus*. *Science*, 311(5768), 1768–1770.

Cottrell, M. T., Moore, J. A., and Kirchman, D. L. (1999). Chitinases from uncultured marine microorganisms. *Applied and Environmental Microbiology*, 65(6), 2553–2557.

Daniel, R. (2005). The metagenomics of soil. *Nature Reviews Microbiology*, 3(6), 470–478.

DeLong, E. F. (2005). Microbial community genomics in the ocean. *Nature Reviews Microbiology*, 3(6), 459–469.

Donadio, S., Monciardini, P., and Sosio, M. (2007). Polyketide synthases and nonribosomal peptide synthetases: The emerging view from bacterial genomics. *Natural Product Reports*, 24(5), 1073–1109.

Donia, M. S., Ruffner, D. E., Cao, S., and Schmidt, E. W. (2011). Accessing the hidden majority of marine natural products through metagenomics. *ChemBioChem*, 12(8), 1230–1236.

Doxey, A. C., Kurtz, D. A., Lynch, M. D., Sauder, L. A., and Neufeld, J. D. (2014). Aquatic metagenomes implicate *Thaumarchaeota* in global cobalamin production. *The ISME Journal*, 9(2), 461–471.

Dufresne, A., Salanoubat, M., Partensky, F., Artiguenave, F., Axmann, I. M., Barbe, V. et al. (2003). Genome sequence of the cyanobacterium *Prochlorococcus marinus* SS120, a nearly minimal oxyphototrophic genome. *Proceedings of the National Academy of Sciences of the United States of America*, 100(17), 10020–10025.

Dunlap, W. C., Jaspars, M., Hranueli, D., Battershill, C. N., Peric-Concha, N., Zucko, J., Wright, S. H., and Long, P. F. (2006). New methods for medicinal chemistry-universal gene cloning and expression systems for production of marine bioactive metabolites. *Current Medicinal Chemistry*, 13(6), 697–710.

Dupont, C. L., Larsson, J., Yooseph, S., Ininbergs, K., Goll, J., Asplund-Samuelsson, J., and Bergman, B. (2014). Functional tradeoffs underpin salinity-driven divergence in microbial community composition. *PLOS ONE*, 9(2), e89549.

Edwards, J. L., Smith, D. L., Connolly, J., McDonald, J. E., Cox, M. J., Joint, I. et al. (2010). Identification of carbohydrate metabolism genes in the metagenome of a marine biofilm community shown to be dominated by Gammaproteobacteria and Bacteroidetes. *Genes*, 1(3), 371–384.

Eloe, E. A., Lauro, F. M., Vogel, R. F., and Bartlett, D. H. (2008). The deep-sea bacterium *Photobacterium profundum* SS9 utilizes separate flagellar systems for swimming and swarming under high-pressure conditions. *Applied and Environmental Microbiology*, 74(20), 6298–6305.

Feller, G., Le Bussy, O., and Gerday, C. (1998). Expression of psychrophilic genes in mesophilic hosts: Assessment of the folding state of a recombinant alpha-amylase. *Applied Environmental Microbiology*, 64, 1163–1165.

Ferreira, A. J. S., Siam, R., Setubal, J. C., Moustafa, A., Sayed, A., Chambergo, F. S. et al. (2014). Core microbial functional activities in ocean environments revealed by global metagenomic profiling analyses. *PLOS ONE*, 9(6), e97338.

Ferrer, M., Beloqui, A., Timmis, K. N., and Golyshin, P. N. (2008). Metagenomics for mining new genetic resources of microbial communities. *Journal of Molecular Microbiology and Biotechnology*, 16(1–2), 109–123.

Ferrer, M., Golyshina, O., Beloqui, A., and Golyshin, P. N. (2007). Mining enzymes from extreme environments. *Current Opinion in Microbiology*, 10(3), 207–214.

Fisch, K. M., Gurgui, C., Heycke, N., van der Sar, S. A., Anderson, S. A., Webb, V. L. et al. (2009). Polyketide assembly lines of uncultivated sponge symbionts from structure-based gene targeting. *Natural Chemical Biology*, 5, 494–501.

Frias-Lopez, J., Shi, Y., Tyson, G. W., Coleman, M. L., Schuster, S. C., Chisholm, S. W., and DeLong, E. F. (2008). Microbial community gene expression in ocean surface waters. *Proceedings of the National Academy of Sciences*, 105(10), 3805–3810.

Fu, C., Hu, Y., Xie, F., Guo, H., Ashforth, E. J., Polyak, S. W., Baoli, Z., and Zhang, L. (2011). Molecular cloning and characterization of a new cold-active esterase from a deep-sea metagenomic library. *Applied Microbiology and Biotechnology*, 90(3), 961–970.

Fu, J., Leiros, H. K. S., de Pascale, D., Johnson, K. A., Blencke, H. M., and Landfald, B. (2013). Functional and structural studies of a novel cold-adapted esterase from an Arctic intertidal metagenomic library. *Applied Microbiology and Biotechnology*, 97(9), 3965–3978.

Fujita, M. J., Kimura, N., Sakai, A., Ichikawa, Y., Hanyu, T., and Otsuka, M. (2011). Cloning and heterologous expression of the vibrioferrin biosynthetic gene cluster from a marine metagenomic library. *Bioscience, Biotechnology, and Biochemistry*, 75, 2283–2287.

Garsoux, G., Lamotte, J., Gerday, C., and Feller, G. (2004). Kinetic and structural optimization to catalysis at low temperatures in a psychrophilic cellulase from the Antarctic bacterium *Pseudoalteromonas haloplanktis*. *Biochemical Journal*, 384, 247–253.

Gilbert, J., Thomas, S., Cooley, N. A., Kulakova, A. N., Field, D., Booth, T. et al. (2009). Potential for phosphonoacetate utilization by marine bacteria in temperate coastal waters. *Environmental Microbiology*, 11, 111–125.

Ginolhac, A., Jarrin, C., Gillet, B., Robe, P., Pujic, P., Tuphile, K., and Nalin, R. (2004). Phylogenetic analysis of polyketide synthase I domains from soil metagenomic libraries allows selection of promising clones. *Applied and Environmental Microbiology*, 70(9), 5522–5527.

Giovannoni, S. J., Bibbs, L., Cho, J. C., Stapels, M. D., Desiderio, R., Vergin, K. L., and Barofsky, D. F. (2005). Proteorhodopsin in the ubiquitous marine bacterium SAR11. *Nature*, 438(7064), 82–85.

Glass, J. B., Yu, H., Steele, J. A., Dawson, K. S., Sun, S., Chourey, K., Pan, C., Hettich, R. L., and Orphan, V. J. (2014). Geochemical, metagenomic and metaproteomic insights into trace metal utilization by methane-oxidizing microbial consortia in sulphidic marine sediments. *Environmental Microbiology*, 16, 1592–1611.

Goulitquer, S., Potin, P., and Tenant, T. (2012). Mass spectrometry-based metabolomics to elucidate functions in marine organisms and ecosystems. *Marine Drugs*, 10, 849–880.

Grindberg, R. V., Ishoey, T., Brinza, D., Esquenazi, E., Coates, R. C., Liu, W. T. et al. (2011). Single cell genome amplification accelerates identification of the apratoxin biosynthetic pathway from a complex microbial assemblage. *PLOS ONE*, 6(4), e18565.

Haefner, B. (2003). Drugs from the deep: Marine natural products as drug candidates. *Drug Discovery Today*, 8(12), 536–544.

Hallam, S. J., Mincer, T. J., Schleper, C., Preston, C. M., Roberts, K., Richardson, P. M., and DeLong, E. F. (2006). Pathways of carbon assimilation and ammonia oxidation suggested by environmental genomic analyses of marine *Crenarchaeota*. *PLoS Biology*, 4(4), e95.

Handelsman, J. (2004). Metagenomics: Application of genomics to uncultured microorganisms. *Microbiology and Molecular Biology Reviews*, 68(4), 669–685.

Handelsman, J., Rondon, M. R., Brady, S. F., Clardy, J., and Goodman, R. M. (1998). Molecular biological access to the chemistry of unknown soil microbes: A new frontier for natural products. *Chemistry & Biology*, 5(10), R245–R249.

Hardeman, F. and Sjoling, S. (2007). Metagenomic approach for the isolation of a novel low-temperature-active lipase from uncultured bacteria of marine sediment. *FEMS Microbiology Ecology*, 59(2), 524–534.

Hayashi, H., Abe, T., Sakamoto, M., Ohara, H., Ikemura, T., Sakka, K., and Benno, Y. (2005). Direct cloning of genes encoding novel xylanases from the human gut. *Canadian Journal of Microbiology*, 51(3), 251–259.

Hildebrand, M., Waggoner, L. E., Liu, H., Sudek, S., Allen, S., Anderson, C., Sherman, D. H., and Haygood, M. (2004). *bryA*: An unusual modular polyketide synthase gene from the uncultivated bacterial symbiont of the marine bryozoans *Bugula neritina*. *Chemistry & Biology*, 11(11), 1543–1552.

Hu, Y., Fu, C., Huang, Y., Yin, Y., Cheng, G., Lei, F. et al. (2010). Novel lipolytic genes from the microbial metagenomic library of the South China Sea marine sediment. *FEMS Microbiology Ecology*, 72(2), 228–237.

Humily, F., Farrant, G. K., Marie, D., Partensky, F., Mazard, S., Perennou, M. et al. (2014). Development of a targeted metagenomic approach to study a genomic region involved in light harvesting in marine *Synechococcus*. *FEMS Microbiology Ecology*, 88(2), 231–249.

Imhoff, J. F., Labes, A., and Wiese, J. (2011). Bio-mining the microbial treasures of the ocean: New natural products. *Biotechnology Advances*, 29(5), 468–482.

Jaeger, K. E., Dijkstra, B. W., and Reetz, M. T. (1999). Bacterial biocatalysts: Molecular biology, three-dimensional structures, and biotechnological applications of lipases. *Annual Reviews in Microbiology*, 53(1), 315–351.

Jaeger, K. E. and Eggert, T. (2002). Lipases for biotechnology. *Current Opinion in Biotechnology*, 13(4), 390–397.

Jeon, J. H., Kim, J. T., Kim, Y. J., Kim, H. K., Lee, H. S., Kang, S. G., Kim, S. J., and Lee, J. H. (2009). Cloning and characterization of a new cold-active lipase from a deep-sea sediment metagenome. *Applied Microbiology and Biotechnology*, 81(5), 865–874.

Jeon, J. H., Lee, H. S., Kim, J. T., Kim, S. J., Choi, S. H., Kang, S. G., and Lee, J. H. (2012). Identification of a new subfamily of salt-tolerant esterases from a metagenomic library of tidal flat sediment. *Applied Microbiology and Biotechnology*, 93(2), 623–631.

Jiang, C., Liu, Y., Meng, C., Wu, L., Huang, J., Deng, J., and Wu, B. (2013). Expression of a metagenome-derived fumarate reductase from marine microorganisms and its characterization. *Folia Microbiologica*, 58(6), 663–671.

Kalyuhznaya, M. G., Martens-Habbena, W., Wang, T., Hackett, M., Stolyar, S. M., Stahl, D. A., Lidstrom, M. E., and Chistoserdova, L. (2009). Methylophilaceae link methanol oxidation to denitrification in freshwater lake sediment as suggested by stable isotope probing and pure culture analysis. *Environmental Microbiology Reports*, 1(5), 385–392.

Kalyuzhnaya, M. G., Lapidus, A., Ivanova, N., Copeland, A. C., McHardy, A. C., Szeto, E. et al. (2008). High-resolution metagenomics targets specific functional types in complex microbial communities. *Nature Biotechnology*, 26(9), 1029–1034.

Kampa, A., Gagunashvili, A. N., Gulder, T. A., Morinaka, B. I., Daolio, C., Godejohann, M., and Andresson, Ó. S. (2013). Metagenomic natural product discovery in lichen provides evidence for a family of biosynthetic pathways in diverse symbioses. *Proceedings of the National Academy of Sciences of the United States of America*, 110(33), E3129–E3137.

Karl, D. M. (2007). Microbial oceanography: Paradigms, processes and promise. *Nature Reviews Microbiology*, 5, 759–769.

Karl, D. M. (2014). Microbially mediated transformations of phosphorus in the sea: New views of an old cycle. *Annual Review of Marine Science*, 6, 279–337.

Kawai, M., Futagami, T., Toyoda, A., Takaki, Y., Nishi, S., Hori, S., and Takami, H. (2014). High frequency of phylogenetically diverse reductive dehalogenase-homologous genes in deep sub seafloor sedimentary metagenomes. *Frontiers in Microbiology*, 5, 80.

Kennedy, J., Flemer, B., Jackson, S. A., Lejon, D. P. H., Morrissey, J. P., O'Gara, F., and Dobson, A. D. W. (2010). Marine metagenomics: New tools for the study and exploitation of marine microbial metabolism. *Marine Drugs*, 8, 608–628.

Kennedy, J., Marchesi, J. R., and Dobson, A. D. (2008). Marine metagenomics: Strategies for the discovery of novel enzymes with biotechnological applications from marine environments. *Microbiol Cell Factories*, 7(1), 27.

Kettler, G. C., Martiny, A. C., Huang, K., Zucker, J., Coleman, M. L., Rodrigue, S., and Chisholm, S. W. (2007). Patterns and implications of gene gain and loss in the evolution of *Prochlorococcus*. *PLoS Genetics*, 3(12), e231.

Kim, S. K. (2013). *Marine Nutraceuticals: Prospects and Perspectives*. Boca Raton, FL: CRC Press.

Kim, T. K. and Fuerst, J. A. (2006). Diversity of polyketide synthase genes from bacteria associated with the marine sponge *Pseudoceratina clavata*: Culture-dependent and culture-independent approaches. *Environmental Microbiology*, 8(8), 1460–1470.

Kimura, M., Wakimoto, T., Egami, Y., Tan, K. C., Ise, Y., and Abe, I. (2012). Calyxamides A and B, cytotoxic cyclic peptides from the marine sponge *Discodermia calyx*. *Journal of Natural Products*, 75(2), 290–294.

Klippel, B., Sahm, K., Basner, A., Wiebusch, S., John, P., Lorenz, U., and Antranikian, G. (2014). Carbohydrate-active enzymes identified by metagenomic analysis of deep-sea sediment bacteria. *Extremophiles*, 1, 11.

Kvist, T., Ahring, B. K., Lasken, R. S., and Westermann, P. (2007). Specific single-cell isolation and genomic amplification of uncultured microorganisms. *Applied Microbiology and Biotechnology*, 74(4), 926–935.

Larsson, J., Celepli, N., Ininbergs, K., Dupont, C. L., Yooseph, S., Bergman, B., and Ekman, M. (2014). *Picocyanobacteria* containing a novel pigment gene cluster dominate the brackish water Baltic Sea. *The ISME Journal*, 8, 1892–1903.

Lee, H. S., Kwon, K. K., Kang, S. G., Cha, S. S., Kim, S. J., and Lee, J. H. (2010). Approaches for novel enzyme discovery from marine environments. *Current Opinion in Biotechnology*, 21(3), 353–357.

Long, P. F., Dunlap, W. C., Battershill, C. N., and Jaspars, M. (2005). Shotgun cloning and heterologous expression of the patellamide gene cluster as a strategy to achieving sustained metabolite production. *ChemBioChem*, 6(10), 1760–1765.

Lozupone, C. A. and Knight, R. (2007). Global patterns in bacterial diversity. *Proceedings of the National Academy of Sciences of the United States of America*, 104(27), 11436–11440.

Luo, H., Benner, R., Long, R. A., and Hu, J. (2009). Subcellular localization of marine bacterial alkaline phosphatases. *Proceedings of the National Academy of Sciences of the United States of America*, 106, 21219–21223.

Mardis, E. R. (2008). The impact of next-generation sequencing technology on genetics. *Trends in Genetics*, 24(3), 133–141.

Martin, M., Biver, S., Steels, S., Barbeyron, T., Jam, M., Portetelle, D., and Vandenbol, M. (2014). Functional screening of a metagenomic library of seaweed-associated microbiota: Identification and characterization of a halotolerant, cold-active marine endo-β-1, 4-endoglucanase. *Applied and Environmental Microbiology*, 80(16), 4958–4967.

Martinez, A., Tyson, G. W., and Delong, E. F. (2010). Widespread known and novel phosphonate utilization pathways in marine bacteria revealed by functional screening and metagenomic analyses. *Environmental Microbiology*, 12, 222–238.

Martinez, A., Ventouras, L. A., Wilson, S. T., Karl, D. M., and DeLong, E. F. (2013). Metatranscriptomic and functional metagenomic analysis of methylphosphonate utilization by marine bacteria. *Frontiers in Microbiology*, 4, 340.

Martiny, A. C., Huang, Y., and Li, W. (2009). Occurrence of phosphate acquisition genes in *Prochlorococcus* cells from different ocean regions. *Environmental Microbiology*, 11, 1340–1347.

Mayer, A., Rodríguez, A. D., Taglialatela-Scafati, O., and Fusetani, N. (2013). Marine pharmacology in 2009–2011: Marine compounds with antibacterial, antidiabetic, antifungal, anti-inflammatory, antiprotozoal, antituberculosis, and antiviral activities; affecting the immune and nervous systems, and other miscellaneous mechanisms of action. *Marine Drugs*, 11(7), 2510–2573.

Medigue, C., Krin, E., Pascal, G., Barbe, V., Bernsel, A., Bertin, P. N., and Danchin, A. (2005). Coping with cold: The genome of the versatile marine Antarctica bacterium *Pseudoalteromonas haloplanktis* TAC125. *Genome Research*, 15(10), 1325–1335.

Methe, B. A., Nelson, K. E., Deming, J. W., Momen, B., Melamud, E., Zhang, X., and Fraser, C. M. (2005). The psychrophilic lifestyle as revealed by the genome sequence of *Colwellia psychrerythraea* 34H through genomic and proteomic analyses. *Proceedings of the National Academy of Sciences of the United States of America*, 102(31), 10913–10918.

Meyerdierks, A., Kube, M., Kostadinov, I., Teeling, H., Glöckner, F. O., Reinhardt, R., and Amann, R. (2010). Metagenome and mRNA expression analyses of anaerobic methanotrophic archaea of the ANME-1 group. *Environmental Microbiology*, 12(2), 422–439.

Mohamed, Y. M., Ghazy, M. A., Sayed, A., Ouf, A., El-Dorry, H., and Siam, R. (2013). Isolation and characterization of a heavy metal-resistant, thermophilic esterase from a Red Sea Brine Pool. *Scientific Reports*, 3, 3358.

Moran, M. A., Buchan, A., González, J. M., Heidelberg, J. F., Whitman, W. B., Kiene, R. P., and Ward, N. (2004). Genome sequence of *Silicibacter pomeroyi* reveals adaptations to the marine environment. *Nature*, 432(7019), 910–913.

Morris, R. L. and Schmidt, T. M. (2013). Shallow breathing: Bacterial life at low O2. *Nature Reviews Microbiology*, 11, 205–212.

Mou, X., Sun, S., Edwards, R. A., Hodson, R. E., and Moran, M. A. (2008). Bacterial carbon processing by generalist species in the coastal ocean. *Nature*, 451, 708–713.

Müller, W. E., Grebejuk, V. A., Thakur, N. L., Thakur, A. N., Batel, R., Krasko, A. et al. (2004). Oxygen-controlled bacterial growth in the sponge *Suberites domuncula*: Toward a molecular understanding of the symbiotic relationship between sponge and bacteria. *Applied and Environmental Microbiology*, 70, 2332–2341.

Nishimura, S., Arita, Y., Honda, M., Iwamoto, K., Matsuyama, A., Shirai, A., and Yoshida, M. (2010). Marine antifungal theonellamides target 3β-hydroxysterol to activate Rho1 signaling. *Nature Chemical Biology*, 6(7), 519–526.

Pace, N., Stahl, D., Lane, D., and Olsen, G. (1985). Analyzing natural microbial populations by rRNA sequences. *American Society for Microbiology News*, 51(1), 4–12.

Palenik, B., Ren, Q., Dupont, C. L., Myers, G. S., Heidelberg, J. F., Badger, J. H., and Paulsen, I. T. (2006). Genome sequence of *Synechococcus* CC9311: Insights into adaptation to a coastal environment. *Proceedings of the National Academy of Sciences of the United States of America*, 103(36), 13555–13559.

Park, S. J., Ghai, R., Martín-Cuadrado, A. B., Rodríguez-Valera, F., Chung, W. H., Kwon, K., and Rhee, S. K. (2014). Genomes of two new ammonia-oxidizing archaea enriched from deep marine sediments. *PLOS ONE*, 9(5), e96449.

Piel, J., Hui, D., Fusetani, N., and Matsunaga, S. (2004). Targeting modular polyketide synthases with iteratively acting acyltransferases from metagenomes of uncultured bacterial consortia. *Environmental Microbiology*, 6(9), 921–927.

Plaza, A., Keffer, J. L., Lloyd, J. R., Colin, P. L., and Bewley, C. A. (2010). Paltolides A–C, anabaenopeptin-type peptides from the palau sponge *Theonella swinhoei*. *Journal of Natural Products*, 73(3), 485–488.

Quinn, J. P., Kulakova, A. N., Cooley, N. A., and McGrath, J. W. (2007). New ways to break an old bond: The bacterial carbon-phosphorus hydrolases and their role in biogeochemical phosphorus cycling. *Environmental Microbiology*, 9, 2392–2400.

Ramaswamy, A., Flatt, P. M., Edwards, D. J., Simmons, T. L., Han, B., and Gerwick, W. H. (2006). The secondary metabolites and biosynthetic gene clusters of marine cyanobacteria. Applications in biotechnology. In *Frontiers in Biotechnology* (eds. P. Proksch and W.E.G. Miller), Horizon Bioscience, Norfolk, England, p. 175.

Ren, D., Bedzyk, L. A., Ye, R. W., Thomas, S. M., and Wood, T. K. (2004). Stationary-phase quorum-sensing signals affect autoinducer-2 and gene expression in *Escherichia coli*. *Applied and Environmental Microbiology*, 70, 2038–2043.

Riesenfeld, C. S., Schloss, P. D., and Handelsman, J. (2004). Metagenomics: Genomic analysis of microbial communities. *Annual Review of Genetics*, 38, 525–553.

Rusch, D. B., Halpern, A. L., Sutton, G., Heidelberg, K. B., Williamson, S., Yooseph, S., and Venter, J. C. (2007). The Sorcerer II global ocean sampling expedition: Northwest Atlantic through eastern tropical Pacific. *PLoS Biology*, 5(3), e77.

Sakurada, T., Gill, M. B., Frausto, S., Copits, B., Noguchi, K., Shimamoto, K., and Sakai, R. (2010). Novel N-methylated 8-oxoisoguanines from pacific sponges with diverse neuroactivities. *Journal of Medicinal Chemistry*, 53(16), 6089–6099.

Santos, A. O. D., Britta, E. A., Bianco, E. M., Ueda-Nakamura, T., Pereira, R. C., and Nakamura, C. V. (2011). 4-Acetoxydolastane diterpene from the Brazilian brown alga *Canistrocarpus cervicornis* as antileishmanial agent. *Marine Drugs*, 9(11), 2369–2383.

Sañudo-Wilhelmy, S. A., Gómez-Consarnau, L., Suffridge, C., and Webb, E. A. (2014). The role of B vitamins in marine biogeochemistry. *Annual Review of Marine Science*, 6, 339–367.

Schattenhofer, M., Fuchs, B. M., Amann, R., Zubkov, M. V., Tarran, G. A., and Pernthaler, J. (2009). Latitudinal distribution of prokaryotic picoplankton populations in the Atlantic Ocean. *Environmental Microbiology*, 11(8), 2078–2093.

Schirmer, A., Gadkari, R., Reeves, C. D., Ibrahim, F., DeLong, E. F., and Hutchinson, C. R. (2005). Metagenomic analysis reveals diverse polyketide synthase gene clusters in microorganisms associated with the marine sponge *Discodermia dissoluta*. *Applied and Environmental Microbiology*, 71(8), 4840–4849.

Schmidt, E. W., Nelson, J. T., Rasko, D. A., Sudek, S., Eisen, J. A., Haygood, M. G., and Ravel, J. (2005). Patellamide A and C biosynthesis by a microcin-like pathway in *Prochloron didemni*, the cyanobacterial symbiont of *Lissoclinum patella*. *Proceedings of the National Academy of Sciences of the United States of America*, 102(20), 7315–7320.

Schofield, M. M. and Sherman, D. H. (2013). Meta-omic characterization of prokaryotic gene clusters for natural product biosynthesis. *Current Opinion in Biotechnology*, 24, 1151–1158.

Sebastian, M. and Ammerman, J. W. (2009). The alkaline phosphatase PhoX is more widely distributed in marine bacteria than the classical PhoA. *ISME Journal*, 3, 563–572.

Seow, K. T., Meurer, G., Gerlitz, M., Wendt-Pienkowski, E., Hutchinson, C. R., and Davies, J. (1997). A study of iterative type II polyketide synthases, using bacterial genes cloned from soil DNA: A means to access and use genes from uncultured microorganisms. *Journal of Bacteriology*, 179(23), 7360–7368.

Shokralla, S., Spall, J. L., Gibson, J. F., and Hajibabaei, M. (2012). Next-generation sequencing technologies for environmental DNA research. *Molecular Ecology*, 21, 1794–1805.

Sogutcu, E., Emrence, Z., Arikan, M., Cakiris, A., Abaci, N., Toksoy Öner, E. et al. (2012). Draft genome sequence of *Halomonas smyrnensis* AAD6. *Journal of Bacteriology*, 194, 5690.

Song, J. S., Jeon, J. H., Lee, J. H., Jeong, S. H., Jeong, B. C., Kim, S. J., Lee, J. H., and Lee, S. H. (2005). Molecular characterization of TEM-type beta-lactamases identified in cold-seep sediments of Edison Seamount (south of Lihir Island, Papua New Guinea). *Journal of Microbiology*, 43(2), 172–178.

Stein, J. L., Marsh, T. L., Wu, K. Y., Shizuya, H., and DeLong, E. F. (1996). Characterization of uncultivated prokaryotes: Isolation and analysis of a 40-kilobase-pair genome fragment from a planktonic marine archaeon. *Journal of Bacteriology*, 178(3), 591–599.

Sunna, A. and Bergquist, P. L. (2003). A gene encoding a novel extremely thermostable 1, 4-beta-xylanase isolated directly from an environmental DNA sample. *Extremophiles*, 7(1), 63–70.

Swan, B. K., Chaffin, M. D., Martinez-Garcia, M., Morrison, H. G., Field, E. K., Poulton, N. J., and Stepanauskas, R. (2014). Genomic and metabolic diversity of Marine Group I *Thaumarchaeota* in the mesopelagic of two subtropical gyres. *PLOS ONE*, 9(4), e95380.

Tang, K., Utairungsee, T., Kanokratana, P., Sriprang, R., Champreda, V., Eurwilaichitr, L., and Tanapongpipat, S. (2006). Characterization of a novel cyclomaltodextrinase expressed from environmental DNA isolated from Bor Khleung hot spring in Thailand. *FEMS Microbiology Letters*, 260(1), 91–99.

Thakur, N., Peroviae-Ottstadt, S., Batel, R., Korzhev, M., Diehl-Seifert, B., Müller, I. et al. (2005). Innate immune defense of the sponge *Suberites domuncula* against gram-positive bacteria: Induction of lysozyme and adaptin. *Marine Biology*, 146, 271–282.

Thomas, T., Egan, S., Burg, D., Ng, C., Ting, L., and Cavicchioli, R. (2007). Integration of genomics and proteomics into marine microbial ecology. *Marine Ecology Progress Series*, 332, 291–299.

Venter, J. C., Remington, K., Heidelberg, J. F., Halpern, A. L., Rusch, D., Eisen, J. A. et al. (2004). Environmental genome shotgun sequencing of the Sargasso Sea. *Science*, 304(5667), 66–74.

Villanueva, L., Bale, N., Hopmans, E. C., Schouten, S., and Damsté, J. S. S. (2014). Diversity and distribution of a key sulpholipid biosynthetic gene in marine microbial assemblages. *Environmental Microbiology*, 16, 774–787.

Walsh, D. A., Zalkova, E., Howes, C. G., Song, Y. C., Wright, J. J., Tringe, S. G. et al. (2009). Metagenome of a versatile chemolithoautotroph from expanding oceanic dead zones. *Science*, 326, 578–582.

Weiner, R. M., Taylor II, L. E., Henrissat, B., Hauser, L., Land, M., Coutinho, P. M. et al. (2008). Complete genome sequence of the complex carbohydrate-degrading marine bacterium, *Saccharophagus degradans* strain 2–40T. *PLoS Genetics*, 4(5), e1000087.

Woyke, T., Xie, G., Copeland, A., Gonzalez, J. M., Han, C., Kiss, H., and Stepanauskas, R. (2009). Assembling the marine metagenome, one cell at a time. *PLOS ONE*, 4(4), e5299.

Xu, M., Xiao, X., and Wang, F. (2008). Isolation and characterization of alkane hydroxylases from a metagenomic library of Pacific deep-sea sediment. *Extremophiles*, 12(2), 255–262.

Yamada, K., Terahara, T., Kurata, S., Yokomaku, T., Tsuneda, S., and Harayama, S. (2008). Retrieval of entire genes from environmental DNA by inverse PCR with pre-amplification of target genes using primers containing locked nucleic acids. *Environmental Microbiology*, 10(4), 978–987.

Yang, J. C., Madupu, R., Durkin, A. S., Ekborg, N. A., Pedamallu, C. S., Hostetler, J. B., and Distel, D. L. (2009). The complete genome of *Teredinibacter turnerae* T7901: An intracellular endosymbiont of marine wood-boring bivalves (shipworms). *PLOS ONE*, 4(7), e6085.

Yooseph, S., Sutton, G., Rusch, D. B., Halpern, A. L., Williamson, S. J., Remington, K., and Venter, J. C. (2007). The Sorcerer II Global Ocean Sampling expedition: Expanding the universe of protein families. *PLoS Biology*, 5(3), e16.

Zeng, R., Xiong, P., and Wen, J. (2006). Characterization and gene cloning of a cold-active cellulase from a deep-sea psychrotrophic bacterium *Pseudoalteromonas* sp. DY3. *Extremophiles*, 10(1), 79–82.

Zhang, C. and Kim, S. K. (2010). Research and application of marine microbial enzymes: Status and prospects. *Marine Drugs*, 8(6), 1920–1934.

Zhang, S. and Bryant, D. A. (2011). The tricarboxylic acid cycle in cyanobacteria. *Science*, 334(6062), 1551–1553.

Zhang, Y., Zhao, J., and Zeng, R. (2011). Expression and characterization of a novel mesophilic protease from metagenomic library derived from Antarctic coastal sediment. *Extremophiles*, 15(1), 23–29.

Index

Printed and bound by CPI Group (UK) Ltd, Croydon, CR0 4YY

01/11/2024

01782603-0017